1st International Symposium of Indonesian Chemical Engineering (ISIChem 2018)

IOP Conference Series: Materials Science and Engineering
Volume 543

West Sumatera, Indonesia
4 - 6 October 2018

Editors:

Hakun Wirawasista Aparamarta
Yi-Hsu Ju
Fusheng Li
Soorathep Kheawhom
Widiyastuti
Himsar Ambarita
Siti Nurkhamidah
Kusdianto
Raden Darmawan

ISBN: 978-1-5108-8922-4
ISSN: 1757-8981

Printed from e-media with permission by:

Curran Associates, Inc.
57 Morehouse Lane
Red Hook, NY 12571

Some format issues inherent in the e-media version may also appear in this print version.

Printed by Curran Associates, Inc. (2019)

For permission requests, please contact the Institute of Physics at the address below.

Institute of Physics
Dirac House, Temple Back
Bristol BS1 6BE UK

Phone: 44 1 17 929 7481
Fax: 44 1 17 920 0979

techtracking@iop.org

Additional copies of this publication are available from:

Curran Associates, Inc.
57 Morehouse Lane
Red Hook, NY 12571 USA
Phone: 845-758-0400
Fax: 845-758-2633
Email: curran@proceedings.com
Web: www.proceedings.com

Table of contents

Volume 543

The 1st International Symposium of Indonesian Chemical Engineering (ISIChem) 2018

4–6 October 2018, West Sumatera, Indonesia

Accepted papers received: 26 April 2019

Preface

Papers

Biochemical technology

Material and nano technology

Chemical Engineering Management

Process System Engineering

Renewable Energy Technology

Separation and purification technology

Water and Waste Water Treatment

IOP Conf. Series: Materials Science and Engineering **543** (2019) 011001 doi:10.1088/1757-899X/543/1/011001

Preface: The 1st International Symposium of Indonesian Chemical Engineering (ISIChem) 2018

Hakun Wirawasista Aparamarta[1], Yi-Hsu Ju[2], Fusheng Li[3], Soorathep Kheawhom[4], Kusdianto[1], Widiyastuti[1], Himsar Ambarita[5], Siti Nurkhamidah[1], Raden Darmawan[1]

*Editor of ISIChem 2018

[1] Chemical Engineering Department, Institut Teknologi Sepuluh Nopember (ITS) Surabaya, Indonesia

[2] Chemical Engineering Department, National Taiwan University of Science and Technology (NTUST), Taiwan

[3] Chemical Engineering Department, Gifu University, Japan

[4] Chemical Engineering Department, Chulalongkorn University, Thailand

[5] Mechanical Engineering Department, University of Sumatera Utara (USU), Indonesia

This proceedings volume contains selected articles from the "1st International Symposium of Indonesian Chemical Engineering (ISIChem) 2018" held in Kryad Bumi Minang Hotel, West Sumatra, Indonesia from October 4-6, 2018, that has been peer reviewed.

Despite the fact that this is only our first international symposium of Indonesian Chemical Engineers, and the sixth for the national one, we are so honored to announce that this year we have abundance of enthusiastic participants from five different countries participating in the conference. In the October 4-6, 2018, 250 research papers have been presented in the ISIChem, covering eleven research areas, such as: Biochemical Engineering, Catalyst and Reaction Engineering, Waste Water Treatment Technology, Renewable Energy Technology, Nano Materials and Nanotechnologies, System Engineering Process, Thermodynamics and Supercritical Technology, Separation and Purification Technology, Food and Pharmaceutical Technology, Chemical Engineering Management, and Chemical Engineering Education. We are also delighted to announce that the selected papers will be published in IOP Conference Series: Materials Science and Engineering.

We would like to take this opportunity to acknowledge the incredible support from our main sponsors: STARBORN (PT. Luas Birus Utama), PT. Semen Indonesia, BPJS Ketenagakerjaan, PT. Pupuk Indonesia, and PTPN X. I would also like to express my sincere gratitude to all the steering and organizing committees from Institut Teknologi Sepuluh Nopember (ITS) and Universitas Bung Hatta (UBH) and all the parties for the hard work and tremendous effort they have dedicated to make this event successful.

Alongside our seminars, another important agenda of the national chemical engineers is also taking place at the same time, namely the Annual General Meeting (Musyawarah Nasional) of the Indonesian Chemical Engineering Education Association (APTEKIM).

We would like to thank again the speakers, presenters, participants, and sponsors, who were responsible for the success of the seminar, and look forward to the second edition of the ISIChem seminar.

Hakun Wirawasista Aparamarta, PhD
Chairman of ISIChem 2018

IOP Conf. Series: Materials Science and Engineering **543** (2019) 011001 doi:10.1088/1757-899X/543/1/011001

The Editors of the Work shall be:

- Hakun Wirawasista Aparamarta, PhD
  Institut Teknologi Sepuluh Nopember (ITS) Surabaya, Indonesia
  hakun2397@gmail.com
- Prof. Yi-Hsu Ju
  National Taiwan University of Science and Technology (NTUST), Taiwan
  yhju@mail.ntust.edu.tw
- Prof. Fusheng Li
  Gifu University, Japan
  lifs@gifu-u.ac.jp
- Prof. Soorathep Kheawhom
  Chulalongkorn University, Thailand
  Soorathep.k@chula.ac.th
- Dr. Widiyastuti
  Institut Teknologi Sepuluh Nopember (ITS) Surabaya, Indonesia
  widi@chem-eng.its.ac.id
- Dr. Eng. Himsar Ambarita
  Universitas Sumatera Utara, Indonesia
  himsar@gmail.com
- Siti Nurkhamidah, PhD
  Institut Teknologi Sepuluh Nopember (ITS) Surabaya, Indonesia
  dst_eureka@yahoo.co.uk
- Dr. Kusdianto
  Institut Teknologi Sepuluh Nopember (ITS) Surabaya, Indonesia
  kusdianto@chem-eng.its.ac.id
- Dr. Eng. Raden Darmawan
  Institut Teknologi Sepuluh Nopember (ITS) Surabaya, Indonesia
  Rdarmawan09@gmail.com

IOP Conf. Series: Materials Science and Engineering **543** (2019) 011002 doi:10.1088/1757-899X/543/1/011002

Peer review statement

All papers published in this volume of *IOP Conference Series: Materials Science and Engineering* have been peer reviewed through processes administered by the proceedings Editors. Reviews were conducted by expert referees to the professional and scientific standards expected of a proceedings journal published by IOP Publishing.

IOP Conf. Series: Materials Science and Engineering **543** (2019) 012001 doi:10.1088/1757-899X/543/1/012001

Utilization of Cocopeat And Goat of Dirt In Marking of Solid Organic Fertilizer To Quality Macro Nutrient (NPK)

E Kurniawan[1*], Ishak[1], Suryani[1]

[1]Chemical Engineering of Departement, Faculty of Engineering, Malilkussaleh University, Batam Street Bukit Indah of Campus, Muara Satu Subdistrict, Lhokseumawe, Aceh 24352, Indonesia.

[*]E-mail : ediekur@yahoo.com

Abstract. Fertilizer is a material that can provide nutrients for plants. Organic fertilizers are fertilizers composed of living creatures that are processed through decomposition by decomposers, such as weathering the remains of plants, animals, and humans. The purpose of this research is to see the effect of fermentation time and influence of EM4 composition on nutrient quality in solid organic fertilizer. The results obtained based on the length of time of fermentation and Volume Efective Microorganism 4 EM4 then obtained the best results at 60 ml EM4 volume and 16th days fermentation time with 0.90% nitrogen content, 0.93% phosphorus, and potassium levels of 1.48%. The longer and more EM4 volume the more nutrients it gets.

1. Introduction

Fertilizer is a material that can provide nutrients for plants, if plants lack even do not have one nutrient element, the growth of these plants will not grow optimally. The availability of nutrients in the soil is not always enough to meet the needs of plants. N, P and K elements are primary nutrients, elements Ca, Mg and S are secondary nutrients. In addition, plants need micro nutrients, namely other important elements needed in small amounts, but determine the development of plants, namely boron (B), chlorine (Cl), copper (Cu), iron (Fe), manganese (Mn). molybdenum (Mo) and zinc (Zn).

Fertilizer is a material that is added to the planting media or plants to meet the nutrient needs needed by plants so that they are able to produce well. Fertilizer material can be either organic or non-organic (mineral) materials. Fertilizers containing raw materials that are needed for plant growth and development in the provision of fertilizers need to be considered for the needs of these plants, so that plants do not get too many nutrients. Too little or too many nutrients can be harmful to plants. Fertilizers can be given through the soil or sprayed on the leaves. One type of fertilizer that became an alternative and began to be popular again after a long time has never been used in the development of organic agriculture, namely organic fertilizer.

Organic fertilizer is a fertilizer composed of living matter which is processed through decomposition by decomposing bacteria, such as weathering of plant, animal and human remains [9]. According to [15], the planting medium serves as a place for plants to lay their roots, a source of nutrients for plants that must be supported by good funding drainage. Planting media provide nutrients and provide moisture for plants that function for growth. Plants will grow well when placed on fertile planting media. Fertility of growing media can be seen from physical aspects which include texture, funderation structure.

Organic fertilizers are fertilizers that are composed of material living things that are processed through a process of decay (decomposition) by decomposing bacteria, such as weathering the remains of plants, animals and humans. Organic fertilizers can be either solid or liquid which is used to improve the physical, chemical and biological properties of the soil. Organic fertilizers will provide many benefits because the basic ingredients of organic fertilizers come from agricultural waste, such as: straw, and rice husks, peanut shells, bagasse, cuttlefish, corn stalks, and other forage materials. Whereas livestock manure that is widely used is cow dung, buffalo, goats, chickens, ducks and pigs. Besides that, with the development of settlements, cities and industries, the basic ingredients of compost are

IOP Conf. Series: Materials Science and Engineering **543** (2019) 012001 doi:10.1088/1757-899X/543/1/012001

increasingly diverse. Materials that are widely used include: feces, liquid waste, municipal waste and settlements. Organic fertilizers are the best soil enhancers compared to other enhancers.

Fertilizer is a material that is added to the planting media or plants to meet the nutrient needs needed by plants so that they are able to produce well. Fertilizer material can be either organic or non-organic (mineral) materials. Fertilizers containing raw materials that are needed for plant growth and development in the provision of fertilizers need to be considered for the needs of these plants, so that plants do not get too many nutrients. Too little or too many nutrients can be harmful to plants. Fertilizers can be given through the soil or sprayed on the leaves. One type of fertilizer that became an alternative and began to be popular again after a long time has never been used in the development of organic agriculture, namely organic fertilizer.

According to [5] cocopeat contains macro and micro nutrients needed by plants including potassium, phosphorus, calcium, magnesium and sodium. In addition, cocopeat can hold water and chemical elements of fertilizer and neutralize soil acidity. Physically, cocopeat has a structure that can maintain the balance of aeration and is crumb so that water and air can easily enter.

One of the planting media that is often used for ornamental plants is cocopeat. Cocopeat has several advantages including :

1. Can store oxygen up to 50%, while soil media can only store oxygen 2-3%. The availability of oxygen in the growing media is needed for root growth.
2. Cocopeat can with stand the water content and chemical elements of fertilizer.
3. Can neutralize soil acidity (acidity of cocopeat fiber 5.8 - 6).
4. Macro and micro nutrients contained in coconut fiber include (K) Potassium, (P) Phosphorus, (Ca) Calcium, (Mg) Magnesium, (Na) Sodium and several other minerals. But of the many nutrient content that is owned by cocopeat, it turns out that the most abundant amount is the element K.

Dolomite is a type of limestone that can also benefit plants and soil. These benefits include:

1. Member of nutrition for plants
2. Helps change the pH of the soil
3. Neutralizing harmful substances that can poison the soil and plants (Al, Fe and Cu)
4. Increasing the effectiveness and efficiency of the soil against absorption of nutrients in it.
5. Maintain the availability of nutrients in the soil.
6. Improve soil structure.
7. Stimulates the formation of fatty substances, carbohydrates and various other nutrients in plants.
8. Helping phosphorus distribution in plants.
9. Helps perfect chlorophyll formation in plants

The advantage of EM4 is that it can be produced in a faster time than natural methods. EM4 contains yeast, photosynthetic bacteria, fungus and lactobacilus sp. EM4 is an agricultural cultivation technology to improve the health and fertility of soil and plants by using beneficial microorganisms for plant growth. While harmful microorganisms that can cause disease can be suppressed. EM4 is able to process or decompose organic materials quickly by fermenting into compost so that it does not cause foul odor but causes fresh aroma.

The quality of bokashi from goat manure will be better if mixing goat manure before the fermentation process begins. Increasing the macro or micro nutrient content will increase and be more complex thanks to the mix of cow manure and goat manure. As a table of composition of macro and micro chemical content between cow and goat manure has advantages in macro elements Nitrogen (N), Potassium (K) and calcium (Ca). However, all that depends on what kind of plants will be fertilized using bokashi made from a mixture of cow and goat manure. in addition to solid goat manure, goat's urine also has very good nutrients to be used as organic fertilizer.

Goat manure is added in cocopeat to add cocopeat nutrition for more decomposition so that it is good for plant nutrition. The selection of goat manure for additional nutrition in cocopeat production because goat manure is harder and denser so that the water content is less and the nitrogen content is greater because the nitrogen element is needed by plants for its growth.

Benefits of goat manure:

IOP Conf. Series: Materials Science and Engineering **543** (2019) 012001 doi:10.1088/1757-899X/543/1/012001

1. Increase the nutrient content needed by plants
2. Increasing crop productivity
3. Stimulates the growth of roots, stems, and leaves
4. Spout and fertilize the soil
5. Provision of macro nutrients (nitrogen, phosphorus, potassium, calcium, magnesium and sulfur) and microstructure such as zinc, copper, cobalt, barium, manganese, and iron, although the numbers are relatively small.
6. Increase soil cation exchange capacity (CEC).
7. Form complex compounds with metal ions which poison plants such as aluminum and iron.

The benefits of solid organic fertilizers include, increasing plant fertility, improving soil chemical conditions, improving soil biological conditions, improving soil physical conditions, safe use for humans, not polluting the environment. There are three macro nutrients in organic fertilizer, namely nitrogen, phosphorus and potassium. The uses of nitrogen are to increase plant growth, rotein levels in the soil, vegetable plants produced by the leaves, the activity of organisms in the soil and helps process the synthesis of amino acids and proteins in plants. The benefits of phosphorus for plants are helps respiration and photosynthesis in plants, helps form nucleic acids, helps the formation of plant seeds and fruit formation, stimulates the development of plant roots so that plants are more resistant to drought and speed up crop harvest time. The content of potassium plays a major role in the process of plant growth, including form and transport carbohydrates in the plant's body, useful as a catalyst in the process of protein formation, regulate various types of activities from mineral elements in plants, neutralize the reactions that are in the cell, especially the reaction of organic amino acids, increase the growth of meristem tissue, adjust the movement of the stomata, increasing the robustness of the plant stem so that it does not easily collapse, increase the level of carbohydrates and also sugar in the fruit so that the fruit has a sweet taste to make the seeds of the plant fuller and denser so that it can be used as superior seeds, improve fruit quality in plants, increase plant resistance from pests and diseases, and improve the development of plant roots.

Organic fertilizers are also called complete fertilizers because they have a complete composition of nutrients, but the amount of each nutrient is relatively low, so it is needed in large quantities. The main advantage of using organic fertilizer is that it can improve the fertility of the chemical, physical and biological soil, in addition to nutrients for plants. Organic fertilizers can spur and increase the microbial population in the soil, far greater than just providing chemical fertilizers. Organic fertilizers are also able to fix the structure and fertility of the soil. No wonder if organic fertilizer can prevent soil erosion. Because the nitrogen content and nutrient content released by organic matter will slowly undergo a process of mineralization. If given continuously, it can help build soil fertility.

Organic fertilizers have several advantages including, increasing water content and can withstand water for sandy conditions, increase resistance to erosion, increase air exchange, the number of pores and the nature of water infiltration for clay conditions, reduce the level of soil surface hardness, contain elements complete macro micro nutrients, safe (environmentally friendly), effective and economical (cheap/easy to get) and eliminate chemical rasidu [14]. While , organic fertilizers also have disadvantages, including, it requires in very large quantities to meet the nutrient requirements of a crop, nutrients contained for similar materials vary widely and are good, both in transportation and use in the field. Similar effect of sewage sludge on the levels of soil acidity was demonstrated by [7] who investigated the effect of fertilization with manure and sewage sludge on soil properties. Others point to the acidifying tendencies of sewage sludge after its introduction to the soil fertilized with it [13] This causes the necessity of further investigations of wastewater and sludge in terms of chemical composition.

Fermentation is the process of producing energy in cells in an anaerobic (without oxygen) state. In general, fermentation is a form of anaerobic respiration. Fermentation is the process of decomposition or alteration of organic matter carried out under certain conditions by fermentative microorganisms. The principle of fermentation is that organic waste material is destroyed by microbes in a range of temperatures and certain conditions, namely fermentation. Fermentation can occur due to the activity of microorganisms found in suitable organic matter, this can cause changes in compounds. Fermentation is also a process of breaking down organic compounds into simple compounds involving microorganisms. Fermentation can occur due to microbial activity due to fermentation on suitable organic substrates.

Fermentation is the process of solving organic compounds into simple compounds involving microorganisms. Fermentation can occur due to microbial activity that causes fermentation on an appropriate organic substrate. The occurrence of fermentation can cause changes in the nature of food as a result of the breakdown of the components of the material. To avoid a competition with food industry, cheap by-products and waste substrates are recommended generally for fermentative production of chemicals [10].

Bio-fermentation is a bioengineering technology that utilizes the metabolism of microorganisms for the production of chemical raw materials. The continuous regulation of the technological parameters of bio- fermentation is the most important means of maintaining biological production of chemical raw materials [2].

The essential factors for the growth of microorganisms are temperature, water, nutrients, pH value, etc. Since the bio-fermentation involves many aspects as physics, chemistry and biology, etc, and it is a typical nonlinear and time-varying reaction process, the existing technologies have many shortcomings in the online monitoring of related metabolic indexes (such as biological concentration, growth rate and metabolic rate) of microorganisms. Currently, the most commonly used method is to manually sample the bio-fermentation and perform offline analysis, however, this method has larger sampling amount and the analysis result is lagging, which severely constrains the development of bio-fermentation technology [11]. In addition, extensive bacterial growth achieved during fermentation of the stillage could bring additional value to the process through utilization of the spent media for animal feed formulae enriched in lactic acid bacteria [1].

The purpose of this study is to see the effect of bio-fermentation time and the effect of EM4 composition on nutrient quality in solid organic fertilizer.

Factors Affecting Fermentation:

1. Microorganisms
2. pH (acidity)
3. Temperature
4. Humidity
5. Number of Microorganisms
6. Material Size

Factors that influence the growth of microorganisms:

1. Supply of Nutrition
2. Time
3. Temperature
4. pH value

2. Materials and Methods

The materials used in this research are cocopeats, goat of dirt, dolomite, EM4 and water. And the tools used in this study are blender, buckets, scales, gloves, plastic bag,knife and fertilizer mold.

Cocopeats are dried in the sun and sifted with a 10 mesh sieve called dust. The following are the stages carried out in this study:

1. Prepared goat samples are then stacked and placed on a place that is not inundated
2. Refined dirt samples
3. Samples of cocopeat as much as 100 grams are taken into the bucket.
4. Added 5 grams of dolomite while watering enough water for moisture then added EM4 of 20.40, and 60 ml.
5. Take the cocopeat which has been inserted into the mold.

IOP Conf. Series: Materials Science and Engineering **543** (2019) 012001 doi:10.1088/1757-899X/543/1/012001

6. Then 5 grams of goat manure is taken and put in a mold that has been filled with cocopeat, then closed again using a mixture that has been mixed.
7. Insert the mold that has filled the fertilizer into the plastic so that no air enters.
8. Leave it for 8, 12 and 16 th days.
9. After finishing the fermentation, the fertilizer is ready to be used.

As for the nitrogen testing method with the Kjeldahl method, phosphorus by spectrophotometry and potassium by the Atomic Absorbsion Spectrophotometry (AAS).

3. Results and Discussion

Table 1. Nitrogen Content Analysis Results Data (%)

Time of Fermentation of Solid Organic Fertilizers (Day)	Volume EM4(ml)		
	20,0	40,0	60,0
8	0,54	0,63	0,76
12	0,64	0,72	0,82
16	0,71	0,82	0,90

The effect of fermentation time on nitrogen content by varying the use of EM4 as a bioactivator for the fermentation process carried out. The lowest Nitrogen content was obtained at 8th day fermentation with 20 ml of EM4 amounting to 0.54%. This is because in the 8th day fermentation period the growth of the initial phase microorganisms is a phase of adjustment between decomposing microorganisms and the substrate described so that there has not been a change in the number of cells that are too significant. This can affect the NPK obtained in fertilizers, because the quality of nutrients in solid fertilizers is influenced by the rate of extraction by microbes, NPK levels will increase with the increasing number of microorganisms where the more microbes that develop, the more substrate described so that the quality of organic fertilizer will be even better [14].

This is because in the 8th day fermentation period the growth of the early phase microorganisms is an adjustment phase between decomposing microorganisms and the substrate described so that there is no significant change in cell number. The best results were obtained on the 16th day with an EM4 volume of 60 ml which was 0.90%. The Nitrogen level has met the quality standards of solid fertilizer, which is at least 0.4%. At 12 th and 16th day fermentation, the Nitrogen levels were 0.64%, and 0.71%, with the same EM4 amount of 20 ml, the Nitrogen levels were increasing, so did the same day with different EM4 volumes of 40 ml , and 60 ml of Nitrogen levels also increase. This is because the existing microorganisms experience an exponential phase which is very rapid cell division [3]. This is because in the 8th day fermentation period the growth of the early phase microorganisms is an adjustment phase between decomposing microorganisms and the substrate described so that there is no significant change in cell number. The effect of application of organic fertilizers on soil reaction and other properties of sorption complex are also essential [4].

Table 2 . Phosphor Content Analysis Results Data (%)

Time of Fermentation of Solid Organic Fertilizers (Day)	Volume EM4(ml)		
	20,0	40,0	60,0
8	0,63	0,74	0,84
12	0,66	0,79	0,87
16	0,71	0,82	0,93

Phosphorus levels obtained are increasing. The phosphorus content obtained is strongly influenced by the length of fermentation time and the volume of EM4 that is varied. The lowest phosphorus content obtained at 20 ml EM4 volume on the 8 th day is 0.63%, this is due to the growth of microorganisms in the initial phase which is the adaptation period, since the inoculation of the medium is carried out during the initial phase where the cell mass can change without changes in cell numbers. Phosphorus in plants plays a role in the formation of flowers, fruits and seeds and plays a role in the transfer of energy in plant cells which cannot be replaced by other elements. The best results were obtained on the 16th day with an EM4 volume of 60 ml which was 0.93%.

This is because in fermentation the growth of microorganisms in the initial phase is a period of adaptation that is since the inoculation of the medium is carried out during the initial phase where the cell mass can change without changes in the number of cells. After the mass change then microorganisms grow into the exponential phase, namely at the volume of EM4 40 ml with a fermentation time of 12th day, where the existing microorganisms develop optimally to the number of cells produced so that the phosphorus content obtained is also increasing. From the Phosphorus level it shows that the phosphorus level has met the standard, which is> 0.10%. The content of Nitrogen in the substrate, the greater the multiplication of microorganisms that overhaul phosphorus will increase, so that the phosphorus content in fertilizers also increases [18].

Table 3. Potassium Content Analysis Results Data (%)

Time of Fermentation of Solid Organic Fertilizers (Day)	Volume EM4(ml)		
	20,0	40,0	60,0
8	0,75	0,92	1,36
12	0,79	1,15	1,42
16	0,83	1,24	1,48

The lowest results were obtained at 20 ml EM4 volume with 8th day fermentation time of 0.75%. This is because in fermentation the growth of microorganisms occurs in the initial phase which is a stage of adjustment between microbes and substrate to be described. During the initial phase, cell mass can change without changes in cell number. After the next mass change microorganisms grow into an exponential phase where the existing microorganisms develop optimally against the number of cells produced by microorganisms so that the levels of phosphorus obtained increase. The best results were obtained on the 16th day and 60 ml EM4 volume of 1.48%. From these levels showed potassium solid organic fertilizer has met potassium standards, namely > 0.20%. The best results were obtained on the 16th day and the volume of EM4 as much as 60 ml which is 1.48%. From these levels indicate potassium solid organic fertilizer has met the standard potassium is > 0.20%. the results obtained were higher compared to the fermentation time and other EM4 volumes. Potassium is used by microorganisms in the substrate material as a catalyst, with the presence of bacteria and its activity will greatly influence the increase in potassium.

4. Conclusions

From the research that has been done it can be taken several conclusions which are :

1. Solid waste of goat and coconut husk waste can be used as a raw material for solid organic fertilizers with
   the addition of some other ingredients.
2. The content of macro nutrients (N, P, K) in solid organic fertilizer was best at 16th day fermentation with EM4 volume of 60 ml that is 0.90% N, 0.93% P and 1.48% K.
3. The contents of macro nutrients (N, P, K) are directly proportional to EM4 volume and fermentation time.
4. EM4 volume and fermentation time have an effect on the macro nutrient content (N, P, K).

Reference

IOP Conf. Series: Materials Science and Engineering **543** (2019) 012001 doi:10.1088/1757-899X/543/1/012001

[1] Aleksandra D-V, Ljiljana M, Svetlana N, Jelena P, Sunčica V-T and Katarina M 2013 Distillery stillage as a new substrate for lactic acid production in batch and fed-batch fermentation. *Chem. Eng. Trans.* **34**, 97-102

[2] Borchiellini R et all 2017 The energy center initiative at politecnico di Torino: Practical experiences on energy efficiency measures in the municipality of Torino. *Int. J. Heat and Tech* **35** (S1) S196-S204

[3] Eustace AI and Dorothy ML 2001 Changes in carbohydrate fraction of cassava peel following fungal solid state fermentation. *J. Food and Tech. in Africa* **6** (3) 101-103

[4] Ewa S-M and Longina S 2013 Properties of sorption complex and humic acids in sandy soils fertilized with sewage sludge. *Chem. Eng. Trans.* **32** 499-504

[5] Fahmi Z.I 2013 Planting Media As External Factors Affecting Plant Growth. Center for Repair and Protection of Surabaya Plantation Plants. Surabaya 1-8

[6] Indriani YH 2004 Making of Compost Light. Penebar Swadaya. Jakarta.

[7] Iżewska A 2007 Effect of fertilization with manure sewage sludge and sludge compost on soil properties (in Polish). Zeszyty Problemowe Postępów Nauk Rolniczych **518**, 85-92

[8] Murbanda L 2004 Solid Organic of Fertilizer. Penebar Swadaya. Jakarta

[9] Murbando LHS 2000 Making of Compost. Penebar Swadaya. Jakarta

[10] Ozalp N and Hyman B 2009 Energy allocation and energy intensity estimates for the U.S. organic chemicals industry. *Chem. Eng. Trans.* **18**, 531-536

[11] Pérez P, Yescas RM, Acata RVG, Caballero VP and López RA 2013 Software sensors design for the simultaneous saccharification and fermentation of starch to ethanol Fuel **110** (4) 219-226.

[12] Ramadhani S 2011 The influence of cocopeat shell fiber addition to ground strainer strain **9** (3) 187-195.

[13] Stańczyk ME, Kępa U and Stępniak L 2007a Variation in the acidity of soils after fertilization with sewage sludge. Polish *J. Environmental Studies* **16** (3B) 470-473

[14] Suriadikarta DA and Setyorini D 2009 Standard quality of organic fertilizer. Ministry of Agriculture. Bogor

[15] Wibowo 2007 Planting Media for Ornamental Plants. Penebar Swadaya. Jakarta

[16] Winarso 2005 Soil Biology and Processing Strategy. Graha Ilmu. Yogyakarta

[17} Yuliarti 2008 Potential compost and fertilizer for organic rice production in inceptisol land. *J. Agrosia* **11** (2) 36-40

[18] Yuli A, Hidayati, Kottelat M, Kartikasari SN and Anthony JW 2011 Liquid fertilizer quality results of processing of beef cattle using saccharomyces cereviceae *Animal Science J.* **11**(2) 104-107

Single cell protein production by a local *Aspergillus niger* in solid state fermentation using rice straw pulp as carbon source: effects of fermentation variables

S D Said[a]*, M Zaki[a], T M Asnawi[a], E Novita[b]
[a]Chemical Engineering Department, Faculty of Engineering, Syiah Kuala University, Darussalam, Banda Aceh, Indonesia
[b]Chemical Engineering Graduate Study Program, Faculty of Engineering, Syiah Kuala University, Darussalam, Banda Aceh, Indonesia

*email: syahiddin@che.unsyiah.ac.id

Abstract. In this study, solid-state fermentation was carried out to produce single cell protein by local *Aspergillus niger*. Rice straw pulp produced from the pretreatment by liquid hot water at temperature 121 °C for 60 minutes of cooking time was used as carbon source. The effects of nitrogen sources, initial medium pH, C/N ratios of substrates and fermentation time on SCP production were evaluated in these experiments. The single cell protein was measured as crude protein content in the fermentation product by the Kjeldahl analysis method. The maximum crude protein of 18.9 % (w/w) contained in the fermentation product occurred at the fermentation condition with C/N ratio 30:1, the initial medium pH 4, temperature 30 °C, moisture content 75 % and relative humidity 95 %.

1. Introduction

Providing animal feed supply for a sustainable food production scenario will be the major concern all over the world, especially in developing countries. Indonesia is one of the biggest agricultural countries producing an abundant amount of rice straw as agricultural residue from paddy field every year. Based on an FAO report [1], the estimated rice straw production amounts about 740.95–1111.42 million tons per year globally while Indonesia itself contributed about 70.84–106.26 million metric tons of rice straw in 2014. This cellulosic biomass provides a low-cost resource with large possibility to support a large-scale production of various biomass-derived products [2];[3]. Rice straw is among major lignocellulosic agricultural residues with a great potential as substrate to produce single cell protein (SCP) with a useful nutrient for animal feed [4]. However, available data on treating rice straw applied as a suitable substrate for SCP production are relatively scarce, therefore, efforts to develop SCP from rice straw still need exploring more information. There are some important aspects affect the SCP production in solid state fermentation (SSF) include to select a pretreatment method with ability to produce a suitable substrate for a selected fungus and to determine the appropriate SSF process conditions for optimum microbial growth such as pH, the moisture content, temperature, C/N ratio, and fermentation time. One of the potential pretreatment methods for preparing a substrate from lignocellulosic materials is liquid hot water (LHW). This simple and low-cost pretreatment method can generate reactive cellulose fibers to make them easily accessible for microbial growth and produce degradation products with little inhibition for a subsequent fermentation process [5]. Therefore, LHW pretreatment of rice straw as substrate for solid state fermentation using local *Aspergillus niger* is a promising cost-effective and environmental-friendly technology for developing countries. *Aspergillus niger* is one species with ability to convert cellulose materials into SCP [6]. Besides, it has an optimum growth temperature in the range of 35–37 °C, hence, it requires less energy for reaching optimal growth temperature in the fermentation process. Unfortunately, the utilization of rice straw pulp as a carbon source for SCP production by local *Aspergillus niger* is still less reported by researchers. This research evaluates the possibility to utilize LHW pretreatment rice straw pulp for SCP production by a local *Aspergillus niger* and the effects of initial pH, initial moisture content, the C/N ratio and fermentation time on crude protein yield.

2. Materials and Methods

2.1 Chemical composition and crude protein content analysis

The chemical composition of rice straw was determined using the method described by Goering and Van Soest [7]. Ash content was measured by thermogravimetric method. The crude protein content was determined by total nitrogen content which measured by Kjeldahl method [8] then multiplying with a traditional conversion factor of 6.25.

2.2 Microorganism and inoculum preparation

The local fungus used in this study was *Aspergillus niger* obtained from Microbiology Laboratory, School of Life Sciences and Technology, Bandung Institute of Technology, Indonesia. It was maintained on potato dextrose agar and sub-cultured periodically on the agar plate for 7 days. The spores produced on the agar were dislodged and suspended in sterile water under aseptic condition. The suspension of 107 spores/mL was inoculated into 200 mL medium in 500 mL Erlenmeyer flasks. The medium consisted of 40 g/L molasses and 7 g/L ammonium sulfate. The flask was incubated at a temperature of 30 °C and agitation speed of 150 rpm in an orbital incubator for 3 days. The incubation broth was used as culture inoculum for SSF.

2.3 Rice straw pretreatments

The rice straw used as substrate was obtained from a local farmer in the Aceh Besar district of Indonesia. It was collected in a plastic bag and placed in a room with an approximate temperature of 30 °C. Rice straw was cut into 2-3 cm, crushed in a blender, and passed through a screen sieve 60 mesh. Furthermore, rice straw was pretreated using LHW in autoclave at a temperature of 121 °C for 60 minutes. Rice straw pulp from autoclave was rinsed with distilled water and dried in an oven at a temperature of 60 °C for 2 days.

2.4 Solid state fermentation

The effects of nitrogen sources, C/N ratio, initial medium pH and fermentation time on the SCP production (determined as crude protein) in SSF were studied. The SSF was performed in each 500 mL Erlenmeyer flask containing 10 g of rice straw pulp at a certain C/N ratio and 75 % moisture content. Nitrogen source and distilled water were added into the flasks to adjust the C/N ratio and moisture content of the solid substrate (rice straw pulp). The flasks were sterilized at 121 °C for 15 minutes. The substrate in the flask was mixed properly using sterile spatula after adding culture inoculums of 10 mL into each flask. Subsequently, all flasks were incubated in the environment chamber at controlled temperature and relative humidity of 30±1 °C and 95±1 % for 12 days. Three types of fertilizer, namely urea, ammonium sulfate and NPK as nitrogen sources were used for evaluating the effect of the nitrogen source on the crude protein yield. The nitrogen source with the highest crude protein content in the fermentation product would be selected for C/N ratio assessment. The C/N ratio of substrate was evaluated at the various C/N ratios of 20:1, 30:1, 40:1, 50:1 and 60:1. The effect of the initial pH of solid substrate was investigated at pH of 3, 4, 5, and 6. The effect of incubation time was investigated at various times, i.e 3, 6, 9, 12, 15 days at a constantly initial C/N ratio 30:1 and pH of 4. Each experiment was carried out two times.

2.5 Determination of initial medium pH
One gram of the sample was mixed with 10.0 mL of distilled water, shaken in an orbital shaker for 10 minutes, and filtered out. Subsequently pH of the filtrate was measured by pH meter (HORIBA pH 1100).

3. Results and Discussion

3.1 Liquid Hot Water Pretreatment of Rice Straw
The liquid hot water (LHW) pretreatment of rice straw was performed in autoclave at temperature condition of 121 °C for 60 minutes. The composition of untreated rice straw and rice straw pulp expressed in percentage was showed in Table 1. LHW pretreatment was able to reduce the lignin content from 11.18 % in rice straw to 7.07 % in rice straw pulp (pretreated rice straw), on the other hand cellulose and hemicellulose content respectively enhanced from 48.29 % and 16.89 % in rice straw to 54.33 % and 21.35 % in rice straw pulp. Rahnama et al., [9] used NaOH solution with low concentration of 0.5 % (w/v) and the ratio of rice straw to solution 1:10 (1 g rice straw /10 mL alkali solution) for delignification process, and revealed decrease of lignin content of rice straw from 9.22 % to 5.91 % after pretreatment with an alkaline solution which conducted in an autoclave at temperature 121 °C for 20 minutes. Referring to the report, LHW pretreatment has a comparable delignification effect to alkaline pretreatment but need longer operation time.

Table 1. Chemical composition of rice straw and rice straw pulp

Samples	Lignin (%)	Cellulose (%)	Hemicellulose (%)	Ash (%)
Rice straw	11.18 %	48.29 %	16.89 %	13.25 %
Rice straw pulp	7.07 %	54.33 %	21.35 %	14.33 %

3.2 Time courses of SCP production
The effect of fermentation time on SCP production was evaluated in solid substrate (mixed rice straw pulp-urea) with the C/N ratios of 20 and 30, and the fermentation condition of the initial medium pH 4, temperature 30 °C, moisture content 75 %, and relative humidity 95 %. The time course of SCP (determined as crude protein) production was shown in Fig. 1. SCP production increased with the time course from 0–12 days and reached the maximum protein production (18.9 %) in 12 days. SCP production began getting constant after 12 days of fermentation. Hamdy [10], explained the similar effect of fermentation time on SCP production by *Aspergillus niger* which was fermented in orange peel media. The total protein in fermentation product yielded by *Aspergillus niger* of 40.1 % was reached at 144 h (6 d) fermentation. The longer fermentation time was needed and less protein content in fermentation product resulted in rice straw pulp fermentation compared to fermentation in orange peel media. It was probably because of higher lignin content remained and lower soluble sugars contained in rice straw pulp.

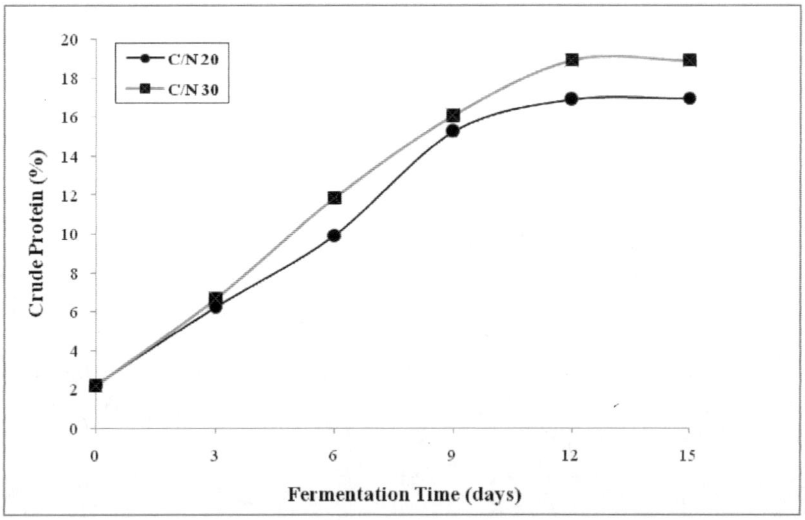

Figure 1. The time course of single cell protein production

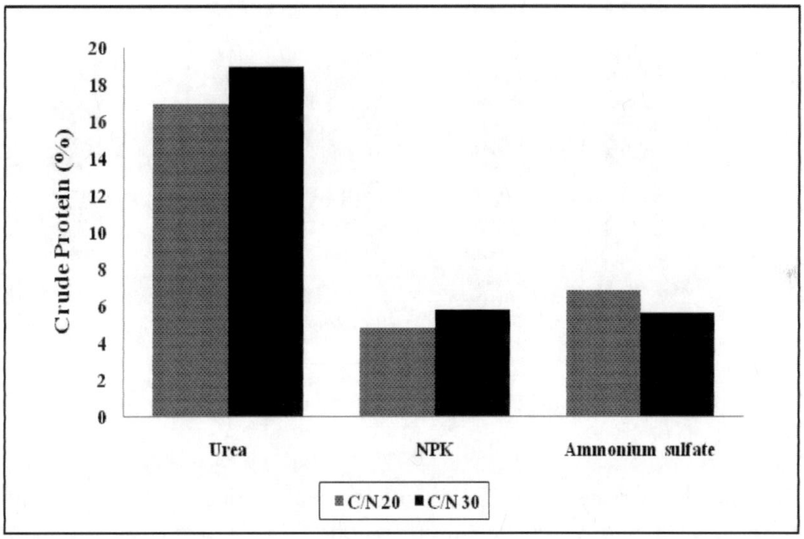

Figure 2. The effect of the various nitrogen sources on SCP production

3.3 Effect of the nitrogen source on SCP production

In order to select the optimum nitrogen source for maximum SCP production, various types of fertilizers which available in a local market (urea, ammonium sulfate and NPK) were used as the nitrogen source in the growth medium. Fig. 2 presented the effect of the nitrogen source on crude protein content in the fermentation product after 12 days fermentation time. The results revealed that the highest protein content (18.9 %, w/w) was produced on solid substrate with urea as nitrogen source. It is presumably that the possible reason urea might have a higher solubility in water compared to those other two nitrogen sources. It gave higher nitrogen concentration on the surface of substrate which was consumed by fungi as nutrient for their growth as the microbial activities take place in or near the surface of substrate. Hence, a higher crude protein yield reached (18.9 %) as compared to other nitrogen sources. The results revealed that urea, a low-cost fertilizer, supported maximum crude protein production.

Oshoma and Eguakun-Owie [11] who studied the effect of different anorganic nitrogen supplement on food waste for the production of *Aspergillus niger* SCP reported that ammonium nitrate and ammonium sulfate addition gave higher protein content.

3.4 Effect of the C/N ratio in the substrate

The effect of the carbon to nitrogen (C/N) ratio of the substrate on a crude protein with urea as a nitrogen source in fermentation product was evaluated. Carbon and nitrogen sources are important factor which will be consumed as nutrient for fungal growth and microbial metabolism during fermentation process. Optimum concentrations of these nutrients were found to increase the growth performance and fungal biomass. Hence, it is necessary to maintain the accurate composition of growth media for effective fermentation process to keep the C/N ratio within the desired range. In Fig. 3 showed that the crude protein content increased with the increasing of the C/N ratio from 10:1 to 30:1, but in contrary, further increasing in the C/N ratio from 30:1 to 50:1 gave a decreasing of crude protein content. The maximum crude protein content (18.9 %) was found at the C/N ratio of 30:1.

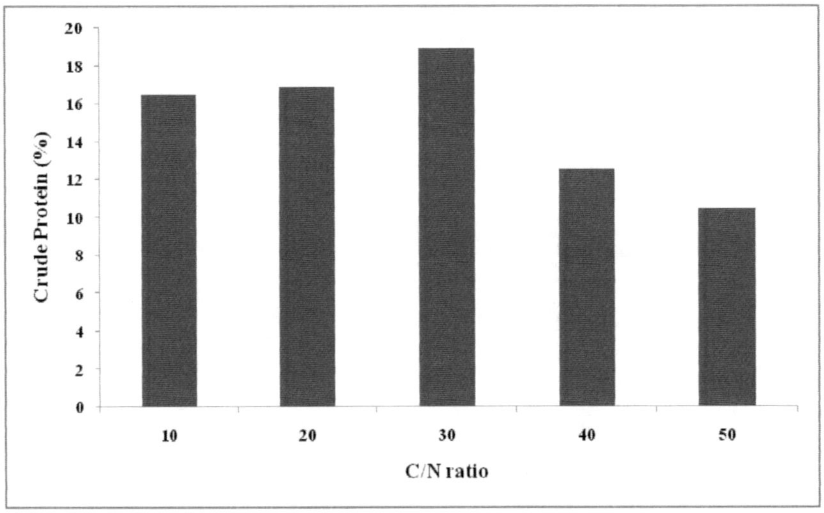

Figure 3. The effect of the C/N ratio on SCP production

A previous investigation also reported a significant enhacement in microbial biomass protein production using corn stover by sequential culture fermentation of *Candida utilis* and *Arachniotus sp.* was observed at 30:1 C/N ratio in 6 % (w/v) corn stover [12]. Thus, determination of appropriate amount of C/N ratio is the vital factor to harvest maximum microbial biomass protein.

3.5 Effect of initial medium pH

The effect of pH was studied by adjusting different ranges of an initial medium pH of the solid substrate in the range of 3–7 on crude protein yield. The results (Fig. 3) shown that crude protein production increased from pH of 3 to 4 and reached maximum crude protein yield at pH 4 (18.9 %, w/w). Further increasing in initial medium pH affected the decreasing of crude protein production. It was observed that the highest crude protein production obtained was at the initial medium pH of 4.This result is in agreement with Maftukhah and Abdullah [13] who used rice straw as substrate for cellulose enzyme production on solid state fermentation by *Aspergillus niger* which found that the maximum enzyme production occurred at pH 4. Sohail et al. [14] also reported that during an exponential phase of growth in solid state fermentation of lignocellulosic substrates, *Aspergillus niger* MS82 produced the highest production of cellulase at initial pH 4 and temperature 35 °C. These results in general also confirmed that initial pH 4 is optimum for *Aspergillus niger* to produce the maximum crude protein or cellulase production in solid state fermentation using cellulosic substrate.

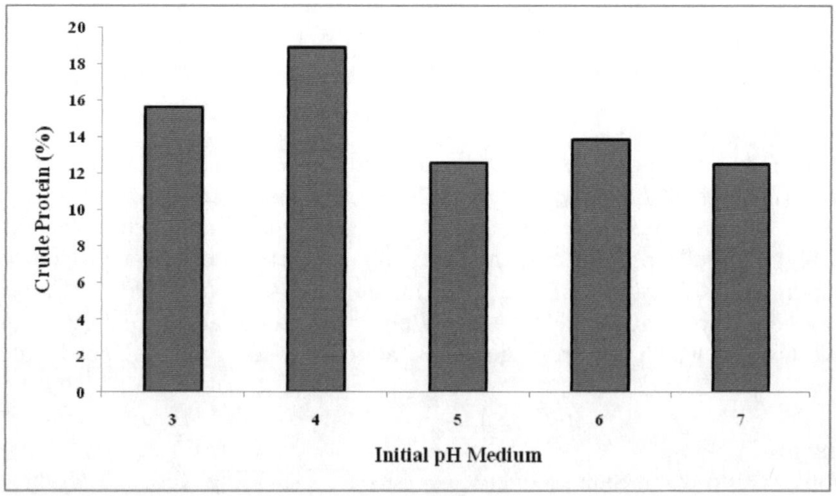

Figure 4. The effect of initial pH medium on SCP production for 12 days at 30 °C and 75 % moisture content

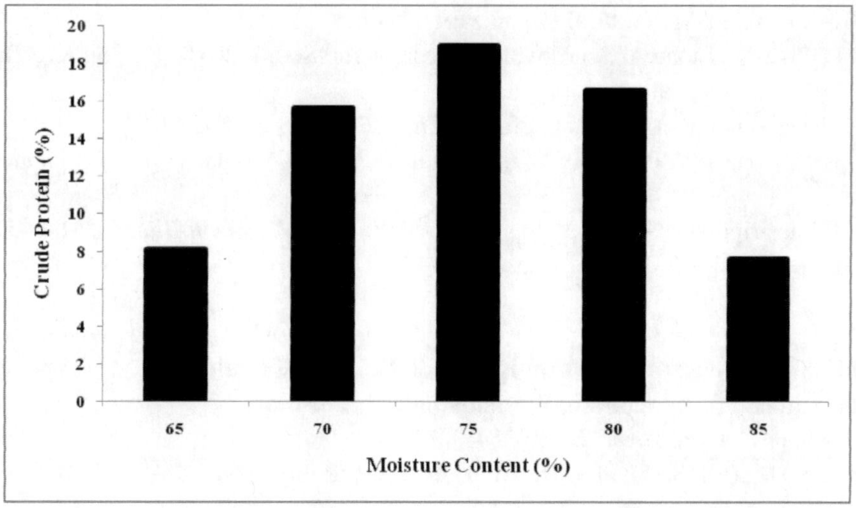

Figure 5. The effect of the initial moisture content on the SCP production for 12 days at 30 °C and relative humidity 95 %

3.6 Effect of moisture content

The initial moisture content of the substrate plays a vital role on microorganisms activities during the crude protein production by *A. niger* cultivated in SSF bioreactor. The moisture contents of substrates were evaluated ranging from 65 % to 85 %. Fig.5 shows the maximum crude protein yield (18.9 %) was obtained at the moisture content of 75 % and tends to decrease at higher moisture content level. Hamdy [10], studied the effect of moisture content in the range of 50 % to 70 %, and revealed that the initial moisture content of orange peels media of 55 % gave the highest protein yield in *A. niger* fermentation product. Water got involved in most vital metabolic activities, and the majority of microbial viable required a moisture content of 70–80 % for reproduction [15]. But the optimum initial moisture content required by the fungus for maximum growth depends on the physicochemical characteristics of the substrate. High moisture content could developed water film which cover the porosity of solid surface and reduce oxygen transfer for the fungal growth. Conversely, low moisture content will limit the fungal growth [16]. This evaluation suggested that the optimal initial moisture content plays an important role for the maximum microbial biomass production.

4. Conclusion

This study showed that it was possible to develop a single cell protein production system by utilizing the solid substrate (mixed rice straw pulp-urea) in a SSF with a local *Aspergillus niger* as bioconversion agent. The present work revealed that *Aspergillus niger* had a good ability to consume the prepared substrate. The highest crude protein content (18.9 %) in the fermentation product occurred at a fermentation condition of C/N ratio 30:1, the initial medium pH 4, temperature 30 ^{o}C, moisture content 75 % and relative humidity 95 %. In addition, LHW pretreatment has ability to reduce lignin content in rice straw to the level in which the pretreated substrate (rice straw pulp) is appropriate for *A. niger* growth.

Acknowledgements

The authors would like to thank Syiah Kuala University for providing Research Grant year 2017 and Microbiology Laboratory, School of Life Sciences and Technology, Bandung Institute of Technology, Indonesia for providing local *Aspergillus niger* is gratefully acknowledged.

References

[1] FAO 2014 Food and Agricultural Organization, Rome.

[2] Masutti D C, Borgognone A, Scardovi F, Vaccari V and Setti L 2015, *J.Chem.Eng Trans.* **43** 487-92.

[3] Pirota R, Delabona P and Farinas C 2014, *J.Chem.Eng Trans.* **38** 259- 64.

[4] Sarnklong C, Cone J W, Pellikaan W and Hendriks W H 2010, *Asian-Aus.J.Animal.Sci.* **23** 680-92.

[5] Imman S, Laosiripojana N and Champreda V 2017, *J App.Biochem.Biotech.* **184** 432–43.

[6] Dhillon G S, Oberoi H S, Kaur S, Bansal S and Brar S K 2011, *J. Ind Crops and Prods.* **34** 1160–7.

[7] Goehring H K and Van Soest P J 1970, *Forage fiber analysis (apparatus, reagents, procedures and some applications)* (Agricultural Handbook No.379, Agricultural Research Service – United States Department of Agriculture, Washington D.C)

[8] Kirk P L 1950, *J. Anal. Chem.* **22** (2) 354-8.

[9] Rahnama N, Mamat S, Shah U K M, Ling F H, Rahman N A A and Arif A B 2013, *J. BioResources.* **8** (2) 2881-96.

[10] Hamdy H S 2013**,** *Roman. Biotech. Letters.* **18** (1) 7929–46.

[11] Oshoma C E and Eguakun-Owie S O 2018, *J. App. Sci. Envi.Management.* **22** (3) 350 –5.

[12] Ahmed S, Ahmad F and Hashimi A S 2010, *Pakistan Journal Botany.* **42** 1225-34.

[13] Maftukhah S and Abdullah A 2018, *MATEC Web of Conferences.* **156** 01010.

[14] Sohail M, Siddiqi R, Ahmad A and Khan S A 2009, *J. New. Biotechnology.* **25** 437-41.

[15] Shahzad F, Abdulah M, Chaudhry A S, Javed K, Bhatti J A, Jabbar M A, Kamran Z, Ahmed F, Ahmed S, Ali A, Irshad I and Ahmad N 2016, *The J. Animal & Plant Sciences.* **26** (2) 309-14.

[16] Farinas C S 2015, *Renew. Sustainable Energy Reviews.* **52** 179–188.

Production of Reducing Sugar from Coffee Pulp Waste Using Mixture of Microorganisms, Enzymes, and Surfactants

T Widjaja[1], N Hendrianie[1], E O Ningrum[2], W H Erliana[1], T Iswanto[1]

[1]Department of Chemical Engineering, Institut Teknologi Sepuluh Nopember Surabaya, Sukolilo, 60111, Surabaya, Indonesia
[2]Department of Industrial Chemical Engineering, Institut Teknologi Sepuluh Nopember Surabaya, Sukolilo, 60111, Surabaya, Indonesia

E-mail: triw@chem-eng.its.ac.id

Abstract. This study has successfully investigated the effect of microorganisms, enzymes, and surfactants mixture to produce a reducing sugar from Coffee pulp waste. The experiment consisted of microbial pretreatment to reduce lignin content followed by comparing the hydrolysis by enzymes and microorganism using a surfactant and without surfactant to get a higher yield of reducing sugar. Pretreatment was conducted by mixture of *Bacillus subtilis* (BS) with *Trichoderma reesei* (TR) in the ratio of 2:1 (v/v) and *Aspergillus niger* (AN) with TR in the ratio of 1:1 (v/v). BS-TR mixture increased the cellulose content to 10.939 % and decreased the lignin and pectin content to 71.261 % and 55.046 %, respectively. Whereas, AN:TR mixture increased the cellulose content to 12.572 % and decreased the lignin and pectin content to 69.941 % and 52.294 %, respectively. Afterward, the result of enzymatic hydrolysis with 3 g of Tween 80 and biological hydrolysis showed increasing of reducing sugar yield of 0.5831 and 0.0341 %, respectively. Further investigation was described as the crystallinity index and the fructose, glucose, and sucrose contents. The addition of both PEG 4000 and Tween 80 as a surfactant in the enzymatic hydrolysis process could significantly increase the concentration of reducing sugar.

1. Introduction

Coffee is the second largest trading commodity in the world including in Indonesia which has a quite high productivity of 739.005 tons in 2015, harvested from the productive coffee plant area of 1,254,382 ha managed by people, government and private sector [1]. Hence, an enormous amount of Coffee Pulp Waste (CPW) was generated, but still without optimized utilization yet and low economic value. Farmers frequently threw away near plantation area or sold the waste at a meager price for cattle feed and compost, where in fact the waste can be utilized to produce more valuable products such as reducing sugar or methane production for biogas [2].

CPW contains 63 % of cellulose, 2.3 % of hemicellulose, 17 % of lignin, 11.5 % of protein, 1.8-8.56 % of tannin, 6.5 % of pectin, and 1.5 % of caffeine [3]. The presence of lignin, pectin, tannin, and caffeine in the CPW can inhibit reducing sugar formation, so it must be eliminated. However, the complex structure of lignin, pectin, hemicellulose, and cellulose contents in the CPW is formed by the covalent bonds, the intermolecular bonds, and the van der Waals force which is hardly broken down by hydrolysis process and also insoluble in the water [4]. Therefore, the pretreatment process is highly recommended.

IOP Conf. Series: Materials Science and Engineering **543** (2019) 012003 doi:10.1088/1757-899X/543/1/012003

Biological pretreatment presents a promising approach to convert lignocellulosic materials into reducing sugar, where it is safer and environmentally friendly. It also does not require substantial energy to degrade lignin from lignocellulosic biomass. Several types of microorganism were reported to degrade lignin and pectin. *Aspergillus niger* can effectively degrade lignin and pectin, *Bacillus subtilis* can degrade pectin, and *Trichoderma reesei* can degrade lignin [5]. In another study, *Aspergillus* and *Penicillium* fungus was known to have a good degradation ability of caffeine [6].
The aims of this study are reducing lignin and pectin level in CPW using *Aspergillus niger, Bacillus subtilis,* and *Trichoderma reesei* to obtain more natural accessible cellulose and hemicellulose; investigating the effect of the enzymatic hydrolysis from pretrated CPW using pure cellulase and xylanase enzymes and culture of *Aspergillus niger* and *Trichoderma viride* on the production of reducing sugar. The selection of the enzymatic hydrolysis method was based on the previous research which states that the enzymatic hydrolysis can produce high pure glucose yield with a specific process, can be carried out in low operating temperature, has low energy consumption and does not cause corrosion [7]. In the enzymatic hydrolysis process, PEG 4000 and Tween 80 as the surfactants was added which are useful for enlarging the conversion value of the biomass substrate and preventing deactivation of cellulase enzymes [8]. The effects of those surfactants on reducing sugar yield was also investigated in the present study.

2. Methods

2.1. Preparation of coffee pulp waste
Coffee pulp waste was obtained from Bangelan, East Java Indonesia. It was dried under the sunlight for approximately 7 days. Then the CPW was mashed using a grinding machine and sieved to get 120 meshes. The smaller substrate particles help the hydrolysis process to make the contact surface between the substrate and the enzymes larger and the lignocellulose content will be easier to be degraded. Before used, the CPW was dried using the oven at 105 °C to reduce moisture content.

2.2. Biological Pretreatment process of coffee pulp
Aspergillus niger, *Bacillus subtilis*, *Trichoderma viride*, and *Trichoderma reesei* were obtained from the Laboratory of Microbiology Engineering, Chemical Engineering Department FTI-ITS. One loop of each microbs was inoculated to 100 ml of sterilized nutrient broth and incubated for 96 h at 30 °C, and 1 ml of the inoculum sample was introduced into the test tube. The sample was diluted until 10 ml and counting chamber analysis method was done to find the log phase of the microorganism.
In the biological pretreatment, the CPW and water was mixed with ratio of 1:4. Thirty milliliter of the mixture of AN and TR with 1:1 (v/v) ratio; BS and TR with 2:1 (v/v) ratio were inoculated into the CPW solution and incubated at 35 °C and pH 5 for 4 d. The ratio between microorganism and CPW was 30:100 (ml/g). The obtained CPW after this pretreatment was used in the next step, the hydrolysis process.

2.3. Hydrolysis
The enzymatic hydrolysis process using cellulose and xylanase enzyme was performed by weighing 1 g of the pretreated CPW and put into an Erlenmeyer flask. 20,552 ml of cellulase and 14,524 ml of xylanase were added into the flask and then heated until 40 °C. Afterward, 0,1 M of citrate buffer with pH 5,5 was added into the enzyme-CPW solution until the total volume reached 30 ml. Then, the surfactant was added from 1 to 3 g. The sample was incubated using incubator shaker at 35 °C for 16 h. The biological hydrolysis process using microorganism was carried out by weighing 1 g of pretreated CPW and introduced it into the Erlenmeyer flask. Three milliliter of the mixture of *Aspergillus niger* and *Trichoderma viride* with various ratio of AN:TV (1:1, 1:2, and 2:1) were inoculated into the flask and 0,1 M citrate buffer solution pH of 5,5 was added until the total volume reached 30 ml. Then, that flask was heated slowly until 30 °C. The surfactan was added into pretreated CPW solution in ratio 1 g per 100 g of CPW. The flask was incubated in the same condition above. Pretreated CPW and 0,1 M citrate buffer solution pH of 5,5 (total volume: 30 ml) was used as control.

2.4. Analysis

Gravimetric method by [9] was used to analyze cellulose, hemicellulose and lignin content. DNS analysis was performed to analyze the reducing sugar before and after the hydrolysis process measured using a spectrophotometer at a wavelength of 540 nm [10]. The content of reducing sugar from the best variable of hydrolysis process was tested by HPLC type Agilent 1100 Series equipped with Agilent 1260 Refractive Index Detector, positive polarity at 35 °C and column of Agilent Zorbax Carbohydrate 4.6 x 150 mm, 5 μm at 45 °C.

The XRD method was used to find the value of Crystallinity Index of lignocellulosic material. XRD will read the Crystallinity Index value of cellulose crystals in the solid sample. The intensity of amorphous cellulose can be seen at a 2θ angle of approximately 18 ° and crystalline cellulose at 22-24 °. The value of Crystallinity Index (CrI) was calculated using the Eq. (1):

$$CrI = \frac{I_{002} - I_{am}}{I_{002}} \times 100\% \quad (1)$$

Yield of reducing sugar was determined using the Eq. (2):

$$Reducing\ Sugar\ Yield = \frac{Reducing\ sugar\ conc. \times total\ volume\ of\ hydrolysate}{(\%cellulose + \%hemicellulose) \times weight\ of\ susbtrate} \quad (2)$$

3. Results and discussion

3.1. The effect of mixed culture pretreatment on lignocellulosic content of CPW

Pretreatment of the CPW was pretreated by mixed culture and single culture as the control. The underlying reason of the selection of such microorganisms was because of brown-rot fungi types such as *Aspergillus niger* usually can degrade cellulose and soft-rot fungi such as *Trichoderma reesei* can degrade cellulose and lignin [11]. Besides, it is also a fungi with the most cellulose degradation activity when compared to other cellulosic fungi [5]. Pretreatment was needed because it could damage the lignin structure so then cellulose can be easily degraded during hydrolysis processs [12].

Table 1. The cellulose content before and after pretreatment using AN:TR and BS:TR

Variable	*Lignin* (%)	Pectin (%)	*Cellulose* (%)	Hemicellulose (%)
BS:TR	1.96	0.98	65.82	24.7
AN:TR	2.05	1.04	67.05	26.18
Initial coffee pulp	6.82	2.18	58.62	21.96

Table 1 shows that there is a decrease of lignin content on pretreated sample of using mixed culture of BS:TR and AN:TR where lignin content of initial coffee pulp is 6.82 % and after pretreatment becomes 1.96 % and 2.05 % for BS:TR and AN:TR, respectively. The pectin content also decreased after the pretreatment process, where the initial pectin content was 2.18 % and after pretreatment became 1.04 % for AN:TR and 0.98 % for BS:TR. The result was also followed by the increase of cellulose and hemicellulose content. After pretreatment, it increased become 67.05 % for AN:TR and 65.82 % for BS:TR. Hemicellulose content after pretreatment increased to 26.18 % and 24.5 % for AN:TR and BS:TR, respectively. Decreasing levels of lignin and increasing of cellulose and hemicellulose levels indicate that the microbial used in this process have damaged the lignin layer and made cellulose and hemicellulose content increasing [12]. From the Table 1, the decreasing of lignin content in pretreatment by AN:TR is higher than the BS:TR. This result was in accordance with previous research conducted by [5] which reported that *Aspergillus niger* effectively degrade lignin and pectin while *Trichoderma reesei* effectively degrade lignin, so this mixed culture has degraded more lignin. Cellulose consists of D-glucopyranose and is bonded with the β-1,4-glycosidic bond. There is a hydroxyl bond in the cellulose macromolecule comprising of several intra- and inter-molecules hydrogen bond resulting in various cellulosic crystalline structure [13]. Since the cellulosic structure is so complex, it was necessary to find out how the crystalline structure of the substrate used by X-Ray Diffraction analysis. The crystalline structure of cellulose is imperfect, most of the cellulose structures that are located in the outer layer (less-ordered) are called amorphous, while the remainder which is located in the inside layer of cellulse (ordered) is called crystalline. The amorphous in the crystalline structure is the easier part of the cellulose substrate that can be degraded by enzymes [13]

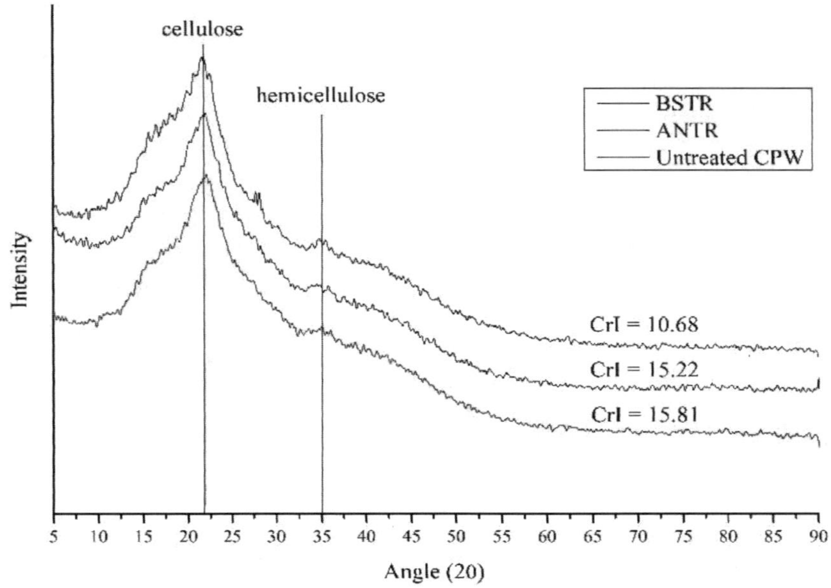

Figure 1. Graph of X-Ray Diffraction of CPW Before and After Pre-treatment using BS:TR and AN:TR

Figure 1 shows the results of XRD analysis for CPW before and after pretreatment. Sample peak sample before pre-treatment and after pre-treatment are shown in Figure 1 where the peak diffraction sample of AN:TR (1:1) at 22.045° and BS:TR (2:1) at 22.033°. Previous research stated that the diffraction peak at $2\theta = 22°$ was the characteristic of cellulose crystalline I. After biological pretreatment, X-ray diffraction analysis with diffraction plane (021) and (002) has increased by the intensity. The result shown a general characteristic of the amorphous form of cellulose which is often found after pretreatment by dissolution and regeneration, where the area of the reflector plane (021) and (002) has increased [14]. From the Figure 1, at unpretreated CPW does not show any diffraction peak at $2\theta = 35^o$ while the pretreatment sample does show the diffraction peak of $2\theta = 35.07^o$. In the hemicellulosic structure, the diffraction peak at $2\theta = 35^o$[15]. This result indicates that the hemicellulose content of the pre-treatment sample has increased. The increase of CrI value showed a loss of amorphous compound from the sample, the cellulose structure has been opened.Coffee pulp waste was obtained from Bangelan, East Java Indonesia. It was dried under the sunlight for approximately 7 days. Then the CPW was mashed using a grinding machine and sieved to get 120 meshes. The smaller substrate particles help the hydrolysis process to make the contact surface between the substrate and the enzymes larger and the lignocellulose content will be easier to be degraded. Before used, the CPW was dried using the oven at 105 ^{o}C to reduce moisture content.

3.2. The effect of enzyme and surfactant mixture towards enzymatic hydrolysis process
Enzymatic hydrolysis of coffee pulp aims to convert cellulose and hemicellulose into reducing sugars using a pure cellulase and xylanase enzyme. PEG 4000 was used to enhance the total reducing sugar after hydrolysis of lignocellulosic biomass [16], while Tween 80 was chosen due to the ability to influence on xylanase adsorption and desorption on/from lignin and occupies a portion of lignin hydrophobic surface [17]. The surfactant mass was designed based on research by [18] which states that non-ionic surfactants which added to the hydrolysis process will produce significant results, but if it is over than five grams, it will not have a significant effect due to the saturated solubility of the surfactant.

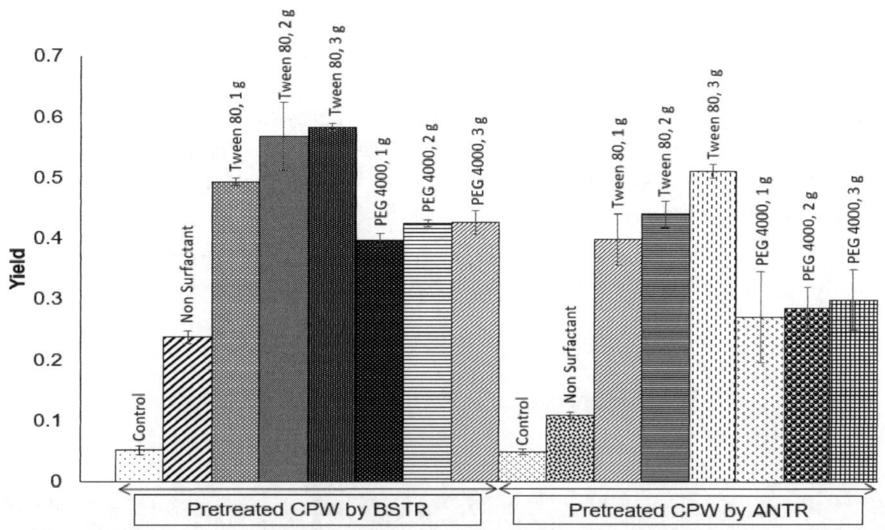

Figure 2. yield of reducing sugar from enzymatic hydrolysis

From the Figure 2, the higher mass of surfactants which has introduced into the sample would give the higher concentrations and yield of reducing sugars. This suggests that surfactants have a positive effect on enzymatic hydrolysis. Compared with non-hydrolysed samples, the other variables has significantly different of reducing sugar content in both with addition of surfactant and non-surfactant. The effects of surfactant addition have shown significant reducing sugar yield especially in the sample resulted from BS:TR pretretment with addition of Tween 80. This indicated that Tween 80 was a better surfactant than PEG 4000. Tween 80 shown the significant result than PEG 4000 because Tween 80 will increase the solubility of enzymes [19] and reduce the enzyme's absorption, so the enzyme can optimally work by encircling the substrate at first. Besides, addition of Tween 80 to the hydrolysis solutioncan increase the temperature by more than 10 °C due to the non-ionic bond formed. This condition will cause the enzyme to work more optimally, but this was not good for the culture that lived at optimum temperature of 35 °C [20].

In prereated CPW by BS:TR, the increase of reducing sugar yield in flask with non-surfactant and addition of 3 g of Tween 80 and 3 g of PEG 4000 was 52.535 %, 91,188 %, and 26.963 %. In the pretreated CPW by AN:TR, the increase of reducing sugar yield in flask with non-surfactant, addition 3 g of Tween 80, and 3 g of PEG 4000 was 49.499 %, 91.814 %, and 41.508 %, respectively.

From the explanation, it could be stated that the mixture of pure enzyme and surfactant might hydrolyze cellulose and hemicellulose in the pretreated CPW to be reducing sugar. The combination of a pure cellulase and pure xylanase enzyme could maximize the process of hydrolysis of the pretreated CPW, where the cellulose was hydrolyzed into glucose-reducing sugar by pure cellulase and hemicellulose was hydrolyzed by the xylanase into the xylose-reducing sugar.

3.3. The effect of microorganisms and surfactant mixture towards biologically hydrolysis process

The biologically hydrolysis has some benefits which are environmentally friendly, low energy requirement does not require chemicals and can avoid sugar degradation, resulting in higher yield [21]. In the present study, hydrolysis process was used *Aspergillus niger* and *Trichoderma viride* with mixture ratio of 1:1; 1:2 and 2:1. The use of *Aspergillus niger* and *Trichoderma viride* refers to previous studies which suggest that *Aspergillus niger* and *Trichoderma viride* are more potentially enzymatic cellulase than other microorganisms [22]. According to research by [23], *Aspergillus niger* is the most suitable microorganism for cultivation on the coffee pulp, where the microorganism can produce pectinase enzymes. *Aspergillus niger* is also known to produce cellulase enzyme with high specific activity [24]. In this study, the use of a mixture of *Aspergillus niger* and *Trichoderma viride* was based on these microorganisms can yield higher hydrolysis yields when compared to their single culture. The effective

addition of *Trichoderma viride* will produce the β-glucosidase enzyme that help *Aspergillus niger* to hydrolyze cellobiose which is an inhibitor of cellulase activity [11].
The biological hydrolysis will produce fungal cellulase using cellulose contained in the substrate as the carbon source [28]. Cellulosic microorganisms can build a synergistic relationship with non-cellulosic species in cellulosic waste, to make the synergistical interaction in cellulose degradation. Generally, endoglucanase and cellobiohydrolase will work synergistically to hydrolyze cellulose and biological hydrolysis will also produce microbial hemicellulase [24].

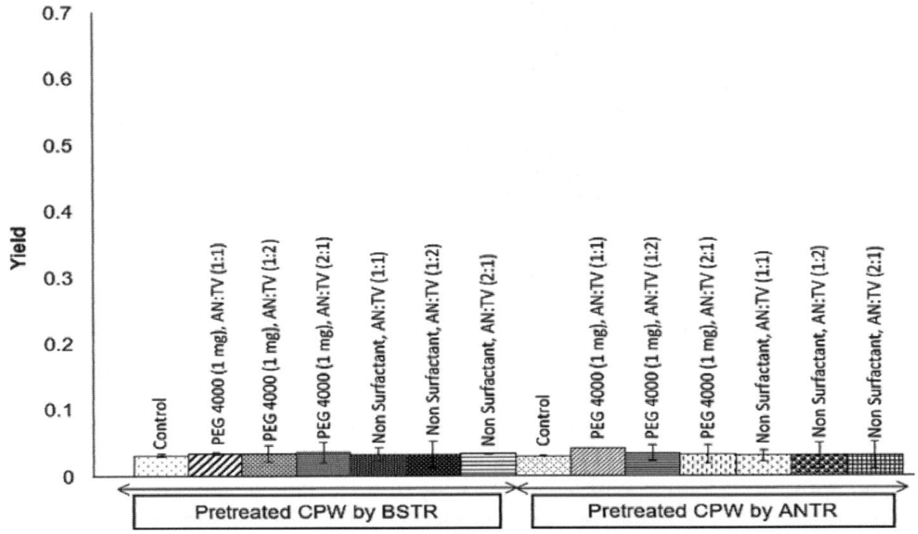

Figure 3. yield of reducing sugar from biological hydrolysis

Figure 3 shows the reducing sugar yield from biological hydrolysis. The yield has higher than the initial yield of pretreated CPW by BS:TR without hydrolysis which amounted to 0.031 (g/g) and 0.029 (g/g) for the pretreated CPW by AN:TR The hydrolysis by AN:TV in 2:1 ratio was resulted the highest reducing sugar yield from pretreated CPW by both of BS:TR (2:1) and AN:TR (1:1) which were 0.033 (g/g) and 0.031 (g/g), respectively. This was consistent with previous research that stated *Aspergillus niger* is the fungi with the highest specific activity to produce cellulase enzyme [24] since the amount of *Aspergillus niger* was higher in the culture, it would result in higher yield of cellulose. The obtained reducing sugar yields from pretreated CPW by AN:TR and hydrolyzed by AN:TV with ratio of 1:1, 1:2, and 2:1 without surfactant addition were 0.0302, 0.03 and 0.0306 (g/g), respectively. While, the reducing sugar yields with addition of surfactants shown a succession of 0.042, 0.034 and 0.032 (g/g) or increase to 39.07 %, 13.33 %, and 4.57 %, respectively. In the pretreated CPW by BS:TR and hydrolyzed by AN:TV with ratio of 1:1, 1:2, and 2:1 without surfactant were 0.032, 0.032 and 0.033 (g/g), respectively. When they were used the surfactant, reducing sugar yield has increased consecutively to 0.034, 0.034 and 0.035 (g/g) or increased 6.25 %, 6.25 %, and 6.06 %. These results were consistent with previous studies which suggest that the use of surfactants may increase delignification and increase the yield by fungal enzyme hydrolysis [25]. The mechanisms of surfactants that can explain their effect on enzymes are: (1) changing the substrate structure becomes more easily degraded by enzymes, (2) stabilizing the enzyme to avoid denaturation [26]. (3) increase the positive interaction between substrate and enzyme [20] and (4) decrease the number of unsupervised enzymes binding to lignin and other molecules involved in cellulase enzyme activity [27].

4. Conclusion

Production of reducing sugar from the coffee pulp waste using biological pretreatment process on a mixture culture between *Aspergillus niger*, *Bacillus subtilis*, and *Trichoderma reesei* followed by hydrolysis process using a mixture of pure cellulase and xylanase enzyme and using verious mixture ratio of *Aspergillus niger* and *Trichoderma viride* was succesfully conducted. The optimum condition was obtained in biological pre-treatment process which the decrease of lignin and pectin content using BS:TR was higher than AN:TR In hydrolysis process, the optimum condition was obtained from pretreated CPW by BSTR with enzymatic hydrolysis and addition of Tween 80 of 3 g which produced yield of reducing sugar was 0.5831 %. Addition of surfactants have shown the significant reducing sugar yield, in case Tween 80 was better than PEG 4000. The increase of reducing sugar yield by addition of 3 g of Tween 80 could reach 91,188 % and 91.814 % in the pretreated CPW by BS:TR and AN:TR, respectively.

Acknowledgements
The writer would like to express gratitude towards the Research and Community Service Foundation of Institut Teknologi Sepuluh Nopember (LPPM-ITS), the financial support from Local Research Fund of ITS with contract number of 133/PKS/ITS/2018.

References
[1] Directorate General of Plantation 2015 Indonesia's Coffee Plantation Statistics
[2] Widjaja T, Iswanto T, Altway A, Shovitri M and Juliastuti S R 2017 *Chem. Eng. Trans.* **56** 1465–70
[3] Corro G, Paniagua L, Pal U, Bañuelos F and Rosas 2013 *Energy Convers. Manag.* **74** 471-81
[4] Ayeni A O, Adeeyo O A, Oresegun O M and Oladimeji T E 2015 *American J. of Eng. Research* **4** 14-19
[5] Shahzadi T, Mehmood S, Irshad M, Anwar Z, Afroz A, Zeeshan N, Rashid U and Sughra K 2014 *Adv. Biosci. Biotech.* **5** 246–51.
[6] Mazzafera P 2002 *Sci. Agri.* **59** 815–21
[7] Sarita C 2013 *Chem. Eng. Trans.* **32** 949–54
[8] Li J, Li S, Fan C, and Yan Z 2012 *Colloids Surf. B. Bio.* **89** 203–10
[9] Datta R 1981 *Bio. Tech.* **23** 2167-70
[10] Miller G L 1959 *Anal. Chem.* **31** 426-28
[11] Sun Y and Cheng J 2002 *Bio. Tech.* **83** 1–11
[12] Mood S H, Hossein G A, Tabatabaei M., Salehi J G, Najafi G H, Gholami M and Ardjmand M, 2013 *Ren. Sust. Energy Reviews* **27** 77–93
[13] Park S, Baker J O, Himmel M E, Parilla P A and Johnson D K 2010 *Biotech.Bio.* 1–10
[14] Xiao W, Yin W, Xia S and Ma P 2012 *Carb. Poly.* **87** 2019–23
[15] Hamada Y, Yoshida K, Asai R, Hayase S, Nokami T, Izumi S and Itoh T 2013 *Green Chem.* **15** 1863
[16] Börjesson J, Engqvist M, Sipos B and Therneld F 2007 *Enzy. Microb. Tech.* **41** 186-95
[17] Zhang K, Pei Z and Wang D 2016 *Bio. Tech.* **199** 21–33
[18] Zhou Y, Chen H, Qi F, Zhao X and Liu D 2015 *Bio. Tech.* **182** 136–43
[19] Li Y, Sun Z, Ge X and Zhang J 2016 *Biotech. Biofuels* **9** 20
[20] Eriksson T, Börjesson J and Tjerneld F 2002 *Enzy. Microb. Tech.* **31** 353–64
[21] Mussatto S I and Teixeira J 2010 *Méndez-Vilas A* **2** 897-907
[22] Ul-Haq I, Javed M M, Khan T S and Siddiq Z 2005 *J. Agri. Bio. Sci.* **1** 241–45
[23] Pandey A, Soccol C R, Nigam P and Brand D 2000 *Bio. Eng.* **6** 153–62
[24] Howard R L, Abotsi E, Jansen V R E L and Howard S 2003 *A. J. Bio.* **2** 602-19
[25] Cao S and Aita G M 2013 *Bio. Tech.* **131** 357-64
[26] Kaar W E and Holtzapple M T 1998 *Biotech. Bioeng.* **59** 419–27
[27] Qing Q, Yang B and Wyman C E 2010 *Bio. Tech* **101** 5941–51
[28] Shin C S, Lee J P, Lee J S and Park S C 2000 *App. Biochem. Biotech.* **84-86** 273-245

Evaluation of Glutathione Production Method using *Saccharomyces cerevisiae*

MTAP Kresnowati[1,2], NA Ikhsan[2], RS Nursa'adah[2], NN Santoso[2], YW Susanto[2]

[1.] Food Engineering Department, Faculty of Industrial Technology, Institut Teknologi Bandung, Jatinangor campus, Jalan Raya Jatinangor, Sumedang, 45363, Indonesia
[2.] Chemical Engineering Department, Faculty of Industrial Technology, Institut Teknologi Bandung, Ganesha campus, Jalan Ganesha 10 Bandung 40132, Indonesia

E-mail: kresnowati@che.itb.ac.id

Abstract. Industrial interest of glutathione, as a pure substance or accumulated in yeast, has been prominent as a result of various applications of glutathione in food and pharmaceutical industries. Glutathione production by fermentation process is easier and more economical compare to production by chemical and enzymatic reaction. This study aims to evaluate environmental condition to the fermentation process producing glutathione in three different strategies. In the following work, stress conditions and cysteine addition coupled with fed-batch fermentation of *Saccharomyces cerevisiae* ITBCC R58 for glutathione accumulation have been investigated. Stress conditions applied on this study included temperature shift from 30 to 50°C and 27°C, pH shift from 5 to 1.2 and 8.8, and osmotic stress by addition NaCl to the solution. Another strategy was to add cysteine as one of glutathione precursors into the fermentation medium. Later, the oxidative state of glutathione harvested was also checked. Osmotic stress showed the best result amongst stress variations applied on this work, however combined cysteine addition with fed-batch fermentation stood out as the best strategy in this study. In all experiments, oxidized-state glutathione was identified, indicating the need of a specific method development to harvest glutathione in reduced state.

1. Introduction

A current expansion of yeast global market has been expected to grow 7.1% between 2016 to 2022 [1] implying that varieties of yeast-derived products are of great interests. Yeast autolysates are commonly used in wide range of application from laboratory uses as an excellent medium for growth of microorganisms to flavor enhancers. Yeast extract was also reported to generate unique flavors, known as umami and kokumi-taste affected by different γ-glutamylcysteine derivatives such as glutathione, a tripeptide γ-glutamyl-L-cysteineglycine [2]. On the other hand, intact yeast cells are widely used as fermentation starter in food industries, for example bread. Glutathione-enriched intact yeast cells are also considered to be able to improve dough extensibility and decrease mixing time in baking [3]. Further, yeast extract containing oxidized form of glutathione has been proven to enhance the gas-retaining capacity of gluten-free rice batter [4]. Glutathione-enriched yeast is priced 4 $/kg which doubles the normal yeast price. As an intracellular product, glutathione in glutathione-enriched yeast benefits from its no-need of purification, hence being very convenient and easy to manufacture.

Pure glutathione finds applications in pharmaceutical and cosmetic industry as a skin whitening agent [5] and scavenger for free radicals and heavy metals thus required in good purity and resulting in the high pricing of glutathione powder. Inside the cell glutathione serves to prevent cell damages caused by reactive oxygen species (ROS) [6]. Other roles of glutathione inside cells range from protecting protein during stress condition [7] to detoxifying heavy metals by generating glutathione conjugates [8]. Intracellular glutathione content in wild-type yeast is varied between 0.1 and 1%-w/w with a ratio of reduced form (glutathione) to oxidized form (GSSG) of 30-100:1 [9].

Industrial interests for glutathione served as motivation for a number of studies aiming to increase glutathione production by yeast cells. Using strain improvement method, it has recently been possible to generate yeast strains containing up to 3.9%-w/w of intracellular glutathione content [10]. Yet, despite promising scientific progresses, target-engineered strains are still hardly accepted in food industry due to authority issues [11]. Other proposed methods are controlling the microenvironment of the yeast cultivation, for example by applying stress condition such as sudden pH change [12], temperature drop [13], osmotic stress [14], magnetic fields [15], high-pressure [16] and H2O2 stress [17], addition of amino acid precursor [18], optimization of growth medium, as well as fed-batch mode to achieve high cell density [19]. In order to get the oxidized form of glutathione, reacting the reduced form of glutathione with oxidizers such as potassium bromate, benzoil peroxide and hydrogen peroxide has been suggested [20].

This study aims at evaluating the potential of local yeast, Saccharomyces cerevisiae ITBCC R58, in producing glutathione. Three different strategies to enhance glutathione production such as applying stress conditions (pH, temperature, or osmotic stress), addition of precursor, and implementing fed batch fermentation configuration were tested. Further the specific state of produced glutathione (whether the reduced or the oxidized state of glutathione) was also analysed.

2. Methods

2.1. Microorganism and Cultivation

Yeast Saccharomyces cerevisiae ITBCC R58 was obtained from the culture collection of Laboratory of Microbiology and Bioprocess Technology, Department of Chemical Engineering ITB. The inoculum was prepared by culturing the yeast at 100 mL YPD media containing glucose 20 g/l, peptone 10 g/l, and yeast extract 5 g/l at 30°C and pH of 5.0 for 16 h on a 150-rpm shaking incubator. The culture was then inoculated into the main fermentation broth. Unless stated otherwise, the medium for glutathione fermentation in batch cultures contained glucose 50 g/l, yeast extract 15 g/l, peptone 5 g/l; salt solution: $(NH_4)_2HPO_4$ 10 g/l, $MgSO_4$ 0,5 g/l; trace element solution: $ZnSO_4$ 10 mg/l, $FeSO_4$ 12 mg/l, $CuSO_4$ 12 mg/l, and $MnSO_4$ 10 mg/l; and vitamin solution: biotin 0,01 mg/L, pantothenic acid calcium salt 0,02 mg/L, and pyridoxine hydrochloride 0,2 mg/l. Main fermentation was carried out in 3 l stirred bioreactor (New Brunswick BioFlo 110) with working volume of 2 l at 30°C and pH of 5.0 using an aeration rate of 1.25 vvm and an agitation rate of 150 rpm. Right after reaching stationary phase, fermentation broth was split into several 250-ml Erlenmeyers with 200-ml working volume in which the three different strategies were applied. This aimed to generate a uniform starting condition for all variations.

2.2. Stress Condition, Cysteine Addition and Fed-Batch Fermentation

Glutathione fermentations were held in 8 different conditions in duplicate. The variations were as follows: low-pH stress from 5.0 to 1.2 by addition of HCl 5 M, high-pH stress from 5.0 to 8.8 by addition of KOH 5 M, low-temperature from 30 to ambient (27°C), high-temperature from 30 to 50°C, osmotic stress by addition of NaCl salt (80 g/l), cysteine addition by adding L- cysteiniumchloride (8 mmol), fed-batch by addition of glucose (50 g/l), and combined fed-batch with cysteine addition by adding glucose (50 g/l) and L- cysteiniumchloride (8 mmol). All variations were applied right after system reached stationary phase and subsequently held for 3 h in which sample was taken every hour.

2.3. Analysis

Biomass concentration was monitored by turbidimetry using spectrophotometer UV/VIS at 560 nm and gravimetry. Glucose and glutathione concentration were measured by using HPLC. For glucose analysis, Aminex HPX-87H column and RID detector were applied, using H_2SO_4 5 mM as the mobile phase. For glutathione analysis, intracellular glutathione was first extracted following the works of [21]. Analysis was conducted by applying μBondapak C18 column and UV detector, using NaH_2PO_4 25 mM as the mobile phase. In order to differentiate the specific state of glutathione, an HPLC analysis method modification of [22] was applied. The same HPLC system was applied, however, the mobile phase used was mixture of water, methanol (2,5 v/v), and phosphoric acid 25 mM, buffered at pH 3.

The measured glutathione concentration in the extract was used to calculate the overall glutathione concentration in the fermentation broth (C_{glu} [=g-glutathione/l]) as well as the intracellular glutathione content ($C_{glu,in}$ [= g-glutathione/g-cell Dry Weight]). The glutathione yield on glucose as the substrate ($Y_{S/glu}$ [= g-glutathione/g-glucose]) and the cell yield on glucose as the substrate ($Y_{S/X}$ [= g-cell Dry Weight/g-glucose]) was calculated following eq 1 and 2, correspondingly.

$$Y_{S/glu} = \frac{C_{glu,t} - C_{glu,0}}{C_{S,0} - C_{S,t}} \quad (1)$$

$$Y_{S/X} = \frac{C_{X,t} - C_{X,0}}{C_{S,0} - C_{S,t}} \quad (2)$$

In which C_S and C_X is the glucose substrate concentration [=g-glucose/l] and the cell concentration [=g-cell Dry Weight/l] in the fermentation broth, whereas the subscripts t and 0 indicates the timing of the samples, final and initial, respectively.

3. Results and Discussion

3.1. Glutathione Production on Different Stress Conditions

Effects of five stress conditions applied on fermentations are shown in Figure 1. The stresses were applied right after system reached stationary phase, and thereby were slightly different for different batches. There was observed a trend of decrease in biomass concentration (Figure1(a-b)) for all variations after cells were exposed to stress conditions. This phenomenon was taken as an indirect marker of stress conditions undergone by cells since for unicellular microorganisms, stress is defined as any conditions or states that could lead to the decrease in cell growth [23].

Figure 1 (c) and (d) show that the applied stress conditions positively affected glutathione production, except for the variations of low pH stress and high temperature stress. High-pH, osmotic, and low-temperature stress enhanced glutathione production and resulted in intracellular glutathione contents which were 56.3, 52.4, and 26.5% higher than initial condition, respectively. However, only the osmotic stress resulted in an increase value of glutathione yield on glucose, from 0.0011 to 0.0023 g glutathione/g glucose while the others did not noticeably improve the glutathione yield. Osmotic stress seems to be a good strategy to enhance glutathione production which in this study, resulted in 1.284%-w/w of final intracellular glutathione content (Table 1).

Glutathione is involved in oxidative stress response and is responsible to fight reactive oxygen species (ROS) by transfer electron via hydrogen for systems which are dependent on thiol such as peroxidase [24]. Even though osmotic stress does not fully associate with oxidative stress response, there has been a study finding that a fraction of oxidative stress response shared the same components with osmotic stress [25].

The high-pH variation resulted in an increase of intracellular glutathione content (Figure1(d)) but not in the glutathione yield (Table 1). This implies that high-pH favors the glutathione accumulation inside yeast cells yet it is not efficient to convert substrate to glutathione. This is probably a result of the extreme condition that cells underwent in high-pH environment that reduced a great number of intact cells. A study performed by [26] showed an increase of glutathione production in *Candida utilis* by 42% when pH was brought up from 5 to 6. In the opposite, a decrease of glutathione production was recognized when pH was shifted from 5 to 4. The reason behinds the enhancement of glutathione production in high-pH condition was explained from a research by [27]. Yeast cells exposed to alkaline stress accumulate ROS in relatively short time. This ROS accumulation induces the expressions of genes involved in oxidative stress such as glutathioneI and glutathioneII that are responsible in catalyzing the biosynthesis of glutathione. Another finding of this current study is the reduction of pH resulted in a slight decrease of glutathione production (Figure1(d)). On the contrary, Nie, et al. [12] reported an increased glutathione production was observed from the fermentation of *C. utilis* with low-pH stress (pH of 1.2). However, almost all intracellular glutathione was extracted out of cells for inexplicable reasons. Regrettably, this phenomenon could not be observed in this current study since we did not analyze extracellular glutathione due to operational limitations.

A slight temperature shift from 30 to 27°C in this study showed an increase in glutathione content (Figure1(d)). This result is in accordance with [13] that applied similar strategy on *Candida utilis*. They found a temperature shift from 30 to 26°C increase glutathione content up to 19.04%. Similar to yeast stress response on alkaline stress, growing yeast cells at a suboptimal temperature also raises the intracellular ROS and promotes an oxidative response involving glutathione [28]. Although heat shock is expected to trigger stress response involving glutathione, we found that high temperature did not seem to be good strategies to increase glutathione accumulation (Figure 1(d)).

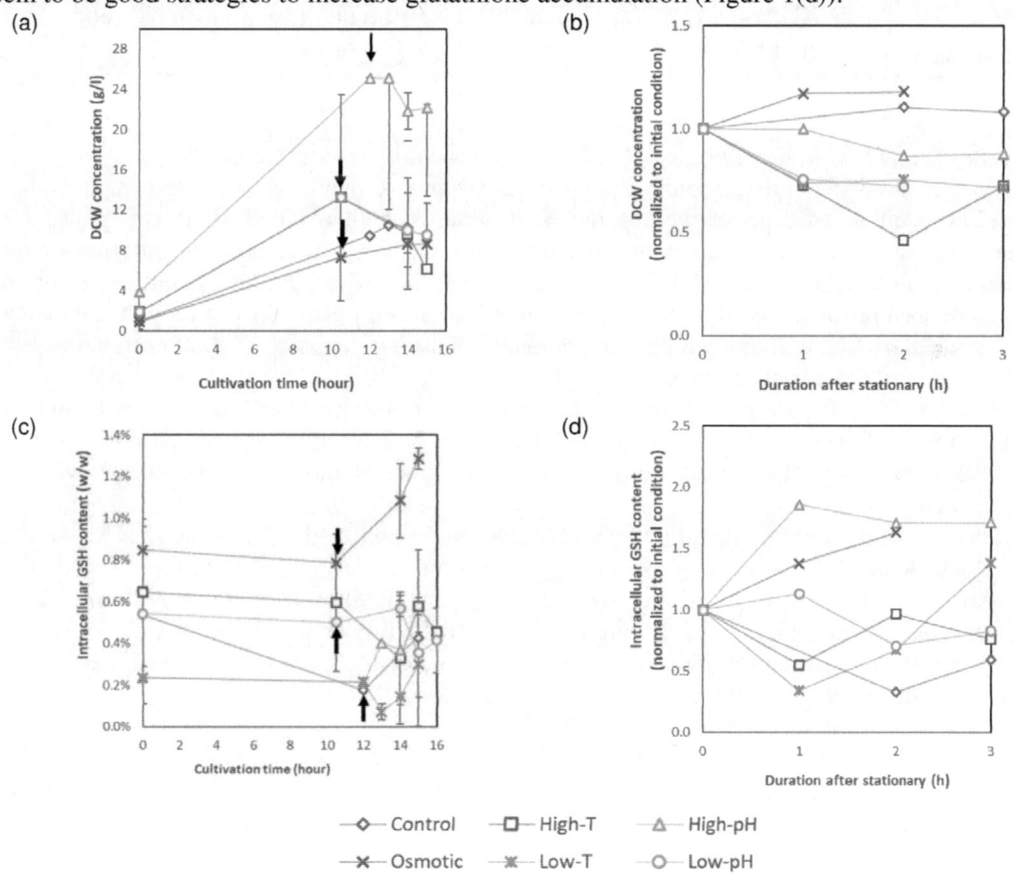

Figure 1 Effects of stress conditions on glutathione production on (a) Cell concentration, (b) Cell concentration relative to initial condition, (c) Intracellular glutathione content (d) Intracellular

IOP Conf. Series: Materials Science and Engineering **543** (2019) 012004 doi:10.1088/1757-899X/543/1/012004

glutathione content relative to initial condition. The black arrows indicate the time when the stress was applied, data are mean values ± SD from duplicate experiments.

Table 1 Overview of batch fermentations with various strategies to improve glutathione content

Variation	Final glutathione concentration [g/l]	Final intracellular glutathione content [%w/w]	Glutathione yield on glucose [g-glutathione/ g-glucose]	Cell yield on glucose [g-cell DW/ g-glucose]
Control	0.0528	0.426	0.0011	0.3675
Low-pH	0.0693	0.370	0.0015	0.4628
High-pH	0.0397	0.415	0.0007	0.1605
High-temperature	0.0337	0.455	0.0010	0.1950
Low-temperature	0.0572	0.299	0.0012	0.3475
Osmotic	0.1112	1.284	0.0023	0.1724
Fed-batch	0.1160	0.912	0.0013	0.1425
Cysteine	0.0781	0.984	0.0080	0.1688
Fedbatch +cysteine	0.1511	1.695	0.0017	0.0984

3.2. Effects of Cysteine Addition and Fed-Batch Mode on Glutathione Fermentation

Effects of cysteine addition and fed-batch mode on glutathione fermentations are shown in Table 1 and Figure 2. Cysteine and glucose were added right after system reached stationary phase, and thereby were slightly different for different batches. Figure 2 shows that cell concentration of fed-batch variation was elevated from 10.73 (control) to 12.72 g DCW/l while the cysteine addition did not affect cell concentration. According to [18], cysteine inhibits cell growth even though it enhances glutathione production. Spike timing which is at stationary phase was appropriately chosen to deter the negative effect of cysteine on cell growth.

Not only did the fed-batch mode improved cell concentration, intracellular glutathione content was also increased up to 16.03%. This overall improvement may be caused by the increase of chemical energy (ATP) supplied by glucose addition that could support glutathione biosynthesis which is dependent on ATP.

An elevation of glutathione production is also observed in cysteine addition strategy. The addition of cysteine alone elevated intracellular glutathione content up to 112.9% higher than initial condition. This result is in harmony with many former studies of cysteine addition in glutathione fermentation. Wei et al. reported an increase of glutathione content 106.5% higher than control due to cysteine addition [18]. A refined explanation of this result is not well understood, yet it is believed that cysteine addition enhances the conversion of γ-glutamylcysteine reaction.

Based on the above results, combined fed-batch and cysteine addition is strongly believed to boost glutathione production way more effective than their individual effect. In this study, this variation resulted in an increase of glutathione production and showed the best result over other variations. glutathione intracellular content is elevated from 0.137% to 1.695%-w/w. However, there is no significant increase in cell concentration although fed-batch fermentation had been applied. This could become some improvements in future studies to improve the method of fed-batch fermentation and timing for cysteine addition to achieve better high cell density as well as the high content of glutathione. Lorentz, et al. [3] successfully performed fed-batch fermentation with RQ control and periodic cysteine spike resulting an increase of intracellular glutathione content from 0.42 (control) to 1.74%-w/w.

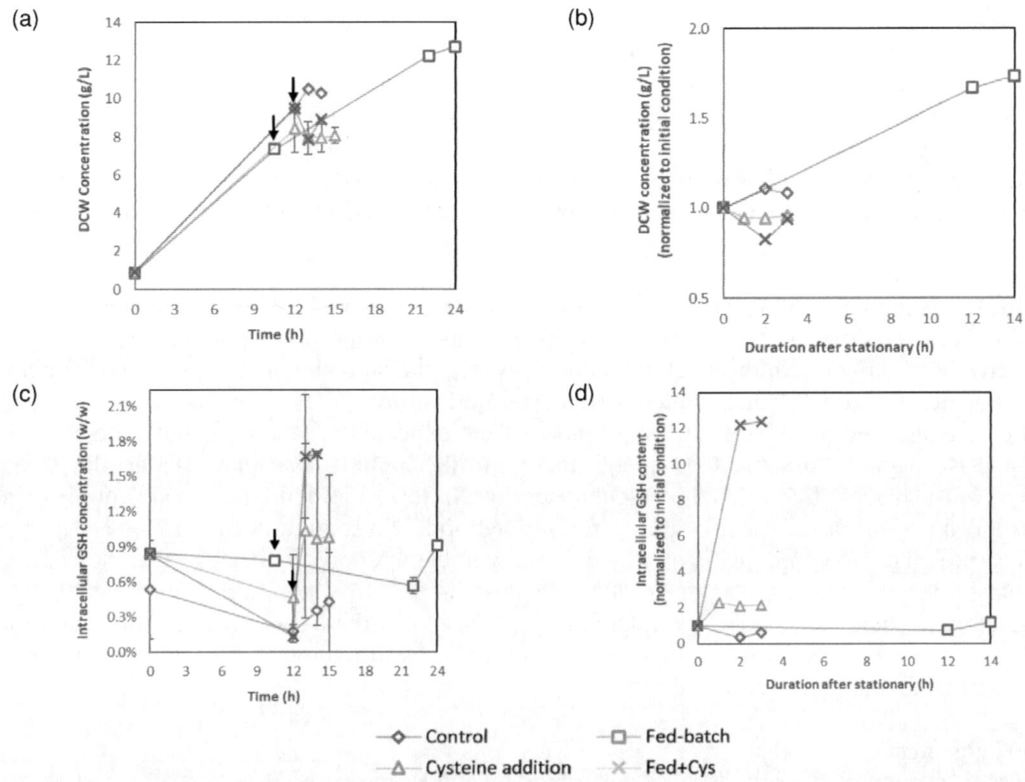

Figure 2 Effects of cysteine addition, fed-batch, and combined cysteine with fed-batch on a) Cell concentration, b) Cell concentration relative to initial condition, c) Intracellular glutathione content, and d) Intracellular glutathione content relative to initial condition. The black arrows indicate the time when the cysteine or glucose was added, data are mean values ± SD from duplicate experiments.

3.3. Glutathione Yield on Substrate at Different Variations

Besides intracellular glutathione content, the glutathione yield on substrate (glucose) is also an important parameter for evaluating performances of each variation in the effort to enhance glutathione production. According to glutathione yield for all variations shown in Table 1, cysteine addition stood out as the best strategy to enhance glucose conversion into glutathione. The yield of control that is only 0.0011 g glutathione/g glucose was brought up to 0.0080 g glutathione/g glucose when cysteine addition was applied. In addition, amongst stress variations, osmotic stress showed a noticeable increase in glutathione yield reaching 0.0023 g glutathione/ g glucose. When the expected outcome is pure glutathione, a strategy that enhances glutathione yield is more favorable since it will improve the fermentation efficiency. However, if the expected product is glutathione-enriched yeast, a variation that improves intracellular glutathione content is more relevant.

3.4. Oxidative State of Glutathione

As aforementioned above, glutathione has two different oxidative states, reduced glutathione (GSH) and oxidized glutathione (GSSG). The reduced state glutathione serves as antioxidant and will transform into the oxidized state after fighting against ROS. However, due to the recycle system in cells, GSSG will transform into glutathione again to maintain a normal ratio between GSH and GSSG which is within a range of 30-100:1. An analysis to determine oxidative state of glutathione has been performed. Only GSSG was recognized in the fermentation product while GSH was completely unidentified. Since there is no oxidizing agent was introduced into the fermentation product that contained glutathione, it is assumed that glutathione was oxidized automatically into GSSG in the presence of oxygen in air. Due to the handling of glutathione during experimentations from glutathione extraction to its analysis, it is highly possible that GSSG is formed from glutathione. Therefore, a better harvesting system is needed to prevent glutathione from getting oxidized.

4. Conclusion

Various stress conditions, cysteine addition, and fed-batch strategy have been applied on fermentation of Saccharomyces cerevisiae in this study. Except for the high pH and high temperature stresses, all strategies have been shown to improve glutathione concentration. The obtained results shed a light of effective strategies for glutathione production since the industrial interest of glutathione, as a pure substance or accumulated in yeast, is still prominent. If the expected product is glutathione-enriched yeast cells, it is favorable to apply the strategy that yield the highest final intracellular glutathione content or give the highest final glutathione concentration. In this case, fed batch mode and cysteine addition stood up as the best strategy, giving final glutathione concentration of 0.151 g/l (1.695%-w/w). If the pure glutathione product is expected, it is favorable to apply the strategy that increase the overall yield of glutathione on substrate. In this case, cysteine addition stood up as the best strategy, giving glutathione yield of 0.0080 g-glutathione/g-glucose. Further, since genetically modified organisms are barely accepted in food industries, uses of a non-modified food grade strain in this study show the potential in glutathione production.

5. Acknowledgement

This research is supported by Laboratory of Microbiology and Bioprocess Technology, Department of Chemical Engineering ITB and Research and Higher Education (Ristekdikti) by Program of Students' Creativity (Program Kreativitas Mahasiswa).

6. References

[1] McWilliams A 2017 Yeasts Yeast Extracts Autolysates and Related Products: The Global Market Wellesley: BCC Research Available online at: wwwbccresearchcom/market-research/chemicals/yeast-yeast-extracts-autolysates-products-chm053chtml Accessed on 7 February 2019

[2] Ueda Y, Yonemitsu M, Tsubuku T, Akako S, Akaguchi M and Miyajima R 1997 Flavor characteristics of glutathione in raw and cooked foodstuffs *Biosci. Biotechnol. Biochem.* **61** 1977-80

[3] Lorentz E, Schmacht M, Stahl U and Senz M 2015 Enhanced incorporation yield of cysteine for glutathione overproduction by fed-batch fermentation of Saccharomyces cerevisiae *J. Biotechnol.* **216** 131–9

[4] Yano H and Fukui A 2015 Glutathione changes physical properties of rice batter without increasing its allergenicity *Report Nat. Food Res. Inst.* **79** 9-17

[5] Sonthalia S, Daulatabad D and Sarkar R 2016 Glutathione as a skin whitening agent: Facts myths evidence and controversies *Indian J. Dermatol. Venereol. Leprol.* **82** 262-72

[6] Jamieson D J 1998 Oxidative stress responses of the yeast Saccharomyces cerevisiae *Yeast* **14** 1511-27

[7] Ayer A, Gourlay C W and Dawes IW 2014 Cellular redox homeostasis reactive oxygen species and replicative ageing in Saccharomyces cerevisiae FEMS *Yeast Res.* **14** 60-72

[8] Herrero E, Ros J, Belli G and Cabisol E 2008 Redox control and oxidative stress in yeast cells *Biochim. Biophys. Acta* **1780** 1217-35

[9] Ask M, Mapelli V, Hock M, Olsson L and Bettiga M 2013 Engineering glutathione biosynthesis of Saccharomyces cerevisiae increases robustness to inhibitors in pretreated lignocellulosic materials *Microb. Cell Fact.* **12** 87
[10] Patzschke A, Streiger M G, Holz C, Lang C, Mattanovic D and Sauer M 2015 Enhanced glutathione production by evolutionary engineering of Saccharomyces cerevisiae strains *J. Biotechnol.* **10** 1719-26
[11] Çakar Z P, Turanli-Yildiz B, Alkim U and Yilmaz Ü 2012 Evolutionary engineering of Saccharomyces cerevisiae for improved industrially important properties FEMS *Yeast Res.* 12 171-82
[12] Nie W, Wei G, Du G, Li Y and Chen J 2005 Enhanced intracellular glutathione synthesis and excretion capability of Candida utilis by using a low pH-stress strategy *Lett. Appl. Microbiol.* **40** 378-84
[13] Wei G Y, Li Y, Du G C and Chen J 2003 Effect of surfactants on extracellular accumulation of glutathione by Saccharomyces cerevisiae *Process Biochem.* **38** 1133–38
[14] Jamnik P, Medved P and Raspor P 2006 Increased glutathione content in yeast Saccharomyces cerevisiae exposed to NaCl *Ann. Microbiol.* **56** 175-8
[15] Santos L O, Alegre R M, Diego C G and Cuellar J 2010 Effects of magnetic fields on biomass and glutathione production by the yeast Saccharomyces cerevisiae *Process Biochem.* **45** 1362–67
[16] Dong Y, Yang Q, Jia S and Qiao C 2007 Effects of high pressure on the accumulation of trehalose and glutathione in the Saccharomyces cerevisiae cells *Biochem. Eng. J.* **37** 226–30
[17] Wang B, Liang G, Zhou Q, Xie J and Mo Y 2010 Combined amino acids modulation with H2O2 stress for glutathione overproduction in Candida utilis *Afr. J. Biotechnol.* **9** 5399-406
[18] Wei G, Wang D and Chen J 2008 Overproduction of glutathione by L-cysteine addition and a temperature-shift strategy *Biotechnol. Bioprocess Eng.* **13** 347-53
[19] Xiong Z, Guo M, Guo Y, Chu J, Zhuang Y and Zhang S 2008 Real-Time Viable-Cell Mass Monitoring in High-Cell-Density Fed-Batch Glutathione Fermentation by Saccharomyces cerevisiae T65 in Industrial Complex Medium *J. Biosci. Bioeng.* **105** 409–13
[20] Finley J W, Wheeler E L and Witt S C 1981 Oxidation of glutathione by hydrogen peroxide and other oxidizing agents *J. Agric. Food Chem.* **29** 404–07
[21] Xiong Z Q, Guo M J, Guo Y X, Chu J, Zhuang Y P and Zhang S L 2009 Efficient extraction of intracellular reduced glutathione from fermentation broth of Saccharomyces cerevisiae by ethanol *Bioresour. Technol.* **100** 1011-4
[22] Rollini M and Manzoni M 2006 Influence of different fermentation parameters on glutathione volumetric productivity by Saccharomyces cerevisiae *Process Biochem.* **41** 1501-05
[23] Ruis H and Schüller C 1995 Stress signaling in yeast *BioEssays* **17** 959-65
[24] Hohmann S and Mager W H 1997 *Yeast Stress Responses* Springer New York USA
[25] Krawiec Z, Bilinski T, Schüller C and Ruis H 2000 Heme synthesis in yeast does not require oxygen as an obligatory electron acceptor *Acta Biochim. Pol.* **47** 1027-35
[26] Liang G, Wang B, Xie J and Mo Y 2009 Novel pH control strategy for glutathione overproduction in batch cultivation of Candida utilis *Afr. J. Biotechnol.* **8** 6337-45
[27] Serra-Cardona A Canadell D Arino J 2015 Coordinate responses to alkaline pH stress in budding yeast *Microb. Cell* **2** 181-96
[28] Garcia-Rios E, Ramos-Alonso L and Guillamon J M 2016 Correlation between low temperature adaptation and oxidative stress in Saccharomyces cerevisiae *Front Microbiol.* **7** 1-11

Utilization of Molasses to Produce Lactic Acid by Using Lactobacillus Delbrueckii and Lactobacillus Plantarum

S Nurkhamidah*, A Altway, Susianto, Y Rahmawati, F Taufany, N Hendrianie, H Ni'mah, I Gunardi, S Zulaikah, E O Ningrum, R D Nyamiati, and A Ramadhani

Department of Chemical Engineering, Institut Teknologi Sepuluh Nopember, Kampus ITS Sukolilo, Surabaya 60111, Indonesia

*E-mail: dst_eureka@yahoo.co.uk

Abstract. Molasses is the final liquid from the process of crystallization of sucrose in a sugar factory. In general this liquid is only disposed to the surrounding environment or used as a fertilizer mixture. However, this fluid still contains glucose that can be fermented into lactic acid. Lactic acid is a water-soluble compound commonly used in food, pharmaceutical and chemical industries. Through the polymerization process, lactic acid changes into polylactic acid (PLA) which is a biodegradable plastic. In general, PLA is made from corn. If PLA can be synthesized from molasses, it can reduce the use of corn as a raw material for making PLA and reduce waste in the sugar industry. This study aims to determine the optimal conditions of lactic acid fermentation by using sugarcane drops from Tulung Agung as a source of glucose and its purification so that it can obtain high levels of lactic acid. The results showed that fermentation with Lactobacillus plantarum could produce more lactic acid with faster fermentation time than the fermentation using Lactobacillus delbrucckii with a concentration of lactic acid 19.68 g/L.

1. Introduction

Lactic Acid (2-hydroxycaboxylic acid) is an organic compound that can dissolve in water and can be found in nature in the form of L - lactic acid [1]. Lactic acid is widely used in industry as a chemical solvent and as a detergent. In addition, lactic acid is also widely used as an additive in cosmetics that serves to maintain moisture. Lactic acid is also used in the food industry as an additional ingredient to give food a sour taste. Another use of lactic acid is when it is in the form of potassium lactate that has properties to control pathogenic bacteria in fish and animal feed products [1]. Lactic acid can be obtained either by the action of fermented microorganisms or chemical synthesis. The fermentation process has the advantage of being more cost-effective [2]. Lactic acid demand continues to grow every year. However, the production of the lactic acid industry does not follow market demand. That happened for several reasons, including the price of raw materials. The substrate used by the industry is glucose and starch, both of which have been processed, which eventually ends the process [3].

Lactic acid fermentation can also be carried out by fungi such as Rhizopus oryzae in aerobic conditions. However, fungi such as Rhizopus oryzae have a slow growth and lower production rates compared to lakat acid bacteria. Therefore, fermentation using lactic acid bacteria is more widely used [4].

According to the product produced by lactic acid bacteria, there are two types of lactic acid bacteria, first homofermentative lactic acid bacteria which are lactic acid bacteria that can produce L-lactic acid or D-lactic acid from glucose with a yield of 85% and above. Whereas, heterofermentative lactic acid bacteria are lactic acid bacteria which only produce L - lactic acid or D - lactic acid from glucose with a 50% yield and the rest are byproducts [5]. In the lactic acid fermentation process, there are several raw materials that can be used as sources of monosaccharides and disaccharides, for example flour and lignocellulose [4]. In addition, lactic acid bacteria also require other nutrients such as vitamin B, amino acids, fatty acids, peptone and yeast extract [5]. The addition of some of these nutrients will increase production costs, but the process is expected to produce purer lactic acid. Lactic acid bacteria can grow under anaerobic conditions and can grow well in the temperature range between 5- 45 °C and in the pH range 5.5 - 6.5 [4]. The polymerization process of lactic acid is initiated by dehydration of the monomer, which generates a prepolymer chains consisting of oligomers and low molecular weight PLA. This process, starting from lactic acid involves three distinct stages: polycondensation, obtaining lactide and ring-opening polymerization [6].

Optical purity of lactic acid is very important for the physical properties of poly-lactic acid (PLA), and pure lactic acid can be polymerized to highly-crystalline poly-lactic acid (PLA) which is suitable for commercial use [7]. As a monomer for the production of poly-lactic acid (PLA) it has become famous for its broad usefulness and beneficial characteristics, such as biodegradability, biocompatibility, elasticity and well-controlled drug release profiles [8]. Poly- lactic acid (PLA) is considered to be one of the most promising biodegradable plastics, mainly due to its high chemical resistance, which is advantageous for the manufacture of fibers and films, while heat resistance is beneficial for the production of many equipments [9]. Molasses is an industrial byproduct of the process of sugar and alcohol and rich in fermented sugar [10], nitrogen and vitamins. This substrate is not expensive and highly available in Brazil, with annual production of 17.9 million tons during the sugar manufacturing process.

The purpose of this study is to determine the optimal conditions of lactic acid fermentation using molasses as a source of glucose by using Lactobacillus plantarum and Lactobacillus delbrucckii bacteria to obtain high levels of lactic acid.

2. Materials and Methods

The main materials used were molasses from waste in the sugar industry. Lactobacillus plantarum and Lactobacillus delbrucckii were used in this study. These strains are efficient lactic acid producing bacteria, which can convert glucose into Lactic acid through the fermentation.

2.1. Starter Preparation

Starter Preparation begins with the addition 10 mL of molasses, 4 gr of urea, and Na-Asetat buffer solution reaches pH 4.5. Finally the pH was set the pH between 6.5-7 by addition of NaOH, a total volume of 100 mL, added Lactobacillus plantarum or Lactobacillus delbrucckii. These were incubated at 37 °C for 24 hours in a incubator.

2.2. Fermentation Lactic Acid

Fermentation begins with the addition of 100 mL molasses, 240 gr urea, and Na-acetate buffer solution reaches pH 4.5. Finally the pH was set to pH 6.5-7 with the addition of NaOH, a total volume of 1 L, put into 5 erlenmeyer, added with a starter of Lactobacillus plantarum or Lactobacillus delbrucckii for 10 mL, incubated at 37 °C for 24 hours in a shaking incubator.

2.3. Fermentation Result Analysis Method

Growth in the amount of microbes is observed through the counting chamber. The fermentation results were autoclaved at 121 °C then reduced sugar test using spectrophotometry after 10 times dilution, and lactic acid test with the addition of 1 mL of fermentation after autoclave, 6 mL of concentrated H_2SO_4, heated 10 minutes, cooled at room temperature, added 100 μL $CuSO_4$ and 200 μL PHF was then measured using spectrophotometry after dilution of 100 times with a wavelength of 560 nm.

3. Results and Discussion

3.1. Analysis of Amount Microbes

Molasses is a waste from the sugar processing process in the sugar industry which means it still contains glucose. Therefore, the level of reducing sugar in molasses were firstly analized before fermenting using the DNS method and obtained a result of 208.73 g/L. In the process of glucose fermentation to lactic acid, there are two types of microorganisms used, namely Lactobacillus plantarum and Lactobacillus delbrucckii. The fermentation was carried out for eight days including the observation of the number of microorganisms, reducing sugars, and lactic acid produced.

The fermentation process was carried out in batches. First, microorganisms Lactobacillus plantarum and Lactobacillus delbrucckii were reactivated and stored in an incubator at 37 °C for approximately 24 hours. Then a starter was made with 0.4 gr of urea added; 1 mL molasses; and added Na-Acetate buffer then set the pH between 6.5-7, the pH is adjusted in such a way because the pH in the fermentation process must be controlled and maintained at a slightly acidic pH close to neutral so that lactic acid bacteria can grow [4]. Then added 1 microbial Lactobacillus plantarum and Lactobacillus delbrucckii which had been activated and incubated at 37 °C for 24 hours, because according to [4] lactic acid bacteria were able to grow at a temperature range of 5-45 °C.

Table 1. Result of Calculation Amount Lactobacillus plantarum and Lactobacillus microorganisms

Time (day)	Amount of Lactobacillus delbrucckii Microbes	Time (day)	Amount of Lactobacillus plantarum Microbes
2	48000000	2	19333333
5	59280000	5	19280000
6	38960000	6	15520000
7	55760000	7	14880000
8	47840000	8	27200000

After the starter was incubated for 24 hours, a fermentation medium was made followed by starter addition into it. The starter medium that had been added to the microbe was stirred evenly and divided into 5 erlenmeyers for fermentation. Fermentation was carried out anaerobically, because bacteria can grow well in anaerobic conditions [4].

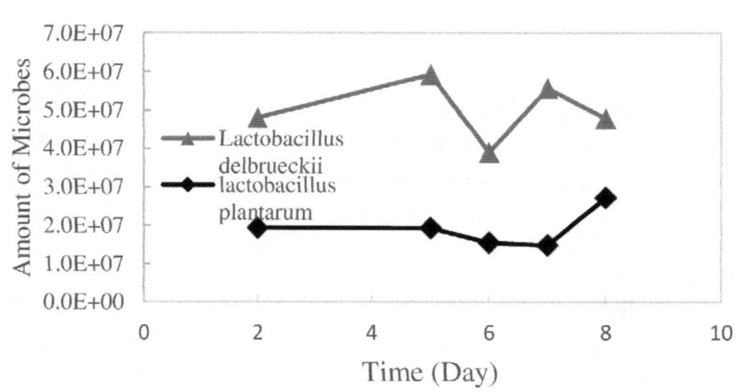

Figure 1. Effect of fermentation time on amount of microbes produced

IOP Conf. Series: Materials Science and Engineering **543** (2019) 012005 doi:10.1088/1757-899X/543/1/012005

The calculation of the amount of microorganisms was carried out on the second day. The observation shows that the amount of microorganisms was still fluctuating. For Lactobacillus delbrucckii, on the 5th day it increased when compared to the second day. However, on the 6 days it decreased and the 7th day rise again and the 8th day declined again. Where in Lactobacillus plantarum, the amount of microorganisms decreased until the 7th day but increase on the 8th day. From Figure 1 it can be seen that the amount of Lactobacillus delbrucckii is more than Lactobacillus plantarum although the number of bacteria added to the starter was the same. According to [4] the commercial fermentation process of lactic acid bacteria takes 3 to 6 days to complete at concentrations between 5-10 wt%.

3.2. Effect of Fermentation Time on Amount of Reduced Sugar

Besides observing amount of microorganisms during the fermentation process, the remaining reducing sugar was also observed. In theory, the amount of reducing sugar will decrease with the increasing of fermentation time, because the sugar has been converted into a product by microorganisms through the fermentation process.

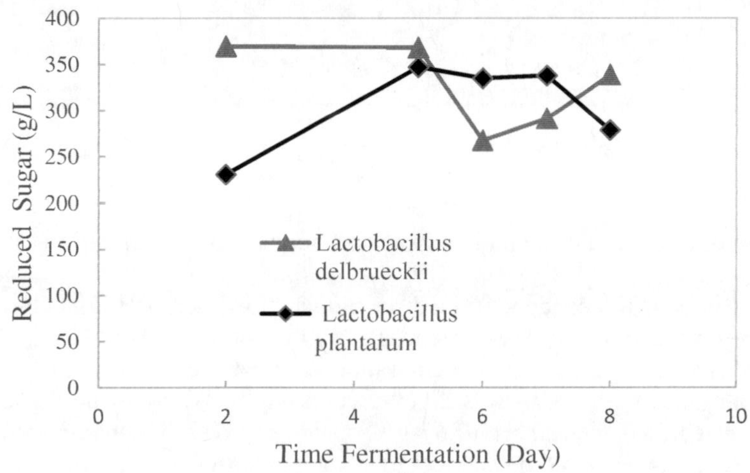

Figure 2. Effect of Fermentation Time on Amount of Reduced sugar

3.3. Analysis Lactic Acid

Table 2 shows the absorbance results of lactic acid fermentation with Lactobacillus delbrucckii and Lactobacillus plantarum by using spectrophotometer analysis.

Table 2. Result Lactid Acid Concentration used Spectrophotometry Analysis

Time(day)	Absorbance Lactobacillus plantarum	Lactid Acid Concentration of Lactobacillus plantarum (ppm)	Absorbance Lactobacillus plantarum	Lactid Acid concentration of Lactobacillus plantarum (ppm)
2	0,119	13,98	0,111	13,29
5	0,154	16,67	0,126	14,52
6	0,143	15,83	0,193	19,67
7	0,178	18,52	0,103	12,75
8	0,205	20,60	0,135	15,21

IOP Conf. Series: Materials Science and Engineering **543** (2019) 012005 doi:10.1088/1757-899X/543/1/012005

The fermentation results were autoclaved at 121 °C then reduced sugar test using spectrophotometry after 10 times dilution, and lactic acid test with the addition of 1 mL of fermentation after autoclave, 6 mL of concentrated H_2SO_4, heated 10 minutes, cooled at room temperature, added 100 μL $CuSO_4$ and 200 μL PHF was then measured using spectrophotometry after dilution of 100 times with a wavelength of 560 nm. After that the absorbance data was calibrated using a calibration curve and obtained the concentration of lactic acid (ppm) in fermentation each microbe.

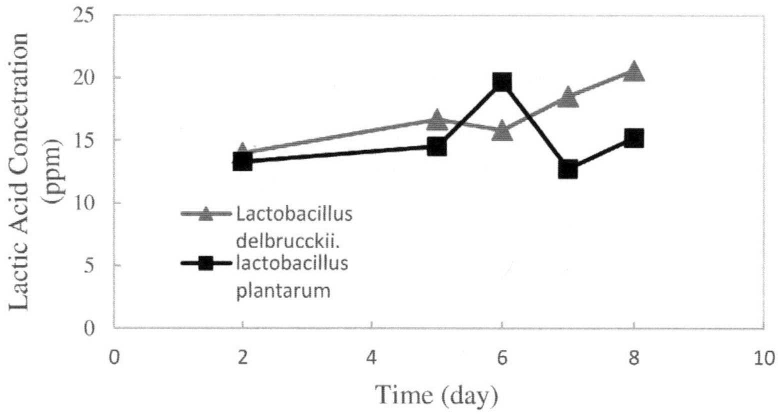

Figure 3. Effect of Fermentation Time on Lactic Acid Concentration

In Figure 3, it can be seen that the concentration of lactic acid for Lactobacillus delbrucckii has an upward tendency, only on the 6th day it experienced a slight decrease. Whereas in Lactobacillus plantarum, the results of fluctuating lactic acid concentration have not been read. The highest lactic acid concentration for Lactobacillus delbrucckii was 20.60 g/L on day 8. Whereas the highest lactic acid concentration for Lactobacillus plantarum is 19.67 g/L on the 6th day. The commercial lactic acid bacteria fermentation process takes 3 to 6 days to complete at a glucose concentration of between 5-10 wt% [4]. Research [11] also the concentration of lactic acid resulting from fermentation using Lactobacillus plantarum was analyzed using HPLC to produce lactic acid concentration of 5 wt%. From the results of lactic acid obtained, it can be concluded that lactobacillus plantarum can produce more lactic acid and a shorter time than Lactobacillus delbrucckii and with fewer microorganisms can see in Table 1. According to research [12] a with lactobacillus plantarum can produce lactic acid more, more quickly and using molasses as a carbon source it was achieved the highest lactic acid yield considering among the alternative sources of carbon, to know 88 %. The productivity using molasses was the highest, even considering the glucose control: 3.17 g L-1 h-1 [13].

3.4 Effect of Observations Time on Amount of Microbes Produced

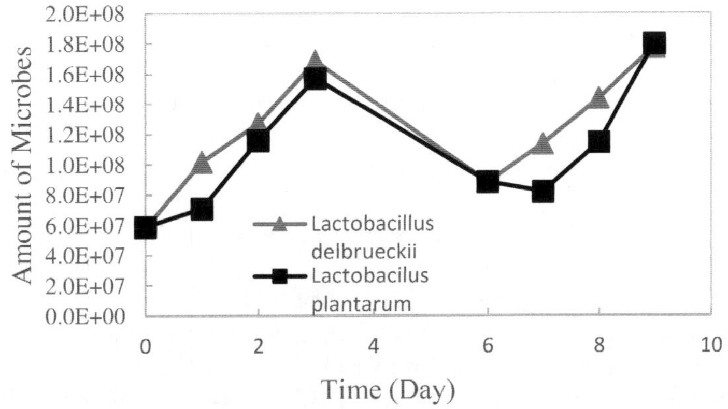

Figure 4. Effect of Observations Time on Amount of Microbes Produced

To ascertain whether the starter with incubation for 24 hours was in the phase log of the growth of microorganisms, then the starter was observed as shown in table 3, the starter observation was done for 9 days.
Starter Preparation started with the addition 10 mL of molasses, 4 gr of urea, and Na-Asetat buffer solution reaches pH 4,5. Finally the pH was set to pH 6.5-7 by addition of NaOH, a total volume of 100 mL, added Lactobacillus plantarum or Lactobacillus delbrucckii. These were incubated at 37 °C.
In Figure 4. Observations show that the phase log of microorganisms in the starter until the 3rd day. Therefore, the next experiment will be fermentation with a starter incubation period for 3 days. With this, it is expected to produce even more lactic acid.

4. Conclusion

From the results of experiments that had been carried out it can be concluded that lactic acid fermentation using molasses as a source of glucose found that the Lactobacillus plantarum bacteria can produce more lactic acid than Lactobacillus delbrucckii with a faster fermentation time, with a concentration of 19.68 g/L for 6 days and phase log of microorganisms in the starter between 0 to 3 days.

Acknowledgments

This work has been financially supported by local funding by Institut Teknologi Sepuluh Nopember (ITS) Surabaya with contract's number 1470/PKS/ITS/2018 to which the authors express their sincerest gratitude.

References

[1] Sin L, Rahmat A and Rahman W 2013 Polylactic Acid (Amsterdam: Elsevier William Andrew)
[2] Silva S S and Mancilha I M 1991 Aproveitamento de Agroindustriais Acido Lático Uma Alternative *Tec. De. Alimen.* **25** 37-40
[3] Bomrungnok W, Sonomoto K, Pinitglang S and Wongwicharn A 2012 Single Step Lactic Acid Production from Cassava Starch by Laactobacillus plantarum SW14 in Conventional Continuous and Continuous with High Cell Density *APCBEE. Proc.* **2** 103-97

[4] Mäki A P, Simakova I, Salmi T and Murzin D 2013 Production of Lactic Acid/Lactates from Biomass and Their Catalytic Transformations to Commodities *Chem. Rev.* **114(3)** 1909-71

[5] Reddy G, Altaf M, Naveena B, Venkateshwar M and Kumar E 2008 Amylolytic Bacterial Lactic Acid Fermentation *Rev. Bio. Adv.* **26(1)** 22-34

[6] Milena S L, Andre L J and Rubens M F *2014* Synthesis and Characterizations of Poly (Lactic Acid) by Ring-Opening Polymerization for Biomedical Applications *Chem. Eng. Trans.* **38** 331-36

[7] Lunt J 1998 Large-Scale Production, Properties and Commercial Applications of Polylactic Acid Polymers and Polymer Degradation Stability **59** 145–52

[8] Djukić V A P, Mojović L V, Vukašinović S M S, Rakin M B, Nikolić S B, Pejin J D and Bulatović M L 2012 Effect of Different Fermentation Parameters on L-lactic Acid Production From Liquid Distillery Stillage *Food Chem.* **134(2)** 1038–43

[9] Tanaka T, Hoshina M, Tanabe S, Sakai K, Ohtsubo S and Taniguchi M 2006 Production of D-lactic Acid from Defatted Rice Bran by Simultaneous Saccharification and Fermentation *Bio. Tech.* **97(2)** 211–17

[10] Lima U A, Aquarone E and Borzani W 1975 Biotecnologia Tecnologia das Fermentações (Brasil: São Paulo)

[11] Andrea K, Patricia F M, Johnatt O, Betania H L, Rubens M F and Maria R W M 2014 Purification of Lactic Acid Produces from Sugarcane Molasses *Chem. Eng. Trans.* **37** 367-72

[12] Regiane A O, Andrea K, Carlos E Y R, Maria R W M and Rubens M F 2017 Hybrid Short Part Evaporation as an Option to Lactic Acid Recovery from Fermentation Broth *Chem. Eng. Trans.* **57** 37-42

[13] Regiane A D O, Rubens M F and Carlos E V R 2016 High Lactic Acid Production from Molasses and Hydrolysed Sugarcane Bagasse *Chem. Eng. Trans.* **50** 307-12

IOP Conf. Series: Materials Science and Engineering **543** (2019) 012006 doi:10.1088/1757-899X/543/1/012006

An experimental study on thermal catalytic decomposition of inulin to 5 hydroxymethylfurfural and levulinic acid and the effect of C6 on furfural during the decomposition process

B A Fachri

Chemical Engineering Department, University of Jember, Jalan Kalimantan 37 Jember, 68121, Indonesia

Email: fachri.teknik@unej.ac.id

Abstract. The attractive biobased platform chemicals, 5-Hydroxymethylfurfural (HMF) and Levulinic Acid (LA), are commonly produced by the conversion of monomeric sugars. Nevertheless, a polymeric sugar such as inulin shows promising. This work reports the study on the thermal catalytic decomposition of inulin to HMF and LA. This work also investigated the effect of furfural during the process. The decomposition reactions were conducted in a batch reactor in a temperature (153-187)°C, an inulin loading (0.03- 0.12) g mL^{-1} and reaction times (18 -74 minutes) using a central composite experimental design. Furthermore, to study the effect of fructose and glucose on furfural, some additional experiments were also performed in the batch system. The decomposition process gained 35 wt% yield of HMF and 13 wt % yield of LA. The HMF and the LA model were statistically formulated and showed a good fit to the experimental data. The fructose particularly plays a role in furfural disappearance during the process.

1. Introduction

5-hydroxymethylfurfural (HMF) and levulinic acid (LA), two of biobased chemical products [1], have been associated to the top 12 biobased chemicals from biomass by the US Department of Energy (DOE) [2], [3], [4]. The HMF is generally synthesized from hexoses by the exclusion of three water molecules in the presence of brönsted acids such as hydrochloric acid and sulfuric acid [5],[6]. The decomposition of both monomeric and polymeric sugar to HMF and levulinic acid in the aqueous medium is typically one spot reaction scheme. This scheme leads to low yield and selectivity due to the formation of insoluble polymers and consecutive reactions. To increase the yield, nowadays, some methods using heterogeneous catalyst and organic solvents are proposed and reported in the literature. Nonetheless, related to catalyst lifetime, catalyst deactivation and down-stream processing, using water gives more benefits than a heterogeneous system and organic solvents [7], [8], [9] [10], [11].

Due to the reactivity, It is stated that d-fructose is preferred and selective feedstock to produce HMF than d-glucose [12]. Thus, the biomass polymers enriched in d-fructose units could be an alternative feed. A promising biomass polymer is an inulin which is an oligosaccharide containing fructose-fructose unit or fructose-glucose unit. The advantages of using inulin are abundant and reactive [6], [13]. Inulin can be easily found in plants, such as Jerusalem artichoke, chicory, and dahlia tubers [14].

This work mainly reports a study on the decomposition of inulin to HMF using water as the solvent in the thermal catalytic process in the absence of a catalyst. Water was chosen as the selected solvent, as it is environmentally friendly, a good solvent for many carbohydrates and it is cheap, nontoxic, and nonflammable [15], [16]. This work is also to report the effect of C6 on furfural during the inulin decomposition.

2. Methods

2.1 Decomposition Process

Inulin from Dahlia tubers was purchased from Acros Organic (Geel, Belgium). Amount of inulin was fed into ampoules (5 mm in ID, 15 cm in length and 1.5 mm in thickness) filled with 4 mL of deionized water. The ampoules were sealed with a torch and placed in a rack in a heating oven (Heraeus Instruments Type UT 6060) at a constant temperature. At different reaction times, an ampoule was taken from the oven and quickly quenched in cold water to stop the reaction. All experiments were conducted in triplicate and the average value is taken. After the reaction, the ampoules were opened and the reaction mixture was taken out and centrifuged for about 10–30 min to remove the solids. The liquid product was diluted with demi water before analysis. A liquid sample was taken and analyzed by HPLC.

2.2 Analysis

HPLC was used to identify and quantify the liquid product from the reactions. The HPLC system consisted of a Hewlett Packard 1050 pump, a Bio-Rad organic acid column Aminex HPX-87H and a Water 410 differential refractive index detector. A sulfuric acid solution (5 mM) was used as the eluent with a constant flow rate of about 0.55 cm3 min-1. The column was operated at 60°C. The HPLC was calibrated with solutions of the pure compounds at a range of concentrations. Using the chromatogram peak area and the external calibration curves, the concentrations of components in the liquid phase were determined.

2.3 Definitions

The yield of HMF and levulinic acid are defined according to equation (1) and (2). Both are reported on a weight basis.

$$y_{HMF} = \frac{C_{HMF} \times M_{HMF} \times V}{W_{in}} \times 100\% \ (wt\%) \quad (1)$$

$$y_{LA} = \frac{C_{LA} \times M_{LA} \times V}{W_{in}} \times 100\% \ (wt\%) \quad (2)$$

Here, C refers to concentration (mol L^{-1}) at a certain time, M is the molecular weight (g mol^{-1}), V refers to reaction volume (L) and W refers to the intake of inulin (g).

3. Results and Discussion

Exploratory experiments on the thermal catalytic conversion of inulin were performed in a range of (153-180°C), using an inulin loading in a range of (0.03- 0.12 g mL^{-1}) and reaction times in a range of (18 -74 minutes). The 24 batch experiments were conducted using a composite design. The data were statistically analyzed and a model was developed to describe the HMF yield versus process conditions like reaction time, inulin intake and reaction temperature.
In this one-spot reactions, since inulin is converted to fructose (94 mol%) and glucose (6 mol%), some compounds are simultaneously produced. HMF, LA, formic acid, acetic acid, and humins were clearly detected during the process. Finally, the possible effects of C6 such as fructose and glucose on furfural during the process were observed.

3.1 Decomposition of Inulin to HMF and LA

During the experiments on the thermal conversion of inulin, some soluble components were detected in the reaction mixtures (HPLC). Three of them, d-fructose, d-glucose and HMF are intermediates in the reaction sequence. The end products are LA, formic acid (FA) and acetic acid (AA). Some brown-black insoluble called humins were identified, which are always formed during the acid catalyzed conversions of carbohydrates in water, either in the monomeric or polymeric form. The typical concentration profile is given in Figure 1 and was performed at 170°C and inulin loading of 0.1 g/mL.

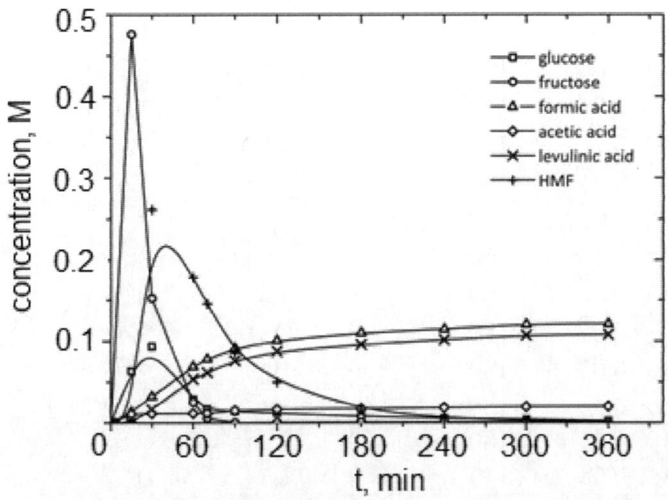

Figure 1. Concentration profile for various compounds during the thermal decomposition of Inulin to HMF

3.2 Effect of Process Condition

To investigate the effect of process variables on HMF, 40 experiments were performed in a batch reactor. Three independent variables which are temperature (150-190°C), inulin intake (0.03-0.12 g /mL) and reaction time (4-74 min) were explored. The yield of HMF was taken as the dependent variable. The results are provided in Table 1.

The highest experimental y_{HMF} is about 35 wt % (entry 6 in Table 1) and was achieved at 180 °C, an inulin intake of 0.05 g/mL and a reaction time of 18 min. At different conditions, the highest y_{LA} is about 13.4 wt % (entry 9 in Table 1) and was achieved at 187 °C, an inulin intake of 0.08 g/mL and a reaction time of 39 min.

The HMF yield as a function of temperature, reaction time and inulin loading was statistically formulated using the Design-Expert 10 software package. The appropriate model equation is given in equation (3). The model accommodates both quadratic and interaction terms. It implies process variables affect the HMF yield. The R-squared of the model is 0.9664. An indication of the good fit between the model and experimental data is given in Figure 2.

Table 1. HMF yield on various reaction conditions

Run	C_{in}, g/mL	$T_{reaction}$, °C	t, min	y_{HMF}, wt %	y_{LA}, wt %
1	0.08	170	39	22.8	4.0
2	0.05	160	18	2.7	0.0
3	0.08	170	39	23.1	3.1
4	0.08	170	39	23.9	3.2
5	0.05	180	18	32.1	5.6
6	0.05	180	18	35.0	6.1
7	0.12	170	39	17.6	6.1
8	0.08	170	39	23.1	3.3
9	0.08	187	39	14.6	13.4
10	0.08	170	39	22.5	1.1

Run	C_{in}, g/mL	$T_{reaction}$, °C	t, min	y_{HMF}, wt %	y_{LA}, wt %
11	0.08	170	39	21.8	1.5
12	0.08	170	39	19.4	1.8
13	0.08	170	39	16,6	2.2
14	0.08	170	74	8.9	1.9
15	0.1	180	18	20.9	5.5
16	0.03	170	39	24.8	3.6
17	0.05	160	60	25.7	3.1
18	0.1	160	60	23.3	3.6
19	0.1	160	18	3.2	0.0
20	0.05	180	60	10.1	9.4
21	0.05	180	60	10.4	9.0
22	0.1	180	60	3.1	9.5
23	0.1	180	60	2.2	9.8
24	0.08	153	39	12.4	0.3

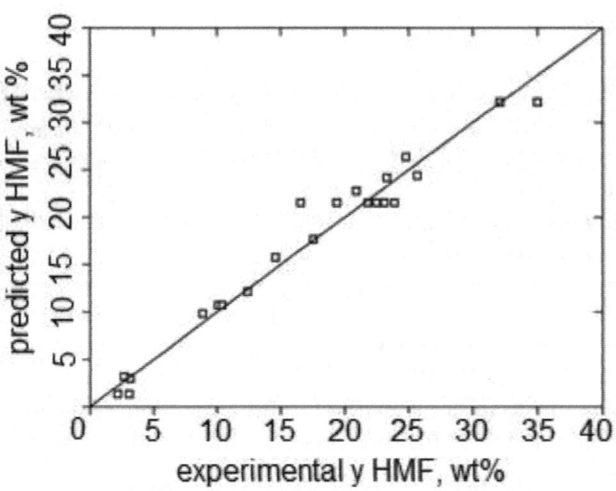

Figure 2. Parity plot between the experimental and modeled HMF yields

$$y_{HMF} = (1466.7)C_{in} + (11.8)T + (9.3)t - (9.2)C_{in}T - (0.05)tT - (0.03)T^2 - (0.009)t^2 - 1223.9 \quad (3)$$

The model predictions at a batch time of 30 min (Figure 3, left), implies that the HMF yield is a function of both the temperature and inulin intakes. Highest yields are obtained at the highest temperature in the range and the lowest inulin intake. The exact temperature dependency of the HMF yield is a function of the inulin intake. At low inulin intakes, the y_{HMF} increases with the temperature, up to a maximum of 35 wt % at 186 °C. At high intakes, a maximum of the y_{HMF} is observed at intermediate temperatures. The reduction of the y_{HMF} at the higher temperatures is likely due to the subsequent reaction of HMF to LA and FA (Scheme 1), leading to a lowering in the y_{HMF}. As expected, this subsequent reaction is occurring to a larger extent at 40 min batch times (Figure 3, right) and the y_{HMF} are considerably lower than for 30 batch times.

IOP Conf. Series: Materials Science and Engineering **543** (2019) 012006 doi:10.1088/1757-899X/543/1/012006

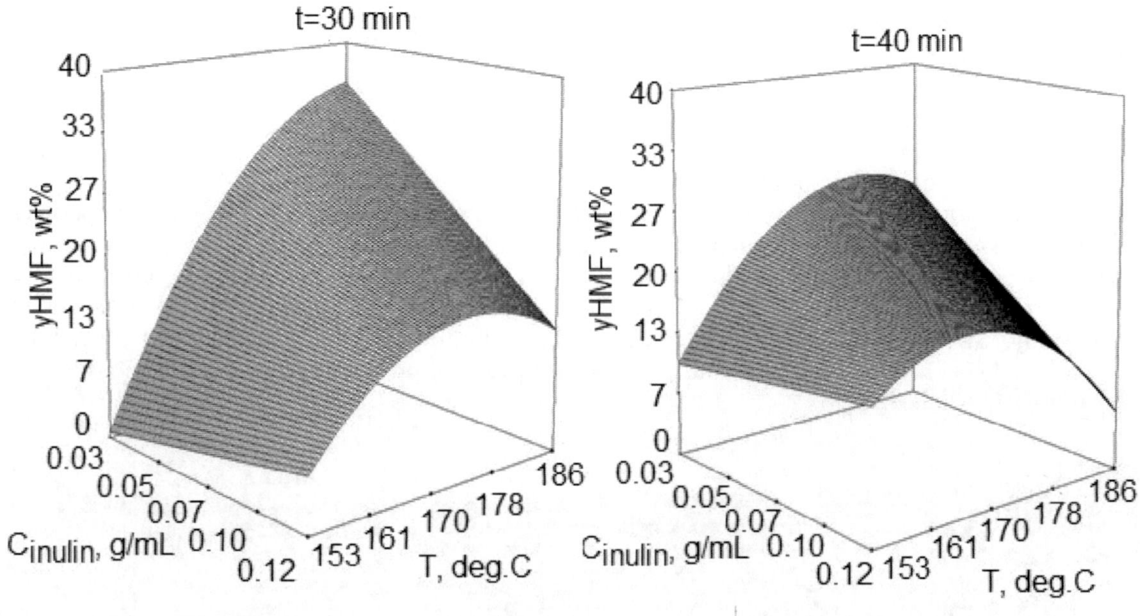

Figure 3. Modeled HMF yield versus temperature and inulin intake at two batch times.

The occurrence and importance of the consecutive reaction are also confirmed by considering the yLA versus the process conditions (Table 1). The y_{LA} was also statistically modeled and is given in equation 4 and Figure 4. The model equation has an R square of 0.9464.

$$y_{LA} = 437.51 - (188.89)C_{in} - (5.41)T + (0.09)t - (0.12)C_{in}T + (0.17)C_{in}t + (5.88.10\text{-}4)Tt + (1422.29)C_{in}^2 + 0.02)T^2 - (1.69.10\text{-}3)t^2 \quad (4)$$

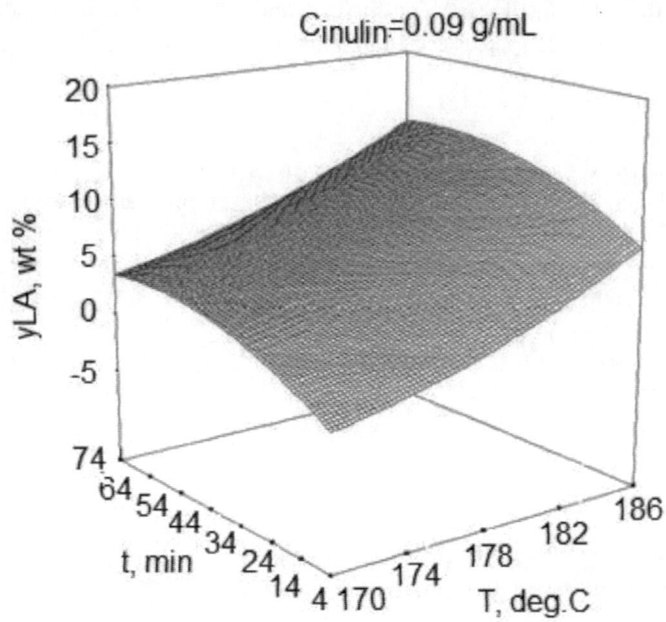

Figure 4. Modeled LA yield versus temperature and reaction time at selected inulin loading

3.3 Effect of C6 on Furfural

Some research has reported furfural as one of the decomposition product of carbohydrate [14], [17], [18], [19]. Conversely, herein this work, furfural is unidentified in the reaction condition window. To check the possible reason, some additional experiments were performed involving fructose, glucose and formic acid. The fructose-furfural and the glucose-furfural interaction were checked as the molecule structure of inulin containing both fructose and glucose. The interaction formic acid-furfural were also checked as the formic acid is able to present the autocatalytic process [20]. All interaction was described in Figure 5. As a result, the interaction between fructose- furfural shows a tremendous effect on the furfural disappearance compared to glucose-furfural and formic acid-furfural (Figure 5 left). To gain insight into the effect of fructose to furfural, the various fructose concentration (0.1-0.4) M was applied in this study. As fructose concentration increases, the furfural concentration dramatically decreases (Figure 5 right). This indicates that the reactivity of fructose may be playing a role in furfural disappearance.

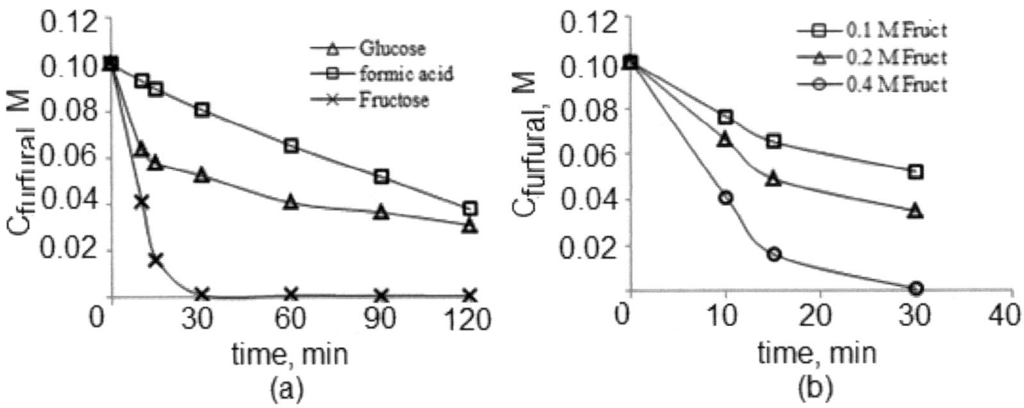

Figure 5. The relationship between furfural and reaction time in the presence of glucose, fructose and formic acid (a) and the relationship between furfural and reaction time on the different fructose concentration (b)

4. Conclusion

This work concludes that the thermal catalytic decomposition of inulin to HMF in water was studied in a wide range of reaction conditions, including variations in temperature, inulin loading and reaction time. The highest HMF yield was 35 wt % (180 °C, inulin intake of 0.05 g mL^{-1} and a reaction time of 18 min). The experimental data were modeled using a statistical approach. The model shows a good fit with the experimental data and allows estimation of the HMF yield as a function of temperature, inulin intake and reaction time.

The C6 such as fructose formed in the reaction strongly indicates playing a role in furfural disappearance during the inulin decomposition.

ACKNOWLEDGMENT

The author would like to thank the Department of Chemical Engineering, the University of Groningen for supporting this research. The author would like also to thank Prof. Hero Jan Heeres for supervising this research.

REFERENCES

[1] Shamsul N S, Kamarudin S K and Rahman N A 2018 *Bioresour Technol.* **247** 821.

[2] Bozell J J and Petersen G R 2010 *Green Chem.* **12** 539.
[3] Lanziano C, Costa M, Moya S F, Barrett D H, Rodella C and Guirardello R 2018 *Chem. Eng. Trans.* **65** 139.
[4] Licursi D, Antonetti C, Fulignati S, Corsini A, Boschi N and Galletti A M R 2018 *Chem. Eng. Trans.* **65** 637.
[5] Pedersen A P, Ringborg R, Grotkjær T, Pedersen S and Woodley J M 2015 *Chem. Eng. J.* **273** 455.
[6] Fachri B A, Abdilla R M, Rasrendra C B and Heeres H J 2016 *Chem. Eng. Res. Des.* **109** 65.
[7] Boisen A, Christensen T B, Fu W, Gorbanev Y Y, Hansen T S, Jensen J S, Klitgaard S K, Pedersen S, Riisager A, Ståhlberg T and Woodley J M 2009 *Chem. Eng. Res. Des.* **87** 1318.
[8] Girisuta B, Janssen L P B M and Heeres H J 2006 *Chem Eng Res Des.* **84** 339.
[9] Yang F, Liu Q, Bai X and Du Y 2011 *Bioresour Technol.* **102** 3424.
[10] Ndayishimiye J and Chun B S 2017 *Biomass Bioenerg.* **106** 1.
[11] Yu I K M and Tsang D C W 2017 *Bioresour Technol.* **238** 716.
[12] Fachri B A, Abdilla R M, Bovenkamp H H V D, Rasrendra C B and Heeres H J 2015 *ACS Sustain. Chem. Eng.* **3** 3024.
[13] Fachri B A, Rasrendra C B and Heeres H J 2015 *Catal.* 5 2287.
[14] Anańina N A, Andreeva O A, Mycots L P and Oganesyan E T 2009 *Pharm. C J.* **43** 157.
[15] Iryani D A, Kumagai S, Nonaka M, Sasaki K and Hirajima T 2013 *Proc. Earth Planet Sci.***6** 441.
[16] Babaei Z, Chermahini N A, Dinari M, Saraji M and Shahvar A 2018 *J. Clean. Prod.* **198** 381.
[17] Qi B, Luo J, Chen X, Hang X and Wan Y 2011 *Bioresour Technol.* **102** 7111.
[18] Rout P K, Nannaware A D, Prakash O, Kalra A and Rajasekharan R 2016 *Chem. Eng. Sci.* **142** 318.
[19] Sweygers N, Harrer J, Dewil R and Appels L 2018 *J. Clean Prod.* **187** 1014.
[20] Ranoux A, Djanashvili K, Arends I W C E and Hanefeld U 2013 *ACS Catal.* **3** 760.

Biodiesel Production Through Waste Cooking Oil (WCO) Esterification Using Sulfated Alumina as Catalyst

M Ulfah[1]*, Firdaus[1], S Octavia[1], H Suherman[2] and Subagjo[3]
[1]Department of Chemical Engineering, Universitas Bung Hatta, Padang, Indonesia
[2]Department of Mechanical Engineering, Universitas Bung Hatta, Padang, Indonesia
[3] Department of Chemical Engineering, Institut Teknologi Bandung, Indonesia

*Email : ulfah_nad@yahoo.com

Abstract. Biodiesel produced from waste cooking oil (WCO) with initial free fatty acids (FFA) of 6.67 % through one-stage reaction (transesterification) and two-stage reaction (esterification followed by transesterification) had been studied. The esterification process was catalyzed by solid catalyst of synthesized sulfated alumina and transesterification reaction was catalyzed by NaOH. The objective of this research is to determine the optimum condition of biodiesel production from WCO using sulfated alumina catalysts by esterification reaction and to compare biodiesel yield using homogeneous catalysts of sulfuric acid with a heterogeneous catalyst of sulfated alumina. The results showed that the optimum conditions for esterification reactions using sulfated alumina catalyst is at the volume ratio of methanol/oil of 1.5, the amount catalyst of 1 wt %/v of oil at 1 hour reaction time. With the optimum condition, the maximum conversions of free fatty acid using a sulfuric acid and sulfated alumina catalyst were 90.4 % and 89.97 % respectively. The yield of biodiesel from waste cooking oil by two-stage transesterification process is 86.67 % and one-stage transesterification is 66.67 %.

1. Introduction

Biodiesel production from WCO is an effective way to reduce the price of biodiesel because of lower raw material cost [1]. WCO price is two to three times cheaper than edible oil [2]. In Indonesia, most of waste cooking oil is derived from palm oil and palm oil demand in 2013 is about 2,210,758 tons with populations of 247,954,018 people [3].

In Canada and United States, each person produces 9 pounds (4.08 kg) of waste cooking oil annually [4]. If WCO production per capita in Indonesia is considered the same as in Canada, the total production of waste cooking oil in Indonesia is estimated 1,012,148 tons. This illustrates that development of biodiesel industries in Indonesia with WCO as a feedstock has a huge opportunity. In addition, utilization of waste cooking oil can also overcome the problem of waste oil disposal and public health [5].

Waste cooking oil typically contains more than 2 % of free fatty acid (FFA). Previous research stated that the maximum conversion and yield of biodiesel from raw material with high FFA content (> 2 %) are using two-stage reaction. The two-stage transesterification system can increase conversion by 30 % compared to one-stage reaction [2]. Issariyakual et al. (2007) [6] reported that double-stage transesterification reaction of biodiesel production from waste cooking oil can reach conversion of 90 %, while one-stage transesterification only achieves 50 %.

Biodiesel production using heterogeneous solid catalysts, as a substitute for homogeneous catalysts provides low production costs. Mbraka and Shank (2006) [7] reported that 88 tons of homogeneous catalysts (NaOH) are required to produce 8000 tons of biodiesel. In contrast, Dossin et al. (2006) [8] used only 5.7 tons of solid heterogeneous catalysts (supported by MgO) to produce 100,000 tons of fatty acid methyl esters (FAME).

The disadvantage of solid heterogeneous catalyst application is in the presence of a mixture of three phases (catalyst, oil and alcohol), thus the reaction is limited by mass transfer. This can be prevented by using supported catalysts which have a specific surface area and large pore size. Gamma alumina has large specific surface area, large pore size and high thermal stability. Ulfah and Subagjo, 2012 [9] has successfully designed manufacturing procedure of gamma alumina with pore diameter of 76 Å and surface area in the range of 150-250 m^2/g, derived from boehmite (AlOOH). The gamma alumina have been applied in industries since 2010 as hydrotreating naphtha catalyst at Pertamina, Inc. in Dumai-Indonesia. Therefore, in this study, the solid acid catalyst is developed based on gamma alumina support.

Kiakalaieh et al. 2013 [2] summarized the utilization of alkali and acid solid catalysts to produce biodiesel from waste cooking oil. The conversion and yield of biodiesel catalyzed by the acid solid catalysts are in the range of 80-98 %. But the WCO esterification catalyzed with the sulfated alumina catalyst has not been published. Previous research indicated that sulfated gamma alumina catalyst can convert free fatty acid in the rubber seed oil more than 85 % [10]. This research will study the optimum esterification reaction condition catalyzed by synthesized sulfated alumina catalysts to produce crude fatty acid methyl ester (FAME) from waste cooking oil and to compare between the performance of synthesized solid acid catalysts and sulfuric acid homogenous-catalyst in the esterification reaction of waste cooking oil.

2. Materials and Methods

2.1 Chemicals and Materials

Waste cooking oil was collected from a fried merchant in Padang-Indonesia. Phosphoric acid (purity of 85 %) used as pretreatment process of waste cooking oil, HNO_3 (purity of 65 %) used for the catalyst preparation and ammonia solution (purity of 25 %) were purchased from Merck (Indonesia). Boehmite used for the catalyst preparation was obtained from Pertamina, Inc., Indonesia. Methanol used for esterification and transesterification, acid sulfuric used as precursor for sulfated alumina (H_2SO_4) catalysts and sodium hydroxide (NaOH) used as alkali transesterification catalysts were purchased from supplier Bratachem, Inc., Indonesia. Technical grade of methanol, H_2SO_4 and NaOH were used in this experiment.

2.2 Catalyst preparation

Sulfated gamma alumina catalyst was prepared in two stages. First stage, boehmite is mixed with water, stirred and then diluted nitric acid (HNO_3 3 % wt) was added. The mixture was stirred to form a paste and subsequently concentrated ammonia and sulfuric acid was added. The weight ratio of boehmite: water: nitric acid was 1:1.67:0.67; and the amount of concentrated ammonia solution and concentrated sulfuric acid added were 0.03 and 0.023 times the weight of boehmite.

The pasta formed was extruded and left overnight at room temperature. Extrusions result was dried at 120 °C for 3 hours and calcinated at 570 °C for 2 hours. This first stage material was denoted as catalyst type I. The catalyst type II was prepared by impregnating industrial grade concentrated sulfuric acid on catalyst type I for 1 hour, screened and followed by drying and calcination. Drying and calcination conditions of type II were the same with type I.

2.3 Catalyst characterization

The chemical composition of both catalysts was identified using XRF analysis by PANalytical's Epsilon 3. The textural properties of catalysts were investigated by Brunauer Emmet and Teller (BET) model using Quantachrome Instrument.

2.4 Waste cooking oil treatment

Firstly, waste cooking oil was filtered to remove the remnants of fried foods. Afterwards, waste cooking oil was heated for 2 hours at a temperature of 100 °C while stirred to remove the water content. Removal of suspended dirt in used cooking oil was done by addition of phosphoric acid 85 % of 0.5 wt % /v of waste cooking oil. The process was conducted for 1 hour at a temperature of 90 °C and atmospheric pressure. The processed waste cooking oil is left for two days and then filtered using a vacuum pump.

2.5 Biodiesel production

2.5.1 Esterification reaction

Heterogeneous and homogeneous (H_2SO_4, 97 %) esterification reactions were conducted in a 500 ml three necked round flask equipped with a reflux condenser. 30 ml of waste cooking oil was taken and stirred at 400 rpm. While stirred, waste cooking oil was until the temperature reached 60 °C. Thereafter, methanol and sulfated alumina catalyst were introduced into the batch reactor containing hot waste cooking oil. The volume ratio of methanol to WCO ranged from 0.9 to 1.8 v/v, and the amount of catalyst varied from 1.0 to 5.0 wt % of WCO. After a certain reaction time (1 and 5 h), the reaction was stopped and the reactor was cooled to room temperature. The product was filtered to separate the catalyst using a vacuum pump. The mixture of remaining methanol, water formed through esterification reaction, and crude methyl ester were inserted into separate funnel and left overnight. The mixture of water and excess methanol were separated and crude methyl ester produced was analyzed to determine free fatty acid content by titration using NaOH solution. This titration method followed method ASTMD-664. The molecular weight of WCO is 280 mol/g of linoleic acid. The conversion of oleic acid into its ester (the catalytic activity) was calculated based on the reduction of the acidity index of the products as compared with the acidity index for the initial oleic acid [11].

2.5.2 Transesterification reaction

Transesterification reaction of crude FAME or WCO (in one step reaction) and methanol was carried out with the same procedure as the esterification reaction. The amount of methanol added was its optimum condition at esterification reaction. The catalyst used is the NaOH catalyst varied by ratio catalyst/crude FAME of 0.5, 1, 3 and 5 wt %/v. The reaction time was set for 1 hour and reaction temperature of 60 °C. The reaction product was separated; it formed two layers, glycerol and excess methanol at lower layer and methyl esters (biodiesel) on the upper layer. The obtained methyl ester was washed using 100 °C hot water to dissolve the residual methanol and salts formed. The washing process was conducted by mixing hot water into methyl ester as much as 30-50 % of the methyl ester volume, then stirred and allowed to stand for approximately one hour until the water and biodiesel separated. Washing process was performed 4-5 times. Biodiesel was heated to 100 °C to remove the remaining water contained.

2.5.3 Product characterization

Biodiesel properties were measured in accordance with ASTM standards, including density, kinematic viscosity, water and sediment, acid number, free glycerol, total glycerol, methyl ester content and color. Biodiesel properties analysis was conducted at Pelita Agung Agrindustri, Inc., Dumai-Pekanbaru, Indonesia.

3. Results

3.1 The physicochemical properties of sulfated gamma-alumina catalyst

The physicochemical properties of sulfated gamma-alumina catalyst ($SO_4^{=}/Al_2O_3$) is presented in Table 1. Catalyst type II was prepared by impregnating concentrated sulfuric acid for 1 hour on catalyst type I which contains 8.821 wt % SO_3. After the deposition of sulfate ion, surface area of catalyst type I was reduced by 30 %. In addition, the increasing of sulfate ion decreasing catalyst pore volume. The addition of more sulfuric acid caused the catalyst structure collapsed during calcination and $SO_4^=$ species covered the catalyst pore. Conversely, the addition of more sulfuric acid increasing catalyst pore diameter. Rane et al.2016 [12], also showed that increased concentration of SO_4^{2-} on Al_2O_3, could increase the diameter of catalyst pores. It was found that the average pore diameter of the type II sulfated alumina catalyst is 63.64 Å which is larger than the triglycerides (or oil) molecule (6 Å) [13]. This beneficial effects, reactants traffic onto catalyst surface, became larger that would minimize diffusion limitations of reactant molecules. Therefore, the reactant molecules can easily penetrate into the interior of the catalyst and most of the active sites will be utilized during the esterification reaction. The result showed that type II sulfated alumina catalyst had bigger pores and contain more SO_4^{2-} compared to type I. Therefore, this study only focusing on type II catalyst performance.

Tabel 1. Physicochemical properties of sulfated alumina catalyst

	S_{BET} ($m^2\ g^{-1}$)	Dp, (Å)	Vp, ($cm^3\ g^{-1}$)	Al_2O_3, wt %	SiO_2, wt %	P_2O_5, wt %	SO_3, wt %
$SO_4^=/Al_2O_3$ type I	295.30	53.31	0.3936	89.487	0.814	0.645	8.821
$SO_4^=/Al_2O_3$ type II	76.387	63.64	0.11	57.957	0.645	0.53	40.715

3.2 The physicochemical properties of sulfated gamma-alumina catalyst

Esterification process was conducted on pretreated waste cooking oil (without drying process) with free fatty acid content (FFA) of 6.77 % catalyzed by developed gamma alumina catalyst. From the process the free fatty acids increased to 7.3 % even though esterification was intended to reduce FFA, according to reaction 1.

$$\text{R-COOH} + \text{R'-OH} \leftrightarrow \text{R-CO-CH}_2\text{-R'} + \text{H}_2\text{O} \quad (1)$$

Based on this result, analysis of water content was conducted and it was found that initial water content in waste cooking oil was 6.61 %. Thus, the preparation stage of waste cooking oil as a feedstock for biodiesel includes: filtering impurities, water evaporation, elimination of suspended dirt and final filtration.

3.3 Effects of esterification reaction variables

To obtain optimum conditions of waste cooking oil esterification reaction using sulfated gamma alumina catalyst, different reaction variables effect on volume ratio of methanol/oil, the amount of catalyst, and reaction time were investigated

3.3.1 Effect of methanol/oil volume ratio

A crucial factor for (trans)-esterification reaction to determine the equilibrium point of the reaction in order to eliminate the effect of the reverse process is methanol-WCO ratio. In this research, the influence of methanol/oil ratio versus FFA conversion was conducted at 4 different ratios of methanol/oil (0.9; 1.2; 1.5 and 1.8), while other parameters, e.g. 1 wt.% loading of the sulfated catalyst, reaction temperature of 60 °C, stirring speed of 400 rpm and reaction time of 1 h (and 5 h) were held constant. The results are presented in Figure 1.

Figure 1 showed that at the methanol/oil ratio of 0.9-1.5 v/v, at 1 hr or 5 hr of reaction time, the maximum conversion of FFA (at steady state) was obtained at the methanol/oil ratio of 1.5 v/v. However, the FFA conversion was decreased if the methanol/oil ratio was increase above 1.5 v/v. This result is similar to those conducted by other researchers. According to Wang et.al (2016) [14], excessive methanol volume more than equilibrium state will reduce WCO concentration and lessen the FFA conversion since the reaction is reversible [11]. Excessive methanol will also result in methanol accumulation that will reduce the catalyst concentration [15]. In addition, water (by-product) obtained during the esterification process of FFAs in WCO feedstock lowers the activity of the catalyst [16].

3.3.2 Effect of methanol/oil volume ratio

Another important factor to improve the rate of the trans-esterification reaction is catalyst loading amount. Figure 2 presents the effect of sulfated gamma alumina catalyst amount on the esterification reaction of waste cooking oil against FFA conversion at a volume ratio methanol/oil of 1.5 v/v and a temperature of 60 °C. At 1 % of catalyst addition, the maximum FFA conversion of 89.97 % was obtained after 1 hour of reaction time. Increasing the catalyst concentration to 3 wt %/v slightly decrease the FFA conversion to 89.45%. However, the reaction time needed was 3 hours. Furthermore, increasing the catalyst concentration into 5% wt/v gave longer reaction time (5 hours) with FFA conversion of 89.76 %. The amount of catalyst from catalyst different sources gave a variation effect of methyl ester convertion [15] [17].

Accordingly, the optimum condition of the waste cooking oil esterification reaction catalyzed by sulfated alumina solid catalyst is a methanol/oil ratio of 1.5 v/v, a catalyst amount of 1 wt % /v of oil and reaction time of 1 hour. Conventionally, FFA esterification reaction generally use homogeneous acid catalyst (particularly sulfuric acid). To compare the performance of sulfated gamma alumina solid acid catalyst, waste cooking oil esterification reaction with concentrated sulfuric acid homogeneous catalysts has been done. The esterification reaction of waste cooking oil using sulfuric acid was carried out at a methanol/oil ratio of 1.5 v/v, the amount sulfuric acid catalyst of 1 % v/v of oil (30 ml of WCO and 0.3 ml of concentrated H_2SO_4), reaction time of 1 hour and the initial FFA content of 6.77 %. Under the optimum esterification condition, the maximum conversion of free fatty acid using the synthesized catalyst and sulfuric acid were 89.97 and 90.4 %, respectively.

Based on table 1, SO_4^{2-} contained in type II catalyst was 40.715 %. According to the optimum condition of esterification reaction mentioned above, catalyst/WCO ratio was obtained at 1 wt %/v (0.3 g type II sulfated alumina in 30 ml of WCO). Thus, it was found that 0.067 ml of H_2SO_4 was deposited on 0.3 g of type II catalyst. Accordingly, sulfated gamma alumina acid solid catalyst has a potential as a substitute for sulfuric acid homogeneous catalyst.

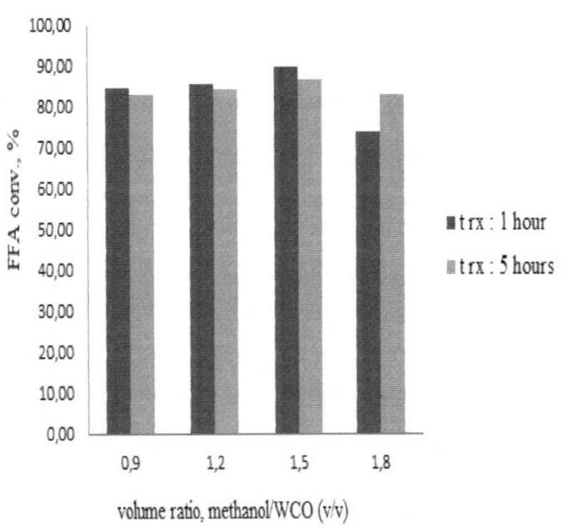

Figure 1. Effect of methanol/oil volume ratio on FFA conversion (T = 60 °C and catalyst 1 wt %/v)

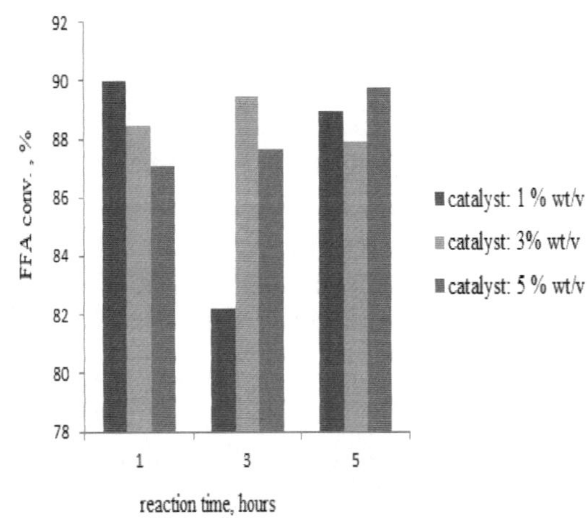

Figure 2. Effect of catalyst amount on FFA conversion (volume ratio methanol/WCO of 1.5, T = 60 °C)

3.4 Biodiesel yield

Crude biodiesel from the esterification reaction in optimum condition was used directly as a reactant in transesterification reaction without washing it. Transesterification process was conducted at a volume ratio methanol/oil of 1.5 v/v, temperature of 60 °C, reaction time of 1 hour and variation of ratio NaOH catalyst to oil at 0.5;1 ;3 and 5 wt % /v. The maximum FFA conversion obtained by using sulfated alumina catalyst was 89.98 %. Thus, the remaining FFA contained in crude FAME was 0.68

%. The results showed that at 0.5 wt %/v of NaOH used, after the reaction was stopped and allowed to stand overnight in separating funnel, the two layers of biodiesel and glycerol was not formed. Transesterification could not take place properly with an insufficient amount of an alkali catalyst loading [18].

At the loading of 5 wt % NaOH catalyst, mixture of methanol and crude biodiesel produced directly formed a solid soap after addition of catalyst. At higher loading of NaOH catalyst, the intensification of mass transfer became more important than increasing the amount of catalyst [19]. Marchetti et.al (2007) [16] reported that when the amount of free fatty acids in the feedstocks exceeds 0.5 %, the use of the alkaline homogeneous technology, which employs sodium hydroxide as catalyst, is not recommended, because a soap making reaction will take place and will consume the catalyst as well as the raw material but also it will make more difficult the downstreaming separation and purification of the biodiesel. This research showed that the biodiesel yield will increase as the NaOH loading increased in the range of 1.0 to 3.0 wt.% and decrease as the catalyst loading exceed 3.0 wt.%. From this experiments 74 % and 86.67 % of biodiesel yield was obtained at the concentration of NaOH catalyst of 1 % and 3 %, respectively.

In this research, the effect of reaction stages of biodiesel production against biodiesel yield were also studied. At one stage system, waste cooking oil with FFA of 6.77 % undergo transesterification process with ratio catalyst to oil of 1 wt %/v, ratio volume methanol/oil of 1.5, temperature of 60 °C and reaction time of 1 hour. The results showed that biodiesel yield with single stage system obtained 66.67 %. Meanwhile, two stages of transesterification process can produced 86.67 % of biodiesel. Issariyakual et al. (2007) [6] reported that more than 90% ester was obtained when two-stage method was used compared to 50 % ester in single stage alkaline catalyst. Issariyakual et al. (2007) [6] was conducted using homogeneous catalyst; esterification stage using concentrated sulfuric acid and transesterification process with potassium hydroxide as catalyst.

3.5 Characteristics of biodiesel product

Table 2 presents biodiesel characteristics from crude palm oil produced by our research's and Pelita Agung Agrindustri, Inc., located in Dumai-Pekanbaru, Indonesia. Most of the key parameters of biodiesel produced found to be met the standards of biodiesel in Indonesia.

Tabel 2. The comparison of biodiesel characteristics produced from ASTM standard, Pelita Agung Agrindustri, Inc. and this research

Key Parameters	Biodiesel			
	Diesel fuel ASTMD975	Crude Palm Oil	Cooking Oil	Unit
Density (40 ^{0}C)	850 – 890	858	857.3	kg m^{-3}
Density (32 ^{0}C)	850 – 890	864.8	863.1	kg m^{-3}
Kinematic viscosity (40^{0}C)	2.3 – 6.0	4.545	4.595	mm^2 s^{-1}
Water and sediment	0.05 max	0.0276	0.0359	% (v/v)
Acid number	0.80 max	0.271	0.383	mg KOH g^{-1}
Free glycerol	0.02 max	0.005	0.015	% (mm^{-1})
Total glycerol	0.24 max	0.2271	0.2571	% (mm^{-1})
Ester methyl content	96.5 min	98.42	98.18	% (mm^{-1})
Color	2.5 max	L 1.0	L 1.0	Color ASTM

4. Conclusions

This study synthesized sulfated alumina catalyst and evaluated its performance in esterification reaction of waste cooking oil. The catalyst performance was compared with homogeneous catalysts of concentrated sulfuric acid. The optimum conditions of waste cooking oil esterification reaction with sulfated alumina catalyst were the volume ratio methanol/oil of 1.5, the amount catalyst/oil of 1 wt

%/v and reaction time of 1 hour. Under the optimum esterification condition, the maximum conversion of free fatty acid using the synthesized catalyst and sulfuric acid were 89.97 and 90.4 %, respectively. Sulfated alumina- acid heterogeneous catalysts has potential as a replacement of sulfuric acid homogeneous catalyst. The maximum biodiesel obtained from waste cooking oil is 86.67 % through two stages while one through stage was only 66.67 %.

References

[1] Filho S C S, Silva T A F., Miranda A C, Fernandesa M P B, Felício H H, Calarge F A, Santana J C C and Tambourgi E B 2014 *Chem. Eng. Trans.* **37** 577-582.
[2] Kiakalaieh A T, Amin N A S and Mazaheri H 2013 *Appl. Energy* **104** 683-710.
[3] Director General of Estate Crops, 2014, The Estate Crop Statistic of Indonesia 2013-2015<ditjenbun.pertanian.go.id/tinymcpuk/gambar/file/statistik/2016/SAWIT%202014-2016.pdf> Accessed in March 2018.
[4] Chetri A B., Watts K C and Islam M R 2008 *Energies* **1** 3-18
[5] Rodríguez D, Riesco J and Malagon-Romero D 2017 *Chem. Eng. Trans.* **57** 679-684.
[6] Issariyakul T, Kulkarni M G, Dalai A K and Bakhshi N N 2007 *Fuel. Proc. Technol.* **88 (5)** 29–36.
[7] Mbaraka I K and Shanks B H 2006 *J. Amer. Oil Chem. Soc.* **83** 79–91.
[8] Dossin T F, Reyniers M F, Berger R J and Marin G B 2006 *Appl. Catal. A: General* **67** 136–48.
[9] Ulfah M and Subagjo 2012 *Reaktor* **14(2)** 151-157.
[10] Ulfah M and Subagjo, 2016 *Reaktor* **16(2)** 109-115.
[11] Nayebzadeh H, Saghatoleslami N, Maskooki A and Vahid B 2014 *Chem. Biochem. Engineering Quarterly* **28(3)** 259-265.
[12] Rane A S, Pudi S M and Biswas P 2016 *Chem. Biochem. Eng.* **30(1)** 33-45.
[13] Maneerung T, Kawi S, Dai J and Wang J H 2016 *Energy. Conv. Manag.* **123** 487–497.
[14] Wang H G, Shi G L, Yu F and Li R F 2016 *Fuel. Proc. Technol.* **145** 9–13.
[15] Gardy J, Osatiashtiani A, Céspedes O, Hassanpour A, Lai X, Lee A F and Rehan M 2018 *Appl. Catal. B: Envir.* **234** 268–278.
[16] Marchetti J M, Miguel V U and Errazu V U 2007 *Fuel* **86** 906–910.
[17] Ramachandran K, Sivakumar P, Suganya T and Renganathan S 2011 *Biores. Technol.* **102** 7289–93
[18] Cayh G and Küsefog S 2008 *Fuel. Proc. Technol.* **89** 118–22.
[19] Patil P D, Gude V G, Reddy H K, Muppaneni T and Deng D 2012 *J. Envir. Protec.* **3** 107-113.

IOP Conf. Series: Materials Science and Engineering **543** (2019) 012008 doi:10.1088/1757-899X/543/1/012008

Kinetic modelling of esterification of glycerol with acetic acid catalyzed by lewatite

L Setyaningsih[1], T Mutiara[1], A Chafidz[1], A Zulkania[1], M A Andrianto[1], N L Farikha[1]

[1]Department of Chemical Engineering, Islamic University of Indonesia, Yogyakarta, INDONESIA

E-mail: lucky.nuzulia@uii.ac.id

Abstract. One of the ester product of glycerol and acetic acid is triacetin which has wide industrial aplication. In this study the esterification of glycerol with acetic acid over Lewatite catalyst to produce triacetin, were carried out. Lewatite is one of ion exchange resin, the use of ion exchange resins as catalyst will not produce waste, not corrosive, high thermal stability and has good selectivity. The esterification reaction of glycerol with acetic acid were performed in a three-neck flask with stirrer (175 rpm) at temperature (90 °C, 100 °C and 110 °C) and atmosphere pressure. A 9.66 % selectivity of triacetin with conversion (77.4 %) was achieved at 100 °C within 90 minutes of reaction time. This research aimed to report experimental data and kinetic modeling of glycerol esterification with acetic acid to evaluate and compare model performance and capability. A kinetic model was developed using a first order pseudo-homogeneous model according to the experimental data. The surface reaction of acetic acid with glycerol was assumed to be rate-controlling. The effect of temperature to rate of reaction constants can be expressed as follows: $k_r = 8.138\ x\ 10^{11} \exp\left(-\frac{57{,}176.47}{RT}\right) \left({}^{1}/_{s}\right)$

1. Introduction

Crude glycerol produced in biodiesel industry need to be purified so can be utilized in cosmetic, food and pharmaceutical industry. However, glycerol purification is high cost, so it is necessary to look for high value derivative product of glycerol [1]. One of the ester product of glycerol and acetic acid is triacetine which has wide industrial application [2]. Triacetins are used as food aroma, solvent in perfume [3], plasticizer for resin [4], and as additive in fuel oil to increase octane number [5]. Some short chain oils such as triacetine, tripropionine, tributyrin have been proved to be good biodiesel additives to improve lubrication and viscosity [6]. In general, these reactions are carried out using homogeneous acid catalysts, which are corrosive, difficulty in separation and generate waste, leading to increased production costs and environmental problems [7]. Hence, the heterogeneous catalysts become alternative catalysts used in esterification reactions because has good physical and chemical properties [8]. The use of ion exchange resins will not produce waste, not corrosive, high thermal stability and has good selectivity [9]. In addition, heterogeneous catalysts were applied in various chemical synthesis reactions, indicating high and sustained catalytic activity [10].

Previous research was conducted experimentally to obtain optimum conditions in esterification reaction of glycerol and acetic acid over Lewatit catalyst, which obtained 3 % weight catalyst and

molar ratio of glycerol reactants to acetic acid was 1: 7. However the development of kinetic models can be relied upon to describe the design and optimization process. This research aimed to report experimental data and kinetic modeling of glycerol esterification with acetic acid to evaluate and compare model performance and capability. The reaction pathway is very important to use to build a kinetic model that can be applied to predict changes in the number of specific products obtained during the reaction process [11].
According to Smith [12], the liquid phase reaction with solid catalyst comprises the following steps:

- Mass transfer of the reagents from the liquid bulk to the catalyst surface
- Reaction at the surface of the catalyst
- Mass transfer of the product from surface of catalyst to the bulk of liquid

In irreversible reactions, the product concentration does not affect the rate of reaction. If the direct reaction and desorption stage are rapid, mass transfer of the product from surface of catalyst to the bulk of liquid can be ignored. Due to excess of acetic acid, the concentration of acetic acid in catalyst surface are assumed to be constant, then in steady state conditions obtained:

$$\frac{dX_A}{dt} = K_0 \frac{3}{C_{A0}} \left(C_{G0} - \frac{1}{3} C_{A0} X_A \right) \tag{1}$$

K_o values can be searched by minimizing the sum of square of errors (SSE) using the Matlab program of calculated acetic acid conversion and actual acetic acid conversion. The value of mass transfer coefficient (kc) can be searched by using equation as follows:

$$SSE = \sum (X_{calculation} - X_{data})^2 \tag{2}$$

The value of mass transfer coefficient (kc) can be searched by using equation as follows:

$$k_c = \frac{S_h \, D_{GA}}{d_p} \, a.m \tag{3}$$

Sherwood (Sh) value is calculated using the following Frossling equation:

$$S_h = 2 + 0.6 \, Re^{1/2} + Sc^{1/3} \tag{4}$$

According to Fogler [13], the overall rate of the reaction is determined by the slowest rate in the process mechanism. By applying the value of mass transfer coefficient kc using (equation 3) and data Ko, the modelling simulation will result the value of kr (constant rate of reaction). If the value of kc » kr indicate that reaction on the surface catalyst control the rate reaction so $Ko \approx kr$, otherwise if the value of kr » kc indicate that the mass transfer of reagents to catalyst is the rate controlling.

2. Methods

2.1. Materials

Lewatite catalyst and glycerol of high purity (96 %) was obtained by CV. Bratachem, Indonesia. Acetic acid (99 %), chloroform, ethanol, natrium thiosulfate and potassium dichromate were purchased from CV. Chemics, Indonesia. All reagent used without any further treatment. Lewatit catalyst was activated by HCl 1 N solution of 200 % volume. The catalyst mixture was stirred using a magnetic stirrer for 30 minutes and rinsed with distillated water to neutral pH.

2.2. Experimental procedure

The esterification reaction of glycerol with acetic acid were performed in a three-neck flask (250 ml) with stirrer (175 rpm) at temperature (90 °C, 100 °C and 110 °C) and atmosphere pressure. 73.31 ml glycerol was placed in the reactor together with catalyst and heated to reaction temperature. After the

IOP Conf. Series: Materials Science and Engineering **543** (2019) 012008 doi:10.1088/1757-899X/543/1/012008

mixture reaches temperature reaction, the 344 mL acetic acid that has been heated in a separate container is inserted into the reactor. To maintain the reaction temperature, the reactor was equipped with a water cooled condenser. Prior the reaction, solution sample was taken for analysis of total acid concentration, free acid concentration, and total glycerol concentration. The reaction product were taken every 15 minutes to analyze the free acid. The reaction was stopped after a reaction time of 90 minutes. The experiment was repeated with other temperature variations. The total acid (Ao) analysis was performed by volumetry using chloride acid, while the free acid (Ab) was analyzed using NaOH, and the total glycerol was analyzed by the FBI-A02-03 method. Calculation of the conversion of reagents based on acetic acid (free acid) were calculated using the following equations:

$$Acetic\ acid\ conversion\ (\%) = \frac{A_0 - A_b}{A_0}\ x\ 100\% \quad (5)$$

3. Result and discussion

3.1. Qualitative analysis

The synthesized triacetin was characterized by Fourier Transform Infra Red Spectrophotometer (FTIR) and Gas Chromatografy Spectrometry (GC). FTIR analysis were intended to determine the presence of triacetin in sample. Figure 1 shows the IR spectra of reaction product. As displayed in Fig. 1 an intense and broad peak at 1724.26 cm^{-1} assignable to the bending vibration a functional group of ester. From (Figure 2) found that the GC spectra of samples have the highest similarity with the triacetin of 98 %. These results prove that product of reaction contain triacetin.

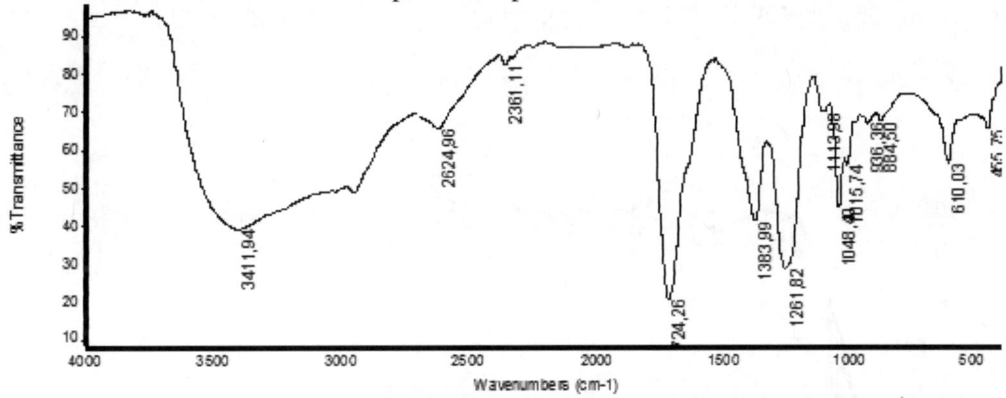

Figure 1. FTIR spectra of esterification reaction product

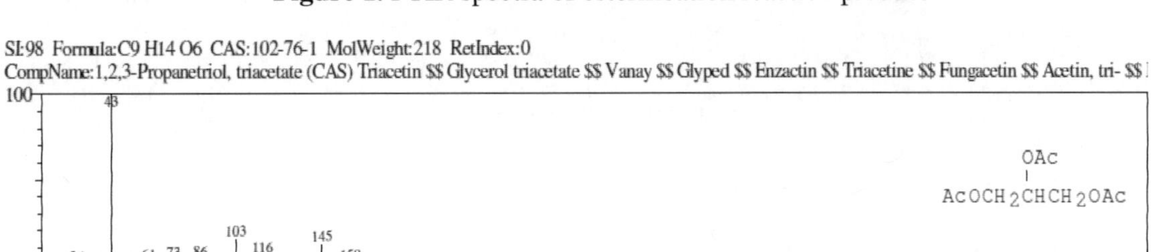

Figure 2. Gas chromatografy spectra of esterification reaction product

3.2. Acetic acid conversion

The effect of temperature on the esterification reaction is studied by reacting acetic acid and glycerol at different temperatures with a fixed initial molar ratio (1:7) and fixed catalyst loading (3% of weight reactant). The effect of reaction temperature is very important to observe for calculating the activation energy of the reaction [14]. Figure 3 indicated that temperature reaction is the significant factor in acetic acid conversion which increasing temperature will increase acetic acid conversion, indicating that mass transfer can be ignored. The highest conversion at the temperature range of 80 °C – 110 °C was reached at 100 °C at value of 77.4 % after 90 minutes of reaction.

Table 1 shows the selectivity of the triacetin obtained by the esterification of glycerol with acetic acid over lewatit catalyst after 90 minutes of reaction. It can be seen that the increase of temperature increase the triacetin selectivity, so triacetin formation was predominant at higher reaction temperature. At temperature 100 °C, the selectivity of 82.86 % to monoacetin, 7.47 % to diacetin and 9.66 % to triacetin. Low selectivity to triacetin can be explained due to this product will be formed through the consecutive nature of glycerol esterification, which can be attributed to the occurrence of nucleophilic attacks by glycerol to form medium asylum ions on the surface of the catalyst and protonation of carbonyl oxygen atoms from acetic acid [15]. This intermediate material is preferably close to the glycerol molecule and subsequently produces successive nucleophilic attacks on the hydroxyl glycerol group. Other reason of low selectivity of triacetin because there is no sufficient time for monoacetin and diacetin convert to triacetin. The increasing reaction temperature up to100 °C the conversion and selectivity increase, while at 110 °C both of conversion and selectivity decrease, this might relate with the maximum operation temperature of Lewatite catalyst (120 °C) resulting in a decrease of catalyst activity. Decrease in catalyst activity might due to the internal diffusion limitations inside the porous of catalyst. Several studies concluded that catalyst activity was predominantly influenced by acidity and textural characteristics of catalyst [16]. Fouling and sintering on surface catalyst were reported as the main causes of catalyst deactivation [17].

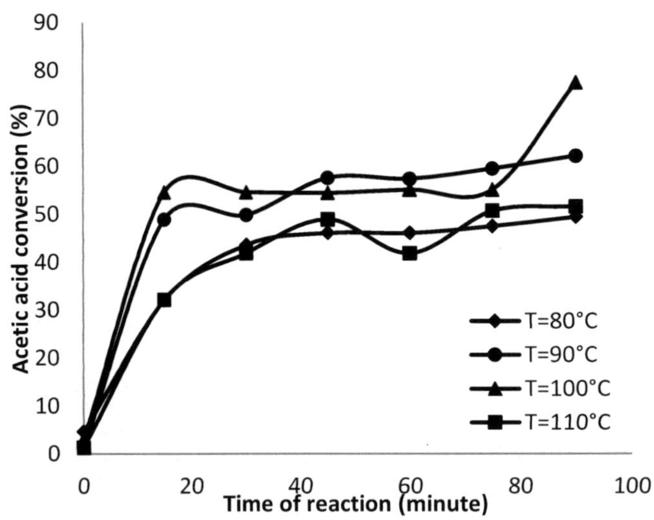

Figure 3. Acetic acid conversion versus time for different reaction temperature

Table 1. Selectivity of monoacetin, diacetin and triacetin at various reaction temperature

Temperature, °C	Monoacetin	Diacetin	Triacetin
90	41.47 %	53.31 %	5.32 %
100	82.86 %	7.47 %	9.66 %
110	49.07 %	38.51 %	4.81 %

IOP Conf. Series: Materials Science and Engineering **543** (2019) 012008 doi:10.1088/1757-899X/543/1/012008

An increase in the reaction temperature effect to the energy of reactant molecule in achieving its activation energy. This causes the collision between reactant molecules occur in high frequency leading to an increase in the reaction rate. The activation energy (E_0) and the parameter constant (k_0) are determined by using the linear regression technique of the Arrhenius equation (Equation 6) by entering the data of the reaction rate constant (k) at different temperature.

$$k = k_0 \exp\left(\frac{-E_0}{RT}\right) \quad (6)$$

3.3. Qualitative analysis

The kinetic study of the esterification of acetic acid and glycerol was performed to obtain wider information of the conversion of chemical species in the reaction mixture. A set of reaction at several temperatures have done to select suitable kinetic model represent behaviour of the reaction. Glycerol esterification could be assumed as irreversible reaction and the kinetic data were fitted to first order pseudo-homogeneous model that considers a constant ratio mole reactant and catalyst loading inthe reaction liquid. The selection of the appropriate model was based on the value of activation energie within adequate range, standard error value and positive value of kinetics parameter. kinetics constants for chemical reactions, adsorption, and the phenomenon of mass transfer are usually influenced by temperature, changes in solubility with temperature depend on the nature of solutes and solvents. Therefore, solubility may not depend on temperature, at least under certain conditions [18]. Although the internal mass transfer in a heterogeneous catalyst might control the reaction rate, it is possible that the rate control step can be neglected for porous support or systems with partially dissolved catalysts. At the liquid-solid heterogeneous reaction, the mass transfer resistance affects the rate resistance of the reaction, which both of resistors can be combined in the overall resistance constant (K_0). The result of the calculation based on the proposed mathematical model is shown in (Figure 4). As shown in (Figure 4), the model used is good enough to describe the process that occurs during the reaction process. To determine a more decisive step between mass transfer and reaction rate, the mass transfer coefficient (kc) and the reaction rate coefficient (kr) value must be known. The kc value is calculated by (Equation 3), thus the value of kr at various temperatures can be determined. The calculation results of kc and kr values are presented in (Table 2).

Table 2. Overall constant, mass transfer constant and rate constant value at various temperature

Temperature		Overall Constant	Mass Transfer Constant	Reaction Rate Constant
		(K_0) x 10^{-3}	(kc)	(kr) x 10^{3}
°C	°K	1/s	1/s	1/s
80	353.15	3.21	38.9	4.53
90	363.15	9.77	40.08	10.11
100	373.15	12.53	40.76	13.57
110	383.15	20.16	40.90	21.15

As seen in (Table 2), the value of Ko is almost equal to kr. This shows that the rate of reaction is a decisive / controlling step in reaction. This is in accordance with previous studies with ion exchange resin catalysts such as Pereira [19] who evaluated kinetics and thermodynamic equilibrium in esterification of lactic acid with ethanol using Amberlyst-15. In esterification of citric acid with ethanol using an ion exchange resin catalyst, the reaction kinetics is determined by chemical reactions [20]. The reaction rate constant of kr is a function of temperature, and is considered to follow the Arrhenius equation as shown in (Figure 4).

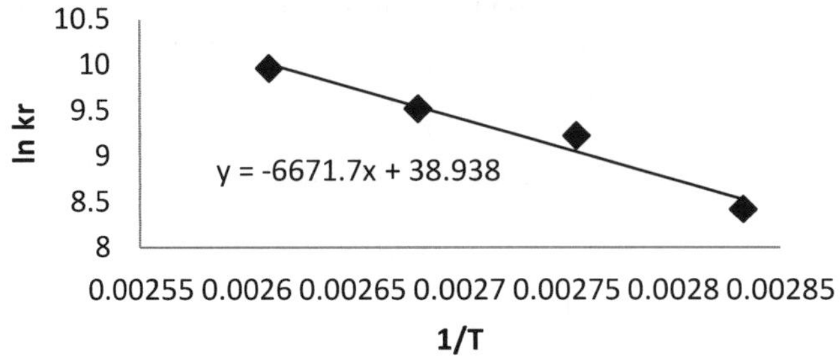

Figure 4. Arrhenius plot of the rate constant (kr)

Regarding experimental kinetic data for esterification of acetic acid with glycerol presented in this study, higher glycerol conversion and selectivity of monoacetin in all reaction can be observed and acetic acid quickly reacts with glycerol to form monoacetin, diacetin and triacetin produced according to the reaction temperature. The From the result of the linearization of the Arrhenius equation in (Figure 4), the effect of temperature to rate of reaction constants can be expressed as follows (equation 7) with an average error of 23.55%. Activation energy and frequency constant were obtained 57,176.47 kJ/mol and 8.138 x 10^{11} 1/s respectively. The adjusted correlation coefficient (R^2) of 0.9645 obtained for Arrhenius model, which indicates good performance of the models in predicting the kinetic behaviour for these esterification reaction investigated and it is in accordance with the assumption made that mass transfer step is negligible for type of Lewatit catalyzed reaction. Therefore the reaction rate in the Lewatit esterification of glycerol seems to be the controlling step.

$$k_r = 8.138\,x\,10^{11}\exp\left(\frac{-57{,}176.47}{RT}\right)\left({}^{1}/_{s}\right) \quad (7)$$

4. Conclusion

The esterification reaction of glycerol with acetic acid was carried out over Lewatite catalyst at different temperatures (80°C, 90°C, 100°C and 110°C) with a fixed initial molar ratio (1:7) and fixed catalyst loading (3% of weight reactant). Among the various temperature reaction, 100°C presented highest conversion (77.4%) with 9.66 % selectivity of triacetin within 90 minutes of reaction time. The model used is good enough to describe the process that occurs during the reaction process. The rate constant of reaction was obtained by plotting temperature of Arrhenius equation. A model with a coefficient of determination of 0.9645 was obtained and the activation energy for triacetine formation using Lewatit catalyst was found to be -57kJ/mol. Mathematical model calculations show the surface reaction is a controlling step in reaction.

5. References

[1] Hunpinyo P and Narataruksa P 2016 Process simulation and costing study for the FT-Liquid fuels production from steam glycerol reforming *Chem. Eng. Trans.* **52** 655

[2] Ferreira P, Fonseca I M, Ramos A M, Vital J and Castanheiro J E 2009 Esterification of glycerol with acetic acid over dodecamolybdophosphoric acid encaged in USY zeolite *Catal. Commun.* **10** 481

[3] Nabeshima H and Ito K 1995 *J.P. Patent* 276787

[4] Hofmann P D E 1985 *Patent* 3512497

[5] Nomura S and Hyoshi T 1995 *J. P. Patent* 203429

[6] Ruiz A E B, Plesu V, Ruiz J B, Tohaneanu M C, Llorens J, Lepinay M and Dineiro C 2015 Experimental study of short chain oils viscosity as biodiesel additives *Chem. Eng. Trans.* **45** 09

[7] Lotero E, Goodwin Jr J G, Bruce D A, Suwannakarn K, Liu Y, and Lopez D E 2006 in *J. Spivey (Ed.)* Catalysis (London: Royal Society of Chemistry)

[8] Gangadwala J, Manker S and Mahajani 2003 Esterification of acetic acid butanol in the presence of ion-exchange resins as catalysts *Ind. Eng. Chem.* **42** 46

[9] Ali S H and Merchant S Q 2006 Kinetics of the esterification of acetic acid with 2-propanol: Impact of different acidic cation exchange resins on reaction mechanism *Int. J. Chem. Kinet.* **38** 593

[10] Zhu S, Zhu Y, Gao X, Mo T and Li Y 2013 Production of bioadditives from grycerol esterification over zirconia supported heteropolyacids *Bioresource Technol.* **130** 45

[11] Hermida L, Abdullah A Z and Mohamed A R 2011 Synthesis of Monoglyceride Through Glycerol Esterification with Lauric Acid Over Propyl Sulfonic Acid Post-Synthesis Functionalized SBA-15 Mesoporous Catalyst *Chem. Eng. J.* **174** 668

[12] Smith J M *Chemical Engineering Kinetics* (Singapore: McGraw Hill International Book Co.)

[13] Fogler S H *Elements Of Chemical Reaction Engineering* 4th Edition (Prentice Hall International Series in the Physical and Chemical Engineering Sciences)

[14] Toor A P, Sharma M, Kumar G and Wanchoo R K 2011 Kinetic study of esterification of acetic acid with nbutanol and isobutanol catalyzed by ion exchanged resin *Chem. Reac. Eng. Catal.* **6(1)** 23

[15] Khayoon M S and Hameed B H 2011 Acetylation of Glycerol to Biofuel Additives Over Sulfated Activated Carbon Catalyst *Bioresource Technol.* **102** 29

[16] Kiakalaieh A T and Amin N A S 2017 Kinetic study on catalytic conversion of glycerol to renewable acrolein *Chem. Eng. Trans.* **56** 241

[17] Atia H, Armbruster U, Martin A 2008 Dehydration of glycerol in gas phase using heteropolyacid catalysts as active compounds *J. Catal.* **258** 71

[18] Ladero M, Gracia M D, Tamayo J J, Ahumada I L D, Trujilo F and Garcia-Ochoa F 2011 Kinetic Modelling of The Esterification of Rosin and Glycerol: Application to Industrial Operation *Chem. Eng. J.* **169** 319

[19] Pereira C S M, Pinho S P, Silva V M T M and Rodrigues A E 2008 Thermodynamics Equilibrium and Reaction Kinetics for the Esterification of Lactic Acid with Ethanol Catalyzed by Acid Ion-Exchange Resin *Ind. Eng. Chem.* **47** 53

[20] Kolah A K, Astana N S, Vu D T, Lira C T and Miller D J 2007 Reaction Kinetics of the Catalytic Esterification of Citric Acid with Ethanol *Ind. Eng. Chem.* **46** 80

IOP Conf. Series: Materials Science and Engineering **543** (2019) 012009 doi:10.1088/1757-899X/543/1/012009

A Kinetics study of 2-ethyl-2-hexenal hydrogenation to 2-ethyl-hexanol over Nickel based catalyst

S Virgana, Subagjo and M L Gunawan*

Department of Chemical Engineering, Faculty of Industrial Technology, InstitutTeknologi Bandung, JalanGanesa No 10, Bandung-40132, Indonesia

*Email: melia@che.itb.ac.id, melialaniwati@gmail.com

Abstract. The 2-ethyl-hexanol was much-needed in the plasticizer industry as a precursor for the synthesis of bis diesters (2-ethyl hexyl) phthalate (DEHP). This compound can be synthesized by oxo reaction, where one of the steps isa consecutive hydrogenation of the 2-ethyl-2-hexenal to 2-ethyl-hexanol. This reaction required metal-based catalyst. This research aimed to develop the formula and procedure of catalyst synthesized and get a kinetics of the reaction. The catalyst was synthesized with a nickel content of 50%-w. The support consists of γ-alumina and SiO_2(cab-osil) in variations of 0%-w up to 100 %-w of cab-o-sil towards support. The activity of synthesized catalysts were tested at 120°C and 30 bars in the fixed bed reactor. The results showed that the best performing catalyst was NiO50-Cab50 with 98.29% 2-ethyl-2-hexenal conversion and selectivity to 2-ethyl-hexanol 87.41 %. By varying the reaction temperature of 100-120°C in the best catalyst, it is found that the consecutive reaction was the first order with the activation energy of 2-ethyl-2-hexenal hydrogenation to 2-ethyl-hexanal was 33.66 kJ/mol and hydrogenation 2-ethyl-hexanal to 2-ethyl-hexanol was 58.39 kJ/mol. This showed that the hydrogenation of C = C bonds was easier than the C = O bond.

1. Introduction

The 2-ethyl-hexanol compound was required in the plasticizer industry as a precursor for the synthesis of bisdiesters (2-ethyl hexyl) phthalate (DEHP). In addition, 2-ethyl-hexanol was used as solvents, oils and additives, paint and coating additives, fuel and fuel additives, agricultural chemicals (non-pesticides), and intermediates. Until now, 2-ethyl-hexanol was produced in the industry through oxo process with synthesis gas and propylene as raw materilas [1]. One of the steps in the oxo process was a consecutive hydrogenation of the 2-ethyl-2-hexenal to 2-ethyl-hexanol via the 2-Ethyl-hexanal or 2--ethyl -2-hexenol as intermediate products. The objective of this research was to get the kinetics model of 2-ethyl-2-hexenal hydrogenation on the best nickel based catalyst, which gives the high selectivity to 2- ethyl-hexanol.

2. Experimental

2.1. Catalyst synthesis

The hydrogenation catalyst consisted of two components, namely Nickel (Ni) active metal as the active phase (main component) and support (γ-alumina and cab-o-sil). As a metal source, Ni was selected by nitrate salt solution. The support was in the form of a mixture of alumina and silica synthesized from AIP2 (Indonesian Patent No: P00201701287 - *granted*) and cab-o-sil. The catalyst was synthesized by peptization method with 50 %-w nickel content and the composition of silica in the support was varied, that was 0 % (NiO50-Cab0), 20 % (NiO50-Cab20), 50 % (NiO50-Cab50), 80 % (NiO50-Cab80), and 100 % (NiO50-Cab100).

2.2. Characterization of catalyst

The physical properties of the catalyst, namely surface area, average pore diameter, and average pore volume were measured using the surface area with BET method and the pore size analyzer of the Quantachrome brand NOVA 320E type with the BET method. The hydrogen chemisorption profile was measured using Chem-BET Pulsar TPR/TPD (Temperature Programmed Reduction/Desorption). Identification of compounds of catalysts were measured using Bruker D8 Advance X-ray diffractometer with Cu anode. Analysis of the composition of the reaction mixture produced using a GC (gas chromatography) equipment type Shimadzu 2010 plus a type of column RTX-5, Flame Ionization Detector, and helium as carrier gas.

2.3.Catalyst activity test

The 2-ethyl-2-hexenal hydrogenation reaction was carried out in an isothermal fixed bed reactor at a temperature of 100-120°C, a pressure of 30 bar. Feed in the form of a mixture of 2-ethyl-2-hexenal and 2-ethyl hexanol with a ratio of 1:4 flowed into the reactor at a rate to provide a LHSV (Liquid Hourly Space Velocity) of 1-1.75 hours^{-1}. Hydrogen gas flowed at a rate of 30 ml/min. Previously the catalyst activation was carried out in situ using a gas with a flow rate of 30 ml/min, 500°C and a pressure of 3-7 bar for 3 hours. The reactor system setup was presented in Figure 1.

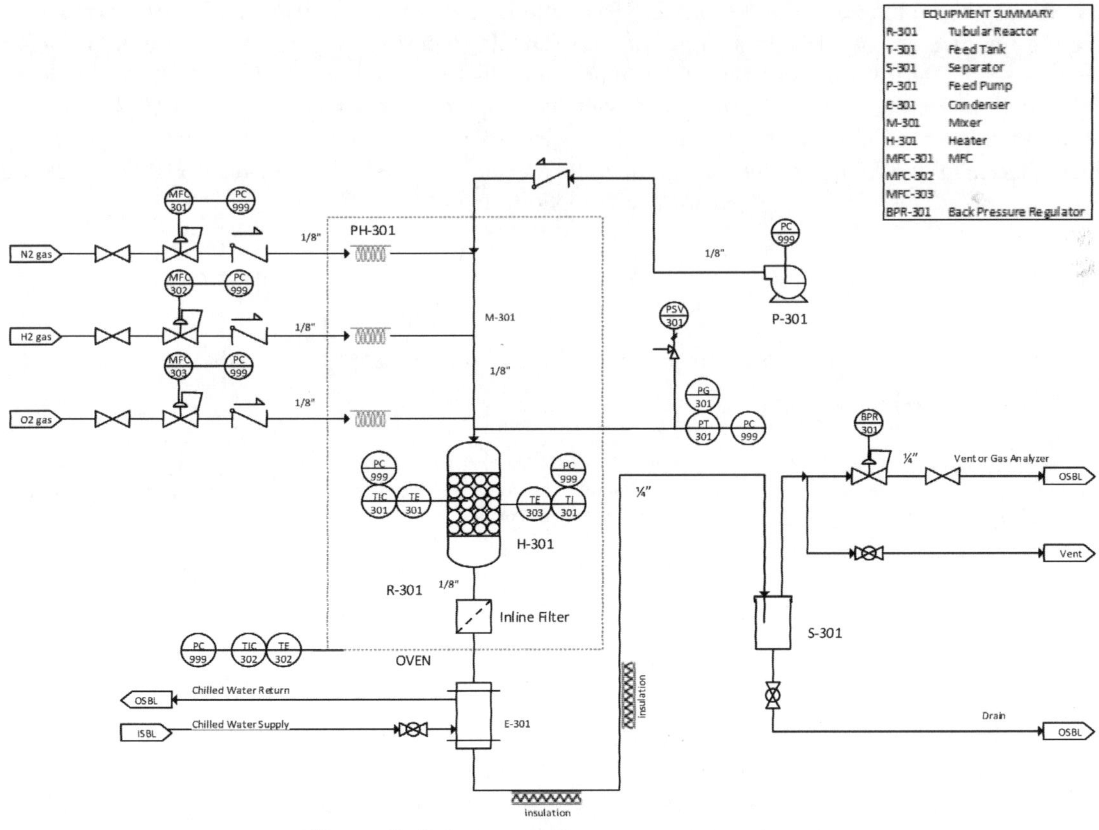

Figure 1. Set up a continuous reactor system

2.4.Kinetics Data Processing

The data obtained was processed using Microsoft Excel and MATLAB soft frames. The parameter values of the rank power law kinetics model were the reaction order, activation energy (Ea) and Arrhenius constant (k_0)

3. Results and Discussion

3.1. Characterization and activity test of catalysts

3.1.1 X-Ray Diffraction method results.

Analysis with the X-Ray Diffraction (XRD) method was used to determine the types of crystals contained in the synthesized catalyst. The diffractogram in Figure 2a showed several forms of the constituent crystal structure of the catalyst in the range of 2θ from 0° to 100°. X-Ray diffractogram catalyst compared with standard data available in JCPDS (Join Committee on Powder Diffraction Standards). From Figure 2a, there was 3 main peaks at diffraction angles (2θ) 37.13°, 43.18° and 62.78°, 75.33°, 79.32°, 95.02° which all show the form of NiO crystals (111), (200), (220), (311), (222), and (400). The diffractogram on the NiO50-Cab50 catalyst was not too different from the NiO diffractogram and did not indicate the peak of the SiO_2 and Al_2O_3 support crystals, probably because the support was completely mixed with NiO when preparing. In addition, sharp NiO crystal peaks can be caused by the occurrence of NiO agglomerations forming larger crystals due to the high temperatures used when calcination [2].

3.1.2 TPR method results.

Analysis of the Temperature Programmed Reduction (TPR) method was used to determine the reduction temperature of the synthesized catalyst. The TPR profile (Figure 2b) on the NiO50-Cab0 catalyst showed the two highest reduction temperature peaks, namely 480°C and 608°C while the NiO50-Cab50 catalyst showed the lowest reduction temperature of 450°C and 520°C. This can be caused by the interaction between NiO and Al_2O_3 which strongly form $NiAl_2O_4$ which was difficult to reduce, thus requiring a higher reduction temperature while the interaction between NiO and SiO_2 was weaker, so it was more easily reduced at lower temperatures.

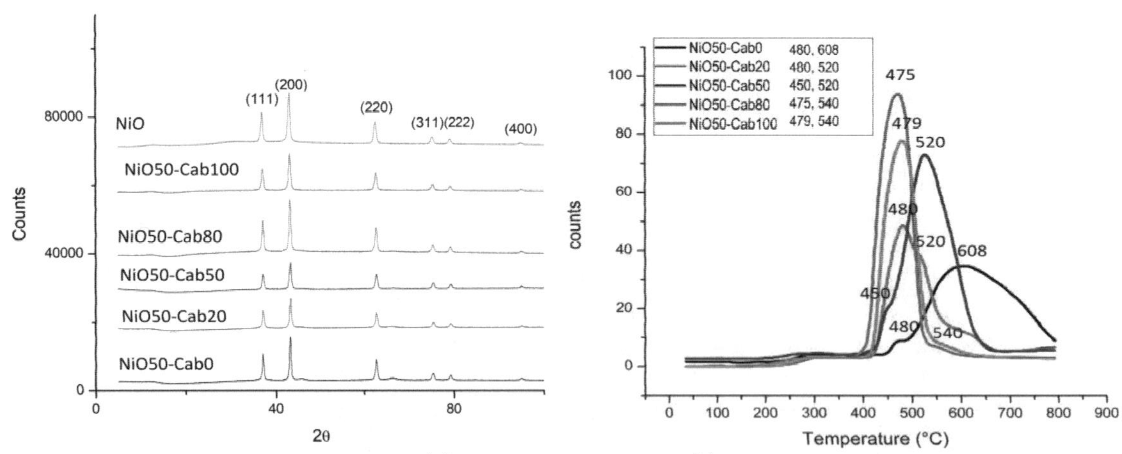

(a) (b)

Figure 2 Results of characterization with the (a) XRD method and (b) TPR method

The peak that appeared on the catalyst with cab-o-sil composition higher or equal to 50 had only one reduction peak, while the catalyst with cab-o-sil composition lower than 50% has more than one reduction peak. The peaks showed different species from NiO: the range 250-350°C showed the crystalline segregation form which had very weak interaction with the support, the range 350-500°C showed NiO which binded to the support, the range 500-750°C showed NiO which was strongly bounded to Al_2O_3 forms nickel aluminate ($NiAl_2O_4$). Species with very strong bonds between NiO and supports were shown at a higher reduction temperature because it was difficult.

3.1.3. BET method results.
Analysis with the BET method was carried out using Quantachrome NOVA's 3200E Surface Area Analyzer by using nitrogen as adsorbate. The results of catalyst analysis with the BET method were shown in Table 1. Characterization results on the catalyst sample indicate that the largest catalyst surface area was NiO50-Cab0. According to Gobara and Hassan [3], the catalyst with SiO_2 support produces the largest catalyst surface area. However, the phenomenon on catalysts with cab-o-sil which was getting higher shows the opposite result. The surface area of the synthesized catalyst with a larger SiO_2 support (higher cab-o-sil), did not produce a larger surface area. Probably, this was due to nickel metal sintering in SiO_2-supported catalysts due to weak Si-Ni interactions [4]. The weak Ni-Si interaction was supported by TPR data which shows the reduction temperature of the NiO50-Cab100 catalyst was lower, so it can be concluded that the interaction was easier to reduce at a lower temperature than the NiO50-Cab0 catalyst which had a higher reduction temperature.

Table 1 The physical properies of catalyst (using the BET method)

No.	Catalyst	Surface Area (m^2/g)	Pore Volume (cc/g)	Pore Size (Å)
1	NiO50-Cab0	188.873	0.22697	42.785
2	NiO50-Cab20	117.153	0.21794	37.206
3	NiO50-Cab50	106.870	0.25144	47.055
4	NiO50-Cab80	103.371	0.23373	45.222
5	NiO50-Cab100	117.510	0.24238	41.253

In the NiO50-Cab80 catalyst, the SiO_2 content was greater than that of Al_2O_3, resulting in more nickel metal deposited in SiO_2 while a weak Ni-Si bond was easily sintered at the time of reduction. This caused a decreased surface area and a larger pore size. Meanwhile, on the NiO50-Cab20 catalyst, the Al_2O_3 content was greater than SiO_2, producing more nickel metal deposited in Al_2O_3. The large surface area affected by the stronger Ni-Al bond was not easily subjected to sintering at the reduction temperature used.

3.1.4 Activity test results
In the activity test in a fixed bed reactor, the conversion data and selectivity were taken when the reaction reaches a steady state. The 2-ethyl-2-hexenal hydrogenation reaction was a consecutive reaction with intermediate products in the form of 2-ethyl-hexanal and 2-ethyl-hexanol final products. Although catalyst conversion was one indicator of catalyst activity, this indicator cannot measure how efficiently the catalyst worked. The performance of the catalyst also depends on selectivity. High selectivity allows few side reactions, so the acquisition of the main product becomes high.

Large pore size was thought to affect the value of selectivity. Based on the BET results, the NiO50-Cab50 catalyst showed the largest pore volume between synthesized catalysts that had the highest selectivity value. Consecutive selectivity values from the largest to smaller values: NiO50-Cab50 > NiO50-Cab100 > NiO50-Cab80 > NiO50-Cab0 > NiO50-Cab20 (Table 2). This sequence also applies to the pore volume measured by the BET method (Table 1).

IOP Conf. Series: Materials Science and Engineering **543** (2019) 012009 doi:10.1088/1757-899X/543/1/012009

Table 2 Performance of catalysts at 120 °C , 30 bar, LHSV 1 hour^{-1}

No.	Catalyst	Conversion, X (%)	Selectivity, S (%)
1	NiO50-Cab20	98.73	16.09
2	NiO50-Cab50	98.29	87.41
3	NiO50-Cab80	73.46	56.34
4	NiO50-Cab100	96.02	66.10

3.2. The Kinetics Model

3.2.1 Internal Diffusion Limitation

The pore diffusion significance can be seen from the Thiele Modulus (Φ) and effectiveness (η). Thiele modulus describes a decrease in the concentration of reactants in the pore of the catalyst [5]

$$\Phi^2 = \frac{k_1 R^2 S_a \rho_p C_{AS}^{n-1}}{D_e} \tag{1}$$

With k_1 being the reaction rate constant, R was the catalyst radius, Sa was the catalyst specific surface area, ρ_p was the catalyst density, C_{AS} was the concentration on the catalyst surface ($\frac{mol}{m^3}$), and D_e was the effective diffusivity in the liquid phase. From the calculation of the Thiele Modulus using equation (1), the value of Φ was 1.387. In this study, the value of the small value of the Thiele Modulus showed the rate of pore diffusion was fast and not as determining stage of reaction. The concentration profile in the catalyst pore, can be calculated by equation (2) and by equation (3) using FlexPDE.

$$\frac{d^2\varphi}{d\lambda^2} + \frac{2}{\lambda}\frac{d\varphi}{d\lambda} - \Phi^2\varphi = 0 \tag{2}$$

$$\varphi = \frac{C_A}{C_{AS}} ; \lambda = \frac{r}{R} \tag{3}$$

Figure 3 showed the distribution of reactant concentrations from the catalyst surface (r = R) to the center of the catalyst (r = 0) at various values Φ. The value of Φ 1.387 occurs only in all parts of the catalyst pore and the reaction reaches the center of the catalyst so that all parts of the catalyst were utilized. This showed that mass transfer in pores was fast and the reaction was not influenced or controlled by internal diffusion so that the study of catalyst kinetics was in a kinetic regime

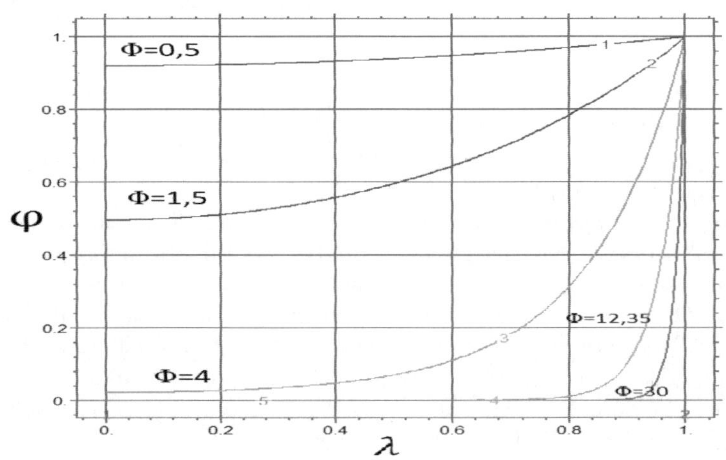

Figure 3 Concentration profiles in the pore catalyst at various values Φ

3.2.2 Determination of Kinetics Parameter

The 2-ethyl-2-hexenal hydrogenation reaction product was identified through GC method analysis. The results of the analysis in the form of chromatograms showed the peaks that appear at certain retention times according to the type of compound at the retention time. Identification was known based on previous research literature data through testing using a standard solution made from a mixture of t-butanal,i-butanal, n-butanal, i-butanol, n-butanol, 2-ethyl-2-hexenal and 2-ethyl-hexanol. The hydrogenation of 2-ethyl-2-hexenal products identified through the GC method analysis are 2-ethyl-hexanal and 2-ethyl-hexanol. The 2-ethyl-hexanal products were known as intermediate products. The 2-ethyl-hexanal product was formed because of the hydrogenation reaction of the C = C double bond in 2-ethyl-2-hexenal.

The hydrogenated 2-ethyl-2-hexenal consecutive reaction rate equation was shown in Equations 4-6:

$$r_{en} = -k_1 . C_{en}{}^{m} \tag{4}$$
$$r_{an} = k_1 . C_{en}{}^{m} - k_2 . C_{an}{}^{n} \tag{5}$$
$$r_{an} = k_2 . C_{an}{}^{n} \tag{6}$$

Optimization of kinetic parameters was done in order m, n, k_1, and k_2 values. Optimization was done by matching the data with a model that provided minimal FOBJ (objective function) values. The results of the optimization of the hydrogenation reaction at 120°C, 30 bar, obtained the value of m and n was 1 and the values of k_1 and k_2 are 1.9866 and 3.9498 ($mol.L.hr^{-1}$) respectively. Distribution of experimental data and optimization results by modeling using MATLAB (Figure 5) according to the mechanism of consecutive reactions in the proposed hydrogenation of 2-ethyl-2-hexenal as Figure 4.

Meanwhile, the determination of reaction kinetics parameters to get the activation energy value (Ea) and Arrhenius constant (k_0) in the 2-ethyl-2-hexenal hydrogenation reaction calculated using Microsoft Excel followed the rank law model using Equations 3-5 at various temperature variations (100-120°C). Experimental data curves on various temperature variations were shown in Figure 5. Data processing methods were performed by distributing ln k_1 and ln k_2 values to 1/T as Equation 7 and the results of the calculations were summarized in Table 3.

$$\ln k_i = \ln k_0 - \left(\frac{E_a}{Rg}\right)\frac{1}{T} \tag{7}$$

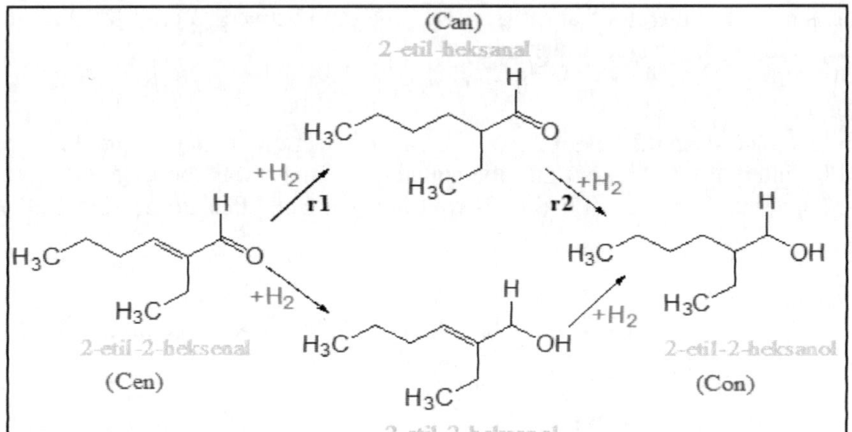

Figure 4 Hydrogenation 2-ethyl-2-hexenal pathway to 2-ethyl-hexanol [6]

Table 3 Kinetics parameter of hydrogenated 2-ethyl-2-hexenal consecutive reactions on NiO50-Cab50

No.	Reaction	Activation energy	Arrhernius contant
1	2-ethyl-2-hexenal $+H_2 \rightarrow$ 2-ethyl-hexanal	Ea_1 = 33.66 kJ/mol	$k_{01} = 2.19 \times 10^2$
2	2-ethyl-hexanal $+ H_2 \rightarrow$ 2-ethyl-hexanol	Ea_2= 58.39 kJ/mol	$k_{02} = 1.70 \times 10^5$

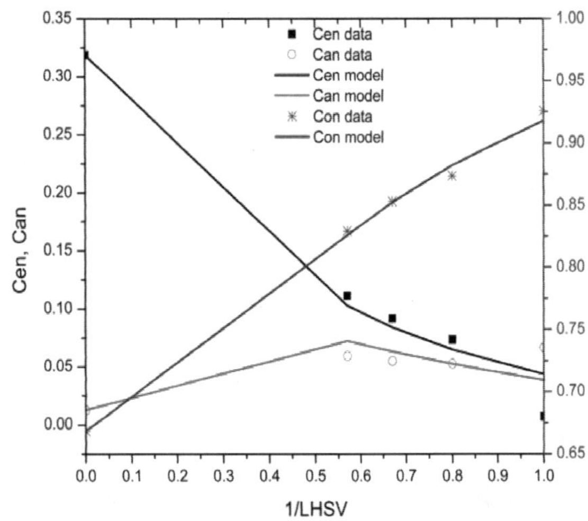

(a)

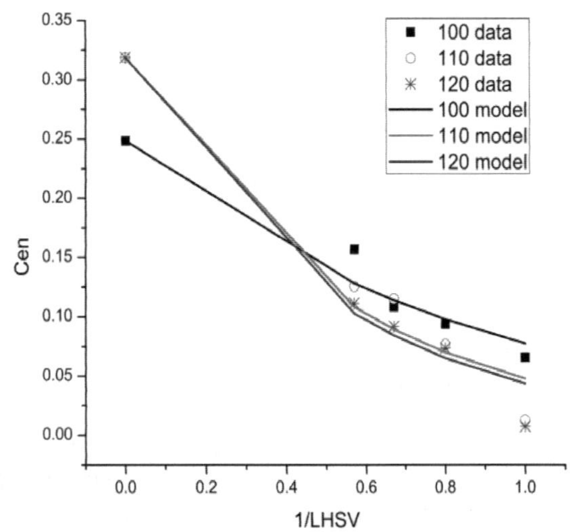

(b)

Figure 5 (a) Product and reactant concentration (en =2-ethyl-2-hexenal; an = 2-ethyl-hexanal; on = 2-ethyl-hexanol) in 2-ethyl-2-hexenal hydrogenation reaction at 120 ° C and 30 bar using NiO50-Cab50 catalyst (b) profile of 2-ethyl-2-hexenal concentration at various temperatures (100 -120°C).

The low value of activity energy showthat the reaction increases rapidly with increasing temperature. The values of $Ea_2 > Ea_1$ showed that the hydrogenation reaction in the C = C double bond to form 2-ethyl-hexanal was easier than hydrogenation in the C = O double bond to form 2-ethyl-hexanol. This mechanism illustrates that there was competition between 2-ethyl-2-hexenal and 2-ethyl-hexanal in occupying the active side of the catalyst [7].

4. Conclusions

The conclusion obtained from this research was the rate of reaction obtained from this experiment was not influenced by internal diffusion so that the equation obtained was the kinetics equation. NiO50-Cab50 catalyst converts 2-ethyl-2-hexenal to 2-ethyl-hexanal faster than converting 2-ethyl-hexanal to 2-ethyl-hexanol.

Acknowledgments

The authors would like to thank LPDP (Lembaga Pengelola Dana Pendidikan) and Institut Teknologi Bandung (ITB) for funding through P3MI for financial support on this research.

References

[1] Agency, U. S. E. P periode 2006 dari Inert reassessment: 2-ethyl-1-hexanol. Chemical Data Reporting, <*https://pubchem.ncbi.nlm.nih.gov/compound/2-Ethylhexanol#section=Food-Additives-and-Ingredients*> accessed 30.07.2017.
[2] Ashok J, Raju, G, ReddyP S, Subrahmanyam, M, and Venugopal A2008*J.Nat. Gas Chem.***17**113–9.

[3] Gobara HM and Hassan S A 2009*Petrol. Sci. Technol.***27** 1555–71.
[4] Guimon C, Auroux A, Romero E and Monzon A 2003*Appl. Catal., A***251**199–214.
[5] Both R, Neaga A O, Festila C and Dulf E H 2009 *J. of Chem. Eng.***15**1–6.
[6] Tuta E and Bozga G 2014 *Rev. Chim.-Bucharest* **65** 603–7.
[7] Smedler G 1989*Can. J. Chem. Eng.***67** 51–61.
[8] Both R, Cormos A M, Agachi P S and Festila C 2013 *J. Comput. Chem. Eng.* **52** 100–11.
[9] Collins D J, Grimes D E and Davis B H 1983*Can. J. Chem. Eng.* **61**36–9.
[10] GallezotP and Richard D 1998*Cat. Rev. - Sci. Eng.***40**81–126

Mathematical modelling of paddy drying using fluidized bed dryer

S Suherman[*] and E E Susanto

Department of Chemical Engineering Faculty of Engineering Diponegoro University
Jl Prof H. Soedarto, SH, Tembalang, Semarang, Indonesia 50275

[*]Email: suherman.mz@che.undip.ac.id

Abstract. Several mathematical models have been developed to illustrate the drying process of crops, and one of them is thin-layer modelling. However, there are still limited information about mathematical modelling in paddy drying. In this paper, thin layer modelling of paddy drying curve on drying using fluidized bed dryer is presented. The drying paddy was done with drying temperature variations of 50, 60, 70, 80, and 90 °C. The initial paddy moisture content is 33 % dry basis. The drying process was done for 1,5 hours. Paddy moisture was measured every 10 minutes using grain moisture meter. The thin layer models used are Newton, Page, Modified Page, Henderson and Pabis, Logarithmic, Two-term, and Wang and Singh models. Model fittings were done using MATLAB 2015 software by sum square error minimization. Experimental results showed that Modified Page model gave the best results, evidenced by the highest average R^2 value of 0.9949 and lowest average RMSE (Root Mean Square Error) value of 0.0631, followed by Page model and Henderson and Pabis model. The plotting of MR (Moisture Ratio) and time at five drying temperatures showed that effective diffusivity (Deff) value ranged from 1.0334×10^{-8} m²/s to 2.1983×10^{-8} m²/s and the value increases as drying temperature increases. The plotting results of ln D_{eff} and 1/T (absolute temperature) showed that the value of diffusivity factor (D_o) is 8.5293×10^{-6} m²/s and activation energy (Ea) is 18.11 kJ/mol.

1. Introduction

Paddy provides around 20 % of total energy per capita and 13 % protein for global population. In Asia, paddy contributes to 35 % energy and 28 % protein, in South America 12 % energy and 9 % protein. Paddy is the main food in several developing countries, contributing to 4000 kJ energy per capita everyday. Starch is the main component of paddy (±75 %) which located in the endosperm, shaped like granules with the size of 3-10 μm. Protein is the second component in paddy (±8 %), located inside the endosperm, shaped like particles with the size of 1-4 μm. Harvested paddy usually has moisture content of 20 % or more (wet basis), therefore it has to be dried before being kept in storage. In tropical countries, paddy is usually dried by direct sunlight until it reaches moisture content of 14 % (wet basis). At this condition, paddy can be stored for 2-3 months. If further storage time is needed, the paddy's moisture content must be lowered to 12 % (wet basis) using dryer [1].

Drying is one of the oldest and most commonly used food preserving methods. The most common drying method used until now is by drying directly under the sun (sun drying). However, recently there are lots of improvement and development on mechanical or semi-mechanical drying methods, which are cheap and effective in handling various adversities of sun drying like the need of wide area to dry the material,

fluctuation of solar radiation, weather change, contamination of microorganism, animal, and other chemical reactions. One of the artificial drying methods developed is Fluidized Bed Dryer (FBD) [2]. Fluidized Bed Dryer is a good alternative for paddy drying due to uniform product quality due to complete mixing and its high drying capacity due to high ratio of air mass to mass of product [3]. The drying rate of paddy in fluidized bed dryer was affected by drying air temperature and bed thickness [4]. The maximum drying temperature in fluidized bed was suggested as 115 °C to reduce moisture to 24-25% (d.b.) for ensuring rice quality [5]. The drying temperature in pulsed fluidized bed should be less than 145 °C for initial paddy moisture content of 28% (d.b.) to maintain rice quality [6].
Several mathematical models have been developed to illustrate the drying process. One of the modelling method commonly used on agricultural drying is thin layer modelling. There are several researches about thin layer modelling, such as by Jafari et al. (2017) about paddy drying in semi-industrial continuous band microwave dryer, Behera and Sutar (2018) about parboiling of paddy, and Li et al. (2016) about changes in moisture effective diffusivity and glass transition temperature of paddy during drying [7-9]. However, study about modelling of paddy drying in FBD has yet to be found. Therefore, the aim of this study is to determine the paddy drying kinetics on FBD by thin layer models and determine the value of activation energy, effective diffusivity, and its relation with temperature.

2. Experimental methods

2.1. Drying process

This study was performed in the Laboratory of Department of Chemical Engineering, Faculty of Engineering, Diponegoro University. The paddy used in this experiment was taken from Sragen region, Middle Java. 10 grams of paddy was put in a porcelain cup, then placed on 105 °C oven to determine the initial moisture content. The weight of the sample was measured every 10 minutes until it reaches constant value. From this test, paddy initial moisture content of 33 % dry basis was obtained. For drying operation, 200 grams of paddy were used. Paddy was inserted in the fluidization column to be dried for 90 minutes at different temperatures of 50, 60, 70, 80, and 90 °C. Every 10 minutes, a small portion of paddy was taken from the fluidization column to measure its moisture content using grain moisture meter.

2.2. Mathematical modelling

Seven drying models were used in this study. The modelling was done using MATLAB software. The definition of Moisture Ratio (MR) is shown in equation (1)

$$\mathrm{MR} = \frac{\mathrm{M}-\mathrm{M_e}}{\mathrm{M_o}-\mathrm{M_e}} \tag{1}$$

Where M is moisture content at certain time, Mo is initial moisture content, and Me is equilibrium moisture content. To determine the accuracy of the models, there are 2 parameters that can be used, namely Root Mean Square Error (RMSE) and determination coefficient (R^2). The calculations of RMSE and R^2 were done using Microsoft Excel. A model is said to be suitable if it has high R^2 value and low RMSE value. The calculations of RMSE and R^2 are expressed in equation (2) and (3) [10].

$$\mathrm{RMSE} = \sqrt{\frac{1}{N}\sum_{i=1}^{N}\left(\mathrm{MR_{exp,i}}-\mathrm{MR_{pre,i}}\right)^2} \tag{2}$$

$$R^2 = \frac{\left[\sum_{i=1}^{N}\left(\mathrm{MR_{exp,i}}-\overline{\mathrm{MR_{exp}}}\right)\left(\mathrm{MR_{pre,i}}-\overline{\mathrm{MR_{pre}}}\right)\right]^2}{\sum_{i=1}^{N}\left(\mathrm{MR_{exp,i}}-\overline{\mathrm{MR_{pre}}}\right)^2\sum_{i=1}^{N}\left(\mathrm{MR_{pre,i}}-\overline{\mathrm{MR_{pre}}}\right)^2} \tag{3}$$

Where Mr_{exp} is experimental Moisture Ratio, Mr_{pre} is Predicted Moisture Ratio, and N is number of observations.

2.3. Effective diffusivity and activation energy

Drying characteristics on falling rate period can be determined using Fick's diffusion equation. With the assumptions of spherical paddy coordinate, uniform initial moisture content distribution, and long drying time, the relation of Moisture Ratio and Effective Diffusivity can be expressed with equation (4) [8]:

$$MR = \frac{6}{\pi^2} \exp\left[\frac{-Deff\,\pi^2}{r^2}\theta\right] \tag{4}$$

Where D_{eff} is effective diffusivity (m^2/s), r is the radius of paddy (m), θ is drying time (s), dan π is a constant. By changing equation (4) into logarithmic form, a new linear equation is obtained, as shown in equation (5).

$$\ln MR = \ln\left[\frac{6}{\pi^2}\right] - \frac{Deff\,\pi^2}{r^2}\theta \tag{5}$$

The value of effective diffusivity (D_{eff}) can be determined by plotting the ln MR data versus drying time (θ). From the slope value obtained, effective diffusivity can be calculated. Furthermore, the relation of effective diffusivity and drying temperature can be illustrated with Arrhenius equation, as shown in equation (6) [11].

$$D_{eff} = D_o \exp\left[-\frac{Ea}{RT}\right] \tag{6}$$

Where D_o is diffusivity constant at infinite drying temperature (m^2/s), R is universal gas constant (8.314 J/mol.K), Ea is activation energy (kJ/mol), and T is absolute temperature (K). The value of Ea and D_o can be determined by plotting ln D_{eff} and 1/T, after linearization of the equation above, as shown in equation (7).

$$\ln D_{eff} = \ln D_o - \frac{Ea}{RT} \tag{7}$$

3. Results and discussion

3.1. Drying curve analysis

The relation of Moisture Ratio and drying time at five different drying temperatures is shown in Figure 1. From Figure 1, it is shown that as drying time increases, the Moisture Ratio decreases, and the higher the drying temperature, the reduction of Moisture Ratio become bigger. As a comparison, on research performed by Kamin and Janaun (2017) about paddy drying in a Laterally Aerated Moving Bed Dryer, at ambient temperature, it took 5 hours to reduce paddy's moisture ratio from 1 to approximately 0,72 [12]. It can be concluded that the whole process of paddy drying were happened on falling rate period and the most important factor that affects paddy drying is moisture diffusion.

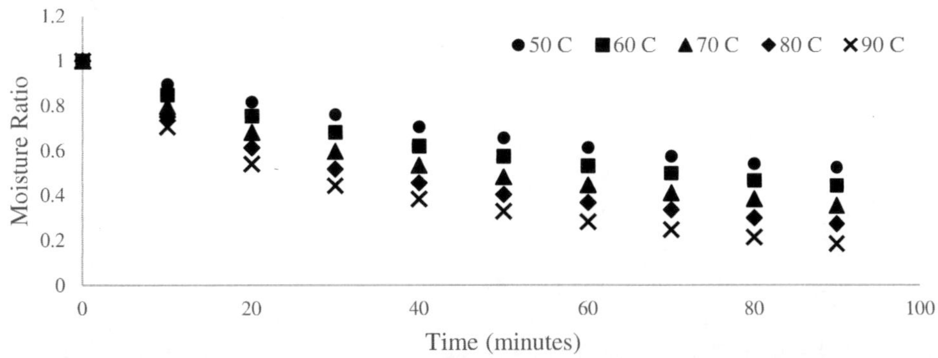

Figure 1. Drying curve of paddy using Fludized Bed Dryer at different temperatures

3.2. Modelling results

Seven models of thin layer drying were fitted with experimental Moisture Ratio data at five different temperatures. The models used are shown in table 1. The model with highest R^2 value and lowest RMSE value is considered the most suitable model to illustrate paddy drying kinetics on FBD. Modelling results shown that the most suitable model is Modified Page model, followed by Page model and Henderson and Pabis Model. The value of model constants used and the value of R^2 and RMSE are shown in table 2 while figure 2 shows the comparison of drying curve between experimental results and modelling results, in this case Modified Page model was used. From figure 2, it can be observed that there is a good agreement between experimental results and Modified Page model.

Table 1. Thin layer drying models

No	Model	Formula	References
1	Newton	MR = exp(-kt)	Ayensu [13]
2	Page	MR = exp(-kt^n)	Menges and Ertekin [14]
3	Modified Page	MR = exp(-kt)^n	White et al. [15]
4	Henderson and Pabis	MR = a exp(-kt)	Kashaninejad et al. [16]
5	Logarithmic	MR = a exp(-kt) + c	Togrul dan Pehlivan [17]
6	Two-term	MR = a exp(-k$_1$t) + b exp(-k$_2$t)	Wang et al. [18]
7	Wang and Singh	MR = 1 + at + bt^2	Wang et al. [18]

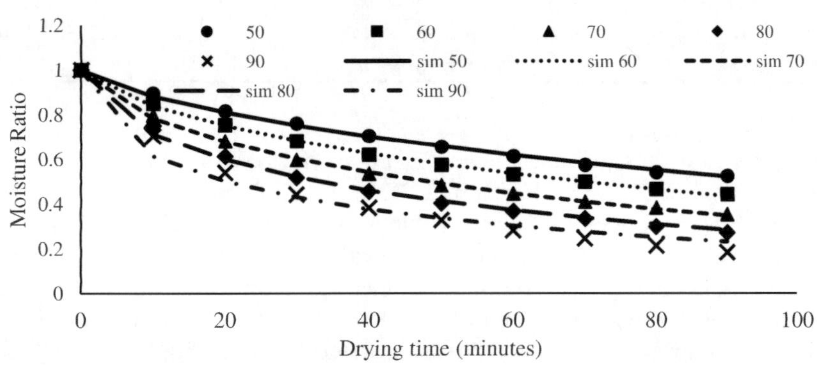

Figure 2. Drying curve comparison between experimental results and prediction results of Modified Page model

Table 2. Thin layer modelling results and value of constants used on paddy drying using FBD

Model	T (^{0}C)	k	n	a	b	c	k_1	k_2	R^2	RMSE
Newton	50	0.0083							0.9898	0.0750
	60	0.0109							0.9784	0.1227
	70	0.0144							0.9700	0.1630
	80	0.0183							0.9687	0.1803
	90	0.0235							0.9759	0.1712
Page	50	0.0272	0.6985						0.9960	0.0408
	60	0.0269	0.7709						0.9980	0.0300
	70	0.0286	0.8241						0.9908	0.0818
	80	0.0293	0.8785						0.9843	0.1208
	90	0.0308	0.9277						0.9840	0.1345
Modified Page	50	0.0062	0.7499						0.9984	0.1119
	60	0.0086	0.711						0.9992	0.0156
	70	0.012	0.6694						0.9990	0.0198
	80	0.0163	0.5978						0.9970	0.0454
	90	0.0237	0.5007						0.9811	0.1227
Henderson and Pabis	50	0.0073		0.9633					0.9873	0.0393
	60	0.0095		0.9489					0.9735	0.0734
	70	0.0116		0.9082					0.9585	0.0867
	80	0.0132		0.8451					0.9471	0.0859
	90	0.0155		0.7878					0.9459	0.0972
Logarithmic	50	0.0089		0.8598		0.1137			0.9911	0.0363
	60	0.0107		0.8154		0.1139			0.9778	0.0483
	70	0.0134		0.7772		0.1114			0.9662	0.0624
	80	0.0161		0.7258		0.1105			0.9603	0.0755
	90	0.0201		0.6901		0.1045			0.9651	0.0942
Two-term	50			0.9125	0.1152		0.0079	0.0221	0.9939	0.0960
	60			0.8766	0.1155		0.0097	0.022	0.9810	0.1066
	70			0.8331	0.1151		0.012	0.0211	0.9660	0.1110
	80			0.7839	0.1112		0.0141	0.0214	0.9562	0.0997
	90			0.7341	0.1099		0.0172	0.0207	0.9557	0.0870
Wang and Singh	50			0.0062	-0.0002				0.6783	0.8202
	60			0.0061	-0.0002				0.5413	0.8455
	70			0.0062	-0.0003				0.3922	1.6431
	80			0.0063	-0.0003				0.5116	1.5674
	90			0.0061	-0.0003				0.5490	1.5393

3.3. *The effective diffusivity and activation energy in paddy drying using FBD*

The effective diffusivity of paddy on drying using FBD at five different temperatures can be determined by plotting ln MR vs time, as shown in figure 3. Table 3 shows paddy's effective diffusivity at different temperatures. From Table 3, it can be observed that the value of effective diffusivity increases as drying temperature increases. Therefore, it can be implied that high temperature will cause water inside the paddy to diffuse to drying air because of the difference in temperature and pressure between drying air and paddy [19]. As a comparison, effective diffusion determination with the same Fick's law on the drying of macroalgae (*Oedogonium sp.*) using solar drying performed by Hammond et al. (2018) shows that, at the drying temperature of 25-60 °C, the optimum effective diffusivity of macroalgae obtained was 5.67 x 10^{-9} m^2/s [20].

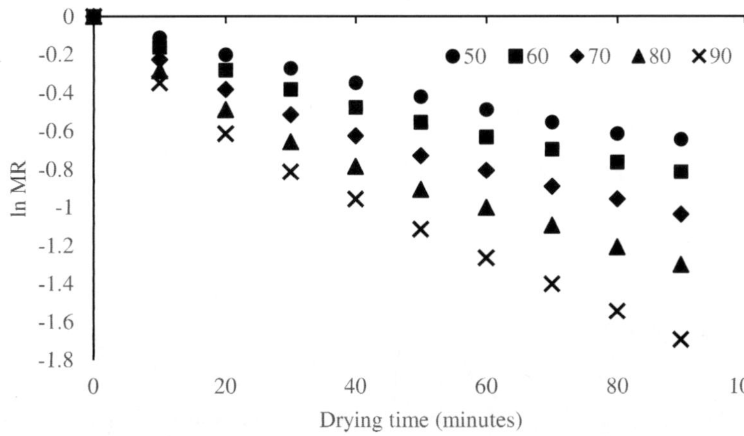

Figure 3. Plotting of ln MR versus drying time

Table 3. Effective diffusivity of paddy dried at different temperatures

No	Temperature (^{0}C)	Effective Diffusivity (m^2/s)
1	50	1.033×10^{-8}
2	60	1.214×10^{-8}
3	70	1.439×10^{-8}
4	80	1.732×10^{-8}
5	90	2.198×10^{-8}

Figure 4 shows the relation of effective diffusivity (ln D_{eff}) and absolute temperature. From Figure 4 it can be implied that the relation of effective diffusivity and absolute temperature is linear, which relates to Arrhenius relation between diffusivity coefficient and temperature [19]. According to Figure 4, the R^2 value (determination coefficient) is 0.9872. From the plotting results, the value of diffusivity constant (D_o) obtained is 8.529×10^{-6} m^2/s and the value of activation energy (Ea) obtained is 18.11 kJ/mol.

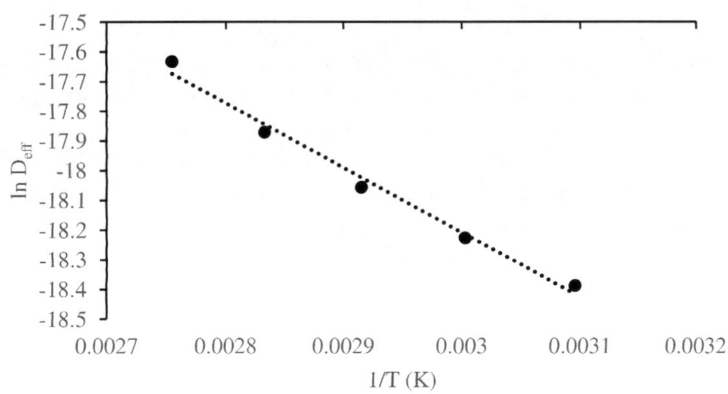

Figure 4. Plotting of ln D_{eff} versus 1/T

4. Conclusions

Modified Page model is the most suitable model to illustrate the kinetics of paddy drying using Fluidized Bed Dryer (FBD). The value of effective diffusivity ranges from 1.033 x 10^{-8} to 2.198 x 10^{-8} m^2/s and will increase as drying temperature increases. The moisture diffusivity of paddy greatly depends on temperature, which follows the Arrhenius relation. From the plotting results of ln D_{eff} and 1/T, the value of diffusivity constant (D_o) obtained is 8.529 x 10^{-6} m^2/s and the value of activation energy (Ea) obtained is 18.11 kJ/mol.

5. Acknowledgement

The authors want to give thanks to Diponegoro University for funding this research in the program Research of International Publication 2018.

References

[1] Prabowo 2006 *Processing and Its Effect on Physico-chemical Properties and Quality of Rice* (Theses, Samarinda: Mulawarman University)

[2] Khalil M S, Salem A Z M, Hassan A A, Gado H M, Alsersy H and Simbaya J 2012 Effects of Sun-Drying and Exogenous Enzymes on Nutrients Intake, Digestibility and Nitrogen Utilization in Sheep Fed Atriplex Halimus Foliages *Anim. Feed Sci. Tech.* **171** 128-35

[3] Soponronnarit S, Prachayawarakorn S and Wangji M 1996 Commercial Fluidized Bed Paddy Dryer *Proc 10th Int Drying Symp, Vol. (A), Krakow, 30 July - 2 August, Poland* 638-44

[4] Tumambing J A and Driscoll R H 1991 Modeling the performance of continuous fluidized bed paddy dryer for rapid pre-drying of paddy *Proc 14th ASEAN Seminar on Grain Post-harvest Tech.* 193-213

[5] Soponronnarit S and Prachayawarakorn S 1994 Optimum Strategy For Fluidized Bed Paddy Drying *Dry. Technol.* **12** 1667-86

[6] Prachayawarakorn S, Tia W, Poopaiboon K and Soponronnarit S 2005 Comparison of performances of pulsed and conventional fluidized-bed dryers *J. Stored Prod. Res.* **41** 479-97

[7] Jafari H, Kalantari D and Azadbakht M 2017 Semi-Industrial Continuous Band Microwave Dryer For Energy and Exergy Analyses, Mathematical Modeling of Paddy Drying and It's Qualitative Study *Energy* **138** 1016-29

[8] Behera G and Sutar P P 2018 A Comprehensive Review of Mathematical Modeling of Paddy Parboiling and Drying: Effects of Modern Techniques on Process Kinetics and Rice Quality *Trends Food Sci. Technol.* **75** 206-30

[9] Li X J, Wang X, Li Y, Jiang P and Lu H 2016 Changes in Moisture Effective Diffusivity and Glass Transition Temperature of Paddy During Drying *Comput. Electron AGR* **128** 112-19

[10] Dhanushkodi S, Wilson V H and Sudhakar K 2017 Mathematical Modeling of Drying Behavior of Cashew in A Solar Biomass Hybrid Dryer *Resource-Efficient Technol.* **3** 359-64

[11] Shen F, Peng L, Zhang Y, Wu J, Zhang X, Yang G, Peng H, Qi H and Deng S 2011 Thin-Layer Drying Kinetics and Quality Changes of Sweet Sorghum Stalk For Ethanol Production As Affected By Drying Temperature *Ind. Crop Prod.* **34** 1588-94

[12] Kamin N H and Janaun J 2017 Drying of Paddy in A Laterally Aerated Moving Bed Dryer at Ambient Temperature *Chem. Eng. Trans.* **56** 877-82

[13] Ayensu A 1997 Dehydration of Food Crops Using A Solar Dryer With Convective Heat Flow *Sol Energy* **59** pp 121-6

[14] Menges H and Ertekin O C 2006 Mathematical Modeling of Thin Layer Drying of Golden Apples *J. Food Eng.* **77** 119-25

[15] White G M, Ross I J and Ponelert R 1981 Fully Exposed Drying of Popcorn *Trans ASAE* **24** 466-68

[16] Kashaninejad M, Mortazavi A, Safekordi A and Tabil L G 2007 Thin-Layer Drying Characteristics and Modeling of Pistachio Nuts *J. Food Eng.* **78** 98-108

[17] Togrul I T and Pehlivan D 2002 Mathematical Modeling of Solar Drying of Apricots in Thin

Layers *J. Food Eng.* **55** 209-16

[18] Wang Z, Sun J, Liao X, Chen F, Zhao G, Wu J and Hu X 2007 Mathematical Modeling on Hot Air Drying of Thin Layer Apple Pomace *Food Res. Int.* **40** 39-46

[19] Suherman, Fajar B, Satriadi H, Yuariski O, Nugroho R S and Shobib A 2012 Thin Layer Drying Kinetics of Roselle *Adv. J Food Sci. Technol.* **4** 51-55

[20] Hammond L, Bai L, Sheehan M and Walker C 2018 Experimental Analysis and Diffusion Modelling of Solar Drying of Macro Algae-Oedogonium sp. *Chem. Eng. Trans.* **65** 427-32

Refining and Physicochemical Test of Tengkawang oil *Shorea stenoptera* Origin Sintang District West Kalimantan

N Hidayat, M A Darmawan, N Intan and M Gozan
Department of Chemical Engineering, Faculty of Engineering, Universitas Indonesia, Depok 16424, Indonesia
E-mail : mgozan@che.ui.ac.id

Abstract. The oil from Tengkawang fruit (*Shorea stenoptera*) has been processed by indigenous people in West Kalimantan as cooking oil and margarine. This research examined the degumming process, neutralization and bleaching of Tengkawang oil to reduce the high acid and peroxide numbers resulted from traditional processes. The oil produced at each stage of the refining process were tested for their physicochemical properties by calculating the acid, iodine and peroxide numbers. Respectively, the values of acid number (mg NaOH/100 g sample), iodine number (g I_2/100 g sample), and peroxide number (mg equivalent O_2/kg sample) for initial Tengkawang fat were 10.14; 32.49 and 10.98; for Tengkawang fat after degumming were 10.25; 28.81 and 10.88); for Tengkawang fat after neutralization were 6.09; 30.52 and 5.57; and for Tengkawang fat after bleaching were 6.14; 31.39 and 5.55. The acid number of Tengkawang oil after neutralization process with variation of volume 5, 7.5 and 10 % were 7.83; 6.14 and 3.12 mg NaOH/100 g sample, respectively. The refining process increased the quality of physicochemical properties of Tengkawang oil to fall into industrial grade.

1. Introduction

One of the export commodities of non-timber forest product group is Tengkawang fruit. Tengkawang fruit is produced by several types of *Shorea* that belong to *Dipterocarpaceae* family [1]. Tengkawang can grow naturally on the mountain slopes at an altitude of 5 - 1.000 mdpl, on all soil types, provided it is not sandy and sufficient sunlight [2]. There are 12 types of Tengkawang protected by PP. 7 of 1999, namely: *Shorea stenoptera, S. gysberstiana, S.pinanga, S. compressa, S. seminis, S. martiana, S. mexistopteryx, S. beccariana, S. micrantha, S. palembanica, S. lepidota*, and *S. singkawang* [3]. The trade name of Tengkawang fruits is Illipé nut or borneo tallow nut [4].

Total consumption of vegetable oil in Indonesia increased by 335 % since 1980 [5]. Tengkawang is one of the potential bio-based sources. The use of bio-based resources for the production of added value products is a key element of sustainable development. The fat content of Tengkawang ranged from 50-70 % [6]. However, the physicochemical properties of Tengkawang fat are influenced by the type, seed size, seed maturity, and storage time [2].

Tengkawang fruit processing into fat can be done in three ways, namely the way of forging, boiling and dissolving (extraction) using organic solvents [6]. The curret practice of extracting Tengkawang fat by indigenous people in Kalimantan is through the seed pressing. This current processing is still traditional so that the yield and quality of processed oil is not optimal [7]. It is characterized by high acid and peroxide number that is 10.14 mg NaOH/100 g sample and 10.98 mg equivalent O_2/kg sample.

Tengkawang oil is reported to have the same type of triglyceride as palm oil [4]. The fat refining stages of palm oil generally consist of degumming, neutralization, bleaching and deodorizing. Steam-stripping is commonly used to reduce the free fatty acid content in the palm oil refining processes to an acceptable level. In this study, we used NaOH in neutralization process to remove free fatty acids from Tengkawang oil.

IOP Conf. Series: Materials Science and Engineering **543** (2019) 012011 doi:10.1088/1757-899X/543/1/012011

Degumming is an early stage purification process whereby the principle is to remove or separate gum (sap or mucus) without reducing the amount of free fatty acids in fat. Neutralization is the process of separating free fatty acids by reacting free fatty acids with alkaline or other reagents to form soap. Free fatty acids it is easily oxidized and can cause rancidity, so that the presence of free fatty acids in palm oil is undesirable [8].

FFA levels of various types of Tengkawang are not much different. FFA levels of fat Tengkawang type *S. pinanga*, the extraction results in an amount of FFA 1.94, after degumming rose to 2.31 and after neutralization fell back to 1.77 [4]. Bleaching is the purification step of removing unwanted substances in oil, which can be done by mixing the oil with a small amount of adsorbents such as absorbent soil, active clay and activated charcoal or with chemicals [6].

Studies of oil extraction from Tengkawang fruit using n-hexane solvent against *S. pinanga, S. stenoptera, S. mexistopteryx,* and *S. parvivolia* showed that oleic acid is the largest content of Tengkawang fat [1]. The highest oleic acid ranged from 31-59 % of the Tengkawang fats produced for the *S. stenoptera*. The study was then continued by extracting oil from seeds and continued with degumming and neutralization of *S. pinanga* and *S. mexistopteryx* [4]. Research on Tengkawang oil production using apit (traditional tool of indigenous people of Kalimantan) on *S. stenoptera* and *S. mexistopteryx* showed that the density and refractive index of both types of Tengkawang were the same, but differed on the iodine number, the saponification number and the FFA content [9].

In this research, the process of refining Tengkawang oil with degumming, neutralization and bleaching of *S. stenoptera* originating from Sintang regency of West Kalimantan was investigated. The purpose of this research is to improve the quality of vegetable oil with low acid number and peroxide number. Thus, it will increase the quality and the economic value.

2. Methods

2.1. Materials

Tengkawang fat was obtained from *S. stenoptera* harvested in 2017 from Sintang District, West Kalimantan. Other chemicals used were KI, NaOH, KOH, $Na_2S_2O_3$, iodine monochloride (ICl), glacial acetic acid, chloroform, alcohol, acetone, indicator of phenolphthalein, starch solution. All chemicals were obtained from Merck, Germany.

2.2. Purification Process

At room temperature the Tengkawang fat is in solid phase. 150 grams of Tengkawang was inserted into 300 mL beaker glass and was heated until it melted perfectly at temperatures 60-70 °C. The melted Tengkawang fat was then added with 20 % phosphoric acid by 1 % (w/w). The mixture was stirred using a magnetic stirrer at 400 rpm.

After 10 minutes of stirring, the mixture was inserted into the separating funnel to wash the remaining phosphoric acid and the impurities. Then, 5 % warm water was added to wash the mixture and was shaken for several minutes. The mixture was allowed to form two phase layers of fat and water. The weight of Tengkawang fat was measured. Then, Tengkawang fat was neutralized with NaOH concentration of ± 1 M added to the mixture with variation of volume 5, 7.5 and 10 % (w/w). The purpose of this variation is to examine the effect of NaOH concentration on the acid number. The mixture was then stirred for 15 minutes at 400 rpm at temperature of 60-70 °C.

After 15 minutes, the soap phase began to separate with the fat of the Tengkawang . The Tengkawang fat was then inserted into the separating funnel to remove the residual soap. Afterward, 5 % of warm water was added to wash the mixture and was shaken for a few minutes to let the mixture to form two phase layers of water and fat. The weight of Tengkawang fat was measured. The bleaching process was carried out by adding 1 % (w/w) of activated carbons into Tengkawang fat and heated at 70 °C for 30 min. After stirring, Tengkawang fat was then separated from the activated carbons.

2.3. Acid Numbers, Iodine and Peroxides

The number of acids, iodine number and peroxide number were examined in duplo at each stage of the refining process. In the acid number test, as much as 2 grams of sample was incorporated into the 100 mL Erlenmeyer. Subsequently, 25 mL of neutral ethanol was added to the Erlenmeyer and shaken to form two layers of ethanol and fat layer followed by the addition of 3 to 4 drops of phenolphthalein indicator. The sample was titrated with 0.1 N NaOH to form a pink solution [10]. The calculation of the numbers of acids was as follows:

$$\text{Acid Number} = \frac{V \times T \times 40}{m} \quad (1)$$

where V is the volume of NaOH required [mL]; T is the standard normality of NaOH; 40 is the molecular mass of NaOH [gram/mol]; and m is the sample weights [gram].

For iodine number test, 1 gram sample was inserted into 250 mL Erlenmeyer with addition of chloroform. *Wijs* solution (iodine monochloride in glacial acetic acid) of 25 ml was added, where iodine as a reagent and glacial acetic acid as an acid catalyst. The mixture was shaken for one minute strongly, and then the mixture was kept in the dark for 60 minutes. KI solution (20%) was added with 100 mL of aquadest and mixed titration with 0.1 N $Na_2S_2O_3$ solution with kanji indicator [10]. Calculations of iodine Numbers were as follows:

$$\text{Iodine Number} = \frac{12.69 \times T \times (V0-V1)}{m} \quad (2)$$

where V_0 is the volume $Na_2S_2O_3$ which required in the blank spreading [mL]; V_1 is the volume $Na_2S_2O_3$ which required in sampling [mL]; T is the standard normality of $Na_2S_2O_3$; and m is the sample weights [gram].

For the test of peroxide number, 2 grams of sample yield was added to 250 mL of Erlenmeyer with chloroform and was shaken until completely dissolved. Then, glacial acetic acid as catalyst and saturated KI solution was mixed for one minute strongly. Aquadest was added as much as 75 mL into mixture and mixed titration with 0.02 N $Na_2S_2O_3$ solution with kanji indicator [10]. Calculations of the Peroxide Numbers were done as follows:

$$\text{Peroxide Number} = \frac{T \times (V1-V0)}{m} \quad (3)$$

The Tengkawang fat was also characterised regarding to Melting Point (SNI.7348:2009), Saponification Number (SNI.01-3555-1998) as well as Monounsaturated, Polyunsaturated and Trans Fat (IUPAC 2, 301 7th edition) standard methods.

3. Results and discussion

The Tengkawang fat from Sintang District was obtained in the form of brownish yellow solid. Analysis of the acid number (mg NaOH/100 g sample), peroxide number (mg equivalent O_2/kg sample) and iodine number (g I_2/100 g sample) was done prior to degumming. The results were 10.14; 10.98 and 32.49, respectively. The values of acid number and peroxide number in Tengkawang Sintang fat sample before the initial degumming were big. Therefore, further treatment was needed to reduce the values of the acid number and peroxide number.

The value of iodine number of sample Tengkawang was small (<35 g I_2/100 g sample). This indicates that the Tengkawang sample treated has resistance to rancidity. The Tengkawang remained in good conditions after 1.5 months of lack of lighting and low temperatures storage. Therefore, the Tengkawang sample had a typical of Tengkawang flavor and did not undergo easily rancid.

The degumming process was carried out by adding a phosphoric acid into melted fat. Phosphoric acid, polybasic organic acids, and hydration with water are usually used to remove phospholipids either individually or in combination. Phospholipids must be removed because they have emulsifying properties. After degumming, the colour of the fat turned to a clear yellow. The Tengkawang oil formed was analysed to determine the acid number, peroxide number and iodine number with the following results as follows 10.25; 10.88 and 28.81, respectively.

These results showed an increase in acid numbers. Although the increase was not too large, the increase in the acid numbers was expected due to the hydrolysis reaction that occurred due to the addition of acid during the degumming process [11]. In the degumming process, there were free fatty acids due to reaction with residual phosphoric acid thus increasing the value of fatty acids and acid numbers. There was an increase in the free fatty acids and acid numbers. The impurities such as phosphatide bonds were detached and dissolved in the aqueous phase (containing residual phosphoric acid).

The degummed oil was neutralized with the addition of NaOH 7.5 % (w/w). During the process of addition and stirring, it formed soap (soap stock). Besides, the glycerol was still mixed in Tengkawang fat. The longer the time of neutralization, the more soaps were formed as more free fatty acids were accumulated. After the stirring process was completed, the soap floated to the surface of the mixture so that the soap layer can be removed and separated. After the soap layer was separated, the oil layer was washed several times with warm water to obtain a clear yellow oil layer. The acid number, peroxide number and iodine number of Tengkawang oil after neutralization were 6.09; 5.57 and 30.52, respectively.

The results showed that the acid and peroxide numbers significantly decreased. In the sample after neutralization there was a decrease in the acid number from 10.25 to 6.09 and in the value of peroxide number from 10.88 to 5.57. The decrease in the acid numbers was the result of acid-base reactions between free fatty acids in the oil with a NaOH base. Decreasing acid numbers improved the quality of oil; a low acid number indicates a longer shelf life [4].

The iodine number increased slightly from 28.81 to 30.52. The higher the iodine number, the more the C = C bonds present in the oil. Fats with lots of unsaturated trace bonds are particularly susceptible to oxidative resistance which can be stimulated at high temperatures and light. In contrast, fats with low iodine indicate a lot of saturated bonds [4].

After the neutralization process, the next step was bleaching using activated carbon. After bleaching, the acid number, peroxide number and iodine number were analyzed. The results showed 6.14; 5.55 and 31.39, respectively. The values of those three numbers were stagnant. In the bleaching process the activated carbon played a role to absorb unwanted dyestuffs without oxidizing the oil, so the color of the fat Tengkawang become darker white than before bleaching.

Table 1. Acid Numbers, Peroxide Numbers and Iodine Numbers of Tengkawang Fat

Sample	Value of acid number (mg NaOH/g fat)	Peroxide Numbers Value (mek O_2/kg)	Iodine Value (gram iodine/100 gram)
Early Tengkawang sample	10.14	10.98	32.49
Tengkawang sample after degumming	10.25	10.88	28.81
Tengkawang sample after neutralization	6.09	5.57	30.52
Tengkawang sample after bleaching	6.14	5.55	31.39

There were variations of iodine number, yet the iodine values contained in Tengkawang sample came within the standard ranges of Tengkawang fats established by SNI No 2903: 2016. The value of acid number was analyzed after bleaching using NaOH ± 1 N as much as 5 and 10 % as can be seen in Table 2.

Table 2. Tengkawang Fatty Acid Numbers with Variation of NaOH Levels

Sample	Value of acid number (mg NaOH/g sample)
Sample + NaOH 5 % (w/w)	7.83
Sample + NaOH 7.5 % (w/w)	6.14

Sample + NaOH 10 % (w/w)	3.12

Table 2 showed that the concentration of NaOH used was inversely proportional to the acid value. The greater the NaOH used, the lower the yield percentage of Tengkawang fat. The triglycerides of the fatty ester in the Tengkawang were broken and saponified. The use of 10 % NaOH (w/w) is effective if there is a system capable of separating the Tengkawang fat and other side-effects such as soap stock and glycerol.

Table 3. Fatty Acid (18:2) Compositions from Different Studies

Butter Name	Fatty Acid (18:2)*	References
Shea Butter	4.3 - 6.65	[12], [13], [14], [15]
Sal Fat	2.2 - 2.8	[12], [16], [17]
Tengkawang	6.65	From this study

*in %

According to the Table 3. The results from our study showed that fatty acid content of Tengkawang was 6.65 %. These results were similar to Shea butter which belongs to the family of *Sapotaceae*, Shea butter come from edible vegetable fat obtained from the shea kernel *Vitellaria paradoxa* which mostly growth in West African and sub-saharan African. This trees also called *Butyrospermum parkii L.* [18]. Casten and Synder, 1985 reported that Shea kernel contains 40–55% valuable fat [19], according to Table 4 the results from our study showed that the monounsaturated and polyunsaturated fat content of Tengkawang was 28.7 % and 2.43 %.

Compared with Sal fat in Table 3, fatty acid content of Tengkawang fat were not similar. Several researchers reported that the variety, climate and regional differences influence the quality of fatty acid compositions in several fat source [18, 20-21]. Sal fat (*Shorea robusta*) is obtained from the seed kernel of sal trees, which grown in India, Malaysia, Borneo, Java, and Philippines.

Approximately 5% of the total forest area in India is occupied by Sal trees. The sal seed kernels contain 19–20% of oil which is known as Sal butter. The iodine value of sal fat is 31–45 and the melting point is 30–36 °C [22]. This properties were similar with Tengkawang fat, the results from our study showed that the melting point and iodine value of Tengkawang butter were 30.4 and 31-35 respectively (Table 1 and 4).

This shows that both tengkawang and sal fat have resistance to rancidity. Based on the triglyceride compositions, both shea butter and sal fat can be used as cocoa butter substitutes in making chocolate [23-24]. The Tengkawang in our study can also be used as a substitute for cocoa butter as the properties come into the properties of commercial cocoa butters from different countries [25].

Table 4. Physicochemicals of Tengkawang oil *Shorea stenoptera* Origin Sintang District West Kalimantan

Parameter	Unit	Result	Test Method
pH	-	12.8	SNI. 01-2891-1992, point 16
Melting Point	°C	30.4	SNI. 7348 : 2009, attachment A.4
Monounsaturated Fat	%	28.7	IUPAC method 2, 301 7th edition
Polyunsaturated Fat	%	2.43	IUPAC method 2, 301 7th edition
Trans Fat	%	<0.005	IUPAC melhod 2, 301 7th edition

Saponification Number	mg KOH/gram	192	SNI. 01-3555-1998, point 7

As can be seen in Table 3 and 4, the physicochemicals of Tengkawang oil *Shorea stenoptera* meet some requirements as an alternative source to substitutes in cocoa butter and confectionary industries. With the right purification process, it is possible for Tengkawang as a potential food source in the future.

4. Conclusion

The value of acid number for Tengkawang fat; Tengkawang after degumming; Tengkawang after neutralization; and Tengkawang after bleaching were 10.14; 10.25; 6.09 and 6.14 (mg NaOH/100 g sample), respectively. The iodine number value of the initial Tengkawang; Tengkawang after degumming; Tengkawang after neutralization; and Tengkawang after each bleaching were 32.49; 28.81; 30.52; 31.39 (g I_2/100 g sample), respectively. The bleaching process was not necessary to reduce the acid number, peroxide number and iodine number. But it was useful to make the fat look clearer. All the samples were in the standard ranges for iodine number of Tengkawang fat according to SNI 2903:2016, i.e 25 - 38. The value of the initial Tengkawang peroxide number; Tengkawang after degumming; Tengkawang after neutralization; and Tengkawang after each bleaching were 10.98; 10.88; 5.57 and 5.55 (mg equivalent O_2/kg sample), respectively. This indicates that the treated Tengkawang sample has resistance to rancidity. The results from the 5; 7.5 and 10 % NaOH neutralization were 7.83; 6.14 and 3.12 (mg NaOH/100 g sample), respectively. Only the 10% neutralization Tengkawang sample had the lowest acid number of 3.12 and came within the standard of SNI 2903: 2016 (maximum of 3.50).ootnotes should be avoided whenever possible. If required they should be used only for brief notes that do not fit conveniently into the text.

5. Acknowledgement

Authors are gratefully thank the financial support by Universitas Indonesia through *Hibah Publikasi Internasional Terindeks untuk Tugas Akhir Mahasiswa* –(PITTA).

6. References

[1] Gusti R E P, Zulnely and Kusmiyati E 2012 *J. Penel. Hasil Hutan* 30 **4** 254 [in Bahasa]
[2] Sunarta S, Widowati and Kasmudjo K 2012 *Laporan Penelitian DPP* [in Bahasa]
[3] Presidential Decree 1999 **7** *Pengawetan Jenis Tumbuhan dan Satwa* [in Bahasa]
[4] Gusti R E P and Zulnely Z 2015 *J. Penel. Hasil Hutan* 33 **1** 61 [in Bahasa]
[5] Nayantakaningtyas J S and Daryanto H K 2012 *J. Manaj. dan Agrib.* 9 **3** 194 [in Bahasa]
[6] Ketaren S 1986 *Pengantar Teknologi Minyak dan Lemak Pangan* (UI Press Jakarta) [in Bahasa]
[7] Fernandes A and Maharani R 2017 *J. of Dipterocarp Ecosys. Res.* 3 **2** 49 [in Bahasa]
[8] Zahrina I, Nasikin M, Krisanti E and Mulia K 2018 *Food Chem.* **240** 490
[9] Maharani R, Fernandes A and Pujiarti R 2016 *AIP Conf. Proc.* **1744** 020051
[10] Indonesia B S N 1998 SNI 01-3555-1998 [in Bahasa]
[11] Sonntag N O 1979 *Bailey's Industrial Oil and Fat Products* **1** 289
[12] Banerji R, Chowdhury A R, Misra G and Nigam S K 1984 *Fett Wiss. Technol.* **86** 279
[13] Jacobsberg B 1977 *Oleagineux* **32** 529
[14] Jatto W O, Yuanfa L, Shan L, Wang X and Aworh O 2010 *Electron. J. Environ. Agric. Food Chem.* 9 **2** 358
[15] Kanematsu H, Maruyama T, Niiya I, Imamura M and Matsumoto T 1978 *J. Jpn. Oil Chem. Soc.* **21** 519
[16] Chaudhuri G, Chakrabarty M M and Battacharyya D 1983 *Fett Wiss. Technol.* **85** 224
[17] Vedaramana N, Puhanb S, Nagarajanc G, Ramabrahmama B V and Velappana K C 2012 *Ind. Crop Prod.* **36** 282
[18] Bail S, Krist S, Masters E, Unterweger H and Buchbauer G 2009 *J. Food Compos. Anal.* **22** 738
[19] Casten J and Synder H E 1985 *Volunteers in Technical Assistance* **40**
[20] Cardi C, Vaillant A, Sanou H, Kelly B and Bouvet J 2005 *Mol. Ecol.Notes* 5 **3** 524

[21] Sanou H, Picard N, Lovett P N, Dembélé M, Korbo A, Diarisso D and Bouvet J M 2006 *Genet. Resour. Crop Evol.* **53** 145
[22] Jahurul M H A, Zaidul I S M, Norulaini N A N, Sahena F, Jinap S, Azmir J, Sharif K M and Omar A K M 2013 *J. Food Eng.* 117 **4** 467
[23] Olajide J O, Ade-Omowaye B I O, and Otunola, E T 2000 *J. Agric. Eng. Res.* **76** 419
[24] Reddy S Y and Prabhakar J V 1989 *Food Chem.* 34 **2** 131
[25] Chaiseri S and Dimick P S 1989 *J. Am. Oil Chem. Soc.* **66** 1771

Processing of Pharmaceuticals for Pulmonary Delivery by Supercritical Technology

F Kurniawansyah[1], R Mammucari[2], N R Foster[3]

[1]Department of Chemical Engineering, Institut Teknologi Sepuluh Nopember, Surabaya 60111, Indonesia
[2]School of Chemical and Biomolecular Engineering, University of Sydney, NSW 2006, Australia
[3]WASM: Mining, Energy & Chemical Engineering, Curtin University, Perth, WA 6845, Australia

*Email: fkurniawansyah@gmail.com

Abstract. There are several advantages of using the human pulmonary pathways for drug delivery. These include large absorption area, more rapid onset of therapeutic effect, reduced dose of bioactive delivery and lower side effects of therapy. Inhalable products should meet several criteria, in particular shape, surface roughness and particle size (approximately 1-5 μm, or 2-3 μm as optimum) to reach the deep region of human lungs (alveolus).
In an attempt to produce such powders, supercritical fluid (SCF) technology has been proposed as an alternative to high energy consumptive yet popular mechanical methods such as milling. In this article, application of SCF technology for inhalable products manufacturing is briefly overviewed. Discussion focuses on the process development and formulation strategies to obtain improved product properties of specifically designed for pulmonary administration.

1. Introduction

The human pulmonary track provides an excellent route for drug delivery as it possesses benefits from large surface area (more than 100 m^2), thin mucosal membrane and good blood supply [1]. Pulmonary delivery has been effective as bioactives can reach the blood system much faster than is in the case for the more common route of oral administration. Degradation due to liver or other digestive fluids is practically absent in pulmonary administration, ensuring more rapid action and less side effect of therapy.
The exploration of the human pulmonary pathway for drug delivery is, in fact, not new in the therapeutic applications. More than 4000 years ago, people started to inhale burnt plant smoke to treat coughing [2]. Modern therapy has explored the pulmonary pathway to treat various diseases such as asthma, diabetes, lung injury and cancer.
The development of pulmonary administration depends on two factors - the drug delivery system (formulation) and the delivery device. Dry powder inhalers (DPI) have been considered more effective compared to that used for liquid based formulations, especially in terms of the delivery system, chemical stability and bioavailability [3]. Particle formulation and design as determined by particle morphology, size, density in particular play critical roles in the creation of dispersed powder aerosols. Particles for pulmonary delivery need to be smaller than 10 μm, 1-5 μm in average, or between 2-3 μm for optimum size, to be able to penetrate biological barriers (cilliata and mucosa) [4, 5]. In addition, fluffy particles in elongated form wrapped by coarse carrier materials tend to have a high fraction of deposition in the

lower part of human pulmonary pathways [3]. Commercial inhalable products usually have at least 20% of fine particle fraction – FPF, or fraction of powder which may reach the lower part of the human lungs (trachea, bronchi, alveoli) [6]. Typical morphology of inhalable powders is presented in Figure 1, which was taken from an image of cyclodextrin co-formulated with curcumin particles, processed with SCF technology (from an unpublished work).

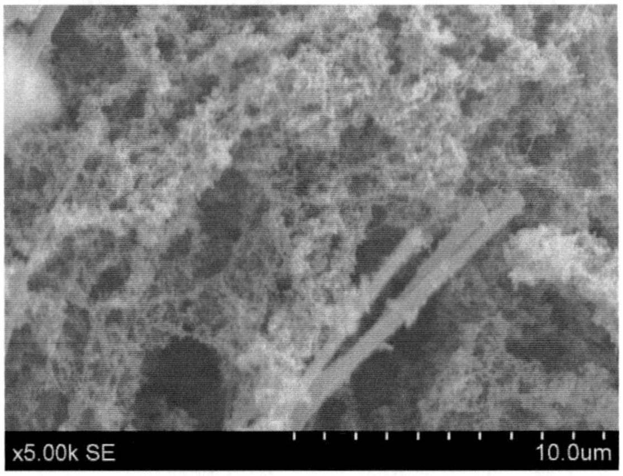

Figure 1. Morphology of curcumin – cyclodextrin powder processed by an SCF-antisolvent technology (FPF 36 %).

Several methods have been used to produce such particles, with milling being the most commonly used technology. Although jet milling is simple to apply, it requires high energy supply in operation. In addition, milling imposes repeated impacts on solid materials which can be detrimental to many organic compounds such as pharmaceuticals [5].

Supercritical fluid (SCF) has been proposed as processing media due to its versatile nature. Vastly diverse properties can be exploited simply by pressure and temperature tuning. SCF technology has been successfully applied to modify powders of consumer products including food and pharmaceuticals. Carbon dioxide, CO2, is by far the most common choice as SCF agent due to its moderate critical points, inflammability, low toxicity and the fact that is relatively inexpensive.

In this article, the application of SCF technology to process particles for pulmonary delivery as inhalable powders is discussed. Emphasis is laid on the development of SCF technology in particle design and formulation in order to enable certain pharmaceuticals to be administered through human pulmonary pathways. Several case studies are presented as evidence of the applicability of SCF technology in this field.

2. Methodology of SCF Processing

The embodiment of SCF extraction methods in the early 1980s spurred the development of further SCF applications in various processing areas, including particle engineering. In particle engineering, SCFs are employed to modify morphology, crystallinity and other physicochemical properties such as mechanical strength and dissolution properties. For respiratory application, the use of SCF provides benefits to improve aerodynamic properties, density and rugosity [7].

In particle processing, SCFs may act differently as processing medium, depending on the materials being processed. CO2 has become the most popular agent selected as SCF due to its inexpensiveness and safety considerations. SCF plays its role as solvent when processed materials are easy to dissolve in the SCF agent. The principle of the so-called rapid expansion of supercritical solutions (RESS) relies primarily on the SCF solvating nature. In a method referred to as particles from gas saturated solution (PGSS), SCF acts as solute, in addition to its function as processing medium. Here in the PGSS, SCF facilitates the softening of materials thus melt solutions are obtained. In both in the RESS and PGSS processes, particles are generated through expansion in which the Joule – Thompson effect is exploited.

When materials have poor compatibility with the SCF agent, the supercritical antisolvent (SAS) method may be selected as alternative strategy. The SAS concept is based on the capability of SCF to rapidly reduce organic solvent solvating power, thus inducing supersaturation thereby resulting in particle formation [8]. Selection of pressure, temperature, SCF: material ratio and feed pre-treatments have been key factors in the production of inhalable powders. Figure 2 is an illustration of the application of SCFs for particle formation.

3. SCF Technology for Pulmonary Powders

The application of SCF to produce inhalable products can be found in a number of articles. Several inhalable composites consisting of pharmaceuticals and nutraceuticals along with various carriers have been processed using SCF technology.
Processing of small molecule bioactives such as steroids, and large biological compounds such as proteins has also been investigated.

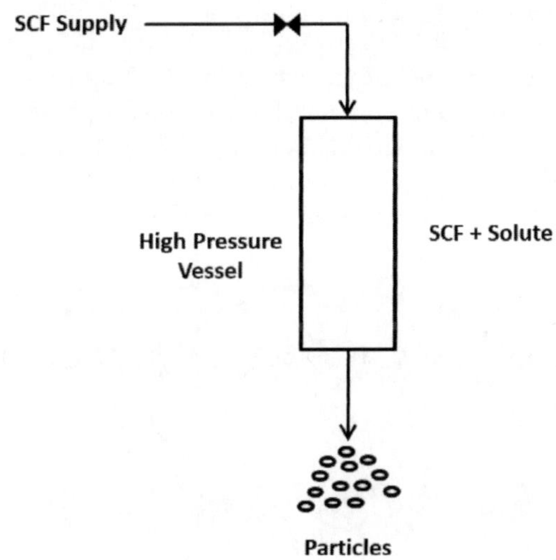

(a)

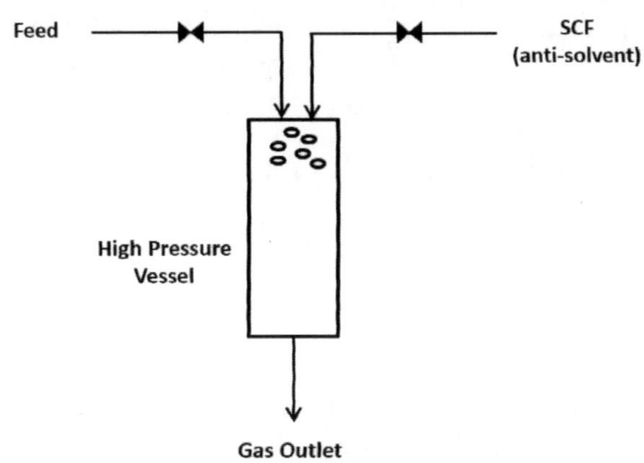

(b)

Figure 2. Schematic diagram of the (a) RESS and (b) SAS processes

Steroids such as budesonide and fluticasone have been processed using semi-continuous SAS in an attempt to improve particle dimensions and wettability. The study was successful as the aerodynamic properties of particles was improved significantly, compared to the products of jet milled processing [9]. Insulin, a robust enzyme to treat diabetic diseases, is an excellent example of how SCF technology is applicable for engineering large molecules for pulmonary delivery. Formulation of insulin is available mainly in suspension form and is delivered mainly via the subcutaneous route. Oral delivery of insulin is hampered by the fact the gastrointestinal fluid is often detrimental to bioactives such as insulin. The exploration of pulmonary pathway has been investigated to achieve increase effectiveness of insulin administration. SCF technology has been used to produce insulin composites, mainly with sugar compounds (or their derivatives) acting as carriers/ excipients. Mannitol and trihalose have been used to formulate SAS inhalable insulin processed in approximately 2 µm particles. Analysis using micro-uniform deposit impactor arrived at 69-70% of estimated particles reaching the deep region of the lungs [10]. In a development of a new method, the semi-batch atomized rapid injection solvent extraction (ARISE) system, insulin as a single compound was precipitated from dimethylsulphoxide (DMSO) solution. Particles of insulin as small as 1 µm with narrow particle size distribution were produced. The ARISE processed insulin was estimated to have 60% of its bulk products capable of reaching the alveolus [7].

Also using the ARISE method, composites of the natural phytochemical, curcumin, has been processed with polymer and cyclodextrin for delivery via the pulmonary route. The ARISE composites of curcumin had respirable fraction of its powder as high as 60%. Estimated deposition of this product, is presented in Figure 3. Further, the SCF processed curcumin composites also exhibited an increased activity towards lung cancer cell model in an in-vitro assessment [11].

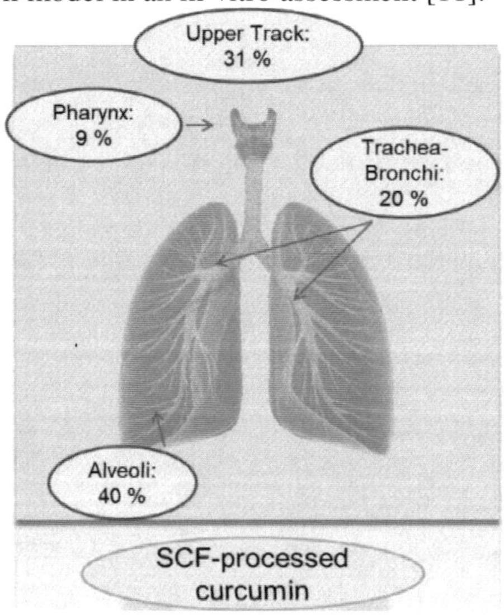

Figure 3. Estimated powder deposition of SCF formulated curcumin in human pulmonary tract (FPF 60%, measurement with Andersen Cascade Impactor, 28.3 L/min) [11].

4. Conclusion

The suitability of Supercritical Fluid technology for the development of pulmonary drug delivery systems has been explored. The versatile properties of SCFs present significant advantages for the processing of particles with very specific criteria, and which are biocompatible. SCFs can act as solvent, solute or anti-solvent depending mainly on the desired product characteristics. Several processing platforms have been developed, mostly confirming the suitability of SCF technology for the production of powders with tailored properties suitable for delivery via the human pulmonary pathway.

References

[1] Chaturvedi N, Solanki H 2013 Pulmonary drug delivery system: review *Int J Pharm* **5**, 7-10.
[2] Labiris N, Dolovich M 2003 Pulmonary drug delivery. Part II: The role of inhalant delivery devices and drug formulations in therapeutic effectiveness of aerosolized medications *Br. J. Clin. Pharmacol.* **56**, 600-612.
[3] Chan HK 2008 What is the role of particle morphology in pharmaceutical powder aerosols? *Expert Opin Drug Deliv*. **5**, 909-914.
[4] Salem H, Katz S 2006 Inhalation Toxicology, *Taylor* - Francis Group Boca Raton.
[5] Rasenack N, Steckel H, Müller BW 2003 Micronization of Anti-Inflammatory Drugs for Pulmonary Delivery by a Controlled Crystallization Process, *J. Pharm. Sci*. **92**, 35-44.
[6] Hickey AJ, Martonen TB, Yang Y 1996 Theoretical relationship of lung deposition to the fine particle fraction of inhalation aerosols *Pharm. Acta Helv.* **71**, 185-190.
[7] Sih R 2008 New Process Development of Dense Gas Technology for The Processing of Pharmaceuticals PhD Thesis-Chemical Science and Engineering University of New South Wales, Sydney.
[8] Jung J, Perrut M 2001 Particle design using supercritical fluids: Literature and patent survey, *J Supercrit Fluids*. **20**, 179-219.
[9] Steckel H, Thies J, Müller BW 1997 Micronizing of steroids for pulmonary delivery by supercritical carbon dioxide *Int J Pharm* **152**, 99-110.
[10] Kim YH, Sioutas C, Shing KS 2008 Influence of Stabilizers on the Physicochemical Characteristics of Inhaled Insulin Powders Produced by Supercritical Antisolvent Process, *Pharm Res* **26**, 61.
[11] Kurniawansyah F, Duong HTT, Luu TD, Mammucari R, Vittorio O, Boyer C, Foster N 2015 Inhalable curcumin formulations: Micronization and bioassay *Chem. Eng. J* **279**, 799-808.

IOP Conf. Series: Materials Science and Engineering **543** (2019) 012013 doi:10.1088/1757-899X/543/1/012013

Modification of Sago Starch for Edible Coating

D Sondari
Research Center for Biomaterials, Indonesian Institute of Sciences, Jl Raya Bogor km 46 Cibinong, Bogor, West Java 16911, Indonesia

E-mail: sondaridewi@gmail.com

Abstract. The edible coating film was prepared using native sago starch and modified sago starch with acetylation. The effects of acetic anhydride (7.5%, 15%, 30%, and 45% w/w) on swelling properties of wheat starch granules, amylose content, lipids content, protein content, a degree of substitution, percent acetylation, and gelatinization temperature were investigated. Modified sago starch and native starch were characterized by Fourier Transform Infrared (FTIR), X-ray Diffraction (XRD), Differential Scanning Calorimetry (DSC), Thermogravimetric Analysis (TGA) and Scanning Electron Microscopy (SEM). As a result, modified sago starch containing 45% w/w acetic anhydride have a degree of substitution 0.12 with acetylation 45%. All modified starch granules exhibited higher amylose content compared to native starch. Higher amylose contents inhibit the extensive granule swelling power, so swelling power decreased while increased amylose content. Edible coating film used were starch concentration 5.0% w/w, and glycerol content 1.0% v/v. Edible coating film made from native and modified sago starch have been studied to hydrophobicity, Fourier transforms infrared (FTIR), water vapor absorption, and scanning electron microscopy (SEM).

1. Introduction

Sago (*Metroxylon sagu Rottb*) variety of crops is one of the promising raw materials for the production of the biodegradable material. Most or around 50% of the world's sago plants are in Indonesia [1]. Sago starch is commonly used as a functional ingredient in food industries such as a thickener, stabilizer, and gelling agent. However, the utilization of sago starch is not optimal because if the sago starch cooked, the pasta is hard, not clear and very sticky. This is a constraint in the application as raw materials for food and non food industries [2-4].. To increase starch added value is to make modifications starch to obtain properties suitable for a particular application [5, 6].

Modification of starch is the treatment given to starch in order to obtain better properties or to change several specific properties. Edible films or coatings could be used as an alternative way of conservation, because of their ability to reduce respiration, reduce moisture, solute migration, and transpiration rate, to maintain firmness and generally delay senescence [7]. Sago starch has great potential to be explored in the development of edible coatings because of the abundance of materials, easy to obtain, low price, edible, and easy to recycle (renewable) and its ease to be physically modified. In addition, the high amylose composition of sago starch that is 27%has potential as a strong film-forming material [8]. The objective in this study was a modification of sago starch using acetylation and characterization of an edible coating.

2. Experimental section

2.1. Sample Preparation
Acetylation was performed according to previous researcher [9]. 200 g of starch (native) were dispersed in 460 g of distilled. Then, acetic anhydride was added drop-wise to the stirred starch suspension at pH kept between 8 and 9 by adding 2% NaOH solution. When the whole amount of acetic anhydride was added, the suspension was further stirred for 15 min. After the reaction, the modified starch was washed with distilled water, dried at a temperature of 20◦C for 48 h. Ten percent (w/v) solutions sago starch were prepared of stirring at 60oC. Glycerol was added as a plasticizer at 0.5-1.25% (w/w of sago). The solutions were then allowed to dry overnight at 45oC. They were then conditioned at 51.4% RH at room temperature (30 ± 2oC) for 48 h prior to any analysis [10].

2.2. Characterization
Amylose content was analyzed by the spectroscopic method, the solution was measured its absorbance by a UV-Vis spectrophotometer at a wavelength of 625 nm and amylose levels (in percent) was determined using the standard curve equation of amylose solution. Swelling power and solubility of starch was determined by previously developed method [11]. Soxhletation method was used to evaluated fat content according to SNI 01-2891-1992. Determination of degree of substitution using previously developed method [12]. The functional groups of native and modified sago starch were analyzed using FTIR at a wave number of 4000-400 cm-1. The morphological analysis of the native and modified sago starch was investigated using a scanning electron microscope (SEM) HITACHI SU3500. X-ray diffraction pattern of native and modified sago starch was obtained by using x-ray diffractometer (PANanalytical, X'pert PRO Multi-purpose X-ray diffractometer. Film moisture content (MC). MC of the film samples was determined gravimetrically by measuring water removed from the initial mass. The film samples were dried at 110°C for 24 h to attain a constant weight [13]. The contact angles of prepared films were assessed by the recommended procedure by previous researcher [14].

3. Results
The effect of addition of acetic anhydride, at different levels, on moisture content (%), ash content (%), protein content (%), lipid content (%), swelling power (%), solubility (%), amylose content (%), acetyl (%) and degree of substitution (DS) of native and modification of sago starch are shown in Table 1.

Table 1. Physicochemical analysis of native sago and modified of sago starch

Proximate test	Native sago	Acetylation modification of sago 7.5%	15.0%	30.0%	45.0%
Moisture content (%)	11.46	12.18	12.83	13.15	13.86
Ash content (%)	1.04	0.72	0.64	0.55	0.51
Protein content (%)	0.016	0.005	0.006	0.011	0.016
Lipid content (%)	1.16	0.68	0.52	0.09	0.20
Swelling power (%)	-	13.23	13.31	13.62	13.56
Solubility (%)	-	1.03	1.18	1.96	1.85
Amylose content (%)	25.85	25.59	28.35	29.41	29.79
Acetyl group (%)	-	7.24	7.72	8.06	7.96
Degree of substitution (%)	-	0.09	0.11	0.11	0.12

The structure morphology native sago starch and sago starch were modified can be seen in Figure 1. Showed that the granule structure of native sago starch is round and oval. The modification acetylation of sago starch brought granule fusion in both starches. The fusion of starch granules is due to increased hydrogen bonds between starch molecules due to the substitution of hydrophilic compounds

IOP Conf. Series: Materials Science and Engineering **543** (2019) 012013 doi:10.1088/1757-899X/543/1/012013

into starch molecules, so that starch molecules combine to produce a fusion of starch granules [15]. [Acetylation modification does not cause damage to the starch granules, but the presence of cracks, scratches and partly occurring granular oval changes result in changes in their specific surface areas.

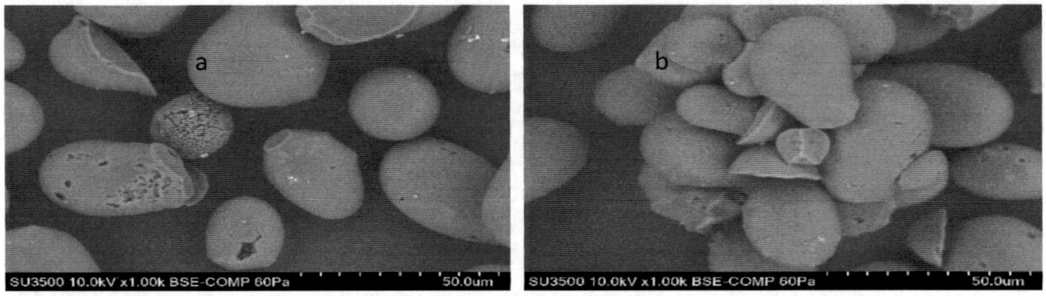

Figure 1. Scanning electron micrographs (SEM) of native (a) and acetylated sago starch (b)

The FTIR analyses were used to study the chemical structural change between native sago and modified sago starch. The chemical structure can be investigated from the change of functional group by means of the absorption peak characteristic.

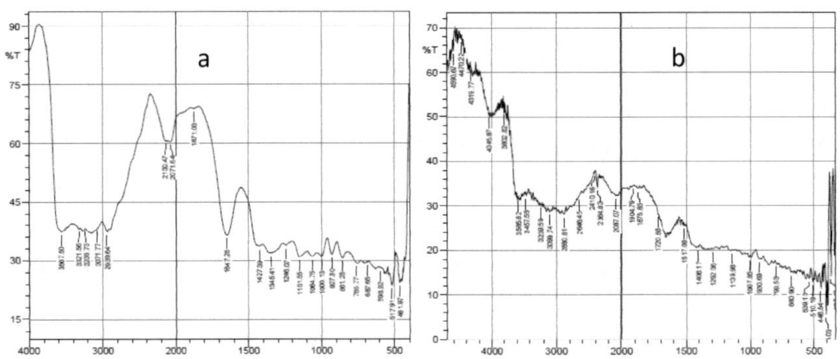

Figure 2. Spectra FTIR of native sago (a) and acetylated sago starch (b)

Native sago starch is composed of amylose fraction (≈30%) and amylopectin fraction (≈70%). Amylose starch is linear molecule consisting of α-1,4- glucopyranose for amorphous region and amylopectin which is α-1 branch fraction, 4 and α-1,6-glucopyranose which have crystalline region.The x-ray pattern for native sago and modification with acetyl in Figure 3.

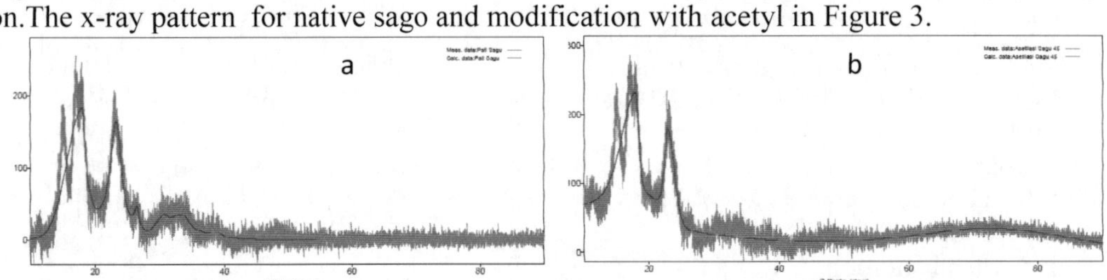

Figure 3. X-Ray patterns of native sago (a) and modified sago starch (b)

The determination of the thermal properties from native sago starch and acetylation was performed using DSC. The analysis results using DSC suggest gelatinization temperature difference where sago acetylation starch has lower gelatinization temperature (Table 2).

Table 2. Gelatinization comparison of native and modified cassava starch

Sample	$C_4H_6O_3$ (ml)	Gelatinization Temperature			Gelatinization enthalpy (ΔHg) (J/g)
		To (°C)	Tp (°C)	Tc (°C)	
Native sago	0	32	102	120	11.6
Acetylated sago	45	40	90	120	5.2

The effect of acetylation of starch on the thermal stability was studied by thermogravimetric analysis (TGA). TGA curves of starch and acetylation starch are shown in Fig. 4. Native sago starch (Fig. 4a) was degraded at 300 °C and ends at 450 °C with a loss percent weight of 95%. Further heating to a temperature of 550 °C produces carbonization and ash formation [16].

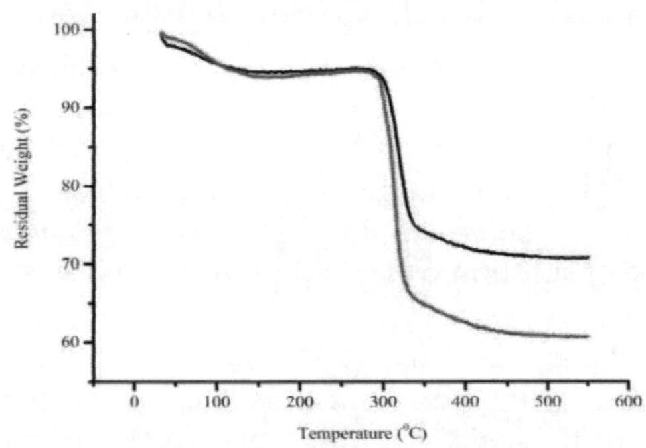

Figure 4. Thermal stability analysis of native (a) and modified cassava starch (b)

The edible film of acetylated sago starch showed higher moisture content (%) than acetylated sago starch (Table 3).

Table 3. Moisture content of native and acetylated sago starch film

Formula	Glycerol (%)	Mc (%)	Contact Angle (°)
Native sago	1	0.16	27.43
Modified of sago (45% acetic anhydride)	1	0.46	60.41
	0.5	0.23	68.75
	0.75	0.29	63.29
	1.25	0.51	50.24

4. Discussion

Native sago starch showed higher ash content than a modified sago starch, but moisture content of sago decreased after acetylation reaction at different levels. The moisture content and ash content of native sago starch and modified sago are in accordance with SNI requirements is for water content

<14.5 and ash content <7. According to Winarno (1989), the level of food with a moisture content of less than 15% is classified as dry food [17]. Fellow (2000) explained the influence factors of the drying process of foodstuffs are the initial moisture content of the material, the humidity of the environment, and the heat transfer [18]. Protein and lipid content decreased with increasing levels of acetyl anhydride as agent acetylation.

The acetylation of modification sago starch increased the swelling power and solubility. This may be due to the substitution of hydrophilic compounds from the acetyl group in the hydroxyl group of sago starch [19]. Swelling power and solubility of oxidation starch increased to 30.0% acetic anhydride concentration. The correlation between solubility and lipid content, high lipid content in sago starch causes lower solubility as report previous researcher [20]. The various factors influenced to swelling power, granule disintegration, and release of amylose depends on the type of starch, starch concentration, temperature, solute, and agitation during heating [21]. Acetylated sago starches had slightly higher amylose contents than native starches, the amylose content increased from 25.85% to 29,79% in sago starch during acetylation with the highest concentration of acetic anhydride. The similar research on the amylose content of starch acetylation which has been previously observed [19]. The presence of acetyl groups in sago starch affects the function of amylose and amylopectin fractions in iodine absorption during amylose testing [19, 22]. The effect of modification using acetic anhydride on acetyl (%) and degree of substitution (DS) of sago starch showed that the increasing of both acetyl and degree of substitution. The degree of substitution (%) increased with increase in the level of acetic anhydride, this research was reported other researcher [23]. . The level of acetic anhydride up to 45.0% with a degree of substitution at 0.12. Similar trends were observed for acetyl (%), acetyl increased with increase in the level of acetic anhydride up to 45.0% (8.06), but the level of acetic anhydride up to 45.0%, the acetyl content decreased (7.96). Acetate starch with a low degree of substitution (<0.1) commonly used in the food industry because it provides structural consistency and stability and starch with a degree of substitution (> 1.0) used as a substitute for thermoplastic cellulose acetate [24].

In Figure 2, shows that modified starch and native starch have a similar functional group structure. The same functional group because the main constituent components of starch are amylose and amylopectin [25]. The intensity of the hydroxyl groups absorption band at range 3000 - 3300 cm^{-1}, and 2,950 cm^{-1} correspond to OH and CH stretchings, while the peaks at 1,647 cm^{-1} and 1,427 cm^{-1} correspond to (OH) and (CH) bendings was observed of the native sago starch. Compared to native starch, starch acetates pressence sorption band at 1,750 cm^{-1} that was attributed to the stretching of the ester carbonyl C=O and indicated the acetylation of starch [26].

Figure. 3 show the C-type Crystallin structure consisting of about 65% of type A and 35% of type B of the crystalline form [8, 27]. According to the Figure 3. Can be known that the intensity of modified sago starch decrease compare to native sago. This result indicated that acetylation reduced the crystallinity of sago starch. The crystalline structure is affected by the intra- and intermolecular hydrogen bonds. In the acetylation process, the hydroxyl groups were replaced partially by the acetyl group. Therefore, acetylation reduce the formation of intermolecular hydrogen bonds and decrease the crystalline structure [24, 28].

The molecular structure of amylopectin, and granule structure can be a control for starch gelatinization (Table 2). Acetylation of starch leads to the opening of double helices and fusion of lamella starch crystals from faster amylopectin which results in decreasing gelatinization temperature [29]. The acetylation reaction with anhydrous acetate results in a decrease in ΔH from the acetylated sago starch. The decrease in ΔH represents the loss of the double helix sequence, and the loss of crystallinity in the granules. The native sago starch has a high degree of substitution and can increase the larger reduction in ΔH. The decrease in ΔH from acetylated sago starch is the result of double helices present in the crystal and non-crystalline regions of the broken granules during the acetylation reaction. The ΔH value of native sago starch shows a lower percentage of regular or less stable crystal structures [29].

As for acetylated sago starch (Fig. 4b) degrades at the same temperature as native sago starch at 300 ^{0}C but ends at higher temperatures at 500 ^{0}C and with a lower percent weight loss of 93%. The high thermal stability of acetylation modified starch is due to the presence of ester groups. The TGA results show that acetylation modification on starch can reduce thermal stability to 330 °C. However, thermal stability increases at temperatures over 330 °C. The degree value of substitution (DS) affects the value of thermal stability. An increase in thermal stability with increased DS is associated with a small number of hydroxyl groups remaining after acetylation. Starch decomposition is the result of inter or intramolecular dehydration reactions of starch molecules with water as the main product of decomposition [24]. The condensation reaction becomes slower after the hydroxyl group is replaced by the acetyl group. Therefore, a high DS has a beneficial effect on the thermal stability of the sample. The edible film of acetylated sago starch showed higher moisture content (%) than acetylated sago starch (Table 3). The presence of glycerol in the film formulation causes an increase in the hydrophilic properties of the edible film. While edible film from native sago starch has a lower water content. Chain of starch molecule is disconnected by acetic anhydride so that its water content value decreases. The water content of the film affects the ability to absorb water of a film. The high water content causes the quality of the film to have a shorter lifetime due to bacterial disorders [30].. Edible films made from native sago and acetylation modifications using glycerol with a concentration of 0.5, 0.75, 1.0 and 1.25 % weight, glycerol as plasticizer to impart the thermoplasticity to the starch film. The quality of the edible film can be seen from the surface hydrophilicity of edible film that using contact angles determination. The results of testing the contact angles edible film of native sago and acetylation modifications can be seen in Table 3. An increase in the concentration of glycerol as a plasticizer decreases the contact angle value of a film due to the hydrophilicity of glycerol. Addition glycerol to impart the thermoplasticity to the starch film, but the addition of excess amount caused the drop contact angle, bringing about the deteriorating mechanical properties of the film [31].

5. Conclusion

The optimum conditions of the acetylation reaction were carried out at a concentration acetic anhydride of 45% with degree of substitution 0.12%. The presence of the ester carbonyl C=O indicated the acetylation of starch has been done. Generally, the physicochemical properties of modified sago starch better than native sago starch.

Conflict of interest

No conflict of interest

Acknowledgment

The authors would like to thank the Ministry of Research, Technology and Higher Education for Program Insentif riset Sistem Inovasi Nasional (INSINAS) 2018.

References

[1] Flach M 1997. Sago palm. In Promoting the conservation and use of underutilized and neglected crops, report No.13, 1–61, Rome, Italy:Institute of Plant Genetics and Crop Plant
[2] Mirmoghtadaie L, Kadivar M, Shahedi M 2009 *Food. Chem.* **116** 709–13.
[3] Cock J H 1982 *Sci.* **218(4574)** 755-62.
[4] Miyazaki M R, Hung P V, Maeda T, Morita N 2006 *Trends in Food. Sci. Technol.* **17** 591-99.
[5] Singh S, Gamlath S, Wakeling L 2007 *Int. J. Food. Sci. Technol.* **42(8)** 916-29.
[6] Huang J, Schols H A, Jin Z, Sulmann E Y, Voragen A G J 2007 *Carbohydr. Polym.* **67** 11–20.
[7] Tezotto-Uliana JV, Fargoni GP, Geerdink GM, Kluge RA 2014 *Postharvest Biol. Technol.* **91** 72–77.
[8] Ahmad F B, Williams P A 1998 *J. Agricul. Food. Chem.* **46(10)** 4060–65.
[9] Fortun T, Juszczak L, Gałkowsk D, Małgorzat B, Aczkowicz, Łabanowska M, Kurdziel, M 2018 *Int. J. Biol. Macromol.* **106** 57–67
[10] Moreno O, Cardenas J, Atarés L, Chiralt A 2017 *Carbohydr. Polym.* **178** 147–158.
[11] Onyango C, Mewa EA, Mutahi AW, Okoth MW 2013 *African. J. Food. Sci.* **7(5)** 80-86.

[12] Shah N N, Vishwasrao C, Singhal R S, Ananthanarayan L 2016 *Food Hydrocol.* **55** 179-188.
[13] Thakura R, Saberia B, Pristijonoa P, Goldinga J, Stathopoulos C, Scarlett C, Bowyer M, Vuong Q 2016 *Int. J. Biol. Macromol.* **93** 952–960.
[14] Rezaei S M, Razavi S H, Hosseini S M H 2010 *J. Biotechnol.* **122** 161–66.
[15] Singh N, Chawla D, Singh J 2004 *Food. Chem.* **86** 601–08.
[16] Qiao L 2016 *Chem. Eng. Trans.* **55** 235-40.
[17] Winarno F G 1989 Kimia Pangan dan Gizi. PT. Gramedia, Jakarta.
[18] Fellow P J 2000. *Food Processing Technology, Principle and Practice.* CRC Press. Cambridge, England Research, Gatersleben/International Plant Genetic Resources Institute.
[19] Betancur D, Chel G L, Canizares H E 1997 *J. Agricul. Food. Chem.* **45** 378-382.
[20] Galliard T, Bowler P 1987. *Starch properties and potential*, In T. Galliard (Ed.), Chichester: Wiley, 57– 78.
[21] Biliaderis C G 1991 *Water relationships in food* In H. Levine, & L. Slade (Eds.) New York: Plenum Press. 251–273.
[22] Whistler R L, Daniel J R 1995 Food. Chem. 69–137. New York: Marcel Decker
[23] Rusman R, Majid R A, Abd Rahman W A W, Low J H 2017 *Chem. Eng. Trans.* **56** 1171-76.
[24] Xu Y, MiladinovV, Hanna M 2004 *Cereal. Chem.* **81(6)** 735–40.
[25] Fleche G 1985. Chemical Modification and degradation of Starch. In G.M.A.V. Beynum and J.A Roels (eds.). Starch Conversion Technology. Marcel Dekker, Inc., New York
[26] Mano J F, Koniarova D, Reis R L 2003 *J. Mater. Sci. Mater. Med.* **14** 127-35.
[27] Zobel H F, Young S N, Rocca L A 1988 *Cereal. Chem.* **66** 443–46.
[28] Colussi R, Pinto V, Halala S, Vanier N, Villanova F, Silva R, Zavareze E, Dias A 2014 *Carbohydr. Polym.* **103** 405– 413.
[29] Colussi R, Mello S, Pinto V, Bartz J, Carlos L, Zavareze E, Dias A 2015 *Food. Sci.Technol.* **62** 1076-82.
[30] Malia S, Victo´ria M, Grossmanna, Maria A, Garcia, Miriam N, Martino E, Zaritzky 2002 *Carbohydr. Polym.* **50** 379–86.
[31] Bangyekan C, Aht-Ong D, Srikulkit K 2006 *Carbohydr. Polym.* **63** 61–71.

IOP Conf. Series: Materials Science and Engineering **543** (2019) 012014 doi:10.1088/1757-899X/543/1/012014

The Effects of Vacuum Evaporation on Amino Acid Contents in Pureed *Aloe Chinensis Baker* Gel using HPLC

T Y Hendrawati[1*], R A Nugrahani [1], A I Ramadhan [2], S Susanty [1], A Siswahyu[3]

[1]Chemical Engineering Department, Faculty of Engineering, Universitas Muhammadiyah Jakarta, Indonesia
[2]Mechanical Engineering Department, Faculty of Engineering, Universitas Muhammadiyah Jakarta, Indonesia
[3]Chemical Engineering Department, Industrial Technology Faculty, Institut Sains dan Teknologi Al Kamal, Jl. Kedoya Raya Nomor 2, Kebon Jeruk, Jakarta, Indonesia

Email : yuni.hendrawati@ftumj.ac.id

Abstract. *Aloe Chinensis Baker* is an aloe species that is frequently utilized in Asia, including Indonesia. Aloe vera contains various compounds in its leaves, such as vitamins, enzymes, and amino acids. The largest component of Aloe vera leaves is water. Vacuum evaporation is considered to be a potential method in the process to reduce the water content while maintaining the integrity of the nutrients. The objective of this research is to reduce the water content in pureed Aloe vera gel to obtain extracts. Evaporation using a rotary vacuum evaporator is conducted at an pressure of 110 mBar and temperature of 40 °C, and a varied time of 30, 60, 90, and 120 minutes. Analyses of amino acid contents are performed with HPLC. The results show that at 120 minutes, the preserved amino acid contents include: L-Isoleucine 29.3 ppm, L-Proline 10.8 ppm, L-Arginine 9.0 ppm, L-Asparagine 243 ppm, L-Threonine 56.9 ppm, L-Leucine 30.9 ppm, L-Methionine 38.4 ppm, L-Histidine 15.4 ppm, Aspartic Acid 24.4 ppm, L-Lysine 108.9 ppm, L-Cysteine 131.0 ppm, L-Alanine 40.2 ppm, and Glycine 230.7 ppm. The resulting Aloe vera extract show the following characteristics, density of 0.99 g/ml and pH of 6.5. Yield of 120 minutes evaporation is 40 %.

1. Introduction

Aloe vera is one of the oldest and well known medicinal plants in the world. Aloe vera gel contains important compounds including 19 out of 20 amino acids needed by human bodies, as well as seven out of eight essential amino acids [1]. One of the species of Aloe vera largely cultivated in Indonesia, especially Potianak, West Kalimantan, is *Aloe chinensis Baker* [2]. In cosmetic and pharmaceutical formulations, Aloe vera extract is needed at a lower water but maintained contents of active compounds [3]. A vacuum evaporation process is hoped to preserve those active compounds, especially amino acids. For organic compounds, the commonly used methods of qualitative and quantitative analysis mainly include such methods of mutual coupling as UV, fluorescence, gas chromatography, liquid chromatography, mass spectrometry, etc. [4]. In previous studies, The HPLC method for determining the blood concentration of nifekalant with Ornidazole as internal standard in chinese people plasma was developed and reported [5]. Quantitative analyses of amino acids are performed in this research using HPLC instruments.

Aloe vera is a very commercial species and its leaves are frequently used in cosmetic and pharmaceutical industries. In the food industry, they are used as a functional food source as well as an ingredient in other products, such as gels in in health drinks. In the cosmetic and hygienic industries, aloe vera becomes one of the main materials in cream, lotion, soap, shampoo, facial wash, and other

IOP Conf. Series: Materials Science and Engineering **543** (2019) 012014 doi:10.1088/1757-899X/543/1/012014

products [6]. In the pharmaceutical industry, it is used in the productions for topical applications, such as balms and non-prescription gels, and also for pills and capsules [7].

Aloe vera leaves contain various compounds such as vitamins, minerals, enzymes, and amino acids [6,8]. The presence of these compounds makes aloe vera a useful plant in many fields including cosmetics, pharmaceuticals, and foods. The largest component in aloe vera leaves is water [9,10]. Vacuum evaporation is hoped to reduce the water contents while maintaining the amount of useful bioactive compounds. In this research, the effects of time, temperature, and pressure during evaporation of pureed aloe vera gel are analyzed. The objective of this study is to reduce the water content in pureed aloe vera gel and produce aloe vera extract.

Amino acids are the main building blocks of proteins and can be categorized into two groups, ie. essential and non-essential amino acids. Essential amino acids cannot be produced in the body, and thus must be added in foods, whereas non-essential amino acids can be produced in the body [11]. Amino acids are carboxylic acids with amino (-NH_2) groups. Amino acids found in proteins possess the –NH_2 group on the carbon atom at the α position from the –COOH group. The types of amino acids, their sequence in proteins, and spatial relationships among them determine the three dimensional structures and biological characteristics of simple proteins. Amino acids are organic compounds with functional groups of carboxyl (-COOH) and amines (usually –NH_2). The carboxyl group is acidic, while the amine group is basic. Amino acids in proteins are connected with peptide bonds, and thus in a dipeptide, there is one peptide bond [12].

All amino acids found in proteins have similar attributes, the carboxyl and amino groups are on the same carbon. The differences lie in the R-groups, in which they vary in structure, size, electrical charge, and solubility in water. Several amino acids have specific reactions based on their R-groups. Using hydrolitic reactions on proteins, 20 amino acids are obtained and can be categorized based on their R-groups. The first group is non-polar amino acids with hydrophobic R-groups, ie. Alanine, Isoleucine, Leucine, Methionine, Phenylalanine, Proline, Tryptophan and Valine. The second group is polar amino acids with no charge on the R-group, which consists of Asparagine, Cysteine, Glutamine, Lysine, Serine, Threonine, and Tyrosine. The third group contains amino acids with positively charged R-group, and the fourth contains amino acids with negatively charged R-groups. From these 20 amino acids, there are eight essential amino acids, and those are Isoleucine, Leucine, Lysine, Methionine, Phenylalanine, Threonine, Tryptophan, and Valine. These essential amino acids cannot be synthesized in human body, and thus must be obtained from food and other supplements. Leucine also has a function in the immune system [13]. Lysine has a function as a precursor in antibodies, strengthens the circulatory system, maintains normal cell growth, forms collagen in collaboration with proline and vitamin C, and lowers excess triglycerides in blood [14]. Amino acids have higher melting points (greater than 200 °C) than carboxylic acids or amines [15].

2. Experimental Method

2.1. Material and Instruments

Aloe chinesis Baker leaves from Pontianak Indonesia, pH meter, Pycnometer, Viscometer, HPLC

2.2. Preparation of sample pureed aloe vera gel

Preparation of Pureed Aloe vera gel:

a. Rind is peeled off aloe vera leaves hygienically with a knife.
b. Aloe vera gel is pureed in a blender.
c. Aloe vera gel is filtered using a filter press.
d. Filtrate from the filtration process is dehydrated with vacuum evaporation with an initial volume of 500 ml.

IOP Conf. Series: Materials Science and Engineering **543** (2019) 012014 doi:10.1088/1757-899X/543/1/012014

e. Evaporation with a rotary vacum evaporator is carried out at a pressure of 110 mBar and a temperature of 40 ^{0}C with a variable time of 0, 30, 60, 90 and 120 minutes.

2.3. Analysis of pureed Aloe vera Gel

Analyses are performed on yield, pH, density, viscosity, and results of HPLC on the quantitative amino acid analyses. Amino acids are analyzed with HPLC. The amino acids analyzed are those mainly freed from proteins through hydrolyses using HCl 6 N. Hydrolysates are dissolved in sodium citrate buffer, and each amino acids are separated using HPLC.

3. Results and Discussion

Yield is calculated as the percentage of volume of evaporated water divided by the initial volume from the evaporation process in a vacuum evaporator at 110 mBar and 40 ^{0}C. The resulting yields at varied evaporation time are presented in the following Table 1.

Table 1. Yields at varied evaporation times (at pressure of 110 mBar and temperature of 40 ^{0}C)

Evaporation time (minutes)	Volume of Evaporated (ml)	Yield (%)
0	500	100
30	350	70
60	250	50
90	200	40
120	150	30

In addition to yields, pH, density, and viscosity are also measured. The results from analyses of pH, density, and viscosity are presented in the following Table 2.

Table 2. pH, density, and viscosity at varied evaporation times (at pressure of 110 mBar and temperature of 40 ^{0}C)

Evaporation time (minutes)	pH (Temp. 25 °C)	ρ (g/ml) (Temp. 25 °C)	η (cP)
0	6.5	1.004	129.6
30	6.5	1.005	133.8
60	6.5	1.003	402.3
90	6.5	0.99	10,474
120	6.5	0.99	13,349

Based on the results in Table 2, pH stays the same at 6,5 as time increases. However, density decreases from 1,005 g/ml to 0,99 g/ml, and viscosity increases from 129.6 cP to 13,349 cP as the evaporation time increases.

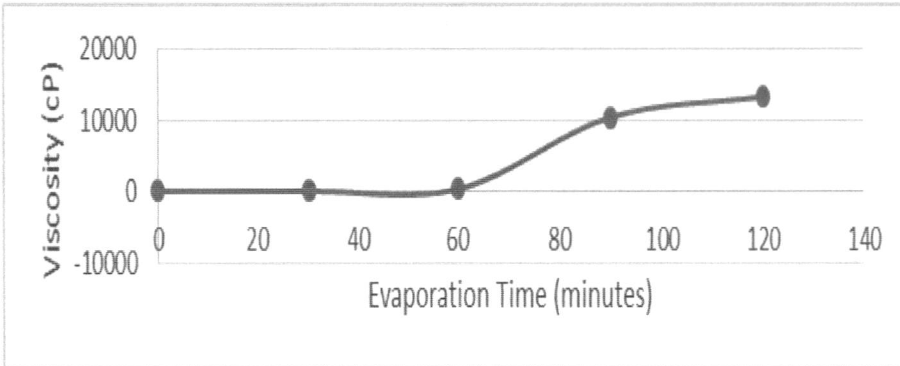

Figure 1. Effects of time of evaporation on Aloe vera gel viscosity in vacuum evaporation process at pressure of 110 mBar and temperature 40 ^{0}C

The analyses of amino acid contents in Aloe vera gel prior to evaporation result in L-Isoleucine 3.72 ppm, L-Valine 6.85 ppm, L-Proline 0.07 ppm, L-Phenylalanine 4.74 ppm, L Arginine 4.81 ppm, L-Threonine 5.68 ppm, L-Leucine 8.53 ppm, L-Serine 6.35 ppm, L-Methionine 1.83 ppm, L-Histidine 5.92 ppm, Aspartic Acid 14.37 ppm, L-Tyrosine 3.34 ppm, L-Lysine 8.27 ppm, Glutamic Acid 14.27 ppm, L-Cysteine 0.02 ppm, L-Alanine 1.09 ppm, and Glycine 7.80 ppm.

The HPLC analysis of amino acid content after 30 minutes of vacuum evaporation results in separation of amino acid components as shown in the chromatogram presented in Figure 2 .

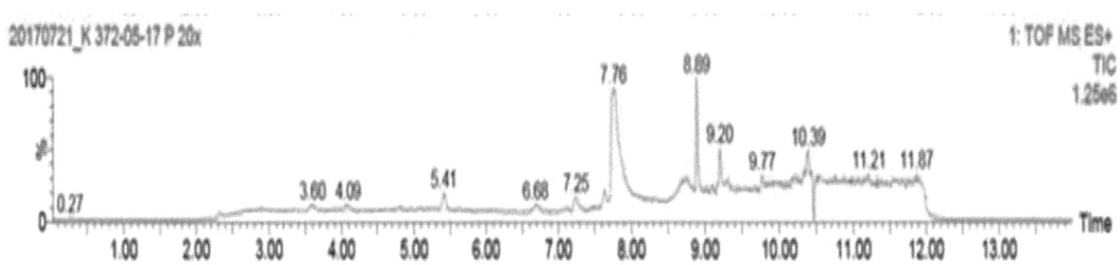

Figure 2. HPLC Chromatogram of Aloe vera gel vacuum evaporated for 30 minutes at pressure of 110 mBar and temperature 40 ^{0}C

In the 30 minutes evaporation. the amino acids that are preserved in the Aloe vera extract are L-Isoleucine 13.4 ppm, L-Valine 455.1 ppm, L-Proline 14.7 ppm, L-Phenylalanine 7.3 ppm, L-Arginine 12.2 ppm, L-Asparagine 73.7 ppm, L-Leucine 118.7 ppm, L-Histidine 21.7 ppm, Aspartic Acid 82.4 ppm, and L-Tyrosine 5.4 ppm. The essential amino acids retained are L-Isoleucine. L-Valine. L-Phenylalanine and L-Leucine.

The HPLC analysis of amino acid content after 60 minutes of vacuum evaporation results in separation of amino acid components as shown in the chromatogram presented in Figure 3.

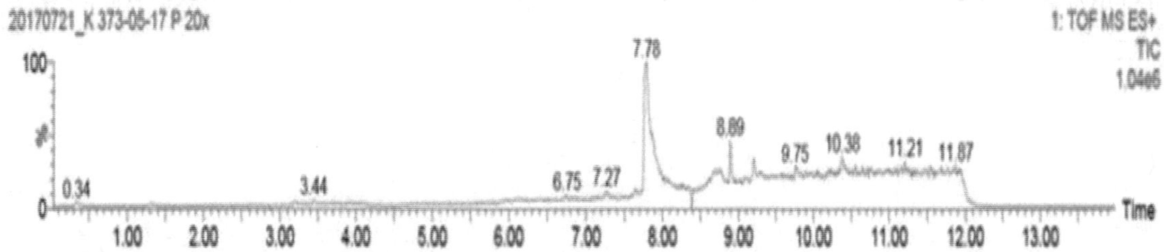

Figure 3. HPLC Chromatogram of Aloe vera gel vacuum evaporated for 60 minutes at pressure of 110 mBar and temperature 40 ^{0}C

In the 60 minutes evaporation. the amino acids that are preserved in the Aloe vera extract are L-Isoleucine 41.2 ppm. L-Valine 16.9 ppm, L-Proline 12.9 ppm, L-Phenyllalanine 3.3 ppm, L-Arginine 10.3 ppm, L -Asparagine 16.8 ppm, L-Threonine 59.7 ppm, L-Leucine 49.6 ppm, L-Histidine 19.3 ppm, and Aspartic Acid 43.9 ppm. The essential amino acids retained are L-Isoleucine. L-Valine. L-Phenylalanine. L-Threonine and L-Leucine.

The HPLC analysis of amino acid content after 90 minutes of vacuum evaporation results in separation of amino acid components as shown in the chromatogram presented in Figure 4.

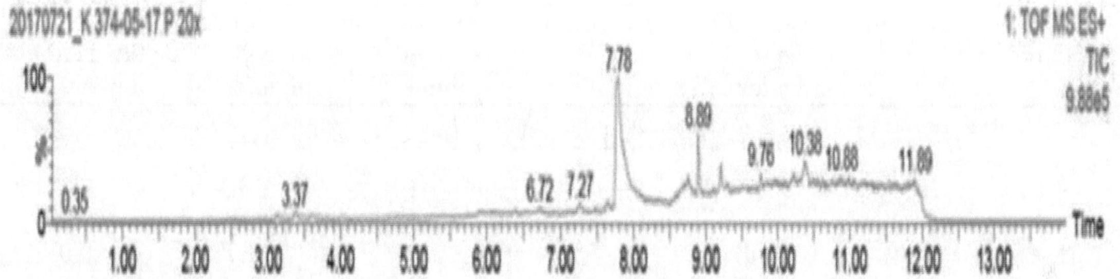

Figure 4. HPLC Chromatogram of Aloe vera gel vacuum evaporated for 90 minutes at pressure of 110 mBar and temperature 40 ^{0}C

In the 90 minutes evaporation. the amino acids that are preserved in the Aloe vera extract are L-Isoleucine 64.3 ppm, L-Proline 12.1 ppm, L-Arginine 10.5 ppm, L-Asparagine 18.5 ppm, L-Threonine 65.1 ppm, L-Leucine 65.8 ppm, L-Histidine 17.5 ppm, Aspartic Acid 39.9 ppm, L-Lysine 112.9 ppm, and L-Cystine 209.9 ppm. The essential amino acids retained are L-Isoleucine. L-Valine. L-Phenylalanine. L-Threonine and L-Leucine and L-Lysine.

The HPLC analysis of amino acid content after 120 minutes of vacuum evaporation results in separation of amino acid components as shown in the chromatogram presented in Figure 5.

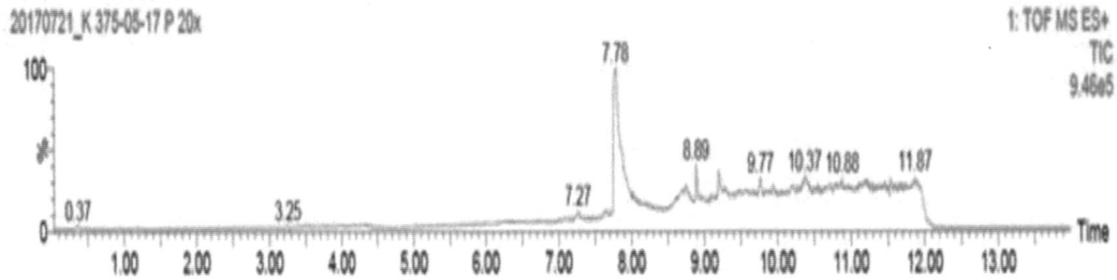

Figure 5. HPLC Chromatogram of Aloe vera gel vacuum evaporated for 120 minutes at pressure of 110 mBar and temperature 40 ^{0}C

In the 120 minutes evaporation, the amino acids that are preserved in the Aloe vera extract are L-Isoleucine 29.3 ppm, L-Proline 10.8 ppm, L-Arginine 9.0 ppm, L-Asparagine 243 ppm, L-Threonine 56.9 ppm, L-Leucine 30.9 ppm, L-Methionine 38.4 ppm, L-Histidine 15.4 ppm, Aspartic Acid 24.4 ppm, L-Lysine 108.9 ppm, L-Cystine 131.0 ppm, L-Alanine 40.2 ppm and Glycine 230.7 ppm. The essential amino acids retained are L-Isoleucine, L-Threonine, L-Leucine, L-Methionine and L-Lysine.

Tabel 3. The results of HPLC analysis on amino acid content of vacuum evaporation of Aloe vera gel at evaporation time 0 minutes until 120 minutes at pressure of 110 mBar and temperature 40 ^{0}C

No	Amino Acid content (ppm)	Evaporation time 0 minutes	Evaporation time 30 minutes	Evaporation time 60 minutes	Evaporation time 90 minutes	Evaporation time 120 minutes
	Essential amino acids					
1	L-Isoleucine	3.72	13.4	41.2	64.3	29.3
2	L-Valine	6.85	455.1	16.9	13.3	0.0
3	L-Phenylalanine	0.07	7.3	3.3	4.7	0.0
4	L-Threonine	5.68	0.0	59.7	65.1	56.9
5	L-Tryptophan	0.0	0.0	0.0	0.0	0.0
6	L-Leucine	8.53	118.7	49.6	65.8	30.9
7	L-Methionine	1.83	0.0	0.0	0.0	38.4
8	L-Lysine	8.27	0.0	0.0	112.9	108.9
	Non Essential amino acid					
9	L-Proline	0.07	14.7	12.9	12.1	10.8
10	L-Arginine	4.81	12.2	10.3	10.5	9.0
11	L-Asparagine	0.00	73.7	16.8	18.5	243.0
12	L-Serine	6.35	0.0	0.0	0.0	0.0
13	L-Glutamine	nil	0.0	0.0	0.0	0.0
14	L-Histidine	5.92	21.7	19.3	17.5	15.4
15	Aspartic acid	14.37	82.4	43.9	39.9	24.4
16	L-Tyrosine	3.24	5.4	0.0	0.0	0.0
17	L-Glutamic acid	14.27	0	0.0	0.0	0.0
18	L-Cysteine	0.02	0	0.0	209.9	131.0
19	L-Alanine	1.09	0	0.0	0.0	40.2
20	Glycine	7.80	0	0.0	0.0	230.7

Polar amino acids include serine, threonine, asparagine, glutamine, histidine and tyrosine. The hydrophobic amino acids include alanine, valine, leucine, isoleucine, proline, phenylalanine,

IOP Conf. Series: Materials Science and Engineering **543** (2019) 012014 doi:10.1088/1757-899X/543/1/012014

tryptophane, cysteine and methionine. In this research, the content remain in Aloe vera gel extract was essential amino acid. The essential amino acid content was changed because the vacuum evaporation conducted at pressure of 110 mBar and temperature 40 ^{0}C showed that the amino acid content increased and then decreased along the evaporation time. This is due to the polar essential amino acid content dissolved in water was evaporated. This can be seen in Figure 6. Hence the effect of evaporation time on hydrophobic essential amino acid content in Aloe vera gel can be seen at Figure 7.

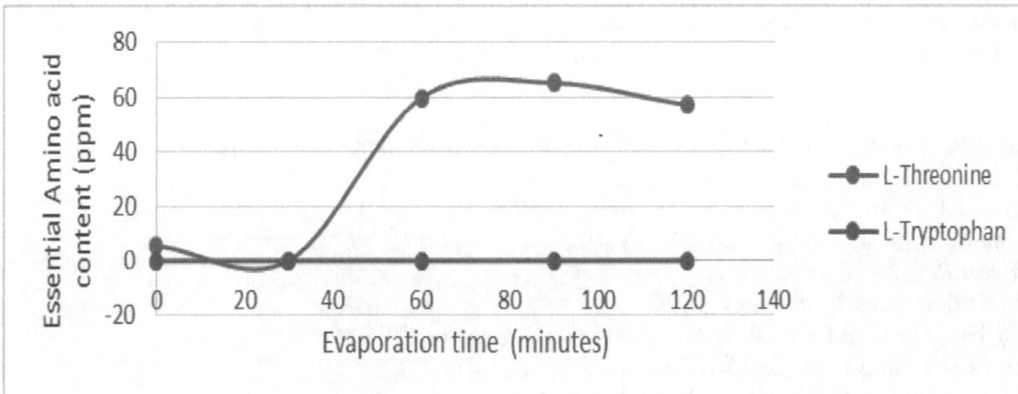

Figure 6. The effect of evaporation time on polar essential amino acid content in Aloe vera gel extract

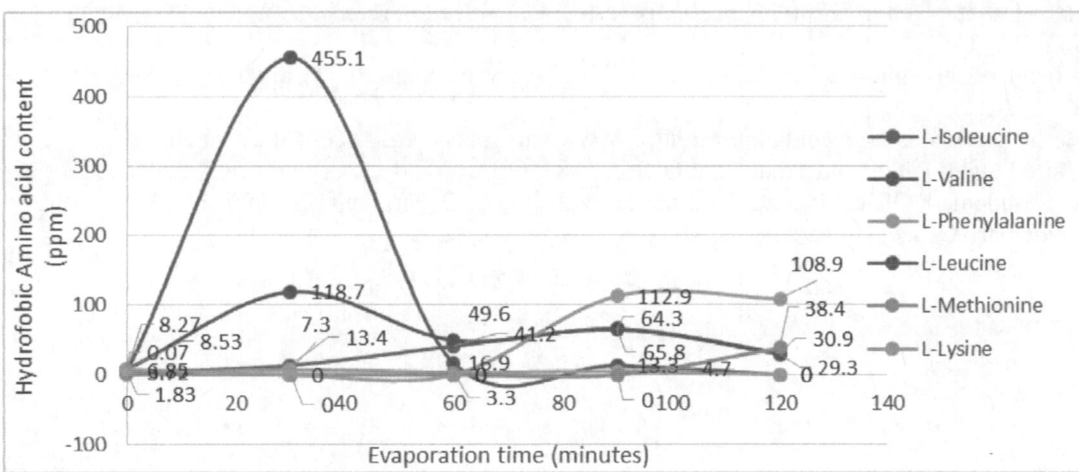

Figure 7. The effect of evaporation time on hydrophobic essential amino acid content in Aloe vera gel extract

4. Conclusion

The processes of vacuum evaporation at the pressure of 110 mBar temperature of 40 °C and variable time of 30, 60, 90 and 120 minutes can maintain the presence of amino acids in the Aloe vera gel extract. Increasing time produces constant pH at 6.5, decreasing densities from 1.005 g/ml to 0.99 g/ml and increasing viscosities from 129.6 cP to 13,349 cP. After 120 minutes of evaporation, the amino acids that are preserved in the Aloe vera gel extract are L-Isoleucine 29.3 ppm. L-Proline 10.8 ppm. L-Arginine 9.0 ppm. L-Asparagine 243 ppm. L-Threonine 56.9 ppm. L-Leucine 30.9 ppm. L-Methionine

38.4 ppm. L-Histidine 15.4 ppm. Aspartic Acid 24.4 ppm. L-Lysine 108.9 ppm. L-Cystine 131.0 ppm. L-Alanine 40.2 ppm. and Glycine 230.7 ppm. The essential amino acids retained are L-Isoleucine. L-Threonine. L-Leucine. L-Methionine and L-Lysine.

Acknowledgment

Thanks to Universitas Muhammadiyah Jakarta and Directorate of Research and Community Service, Directorate General for Research and Development, Ministry of Research and Technology Higher Education on the Research Grant PTUPT in 2017-2019 contract number 03/E/KPT/2018, 03/01/2018.

References

[1] Rajeswari R, Umadevi M, Sharmila R C, Pushpa R, Selvavenkadesh S, Sampath K P and Debjit B 2012 *J. Pharmaco. Phytochem.* **1** 118 – 24.
[2] Hendrawati T Y 2015 *J. Eng. Sci. Tech.* **2015** 47-59
[3] Boudreau M D and Beland F A 2006 *J. Environ. Sci. Health Part C: Environ. Carcinog. Ecotoxicol. Rev.* **24** 103-54.
[4] Qiong H 2017 *Chem. Eng. Trans. 62* 439 – 44.
[5] Xin X, Changbin L and Yali W 2015 *Chem. Eng. Trans.* **46** 1375-80.
[6] Hamman J H 2008 *Molecules* **13** 1599-16.
[7] Eshun K and He Q 2004 *Food Sci. Nutr.* **44** 91-96.
[8] He Q, Changhong L, Kojo E and Tian Z 2005 *Food Control* **16** 95-104.
[9] Grindlay D and Reynolds T 1986 *J. Ethnopharmacol.* **16** 117-51.
[10] Josias H and Hamman 2008 *Molecules* **13** 1599-16.
[11] Sitompul S 2004 *Buletin Teknik Pertanian* **9** 33-37.
[12] Hans D J and Nobert S 2008 *Peptides from A-Z A Concise Encyclopedia* Wiley VCH Verlag GmbH & Co. KGaA.
[13] Edison T 2009 *Amino acid: Esensial for our bodies*. http://livewellnaturally.com, Accessed 27.07.2018.
[14] Harli M 2008 *Asam amino esensial* http://www.suparmas.com. Accessed 27.07.2018.
[15] United States Patent and Trademark Office, 1989, Official Gazette of the United States Patent and Trademark Office: Patents, Volume 1108, 3-4, U.S. Department of Commerce, Patent and Trademark Office, University of California.

IOP Conf. Series: Materials Science and Engineering **543** (2019) 012015 doi:10.1088/1757-899X/543/1/012015

The utilization of gadung (dioscorea hispida dennst) starch for edible coating making and its tomato packaging

P N Alam[1], Mukhlishien, H Husin, T M Asnawi, Santia and A Yustira
Department of Chemical Engineering, Syiah Kuala University,
Darussalam, 23111, Banda Aceh, Indonesia

E-mail: pocutnurulalam@unsyiah.ac.id

Abstract. An edible coating is one method that could be utilized in preserving and coating on fruits. The utilization of gadung (Dioscorea hispida Dennst) starch as a raw material in edible coating making can be used as an alternative renewable raw material due to of high starch content and its potential still not been utilized optimally. This study has been evaluated the influences of hydrocolloids and plasticizers contents of the mixture to produce proper tensile strength and elongation of the coating produced. The quality and characterization of edible coating have been evaluated using weight loss test, thickness test, elongation test, tensile strength test and oxygen permeability test. The gadung starch edible coating was shown to be effective in tomato fruit preserved until 21 days. Mechanical properties of edible coating were obtained the tensile strength ranging from 5.76 to 7.03 kgf.mm-2, elongation ranging from 13.57 - 38.56%, thickness ranging from 0.01 - 0.05 mm and optimum oxygen permeability of 3.739 g.m-2.h-1 by adding of the CMC of 1% and glycerol of 10%.

1. Introduction

Fruit is an important source of nutrients and energy for human life. The problems faced by fruit traders and consumers is on how to maintain longer its freshness and quality [1]. The decreasing of fruits quality can be affected by oxygen and water content, light exposure and storage time. To overcome the degradation of fruit quality in order to prolong its shelf life before consumption can be carry out by cooling, packaging with plastic and storage in atmospheric conditions [2]. Among disadvantages of packaging with plastic would damage the fruits due to its characteristic might could entrap water vapor and instable if excessive exposed to heat [3]. The best way to maintain the quality of fruits is to use edible coating [4, 5].

An edible coating is a method used to extend shelf life and maintaining the quality of fruits at room temperature. An edible coating is a thin layer that aims to provide a selective resistant of mass transfer so that fruits can maintain weight loss [6]. Edible coatings can be made from three different types of materials: hydrocolloids, lipids, and composites. Edible coating layer provides an alternative as a packaging material that lesser impact on environmental pollution due to its safer and biodegradable materials and low cost [4].

One of the most effective ways to retain fruit shelf life is to packaging the fruits with edible coating [7]. The edible coating is safe for consumption. In this study the edible coating layer is made from starch of gadung (Dioscorea hispida Dennst) due to of its starch is easily biodegraded into harmless compounds. In Indonesia there are various plants producing starch such as cassava, gadung, taro, rice, potato, sorghum, banana and others [5]. Therefore, this research utilizes gadung as raw material for edible coating production because gadung contains starch is rarely consumed and by society due to complicated processing process. The utilization of hydrophilic plasticizers could increase edible coating solubility. The most widely used plasticizer in edible coating making is glycerol. This study proposed the preparation of edible coating from gadung starch to study the potential utilization of gadung starch as edible coating in order to maintain the shelf life and the quality of fruits.

2. Experimental section

2.1. Materials
The material used in this research is gadung starch tubes, liquid glycerol, aquadest, silica gel and Carboxymethyl Cellulose (CMC). The tool used in this research is 250 ml glass, 50 ml measuring cup, blender, spatula, hot plate, mercury thermometer, oven dryer, magnetic stirrer, knife, spatula and stopwatch.

2.2. Preparation of gadung starch
Methodology This study used a complete randomized design (RAL) experimental design method. The variables evaluated were variation of hydrocolloid concentration of CMC (A) consisting of 4 levels i.e 1, 3, 5 and 7% (b/v) and concentration factor of plasticizer ie glycerol (B) consisting of 4 levels ie 10, 15, 20 and 25% (w/v). So the combination of treatment is 4 x 4 = 16 by using 2 replications, so that in 32 units obtained the experiment. Gadung starch was extracted from a local variety gadung which is a commonly grown in forest in Aceh Province, Indonesia. Freshly harvested tomato (Solanum lycopersicum L.) fruit at the mature-green stage of ripening were purchased from local market of Banda Aceh. The fruits were chosen for uniformity in color, size, and absence of blemishes, mechanical damages and fungal infection. Before coating, tomatoes were thoroughly washed with a 0.05% sodium hypochlorite solution (Product Code: 425044 Sigma-Aldrich) and air-dried. Sorbitol (Product Code: S1876), glycerol (Product Code: G9012), and all other chemicals used were of analytical reagent grade and purchased from Sigma (Sigma-Aldrich GmbH, Sternheim, Germany).

2.3. Isolation of starch
Gadung starch was extracted by blending with 0.16% sodium metabisulphite solution and the resultant slurry was screened through 100 mesh sieve followed by 300 mesh sieve. The supernatant was decanted and the starch layer was resuspended in distilled water and then centrifuged at 8000 rpm for 5 min. The brownish layer was discarded; the under layer was resuspended in distilled water and recentrifuged at 8000 rpm for 5 min. The resultant starch was finally air dried at 45 ◦C for 12 h and then pulverized. The extracted starch had approximately 98.5% purity (fat 0.22%, fiber 0.53%, protein 0.42%).

2.4. Coating application
The selected tomatoes were divided into four groups each contained thirty fruits and dipped in ascorbic acid and citric acid for 5 m. The coating slurry was prepared by gelatinizing the gadung starch (4.0 g/100 mL solution) in distilled water with constant stirring on hot plate. Glycerol was then added in the gelatinized slurry and again stirred for 30 min. Tomatoes were dipped separately in coating slurry for 30 s, air dried and finally stored at 25 ◦C and 60% RH for 20 days in a controlled atmosphere chamber. The uncoated fruits were used as control. Quality assessment of each sample was evaluated after every 4 days.

2.5. Film preparation
Gadung starch films were prepared by casting technique. Aqueous suspension of 2% gadung starch was gelatinized at 85°C for 30 min on a hot plate followed by adding of Carboxymethyl Cellulose

(CMC) with continuous stirring. Glycerol as plasticizer was added (10, 15, 20, and 25% of starch weight) to the gelatinized suspensions and again stirred for 30 min. The concentration of plasticizers was selected based on preliminary trials to obtain smooth and flexible films. The air bubbles formed during heating were removed by means of sonnication at reduced pressure. The slurries were then casted on acrylic trays and dried overnight at 55°C for 12 h. Films were removed carefully from trays and conditioned in a desiccator at 58% relative humidity (RH) for 48 h prior to further testing.

2.6. Edible Film Testing

Analysis of tomato fruit is weight loss, and organoleptic test. While the characterization test of Edible Coating includes film thickness, elongation, tensile strength, and oxygen permeability.

3. Results and discussion

3.1. Weight loss

All coated and uncoated tomato samples were weighed at the beginning of the experiment and after every 3 days for upto 21 days. In this study tomato coated edible coating with different concentrations of CMC and Glycerol and observed tomatoes for 21 days. Weight loss was determined by taking their initial and final weights differences and expressed as percentage. It can be seen in Figure 1 the decrease in weight loss that occurs with different levels of CMC and Glycerol.

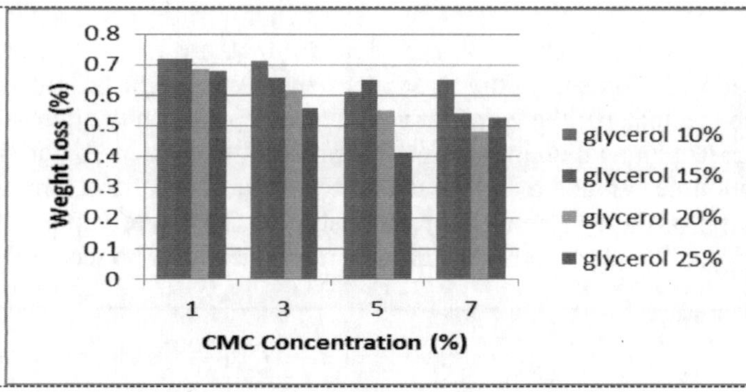

Figure 1. Experiment set-up and apparatus

Figure 1 shows that the greater concentration of glycerol causes less weight loss value. Weight loss with a concentration of 25% glycerol is greater than weight loss with a concentration of 10%. As well as on the increase in glycerol concentration, Figure 1 also shows an increase in CMC concentration causing decreasing weight shrinkage values. The shrinkage value of successive weights decreases with the addition of glycerol. The highest shrinkage value at 10%, second and third concentrations of glycerol at concentrations of 15% and 20%, and the smallest at 25% concentration. The shrinkage value of the higher weights indicates the weight loss in the tomato fruit is greater which means the loss is also greater. With the addition of fruit glycerol will be coated so that the inhibition of the greater microbial and fruit rotation will be longer [8]. According to Nawab et al., the shrinkage of tomato weight will increase during storage, this is because tomato is a climacteric fruit that has a level of respiration that continues to increase along with the maturation of fruit [9].

3.2. Edible coating thickness

The film thickness will affect the oxygen permeability /vapor transmission rate and tensile strength of the film's edible. The addition of CMC and glycerol concentrations also affects the thickness of the edible film, wherein the bonds formed between hydrocolloids and glycerol with the starch tuber starch increase the thickness. The relationship between CMC and glycerol concentrations in edible films with the resulting edible film thickness can be seen in Figure 2.

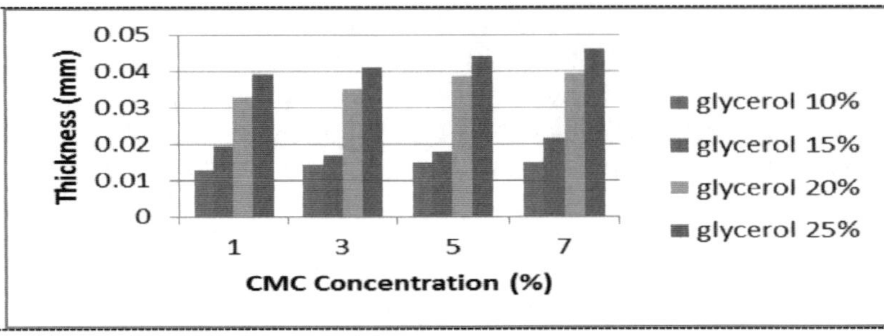

Figure 2. Glycerol and CMC concentration correlation to Elongation of edible film

Figure 2 shows that the addition of CMC and glycerol concentrations tends to produce an average of different thicknesses. The addition of glycerol causes edible films to increase in thickness. The average thickness of edible film with CMC concentration variations were 1, 3, 5 and 7% and the variation of glycerol concentrations were 10, 15, 20 and 25% were between 0.01-0.05 mm. The average standard of edible film thickness is less than 0.25 mm. The thickness of this edible film is relatively thinner than some of the previous studies.

3.3. Elongation

Elongation is the mechanical properties of edible films that are closely related to the physical properties of the film. Elongation shows the maximum length change of the film when it gets tensile until the movie is disconnected [10]. Percent elongation also depends on the ingredients mixed into the edible film mixture. Percent elongation indicates the film's ability to extend before being disconnected. The higher the elongation percentage of a film the better the quality of the film. The data of the test results can be seen graph of the concentration relationship of CMC (carboxyl methyl cellulose) with the concentration of glycerol to elongation on edible film produced can be seen in Figure 3.

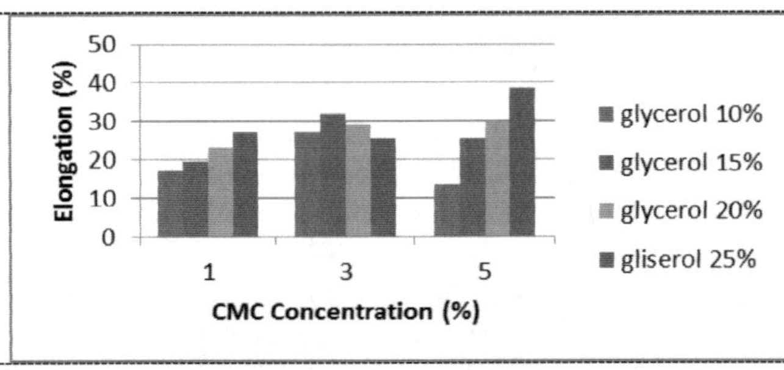

Figure 3. Glycerol and CMC concentration correlation to Elongation of edible film

Based on Figure 3 the addition of CMC with concentrations of 1, 3 and 5% and plasticizers of glycerol at concentrations of 10, 15, 20 and 25% averaged 13.57 - 38.56%. The optimum elongation percentage obtained at the addition of 5% CMC concentration and 25% glycerol concentration is 38.56%, because in this composition the most additions of plasticizer and CMC (carboxyl methyl cellulose) mixture adds moisture to the film causing the film to be elastic. The addition of CMC and plasticizer is needed to overcome the fragility caused by molecular intermolecular forces on the main raw material. Without CMC and plasticizer, films made from starch will be fragile(Thakur et al., 2018). Increased concentrations of glycerol will lead to an increase in elongation percentage rather than the film.

3.4. Tensile strength

Tensile strength is the maximum force that occurs in the edible film during the measurement until the film lasts before breaking or tearing. Measurement of tensile strength is intended to determine the amount of force required to achieve maximum attraction in the area of the film lengthening or stretching.

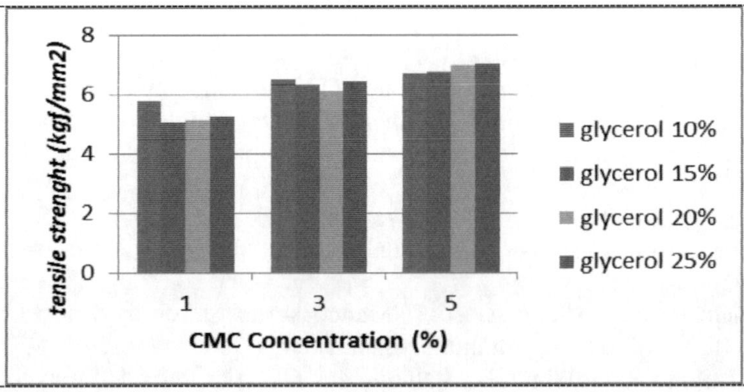

Figure 4. Glycerol and CMC concentration correlation to Tensile Strength of edible film

4. Conclusion

A new non-conventional source of starch from gadung (Dioscorea hispida Dennst) starch was used to prepare edible coating films and the influence of different plasticizers on its functional properties was investigated. Gadung films plasticized without glycerol and CMC formed a fragile film indicating that glycerol was appropriate plasticizer for gadung films. The edible films produced with thickness ranging from 0.01 - 0.05 mm plasticized with glycerol has a good mechanical characteristic with a higher elongation, tensile strenght and lower oxygen permeability with increasing the concentration of glycerol and CMC. The optimum elongation percentage obtained at 38.56% with the addition of 5% CMC and 25% glycerol concentration. The films plasticized with this amount concentration of CMC (5%) and glycerol (25%) also showed a higher tensile strength, it was 7.03 kgf/mm2. Films plasticized with glycerol concentration of 10% combined with CMC 1% gave a lower oxygen permeability values of 3.739 g/m2/h than that of at higher plasticizers concentrations. It could be proved to be a good barrier for water vapors. The findings of this study will facilitate the use of gadung as an alternative starch source for edible coating making. It will also provide foundation for further developments of edible packaging from gadung starch. These results suggest that gadung edible coating can be used effectively to prolong the shelf life of tomato until 21 days.

Acknowledgement

The authors are grateful to the Syiah Kuala University of the Republic of Indonesia for the financial support and contribution through the Research Grant for the Chemical Reaction Engineering Laboratory, Syiah Kuala University year 2018.

References

[1] Miao, J, 2018, RFID-Based Key Technology for Fresh Food Quality Inspection *CET* **66** 1363-68
[2] Shin, S.-H, Chang Y, Lacroix M and Han J, 2017, Control of microbial growth and lipid oxidation on beef product using an apple peel-based edible coating treatment. *LWT* **84** 183-8
[3 Acevedo-Fani, A, Soliva-Fortuny, R. and Martín-Belloso O, 2017, Nanoemulsions as edible coatings. *Curr Opin Food Sci.* **15** 43-9

[4] Saberi B, Golding JB, Marques JR, Pristijono, P, Chockchaisawasdee S, Scarlett CJ and Stathopoulos CE, 2018, Application of biocomposite edible coatings based on pea starch and guar gum on quality, storability and shelf life of 'Valencia' oranges. *Postharvest Biol. Technol.* **137** 9-20

[5] Tavassoli-Kafrani E, Shekarchizadeh H and Masoudpour-Behabadi M, 2016, Development of edible films and coatings from alginates and carrageenans. *Carbohydr. Polym.* **137** 360-74

[6] Takahashia, GCdS, Barbosaa, HD, Bergamascoa RdC, Madronaa GS, Tonona LAC, Yamashitab F and Scapim MRdS, 2017, Development and Active Biodegradable Film Evaluation Incorporated with Oregano Essential Oil and Nanoclay *CET* **57** 403-8

[7] Feng Z, Wu G, Liu C, Li D, Jiang, B and Zhang X, 2018, Edible coating based on whey protein isolate nanofibrils for antioxidation and inhibition of product browning. *Food Hydrocoll.* **79** 179-88

[8] Limchoowong N, Sricharoen P, Techawongstien S and Chanthai S, 2016, An iodine supplementation of tomato fruits coated with an edible film of the iodide-doped chitosan. *Food Chem.* **200** 223-9

[9] Nawab A, Alam F and Hasnain A, 2017, Mango kernel starch as a novel edible coating for enhancing shelf- life of tomato (Solanum lycopersicum) fruit. *Int. J. Biol. Macromol.* **103** 581-6

[10] Thakur R, Pristijono P, Golding JB, Stathopoulos CE, Scarlett CJ, Bowyer M, Singh SP and Vuong QV, 2018, Development and application of rice starch based edible coating to improve the postharvest storage potential and quality of plum fruit (Prunus salicina). *Sci. Hort.* **237** 59-66

IOP Conf. Series: Materials Science and Engineering **543** (2019) 012016 doi:10.1088/1757-899X/543/1/012016

Innovation of Cooking Juice in Process Production of Red-Cane Sugar in Lawang Village of Agam District

Firdaus, E Praputri, E Sundari and M Ulfah
Department of Chemical Engineering, Universitas Bung Hatta, Padang, Indonesia

*Email: firdausftimtdr@gmail.com

Abstract. Cane refining activities which produce red-cane sugar has traditionally been the primary source of income in Lawang villagers of Agam district since hundred years ago. Three main weaknesses found in the process traditionally conducted are lower evaporation rate caused in longer time production and need more fuel, lower product quality caused in lower price, and flue gas pollution. Heat transfer mechanisms involved as long as evaporation takes place in traditional cooking process of juice, namely conduction, convection, and radiation, are not well evaluated yet. Meanwhile, falling film evaporation is not found and gummy cakes formed and accumulated in the surface of basket made of woven bamboo placed in the juice pot. The influence of evaporator like device made of a series of diameters of aluminium circular plates to evaporation process has been studied in traditional cane refining in Lawang village. The device was introduced to the juice pot which significantly faster evaporation process, namely reached about 45% faster then what usually run by villagers. The influences of introduction of lids of aluminium plates as well as any variation in circular plates to faster evaporation process had also been studied. It showed that the numbers of circular plates in a series and introduction of lid of aluminium plate contributed dominantly to faster evaporation process till about 56.67% faster, while the incline of circular plates in a range of 0 to 20 degree and the existence of some holes in the inner plates for circulation of juice cooking are not significantly shorten the time needed for evaporation process. It was also found in this study that falling film evaporation occurred in the evaporator like device introduced that contributed to faster evaporation process. The special channel added at the outsider of circular plate could play a role to accumulate gummy cakes formed.

1. Introduction

ASEAN free trade era compels people tobe higher in efficiency and productivity, especially villagers in West Sumatera who are economically heavily dependent on micro and small enterprises. Many micro and small enterprises in West Sumatera use traditional stove having common characteristics that are wasteful fuel, need longer time for cooking, and unhygienic in production process [1]. Renewable energy from biomass is supplied and converted by traditional open combustion, which is very inefficient, in these stoves. Generally, the thermal efficiency of these local stove is lower than about 20% [2]. As tens of thousands of these traditional stove use firewood fuel then they should contribute to deforestation impact [1]. Multi Effect Evaporator application, in order to achieve a more efficient process in the sense of energy use, is also not introduced in these local stoves [3]. While, fire protection is not implemented [4].

Cane refining activities which produce red-cane sugar in West Sumatera are centred in Agam and Tanah Datar districts. Lawang is a village (nagari) in Agam district as the centre of red-cane sugar production. Headman of Lawang village said that there were about 300 cane refinings there. Less papers had

IOP Conf. Series: Materials Science and Engineering **543** (2019) 012016 doi:10.1088/1757-899X/543/1/012016

discussed traditional red-cane sugar production. The key process in production of red-cane sugar from the juice is to vaporize water to paste phase follows by cooling the paste to form mold crystals [5]. There are four sets of pot in the stove can be found in Lawang to vaporize the juice with the height of 40, 55, 70, and 85 cm respectively. The hot flue gas flows naturally from the combustion chamber at the first pot to the fourth one. It needs more or less 5 hours for cooking cane juice to get the first pot product of caramel ready to mold to become red-cane sugar. One pot of juice (around 50 kilograms of juice per pot) gives 7 to 9 kilograms of red-cane sugar. Three modes of operation for milling sugar juice are find in this Lawang village, namely buffalo-powered mill, motor to replace buffalo power mill, and engine-powered mill with product capacity reach around 40, 60, and 120 kilograms respectively.
This paper attempted to design and construct evaporator like devices introduced to the juice pot using aluminium plate material and to evaluate their contribution to faster evaporation process, minimizes fuel need, collecting gummy cakes formed, and shortening total time work needed.

2. Methods

This study consists of four main steps, namely field survey, design, construction, and introduction evaporator like device to cane juice pot in testing cooking process innovation. The field survey was conducted in Lawang village to collect dates about existing condition on cane refining process applied by villagers for state comparison need. The evaporator like devices were designed in the form of circular plates to enhance heat transfer and evaporation area. Then, they were constructed using aluminium plates material [6]. The influences of number of circular plates in a unit introduced to the juice pot, namely 4 and 9 circular plates, on their contribution to faster evaporation process, minimizes fuel need, and shortening total time work needed were to be evaluated. The height of the circular plate is set to about 45 cm above the initial level of juice in the pot and the biggest diameter one is 56 cm. The influences of the introduction of lid made of aluminium plates as well as any variation in circular plates, namely the thickness, the inclined of plates, and the existence of some holes in inner circular plates, were also be studied. Handmade special pocket in the outer of circular plate is also introduced to collect gummy caked formed as long as cooking process takes place. Combustion and the end of cooking process were determined as usually be done by villagers in cane refining. Cooking time equal to 5 hours to get the first pot product of caramel ready to mold to become red-cane sugar was set to be state comparison. Temperatures at any position in the stove were measured by infrared-thermometer [7]. All test are conducted in the first pot.

3. Results

Simple sketches of evaporator like device are given in Figure 1. Its construction introduced to the cane juice pot compared to the existing pot without evaporator like device belong to villager is given in Figure 2. Falling film takes place in evaporator like device and contributes to evaporation process. Meanwhile, gummy cakes formed as long as cooking process were accumulated at the outsider edge of circular plate. The results of cooking test of cane juice by introducing the evaporator like devices to juice pot are given in Table 1.

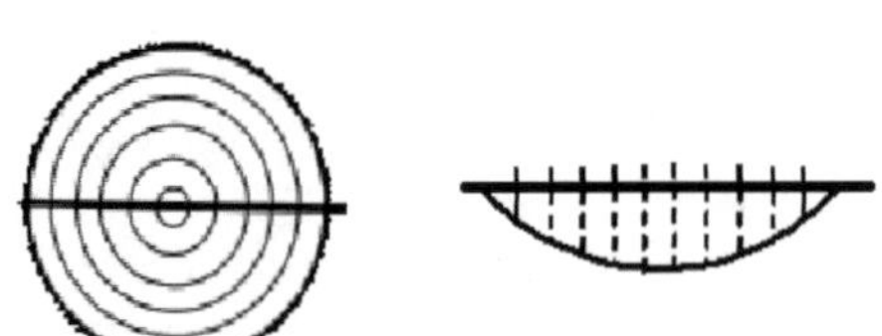

Figure 1. Sketches of Evaporator like device; (a) top and (b) side views

Figure 2. a) Evaporator like device introduced to cane juice pot and b). the existing one

Table 1. The Results of Cooking Test of Cane Juice by Introducing Evaporator Like Device to Juice Pot

Parameters	Variation in distances between plates (0.7 mm thick), cm			
	4	9	9	9
Cooking time in first pot, hours	2.75	3.15	3.15	3.80
Evaporation rate, liters/hour	15	12	12	9
Chamber temp. °C	594	500	500	460
Cooking juice temp. °C	96	93	93	90
Outest circular plate temp. °C	47	42	42	37
Outer chamber wall temp. °C	47	46	46	36.5
Early juice temp. °C	24.3	24.3	24.3	24.3
Environment temp. °C	20.6	20.6	20.6	20.6

3.1 Falling Film Phenomena

Falling film phenomena seems to occur on any variation of circular plates evaporator like device introduced to cane juice. This phenomena takes place on the surface of circular plate as the boiled juice is falling down it. It makes evaporation process going faster by addition of the mechanism of evaporation [Coulson, 2005]. As long as cooking takes place, the boiled juice molecules conducts a series of re-circulation by flowing upward and falling down the circular plates. This re-circulation motion of boiled juice takes place by thermosyphon effect based on natural convection [Coulson, 2005] as well as capillary action which occurs because of intermolecular forces between the liquid and surrounding solid surfaces.

In the annular between circular plates the hot molecules will move upward while the cold ones will fall down. This re-circulation will going on and boiled juice move upward through annular then falling as the film of fluid on the outer surface of plate. The molecules of water vaporizes from the boiled juice as it both flowing upward in the annular and falling down the surface of plates. The closer the the distances between plates to make annular the more rapid re-circulation of boiled juice to occur and the faster evaporation process will going on.

Ignoring any variation on others parameters, the results of cooking test in Table 1 show that evaporation rate increase significantly to 15 liters/hour as variation in distances between circular plates equal to 4 cm compare to 12 liters/hour when distances between circular plates equal to 9 cm. The increase in evaporation rate due to closer circular plates which make the smaller annular where a series of re-circulation of boiled juice takes place more rapidly. This phenomena is not found in traditional cooking and it perhaps plays the main role in shortening cooking time.

IOP Conf. Series: Materials Science and Engineering **543** (2019) 012016 doi:10.1088/1757-899X/543/1/012016

In traditionally cooking juice in Lawang village, re-circulation of the boiled juice flowing upward and falling down takes place by thermosyphon effect only on the surface of a basket made of woven bamboo placed in the first pot of cane juice. There is no capillary action exist here. The villagers placed the basket in their pot to prevent the squirt of juice as it boiled as well as to trap gummy cakes formed as long as cooking process takes place to control the dirt content of product of the red cane sugar. They did not take care of falling film phenomena.

As an isolation material, woven bamboo certainly obstructs falling film evaporation to occur. Conversely, cooling effect happened to the boiled juice due to contact with the surface of basket which restrained the evaporation process to occur. Then evaporation process in traditional cooking due to vaporize of water to air from the boiling juice only without contribution of any surface area addition.

3.2 Number of Circular Plates and Holes in Inner Plates

The number of circular plates for each unit of evaporator like devices used is dictated by the distances between the circular plates which were set up to 4 and 9 cm. Consequently, the number of circular plates of the two units will be difference, one consists of 9 circular plates and the other one has 5 circular plates. As cooking time more or less 5 hours was taken to be state comparison then evaporator like devices made of aluminium plate 0.7 mm thick seem to give significant contribution to shorten cooking time as it enlarge the surface area for heat transfer and evaporation as well as for a series of re-circulation of boiled juice to occur [8].

The result of cooking test in Table 1 show that cooking time for 4 cm distance between circular plates in one unit was noted to become 2.75 hours and for 9 cm distance one it was noted to 3.15 hours. Then, the reduced cooking time is within the range of 23.33 to 45%. It is clear that the first unit gives better performance as it has larger total surface area with 9 circular plates where evaporation will takes place and the other one has only 5 circular plates.

The influence of the holes exist in inner circular plates were studied for 9 mm thick plates. The existences of some holes by diameter around 1.5 cm in each inner plates were aimed to serve the juice in order to move on horizontal and radial direction between annular in inner plates and to produce agitation effect. However, it seems that they give no contribution to evaporation process and to shorten cooking time.

In addition, temperature drop in chamber from 500 to 460 °C followed by slightly reduced in cooking juice temperature, namely from 93 to 90 °C and caused cooking time 40 minutes longer. It can be seen in Table 1 for evaporator like device with 9 cm distance between circular plates that it needs cooking time of 3 hours and 50 minutes, or it is still 23.33% faster than state comparison. As firing in traditional chamber is conducted manually, and using mixing of biomass fuel of bagasse, bamboo, and firewood then temperature drop is often to occur.

Measurement of low temperature on the outer chamber wall as well as on the outer circular plate showed that heat loss through wall has relatively been very small. It means that much heat losses through hot flue gas flow in traditional cooking cane juice done by villagers [1]. Meanwhile, the temperature of the outer circular plate which was only reach 47 °C indicates that the plates was only heated by heat transfer from the boiled juice.

It is seemed that the mechanism of conduction of heat transfer from the pot to the circular plates was not exist as they were not be joint mechanically and to be separated by film of fluid of juice. Suppose it is heated by mechanism of conduction from the pot, its temperature certainly will higher than the temperature of boiled juice. Then, heat will be transferred from the plate to juice and evaporation process will take place faster.

Well design and construction of integrated pot and evaporator like device will make possible heating of the plates by conduction mechanism, namely heat from the flue gas in the chamber will be transferred to the pot as well as to the circular plates connected to the pot. This will certainly increase the temperature of the plates and make evaporation process will be going faster.

The boiled cooking juice temperature which was measured to be 96 °C or less showed that surface area of evaporation served by circular plates play a role for mass transfer of molecules of water from the juice body to the air. The larger surface area of evaporation the more rapid water evaporates from the juice, and less working time needed.

Comparison of cooking time with and without introduction of circular plates evaporator like device is given in Table 2. It is clear that introduction of circular plates are very help full to reduce working time, and consequently to save fuel need significantly. Therefore, workers can either increase their product capacity or reduce their working time.

Table 2. Comparison of Cooking Time with and without Circulate Plates Evaporator Like Device

Capacity, pots of juice	With circular plates, time	Without circular plates, time	Reduced time work, %
10	06.00-15.00	06.00-18.20	27.03
7	07.50-15.00	07.50-17.00	21.82
6	06.00-11.30	06.00-14.00	31.25

3.3 The Thick of Plate and Mode of Lid

The influences of plate thick and modes of lids to cooking parameters' values are given in Table 3. Plate thick slightly influences the cooking parameters values; cooking time with 0.7 mm thick plate of evaporator like device is 25 minutes faster than 0.9 mm thick one. It is 33.33% faster than cooking without evaporator like device. The identical value is 25% with 0.9 mm thick plate.

Table 3. The Influences of Plate Thick and Modes of Lids to Cooking Parameters

Parameters	Plate thick (9 cm plates distance), mm		Mode of lid (9 cm plates distance, 0.7 mm thick)	
	0,9	0,7	Large one hole lid	Small many holes lid
Cooking time, hours	3.75	3.3	2.15	2.5
Evaporation rate, liter/hour	10	12	18	16
Chamber temp. °C	505	500	587	574
Cooking juice temp. °C	91	93	97	92
Outest circular plate temp. °C	44	42	68	65
Outer chamber wall temp. °C	44	46	48	46
Early juice temp. °C	24,3	24,3	24,3	24,3
Environment temp. °C	20,6	20,6	20,6	20,6

The rate of heat transfer by conduction mechanism is influenced by the thick of plate [Coulson, 2005]. This finding is in consistence with the value of cooking juice temperature, where 0.7 mm thick plate gives slightly higher temperature than 0.9 mm one, namely 93 compare to 91 0C. However, the evaporator like device made of 0.7 mm thick of plate is lighter and not comfortable in used. The villager has recommended to use heavier one mode of 0.9 mm thick of plate.

The consistence value is also due to evaporation rate, where 0.7 mm thick plate gives the value of 12 liters/hour compared to 0.9 mm thick plate which gives 10 liters/hour. Introduction of lid on the evaporator like device influence significantly the performance of device, such as on cooking time, evaporation rate, cooking juice temperature, and the temperature of the outer circular plate. However, chamber temperature for this test are around 20% higher than averages.

Large centered one hole lid introduce on evaporator like device reduces cooking time to become 2 hours and 10 minutes, or 56.67% faster than what usually run by villagers. Meanwhile, small distributed many holes lid gives lower performance than the large centered one hole lid, where the cooking time reduce only to become 2 hours and 35 minutes, or 48.33% faster compare to what usually run by villagers.

The lid on top of evaporator like device prevent heat release freely, then it was accumulated in the device. Consequently, circular plate temperature and evaporation rate increase significantly [8]. For large centered one hole lid as an example, circular plate temperature is 68 compare to 42 °C and evaporation rate is 18 comp are to 12 liters/hour.

3.4 The Incline of Circular Plate

The influences of the inclines of circular plate to cooking parameters' value are given in Table 4. Generally, the incline in circular plate between 0 to 200 gives no influence on cooking parameters value. Evaporation rate was relatively constant in 12 liters/hour. It can also means that falling film phenomena is not influence by the incline of circular plates. Slightly higher in cooking juice temperature triggered by higher temperature chamber was not accompanied by the increase in evaporation rate or cooking time. Cooking time even slightly longer for inclined plate than the vertical one. In contrast with higher difficulty in its construction, inclined circular plate evidently gives no contribution to reduce cooking time.

Table 4. The Incline of Circular Plate to Cooking Parameter

Parameters	0° inclined (9 cm plate distances, 0.7 mm thick)	10° inclined (9 cm plate distances, 0.7 mm thick)	20° inclined (9 cm plate distances, 0.7 mm thick)
Cooking time, hours	3.33	3.67	3.65
Evaporation rate, liters/hour	12	12	12
Chamber temp. °C	500	555	567
Cooking juice temp. °C	93	96	96
Outer circular plate temp. °C	42	46	46
Outer chamber wall temp. °C	46	47	47
Early juice temp. °C	24,3	24,3	24,3
Environment temp. °C	20,6	20,6	20,6

3.5 Special Pocket to Collect Gummy Cakes Formed

Gummy cakes from organics and mineral contents formed in boiled juice occur as long as cooking process takes place. Replacing the basket made of bamboo by evaporator like device made of aluminum plates give impact on the increase of dirt content on product of red-cane sugar which reaches 7% or higher. The aluminum circular plates did not play role to trap the gummy cakes as could be done by basket made of woven bamboo. To collect the gummy cakes the evaporator like device needs to be equipped with special pocket.

So, the special aluminum pocket was placed around the circular plate to collect the gummy cakes formed has also been introduced in this evaporator like device. It should give the result as good as traditional cook used a basket made of bamboo except its construction is not qualified. Unfortunately, as handmade special pocket is not qualified in construction then the dirt content of red-cane sugar product found in this study is not less than about 5.4%. However, this device make the cooking process more comfortable as the worker is not used to collect the gummy cakes many times as long as the cooking process take place.

Indeed, the dirt content of high quality product of red-cane sugar is lower than 3%. It certainly needs some treatment to achieve this standard, namely to clean the cane from sand, dust, and others dirt sources and to filter cane juice before cooking. Unfortunelly vilagers is commonly not interest to do such treatment.

4. Conclusion

Innovation of cooking juice in process production of red-cane sugar in Lawang Agam district has been conducted by introducing evaporator like device made of aluminium plates in the first pot of stove. Falling film phenomena occur on any variation tested in this study and contributes to make faster the cooking process, to save fuel need, and to shorten total time work needed. Circular plate evaporator like device equipped with lid made of 0.7 mm thick aluminium plate introduced to the juice pot has successfully reduced cooking time till 56.67% faster than what had usually run by cane refinery workers in Lawang village. The distances between circular plates in one unit, and consequently numbers of circular plates in one unit, contribute significantly in reducing cooking time, namely 4 cm distance between circular plates reduces cooking time up to 45% and 9 cm one reduce 23.33%. The thick of aluminium plate gives difference results in reducing cooking time, namely 25% for 0.9 mm thick and 33.3% for 0.7 mm one. However, some holes made on the inner circular plates and the incline of circular plates give no contribution to shorten cooking time. Gummy cake formed as long as cooking process take place can be collected in special channel in the device to make the cooking process more comfortable.

Acknowledgment
The authors are grateful to Bung Hatta University Rector and the guardian of Lawang village for sustained encouragement and facilitate throughout the course of this study. Thanks are due to Yunita Amri, Shellia Harlita, Novri Fatmoheri, Asgumber Lindo, Muhammad Teguh Aditama, Jefri Wahyudi, Sari Andela and Arsy Annisa Phoetry for their help in conducting the experiments.

References

[1] Firdaus, Dahlan M H and Faizal M 2013, *Proc. Int. on Environmental Engineering* (Institut Teknologi Sepuluh Nopember, Public Health Program Study, Medical Faculty, Udayana University, Bali)
[2] Katoch S S and Bhat I K 2010 Clean energy technology approaches of mitigatinggreenhouse gases *Chem Eng Trans* **21 607-12**
[3] Cortes M G 2010 *Chem Eng Trans* **21** 277-82.
[4] Wehmeier G and Mitropetros K 2016 *Chem Eng Trans* **48,** 259-264
[5] Novi P 2016 *JNEP* **3** 17 – 20
[6] Haslego C, Laval A, Polley G 2002 *Designing Plate and Frame Heat Exchanger*, www.PINCTECHNOLOGY.COM
[7] Firdaus, Dahlan, M. H., Faizal, M., Kaprawi, Arita, A., 2015, *Journal of Sustainable Development,* **8**
[8] Coulson 2005 *Chemical Engineering Design* ed. 6 (Elsevier Butterworth-Heinemann).

IOP Conf. Series: Materials Science and Engineering **543** (2019) 012017 doi:10.1088/1757-899X/543/1/012017

Calcium Soap from Palm Fatty Acid Distillate for Ruminant Feed: The Influence of Initial Mixing Temperature

L A Handojo*, A Indarto, D Shofinita, M R Saadi, D Yulistia, and F I Hasyyati
Department of Chemical Engineering, Faculty of Industrial Technology, Institut Teknologi Bandung, Labtek X 2nd Fl., Jl. Ganesha 10, Bandung 40132, Indonesia

Email : lienda@che.itb.ac.id

Abstract. The breastfeeding dairy cows often go through metabolic disturbances and low milk production due to the use of stored body fat and minerals to assist milk production. Those problems can be solved by feeding feed supplement to the dairy lactating cows. Calcium soap is a well-known lactating dairy cattle feed supplement, which gives energy in the form of protected fat so it can increase milk productivity of dairy cattle. This research studied the use of PFAD (Palm Fatty Acid Distillate) as the free fatty acids source and calcium oxide as a calcium source to produce calcium soap using modified fusion method with water as the reaction catalyst. This study evaluated the initial mixing temperature of PFAD as the parameter affecting the conversion of PFAD, which was determined by the product acid value. The result showed that, at the same mole ratio of calcium source to PFAD, the lower initial mixing temperature gave higher reaction conversion of PFAD. The recommended initial temperature for mixing was found to be 60 ^{o}C, which gave the acid value of the product in the range of 2.96 – 3.14 mg KOH/g sample for mole ratio of CaO/PFAD 1 – 1.15.

1. Introduction

In the early days of breastfeeding, dairy cows are often malnourished because of insufficient feed intake. In this condition, dairy cows have to use stored body fat and minerals from the body to support milk production. This often leads to metabolic disturbances and low milk production [1]. The addition of fat supplements directly to the cows can increase the energy density of feed, but high-fat content of feed can create the decrease of fiber digestibility and milk fat percentage [2]. According to Chalupa et al. [3], to avoid such adverse effects, the salt of long chain fatty acids (such as calcium soap) as an animal feed supplement is a good alternative. The addition of calcium soap as a mixture of dairy cattle feed can increase the yield of milk, the composition of fat in milk, as well as the gross energy of cow's milk [4]. In Indonesia, this supplement can be used to solve the problem of low milk production, especially in fulfilling the projected consumption of cow milk for consumption which tends to increase by an average of 4.1% per year in 2017 – 2020 [5].

Crude Palm Oil (CPO) contains unwanted impurity components. CPO is generally purified for sale as Refined Bleached Deodorized Palm Oil (RBDPO). RBDPO is palm oil which has undergone a purification process to eliminate free fatty acids and purification to remove color and odor. The elimination of free fatty acids (FFA) may be performed in industrial practice using different units for neutralization with caustic, washing and drying of the fuel. The method of dry washing has been done by using modified potato and cassava starch as adsorbents [6]. The side product of the process is Palm Fatty Acid Distillate (PFAD) which is obtained as much as 5 % of the weight of palm oil. The main component of PFAD is free fatty acids that can reach around 80 % [7].

Many types of research are focusing on utilizing this by-product to increase its value. One of them is further processing through an esterification reaction with glycerol and lipase catalyst to produce

IOP Conf. Series: Materials Science and Engineering **543** (2019) 012017 doi:10.1088/1757-899X/543/1/012017

monoacylglycerol and diacylglycerol, then monoacylglycerol and diacylglycerol can be used as emulsifiers in food or non-food products such as cosmetics and drugs [8]. The utilization of PFAD also has great potential in biodiesel industries [9]. PFAD can also be utilized as raw material for producing calcium soap, a kind of ruminant feed supplement [10]. The production of calcium soap from PFAD is expected to be profitable in the palm oil industry.

There are three basic methods used in the manufacturing process of calcium soap from fatty acids, namely the double decomposition method, the modified fusion method, and the metal-acid reaction method. Modified fusion methods are generally chosen for the production of calcium soap. Calcium oxide is reacted directly by melting free fatty acids using a catalyst in the form of water [11]. This method is more economical than other methods and has been used in the manufacture of industrial-scale calcium soap. The chemical reaction involved in the modified fusion method is as follow.

$$MeO + 2RCOOH \rightarrow Me(OOCR)_2 + H_2O$$

or

$$Me(OH)_2 + 2RCOOH \rightarrow Me(OOCR)_2 + 2H_2O$$

The "Me" in the chemical reaction above can be replaced by any divalent metals in the modified fusion method. Certain divalent metals such as calcium, copper, magnesium, and zinc have long been recognized as beneficial mineral nutrients for humans and certain companion animals and livestock, such as ruminants, horses, dogs, cats, rabbits, hamsters, birds, fish, and the like. For example, the mineral calcium not only builds and strengthens bones and teeth, but it also maintains a normal heartbeat and regulates blood pressure. It is also essential for the healthy functioning of the nervous system. [12]

This study aims to evaluate the production of calcium soap from PFAD and a certain calcium source for ruminants feed, specifically the effect of initial mixing temperature on the conversion of the reaction. The acid value of the calcium soap was chosen as the quantitative parameter to be observed. In addition, the effect of initial mixing temperature on saponification reaction time was also observed.

2. Experimental section

2.1. Materials

Materials used in this study were technical grade calcium source and PFAD as the source of fatty acid. The PFAD was obtained without any further treatment from PT Tunas Baru Lampung Tbk – Sidoarjo, a CPO refining plant in Indonesia.

2.2. Calcium soap preparation

Several calcium soap samples were prepared using the modified fusion method with two different operating variables, initial mixing temperature, and reactants mole ratio. The PFAD was melted to various temperatures (60-90 °C), then calcium source powder was added with the mole ratio of CaO/PFAD that ranged from 1 to 1.4. Hot water was added to the mixture immediately after the homogeneous phase was achieved.

2.3. Acid value determination

For all experiments, the analysis of reaction conversion was focused on acid value determination of the calcium soap product. The analysis was done according to the standard method described in ISO 660:1990. One gram of calcium soap was dissolved in ethanol at 60 °C for 10 min. The resulting mixture was titrated with 0.01 N potassium hydroxide using phenolphthalein as an indicator. The acid value in the product was calculated with the following equation:

$$\text{Acid value} = \frac{56.1 \cdot c \cdot V}{m}$$

Where c is the exact concentration of the potassium hydroxide solution used (mol/L), V is the volume of the potassium hydroxide solution (mL), and m is the mass of the test portion (grams).

3. Results and discussion

The modified fusion method is similar to that of the fusion method, except the addition of a small amount of water to the mixture of molten fatty acid and oxide or metal hydroxide. The added water acts as a catalyst and causes reactions to occur at lower temperatures and faster reaction [13]. In the modified fusion reaction, the usually high reaction temperature can be reduced up to 90-100 °C [14]. However, in the production of calcium soap, the lowest initial mixing temperature may be the same as molten PFAD temperature, which seems to be 60 °C [15]. The reaction can reach the saponification temperature with the help of the exothermic reaction between calcium oxide and water, which produces heat. Therefore, the reaction mixture in the modified fusion reaction does not need energy as much as the usual saponification reaction.

The experiment in this study varied the stoichiometry reactant mole ratio of CaO/PFAD in stoichiometry and the initial mixing temperature. All saponification reaction with different initial mixing temperatures and reactants mole ratios gave similar trends. Therefore, only two sets of data samples were selected to be presented, the variation of the mole ratio of reactants 1 and 1.15. The change of reaction temperature with respect to reaction time was plotted for each initial mixing temperature, as shown in Figure 1 and 2 for mole ratio 1 and 1.15 respectively. The plot was terminated as the temperature reached its peak.

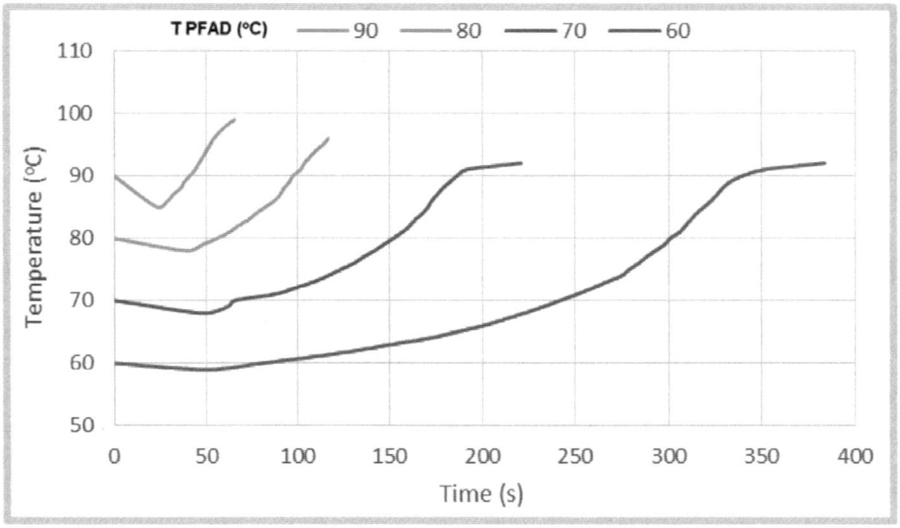

Figure 1. Reaction temperature profile as a function of reaction time at various initial mixing temperature and the mole ratio of calcium source to PFAD = 1

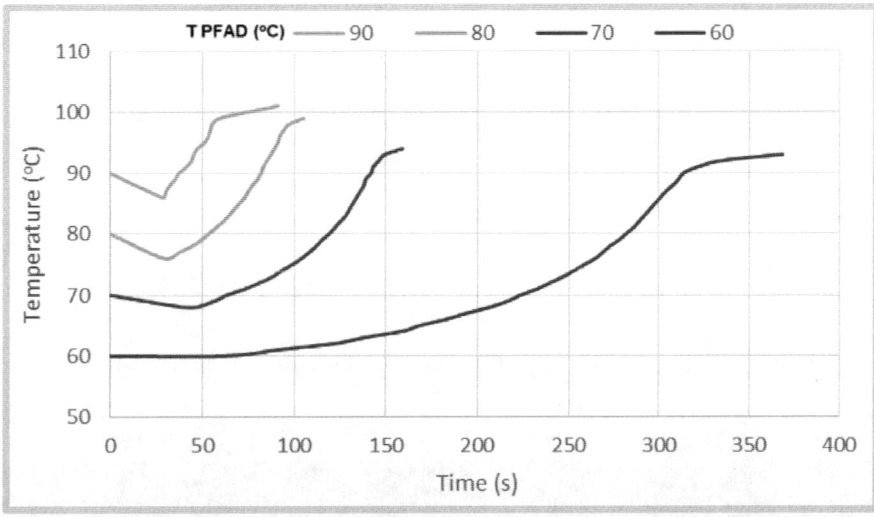

Figure 2. Reaction temperature profile as a function of reaction time at various initial mixing temperature and the mole ratio of calcium source to PFAD = 1.5

From Figure 1 and 2, the effect of various initial mixing temperature on the time needed for the system to reach the peak temperature can be observed. The peak temperatures of reaction mixture made with initial mixing temperature of 60, 70, 80, and 90 °C differed with the range between 90-100 °C, which were different from the experiment conducted by Scott [14]. The author found that the modified fusion reaction initiated at 90-100°C and reached the peak temperature of 100-200 °C. This may be caused by the different operating conditions such as mixing equipment. Just like the experiment by Scott, the experiment in this research also showed that the reaction can be said to end when the mixture reached its peak temperature.

Both figures, Figure 1 and 2, show the increasing temperature during the experiment. To more easily understand the stages in the reaction of calcium soap production, the reaction itself can be divided into four stages. The stages are depicted in form of zones in Figure 3. The reaction stages that represent the four zones are shown in Figure 4.

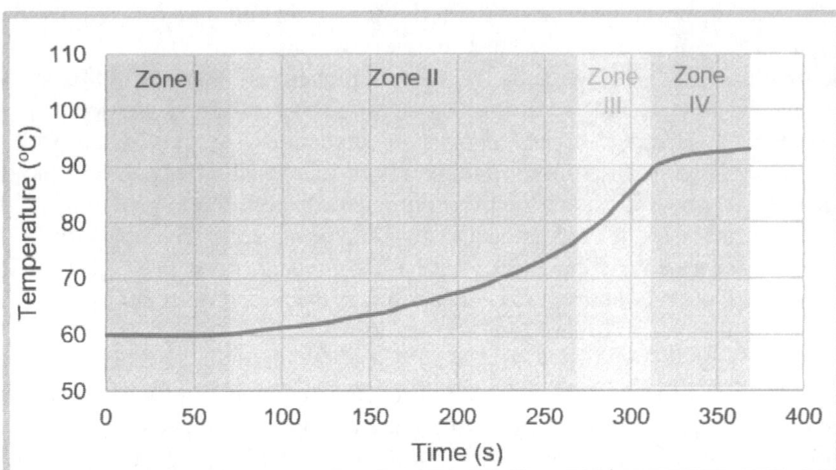

Figure 3. Reaction temperature profile as a function of reaction time at 60 °C mixing temperature and mole ratio of calcium source to PFAD = 1.15

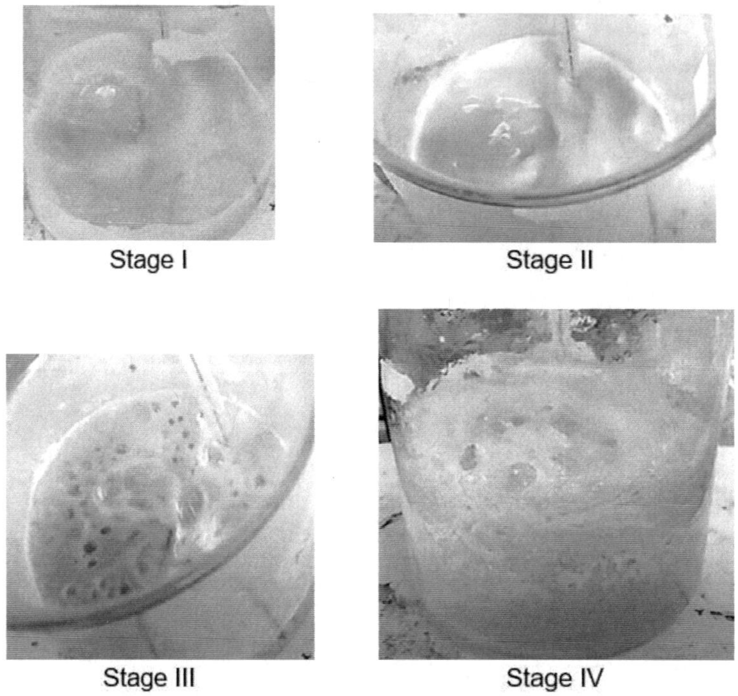

Figure 4. Images of PFAD saponification reaction stages

Zone I (highlighted blue) shows that at the beginning of the reaction, the mixture underwent temperature adjustment after all raw materials were mixed. It can be observed for the samples which were made with various initial mixing temperature above 60 °C that the temperature of the mixture fell slightly and then rose to the specified mixture temperature. Meanwhile, with the initial mixing temperature of 60 °C, the mixture temperature remained the same. Zone II (highlighted red) shows that the mixture temperature starts to rise. This was due to an exothermic reaction from dissolving calcium source with water. At the time the mixture reached the optimal temperature for saponification reaction, which is shown by zone III (highlighted yellow), the reaction took place while continuing to raise the mixture temperature more rapidly than that of in zone II. Zone IV (highlighted green) shows the reaction temperature as it reached its peak. It can be seen that the higher the initial mixing temperature, the higher the peak temperature values reached by the reaction mixture.

Stage I shows that at the beginning of the reaction there were no changes in the appearance of the mixed raw materials, the changes that occur only changes in temperature which depended on the initial mixing temperature used. There were no changes in appearance in stage II as well, however as mentioned in the prior paragraph, the mixture temperature increased slowly during the dissolution of calcium source and water. Stage III shows the beginning of the saponification reaction. At the beginning of stage III, the mixture started to expand while forming a more viscous mixture. As the mixture expands, the temperature also kept increasing until it reached its peak in stage IV. After the peak temperature was reached, the mixture will stop expanding as the temperature dropped. At the end of the reaction, the mixture could expand to four times the initial volume.

Other than peak temperatures, the reaction times are also shown in Figure 1 and 2. Both Figure 1 and 2 show that higher initial mixing temperature boosted the reaction time needed to reach the peak temperature significantly. This is shown by the higher the initial mixing temperature the greater the slope of the graph. This means that the temperature of the reaction mixture increases faster. However, it was found in the experiment that the reaction time did not ensure the same reaction conversion. The faster the reaction time meant the less time for the mixing, meanwhile mixing was important in this experiment. To ensure a high reaction conversion in the process, the raw materials had to be well mixed during the reaction. Just as Burgos and Bailey [16] proposed, a disadvantage with known methods of preparing such ruminant feedstuff is that they may not provide adequate mixing of the fatty acids, calcium oxide, and water, resulting in pockets of unreacted chemicals. Such unreacted chemicals make the mixture unstable. If such unreacted chemicals later come in contact with water, they will cause an exothermal reaction that releases a lot of heat.

The products were then analyzed by determining the acid value. Products with lower acid value were known to achieve higher reaction conversion. All samples produced with different initial mixing temperatures and reactants mole ratios gave similar trends of acid value. Therefore, only two sets of data samples were selected to be presented. The acid values for the variation of the mole ratio of calcium source and PFAD 1 and 1.15 using all the initial mixing temperature are shown in Figure 5.

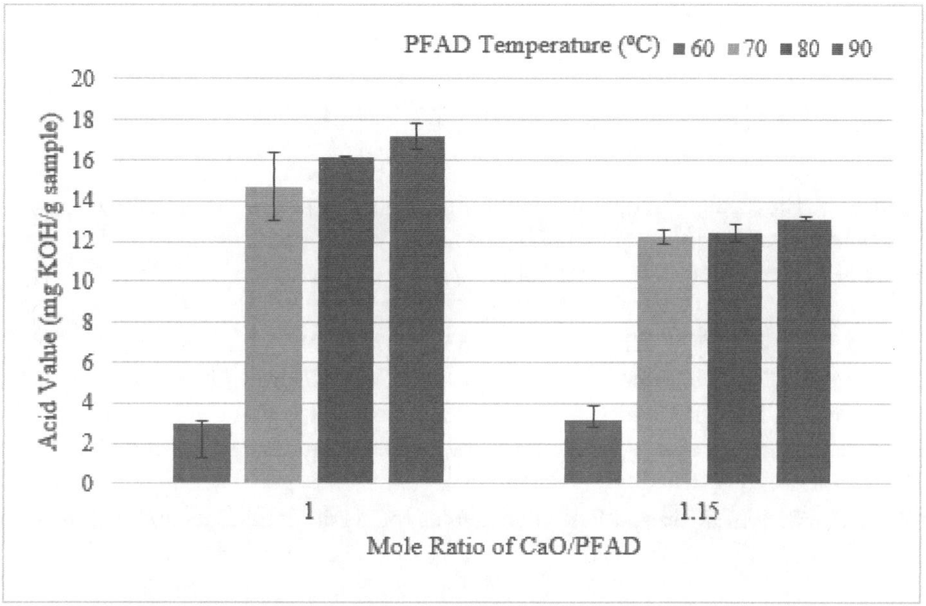

Figure 5. The acid value of the products as a function of CaO to PFAD mole ratio and initial mixing temperature

The low acid value indicates high reaction conversion because the amount of free fatty acids remaining in the calcium soap are lower. Based on the results, the higher the mole ratio of the reactant, the lower the acid value of the product or the higher the reaction conversion. The larger mole of the reactant meant more amount of calcium source was added to a certain amount of PFAD. The use of excess amounts of calcium source could ensure that almost all free fatty acids reacted with calcium source to form calcium soaps. However, a lower ratio was preferred to reduce excess amount of calcium in the product, which may result in the inhibition of the absorption of other minerals such as zinc and phosphorus in dairy cow digestion [17]. The obtained profile was consistent with the previous statement by Handojo et al. [18].

In contrast to the reactant mole ratio, lower product acid values were obtained at lower initial mixing temperatures. This trend also appeared on every data sets of the same reactant mole ratio in this research, which may be because of the shift equilibrium reaction. Furthermore, it is depicted in Figure 5 that generally, the reaction conversion was not significantly influenced by adjusting the temperature at 70, 80, and 90 °C. On the other hand, the increase in acid value was seemingly drastic at the temperature from 60 to 70 °C. The high temperature of the mixture may hinder the reaction process since the reaction itself is exothermic. Another possibility was that the higher the temperature of the mixture, the easier the water evaporates. This may cause a problem to the reaction since an amount of water (as a catalyst) is needed to dissolve the calcium source to facilitate the reaction in the liquid phase. Thus, it can be concluded that for this process, the lowest possible initial mixing temperature is preferred, which is 60 °C (PFAD has melted completely).

4. Conclusion

The best operating condition for producing calcium soap is required to achieve the highest reaction conversion (lowest acid value). The initial mixing temperature is one of the variables that affect the saponification reaction of calcium soap. This study found that the higher the initial mixing temperature, the higher the peak temperature values reached and also the less time needed to complete the reaction. However, the lowest possible initial mixing temperature is recommended to achieve a higher reaction conversion (the lower acid value), which was found to be 60 °C.

Acknowledgement

The authors would like to express their sincere gratitude to BPDPKS Indonesia for the 2017 research grant that enables the authors to conduct this research and PT Tunas Baru Lampung Tbk for providing PFAD for this research.

5. References

[1] Kronfeld D S 1982 Major metabolic determinants of milk volume, mammary efficiency and spontaneous ketosis in dairy *Obs. J. Dairy. Sci.* Vol. **65** pp. 2204-2212

[2] Palmquist D L and Jenkins T C 1980 Fat in lactation ration, review *J. Dairy. Sci.* Vol. **63** pp. 1-14

[3] Chalupa W, Rickabaugh B, Kronfeld D S and Sklan D 1984 Rumen fermentation in vitro as influenced by long chain fatty acid *J. Dairy. Sci.* Vol. **67** pp. 1439-1444

[4] Purushothaman S, Kumar A and Tiwari D P 2008 Effect of feeding calcium salts of palm oil fatty acids on performance of lactating crossbred cows, asian-aust *J. Animal. Sci.* Vol. **21(3)** pp. 376-385

[5] Pusat Data dan Sistem Informasi Pertanian/Center for Agriculture Data and Information Systems 2016 *Outlook Susu: Komoditas Pertanian Subsektor Peternakan/Milk Outlook: Agricultural Commodities In Livestock Subsector* (Indonesia: Ministry of Agriculture Indonesia) ISSN: 1907-1507 Available online at http://epublikasi.setjen.pertanian.go.id/epublikasi/outlook/2016/Peternakan/OUTLOOK%20SUSU%202016/files/assets/common/downloads/OUTLOOK%20SUSU%202016.pdf Accessed on May 2 2018

[6] Ismaila A, Saimon N N, Jusoh M and Zakaria Z Y 2017 Adsorption of free fatty acid in biodiesel from palm fatty acid distillate using KOH-activated starch *Chem. Eng. Trans.* Vol. **56** pp. 619-624

[7] Pitoyo 1991 *Pemisahan Tokoferol dari Distilat Asam Lemak Minyak Sawit (Separation of Tocopherol from Palm Oil Fatty Acid Distillate)* Master's Thesis Graduate Program (Yogyakarta, Indonesia: Universitas Gadjah Mada)

[8] Elizabeth B 1997 Monoglycerides in food system: current and future uses *J. Food Tech.* Vol. **51 No. 8**.

[9] Saimon N N, Thavil S, Jusoh M, Ngadi N and Zakaria Z Y 2018 Preparation of Ti-SO3H/ICG using microwave-assisted for esterification of palm fatty acid distillate to methyl ester *Chem. Eng. Trans.* Vol. **63** pp. 463-468

[10] Handojo L A, Indarto A, Shofinita D, Meitha A, Nabila R, Triharyogi H, Hakim M and Saadi M 2018 Calcium soap from palm fatty acid distillate for ruminant feed: $Ca(OH)_2$ as calcium source *MATEC Web of Conferences* 159 02062

[11] Perez E P 2009 *Method for Producing Calcium Soaps for Animal Feed* US Patent No. 20090220638A1

[12] Luchini D N, Strohmaier G K, Frederiksen E D, Hawkes J G and Dull B J 2012 *Polyunsaturated Fatty Acid Monovalent and Divalent Metal Salt Synthesis* US Patent 8 093 415 B2

[13] Blachford J 1982 *Manufacture of Metallic Soaps* US Patent 4 316 852

[14] Scott L, Monrovia, Strachan H D, Monte E and McCloskey C M 1974 *Manufacture of Metallic Soaps* US Patent No. US3803188A

[15] Handojo L A, Indarto A, Shofinita D, Meitha A, Nabila R, Triharyogi H and Kevin L 2018 Calcium soap from palm fatty acid distillate for ruminant feed: the influence of water temperature *IOP Conference Series: Earth. Envi. Sci.* 141 012010

[16] Burgos A and Bailey W 2007 *Apparatus for Use in Making Ruminant Feedstuff* US Patent 2007/0092617 A1.

[17] NRC 2001 *Nutrient Requirements of Dairy Cattle, 7th Rev. Ed.* (Washington D.C: National Academy Press-Subcommittee on Dairy Cattle Nutrition, Committee on Animal Nutrition, Board on Agriculture and Natural Resources) ISBN 0-309-06997-1 Available online at http://www.nap.edu/openbook.php?isbn=0309069971 Accessed on February 14 2018

[18] Handojo L A, Indarto A, Shofinita D, Meitha A, Nabila R and Triharyogi H 2018.Calcium soap from palm fatty acid distillate (pfad) for ruminant feed: quality of calcium source *MATEC Web of Conferences* 156 02007

Effect of Extraction Time on *Averrhoa bilimbi* Leaf Ethanolic Extracts Using Soxhlet Apparatus

H Niawanti, Y S Lewar and N N Octavia

Department of Chemical Engineering, Engineering Faculty, University of Mulawarman, Sambaliung Street No. 9, Kampus Gunung Kelua, Samarinda, Indonesia

E-mail: niawanti@ft.unmul.ac.id

Abstract. *Averrhoa bilimbi* leaf extract is used as herbal medicine for various disorders treatment by communities. The causes of coronary disease such as cholesterols and triglycerides are widely established by *Averrhoa bilimbi* leaf extracts due to the presence of bioactive compounds such as glycoside, tannins, flavonoids and phenolic compounds. This study extracted *Averrhoa bilimbi* leaf using Soxhlet apparatus and 500 mL ethanol 96 % as a solvent extraction. Ethanol was used as solvent due to the solubility of bioactive compounds, non-toxic and biodegradable. The leaves were collected, chopped and dried at 60 °C until reached a constant weight. The moisture content of leaf was obtained 48.67 % at 24 hours of drying. The objective was to examined extraction time effect on the yield of leaf ethanol extracts. Extraction was carried out at constant temperature 70 °C and the time was varied at 60, 120, 240, 360 and 480 minutes. The highest yield and total flavanoid obtained for 13.77 % 13132.52 mg/L at 240 minutes. The yield of extraction was reduced at 360 minutes due to the optimum extraction time had been reached. In spite of that, this work indicate that the longer of extraction time, the more bioactive compound was extracted.

1. Introduction

Plants are the source of herbal medicine that has been used by civilization since a long time ago, before the discovery of chemicals drugs. The example of using plants as medicine is the treatment of *Diabetes mellitus* disease. *Diabetes mellitus* is a metabolic disease characterized by hyperglycemia and it is caused by insulin secretion damage, insulin action, or both. Chronic hyperglycemia of diabetes is associated with long-term damage, dysfunction, and the failure of different organs, especially the eyes, kidneys, nerves, heart and blood vessels [1]. Prior to the discovery of insulin, various herbs were used as a cure for diabetes. After the introduction of insulin as a diabetes drug, the used of herbal medicine is increasingly abandoned or even rejected by some communities, but others still continue to utilize herbal plants as a conventional treatment solution [2].

The Indonesian people used herbs as a medicine since a long time ago even though it was not based on the knowledge about the content of any compounds in plant [3], for example is the use of *Averrhoa bilimbi* fruit and leaves to overcome various dangerous diseases such as *Diabetes mellitus*, hypertension, inflammation and bacterial infections [4]. The result of phytochemical analysis show that *Averrhoa bilimbi* fruit extracts contain phenol, flavonoid and tannin which are useful as antioxidants. Fruit extract containing total phenolic compound around 50.23-68.67 mg GAE/g extract was obtained by fifthteen days of maceration using methanol as solvent [5]. The previous research by Pushparaj et al. [6] show that *Averrhoa bilimbi* leaves extract successfully reduced 50 % of blood glucose levels in diabetic rats and 130 % of triglyceride levels in rats bloods.

Many researchers developed the utilization of *Averrhoa bilimbi* leaf extract in Indonesia. Most of them studied about the benefits of extracts for human health e.g. measuring of *Averrhoa bilimbi* extract in vitro effectivity to againts *Escherichia coli* and *Staphylococcus aureus* growth [7], evaluating of *Averrhoa bilimbi* leaves and *Catharanthus roseus* combination as hypoglycemic agent [8], evaluating of *Averrhoa bilimbi* extract to mice physical ability against cigarette smoke-induced [9] and measuring of *Averrhoa bilimbi* leaf extract antibacterial activity [10]. Ethanol and water were used as solvent in their extraction process to attracted bioactive compound. However, water is only effective to extract polar and hydrophilic bioactive compounds but less effective to extract non-polar and hydrophobic ones [11].

Nair et al. [12] studied about the anticancer activity of *Averrhoa bilimbi* leaf and fruit extract. Extraction was processed in soxhlet apparatus using several solvents i.e. petroleum ether, chloroform, ethyl acetate, ethanol 80 % and water. The result show that ethanol 80 % was the optimum solvent by producing only 47 ± 1.2 µg/g plant of total flavonoids. *Averrhoa bilimbi* leaves extract contain flavonoids, phenolic compound and glycosides which are not only reduced blood glucose levels, but also reduced the levels of triglyceride. Triglyceride is a substance which cause coronary heart disease in human body. Therefore, ethanol 96 % was used as extraction solvent in this work due to the result of previous research which reported that ethanol is best solvent for flavonoid isolation. This study aims to determine the effect of extraction time on the yield of *Averrhoa bilimbi* extract and the total content of flavonoids extract.

2. Methods

2. 1 Materials

Averrhoa bilimbi leaf was collected from Penajam, East Borneo, Indonesia. Ethanol 96 % was purchased from One Med, $AlCl_3$ and Quarcetin standard were purchased from Sigma Aldrich. All other chemicals used were obtained from commercial sources.

2.2 Averrhoa bilimbi leaf extraction

Averrhoa bilimbi leaf was extracted using soxhlet apparatus and 500 mL of ethanol 96 % as solvent. Before the extraction process, the leaf was dried at 60 °C. Dry leaf sample was chopped and crushed until the coarsely powdered formed. 25 gram of *Averrhoa bilimbi* leaves powder were placed in filter paper and inserted into soxhlet apparatus. The operating temperature at 70 °C under atmospheric pressure and the extraction time were varied at 60, 120, 240, 360 and 480 minutes. The extract was purified using distillation apparatus at 80 °C for removing ethanol content. Then, the extract was analyzed by spectrophotometer to measure the total flavonoid content. The moisture content and extraction yield also calculated by using equation that can be seen below:

$$\text{Moisture content (\%)} = \frac{Mass\ of\ fresh\ leaves - Mass\ of\ dry\ leaves}{Mass\ of\ fresh\ leaves} \text{ x } 100\ \% \quad (1)$$

$$\text{Extract yield (\%)} = \frac{mass\ of\ extract\ (g)}{mass\ of\ dry\ leaves\ sample\ (g)}\ x\ 100\ \% \quad (2)$$

2.3 Total flavanoid analysis

Total flavonoids content in this work was determined using UV-Vis Spectrophotometer V-550. The calibration curve was obtained by using $AlCl_3$ reagents with quarcetin standard 95 % (Sigma Aldrich). A straight line passing through the origin was obtained at a wavelength of 416 nm (R^2 = 0.9666). The *Averrhoa bilimbi* extract (0.1 gram) was dissolved in 1 mL of distilled watert to make a solution. Then, 0.5 mL of extract solution was dissolved into the mixture of 1.5 mL methanol, 0.1 mL $AlCl_3$

IOP Conf. Series: Materials Science and Engineering **543** (2019) 012018 doi:10.1088/1757-899X/543/1/012018

(10 %) solution, 0.1 mL CH_3COONa 1 M, and 2.8 mL distilled water. The mixture of solution was placed at room temperature (± 30 °C) for 30 minutes. After 30 minutes, the solution was dilute in distilled water. Sample was placed in 1 cm cuvette and the spectrophotometer was operated at 1 nm bandwidth and 1 nm data pitch. The blank solution consist of 1.5 mL methanol, 0.1 mL $AlCl_3$, 0.1 mL CH_3COONa, and 2.8 mL distilled water. The calibration curve of quarcetin standard shown in **Figure 1.**

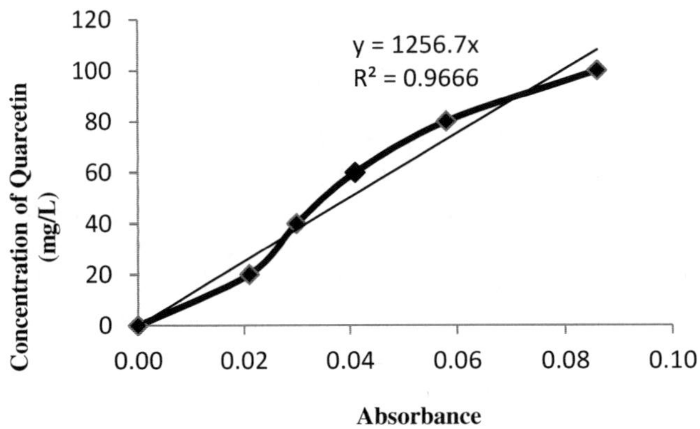

Figure 1. Calibration curve of quarcetic standard

3. Results and Discussion

3.1 Extraction time effect on the yield of Averrhoa bilimbi extract

Averrhoa bilimbi leaf contain flavonoid compounds which is useful as natural antioxidants. Moisture content of *Averrhoa bilimbi* leaf was obtained for 48.66 %. This work used ethanol as a solvent to isolate flavonoids from the leaf. Flavonoid compound was well dissolved in ethanol due to its polarity. Ethanol is composed of a polar -OH group and a non-polar CH_2-CH_3 group, the non-polar property made flavonoid is optimally extracted by ethanol from *Averrhoa bilimbi* leaf. In spite of solvent selection, extraction time also influence the result of extraction process. The longer of extraction time will provide the longer of contacting time between solvent and the substance to be dissolved. The results of this study indicated that the extraction time affects the yield of the extract. **Figure 2** is show the relationship between extraction time and extract yield at the end of extraction process.

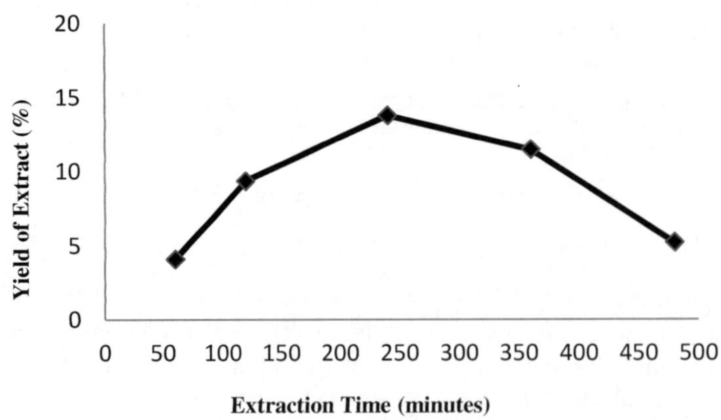

Figure 2. Extraction time effect on the yield of extraction

Figure 2 shows that the longer of extraction time resulted the higher yield of *Averrhoa bilimbi* extract. The highest extract yield was 13.77 % at 240 minutes of extraction time. However, the result of extract yield at 360 and 480 minutes of extraction time was decreased, it was due to the natural antioxidants may be substantially loss during the longer of heating in soxhlet extraction process. The fact that most of the bioactive compounds are relatively unstable when exposed to heat. The previous research by Rawson et al. [13] found that most of the bioactive compound in plants for example flavonoid, phenolic, antioxidant, carotenoid and vitamin C will change their chemical and physical structure also loss the bioavailability due to the thermal degradation process when exposed to high temperature. Another research by Arancibia-Avalia et al. [14] reported that heating process caused the dropped in flavonoids content.

3.2 Extraction time effect on the total flavonoids content

Flavonoids can be found as a group of polyphenol compounds. In general, flavonoids are divided into 4 groups namely flavones, flavonols, flavanones and flavanonols. *Averrhoa bilimbi* based on previous studies contain flavone and flavonol [15]. Quercetin and kaempferol are included in the flavonol class, hence the flavonoid analysis using the quarcetin standard. It is well dissolved in an ethanol solvent due to these class of flavonoids belong to a less polar species.

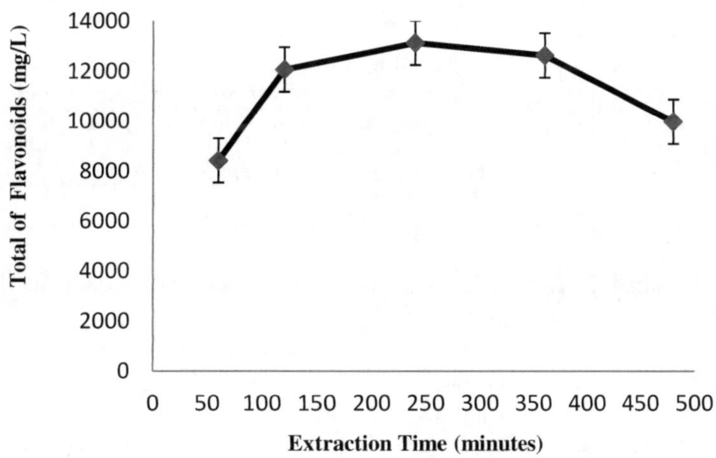

Figure 3. The effect of extraction time on total flavonoids

Based on **Figure 3**, it was found that total flavonoid content at 60, 120, 240, 360 and 480 minutes are 8419.89, 12064.32, 13132.52, 12629.84, 927.3 mg/L, respectively. The highest total flavonoids concentration was obtained at 240 minutes for 13132.52 mg/L. Overall, the total flavonoid content increased due to the longer of extraction time, but when extraction time running at 360 and 480 minutes the total flavonoid was decreased. This is in accordance with the results obtained on the effect of extraction time on the yield of *Averrhoa bilimbi* extract.

4. Conclusion

The highest yield of extraction and total flavonoid content were obtained at 240 minutes extraction time using ethanol 96 % as a solvent. The result of this research represent the effect of extraction time on total flavonoid and the yield of extraction. Despite of that, this work also shows that *Averrhoa bilimbi* leaves contain flavonoid as bioactive compound that can be used as raw material for medicine production.

Acknowledgements
The financial support of Department of Chemical Engineering, Engineering Faculty, University of Mulawarman is gratefully acknowledged.

References

[1] American Diabetes Association 2014 Diagnosis and classification of diabetes mellitus **37** 81-90.
[2] Swanston-Flatt S K, Day C, Bailey C and Flatt P R 1990 Traditional plant treatments for diabetes studies in normal and streptozotocin diabetic mice *Diabetologia* **33** 462-64.
[3] M Sangi, Runtuwene M R J, Simbala H E I and Makang V M A 2008 Analisis fitokimia tumbuhan obat di kabupaten minahasa utara *Chem. Prog.* **1** 47–53.
[4] Gunawan , Cordero A and Paano A 2013 Structure elucidation of two new phytol derivatives , a new phenolic compound and other metabolites of Averrhoa bilimbii *DLSU Res. Congress* 1–8.
[5] Hasanuzzaman M, Ali R M, Hossain M, Kuri S and Islam M S 2013 Evaluation of total phenolic content, free radical scavenging activity and phytochemical screening of different extracts of Averrhoa bilimbi (fruits) *Int. Curr. Pharm. J.* **2** 92–96.
[6] Pushparaj P, Tan C H and Tan B K 2000 Effects of averrhoa bilimbi leaf extract on blood glucose and lipids in streptozotocin-diabetic rats *J. Ethnopharmacol.* **72** 69-76.
[7] Widhianto E K, Elmarda R V and Rakhmawatie M D 2017 Effectivity in vitro of Averrhoa bilimbi ethanolic extract againts Escherichia coli and Staphylococcus aureus growth http://www.researchgate.net/publication/320893786.
[8] Sutrisna E M and Sujono T A 2015 The combination of belimbing wuluh fruit (*Averrhoa bilimbi*) and leaves of tapak dara (Catharanthus roseus G.) from Indonesia as a candidate hypoglycemic agents and thin layer chromatography profiles *Biomed. Pharmacol. J.* **8** 39-46.
[9] Samudra A G, Yanti S and Wijayanti D A 2017 The effect of star fruit (*Averrhoa bilimbi* L.) to physical ability of mice (Mus musculus) against cigarette smoke-induced *J. Pharm. Med. Sci.* **2** 41-43.
[10] Aryantini D and Sari F 2017 Uji aktivitas antibakteri fraksi aktif terstandar flavonoid dari daun belimbing wuluh (*Averrhoa bilimbi* L.) *J. Wiyata* 143-50.
[11] Zullaikah S, Rachmaniah O, Utomo AT, Niawanti H and Ju Y H 2018 Green separation of bioactive natural products using liquefied mixture of solid *Green Chemistry* vol **1** ed Hosam E D M S and Koller M (London: InTech Open) chapter 2 pp 17-38.
[12] Nair M S, Soren K, Singh V and Boro B 2016 Anticancer activity of fruit and leaf extracts of *Averrhoa bilimbi* on MCF-7 human breast cancer cell lines: a preliminary study Austin J. Pharmacol. Ther. **4** 1082.
[13] Rawson E S, Stec M J, Frederickson S J and Miles MP 2011 Low-dose creatine supplementation enhances fatigue resistance in the absence of weight gain *Nutrition* **27** 451–55.
[14] Avila P A, Toledo F, Werner E, Suhaj M, Leontowicz H, Leontowicz M, Ayala A.L.M and Gorinstein S 2011 Partial characterization of a new kind of Chilean Murtilla-like berries *Food Res. Int.* **44** 2054–62.
[15] Hamdanah S, Anam S and Jamaluddin 2015 Isolation and identification of flavonoid compound from ethanol extract of strafruits (Averrhoa bilimbi L.) by using spectrophotometry uv-vis method *GALENIKA J. Pharm.* **1** 22-34.

IOP Conf. Series: Materials Science and Engineering **543** (2019) 012019 doi:10.1088/1757-899X/543/1/012019

Minerals Improvement of Soy Milk by Addition of Drumstick Tree (*Moringa oleifera*) Extract

O Rachmaniah[1,4], S Kusumahati[2], E O Ningrum[3], F Kurniawansyah[1], S Zulaikah[1], H Ni'mah[1], S Soeprapto[3], and S Soeprijanto[3]

[1] Department of Chemical Engineering, Institut Teknologi Sepuluh Nopember, Kampus ITS Sukolilo, Surabaya 60111, Indonesia

[2] Small-Medium Sized Enterprise of Soy Milk, Rungkut, Surabaya 60293, Indonesia

[3] Department of Chemical Industrial Engineering, Institut Teknologi Sepuluh Nopember, Kampus ITS Sukolilo, Surabaya 60111, Indonesia

E-mail : orchidea@chem-eng.its.ac.id

Abstract. Soy milk is a traditional beverage widely consumed in Asian countries, one of which is Indonesian. Soy milk has a balanced nutrient combination which is more or less similar to cow milk, excluding its vitamins (vitamins A and B2) and minerals (K, P, and Na). Behind its limitation, soy milk is free of cholesterol, lactose, gluten, and rich of phytochemical compounds. Soy milk has been consumed as an alternative to milk for people with lactose-intolerant problem, and for vegetarian as well. Therefore, a fortified soy milk using water extract of drumstick tree (*Moringa oleifera*) leaves was developed. A water extract of pandan leaves (*Pandanus amaryllifolius*) was also added to minimize the strong aroma of leaves/sepat from the *M. oleifera* extract and to appeal the appetite. Hedonic ranking test as an organoleptic testing method was conducted to identify consumer preferences. The test revealed the soy milk with low addition of leaves *M. oleifera* extract (0.26 g/L) is correspondents' more preferable options than higher concentration of extract, i.e. 0.50 and 0.74 g/L. Minerals contents of Ca and Fe were increased to 9% and 32%, respectively, with the addition of 0.26 g/L of the extract. Unfortunately, no vitamins B and C were detected in all variables of added extract concentrations to soy milk.

1. Introduction

Micronutrients are essential for the human body to perform physiological functions. Micronutrient deficiencies are critical especially in pregnant women and young children, as it affect fetal and child growth in cognitive development (such learning disabilities, mental retardation), and resistance to infection. Therefore, nutritional interventions, able to ensure the fulfilment of the adequacy of micronutrients, need to be developed, especially for poor and rural areas. Empowering the diversity of local sources to improve the micronutrient status of the community is one of the answers.

In 2006, the Ministry of Health of the Republic of Indonesia that malnutrition in Indonesia is still a major problem and may cause death, especially in high risk groups of infants and toddlers. Malnutrition cases found in Indonesia are quite high, reached out 1,401 cases. Recently, there have been extraordinary outbreaks of malnutrition in Papua, especially in Asmat District. Therefore, local food fortification in rural areas is one of the strategies to overcome the problem.

Food fortification programs have proven their effectiveness in eliminating pellagra, beriberi, ariboflavinosis, rickets, and goiter. It can also be made by adding key nutrients to foods, such as flour and bread which are economically priced and widely consumed by local population. One of the nutrient-rich sources is *Moringa oleifera. Moringa oleifera* leaves are reported to contain 7 times as much vitamin C as oranges, 10 times as much vitamin A as carrots, 17 times as much calcium as milk, 9 times as much protein as yoghurt, 15 times as much potassium as bananas and 25 times as much iron as spinach [1]. It is also rich in phytosterols like stigmasterol, sitosterol, and cholesterol which are precursors of hormones. It is used to treat malnutrition in children younger than 3 years [2].

The use of *M. oleifera* as a food fortificant is on the increase. For instance, both fresh and dried Moringa leaves are included in meals in African countries such Ghana, Nigeria, Ethiopia, East Africa, and Malawi [3]. The potential use of *M. oleifera* in food preparation such in making amala (dough made from yam and plantain (*Musa paradisca*) flour), bread [4], cake [5], yoghurt [6], jelly [7], and drinking tea preparation [8] were reported. *M. oleifera* is well known as kelor leaves in Indonesia. One of the traditional recipes from Indonesia, uses processed *M. oleifera* leaves in soup and other processed food. Utilization of M. *oleifera* such as natural coagulant [9] and for biodiesel production [10] were also reported.

Tofu and soy milk, the main products of soy bean, are favorites especially in Asian countries. Soy milk is widely used as substitute for milk instead of cow milk due to possible allergenic problems as well as for vegetarian [11]. Soy milk has a balanced nutrient combination which is more or less similar to cow milk, excluding its vitamins (vitamins A and B2) and minerals (K, P, and Na) content. It contains no cholesterol or lactose and only small quantities of saturated fatty acids [12]. In fact, it must be a healthier protein source than milk because it does not contain cholesterol and lactose [13]. Thus, soy milk is a good candidate for nutritional sources especially for rural areas with less expensive prices and more affordability. Combining extracts of *M. oleifera* leaves as a food fortificant to increase the nutritional value of soy milk is a good strategy to raise an alternative food sources to combat malnutrition, especially among children, infants, and lactating mothers in Indonesia.

2. Experimental Section

2.1. Materials

All the ingredients for soy milk production was bought from the local market in Rungkut, Surabaya, Indonesia. The high quality of soy bean (FNXX brand) was used as a raw ingredient. The Soy milk was processed and produced in collaboration with a small-medium sized enterprise (Usaha Kecil Menengah, UKM) of Soy Milk, Rungkut, Surabaya, Indonesia using a conventional method. While all the chemicals for analysis were provided by Laboratorium Dasar Bersama (LDB), Universitas Airlangga, Surabaya, Indonesia.

2.2. Extraction of Leaves of Drumstick Tree (Moringa oleifera)

Fresh leaves of drumstick tree (*Moringa oleifera*) were supplied from Jemur Wonosari, Wonocolo, RT 6, RW 4, Surabaya, Indonesia. Subsequently after the leaves had been received, 100 g of the fresh leaves was washed with water, and chopped with a kitchen blender using 100 mL of water addition (1:1, w/v). The obtained extract was filtered twice using cloth filter. The obtained filtrate was mixed. The solid residue from the first step of filtration was denoted as a rough cake, and the residue from the second step as a fine cake. All the obtained filtrate and residues were put into freeze dryer for 3 days.

2.3. Soy Milk Production

Sorted soybeans (500 g) were soaked in water (3 L) for 10-12 hours. Afterwards, the soybeans were cleaned, washed, and drained. Subsequently, the soybeans were blended with water at 1:10 (w/w), boiled at 100 ^{o}C for 30 minutes, and filtered to separate the pulp from the juice of the soybeans. The produced soy milk was reheated, and sugar was added. A preservative (0.2 g/L) was added as optionally. Furthermore, the extract of *M. oleifera* leaves was added to the soy milk according to the pre-determined variables. The soy milk was immediately packed (330 mL capacity) and should be kept in cold temperature ca. 4-5 °C for further consumption and analysis.

2.4. Analysis of Minerals

The analysis of all the minerals of Ca, Mg, and Fe were conducted using Atomic Absorption Spectroscopy (AAS) with dry destruction method. The sample was accurately weighed (in Duplo), ca. 10±0.0001 g, in curs porcelain. Sample was ashes in muffle furnace at 450-500 °C overnight. Subsequently, after the sample temperature reached room temperature, 2 mL of concentrated HNO_3 was added till all of the sample was dissolved and homogenized carefully in fume hood. The vaporization process of the sample was slowly done using hot plate or water bath until the sample was dried; directly put it in to the muffle furnace at 450-500 °C for an hour. Make sure that all the volatile carbon was vaporized by repeating the addition step of HNO_3 (two or three times). The ash was transfer quantitatively to 25 mL volumetric flask, and La was added until its concentration reached 0.1 or 1% La; add 1% (v/v) HNO_3 up to the volumetric mark. Finally, the filtration was done and the filtrate was analyzed using polyethylene bottle at AAS.

2.5. Organoleptic Test of The Fortified Soy Milk

The consumer acceptance was tested using an organoleptic test form which was distributed and filled by 75 respondents in the age range of 20-30 years. Each respondent assess 4 types of corresponding fortified soy milk including a Blanco of soy milk (soy milk without any addition of extract leaves of *M. oleifera*) and soy milk with addition of 0.24, 0.5, and 0.74 g/L of filtrate. The assessments of the consumer preferences was scored between 0 and 7 with 0 (neutral/undefined), 1 (extremely dislike) to 7 (I extremely like). The detailed scores as follows: 0 = zero/undefined, 1 = extremely dislike, 2 = I dislike, 3 = barely dislike, 4 = moderately like, 5 = I like, 6 = I like very much, and 7 = really like.

One way-ANOVA analysis was applied to analyse the data of the consumers' overall acceptance of organoleptic test for soybean milk fortified using the extract of *M. oleifera* leaves with p value <0.05. Meanwhile for the homogeneity test, a Levine statistics (based on mean) with p value> 0.05 was applied. A sample of dataset were consisted of the following 14 attributes (colour, texture, uniformity, viscosity, appearance, tastefulness, pleasantness, sweetness, bitterness, sourness, saltiness, aroma of leaves/sepat, fresh, and texture) and 4 types of samples (blank of soy milk, soy milk with addition of 0.24, 0.5, and 0.74 g/L of filtrate).

2.6. Analysis of Vitamin C

Vitamin C analysis was conducted using high performance chromatography (HPLC) with ascorbic acid as a standard for Vitamin C and tartaric acid as an internal standard. A sample of 2.0 g was accurately weighed and diluted up to 50 mL using methanol p.a. in HPLC grade addition in corresponding volumetric flask. A sample was filtered with 0.2 μm of nylon membrane syringe filter. Then this sample was ready to be injected into HPLC-DAD at 260 nm using the LiChrospher 100 RP-18 HPLC column (5 μm, 250 x 4 mm ID). A series of Agilent HPLC 1100 equipped with a PDA auto sampler and detector was used. An isocratic mobile phase system with KH_2PO_4 composition (0.02 M): methanol = 40:60 (v/v) is used at a flow rate of 0.6 mL/min, injection volume = 10 μL, 22 °C column temperature, and total running time = 10 minutes. Each sample was made in Duplo and triplet analysis.

2.7. Analysis of Vitamin B

The sample was accurately weighed and diluted up to 10 mL using methanol p.a. in HPLC grade addition in corresponding volumetric flask. The sample solution was further ultra-sonicated for 30 minutes and 15 minutes in vortex. It was subjected to stand for about 1 hour; filtered with 0.2 μm of nylon membrane syringe filter. Then it was ready to be injected into the HPLC at 264 nm. A similar HPLC systems was used (subsection 2.4 analysis of vitamin C). The Luna C8 HPLC column (100 Å, 150 x 4.6 mm ID) was used with mixture of mobile phase as follow: acetonitrile (eluent A) and KH2PO4 (pH = 3) (eluent B) at a flow rate of 0.5 mL/min, injection volume = 10 μL, 25 °C column temperature, and total running time = 25 minutes. Each sample was made in Duplo and triplet analysis. Gradients of the mobile phase are as shown in Table 1.

Table 1. Gradients of the HPLC mobile phase system for analysis of vitamin B

Time	Mobile phase composition (v/v)	
(min)	Eluent A	Eluent B
7	10	90
8	40	60
10	30	70
15	40	60
20	10	90

3. Results and Discussion

Consumer acceptance generally depends on the attributes were tested at the organoleptic or hedonic test (subsection 2.5). Test was conducted to determine the quality of the fortified soy milk with extract leaves of *M. oleifera,* i.e consumer preferences. The test sheet dividing sensory parameters in three categories: appearance, smell, and flavour which is very important for the food industry. The appearance/sight categories were based on colour, texture, uniformity, viscosity, and appearance. The smell categories were based on the degree of tastefulness and pleasantness. The flavour categories included sweetness, bitterness, sourness, saltiness, aroma of leaves/sepat, freshness, and texture.

3.1. Organoleptic Test Analysis of Produced Soy Milk

Data analysis of the organoleptic dataset of the fortified soy milk with filtrate of *M. oleifera* leaves showed fortified soybean milk was not superior to soy milk in all categories of appearance/sight, smell, and flavor (Figure 1). Increasing the added concentration of filtrate of *M. oleifera* leaves affected the scores of color and texture (appearance category); tastefulness and pleasantness (smell category); and freshness and texture (flavor category). The extract addition did not affect the sweetness of the soy milk but tended to decrease the score of the tastefulness of the soy milk, especially when high concentrations of filtrate was added, i.e. 0.74 g/L. Overall, organoleptic test results suggested that an 0.24 g/L concentration of *M. oleifera* leaves be acceptable to be added to the soy milk without interfering the tastefulness of the consumer.

One way-ANOVA analysis of the overall acceptance of organoleptic test for soy milk fortified with filtrate of the extract of *M. oleifera* leaves, at concentration of 0.24 and 0.50 g/L, gave a significance of 0.008 (p value <0.05). There was a significant difference in the overall acceptance of the organoleptic test between the compared sample groups (Blanco, soy milk fortified with extract of *M. oleifera* leaves at filtrate of concentration 0.24 g/L (F2) and 0.50 g/L (F1)). While the homogeneity test using Levine statistics (based on mean) gave p value 0.226 (p value> 0.05), meaning that there was no significant variation between the groups of compared samples (results are not shown here). In other words, the data used were uniform. A post hoc test-multiple comparison (Tukey HSD) was also applied to find out which sample group was superior (Table 2). The F1 sample and Blanco gave significant differences on the overall acceptance results (Table 2) indicated by the different locations of F1 (in column 1) and B (in column 2). The overall acceptance for F1 and F2; however was not significantly different. The acceptance is similar for F2 and B. Overall, respondents prefer Blanco with a hedonic value scale of 5.26 (medium). The Blanco was superior to the F2, with a value of 4.93 (Table 2).

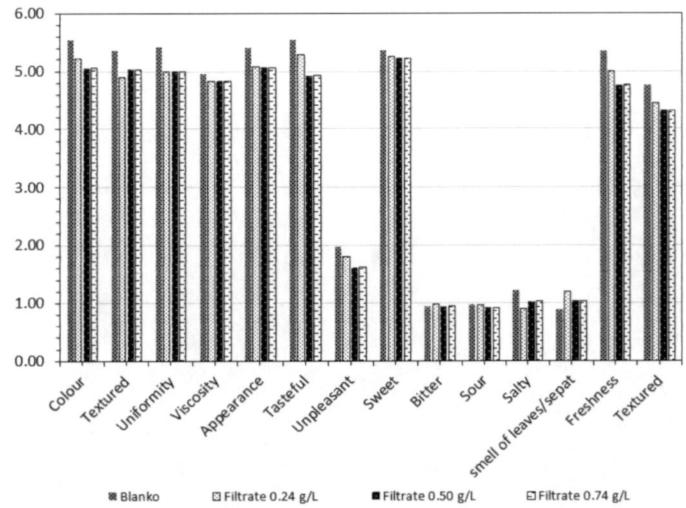

Figure 1. Organoleptic test results of the soy milk fortified with the extract of *M. oleifera* leaves (filtrate) at concentration level of 0.24, 0.50, and 0.74 g/L

Table 2. Overall acceptance of compared samples of soy milk

Sample	N	α = 0.05	
		1	**2**
Soy milk fortified with filtrate of the *M. oleifera* leaves 0.50 g/L (**F1**)	69	4.81	
Soy milk fortified with filtrate of the *M. oleifera* leaves 0.24 g/L (**F2**)	69	4.93	4.93
Blanco (Soy milk no fortification) (**B**)	69		5.26
Significantly		0.714	0.065

Recapitulation of the organoleptic test of the respondent to the fortified soybean milk with fine cake of the *M. oleifera* leaves showed that fortified soy milk could not beat the Blanco in all categories (appearance/sight, smell, and/or flavor) (Figure 2). Increasing the concentration of *M. oleifera* leaves (fine cake) will further lower the score to dislike category. On all aspects of the assessment, the fortified soy milk with fine cake of the *M. oleifera* leaves at the concentration of 0.24 g/L yielded a better test value than the one with the higher concentrations (0.5 and 0.74 g/L). The effect of addition of the *M. oleifera* leaves (fine cake) in soy milk yielded more significant compared to the filtrate addition (Figure 1). Hence, the filtrate of *M. oleifera* leaves was the preferred addition to soy milk compared to the fine cake of *M. oleifera* leaves.

The addition of a rough cake of the *M. oleifera* leaves as well as the addition of fine cake of the *M. oleifera* leaves to the soy milk was also significant (result not shown). The addition of both fine and rough cake greatly affects the assessment of appearance/sight and taste categories. The fine cake is more preferable than the rough cake. The filtrate and fine cake of 0.24 g/L of the *M. oleifera* leaves can be used as the soy milk fortificant with acceptable organoleptic test result (Table 3). However, the filtrate of the *M. oleifera* leaves 0.24 g/L was selected as the most preferred fortificant over all other extract types of *M. oleifera* leaves. Thus, concentration apparently lower than standardized addition portion of Moringa powder with classifications of acceptability "liked moderately" and "liked very much", ca. 5-7% [14].

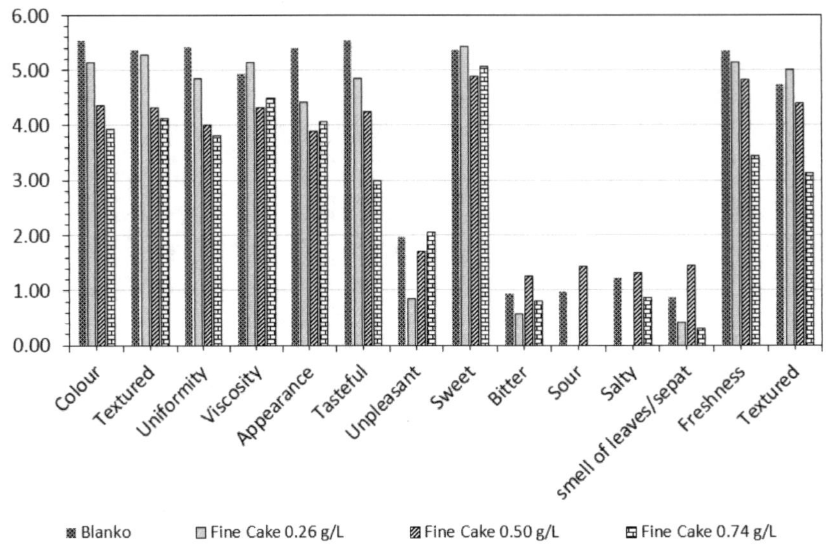

Figure 2. Organoleptic test results of the soy milk fortified with the extract of *M. oleifera* leaves (fine cake) at concentration level of 0.24, 0.50, and 0.74 g/L.

Table 3. Organoleptic score of Blanco, fortified soy milk with filtrate and fine cake (0.24 g/L).

			Fortified Soy milk (at 0.24 g/L)	
		Blanco	**Filtrate**	**Fine cake**
Appearance/ Sight Test	Color	5.54	5.23	5.14
	Texture	5.37	4.89	5.29
	Uniformity	5.42	4.99	4.86
	Viscosity	4.94	4.83	5.14
	Appearance	5.41	5.07	4.43
Smell Test	Tastefulness	5.54	5.29	4.86
	Pleasantness	1.97	1.80	0.86
Flavor Test	Sweetness	5.37	5.26	5.43
	Bitterness	0.94	0.99	0.57
	Sourness	0.97	0.97	0
	Saltiness	1.23	0.90	0
	aroma of leaves/langu	0.89	1.20	0.43
	Freshness	5.35	5.00	5.14
	Texture	4.75	4.44	5.000
	average score	**3.84**	**3.63**	**3.37**

3.2. Nutrition Analysis of the Produced Soy Milk

To find the best product of the soy milk, the nutrition analysis of the produced soy milk was also conducted, focusing on vitamin B and C and minerals (Ca, Mg, and Fe). Moyo et al. [15] reported vitamin C content of *M. oliefera* leaves was 34.78-40.64 mg/100 g, while 0.06 mg/100 g of vitamin B1 (thiamine), 0.05 mg/100 g of vitamin B2 (riboflavin), 0.8 mg/100 g of vitamin B3 (niacin), 220 mg/100 g of vitamin C, 440 mg/100 g of calcium, 7 mg/100 g of iron, and Mg content of 0.5% weight of dry leaves of *M. oliefera*. Lower minerals composition from *M. oleifera* leaves flour was reported by

Teixeira et al. [16]. They contained 2.97 mg/100 g of calcium, 103.12 mg/100 g of iron, 1.9 mg/100 g of magnesium, 1.58 mg/100 g of zinc, 4.16 mg/100 g of potassium, and 3.38 mg/100 g of copper.

Mineral contents of fortified soy milk were shown in Table 4. The Blanco of soy milk itself contained more Mg and less Ca and Fe. The fortified soy milk with filtrate of *M. oleifera* leaves (no preservative addition) yielded higher content of minerals than the fortified soy milk with fresh extract of *M. oleifera* leaves. The preservative, however, cannot store the minerals.

Table 4. Minerals in the Fortified Soy Milk

			Type of Fortified Soy Milk		
	Soy Milk (Blanco)		freeze dry	fresh extract (0.26 g/L)	
Minerals	**no preservative**	**+ preservative**	**+ filtrate (0.26 g/L)**	**no preservative**	**+ preservative**
Ca (ppm ± RSD)	5.12 ± 0.4	2.53 ± 3.1	5.58 ± 4.4	2.84 ± 1.4	2.35 ± 0.3
Mg (ppm ± RSD)	152.30 ± 4.2	80.00 ± 3.6	144.70 ± 7.0	75.80 ± 0.1	82.50 ± 1.8
Fe (ppm ± RSD)	4.13 ± 1.8	1.27 ± 0.59	6.09 ± 1.6	1.35 ± 1.41	1.25 ± 1.51

No vitamins B and C were detected in all fortified soy milk including it Blanco itself. Soymilk fortified by a fresh filtrate of *M. oleifera* (without freeze dried process) was also analyzed. Preservation addition to the soy milk still could not preserve both of vitamins B and C. No vitamins were detected in all soy milk samples though the limit of detection in the analysis of both vitamins was very low, ca. 0.0125% w/w. Vitamins B and C were absent presumably due to the high temperature process of soy milk production, ca. 100 ^{o}C, 30 minutes (subsection 2.3). This temperature might be damaging the vitamins. However, this step is needed to sterilize the milk. Therefore, a conventional method is not appropriate to preserve the vitamins of the extract of *M. oliefera* leaves.

4. Conclusions

The filtrate of *Moringa oleifera* leaves extract at the concentration of 0.24 g/L is the most preferable fortificant for the soy milk, based on the organoleptic test. Water extraction of *M. oleifera* leaves as well as freeze dried processed cannot preserve the vitamins B and C in *M. oleifera* leaves. Hence, the vitamins B and C in the soy milk cannot be substituted either. Fortified soy milk with filtrate of *M. oleifera* leaves (0.24 g/L) increase the minerals content of Ca and Fe to 9% and 32%, respectively. More work is still needed, though, to find the suitable extraction process of *M. oleifera* leaves, especially in the process to preserve the high content of vitamins of *M. oleifera*. This work will eventually result in more competitive soy milk products, in both quality and price.

Acknowledgment

The authors are grateful to the Ministry of Research, Technology, and Higher Education of the Republic of Indonesia for their financial support via Research-Based Community Development Program with Local Fund from ITS (Program Pengabdian Masyarakat berbasis Penelitian, Dana Lokal ITS) No. 1198/PKS/ITS/2018.

References

[1] Rockwood J L, Anderson B G and Casamatta D A 2013 *Int. J. Phytoteraphy Res.* **3** 61-71
[2] Mutiara K T, Harijono, Estiasih T and Endang S W 2013 *J. Basic App. Sci. Res.* **3** 430-4
[3] Agbogidi O M and Ilondu E M 2012 *J. Biol. Innovations* **6** 156-167
[4] Chinma C E, Abu J O and Akoma S N 2014 *J. Food Processing and Preservation* **38** 721-7
[5] Kolawole F L, Balogun M A, Opaleke D O and Amali H E 2013 *Agrosearch* **13** 87-93
[6] Kuikman M and O'Connor C P 2015 *J. Food Res.* **4**, 165-71

[7] Yulianti R 2018 *Moringa oleifera lam as a source of vitamin C and β-carotene in jelly drink* (Bogor: Institut Pertanian Bogor)
[8] Sugiyanto AK 2016 *Nutritional content of Moringa oleifera leaves based on leaf position and brewing temperature* (Bogor: Institut Pertanian Bogor)
[9] Camacho F P, Bongiovani M C, Silva M O, Coldebella P F, Amorim M T S P and Bergamasco R 2015 *Chem. Eng. Trans.* **43** 1123-28
[10] Camacho F P, Bongiovani M C, Nishi L, Rezende D, Silva M F and Bergamasco R 2015 *Chem. Eng. Trans.* **43** 2233-38
[11] Farnworth E R, Mainville I, Desjardins M P, Gardner N, Fliss I and Champagne C 2007 *Int. J. of Food Microbiol.* **116** 174-81
[12] Je-Ruel L and Chin-Wen L 2000 *Food Microbiol. Safety* **40**, 716-19
[13] Wang J, Wu R, Zhang W, Sun Z, Zhao W and Zhang H 2013 *J. Dairy Sci.* **96**, 5603-24
[14] Mamta, Dunkwal V, Goyal M and Monika 2017 *J. Pharmacog. Phytochem.* **6**, 1060-63
[15] Moyo B, Masika P J, Hugo A and Muchenje V 2011 *African J. Biotechnol.* **10** 12925-33
[16] Teixeira E m B, Carvalho M R B, Neves V A, Silva M A and Arantes-Pereira L 2014 *Food Chem.* **147** 51-4

IOP Conf. Series: Materials Science and Engineering **543** (2019) 012020 doi:10.1088/1757-899X/543/1/012020

Defatted Rice Bran a Byproduct of Oil Extraction with Ultrasonic Method for Protein Supplement in Cassava-Flour Biscuits

R A Nugrahani[1], N H Fithriyah[1], and Nelfiyanti[2]

[1]Chemical Engineering Department, Engineering Faculty, Universitas Muhammadiyah Jakarta, Kampus B Jl. Cempaka Putih Tengah 27 Jakarta, 10510, Indonesia
[2]Industrial Engineering Department, Engineering Faculty, Universitas Muhammadiyah Jakarta, Kampus B Jl. Cempaka Putih Tengah 27 Jakarta, 10510, Indonesia

E-mail: ratri.ariatmi@ftumj.ac.id

Abstract. One of the components in rice bran is oil, which is mainly used for cooking. Because of its high uses, there is also an increase in the supply of the defatted rice bran. Currently, it is used generally as a supplement for livestock feed. Thus, there is a need to diversify their applications to increase protein content in other food sources. The objective of this study is to extract rice bran oil with ultrasonic method, characterize defatted rice bran, and as a supplement in cassava flour biscuits. The method of this research is the stabilization of rice bran by roasting for 10 minutes, extractions of rice bran with n-hexane solvent: 1:4; 1:5, 1:6; 1:7; 1:8, 10-minute ultrasonic technique to remove the oil content, and protein analyses of defatted rice bran. Defatted rice bran is used as supplement for cassava flour biscuits. Simple organoleptic tests were conducted based on respondents' preference. Results showed that there was a significant correlation between rice bran: solvent ratios and protein contents in defatted rice bran. The highest protein content was found at 18.31% in rice bran: solvent ratio 1: 5. The most preferred by respondents in its application for cassava flour biscuits was 13.6%.

1. Introduction

Currently Indonesia still imports wheat flour as one of raw materials for food products. Asosiasi Produsen Tepung Terigu (APTINDO – Wheat Flour Producer Association) reported that there was a six percent increase of wheat flour import in 2016 compared to that in 2015 at around 8.3 million tonnes. National demand for wheat flour in 2017 was 8.79 million tonnes [1]. To reduce dependency on wheat flour import, one of the alternative flours that can be used is the flour of cassava (Manihot esculenta Crantz), which is already frequently used in flour-based food products. Proximate analyses by [2] revealed that 100 g of cassava contains 0.3-3.5 grams of proteins, 25.3-35.7 of carbohydrates, and 0.1-3.7 grams of fiber. While, according to [3], functional nutrition containing proteins and fibers can decrease blood cholesterol level and protect from colon cancer, cardiovascular diseases, constipation, and obesity.

The biggest disadvantage of cassava flour is its low protein contents, and thus the need to enrich it with several sources of protein such as cereal, legumes and bran [3]. The advantages of cassava flour are the

IOP Conf. Series: Materials Science and Engineering **543** (2019) 012020 doi:10.1088/1757-899X/543/1/012020

absence of cereal flavor and several characteristics needed in some applications, such as lower gelatinization and higher viscosity [4].

Rice bran is a byproduct of rice dehulling process. Its availability in Indonesia, as an agrarian country, is significantly abundant. Currently, the use of rice bran is mainly limited to livestock feed. Rice bran contains less than 90% triglycerides, comprising saturated and unsaturated fatty acids, and also antioxidants, that can be used as cooking oil and flavoring agent.

Before being extracted, rice bran has to be stabilized to prevent an increased free fatty acid (FFA) content, and extraction results can be purified by degumming [5]. In addition to cooking oil, rice bran oil can be used as oleochemical, such as base oil for uses as inhibitor and surfactant [6], [7].

Several extraction methods of rice bran oil, that are frequently used following stabilization stage by steaming, include pressing with expeller machine [8]. Another researcher used hydraulic press machine to separate the oil [9]. Rice bran that had been stabilized in a microwave for three to five minutes at 120°C was extracted using n-hexane at 50°C for eight to nine hours [10]. Ultrasonic method using low frequency sound wave around 20-50kWz for 20-40 minutes reduced extraction time by 70% compared to that of conventional hydrodistillation [11]. Rice bran oil extraction using a 24-hour maceration method in methanol and hexane solvents yields a byproduct called defatted rice bran. Previous research have been done by [12] about the extraction of cherry seed oil. The supercritical carbon dioxide showed an efficient process and a green method. Other related research have been done by [13] is about Larix Decidua Bark extraction. The highest yield is obtained by microwave with solid/liquid ratio of 1:8 (w/v).

Byproduct of rice bran extraction is defatted rice bran that can be used to increase the contents of fiber and protein, antioxidant characteristics, and storage stability in bread by adding 15% of defatted rice bran to wheat flour [14].

The objectives of this study is to extract oil from rice bran using the ultrasonic method, characterize the resulting byproduct (defatted rice bran), and test its application as a protein supplement in cassava flour biscuits.

2. Methods

2.1. Materials and Equipment

The materials used in this study are rice bran, n-Hexane, cassava flour, margarine, palm sugar, egg yolk, vanilla, and baking soda. The equipment used is a Wisd™ ultrasonicator with frequency at 20 kHz, a Rotary evaporator Buchi, and a Sharp™ oven.

2.2. Procedures

The study was began with the rice bran stabilization, which is to roast for 10 minutes. The method in rice bran oil extraction was using several ratios of rice bran that has been stabilized : n-hexane 1:4; 1:5, 1:6; 1:7; 1:8 (w/v) and ultrasonication for 10 minutes. Defatted Rice bran and the mixture of the oil extract and solvent are separated using a vacuum filter. The next step, n-hexane residue is separated from the oil extract using Rotary evaporator. Protein analyses in defatted rice bran was conducted with Kjeldahl methodology. The Kjeldahl method is the total nitrogen content analysis. The sample is mineralized with concentrated sulfuric acid and catalyst (copper and potassium sulfate) to convert the nitrogen into ammonium sulfate. Ammonia from ammonium sulfate is replaced with concentrated sodium hydroxide, then ammonia is distilled and collected in a solution of boric acid and titrated with hydrochloric acid. The protein content is determined by multiplying the nitrogen content with the NCF (Nitrogen conversion factor) [15]. Defatted rice bran sample with the highest protein content was use as a supplement in cassava flour biscuits, which were assessed in taste, color organoleptic tests to determine respondents' preference.

3. Results and Discussion

3.1. Oil Extractions and Protein Content Analysis in Defatted Rice Bran

Rice bran contains oil that has many kinds of bioactive compounds, including antioxidants. Rice bran oil that has been further separated from n-hexane using Rotary evaporator can be put in many uses, such as cooking oil, flavoring, and cosmetic additives. Extraction with ultrasonication in n-hexane solvents was performed in rice bran that had been stabilized by roasting for 10 minutes. The variation of weight to volume ratios between rice bran to solvent was 1:4; 1:5; 1:6; 1:7; 1:8. The residue of n-hexane in the defatted rice bran was separated by oven. Figure 1 shows the ultrasonic extraction equipment of rice bran oil and Figure 2 shows the defatted rice bran.

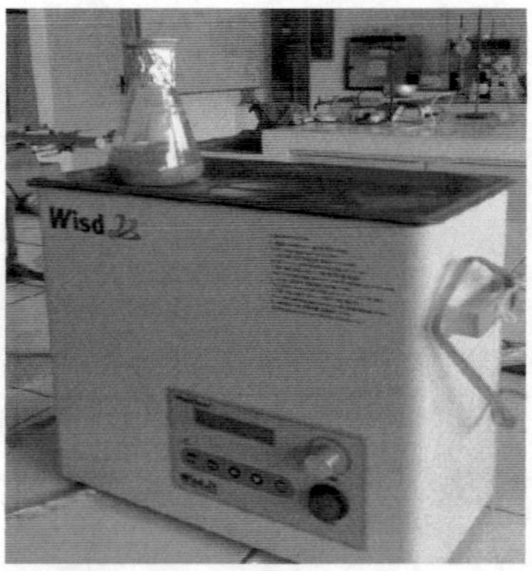

Figure 1. The Ultrasonic extraction equipment of rice bran oil and Figure

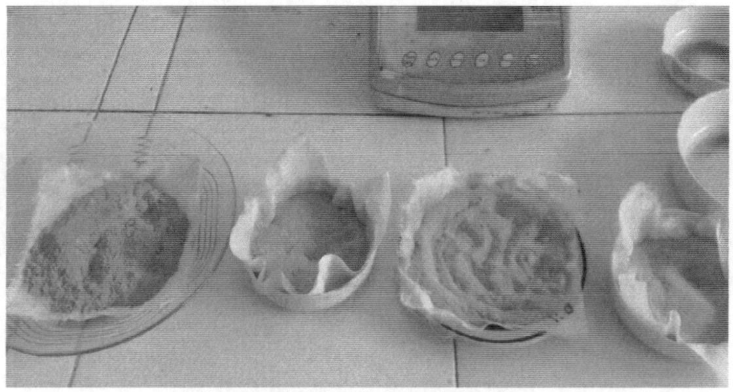

Figure 2. The defatted rice bran

Table 1 shows the protein contents of defatted rice bran resulted from rice bran oil extraction using ultrasonic method.

Table 1. Protein contents of defatted rice bran resulted form rice bran oil extraction using ultrasonic method at different solvent ratios.

Rice bran : Solvent (w/v)	Protein (%)
1:4	17.23
1:5	18.31
1:6	17.84
1:7	18.19
1:8	18.05

Based on Table 1, the protein contents of defatted rice bran increase with increased ratios of rice bran and n-hexane solvent (w/v) from 1:4 to 1:5 in the rice bran oil extraction processes using ultrasonic method. The highest protein content was found at rice bran and solvent ratio (w/v) of 1:5. Then the protein contents of defatted rice bran decrease with increased ratios of rice bran and n-hexane solvent (w/v) from 1:5 to 1:8.

Hexane is one of the nonpolar solvents and better suited to oil extraction, while the polar solvents are known to extract free fatty acid (FFA) and other product such as protein [16]. It was expected that the higher the ratio of rice bran and solvent, the higher oil and the lower protein is extracted, which leads to the increase in protein content of defatted rice bran. In previous research, [17] concluded that protein stable in nonpolar solvents such as cyclohexane. Based on [18], protein decomposition has occurred in defatted soybean protein in n-hexane solvent, especially at high temperatures. The separation of n-hexane residue in the oven caused the denaturation of the protein.

3.2. Correlation Analyses Between Rice Bran to Ratio and Protein Content in Defatted Rice Bran
Regression analysis and R^2 tests were performed on the data to determine the correlation between rice bran and solvent ratios in ultrasonic extraction and protein contents in defatted rice bran. Figure 3 shows this correlation.

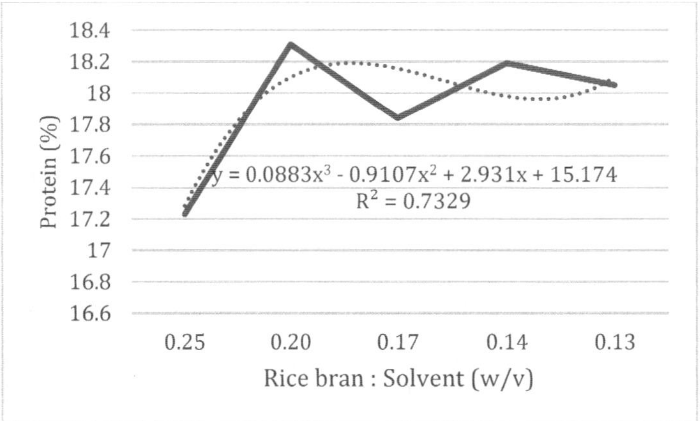

Figure 3. Protein contents in defatted rice bran (%) as a function of rice bran : solvent ratios (w/v)

The result shown in Figure 3 describe that the higher the solvent: rice bran ratio, the higher the protein content in the defatted rice bran, with the optimal ratio of solvent to rice bran at five. The more oil is extracted from the rice bran, the higher protein content is retained in the defatted rice bran. However, when the ratio value is greater than five, the protein content slightly decreases. The correlation between the solvent: rice bran ratio in x and protein content in rice bran in y can be expressed in the following formula: $y = 0.0883x^3 - 0.9107x^2 + 2.931x + 15.174$, with an R^2 value of 0.7329. This model is the equation that best suits to the real data. This result shows that there is a correlation between the ratio of solvent: rice bran and protein contents in defatted rice bran.

3.3. Analysis of Respondents' Preferences to Different Biscuit Recipes
The test for the applications of defatted rice bran was conducted by using it as a supplement for biscuits. The biscuits also has cassava flour, margarine, palm sugar, egg yolk, vanilla, and baking soda as its main ingredients. Furthermore, the biscuits has addition of defatted rice bran for recipe 1 and both defatted rice bran and rice bran for recipe 2. Those formulas are prepared at 140°C in the oven for 25 minutes, as shown in Figure 4.

IOP Conf. Series: Materials Science and Engineering **543** (2019) 012020 doi:10.1088/1757-899X/543/1/012020

Figure 4. The cassava flour biscuit preparation with the supplementation of defatted rice bran

The results of preference tests from 10 respondents are presented in Table 2.

Table 2. Results of taste organoleptic tests based on respondents' preference

Ingredients	Composition in Biscuit		Respondents' Evaluation	
	Recipe 1	Recipe 2	Recipe 1	Recipe 2
Defatted rice bran flour	60 g	40 g		
Rice bran flour		20 g		
Cassava flour	150 g	150 g		
Palm sugar	100 g	100 g	less preferred	more preferred
Margarine	100 g	100 g		
Egg yolk	1 sp	1 sp		
Vanilla	1 tsp	1 tsp		

The results in Table 2 show that biscuits with 40% defatted rice bran in Recipe 1 were less preferred than those of 13.33% in Recipe 2. This result shows equal to the [19], in which the best-tasted biscuits is the one with addition of 10-15% w/w rice bran in wheat flour. Then, the final products of the biscuit preparations with recipe 1 and recipe 2 are shown in Figure 5a and Figure 5b.

a). Biscuit with Recipe 1 | **.b).** Biscuit with Recipe 2

Figure 5. The cassava flour biscuit with Defatted rice bran flour

While the result of the organoleptic color test in Figure 4a and 4b shows that recipe 2 biscuit with the addition of rice bran in defatted rice bran has darker color compared to recipe 1 biscuit. This is accordance to [19] that the evaluation results of biscuit sensory shows that the score of biscuit color decrease significantly with the increase of rice bran.

4. Conclusion

One of the components in rice bran is oil, which is mainly used as cooking oil and main ingredient of oleochemical products. It can also be processed further to produce active compounds for health and cosmetic products. Because of the high benefits and uses of oil and extractions of rice bran, therefore there is also an increase in the supply of defatted bran. Extraction of rice bran oil with an ultrasonic process resulted in defatted rice bran. Defatted rice bran contains proteins and can be utilized as a supplement in other food materials that have low protein contents, such as cassava flour. The highest protein content was 18.31% and was obtained from the byproduct of rice bran oil extraction with rice bran: n-hexane solvent ratio at 1:5. The more oil is extracted from the rice bran, the higher protein content is retained in the defatted rice bran. However, when the ratio value is greater than five, the protein content slightly decreases. In its application in cassava flour biscuits, the composition of 13.33% defatted rice bran was the most preferred by respondents. The cassava flour biscuits with 40% defatted rice bran in Recipe 1 were less preferred than those of 13.33% in Recipe 2. The result of organoleptic test shows that the biscuit's color turns darker when there is an increase in supplementation of defatted rice bran.

5. Acknowledgement

The author would like to express a profound gratitude for the grant awarded by the Ministry of Research, Technology, and Higher Education of the Republic of Indonesia, in accordance with 2018 Contract Number: 006/KM/PNT/2018, 6 March 2018.

6. References

[1] Laoli N and Caturini R 2017 Impor gandum 2017 diprediksi tembus 8,79 juta ton (industri.kontan.co.id)
[2] Montagnac J A Davis C R Tanumihardjo S A 2009 *Compr. Rev. Food Sci. F.* **8** 181-194.
[3] Jisha A Padmaja G Sajeev M S 2010 *J. Food Qual.* **33** 79-99.
[4] Demiate I M and Kotovics V 2011 *Food Sci.Technol.* **31** 388-397.
[5] Prasad R.B.N., 2016. Rice Bran Oil Processing and Value Added Products, CSIR-Indian Institute of Chemical Technology, India (ofievents.com/india/contentimages/advertising/DrRBNPrasad.pdf)
[6] Nugrahani R A Redjeki A S Mentari Y and Hasanah M 2017[a] *Adv. Sci. Lett.* **23** 5720–5722.
[7] Nugrahani R A Redjeki A S Teresa Y and Hidayati N 2017[b] *J. Chem. Technol. Metall.* **52** 797-802.
[8] Matouk A M El-Kholy M M Sadany M Hendawy Y T 2009 *Misr J* Ag *Eng.* **26** 324- 342.
[9] Husain SM 2015 Newer Processing Methods for the Extraction and Refining of Rice Bran Oil Global Rice Bran Oil Conference Trident Hotel Mumbai India 5.
[10] Bopitiya D and Madhujith T 2014 *Trop. Agric. Res.* **26** 1-11.
[11] Nora F M D and Borges C D 2017 *Ciênc. Rural* **47** 1-9.
[12] Straccia M C Siano F Coppola R La Cara F and Volpe M G 2012 *Chem. Eng. Trans.* **27** 395.
[13] Silleroa L Pradob R and Labidia J 2018 *Chem. Eng. Trans.* **70** 1369.
[14] Sairam S Gopala Krishna A G Urooj A 2011 *Food Sci.Technol.* **48** 478–483
[15] Maubois J L and Lorient D 2016 *Dairy Sci. Technol.* **96** 15–25.
[16] Eftthymiopoulos I and Mills-Lamptey B 2018 *Ind. Crop. Prod.* **119** 49-56.

[17] Pace C N Trevino S Prabhakaran E and Scholtz J M 2004 *Philos. Trans. Royal Soc. B.* **359** 1225-1235.
[18] Fukushima D 1969 Denaturation of Soybean Proteins by Organic Solvent Presented at the 53rd Annual Meeting Washington DC 157.
[19] Younas A Bhatti M S Ahmed A and Randhawa M A 2011 *Pak. J. Agr. Sci.* **48** 129-134.

IOP Conf. Series: Materials Science and Engineering **543** (2019) 012021 doi:10.1088/1757-899X/543/1/012021

Enhanced Extraction of Phenolic Compounds from *Moringa Oleifera* Leaves Using Subcritical Water Ethanol Mixture

S Zullaikah[1,*], R Y Naulina[1], P Meinawati[1], K Fauziyah[1], M Rachimoellah[1], O Rachmaniah[1], S Nurkhamidah[1], N M I P Suari[1] and E N Prasetyo[2]

[1]Department of chemical engineering, Institut Teknologi Sepuluh November (ITS), Kampus ITS Keputih Sukolilo, Surabaya 60111, Indonesia

[2]Department of Biology, Institut Teknologi Sepuluh Nopember (ITS) Kampus ITS Keputih Sukolilo, Surabaya 60111 Indonesia

Email: szulle@chem-eng.its.ac.id, szullehakim@gmail.com

Abstract. A *Moringa Oleifera* leaves are rich of phenolic compounds that have the ability to serve as antioxidants. Phenolic compounds as an antioxidant can stabilize free radicals with complementing the electron deficiency of free radicals. Conventional method (maceration) can extract phenolic compounds from *Moringa Oleifera* leaves. However, this method is known as time consuming, inefficient and using solvent that is non-environmentally friendly. In this study, the extraction of phenolic compounds from *Moringa Oleifera* leaves was enhanced using a subcritical water-ethanol mixture. Fresh and dried *Moringa Oleifera* leaves were extracted using ethanol solution with different ratio and ratio of dried leaves and ethanol solution at 200 °C for 15 min. The highest amount of phenolic compounds of 87.11 ± 0.81 mg GAE/g of dried leaves with flavonoid contents of 75.79 ± 0.73 mg/g of dried leaves, antioxidant activity of 88.75 ± 0.93 mg vitamin C/L and yield of extracts of 32.73 % ± 0.08 were obtained under following operation conditions: ratio of dried leaves 0.13:20 (g/mL) and ethanol solution 96 %. The utilization of ethanol solution as a solvent on the subcritical condition was able to enhance the amount of total phenolic compounds, flavonoid content, antioxidant activity and yield of extract.

1. Introduction

Worldwide consumption of herbal medicines has markedly increased. According to the Secretariat of the Convention on Biological Diversity, global sales of herbal products were estimated to be the US $60 billion in 2000. In 2008, the global market for herbal remedies was about the US $83 billion with a steady growth rate ranging between 3 % - 12 % per year [1]. A Moringa leaf is one of the plants that can be used as herbal medicine. High nutrients and antioxidants cause this plant is called the miracle of a tree. This magic tree is not used only as medicine but can also be used as a nutritional supplement. Based on [2], Moringa leaves can be an alternative food to reduce malnutrition, especially in children and toddlers, because it has a high concentration in protein, potassium, and iron. It has benefits in the health field and good source of natural antioxidants because of the content of various types of antioxidant compounds such as ascorbic acid, flavonoids, phenolics and carotenoids[3],[4]. Moringa Oleifera is a plant that thrives and wild in Indonesia. These plants include in the Moringaceae family of the single genus Moringa. The genus Moringa is thought to include 13 species of which 11 are from

Africa [5]. The growth of herbal plants as drugs triggers to increase the production of extracts of natural materials. It has developed several methods in extracting natural materials ranging from conventional to modern. Has been described in the literature [6], there are methods of extracting natural materials like maceration, decoction, percolation and Soxhlet extraction. However, in the application they need a lot of solvent and long time for extraction. An alternative method to increase the extraction of phenolic compounds from Moringa Oleifera leaves by using subcritical water extraction. This method is time efficient and water is using as solvent, therefore is environmentally friendly. Besides that, issue about design and economic potential for the extraction of active components from herbs to be another reason to enhance the product [7]. Subcritical water extracts rapidly with the use of low-pressure high temperatures thus avoiding the degradation of volatile and thermo labile compounds. Subcritical water is water at a temperature of 100 – 374 °C under mild pressure to keep water in the liquid condition, and the dielectric constant of water also changed under this condition. It changes the nature of water that was originally polar to non-polar [8]. In our previous work on extracting phenolic compounds using subcritical water extraction resulted a total yield of 30.70 % at T = 200 °C for 15 min. with total phenolic compounds of 48.73 mg tannic acid/g extract and antioxidant activity of 45.86 mg vitamin C/L. Since phenolic Compounds have low polarity, solvent extractions are the most typically procedures used for extraction of plant materials due to their ease of use, efficiency, and wide applicability [9]. In this work, a mixture of water and ethanol was used to enhance extraction of phenolic compounds from Moringa oleivera leaves. The objective of this work was to study the effect of sample to solvent ratio using dry Moringa leaves and fresh Moringa leaves and ethanol solution concentration on the yield of extract, total phenolic compounds, antioxidant activity and flavonoid content using subcritical water-ethanol mixture.

2. Experimental section

2.1 Materials

Fresh moringa leaves with water content of 74.55 % was obtained from yard around campus. The dry moringa leaves was obtained by drying in a oven at 60 °C for 24 h. The dry moringa leaves were reduced in size used a blender and fresh moringa leaves used a mortar. Aquades, ethanol 96 %, methanol, DPPH (2, 2-diphenyl-1-picryl-hydrazyl), vitamin C (ascorbic acid), Na_2CO_3, Follin-Ciocalteu reagent, quercetin, $AlCl_3$, CH_3COONa, HCl were obtained from commercial sources.

2.2 Subcritical Extraction

Extraction of phenolic compounds from moringa oleifera leaves using subcritical water ethanol mixture was carried out in a batch hydrothermal reactor (V = 45 mL) made by stainless steel tube (SUS 316) with a Swagelok fitting. Generally, a certain ratio of fresh/dried leaves of moringa oleifera and solvent (ethanol solution) were charged into reactor. The reactor was heated in a furnace at temperature of 200 °C (time to reach from ambient temperature to T = 200 °C was 20 min.) and extraction was carried for 15 min. Increasing temperature in the reactor was followed with increasing of vapor pressure. The ratio of fresh leaves to ethanol solution (0, 35, 70, 80 and 96 %, (v/v)) is 3 :18; 0.5 : 20 and 1 :20 (g/mL), while the ratio of dried leaves to ethanol solution is 0.7635 : 18; 0.1273 : 20 and 0.2545 : 20 (g/mL). The reactor is then quenched by direct insertion into cold water; the extract (liquid phase) is collected by filtered with filter paper. The extract is separated from the solvent by a rotary evaporator. To remove a residual solvent, the extract is placed in the oven at 60 °C to obtain constant weight.

2.3 Total Phenolic Compounds

Total phenolic compounds was analyzed using Folin-Ciocalteau reagents according to [6].

2.4 Antioxidant Activity

Free radical scavenging activity was measured by using DPPH (2, 2-diphenyl-1-picryl-hydrazyl) according to[10], [11] and [12]

2.5 Flavonoid Content

Total flavonoid content was analyzed according to [13].

3. Results and discussions

3.1 Effect of concentration of ethanol solution at different ratio of fresh/dried leaves and solvent on the yield of extract

Moringa leaves contain phenolic compounds that are useful as natural antioxidants. In this work, to enhance extraction of phenolic compounds was carried out by using subcritical water-ethanol mixture. The ratio used was maintained to the volume reactor of 21 ml in order to maintain the vapor space since increasing temperature in the reactor was followed with increasing of vapor pressure. For concentration of ethanol solution of 0 % showed that the solvent used was water and this condition is called subcritical water. Effect of concentration of ethanol solution at different ratio of fresh/dried leaves and solvent on the yield of extract was shown in figure 1.

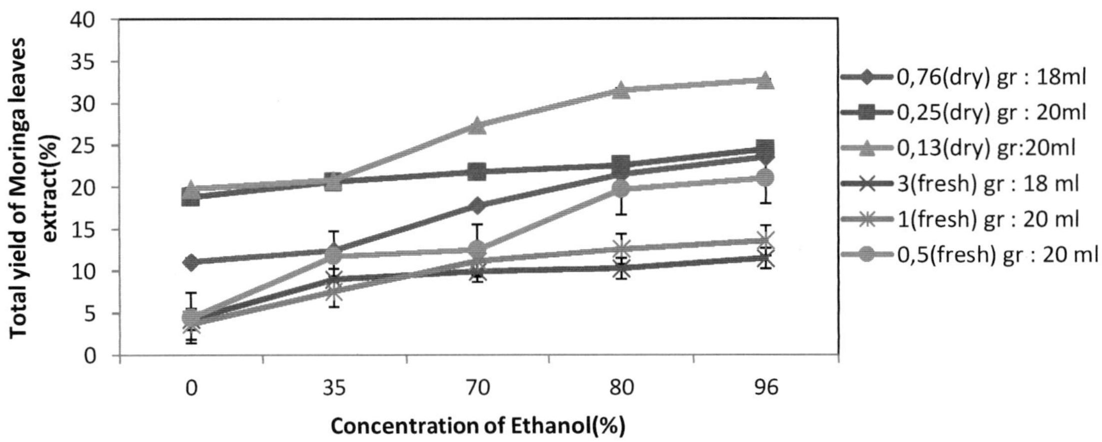

Figure 1. Effect of concentration of ethanol solution at different ratio of fresh/dried leaves and solvent on the yield of extract

From figure 1, it can be seen that the highest yield of extract was obtained using concentration of ethanol solution of 96 % for both fresh and dried leaves. Highest yield of extract from fresh leaves is 21.01 ± 0.75 at a ratio of 0.5 g: 20 mL and dried leaves of 32.73 ± 0.08 at ratio 0.13 g: 20 mL. The extract resulted from dried leaves is higher than that of fresh leaves. This shows that the ratio and concentration of ethanol solution used in the extraction has affected the yield of extract. This is consistent with literature [14] which states that the amount of the extract decreases with decreasing ratio of solid to solvent. The greater of the concentration of ethanol solution will increase the yield of extract. This is due to higher concentration of ethanol solution lowering solvent polarity, and lowering solvent polarity increasing extractable ability of total extract that is composed of phenolic and flavonoid compounds [9]. In accordance with the principle of mass transfer, the orientation of the thrust force during mass occurs gradient compilation of greater solids concentration [15]. The yield of extract from fresh leaves enhanced by 37.71 % by increasing concentration of ethanol solution from 0

% (solvent is water) to concentration of ethanol solution of 96 %, while using dried leaves the yield of extract enhanced by 60.50 %.

3.2 Effect of concentration of ethanol solution at different ratio of fresh/dried leaves and solvent on the total phenolic compounds (TPC)

The influence of concentration of ethanol solution at different ratio of fresh/dried leaves and solvent on the TPC (figure 2) shows the same trend as the yield of extract, where the greater the concentration of ethanol solution, the higher the TPC concentration. TPC composed of flavonoid, where it belongs to flavonol such as quercetin and kaempferol. From figure 2, the highest TPC was obtained using concentration of ethanol solution 96 % at a ratio of dried leaves and solvent of 0.1273 g: 20 mL, with TPC concentration of 87.11 ± 0.81 mg GAE / g dried leaves. Phenolic compounds are less polar compounds; therefore organic solvents such as water and ethanol under subcritical conditions are particularly suitable for extracting. Extraction of moringa leaves using water (concentration of ethanol solution = 0 %) show that the content of phenolic compounds extracted is lower compared using concentration of ethanol solution 96 %. This is because the water polarity index is larger than ethanol 96 %, the lower polarity index (nonpolar solvent) is an effective solvent to extract TPC.

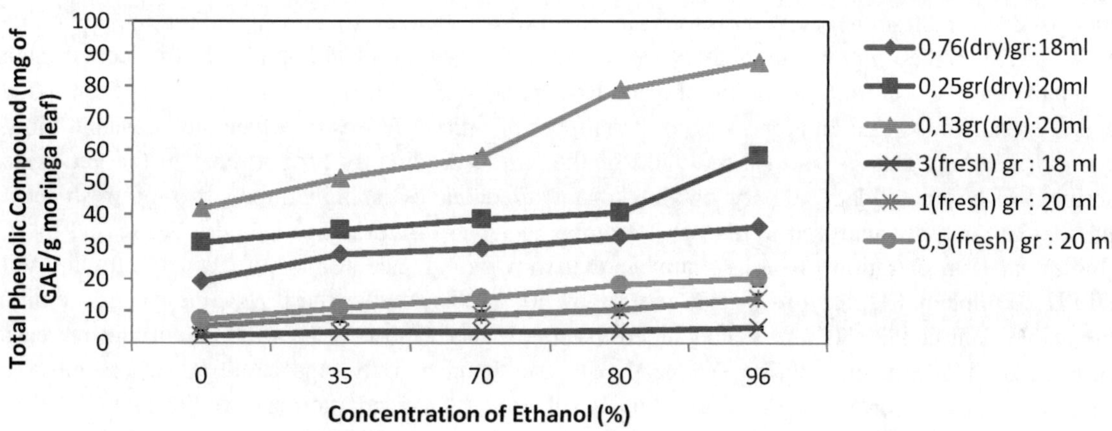

Figure 2. Effect of concentration of ethanol solution at different ratio of fresh/dried leaves and solvent on the TPC (Total Phenolic Compound)

3.3 Effect of concentration of ethanol solution at different ratio of fresh/dried leaves and solvent on the antioxidant activity

Antioxidants are any substances that when present at low concentrations versus oxidized substrate significantly delay/prevent substrate oxidation. Activity is measured by counting the amount of DPPH(2,2-diphenyl-1-picryl-hydrazyl) light intensity reduction equivalent to DPPH concentration reduction. Dyeing is produced by reacting the DPPH(2,2-diphenyl-1-picryl-hydrazyl) molecule with the hydrogen atom released by the component molecule of the sample to form the DPPH (2,2-diphenyl-1-picryl-hydrazyl) compound and causing the DPPH (2,2-diphenyl-1-picryl-hydrazyl) color decay from purple to yellow [16]. Effect of concentration of ethanol solution at different ratio of fresh/dried leaves and solvent on the antioxidant activity was shown in figure 3.

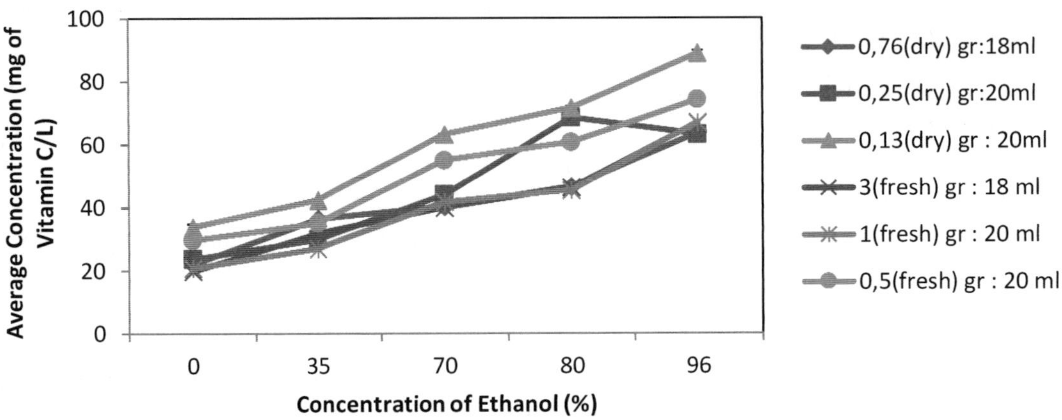

Figure 3. Effect of concentration of ethanol solution at different ratio of fresh/dried leaves and solvent on the antioxidant activity

Increased concentration of ethanol solution and ratio of leaves to solvent affect the antioxidant activity. The highest antioxidant activity (88.75 ± 0.93 mg vitamin C/L) obtained using concentration of ethanol solution of 96 % and ratio of dried leaves and solvent of 0.1273 g: 20 mL. The antioxidant activity tends to increase in proportion to the number of ratio of leaves to solvent and concentration of ethanol solution. This is because the yield of the extract is directly proportional to the antioxidant activity. The antioxidant activity obtained from dried leaves is higher than that of fresh leaves. Antioxidant activity analyzed by DPPH(2,2-diphenyl-1-picryl-hydrazyl) method is expressed by IC_{50} (Inhibitory Concentration). IC_{50} is a number that shows the concentration of extract that can inhibit DPPH(2,2-diphenyl-1-picryl-hydrazyl) activity by 50 %. The smaller the IC_{50} value means the higher the antioxidant activity. It can be concluded that the higher ratio of leaves to solvent and increasing concentration of ethanol solution enhanced the antioxidant activity, and resulting in less inhibition percentage. IC_{50} for dried leaves extracted using subcritical water-ethanol mixture (96 %) at 200 °C for 15 min are 181.45, 119.34, and 106.48 for ratio of dried leaves of 0.7635 g: 18 mL, 0.2545 g: 20 mL, and 0.1273 g: 20 mL, respectively. While IC_{50} for fresh leaves extracted using subcritical water-ethanol mixture (96 %) at 200 °C for 15 min are 166.91, 165.40, and 136.09 87 for ratio of fresh leaves of 03 g: 18 mL, 1 g: 20 mL, and 0.5 g: 20 mL, respectively.

3.4 Effect of concentration of ethanol solution at different ratio of fresh/dried leaves and solvent on the flavonoid content

Flavonoids are defenilpropane which is widely found in plants. The immediate family members of flavonoids include flavones, isoflavones, and the 2,3-dihydroderivatives of flavones, which are flavanones, it interconvertible with the isomeric chalcones. Flavanones undergo a series of transformations affecting the heterocyclic C ring to give rise to other family members of flavonoids, including anthocyanins and catechins. Some flavonoids have several fungi as antilipoperoxidant, antitumoral, antiplatelet, antiischemic, anti-allergic, and anti-inflammatory activities. There are also useful enzymes, such as lipoxygenase, cyclooxygenase, monooxygenase, xanthine oxidase, mitochondrial succinoxidase and NADH-oxidase, phospholipase A, and protein kinases. This biological function is a form of antioxidant property [17]. The principle method used for the determination of flavonoid content is when aluminum chloride forms stable acid complex with carbonyl group at C4 and hydroxyl at C3 (flavonol) and C5 in flavonols and flavones, in addition to forming a labile acid complex with hydroxyl at ortho position in ring A or B flavonoids [4]. Total flavonoids extracted with subcritical water had lower levels and shown by figure 4.

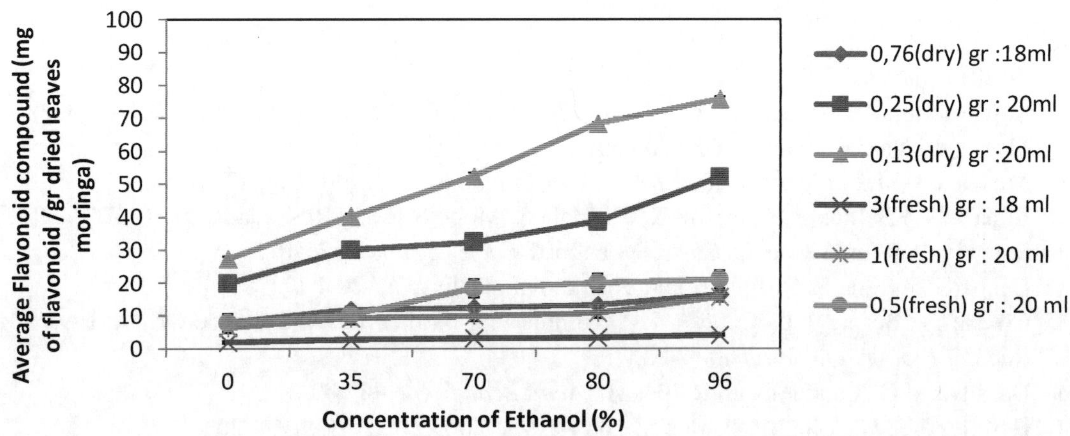

Figure 4. Effect of concentration of ethanol solution at different ratio of fresh/dried leaves and solvent on the flavonoid content

Figure 4 shows that by using higher concentration of ethanol solution and higher ratio of leaves to solvent extracted higher flavonoid content. The highest flavonoid content (75.79 ± 0.73 mg quercetin/ g of dried leaves) was found using dried leaves with ratio of leaves to solvent of 0.1273 g: 20 mL.

4. Conclusions

In this work the yield of extract, total phenolic compounds, antioxidant activity and flavonoid content of Moringa oleifera leaves was extracted using subcritical water-ethanol mixture. The highest yield of extract, total phenolic compounds, antioxidant activity and flavonoid content of 32.73 ± 0.08 %, 87.11 ± 0.81 mg GAE / g of dried leaves, 88.75 ± 0.93 mg of vitamin C/L with IC_{50} value of 119.34 μg / mL, and 75.79 ± 0.73 mg quercetin / g of dried leaves, respectively was obtained using concentration of ethanol solution 96 % and ratio of dried leaves to solvent of 0.1273 g: 20 mL. The yield of extract from fresh leaves enhanced by 37.71 % by increasing concentration of ethanol solution from 0 % (solvent is water) to concentration of ethanol solution of 96 %, while using dried leaves the yield of extract enhanced by 60.50 %. From this data we can conclude that the utilization of ethanol solution as a solvent on the subcritical condition was able to enhance the amount of the yield of extract, total phenolic compounds, antioxidant activity as well as flavonoid content.

Acknowledgments

This research was financially supported by Ministry of Research, Technology and Higher Education and Institut Teknologi Sepuluh Nopember (ITS).

References

[1] Zhang J, Barbara Wider, Hongcai Shang, Xuemei Li and Edzard Ernst 2012 *Complementary Therapis in Medicine* **20** 100

[2] Jongrungruangchok S, Bunrathep S and Songsak T 2010 *J Health Res* **24** 123
[3] Anwar F, Latif S, Ashraf M and Gilani A H 2006 *Phytother* **21** 17
[4] Makkar H P S and Becker K 1996 *Animal Feed Science Technology* **63** 211
[5] Falowo B A, Mukumbo F E, Idamokoro E M, Lorenzo J M, Afolayan A J and Muchenje V 2018 *Food Research International* **106** 317
[6] Vongsak B, Sithisarn P, Mangmool S, Thongpraditchote S, Wongkrajang Y and Gritsanapan W 2013 *Industrial Crops and Products* **44** 566
[7] Kamaruddin M J, Abdul Hamid S R, Ali Othman S I, Hisham Zainal Alam M N, Ahmad Zaini M A and Zakaria Z Y 2018 *CET* **63** 601
[8] Mc Cune L M and Johns T 2002 *Journal of Ethnopharmacology* **82** 197
[9] Elma S N, Khalilah B, Dzulfadli R, Shamala S, Mohd S H and Roshada H 2018 *CET* **63** 817
[10] Soto Ayala R and Luque de Castro M D 2001 *Food Chemistry* **75** 109
[11] Deepa G, Ayesha S, Nishtha K and Thankamani 2013 *IFRJ* **20** 4 1711
[12] Cotelle N, Bernie J L, Catteau J P, Pommery J, Wallet J-C and Gaydou E M 1995 *Free Radical Biology and Medicine* **20** 35
[13] Da Silva M C A and Paiva S R 2012 *An Acad Bras Ciènc* **84** 609
[14] Herodez S S, Hadolin M, Skerget M and Knez Z 2003 *Food chemistry* **80** 275
[15] Pinelo M, Rubilar M, Jerez M, Sineiro J and J.N Maria 2005 *J Agric Food Chem* **53** 2111
[16] E D Daffodil, V R Mohan and P Jegadeeswari 2014 *Int J Pharm Sci* **6** 747
[17] G Cao, Sofic E and Prior R. L 1997 *Free Radical Biology and medicine* **22** 749

The Influence of Strong Reduced Water and Fresh Coconut Oil in Viscosity of Liquid Hand Soap

E Sulistiawati[1], E Astuti[1], I Santosa[1]

[1] Departement of Chemical Engineering, Faculty of Industrial Tehnology, Universitas Ahmad Dahlan, UAD Campus 4, Jl. Kolektor Ringroad Selatan, Tamanan, Banguntapan, Bantul, Yogyakarta, Indonesia.

Email: endahsulistiawati@che.uad.ac.id

Abstract. Innovation of liquid hand soap continues to be done, among others, using natural local raw materials. This research used strong reduced water (SRW), an alkali produced from a water ionizer machine (with enhancer Sodium Chloride solution), which is eco-friendly and comfortable on the skin. The study also used fresh coconut oil (FCO), was produced by drying process and zero waste. The objective of this research was to determine the effect of utilization of SRW and FCO in the making process of liquid hand soap, which gave good viscosity. Fresh coconut oil (FCO) was reacted with SRW (pH 11.5) for 30 minutes at room temperature (30 ºC), then mixed with sodium lauryl sulfate, sodium sulfate, other additives such as dyes, fragrances, electrolyzed oxydized water (EOW, pH 2.5), Ethylenediaminetetraacetic acid (EDTA) and distilled water. EOW was produced from the ionizer machine too. After saponification process was completed (about 24 hours), the viscosity and density of the liquid soap was measured. The better composition of FCO-SRW yielding good viscosity of liquid soap was 30 to 50 ml based on 500 ml distilled water production. The viscosities of this product were 982.55 to 1,782.25 centipoises, density of 1.0049-1.1175 g/ml, and pH 7. The using of reaction product of fresh coconut oil (FCO) and strong reduced water (SRW) can increase the viscosity of liquid hand soap.

1. Introduction

Soap is produced by saponification process. The forms of soap can varyies, among others: solid soap, soft soap, liquid soap, and powder soap. If an alkali used is sodium hydroxide, then the result is a solid soap. If potassium hydroxide is used, the result is soft soap. Sodium hydroxide is not eco friendly, and not comfortable on the skin. Sodium hydroxide of 10 % concentration was used to produce detergents [1]. This study used alkali from a water ionization machine, instead of sodium or potassium hydroxide, and applied to the manufacture of liquid hand soap.

In this study, an alkali used was not NaOH or KOH, but it used Strong Reduced Water (SRW). Raw water (pH.7) was fed into the ionizer machine, with the help of enhancers (solution which contains Sodium chloride salt); it produced SRW (pH 11.5) and Electrolyzed Oxydized Water (EOW, pH.2.5). The SRW can remove fat and dirt on the equipment. The EOW is different from hydrochloric acid (HCl) or sulfuric acid (H_2SO_4), which is not corrosive to skin and mucous membrane. The SRW and EOW is environmentally friendly and relatively inexpensive. The EOW has been tested and used as a desinfectant in food industry and other applications [2]. It was studied the disinfection efficacy of the

IOP Conf. Series: Materials Science and Engineering **543** (2019) 012022 doi:10.1088/1757-899X/543/1/012022

combination of strongly acidic electrolyzed water and alkaline electrolyzed water on fresh-cut cilantro [3]. Electrochemically activated aqueous solutions was used in the manufacture of fur materials. The benefit of the water used was safe, the process was accelerated three fold then soaked in chemicals such as formalin and silicofluoride [4].

One of the vegetable oils can be used as raw materials for making soap is coconut oil. Several studies on the manufacture of coconut oil and its processing have been carried out. Coconut oil can be made in a wet or dry process, which dry process saves more energy. In this process, Fresh Coconut Oil (FCO) was produced from dry process using a batch modified screw mixed dryer. The dryer was operated at temperature of 80 °C to 100 °C [5].The Virgin Coconut Oil (VCO) extract contains phytochemicals such as phenolic compounds and also shows high antioxidants activity that might induce fibroblast cell growth and proliferation to enhance the wound closure activity. The optimum VCO concentration obtained in this study was 1.0 mg/mL to give the highest wound closure about 56.89 ± 2.35 % after 24 h [6].

The rheological properties may deviate from the nominal ones, because of process disturbances, and they can be adjusted by varying the ingredients flow rates which have a major impact without affecting excessively detergent characteristics. A point control for the viscosity curve of the non-Newtonian detergent is implemented, where the objective is to maintain the viscosity curve inside a specific range of values which represent the target [7]. It was simulated the mixing or demixing process of a quiescent binary mixture with a composition-dependent viscosity which is instantaneously brought from the two-phase (resp. one-phase) to the one-phase (resp. twophase) region of its phase diagram. The theoretical approach follows a standard diffuse-interface model of partially miscible regular binary mixtures wherein convection and diffusion are coupled via a non equilibrium capillary force, expressing the tendency of the phase-separating system to minimize its free energy. For a phase-separating system, at a fixed value of the fluidity coefficient (with the continuous phase viscosity taken as a reference), the separation depth and the characteristic length scale of single-phase microdomains increase monotonically for increasing values of viscosity ratio. However, for a mixing system the attainment of a single-phase equilibrium state by coalescence and diffusion is retarded by an increase in viscosity ratio at a fixed fluidity for the dispersed phase [8]. Viscosity is one of quality component of liquid soap. Indonesian National Standards (SNI) sets a minimum 500 centipoises dan a maximum 20,000 centipoises. The purpose of this study was to determine the effect of utilization of fresh coconut oil (FCO) and strong reduced water (SRW) on physical properties, especially the viscosity of liquid hand soap.

2. Methods

2.1. Materials

Strong reduced water (SRW or strong alkaline electrolyzed water), pH 11.5, was produced from mineral water (drinking water grade, from local, Kaliurang,Yogyakarta, Indonesia) and enhancer solution (sodium chloride solution, Enagic Co., Japan) through water ionizer machine (Leveluk SD501, Enagic Co., Japan), as shown in Figure 1. SRW contained sodium 3493.45 mg/liter. Fresh coconut oil, was produced from fresh coconut meat (from local market Yogyakarta, Indonesia). The fresh coconut meat was grated, dried at 80-100 °C using drying machine (modified screw batch drier) until zero water content, then pressed. It was zero waste process, because the residue can be processed as coconut fluor. Sodium lauryl ether sulfate (technical), sodium sulfate (technical), cocoamide di ethanol amine (technical), citric acid (technical), Ethylene diamine tetra acetic acid (EDTA, technical), fragrance, propylene glycol (PG, technical), dyes (cosmetics grade), and distilled water. All of those materials were purchased at local market, Yogyakarta, Indonesia. It also used electrolyzed oxidized water (EOW, pH 2.5) contained of sodium (Na) 747.29 mg/liter for substituting of citric acid.

Figure 1. Water ionizer machine for producing strong reduced water and strong electrolyzed oxidized water.

2.2 Methodology

2.2.1 Reaction of Fresh coconut oil and Strong reduced water

Fresh coconut oil (FCO) 100 ml and strong reduced water (SRW) 300 ml were mixed in a basin. The reaction was carried out for 1 hour, stirring speed 1000 rpm, and reaction temperature of 50 °C. After that, the product of FCO-SRW was placed into the separation funnel, waited for separating. The bottom layer (380 ml) was used for manufacturing of liquid hand soap, and had viscosity of 0.56 centipoises.

2.2.2 Liquid handsoap production

Experiments of liquid hand soap producing have been conducted in the laboratory of Chemical Engineering Departement, Faculty of Industrial Technology, Universitas Ahmad Dahlan, Yogyakarta, Indonesia. General or standard formula was applied as controls. Sodium lauryl sulfate 200 g, sodium sulfate 150 g, cocoamide diethanol amine 50 ml, citric acid 50 g, EDTA 10 g, fragrance 10 ml, propylene glycol 10 ml, dye (cosmetics grade) 8 drops, and distilled water 1 liter. The mixture was poured into bottles (as shown in Figure 2) and waited for 24 hours. After saponification process had been completed, the hand soap was measured its viscosity and density.

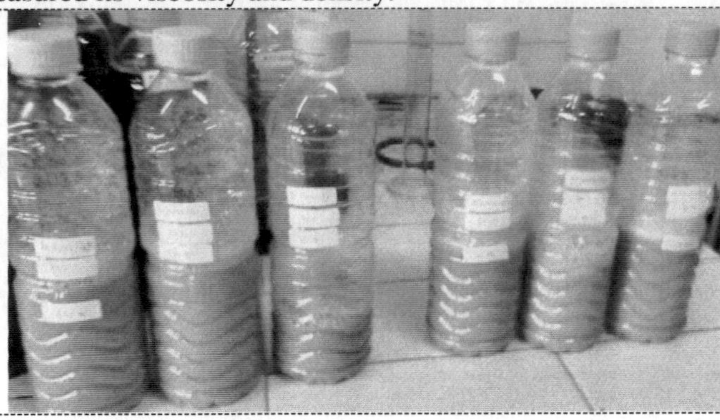

Figure 2. Liquid hand soap before saponification process completed

2.2.3 Density and viscosity

The density of the sample was determined by using a 10 ml pycnometer at room temperature.

Viscosity: Ostwald viscometer was used at room temperature of 30 °C, the viscosity was computed by equation (1).

$$\mu_s = \frac{\rho_s . t_s}{\rho_{aq} t_{aq}} . \mu_{aq} \quad (1)$$

where
μ_s is the viscosity of soap, centipoises
μ_{aq} is the viscosity of distilled water, centipoises
ρ_s is the density of soap, g/ml
ρ_{aq} is the density of distilled water, g/ml
t_s is the time required of soap in Ostwald viscometer, minutes
t_{aq} is the time required of distilled water in Ostwad viscometer, minutes.

3. Results and discussions

It was studied to substitute cocoamide diethanol amine by the reaction result of commercial coconut oil (CO) and strong reduced water (SRW). The viscosity and density of liquid hand soap resulted from CO-SRW are shown in Table 1.

Table 1. Effect of CO-SRW on viscosity and density of liquid hand soap (distilled water 500 ml, citric acid 22.56 g)

CO-SRW, ml	Viscosity, centipoises	Density, g/ml
25	94.59	1.1757
50	95.54	1.1056
S	543.25	1.0172
Y	885.09	1.0115

After the liquid soap making was completed, the average viscosity and density was measured, 95.06 centipoises and 1.1407 g/ml, respectively. National standard of Indonesia sets 500 to 20,000 centipoises for viscosity, and 1.010 to 1.100 g/ml for density. So, the results of hand soap for standard formula was not accordance with national standard. It was also measured the viscosity and density of commercial liquid hand soap S and Y (from local market at Yogyakarta, Indonesia). The S branded had viscosity 543.25 centipoises and density 1.0172 g/ml. The Y branded had viscosity 885.09 centipoises and density 1.0115 g/ml.

The fresh coconut oil and strong reduced water with the same amount was mixed, and the result of hand soap can be seen in Figure 3. It can be seen from Figure 3 that the using of fresh coconut oil increased sharply the liquid hand soap viscosity. This can be happened because FCO is more soluble in SRW compared to commercial coconut oil, so the mixture is more homogeneous, and the reaction is more completed. The viscosity of reaction product of CO-SRW mixture was 11.23 centipoises, whereas the FCO-SRW mixture was 0.56 centipoises.

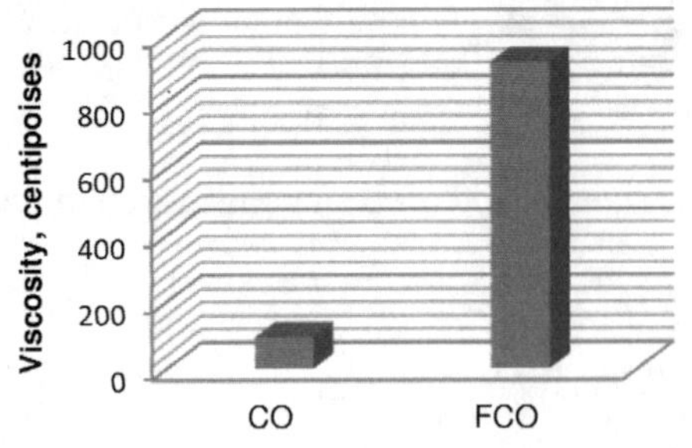

Figure 3. Comparison of CO and FCO on viscosity and density of liquid hand washing soap

IOP Conf. Series: Materials Science and Engineering **543** (2019) 012022 doi:10.1088/1757-899X/543/1/012022

The composition of FCO-SRW was made varies from 10 ml to 70 mililiters, which the physical properties and pH of liquid hand soap can be seen in the Table 2.

Table 2. Effect of FCO-SRW on viscosity, density, and pH of liquid hand soap (distilled water 500 ml, EOW = 50 ml)

FCO-SRW, ml	Viscosity, centipoises	Density, g/ml	pH
10	748.83	1,127.0	7.0
20	606.19	1,121.1	7.0
30	982.55	1,117.5	7.0
40	1,001.62	1,109.6	7.0
50	1,075.90	1,107.3	7.0
60	466.44	1,121.4	7.0
70	467.23	1,123.3	7.0

In Table 2, the composition FCO-SRW 10 ml to 50 ml gave the viscosity of liquid soap was above 500 centipoises, but at 60 ml to 70 ml the viscosity dropped under 500 centipoises. It could be caused of the excess of FCO-SRW, which decreased the mixture viscosity. The individual viscosity of FCO-SRW was 0.56 centipoises. After six months of storage, the fresh coconut oil was tested to make liquid hand soap. The composition of FCO-SRW was made varies from 10 ml to 90 mililiters, which the physical properties and pH of liquid hand soap can be seen in the Table 3. From the Table 2 it shows that the highest viscosity was achieved for fresh at 50 ml (1,075.90 centipoises), and after 6 months storage at FCO-SRW 30 ml (1782.25 centipoises). The density of liquid soap at this composition was also close to the national standard of Indonesin Government (SNI) 1.0049 g/ml. National standard for liquid soap density is 1.010 to 1.100 g/ml. So, fresh coconut oil can be stored for several months and can be used to produce the relative same quality of liquid hand soap.For the composition FCO-SRW 10 ml to 40 ml (Table 3) the viscosity of liquid soap was above 1,000 centipoises, but for 50 ml to 90 ml the viscosity decreased sharply under 500 centipoises.

Table 3. Effect of FCO-SRW (after FCO 6 months storage) on viscosity, density, and pH of liquid hand soap (distilled water 500 ml, EOW = 50 ml)

FCO-SRW, ml	Viscosity, centipoises	Density, g/ml	pH
10	1143.31	1.1178	7.0
20	1192.52	1.1071	7.0
30	1782.25	1.0049	7.0
40	1391.08	1.1095	7.0
50	439.31	1.1099	7.0
60	285.41	1.0957	7.0
70	194.26	1.1027	7.0
80	60.23	1.1063	7.0
90	50.35	1.1065	7.0

The liquid soap is Non newtonian fluid, is made from the mixtures of vegetable oil, alkali, surfactant (sodium lauryl sulfate), sodium sulfate, and others. The composition of FCO-SRW must be appropriate, 10 ml to 40 ml, in order to obtain a good viscosity (above 1,000 centipoises). The addition of 50 ml and more made the viscosity of the mixture decreased, it could be caused of excess FCO-SRW. The optimum

composition makes the mixture can be reacted completely, and the reaction of saponification is also completed. It was reported the viscosity of the mixture was up three times larger than that of pure liquids, with the maximum viscosity corresponding to either a 50 %-50 % (type 1 mixture), or a 25 %-75 % composition (type 2 mixture), to better emulate the behavior of real mixtures [9]. However, in this case, it must be found a certain composition to make a good formula of liquid hand soap.
The next investigated step was to vary the amount of electrolyzed oxydized water (EOW), to substitute the citric acid, at 25 ml FCO-SRW (Table 4). The viscosity of liquid soap was fluctuated, at 100 ml EOW the viscosity of liquid soap was maximum, but at 20 ml was minimum (164.82 centipoises). It need to be studied further about this cause. The density was vary 1.1246 to 1.1620 g/ml. The pH of liquid soap for 100 ml EOW was 6.5.

Table 4. Fresh coconut oil, the effect of electrolyzed oxydized water (EOW, pH 2.5) on viscosity, density, and pH of liquid hand soap (distilled water 500 ml, FCO-SRW = 25 ml)

EOW, ml	Viscosity, centipoises	Density, g/ml	pH
10	574.08	1.1620	7.0
20	164.82	1.1271	7.0
30	742.18	1.1533	7.0
40	761.33	1.1270	7.0
50	1157.43	1.1246	7.0
100	1404.26	1.1256	6.5

The FCO after 6 months of storage was also evaluated to produce liquid hand soap (Table 5). For the certain FCO-SRW 30 ml, the EOW was varied 10-100 ml, so it can influences the pH of liquid soap, as shown in Table 5. At the EOW 10 to 50 ml, the viscosity was above 1000 centipoises, but at 60 to 100 ml was fluctuated. The pH of liquid soap decreased for EOW 70 ml to 100 ml, the soap was acid (pH = 6.5), because the excess of EOW. The mixture of acid and alkali could be neutral, acid or base, depends on the amount of each compound. The Kay's mixing rule is a quick method for estimating these physical properties, since it takes into account just the system composition and the physical properties of the pure compounds [10].

Table 5. Effect of electrolyzed oxydized water (EOW, pH 2.5) on viscosity, density, and pH of liquid hand soap (distilled water 500 ml, FCO-SRW (after 6 months storage) = 30 ml)

EOW, ml	Viscosity, centipoises	Density, g/ml	pH
10	1668.27	1.1067	7.0
20	2335.46	1.1030	7.0
30	1321.98	1.0147	7.0
40	1656.63	1.0977	7.0
50	1682.23	1.0049	7.0
60	454.63	1.0873	7.0
70	386.06	1.0947	6.5
80	631.57	1.1023	6.5
90	1441.10	1.1107	6.5
100	78.32	1.0947	6.5

4. Conclusion

The utilization of reaction product of fresh coconut oil (FCO) and strong reduced water (SRW) can improve the viscosity of liquid hand soap. The FCO after six months storage can be applied to make the same quality of liquid hand soap. The better composition of FCO-SRW yielding good viscosity of liquid soap was 30 to 50 ml, and electrolyzed oxidized water (EOW) 30-50 ml based on 500 ml distilled water production. The viscosities of this product were 982.55 to 1,782.25 centipoises, density of 1.0049-1.1175 g/ml, and pH 7. The using of reaction product of fresh coconut oil (FCO) and strong reduced water (SRW) can increase the viscosity of liquid hand soap.

Acknowledgement
The authors are grateful to Universitas Ahmad Dahlan for the financial support, and also M. Tamrin, Eko Susilowati, Donianto, M. Solihin Akbar, Mita Prastika, and Asri Dwi Sakinah as technical staffs.

References

[1] Taris A, Grosso M, Zonfrilli F, Guida V 2015 *CET* **43** 1549-1554.
[2] Huang Y, Hung Y, Hsu S, Huang Y, Hwang D 2008 *Food Control* **19(4)** 329-345.
[3] Hao J, Li H, Wan Y, Liu H 2015 *Food Control* **50** 699-704.
[4] Danylkovych AG, Lishchuk VI, Romaniuk OO 2016 *Springer Plus* **5** 214.
[5] Sulistiawati E, Santosa I, Aktawan A 2018 Mater. Sci. Eng.: IOP Conf. Ser.: **403** 012012.
[6] Ahmad Z, Sarmidi M R, Hasham R 2017 *CET* **56** 1657-1662.
[7] Mei R, Gosso M, Tronci S, Baratti R, Corominas F 2017 *CET* **57** 1099-1104.
[8] Lamorgese A, Mauri R, 2017 *CET* **57** 1225-1230.
[9] Galletti C, Brunazzi E, Mauri R *CET* **43** 1645-1650.
[10] Florido PM, Lobo DPS, Pinto CN, Rodrigues CEC, Gonçalves CB 2017 *J. Chem. Thermodynamics* **113** 198–212.

IOP Conf. Series: Materials Science and Engineering **543** (2019) 012023 doi:10.1088/1757-899X/543/1/012023

Heat Transfer Evaluation on Microwave Sterilization: Case Study of Oil Palm Fruit

M Sarah[1,2]

[1]Departement of Chemical Engineering, Universitas Sumatera Utara, Padang Bulan Medan, Indonesia
[2]Sustainable Energy and Biomaterial Centre of Excellent, Universitas Sumatera Utara

Email: mayasharid@yahoo.com or Maya3@usu.ac.id

Abstract. Microwave energy can be used for oil palm fruit bunch sterilization. It penetrates onto the core of oil palm fruit bunches and generates heat inside out. This sterilization aims to inactivate lipase that influences palm oil quality. This study evaluates effect of dielectric properties power developed per unit volume and interior temperature distribution on heating rate during heating process. The experiment carried out at various combinations of microwave power and the oil palm bunches portion in microwave oven. This study observed that distribution of interior temperature in a single oil palm fruit was influenced by power density and fruit's diameter. Greater power density provides high microwave energy per kg sample to attain greater temperature of microwave heating and vice versa. Distribution of interior temperature in the fruit with the thickness (r) of 6 mm and above indicated non-uniformity heating. Furthermore, interior temperature of fruitlet with r = 4.5 mm indicated distribution of temperature from endocarp to exocarp, meanwhile simulation temperature on smaller fruit diameter (r = 3 mm) indicated thermal runway effect after 5 min heating duration.

1. Introduction

Every palm fruit bruises during harvesting and transportation to palm oil milling facilitate the release of free fatty acid by lipase with the presence of water. Sterilization is compulsory in the mill to protect palm oil quality through inactivation of 90 % lipase activity at certain time and temperature. Fundamentally, lipase activity can be reduced through heating process at temperature above 50 ^{o}C. One of the heating method for oil palm sterilization is microwave irradiation. This method has several advantages because it can either carried out very fast at high temperature or proceeded slower at lower temperature. Utilization of microwave energy to enhance chemical processes such as sterilization, extraction and demetallization reported by several authors elsewhere [1, 2]. Opportunity to utilize microwave energy for heating palm fruit firstly introduced by Tan (1980) who investigated dielectric properties and moisture loss of fresh oil palm fruit during heating process [3]. Ever since many studies reported utilization of microwave energy for oil palm fruit sterilization with respect to stripping efficiency after sterilization [4], crucial time and temperature for sterilization [5] and quality of extracted oil [6, 7]. Over all studies on heating oil palm fruit with microwave energy can be classified into three parts: (1) the study of the opportunities for the use of microwave energy to heat the oil palm fruit, (2) the study of the quality of palm oil after heating with microwave energy, and (3) the study of the ability of microwave heating to facilitate the release of oil palm fruit from their stems. All those previous studies reported potency of microwave irradiation to sterilize palm fruit at laboratory scale. Sarah (2018) also reported microwave irradiation offered various modification on mode of operation to obtain palm oil with high carotenoids content (Red Palm Oil) or on the contrary produce palm oil without any carotenoids [8]. Conventional palm oil milling produces crude palm oil that refined further to remove as much as carotenoids content in palm oil to satisfy costumer. In facts carotenoids is very important to human health because it contains β-carotene that acts as pro vitamin A in human living. The fact β-carotene destroyed during sterilization or heating process due to high temperature or long period of exposure indicates sterilization time and temperature are very crucial to oil palm sterilization by microwave irradiation. Both indicates significant heat transfer process occurs during sterilization from microwave energy as a source of energy into thermal energy.

This study evaluates mechanism and effect of microwave energy transformation into thermal energy at microscopic level. Heat generation occur prior of heat transfer as electromagnetic waves applied into oil palm fruit sample. Interaction between microwave, polar water molecules and charged ions in oil palm fruit causes water molecules in oil palm fruit to constantly rotate and couple with electromagnetic field. Molecular friction resulting from dipolar rotation of water molecules generates heat and resulting rise on fruit temperature in kernel or endocarp. Furthermore, heat transfer occurs when heat flows from kernel or endocarp into exocarp of oil palm fruit.

Heat generation in the concept of microwave heating is indicated by an increase in temperature of the material per unit of time, and heat transfer from the interior of material into the exterior or surface. This is indicated by the difference in temperature of the interior and exterior of material, where the interior temperature is higher than the temperature of the exterior. At time the microwave energy is absorbed by dielectric material, rate of temperature increases depends upon number of distinct parameters, such as frequency, dielectric loss factor and electric field. The rate of heating of material expressed by power equation [9, 10, 11]. Energy produced per unit volume of the microwave heating of oil palm fruit can be estimated using Eq. (1) where $\boldsymbol{P_v}$ represent the energy developed per unit volume (Wm^{-3}), f is frequency (Hz), ε_o is vacuum permitivity ($F\ m^{-1}$), E is electric field strength inside the load ($V\ m^{-1}$) and $\boldsymbol{\epsilon}''$ is dielectric loss factor.

$$\boldsymbol{P_v = 2\pi f \varepsilon_o E^2 \epsilon''} \quad (1)$$

Energy transfer with respect to Eq. (1) is highly dependent on the dielectric properties of oil palm fruit and frequency of the microwave oven. Meanwhile, heat transfer in material during microwave heating occurs from the interior, where generation of heat took place, into the exterior, by dissipating heat. Heat transfer equation can be used to evaluate distribution of interior temperature during heating process with power absorbed as source term in heat transfer equation [12, 13, 14, 15]. Eq. (2) can be used to estimate the interior temperature distribution within a single oil palm fruit where k is the thermal conductivity ($W\ m^{-1}K^{-1}$), ρ is density ($kg\ m^{-3}$), c_p is specific heat capacity ($J\ kg^{-1}\ K^{-1}$), T is the temperature (K), t is the time (s) and $\boldsymbol{P_v}$ is energy developed per unit volume ($W\ m^{-3}$).

$$\boldsymbol{\rho c_p \frac{\partial T}{\partial t} = k\nabla^2 T + P_v(x, y, z, t)} \quad (2)$$

Eq.(2) solved numerically with respect to boundary as given in Eq.(3) [12, 13, 14, 15] where h_c is convection heat transfer coefficient over a surface and T_a is temperature of air inside the microwave oven (°C). The initial temperature of air and fruit inside the microwave oven was assumed constant at 25 °C.

$$\boldsymbol{-k\nabla T = h_c(T - T_a)} \quad (3)$$

2. Materials and Methods

2.1. Materials

Materials used in this study is oil palm fruitlets *Tenera* variety taken from oil palm tree around the campus. The fruitlets sorted with respect to their size (big, medium and small sizes) and were put in the dry environment. This study utilized domestic microwave oven (Sharp R-958A series) with frequency of 2450 MHz and maximum allowance power of 800 W. Sample of palm fruit at various mass (0.5 kg to 1.5 kg) exposed by microwave energy at various power level (medium, medium high, high respectively). Data logger (Pico Temperature Data Logger PT 104) was installed to measure temperature of fruit during irradiation period. Dielectric properties of sterilized palm fruit were measured utilizing dielectric probe (8710-2038) and computer controlled (ENA Series Network Analyzer Agilent Technologies), at frequency range of 300 kHz to 20 GHz. Dielectric properties obtained was then used to evaluate heat transfer in a fruitlet during irradiation period.

2.2. Methods

This study evaluated heat transfer with respect to Eq (2). Several authors reported their simulation work using similar equation [12, 13, 14, 15]. In this study, Eq (2) was modified into Eq (4). For one dimensional body, partial diffential equation from Eq. (2):

$$\rho c_p \frac{\partial T}{\partial t} = k\nabla^2 T + P(r, y, z) \tag{4}$$

Rearrangement of Eq. (4):

$$\frac{\partial T}{\partial t} = \left(\frac{k}{\rho c_p}\right)\left(\frac{\partial^2 T}{\partial r^2}\right) + \frac{P}{\rho c_p} \tag{5}$$

Finite Difference Method (semi discretization, central different) was used to solve Eq. (5). With respect to Eq. (3) boundary for Eq. (5) is:

$$-k\frac{\partial T}{\partial r} = h_c(T - T_{a)} \tag{6}$$

Power from microwave that absorbed by dielectric of oil palm fruits per unit volume calculated using Eq. (1). while electric field could be estimated numericaly by using Eq (7) [16].

$$E_z = \frac{1}{2\pi}\frac{1}{\ln(b/a)}\int_a^b\int_0^\pi \cos\phi' \frac{\exp(-jkR')}{R'} . \left[\frac{1}{\rho} - \left(jk' + \frac{1}{R'}\right)\frac{(\rho - \rho' \cos\phi'}{R'}\right] d\phi' d\rho' \tag{7}$$

Term R' and k' from Eq. (7) are: $R' = [(z - z')^2 + \rho^2 + \rho'^2 - 2\rho\rho' \cos\phi']^{\frac{1}{2}}$ and $k' = \frac{2\pi f}{c}\sqrt{(\epsilon_r' - \epsilon_r''}$ respectively.Temperature distribution inside mesocarp was modeled into radial direction at various combination of mass and power. The endocarp was represented by r=0 while exocarp depends on thickness of mesocarp. inner mesocarp. Heat transfer within the fruitlets was evaluated using Matlab software R12a series.

3. Results and Discussions

3.1 Potency of palm fruit properties for heat transfer process during irradiation

Dielectric properties of oil palm fruit figure out the potency of fruit to be heated or irradiated by microwave energy. Dielectric constant shows the ability of palm fruit absorbs the microwave energy while dielectric loss factor informs conversion of microwave energy into thermal energy. Ratio between two properties become an indicator for heat transfer efficiency. Overall, the mean value of dielectric properties of fresh fruit and heated fruit from triplicate measurement is shown in Table 1.

Table 1. Dielectric properties, relative permitivity and loss tangent of various fruit categories

Fruit categories	Power level	Dielectric constant	Dielectric loss factor	Relative permitivity	Loss tangent
Fresh fruit	-	14.155±1.797	4.162±0.376	14.754<-16.385°	0.294±0.014
Heated fruit	Medium	15.896±1.938	6.030±0.580	17.001<-20.774°	0.379±0.027
	Medium High	10.523±0.316	3.010±0.216	10.945<-15.963°	0.369±0.042
	High	14.042±1.222	4.467±0.580	14.095<-4.974°	0.286±0.011

Dielectric constant always greater compared to dielectric loss factor. This study did not observe significant variability on dielectric properties of fresh ripe fruit and ripe fruit (of heated fruit) while dielectric constant and dielectric loss of heated fruit resulted from this study vary according to degree of ripeness of the fruit. For ripe fruit, the dielectric constants ranged between 10.285 to 17.662, while dielectric loss factors ranged between 2.811 to 7.156.

Dielectric constant and dielectric loss factor of medium high treatment was observed lower as compared to dielectric values of others treatment. Variability of dielectric properties of heated fruit in this study was due to fruit characteristic and did not influence by the power. Dielectric loss factor of the study ranged between $10^{-2}<\varepsilon''<5$. Value of dielectric loss factor meets the theoretical requirements for dielectric heating process. According to Metaxas and Meredith (1988), dielectric loss factor of less than 10^{-2} requires the strength of the very high electric field to ensure a reasonable level of temperature rise in the material. Meanwhile, dielectric loss factor greater than 5 might cause problems on the depth of penetration, because the power of the microwave absorption materials very high. Most of the electromagnetic energy is absorbed within a first few mm, leaving the internal parts slightly affected, thus the cause of non-uniformity of heating [17].

Ratio between dielectric loss factor and dielectric constant was expressed as loss tangent (tanδ), which explained potency of oil palm fruit to generate heat from microwave energy. Based on the loss tangent, only 29.4 % of the electromagnetic energy absorbed by the sample of oil palm fresh fruit has the potential to be converted into thermal energy. Meanwhile for sample ripe oil palm fruit, which has been heated at medium power level, approximately 37.9 % of the amount of electromagnetic energy that is absorbed by the oil palm fruit has been converted into thermal energy. Overall, palm fruits show the potency to be heated by microwave energy with respect to their dielectric properties values. Dielectric properties data obtained were used to estimate interior temperature distribution during heating as describes in the following discussion.

3.2 Effect of microwave power and sample thickness to interior temperature distribution

Heat transfer evaluation in this study carried out with respect to Eq. (5) to Eq. (7) and the effect of microwave power to heating process. Actualy investigation on interior temperature distribution in this study is very complex due to the heating process utilized 0.5 to 1.5 kg sample with irregular shape and not uniform in size. To conduct simulation, it was assumed oil palm fruit is spherical elongated in shaped which has 2 diameters: the equatorial diameter (d_e) and polar diameter (d_p) as shown in Figure 1. There were several fruitlets size in this study with d_e and d_p combination of 2.5 cm and 4 cm (fruit 1); 2.5 cm and 3 cm (fruit 2); 2 cm and 3 cm (fruit 3); and 1.5 cm and 3 cm (fruit 4) respectively. For kernel, the average kernel in this study has d_e ranged from 1 into 1.5 cm, and d_p ranged from 1.5 into 2.5 cm. Total thickness of mesocarp (for both sides of the kernel) ranged from 0.5 to 1 cm (in the equatorial direction) and 1 to 2 cm (in the polar directions). Simulation in this study was carried out in accordance with the thickness of mesocarp. Mesocarp thickness on each side of the mesocarp can be expressed as the radius of the mesocarp, each of 2.5 to 5 mm (equatorial direction) and 5 to 10 mm (polar direction). Based on that, the simulation of temperature distribution in the oil palm fruit is done in a radius of 3, 4.5 and 6 mm each represent fruit size from small, medium and big sizes respectively.

Figure 1. The size of a single oil palm fruit (fruitlet) in palm fruit sample in this study

Eq. (6) is solved numerically by derivative approximation, by using finite differences method with central difference approach at both time and space derivative at various position [18]. Scheme of single oil palm fruit for discretization process is shown in Figure 2.

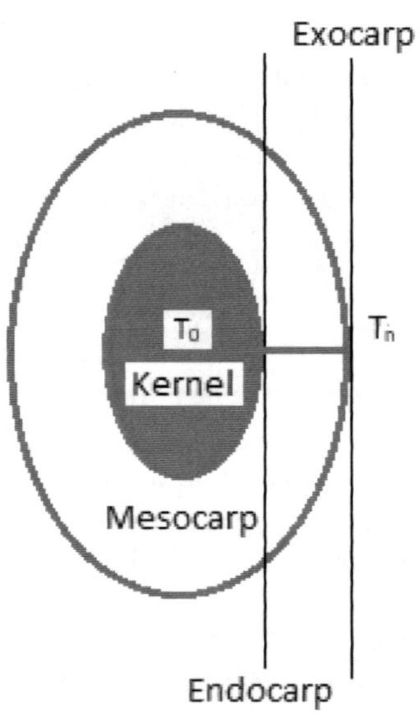

Figure 2. Scheme of single oil palm fruit

Semi discretization of transient heat transfer equation to evaluate interior temperature at endocarp, mesocarp, and exocarp are given in Eq. (8), Eq. (9) and Eq. (10) respectively.

Model for interior temperature distribution at endocarp:

$$\frac{dT_1}{dt} = \frac{2k}{\rho Cp}\left(\frac{T_2-T_1}{(\Delta r)^2}\right) + \frac{P}{\rho Cp} \quad (8)$$

Model for interior temperature distribution in the mesocarp if radial distances is divided into n section, thus for i = (n-1):

$$\frac{dT_{n-1}}{dt} = \frac{k}{\rho Cp}\left(\frac{T_n - 2T_{n-1} + T_{n-2}}{(\Delta r)^2}\right) + \frac{P}{\rho Cp} \quad (9)$$

Model for interior temperature distribution at exocarp:

$$\frac{dT_n}{dt} = \frac{2k}{\rho Cp}\left(\frac{T_{(n-1)} - T_n}{(\Delta r)^2}\right) + \frac{P}{\rho Cp} \quad (10)$$

The results in Figure 3 show temperature distribution of the model in mesocarp fits with temperature of mesocarp obtained from experimental data at a radial distance of 4.5 mm.

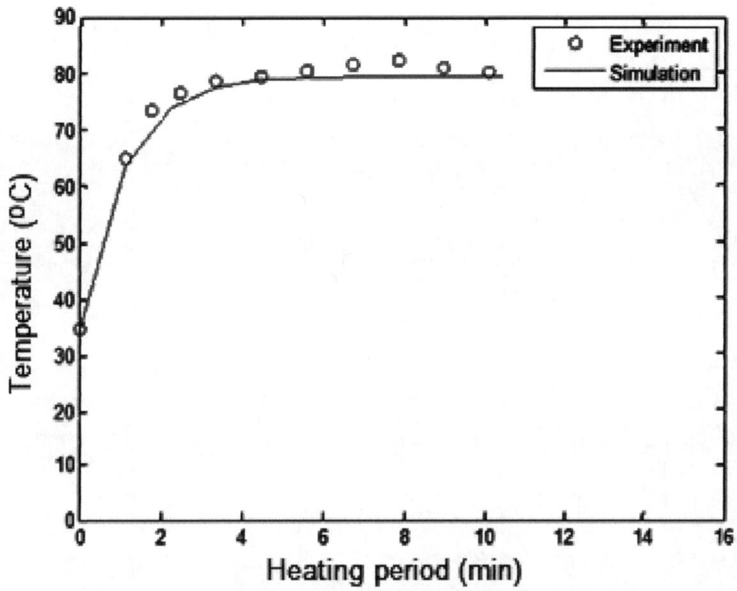

Figure 3. Curve fitting in this study at r=4.5 mm

Figure 4 indicates mathematic model of equation from this study could be used to estimate temperature distribution of oil palm fruit. Similar trend model of time-temperature distribution on microwave heating of ceramic material using different methods reported by others studies elsewhere [12, 13, 15, 19]. Figure 4 shows the temperature distribution in the oil palm fruit at kernel, endocarp, mesocars and exocarp. This simulation was performed at various power levels, and weight sample of oil palm fruit bunches. The temperature distribution was modelled at various thickness of mesocarp: r=3 mm (small fruitlet), r=4.5 mm (medium fruitlet), and r=6 mm (big fruitlet). The simulation shows temperature increased during the first 3 min. Temperature at the centre of the fruit is the highest during initial heating process, and the temperature will decrease along with an increase in the radial direction distance, from the centre towards the fruit exocarp. The lowest temperature is the temperature at the surface of the oil palm fruit (exocarp). It shows the process of heat generation takes place from the kernel, and so on heat to move toward the direction of endocarp, mesocarp and exocarp. However, after 3 min, the temperature in the kernel becomes constant, while the temperature in some hot spots in the mesocarp and exocarp showed different responses.

Figure 4 shows evaluation resulted from 0.5 kg sample that shows thermal runway for small fruitlet (r=3 mm) occurred at some hotspots, excluded at the exocarp. The thermal runway occurred from kernel into outer layer mesocarp after 3 to 5 min. Runway effect is uncontrolled rise in temperature in a material which cause damage of material [17].

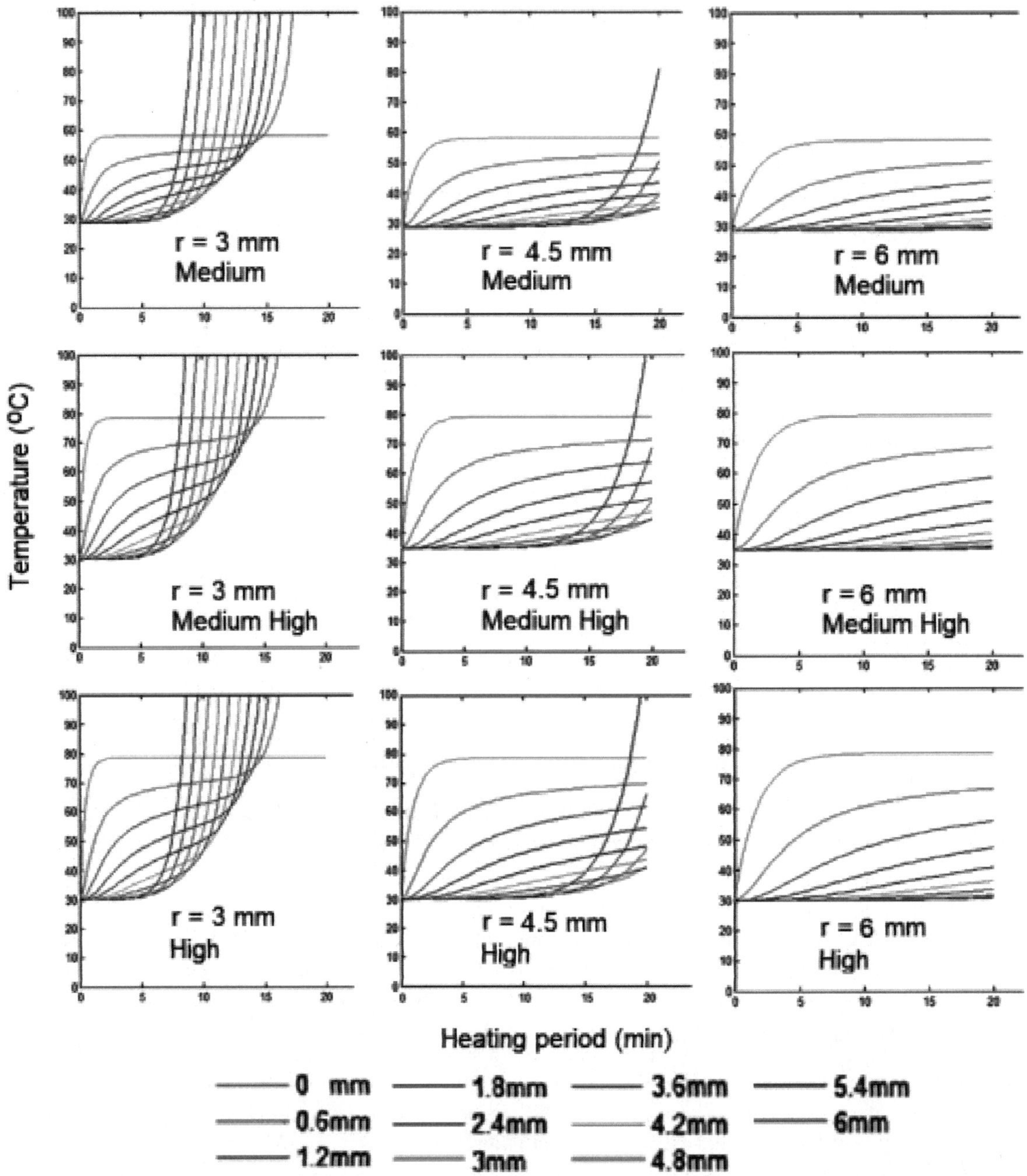

Figure 4. Temperature evaluation on 0.5 kg palm fruit sample at various size of fruitlets and power

Thermal runway indicated by the increase in local temperature exponentially arising from the absorption of microwave energy that is effective at higher temperatures in the oil palm fruit, especially in the endocarp. This is according to Clemens and Saltiel (1996) is due to an increase in the electrical conductivity of the oil palm fruit. It is also reported by Clemens and Saltiel (1996) who reported the same phenomenon when performing studies on ceramic heating in a microwave oven [20]. Thermal runway effect from microwave heating of medium size fruitlet (r=4.5 mm) occurred only in some hotspot location (radial distances were 0; 0.45; and 0.9 mm respectively). On the contrary no thermal runway effect occurred on microwave heating of big size fruitlet. Increment of microwave power from medium to medium high and high shows similar phenomenon.

Figure 5 shows temperature evaluation for 0.5 to 1.5 kg sample at high microwave power. Increment weight of the sample from 0.5 kg to 1 kg and 1.5 kg respectively, lowering the maximum temperature from 78.85 °C and 76.77 °C (for 0.5 and 1 kg respectively) into 71.53 °C (1.5 kg). This was due to power density variability obtained from ratio power to mass sample. Small sample gained higher temperature due to higher power density applied. Overall, excluded diameter of fruitlets, no significant parameters affect the temperature or heat transfer inside the palm fruit during irradiation period.

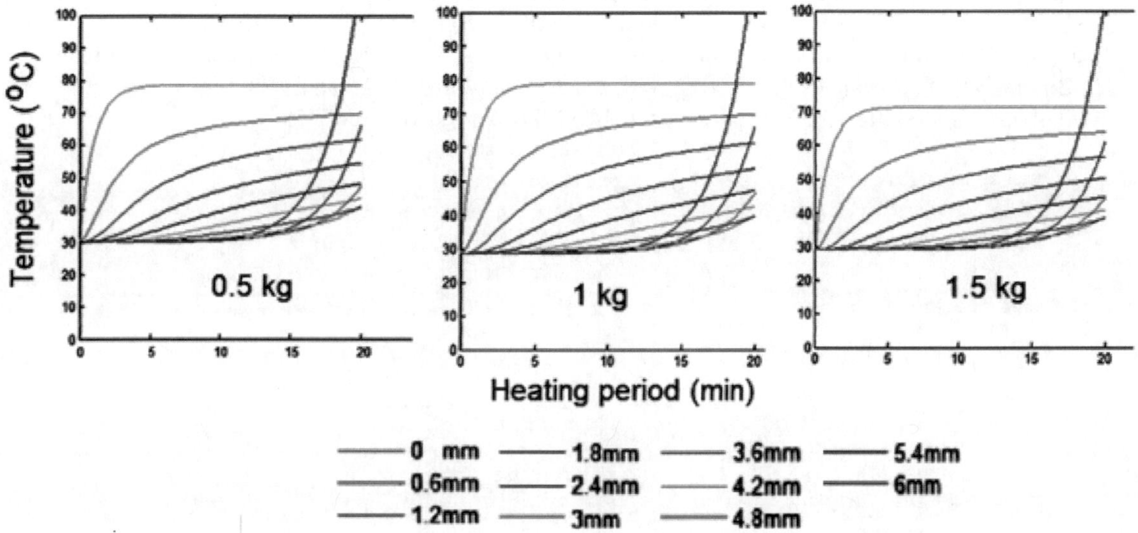

Figure 5. Temperature evaluation of sterilization with high microwave power at various mass

4. Conclussion

Microwave energy shows significant potency for sterilization of palm fruit with respect to their dielectric properties. Interior temperature in a single palm fruit was distributed inside out and it less influenced by power density but fruit's diameter. Greater power density provides high microwave energy per kg sample to attain greater temperature of microwave heating and vice versa. Distribution of interior temperature in the fruit with the thickness (r) of 6 mm and above indicated non-uniformity heating. Furthermore, interior temperature of fruitlet with r = 4.5 mm indicated distribution of temperature from endocarp to exocarp, meanwhile simulation temperature on smaller fruit diameter (r = 3 mm) indicated thermal runway effect after 5 min heating duration.

5. References

[1] Jusoh Y M, Idris A A, Khairuddin N, Zaidel D N A, Hashim Z N, Mahmooda N A, Zakaria Z Y and Muhamad I I 2018 *Chem. Eng. Trans.* 63 p 541-46
[2] Shang H, Shi J, Li J, Liu Y and Zhang W 2014 *Chem. Eng. Trans.* 39 p 511-16
[3] Tan 1981 United Nations Educational, Scientific and Cultural Organization.
[4] Sukaribin N and Khalid K 2009 *Ind. Crops and Prod* 30(2) p 179-183
[5] Sarah M and Taib M R 2017 *Proc. Int. Conf. on Chemical Materials and Process* vol 1879 (Beijing:China/AIP) p 05000 1-7
[6] Chow M C and Ma A N 2007 *J Microwave Power & Electromagnetic Energy* 40(3) p 165-173
[7] Cheng S F, Nor L M and Chuah C H 2011 *I Ind. Crops and Prod* 34(1) p 967-71
[8] Sarah M 2018 *ARPN J. Eng. App. Sci.* 13 p 1009-14
[9] Ryynanen S 1995 *J. Food. Eng.* 26 p 409-29

[10] Khraisheh M, Cooper T and Magee T 1997 *J. Food. Eng.* 33 p 207-19
[11] Meredith R 1998 *Engineers' Handbook of Industrial Microwave Heating* (London: The Institution of Electrical Engineers)
[12] Zhang H and Datta A K 2001 *AIChE Journal 47*(9) 12
[13] Geedipalli S S R, Rakesh V and Datta A K 2007 *J. Food. Eng.* 82 p 359-68
[14] Dev S R S, Gariepy Y, Orsat V and Raghavan G S V 2010 *Prog. in Electromagnetics. Res. M* 13 p 229-43
[15] Navarrete A, Mato R B and Cocero M J 2012 *Chem. Eng. Sci.* 68(1) p 192-201
[16] Metaxas A C and Meredith R J 1988 *Ind. Micro. Heating* (London: Peter Peregrinus Ltd.)
[17] Tsai L L1972 *IEEE Trans. Antennas and Prop.* AP-20(5) p 569-76
[18] Riggs J B 1988 *Intro. Num. Methods. Chem. Eng.* (Texas USA: Texas Tech University Press)
[19] Klinbun W and Rattanadecho P 2012 *App. Math. Modelling* 36(2) p 813-28
[20] Clemens J and Saltiel C 1996 *Int. J. Heat. Mass. Trans.* 39(8) p 1665-75

IOP Conf. Series: Materials Science and Engineering **543** (2019) 012024 doi:10.1088/1757-899X/543/1/012024

Iodization of Local Salt Based on Purification Technique using Saturated Brine Washing Method

T Widjaja[1*], A Altway[1], I Gunardi[1], L Pudjiastuti[2], D F Nury[1], A F Nabila[1], N Prasetyawati[3], M Saifulloh[3], Noviyanto[4]

[1]Department of Chemical Engineering, Institut Teknologi Sepuluh Nopember Surabaya, Sukolilo, 60111, Surabaya, Indonesia
[2]Technical Implementation Unit of Social Sciences and Humanities, Institut Teknologi Sepuluh Nopember Surabaya, Sukolilo, 60111, Surabaya, Indonesia
[3]Department of Industrial Chemical Engineering, Institut Teknologi Sepuluh Nopember Surabaya, Sukolilo, 60111, Surabaya, Indonesia
[4]Department of Mathematics Education, STKIP PGRI Sumenep, Sumenep, 69451, Indonesia

E-mail: triw@chem-eng.its.ac.id

Abstract. Based on Indonesia National Standards (SNI), Indonesian local-salt is currently unavailable to produce high-quality salt. This study was aimed to investigate the purification process of local salt that has been carried out by local salt farmers in Sumenep, Madura Island, Indonesia and to determine the highest level of NaCl content from local salt using a washing method to make iodized salt. Purification process involves washing technique using a nearly saturated brine. The research was conducted in various particle-size (10-20 mesh and 20-35 mesh), salt/brine ratio (1:20, 1:40, 1:60), duration (10, 30, 60 minutes) and stirring speed (100, 200, 300 rpm). The highest NaCl content of 94.45 % was obtained at the washing process using the salt of 20-35 mesh at 1:60 ratio with washing time of 30 minutes at a speed of 100 rpm. The result of the process shows a decrease of Calcium and Magnesium levels by 95.56 %. In iodization process, the determination of the highest NaCl concentration was carried out using washing brine method and spray mixing injection was also done to develop a stable salt formulation. The optimum stability of KIO_3 in salt was observed as a function of time.

1. Introduction

Salt is one of the primary commodities in Indonesia, which has a strategic role that staple for both consumption and industrial needs. However, the quality of salt produced by salt farmers who are still mainly using conventional technology (utilizes solar thermal energy) has an average sodium chloride of 85-89 % of sodium chloride content. The salt produced also contains impurities such as $MgSO_4$, $CaSO_4$, $MgCl_2$, KCl and soil impurities[1]. This local salt could not be consumed directly both for consumption and industrial needs because its NaCl content is still under Indonesian National Standard (SNI) Number. 01-3556-2000 for consumption needs and Indonesian National Standard (SNI) Number 06-0303-1989 for industrial needs[2]. Those salt quality are still below to Indonesian National Standard (SNI) 3556-2016 standard for consumption salts, National Standardization Agency of Indonesia SNI 0303-2012 standard for industrial salts 94.7 % and 98.5 % (dry base), respectively. Cooking salt is salt which is used for consumption with NaCl content of at least 94.7 % calculated from the dried base, included maximum water content 7 %, Iodine as KIO_3 minimum 30 ppm. While industrial salt is usually used as raw material or supporting industrial activities, auxiliary materials for petroleum industry needs, textile, pharmaceutical and CAP (Chlor-Alkali Plant) with a minimum of 97 % NaCl content (dry base), impurities such as: Calcium 0.1 %, 0.05 % Magnesium, Sulphate maximum of 0.2 %, and maximum water content of 2.5 %, (% w/w) [3].

To obtain a good quality salt, the impurities should be precipitated. Salt which produced using the conventional methods still could not meet the standards of salt both for consumption and industrial determined by the Indonesian National Standard (SNI). Based on the quality ofsalt production requirements, there are several methods of purifying in order to fulfilled the requirements, included chemical and physical method. There are some salt purification technologies namely brine washing [4], evaporation (recrystallization) [5], and chemical addition [1]

The purification process of salt is done using a washing technique and evaporation (recrystallization). Washing technique is done with a nearly saturated brine while evaporation (recrystallization) process. The evaporation process is one method to purify (purification) a solid material from its impurity through the dissolution process and crystallization. The evaporation process based on the solubility of the material in a solvent in which the solubility of the material will increase as the temperature rises and vice versa. The impurities may have different properties where the solubility of the impurities will be low at high temperature and vice versa. This process requires considerable heat energy to be able to evaporate the remaining water and recrystallize the purified salt [6]. The use of a nearly saturated solution was aimed to dissolve the impurities and other impurities to obtain salt with higher sodium chloride (NaCl) content. Salt can reach up to 5 % or NaCl levels reached from 91 to 96 % using this washing method. conducted the comparison of the physical method using both recrystallization and washing method in resulting salt with the highest NaCl content. The purity of sodium chloride could be enhanced up 99.21 % from the re-crystallization method, while the washing method enhanced up to 96.64 % using the particle size of 20-30 mesh. The result showed that washing method is more efficient than the re-crystallization method because the re-crystallization method requires heat energy and could be used both for industrial and consumption needs.

The iodized local salt was done using the spray system mixing. The mechanical iodization process of salt can be carried out in three ways, namely the first method, dry mode (solid-solid mixing). The second method wet way of drying (mixing of liquid-solid) where the salt is mixed with iodine-containing liquids by dripping or spraying in this process. This method further guarantees the homogenization of the results, but the water content in the salt will increase. The third method is the wet (liquid-liquid mix). In this process, the mixing is carried out during the crystallization process, in which iodine increases in the crystallization process. The resulting results are evenly distributed but require a more significant cost and higher technology [7].

The objective of this research is to determine the most effective and efficient method to make iodized salt from the local's salt. Salt crystals in varied coarse size of 10-20 mesh and 20-35 mesh, were introduced into brine with concentration 23° Be and the ratio of salt and brine solution used 1:20, 1:40, and 1:60. The solution was stirred for 10, 30, and 60 minutes, as well as at stirring speeds of 100, 200 and 300 rpm. These variables were tested to get lower Calcium and Magnesium level and increased NaCl concentration to 94.7 % as same as the National Standards Indonesia (SNI). From the best variation, after the improved quality of salt generated by the washing process, combined with iodized salt by the process of spray mixing method in which the homogeneity and stability of iodine in iodized salt as a function of time.

2. Experimental section

2.1. Raw material

Local-salt was kindly obtained from the local salt farmer in Sumenep, Madura island, Indonesia. Aquadest, sulfuric acid, potassium iodate, sodium thiosulfate, starch indicator were purchased from

Merck. Sharp EM-S53-WH Standing Mixer was used for iodization and purchased from local market Surabaya, Indonesia.

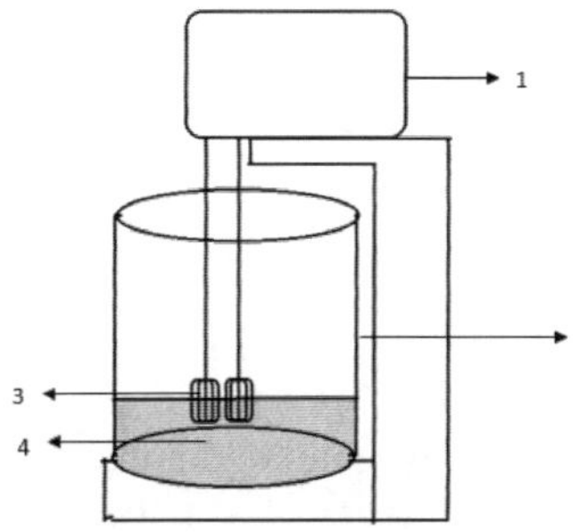

Figure 1. Mixer set-up, (1): Motor; (2): Vessel; (3): Mixer; (4) Salt

2.2. Salt preparation

Salt was firstly determined gravimetrically. Salt was weighed and then dried at 110 °C to reduce its water content of the salt solid. After the drying process, salt was varied by size, coarse size, 10-20 mesh and 20-30 mesh, respectively. The impurities can be dissolved when salt solids contacted with the brine solution. Chemical composition of local salt as the raw material in brine washing method resulting in 82.335 % (dry base) using Argentometric method. All paremeters, such as Calcium 0.235 %, Magnesium 1.104 %, Sulphate 1.411 % were analyzed using titrimetric. Impurities was analyzed by gravimetric.

2.3. Brine washing method

The process was conducted in a batch laboratory scale. Brine solution was used for washing the solution. Brine solution contains 98.5 % NaCl content. The brine washing method is an extraction process or a separation of a component which is in a solid phase by a liquid phase as a solvent (Wu et al., 2017). In this case, the salt is the solid phase and the salt solution as the solvent (Wu et al., 2017). The performance of brine washing method was influenced by the concentration of brine solution which the 23 °Be have been measured by Baume meter. According to (Sedivy, 1996), the brine solution includes a saturated solution at 24.6 °Be and becomes a salt crystal at 25 °Be - 30 °Be. Salt crystals in varied of coarse size, 10-20 mesh, and 20-35 mesh was contacted on brine with concentration 23 ° Be with the ratio of salt and brine solution used 1:20, 1:40, 1:60, time of stirring 10, 30, and 60 min., as well as stirring speeds of 100, 200, and 300 rpm is constant. After the washing process, the filtration separation process is performed. The next stage performs a complexity titration analysis to determine the content of Ca and Mg after washing with brine. The initial content of Ca and Mg obtained were 3.6072 and 2246.392 milligrams using titration, respectively.

2.4. Iodization

Local salt was used to prepare the fortified salt sample. Duration of 1,2,3,4,5, 10 and 15 minutes, stirring speed of 100 rpm, and particle size of salt 20and 60 mesh was carried out in the present study. Salts were weighed and mixed in a static mixer using a stirring speed of 100 rpm. 1 ml of potassium iodate was sprayed with mixed salt. Iodized salt samples were stored in a desiccator at 25 °C for 90 d. Iodine content analysis was estimated as described by iodometric titration method in Indonesian National Standard (SNI) number:01-3556 which can be titrated by Sodium Thiosulphate using starch as color indicator.

2.5. Statistical analysis

Data from chemical assays were obtained for once and examined as the homogeneity of KIO_3 in salt. Stability KIO_3 in salt was observed as function time every 10 d (days) once. The results of this observation will be obtained the response as index homogeneity.

3. Results and discussion

NaCl levels after the washing process are presented in Table 1. From the best variation, i.e., in the size of 20-35 mesh,100 rpm, 1:60 ratio with 30 minutes washing time, salt quality improvement of 12 %, from the initial salt before washing 82.22 % to 94.45 %. Salt produced by quality 94.45 %, combined with the process of iodized salt by mixing spray method in which the homogeneity and stability of iodine in iodized salt as a function of time.

3.1. Decreasing Calcium and Magnesium in varied particle size

By using the washing method, the impurity content can be reduced because ions Ca^{2+} and Mg^{2+} can dissolve in the brine solution. In this study, the concentration of brine solution used is 23°Be which means the solution contains 23 % salt and 77 % water. Water will dissolve Ca and Mg so that there is a solvation process. In the solvation process, the ion or molecule is surrounded by a solvent molecule having a particular arrangement. So the reaction that occurs when the washing process of salt solids as follows:

$$2H_2O + Ca^{2+} \rightarrow Ca\,(OH)_{2(aq)} + H_{2(g)} \quad (1)$$

$$2H_2O + Mg^{2+} \rightarrow Mg\,(OH)_{2(aq)} + H_{2(g)} \quad (2)$$

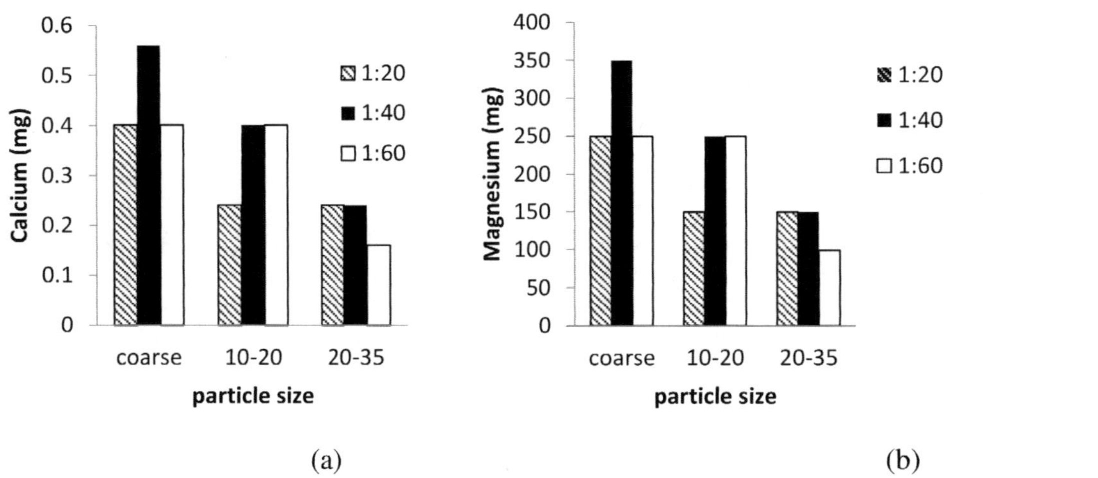

Figures 2. (a) Calcium content; (b) Magnesium content after washing process, varied particle size

From Figures 2 (a) and (b), it can be seen that decrease in Calcium and Magnesium in various particle size variations of the best variation at washing time of 30 minutes and 100 rpm. Washing time becomes one of the factors determine the quality of salt on salt refining with washing method. The longer the wash time than the longer the salt contacts with the brine solution, so impurities present in the salt crystals can be reduced more lot and impurities left in salt crystals a little more. Besides, levels of Magnesium which was obtained tend to be larger than Ca because the concentration of Magnesium in salt crystals was higher than Calcium so that the difference of concentration of Mg with brine was also more significant [8].

One factor is the low purity of the salt due to the presence of impurities in the salt of $Ca^{2+}$and Mg^{2+}as the largest impurities [1]. In salts, Ca^{2+} and Mg^{2+} ions in the form of $MgSO_4$, $MgCl_2$, $CaSO_4$, $CaCl_2$. With the solubility of these compounds affects the salt purity itself, because with a small solubility it will be difficult to be removed or dissolved. The decrease of Ca^{2+} and Mg^{2+} was influenced by several factors such as salt size, washing time, the salt ratio with brine, and stirring rate. Data were obtained by using titration. From the results, at a speed of 100 rpm and 30 minutes washing time gave the best data.

IOP Conf. Series: Materials Science and Engineering **543** (2019) 012024 doi:10.1088/1757-899X/543/1/012024

Table 1. NaCl levels after washing using AAS analysis

Particle size (mesh)	Ratio (salt/brine)	Analysis result (%)	Particle size (mesh)	Ratio (salt/brine)
	1:20	88.12		1:20
coarse	1:40	93.18	coarse	1:40
	1:60	88.2		1:60
	1:20	91.93		1:20
10-20	1:40	90.65	10-20	1:40
	1:60	90.65		1:60
	1:20	90.65		1:20
20-35	1:40	89.38	20-35	1:40
	1:60	94.45		1:60

Also, the solubility of Mg is much higher than that of Ca solubility. One factor is the low salinity of salt due to the presence of impurities in the salt of the compound Ca^{2+} and Mg^{2+} as the most considerable impurities in the salts of Ca^{2+} and Mg^{2+} ions in the form of $MgSO_4$, $MgCl_2$, $CaSO_4$, $CaCl_2$. At smaller particle sizes, Calcium and Magnesium tend to move more easily from within the salt crystals to the brine.

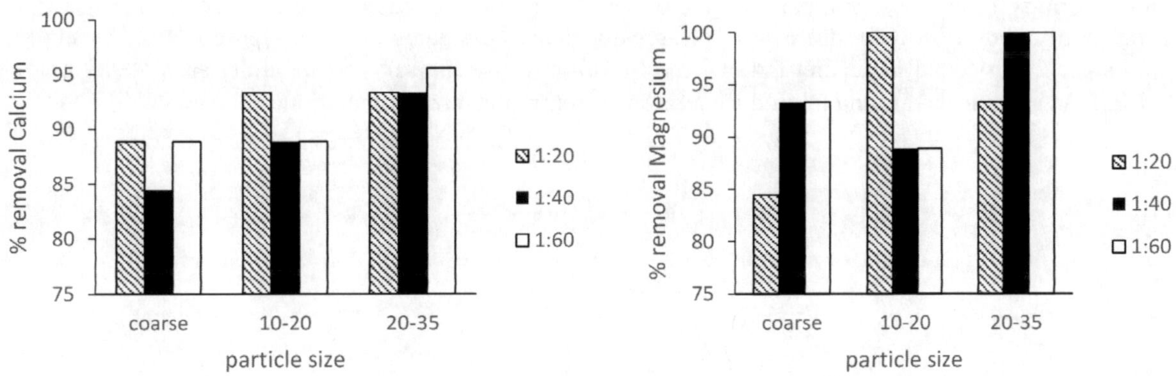

(a) (b)

Figures 3. (a) Removal Calcium; (b) Magnesium in varied particle size

From Figures 3, particle size 20-35 gave a significant decrease results. In Figures 2(a), Calcium content 3.6072 to 0.160 milligrams produce a percentage of removal Calcium 95.556 %, where the percentage decrease in Calcium and Magnesium can be seen in Figures 2 (a) and (b), respectively. The largest decrease in Magnesium also occurs in particle size of 20-35. At 20-35 mesh, a large decrease of Calcium and Mg compared to the size of coarse and 10-20 mesh. This is due to the reduction of size so that impurities in the salt can go out into the salt surface and dissolve with brine. Ratio (Salt/brine) also affects the salt quality resulting from the washing process. In Figures 3 (a) and (b) show that the ratio is at 1:60, for each Calcium and Magnesium a significant decrease in coarse size10-20 mesh, as well as the size of 20-35 mesh. On size 10-20 mesh with variable 60 minutes, a ratio (salt/brine) which can result in a decrease in Calcium and Magnesium % the highest amounted to 93.333 % occurred in the variable comparison 1:60. The higher the ratio (salt/brine), the more brine solution used so that impurities that can be removed from the salt crystals are also more numerous. This caused the Calcium and Magnesium in the washing salts to decrease, and the salt becomes purer with higher NaCl content. Comparison of salt and brine also determine how effective and efficient washing process performed. By NaCl levels of 94.45% using AAS analysis after saturated washing brine process, the salt could be formed as both for consumption and chemical needs.

3.2. Homogeneity of the KIO3 as function of time

Salt iodization was done using spray mixing. On the highest NaCl content of 94.45 %, salt iodization was done using spray mixing method and iodine content determination analysis using iodometric titration method. KIO_3 KIO_3 as much as 1 ml (% v / v) sprayed mechanically into the salt to be iodized using the variable rotation speed specified.

In this study, an iodized product was analyzed to determine KIO_3 levels in (ppm). It is then calculated using a homogeneity index to compare each variation of a predetermined operating condition.

$$homogeneity\ index = \left\{1 - \sqrt{\frac{\Sigma(y_i - \bar{y})^2}{\Sigma y_i}}\right\} x\ 100\% \tag{3}$$

Figure 4 shows the best variables of iodized result that is on the particle size of 20 and 60 mesh. In this case, the smaller the particle size indicates the more homogeneous KIO_3 present after injection with the spray system in the salt. In this case, time affects the homogeneity index of the resulting KIO_3. At (t) 5 min, the homogeneity index shows 85.13 for the particle size of 20 mesh and 88.62 at 60 mesh particle size. At (t) 6 to 15 minutes the homogeneity index shows more uniformity in both 20 and 60 mesh. The more serious issue arises with particle size of salt that has a large range of grain size. This causes an uneven distribution of KIO_3 due to surface area differences. As can be seen in Figure 4, at 20 mesh the homogeneity is found in smaller than 60 mesh. This because the salt will naturally separate by grain size. Fine salt may be ground to produce a more uniform grain size distribution.

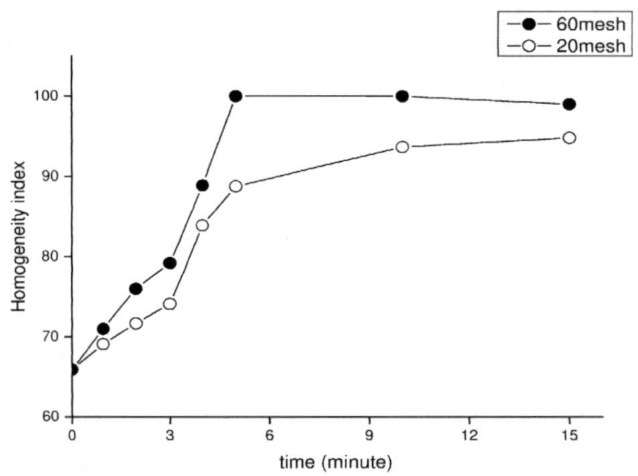

Figure 4. Homogeneity index at 20 and 60 mesh of particle size

3.3. Stability of KIO_3 as function of time

Table 2 shows the concentration of KIO_3 during 90 days stored in desiccators at 25°. It was observed that the stability of KIO_3 did not show the difference, apparently for 90 d (days). In summary, spray solutions of 1 ml KIO_3 are suitable for salt fortification as they retain the micronutrients and remain in solution for a minimum 4 months. These solutions may be used to fortify salt which retains sufficient micronutrient stability. As a result, impure iodized salt can lose most of the added iodine during extended storage. Therefore, the suggestion of 9 to 12 months storage should be investigated.

Table 2. Stability of KIO_3 at 90 d (days)

Time (days)	10	20	30	40	50	60	70	80	90
KIO_3 concentration (ppm)	80.1	80.1	80.1	80.1	80.1	80.1	80.1	80.1	80.1

4. Conclusion

The development of a novel, saturated brine washing method to make iodized local salt using spray mixing has been reported. The results of saturated brine washing method could enhance the NaCl content of 94.45 %, which was obtained at brine washing process of particle size 20-35 mesh at (Salt/brine) 1:60 ratio with washing time of30 minutes at a speed of 100 rpm. This result of the process could decrease Ca and Mg by levels 95.56 %. The stability of KIO_3 local salt samples using brine washing method was confirmed stable for 90 d stored in the desiccator at 25 °C.

Acknowledgement

The authors would like to express their most profound gratitude to The Institute for Research and Community Services (LPPM) of Institut Teknologi Sepuluh Nopember Surabaya, Indonesia for funding this research with grant number [1457/PKS/ITS/2018]. The authors would also like to express special thanks to all member of Biochemistry Technology and Chemical Reaction Engineering laboratory for their unconditional help and support during the research process.

References

[1] Ihsan D and Djaeni M 2002 Improving public salt quality by chemical treatment *J. Coast. Dev.* **5** 111–16.
[2] National Standardization Agency of Indonesia SNI 0303-2012 standard for idnustrial salts
[3] Anwar M A, Bakar A, Ahmad A, and Harun M 2017, *Chem. Eng. Transc.* **56** 235–40.
[4] Sumada K, Dewati R and Suprihatin 2018 *J. Phys.: Conf. Ser.* **953** 012214
[5] Chenli Z, Lian F, and Ma L 2016 *5th Int. Conf. on Measurement, Instrumentation and Automation* pp. 105–109.
[6] Caecilia P, Ketut S, Yustina N and Prasetyo H 2016 *MATEC Web Conf.* **58** 01022 10–13.
[7] Li Y O, Diosady L L and Wesley A S 2010 *J. Food Eng.* **99** 232–38.
[8] Clar C, Oxon D and Wu T 2002 *Endocrinolog. Metab. Clinics North America* **31** 681-98.

Thermal stability and reusability of home-made co-immobilized lipase from *Mucor miehei* in polyurethane foam for the production of bio-flavor

D Moentamaria, M Muharja, T Widjaja, A Widjaja
Department of Chemical Engineering- Institut Teknologi Sepuluh Nopember, Surabaya 60111, Indonesia

E-mail: arief_w@chem-eng.its.ac.id

Abstract. In this study, polyurethane foam (PUF) was used as a matrix to immobilize *M. miehei* lipase through a modified substance coating consisting of lecithin, gelatine, polyethylene glycol (PEG), $MgCl_2$, known as co-immobilized. Co-immobilized lipase application was carried out on bio-flavor manufacturing through hydrolysis and esterification reactions. Fatty acid was gained from hydrolysis of glyceride coconut oil (CNO) containing main components of C8 to C16. Furthermore, esterification of fatty acids and citronellol from citronella oil was employed. The remaining triglycerides (TAGs) in hydrolysis were analysed using LC-MS/MS. While bio-flavor was analysed using GC-FID. Maximum bio-flavor production was obtained at 40 °C and fatty acids lauric acid/citronellol mole ratio of 1:5. The remaining activity of free lipase, immobilized lipase on PUF, and immobilized lipase on co-coated PUF at 50 °C for 50 min was 0 %, 34 %, and 54 %, respectively. Immobilized lipase on co-coated PUF was able to reuse up to 5 cycles. The maximum conversion of bio-flavor was attained using lipase on modified co-immobilized. The use of co-immobilized lipase of the non-commercial source used in this study could be a promising method for a zero-waste bio-flavor production in a mild and solvent-free process to support low-cost and green technology.

1. Introduction

These Bio-flavor as an ester produced from plant extracts is rare in commercial scale production [1]. Until now, many food industries use artificial flavor as a food additive which is carried out through organic chemical synthesis. Even though, bio-flavor can be produced as an ester through an enzymatic reaction using lipase [2].

Lipases (triacylglycerol EC 3.1.1.3 hydrolases) are enzymes used for the reaction of lipid hydrolysis to fatty acids and glycerol. Lipase has the ability to catalyse several reactions such as esterification (alcohols and carboxylic acids), transesterification (esters and alcohols), interesterification (esters and acids) and transfer of acyl groups from esters to other nucleophiles (e.g. amines and thiols) [3–5]. Commonly, lipases are produced from a variety of microorganisms, but yeast and fungi are the primary sources widely used in industry. Commercial lipases obtained from *M. miehei* are able to catalyse the CNO which consist of lauric acid (C12) and myristic (C14) about 60% to produce ester flavor [6–9]. However, the low stability of free enzymes, especially the effect of temperatures above 40 °C for a long operational time, causes the use of this enzyme on a limited industrial scale [10].

There are three ways to increase enzyme stability. According to Cui et al. [6], *Y. lipolytica* lipase Lip2 was stabilized with improved thermal stability by adding a material for protective lipase and immobilizing various matrixes. PUF was widely utilized as a lipase immobilization matrix because of commercially inexpensive, inert, and excellent mechanical properties for high resistance. Nyari et al. [12] used PUF as a matrix for immobilizing lipases in the esterification reaction of alcohols and carboxylic acids.

The use of immobilization techniques is a solution to the low stability of free enzymes through various carrier binding or crosslink and encapsulation methods in the inorganic/organic polymer matrix [13]. The immobilization of enzyme on various solids supports through physical and chemical mechanisms was utilized to recover, reuse and increase temperature stability (thermal, operational and storage) of the enzyme [14]. This biotechnology process can be economically an option for the development of enzyme applications [12,15]. Enhancing enzyme stability can also be prepared by adding material support such as glutaraldehyde or a combination of lecithin-gelatine / lecithin-PEG / gelatine-PEG and the addition of metal ions as cofactors so as to improve enzyme performance [16–18]. Based on literature review, the combination of lecithin-gelatine-PEG-cofactor Mg has never been studied related increasing stability and reusability for immobilization of lipase on the bio-flavor hydrolysis from CNO.

This study aims to prepare co-immobilized coating on PUF matrix, followed by covalent and crosslinking methods, and study thermal stability and reusability of immobilized lipase. The use of lipase immobilization in PUF matrix was applied to produce bio-flavor from CNO through hydrolysis process producing fatty acids then reacted with citronellol in the esterification reaction.

2. Experimental section

2.1. Materials

CNO was collected from the local Permata Agrindo Pendowoharjo (Sewon, Bantul, Indonesia) under the commercial name Laitco. The stock of *M. miehei* was cultured in Biochemical Technology Laboratory, Department of Chemical Engineering, Sepuluh Nopember Institute of Technology, Surabaya. Co-immobilized consisting of gelatin, lecithin, PEG, $MgCl_2$ was purchased from Merck (Germany).

2.2. Production of crude lipase from M. miehei through solid state fermentation

Media consisting of KH2PO4, FeSO4.7H2O, olive oil, palm oil, dried coconut grout (solid), and water was sterilized at 121 °C for 15 min. *M. miehei* culture stock was inoculated in the media and incubated through solid state fermentation at 37 °C for 5 days. Furthermore, crude lipase obtained by extraction was added phosphate buffer- solid media ratio of 4:1 (wt/wt), then incubated at 37 °C, for 135 min, 150 rpm. Crude/free lipase as a supernatant was filtrated and analysed the activity (U/ml) [19,20].

2.3. Assay of lipase activity

The activity of crude lipase was determined using olive oil as a substrate. Olive oil and Gum Arabic were emulsified for 2 min. Next, 5 ml of emulsified olive oil was mixed with 2 ml of 0.1 M phosphate buffer (pH 7) and 1 ml of enzyme suspension. The mixture was incubated at 37 °C for 30 min. After incubation, the reaction was stopped by adding acetone-ethanol of 1:1 (v/v) and the free fatty acid (FFA) was titrated by 0.05 M NaOH. One unit of lipase activity is defined as a number of enzymes capable of liberating 1 µmol of fatty acids per minute [21]. Lipase activity was calculated by Equation (1), where Vs and Vb are the volume of NaOH needed to titrate the sample and blank (ml), respectively. The value of 1000 is the conversion from mmol to µmol and t is the reaction time (30 min).

$$Lipase\ activity\ (U/ml) = \frac{(V_s - V_b) x\ N\ NaOH\ x\ 1000}{t} \quad (1)$$

Matrix activity was determined in the same way as above, but a 1 ml crude sample was replaced with 1 g of PUF co-immobilized lipase. Hence, the co-immobilized lipase-PUF activity has units of U/g matrix

2.4. *The procedure of lipase immobilization*

2.4.1. *Co-immobilized coating*

Co-immobilized was prepared by making a mixture of solutions consisting of lecithin (50 g/l), gelatine (50 g/l), $MgCl_2$ (10 g/l), and PEG 6000 (20 g/l) with the same weight composition ratio. PUF was made by reacting isocyanates and polyols with the same volume ratio, then cut to size of 5 mm x 5 mm x 5 mm. The co-immobilized-coated PUF was prepared by soaking PUF in co-immobilized with a ratio of 1:20 (wt/wt) and then stored at 40 °C.

2.4.2. *Immobilization of lipase on PUF*

Immobilization of lipases on PUF was carried out by the following steps. First, physical adsorption on PUF was done by soaking 1 g of plain PUF into free lipase for 24 h at room temperature, washed and stored. Second, 20 gr of co-immobilized consisting of dispersed lecithin, gelatine, PEG, $MgCl_2$ was dispersed cross-linking through the reaction of each material. Lastly, covalent lipase binding in PUF was activated by co-immobilized.

2.5. *Hydrolysis and esterification reaction*

2.5.1. *Hydrolysis of CNO by immobilized lipase of M. miehei*

Hydrolysis of CNO by lipase immobilized on PUF produced FFA and glycerol. The hydrolysis reaction was carried out in a three-neck flask. CNO of 10 g and water of 50 g was mixed and added 26.24 U of immobilized lipase on PUF. The reaction was maintained at 40 °C during 900 min. The product of hydrolysis reaction consisting of two layers, the fatty acid and glycerol layer, was separated using a separating funnel. Chemical properties of the remaining TAGs were determined by LC-MS/MS.

2.5.2. *Esterification of FFA and citronellol by immobilized lipase from M. miehei*

The FFA produced from hydrolysis reaction of CNO was used as the substrate for the further esterification reaction. The esterification reaction was carried out in a batch process using erlenmeyer containing FFA, citronellol and immobilized lipase. The substrates of FFA and citronellol were prepared in a molar ratio of 1:1 and the immobilized lipase was added in 6 U. The reaction was maintained at 40 °C for 5, 10, 15, 20, and 25 h. The product of esterification consisting of two layers, bio-flavor as ester and water, were separated by a separating funnel.

2.6. *Thermal stability and reuse analysis*

Stability of free lipase, immobilized lipase on PUF, immobilized lipase on co-coated PUF were determined through thermal stability test at an incubation temperature of 50°C. During the incubation period, the remaining activity was analysed every 10-minute interval and defined by the Equation (2).

$$\text{Remaining activity (\%)} = \frac{[(initial\ activity) - (activity\ after\ storage)]}{initial\ activity}\ x\ 100 \qquad (2)$$

The relative percentage of hydrolysis was calculated by the Equation (3), where n denotes to a number of cycles.

2.7. *Analytical method*

The remaining TAGs from the CNO hydrolysis reaction were analysed using LC-MS/MS. ACCELLA 1250 LC-MS/MS (Thermo Scientific-Waltham, MA USA) was equipped by Hypersil Gold column (50 mm x 2.1 mm x 1.9 μm), vacuum degreaser, quartener pump, thermostatic autosampler controlled by a personal computer through x-calibre program 2.1. The mobile phase of 2 μL was employed at a flow rate of 300 μL/min. The column and autosampler compartment were controlled at 30 °C and 16 °C, respectively. The use of MS/MS Triple Q TSO QUANTUM ACCESS MAX mass spectrometer with Electrospray Ionization was controlled by TSQ Tune software and operated in positive mode.

IOP Conf. Series: Materials Science and Engineering **543** (2019) 012025 doi:10.1088/1757-899X/543/1/012025

Bio-flavor was examined using GC-FID (HP 5890, Santa Clara-California, United States). GC FID was equipped by HPL 608 column in dimension of 30 m length x 0.53 mm i.d. x 0.5 μm film thickness. The temperature of the GC-FID oven was adjusted to 125 °C for 3 min elevating of 7.5 °C/min up to 250 °C. The temperature of the injector and detector were kept at 255 °C and 275 °C, respectively. Conversion of citronellol was determined based on the area under the curve [22,23].

3. Results and Discussion

3.1. Effect of thermal stability free and immobilized activity on hydrolysis reaction

The operating conditions of the hydrolysis and esterification reactions using co-immobilized lipase in the PUF matrix were carried at 40 °C and pH 7 based on an optimum condition of previous work by Patel et al. [24]. The thermal stability of the lipase activity was examined at 50 °C from free lipase, immobilized lipase on PUF, and immobilized lipase on co-coated PUF in the hydrolysis of olive oil [6].

Figure 1 demonstrated the remaining activity of the enzyme on hydrolysis using free and immobilized lipase. As presented in Figure 1, in the temperature above the optimum condition, the thermal stability of the three immobilization methods shows different performance. Free lipase has lower stability compared to the others. At 50 min of hydrolysis, the remaining activity of each free lipase, immobilized lipase on PUF, immobilized lipase on co-coated PUF was 0 %, 34 %, 54 %, respectively. The composition of co-immobilized would affect stability remarkably. These results are owing to gelatine and lecithin as emulsifiers are dispersed into small particles which can penetrate the substrate entry into the immobilized lipase on co-coated PUF. Moreover, PEG would reinforce the polymerization of gelatine lecithin through its hydroxyl bond and $MgCl_2$ as a cofactor metal ion [18]. Perez-Miller and Hurley [25] also showed the covalent bonding between Mg and phospholipid owned by lecithin.

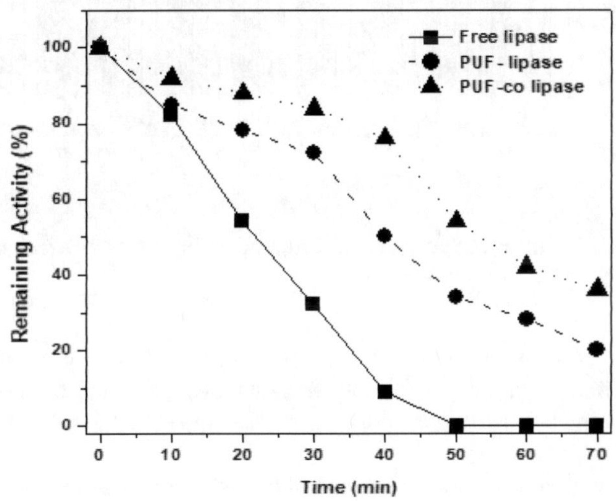

Figure 1. Thermal stability of free and immobilized lipase on PUF

3.2. Reusability on hydrolysis reaction

Generally, CNO contains saturated fatty acids in the form of 92 % TAGs and 0.1-0.3 % FFA. Most of the TAGs (about 70 %) are saturated fatty acids known as medium chain fatty acids (MCFA), with the lauric acid content of 45-56 % [9]. the FFA content before and after the hydrolysis process which analysed using the titration method were 0.2 and 0.85 %, respectively. Residual TAGs analysis using LCMS/MS was observed as TAGs in the form of C8, C8, C10; C8, C10, C10 or C8, C8, C12; 3×C10; C10, C10, C12; C10, C12, C12 or C10, C10, C14; 3×C12, C12, C12; C14, C12, C14, C14 or C12, C12, C16; 3×C14; C14, C14, C16; C14, C16, C16 or C14, C14, C18.

TAGs, also known as lipids, need a lipase biocatalyst to dissolve into the water on the hydrolysis reaction. The surface actives work on the interface area of water and lipid phase. Immobilization of lipase was done by combined encapsulated-coating, crosslinking and covalent binding to increase the enzyme performance. The advantages of using immobilized enzyme techniques are due to highness of stability and reusability. In this immobilization technique, lipase was bounded to the solid matrix PUF facilitating separation between the product and immobilized lipase directly [26].

The hydrolysis reaction was carried out on the water-oil ratio of 5:1 (w/w) using immobilized lipase-PUF resulting in FFA and glycerol (see Figure 2). Immobilized lipase-PUF in this study can be reused up to 5 times. As shown in Figure 2, the remaining TAGs show an unreacted composition of fatty acids. As the example in reuse 1, from mass spectra analysis shows BM 527 which is identified as C8, C10, C10 or C8, C8, C12, while reuse 4 is identified as C14, C14, C16.

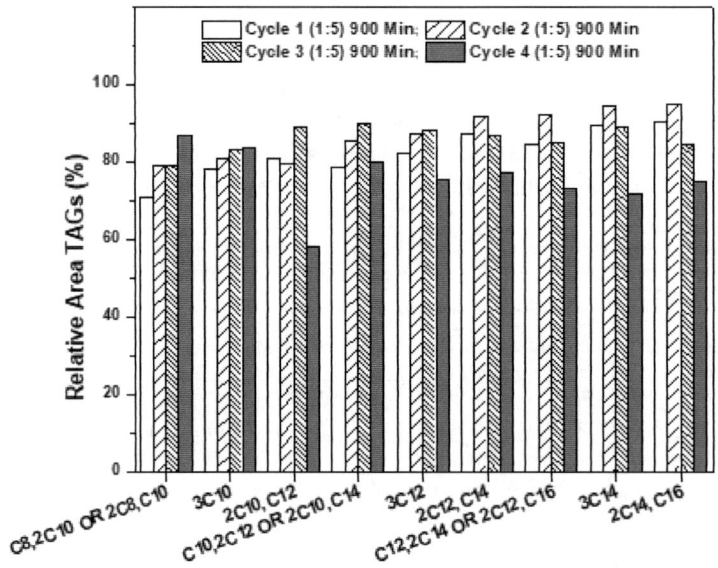

Figure 2. The relative area of fatty acids in immobilization of lipase for CNO hydrolysis

In this study, after completing the hydrolysis reaction, the enzyme was separated from the product then immediately reused in the reaction with the same variables to compare. The immobilized enzyme was washed without solvent before reuse. This step was confirmed by Nyari et al. [12] that using washing stages without hexane solvents.

Figure 3 exhibits the relative conversion of hydrolysis (%) on the reuse of immobilized lipase. As illustrated in Figure 3, on the cycles number of 2 and 3, the relative conversion decreased from 17.48% to 12.65%. These results were confirmed by another previous work [24,27]. However, on the number of cycles of 4 and 5, relative conversion increased gradually. This phenomenon was also attained by Ferraz et al. [20] using home-made immobilized lipase from *P. crustosum* for the esterification of geraniol laurate. In the study, it was found that the percentage of relative activity and conversion tends to fluctuate with increasing the number of cycles. This trend was related to the commercial and non-commercial use of lipases from various sources and reaction systems.

The utilization of commercial lipases commonly presented the consistency of decreasing the percentage of conversion and enzyme activity during a number of cycles. In the immobilization of commercial lipase, enzyme activity efficiency decreases with increasing number of reuse [4]. As revealed by Mohammadi et al. [27] on the study of the hydrolysis of fish oil using immobilized lipase R. miehei, the remaining activity decreased from 100% to 83% after 5 cycles. Moreover, Patel et al. [24] also showed a decrease in residual esterification percentage from 100% to 50% on the production of ethyl caprylate after 29 times cycles.

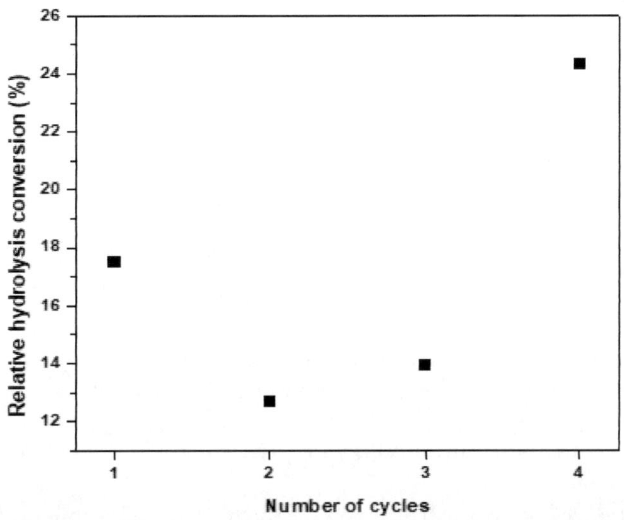

Figure 3. Relative hydrolysis conversion (%) of the immobilized lipase reuse

3.3. Application immobilized lipase in esterification for bio-flavor production

Immobilized lipase on PUF which has been examined the stability and reusability performance was applied to the esterification reaction with the FFA substrate produced by hydrolysis (in section 3.2) and citronellol, resulting in bio-flavor. The content of citronellol before and after the esterification process was analysed by GC-FID to determine the conversion percentage. The conversions of citronellol (%) in producing bio-flavor after 5, 10, 15, 20 and 25 h were 32.5, 42.6, 46.4, 51.9 and 52.1, respectively.

The results of the optimum bio-flavor conversion by employing immobilized lipase and free lipase were 52.1 and 44.4 %, respectively. Applying immobilized lipase provides a higher conversion than the free lipase because, after immobilized, the enzyme will be a restructuring of the "lid" lipase structure yielding configuration changes. This configuration changes may be due to contact of the enzyme into the interfacial region which causes hydrophobic exposed outside and the hydrophilic inside (becomes an open form) leading to high contact between the enzyme and substrate [4,24]. As a comparison, the highest yield produced by the esterification reaction using immobilized lipase enzyme was 85 % for 48 h (higher than the yield of this study). However, this study uses non-commercial *M. miehei* lipase, while Patel et al. [24] used commercial lipase *C. rugosa* so that giving different performance.

4. Conclusion

Lipase immobilized on co-coated PUF gave the highest thermal stability compared to immobilized on PUF only and free lipase. The immobilized lipase could retain its hydrolytic activity after 5 times cycles without organic solvent washing. The immobilized enzyme prepared in this study also gave better performance than free lipase in the production of bio-flavor.

Conflict of interest

No conflict of interest.

Acknowledgments

The authors thank the Ministry of Higher Education for Fundamental Research Grant (205/SP2H/LT/DRPM/III/2016) funded by the Indonesian Government.

References

[1] Gao Z, Chu J, Jiang T, Xu T, Wu B and He B 2018 Lipase immobilization on functionalized

mesoporous TiO2: Specific adsorption, hyperactivation and application in cinnamyl acetate synthesis *Process Biochem.* **64** 152–9

[2] Zhong N, Chen W, Liu L and Chen H 2019 Immobilization of Rhizomucor miehei lipase onto the organic functionalized SBA-15: Their enzymatic properties and glycerolysis efficiencies for diacylglycerols production *Food Chem.* **271** 739–46

[3] Andrade T A, Errico M and Christensen K V 2017 Castor oil transesterification catalysed by liquid enzymes: Feasibility of reuse under various reaction conditions *Chem. Eng. Trans.* **57** 913–8

[4] Gupta S, Bhattacharya A and Murthy C N 2013 Tune to immobilize lipases on polymer membranes: Techniques, factors and prospects *Biocatal. Agric. Biotechnol.* **2** 171–90

[5] Paludo N, Alves J S, Altmann C, Ayub M A Z, Fernandez-Lafuente R and Rodrigues R C 2015 The combined use of ultrasound and molecular sieves improves the synthesis of ethyl butyrate catalyzed by immobilized Thermomyces lanuginosus lipase *Ultrason. Sonochem.* **22** 89–94

[6] Cui C, Tao Y, Li L, Chen B and Tan T 2013 Improving the activity and stability of Yarrowia lipolytica lipase Lip2 by immobilization on polyethyleneimine-coated polyurethane foam *J. Mol. Catal. B Enzym.* **91** 59–66

[7] Molina G, Pimentel M R, Bertucci T C P and Pastore G M 2012 Application of fungal endophytes in biotechnological processes *Chem. Eng. Trans.* **27** 289–94

[8] SÁ A G A, Meneses A C de, Araújo P H H de and Oliveira D de 2017 A review on enzymatic synthesis of aromatic esters used as flavor ingredients for food, cosmetics and pharmaceuticals industries *Trends Food Sci. Technol.* **69** 95–105

[9] Sun J, Yu B, Curran P and Liu S Q 2013 Lipase-catalysed ester synthesis in solvent-free oil system: Is it esterification or transesterification? *Food Chem.* **141** 2828–32

[10] Qi H, Du Y, Hu G and Zhang L 2018 Poly(carboxybetaine methacrylate)-functionalized magnetic composite particles: A biofriendly support for lipase immobilization *Int. J. Biol. Macromol.* **107** 2660–6

[11] Pires-Cabral P, da Fonseca M M R and Ferreira-Dias S 2010 Esterification activity and operational stability of Candida rugosa lipase immobilized in polyurethane foams in the production of ethyl butyrate *Biochem. Eng. J.* **48** 246–52

[12] Nyari N L D, Fernandes I A, Bustamante-Vargas C E, Steffens C, De Oliveira D, Zeni J, Rigo E and Dallago R M 2016 In situ immobilization of Candida antarctica B lipase in polyurethane foam support *J. Mol. Catal. B Enzym.* **124** 52–61

[13] Xie W and Huang M 2018 Immobilization of Candida rugosa lipase onto graphene oxide Fe3O4nanocomposite: Characterization and application for biodiesel production *Energy Convers. Manag.* **159** 42–53

[14] Wan D, Tian L, Li X, Li B and Zhang Q 2018 A versatile strategy for enzyme immobilization: Fabricating lipase/inorganic hybrid nanostructures on macroporous resins with enhanced catalytic properties *Biochem. Eng. J.* **139** 101–8

[15] Zhao X, Qi F, Yuan C, Du W and Liu D 2015 Lipase-catalyzed process for biodiesel production: Enzyme immobilization, process simulation and optimization *Renew. Sustain. Energy Rev.* **44** 182–97

[16] El-Ghaffar M A A and Hashem M S 2010 Chitosan and its amino acids condensation adducts as reactive natural polymer supports for cellulase immobilization *Carbohydr. Polym.* **81** 507–16

[17] Liang Y, Bar-Shir A, Song X, Gilad A A, Walczak P and Bulte J W M 2015 Label-free imaging of gelatin-containing hydrogel scaffolds *Biomaterials* **42** 144–50

[18] Xue J and Zhong Q 2014 Blending lecithin and gelatin improves the formation of thymol nanodispersions *J. Agric. Food Chem.* **62** 2956–62

[19] Balaji V and Ebenezer P 2008 Optimization of extracellular lipase production in Colletotrichum gloeosporioides by solid state fermentation *Indian J. Sci. Technol.* **1** 1–8

[20] Ferraz L I R, Possebom G, Alvez E V, Cansian R L, Paroul N, de Oliveira D and Treichel H 2015 Application of home-made lipase in the production of geranyl propionate by esterification of geraniol and propionic acid in solvent-free system *Biocatal. Agric. Biotechnol.* **4** 44–8

[21] Lopes D B, Fraga L P, Fleuri L F and Macedo G A 2011 Lipase and esterase: to what extent can

this classification be applied accurately? *Ciência e Tecnol. Aliment.* **31** 603–13

[22] Garlapati V K and Banerjee R 2013 Solvent-free synthesis of flavour esters through immobilized lipase mediated transesterification *Enzyme Res.* **2013** 1-6

[23] Sharma A, Chaurasia S P and Dalai A K 2013 Enzymatic hydrolysis of cod liver oil for the fatty acids production *Catal. Today* **207** 93–100

[24] Patel V, Gajera H, Gupta A, Manocha L and Madamwar D 2015 Synthesis of ethyl caprylate in organic media using Candida rugosa lipase immobilized on exfoliated graphene oxide: Process parameters and reusability studies *Biochem. Eng. J.* **95** 62–70

[25] Perez-Miller S J and Hurley T D 2003 Coenzyme isomerization is integral to catalysis in aldehyde dehydrogenase *Biochemistry* **42** 7100–9

[26] Mathpati A C, Badgujar K C and Bhanage B M 2016 Kinetic modeling and docking study of immobilized lipase catalyzed synthesis of furfuryl acetate *Enzyme Microb. Technol.* **84** 1–10

[27] Mohammadi M, Habibi Z, Dezvarei S, Yousefi M and Ashjari M 2015 Selective enrichment of polyunsaturated fatty acids by hydrolysis of fish oil using immobilized and stabilized Rhizomucor miehei lipase preparations *Food Bioprod. Process.* **94** 414–21

Synthesis of nickel – hydroxyapatite by electrochemical method

A Nur[1], A Jumari[1], A W Budiman[1], A H Wicaksono[1], A R Nurohmah[1], N Nazriati[2] and F Fajaroh[2]

[1]Chemical Engineering Department, Faculty of Engineering, Sebelas Maret University, 57126 Surakarta, Indonesia
[2]Chemistry Department, Faculty of Mathematics and Science, State University of Malang, 65111 Malang, Indonesia

Email : adriannur@staff.uns.ac.id

Abstract. The electrochemical method in two chambers with the bipolar membrane has been used to produce the nickel – hydroxyapatite particle. Hydroxyapatite is used as the supported catalyst for nickel catalyst. In this work, in-situ electrochemical synthesis in two chambers with bipolar membrane has been used in the formation of nickel-hydroxyapatite. A simple electrolysis cell from acrylic was used. The anode and the cathode were set parallel with a distance of 3 cm and immersed in the solution at a depth of 2 cm. The anode and the cathode were connected to a DC power supply. The particle was produced in the electrolysis cell. In the anode chamber, the pH solution decreases for the time of electrolysis while in the cathode chamber the pH solution increases. The nickel – hydroxyapatite particle is formed in the cathode chamber. There is no particle in the anode chamber. At the lower current density, the first formed is brushite. At the higher current density, brushite converts to HA. The higher current density the greater the chances of HA formation. The increase in the initial concentration of $NiCl_2$ accelerates the increase of the pH of the electrolysis solution.

1. Introduction

Hydroxyapatite (HA) is a very attractive calcium phosphate because of its many applications. HA has been used as the coating for metallic implants [1, 2], as the bone filler [3], bone tissue engineering [4], and drug delivery [5]. In addition, HA may be used as a supported catalyst. Hydroxyapatite has attractive properties as a supported catalyst such as the ability of ion exchange, adsorption ability, acid-base properties, not toxic, and good thermal stability. The several methods for HA synthesis methods that have been commonly used are hydrothermal [6], emulsion [7, 8], mechanochemical [9, 10] and chemical precipitation [11, 12]. Among several methods of HA synthesis, electrochemical method is the most promising method. The advantages of the electrochemical method are the ease of controlling the product by regulating the current or potential system. The electrochemical methods for HA synthesis have been developed using the bipolar membranes. The bipolar membrane is used to separate the cell into two chambers, namely the anode chamber and the cathode chamber. The formation of HA is very effective in the cathode chamber because the cations (OH^- ions) still around the electrode and react to form hydroxyapatite. HA can be formed at low current density [13, 14]

Nickel is one of the potential candidates that exhibit good catalyst properties. Nickel as a catalyst has been widely used in some industries such as for the production of hydrogen [15] and syngas [16]. The support of the catalyst is required to increase the catalyst dispersion and prevent the catalyst being lost out with the gas bubbles.

The impregnation of Ni catalyst to HA as a supported catalyst by the electrochemical method in two chambers and the bipoar membrane becomes an interesting challenge. The electrosynthesis of HA with the bipolar membrane, HA was produced in the cathode. In the other side, the Ni^{2+} ions were oxidized in the anode to produce Ni and NiO. Synthesis of HA electrochemically requires the solution in the high pH because at low pH it tends to form brushite [17]. The OH^- ions produced by water reduction at the cathode is very important in HA formation [18]. The addition of nickel to the electrolysis solution of HA causes a decrease in pH. This condition causes the formation of HA getting slower.

In this work, in-situ electrochemical synthesis in two chambers with bipolar membrane has been used in the formation of nickel-hydroxyapatite. The novelty of this research is the electrochemical synthesis in situ of nickel-hydroxyapatite as catalysts in the two chamber and the bipolar membrane. The optimal conditions of the electrolysis process have been studied i.e. the current density, the electrolysis time, and the initial concentration of nickel.

2. Experimental

The electrochemical cell is separated into 2 chambers i.e. the anode chamber and the cathode chamber using a bipolar membrane (Figure 1). The bipolar membrane used is Fumasep FBM from FUMATECH BWT GmbH. It consists of an anion layer and a cation layer. On the borderline between anion membrane (AEM) and cation membrane (CEM) water is split into OH^- and H^+ ions when exceeding a potential difference of approximately 0.8 V. The bipolar dimension is 2×5 cm and 0.2 – 0.25 mm in thickness.

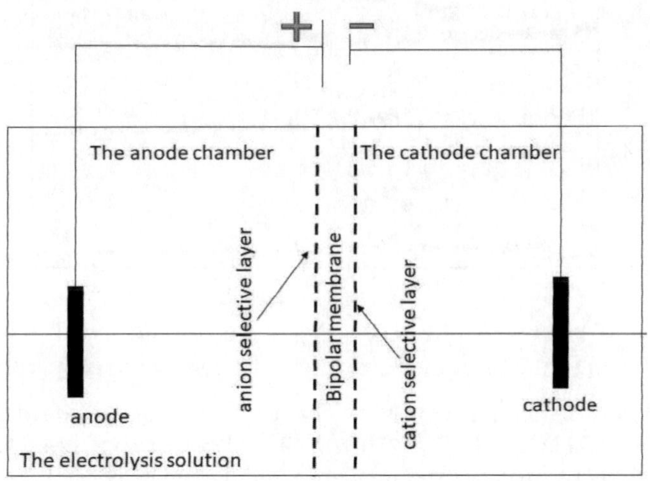

Figure 1. A schematic presentation of the electrolysis cell for synthesis of Ni-HA

The electrolysis solution for the electrosynthesis of HA was obtained from $Na_2H_2EDTA.2H_2O$ (Merck, reagent grade), KH_2PO_4 (Merck, reagent grade) and $CaCl_2$ (Merck, reagent grade). The concentration relationship of Ca^{2+}/EDTA/PO_4^{3-} is 0.25/0.25/0.15 M. The $NiCl_2$ (Merck, reagent grade) was dissolved in the electrolysis solution at the initial concentration of 0.2 to 1 M. All chemicals were used as supplied with no further treatment. Two carbon electrodes with dimensions of (2×1) cm were used as the anode and the cathode. A simple electrolysis cell from acrylic was used. The anode and the cathode were set parallel with a distance of 3 cm and immersed in the solution at a depth of 2 cm. The anode and the cathode were connected to a DC power supply (Zhaoxin PS-3005D). The bipolar membrane was installed between the anode and the cathode. The electrosynthesis was performed for 30 to 60 minutes at 80 to 240 mA/cm^2 at room temperature. The ultrasonic cleaner was used in the

IOP Conf. Series: Materials Science and Engineering **543** (2019) 012026 doi:10.1088/1757-899X/543/1/012026

electrolysis cell to void agglomeration. The pH of the electrolysis solution was measured by the portable pH meter (ST300 Ohaus).

The particle was produced in the electrolysis cell. The particle was separated from the solution by filtration. The particle was washed with demineralized water and dried in the oven at 110°C overnight. The X-ray diffraction (XRD) pattern of the particles was detected using an X-ray diffractometer (Shimadzu 6000). Fourier transform infrared spectroscopy (FTIR) (Shimadzu 8400s) was performed to evaluate the functional groups of specimens. The FTIR spectra were obtained over the region 500-4,000 cm^{-1}.

3. Result and Discussion

3.1. The current density

The study on the effect of the current density on the particle produced was done at the initial concentration 0.5 M $NiCl_2$ for 1 hour the electrolysis time. The green particle was obtained by the electrochemical method of the bipolar membrane under the condition described above. These particles were found in the cathode chamber and no particle in the anode chamber. The XRD patterns of the particle produced were shown in Figure 2.

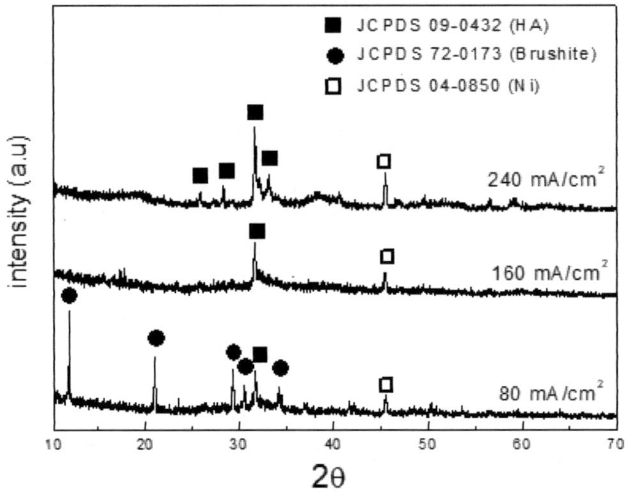

Figure 2. The XRD patterns of the particles produced at various the current density

At the current density 80 mA/cm^2, there are five characteristic peaks matched the standard pattern of brushite (JCPDS 72-0713). They are 11.71° (020), 20.99° (12-1), 29.35° (14-1), 30.53° (121), 34.21° (150). The peak at 31.71° (211) matches the standard pattern of hydroxyapatite (JCPDS 09-0342) and the peak at 45.45° match the standard pattern of Ni (JCPDS 04-0850). At the current density 160 mA/cm^2, the characteristic peaks of the brushite disappear and the peaks of HA and Ni still appear. At the higher current density (240 mA/cm^2), there are four characteristic peaks matched the standard pattern of HA. They are 25.99° (002), 28.37° (210), 31.73° (211), and 33.23° (300). At the lower current density, the first formed is brushite. At the higher current density, brushite converts to HA. The higher current density the greater the chances of HA formation.

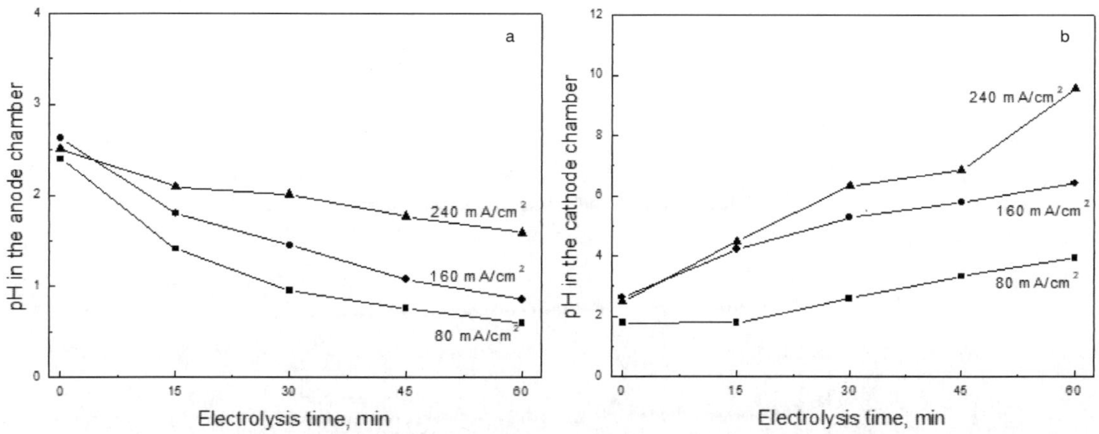

Figure 3. The profile of the solution pH for the electrolysis time at various the current density at the anode chamber (a) and the cathode chamber (b).

The profile of the pH solution for the electrolysis time at the anode chamber and the cathode chamber was shown at Figure 3. In the anode chamber, the pH solution decreased for the time of electrolysis. The decrease in the pH of the solution at the anode chamber is due to the formation of H+ ions in the anode as the result of water oxidation.

$$2H_2O_{(l)} \rightarrow O_{2(g)} + 4H^+_{(aq)} + 4e^- \quad (1)$$

The formation of O_2 was shown by the gas bubble formed in the anode for the electrolysis time. The lower the current density the lower the pH of the solution. At the condition of the pH is very low the solution can damage the bipolar membrane. No particle was formed in the anode chamber because of low the solution pH. In the cathode chamber, the pH solution increased for the time of electrolysis. The increase in the cathode chamber is due to the formation of OH^- ions in the cathode as the result of water reduction.

$$2H_2O_{(l)} + 2e^- \rightarrow H_{2(g)} + 2OH^-_{(aq)} \quad (2)$$

The formation of H_2 was confirmed by the gas bubble formed in the cathode for the electrolysis time. The higher the current density the higher the pH of the solution. At 80 mA/cm^2, the final pH solution was 4. This pH causes the calcium phosphate to consist of brushite and HA. At 160 and 240 mA/cm^2, the final pH solution was more than 4. The HA was formed at this pH or higher.

3.2. The initial concentration of $NiCl_2$

The influence of the initial concentration of $NiCl_2$ on the particle produced was performed at 240 mA/cm^2 for 30 minutes. The initial concentrations of $NiCl_2$ were 0.2 M, 0.5 M, and 1 M. Figure 4 shows the XRD patterns of particle produced for various of the initial concentration of $NiCl_2$.

The characteristic peaks of HA are shown at the XRD patterns of the particles produced at the 0.2, 0.3, and 0.5 M initial concentration of $NiCl_2$. The characteristic peaks of HA are 25.99° (002), 28.37° (210), 31.73° (211), and 33.23° (300). The characteristic peaks of Ni appear at 45.5°.

The crystallite sizes of HA and Ni were calculated using the Scherrer Formula, which is presented as

$$D = \frac{K\lambda}{\beta \cos\theta} \quad (3)$$

IOP Conf. Series: Materials Science and Engineering **543** (2019) 012026 doi:10.1088/1757-899X/543/1/012026

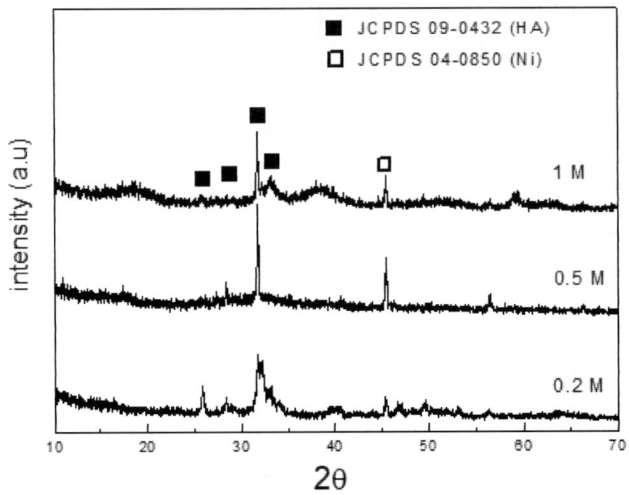

Figure 4. The XRD patterns of the particles produced at various the initial concentration of $NiCl_2$

In this formula, K is the shape coefficient for reciprocal lattice point (assuming $K = 0.9$), λ the wavelength of the X-ray radiation (0.1540 nm), β is full width at half maximum (FWHM) of the peak, and θ is the diffraction angle. The crystallite size of the HA and Ni at several of the initial concentration of $NiCl_2$ is shown in Table 1. The higher the initial concentration of $NiCl_2$ the higher the crystallite size of HA and Ni particles.

Table 1. The Crystallite size of the particle produced at various the initial concentration of $NiCl_2$.

The initial concentration of $NiCl_2$ (M)	The Crystallite Size (nm)	
	HA ($2\theta = 31.7^o$)	Ni ($2\theta = 45.5^o$)
0.2	52.5	36.4
0.5	59.9	43.7
1	59.9	54.7

3.3. *The electrolysis time*

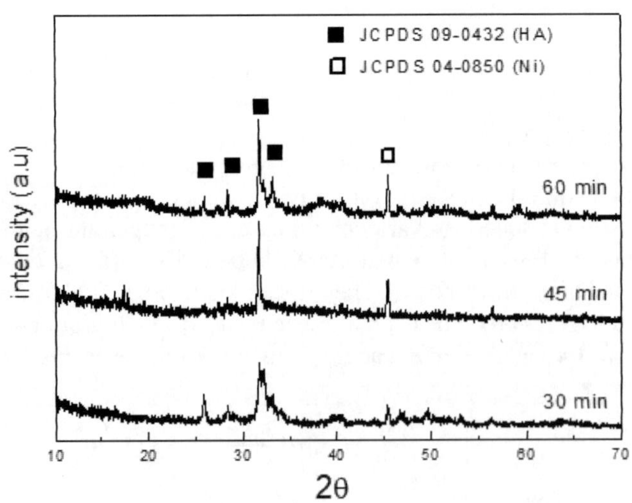

Figure 5. The XRD patterns of the particles produced at various the electrolysis time.

The study of the electrolysis time was done at 0.5 M the initial concentration of $NiCl_2$ and 240 mA/cm$_2$ for 30, 45, and 60 minutes the electrolysis time. The particle produced was analyzed by X-ray diffractometer. The XRD patterns of the particle produced at various the electrolysis time shown in Figure 5. The HA particle was produced at 30 min the electrolysis time. In the high current density (240 mA/cm^2), the formation of HA at the bipolar membrane was very effective. Both the high current density and the bipolar membrane caused the formation of HA was very fast.

Figure 6 show the profile of solution pH for the electrolysis time. In the anode chamber, the pH solution decreases for the time of electrolysis while in the cathode chamber the pH solution increase. The pH of the solution increases fast in 15 minutes. This condition causes the formation of HA is fast. In the anode chamber, the pH of the solution is very low. The low of the solution pH cause there is no particle formed.

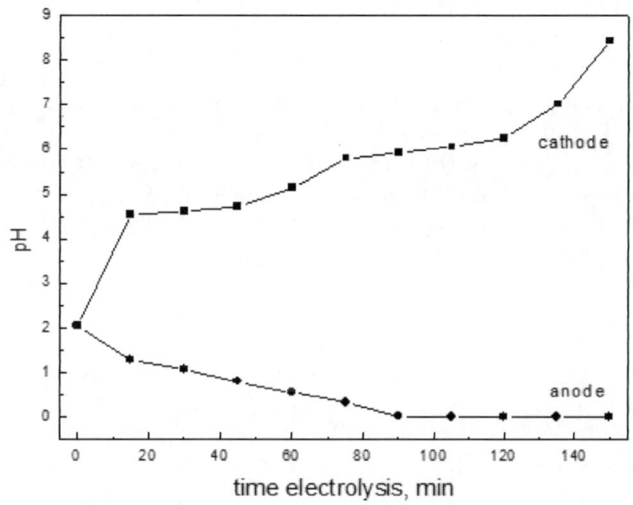

Figure 6. The profile of the solution pH for the electrolysis time.

4. Conclusions

The electrochemical method in two chambers with the bipolar membrane can produce the nickel – hydroxyapatite particle. In the anode chamber, the pH solution decreases for the time of electrolysis while in the cathode chamber the pH solution increases. The nickel – hydroxyapatite particle is formed in the cathode chamber. There is no particle in the anode chamber. At the lower current density, the first formed is brushite. At the higher current density, brushite converts to HA. The higher current density the greater the chances of HA formation. The increase in the initial concentration of $NiCl_2$ accelerates the increase of the pH of the electrolysis solution.

Acknowledgments

The authors would like to thanks to The Ministry of Research and Higher Education of the Republic of Indonesia for Research Grant of PDUPT 2018.

References

[1] Furko M, Havasi V, Kónya Z, Grünewald A, Detsch R, Boccaccini A R and Balazsi C 2017 *Bol la Soc. Esp. Ceram.* **57** 55-65
[2] Harun W S W, Asri R I M, Alias J, Zulkifli F H, Kadirgama K, Ghani S A C and Sharifuddin J H M 2018 *Ceram. Int.* **44** 1250-68
[3] Kolanthai E, Ganesan K, Epple M and Kalkura S N 2016 *Mater. Today Commun.* **8** 31-40
[4] Farokhi M, Mottaghitalab F, Samani S, Shokrgozar M A, Kundu S C, Reis R L, Fatahi Y and Kaplan D L 2018 *Biotechnol. Adv.* **36** 68-91

[5] Kim H, Mondal S, Bharathiraja S, Manivasagan P, Moorthy S and Oh J 2018 *Ceram. Int.* **44** 6062-71
[6] Bensalah H, Bekheet M F, Younssi S A, Ouammou M and Gurlo A 2018 *J. Environ. Chem. Eng.* **6** 1347-52
[7] Ezekiel I, Kasin R S, Ismail Y M B and Noor A F M 2018 *Ceram. Int.* **44** 13082-9
[8] Xiao Q, Zhou K, Chen C, Jiang M, Zhang Y, Luo H and Zhang D 2016 *Mater. Sci. Eng. C* **69** 1068-74
[9] Bulina N V, Chaikina M V, Prosanov I Y, Dudina D V and Solovyov L A 2017 *J. Solid State Chem.* **252** 93-9
[10] Youness R A, Taha M A, Elhaes H and Ibrahim M. 2017 *Mater. Chem. Phys.* **190** 209-18
[11] Guesmi Y, Agougui H, Lafi R, Jabli M and Hafiane A. 2018 *J. Mol. Liq.* **249** 912-20
[12] Yelten-yilmaz A and Yilmaz S 2018 *Ceram. Int.* **44** 9703-10
[13] Nur A, Jumari A, Budiman A W, Nazriati N, Fajaroh F, Fariza J H and Anisa T L 2018 *MATEC Web of Conf.* **156** 05015
[14] Nur A, Nazriati N, Jumari A, Santi E F N and Asiffa S Y 2017 *MATEC Web of Conf.* **101** 03004
[15] Lukman H, Zahira Y, Manal I and Daud W R W 2010 *Key Eng. Mater.* **447-448** 770-4
[16] Jun J H, Lee T J, Lim T H, Nam S W, Hong S A and Yoon K J 2004 *J. Catal.* **221** 178-90
[17] Nur A, Setyawan H, Widjaja A and Lenggoro I W 2014 *BCREC* **9** 168-75
[18] Nur A and Setyawan H. 2016 *J. Chem. Eng. Jpn.* **49** 144-51

The Effect of Total Sulfur Content and Total Acid Number (TAN) on Small Scale Oil Refinery

H Devianto[1], M Eviani[2], R A Rahardi[1], F Yusupandi[1]
[1] Department of Chemical Engineering, Bandung Institute of Technology, Ganesha 10, Bandung, Indonesia
[2] Ministry of Energy and Mineral Resources Republic of Indonesia, Research and Development Agency for Energy and Mineral Resources, Research and Development Centre for Oil and Gas Technology, LEMIGAS, Ciledug, Jakarta Selatan, Indonesia

E-mail : hardev@che.itb.ac.id

Abstract. Corrosion is one of major problems in refinery industry which leading to mechanical failure of equipments. Corrosion mechanism and rate depend on the corrosive substance, temperature, and material of equipment. Refinery feedstock commonly is a blending of different oils contains different corrosive substance. One of corrosive substance, Sulfur Content and TAN, can cause thinning of material surface at high temperature. This research aims to study the effect of sulfur content and TAN on corrosion mechanism and rate of equipment and piping in small scale refinery. Corrosion mechanism and rate for several refinery feedstocks, containing various sulfur content and TAN, were determined using API-581RP. Mass and energy balance for small scale refinery consists of distillation unit, sulfur removal unit, and platformer unit was done based on process simulator. Materials concerned in this research were carbon steel, alloy steel, and stainless steel. The result showed most of the mechanism in distillation column was thinning high temperature sulfidic/naphthenic corrosion while in sulfur removal and platformer were thinning high temperature H_2S/H_2 corrosion. The change of sulfur and acid content caused change corrosion rate in non-linear way. These results can be used as prediction for future corrosion assessment based on sulfur and TAN composition.

1. Introduction

Corrosion is major problems in industry especially oil refinery [1]. Corrosion is natural phenomenon and cause material depletion of equipment because of its interaction with its environment [2]. Corrosion can cause many casualties and economical loss, even endanger worker safety [3]. There are many refinery explosions caused by corrosion in one of the equipment which lead to leaking or explosion [4]. Corrosion analysis is needed to prevent major accident caused by corrosion [5]. Prevention of corrosion is key point to assure process safety in the refinery and petrochemical industry [6]. Corrosion analysis can be used to predict lifetime of equipment and prepare the replacement to be installed at plant turnaround [7]. Corrosion analysis can be done by using API 581 to predict corrosion mechanism and rate based on impurities, temperatures, and equipment materials [8].

In general, small scale refinery consists of three main units: distillation, desulphurization unit, and platformer [9]. Distillation unit is used to separate raw crude into several product; naphtha, kerosene, diesel, etc. Distillation unit has impurities such as sulfur organic, TAN, salt, and water which sulfur

IOP Conf. Series: Materials Science and Engineering **543** (2019) 012027 doi:10.1088/1757-899X/543/1/012027

organic and TAN cause thinning at high temperature while salt can cause localized corrosion such as pitting even at room temperature [10]. Desulphurization unit is used to convert organic sulfur into H_2S and lower the corrosion potential. In Desulphurization unit beside sulfur and TAN, impurities such as H_2S, hydrogen, ammonia, water can also cause localized and general corrosion [11]. Platformer is used to increase gasoline octane number by reforming lower octane number hydrocarbons into higher octane number using catalyst. This process is done at high temperature in environment rich of hydrogen. This can cause high temperature hydrogen attack, which is one types of corrosion that can be dangerous [12].

The impurities in small scale refinery units are very diverse and complicated. Small change in hydrocarbons or impurities content can cause major change of corrosion severity [13]. Thus, a study to predict the effect of impurities or hydrocarbons composition to corrosion in small scale refinery must be done. This study aim to analyze the effect of different crude, in which contains different impurities, to corrosion mechanism and rate in small scale refinery plant. The output hopefully can be used as basic prediction for future corrosion in refinery.

2. Experiment Methods

The methodology for this study is shown by figure 1. Initially, by using crude assay, mass and energy balance of small scale refinery plant is calculated. From mass and energy balance, the data needed for corrosion analysis can be extracted. In corrosion analysis, corrosion mechanisms and rates are checked for individual equipment. Each corrosion rate is labeled by different colors to indicate how severe the damage is. Then every analysis for equipment is combined into corrosion mapping followed by evaluation.

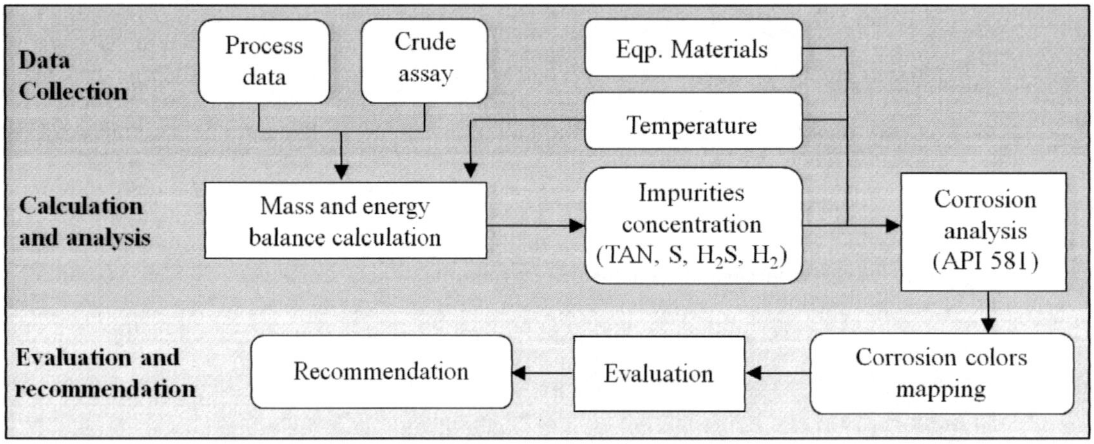

Figure 1. Methodology for this study

This study is done by using three types of crudes: crude that contains most heavy hydrocarbons (Crude A); crude that contains most light hydrocarbons (Crude B); and mixture of both Crude A and Crude B in ration 50:50 (Crude C). Each crude contains different sulfur and TAN content, which shown in table 1. Crude B which contains most light hydrocarbons has higher sulfur and TAN content than the others. The mass and energy balance calculation is done for three crude assays. Mass and energy balance calculation is carried out by using Process Simulation Software.

Table 1. Sulfur and TAN content in three types of crudes

Impurities	Crude A	Crude B	Crude C
Sulfur content (%-wt)	0.31	1.56	0.94
TAN content (mg KOH/g)	0.04	0.13	0.09

Process Simulation Software used crude assay as input to construct different types of hydrocarbons based on its boiling point. The impurities such as sulfur and TAN were added by using compound model such as carboxylic acid or sulfur organic. For this study, the compound models used for sulfur are thionaphthene (boiling point 219°C) as light sulfur and 1-octadecanethiol (boiling point 360°C) as heavy sulfur. For TAN, the compound models used are pentadecanoic acid ($C_{15}H_{30}O_2$) and eicosanoic acid ($C_{20}H_{40}O_2$). Those compounds are used because sulfur and TAN content are much concentrated in two boiling points; around 200 – 230°C and 320 – 360°C [14].

Small scale refinery consists of three main units: Crude Distillation Unit or CDU, Naphtha Hydrotreater or NHT, and Platformer or PTU [9]. Process flow diagrams of each unit are shown in figure 2. Each units consist of different number of equipment which shown in table 2. There are three types of equipment material used in each unit; there are carbon steel, stainless steel, and low alloy carbon steel.

Table 2. Number of equipment contained in each unit

Equipment Types	Number of Equipments		
	CDU	NHT	PTU
Columns	4	1	1
Reactor	0	2	3
Heat Exchanger	2	10	7
Vessels	24	2	3
Furnace	1	1	3
Total Equipment	31	16	17

Corrosion analysis is based on API 581. In API 581, determining corrosion mechanism and rate needs concentration and types of impurities in each stream, equipment materials, and temperature. In this study which mass and energy balance is done by simulation, there are only three types of corrosion can be determined. First, the impurities are TAN and sulfur organic which cause Thinning High Temperature Sulfidic/Naphthenic Acid Corrosion at high temperature (>400°F). Second, the impurities are H_2S and H_2 which cause Thinning High Temperature H_2S/H_2 Corrosion at high temperature (>400°F). Last, hydrogen gas at high temperature (>400°F) in carbon steel or low alloy steel can cause High Temperature Hydrogen Attack (HTHA).

Corrosion mapping severity level is based on four colors; green for low, blue for moderate, yellow for high, and red for severe. Based on API 581, boundary limits of corrosion rate for each mapping colors are shown in table 3.

Table 3. Number of equipment contained in each unit

	Colors	mpy	mmpy
Low	Green	< 2	< 0,05
Moderate	Blue	2–20	0.05–0.51
High	Yellow	20–50	0.51–1.27
Severe	Red	> 50	> 1.27

IOP Conf. Series: Materials Science and Engineering **543** (2019) 012027 doi:10.1088/1757-899X/543/1/012027

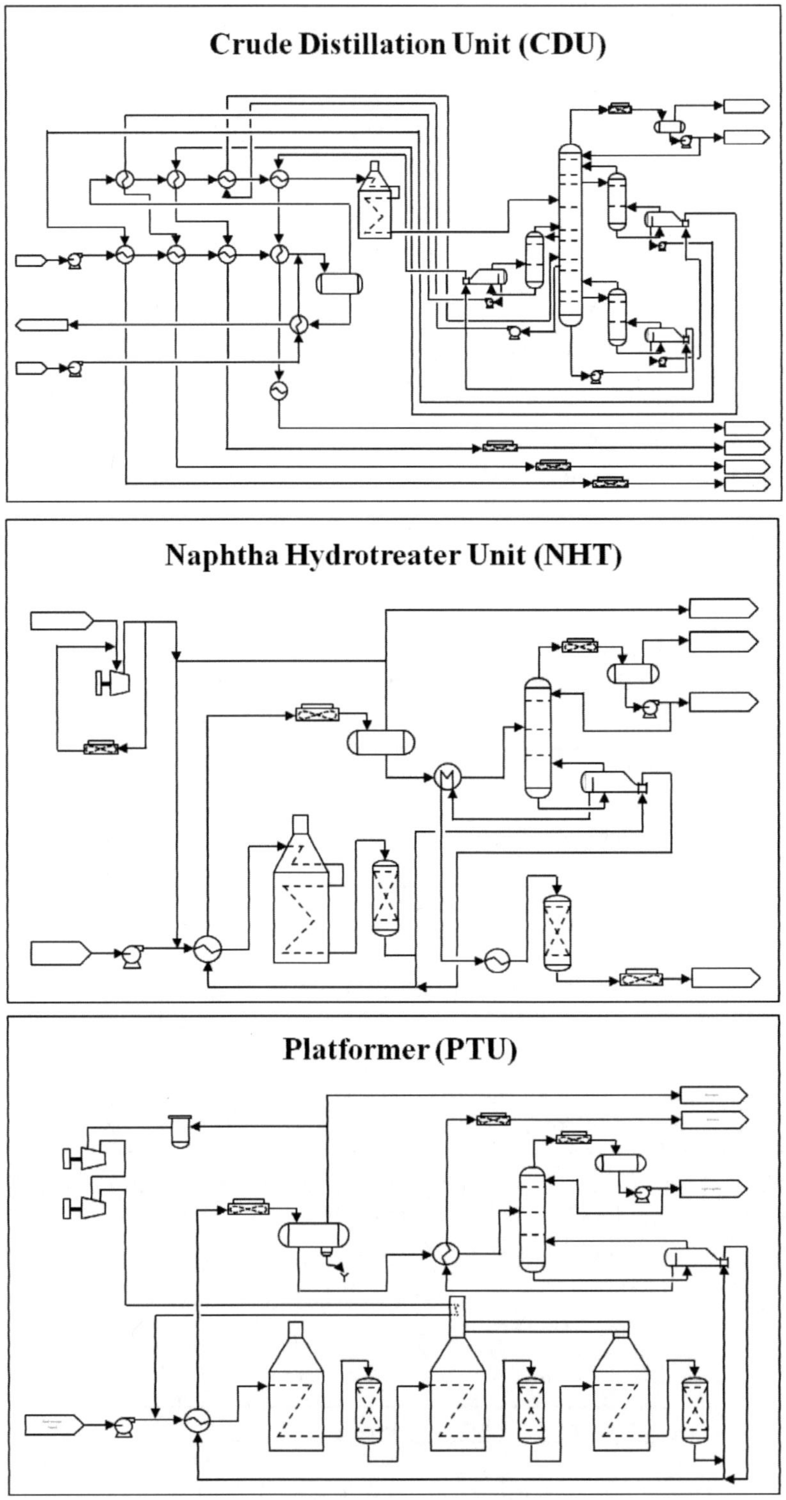

Figure 2. Process flow diagrams

3. Results and Discussion

3.1. Crude assay analysis for early prediction

Figure 3 shows hydrocarbon composition inside each crude based on its boiling points. Crude A contains more hydrocarbons with boiling points above 300°C while Crude B contains more hydrocarbons with boiling points below 200°C. Thus, Crude A is categorized as heavy crude and Crude B is categorized as light crude. Crude C is a mixture between Crude A and B in 50:50 ratio, thus the composition is rather even between each boiling point.

It is expected, that Crude A will have heavier product, such as diesel and atmospheric residue rather than light product such as naphtha or kerosene. This means, in distillation column the flow of hydrocarbons is mostly headed toward bottom columns. While in Crude B, the flow of hydrocarbons is mostly headed toward top columns. Sulfur and TAN content is distributed mostly towards bottom columns, thus near bottom columns sulfur and TAN content should be higher [15]. The concentration of sulfur and TAN in Crude A at bottom columns should be lower than Crude B because Crude A has more heavy hydrocarbons which can dilute sulfur and TAN concentration. This means, Crude A should have lower corrosion severity than Crude B. In addition of that, Crude B contains more sulfur and TAN than Crude A. In Crude C, because the hydrocarbon composition is spread evenly, corrosion severity became more unpredictable. The corrosion severity will depend strongly to sulfur and TAN concentration which tied to hydrocarbon movement inside columns.

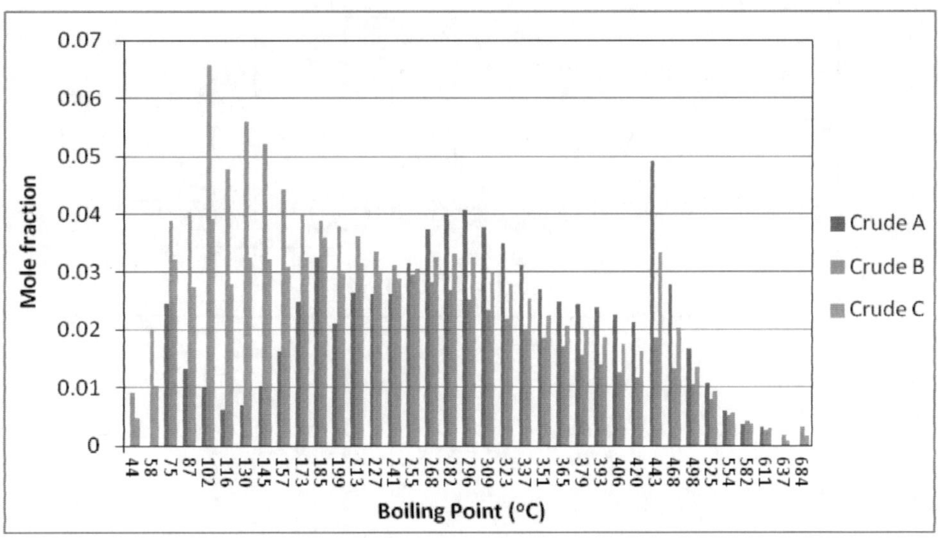

Figure 3. Hydrocarbons composition based on boiling point for each crude assays

3.2. Corrosion Mapping Results and Analysis

The corrosion mapping results for each unit are shown in figure 4 – 7.

IOP Conf. Series: Materials Science and Engineering **543** (2019) 012027 doi:10.1088/1757-899X/543/1/012027

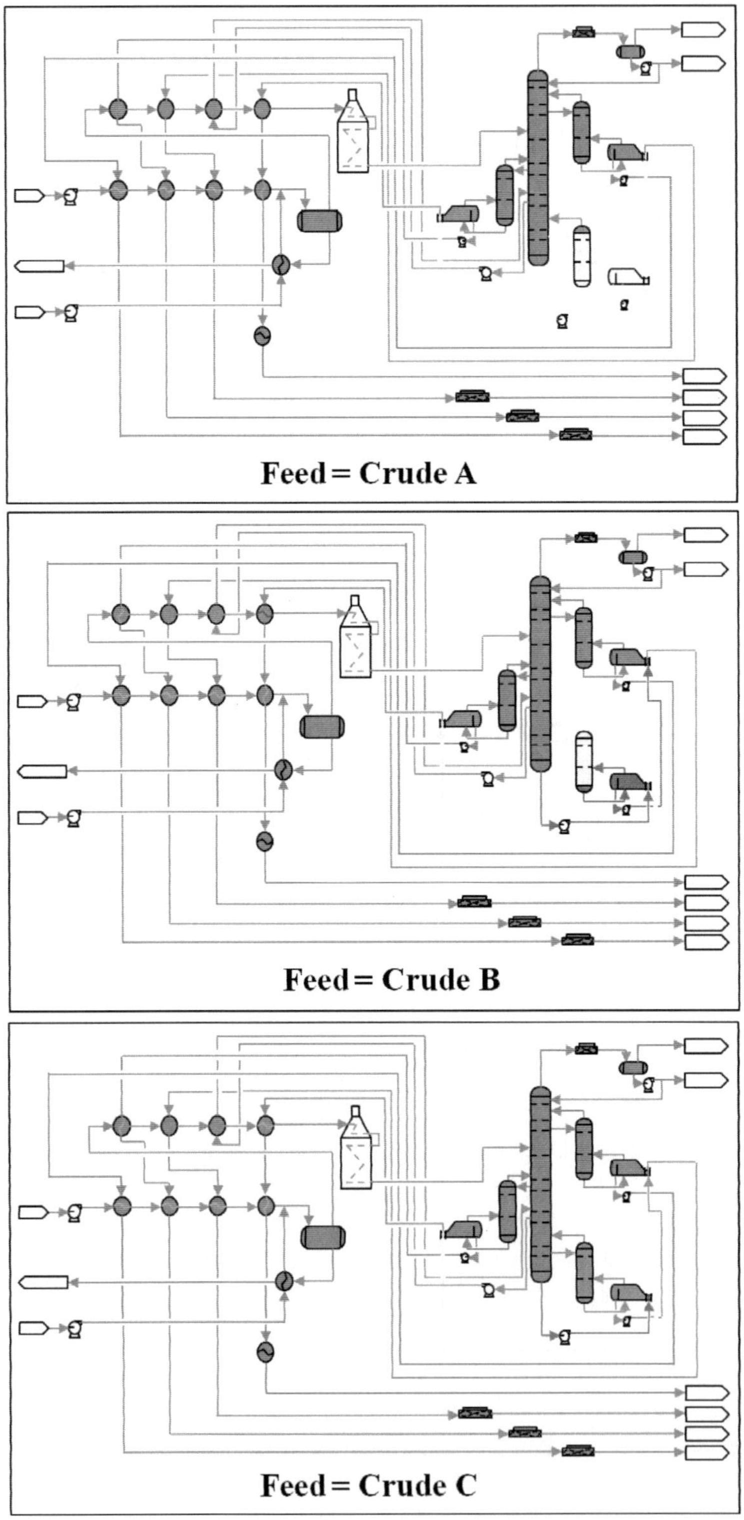

Figure 4. Corrosion mapping of Crude Distillation Unit (CDU)

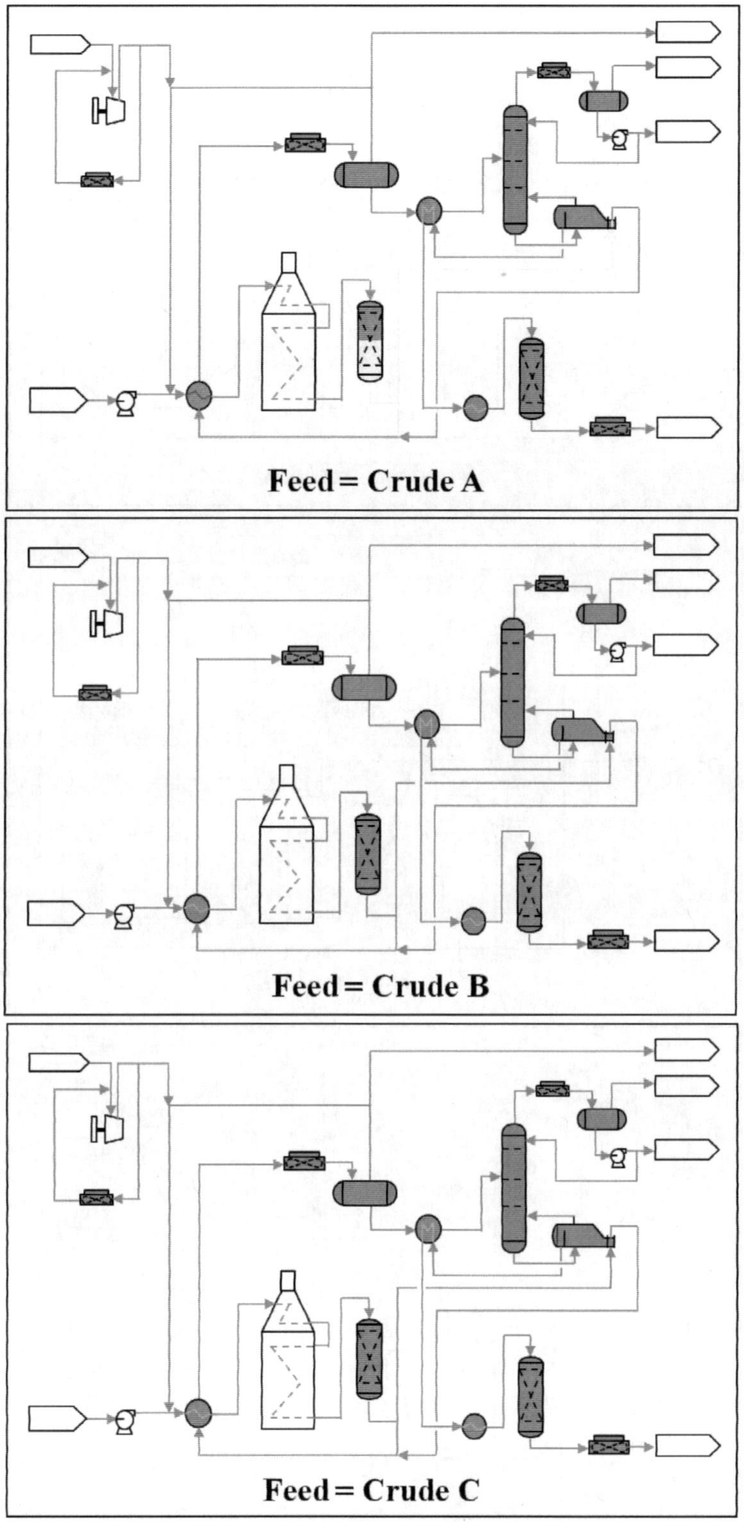

Figure 5. Corrosion mapping of Naphtha Hydrotreater (NHT)

IOP Conf. Series: Materials Science and Engineering **543** (2019) 012027 doi:10.1088/1757-899X/543/1/012027

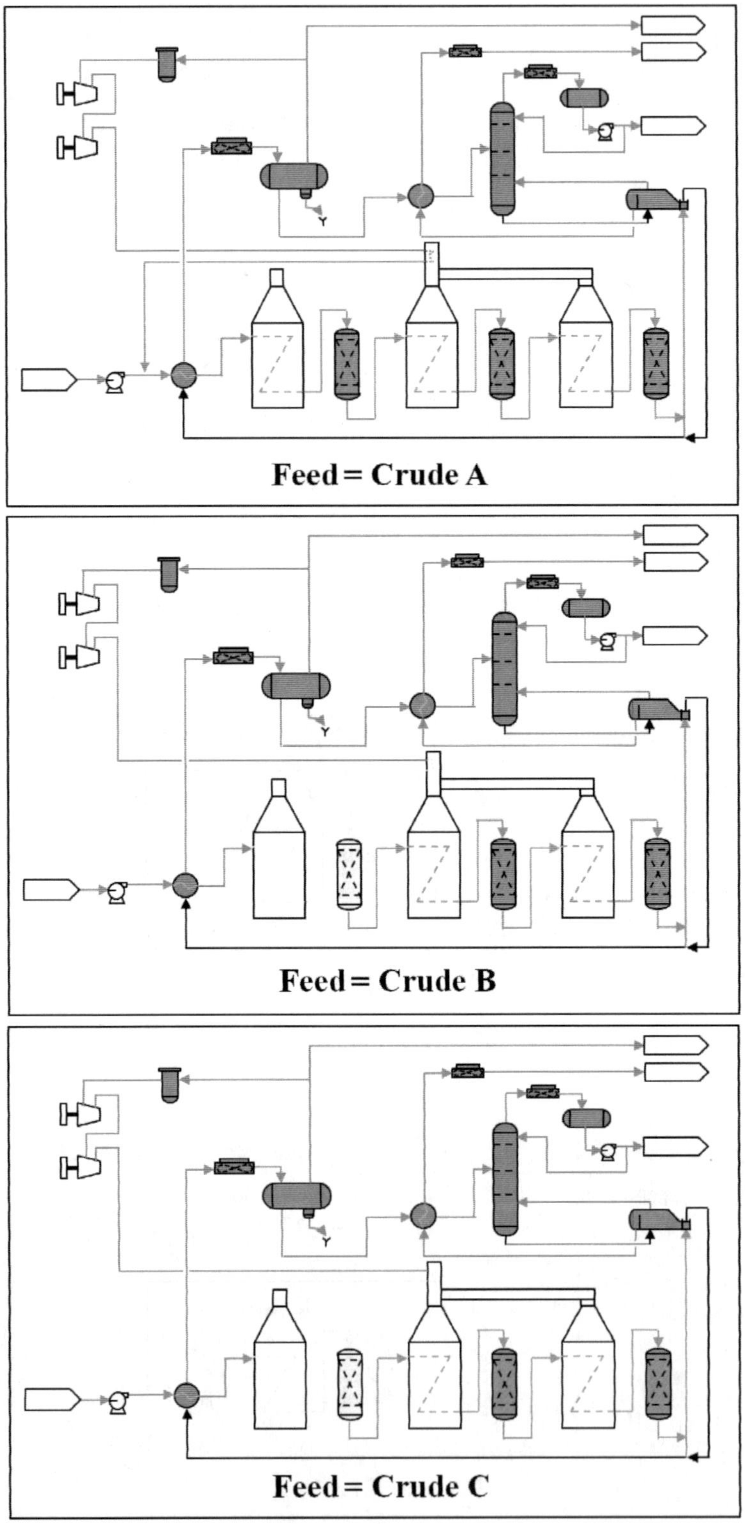

Figure 6. Corrosion mapping of Platformer (PTU)

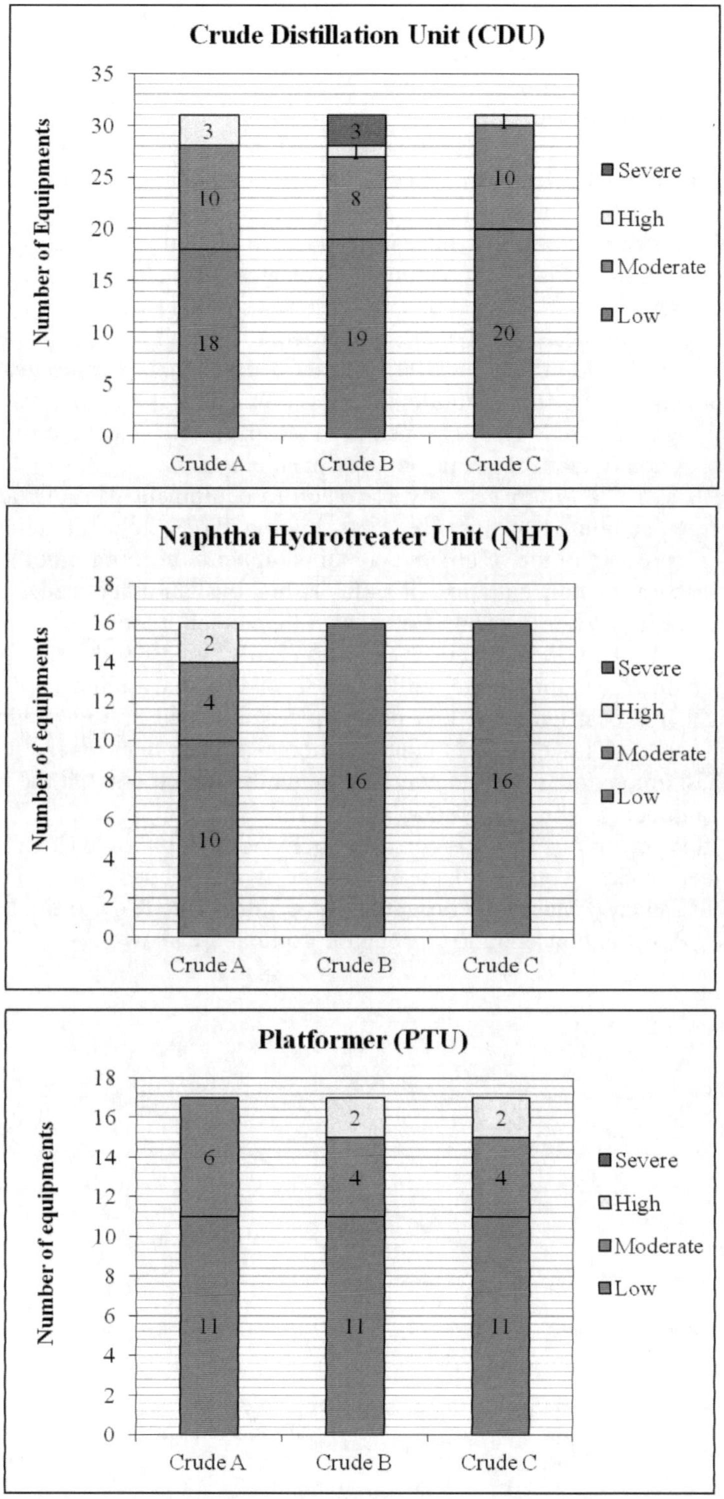

Figure 7. Number of equipment damaged by corrosion and its severity

Between these three units, Crude distillation Units or CDU have the highest number of equipment that damaged by corrosion. This happens because CDU is the first unit that raw crude enters. The impurities in CDU are higher than any other units. NHT has less corrosion damage because NHT feed comes from heavy naphtha product around the top column of CDU. Sulfur and TAN are mostly concentrated in bottom columns [15]. Thus, this heavy naphtha contains very little sulfur and TAN to cause significant corrosion. PTU feed also comes from NHT, thus the corrosion damage is also low.

In CDU, the corrosion mechanism happened is only Thinning High Temperature Sulfidic/Naphthenic Acid Corrosion, while in NHT only Thinning High Temperature H_2S/H_2 Corrosion, and PTU only High Temperature Hydrogen Attack. Thinning corrosion due to sulfidic/naphthenic acid is only happened at CDU because in NHT sulfur organic is converted into H_2S. Conversion of sulfur organic into H_2S in NHT process is usually very high, around 90-99% [16]. Thus, resulted no sulfur carried throughout NHT and PTU. However, H_2S produced from the reaction also will cause thinning at high concentration in NHT. In PTU, the reaction is conducted in high temperature (500°C) which raised HTHA probability to happen [17]. Because only HTHA happened, the corrosion damage should relatively same between each crude.

In NHT, Crude B and C do not cause any corrosion to equipment while Crude A does. Crude A contains heavier hydrocarbons which cause most hydrocarbons concentrated in bottom column product. This causing product in top of column contains higher concentration of sulfur and TAN than the other crudes. Because the concentration of sulfur is higher than other crudes, then the reaction of sulfur to H_2S will also give higher concentration of H_2S than other crudes.

On the contrary, in CDU, Crude B causes corrosion with highest severity. This happens because two things. First, crude B contains more sulfur and TAN than any other crude. Second, Crude B contains mostly light hydrocarbons thus less heavy hydrocarbon in bottom columns product. Fewer heavy hydrocarbon make sulfur dan TAN concentration is higher than other crudes. This cause the severity is the highest among other crudes which is shown by red colors. On the other hand, by using Crude A, several equipments, which colored as red in Crude B, are colored as yellow.

The interesting thing from this result is corrosion behavior using Crude C. Crude C has the lowest severity among other crudes. This could mean that the hydrocarbon is separate evenly in top and bottom column in which case cause the concentration of sulfur and TAN to the lowest. To prove this hypothesis, in figure 8, sulfur concentrations along the column are shown.

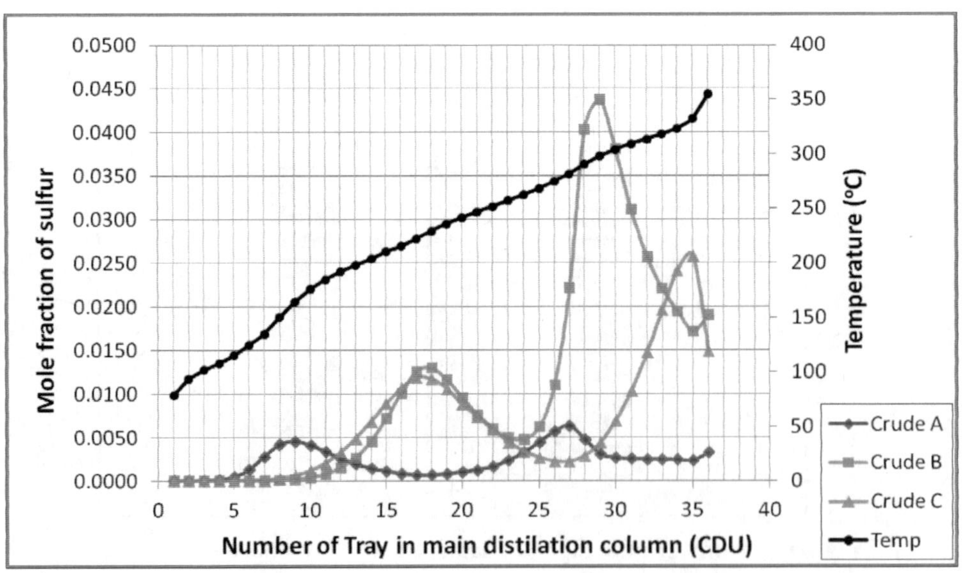

Figure 8. Sulfur concentrations along the main distillation column in CDU

IOP Conf. Series: Materials Science and Engineering **543** (2019) 012027 doi:10.1088/1757-899X/543/1/012027

Figure 8 shows the concentration of sulfur at near bottom column in crude B is the highest among the other. This proved the discussion before that heavier hydrocarbon make concentration sulfur and TAN at bottoms columns and vice versa. Meanwhile, in Crude C, sulfur concentration is separated and concentrated near two points in tray number 17 and 35. Figure 8 also shows temperature profile along the column. It shows that tray number 17 temperatures is around 220°C and tray number 35 is around 330oC. According to literature, sulfur and TAN content are usually concentrated in two boiling points; around 200–230°C and 320–360°C [18]. This means because Crude C has hydrocarbons composition even at boiling point 200 until 360°C, the sulfur movement follows the common average temperature of light sulfur and heavy sulfur. The separation is very clear due to this study only modeled sulfur as two kinds; heavy and light. Thus the heavier sulfur will go to hydrocarbons around 330°C and the lighter goes to hydrocarbons around 220°C.

This phenomenon clearly can be seen in corrosion mapping (figure 4). Sub-column distillation and reboiler near the bottom columns of main distillation has distinguished color in each crude. In Crude A, the color is mostly yellow, while in Crude B is mostly yellow and red, and in Crude C is mostly blue. This means, the corrosion damage in Crude A is high, Crude B is severe, and Crude C is moderate. This sub-column has feed comes from the 27th tray of main distillation column. From Figure 7, in tray 27 the sulfur concentration for Crude C is the lowest than other crudes. This proves that different types of hydrocarbon can effect strongly to corrosion severity.

4. Conclusion

This study presents corrosion analysis for small scale refinery. The results showed that the change of sulfur and TAN composition on crude oil could cause significant change in corrosion rate in nonlinear way especially in Crude Distillation unit which has the highest number of equipment damaged by corrosion with the most severity. These results can be used as prediction for future corrosion assessment based on sulfur and TAN composition.

5. References

[1] Cala J S, Miranda C M, Catano D L, Pena D Y and Klapper H S 2017 *Chem. Eng. Trans.* **57** 1435–40.

[2] Landolt D 2007 *Corrosion and Surface Chemistry of Metals* CRC Press Taylor & Francis Group Boca Raton FL.

[3] Kim J, Park S R and Moon I 2009 *10th Int. Symp. Proc. Sys. Eng. PSE2009* **27** 1839–1844.

[4] Chettough S, Hamzi R and Benaroua K 2016 *J. Loss Prev. Process Ind.* **41** 186–93.

[5] Wu W, Cheng G and Zhou HH 2013 *Eng. Fail. Anal.* **32** 23–34.

[6] Alviz A, Kafarov V and Merino L 2017 *Chem. Eng. Trans.* **61** 1315–20.

[7] Tak K and Kim J 2018 *Comp. Chem. Eng.* **117** 97–104.

[8] Shishesaz M R, Bajestani M N, Hashemi S J and Shekari E 2013 *Int. J. Press. Vess. Pip.* **111-112** 202–8.

[9] Andrews A, Lattanzio R K, Pirog R, Warner J D and Yacobucci B D 2014 *Cong. Res. Serv. Report* **R43682** 1–19.

[10] Loto R T 2017 *J. Mat. Res. Tech.* 6 203–12.

[11] Groysman A 2017 *Corrosion Problems and Solutions in Oil Refining and Petrochemical Industry* Springer Switzerland.

[12] Schuyten J 1947 *Mat. Performance* **4** 13–9

[13] Benziane M M, Said B B and Boudouani, N 2017 *J Pet. Sci. Eng.* **158** 672–9

[14] Eriazy M R, Nasimi N and Roomi Y A 1999 *Ind. Eng. Chem. Res.* **38** 4507–12.

[15] Andari M K, Bchbehani H and Stanislaus A 2007 *Fuel Sci. Tech. Int.* **14** 939-61.

[16] Bose D 2015 *World Scientific News* **3** 99–111.

[17] Djukic M B, Zeravcic V S, Bakic G M, Sedmak A and Rajicic B 2015 *Eng. Fail. Anal.* **58** 485–98.

[18] Ali M F, Hasan M U and Saleem M 1981 *Mid. East Oil Tech. Conf. Soc. Pet. Eng.* **9583** 51–3.

IOP Conf. Series: Materials Science and Engineering **543** (2019) 012028 doi:10.1088/1757-899X/543/1/012028

Comparative Characterization of Chitosan/Gelatin/Geothermal Silica Biocomposites in Two-Dimensional Film and Three-Dimensional Scaffold Forms

Y Kusumastuti[1,*], F Larasati[1], D R Gunawan[1], M Najmina[1], N R E Putri[1], H T B M Petrus[1], and M Kobayashi[2]

[1]Chemical Engineering Department, Universitas Gadjah Mada, Jl. Grafika No. 2, Kampus UGM, Yogyakarta 55281, Indonesia
[2]Division of Materials Science, Nara Institute of Science and Technology, 8916-5 Takayama, Ikoma, Nara 630-0192, Japan

E-mail: yuni_kusumastuti@ugm.ac.id

Abstract. Scaffold, a template resemble an extracellular matrix, contributes a necessary part in tissue engineering to accommodate the growth of cells. In the development of scaffold made from organic materials such as chitosan and gelatin, researchers have done various ways to modify its properties and one of them is by incorporate it with inorganic materials. This research explored the potential of silica derived from geothermal power plant waste as a biocomposite material for scaffold. Biocomposites with two-dimensional (2-D) film form were prepared by simple drying process at room temperature and ambient pressure, while three-dimensional (3-D) scaffold form were fabricated by freeze-drying. The obtained biocomposites were characterized by Fourier Transform Infra-Red (FTIR) spectroscopy and Scanning Electron Microscopy (SEM). In addition, swelling and degradation tests were also performed on the films and scaffolds. The results showed that there are interactions between each component in chitosan/gelatin/geothermal silica biocomposites and the addition of geothermal silica decreases the swelling and degradation rates of the biocomposites. These results indicate that geothermal silica has a high potential to be used as an additive for controlling the physical properties of chitosan/gelatin scaffolds.

1. Introduction

In the regeneration process of damaged tissue, implanted scaffolds have an essential role. Tissue damage causes the loss of cells and native extracellular matrix (ECM). In order to support cell growth in the healing process, scaffolds should be properly designed to resemble the functions of ECM and provide a supportive environment [1]. A great number of 2-D films and 3-D scaffolds have been developed utilizing a wide range of biomaterials to promote tissue regeneration. The biomaterials have to possess the capability to support cell attachment and nutrient supply. In addition, they need to show good biocompatibility and biodegradability [2].

Scaffolds are made of natural or synthetic materials, or a combination of both in hydrogel [3], composite [4], thin film [5] or lyophilized forms [6]. Recently, development of organic scaffolds incorporating inorganic materials has gained much interest. The purpose of this blending is to overcome the certain limitations of stand-alone organic scaffolds; such as, high degradation rates and overly high solubility in physiological conditions [7].

One common organic-based scaffold is chitosan/gelatin. Chitosan is a natural polysaccharide derived from chitin, a major component in the exoskeleton of crustaceans. Chitosan is obtained by the partial deacetylation of chitin, and is a promising biomaterial for various medical applications; such as, bone grafting, wound dressing, and drug delivery systems, due to its low cost, resource abundance, anti-microbial activity, low toxicity, biocompatibility, and biodegradability [8]. The structure of chitosan resembling glycosaminoglycan and the polysaccharide's hydrophilic nature makes it an attractive material as a scaffold for tissue engineering; especially since the monomeric unit of chitosan, N-acetyl glucosamine, is an essential compound in wound healing [9].

Gelatin, an anionic natural polymer, is one of the blending components for chitosan to improve the biological activity of the scaffolds by improving cell adhesion and migration via the Arg-Gly-Asp (RGD)-like sequence [10]. Chitosan and gelatin form polyelectrolyte complexes that increase the bond stability between the components [11]. On the other hand, gelatin is known to have a hydrophilic nature that indirectly affects the degradation rate of the resulting scaffold. In order to control the degradation rate of chitosan/gelatin scaffolds, methods of forming blends with other polymers have been studied extensively.

Silica sludge or geothermal silica with more than 50 % amorphous silica content in amorphous structure is obtained as an unutilized side product of PT. Geo Dipa Energi Dieng geothermal power plant. Amorphous silica has bioactive characteristics and has an ability to assist the bone-forming process by stimulating cell mineralization [12]. The purpose of this research is to explore the viability of using the geothermal silica as a silica source to modify the properties of chitosan/gelatin scaffolds. In particular, the research focuses on the fabrication of 2-D films and 3-D scaffolds to better understand the fundamental characteristics of the biocomposites.

2. Experimental section

2.1 Materials

Chitosan powder was purchased from CV. Naturindo Perkasa. Gelatin was purchased from a local supplier. Geothermal sludge was provided by PT. Geo Dipa Energy, Dieng, Central Java, Indonesia. Acetic acid (CH_3COOH) 98 %, hydrochloric acid (HCl) 37 %, and sodium hydroxide (NaOH) pellets were purchased from Merck.

2.2 Procedures

Table 1. Composition of chitosan/gelatin/geothermal silica blended scaffolds

2-D films		3-D Scaffolds	
Sample	Chi/Gel/Si Ratio	Sample	Chi/Gel/Si Ratio
A1	1:1:0	B1	1:1:0
		B2	1:1:0.5
A2	1:1:1	B3	1:1:1
		B4	1:1:1.5
A3	1:1:2	B5	1:1:2
		B6	1:1:2.5

2.2.1 Geothermal Silica Purification

The purification of geothermal silica was conducted based on previous experiment with slightly modification [4, 13, 14]. Silica sludge was washed with demineralized water, dried in an oven at 60 °C, subjected to size reduction by using a mortar grinding machine, and sieved through a 60-mesh sieve. The purified silica was prepared by alkali extraction followed by acidification steps as described below. Twenty grams of the dried silica sludge was dissolved in 800 ml of 1.5 N NaOH solution, gradually heated to 90 °C, and stirred at 300 rpm for hour. The solution was left to cool down and subsequently filtered through filter paper. The filtered sodium silicate solution was titrated with 2 N HCl to form silica gel. The gel was stored for 18 hours, filtered, and continuously washed until reaching pH 7. Finally, the neutralized gel was dried in an oven at a temperature of 60 °C prior to subsequent use.

2.2.2 Preparation of Scaffolds

In order to fabricate 2-D and 3-D scaffolds, chitosan (1 % w/v) and gelatin (1 % w/v) were dissolved in 60 mL of 1 % acetic acid solution in water. After a gel was formed, silica powder (1 % w/v) was added to the solution and stirred for 2 hours at ambient temperature until the solution was homogeneously mixed. The 2-D films were fabricated by drying the solution at room temperature and ambient pressure in an 8.5 cm diameter polyethylene petri dish until a thin layer of film was formed. For 3-D scaffold preparation, the polymer mixture was poured into a 9.5 cm diameter glass petri dish and chilled overnight at -20 °C. It was subjected to freeze-drying until dry biocomposite sponges were obtained. These were stored in a vacuum container prior to subsequent use. A series of blended biocomposites were prepared by changing the amounts of silica in the chitosan/gelatin/silica blend within the range of 1:1:0 to 1:1:2.5 according to the ratios shown in Table 1.

2.2.3 Characterization

The components of geothermal silica were analyzed with an Energy Dispersive X-ray (EDX) instrument (Shimadzu Co., Japan) by detecting X-ray spectra emitted from materials as a response to bombardment with a focused electron beam. The molecular interaction between components in a scaffold was observed with Fourier Transform Infra-Red (FTIR) spectrometer (ABB MB3000) by using KBr method. FTIR measurements were performed in the spectra range between 4000 and 400 cm^{-1}. Morphology and porosity of the scaffolds were observed by Scanning Electron Microscopy (SEM: JEOL JSM-6510 LA) after coating the specimens with a platinum layer.

2.2.4 Swelling Study

The scaffold swelling test was performed by soaking a dry sample of known weight into pure water or Phosphate Buffered Saline (PBS) at room temperature for one hour. The sample was then removed and water on the sample surface was removed by absorbing with tissue paper. The degree of swelling was calculated using the Eq. 1:

$$\text{Degree of Swelling} = \frac{W_s - W_d}{W_d} \; x \; 100\% \tag{1}$$

Where, W_d and W_s are the weights of dry and swollen samples, respectively. All data are the average of triplicate samples.

2.2.5 Degradation Test

The biocomposite degradation test was similar to the swelling test, but instead of 1 hour, the samples were soaked in pure water (H_2O) or PBS for 4 days and weighted every 24 hours. The degradation ratio was calculated using the Eq. 2 [4]:

$$\text{Degradation ratio} = \frac{W_0 - W_1}{W_0} \; x \; 100\% \tag{2}$$

where, W_0 and W_1 are the weights of initial and final samples, respectively. All data are the average of triplicate samples.

3. Results and discussion

3.1 Silica Purification

The elemental composition of purified silica was analyzed by EDX Spectroscopy. The results are shown in Table 2. The result of the EDX analysis indicates a relatively high purity of silica (95.58 %) after several steps of the washing process. Thus, geothermal silica has potential as a blending component for biocomposites [13].

Table 2. Elemental composition of purified geothermal silica by EDX Analysis

Element	Atomic %
Si	95.58
S	1.63
Al	1.39
Fe	0.91
Ca	0.38
K	0.08
Cu	0.02
Zn	0.01

3.2 Morphology

The surface morphology of chitosan/gelatin/silica 2-D films and 3-D scaffolds was observed by SEM as shown in Figure 1. The SEM result revealed that the 2-D films show a rough surface with homogeneous silica distribution, and no interconnected pore was observed. The surface roughness of the films has the tendency to increase along with the amount of silica contained. Silica attached on the film surface are apparent and may provide better properties for cell adhesion on the films. Silica attached on the film surface may assist bone forming process by stimulating cell mineralization [12].
Contrastingly, interconnected pores formed in the 3-D scaffolds. By increasing the amount of silica in biocomposites, the pore size decreases. Pore diameters in chitosan/gelatin/silica=1:1:2 are in the range around 100 to 150 μm. The pore size of scaffold could affect cell penetration and distribution through the scaffold which is important to promote better vascularization and tissue formation [15]. Physically, higher composition of silica for the equal total mass would produce a more fragile biocomposite due to smaller number of silica binding polymers. The interconnected pores contribute to the nutrient transfer and cell infiltration: essential roles expected for tissue engineering applications. By changing the amount of silica contained in scaffolds, optimal morphology and physical properties can be achieved.

3.3 Chemical structure of the biocomposites

FTIR spectra of the 1:1:1-chitosan/gelatin/geothermal silica composite along with its pure components are shown in Figure 2. Peaks at 1634 cm^{-1} and 1557 cm^{-1} in the spectrum of chitosan are assigned to the presence of $NHCOCH_3$ (amide I) and $-NH_2$ groups (amide II), respectively. Peaks at wavenumbers 2884 cm^{-1} and 3400-3500 cm^{-1} represent $-CH_3$ and –OH groups. A tapered peak at 1411 cm^{-1} indicates symmetrical deformation and a peak at 1090 cm^{-1} indicates the vibration of amine groups. In contrast, the characteristics of gelatin are evident by the peaks at 3294 cm^{-1}, 1500-1200 cm^{-1}, and 1550 cm^{-1} corresponding to N-H group, CH_2 deformation, and the deformation of N-H by alpha helix, respectively. A silanol (Si-OH) group in silica spectra is shown at the peak of 1635 cm^{-1}, while the existence of Si-O-Si is shown at 1100 cm^{-1}. The interaction of silica with chitosan and gelatin leads to the friction on N-H group that overlap with –OH group at the peak of 3411 cm^{-1}. Meanwhile, a peak at 647 cm^{-1} demonstrates the Si-O-C bond of the blended scaffold [15].

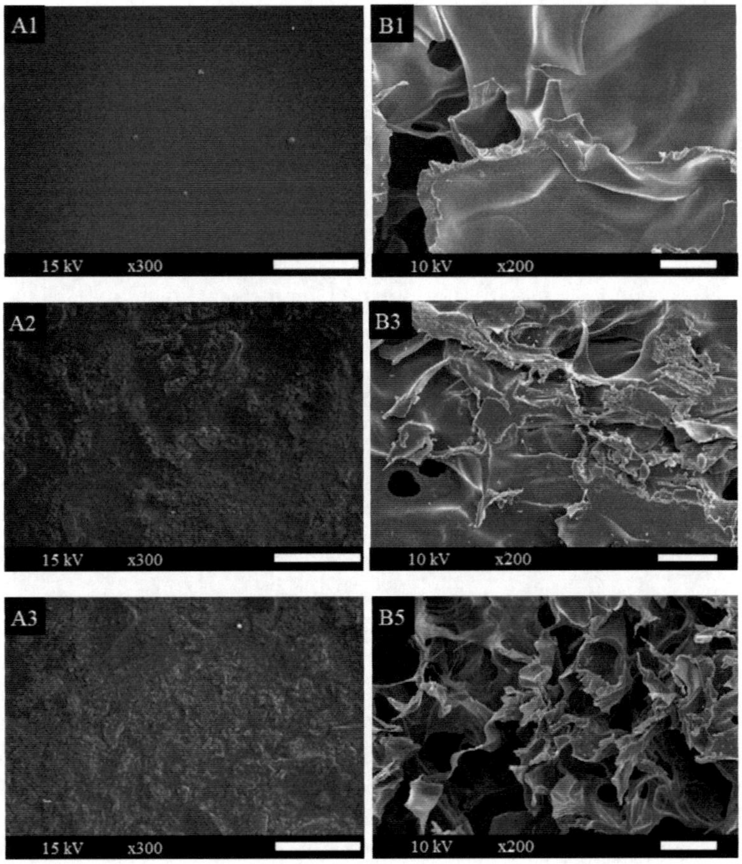

Figure 1. SEM images of 2-D films of chitosan/gelatin/silica = 1:1:0 (A1), 1:1:1 (A2), 1:1:2 (A3) and 3-D scaffolds of chitosan/gelatin/silica = 1:1:0 (B1), 1:1:1 (B3), and 1:1:2 (B5). Scale bar: 100 μm

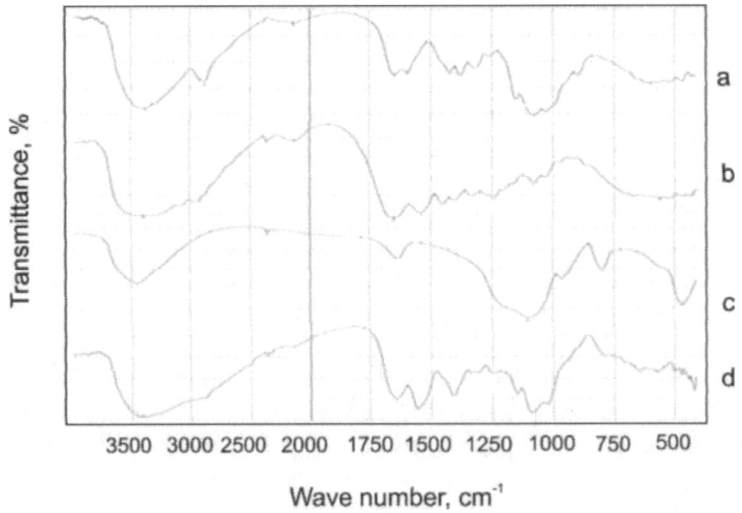

Figure 2. FTIR spectra of (a) chitosan, (b) gelatin, (c) geothermal silica, and (d) 1:1:1 blended biocomposite

3.4 Swelling properties

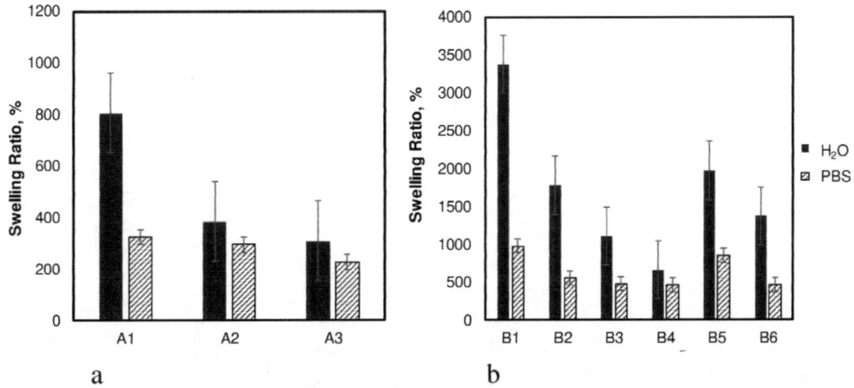

Figure 3. The degree of swelling in H_2O and PBS versus silica composition in 2-D films (a) and 3-D scaffolds (b). Silica content is 0 (A1, B1), 0.5 (B2), 1 (A2, B3), 1.5 (B4), 2 (A3, B5), and 2.5 (B6) as shown in Table 1.

The degree of swelling versus the composition of geothermal silica in 2-D films and 3-D scaffolds are shown in Figure 3. The swelling test was performed for 1 hour to determine the absorption ability of biocomposites in H_2O or PBS. The addition of silica to the biocomposites was found to considerably reduce the degree of swelling. This decreasing trend can be ascribed to the interaction between silica and chitosan's positive charge that reduces the number of free amine groups in chitosan. The amine group can form hydrogen bonds with H_2O as the swelling media, thus the reduction of free amine groups leads to a lower degree of swelling.
Meanwhile, the difference in the degree of swelling in H_2O and PBS is caused by a higher ion concentration of PBS that leads to lower osmotic pressure, resulting in a lower absorption capability of the biocomposites. This low swelling ratio is preferred in the application of tissue engineering to avoid increasing mean pore size of the 3-D scaffolds.

3.5 Degradation rate

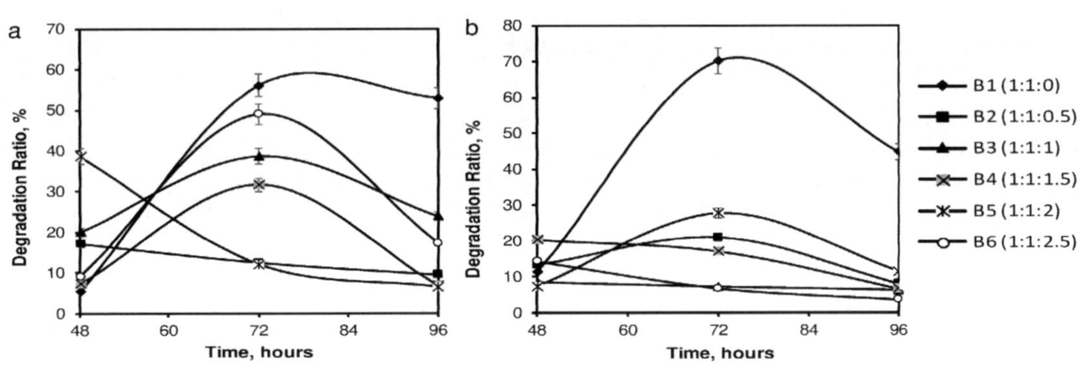

Figure 4. The degradation ratio of 3-D scaffolds as a function of time in (a) H_2O and (b) PBS

The results of the degradation test are shown in Figure 4. A significant difference was observed in the degradation rate of the scaffolds with silica incorporated compared to the scaffold without silica. The addition of silica significantly decreased the degradation rate of the scaffolds. The maximum degradation ratio was achieved after 72 hours of immersion.

4. Conclusion

In this study, blended scaffolds were synthesized using chitosan, gelatin and silica. The addition of silica from a geothermal source was aimed at controlling the degradation rate of the scaffolds. FTIR analysis showed interactions between three blended components, while SEM analysis of the 3-D scaffolds showed the formation of interconnected pores that play important roles in tissue engineering. The results of swelling and degradation analyses indicated that the addition of silica reduces the degree of swelling leading to scaffolds with a low degradation rate. In conclusion, geothermal silica has high potential for improving the properties of organic scaffolds, particularly by controlling the degradation rate of the scaffolds.

Acknowledgement

The authors gratefully acknowledge a partial financial support provided by The Global Collaboration Program of Nara Institute of Science and Technology sponsored by MEXT, Japan. English proofreading by Leigh McDowell is also acknowledged.

References

[1] Pavia F C, Rigogliuso S, La Carrubba V, Mannella GA, Ghersi G, and Brucato V 2012 *Chem. Eng. Trans.* **27** 409-414.
[2] Patricio T, Gloria A, Bartolo P 2013 *Chem. Eng. Trans.* **32** 1645-1650.
[3] Kusumastuti Y, Shibasaki Y, Hirohara S, Kobayashi M, Terada K, Ando M, Tanihara M, 2017a *J. Tissue Eng. Regener. Med.* **11** 869-876.
[4] Kusumastuti Y, Kobayashi M, Purwaningtyas F Y, Petrus H T B M, Putri N R E, Budhijanto, Tanihara M 2018 *J. Eng. Sc. Technol.* **13** 3500-3515.
[5] Sodha S, Wall K, Redenti S, Klassen H, Young M J, Tao S L 2011 *J Biomater Sci Polym Ed* **22** 443-456.
[6] Sanad R A, Abdel-Bar H M 2017 *Carbohydr Polym.* **173** 441-450.
[7] Dorati R, Colonna C, Tomasi C, Genta I, Bruni G, Conti B 2014 *Mater. Sci. Eng. C* **34** 130-139.
[8] Kong M, Chen X G, Xing K, Park H J 2010 *Int. J. Food Microbiol.* **144** 51-63.
[9] Ahmed S, Ahmad M, Jayachandran M, Qureshi M A, Ikram S 2015 *Immunochem Immunopathol* **1** 1000106.
[10] Xia W, Liu W, Cui L, Liu Y, Zhong W, Liu D, Wu J, Chua K, Cao Y 2004 *J. Biomed.l Mater. Res. B* **71** 373-380.
[11] Yin Y J, Yao K D, Cheng G X, Ma J B 1999 *Polym. Inter.* **48** 429-432.
[12] Vallet-Regí M and Balas F 2008 *Open Biomed. Eng. J.* **2** 1-9.
[13] Kusumastuti Y, Petrus H T B M, Yohana F, Buwono A T and Zaqina R B 2017 *AIP Conf. Proc.* **1823** 020127.
[14] Purwaningtyas F Y, Kusumastuti Y, Petrus H T B M, Budhijanto 2018 *Proc. of the 15th Int. Conf. on QIR* (Jakarta: Faculty of Engineering Universitas Indonesia) pp. 627- 634.
[15] Kavya K C, Jayakumar R, Nair S, and Chennazhi K P 2013 *Inter. J. Biol. Macromol.* **59** 255-263.

IOP Conf. Series: Materials Science and Engineering **543** (2019) 012029 doi:10.1088/1757-899X/543/1/012029

CNT Synthesis from POME by Pyrolysis using Tubular Furnace

I Nurdin*, H Devianto, P Widiatmoko, T Prakoso, N F Nugraha1, S Srimurti
Department of Chemical Engineering, Bandung Institute of Technology, 40132, Bandung, Indonesia

E-mail: isdi@che.itb.ac.id

Abstract. The utilization of the palm oil waste, such as Palm Oil Mill Effluent (POME), is desirable to reduce environmental problems and increase its value as a source of biomass. POME can be processed to make Carbon Nanotube (CNT), a cylindrical carbon allotrope composed of graphene sheets. CNT, since its nanosized and high electrical conductivity, is well-suited to be used as electrodes in energy storage device such as battery. This study focuses on CNT synthesis from POME by pyrolysis process at 900°C using ferrocene as a catalyst in flowing nitrogen atmosphere. The synthesis of CNT consists of preparation of raw material (POME), synthesis of resin oil, pyrolysis, and characterization of CNT. The result shows that this process yields 0.28 g CNT (from 15 g POME, using 30 weight-% Ferrocene) whose diameter varies between 26-171 nm, the surface area is 404.62 m^2g^{-1}. From the XRD analysis, a C (002) is detected at 2θ of 29.098° with intensity of 3.173 Å, slightly under perfect graphite intensity (3.354 Å). It can be concluded that CNT can be produced from POME using pyrolysis process; an affordable yet applicable way on industrial scale to produce a good quality of CNT.

1. Introduction

Indonesia is the largest palm oil producer in the world with capacity of 31.4 million tons per year [1]. As consequence, abundant wastes are generated. Palm oil mill effluent (POME) is the liquid waste which represents 61% of total waste. From each ton of processed palm oil fresh fruit, 0.7 – 1 m^3 POME is produced [2]. Palm oil mill has been regarded as a profit-making industry for the past decades. Besides revenue from the palm oil production itself, the abundance of biomass could generate high economic return to the palm oil mill by converting it to value added products [2]. POME contains nitrogen, phosphate, calcium, potassium, carbon and magnesium, so it can be processed biologically to a fertilizer and biogas [3]. However, to process POME to biogas, vast investment, especially huge area is needed. Because of that, in Indonesia, biogas from POME is still not commercially applicable. Nevertheless, POME has a prospect to become the raw material of valuable carbon nanotubes (CNT).

Today, carbon nanotubes (CNT) can be derived from organic liquid such as, turpentine oil, neem oil, and even organic wastes, like Bio-oil PCB and waste cooking oil. POME is a liquid waste which is rich of organic compounds, so it has potential as a carbon source to produce CNT.

CNT is a cylindrical allotrope of carbon with nanometric size. CNTs have high electrical conductivity of 10^7 $S.cm^{-1}$. Application of CNTs in energy storage such as supercapacitor produce specific capacitances of 12-120 $F.g^{-1}$, energy density of 0.5-40 $Wh.kg^{-1}$and power density of 30-100 $kW.kg^{-1}$ [4]. Combination of graphene and CNT will prevent graphene layer attachment, thus increasing capacitance [5].

IOP Conf. Series: Materials Science and Engineering **543** (2019) 012029 doi:10.1088/1757-899X/543/1/012029

2. Methods

2.1. Materials

CNT was produced from Palm oil mill effluent (POME) which was collected from palm oil mill of PT. Perkebunan Nusantara VIII in Bogor. The POME can also be described as wastewater rich in organic carbon with a colloidal suspension of 95 - 96% water, 0.6 - 0.7% oil, and 4 - 5% total solids, including 2 - 4% suspended solids [6]. Gas Chromatography-Mass Spectroscopy (GC-MS) analysis of POME used in this research, gives characteristic as shown in Table 1.

Table 1 Retention data of compounds from POME

No	Compound	Receptor	Area
1	Cyclohexanecarboxylic acid	$C_7H_{12}O_2$	58.39
2	Hexanoic acid	$C_6H_{12}O_6$	14.74
3	1,3,5-Trimethyl-6-oxabicyclo	$C_{10}H_{16}O_2$	5.53
4	Furanone	$C_6H_{10}O_3$	1.40

Regulation of Minister of Environment and Forestry No. 5 of 2014 about standard specification of palm oil waste, limit its pH of 6-9, BOD of 100 mg L^{-1} max, COD of 350 mg L^{-1} max etc. While POME used in this research has pH of 3-6, BOD of 15,600 mg L^{-1} and COD of 25,000 mg L^{-1}. This means that the POME used in this research does not comply the standard specification. According to its large amount, this waste will potentially damage the environment, so it should be utilized to produce a valuable substance such as CNT.

Formaldehyde water solution used to polymerize POME and ammonium hydroxide as catalyst. Pyrolysis of the resin used ferrocene as catalyst in nitrogen flow. Ferrocene are frequently used as precursors to prepare carbon nanostructures, because they cannot only act as carbon source, but also give rise to small metal clusters as catalyst.

2.2. CNT Production

Figure 1 shows the process flow diagram of CNT synthesis from POME. First of all, POME has to be polymerized with 37% formaldehyde solution with weight ratio of 5:6.5, using ammonium hydroxide as catalyst at 95°C for 4 hours. The resin formed is then heated at 60°C for 2 hours and at 120°C for 12 hours. After that, the resin was ground to fine powder. The fine resin powder is then mixed with 20-30%-w ferrocene, homogenized with ethanol at 50°C until the mixture is uniform and all ethanol is evaporated. Afterward, the mixture is loaded in a ceramic boat and placed in the tubular furnace. The pyrolysis process consists of 2 steps; the first step is conducted at 200°C during 1 hour with heating rate of 5°C/min in natural atmosphere, and the second is run at 900°C for 1 hour with a heating rate of 3°C/min in flowing nitrogen atmosphere.

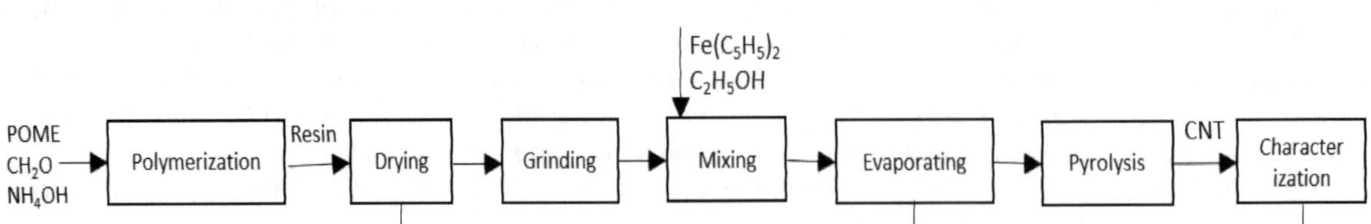

Figure 1 Process flow diagram of CNT synthesis

2.3. *Characterization*

The CNT produced by pyrolysis is characterized as follows: the chemical composition is analyzed using FT-IR, its morphology is analyzed using SEM and TEM, its graphitic characteristic is determined using Raman spectroscopy, its crystallinity is determined using XRD, while its surface area is analyzed using Nitrogen adsorption-desorption analysis with BET calculation method.

The electrochemical performance of CNTs was tested by Cyclic Voltammetry (CV) and Electrochemical Impedance Spectroscopy (EIS) methods. Electrochemical measurements were carried out using Gamry V3000 potentiostat – galvanostat, by means of an asymmetrical two-electrode configurations with 6 M KOH solution as electrolyte at room temperature. The negative electrode of the asymmetrical cell was prepared by mixing CNT with polyvinylidene fluoride (PVDF), which is coated on 1 cm2 stainless steel mesh and dried at 80°C for 1 day. The positive electrode was prepared by mixing NiO with PVDF, which is coated on 1 cm^2 stainless steel mesh and dried at 80°C for 1 day. Nafion 212 impregnated with the electrolyte, was placed between both electrodes.

3. Results and Discussion

The morphology of CNT made from POME using ferrocene as catalyst by pyrolysis at 900ºC, was identified using SEM and TEM. There are 4 samples of CNT produced with different catalyst concentration (0%, 20%, 25% dan 30% wt). The SEM observation results are shown on Figure 2. Figure 2a shows the pyrolysis product formed without ferrocene; it reveals amorphous carbon only. This Figure confirms that carbon nanostructure cannot be formed without catalyst.

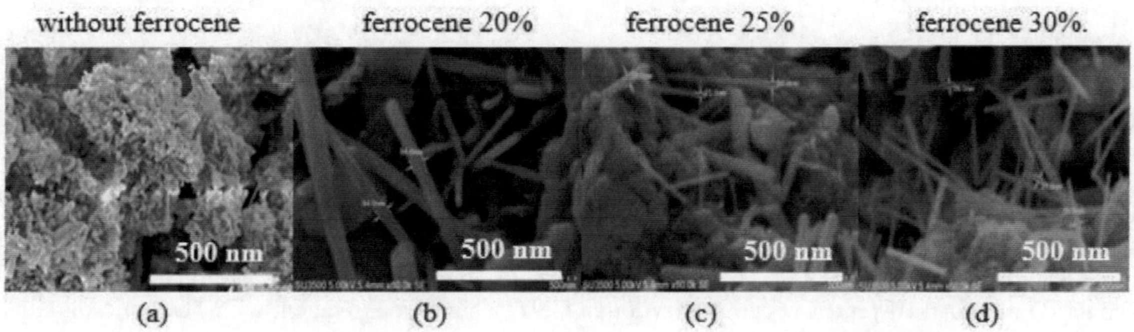

Figure 2 SEM micrographs of CNT with different catalyst concentration

Figure 2b presents the SEM observation of pyrolysis product using 20%-wt. ferrocene, which indicate the existence of CNT of 84-94 nm diameters. The CNT with smaller diameter will give larger capacitance in supercapacitor [7]. As shown on Figure 2b, 2c and 2d, increasing the mass percent of ferrocene (up to 30%), will increase the amount of CNT obtained. This assumption is supported by Table 2 which presents the alteration of CNT yield due to variation of ferrocene concentration.

Table 2 Carbon yield of pyrolysis (mass ratio of carbon product and initial POME) as ferrocene concentration changed

%-wt. ferrocene	20	25	30
%-wt. C in product	2.02	2.17	1.87

As the mass percent of ferrocene increased, the number of nuclei initiator become larger, and increase the number of carbon nanotubes. The growing mechanism of CNT during the synthesis can be proposed as follows: nanometer iron produced from ferrocene decomposition accumulates carbon on its surface with the formation of Fe–C solid solution. After a very short period of time, the Fe–C solid solution reached a supersaturated point and carbon atom started to nucleate and grow in a nanotube shape During the slow growth, the shape of the Fe catalyst may be changed, e.g. from nanoparticle to nanorod by cylinder carbon formation around it. If the catalyst is reduced before the decomposition of resin, the size of the Fe catalyst does not increase during the growth of the nanotubes. In this condition, the decomposition of pyrolysis oil-based resin will result in the formation of nanotubes free of metal [8]. TEM micrographs of CNT formed with 30%-wt. ferrocene reveals some single open-end tubes, as shown on Figure 3. Ferrocene concentration more than 30%-wt. didn't significantly change the yield of CNT [9].

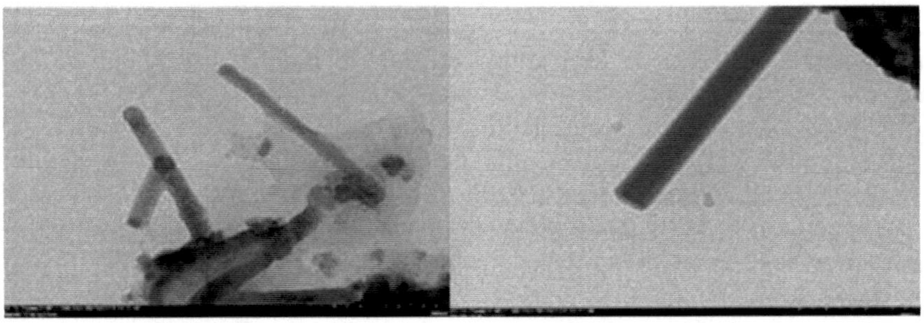

Figure 3 TEM micrographs of CNT with 30%-wt. ferrocene

Raman spectroscopy provides information about the structure and the presence of disorder in the sample. Figure 4a shows the representative Raman spectra of a CNT formed with 30%-wt. ferrocene. In the Raman-shift range of 500–2000 cm^{-1}, two peaks are observed at 1350 cm^{-1} and 1582 cm^{-1} corresponding to D and G bands, respectively. The G band oriented pyrolytic graphite and suggests the CNT are composed of crystalline graphitic carbon. The higher intensity of the G band peak indicates the higher degree of graphitization/crystallinity. The D band oriented disorder in the sp2-hybridized carbon and indicates lattice distortions in the curved tube ends. The intensity ratio of G and D peaks (IG/ID) is used to characterize the purity of CNT. Higher IG/ID ratio indicates a higher degree of graphitization [10]. The IG/ID ratio of 1.01 represents graphitization of CNT. This is in agreement with TEM micrographs of CNT which show a hollow form.

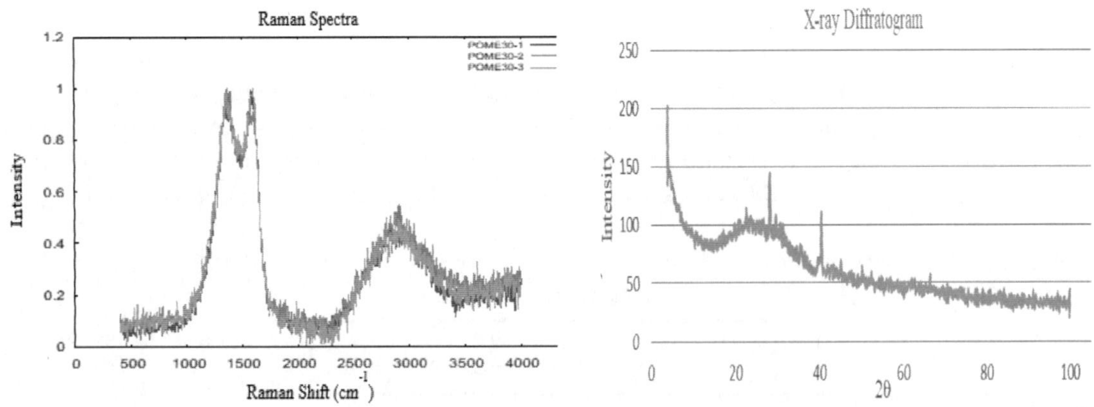

(a) (b)

Figure 4 Raman spectra & X-ray diffractogram of CNT using 30% wt of ferrocene

IOP Conf. Series: Materials Science and Engineering **543** (2019) 012029 doi:10.1088/1757-899X/543/1/012029

Figure 4b shows the typical XRD pattern of CNT sample. The peaks at 2θ of 29.098° with intensity of 3.173 Å, slightly lesser a perfect graphite intensity (3.354 Å) are indexed to be the (00.2). The presence of the (00.2) peak in the XRD spectra indicates the concentric cylindrical nature of the CNT, or the multi-walled CNT (MWCNT). The nature of surface groups of resin-oil was studied by FT-IR and the results are shown in Figure 5a. The detectable transmission band at 3413 cm^{-1} represents the stretching vibration of O-H bond in alcoholic or phenolic compounds. The peak at wave numbers 2800-3000 cm^{-1} are assigned to alkane stretching vibration. The peak at wave number 1600 cm^{-1} is assigned to stretching vibration of C-O bond in aldehyde, ketone or carboxylic groups. While the peak at wave number 1300 cm^{-1} is assigned to stretching vibration of C-N bond in amine group probably present from NH_4OH addition.

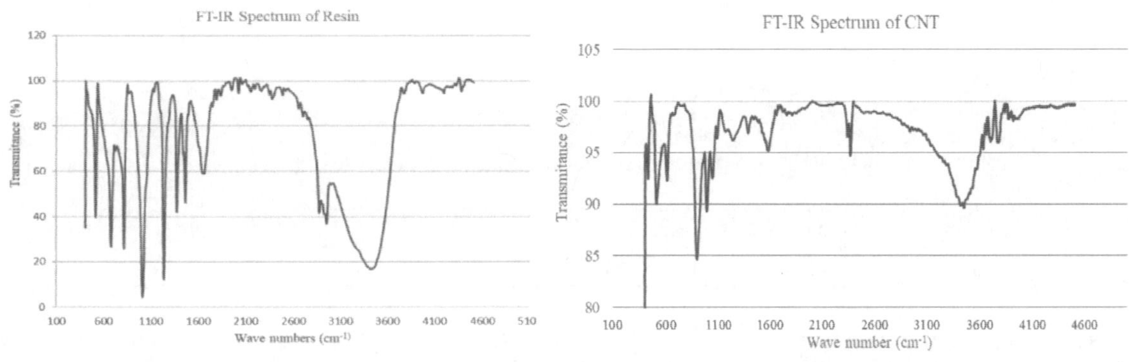

(a) (b)

Figure 5 FT-IR spectrum

FT-IR spectrum of CNT is depicted on Figure 5b. The large peak at wave number ~3430 cm^{-1} is assigned to stretching vibration of hydroxylic groups attached on CNT surface [11]. Transmission band at wave number 3728 cm^{-1} can be interpreted as stretching vibration of free hydroxylic groups, while the peak at wave number 3425 cm^{-1} represents the stretching of O-H bond in carboxylic and alcoholic groups. The peak at wave number 2380 cm^{-1} can be interpreted as stretching of C=O bond in carboxylic groups. The broad peak at wave numbers 1442–1736 cm^{-1} can be assigned to CNT, and the peak at wave number 1372 cm^{-1} is interpreted as stretching of $N\text{-}CH_3$ bond. These identified bonds indicate the specific framework of CNT [12]. Therefore, it can be concluded that CNT is formed from cyclohexane carboxylic acid in POME.

The surface area and porosity of CNT were determined using nitrogen adsorption-desorption with BET calculation method. The surface area of CNT made from POME is 404.62 m^2g^{-1}, while commercial CNT has surface area between 120 and 500 m^2g^{-1} [7]. The cyclic polarization of the asymmetrical cell made of produced CNT gives cyclic voltammograms that can be used to calculate the cell capacitance. The cell capacitance can be calculated based on cyclic voltammogram data, following this equation [4]:

$$C = \frac{\sum|I|.\Delta t}{m.\Delta V} \qquad (1)$$

Where C is capacitance (F/g), I is electric current (Amperes), Δt is time elapsed (second), m is cell mass (grams) and ΔV is voltage difference (Volt). The voltammograms obtained with various scan rate are presented on Figure 6, while the capacitance data are depicted on Table 3.

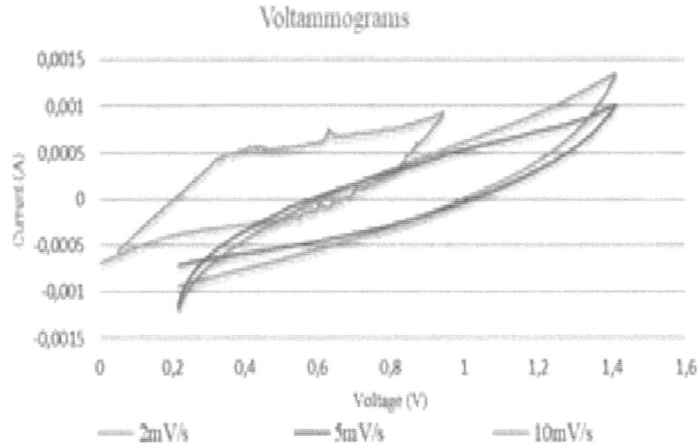

Figure 6 Voltammograms of asymmetrical cell of CNT with various scan rate

Table 3 Asymmetrical cell capacitances (F/g carbon)

Scan rate (mV/s)	2	5	10
Cell capacitance (F/g)	26.09	17.04	9.87

In general, increasing scan rate will reduce the cell capacitance. This phenomenon reveals that at higher scan rate, ionic diffusion in pores become faster and induces more ionic collisions. Too large number of ionic collisions will produce an irregular electrical double layer which has a lower capacitance [13-16] The application of Electrochemical Impedance Spectroscopy on asymmetrical cell of CNT provides the Nyquist curve which represents the relation between real and imaginary impedances, as depicted on Figure 7a. The Nyquist plot gives the information of the cell components' resistance as well as its capacitance that allow designating the cell configuration. Such cell configuration is presented as an equivalent electric circuit on Figure 7b.

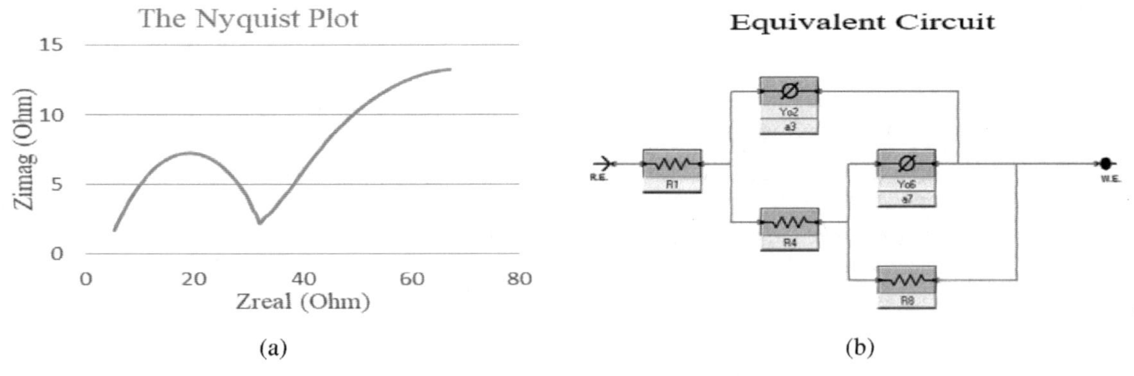

Figure 7 The Nyquist plot of asymmetrical cell made of CNT and its equivalent circuit

This Nyquist plot shows a large resistance of charge transfer as well as of mass transfer, while the equivalent circuit displays a normal equivalent circuit of an asymmetrical cell, despite its resistances and capacitances values. These phenomena reveal that CNT can be used as the electrode of an electrochemical device, but should be well-ordered to obtain very low resistances and high capacitances.

4. Conclusion

CNT can be produced from POME using pyrolysis process at 900°C in flowing nitrogen atmosphere, using Ferrocene as catalyst. All morphology and physical examinations results confirm the CNT formation. The CNT yield increases with the concentration of ferrocene. The capacitance of CNT cell is still low and its resistance is still high due to the cell construction which is not well-organized.

Acknowledgement
This work is supported by Badan Pengelola Dana Perkebunan Kelapa Sawit during the fiscal year 2016.

References
[1] Central Bureau of Statistics 2016 *Indonesian Palm Oil Statistics*, (Indonesia: BPS)
[2] Rahayu A S *et al* 2015 *POME to biogas -project development in indonesia,* (Indonesia: Winrock International)
[3] Deublein D and Steinhauser A 2008 *Biogas from waste and renewable resources* (Weinheim-Wiley VCH)
[4] Quang N *et al* 2017 *Phys. Status. Solidi.* **214** 1600852
[5] Hendaryati D D and Arianto Y 2016 *Indonesian Plantation Statistics – Palm Oil* (Indonesia: Directorate General of Estate Crops)
[6] Ahmad A L Chong M F and Bhatia S 2009 *J. Hazard. Mater.* **171** 166
[7] Yu A Chabot V and Zhang J 2013 *Electrochemical supercapacitors for energy storage and delivery Fund. Appli* (Boca Raton: CRC Press) p 373
[8] Quan C Li A Gao N 2010 *J. Hazard. Mater.* **179** 911
[9] Parshotam H 2007 *The use of ferrocene and camphor for the synthesis of carbon nanotubes using catalytic chemical vapor deposition*. (South Africa: PhD thesis-University of Johannesburg).
[10] Awasthi K Kumar R Tiwari R and Srivastava O N 2010 *J. Exp. Nanosci.* **5** 498
[11] Vinayan B P *et al* 2012 *App. J. Mat. Chem.* **22** 9949
[12] Robati D 2013 *J. Nanostructure. Chem.* **3** 55
[13] Taurino I *et al* 2012 *Surf. Sci.* **606** 156
[14] Zhang Y *et al* 2017 *Industrial applications of carbon nanotubes chapter* 10 277
[15] Kumar S *et al* 2018 *Prog. Energy Combust. Sci.* **64** 219
[16] Termehyousefi A *et al* 2014 *BioMed Res. Int.* **2014** 691537
[17] Kerdnawee K *et al* 2017 *KONA Powder Part. J.* **34** 24
[18] Liu Q *et al* 2008 *Carbon* **46** 1892
[19] Mitrichev I I Vasilenko V Scherbakov A and Koltsova E M 2016 *Chem. Eng. Trans.* **52** 1003
[20] Nasrullah M *et al* 2017 *Water. Res. Ind.* **17** 7
[21] Saberi D Sheykhan M Niknam K and Heydari A 2013 *Cat. Sci. Tech.* **3** 2025
[22] Sarno M *et al* 2017 *Chem. Eng. Trans.* **57** 1519
[23] Yan Y *et al* 2015 *Chem. Soc. Rev.* **44** 3295
[24] Zhang H Y *et al* 2015 *Pigm. Resin. Technol.* **44** 7
[25] Hernández V J *et al* 2015 *Chem. Eng. Trans.* **45** 1183

IOP Conf. Series: Materials Science and Engineering **543** (2019) 012030 doi:10.1088/1757-899X/543/1/012030

Formation of Carbon Fibres From Polymer Poly(vinyl alcohol)/Acetylene Black using Electrospinning Method

R A Yuniar[1], W Widiyastuti[1,*], H Setyawan[1], H Purwaningsih[1], S Machmudah[1] and D Anggoro[2]

[1]Department of Chemical Engineering, Institut Teknologi Sepuluh Nopember, Kampus ITS Sukolilo, Surabaya 60111, Indonesia
[2]Department of Physics, Institut Teknologi Sepuluh Nopember, Kampus ITS Sukolilo, Surabaya 60111, Indonesia

*E-mail: widi@chem-eng.its.ac.id

Abstract. Carbon fibres are important materials due to their excellent properties. Carbon fibres were formed by blending Poly(vinyl alcohol) (PVA) and a high conductivity material of Acetylene black (AB). Electrospinning technique was used to fabricate PVA and AB solutions in demineralized water. Electrospinning is an efficient technique for the formation of polymer nanofibers. The effects of solution concentration and AB contents on morphological appearance and diameter of those as-spun fibres were investigated by scanning electron microscopy (SEM). Iodine treatment on the fabricated PVA/AB composite was successfully achieved in order to build up the carbonization process and retained their original fibrous structure. The mass of iodinated nanofibers at 80 °C for 24 h increased about 130 %. The mechanism of iodine uptake by electrospun PVA/AB fibres have been obtained by thermal behaviour and its morphological characteristics. It was found that iodine treatment played an important role in changing the morphological structure during the carbonization process. With the iodine treatment of PVA/AB precursor fibres successfully converted into thin carbon fibres and SEM analysis confirmed that the diameter of fibres was about 290 nm. The resulted pure fibres and iodine treatments show guarantee for production carbon fibres.

1. Introduction

Nano-scale technology is called nanotechnology has changed the paradigm and the way of view of technology because the material is designed and arranged in the order of atomic per atom or molecules, so that it is found a material that has special superior properties to the existing material. Rapid developments in the application of nanotechnology in the industrial world indicate that the world is currently leading to the revolution of nanotechnology, one of which is nanofiber. Nanofiber has a large surface area to the ratio of the volume of the solution and has a large absorption with small pore size [1]. Nanofibers have the advantage of physical characteristics that can be utilized in various aspects such as healthcare, biotechnology, energy storage application and textile [2]. There are several techniques to produce nanofibers, such as drawing [3], template synthesis [4], phase separation [5] and electrospinning technique was the most frequently used [6].

Among several techniques, electrospinning is a simple and efficient technology for the formation of electrostatic polymer nanowires utilizing electric power to produce fibres of nano-sized diameters. This

technique has received much attention due to the flexibility of the materials used and its ability to produce fibers in the submicron range. Electrospinning is the only technique available for the production of fibers with very small diameters [7]; hence electrospinning can be a main alternative technique.
Carbon is a material widely used as an electrochemical support because it has high conductivity, thermal and chemical stability and surface characteristics that can be changed [8]. Carbon nanofibers produced from electrospinning technique have attracted due to their superior characterization such as high-modulus, high-resistance [9], flexibility and their special morphology nanomaterials [10]. Today the development of carbon materials is directed towards nano meter-sized carbon by modifying the pore structure and surface [11].
Carbon fibers from polymer/organic or polymer/inorganic composite have been developed by electrospinning cause of their potential properties such as mechanical and electrical conductivity. The composites of PVA/lignin [12] and PVA/Carbon black [13] have been investigated as electrode material. Their investigations claimed having good electro-properties. Poly(vinyl alcohol) (PVA) has a fairly high carbon content and has a hydroxyl group in the polymer chain so that PVA is great for precursor in carbon material. PVA is a resin made from the incorporation of vinyl acetate molecules (polymerization), by partial hydrolysis of the catalyzing ester. The physical properties of the polymerization product depend on the degree of polymerization and degree of hydrolysis. PVA fibers have good tensile and bending strength [14]. Carbon fibers from electrospun PVA nanofibers when heated at a higher temperature than their melting point will result in low carbon. Therefore, PVA percussion requires high thermal stability between 100-290 °C [15]. On the other hand, iodine can be used for PVA stabilizer in order that at high temperature carbonization, PVA kept on it structure and hard to melt. This is due to that the addition of iodine treatment can modify the bonds in PVA hydrocarbons into double bonds such as polyene [16]. Some studies have compared halogen compounds such as bromine to iodine. Iodine treatment has resulted more superior than bromine due to bromine treatments is only non-selective while using iodine treatment selectively to aromatics. Thus, iodine is good stabilizer suited to treat a material for carbonization.
According to previous studies, iodine treatments have no yet been examined toward PVA/Acetylene black composite. Thus, in this paper, electrospinning of PVA blended with acetylene black (AB) dissolved in distilled water to produce precursor fibers was investigated. The effects of iodine treatment on PVA/AB composite electrospun fibers are then investigated in order to produce carbon nanofibers.

2. Materials and Methods

2.1. Spin Dope Solution

PVA powder was used technical grade, acetylene black powder and iodine crystal was purchased from SAP chemicals. Spin dope PVA solution were prepared by the addition PVA powder (concentration ranging between 5 and 15 % w/v with 5 % w/v increment) which dissolved in distilled water at 85 °C under constant stirring for an hour. In order to blend solution of PVA and AB were prepared by adding a weighed amount of AB in PVA aqueous solutions followed by sonication at 50 °C for 2 hours. After sonication, the mixture was mechanically stirred at 85 °C for 30 minutes then cooling down to room temperature. AB was varied amount 0, 5, 10 and 15 wt. % based on the weight of PVA powder to obtain the effect of AB content in PVA solution.

2.2. Electrospinning of nanofibers

The polymer solution was carried out in an OneMed plastic 50 ml syringe with needle an inner diameter 0.6 mm. The steady flow rate was maintained by the ColeParmer Instrument Co. syringe pump at 1 ml per hour. As shown in Figure 1, a high voltage power supply was connected the needle and the collector. The applied voltage at 10 kV was applied to the polymer solution using 10 kV was applied onto a collection device. Collection device was wrapped by a square 15 cm x 30 cm of aluminium foil around rotating cylinder. The distance from needle to collector was 10 cm.

2.3. Iodine treatment

An electrospun nanofibers PVA/AB were peeled off from aluminium foil on cylinder collector placed into a sealed glass jar. The glass jar was contained iodine crystal placed in an oven at 80 °C for 24 hours to vaporize the iodine. After iodine treatment, sealed glass jar from the oven was removed and then was cooled at room temperature. For a few moments the remaining iodine vapour in the jar recrystallized after which electrospun nanofibers can be removed from the jar. The samples of PVA/AB iodine treatment and non-treated was carbonized in a Lindberg tubular furnace with heating operation: (i) 30 – 200 °C; (ii) 200 – 400 °C; (iii) 400 – 600 °C, each step holding for 15 minutes. (iv) 600 – 800 °C, the temperature was held for an hour. A constant N_2 flow was maintained during carbonization. For their morphology, the samples were observed using FEI-Inspect S50 Scanning Electron Microscope (SEM). ImageJ software was used for analysing images and calculating fibre diameters from at least 40 randomly selected.

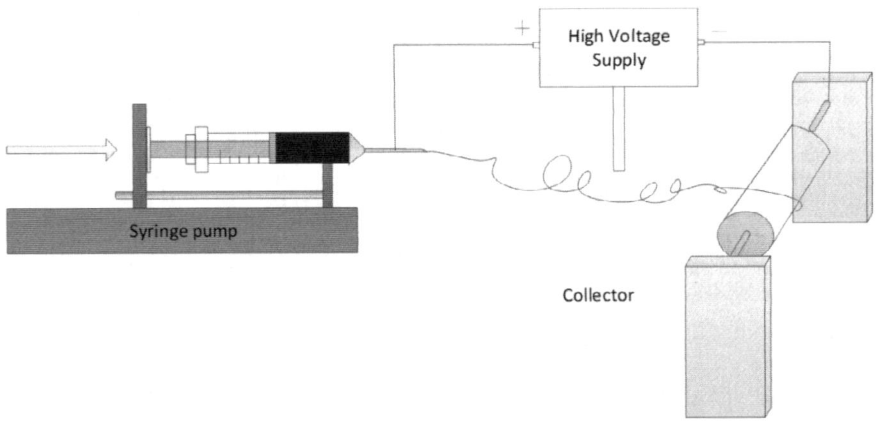

Figure 1. Schematic of the electrospinning process

3. Results and Discussion

3.1. Nanofibers of PVA

Many factors or parameters affected the morphology of the nanofiber, for example electrospinning process, the distance between the needle to collector, viscosity, flowrate, applied voltage [6]. Ambient parameters such as temperature [17] and relative humidity play role to the morphology nanofibers [18]. Figure 2 shows the SEM images that described the morphological as spun nanofibers from various concentrations. Electrospinning of the 5 and 10 % w/v PVA resulted in the formation of beaded fibres, because the solution has low viscosity.

Figure 2. SEM images of electrospun PVA fibres with concentrations: (a) 5, (b) 10, and (c) 15 % w/v

IOP Conf. Series: Materials Science and Engineering **543** (2019) 012030 doi:10.1088/1757-899X/543/1/012030

Figure 2 shows the results of SEM analysis of PVA concentration variables, where Figure 2a shows the formation of relatively non-homogeneous fibres both in shape and size at a concentration of 5 % w/v. It was observed that there were large and oval beaded in the middle of the fibres and at the vertex between fibres. The concentration of PVA 10 % w/v produces relatively homogeneous fibres in shape and size as shown in Figure 2b. However, it can still be observed that there are beaded at several points along the fibre and fibre point vertices. If the concentration is added up to 15 % w/v, it produces fibres that are finer morphologically and more homogeneous. There is no observation of beaded along the fibres and knot points. The concentration and the viscosity of the solution is too low hence the charged liquid could not form fibres instead of droplets [13]. At the concentration of 10 % w/v PVA, the number of beaded fibres decreased as the viscosity increased. The lower viscosity of solution, the lower viscoelastic force because the coulomb repulsion force worked. Increasing the concentration of solution led to the uniform fibres without beads. This is due to the solution has low viscosity that increased the viscoelastic force and it may cause the coulomb repulsion force broke the charged jet much uniform. Increasing PVA concentration up to 15 % w/v resulted smooth nanofibers and uniform in size about 220 nm, as shown in Figure 2c. The fibre of 15 % w/v PVA had oriented-nanofibers. In this composition, solution had optimum viscosity correlated with applied voltage and distance of needle to collector.

3.2. Nanofibers of PVA/AB composite

The purpose of Acetylene Black added to the PVA solution is to increase carbon sources because PVA has low carbon yield when exposed to high temperatures and is expected to increase the conductivity of carbon fibres. Figure 3 shows the results of SEM observations on 15 % w/v PVA samples added with concentration variables AB. The addition of 5 wt. % has been able to change the morphology of PVA fibres to be larger and not smooth on the surface. The shape of the fibre also changes from straight to slightly curly-fibre. Increasing of AB 10 wt. % produces fibre that is relatively heterogeneous in size because there are large and many small sized fibres. While the fibre surface is no longer smooth. While the addition of AB to 15 wt. % produces a fibre that is not homogeneous in size and morphology. An increase in AB addition of up to 15 wt. % causes changes in the morphology and size of PVA/AB composite fibres.

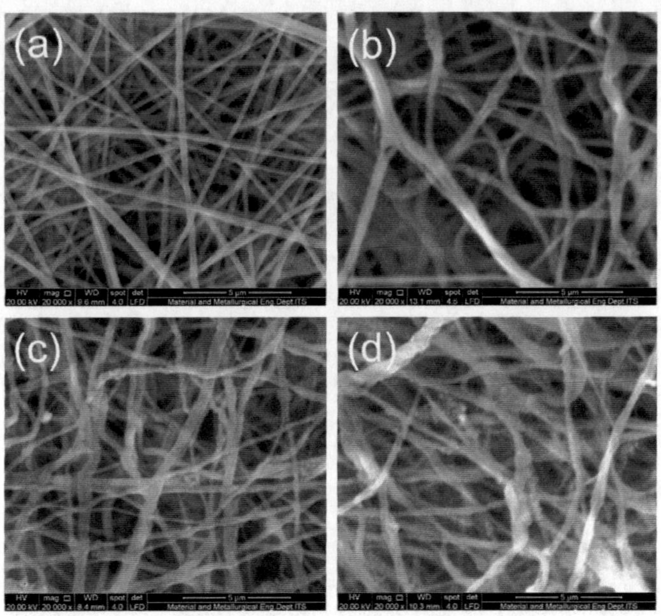

Figure 3. SEM images of electrospun PVA/AB fibres with various AB contents: (a) 0 wt. %, (b) 5 wt. %, (c) 10 wt. %; (d) 15 wt. %

IOP Conf. Series: Materials Science and Engineering **543** (2019) 012030 doi:10.1088/1757-899X/543/1/012030

The homogenous of particles in solution is the important factor in electrospinning polymer/inorganic composite. Sonication is an effective way for mixing polymer/inorganic composite [13]. The electrospinning of PVA/AB composite resulted grey coloured fibre mats. On the other hand, only PVA fibre mats had white colour. The grey colour of 15 wt. % AB was darker than others sample.
Figure 3 shows the SEM images of PVA/AB nanofibers with various AB amount of 0, 5, 10, and 15 wt. % to PVA powder weight with 15 % w/v. The average diameter of PVA/AB composite is almost similar for all samples at around 280 nm. Sample with 5 wt. % AB had smoother fibre than that of higher wt. % AB. Non uniform surface morphologies were found for 10 and 15 wt. % AB. The same phenomenon was also observed for polyacrylonitrile (PAN)/silica composite in which increasing the silica content exhibited more non uniform surface fibres [19].

3.3. Effect of iodine treatment on PVA/AB nanofibers
Figure 4a shows that the nanofiber structure was broken when the origin PVA/AB nanofiber was heated at temperature 200 °C under N_2 atmosphere. Increasing the carbonization temperature at 300 °C led to the further broken fibre structure becoming nanosized spherical particles as shown in Figure 4b. Furthermore, as depicted in Figure 4c, increasing the carbonization temperature at 450 °C made the particle size was smaller. PVA is one of the higher carbon content among vinyl alcohol polymers but on the other hand PVA has low carbon yield due to it melts before carbonization temperature [16]. The PVA thermally decomposed into the volatile low molecular weight molecules. The chains of PVA were pyrolyzed during carbonization. Therefore, PVA needs pre-treatment or additive to produce sustainable form of fibre and high carbon yield [20].

Figure 4. SEM images PVA/AB mats without iodine treatment after carbonization at temperature: (a) 200 °C; (b) 300 °C; (c) 450 °C

Iodine treatment can be considered to take over losing of carbon yield followed by carbonization. The iodine absorption of the PVA/AB mats were investigated through carbonization the samples with treated and untreated iodine. The weight of PVA/AB mats before and after iodine treatment were 0.541 and 1.242 gram, respectively. The 1:4 ratio of iodine crystals to PVA/AB mats was added into sealed glass jar. After 24 h of treating with vapour of iodine crystals, the colour of PVA/AB mats was changed from a grey to dark brown and the weight of PVA/AB mats increased about 50 wt. %.

As shown in Figures 5a and 5b, SEM images of PVA/AB mats prepared with iodine treatment at 80 °C for 24 hours for before and after carbonization, respectively, have almost the same structure fibre with non-homogeneous fibre diameters. Iodine treatment has a significant effect to sustain nanofiber structures after carbonization. A similar effect was investigated for pure of PVA nanofibers [20]. They found that during carbonization with iodine treatment for 24 hours, nanofibers can maintain the original nanofibers without melting. It can be explained that during the carbonization, PVA turned into a stable cyclic aromatic structure with intermolecular cross-linked resulted polyene structure by dehydration

IOP Conf. Series: Materials Science and Engineering **543** (2019) 012030 doi:10.1088/1757-899X/543/1/012030

PVA. Conversely, if the iodine process was carried out only 3-10 hours, the neat PVA entirely melted to form a solid layer and cracking in the structure of morphology.

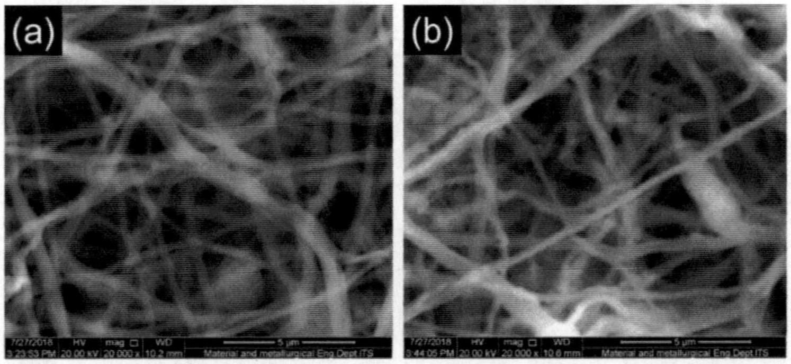

Figure 5. SEM images PVA/AB mats after iodine treatment: (a) before and (b) after carbonization

4. Conclusion

Effective methods have been found to fabricate carbon fibres from PVA/AB composite nanofibers using electrospinning techniques. The average diameter of the as-spun pure PVA was around 220 nm. The effect of AB content on PVA/AB composite influenced the diameter and their morphological structure. The average diameter of the PVA/AB nanofibers was around 280 nm. In the process of carbonization at 800 °C with iodine treatment for 24 hours played a very important role to maintain the nanofiber structure without melting. The ratio of nanofibers of PVA/AB mats to iodine crystals used in iodine treatment was 1:4 which resulted in diameter after carbonization at around 300 nm.

Acknowledgments

Research Grant sponsored by Directorate of Research and Community Services, Directorate General of Strengthening Research and Development, Ministry of Research Technology and Higher Education, Republic Indonesia (No. 128/SP2H/PTNBH/DRPM/2018) is gratefully acknowledged.

References

[1] Liu W and Adanur S 2009 *Text. Res. J.* **80** 124–34
[2] Ramakrishna S, Fujihara K, Teo W, Yong T and Ramaseshan R 2006 *Mater. Today* **9** 40–50
[3] Ondarçuhu T and Joachim C 1998 *Europhys. Lett.* **42** 215–20
[4] Feng J J 2002 *Phys. Fluids* **14** 3912–3926
[5] Ma P X and Zhang R 1999 *J. Biomed. Mater. Res.* **46** 60–72
[6] Bhardwaj N and Kundu S C 2010 *Biotechnol. Adv.* **28** 325–47
[7] Dersch R, Steinhart M, Boudriot U, Greiner A and Wendorff J H 2005 *Polym. Adv. Technol.* **16** 276–82
[8] Huggins T M, Pietron J J, Wang H, Jason Z and Biffinger J C 2015 *Bioresour. Technol.* **195** 147–53
[9] Setton R, Bernier P and Lefrant S 2002 *Carbon Molecules and Materials* (London and New York: Taylor and Francis Inc) p 489
[10] Tanabe Y and Yasuda E 2000 *Carbon N. Y.* **38** 329–34
[11] Ma X, Elbohy H, Sigdel S, Lai C, Qiao Q and Fong H 2016 *RSC Adv.* **6** 11481–7
[12] Lai C, Zhou Z, Zhang L, Wang X, Zhou Q, Zhao Y, Wang Y, Wu X F, Zhu Z and Fong H 2014 *J. Power Sources* **247** 134–41
[13] Chuangchote S, Sirivat A and Supaphol P 2007 *Nanotechnol.* **18** 145705
[14] Du E, Dong S and Sun J 2018 *Chem. Eng. Trans.* **66** 1057–62
[15] Bin Y, Chen Q, Nakamura Y, Tsuda K and Matsuo M 2007 *Carbon N. Y.* **45** 1330–9
[16] Donnet J, Wang T K, Rebouliat S and Peng J 1998 *Carbon fibers* (New York: Marcell Dekker Inc.)

[17] Pelipenko J, Kristl J, Janković B, Baumgartner S and Kocbek P 2013 *Int. J. Pharm.* **456** 125–34
[18] Fuenmayor C A, Mascheroni E, Cosio M S, Piergiovanni L, Benedetti S, Ortenzi M, Schiraldi A and Mannino S 2013 *Chem. Eng. Trans.* **32** 1771–6
[19] Ji L, Saquing C, Khan S A and Zhang X 2008 *Nanotechnol.* **19** 085605
[20] Fatema U K, Uddin A J, Uemura K and Gotoh Y 2011 *Text. Res. J.* **81** 659–72

IOP Conf. Series: Materials Science and Engineering **543** (2019) 012031 doi:10.1088/1757-899X/543/1/012031

Nano Carbon Produced by Advanced Mild Hydrothermal Process of Oil Palm Biomass for Supercapacitor Material

T D Larasati, T Prakoso, J Rizkiana, H Devianto, P Widiatmoko, I Nurdin
Chemical Engineering Department, Faculty of Industrial Technology, Bandung Institute of Technology, Indonesia 40132

E-mail: tirto@che.itb.ac.id

Abstract. Palm oil waste is an abundant biomass resource in Indonesia that is not well utilized as a raw material for producing high value-added materials. One product that is possible to be manufactured from oil palm biomass is activated nano carbon. Characteristics of high porosity and good conductivity will lead the activated nano carbon suitable as an electrode of supercapacitor material. Preparation of activated nano carbon is conducted by two main steps namely carbonization and activation. In this research, carbonization is done by hydrothermal process while activation is done by physico-chemical activation. This study focused on manufacturing activated nano carbon from empty fruit bunches of oil palm with hydrothermal carbonization for supercapacitor applications. The activation of mesopore area of the carbon has been taken place by using $ZnCl_2$ and $CaCl_2$ as a chemical activating agent during the hydrothermal process. The activated nano carbon produced from empty fruit bunches has a surface area of 41-1571 m^2/g and a pore size of 2.0–6.8 nm. In this study activated nano carbon is used as a working electrode in symmetric hybrid supercapacitor.

1. Introduction

Indonesia is one of the largest palm oil producer in the world with palm oil production reaches 31 million tons in 2015. The produced of palm oil is also produced a high amount of palm oil waste. One product that is possible to be manufactured from oil palm biomass is activated nano carbon. Activated nano carbon is an amorphous carbon material with high degree of porosity and an extended surface area [1]. Activated nano carbon preparations from wastes is vital for the regional economy when products of high value are obtained from low cost materials. These bring about the solution to problem of waste management [2]. Activated carbon is usually used as an adsorbent, energy storage aplication, synthesis carbon, catalyst support, etc [3]. One of interesting energy storage aplication is supercapacitor. Porous carbons are used as electrode material in supercapacitor because of high electrical condutivity, low cost and commercially available [4]. The pore size of activated nano carbon which is suitable for capacitor electrode is in the mesopore range (2-50 nm) [5]. Different raw material type and activation procedure has an important role in the physico-chemical characteristics of activated carbon e.g. surface area, porosity, and pore distribution,and surface functional groups which are all important properties for supports [6].

Preparation of activated nano carbon is conducted by two main steps namely carbonization and activation. The carbonization process is commonly conducted by pyrolysis method by heating the raw material at temperature range 300-600 °C in the total absence of oxygen condition and decomposes biomass into gas, liquid (bio-oil) and solid (bio-char). Another carbonization process which can be used is a hydrothermal method. Hydrothermal carbonization (HTC) is a thermo-chemical process for recovery and valorization of biomass and organic waste that uses sub-critical liquid water as a reaction medium at mild temperature (180-250 °C) for converting feedstock into value added products [7]. The advantage of the hydrothermal process due to it used lower energy than pyrolysis process because the operation temperature is lower than pyrolysis process [8]. Study on activated nano carbon synthesis using biomass precursors has been done previously. Laginhas et al. (2016) have made activated carbon from chitosan by hydrothermal treatmant at a temperature of 200 °C with surface area of about 852 m^2/g. This research focused on manufacturing activated nano carbon from empty fruit bunches of oil palm with hydrothermal carbonization for supercapacitor applications by adding pre-treatmant deashing raw material.

2. Experimental

2.1. Materials
Empty fruit bunch (EFB) was chosen as biomass samples. The sample was gathered from PT. Perkebunan Nusantara VIII in Bogor. EFB was cleaned using water and dried. After that, dried biomass is crushed using disc mill and sieved into coarse granules (35-80 mesh). The sample has been subjected to two different types of treatments in this study. The pretreatment processes are deashing and without deashing raw material. The pretreatment is leaching with 1M HF solution for an hour [9]. The samples were filtered off and washed with distilled water. The process of washing was continued until the wash-water remained neutral. Finally, the samples were dried at 105 °C in oven. Zinc chloride and calcium chloride was used as chemical activating agent.

2.2. Hydrothermal
EFB with the specification ratio of biomass to water (1:3, 10 g biomass in 30 mL water) was fed into hydrothermal reactors containing water mixed with activating agent. The ratio of activating agent to the biomass was 2:1. The samples were held at the desired temperature (225, 250 and 275 °C) for 60 min. The obtained biochar was then filtered and dried at 105 °C for 24 hours.

2.3. Activation
The biochar which recovered from hydrothermal process loaded onto crucible and positioned inside the tubular furnace. The temperature was ramped to 800 °C at a rate of 10 °C/min under nitrogen atmosphere at a flow rate of 50 mL/min. After the temperature was reached, N_2 was replaced by CO_2 at a flow rate of 40 mL/min for two hours. Afterwards the furnace was cooled to room temperature in the presence of N_2. The product was added with HCl solution 0,1M and stirred for 30 minutes, then washed with demineralized water until rinsed water remained neutral. The product was dried at 105°C for 24 hours to obtained activated nano carbon.

2.4. Characterizations
Ash content of EFB determined by gravimetric analysis. FTIR analysis was conducted to get the qualitative and quantitative estimation of OFGs. The morphology of raw material and product was obtained by SEM. The surface area and pore diameter of activated nano carbon were calculated by applying the Brunauer-Emmet-Teller (BET) model using a gas sorption Quantochrome Instruments Nova 3200e.
Samples treated with chemicals used were named as AC-$ZnCl_2$ (activated nano carbon with $ZnCl_2$ as activating agent) and AC-$CaCl_2$ (activated nano carbon with $CaCl_2$ as activating agent).

2.5. Electrochemical Characterizations
The electrochemical analysis was conducted to observed activated nano carbon performance as a supercapacitor cell. Cyclic voltammetry, galvanostatic charge/discharge, and electrochemical

impedance spectroscopy were examined using Gamry V3000. Electrochemical analysis was performed by assembling in a symmetrical two-electrode configuration. The working electrodes were prepared from activated nano carbon which placed in 1 cm^2 of stainless steel mesh. Nafion 212 was sandwiched between two electrodes and then infiltrated with electrolyte solution (6 M KOH).

3. Results and Discussion

3.1. Effect of deashing treatments on empty fruit bunch-sample analysis

The proximate analysis of empty fruit bunch are reported in Table 1. Based on Table 1, it was the common characteristic of empty fruit bunch that growth in Indonesia.

Table 1. Proximate analysis of empty fruit bunch (EFB)

Sample	Moisture (wt%)	Volatile matter (wt%)	Fixed carbon (wt%)	Ash (wt%)
EFB	5.47	73.17	17,17	4.19

Fixed carbon and volatile matter content in empty fruit bunch is high so that this biomass suitable as a precursor material for activated nano carbon [10]. The deashing treatments is associated with the mass reduction in the original biomass feedstock. The ash-forming mineral content in biomass is removed by the use of a strong acid catalyst [11]. The percentage ash in the untreated and the treated EFB are presented in Table 2. It shows that treatment with HF can decrease the ash levels about 40%.

Table 2. Percentage of ash in treated EFB

	Untreated EFB (%)	Treated EFB (%)
Ash	4.19	2.86

3.2. Effect of hydrothermal treatment temperature

Table 3 shows the result of mass yield of activated nano carbon after hydrothermal carbonization and activation process. From Table 3, mass yield of activated nano carbon increase when temperature during hydrothermal increase from 225 °C to 275 °C.

Table 3. Mass yield of activated nano carbon

Activating agent	Mass yield of AC untreated EFB (g)			Mass yield of AC treated EFB (g)		
	225 °C	250 °C	275 °C	225 °C	250 °C	275 °C
$ZnCl_2$	8.857	8.940	9.594	9.110	9.588	9.808
$CaCl_2$	7.362	7.560	7.981	8.179	8.335	8.933

Higher temperature leads to extensive dehydration and an increase in the degree of condensation of the biochar. The effect of hydrothermal carbonization temperature (225, 250 and 275 °C) on the formation of OFG on EFB based biochar was observed that irrespective of temperature employed there was an increase in the OFG content in biochar compared to raw material [12].

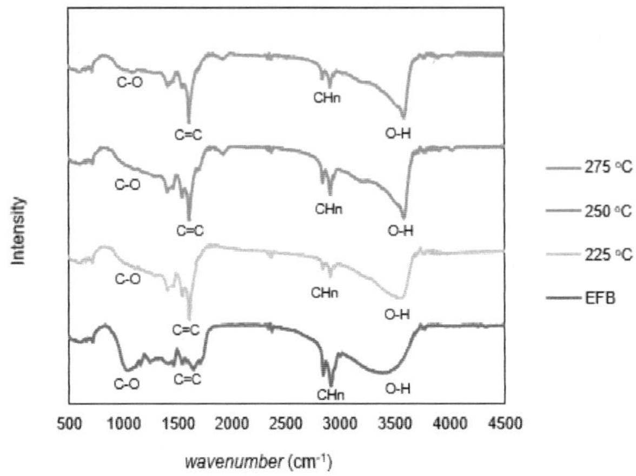

Figure 1. FT-IR spectrum of empty fruit bunch and biochar

Figure 1 shows FTIR spectrum from EFB and biochar. The peak around 1600 cm^{-1} attribute to aromatic groups (lignin and aromatic groups) which is formed during carbonization [13] where the intensity of bonds is not much different for the three samples. The intensity of OH-bonds decreased after the hydrothermal process. It is due to dehydration of biomass occurred so that water molecules within the biomass removed [14]. The intensity of the peak which is around ~3500 cm^{-1} shows stretching vibration of OH-bonds (alcohol, phenol, and carboxyl groups).

3.3. Scanning Electron Microscopy (SEM) *and* BET *Surface Area*

Figure 2 shows the morphology of EFB and activated nano carbon after hydrothermal carbonization and activation process. Figure 2a indicate that EFB have a rough surface and closed pores. More pores are opened on EFB structure creating holes along the EFB surface, so that enhanced the porosity and the surface area of activated nano carbon.

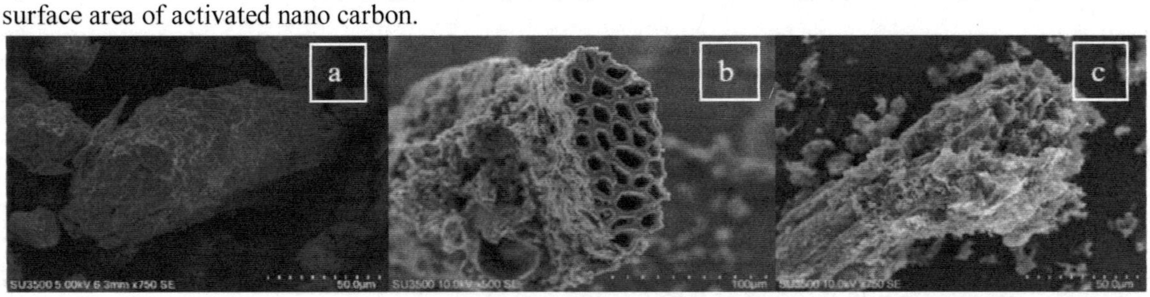

Figure 2. SEM result of (a) empty fruit bunch, (b) AC-$ZnCl_2$, and (c) AC-$CACl_2$

The BET surface area of activated nano carbon which is produced at different process parameters is shown in Table 4. The surface area of activated nano carbon sample increase when temperature increase from 225 °C to 275 °C. This may happen because some pores which are formed by the decomposition of organic components in materials increase when temperature of the hydrothermal process is increase. AC-$ZnCl_2$ shows the highest BET surface area. This is due to more effective dehydration in the addition of $ZnCl_2$ giving rise to higher OFG content and better activation. Likewise, $ZnCl_2$ which acts as an activating agent removes the oxygen and hydrogen content in biomass by forming water and could enhance the porosity of the material [15].

Table 4. Surface area of activated nano carbon

Activating agent	BET Surface area untreated EFB (m^2/g)			BET Surface area treated EFB (m^2/g)		
	225 °C	250 °C	275 °C	225 °C	250 °C	275 °C
$ZnCl_2$	740	1048	1229	1006	1240	1571
$CaCl_2$	41	406	528	268	454	723

Surface area of activated nano carbon which produced with treated EFB is higher than activated nano carbon which produced with untreated EFB. The treatment with HF is more effective to remove some of the ashes contained in EFB, because HF is known to be effective to remove compounds such as carbonates, oxides and silica [16]. Average pore size diameter of activated nano carbon shown in Table 5.

Table 5. Activated nano carbon pore size

Activating agent	Average pore size untreated EFB (nm)			Average pore size area treated EFB (nm)		
	225 °C	250 °C	275 °C	225 °C	250 °C	275 °C
$ZnCl_2$	2.18	2.01	2.50	2.63	2.94	2.13
$CaCl_2$	2.88	2.79	4.97	6.81	5.53	5.63

The AC-$CaCl_2$ has bigger pore size than AC-$ZnCl_2$. The contact between activating agent and EFB is more intense and resulted the porosity of the activated nano carbon enhanced. Activating agent to the raw material ratio which is used in this research drive the micropore formation and pore widening simultaneously. Even though activated nano carbon has mesoporous size, it close to micropore size [17]. The different activating agent is used for activated nano carbon manufacturing has also affected the character of activated nano carbon characteristics [18].

3.4. Electrochemical characterization

Electrochemical analysis was performed to analyze the performance of activated nano carbon while was applied in supercapacitor cell. Figure 3a shows the voltammogram has a tendency to form a slope and not in rectangular shape, that indicates charging-discharging process through high resistance inside the supercapacitor cell [19]. Furthermore, from the voltammogram was suspected that corrosion reaction occurred in electrode due to the dielectric material used was too thin.

Figure 3b shows galvanostatic charge/discharge of activated nano carbon has a triangle and symmetric shape which indicate the supercapacitor cell has good capacitive performance [20]. This result shows that the capacitance of supercapacitor cells made from AC-$CaCl_2$ is greater than AC-$ZnCl_2$ which is suitable with the results from calculations performed by cyclic voltammetry (CV) analysis

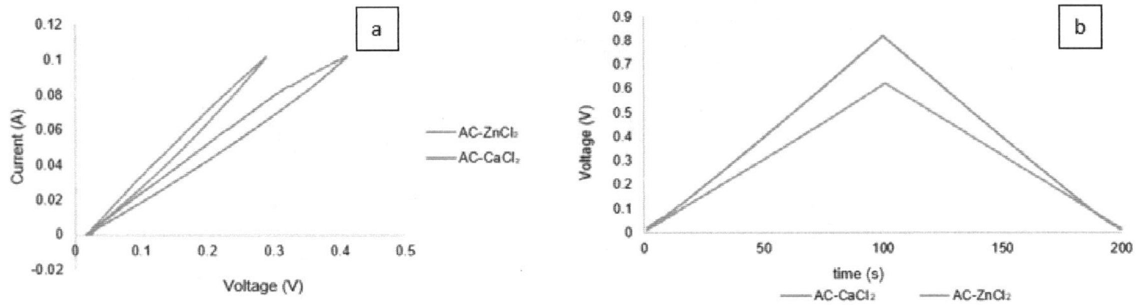

Figure 3. (a) Cyclic voltammogram and (b) Galvanostatic charge/discharge diagram

The specific capacitance of the produced activated nano carbon can be calculated from voltammogram according to the following Equation (1): [21]

$$C= \sum|I|.\Delta t \ / \ m \ \Delta V \qquad (1)$$

with $\sum|I|.\Delta t$ is the area of the current $|I|$ (A) against time (s); m (g) is the mass of active material in the electrode; and ΔV (V) is the potential window.

Table 6. Specific capacitance of activated nano carbon electrode

Material	Capacitance (F/g)
AC-$ZnCl_2$	2.452
AC-$CaCl_2$	4.305

The results of specific capacitance in Table 6 show that the specific capacitance of AC-$CaCl_2$ electrode is greater than AC-$ZnCl_2$ electrode. The results show the contradiction to the average pore size of activated nano carbon. The capacitance is not affected by the surface area of materials. The lower specific capacitance could be caused due to imperfect interphase connection, cell resistance and cell sealing system not optimal [22]. This imperfect cell sealing system could cause the presence of reactive and unstable oxygen functional groups which can interfere the ion transfer process [23]. In addition, side reactions between carbon, electrolyte and impurities could also decrease cell capacitance [24].

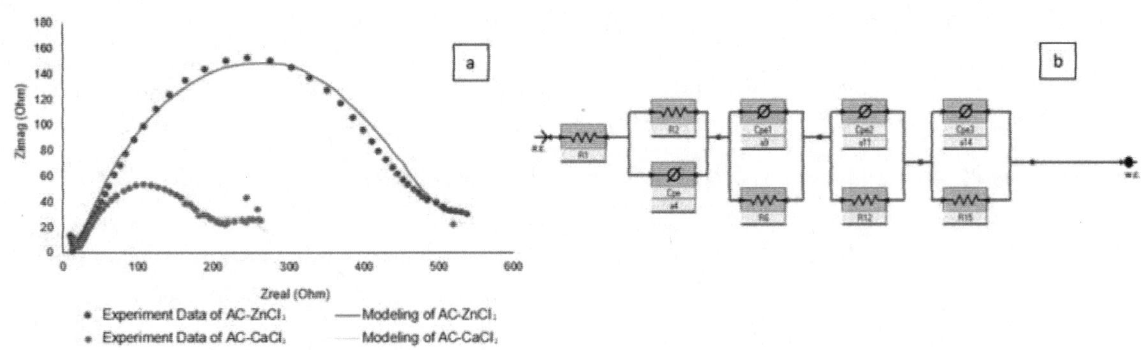

Figure 4. (a) Nyquist Curve and (b) Equivalent circuit of Supercapacitor from activated nano carbon

The result of EIS analysis is represented in a Nyquist curve which is modelled with equivalent circuits to obtain the impedance value of the supercapacitor cell. Figure 4a and Figure 4b show the Nyquist curve and equivalent circuit, respectively.
The Nyquist curve of activated nano carbon supercapacitor has two semicircle curve which indicates cell impedance is affected by charge transfer and mass transfer.

Table 7. Impedance parameters of carbon electrode

Parameter	Value	Unit	Parameter	Value	Unit
R1	1.13 x 10^{-1}	Ohm	Cpe2	5.55 x 10^{-4}	S*s^a
R2	19.04	Ohm	a11	5.76 x 10^{-1}	
Cpe	6.80 x 10^{-8}	S*s^a	R12	128.3	Ohm
a4	9.83 x 10^{-1}		Cpe3	2.37 x 10^{-2}	S*s^a
R6	57.51	Ohm	a14	6.47 x 10^{-1}	
Cpe1	3.41 x 10^{-4}	S*s^a	R15	80.12	Ohm
a9	8.13 x 10^{-1}				

Table 7 shows the parameters of the equivalent circuit modelling against the experimental data. Ohmic resistance (R1) is a parameter which is related to the electrolyte resistance. Based on the EIS analysis, the interphase resistance of activated nano carbon supercapacitor causes the capacitance of supercapacitor cell becomes smaller which is suitable with the results from calculations performed by cyclic voltammetry (CV) analysis.

4. Conclusion

Activated nano carbon as an electrode material of supercapacitor can be obtained from palm oil waste through the hydrothermal carbonization and activation process. Increasing temperature during the hydrothermal carbonization process could increase the surface area of activated nano carbon. The capacitance of supercapacitor is 4.305 F/g for activated nano carbon using $CaCl_2$ as activating agent (AC-$CaCl_2$).

Acknowledgement
This work is supported by Badan Pengelola Dana Perkebunan Kelapa Sawit, funding year of 2018.

References

[1] Cencen F and Aktas O 2012 *Activated Carbon for Water and Wastewater Treatment* Integration of Adsorption and Biological Treatment (Germany: Wiley-VCH).
[2] Nor N M, Lau L C, Lee K T and Mohamed A R 2013 Synthesis of activated carbon from lignocellulosic biomass and its applications in air pollution control-A review *J. Environ. Chem. Eng.* **1(4)** 658-66.
[3] Frackowiak E, Abbas Q, and Béguin F 2013 Carbon/carbon supercapacitors *J. Energy Chem.* **22(2)** 226-40.
[4] Marsh H, and Rodriguez F 2006 *Activated Carbon Elsevier Science & Technology Books.*
[5] Jain A, Xu C, Jayaraman S, Balasubramanian R, Lee J Y and Srinivasan M P 2015 Mesoporous activated carbons with enhanced porosity by optimal hydrothermal pre-treatment of biomass for supercapacitor applications *Microporous Mesoporous Mater.* **218** 55-61.
[6] Lahti R, Bergna D, Romar H, Tuuttila T, Hu T and Lassi U 2017 Physico-chemical properties and use of waste biomass-derived activated carbons *Chem. Eng. Trans.* **57** 43-48.
[7] Puccini M, Stefanelli E, Hiltz M, Seggiani M and Vitolo S 2017 Activated carbon from hydrochar produced by hydrothermal carbonization of wastes *Chem. Eng. Trans.* **57** 169-74.

[8] Tekin K, Karagöz S, and Bektaş S 2014 A review of hydrothermal biomass processing *Renew. Sustain. Energy. Rev.* **40** 673-87.

[9] Pottathil R, Bowers J S Jr and Havemann G D 2012 Method of ash removal from biomass US Patent 2012 0009660.

[10] Lua A C, Lau F Y, and Guo J 2006 Influence of pyrolysis conditions on pore development of oil-palm-shell activated carbons *J. Anal. Appl. Pyrolysis.* **76(1)** 96-102.

[11] Das P, Ganesh A, Wangikar P 2004 Influence of pretreatment for deashing of sugarcane bagasse on pyrolysis products *Biomass Bioenergy* **27** 445-57.

[12] Jain A, Balasubramanian R, and Srinivasan M P 2015 Tuning hydrochar properties for enhanced mesopore development in activated carbon by hydrothermal carbonization *Microporous Mesoporous Mater.* **203** 178-85.

[13] Nizamuddin S, Kumar J, Subramanian N, Sahu J N, Ganesan P, Mubarak N M and Mazari S A 2015 Synthesis and characterization of hydrochars produced by hydrothermal carbonization of oil palm shell *Can. J. Chem. Eng.* **93(11)** 1916-21.

[14] Yuliansyah A T, Hirajima T, Kumagai S, and Sasaki K 2010 Production of solid biofuel from agricultural wastes of the palm oil industry by hydrothermal treatment *Waste Biomass Valorization* **1(4)** 395-405.

[15] Jain A, Jayaraman S, Balasubramanian R and Srinivasan M P 2014 Hydrothermal pre-treatment for mesoporous carbon synthesis: enhancement of chemical activation *J. Mater. Chem. A* **2(2)** 520-28.

[16] Ahmad F, Daud W M A W, Ahmad M A and Radzi R 2013 The effects of acid leaching on porosity and surface functional groups of cocoa (Theobroma cacao)-shell based activated carbon *Chem. Eng. Res. Des.* **91(6)** 1028-38.

[17] Ahmadpour A and Do D D 1997 The preparation of activated carbon from macadamia nutshell by chemical activation *Carbon* **35(12)** 1723-32.

[18] Jain A, Balasubramanian R and Srinivasan M P 2016 Hydrothermal conversion of biomass waste to activated carbon with high porosity: A review *Chem. Eng. J.* **283** 789-805.

[19] Thambidurai A, Lourdusamy J K, John J V and Ganesan S 2014 Preparation and electrochemical behavior of biomass based porous carbons as electrodes for supercapacitors - a comparative investigation *Korean J. Chem. Eng.* **31(2)** 268-75.

[20] Abioye A M, Noorden Z A and Ani F N 2017 Synthesis and characterizations of electroless oil palm shell based-activated carbon/nickel oxide nanocomposite electrodes for supercapacitor applications *Electrochim. Acta* **225** 493-502.

[21] Elmouwahidi A, Zapata-Benabithe Z, Carrasco-Marín F and Moreno-Casti C 2012 Activated carbon from KOH-activation of argan (Argania spinosa) seed shells as supercapacitor electrode *Bioresour Technol* **111** 185-90.

[22] Widiatmoko P, Devianto H, Nurdin I and Yandra R E 2016 The effect of carbon nanotube composite addition on biomass-based supercapacitor *J. Eng. Technol. Sci.* **48(5)** 597-613.

[23] Gu W, Peters N and Yushin G 2013 Functionalized carbon onions, detonation nanodiamond and mesoporous carbon as cathodes in Li-ion electrochemical energy storage devices *Carbon* **53** 292-301.

[24] Puthusseri D, Aravindan V, Madhavi S and Ogale S 2014 3D micro-porous conducting carbon beehive by single step polymer carbonization for high performance supercapacitors: the magic of in situ porogen formation *Energy Environ. Sci.* **7(2)** 728-35.

IOP Conf. Series: Materials Science and Engineering **543** (2019) 012032 doi:10.1088/1757-899X/543/1/012032

Increasing yield of graphene synthesis from oil palm empty fruit bunch via two-stages pyrolysis

P Widiatmoko, I F Sukmana, I Nurdin, T Prakoso and H Devianto
Institut Teknologi Bandung, Jl. Ganesha 10, Bandung, 40132, Indonesia

E-mail: pramujo@che.itb.ac.id

Abstract. Graphene is a 2D hexagonal lattice structure of sp^2 carbon atoms which has been acknowledged for its superior electrical, mechanical, and thermal properties. Production of graphene in large scale and low cost are attracting topic in recent years. Previous study shows that production of graphene from biomass via pyrolysis has low yield of graphene. In this study, we produced graphene sheets from oil palm empty fruit bunch via two-stages pyrolysis to increase the yield. The produced graphene sheets were characterized by scanning electron microscopy, transmission electron microscopy, Raman scattering, and X-ray diffraction. Surface properties (i.e. effective surface areas, pore volumes, and pore size distributions) were studied by nitrogen adsorption–desorption measurements. Effect of the first stage temperature of pyrolysis on the yield, structure, and properties of graphene has been investigated. Our result indicated that two-stages pyrolysis could increase the yield of graphene up to 70%. Graphene sheets shows favorable features of nanosheet frameworks (4–10 atomic layers) and high surface area (870 $m^2\ g^{-1}$).

1. Introduction

Indonesia is well known as the largest producer and exporter of palm oil in the world. The directorate general of plantation (Indonesia) recorded that in 2015, Indonesia produces 31 million tons of crude palm oil (CPO). Production of the CPO increased from 5.39 to 8.42% per year in term of 2010 to 2015. Processing oil palm fresh fruit bunch (FFB) into CPO result in abundant biomass by-products such as empty fruit bunch, oil palm kernel, and palm oil milling effluent. The produced CPO is only about 30% of the FFB mass, while the rest will become waste of biomass. High content of organic material in the by-product of oil palm milling may harmful to environment. Therefore, proper treatment is required. Recently, biomass is considered as a promising source of energy or biofuels, chemicals, and bio-based products which will reduce the dependency on fossil fuels [1]. Charcoal and activated carbon are industrial example of the biomass-based products. Another promising high value product from the biomass is graphene.

Graphene is a thin sheet of carbon which has high conductivity, large surface area [2]. The graphene has been widely applied in nanoelectronic and nanocomposite components such as transistors, solar cells, molecular sensors, and lithium ion batteries. In addition, graphene can also be utilized as a sturdy electrode material for supercapacitors [3]. Supercapacitor is an emerging technology on electrical energy storage. The supercapacitor utilizes large electrode surface and electrical double layer which allows it to possess greater energy density than the conventional capacitors as well as much greater power density than the batteries [4].

Study on graphene synthesis using biomass precursors has been done previously. Shams et al. [5] have made graphene from dried camphor leaf precursors by pyrolysis method at a temperature of 1200°C

IOP Conf. Series: Materials Science and Engineering **543** (2019) 012032 doi:10.1088/1757-899X/543/1/012032

with yield of about 0.8%. Graphene has also been produced from palm empty fruit bunches (EFB) and palm kernel shells by pyrolysis method at 900°C with yield of fixed carbon is about 27%. Higher yield of graphene, therefore, is desired from the process. In this study, we utilize two-stages pyrolysis for obtaining higher graphene yield. The idea is to exploit different decomposition temperature of cellulose, hemicellulose, and lignin as main structure of the EFB.

2. Methods

2.1. Graphene synthesis
Dried EFB was crushed using ball mill and then sieved into coarse granules (170-200 mesh). The prepared raw material was mixed with 3M $FeCl_3$ as a catalyst, urea as a nitrogen source with ratio of 1: 1 and $ZnCl_2$ as activating agent with ratio of 2: 1, all respect to the raw material. The mixture was stirred and heated on a hot plate (80°C) for 2 hours. Furthermore, the mixed material is dried in the oven for 2 hours. The mixed material then put into a tubular furnace for carbonization and activation. Tubular furnace is conditioned under nitrogen atmosphere, heated to 250°C within 45 minutes. After reaching the desired temperature, the temperature is maintained for 60 minutes. This is the first stage. Afterward, the temperature was increased to graphitization stage at 900°C and kept the temperature for another 90 minutes under flowing nitrogen. This is the second stage of pyrolysis. The sample was then denoted as G250. Experiments were repeated with the first stage of pyrolysis temperature varied at 350°C and 450°C, denoted as G350 and G450 respectively. Holding time for all treatments are the same. The product was added with HCl 2M and stirred for 30 minutes. Consecutively, the solutions were washed with demineralized water several times until rinsed water reached pH 7. The product then was dried at 105°C for 24 hours to obtain graphene.

2.2. Characterization
Fixed carbon in product was studied using gravimetric analysis. Electron microscopy analysis (SEM and TEM) was performed to study the graphene morphology. Nitrogen adsorption-desorption analysis with Brunauer–Emmett–Teller (BET) calculation method was performed to determine the surface area of the products. Crystallinity of the product was also studied using XRD analysis. To confirm graphitic characteristic of the produced graphene, Raman spectroscopy analysis was performed.

2.3. Electrochemical analysis
Asymmetrical two-electrode configurations were used to evaluate the capacitive performance of the produced graphene. The negative electrodes were prepared by mixing the produced graphene with polyvinylidene fluoride (PVDF) at a weight ratio of 90:10. The above materials were coated onto a 1 cm^2 stainless steel mesh current collector and dried at 80°C for 8 h. The positive electrodes were prepared by mixing nickel oxide with polyvinylidene fluoride (PVDF) at a weight ratio of 90:10. The materials were coated onto a 1 cm2 stainless steel mesh current collector and dried at 80°C for 8 h. Nafion 212 was sandwiched between two electrodes. The test was performed with KOH 6M aqueous electrolyte solution under ambient conditions using Gamry V3000. Electrochemical characteristic of samples was studied using cyclic voltammetry (CV) and electrochemical impedance spectroscopy (EIS) techniques.

3. Results and discussions
The pyrolysis in this experiment consists of two stages, with variation on the temperature of the first stage i.e. 250, 350, and 450°C, followed by the second stage pyrolysis at 900°C. The pyrolysis products were then immersed in HCl 2M for 2 hours to remove Fe impurities. The acid residue was rinsed with aqua dm. The product was then dried at 105°C temperature. To analysis fixed carbon content in this dried sample, the sample was heated at 800°C for 6 hours in the furnace. The remaining unburned sample is ash. The fixed carbon content calculated from the weigh difference between the dried sample and

remaining ash. Result of dried pyrolysis product and fixed carbon content after pyrolysis process are listed in Table 1. To be noted that biomass precursor weight is 3 g.

Table 1. Yield of pyrolysis product and fixed carbon.

Sample	G250	G350	G450
Char (gram)	0.56	0.49	0.37
Fixed Carbon(%w/w)	79.9	90.01	63.37

The results show that temperature of first stage pyrolysis affects to the yield of dried pyrolysis product. The yield decreases with increasing the pyrolysis temperature due to increased loss of condensable material [6]. During thermal decomposition process, biomass components decompose at different temperature ranges. In the first stage (below 200°C), a weight loss of about 10% is caused by loss of water molecules (dehydration). Furthermore, hemicellulose decomposes in range of 150-350°C, cellulose at 275-350°C, whereas the lignin in range of 250-500°C [7]. EFB contains 50.7% lignin [8]. The maximum decomposition of lignin occurs in the temperature range of 350-450°C. Lignin pyrolysis produces more aromatics and char than produce by cellulose. Pyrolysis of the lignin typically produces about 55% char [9]. Therefore, in order to obtain high graphene yield from pyrolysis of EFB, pyrolysis should be carried out at temperature decomposition of lignin. The idea is proved by the results of this study where the highest yield is obtained at first stage pyrolysis temperature of 350°C.

Morphology of graphene resulted from this pyrolysis was analyzed using electron microscopy. Figure 1 shows the SEM and TEM images of the pyrolysis product from EFB with the ratio $ZnCl_2$:biomass of 2:1 at various first stage pyrolysis temperature. It is concluded that the temperature of first stage pyrolysis, with advantage on increasing the yield, has insignificant effect on produced graphene. Graphene formation mainly occurs at higher temperatures of second stage pyrolysis. According to Zhang et al. [10], Fe_3C was formed gradually at 700°C and changed to Fe. At higher temperatures, amount of Fe_3C that turns into Fe will increase. As Fe_3C changes to Fe, the activated carbon atoms in Fe_3C layer will diffuse outward to form graphene sheet.

Layer morphology is studied using the TEM analysis in Figure 1(d). The produced graphene seems consist of several layers of graphene. The layers are caused by buildup and assembly of formed template. The few-layers or multilayers graphene-like nanosheets are stacked and folded together, which is beneficial for rapid ion and electron transport during the charge–discharge process of supercapacitors [11].

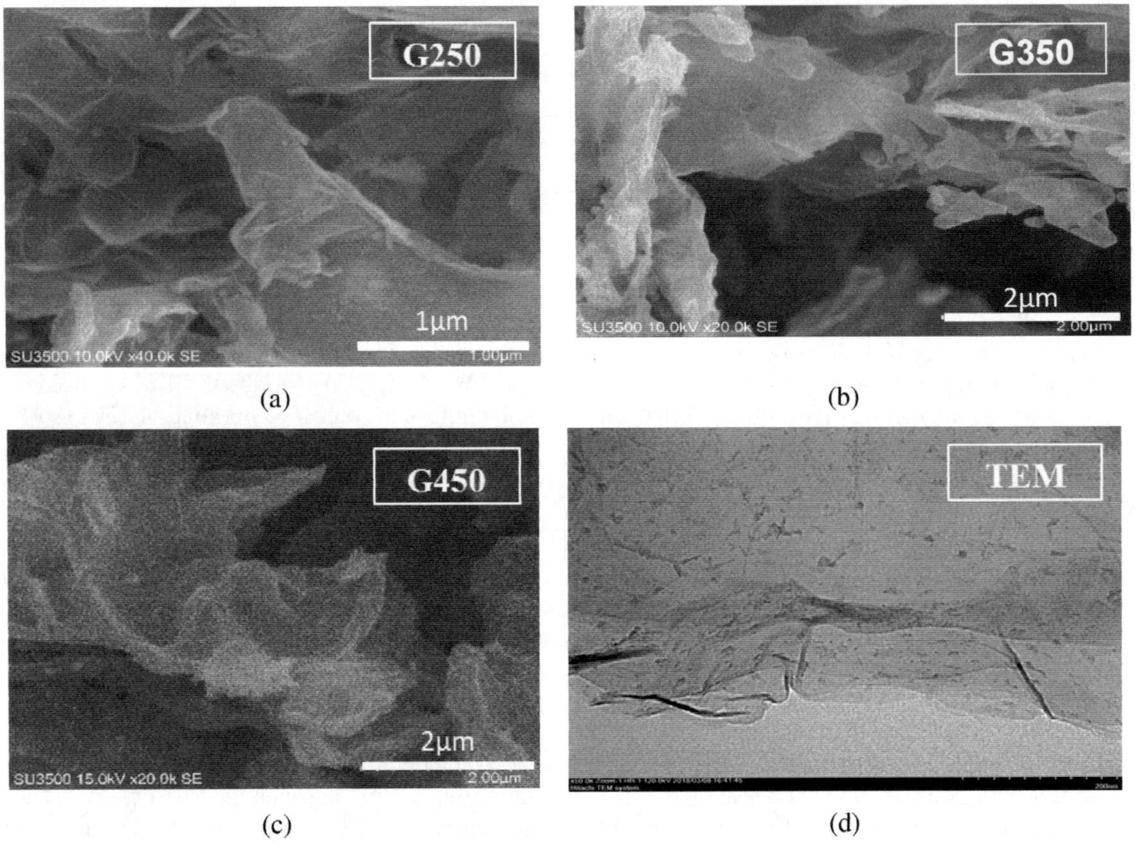

Figure 1. SEM and TEM images of graphene.

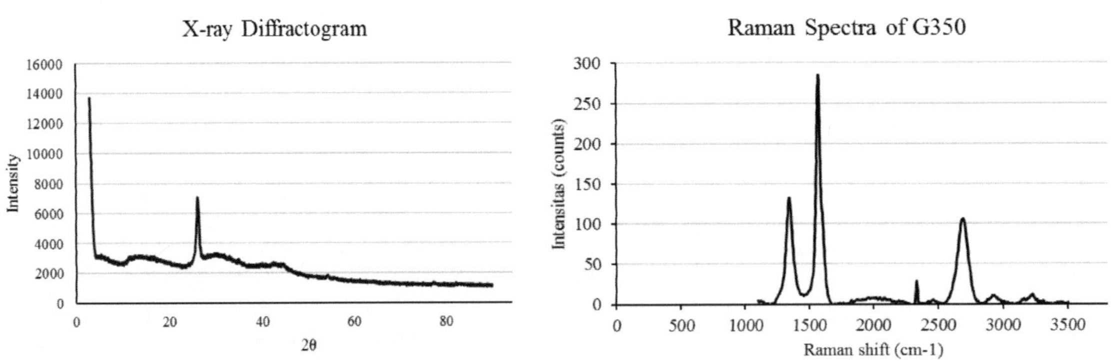

(a) (b)

Figure 2. (a) X-ray diffractogram and (b) Raman spectra of G350.

The XRD analysis as shown in Figure 2(a) shows a peak at 26° which indicates graphite structure [11]. The peak on the widened XRD spectrum shows the amorphous carbon structure generated by the destruction process of the initial graphite structure during the pyrolysis process. The resulting carbon structure is an intermediate structure between graphite and amorphous structures called turbostatic structures or random layered lattice structures [12]. The distance between the pseudographitic layers

(d_{002}) calculated using *Match* software is in range of 0.337 - 0.342 nm, in accordance with results of HRTEM analysis.

Further, degree of product graphitization is calculated based on the equation (1) [3]:

$$G = \frac{0,3440 - d_{002}}{0,3440 - 0,3354} \times 100\% \quad (1)$$

Value of 0.3440 in the equation refers to interlamellar distance of the turbostatic graphite (in nm) and 0.3354 is interlamellar distance of a single graphite crystal (in nm). From the calculation, the degree of product graphitization reached 77%. The high degree of graphitization of multiple layers graphene indicates good electrical conductivity [3].

Raman spectroscopy analysis was performed to study further on the produced graphene. The Raman spectra in Figure 2(b) shows the peaks corresponding to the regular graphite lattice mode vibrations. The D-Band is corresponding to the A1g mode which is the vibration of the carbon atoms in the planar termination of the irregular graphite. The G-band determines the E2g mode of 2D graphite, which corresponds to the bonding vibrations of the sp^2 carbon atoms in the hexagonal lattice. Intensity ratio of I_G / I_D is proportional to the crystalline degree of the carbon material. The G and D peak is at a wave number of about 1573 cm^{-1} and 1346 cm^{-1}, respectively. The intensity of G and D peaks depends on the percentage of the graphite domain containing the sp^2-carbon layer of hybridization in the formed graphene [13]. Based on the Raman spectra, the produced graphene show a regular graphitic character because intensity of G peak is much higher than the D peak. The I_G/I_D value of 2.14 suggests high graphitization degree of the product. The peak in 2690 represents the 2D band. Position and height profile at the 2D peak indicate layering on graphene. Monolayer graphene has 2D peak at 2679 cm^{-1}. The shifting of 11 cm^{-1} to higher wave numbers indicates that our product has 2-4 layers of graphene, in accordance with the results of TEM analysis [13].

Surface area analysis using BET calculations was performed to determine the surface area and pore distribution of graphene. These parameters are important when the carbon material applied as electrode of supercapacitor. The result in Table 2 shows that graphene derived from EFB has significant increment surface area at first stage pyrolysis temperature of 350°C. According to [14] and [15], the surface area decreases at high temperatures treatment due to collapse of micropore into bigger pore. In addition, at high temperatures, the carbon structure shrinks as volatile substances evolve from the structure.

Table 2. Surface area of graphene.

Sample	Surface area (m^2 g^{-1})
G250	290
G350	880
G450	264

Important to note that the graphene surface area can be increased significantly using this two-stage pyrolysis method. In the previous research, surface area analysis of pyrolysis (single stage) EFB at 900°C using $FeCl_3$ 3M catalyst with activating agent $ZnCl_2$ is 63.12 m^2 g^{-1}. In addition, one stage pyrolysis of EFB performed by Nasir [16] at 900 ^{0}C obtained surface area of EFB is 117 m^2 g^{-1}.

Our product of graphene is aimed for supercapacitor. Therefore, electrochemical analysis was performed to test the performance of produced graphene as supercapacitor electrode. Cyclic voltammetry (CV) and electrochemical impedance spectroscopy (EIS) were performed. Cyclic voltammetry was used to observe cell characteristics and capacitance. Supercapacitor cells were prepared for size of 1 cm x 1 cm. The dried product of pyrolysis, acting as a negative electrode, was mounted on a stainless-steel mesh. Nickel oxide, as a positive electrode, was placed on a carbon sheet with a 6M KOH electrolyte. These two electrodes were separated using Nafion 212 membrane.

IOP Conf. Series: Materials Science and Engineering **543** (2019) 012032 doi:10.1088/1757-899X/543/1/012032

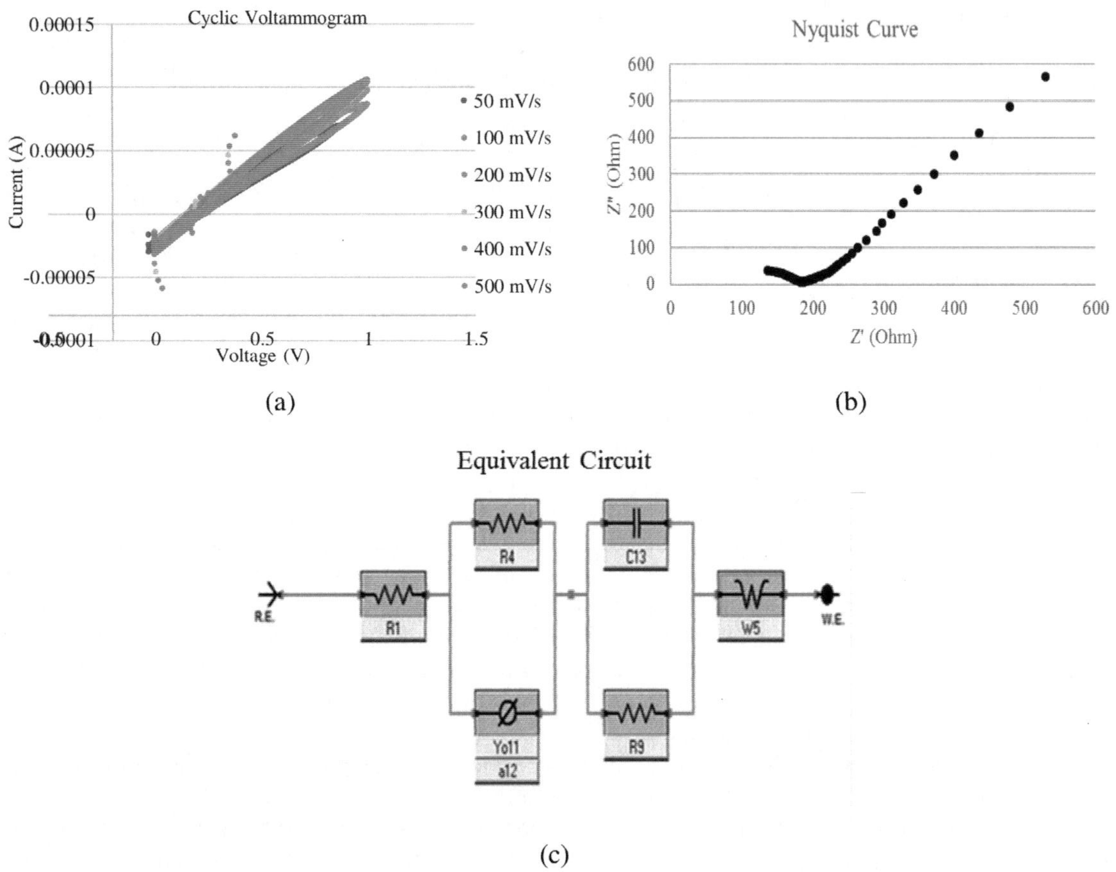

Figure 3. Cyclic voltammogram, nyquist curve and equivalent circuit of supercapacitor from produced graphene.

Under scan rate of 10 mV s^{-1}, a non-ideal CV curves (not quasi-rectangular) was obtained as in Figure 3(a). The curve slope increases with cycle number, indicate the existence of leak current. This leak current may occur due to side reactions between the electrolyte and a functional group in the sample and can be minimized by electrolyte selection [17]. Shape of the curve elongates with increasing scan rates from 50 to 500 mV s^{-1}, result in higher current. This indicates ion transfer barrier from the electrolyte to the electrode surface [18]. Specific capacitance of the produced graphene was calculated from the voltammogram according to following equation:

$$C = \sum|I|.\Delta t \ / \ m \, \Delta V \quad (2)$$

with $I.\Delta t$ is the area of the curve I (A) to t (sec), m (gram) is the mass of active ingredient in the cell, ΔV (V) is operating range of cell voltage. Obtained capacitance for G350 sample is 0.054 F g^{-1}. This low capacitance is supposed due to cell resistance. Therefore, electrochemical impedance spectroscopy (EIS) analysis to the cell was performed.

The EIS result is shown in a Nyquist plots (Figure 3(b)). Equivalent electrical circuit model in Figure 5c was used to describe electrochemical impedance in the cell. Ideal supercapacitor shows a 90° straight line in low frequency as the capacitance is dominant. Good supercapacitor indicated by Warburg impedance close to the 90° straight line. In this study, however, Nyquist plot show a straight line of 45° in low frequency as supposed in the conventional capacitors. This indicates a considerable resistance value of active species diffusion from electrolyte to electrode. In medium frequency, there are electrolyte interphase resistance within the pores of electrodes with domino effect. In high frequency, the real axis intercepts are equivalent series resistance (ESR). Semicircle diameter of the impedance loop represents the charge transfer resistance in the electrode material. The equivalent electrical circuit shown in Figure 3(c) is modeled by the Schrödinger coupled nonlinear equation (CNLS). The whole capacitor circuit is constituted by R1, R4, CPE, C13, R9 and W5. R1 is the sum of the contact resistance and material resistance (electrode, electrolyte and separator). W5 is the Warburg resistance which is related to the ion diffusion/transport in the electrolyte. The resistance parameter of R4 and R9 are interface resistances for mass transfer of electrolytes into the electrode pores [11]. The capacitive element of C13 is related to capacitors formed during charge-discharge process and CPE is not ideal capacitor. Parameters of the equivalent electrical circuit model are presented in Table 3.

Table 3. Impedance parameters of graphene cell

Parameter	Unit	Value	Parameter	Unit	Value
R1	Ohm	9.395×10^{-5}	Yo11	S*s^a	3.952×10^{-3}
R4	Ohm	186.1	a12		5.781×10^{-1}
W5	S*s^(1/2)	1.03×10^{-2}	C13	F	4.959×10^{-2}
R9	Ohm	1.26×10^{3}			

The interface resistance of R4 and R9 are high. Graphene which is a hydrophobic material has low interaction with aqueous based electrolyte, increasing interface resistance. Surface modifications by hydrophilizing agents are proposed to make the graphene surface more hydrophilic [19]. This modification is still being studied further in our laboratory. Internal resistance in the electrode paste matrix can be reduced by increasing percolation. Addition of 10 to 20% conductive carbon filler is proposed. Current collectors and conductive pastes contribute to the weight that could not be ignored in the device. Any non-capacitive weights in cells reduce device performance. As a result, the additive paste used should be as minimal as possible and optimal, as well as the current collector used should be thin and light [20]. Proper cell sealing is also a key component of cell assembly. Evolution of gases from electrolyte degradation, corrosion, and oxidation of electrode and packaging surfaces could be a problem [21]. A tight seal will prevent water and gas molecules from entering the cell. Another major reason for proper sealing is to prevent shunt resistance between electrodes and adjacent cells. The shunt current significantly increases self-discharge and reduces the efficiency of the device. Bad seals will be damaged over time which causes degradation and short circuit [22]. Heat can be applied to make the plastic shrink and tightly seal the cell.

4. Conclusion

Yield of graphene synthesis from oil palm empty fruit bunches can be increased by using two-stage pyrolysis method. In addition, the pyrolysis temperature at the first stage of heating affects surface area and fixed carbon in the product. The highest graphene yield was obtained when the first pyrolysis temperature at 350°C, which is about 70%.

5. Acknowledgement

This work is supported by Badan Pengelola Dana Perkebunan Kelapa Sawit, funding year of 2018. We deeply thank Prof. Wuled Lenggoro, Ferry Faizal, and Yuria Watanabe for insight and helpful supports.

6. References

[1] Hernández M A, Romero J, Jaime C, Pulido J L and Cabeza I O 2017 *Chem. Eng. Trans.* **58** 541-6
[2] Castro N, Guinea A H, Peres N M R, Novoselov K S and Geim A K 2009 *Rev. Mod. Phys* **81** 109-162
[3] Chen F, Yang J, Bai T, Long B and Zhou X 2016 *J. Elec. Chem.* **768** 18-26
[4] Tan Y B and Lee J M 2013 *J. Mat. Chem. A* **1** 14814-43
[5] Shams S, Zhang L Z, Hu R, Zhang R and Zhu J 2015 *Mat. Lett.* **161** 476–9

[6] Giudicianni P, Stefania P, Grottola C M, Stanzione F Faugno S, Fagnano M, Fiorentino N and Ragucci R 2017 *Chem. Eng. Trans.* **5** 500-5
[7] Chen W and Kuo P A 2010 *Energy* **35** 2580-6
[8] Chang G, Huang Y, Xie J, Yang H, Liu H, Yin X and Wua C 2016 *Ener. Conv. Man.* **124** 587-97
[9] Soltes E J and Elder T J 1981 *Pyrolysis* (New York: CRC Press)
[10] Zhang S, Zeng M, Li J, Xu J and Wang X 2014 *J. Mat. Chem. A* 2014 **2** 4391-7
[11] Sun L, Tian C, Li M, Meng X, Wang L, Wang R, Yin J and Fu H 2013 *J. Mat. Chem. A* **1** 6462-70
[12] Manoj B and Kunjomana A G 2015 *IOP Conf. Ser.: Mater. Sci. Eng.* **73** 012096
[13] Sahu V, Grover S, Tulachan B, Sharma M, Srivastava G, Roy M, Saxena M, Sethy N, Bhargava K, Philip D, Kim H, Singh G, Singh S K, Das M and Sharma R K 2015 *Elec. Acta.* **160** 244–53
[14] Kaghazchi T, Kolur,N A and Soleimani M 2010 *J. Ind. Eng. Chem* **16** 368–374
[15] Yang T and Lua A C 2006 *Mat. Chem. Phys.* **100** 438–44
[16] Nasir S, Hussein M Z, Yusof N A and Zainal Z 2017 *Nanomater.* **7** 1-18
[17] Gu W, Peters N and Yushin G 2013 *Carbon* **53** 292-301
[18] Yu A, Chabot V and Zhang J 2013 *Electrochemical Supercapacitors for Energy Storage and Delivery* (New York: CRC Press) p 66
[19] Akkarachanchainona N, Rattanawaleedirojn P, Chailapakula O and Rodthongkumb R 2017 *Talanta* **165** 692-701
[20] Demarconnay L, Raymundo-Piñero E and Béguin F 2010 *Electrochem. Commun.* **12** 1275-8
[21] Simon P and Gogotsi Y 2008 *Nat. Mater.* **7** 845–54
[22] Kotz R and Carlen M 2000 *Elec. Acta* **45** 2483-98

IOP Conf. Series: Materials Science and Engineering **543** (2019) 012033 doi:10.1088/1757-899X/543/1/012033

Study on corrosion behavior of storage tanks filled with biodiesel and the blends

L N Komariah[1], T K Dewi[1], and C Ramayanti[2]
[1]Department of Chemical Engineering Faculty of Engineering Universitas Sriwijaya, Indralaya 30662, Indonesia
[2]Department of Industrial Engineering Faculty of Engineering Universitas Singaperbangsa, Karawang 41361, Indonesia

E-mail : leilynurul@unsri.ac.id; tkdewi@yahoo.com; cindi.ramayanti@ft.unsika.ac.id

Abstract. Corrosion is one of rising problem in the use of biodiesel. Improper storage of biodiesel may cause corrosion to occur faster and give a worse damage effect. Biodiesel is commonly stored longer in B100 (neat biodiesel). In this work, the corrosion characteristics of three different material of tanks for B100 and the blends (B20) are examined. Those metallic materials used are Stainless Steel 304 (SS), Carbon Steel (CS), and Galvanic Steel (GS). The fuel has been monitored every 45 days in 135 days. The corrosion occurrence related to some changes in the physical properties of the fuel stored. It is also identified through the change in tanks' thickness, weight, and the morphology of metal tank surfaces. The result shows a great change in biodiesel physical properties occurred in GS tank, where the BXX acid value increased by 25.39% and the water content changed to 43.18%. While, biodiesel which stored in SS tank performed best with minor change in base and wall thickness with the change of under 1% and the weight change was only 0.02%. In a certain of time, the morphological examination shows CS as the recommended material for biodiesel storage due to uniform corrosion performance.

1. Introduction

1.1. *Corrosion in Fuel Tank*

Corrosion is a process of the degradation of construction material due to electrochemical reaction with its environment. It is considered a very serious problem as it causes damages to properties and environment, in addition to high costs of mitigation of engine failure. Corrosion could also be problem in the fuel oil storage. Corrosion in fuel oil storage tanks occurs due to the contact of stored oil with the tank during its use. Even if non-polar hydrocarbon in fuel oil has been known will not cause corrosion, yet corrosion may occur due to a number of initiators. Contributors for corrosion to occur may come from specific constituents in the fuel oil, e.g. sulfur compound, organic acid and in-organic acid soluble in water or oil. Among any others, are active sulfur (sulfur element, hydrogen sulfide) and water. Previous studies have observed the corrosion of biodiesel in in specified material of tanks, but it was not thoroughly studied the characteristics of corrosion and its relation to changes in physical properties of the fuel stored that may occurred.

Fuel oil for industrial use is generally stored in tanks made from metal. It had been convinced that the only metallic materials compatible and recommended being used with biodiesel. On the other hand, metals are thermodynamically unstable under normal conditions and it is natural for the system (comprising metals and their alloys) to achieve a form with a better thermodynamic stability [1]. The stainless steel and aluminum were strong recommended materials [2]. Storage tank made from metal intended to store fuel oil for long period (more than 10 years) should consider the chemical stability of the stored fuel. There are different corrosion zones in the storage tank depends on position and the adjacent area [1]. Almost all part of this zone is in contact with water, soluble salt, organic and/or inorganic precipitate. The organic precipitate consists of hydrocarbon and biofilm, and the inorganic deposit may consist of corrosion products, salt and sand [3]. The behaviour of potential corrosion occurred in each tank of biodiesel is important to be studied in order to find the best metals materials for storing the fuel and its blends.

1.2. Biodiesel and its stability during storage

Biodiesel can be used as fuel in mixture with petro-diesel. It may directly use in engine without modification, particularly in low blends (<20%). Due to the force for using biodiesel in mixture with petro-diesel as fuel oil, equipment related with this fuel oil should be prepared as to prevent failure or equipment components' damage whilst they are in operation, including potential corrosion in storage tanks. The change of biodiesel characteristics will give significant effect to the performance and lifetime of storage tanks. Air and atmospheric oxygen may dissolve in fuel during its storage. These substances may diffuse to storage tank during distribution. With the change of temperature which is fluctuate between day and night, water and its contaminants may be separated from fuel stored in the tank, forming thin film of electrolyte. Normally, this film is formed in tank wall and moving downward. This causes electrochemical reaction to occur.

Biodiesel is composed from molecules with double bond, causing it to have high reactivity against oxygen, especially when it is contacted with air. Consequently, biodiesel storage at certain period of time may cause degradation of fuel physical properties [4, 5]. Biodiesel can hold many times more dissolved water than conventional diesel fuel. The presence of water in biodiesel may induce the proliferation of microorganisms and increase corrosiveness of biodiesel [3]. Various process may occur during storage of biodiesel, including oxidation, condensation and polymerization.

Detected contaminants during long period of fuel storage include corrosion products, organic particle and deposit. Parts of the dirt like materials tend to adhere to tank's bottom and side wall, even though major parts were existed in the fuel and present in the machine room at operation. Components and certain impurities (iron oxide, sulphide) are also affect the burden given by the fuel properties, it may cause an increase in surface wear of the main machine. Biodiesel encounters degradation via vapour absorption, auto oxidation and microbial attack during storage.

Corrosion is higher with biodiesel than petro-diesel fuel. The rate of corrosion is influenced by temperature, water content, microbial growth, and type of feedstock used for synthesis of biodiesel [6]. Wu et al., [7] were tested biodiesel storage for 52 weeks at temperature 4 °C and 20 °C. Degradation of 40% of the properties was occurred at storage temperature of 40°C. The chemical with high triglyceride content, if stored at room temperature will be hydrolysed and its FFA content may increase to 76% within 6 months storage.

Biodiesel oxidation is caused by unsaturated fatty acid chain and the presence of double bond in its molecule, which may cause increase in its reactivity to oxygen, especially when placed in container which allow contact with air or water [4]. In some case, biodiesel oxidation was known may cause formation of deposit and corrosive materials [8].

Generally, fuel oil is stored using common method by preventing it from moisture, sunlight, the change in temperature and pollutants. With the same storage style, Terry (2005) has taken note that storage of B100 and B20 shows the occurrence of oil quality change shown by the change of acid number, cetane number and viscosity in the period of 6 months. Contamination of water in BXX may cause biodiesel more aggressive in increasing electric conductivity, accelerate oxidation process, so that corrosion process also occurs quicker [9]. Corrosion attacks for metal surfaces are comparatively more in biodiesel than that in diesel fuel [10].

There are many types of metals used as materials for fuel tank, such as stainless steel, mild or galvanic steel, carbon steel, etc. Stainless steel (SS) is a highly alloyed, low-carbon steel with a high chromium content (about 11%). In more aggressive condition, the basic type of SS may corrode. There are most common forms of corrosion in metals, such as general corrosion, crevice corrosion, pitting corrosion, stress corrosion cracking, intergranular cracking and galvanic corrosion [11]. Pitting is one of the most destructive types of corrosion.

2. Methods

The equipment and materials used in this research are storage tanks made from Stainless Steel SS 304, Carbon Steel and Galvanic Steel. The focus of this research is to observe the effect of variation of construction material for the tanks to corrosion occurrence in the side wall of the of the tank and the bottom of the tank. The fuel used in this test is biodiesel (FAME). The samples of biodiesel are stored for 90 days at room temperature, in two storage conditions, that are inside and outside door. The samples are stored and prevented from direct contact with air (closed), and the properties of biodiesel was analysed. Test parameters for biodiesel properties are viscosity, density, acid value and water content. Biodiesel (B100) was taken from PT Sumiasih Oleochemicals, Bekasi, East Java, which is compatible with the standard of EN 14214. Biodiesel used is based on CPO, especially from RBD Stearin.

Table 1. Parameters and analysis method of the fuel sample

Parameters	**Density, at 15 C**	**Viscosity, at 40 C**	**Acid value**	**Water content**
Unit	kg/m^2	mm2/sec	mg KOH/g	ppm
Analysis Method	ISO 3675/ASTM D5453	ASTM D445	ASTM D664	ASTM D6751

It was assumed that the fuel has been stored for more than 3 months before it is used for this work. Analysis to sample of B100 carried out for this work involved parameters of viscosity, density, acid value, and water content. The detection of corrosion characteristics on the metal tanks was conducted using approximation of dimension changes and the weight loss. For visual examination of metal surface (tank), equipment of Celestron Digital Microscope Imager with magnification to 500x has been employed.

3. Result and Discussion

As has been reported by previous researchers in metal tank, biodiesel properties have also encountered change significantly during storage time [12]. In this work, the properties of stored biodiesel has changed in any type of metal fuel tanks. From the measurement of biodiesel properties, its changes may correlate to the corrosion of tanks' materials as presented in Table 2

Table 2. Physical properties of biodiesel B100 after observation for 135 days

Physical properties of B100		Storage Tank1 (Stainless Steel 304)		Storage Tank 2 (Carbon Steel)		Storage Tank 3 (Galvanic Steel)	
		T_0	T_{90}	T_0	T_{90}	T_0	T_{90}
Viscosity	mm2/s	3.700	3.850	3.800	4.060	3.700	3.870
Density	kg/m3	0.852	0.876	0.852	0.894	0.852	0.876
Acid value	mgIOD/g	0.575	0.672	0.575	0.716	0.575	0.721
Water Content	ppm	945.600	1298.900	945.600	1363.400	945.600	1353.900

High change in viscosity and density occurs in biodiesel stored in tank of Carbon Steel as compared with the ones stored in Stainless Steel and Galvanised Steel tanks, as seen in Figure 1a.

The increase in viscosity and mass density of the oil stored in tank of CS indicates that tank material of CS is more prone to molecular interaction with oil. In this case, the viscosity and mass density were connected with the formation of corrosion products, including dissolved sediment. As known, oxidation process occurred during fuel storage may cause the formation of free fatty acid, isomerization of double bond (change of cis- into trans-isomers), saturation and production of higher molecular weight substances. In Figure 2a., the same pattern of changes in acid value and water content in those three materials.

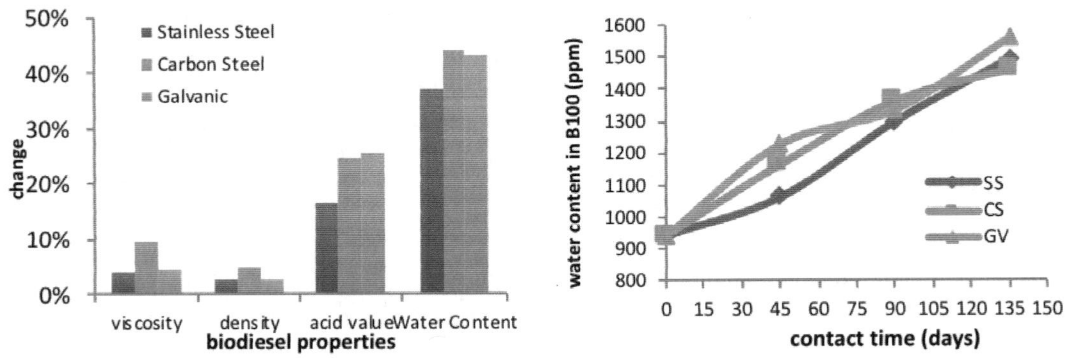

(a) *(b)*

Figure 1. (a) Percent change of biodiesel properties during storage in metal tanks, 1(b) The change in water content of B100 after storage time

High acid value in biodiesel reflects high free fatty acid comes from oil or fat hydrolysis, or from bad treatment process. Acid value of biodiesel tends to increase with storage time, which is connected to the increase of hydroperoxide formed which subsequently oxidized further to form acid. Acid may form if there is water to cause ester hydrolysis into alcohol and acid. A study made by [13] reported that metals (mild steel) is oxidised at a higher rate at high temperature which causes oil to desorb from the surface of mild steel. It has proven that there are strong relation and interactive effects between acid condition with metals corrosion. As seen on Figure 1b., the water content of each sample in three tanks were increased by the increase of contact time. The samples taken from galvanic tank was performed the highest content in water content compared to stainless steel and carbon steel tank. However, high water content in oil fuel samples stored in three different types of tank material is not identical to the corrosion profiles that occurs on each tank. As stated by [3] the effect of water content on corrosion rate is not significant but It does give effect on corrosion process. The corrosion products and some impure precipitates are easily to gather and stick to the wall and/or bottom of fuel tank under the static condition. It performed as protective film and hindered the diffusion of corrosive medium or water molecules to the metal surface and dissolved diffusion of metal ions, and meanwhile caused anodic polarization and cathodic polarization [14].

3.1 The Change of Dimension of Metal Tanks

The change of tank dimension analysed are thickness of tanks' side wall (mm) and the tanks' bottom wall (mm) or base which made from SS, CS and GS. The test is carried out after 135 days exposure with using digital depth gauge for thickness measurement. In Figure 2, it showed the change in dimension considering the base thickness, wall thickness and weight of each tank materials in 135 days of observation.

In general, each metal tank showed a reduction in the thickness of the base and the tank wall. It is detected after the oil has been removed from the tank and some corrosion products have been eroded (the outer most layer of the metal tank). After 135 days of storage, the changes that occurred on the tank wall are a thickness reduction. Galvanic Steel seemed performing the highest base thickness decrease, while the acceptable degree of decrease in industrial application is 0.07%. Carbon Steel shows a little decrease in wall thickness same as indicated by the GS. The position of measurement is crucial to indicate the changes in thickness, it is due to some corrosion products is adhered to the wall or collected at the base or bottom of the tank. The layers of metal tanks eroded in oil and dissolved, then it discharged during tank observations.

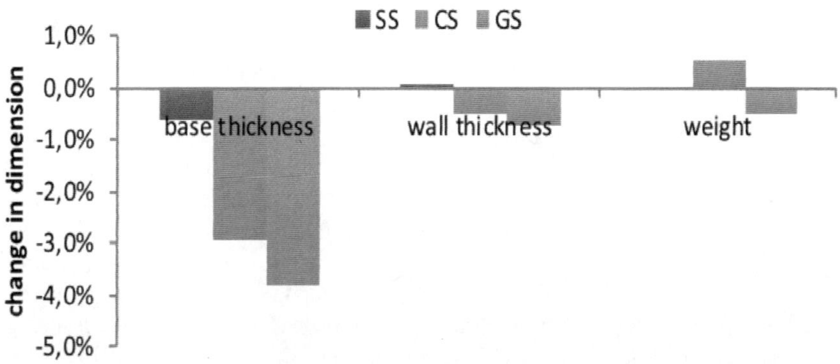

Figure 2. Change in tank dimension

Oxygen content in biodiesel exposed to metal surface of storage tank may be increased with time. It can be indicated with the extreme increases on the fuel' water content after 135 days of storage. Oxygen presents in the form of compounds namely iron carbide or iron oxides. Subsequent formation of oxide compounds and their dissolution seems to degrade the metal surfaces [10].

In some other condition it may presents as corrosion product so the metal thickness will be changed. In the tank wall section, CS and GS tank thickness decreases while SS tank tends to increase. The corrosion products formed on GS and CS tanks tend to erode or crush the surface of the tank wall. Furthermore, carbon steel is susceptible to stress corrosion cracking (SCC) in fuel under particular conditions [3].

3.2 Weight Loss Analysis of the Metal Tanks

Weight analysis is used to measure when there is any change in tanks' weight using analytical balance. After examined for 135 days, the over-all weight of the tanks has decreased by 0.5-0.8%, except for GS tank as seen in Figure 2. The increase in weight of GS tank is caused by adhered material to the bottom of the tank which comes from tank's wall corrosion products when the tank is contacted with B100. Moreover, the material found at the bottom of the tank may also come from the degradation of B100 entrapped in corrosion products, since the oil has been stored for long and it has encountered instability of its physical properties. Best materials for the storage tank from weight loss examination come from SS or CS, excluding GS due to its inconsistent sequence. The corrosion processes were accelerated by bubbling with air, leading to a slight weight loss in the tested metals, particularly in galvanic tank.

3.3. Visual Analysis at Metal Tanks' Surface

Three sample tanks of the same materials, SS, CS and GS are also observed for more than one year exposed to the environment. Various morphological changes of the tanks' surface were detected as corrosion to take place. Morphological view of tanks' surface made from CS, GS and SS were shown in Figure 3-5.

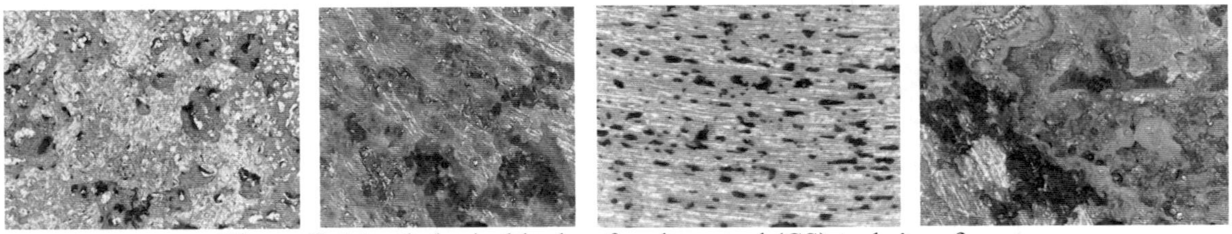

Figure 3. Morphological looks of carbon steel (CS) tanks' surface

From Figure 3, it is obvious seen that CS shows uniform corroded metal surface. The black deposit is also found in the surface. It caused by the reaction between metal oxides and fatty acids to form salts, which strongly stick to the metal surfaces [1]. These appearances lead to conclusion that CS is the better material for biodiesel storing, since it has the most uniform morphological surface, which preventing it of having localized corrosion during its usage, hence give predictable corrosion rate and thus the lifetime of the tank. As reported by Ortega, carbon steel showed passive behaviour in B100 and biodiesel-diesel blends, but active uniform corrosion in diesel. Stainless steel remained passive in all exposures while other metals was susceptible to pitting corrosion [2].

The bottom surface of GS tank performed many various of metal degradation rather than the other tank' surface as seen in Figure 4. In one side of the GS surface, it presented the uniform corrosion but in other place it may detected a shallow but wide through pits which encompassed as local corrosion.

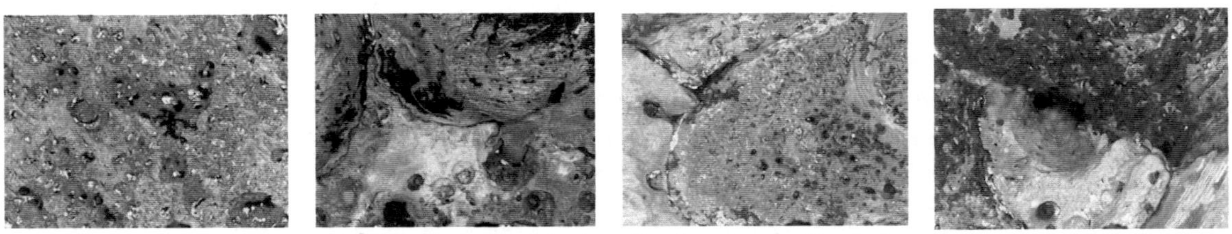

Figure 4. Morphological looks of galvanic steel tanks' surface

While the corrosion of each tank materials in the coupon exposures may possibly occurred due to the extreme environment conditions, such as high humidity and uncontrolled oxidation. Corrosion of metals in biodiesel may not fundamentally different from corrosion of metals in any aqueous solutions. Similar variables (dissolved water and oxygen, acids, bases, ions, microorganisms, temperature, fluid regime and velocity) can affect corrosion in biofuels [3]. The most problems with biodiesel occur due to its low stability on oxidation and its high solvency, tendency to absorb water which accelerate the corrosion process.

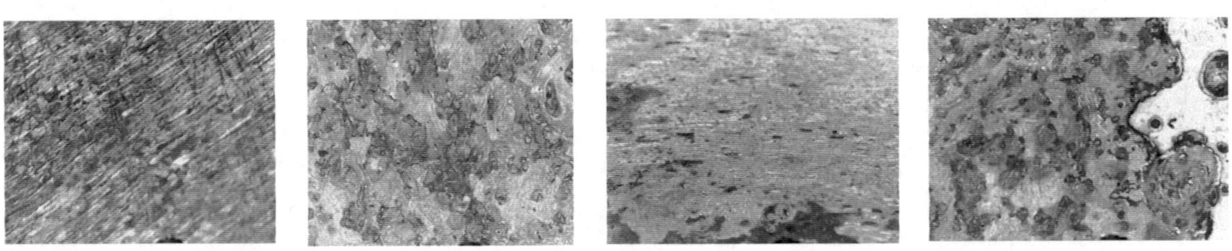

Figure 5. Morphological looks of stainless steel tanks' surface

The corrosion profile on SS tank is shown in Figure 5. It seems that SS has the least corroded surface than the other two metals. However, at high magnification of morphology imaged, the surface of SS was less attacked but still corroded in different pattern. After a certain of time, it is found that a pitting corrosion may be occurred. It can be seen from the narrow and deep pits found in the surface.

Corrosion in SS is more potential to occur in extreme environmental conditions such as low pH, moderate temperature, and high chloride concentration. The electrochemical reaction and oxidation that triggered by biodiesel properties may aggravate the degradation of metals. In term of the type of corrosion detected on the metal surface, from this work it suggests that tank made of SS is no longer recommended for storing biodiesel due to its prone to localized (pitt) corrosion. It may cause a leakage quicker than other type of metals.

4. Conclusion

During storage in the metal tanks, biodiesel encountered some changes in several parameters of its physical properties. Moisture content at each tanks' type has change at the highest with the increase of storage time. The limit of corrosion resistance of any metals depends on its constituent elements when exposed to corrosive environment and corrosive fluid. Biodiesel is recognized as more corrosive than traditional petro-diesel. The fatty acid methyl esters are readily hydrolyzed by microbes, which transform it into organic acids and highly corrosive hydrogen sulfide. Corrosion at biodiesel storage tank with various metal tanks' material show that the damage due to corrosion is the most at material of SS due to localized corrosion. The CS tank was performed more uniform corrosion. The mechanism of corrosion on tank beds and walls seems to be not the same in each type of metal. It may be caused by resistance of each materials to environmental factors. The properties of fuel oil and its aggressiveness may contribute to the corrosion behaviour depends on the presence of water, acid, and other constituents, particularly in relation to cause degradation and oxidation in biodiesel.

Acknowledgments

Palm based Biodiesel and some of test materials are assisted by PT Sumiasih Oleochemicals, Bekasi West Java, Indonesia. The research was conducted well with full support from Laboratorium of Energy Development and Wastewater Treatment Faculty of Engineering Universitas Sriwijaya.

References

[1] Ziółkowska M and Wardzińska D 2015 *Corrosiveness of Fuels During Storage Processes Storage Stability of Fuels* (Intech-Open)
[2] Ortega M A C, Alpírez G M, Eliezer A, González C G, Román J C, Pelayo L J P, Bautista J R A 2013 Materials technological challenges for the biodiesel industry development in Mexico *Materials and Processes for Energy: Communicating Current Research and Technological Development* (Formatex)
[3] Groysman A 2015 *Corrosion in Systems for Storage and Transportation of Petroleum Products and Biofuels* (NACE International)
[4] Kapilan N, Babu T P and Reddy R P 2009 Technical Aspects of Biodiesel and its Oxidation Stability *International Journal of ChemTech Research* **1** (2) 278-282
[5] Bouaid A, Martinez M, Aracil J 2007 Long storage stability of biodiesel from vegetable and used frying oils *Fuel* **86** (16) 2596–2602
[6] Singh B, Korstad J, Sharma Y C 2012 A critical review on corrosion of compression ignition (CI) engine parts by biodiesel and biodiesel blends and its inhibition *Renewable and Sustainable Energy Reviews* **16** 3401– 3408
[7] Wu Y P, Lin Y F and Ye J Y 2011 *Biodiesel-Quality, Emissions and By Products* (InTech open) 71-88.
[8] Terry B 2005 *Impact of Biodiesel on Fuel System Component Durability* (The Associated Octel Company Limited Coordinating Research Council The Associated Octel Company Limited)
[9] Hodam R 2008 *Biodiesel: Compatibility and Regulation* (National Tanks Conference)

[10] Fazal M A, Haseeb A S M A, Masjuki H H 2011 Effect of temperature on the corrosion behavior of mild steel upon exposure to palm biodiesel *Energy* **36** 3328-3334
[11] British Stainless Steel Association 2016 *Making the Most of Stainless Steel*, www.bssa.org.uk/publication accessed 24.10.2018.
[12] Komariah L N, Marwani, Aprisah S, Rosa Y S L 2017 Storage tank materials for biodiesel blends; the analysis of fuel property changes *Matec Web of Conferences* **101** 02012
[13] Choon O C, Karim K A 2017 Inhibitory Effect of Red Onion Skin Extract on the Corrosion of Mild Steel in Acidic Medium *Chemical Engineering Transactions* **56** 913-918
[14] Yin Z, Wanga Z and Li X 2017 Optimization of Anti Corrosion Technology for Petroleum Bit Chem. Eng.Trans. **62** 157-162

Characterization of Zeolite A from Coal Fly Ash Via Fusion-Hydrothermal Synthesis Method

W Wulandari, T Paramitha, J Rizkiana and D Sasongko

Department of Chemical Engineering, Faculty of Industrial Technology, Institut Teknologi Bandung, Bandung, 40132, Indonesia

E-mail : winny@che.itb.ac.id

Abstract. Zeolite A has been successfully synthesized from coal fly ash by using fusion followed by hydrothermal method. This paper describes the characterization of zeolite A. The effects of hydrothermal time, Si/Al molar ratio, and alkalinity in the converting coal fly ash to zeolite A were also investigated. The coal fly ash was obtained from a local power plant at East Java, Indonesia and contained major oxides such as SiO_2 (18.60 wt%), Al_2O_3 (7.18 wt%), Fe_2O_3 (40.20 wt%), CaO (25.20 wt%). The fusion hydrothermal method consists of the following steps: pre-treatment, fusion of coal fly ash with sodium hydroxide, aging, and hydrothermal process. The synthesized material was characterized by using X-Ray Diffraction (XRD), Scanning Electron Microscope (SEM), Fourier Transform Infrared (FTIR), and specific surface area analytical method. The results show that the products contain zeolite A as the major phase, while the highest specific surface area of zeolite A is 37.121 m^2/g. It implies that zeolite A as a higher value added product can be obtained from a solid waste/by-product of power plant, which has wide range applications, including for ion exchange and heavy metal adsorbent from waste water.

1. Introduction

The majority of power generation is currently supplied from fossil fuel, especially coal. Based on World Energy Council 2016, coal-based thermal power plants supply forty percent of world's electricity and is forecast to increase over the next three decades [1]. The combustion of coal not only generates energy, but also generates a large amount of by-products such as coal fly ash contributing environmental problem [2]. In refer to coal fly ash, utilization of coal fly ash is the key to reduce the significant amount of coal fly ash. Numerous studies have reported the utilization of this material as the raw material for zeolite synthesis due to their silica and alumina contents [3].

Zeolite is a crystalline material composed of aluminosilicates with a three-dimensional network of silicon and aluminium tetrahedral linked by oxygen atoms [4]. Various types of synthetic zeolites are zeolite A, zeolite X, ZSM-5, and others. Among the different types of synthetic zeolites, zeolite A has attracted attention to researchers due to its properties. Some of the noteworthy properties of zeolite A include high cation exchange capacity (CEC), high porosity and specific surface area, as well as high affinity towards polar molecules [5]. As a result, zeolite A can be applied as a low-cost adsorbent for pollutant removal from wastewater.

Two common methods for converting coal fly ash to zeolite are hydrothermal method (direct method) and alkaline fusion followed by hydrothermal method. For hydrothermal method, it has small fraction of reacted coal fly ash and low zeolite yields [6]. Compared to hydrothermal method, alkaline fusion plays an important role in converting coal fly ash into large amounts of silicate and aluminosilicate which are soluble in alkaline solutions, thus the purity of the final product is relatively high [7].

In this paper, coal fly ash was used to synthesize zeolite A by the alkaline fusion followed by hydrothermal method. Several synthesizing factors that affect the type of zeolite product were investigated including hydrothermal time, Si/Al molar ratio, and the alkalinity of solution.

2. Experimental section

Coal fly ash was obtained from PT YTL Paiton East Java, Indonesia. It was pre-treated using hydrochloric acid technical grade. Sodium hydroxide and sodium aluminate were used in zeolite synthesis. Analytical grade of sodium acetate, ammonium hydroxide, acetic acid, and 99% isopropyl alcohol were used in the determination of the CEC.

The experiments were conducted in two steps including pre-treatment of coal fly ash and zeolite synthesis. In the pre-treatment step, coal fly ash was roasted at 850°C for 2 h. Then, it mixed with 10% of hydrochloric acid solution (ratio of 1 g : 25 ml) in a 500 ml three-neck flask. The mixture was heated at 80°C for 1 h and stirred constantly at 300 rpm. After that, pre-treated coal fly ash recovered by filtration, repeatedly washed with distilled water, and dried at 100°C for 24 h. For the main process, the zeolite synthesis step consists of fusion, aging, and hydrothermal. In fusion process, 10 g of pre-treated coal fly ash were mixed with 12 g of sodium hydroxide. The mixture was put into crucible nickel and left to fuse at 550°C for 1 h. The sintered granules of fused coal fly ash were cooled down, grinded, while sodium aluminate was added. To adjust the alkalinity and Si/Al molar ratio, water and sodium aluminate were added at different amounts. Furthermore, the mixture was shaken at 30°C for 16 h in incubator shaker for aging process and subsequently heated to temperature of 100°C during 1 h for hydrothermal process. After hydrothermal step was complete, the suspension formed was filtered and zeolite was washed repeatedly until pH of solution reached 9-10. Then, zeolite was dried at 105°C for 16 h.

The chemical composition of coal fly ash and pre-treated coal fly ash were determined by X-ray fluorescence equipment (Rigaku Supermini 200). To analyze the specific surface area, synthesized zeolites were analyzed by BET equipment (Quantochrome Instruments Nova 3200e) by applying the BET equation to the nitrogen adsorption data. Fourier Transform Infrared (Bruker Platinum-ATR) was used to identify functional groups contained in the synthesized zeolites. The morphology of synthesized zeolite was observed by scanning electron microscopy (Hitachi SU3500). The cation exchange capacity was determined using Method 9081.

3. Results

3.1. Pre-treatment of coal fly ash

As observed in Figure 1, the coal fly ash used in this study consists of amorphous phase and some amounts of crystalline phases such as quartz, mullite, calcium oxide, hematite, and magnetite, which is also confirmed from its chemical composition (Table 1). According to Molina and Poole (2004), Fe-bearing of coal fly ash may be decomposed during fusion process and these element are incorporated into the structure of zeolite [8]. Meanwhile, high CaO content may promote the formation calcium silicate hydrate phases [9]. In this work, pre-treatment process by hydrochloric acid successfully reduce the iron oxides and some of the alkali oxide, as shown in Table 1.

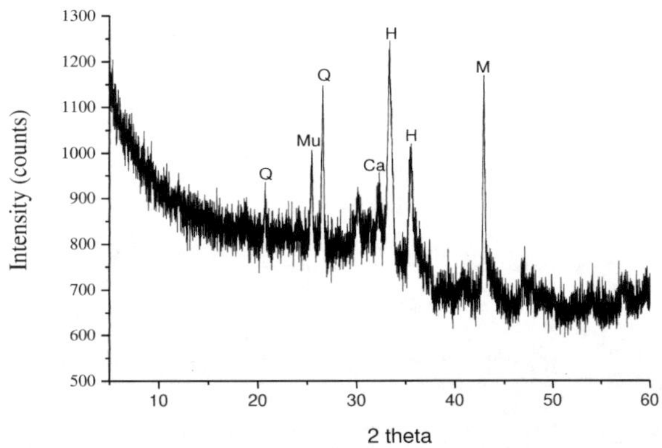

Figure 1. XRD pattern of coal fly ash. Q: quartz, Mu: mullite, Ca: calcium oxide, H: hematite, and M: magnetite.

Table 1. Chemical composition of coal fly ash and pre-treated coal fly ash.

Component	Coal fly ash (% mass)	Pre-treated coal fly ash (% mass)	Component	Coal fly ash (% mass)	Pre-treated coal fly ash (% mass)
SiO_2	18.60	65.60	SO_3	2.23	0.42
Al_2O_3	7.18	5.25	SrO	0.45	0.12
Fe_2O_3	40.20	17.30	BaO	1.04	0.91
MnO	0.45	0.14	Na_2O	0	0.11
MgO	1.17	1.37	ZrO_2	0	0.11
CaO	25.20	4.48	NiO	0	0.03
K_2O	1.54	2.29	ZnO	0	0.06
TiO_2	1.46	1.12	Rb_2O	0	0.02
P_2O_5	0.48	0.63			

The morphological characteristics of coal fly ash and pre-treated coal fly ash is shown in Figure 2 and 3. After the pre-treatment process, the surface of pre-treated coal fly ash shows noticeable changes. It is observed that the hydrochloric acid rapidly disintegrates the outer layer of coal fly ash, where the leaching of some elements may take place.

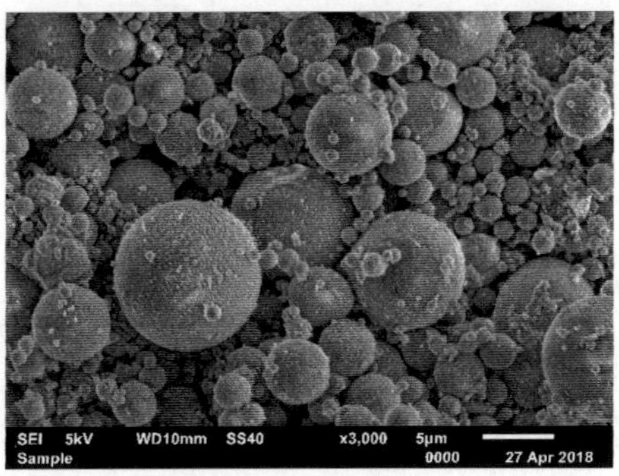

Figure 2. SEM micrograph of coal fly ash.

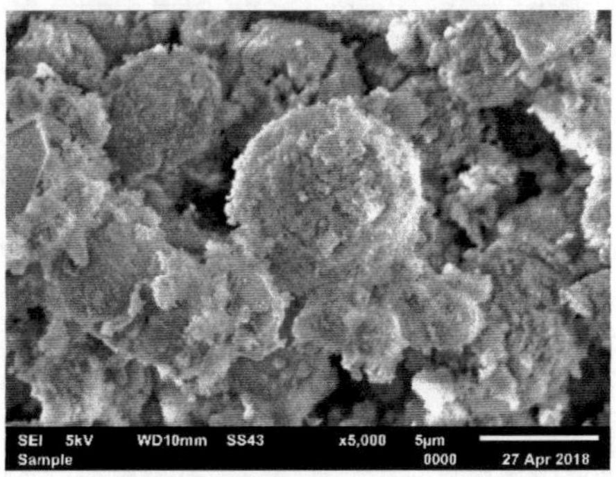

Figure 3. SEM micrograph of pre-treated coal fly ash.

3.2. Effect of hydrothermal time

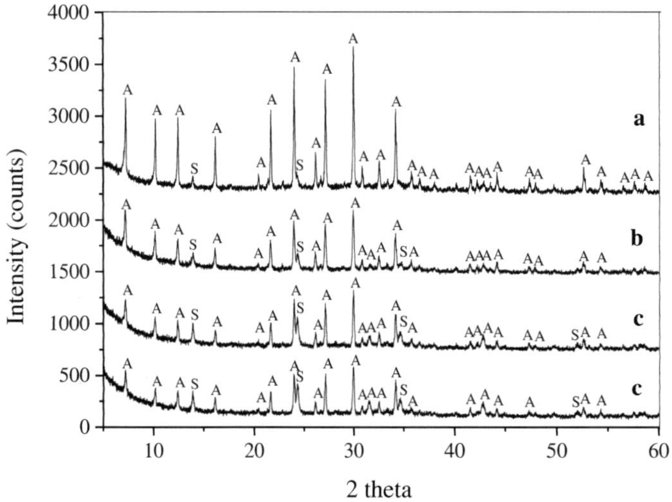

Figure 4. XRD patterns of synthesized zeolites obtained by variation of hydrothermal time (a) 1 h, (b) 3 h, (c) 5 h, and (d) 7 h. A: zeolite A, S: hydroxysodalite.

Figure 4 illustrates the XRD patterns of synthesized zeolites from coal fly ash at Si/Al molar ratio of 0.50 during the hydrothermal time of 1, 3, 5, and 7 h. According to the XRD patterns, all synthesized zeolites contain zeolite A as the major constituent phase, whereas hydroxysodalite is found as the minor phase. Hydroxysodalite is a denser phase and has small pore size of 2.8 Å [10]. In addition, XRD patterns indicate that intensities of the hydroxysodalite gradually increase with the increasing hydrothermal time. These results are likely to reflect the fact that zeolite A as metastabile phase is decomposed gradually with an increase the hydrothermal time results in the supersaturation of liquid phase with aluminosilicate anions. Furthermore, it promotes the formation of hydroxysodalite nuclei [11]. Thus, the formation of zeolite A requires shorter hydrothermal time and it is considered that the stages of transformation are coal fly ash – zeolite A – hydroxysodalite.

Table 2. Properties of synthesized zeolite obtained at varied hydrothermal time.

Hydrothermal time (h)	Degree of crystallinity (%)	CEC (meq/g)	Specific surface area (m^2/g)
1	92.85	4.64	26.583
3	90.61	4.23	24.920
5	89.80	4.01	24.259
7	87.09	3.62	24.101

Based on Table 2, the degree of crystallinity reduce with an increase of hydrothermal time. This behaviour occurs due to short induction and nucleation periods as well as the rapid formation of zeolite A, especially in the system with high alkalinity. Subsequently, zeolite A dissolve in alkaline solution and result in a decreasing degree of crystallinity. In addition, the CEC and specific surface area reduce from 4.64 meq/g to 3.62 meq/g and 26.583 m^2/g to 24.101 m^2/g when the hydrothermal time is increased from 1 h to 7 h.

3.3. Effect of Si/Al molar ratio

Si/Al molar ratio is one of important aspects in determining a pure zeolite A. After conducting the fusion process, sodium aluminate is added to increase its concentration in liquid phase and to adjust the Si/Al molar ratio at 0.50, 0.625, and 0.75. Figure 5 illustrates the XRD patterns of synthesized zeolites which are obtained by variation of Si/Al molar ratio at hydrothermal time of 1 h. In all cases, the peaks of hydroxysodalite are detected. The intensities of hydroxysodalite are varied and are showed in Table 3.

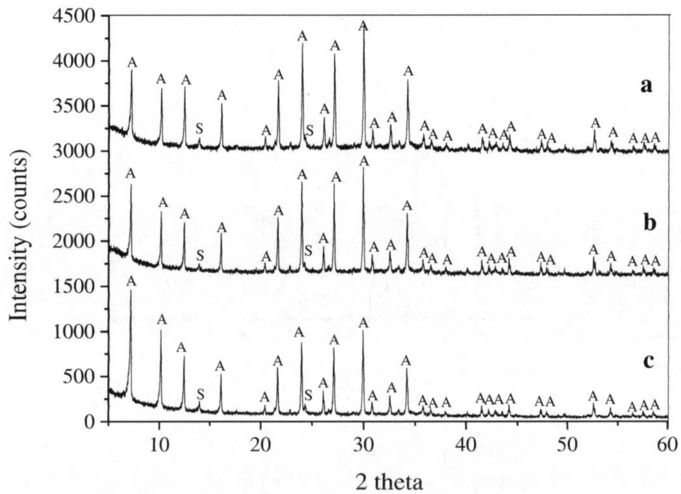

Figure 5. XRD patterns of synthesized zeolites obtained by variation of Si/Al molar ratio (a) 0.50, (b) 0.625, and (c) 0.75. A: zeolite A, S: hydroxysodalite.

Table 3. Intensities of hydroxysodalite at varied Si/Al molar ratio.

2 theta	0.50	0.625	0.75
13.96	290	227	193
24.42	253	189	139

According to Murayama et al. (2002), amount of Na^+ ions in the alkali solution controlled the crystallization rate [12]. At Si/Al molar ratio of 0.50, the Na^+ ions is at most because sodium aluminate introduced in alkali solution is at most too. It can be inferred that crystallization rate of synthesized zeolite with Si/Al molar ratio of 0.50 is the fastest and result in the highest intensities of hydroxysodalite.

Table 4. Properties of synthesized zeolites obtained at varied Si/Al molar ratio.

Si/Al molar ratio	Degree of crystallinity (%)	CEC (meq/g)	Specific surface area (m^2/g)
0.50	92.85	4.64	26.583
0.625	93.95	4.63	31.797
0.75	92.87	5.05	37.530

The degree of crystallinity, CEC, and specific surface area of the synthesized zeolites are shown in Table 4. The degree of crystallinity slightly increase with an increase in the Si/Al molar ratio ranging from 0.50 to 0.625. Furthermore, the degree of crystallinity decrease at Si/Al molar ratio of 0.75. From these results, it is considered that the degree of crystallinity increase with increasing the amount of Na^+ ions up to a certain amount and then it indicates a decreasing tendency. However, the highest CEC value and specific surface area are obtained at Si/Al molar ratio of 0.75. This phenomenon describes that the main factor affected the CEC value and specific surface area is the purity of the synthesized zeolite A [13].

3.4. Effect of alkalinity of solution

Alkalinity is defined in terms of the H_2O/Na_2O molar ratio. Figure 6 shows the XRD patterns of synthesized zeolites at different H_2O/Na_2O molar ratio of 26.26 and 39.04. All samples were synthesized at hydrothermal time of 1 hour. XRD patterns clearly reveal that relatively pure zeolite A without hydroxysodalite phase is detected at H_2O/Na_2O molar ratio of 39.04. Thus, it is considered that hydroxysodalite is not form in lower alkalinity and shorter time.

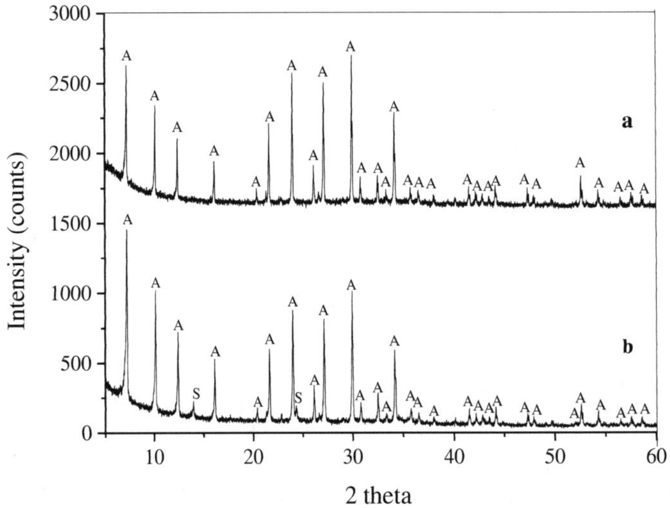

Figure 6. XRD patterns of synthesized zeolites obtained by variation of alkalinity of solution (a) 39.04 and (b) 26.26. A: zeolite A, S: hydroxysodalite.

Table 5 shows the degree of crystallinity, CEC, and specific surface area. In this study, the degree of crystallinity and specific surface area are not vary greatly with an increase of H_2O/Na_2O molar ratio. Meanwhile, CEC value at H_2O/Na_2O molar ratio of 39.04 is higher than 26.26. The results of CEC indicate that relatively pure zeolite A without hydroxysodalite has good capability to absorb metals than zeolite A with impurities hydroxysodalite.

Table 5. Properties of synthesized zeolites obtained at varied alkalinity of solution

H_2O/Na_2O molar ratio	Degree of crystallinity (%)	CEC (meq/g)	Specific surface area (m^2/g)
26.26	92.87	5.05	37.530
39.04	92.52	5.14	37.121

The XRD pattern of synthesized zeolite at H_2O/Na_2O molar ratio of 39.04 is in good agreements with the SEM and FTIR as shown in Figure 7 and Figure 8. Based on Figure 7, it can be seen that relatively pure zeolite A without hydroxysodalite phase is cubic-shaped which corresponds to the typical shape of zeolite A [14]. In addition, there are no spherical particles in relatively pure zeolite A without hydroxysodalite phase, indicating the conversion of coal fly ash to zeolite A is high.

Figure 7. SEM micrograph of relatively pure zeolite A without hydroxysodalite phase.

Based on Figure 8, IR spectra clearly shows zeolite A characteristic bands, indicating the band at wavelength 462.92 cm^{-1}; 549.71 cm^{-1}; 667.37 cm^{-1}; 1,006.84 cm^{-1}; 1,656.85 cm^{-1}; and 3,412.08 cm^{-1}. The band at 549.71 cm^{-1} is assigned to external vibration of double four-rings, which is characteristic of zeolite A. The internal symmetric stretching vibration is indicated at 667.37 cm^{-1}, the internal asymmetric stretching is shown at 1,006.84 cm^{-1}, and the internal linkage vibration of the TO_4 (T = Si or Al) is observed at 462.92 cm^{-1}. In addition, there are two well-defined band at 1,656.85 cm^{-1} and 3,412.08 cm^{-1} associated with the presence of water content and hydroxyl groups in the zeolite framework [15].

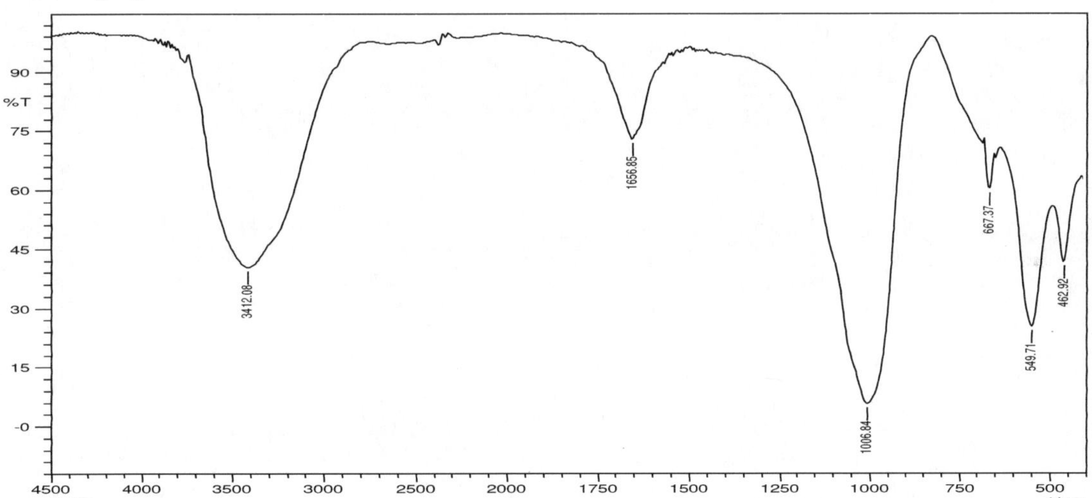

Figure 8. FTIR of relatively pure zeolite A without hydroxysodalite phase

4. Conclusion

A relatively pure zeolite A without hydroxysodalite is successfully synthesized from coal fly ash by alkaline fusion followed by hydrothermal method. The optimum hydrothermal time are 1 h with Si/Al molar ratio of 0.75 and H_2O/Na_2O molar ratio of 39.04. The important aspects are the prolongation of hydrothermal time and higher alkalinity can result in unwanted product: hydroxysodalite phase. A relatively pure zeolite A without hydroxysodalite has degree of crystallinity of 92.52 %, CEC of 5.14 meq/g, and specific surface area of 37.121 m^2/g.

Acknowledgement

The authors would like to thank Program Penelitian, Pengabdian kepada Masyarakat, dan Inovasi (P3MI) 2017 Institut Teknologi Bandung for funding in this research.

References

[1] Liam M, Fernando Z, Roland L, Sandro S, Abhasha M, Mifathul H, Chris B, A. O. Yusuf, Lidia G and Mark M 2016 *World Energy Resources* 2016 (London World Energy Council 2016) p 15
[2] Rafieizonooz M, Salim M R, Hussin M W, Mirza J, Yunus S M and Khankhaje E 2017 *Chem. Eng. Trans.* **56** 439-44.
[3] Izidoro J C, Fungaro D A, Abbott J E and Wang S 2013 *Fuel* **103** 827-34.
[4] Kongnoo A, Tontisirin S, Worathanakul P and Phalakornkule C 2017 *Fuel* **193** 385-94.
[5] Golbad S, Khoshnoud P and Abu-Zahra N 2017 *J. Minerals Mater. Charact. Eng.* **5** 237-51.
[6] Belviso C 2017 *Prog. Energ. Combust. Sci.* 1-27.
[7] Ruen-ngam D, Rungsuk D, Apiratikul R and Pavasant P 2009 *J. Air & Waste Manage. Assoc.* **59** 1140-47.
[8] Molina A and Poole C 2004 *Miner. Eng.* **17** 167-73.
[9] Rios C A and Williams C D 2008 *Fuel* **87** 2482-92.
[10] Ojumu T, Plessis P D and Petrik L 2016 *Ultrason. Sonochem.* **31** 342-49.
[11] Liu X D, Wang Y P, Cui X M, He Y and Mao J 2013 *Powder Technol.* **243** 184-93.
[12] Murayama N, Yamamoto H and Shibata J 2002 *J. Chem. Technol. Biotechnol.* **77** 280-86.
[13] Fotovat F, Kazemian H and Kazemeini M 2009 *Mater. Res. Bull.* **44** 913-17.
[14] Ameh A E, Fatoba O O, Musyoka N M and Petrik L F 2017 *Powder Technol.* **306** 17-25.
[15] Gougazeh M and Buhl J Ch 2014 *J. Assn. Arab Univ. Basic Appl. Sci.* **15** 35-42.

IOP Conf. Series: Materials Science and Engineering **543** (2019) 012035 doi:10.1088/1757-899X/543/1/012035

Natural Rubber Powder Production from Latex

A Z Abidin, N R Winartha, I Rahardi, D A Trirahayu
Chemical Engineering Program, Faculty of Industrial Technology, Institut Teknologi Bandung, Jl. Ganesa 10 Bandung, Indonesia 40132

E-mail : zainal@che.itb.ac.id

Abstract. Natural rubber-based products are widely used in everyday life. There are three types of upstream processed natural rubber, that are crumb rubber, conventional rubber, and concentrated latex. Latex is easily damaged by clotting and contains 35-40% w/w of water so that the packaging volume is relatively large and inefficient in transportation. Latex also contains ammonia as an anti-clotting agent so the packaging must be special to prevent unpleasant odour. This study aims to create a pilot plant of the spray drying latex from simulation results then test the performance, evaluate the performance test results, as well as making improvements to the production of rubber powder. The steps of this research are data collection for spray dryer design, design verification, construction of spray dryer, spray dryer appliance testing, and characterization of rubber powder produced. The heating system used in the spray dryer is LPG gas-fired furnace. Testing is done by varying the spray dryer drying air temperature of 120°C, 150°C, and 180°C, the water content of the feed latex 70% and 80% w/w, as well as non-stick material such as talc powder and MgO. Rubber powder characterized by water content with digital tools moisture meter and analysed spectra with Fourier Transform Infrared Spectroscopy (FTIR). The results showed the optimum operating conditions for the production of natural rubber powder is drying air temperature of 150°C and the water content of the feed latex 80% w/w. Natural rubber powder produced contains 10 to 13% w/w water but still sticky so that the powder grains stick to each other. The FTIR spectrum analysis of natural rubber powder products show similarity with bonding group of polyisoprene natural rubber compound. Natural rubber powder product can also be used as a substitute for crumb rubber.

1. Introduction

Natural rubber derived from the rubber tree sap is currently widely used in everyday life. Latex is concentrated liquid and contains about 35-40% water. This will affect transportation costs that become less efficient because it is used to transport water as well. In addition, the required volume of packaging is also greater and need special handling because of latex late contain ammonia as an anticoagulant material so that latex does not clot and become damaged. Reduction of water content and activity has a positive effect on product quality [1]. In dry conditions, the packaging and transportation process becomes easier and can be stored longer. Latex in the form of liquid contains proteins that are vulnerable to overgrown by microbes. However, it can be minimized if the rubber does not contain water or solids as powder using technology to make the latex into powder by spray drying. Laboratory of Polymer and Membrane Technology - Institut Teknologi Bandung has simulated spray drying technology for the production of natural rubber latex powder with the help of ANSYS FLUENT software. The simulation results obtained in the form of equipment design, operating conditions, and prediction of dry rubber powder products.

To complete the previous research, the present study will generate what has been simulated and perfecting it. Fresh or concentrated latex still has high water contain that make the transportation process becomes inefficient. In addition, the use of ammonia as an anticoagulant material are volatile and has a characteristic odour that makes latex should be packed in particular way. Thus, the rubber in the form of powder becomes a solution to improve the transportation efficiency and handling of latex from plantation to finished goods processing plant. The purpose of this research is to build pilot plant based on simulation results, to test the performance of the pilot plant, to evaluate the performance test results, and making improvements to the production of rubber powder so it can be managed well. The benefit of this research is to obtain the optimum condition and efficient operation of the spray drying process to produce powder in the form of durable rubber and easily dissolved again with water so that the storage, transport, and use in downstream rubber industry becomes more convenient and flexible.

In several studies, there are some works that focused on the possibility to obtain ultrafine rubber powder, particles from micrometer to nanometer dimensions, based on rubber lattices, like styrene-butadiene, acrylonitrile-butadiene and natural rubber. According to Paiva et al, SBR rubber powder prepared by the chemical modification of latex with methyl methacrylate and colloidal silica followed by spray drying had dispersions without any additional treatment [2]. Preparing rubber nanoparticles using conventional methods is difficult, but rubber powder can be prepared from irradiation and spray-drying of rubber lattices [3].

Spray drying is the process of contacting the atomized liquid feed with a gas temperature higher than the liquid [4]. The spray drying process fluid feed change into a dry product from the solvent. Fluid feed nozzle produces atomized using granules are contacted with a medium hot dryer. The commonly used drying medium is air. The heat of the drying medium is used to evaporate the fluid solvent so as to produce the product of dry powder with a temperature that is not too high. The drying time of the granules is relatively short compared to other drying processes. According to Mujumdar (2006) spray drying process consists of three stages, atomization, evaporation, and separation.

In general, the design of the chamber depends on the type of atomizer and liquid-air contact system selected [5]. Selection of an atomizer and liquid-air contact system is determined by the desired characteristics of the dry product and speed of production. Product specifications were obtained from small-scale testing results in the laboratory. However, product specifications are sometimes in the literature. Altitude chamber defined droplet diameter and the temperature difference between the air dryer and the particles product. The air dryer used is high temperature air. Air heating system (water heater) is an important support tools in the spray drying process. Use blower air flows toward the air heating system before entering the chamber. Air heating can be done directly or indirectly using steam, fuel oil, gas, electricity, and heating fluids.

2. Methods

The research was conducted in five major stages, namely the collection of data related to the design of the spray dryer simulation results, design verification simulation results, the construction process of a spray dryer, testing tools spray dryer that was created using baits fresh latex and perform characterization of the powder rubber produced. Flow diagram of the major stages of the experiment can be seen in Figure 1.

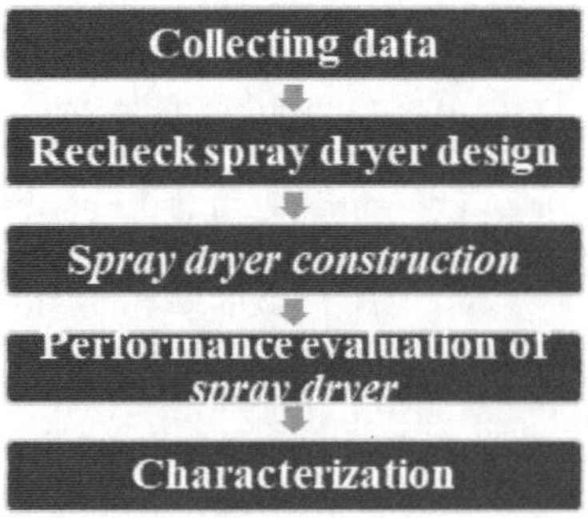

Figure 1. Implementation phase of research

The first stage in this study is data collection. The data collected from the results of previous research that has been conducted at the Laboratory of Polymer Technology and Membrane which was the simulation of spray drying of natural rubber latex. The second stage after the data collected is data verification. Data verified by conducting a literature review of the scientific journals and textbooks related to natural rubber and spray drying technology as well as perform repeated calculations. The third phase is the construction of a spray drying equipment. The fourth stage is a performance test of constructed spray drying. The test is carried out by varying the temperature of the drying air and the natural rubber latex feed water content as well as the non-sticking material. Natural rubber powder is stored in a cup every 10 minutes for 3 times or until 30 minutes. The rubber powder produced will be analysis for the water content using moisture meter. qualitatively observed stickiness and coagulated, also spectrum analysed with Fourier Transform Infrared Spectroscopy (FTIR).

2.1. Main Equipment

Detail specifications of main equipment are presented in Table 1.

Table 1. Specification of the main experimental equipment

Equipment	Specification
Feed tank	60 liter
Pump	Gear pump Oriental Koshin GL-25-5, outlet diameter 1", power 1,5 kW, pressure 5 bar, capacity 58 L/min
Pressure nozzle	Pressure nozzle 0,3 mm with capacity 0,86 mL/s at 4 bar
Drying Chamber	Upper diameter 33 cm, height 64 cm cylinder section, cone height 74 cm, and bottom diameter 7 cm
Heater	LPG gas-fired furnace that can heat the air up to 200°C
Blower	Chuanfan Cx-75A. Outlet 4", capacity 1,22 Nm^3/min

The main equipment scheme is presented in Figure 2.

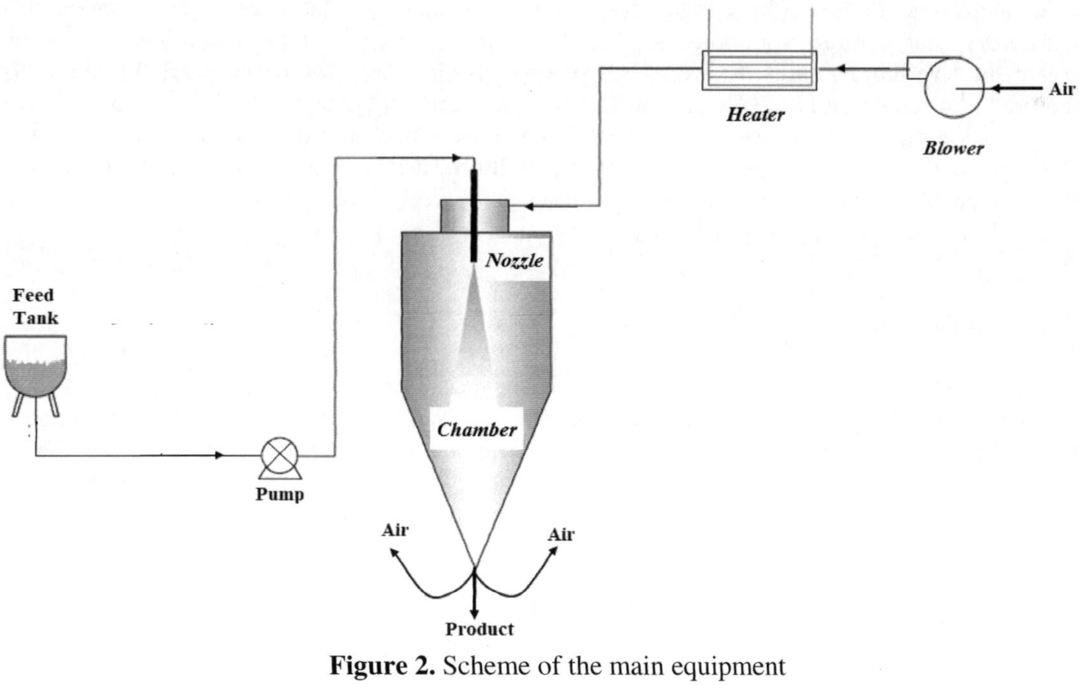

Figure 2. Scheme of the main equipment

IOP Conf. Series: Materials Science and Engineering **543** (2019) 012035 doi:10.1088/1757-899X/543/1/012035

2.2. Equipment and Material

Equipment and materials in this experiment are presented in Table 2.

Table 2. Equipment and material experimental

Equipment	Material
Bucket bait tank	Latex
Pump	Water
Pressure nozzle	Air
drying Chamber	LPG gas
Furnace	Formic acid
High pressure regulator	Talak
High pressure cooker	MgO Powder
Blower	
pressure gauge	
Thermocouple	
Shell	
Stainless steel pipe	
Pipe insulator	
Oven	
Evaporation plate	
Spatula	
Analytical balance	
Digital moisture meter	
Digital Anemometer	

The experiment was conducted by varying drying air temperature and latex feed water content. The selected air dryer temperature variations were 120, 150, and 180oC. The three variations are selected because of the five temperatures the water is estimated to have been well vaporized. Variations of selected feed water content are 70%, and 80%. Both variations are selected because with the water content is not too thick latex, viscosity is not too high. This experiment using a pressure nozzle that cannot be operated for a fluid with high viscosity. In addition, variations are made by adding non-stick ingredients such as talc powder, and MgO. The variables are kept constant throughout the experiment is the feed flow rate of natural rubber latex exit nozzle of 0.86 mL/s and drying air flow rate of 1.22 Nm3/min.

3. Results and Discussion

3.1. Furnace Performance

The furnace fuel is LPG gas. Test performance of the furnace is done by heating the air conditioning as much as 1.22 Nm3/min by varying the drying air temperature of 120, 150, and 180°C. The air conditioning temperature furnace entrance 27°C. LPG as a fuel has a lower heating value (LHV) 46.04 MJ / kg. LPG gas needs and the efficiency of the furnace for each variation of the drying air temperature is presented in Figure 3.

IOP Conf. Series: Materials Science and Engineering **543** (2019) 012035 doi:10.1088/1757-899X/543/1/012035

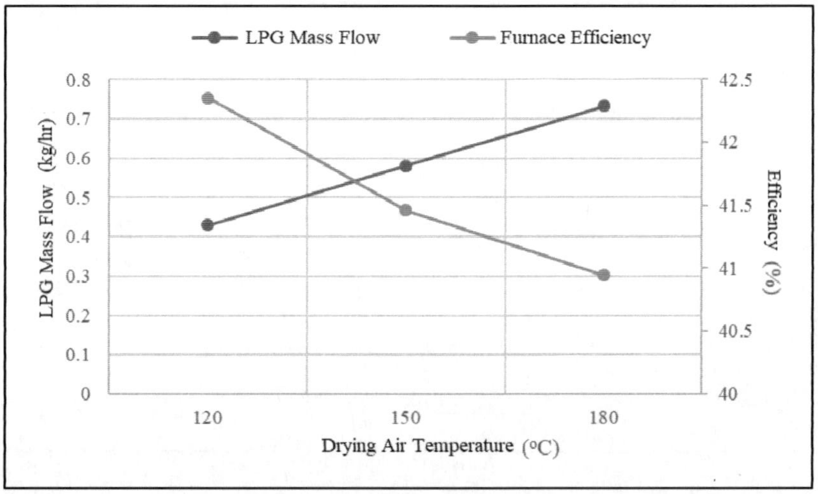

Figure 3. LPG gas requirements and furnace efficiency for each variation of dryer air temperature

Furnace efficiency is 41 to 42.3%. The efficiency value represents the absorbed energy to increase the drying air temperature of the overall heat supply. The value is not up to 50% meaning half of the energy supplied to the furnace is wasted or unused. This can be evidenced by the high temperature of the furnace exhaust gas. Furnace exhaust gas has a high temperature in the range of 230 up to 380°C. The high temperature exhaust gas can be utilized by installing an economizer so as to improve the efficiency of the furnace and is no longer much energy is wasted.

3.2. Spray Drying Performance with Natural Rubber Latex Feed

Natural rubber latex is supplied from PTPN VIII Jalupang farm, Subang in concentrated latex form with dry rubber content (KKK) about 60% -w/w or moisture content about 40% -w/w. The test results of concentrated latex water content showed a water content of 39.38% -w/w. The concentrated latex was diluted to 70 and 80% w/w of water to be fed the spray drying process. The test results density of PTPN VIII concentrated latex or latex diluted presented in Table 3.

Table 3. Natural rubber latex density test results

Type of Latex	Concentred Latex	Latex 70%-w/w Water	Latex 80%-w/w Water
Density (kg/m^3)	943.70	974.08	981.02

Operating temperature for drying air entered the chamber optimum based on Figure 5 is 150°C. At temperatures of drying air entered the chamber at 120°C, water content of rubber powder products is still high at above 20% w/w water. Operating temperature 150°C can be achieved with fuel consumption of LPG is lower than the operating temperature of 180°C. At temperatures of drying air entered the chamber of 150 and 180°C, levels water product powder rubber that generated could it says same, that is amount 10 - 13.5% w/w, so that the selected optimum operating temperature is 150°C.

The optimum latex feed water content based on Figure 5 is 80% w/w. Rubber with 80% w / b latex feed water content is easier to pump. Rubber latex feed moisture content of 70% w/w has a rubber content of more than bait latex with water content of 80% w/w. It causes more quickly formed a layer of rubber on the gear pump to the motor no longer strong enough to turn the gear pump and gear pump must be in maintenance. Maintenance gear pump is done by soaking it with turpentine oil so that a layer of rubber on the gear pump oil dissolved in turpentine. The experimental results show that 70% b / b water latex bait can run up to 4 flights before motor of gear pump no could rotate gear pump again, while attack bait Latex 80% w/w water can run up to 5 attacks before the motor gear pump gear pump cannot play anymore.

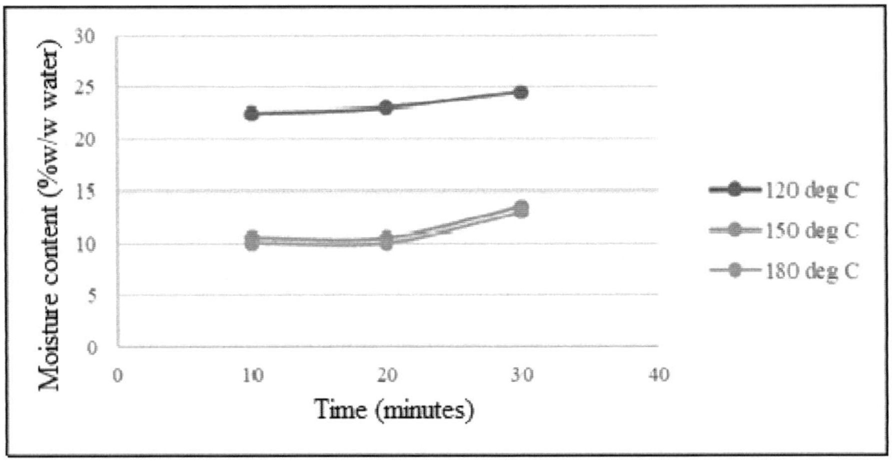

(a)

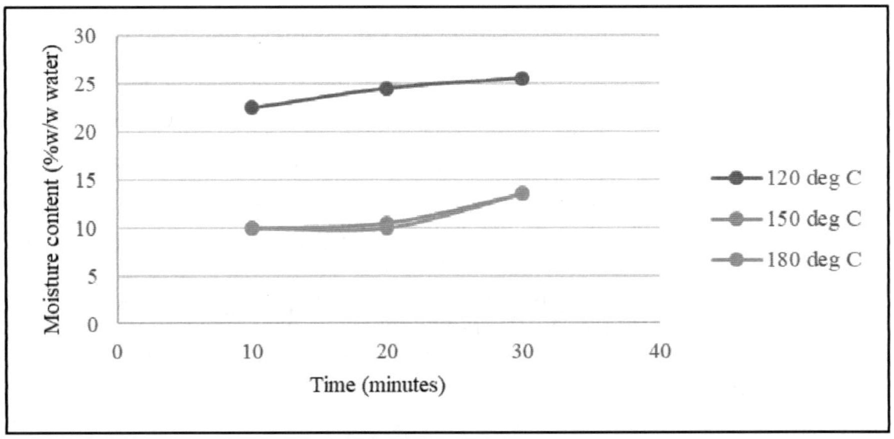

(b)

Figure 4. The water content of the rubber powder product at a temperature of 120, 150, and 180°C to the moisture content of feed (a) 70% and (b) 80% w / w water

Operating temperature for drying air entered the chamber optimum based on Figure 5 is 150˚C.At the entrance chamber drying air temperature of 120oC, moisture content of rubber powder products is still high at above 20% w/w water. Operating temperature 150˚C can be achieved with fuel consumption of LPG is lower than the operating temperature of 180˚C. At temperatures of drying air entered the chamber at 150 and 180˚C, the water content of powder rubber products produced can be said to be equal, at 10 - 13.5% w / w, so that the selected optimum operating temperature is 150˚C. The water content of latex optimum feed based on Figure 5 is 80% w/w.

Rubber with 80% -w/w latex feed water content is easier to pump. Rubber with 70% -w / w latex feed water content has more rubber content than latex feed with 80% -w / w water content. It causes more quickly formed a layer of rubber on the gear pump until the motor no longer strong enough to turn the gear pump and gear pump must be in maintenance. Maintenance gear pump is done by soaking it with turpentine oil so that a layer of rubber on the gear pump dissolved in turpentine oil. The results showed 70% attack latex feed w / w water can run up to 4 attacks before the motor gear pump cannot turn the gear pump again, while the decoy attack latex 80% w/w water can run up to 5 attacks before the motor gear pump cannot turn the gear pump again.

When the spray drying process takes place, the heat energy carried by the drying air is used to evaporate the solvent from the feed. This is indicated by the change in the temperature profile throughout the chamber. Temperatures along the chamber during the spray drying process lasts less than the temperature throughout the chamber when only flowed air dryer only. The temperature profile chamber to a temperature of 150°C while before the feed is sprayed latex and latex sprayed when the bait is presented in Figure 5.

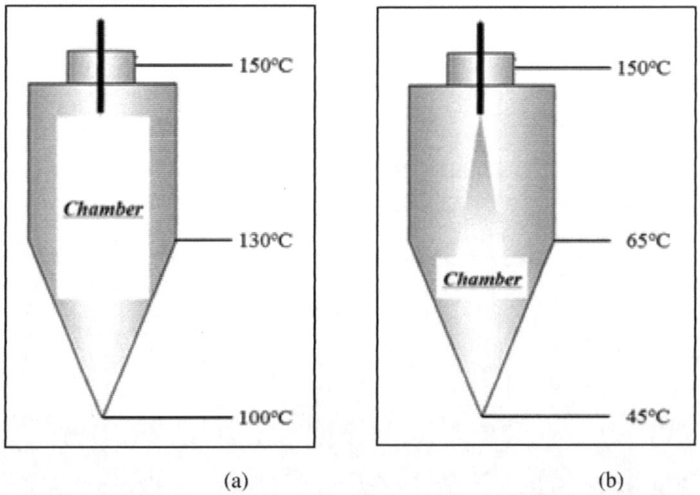

Figure 5. The temperature profile of chamber (a) before the feed latex sprayed, it contains only air dryers and (b) after the feed latex 80% w / w water sprayed

The optimum operating conditions of spray dryer from the experimental are the input drying air temperature enter chamber at 150°C and the feed latex water content of 80% w/w was used for non-stick material rubber powder. The 2% -w/w mixed-in rubber sticky material into the latex feed is varied, namely talc powder and MgO powder. Powder products of rubber latex from spray drying results that have been mixed with non-stick agent talc and MgO powder is still sticky. However, the addition of non-stick material does not degrade the performance of the spray drying. Performance of spray dryer with a non-stick material variation talc powder and MgO powder is presented in Figure 6.

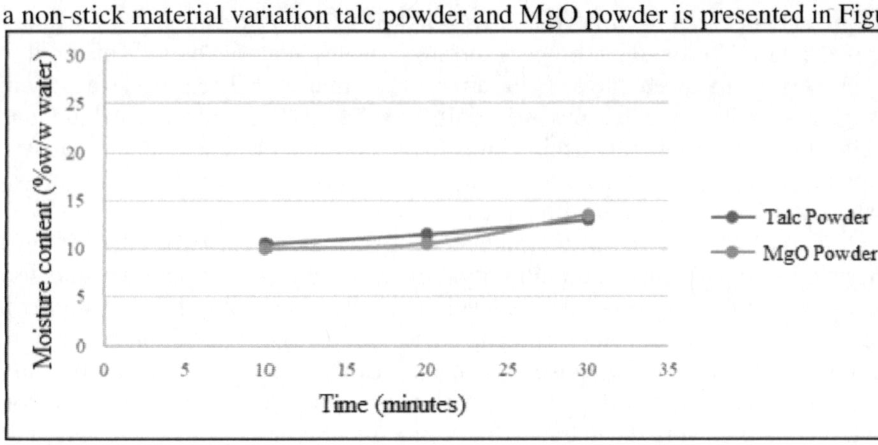

Figure 6. Performance spray drying a mixture of natural rubber latex with non-stick material

3.3. Analysis of Natural Rubber Powder Spectrum

The result of the FTIR spectrum analysis yield curves percent transmittance against wave numbers. Each bonding group has a unique wave number that distinguishes it from other clusters of bonds. The result of the FTIR spectrum analysis for natural rubber powder samples are presented in Figure 7.

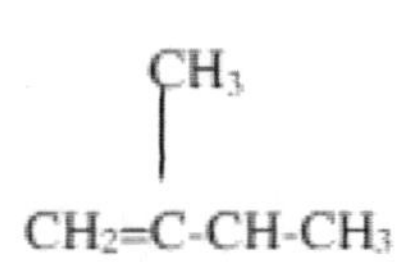

Figure 7. The chemical structure of natural rubber monomer

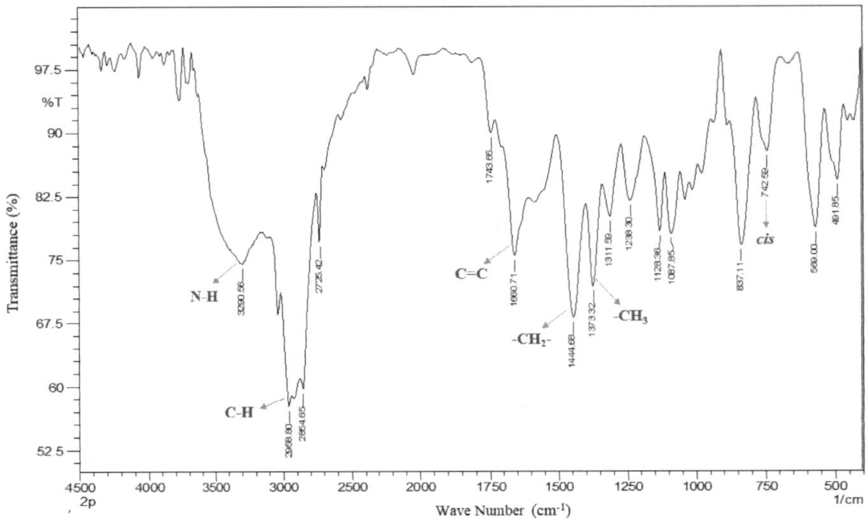

Figure 8. FTIR spectrum of natural rubber powder samples

Characteristic of natural rubber is having unsaturated C = C bond that forms the position of cis C = C bond that forms the cis position read on the FTIR spectrum. Other groups such as the CH bond, -CH2 -, and -CH3 existing in the structure of natural rubber are also accepted FTIR spectrum. The FTIR spectrum of natural rubber powder also indicates that there is a NH group. The NH group can be derived from a protein contained in natural rubber latex or may also be caused by spray drying feed used is concentrated latex containing ammonia. Thus, the true natural rubber powder is a natural rubber compounds and prove that the spray drying process does not damage the rubber compound.

3.4. Comparative Study between Natural Rubber Powder and Crumb Rubber

Crumb rubber is dry rubber processing process through various stages. The raw material comes from the crumb rubber latex is processed into a coagulum using formic acid. The processing of rubber crumb rubber itself is the processing of rubber raw materials (in solid form) by means crushing, blending, and drying which aims to get dry rubber. Crushing aims to obtain rubber crumbs that are ready to be dried, the drying is done by using heating for 3 hours with the temperature further 122°C.Tahap is pressing to get the bale-bale of dried rubber crumb with a press machine.

The impurities contained in the crumb rubber is very damaging properties for the finished goods, especially rubber bending resistance and durability. These properties are important in determining the quality of motor vehicle tires, so the higher the level of crumb rubber impurities, the lower the quality. Production process of crumb rubber requires many stages of process, material like used of formic acid, tools for obtaining crumb rubbers, drying using heaters, pressing tools, and other supporting tools.

Latex preservation process used in dry form by using spray drying technology is more effective and efficient when compared with the manufacturing process of crumb rubber. The resulting product is non-stick so that the rubber powder fused together to form a thin sheet with low moisture content makes the product more durable and more flexible in its use. Bonding groups in the crumb rubber is the same with groups bonding natural rubber powder that is detectable group C = C who plays cis. The cluster is characteristic of the polyisoprene natural rubber compound. The FTIR spectrum of crumb rubber are presented in Figure 9.

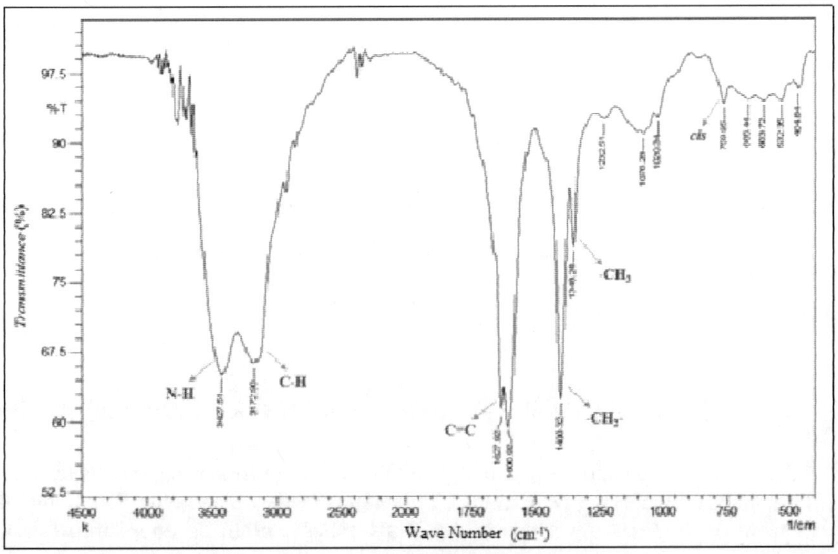

Figure 9. The FTIR crumb rubber spectrum

4. Conclusion

It can be concluded that the optimum operating conditions for natural rubber latex spray drying is the drying air temperature of 150°C entered the chamber and feed moisture content of 80% latex w/w. The resulting rubber powder product is actually a compound of natural rubber, natural rubber compound that does not react and form other compounds not damaged as a result of the spray drying process. The non-stick material of talc powder and MgO powder mixed into the natural rubber latex bait is incapable of making the natural rubber powder product non-stick. However, the mixture does not degrade the performance of spray drying.

References

[1] Zbincinsky I and Zietara R 2004 CFD *Model of Counter Current Spray drying Process* Proceedings of the 14th International Drying Symposium (IDS) Sao Paulo, Brazil
[2] Paiva L B de, Oliveira A M de, and Gavioli R R 2014 *Powder Technol.* **264** 507-13.
[3] Li D, Xia H, Peng J, Zhai M, Wei G, Li J and Qiao J 2007 *Rad. Phys. Chem.* **76** 1732-35
[4] Shabde V 2006 *Optimal Design and Control of a Spray drying Process that Manufactures Hollow Micro- Particles* Disertation Texas Tech University
[5] Mujumdar A S 2006 *Handbook of Industrial Drying* (CRC : Taylor and Francis Group)

IOP Conf. Series: Materials Science and Engineering **543** (2019) 012036 doi:10.1088/1757-899X/543/1/012036

Melt Rheological Properties of Polypropylene/Clay Nanocomposites Prepared from Masterbatch: Effect of Nanoclay Loadings and Reprocessing

A. Chafidz[1], S. Ma'mun[1,*], D. S. Fardhyanti[2], I. Kustiningsih[3], A. Hidayat[1]

[1]Chemical Engineering Department, Universitas Islam Indonesia, Yogyakarta 55584, Indonesia
[2]Chemical Engineering Department, Universitas Negeri Semarang, Semarang 50229, Indonesia
[3]Chemical Engineering Department, Universitas Sultan Ageng Tirtayasa, Cilegon 42435, Indonesia
E-mail: sholeh.mamun@uii.ac.id

Abstract. In the present work, polypropylene (PP) and nanoclay in the form of masterbatch were melt compounded using a co-rotating twin screw extruder to fabricate polypropylene (PP)/clay nanocomposites. PP/clay nanocomposites samples with four different nanoclay loadings (i.e. 0, 5, 10, and 15 wt%) were prepared. Additionally, to study the effect of reprocessing on the properties of the nanocomposites, the melt compounding process was carried twice/recycle (i.e. 1st cycle and 2nd cycle). The surface morphological of the nanocomposites and their melt rheological behavior were investigated using a Scanning Electron Microscopy (SEM) and an oscillatory rheometer, respectively. The SEM micrographs showed that the nanoclay was well distributed in the polypropylene matrix for both 1st and 2nd cycles. While, the dispersion state of the nanoclay (either intercalated or exfoliated) could not be observed due to the limitation of the apparatus. Additionally, the melt rheological analysis results showed that the complex viscosity, $|\mu^*|$ of all nanocomposites samples (for both 1st and 2nd cycles) were higher than the neat PP and the $|\mu^*|$ also increased with the increase of nanoclay loadings. However, the 2nd cycle showed a slightly higher improvement of $|\mu^*|$ compared to the 1st cycle, and the improvement of $|\mu^*|$ was more prominent at lower angular frequencies (e.g. 0.1, 1, 10 rad/s) than at higher ones (e.g. 100, 500 rad/s).

1. Introduction

Nanotechnology has become a new trend of research in the world, which could solve problems of the world (e.g. energy, water, food, etc.), and thus affect the economies and societies of the world over the next decades. One of nanotechnology applilcation is polymer nanocomposite (PNC). It is a relatively new class of composites materials that consist of nano-size materials incorporated in polymer matrix [1]. The PNCs have been widely investigated and developed for many applications in the last two decades [2, 3]. It was due to their improved mechanical (e.g. tensile modulus and strength), thermal stability, reduced gas permeability and flammability, etc. as compared to their conventional composites materials [4]. Among the various nanomaterials used for the preparation of nanocomposites, two of the widely investigated nanomaterials are carbon nanotubes (CNTs) and nanoclay or layered silicates, which are one- and two- dimensional nanomaterials, and therefore have a large aspect ratio [5]. The focus of the current work is the nanoclay materials, which has been found to be effective reinforcing nanomaterials due to their availability, economic competitiveness, high aspect ratio, low density, lamellar structure, high specific surface area, etc [6, 7]. Among various types of clay, montmorillonite

(MMT) type of clay has been extensively used in the fabrication of clay based polymer nanocomposites. The aspect ratio of the MMT clay approximately one nm thick and hundreds nm long [2].

Polypropylene (PP) is one of most important and widely used thermoplastic polymers for the production of versatile products in many applications, e.g. automotive, electronics, packaging, etc. It is due to its good mechanical properties, low density and cost, good processability and recyclability [4]. Therefore, fabrication of PP/clay nanocomposites has become an interesting topic of research. The work on these nanocomposites was first started at Toyota Research Laboratory, and until now, the studies about these nanocomposites have been carried out by many researchers [2]. In general, there are three methods to fabricate polymer nanocomposites, which are in-situ polymerization, solution intercalation, and melt compounding/blending [3, 4, 8]. The later one is the most favored and used method to fabricate polymer nanocomposites. It is because this method is simple, use no-solvent, and thus low cost and environmentally friendly. Moreover, this method is very compatible and suitable with existing industrial processes, such as extrusion and injection molding, and thus make the nanocomposites easy to be commercialized [3, 4].

Recently, the use of masterbatch has become an attractive alternative method to produce polymer nanocomposites via melt compounding compared with the use of bulk nanomaterials. Because, this method is a dust free, thus it has low health and safety risks compared to the use of bulk nanomaterials, as well as easy to handle. Additionally, the use of this method could eliminate the dispersion and formulation difficulties [5, 8]. Nevertheless, the number of literatures that studied the use of nanoclay based masterbatch to produce PP/clay nanocomposites were still limited, and thus it becomes an interesting topic of research. In the current work, PP/clay nanocomposites has been fabricated via melt compounding method using a commercial nanoclay based masterbatch. The prepared nanocomposites were characterized for their morphological and melt rheological properties. The objectives of this work was to effect of nanoclay loadings and reprocessing on the morphological and rheological properties of the prepared nanocomposites.

2. Experimental

2.1. Materials and nanocomposites fabrication

PP/clay nanocomposites were fabricated by melt compounding nanoclay masterbatch pellets (concentration of 50 wt% of nanoclay) with polypropylene (PP) pellets. The masterbatch (product name: NanoMax) was obtained from Nanocor, USA. Whereas, the PP (product name: PP570, MFI 8 g/10 min) was obtained from SABIC, Saudi Arabia. Based on the information provided by the supplier, the nanoclay material is organophillic montmorillonite (MMT) or organoclay, which has been modified with dimethyl-dihydrogenated tallow ammonium. The polymer carrier is PP, which has also been compatibilized with maleic anhydride (MA). The melt compounding process was done in a co-rotating twin screw extruder (TSE). Details about the extruder, preparation step, and processing conditions have been explained in our previous work [4]. The melt compounding process was carried out twice. The first melt compounding process was called 1st cycle, and the pellets obtained from the 1st cycle was then recycled/reprocessed again using the same procedure to get the 2nd cycle nanocomposites pellets. The nanocomposites obtained from the 1st cycle were referred to as NC-0-I, NC-5-I, NC-10-I, and NC-15-I for nanoclay loadings of 0, 5, 10, and 15 wt%, respectively. Whereas, the nanocomposites obtained from the 2nd cycle were referred to as NC-0-II, NC-5-II, NC-10-II, and NC-15-II. Afterward, all the nanocomposites pellets (i.e. 1st and 2nd cycle) from the extruder were molded into a set of ASTM standard samples for testing by using an injection molding machine. Details of the injection molding machine and processing conditions have been mentioned in our previous work [4].

2.2. Sample characterization

Morphological characterization is carried out by utilizing a Scanning Electron Microscopy (SEM). The SEM used in this work is a JEOL JSM-6360A, Japan. Molded bar samples were cryogenically-fractured prior to SEM examination. To examine using SEM, first the fractured samples were finely coated with a gold. The coating can minimize the overcharging and overheating of the samples, and also can improve the signal as well as the surface resolution. All the samples were analyzed by the SEM at voltage of 15 kV. Additionally, an oscillatory rheometer instrument model AR G2 (TA Instruments, USA) was used to investigate the effect of nanoclay loadings and reprocessing on the melt rheological characteristic of the PP/clay nanocomposites. All the rheological analyses were done using frequency sweep procedure with parallel plates fixture (dia. 25 mm and gap 2 mm). The following test conditions were used: constant temperature of 260°C; oscillation stress of 500 Pa; angular frequency range of 0.1 - 628.3 rad/s. The samples used for this test were disk-disk like sample from the injection molding machine. Prior to the test, the sample was conditioned first at constant temperature of 260°C until the samples were completely melted. Afterward, the excess of the molten nanocomposites samples around the parallel plates was cleaned to ensure the validity of the analysis. One of the data generated and studied in the current work was complex viscosity $|\mu^*|$ vs angular frequency (ω). Figure 1 shows a photograph of an oscillatory rheometer with a parallel plates geometry. As seen in the figure, the excess of the molten nanocomposites samples is cleaned prior to the melt rheological test.

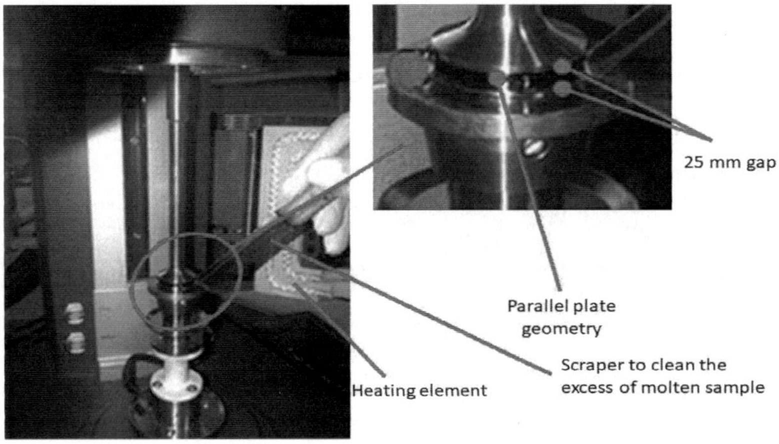

Figure 1. Photo of an oscillatory rheometer with a parallel plate fixture. (During cleaning proses of the excess molten sample prior to the frequency sweep test

3. Results and discussion

The surface morphology of the PP/clay nanocomposites samples for both 1st and 2nd cycle are shown by scanning electron microscopy (SEM) images depicted in Figures 2a-b and Figures 2c-d, respectively. The distribution of nanoclay platelets in the PP matrix could be analyzed by observing the SEM images. The SEM images shown in Figure 2 were only for nanocomposites samples with nanoclay loading of 5 wt%, i.e. NC-5-I (1st cycle) and NC-5-II (2nd cycle). In general, the nanoclay platelets were showed by the white needle-like in the dark area of PP matrix (as shown by the arrows). This has been confirmed by an EDX analysis of the SEM, which has been reported in our previous work [9]. As seen in Figure 2a (SEM magnification of 2,000X), the nanoclay platelets of the 1st cycle were fairly well distributed in the PP matrix. This distribution of nanoclay platelets were looked more clear at higher SEM magnification, i.e. 10,000X (see Figure 2b).

Whereas, for the nanocomposites 2n cycle, the distribution of nanoclay platelets in the PP matrix were quite better than that of the 1st cycle, which can be observed in Figures 2c and 2d. This results could be attributed to the fact that during the recycle or melt compounding of PP/clay nanocomposites (1st cycle), the nanoclay platelets that already distributed in the PP matrix were redistributed again by the shear stress induced by the twin screw extruder, and thus became much better distributed than the 1st cycle. Nevertheless, the dispersion state of the nanoclay (either intercalated or exfoliated) could not be

IOP Conf. Series: Materials Science and Engineering **543** (2019) 012036 doi:10.1088/1757-899X/543/1/012036

observed in these SEM images. To further study the dispersion state of the nanoclay, better microscopy technique such as Transmission Electron Microscopy (TEM) or Atomic Force Microscopy (AFM) are needed. Additionally, for the rest of the nanocomposites samples, i.e. NC-0, NC-10, and NC-15 for both 1st and 2nd cycle also showed the similar trend, but they were not presented in this paper.

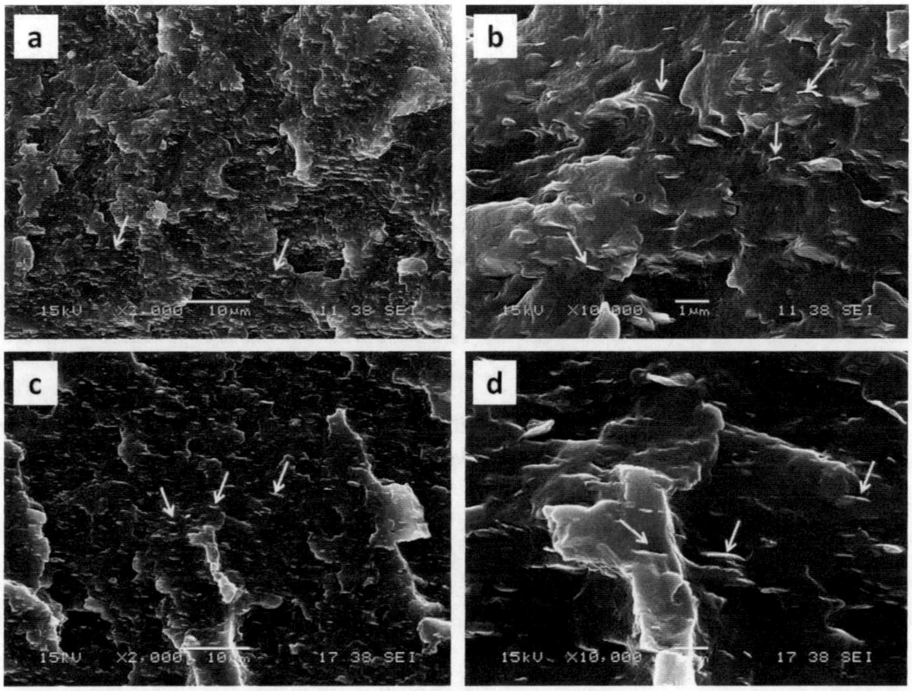

Figure 2. SEM micrographs of a) NC-5-I (2,000X); b) NC-5-I (10,000X); c) NC-5-II (2,000X); and d) NC-5-II (10,000X)

Additionally, the melt rheological analysis of the PP/clay nanocomposites for both 1st and 2nd cycle samples were also carried out using an oscillatory rheometer instrument. The melt rheological analysis results were presented as the plot of complex viscosity, $|\eta^*|$ vs angular frequency (ω), as shown in Figure 3. As seen in the figure, the nanocomposites 1st cycle and 2nd cycle have similar trends. As seen in the figure, the complex viscosity of the nanocomposites for both cycles increased with increasing nanoclay loadings at all angular frequencies. This indicates that the incorporation of nanoclay in the PP matrix has enhanced the complex viscosity of the nanocomposites [10]. This improvement of the $|\eta^*|$ was likely due to the interaction of nanoclay platelets with the PP matrix that restricted the mobility of the molten PP chains. The higher the nanoclay loadings, the more hindered mobility of the molecular PP chains, and the higher the $|\eta^*|$ values. Additionally, it was also in accordance with the literature's reports, which mentioned that the distribution and dispersion of nano-materials in the polymer matrix could affect the melt rheological behavior of the nanocomposites [5, 11]. The better the distribution and dispersion of the nano-materials, the better the interaction between the nano-materials and the polymer matrix. Therefore, the $|\eta^*|$ of the 2nd cycle (indicated by solid lines) was slightly higher than that of the 1st cycle (indicated by symbols). It was likely due to the better distribution of the nanoclay in the PP matrix compared to the 1st cycle.

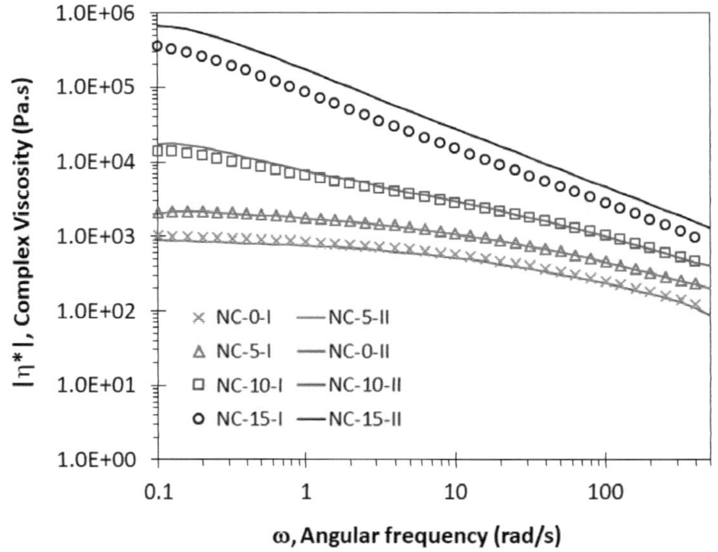

Figure 3. The plot of complex viscosity |η*| vs angular frequency (ω) for PP/clay nanocomposites at different nanoclay loadings for 1st cycle (symbols) and 2nd cycle (lines)

Furthermore, to observe effect of nanoclay loadings and reprocessing on the melt rheological characteristics of the nanocomposites, several representative complex viscosities of PP/clay nanocomposites (1st cycle and 2nd cycle) are listed in Table 1 and also plotted in Figure 4. As seen in the table and figure, the |η*| of both nanocomposites (1st and 2nd cycle) were all higher than the neat PP, and increased with the increasing nanoclay loadings. Additionally, as mentioned in the previous paragraph, the nanocomposites 2nd cycle showed a slightly higher improvement of |η*| compared to the 1st cycle. Additionally, as seen in Table 1 and Figure 4, the improvement of |η*| was more prominent at lower angular frequencies (e.g. 0.1, 1, 10 rad/s) than at higher ones (e.g. 100, 500 rad/s). For example, at angular frequency of 0.1 rad/s, for the PP/clay nanocomposites 1st cycle, the increases of complex viscosity compared to the neat PP (i.e. NC-0-I) were about 101.38%; 1255.77%; and 34,423.23% for NC-5-I; NC-10-I; and NC-15-I, respectively. Whereas, at angular frequency of 500 rad/s, for the PP/clay nanocomposites 1st cycle, the increases of complex viscosity compared to the neat PP (i.e. NC-0-I) were about 97.34%; 292.49%; and 626.59% for NC-5-I; NC-10-I; and NC-15-I, respectively.

Table 1. Representative complex viscosity of PP/clay nanocomposites (1st cycle and 2nd cycle)

Ang. Freq. (rad/s)	PP/clay nanocomposites - 1st cycle				PP/clay nanocomposites – 2nd cycle			
	NC-0-I	NC-5-I	NC-10-I	NC-15-1	NC-0-II	NC-5-II	NC-10-II	NC-15-II
0.1	1022.5	2059.1	13862.7	353000.0	895.5	2007.8	17045.0	666000.0
1	846.7	1778.0	6499.7	85060.5	746.2	1739.1	7547.2	172000.0
10	567.4	1101.4	2811.3	15000.8	512.3	1071.2	2936.4	27733.3
100	250.5	458.3	1027.5	2787.0	232.5	448.1	1001.4	4585.4
500	105.3	207.8	413.3	765.1	87.5	202.5	408.8	1270.7

The similar results also found for the PP/clay nanocomposites 2nd cycle. For example, at angular frequency of 0.1 rad/s, the increases of complex viscosity compared to the neat PP (i.e. NC-0-II) were about 124.21%; 1803.41%; and 74,271.86% for NC-5-II; NC-10-II; and NC-15-II, respectively. Whereas, at angular frequency of 500 rad/s, for the PP/clay nanocomposites 2nd cycle, the increases of complex viscosity compared to the neat PP (i.e. NC-0-II) were about 131.43%; 367.2%; and 1352.23% for NC-5-II; NC-10-II; and NC-15-II, respectively. As seen from the results, at low angular frequencies, the presence of nanoclay platelets on the melt rheological properties of the nanocomposites was more prominent. The interaction between the matrix and the nanoclay resulted in considerable improvement of complex viscosity of the nanocomposites. Whereas, at higher angular frequencies, the nanoclay layers have oriented due to the shear induced by the rheometer apparatus, thus the effect of interaction between the matrix and the nanoclay on the complex viscosity of the nanocomposites was much lower.

Additionally, as seen in Figure 4, the |η*| of the 2nd cycle (indicated by dashed lines) was higher than that of the 1st cycle (indicated by solid lines). The differences of |η*| were more prominent at higher nanoclay loading (e.g. 15 wt%) than at neat PP (i.e. 0 wt%) and at lower nanoclay loadings (e.g. 5, 10 wt%). Whereas, the effect of angular frequencies on the differences of |η*| between 1st and 2nd cycles for all nanoclay loadings was not significant. As seen in Table 1 and Figure 4, the differences of |η*| between 1st and 2nd cycles at 15 wt% of nanoclay loading (i.e. NC-15-I vs NC-15-II) were about 313,000; 86,939; 12,732; 1798; and 505 Pa.s for angular frequencies of 0.1; 1; 10; 100; and 500 rad/s, respectively.

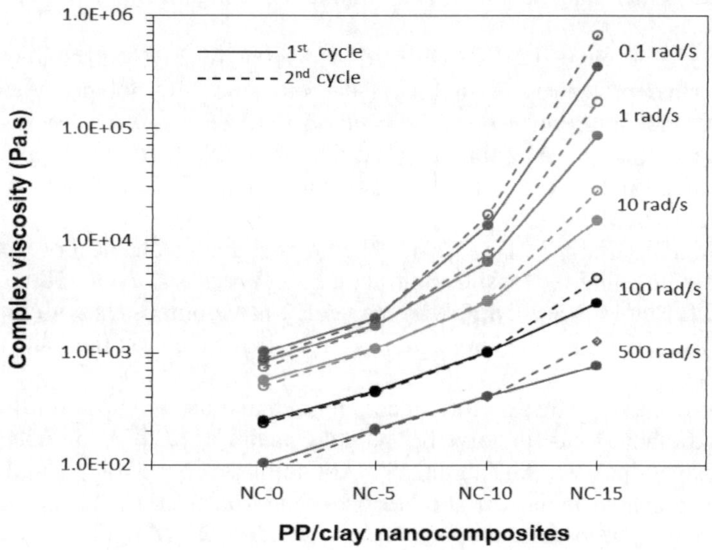

Figure 4. The plot of representative complex viscosity |η*| at several angular frequencies (ω) for PP/clay nanocomposites at different nanoclay loadings for 1st cycle (solid lines) and 2nd cycle (dashed lines)

4. Conclusion

The effects of nanoclay loadings and reprocessing on the morphology and melt rheological properties of the PP/clay nanocomposites have been investigated. In order to study the reprocessing effect on the properties of PP-nanoclay composites, melt compounding was carried out twice/recycle (i.e. 1st cycle and 2nd cycle). Morphological study on the nanocomposites revealed that all the nanocomposites samples (both 1st cycle and 2nd cycle) showed good distribution of nanoclay platelets in PP matrix at all nanoclay loadings. In melt rheological behavior, all the nanocomposites samples for both 1st cycle and 2nd cycle exhibited higher complex viscosity than the neat PP. The complex viscosity of the nanocomposites increased with increasing nanoclay loadings at all angular frequencies. However, the improvement of |η*| was more prominent at lower angular frequencies (e.g. 0.1, 1, 10 rad/s) than at higher ones (e.g. 100, 500 rad/s). For example, at angular frequency of 0.1 rad/s, for the PP/clay nanocomposites 1st cycle, the increases of complex viscosity compared to the neat PP (i.e. NC-0-I) were about 101.38%; 1255.77%; and 34,423.23% for NC-5-I; NC-10-I; and NC-15-I, respectively. Whereas, at angular frequency of 500 rad/s, for the PP/clay nanocomposites 1st cycle, the increases of complex viscosity compared to the neat PP (i.e. NC-0-I) were about 97.34%; 292.49%; and 626.59% for NC-5-

I; NC-10-I; and NC-15-I, respectively. Additionally, the 2nd cycle showed a slightly higher improvement of |η*| compared to the 1st cycle. The differences of |η*| were more prominent at higher nanoclay loading (e.g. 15 wt%) than at neat PP (i.e. 0 wt%) and at lower nanoclay loadings (e.g. 5, 10 wt%). Whereas, the effect of angular frequencies on the differences of |η*| between 1st and 2nd cycles for all nanoclay loadings was not significant.

References

[1] Martínez-Colunga J G, Sánchez-Valdés S, Ramos-deValle L, Muñoz-Jiménez L, Ramírez-Vargas E, Ibarra-Alonso M C, Lozano-Ramirez T, and Lafleur P G 2014 Simultaneous polypropylene functionalization and nanoclay dispersion in PP/Clay nanocomposites using ultrasound *J. Appl. Polym. Sci.* **131** 40631-42.

[2] Lee S, Yoo J, and Lee J W 2015 Water-assisted extrusion of polypropylene/clay nanocomposites in high shear condition *J. Ind. Eng. Chem.* **31** 317-22.

[3] Zhang G, Wu T, Lin W, Tan Y, Chen R, Huang Z, Yin X, Qu J 2017 Preparation of polymer/clay nanocomposites via melt intercalation under continuous elongation flow *Compos Sci Technol.* **145** 157-64.

[4] Chafidz A, Latief F H, Samad U A, Ajbar A, Al-Masry W 2016 Nanoindentation creep, nano-impact, and thermal properties of multiwall carbon nanotubes-polypropylene nanocomposites prepared via melt blending *Polym-Plas. Technol.* **55** 1373-85.

[5] Chafidz A, Kaavessina M, Al-Zahrani S, Al-Otaibi M N 2016 Rheological and mechanical properties of polypropylene/calcium carbonate nanocomposites prepared from masterbatch, *J. Thermoplast. Compos.* **29** 593-22.

[6] Wang M, Fan X, Thitsartarn W, He C 2015 Rheological and mechanical properties of epoxy/clay nanocomposites with enhanced tensile and fracture toughnesses *Polymer* **58** 43-52.

[7] Adak B, Joshi M, Butola B S 2018 Polyurethane/clay nanocomposites with improved helium gas barrier and mechanical properties: Direct versus masterbatch melt mixing route *J. Appl. Polym. Sci.* **135** 46422.

[8] Pokharel P, Lee D S 2014 High performance polyurethane nanocomposite films prepared from a masterbatch of graphene oxide in polyether polyol *Chem. Eng. J.* **253** 356-65.

[9] Chafidz A, Kaavessina M, Al-Zahrani S, Al-Otaibi M N 2014 Polypropylene/organoclay nanocomposites prepared using a Laboratory Mixing Extruder (LME): crystallization thermal stability and dynamic mechanical properties, *J. Polym. Res.* **21** 483-95.

[10] La-Mantia F P, Scaffaro R, Ceraulo M, Mistretta M C, Dintcheva N T, Botta 2016 A simple method to interpret the rheological behaviour of intercalated polymer nanocomposites, *Compos. Part B: Eng.* **98** 382-88.

[11] Choi H J, Kim S G, Hyun Y H, Jhon M S 2001 Preparation and Rheological Characteristics of Solvent-Cast Poly(ethylene oxide)/Montmorillonite Nanocomposites *Macromol. Rapid Comm.* **22** 320-25.

The Adsorption Behaviors of Thermosensitive poly(DMAAPS) Grafted Onto EVA Porous Support

E O Ningrum[1]
[1]Department of Industrial Chemical Engineering, Faculty of Vocational Studies, Institut Teknologi Sepuluh Nopember, Kampus ITS Sukolilo, Surabaya, 60111, Indonesia

eva-oktavia@chem-eng.its.ac.id

Abstract. The adsorption ability of poly(DMAAPS) grafted onto ethylene vinyl acetate (EVA) porous support prepared using various polymerization times was investigated at a variety of temperatures and permeated sample volumes. The permeability coefficient through Poly(DMAAPS) was also investigated. The poly(DMAAPS) was successfully grafted onto a EVA porous support through graft polymerization in water containing DMAAPS monomer with polymerization time of 4, 6 days, and 7 days. The permeability coefficient of Poly(DMAAPS) onto a EVA porous support decreased with increasing the temperature. Additionally, the adsorption amount of cation increased as permeated volume increased.

1. Introduction

The biggest challenge confronting wastewater produced by industrial is about the removal of the heavy metal ions. This topic becomes the point of interest since the heavy metal ions causing water pollution is non biodegradable substance and bio-accumulate over a period of time in human tissues. When it is consumed above the bio-recommended limits. it will give biotoxin effects on the human body which induces various diseases [1]. The treatments including physical and chemical such as precipitation, neutralization, coagulation/flocculation, ion exchange, and complexation which involve the formation of a complex or chelating agent have been practiced to remove the heavy metals; each has its merits and limitations in its application [2,3]. Several innovative approaches have been considered to overcome this limitation and among these, the use of polymer grafted onto supporting substrate is promising.

The change generally happens in the environment with sensitive polymers and hydrogel ions is the changes in pH or temperature. It also includes water polymer precipitation or hydrogel collapse completed with the expulsion of water large fraction in gel pores [4–6]. Such polymer can be grafted or absorbed onto the other polymer surface presents or porous membrane pores when it is stimulated environmentally which leads to the changes of wettability or permeability [5].

A research was conducted to study the molecular imprinted thermosensitive gels grafted onto PP substrate by plasma-initiated graft polymerization [7]. The thermosensitive N-isopropylacrylamide microgel adsorbent was prepared by emulsion polymerization using anionic polymerizable surfactant and the molecular imprinting technique using Cu(II) ions as the target ion. These microgels were suspended stably even when the temperature of the suspension exceeded the transition temperature of poly(NIPAM) which is about 32 ^{o}C and the adsorption amount of Cu(II) ions on the microgel at the shrinking state of 50 ^{o}C was quite large compared to the swollen state of 10 ^{o}C. Although the adsorption of these microgels was quite rapid compared to the grinding gels, however it is difficult to handle in the industrial application due to the small size of the particle.

In this study, we intend to employ the DMAAPS polymer grafted onto EVA porous support in order to regenerate the heavy metal ions from salt solution. This method could overcome the adsorption

IOP Conf. Series: Materials Science and Engineering **543** (2019) 012037 doi:10.1088/1757-899X/543/1/012037

limitations in the use of gel adsorbent. In this research, the adsorption and desorption mechanism of DMAAPS polymer could be simply controlled by tuning the temperature of the system. By grafting DMAAPS polymer onto EVA porous support and after thin film is contacted with a salt solution containing metal ions, the ions directly approach and bind to the opposite charge of betaine polymer DMAAPS without any hindrance. In this system, the ability of ions to infiltrate into the network as occurred in gel does not need to be considered. At temperatures above its upper critical solution temperature (UCST), adsorption occurs due to the polymer swells on the surface-grafted from of EVA porous support leading to the attachment of metal ions into the betaine DMAAPS. At temperatures below its critical solution temperature (UCST), the polymer collapse, the intrachain and/or interchain interactions between the side chains of the polymer are strong and betaine is forced to release the binding with ions and desorption would take place.

Based on the previous reports, most work has been done on the swelling behavior of gels and thermosensitive microgel. This paper reports the study on the ability of DMAAPS polymer grafted onto EVA porous support in recovering and regenerating various kinds of heavy metal ions. The adsorption properties of ground DMAAPS gel are also reported as a comparison.

2. Experimental Section

2.1 Materials

The material used in this work such as *N,N*-dimethylaminopropylacrylamide (DMAPAA), 1,3-propanesultone (PS) was obtained from Kohjin Co. Ltd and purified through reduced distillation. Additionally, 1,3-propanesultone (PS) was used as received without further purification. Other reagents such as acetonitrile and acetone were purchased from Sigma-Aldrich Co. (USA).

2.2 Synthesis of DMAAPS monomer

The DMAAPS monomer was synthesized through the ring-opening reaction of *N,N*-dimethylaminopropylacrylamide (DMAPAA) and 1,3-propanesultone (PS) [8–11]. A mixture of PS (75 g, 11. 4 M) and acetonitrile (75 g) was added by a dropwise to the mixture of DMAPAA (100 g, 5.1 M) and acetonitrile (200 g) with continuous stirring at 30 °C for 90 minutes. After the completion of the addition, the mixture was then stirred at 30°C for 6 h and allowed to stand for two days. The precipitated white crystal was collected by filtration and washed several times with 500 ml of acetone. Finally, the crystal obtained was dried under reduced pressure for 24 h. The structure of DMAAPS is shown in Figure 1.

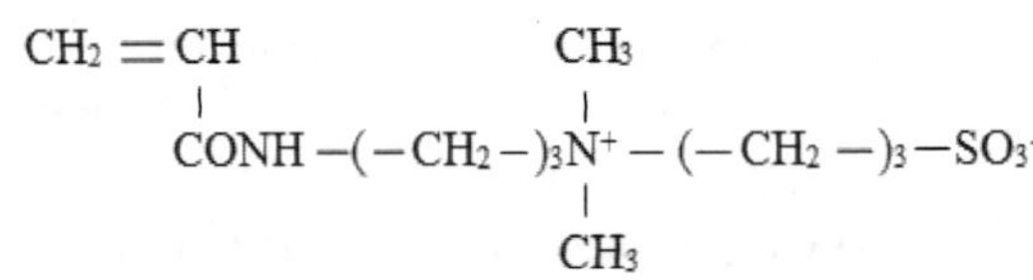

Figure 1. Chemical structure of DMAAPS.

2.3 Plasma and peroxide-initited graft polymerization

As previously reported by Tokuyama et.al, the procedure of the graft polymerization consists of three steps. A piece of ethylene vinyl acetate (EVA) porous support with size of 20x20x0.82 mm to be grafted was treated with Ar plasma in a reactor in order to promote the graft-polymerization of the monomer. Initially, air in the reactor was displaced with Ar gas and evacuated to approximately 10 Pa, and the radio-frequency-plasma treatment was then started for the activation for 60 s (power input 40 W and 13,56 MHz). It was followed by treating the porous support with O_2 gas for 60 s to produce peroxide on the surface of activated EVA porous support. Activated EVA sample was then contacted with a graft monomer solution of the desired concentration, e.g., solution containing monomers of DMAAPS 1000 mol/m^3. The graft polymerization was carried out under nitrogen atmosphere for four days at 80 °C. Finally, the grafting polymers product was washed by water and dried to calculate the amount of grafting as the mass ratio of the grafted polymer to the initial EVA. The plasma initiated graft polymerization apparatus is presented in Figure 2.

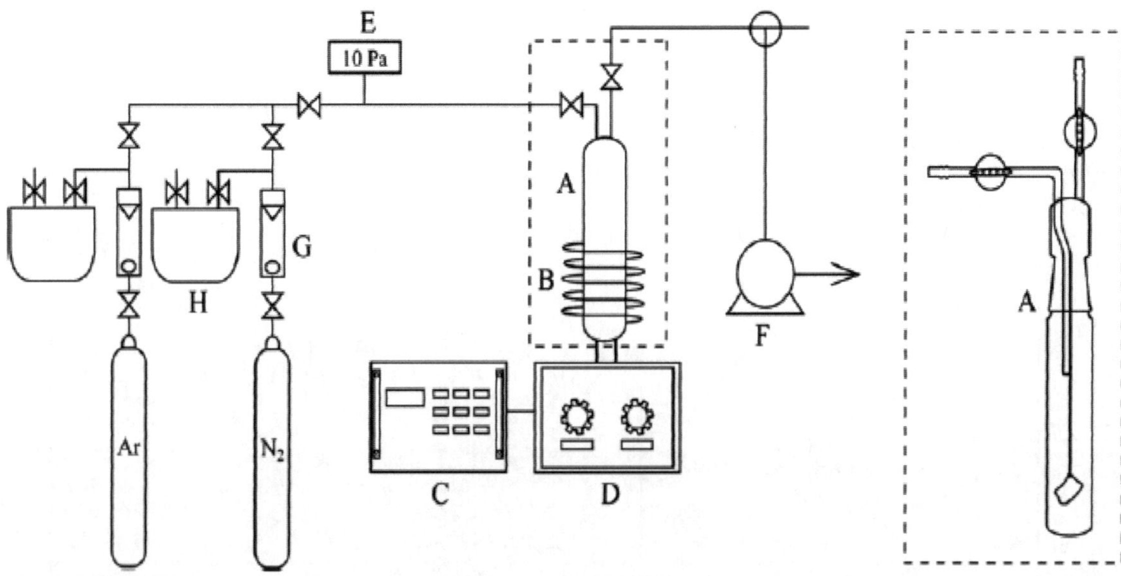

Figure 2. Apparatus of plasma-initiated graft polymerization: (A) reactor, (B) inductively-coupled coil, (C) radio-frequency power source, (D) matching unit, (E) vacuum gauge, (F) vacuum pump, (G) flowmeter and (H) gas reservoir.

Table 1. Polymerization condition of DMAAPS graft

		(mol/m^3)
Monomer	: *N,N*-dimethyl(acrylamidopropyl)ammonium propane sulfonate (DMAAPS)	400, 1000
Accelerator	: *N,N,N',N'*-tetramethylethylenediamine (TEMED)	4
Temperature	80 °C	
Solvent	water	
Polymerization time	4 days	

2.4 Measurement of permeability coefficient and concentration of permeated solution

The permeability through the porous material grafted with poly(DMAAPS) was evaluated by the permeability coefficient *k* defined by the following equation.

$$Q = Ak\frac{\Delta P}{\mu L} \tag{1}$$

where, Q is the flow rate, μ: viscosity, L: the thickness of the porous material, A: the cross sectional area of the porous material, and ΔP: the pressure difference between the both ends of the porous material. Pressure was conducted by an aqueous solution column on the porous material, and the flow rate was measured by the change in the solution level in the column with time. First, the salt solution of 50 °C was permeated through the porous material to form the open-chain conformation of the grafted polymer. Then, the salt solution of the temperature was permeated.

2.5 Adsorption Experiment

The metal ion adsorption experiment was performed by the similar method as the permeability mentioned above of the desired. The salt solution concentration was permeated through the porous material grafted with poly(DMAAPS) and the permeated samples solution was collected at the desired volume. When the desired volume was reached, the adsorption process was stopped and washed EVA porous support by HCl 0.01 M, followed rinsed several time by water. The permeate concentration and initial feed concentration of nitrate solution were measured by an ion chromatography. The adsorption amount was calculated from the concentration difference between the feed solution and permeated solution per gram of the grafted polymer.

3. Result and Discussion

3.1 Effect of polymerization time on the permeability of water

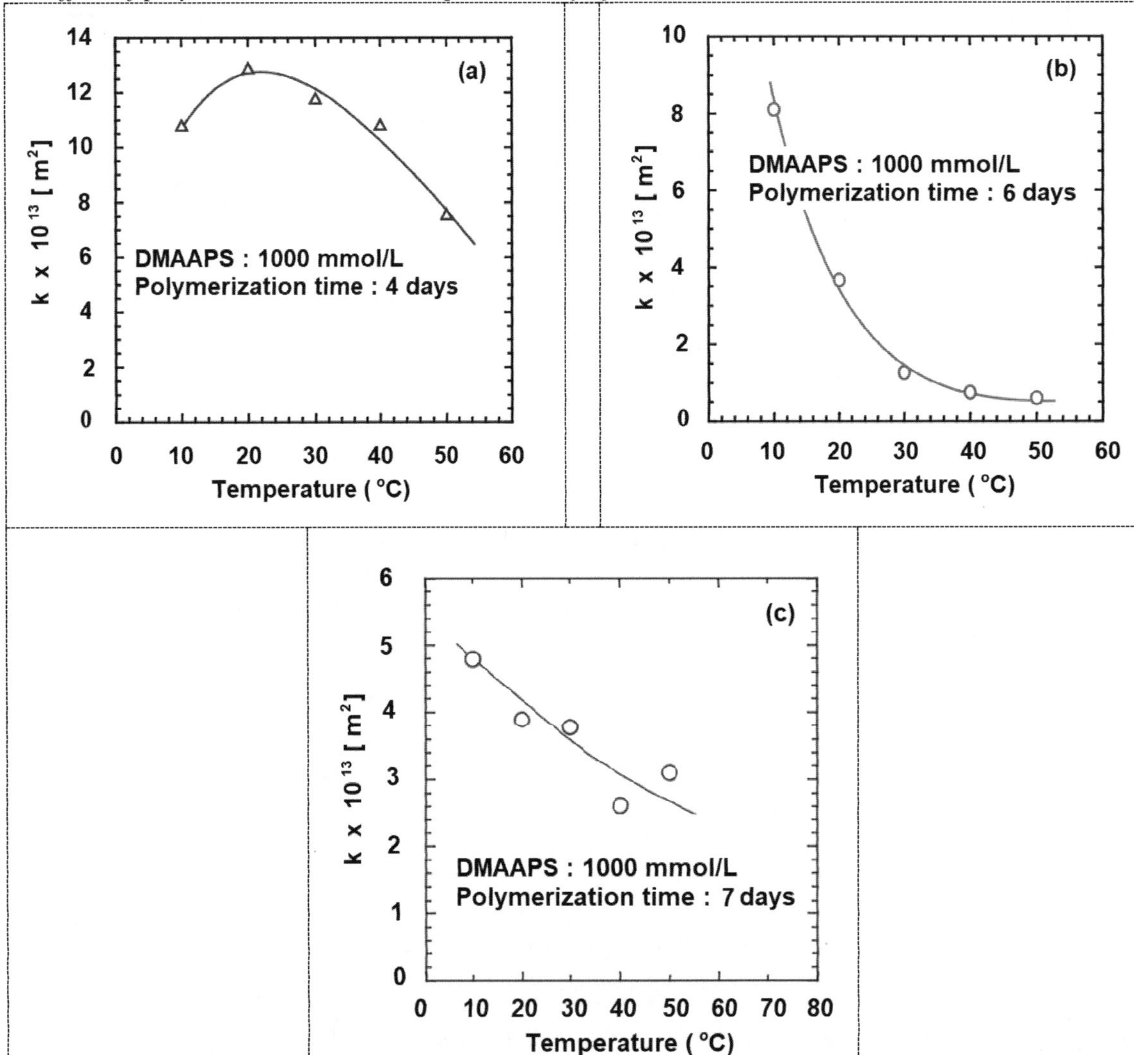

Figure 3. Permeability coefficient through poly(DMAAPS) grafted onto a EVA porous support in water containing DMAAPS monomer of 1000 mol/m^3 with polymerization time of (a). 4 days (b). 6 days (c). 7 days at various temperatures.

Figure 3 shows the permeability coefficient through poly(DMAAPS) grafted onto the EVA porous support polymerization time of (a) 4 days, (b) 6 days, (c) 7 days at various temperatures. As shown in Fig. 3 (a), it can be seen that in 4 days polymerization, permeability coefficient increased first and the value reached maximum and further increasing the temperature, permeability coefficient then decreased. However, in 6 days, the polymerization permeability coefficient decreased along with the increasing the temperature. This phenomenon was considered to be due to the factor affecting the permeability coefficient in polymerization time of 4 days which is only the viscosity of water. However, in polymerization time of 6 days, the viscosity of water and polymer characteristic at various temperatures strongly affect the permeability coefficient. As shown in Fig.3 (c), it can be seen that in 7 days of polymerization, permeability coefficient decreased along with the increasing temperature. This phenomenon was considered to be due to the factor affecting the permeability coefficient in i.e., the viscosity of water and polymer characteristic at various temperatures strongly affect the permeability coefficient

3.2 Effect of polymerization time on the adsorption amount

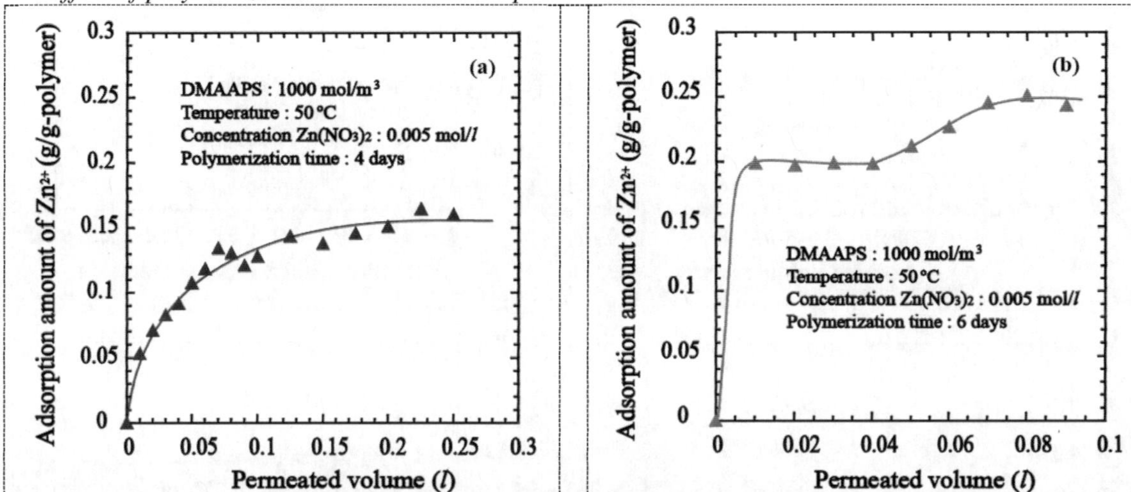

Figure 4. Adsorption amount of nitrate solution by poly(DMAAPS) grafted onto EVA porous support in water containing DMAAPS monomer 1000 mol/m^3 at 50 °C with polymerization time of (a). 4 days (b). 6 days. The porosity and pore size of EVA is 80 % and 20μm.

In this work, the effect of polymerization time in graft polymer solution on adsorption amount was studied. The adsorption amount of nitrate solution by poly(DMAAPS) grafted onto the EVA porous support in water containing DMAAPS monomer 1000 mol/m^3 at 50 °C with polymerization time of (a). 4 days (b) 6 days are shown in Figure 4. From both figures, we can conclude that the poly(DMAAPS) was successfully grafted onto the EVA porous support by graft polymerization in water solution. It was also found that in polymerization time of 4 days and 6 days adsorption amount shows the same tendency. Adsorption amount increased as permeated volume increased, however adsorption amount in 6 days is higher than 4 days of polymerization. This can be considered due to the graft ratio in 6 days polymerization which is higher than the 4 days of polymerization. The grafting ratio in 4 days was found to be 0.043 g-polymer/g-support, however for 6 days was 0.0833 g-polymer/g-support.

IOP Conf. Series: Materials Science and Engineering **543** (2019) 012037 doi:10.1088/1757-899X/543/1/012037

3.3 Effect of permeated sample volume on adsorption amount

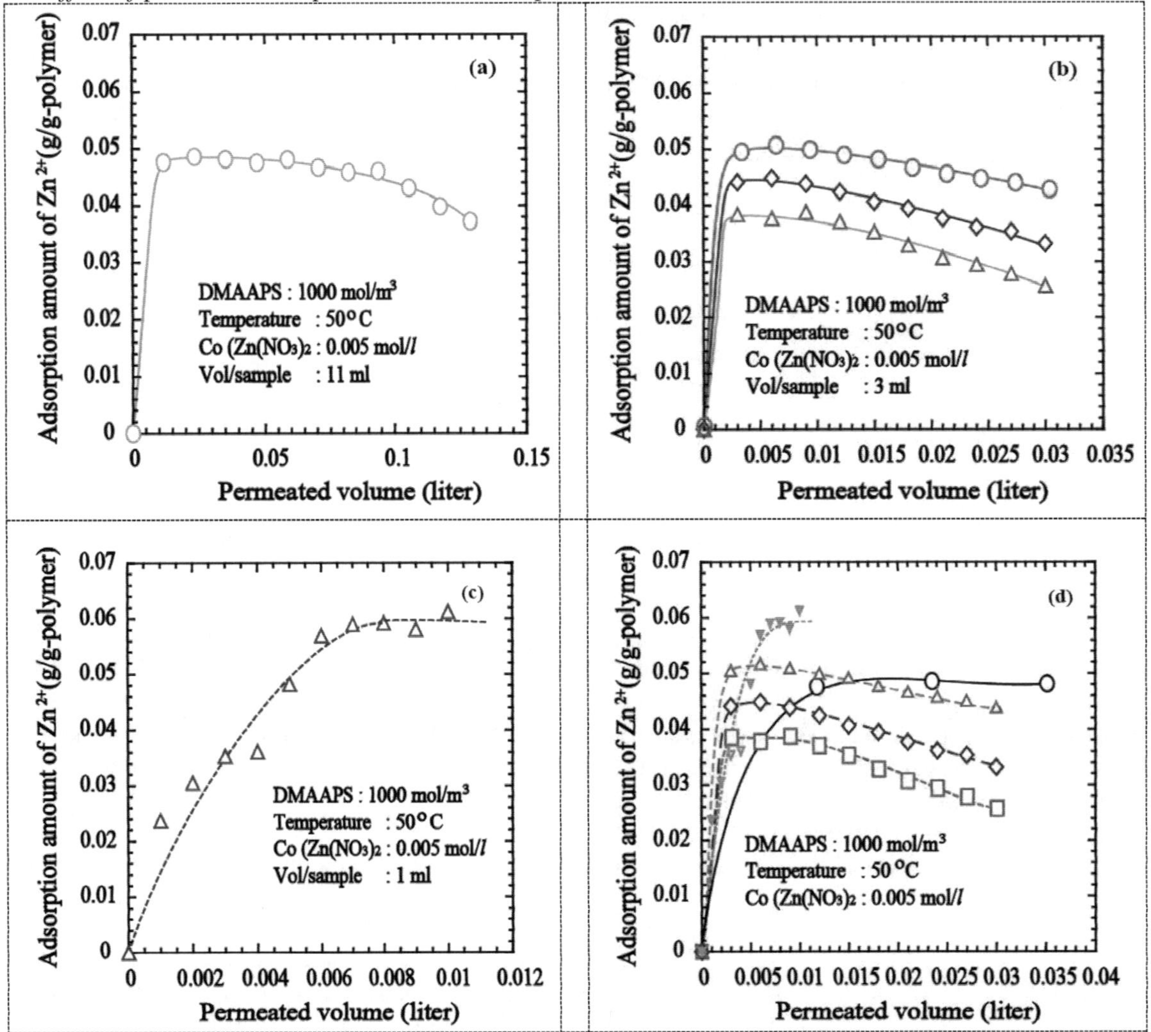

Figure 5. Adsorption amount of nitrate solution by poly(DMAAPS) grafted onto a EVA porous support in water containing DMAAPS monomer of 1000 mol/m^3 in the permeated solution of (a). 11 ml; (b). 3 ml (c). 1 ml; (d) various volume comparation by plasma and peroxide-initiated grafting at 50 °C.

Figure 5 shows the adsorption amount of DMAAPS polmer grafted onto a EVA porous support by graft polymerization in water containing DMAAPS monomer of 1000 mol/m^3. As shown in Fig.5 (a), it can be seen that in 11 ml of permeated solution, the adsorption amount increased drastically in the first sample obtained and the value remained constant with further increasing the permeated solution. However, when the permeated volume reached 100 ml, the adsorption amount gradually decreased. This phenomenon was also observed when changing the permeated solution volume up to 3 ml. On the contrary, the adsorption amount increased significantly when the permeated solution volume was 1 ml.

4. Conclusions

The poly(DMAAPS) was successfully grafted onto a EVA porous support through graft polymerization in water containing DMAAPS monomer of 1000 mol/m^3 at 50 ^{o}C with polymerization time of 4, 6 days, and at 70 ^{o}C with polymerization time of 7 days. However, the amount of polymer grafted on the porous support was too small. The permeability of poly(DMAAPS) onto a EVA porous support with polymerization time of 4 days depends only on the viscosity of water and of polymerization time of 6 and 7 days strongly depends on the thermosensitive characteristic of DMAAPS. The adsorption amount of poly(DMAAPS) was successfully grafted onto the EVA porous support by graft polymerization in water containing DMAAPS monomer of 1000 mol/m^3 in sample volume of 3 ml and 1 ml.

References

[1] Duruibe J O, Ogwuegbu M, Egwurugwu J 2007 *Int. J. Phys. Sci.* **2**(5) 112–8
[2] Rengaraj S, Yeon K-H, Moon S-H. 2001 *J. Hazard. Mater.* **87**(1-3) 273–87
[3] Vibhandik A D, Marathe K V 2014 *Front. Chem.Sci. Eng.* **8**(1) 79–86
[4] Winnik F M. 1990 *Macromolecules* **23**(1) 233–42
[5] Winnik F M, Ottaviani M F, Bossmann S H, Pan W, Garcia-Garibay M, Turro N J 1993 *Macromolecules* **26**(17) 4577–85
[6] Vesterinen E, Tenhu H, Dobrodumov A 1994 *Polymer* **35**(22) 4852–6
[7] Kanazawa R, Mori K, Tokuyama H, Sakohara S 2004 J. *Chem. Eng. Jpn.* **37**(6) 804–7
[8] Ningrum E O, Purwanto A, Ni'mah H, Sumarno, Dewitasari D I, Mulyadi E O 2017 *AIP Conf. Proc.* **1840**(1) 090007
[9] Ningrum E O, Purwanto A, Mulyadi E O, Dewitasari D I, Sumarno S 2017 *Indones. J. Chem.* **17**(3) 446–52
[10] Ningrum E O, Ohfuka Y, Gotoh T, Sakohara S 2015 *Polymer* **59** 144–54
[11] Ningrum E O, Murakami Y, Ohfuka Y, Gotoh T, Sakohara S 2014 *Polymer* **55**(20) 5189–97

IOP Conf. Series: Materials Science and Engineering **543** (2019) 012038 doi:10.1088/1757-899X/543/1/012038

Melt Rheological Behavior of High Density Polyethylene/Poly(Vinyl Alcohol) Fiber Composites Prepared via Melt Compounding Method

A. Chafidz[1], L. Setyaningsih[1,*], N. Indah[1], W. D. P. Rengga[2], Haryanto[3], J. Suhartono[4]

[1]Chemical Engineering Department, Universitas Islam Indonesia, Yogyakarta 55584, Indonesia
[2]Chemical Engineering Department, Universitas Negeri Semarang, Semarang 50229, Indonesia
[3]Chemical Engineering Department, Universitas Muhammadiyah Purwokerto, Purwokerto 53182, Indonesia
[4]Chemical Engineering Department, Institut Teknologi Nasional, Bandung 40124, Indonesia
*E-mail: lucky.nuzulia@uii.ac.id

Abstract. In this work, high density polyethylene (HDPE)/poly(vinyl alcohol) (PVA) fiber composites have been fabricated via melt compounding by employing a twin-screw extruder. The resulted composites samples of four different PVA fiber loadings (i.e. 0, 5, 10, and 20 wt%) were then characterized using an oscillatory rheometer to investigate the effect of PVA loadings on their melt rheological behavior (e.g. storage modulus, loss modulus, complex viscosity). Additionally, the surface morphology of cryo-fractured surface of the composites were also investigated by using a scanning electron microscopy (SEM). The SEM micrographs showed that PVA fibers were perfectly embedded and well blended in HDPE matrix. Additionally, the melt rheological analysis results showed that the complex viscosity of all nanocomposites samples (i.e. PVAC-5, PVAC-10, and PVAC-20) were higher than that of neat PP (i.e. PVAC-0) and increased with the increase of nanoclay concentrations. Moreover, to further study the relationship between the PVA fiber loadings and complex viscosity of the composites, a well-known Carreau-Yasuda equation was employed to model the complex viscosity data from the rheological test. It was found that the equation fitted well the rheological test data.

1. Introduction

It is well-known that plastics have been widely used in the world and become part of our life. It was due to their many advantages (e.g. low density, low price, good mechanical properties, various applications, easily to process into any kind of forms) compared to other materials (e.g. metals, ceramics, wood, etc). High density polyethylene (HDPE) is considered as one of the most widely and frequently used thermoplastic polymers. It has been used in many applications, e.g. packaging, boxes, household goods, piping, etc. It is because HDPE is low cost, low density, good chemical and corrosion resistance, easy to be processed and recycled [1]. Moreover, its properties could be enhanced by blending and filling with reinforcing materials, i.e. micro-/nano-fillers, fibers, etc [2].

In other hand, poly(vinyl) alcohol (PVA) is a biodegradable type of thermoplastic polymers made from vinyl acetate. PVA fibers can be considered as one of fibers that have high reinforcement performance and have demonstrated good results in terms of tensile strain capacity, toughness, and crack bridging properties. Additionally, these fibers also suitable for many applications due to their high resistance to UV, chemicals/alkali, abrasion, and fatigue [3]. Recently, PVA fibers have been widely used in

construction applications, such as in fiber reinforced cementitious composites (FRCC) [4, 5]. The FRCC at least consists of three phases, which are: cement paste/matrix, fiber, and aggregates. The incorporation of fibers in FRCC may increase the energy absorption/toughness, strength, and/or strain of the FRCC [3]. Additionally, PVA fiber can also be used to fabricate polymer/PVA fiber based composites. There have been several literatures about polymer/PVA fiber based composites. Wu et al. [2] studied the preparation and characterization of HDPE/PVA. They found that the mechanical properties of HDPE increased with the addition of PVA fiber. Nevertheless, there have been limited number of literatures that studied PVA fiber filled polymer based composites, and thus it becomes an interesting topic of research.
One of the most preferred and widely used method to fabricate polymer composites is melt compounding. It is because this method is quite simple, low cost, and "green" method since no-solvent is used. Furthermore, this method is very suitable with the plastic industrial machines (i.e. extruder, injection molding), which makes the industrial scale production and commercialization of the polymer composites becomes easier [6, 7]. One of the key factors to fully understand the internal morphology/structure and processability (especially via extrusion/injection molding) of the composites is by studying their melt rheological properties. To the best of our knowledge, the number research studies regarding with the melt rheological studies of HDPE/PVA fiber composites are still limited, and thus it becomes an interesting topic of research. The present work focuses on the fabrication of HDPE/PVA fiber composites and characterization their morphology and melt rheological propertiesof the prepared nanocomposites.

2. Experimental

2.1. Materials
High density polyethylene (HDPE)/Polyvinyl alcohol (PVA) fiber composites were fabricated by melt compounding HDPE pellets (product name: M-300054, from SABIC, Saudi Arabia) with PVA fibers (product name: RECS15, from Kuraray, Japan). The HDPE has density of 954 kg/m^3 and melt flow index (MFI) of 30.0 g/10 min (low melt viscosity). Whereas, the PVA fiber has diameter 0.04 mm, thickness of 15 dtex , cut length of 8 mm, specific gravity of 1.3. Figures 1a and 1b show the appearances of HDPE pellets and PVA fibers, respectively. Whereas, Figure 1c shows a scanning electron microscopy (SEM) image of single of PVA fiber.

Figure 1. a) Photograph of HDPE pellets; b) photograph of PVA fibers; and c) SEM image of a single PVA fiber

2.2. Composites preparation
The melt compounding process was carried out in a twin-screw extruder (TSE) (Farrell FTX-20, UK) at screw speed of 17 rpm and die temperature of 200 °C. The melt composites from the die was then cooled down in waterbath, dried (using air blower), pelletized (by cutting machine), and put in oven for moisture removal. After completely dried, the composites pellets were then molded into ASTM standard samples using an injection molding machine (Super Master, SM120, from Asian Plastic Machinery Co., China). The HDPE/PVA fiber composites were fabricated at four different PVA loadings, i.e. 0, 5, 10,

20 wt%, which correspond to PVAC-0, PVAC-5, PVAC-10, PVAC-20, respectively. Figures 2a and 2b exhibit the appearances of twin-screw extruder (TSE) and injection molding machine used in this work.

2.3. Composites characterization

The surface morphological study was done by using a Scanning Electron Microscopy (SEM), JEOL JSM-6360A, Japan. The samples used for the SEM analysis were cryo-fractured molded bars. Prior to the SEM examination, the cryo-fractured bar samples were coated with a fine layer of gold to minimize the overcharging/overheating of the samples, and also to improve the signal and resolution. The composites samples were examined by the SEM at 15 kV and magnification of 200X. Additionally, the effects of PVA fiber loadings on the melt rheological behavior of HDPE/PVA fiber composites were examined by the use of an oscillatory rheometer ARG2 (from TA Instrument, USA). The melt rheological test was performed under frequency sweep procedure using parallel plates fixture. The testing conditions were as follow: strain of 0.01 %; fixed temperature of 150 °C; angular frequency range of 0.05 - 500 rad/s. The samples used in the melt rheological test were disk-like composites samples obtained from the injection molding. Prior to the melt rheological test, the disk-like sample was put in between the parallel plates, and conditioned at fixed temperature of 150 °C until the whole composites sample was completely melted. Before starting the analysis, the excess of samples melt around the parallel plates was removed using a scraper. The data generated from the melt rheological test was complex viscosity $|\eta^*|$ vs angular frequency (ω). Figures 2c and 2d exhibit the appearances of Scanning Electron Microscopy (SEM) and oscillatory rheometer apparatus used in this work.

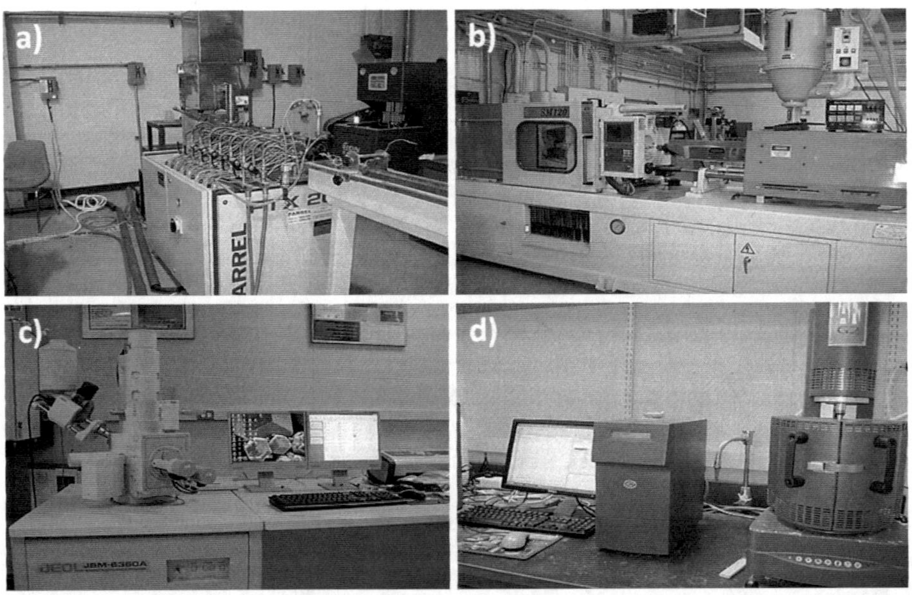

Figure 2. Photograph of a) Twin-screw extruder (TSE); b) Injection molding machine; c) Scanning Electron Microscopy (SEM); d) Oscillatory rheometer

3. Results and discussion

The surface morphology of the HDPE/PVA fiber composites has been examined by a Scanning Electron Microscopy (SEM). Figures 3a and 3b show the SEM images of HDPE/PVA composites, i.e. PVAC-0 and PVAC-10, respectively. The figure shows that the PVA fibers were well blended and embedded in the HDPE matrix (shown by the arrows). It was also observed that the fibers entangled one with the others. Additionally, it seemed that there were less interactions between the HDPE matrix and the PVA fibers (shown by the circle dashed-line). The SEM images of the other samples were similar but not presented in this paper.

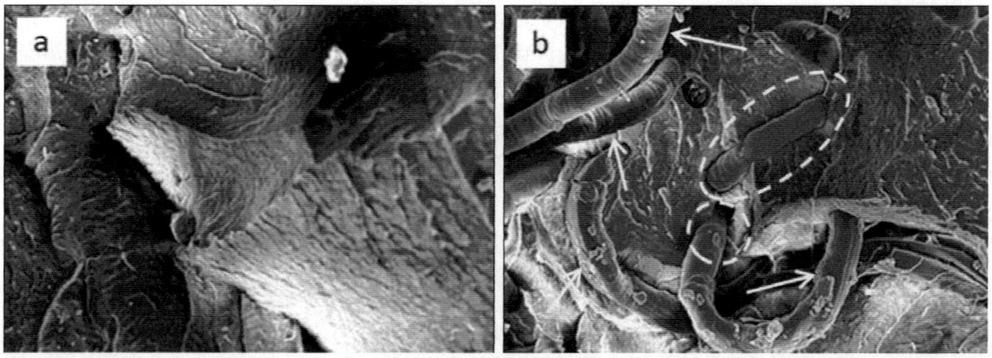

Figure 3. SEM images of cryo-fractured surface of HPDE/PVA fiber composites: a) PVAC-0 and b) PVAC-10

Additionally, the melt (state) rheological analysis of the HDPE/PVA fiber composites was carried out under torsion mode in parallel plate geometry using frequency sweep test. This test was performed at a constant temperature of 150 °C; constant strain of 0.1 %, and angular frequency range of 0.05 and 500 rad/s. Figure 4 shows the complex viscosity $|\eta^*|$ as function of angular frequency (ω) of HDPE/PVA fiber composites measured at fixed temperature of 150 °C. As seen in the figure, the complex viscosity of the composites were all higher than the neat HDPE. Furthermore, the improvement of complex viscosity was more significant at higher PVA loadings (i.e. PVAC-10 and PVAC-20) than at the lower one (PVAC-5).

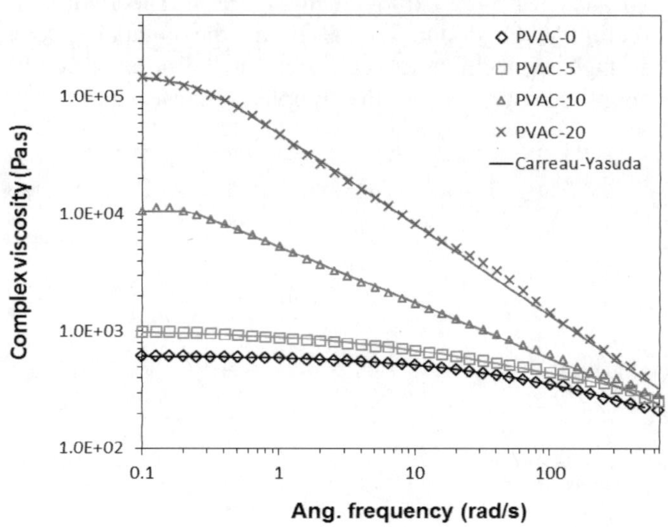

Figure 4. Complex viscosity $|\eta^*|$ as function of angular frequency (ω) for HDPE/PVA fiber composites at different PVA fiber loadings

Additionally, several complex viscosities of the composites samples (at several angular frequencies) are given in Table 1 and also plotted in Figure 5. Using these table and figure we could further observe the effect of PVA fiber loadings as well as the angular frequency on the complex viscosity of the HDPE/PVA fiber composites. As shown in Figure 5 and in accordance with Figure 4, the complex viscosity considerably increased with increasing PVA fiber loadings. The increase of complex viscosity $|\eta^*|$ was most likely due to the dispersion of the PVA fiber in the HDPE matrix, as well as the interaction between them, which resulted in the restricted mobility of HDPE molecular chains in the molten state.

Additionally, the rheological properties of composites materials could also be affected by the frictional interaction among the fillers as well as the interaction between the polymer chains and the fillers [8].

Table 1. Representative complex viscosity of HDPE/PVA fiber composites

Angular Frequency (rad/s)	Complex viscosity of HDPE/PVA fiber composites (Pa.s)			
	PVAC-0	PVAC-5	PVAC-10	PVAC-20
0.1	621.1	1004	10,630	1.49E+05
1	592.3	882.7	5,304	47,980
10	519.4	688.7	1,725	8,235
100	360.6	453.3	635.6	1,441
500	225.3	276.4	374.5	693.8

Furthermore, the increase of complex viscosity was prominent at lower angular frequencies (i.e. 0.1 – 10 rad/s). For example, at angular 100 rad/s the increases of complex viscosity of the composites were only about 0.26; 0.76; 3 times fold of the neat PP, for PVAC-5, 10, 20, respectively. Whereas, at angular frequency of 1 rad/s, the increases of complex viscosity were about 0.5; 8; 80 times fold of the neat PP, for PVAC-5, 10, 20, respectively. Furthermore, at angular frequency of 0.1 rad/s, the complex viscosity of the composites dramatically increased by 0.6; 16; and 240 times fold of the neat PP, for PVAC-5, 10, 20, respectively. At lower angular frequencies (i.e. 0.1 – 10 rad/s), the movement of HDPE molecular chains were very limited, due to the presence of PVA fiber and the interaction between both of them, and thus the complex viscosity was high. Whereas, when the angular frequency of the test increased, the HDPE molecular chains started to move (i.e. less restricted) regardless the presence of PVA fiber, and thus the complex viscosity started to decreased rapidly. This explanation was in accordance with the curve shape of PVAC-10 and PVAC-20 exhibited in Figure 4. The figure shows rapid decline of complex viscosity for both PVAC-10 and PVAC-20 as the angular frequency increased. This phenomenon was called "high shear-thinning" behavior. In the other hand, PVAC-0 and PVAC-5 showed "low shear-thinning" behavior, because the complex viscosity decrease gradually [8].

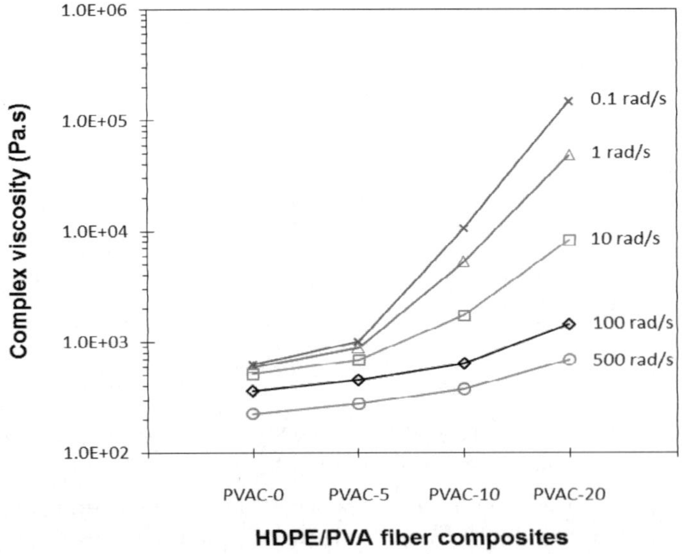

Figure 5. Representative complex viscosity |η*| of HDPE/PVA fiber composites at different angular frequencies

Additionally, to further study the complex viscosity |η*| behavior as function of angular frequency (ω) of the HDPE/PVA fiber composites, we have fitted the complex viscosity |η*| data from the frequency sweep test results to the Carreau-Yasuda equation [9, 10], as shown in Eq (1). Figure 6 shows the schematic diagram of the Carreau-Yasuda equation.

$$\eta^*(\omega) = \eta_0\left[1 + (\lambda\omega)^a\right]^{(n-1)/a} \quad (1)$$

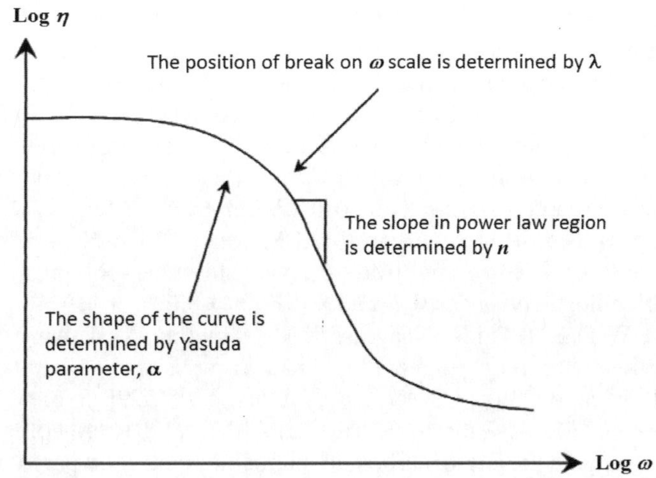

Figure 6. Schematic diagram of Carreau-Yasuda model

Where, η_0 is the zero shear viscosity (or complex viscosity, in this study), λ is the time constant (or cross-over time) and the inverse $1/\lambda$ is the critical value of shear rate (or angular frequency, in this study) at which the complex viscosity begins to decrease. The slope in the power-law region is given by (n-1). Equation (1) reduces to the Newtonian fluid behavior for $n = 1$ and describes shear-thinning behavior when $n < 1$, while a is the Yasuda parameter which is attributed to the amplitude of the region of transition between η_0 and the power law region [9, 10]. All parameters, η_0, λ, n, and a obtained from fitting complex viscosity data are listed in Table 2. As noticed from Figure 4, Carreau-Yasuda model fitted the complex viscosity data of the composites satisfactorily.

Table 2. Parameters with respect to the Carreau-Yasuda model for HDPE/PVA fiber composites

Sample	η_0	λ	n	A
PVAC-0	621.1	0.02	0.61	0.64
PVAC-5	1,004	0.24	0.75	0.68
PVAC-10	10,630	4.39	0.52	11
PVAC-20	1.49E+05	4.31	0.23	2.74

4. Conclusion

In the present work, HDPE/PVA fiber composites have been prepared via melt compounding method using a co-rotating twin screw extruder. The prepared composites samples were then characterized to study the effect of PVA fiber loadings on morphology and melt rheological properties of the composites. Morphological study revealed that all the composites samples showed that the PVA fibers were well blended and embedded in the HDPE matrix. Though it seemed that there were less interactions between the HDPE and PVA fibers. In the melt rheological analysis, the complex viscosity of all the composites samples increased with increasing PVA fiber contents at all angular frequencies. However, the improvement of $|\eta^*|$ was more prominent at lower angular frequencies (e.g. 0.1-10 rad/s) than at higher ones (e.g. 100, 500 rad/s). Additionally, composites samples PVAC-10 and PVAC-20 exhibited a "high shear-thinning" behavior as the angular frequency increased. In other hand, the PVAC-0 and PVAC-5 demonstrated a "low shear-thinning" behaviour.

IOP Conf. Series: Materials Science and Engineering **543** (2019) 012038 doi:10.1088/1757-899X/543/1/012038

References

[1] Low J H, Andenan N, Abdul Rahman W A, Rusman R, Majid R A 2017 Evaluation of rice straw as atural filler for injection molded high density polyethylene bio-composite materials *Chem. Eng. Trans.* **56** 1081-86

[2] Wu S, Ji G, Shen J 2003 Study on ultraviolet irradiation modification of high-density polyethylene and its effect in the compatibility of HDPE/PVA fibre composites, *Mater. Lett.* **57** 2647-50.

[3] Ahmed S F U, Mihashi H 2011 Strain hardening behavior of lightweight hybrid polyvinyl alcohol (PVA) fiber reinforced cement composites, *Mater. Struc.* **44** 1179-91.

[4] Xu B, Toutanji H.A, Gilbert J 2010 Impact resistance of poly(vinyl alcohol) fiber reinforced high-performance organic aggregate cementitious material, Cement and Concrete Research **40** 347-51.

[5] Nematollahi B, Sanjayan J, Ahmed Shaikh FU 2015 Tensile strain hardening behavior of PVA fiber-reinforced engineered geopolymer composite *J. Mater. Civil Eng.* **27** 04015001.

[6] Chafidz A, Latief, F H, Samad U A, Ajbar A, Al-Masry W 2016 Nanoindentation creep, nano-impact, and thermal properties of multiwall carbon nanotubes-polypropylene nanocomposites prepared via melt blending, *Polym-Plast Technol.* **55** 1373-85.

[7] Chafidz A, Rengga W D P, Khan R, Kaavessina M, Almutlaq A M, Almasry W, Ajbar A 2017 Polypropylene/multiwall carbon nanotubes nanocomposites: Nanoindentation, dynamic mechanical, and electrical properties, *J Appl. Polym. Sci.* **134** 45293-316.

[8] Chafidz A, Kaavessina M, Al-Zahrani S, Al-Otaibi, M N 2016 Rheological and mechanical properties of polypropylene/calcium carbonate nanocomposites prepared from masterbatch *J. Thermoplast. Compos.* **29** 593-622.

[9] Choi H J, Kim S G, Hyun Y H, Jhon M S 2001 Preparation and Rheological Characteristics of Solvent-Cast Poly(ethylene oxide)/Montmorillonite Nanocomposites *Macromol. Rapid Comm.* **22** 320-25.

[10] Sorrentino L, Berardini F, Capozzoli, M R, Amitrano S, Iannace S 2009 Nano/micro ternary composites based on PP, nanoclay, and $CaCO_3$, *J Appl. Polym. Sci.* **113** 3360-67.

IOP Conf. Series: Materials Science and Engineering **543** (2019) 012039 doi:10.1088/1757-899X/543/1/012039

Fabrication of porous hydroxyapatite/chitosan composite bodies through dip-coating method

A Fadli, E Saputra, Komalasari and Arosyidin

Chemical Engineering Department, Faculty of Engineering, University of Riau
Jln. HR. Subrantas, Pekanbaru, Riau 28293, Indonesia

E-mail: fadliunri@yahoo.com

Abstract. Porous hydroxyapatite have been attracting considerable attention for bone surgery. Relatively low mechanical strength of porous hydroxyapatite hampers its application for permanent bone implant. The aim of this work was to study the effect of chitosan concentration and time of coating on the physical, chemical and mechanical properties of porous hydroxyapatite/chitosan composites prepared using dip coating technique. Firstly, chitosan powder was dissolved into acetic acid with a concentration of 0.50 %, 0.75 % and 1.00 % (w/v). Subsequently the porous hydroxyapatite bodies were coated with chitosan using dip-coating method at dip time during 30, 45, 60 min. The composites bodies were dried in room temperature for 24 h. The hydroxyapatite/chitosan bodies with compressive strength in the range of 1.52 - 4.05 MPa, porosity of 75 – 78 %, pore size of 22.8 - 26.61 μm, and density of 0.71 - 0.78 g/cm3 were obtained. The increase of chitosan concentration from 0.50 % to 0.75 % can decrease the porosity from 78 % to 76 % and pore size from 28.25 μm become 22.88 μm. The compressive strength of the composites increased with the coating time.

1. Introduction

Patients with bone damage such as osteoporosis increase annually. Women over the age of 50 years and one in between five men over 50 years suffering from osteoporosis [1]. Therefore, the handling of bone damage is very important considering that bone have a role as human body support. Implantation is the replacement of body tissues by using compatible materials [2]. Implant treatment has a high success rate that has been rated as high as 95 to 99% [3]. There are three material groups can be usually used as implant i.e. allograft, xenograft and autograft [4]. But, availability of natural materials is very limited. Therefore, it has been a lot of research to find a synthetic material that compatible of the human body. One of the applied implantation methods is a porous material [5]. The pore size of 100 μm creates the bone tissue grow and connect to the entire surface of the material, in this way the implant serves as a structural bridge or scaffold for bone formation. The advantage of porous implants is that mechanical stability increases with the growth of bone tissue in the material.
Hydroxyapatite [$Ca_{10}(PO4)_6(OH)_2$] a ceramic material that has the physical and chemical similarity to the minerals contained in the bones and teeth of humans so that biocompactible and can be used in orthopedic surgery [6]. Most of the resulting porous hydroxyapatite is still deficient in bone implant applications because the poor compressive strength [5]. To solve the problem, porous hydroxyapatite can be composited with polymer material for use as bone substitutes. Use of polymers as the composite porous hidoroksiapatit aims to become harder and not easily brittle [7]. Polymer that can be used porous hydroxyapatite composite is chitosan. Chitosan contains glucosamine and N-

acetylglucosamine unit 1-4 glycosidic long chain [8]. Chitosan is a polymer that is considered suitable for composites with ceramics because of the nature of the chitosan that is biocompatible, biodegradable, osteoconductive and can reduce the brittle properties of calcium phosphate compounds-chitosan as bone substitution biomaterials. Chitosan can be produce from shrimp waste and the precentages of deacetylation obtained varied between 80.0–81.8 % near the degree of deacetylation of the commercial chitosan [9].
Composite alumina with potassium phosphate and poly vynil alcohol as binder have been done using a coating method that produces alumina ceramics with a porosity level of 65 % and a compressive strength value of 3.0 MPa [5]. In addition, composite hydroxyapatie was produced a scaffold of bovine hydroxyapatite with poly vynil alcohol coating where the optimum compressive strength of 3.2 MPa in 3 % poly vynil alcohol [10]. Also, composite was produced hydroxyapatite/poly-l-lactid acid (PLLA) scaffold were prepared via Thermally Induced Phase Separation (TIPS) by using a ternary PLLA- dioxane-water solution (polymer-solvent-nonsolvent) [11]. The obtained results demonstrated the possibility to analyze the internal distribution of the particles via a non-invasive technique; furthermore a homogeneous hydroxyapatite content and a uniform size distribution in the whole sample was observed, which allows one to assess that the scaffold production technique does not create neither a filler gradient nor a particle sedimentation. In this paper reports the effect of chitosan concentration and coating time on compressive strength, pore size, porosity and density of hydroxyapatite/chitosan body as composite substitution of bone.

2. Materials and methods

2.1. Preparation of porous hydroxyapatite bodies

The foaming protein starch consolidation method was used to produce porous hydroxyapatite bodies [12]. Slurry was made by mixing 24 g of hydroxyapatite (Sigma Aldrich, German), 3 g of starch, 8 g Darvan (R.T Vanderbilt, USA), with 28 g of egg yolk in a glass beaker. The slurry was stirred at 200 rpm for 3 hours. Furthermore, the slurry was poured into a stainless steel mold and heated at 180 ^{O}C for 1 hour. The yolk were removed by burn-out in a furnace at 600 ^{O}C for 2 hours. Then heating was continued up to 1,200 ^{O}C ended by 2 hours holding time at the temperature.

2.2. Process of chitosan coating

Chitosan solution was prepared by dissolving chitosan powder (Sigma Aldrich, German) with concentration 0.50 %, 0.75 %, 1.00 % w/v into 2 % v/v acetic acid solution (Merck, Germany). Then the mixture was stirred at room temperature until a homogeneous solution. Coating process of porous hydroxyapatite with chitosan was conducted using dip-coating technique during 30, 45 and 60 min dip time. Furthermore, porous hydroxyapatite/chitosan composite bodies were dried at room temperature for 24 hours.

2.3. Analysis method

Analysis of morphological of the composite bodies were conducted using SEM (JSM-6390A, Japan). The XRD (PanAnalytical, German) analysis was used to determine the composite crystalline phases whereas the compressive strength was test using universal testing machine. Composite density was measured by comparing the weight with the composite volume and porosity was calculated using the following equation 1.

$$Porosity\,(P_t) = 1 - \frac{\rho_s}{\rho_d} \times 100\% \qquad (1)$$

Where ρ_s is sample density and ρ_d the theoretical density 3.14 g/cm^3.
Pore size distribution of porous bodies were analysed using ImageJ 1.5 j8 application from Wayne Rasband National Institute of Helath USA [13].

IOP Conf. Series: Materials Science and Engineering **543** (2019) 012039 doi:10.1088/1757-899X/543/1/012039

3. Result and discussion

3.1. Characteristics of porous hydroxyapatite bodies

Porous hydroxyapatite/chitosan composite bodies obtained after drying process were performed resulted in samples without significant deformation which indicated good homogeneity of the materials. Figure 1 shows three porous hydroxyapatite/chitosan composite samples with cylindrical shape.

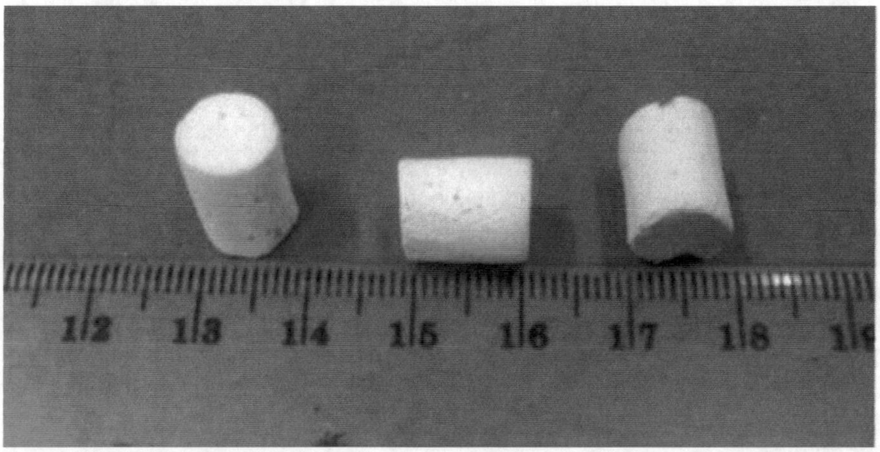

Figure 1. Hydroxyapatite/chitosan composite bodies with cylindrical shape

Coating time and chitosan concentration influence the density, porosity and pore size of the composite bodies. Figure 2 shows the effect of chitosan solution concentration on the density of hydroxyapatite bodies. Density plays an important role in determining the microstructure of the dried composite bodies. Figure 2 shows the density of porous hydroxyapatite before and after coated with chitosan.

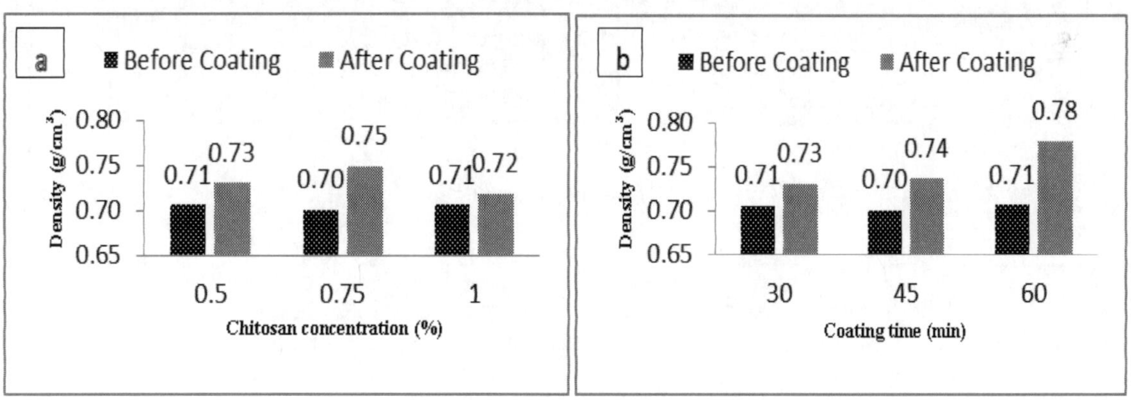

Figure 2. Density of composite bodies versus (a) chitosan concentration and (b) coating time

The density of composite body was found to increase from 0.73 g/cm3 at 0.50 % chitosan concentration to 0.75 g/cm3 at 0.75 % concentration. However, the density decrease again when the concentration increased to 1.00 %. It is believed that high viscosity chitosan solution resulted in thick chitosan layer at surface body of the composite, so that the density also increase. But, when the concentration increase to 1.00 % resulted in higher viscosity too. It will make the chitosan solution can not coat the surface of hydroxyapatite bodies in the pores, moreover addition of chitosan concentration produce high density of bodies. Similarly, longer coating time leads to higher density of the hydroxyapatite bodies and corresponding thicker chitosan layer at hydroxyaptite bodies [5].

IOP Conf. Series: Materials Science and Engineering **543** (2019) 012039 doi:10.1088/1757-899X/543/1/012039

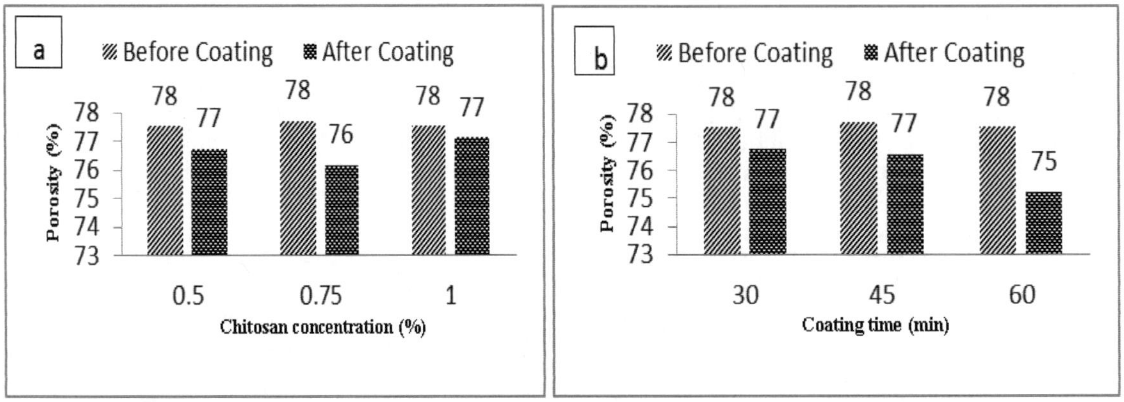

Figure 3. Porosity of hydroxyapatite bodies at different (a) chitosan concentration and (b) coating time

Porosity is considered to have a significant impact on compressive strength of hydroxyapatite bodies. Figure 3 shows the effect of chitosan concentration and coating time of hydroxyapatite bodies porosity. The control of density and porosity not only by the chitosan concentration but also by managing coating time. Porosity of hydroxyapatite before chitosan coated on the body wall surface is 78 %. Increasing coating time can decline the porosity of the composite body. While varied chitosan concentration will make the fluctuation of hydroxyapatite porosity [5]. Physical Porperties of Porous Hydroxyapatite Bodies.

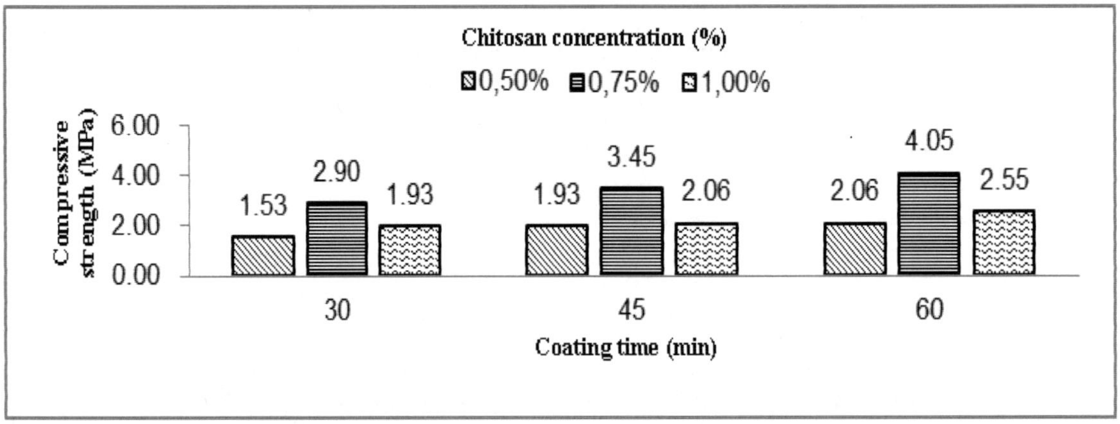

Figure 4: Compressive strength of hydroxyapatite bodies

Many studies have been made to establish a relationship between strength and microstructural properties such as pore structure and pore size distribution. The strength of a porous body is strongly affected by the strength of the body wall and the surface flaws on the wall. Figure 4 shows the effect of chitosan concentration and coating time on compressive strength of the hydroxyapatite/chitosan bodies. The compressive strength value of porous hydroxyapatite before coated is 1.43 MPa and it increase when chitosan coated on the body. The highest compressive strength of the obtained composite is 4.05 MPa at 0.75 % chitosan concentration and 60 min coating time. Compressive strength of composite body at 1.00 % chitosan concentration decreased compared to the 0.75 % concentration. It can be explained that pore size and and pore distribution are the dominant factors in controlling the strength of material mechanisms, so that the chitosan layer distribution can be evenly distributed throughout the pore and surface of hydroxyapatite bodies [5].

IOP Conf. Series: Materials Science and Engineering **543** (2019) 012039 doi:10.1088/1757-899X/543/1/012039

3.2 Micro and macrostructure of hydroxyapatite body

Figure 5 shows the morphology of the hydroxyapatite bodies coated with chitosan 0.50 % and 0.75 %. The figure depict a layer of chitosan seen starting to form but still imperfect in which there is a layer of chitosan ruptures. In contrast to Figure 5b chitosan layer is formed not rupture but distribution still uneven.

Furthermore, from Figure 5c and 5e show the morphology of bodies hydroxyapatite coated with chitosan 0.75 % for 30 min. Analysis result showed that the chitosan layer is formed evenly. Layers are not rupture and can coat all of bodies but the grain size on the bodies still visible. This indicates that the layer of chitosan coat the bodies is not perfect (Figure 5e).

Similarly, Figure 5d and 5f, show chitosan layer are evenly and thick to the hydroxyapatite bodies. In addition, the grain size on the bodies begin invisible (Figure 5f). This indicates the layer of chitosan coated throughout the bodies so, compressive strength of the bodies can achieve great point. The unevenness of chitosan layers can be due to uneven distribution of pore, porosity and density of hydroxyapatite bodies. Other causes are the density and viscosity chitosan solution. The lower density of solution formed the layer of chitosan becomes thinner and if the density solution is higly then chitosan layer formed thicker, higly density makes difficult solution to coat all the pores. And this phenomena causes the chitosan layer is formed unevenly.

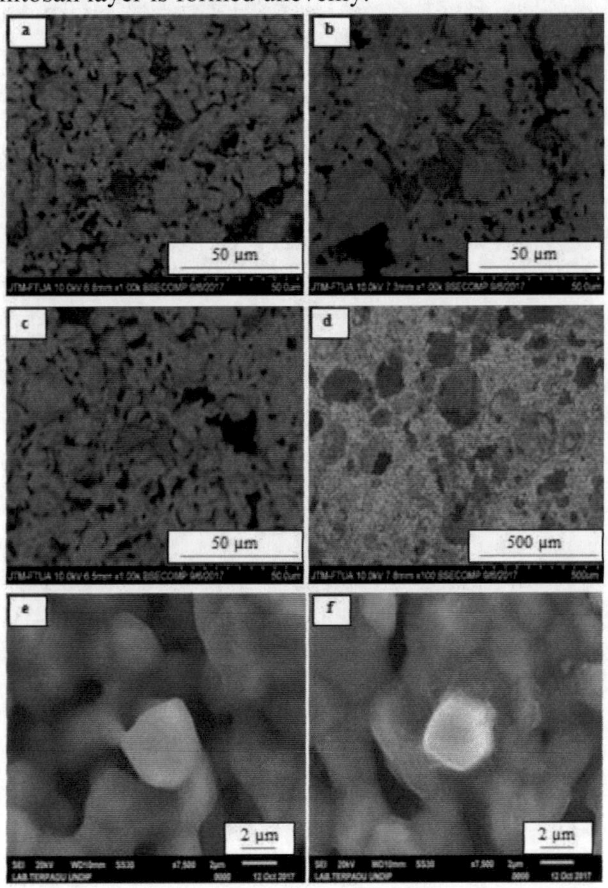

Figure 5. Macro and microstructures of hydroxyapatite/chitosan composite for (a) 0.50 %, 30 min; (b) 0.50 %, 60 min; (c, e) 0.75 %, 30 min; (d, f) 0.75 %, 60 min.

Pore size of hydroxyapatite before composited with chitosan was 28.25 µm. Pore size distribution of hydroxyapatite bodies before the coating was 1 - 20 µm with a pore number of 111 pores. The pore size distribution of obtained composite body was 25.56 - 26.62 µm and 22.88 – 23.14 µm at chitosan concentration of 0.5 % and 0.75 %, respectively.

Table 1. Pore size distribution of composite bodies.

Pore Size (μm)	Number of Pore				
	Before Coating	0.50 % chitosan		0.75 % chitosan	
		30 min	60 min	30 min	60 min
1 – 20	111	106	114	129	242
21 – 100	91	46	69	102	76
> 100	6	6	6	2	9
Total	208	158	189	233	327

3.3. XRD analysis

The chemical composition and crystallinity of hydroxyapatite bodies were investigated using XRD analysis. XRD diffractogram of the composite bodies at various concentration of chitosan and coating time shown in Figure 6. The presence of hydroxyapatite and chitosan is noticeable. The hydroxyapatite peaks is obtained at an angle of 2Θ i.e. 25.8287°; 31.7259°; 32.1370° and 32.8529°, while at 2Θ i.e. 29.3467° and 34.0105° are the chitosan peaks.

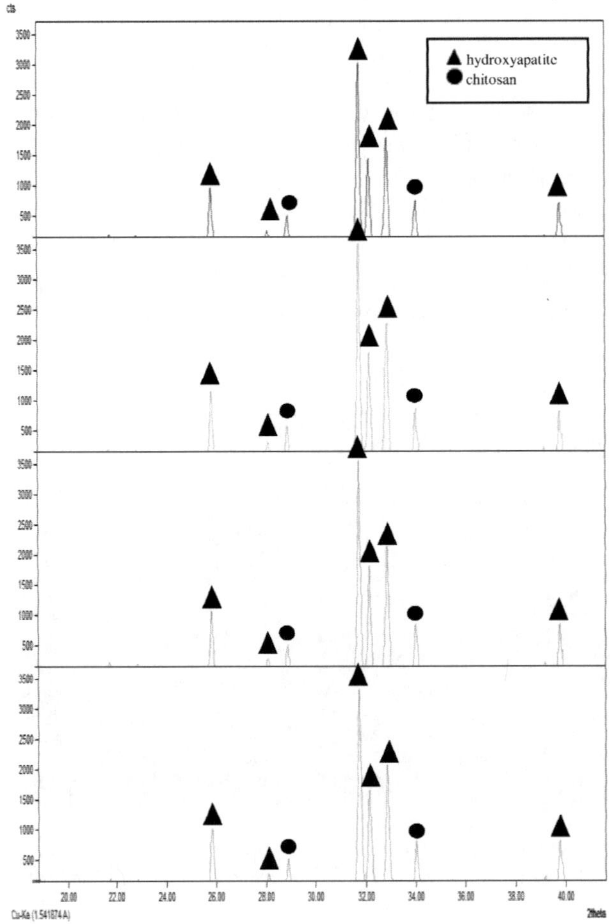

Figure 6. XRD analysis of hydroxyapatite body with chitosan concentration and coating time of (a) 0.50 %, 30 min, (b) 0.50 %, 60 min, (c) 0.75 %, 30 min, (d) 0.75%, 60 min.

Diffractogram at various concentrations of chitosan and coating seen in Figure 6. The results of data processing shows the position of the hydroxyapatite and chitosan peaks. The hydroxyapatite peaks is obtained from figure. 6 a with an angle of 2Θ i.e. 25,8287°; 31,7259°; 32,1370° and 32,8529° with intensity peak 2929,68 cts. figure 6 b shown peaks of hydroxyapatite 25,8420°; 31,7277°; 32,1478° dan 32,8647° with the intensity peak 3743,55 cts. While from the figure 6 c hydroxyapatite peaks formed i.e. 25,8300°; 31,7079°; 32,1288° and 32,8480° with the intensity peak 3478,81 cts. Furthermore, from Figure 6 d peaks of hydroxyapatite formed 25,8398°; 31,7360°; 32,1533° and 32,8719° with the intensity peak 3311,75 cts.
The chitosan peaks from figure 6 a with angle 2Θ i.e. 29.3467° and 34.0105°. Figure 6 b shows that chitosan peaks 29.3623° and 34.0162°. While from figure 6 c chitosan peaks formed i.e. 28.0684° and 33.9957°. Furthermore figure 6 d chitosan peaks formed that is 28.1043° dan 34.0230°.
However, the addition of chitosan concentration from 0.50 % to 0.75 % (Figure 6) and coating time from 30 min to 60 min (Figure 6) are not significantly influence crystallinities of because low amount of chitosan compared to hydroxyapatite content.

4. Conclusion
The porous hydroxyapatite/chitosan composite bodies were successfully prepared via dip coating technique. The control of properties of the composites bodies not only by varying chitosan concentration but also by managing the coating time. As the coating time increased, the compressive strength of the composites increased as well and was found in range of 1.43 MPa to 4.03 MPa. If coating time was increased that made chitosan particle fill in pores but coating time overlong made pore enclosed and uneven. While chitosan concentration affect the porosity and pore size. The porosity of composite decreased with addition of chitosan concentration and coating time with pore size in the range of 22.88 – 28.25 μm.

5. Acknowledgments
Authors are thankful to the Ministry of Research, Technology and Higher Education of the Republic of Indonesia (Kemenristekdikti) for providing financial support to undertake this research work.

6. References
[1] Laurencin C T, Ambrosio A M A, Borden M D, Cooper J A 1999 *Annu. Rev. Biomed. Eng.* **1** 19-46.
[2] Kuo T C, Lee B S, Kang S H, Lin F H, Lin C P 2007 *J. Endod.* **33(4)** 451-54.
[3] Bowen A A, Pascua G M T, Nasimi A 2007 *Med. Oral Patol. Oral Cir. Bucal.* **12** 323-30.
[4] Sopyan I, Mel M, Ramesh S, Khalid K A 2007 *Sci. Technol. Adv. Mater.* **8** 116-23.
[5] Costa H S, Mansu A A P, Pereira M M, Mansu H S 2006 *J. Mater. Sci.* **43** 510-24.
[6] Uchida A, Araki N, Shinto Y, Yoshikawa H, Ono K, Kurisake E 1990 *J. Bone. Joint. Surg. Br.* **72(2)** 298-302.
[7] Taguchi T, Kishida A, Akashi M 1999 *J. Biomat. Sci-Polym. E* **10(8)** 331-39.
[8] Onishi H, Kushitani S, Yasukawa E, Iwaki H, Hench L L, Wilson J, Tsuji E, Sugihara T 1997 *Clin. Orthop. Relat. Res.* **334** 316-25.
[9] Bonfante-Alvarez H, Avila-Montiel G D, Herrera-Barros A, Torrenegra-Alarcon M, Gonzalez-Delgado A D 2018 *Chem. Eng. Trans.* **70** 1969-74.
[10] Alva E T, Punto D, Endang T W 2012 *J. Teknosains* **2** 126-34.
[11] Vitrano I, Carfi Pavia F, Conoscenti G, Lombardo M, La Carrubba V, Brucato V 2018 *Chem. Eng. Trans.* **64** 193-98.
[12] Fadli A, Sopyan I 2009 *Mater. Res. Innovations* **13** 327-29.
[13] Haeri M, Haeri M 2015 *J. Open Source Softw.* **3(1)** 1-4.

The utilization of bagasse fly ash for mesoporous silica synthesis

C W Purnomo[1,2], S K Wirawan[2] and H Hinode[3]
[1] Agrotechnology Innovation Center (PIAT), Universitas Gadjah Mada, Kalitirto, Tanjungtirto, Berbah Sleman, Indonesia.
[2] Chemical Engineering Department, Universitas Gadjah Mada, Jl. Grafika No. 2 Yogyakarta 55281, Indonesia.
[3] Department of Transdisciplinary Science and Engineering, Tokyo Institute of Technology 2-12-1 Ookayama, Meguro, Tokyo 152-8550, Japan.

E-mail : chandra.purnomo@ugm.ac.id

Abstract. Natural silica sources such as from biomass ashes provide an alternative cheap silica precursor for synthesizing advanced materials. Mostly, the silica content in the biomass is in the form of amorphous structure which is suitable to be used directly into the material formation. Bagasse fly ash (BFA) contains carbon and silica in a significant amount. The silica content can be recovered by a simple NaOH fusion method followed by water dilution to produce sodium silicate solution. Then, the obtained solution can be used for various material preparations such as zeolites and mesoporous silica materials. It was observed that the alkaline dissolution method is not only extracting silica but also other content such as Alumina and trace metals. Two different type of mesoporous silica, MCM-41 and SBA-15, have been synthesized by incorporating suitable template and sol-gel hydrothermal method using extracted silica source. The synthesized materials were characterized using TEM and XRD to confirm the morphology and crystal structure. Gas adsorption experiment of n-hexane has been done to examine the adsorption behaviour of the two prepared mesoporous silica. The results suggest that low cost natural silica sources can be an alternative for advanced material preparation.

1. Introduction

Recent trend in material science is utilizing abundant solid waste materials such as biomass [1], manure [2], coal ash [3] and bagasse fly ash [4] to create novel products. In terms of fly ash, there are two major source of this waste which are coal and biomass combustion system. Bagasse fly ash (BFA) is a solid residue from bagasse burning system in sugar industries. The main content inside the ash are silica and carbon which open many possibilities in utilization. The carbon rich particles can be easily separated from silica rich ash fraction by mechanical sieving or floatation method [5]. The carbon fraction then can be used for activated carbon preparation using chemical activation [5] and chemo-physical activation [6] while the silica rich particles can be used for preparing synthetic zeolites [7]. Moreover, a specific zeolite type can be grown on the carbon particles for creating a unique carbon-zeolite composite from BFA [8]. To increase the flexibility and reliability in preparing silica based material, the silica content inside the ash should be extracted to get liquid (supernatant) or powder precursor.

Ordered mesoporous silica is one of novel materials that commonly synthesized from expensive pure silica precursors such as TEOS, fumed silica and water glass. Recently the greener precursors obtained from natural or solid waste resources for mesoporous silica preparation are intensively studied. In the early attempt of using other source of silica than the commercial ones, coal fly ash was the potential option due to its abundant availability [9], high silica content [10] and can be processed by green approach [10]. Up to now, many different silica sources have been used that can be categorized into three large groups. The first group is from agricultural residues such as rice husks [11], wheat straws [12], cane leaves [13], stems [14], bagasse [15] and weeds [16]. The second group is from mineral processing related activities such as mine tailing [17], iron ore tailing [18], clays [19], slag [20], and geothermal sludge [21]. And the third group is from urban waste such as LCD-TFT electronic waste [22] and municipal solid waste residue [23]. The type of prepared mesoporous silica were diverse from highly ordered such as MCM-41, SBA-16, and SBA-15 to less ordered solid structure such as silica gels and amorphous aluminosilicate.

From the three silica source groups, the precursor from agricultural waste is considered lower in the content of impurities (heavy metals), easier extraction process of amorphous natural silica and renewable resources. However in general, the problem of this resource is the collection since the agricultural sites is usually widespread. Unless the agricultural products are brought to a large post harvesting industry such as cane which is collected and transported to cane sugar industries. By this consideration, bagasse fly ash (BFA) side product of bagasse (pressed cane) burning in sugar industry can be a potential candidate for silica based advanced materials preparation in a mass production.

In here, silica extraction using NaOH fusion method has been performed to obtain sodium silicate solution from BFA. This method can ensure a better extraction efficiency of silica content than NaOH dissolution method under room condition [5]. The obtained solution contains high silica content that sufficient as a chemical precursor for preparing highly ordered mesoporous silica i.e. MCM 41 and SBA 15. The usage of this resources for preparing highly ordered materials is still rare to be found.

2. Methods

The Bagasse fly ash was collected from Indonesian sugarcane industry which its properties has been reported before [5]. To extract the silica content, BFA was mixed with sodium hydroxide (1 : 1.2, BFA : NaOH by weight) and heated at 673 K for 1 h to obtain a fused mass using closed container (hydrothermal reactor), which was cooled to room temperature and milled for several minutes. The obtained fused powder was mixed with water in a weight ratio of 1:4 and aged for 1 day at room temperature with agitation. The resultant suspension was then filtered to obtain the supernatant. The concentrations of Si, Al and Na in the supernatant measured by ICP method (Perkin Elmer Optima 8300) are 27,000, 1,700 and 7,000 ppm, respectively.

In a typical MCM-41 synthesis, 0.433 g of $C_{16}H_{33}(CH_3)_3NBr$, 7.5 ml water and 0.35 g aq. NH_4OH were combined with 19 ml of the supernatant and the mixture hydrothermally treated at 97 °C for 3 days. The synthesis pH was adjusted at about10.25 by using 1 M acetic acid solution before hydrothermal treatment. Further details about such hydrothermal apparatus can be found elsewhere [24].

For synthesizing of SBA-15, the surfactant used in the synthesis was triblock polymer $(EO)_{20}(PO)_{70}(EO)_{20}$. SBA-15 was prepared by using 1.8 g of triblock polymer dissolved in 75.0 ml aq. HCl (2 M), and then 20.0 ml of the supernatant was added at room temperature under vigorous stirring. To this reaction mixture, 5.0 ml of 35% HCl and 25.0 ml of distilled water was quickly added. The resulting gel was aged at room temperature for 1 day and subsequently heated at 90 °C for 2 days. All samples were calcined at 550 °C for up to 6 h. Then, MCM-41 and SBA-15 samples were characterized using XRD (Rigaku, Multiplex) for determining the ordered pore structure of the solid, and TEM (JEOL) to observe the pore structure and morphology of the materials, and also Surface Area Analyzer (Quantachrome, Autosorb 1) to understand the porosity of the materials. Beside common characterization methods, the samples adsorption capacity was studied using vapour (n-hexane) adsorption method at room temperature by a self-assembled gravimetric adsorption apparatus.

IOP Conf. Series: Materials Science and Engineering **543** (2019) 012040 doi:10.1088/1757-899X/543/1/012040

3. Result and discussion

The MCM-41 and SBA-15 prepared from silicate solution have been characterized by nitrogen adsorption, SEM and XRD. The results displayed in Figure 1 are nitrogen adsorption isotherm profile and PSD (pore size distribution) of the materials. The capillary condensation of MCM-41 appears at about P/Po =0.35 while for SBA at about P/Po = 0.7 which are typical for those materials [25]. This steep increase and decrease (hysteresis loop type H1) of the adsorbed nitrogen volume indicates the narrow pore size in a mesoporous range. Higher P/Po of the hysteresis loop belongs to the higher pore size materials (SBA 15) as shown also in the PSD (Figure 1.d). The detailed porosity of SBA-15 and MCM-41 is listed in Table 1.

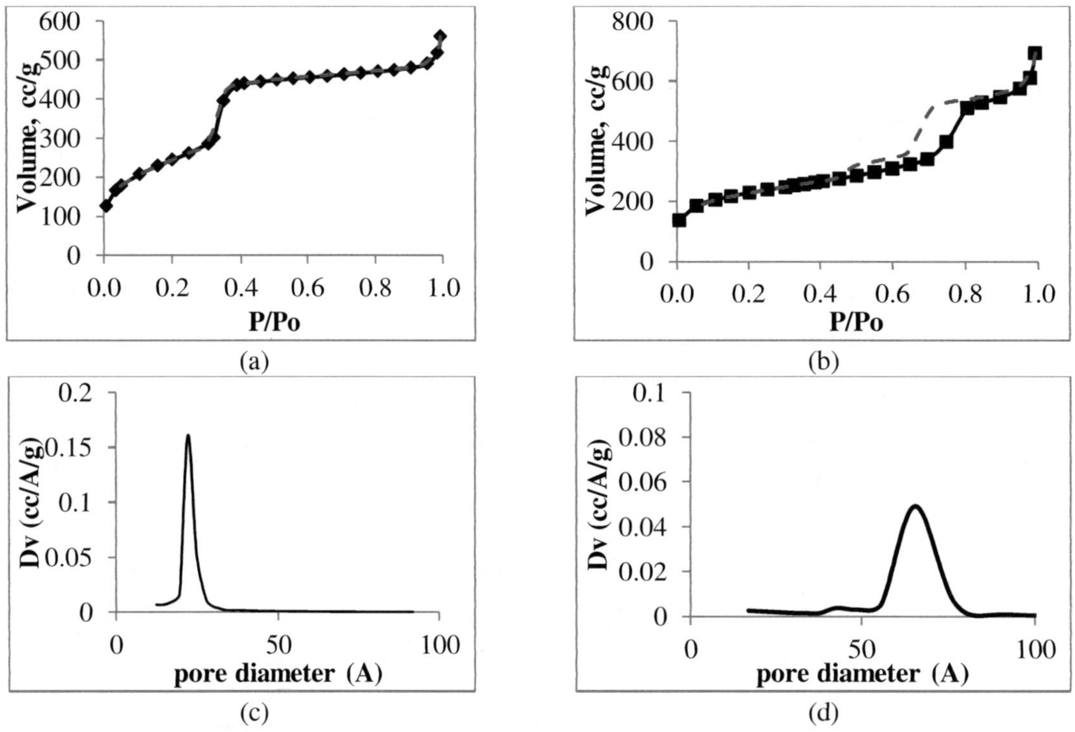

Figure 1. N_2 adsorption (black line) and desorption (red dashed line) isotherms of MCM-41 (a) and SBA-15 (b) and pore size distribution of MCM-41 (c) and SBA-15 (d).

From the TEM image in Figure 2, well ordered pore structure can be observed which confirm the suitability of the preparation method. From the images, it is shown that the pore inside SBA-15 is larger than in MCM-41 which is similar to PSD from the nitrogen adsorption results above. Thicker pore wall of SBA-15 than MCM-41 is also can be observed which suggest the suitability of this material for aqueous application. In brief, all these characterization results are supported each other such as BET pore size distribution and SEM image which shows ordered pore structure with similar pore size.

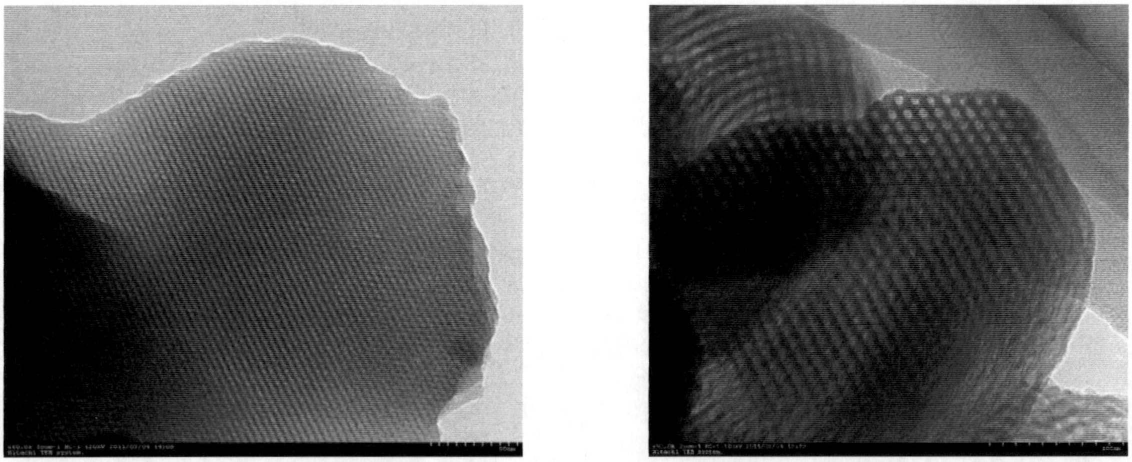

(a) (b)

Figure 2. TEM image of MCM-41 (a) and SBA-15 (b).

The ordered pore structure of mesoporous silica has also been confirmed by XRD analysis as shown by Figure 3. The XRD profile has a good agreement with early report [26]. The pattern basically consists of a single intensive peak at very low angle and followed by weak peaks at higher 2θ degree. These peaks can be indexed on a hexagonal unit cell.

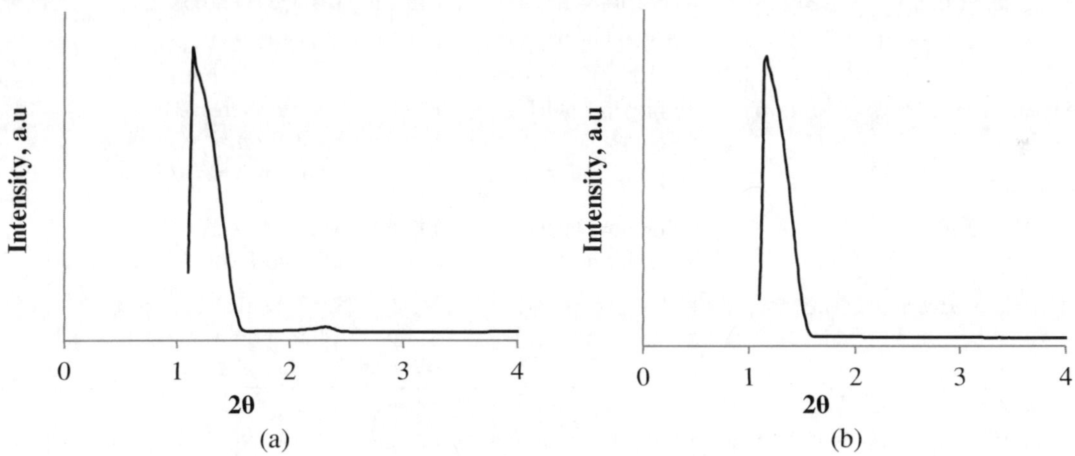

(a) (b)

Figure 3. XRD pattern of MCM 41(a) and SBA 15 (b).

The porosity data of mesoporous silica is provided in Table 1. From the data, high surface area materials have been successfully obtained with most of the pore volume is in the mesopore range. It is indicated that the sodium silicate extract from BFA has reasonable purity and contain no inhibitory substance for mesoporous silica formation.

IOP Conf. Series: Materials Science and Engineering **543** (2019) 012040 doi:10.1088/1757-899X/543/1/012040

Table 1. Porosity of synthesized mesoporous silica material.

Sample	Vol.microporous (cc/g)	Porosity Vol. Mesoporous (cc/g)	Surface area BET (m^2/g)
MCM-41	0.30	0.56	885
SBA-15	0.37	0.50	775

In the literature, some synthesis schemes variation have been implemented using several agricultural waste as the silica precursor. A summary of the experiments is provided in Table 2. From the sample's porosity point of view, it is indicated the most residue are suitable to be used for silica precursor. The results in this study is also comparable with the high porosity samples in the table.

Table 2. Previous synthesis of mesoporous silica results from agricultural residues.

Raw materials	Silica type	Extraction route	Template	Optimum properties	Ref.
Rice husk	na	Boiled in NaOH solution	lignin	surface area: 471.7 m^2/g	[11]
Wheat straw ash	MCM 41	Boiled in NaOH solution	CTAB	Surface area: 1311.6 m^2/g total pore vol.: 1.053 cc/g	[12]
Sugarcane leaf ash	SBA-15	Sonication in heated NaOH solution	P123	surface area: 632 m^2/g. total pore vol.: 1.232 cc/g	[13]
Stem cane ash (SCA)	SBA 16	Boiling in NaOH solution and crystallization	F127	surface area: 316.5 m^2/g total pore vol.:: 0.3 cc/g	[14]
Bagasse ash	na	Boiling in NaOH solution	PEG	surface area: 656 m^2/g total pore vol.: 0.8 cc/g	[15]
Sedge (Carex riparia) ash	MCM 41	Acid refluxing and leaching	CTAB	Surface area: 1174 m^2/g total pore vol.: 0.98 cc/g	[16]

Vapour isotherm curves of the two mesoporous silica are provided in Figure 4 which are typical of type IV adsorption–desorption isotherms. The vapour uptake process can be described as several steps which are monolayer formation at low relative pressure, then multilayer formation followed by capillary condensation of mesopore filling indicated by high increase of adsorbed vapour uptake and then external surface adsorption represented by very low uptake to the highest achievable partial pressure. In terms of hysteresis loop, the phenomenon is observable only for SBA-15 while for MCM-41 the adsorption and desorption is almost reversible. This can be a clue of larger pore sixe existence in SBA-15. At low partial pressure (less than 0.5) the uptake capacity of MCM-41 is superior to SBA-15, this could be due to small pore size of this material which attract hexane molecule stronger than lager pore size. However at higher pressure, the larger pore size in SBA 15 can be filled with more hexane molecule thus the adsorption capacity surpassing the capacity of MCM-41.

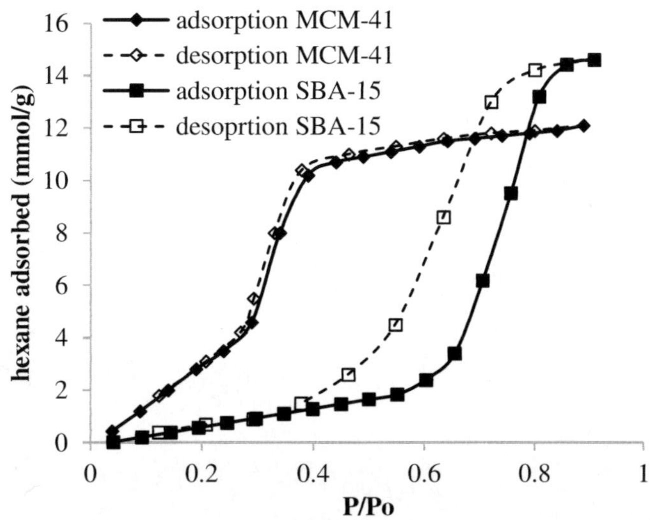

Figure 4. N-hexane adsorption-desorption isotherm of MCM-41 and SBA-15.

4. Conclusions

High ordered mesoporous silica materials have been successfully prepared from alkali fusion supernatant of BFA. The prepared samples exhibited highly ordered hexagonal structure, high surface area and narrow pore size distributions. The successful attempt indicated that the silicate extract is suitable to be used as the precursor. The other impurities such as aluminate and trace metal elements inside the supernatant are not hindering the material synthesis. This finding opens a wide possibility in utilizing waste to produce advanced material such as catalysts and adsorbents.

5. References

[1] Oktaviananda C, Rahmawati R F, Prasetya A, Purnomo C W, Yuliansyah A T and Cahyono R B 2017 AIP Conf. Proc. **1823** 020029
[2] Purnomo C W, Indarti S, Wulandari C, Hinode H and Nakasaki K 2017 *Chem. Eng. Trans.* **56** 1531–36
[3] Rafieizonooz M, Salim M R, Hussin M H, Mirza J, Yunus S M and Khankhaje E 2017 *Chem. Eng. Trans.* **56** 439–44
[4] Purnomo C W, Respito A, Sitanggang E P and Mulyono P 2018 *Environ. Techol. Innovation* **10** 275 – 80
[5] Purnomo C W, Salim C and Hinode H 2011 *J. Anal. App. Pyrol.* **91** 257–62
[6] Purnomo C W, Salim C and Hinode H 2012 *Fuel Process. Techol.* **102** 132–39
[7] Purnomo C W, Salim C and Hinode H 2012 *Microporous Mesoporous Mat.* **162** 6–13
[8] Purnomo C W 2013 *J. Porous Mat.* **20** 1305–13
[9] Misran H, Singh R, Begum S, Yarmo M A 2007 *J. Mat. Process. Technol.* **186** 8–13
[10] Hui K S and Chao C Y H 2006 *J. Hazard. Mat.* B **137** 1135–48
[11] Qu Y, Tian Y, Zou B, Zhang J, Zheng Y, Wang L, Li Y, Rong C and Wang Z 2010 *Bioresource Technol.* **101** 8402–05
[12] Ma Y, Chen H, Shia Y and Yuan S 2016 *Mat. Res. Bull.* **77** 258–64
[13] Arumugam A and Ponnusami V 2015 *J. Sol-Gel Sci. Technol.* **74** 132–42
[14] Azizi S N, Ghasemi S and Yazdani-Sheldarrei H 2013 *Int. J. Hydrogen Energy* **38** 12774-85
[15] Rahman N A, Widhiana I, Juliastuti S R and Setyawan H 2015 *Colloids Surf. A* **476** 1–7
[16] Ghorbani F, Younesi H, Mehraban Z, Celik M S, Ghoreyshi A A and Anbia M 2013 *J. Taiwan Inst. Chem. Eng.* **44** 821-28
[17] Kim S, Han Y, Park J and Park J 2015 *Int. J. Mineral Process.* **140** 88–94

[18] Yang G, Deng Y, Ding H, Lin Z, Shao Y and Wang Y 2015 *Appl. Clay Sci.* **111** 61–66
[19] Qoniah I, Prasetyoko D, Bahruji H, Triwahyono S, Jalil A A, Suprapto, Hartati and Purbaningtias T E 2015 *Appl. Clay Sci.* **118** 290–94
[20] Li C, Qiao X and Yu J 2016 *Mat. Lett.* **167** 246–49
[21] Muljani S, Setyawan H, Wibawa G and Altway A 2014 *Adv. Powder Technol.* **25** 1593–99
[22] Lin L Y and Bai H 2013 *Microporous Mesoporous Mat.* **170** 266–73
[23] Liu Z S, Li W K and Huang C Y 2014 *Waste Management* **34** 893–900
[24] Purnomo C W, Castello D and Fiori L 2018b *Appl. Sci.* **8** 331–47
[25] Zhao D, Huo Q, Feng J, Chmelka B F and Stucky G D 1998 *J. Am. Chem. Soc.* **120** 6024–36
[26] Kresge C T, Leonowicz M E, Roth W J, Vartuli J C and Beck J S 1992 *Nature* **359** 710–12

Mixing Solvent's Effect to the Characterization and Performance Cellulose Acetate/Polyethylene Glycol Membrane

S Nurkhamidah*, Y Rahmawati[1], F Taufany, Y E Fahrizal and I H Ramadhan

Department of Chemical Engineering, Institut Teknologi Sepuluh Nopember, Kampus ITS Sukolilo, Surabaya 60111, Indonesia

*E-mail: dst_eureka@yahoo.co.uk

Abstract. The need for clean water continues to increase. Therefore, it is necessary to find an alternative for the get of clean water. One way is desalination process using membrane. Cellulose Acetate/Polyethylene Glycol (CA / PEG) membrane is one type of membrane that can be used for desalination process. Several studies on the manufacture of CA / PEG membranes have been performed, but have not yet produced a good performance membrane. Therefore in this research was synthesize CA/PEG/ mixing solvent membrane using *Acetone* – THF/ *Acetone* – DMF as the solvents at ratio 100-0, 90-10, 70-30, 50-50 (%volume) each and PEG as the additive vary PEG 200 and 400. CA/PEG/ mixing solvent membrane characteristics are tested. The results showed that the hydrophilicity of *mixing solvent* method using *Acetone-DMF* (50:50) increase water content up to 375.86% and *Acetone-THF* (50:50)'s water content increase up to 177.10% with additive PEG 200, and for PEG 400 as additive showed that the hydrophilicity of *mixing solvent* method using *Acetone-DMF* (50:50) increase water content up to 503,86% and *Acetone-THF* (50:50)'s water content decrease from 173.28 % till 119.23%, as THF content increase.

1. Introduction

On the earth surface consist of 2/3 sea water and 1/3 land. All water resources in the world not consumeable, just 3 % are consumeable and 97 % not consumeable such as seawater, 3 % of it consist of 2 % in glasier form and 1 % use for potable water. UNESCO at world water development report state that water in the world are used for household need, industrial need and agricultural, from All of this fact it can conclude that potable water resource for human life so crucial. Water needed in 2050 was increase till 55 %, from ths prediction will be human motivation for develop and researching water purification technology and fulfil human need for fresh water [1].

One factor caused water not consumeable is total dissolve solid value exceed the limit of consumeable water. One of technology that develop for separate water from impurities as salt was membrane desalination method. Along with the different polymeric materials that are used for the preparation of phase-inversion membranes. Membranes have gained an important consign in chemical technology and have been used in a wide range of applications, such as the production of high-quality water, removal or recovery of toxic or valuable components from various industrial effluents, and applications in the food and pharmaceutical industries. With the advent of membrane technology,

separation, concentration, and purification have become industrially viable unit operations because of the high efficiency of separation, low energy of operation, simplicity of operation with modern compact modules [1].
Cellulose acetate (CA) is very widespread, with characteristics such as good toughness, high biocompatibility, good desalting, high potential flux, and relatively low cost. Thus, it has been widely used for reverse osmosis, microfiltration, and gas separation, Also, CA membranes have excellent hydrophilicity, which is very essential in the minimization of fouling [2].

2. Materials and Methods

2.1. Material

The materials used in this experiment was Cellulose acetate (CA, *Mw* of 30,000 Da) polymers were purchased from Sigma Aldrich. The three solvents, namely, Tetrahidrofuran (THF) and acetone (both with an analytical purity of 99 %); N, N-Dimethyl formamide (DMF) with 99.5 % purity; polyethylene glycol 400 were all obtained from Sigma Aldrich. The coagulation bath used throughout this experiment was DI water purchased in Chemical Store klampis. NaCl with *Mw* of 58,5 from Sigma Aldrich.

2.2. Membrane preparation

Different concentrations of CA/PEG/Acetone: THF and CA/PEG/Acetone: DMF solutions with constant ratio of CA/PEG 200 (20% wt / 80% wt) and with constant ratio of CA/PEG 200 (20% wt / 80% wt) were prepared. The compositions of all prepared membranes have been presented in **Table 1**. All the membranes were prepared by the casting solution technique and the phase inversion method. For preparation of membranes, at first, homogenous solutions were prepared in Acetone: THF with different ratios (100:0, 90:10, 70:30 and 50:50) by stirrer at 3000 rpm for 12 h. The obtained homogeneous solutions were casted onto glass plates by using glass stck with constant thickness of 200 microns. Then, prepared casting films were dipped into deionized water as non-solvent for 15 minutes. In this step exchanging of solvent and non-solvent leads to membrane formation. After completing phase separation step and finishing membrane formation, the membranes were kept in two paper sheets at room temperature (25± 2°C) for one day to remove water in the membrane. The preparation of the membrane adopted from [3].

Table 1. Membrane code

Membrane with additive PEG 200	*Solvent (%v : %v)*	*Membrane with additive PEG 400*	*Solvent (%v : %v)*
M1	Acetone (100)	M8	Acetone (100)
M2	Acetone : DMF (90 : 10)	M9	Acetone : DMF (90 : 10)
M3	Acetone : DMF (70 : 30)	M10	Acetone : DMF (70 : 30)
M4	Acetone : DMF (50 : 50)	M11	Acetone : DMF (50 : 50)
M5	Acetone : THF (90 :10)	M12	Acetone : THF (90 :10)
M6	Acetone : THF (70 : 30)	M13	Acetone : THF (70 : 30)
M7	Acetone : THF (50 : 50)	M14	Acetone : THF (50 : 50)

2.3. Water content

Water content of the membranes was obtained after soaking membranes in water for 24 h and the membranes were weighed followed by mopping it with blotting paper. The wet membranes were placed in oven at 75 °C for 48 h and the dry weights of the membranes were determined [3]. The percent of water content (WC) was calculated using the equation :

$$water\ content = \left(\frac{Wwet - Wdry}{Wdry}\right) x100\% \qquad (1)$$

The term "water cotent" is water that can be filling in void volume of the membrane

2.4. Scanning electron microscopy (SEM)

Membrane were snapped under liquid nitrogen using pinset to give a generally consistent and clean cut. The membranes were then sputter-coated with thin film of gold. The membranes were mounted on brass plates with double-carbon tape in a lateral position. Cross-sectional images of the membranes were obtained. SEM analysis measured by Bruker EDS made in Germany.

2.5. Fourier Transform Infrared (FTiR)

In order to know OH bonding in the various membrane with different concentration of mixing solvent and PEG as additive, FTiR spectra of all of the sample was obtained by spectrophotometer. Range for Fourier Transform Infrared test was 0 cm^{-1} untill 4000 cm^{-1}. The type of bonding observed in this experiment was OH bonding stretching mode, in peak between 3600 cm^{-1} till 3800 cm^{-1}, peak area measured by ImageJ.

3. Results and Discussion

3.1. Scanning electron microscopy (SEM)

Miscibility or solubility parameter between solvent and nonsolvent is one of the most important parameters in membrane formation. If miscibility between solvent and non-solvent is high (difference of solubility parameter between solvent and non-solvent is low), the non-solvent is more able to diffuse in casting film and in this case, solvent leaves casting solution faster. Slower diffusion rate lead to membrane finger-like formation [4] Therefore exchanging of solvent and non-solvent in coagulation bath happens faster and consequently, instantaneous demixing happens [3]. This leads to formation of porous fracture surface. The values of solubility parameters (in this paper it define with RHSP value) of acetone/THF mixture and acetone/DMF mixture with different ratio of acetone :THF and acetone : DMF as mixing solvents and water as non-solvent have been shown in **Table 2** and **Table 3**. SEM image of cross sectional area in Figure 1 shows that CA/PEG membrane with different mixing solvent ratio exhibits finger-like structure with different pore size.

Table 2. Types of solubility of all component

Component	*δp(disperse)*	*δh (hidrogen)*	*δp (polar)*	*Reff*
THF	5.45	8	6.1	[3]
Acetone	15.5	7	10.4	[3]
DMF	17.4	11.3	13.7	[6]
Cellulose	7.9	3.5	6.3	[6]
Water	15.6	16	42.3	[6]

The value of RHSP s-p will decrease as THF in mixing solvent ratio increase, this lead to more better interaction as THF ratio increase in the solvent, then it made longer phase separation process. The value can make this prediction because This calculation related to cohesive energy density of energy that required for electron movement from functional group of polymer to solvent [5]. The effect of DMF was opposite as DMF ratio increase RHSP (S-W) ratio was increased, this result lead to solvent difuse more easly in water in coagulant bath, then it lead to increasement of membrane porosity. For membrane CA/PEG 200 mixing solvent acetone : THF and acetone : DMF with ratio of 70 : 30 has the same symmetric membrane morphology, it can be due to very fast phase inversion caused by very thin membrane thickness.

Table 3. RHSP (s-p) and RHSP (s-w) of mixing solvent

Membrane	*RHSP (s-w)*	*RHSP (p-s)*	*Reff*
M1,M8	34.4	12.7	[6]
M2,M9	32.7	16.6	[6]
M3,M10	31.8	17.7	[6]
M4,M11	31.06	18.9	[6]
M5,M12	33.6	14.1	[6]
M6,M13	34.8	10.3	[6]
M7,M14	36.5	6.8	[6]

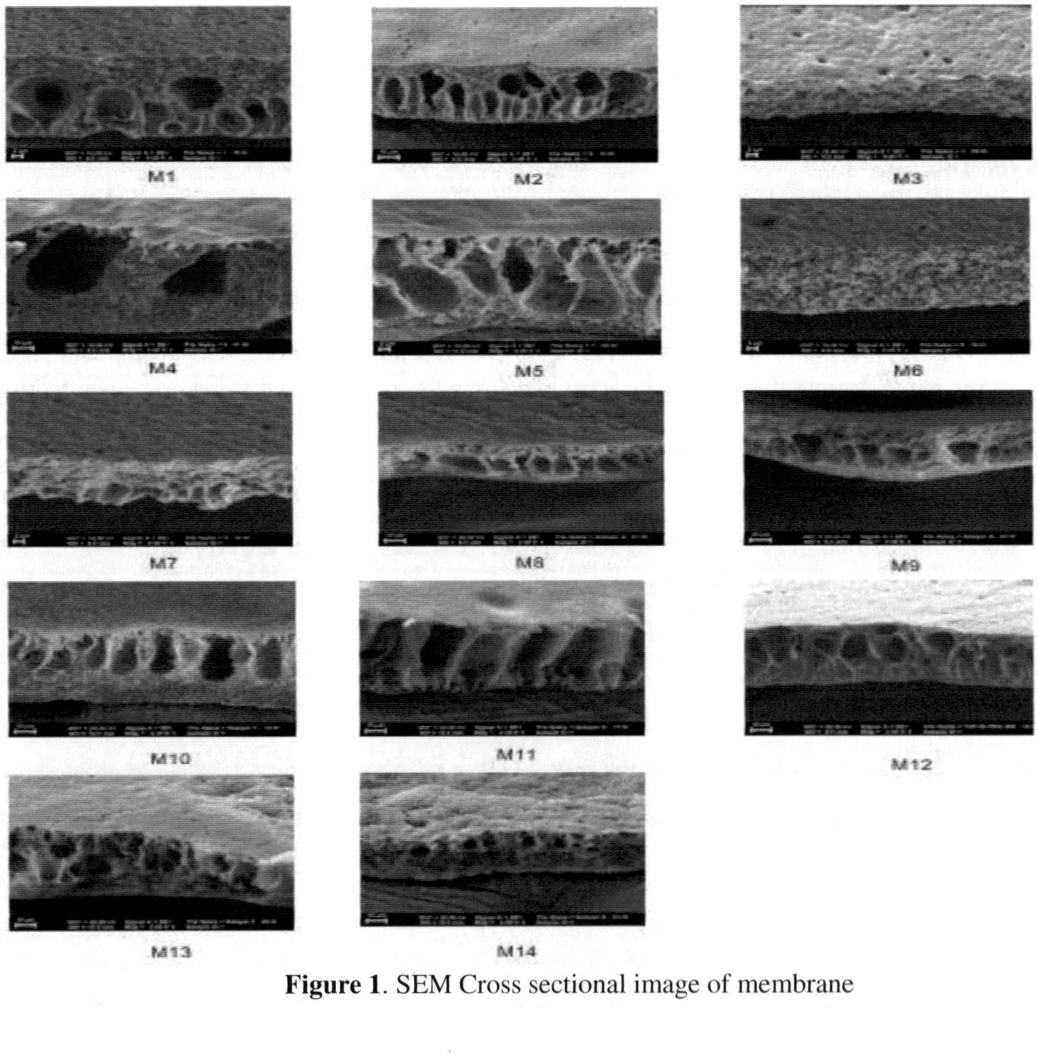

Figure 1. SEM Cross sectional image of membrane

3.2. Water content

The Water Content is related to the membrane hydrophilicity property . Water content of each membrane is calculated using **Eq. (1).** As **Table 4** shows, Water Content for membrane by using pure acetone is equal to 113.18 %, whereas it for the membrane with solvent ratio acetone : DMF (50 : 50) is equal to 375.86 %. It is obvious that water content is increased by increasing the weight percent of DMF amount in Ratio solvent, it is due to increasing finger – like pore size of membrane, as DMF solvent ratio increase. This result can increase porosity of membrane, the caused is phase inversion between bimodal and spinodal curve and result regular pore in membrane selection [7], the result of membrane with PEG 400 additive has the same pattern as membrane with PEG 200 additive. But membrane with PEG 400 as additive has more porous than membrane that use PEG 200 as additive its due to PEG 400 has more glycol content than PEG 200 glycol can be use for cutter the chain C-C in cellulose [8].

Table 4. Showed The Water Content Value of Mixing Solvent Membrane

Membrane with additive PEG 200	*Water content (%)*	*Membrane with additive PEG 400*	*Water content (%)*
M1	113.1	M8	198.7
M2	142.8	M9	190.6
M3	269.4	M10	380.6
M4	375.8	M11	468.6
M5	75.34	M12	173.2
M6	90.3	M13	162.5
M7	177.1	M14	119.2

3.3. FTIR

FTIR result shows the increasing of –OH group with the addition of THF and DMF up to 30 % volume for membrane with PEG 400 as additive and the increasing of it contents as shown by the increasing of area in the wavenumber of 3300 cm-1 as shown in the **Figure 2** and **Figure 3**. The increasing of –OH group up to 30% volume and decrease at 50 % volume of DMF as well as THF due to . This result shows that hydroxyl group increases up to the addition of THF and DMF content up to 30% volume solvent. The increasing of hydroxyl group in addition is caused by auto-oxidation of THF in the air [9], this reaction can lead to enrichment of O-H bonding in membrane and for DMF, the functional group in DMF have possibility in forming hydrogen bond with another DMF ring. And O-H in different chain in Cellulose acetate. **Figure 2** show the result of FTiR test. The trend of hydroxyl grup in mixing solvent between THF : Acetone has similar result both PEG 200 and PEG 400 as additive. But for PEG 400 the increment not as wide as PEG 400, this can be due to formation of new hydrogen bonds between O–H groups of CA and the C=O groups of PEG [6], this can make PEG 400 has better interaction than PEG 200. The most abundant amount of hydroxyl grup in PEG 200 addition was membrane with 30 % volume, this is due to formation of hydroxyl group in THF vapour phase with oxygen. The values of OH area of FTiR image are shown in **Table 5**. The product contain superoxide, and have free radical properties. In final solution, the product has condense and reacted in free branch of cellulose acetate [9].

Table 5. The OH Area Value of Mixing Solvent Membrane

Membrane with additive PEG 200	*OH Area (cm^2)*	*Membrane with additive PEG 400*	*OH Area (cm^2)*
M1	1.46	M8	1.86
M2	1.66	M9	1.45

M3	1.29	M10	1.37
M4	1.67	M11	1.23
M5	1.32	M12	1.50
M6	2.02	M13	1.85
M7	1.32	M14	1.25

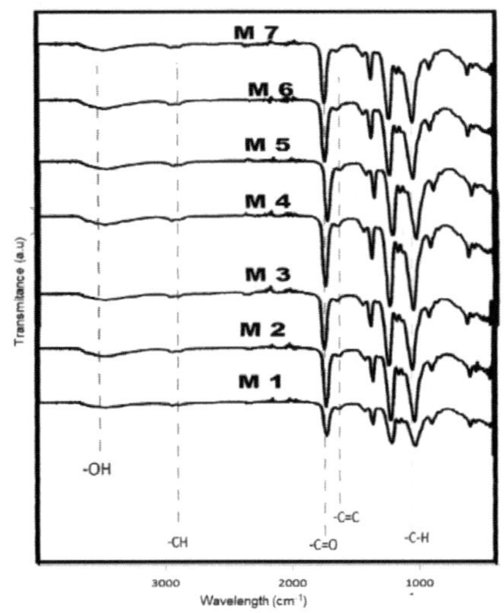

Figure 2 . FTiR graph for CA/PEG200/ Acetone : DMF and Acetone : THF

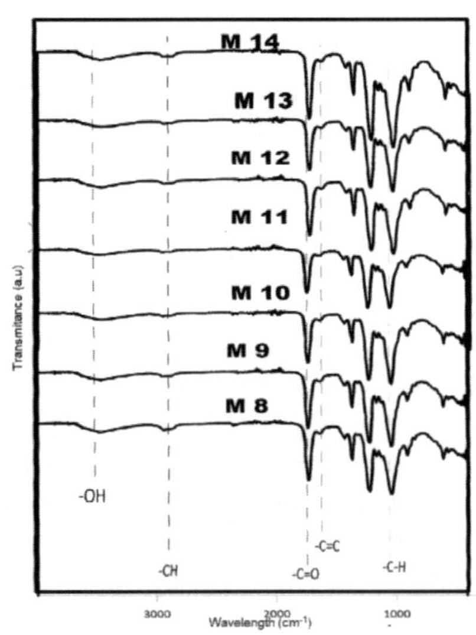

Figure 3 . FTiR graph for CA/PEG400/Acetone : DMF and Acetone : THF

4. Conclusion

In this paper, CA nanofiltration membrane with low concentration of polymer (20% v: 80% v solvent) was prepared with pure acetone and different ratios of Acetone :THF and acetone : DMF as mixing solvents (90:10, 70:30 and 50:50) to investigate the effect of addition of THF with low micibility to non-solvent (deionized wateras coagulation bath). The results of experiments confirm that addition of THF to casting solution leads to decreasing of water content and porosity of the membrane due to decreasing of RHSP (p-s) and lead to increasing phase separation time of casting film in coagulation bath. The optimum ratio of Acetone :THF was found to be 50:50 for additive PEG 200 and 90 : 10 for PEG 400 in terms of water content. Then optimum ratio of Acetone :DMF was found to be 50:50 for both additive PEG 200 and 400 in terms of water content. For optimum hydroxyl grup based FTiR The optimum ratio of Acetone :THF was found to be 70:30, Acetone : DMF 50:50 for additive PEG 200 and Acetone ; DMF 50 : 50 and Acetone : THF 70 :30 for PEG 400

Acknowledgments
The authors gratefully acknowledge Directorate of Research and Community Service Directorate General of Research and Development Strengthening, Ministry of Research, Technology and Higher Education in accordance with the Research Funding Agreement Budget Year 2018 with contract number: 926/PKS/ITS/2018 for the financial support during this research.

References

[1] R W Baker 2004 *Memb. Technol. Appl.*, 2nd edition (Chichester : John Wiley & Sons Ltd) p 14
[2] R. Abedini, S M Mousavi, and R Aminzadeh 2011 *Desal.* **277** 40-45
[3] E Bagheripour, A R Moghadassi, S M Hosseini 2014 *Arab. J. of Chem.* DOI : 10.1016
[4] Jing Ying Chuah, Kok Chung Chong, Soon Onn Laia, Woei Jye Laub, Sze Shin Lee, and Hang Meng Ong 2018 *Chem. Eng. Trans.* **63** 697-7702
[5] Lucia Baldino, Stefano Cardea, and Ernesto Reverchon, 2017 *Chem. Eng. Trans.* **57** 1873-1878
[6] Muddassir Ali, Muhammad Zafar, Tahir Jamil, and Muhammad Taqi Zahid. Butt 2011 *Desal.* **270** 98-104
[7] Mulder M 1996 *B. Princip. Memb. Technol.* (dotchreckt : Cuwer Academic Publisher) p 112
[8] Rajai H Atalla 2010 *Cellulose Sci. Int.* (USA:Elsevier) p 552
[9] Hiroshi Matsubara, and Syouhei Suzuki. 2015 *Org. Biomol. Chem.* DOI: 10.1039/c5ob00012b
[10] Kibrom Alebel Gebru, and Chandan Das 2017 DOI : 10.1016/j.cjche.2016.11.017

IOP Conf. Series: Materials Science and Engineering **543** (2019) 012042 doi:10.1088/1757-899X/543/1/012042

Optimization of LNG Logistics System to Meet Gas Supply at Gresik LNG Receiving Terminal

M Raharjo and M Sudibandriyo*
Chemical Engineering Department, Faculty of Engineering, Universitas Indonesia, Depok 16424, Indonesia

* E-mail : msudib@che.ui.ac.id

Abstract. The construction of Gresik LNG receiving terminal is aimed to fulfil the 109 MMSCFD (0.87 MTPA) gas requirement for combined-cycle power plant (PLTGU) Jawa-3 which is integrated with LNG Gresik Terminal. There are several potential LNG plants in Indonesia and abroad that can be the source of gas for Gresik LNG Terminal. Each of these LNG plants has varying gas price and distance to Gresik. The modelling of the LNG logistics system for Gresik LNG Terminal was built to illustrate the LNG supply chain from the LNG plants to Gresik LNG Terminal and several supply scenarios will be proposed. The built model is linear and optimized by using linear programming. Linear programming involves the determination of objective functions, decision variables and constraints. The optimization of the logistics system aims to obtain the cheapest gas supply cost for Gresik LNG receiving terminal. The result shows that the cheapest gas supply cost is obtained through direct supply scenario with gas source from domestic and abroad with combination of delivery from Bontang equal to 0.34 MTPA (40 %) with 12 shipments, Tangguh equal to 0.26 MTPA (30 %) with 9 shipments, and Bintulu Malaysia equal to 0.26 MTPA (30 %) with 9 shipments.

1. Introduction

The installed capacity of power plants in Java-Bali electrical system by the end of 2015 is 33,825 MW consisting of various types of power plants. Gas-fired generation composition in 2015 is 24.8 % and in 2025 it is projected to be 29.4 %. One of the gas-fired power plant development projects included in the Power Plant Development Project Plan (RUPTL) 2016-2025 as well as the 35,000 MW Project is combined-cycle power plant (PLTGU) Jawa-3 with a capacity of 800 MW which is located in Gresik East Java [1]. Gas for power plant fuel is taken from LNG terminal which is integrated with PLTGU Jawa-3. The LNG source for Gresik LNG Terminal is obtained from various LNG plants which are located around Gresik. LNG is transported from an LNG plant to Gresik LNG Terminal using a tanker. Several scenarios for the selection of LNG supply sources and optimization of the logistics system are required to obtain the most optimal results in terms of technical and economic. Most existing supply chain design and optimization studies rely on centralized models [2]. All the decisions can be implemented successfully to pursue a single objective for gas supply chain optimization [3]. The linear programming approach involving multiple stakeholders with uncertainties is also successfully applied for natural gas supply chain optimization [4]. Based on these results, the linear programming is an appropriate approach for optimizing the LNG supply chain in our study.

IOP Conf. Series: Materials Science and Engineering **543** (2019) 012042 doi:10.1088/1757-899X/543/1/012042

2. Methods

The steps taken in conducting research on logistics system optimization to fulfil gas supply in Gresik LNG Terminal are: data collection, system boundary determination, logistic supply chain calculation, objective function determination, decision variable, constraints and optimization calculation using Solver from Microsoft Excel software.

2.1. LNG carrier selection

The volume gas required for power generation in Terminal Gresik is 109 MMSCFD or equivalent to 0.87 MTPA of LNG. The LNG carrier selection that will be used to send LNG from the liquefaction plant to the LNG Receiving Terminal is influenced by the condition of the port waterway where the terminal is located. The influential factors are the depth of the waters where the ship is leaning and the ship maneuvering area available in the waterways. The Gresik LNG Receiving Terminal is located at the APBS (West Surabaya Watershed) which has a water depth of 13 m and a turning radius of 900 m [5], so that the LNG carrier used should not have a draft exceeding 13 m and a rotating radius of the ship is not exceeding 900 m.

2.2. Logistics chain modelling

In this study the LNG supply chain under review is from a liquefaction plant, transported by an LNG carrier and an LNG receiving terminal as shown in Figure 1 below.

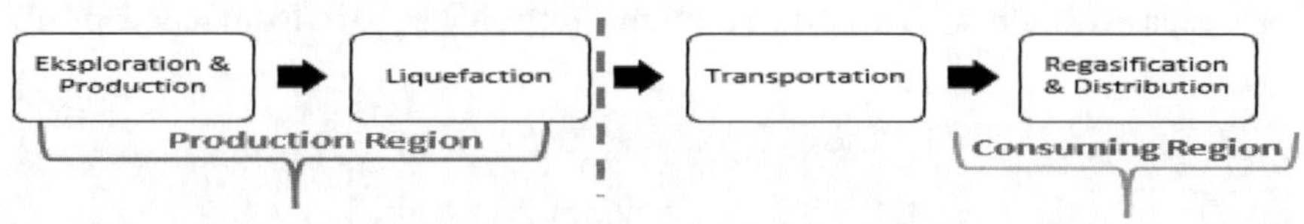

Figure 1. LNG supply chain model

The economic model or LNG project structure under consideration is the merchant model with the LNG supply chain cost calculated starting from the LNG price at the liquefaction plant where the upstream gas exploration and production costs as well as the liquefaction costs of the gas are considered as the cost of gas supply from the production region to the point at the nearest port of the liquefaction plant. Furthermore, transportation costs are calculated on the basis of the chartering a vessel scheme and the subsequent chain cost of LNG regasification at LNG Gresik receiving terminal and gas distribution to PLTGU Jawa-3.

2.2.1. Calculation from production region cost

In general the LNG price, production region cost, C_P is formulated by equation 1 below [6]. LNG price is FOB (free on board) price.

$$C_P = A.P_O + B \tag{1}$$

For domestic gas purchase contracts, P_O refers to ICP (Indonesian crude price), while for LNG purchase contracts in Asia and Australia generally refers to JCC (Japan crude cocktail) or Brent crude oil price. A and B are coefficients with certain values that have been previously set.

2.2.2. Calculation of transportation cost

LNG shipments from liquefaction plant to destination LNG receiving terminals are carried out using LNG carrier in the form of tanker ships using charter a vessel schemes. LNG shipping cost from liquefaction plant to LNG receiving terminal by using time charter, C_T can be formulated by the following equation:

$$C_T = C_{fixed} + C_{variabel} \quad (2)$$

Transportation cost consists of 2 components, that is:

1. Fixed cost, C_{fixed} which is the charter cost.
2. Variable cost, $C_{variabel}$ which consists of 4 components:
   a. Fuel cost. Ships with a capacity of 160,000 m^3 with steam propulsion require fuel of 0.15 % boil-off LNG per day plus fuel oil of 48.1 tonnes of HFO per day while ships with DFDE propulsion require 0.10 % fuel LNG boil-off per day;
   b. Port charge, consisting of an unloading cost of € 16,988 per ship (fix cost) and € 35 per GWh (variable cost) when using a third party port facility;
   c. Insurance, which is 12 % of operating costs;
   d. General cost, which is 16 % of operating costs.

The LNG delivery method used in this study is single round. In this method, LNG shipments from each liquefaction plant use different ships and each LNG plant will use one ship for one purpose. To find out the ship's round trip time, LNG carrier capacity used and storage tank volume at LNG receiving terminal is calculated using equations developed by Nikolaou [7].

2.2.3. Calculation of regasification cost

The cost of LNG regasification at the LNG receiving terminal, C_R is formulated by the following equation:

$$C_R = CAPEX + OPEX \quad (3)$$

Regasification cost consists of 2 components, that are:

1. Investment costs (CAPEX), the average for onshore LNG terminal in 2015 is US$ 242 per tonne, 2016 is US$ 334 per tonne, 2017 is US$ 212 per tonne and estimated for year 2018 is US$ 285 per tonne [8];
2. Operational cost (OPEX), influenced by regasification technology used and assumed 4 % of CAPEX [9].

After each cost in each chain is calculated, the total cost of the overall gas supply is formulated by the following equation:

$$C = C_P + C_T + C_R + C_D \quad (4)$$

The cost of gas distribution, C_D can be ignored due to the position of power plant integrated with the LNG terminal so that the equation 4 above becomes:

$$C = C_P + C_T + C_R \quad (5)$$

2.3. Determining objective functions of the model

The illustrations of the problems in this study are shown in Figure 2 below.

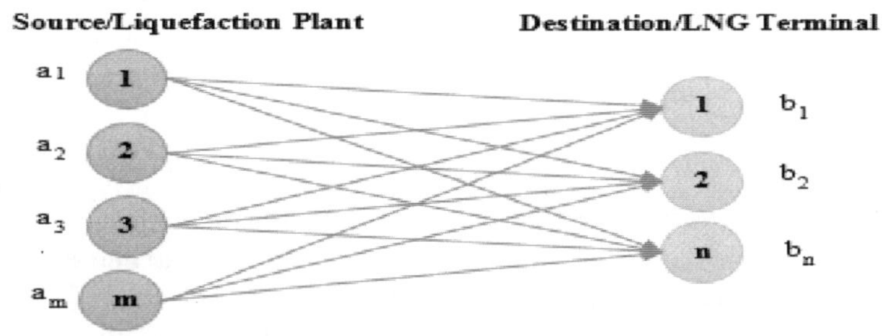

Figure 2. Illustration of supply model problem

The objective function to minimize the total cost of gas supply from the LNG plant to the LNG (Z) terminal is as follows:

$$Z = \sum_{\substack{1 \le i \le m \\ 1 \le j \le n \\ 1 \le t \le T}} C(i,j,t)\, X(i,j,t) \tag{6}$$

The total cost of gas supply, Z (US$) is a function of the entire gas supply chain cost, C (US$ per MMBtu) and the amount of LNG shipped, X (MMBtu).

2.4. Determining decision variable

The decision variable is a variable that can be changed and controlled in order to produce the desired result. In this study the decision variable is:

$X_{i,j,t}$ is a variable showing the ability of LNG delivery (LNG volume delivered) by source *i* to delivery destination *j* within certain period of time *t*, where i = 1, 2…, m, j = 1, 2…, n and t = 1, 2…, T.

2.5. Determining constraints

The decision variable is the objective function to be optimized to get the optimum result is limited by the following constraints:

a. The uncommited capacity of the supply source (liquefaction plant), **a** is at least equal to the amount of LNG delivered, **X**:

$$\sum_{j=1}^{n} X_{i,j} \le a_i \tag{7}$$

b. The LNG terminal's regasification capacity, **b** is at least equal to the amount of LNG delivered, **X**:

$$\sum_{i=1}^{m} X_{i,j} \le b_j \tag{8}$$

c. The gas consumption rate, **q** is at least equal to the terminal regasification capacity, **b**:

$$b_j \ge q_{c,j} \tag{9}$$

d. The number of LNG shipments with LNG carriers sent from source i, $\mathbf{F_{actual}}$ must not exceed the maximum number of per year that the LNG carrier can carry, $\mathbf{F_{max}}$:

$$F_{i,actual} \le F_{i,max} \tag{10}$$

IOP Conf. Series: Materials Science and Engineering **543** (2019) 012042 doi:10.1088/1757-899X/543/1/012042

e. X value should not be negative:

$$X_{i,j} > 0 \quad \forall i,j \tag{11}$$

2.6. Logistics supply scenario

The logistics supply scenarios under review is the combination of direct supply of gas sources at domestic and abroad with a long term contract of 20 years. In this scenario gas sources for Gresik LNG Terminals are obtained from domestic liquefaction plants: Bontang, Masela and Tangguh coupled with a liquefaction plant from abroad: Withnell Bay in Australia, Pluto in Australia, Gorgon in Australia, Lumut in Brunei and Bintulu in Malaysia. The gas supply quotas from each of the above plants are assumed as shown in Table 1 below.

Table 1. List of gas sources at domestic and abroad for Gresik LNG Terminal

No.	Liquefaction Plant	Distance Miles	Allocation for PLN Bcf/year	LNG Price US$/MMBtu	Remark
1	Bontang	573	15.91	7.88	
2	Tangguh	1,290	12.53	8.05	
3	Masela	925	32.71	8.40	Operation in 2026
4	Withnell B	985	26.41	9.73	
5	Pluto	986	13.50	9.73	
6	Gorgon	1,006	19.16	9.73	
7	Lumut	1,382	10.60	8.40	
8	Bintulu	1,495	14.85	8.05	

2.7. Key assumptions

The key assumptions used as the basis for calculating the gas supply cost are as follows:

1. The price of crude oil refers to the estimated price of Brent crude up to 2050 issued by the EIA (US Energy Information Administration);
2. The LNG carrier's charter rate for the Asian market is 80,000 US$ per day with a capacity of 160,000 m^3 with a 5-year contract and the charter rate is assumed to increase by 10 % every 5 years;
3. Average carrier LNG velocity of 10.5 knots with steam propulsion steam type;
4. LNG carrier fuel prices for diesel oil type of IFO 380 is 382.5 US$ per tonne and fuel prices from boil- off gas is 8.40 US$ per MMBtu. The increase in fuel prices follows crude oil prices;
5. The fuel consumption LNG carrier with capacity 160,000 m^3 is 145 tonnes per day LNG equivalent, is a combination of LNG boil-off and diesel oil type IFO 380;
6. LNG boil-off by 0.15 % per day and heel cargo at 4 % of shipload. Cargo heel is required to keep the LNG storage tank in storage compartment cool until the vessel returns to its original location.

3. Result and Discussion

3.1. LNG carrier selection

To make a single voyage from all liquefaction plant to Gresik LNG Receiving Terminal requires 1 unit of vessel with capacity of 65,000 m^3, whereas when using a vessel with a capacity of 27,500 m^3 requires 2 units of ships, except from Bontang liquefaction plant requires 1 unit of ship and from Bintulu liquefaction plant requires 3 units of ship. As depicted in Figure 3, transportation cost is directly proportional to the mileage of ships and ships with larger capacity having smaller transportation costs than smaller capacity vessels.

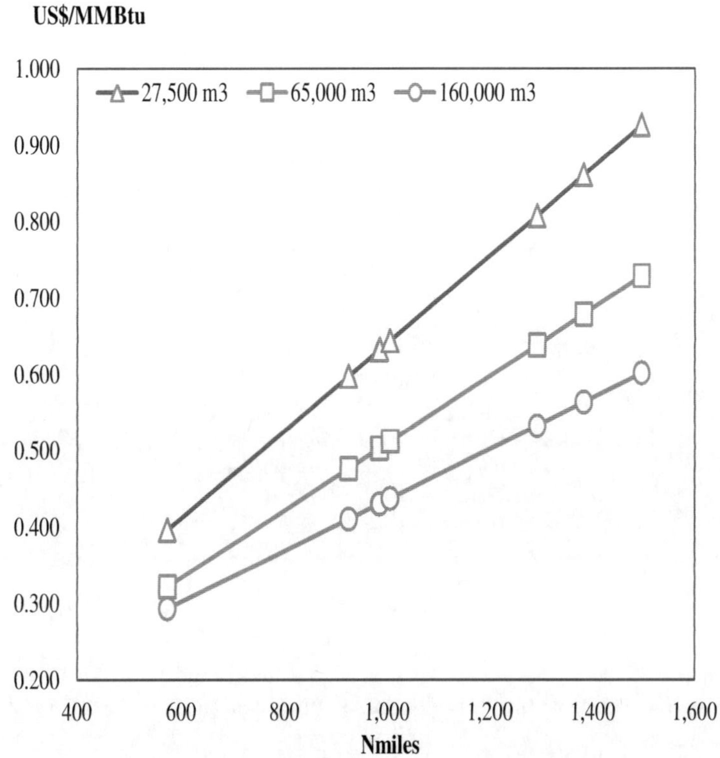

Figure 3. The relationship among the shipping distance, vessel capacity and transportation cost

3.2. Gas supply cost

The calculation is based on the use of LNG carrier with capacity of 65,000 m^3. The selling price of LNG (FOB price) and the price of IFO 380 oil diesel in the period 2019 - 2038 are adjusted to the crude oil price according to Brent crude oil price estimation data for the period of 2016 - 2050 issued by EIA (US Energy Information Administration). From the table, the lowest gas supply costs are obtained from the Bontang liquefaction plant, followed by the Tangguh liquefaction plant, the Bintulu liquefaction plant in Malaysia, the Masela liquefaction plant, the Lumut liquefaction plant in Brunei and the last liquefaction plants in Australia. The smallest gas supply cost is obtained from the Bontang liquefaction plant because it has the cheapest FOB LNG price and the closest delivery distance, followed by Tangguh and Bintulu. The largest gas supply cost is obtained from a liquefaction plant in Australia because although transportation costs are lower than Tangguh, Lumut and Bintulu but are unable to cover higher LNG (FOB) prices. The LNG (FOB) price of the liquefaction plant is the largest component of gas supply cost, which is about 86 % on average, followed by LNG regasification cost component at LNG receiving terminal of 9 % and last 5 % transportation cost component.

Gas supply costs follow world crude oil price, especially in LNG price from liquefaction plant and transportation cost. The relationship between the cost of gas supply and the reference crude oil price is shown in Figure 4. It shows that the increase in gas supply costs is proportional to the crude oil price.

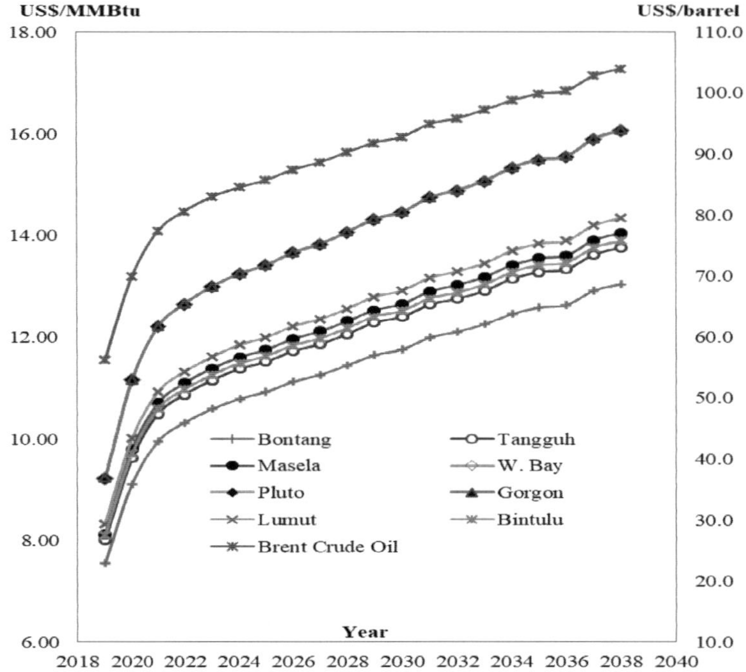

Figure 4. The relationship between gas supply cost and crude oil price

3.3. Result of optimization gas supply

From the optimization result obtained gas supply to Gresik LNG Receiving Terminal as follows:

- 2019
  The Gresik LNG Receiving Terminal starts to operate gradually and the gas requirement of 9.97 Bcf is fully met by Bontang refinery with 7 shipments;
- 2020 – 2038
  The Gresik LNG Receiving Terminal is fully operational and gas supply is obtained with a combination of supply from Bontang liquefaction plant of 15.79 Bcf per year with delivery 12 times per year, Tangguh liquefaction plant of 11.84 Bcf per year with delivery of 9 times per year and Bintulu liquefaction plant of 11.84 Bcf per year with shipment 9 times per year.

The total cost of gas supply for 2019 - 2038 is shown in Table 2.

Table 2. The results of gas supply optimization for Gresik LNG Receiving Terminal for long-term contracts

Year	Crude Oil $/bbl	Supply Cost Mio $	Year	Crude Oil $/bbl	Supply Cost Mio $	Year	Crude Oil $/bbl	Supply Cost Mio $
2019	56.3	71.63	2026	87.5	468.16	2033	97.2	515.62
2020	70.0	383.97	2027	88.7	473.79	2034	98.7	524.78
2021	77.4	418.96	2028	90.3	481.60	2035	99.9	530.11
2022	80.5	434.02	2029	91.8	490.26	2036	100.3	532.24
2023	82.9	445.35	2030	92.8	495.06	2037	102.8	543.86
2024	84.5	454.18	2031	94.9	504.73	2038	103.9	549.29
2025	85.7	459.79	2032	95.8	509.32			

4. Conclusion

The minimum gas supply costs are obtained under the direct supply scenario of the domestic liquefaction plant plus the supply from abroad liquefaction plant. This is because the uncommitted capacity of the domestic liquefaction plant is not able to meet the gas requirements for PLTGU Jawa-3. The minimum gas supply cost is obtained with the following supply combinations:

- Bontang liquefaction plant: gas shipments of 15.79 Bcf or 40 % supply and shipped 12 times per year;
- Tangguh liquefaction plant: gas shipments of 11.84 Bcf or 30 % supply and shipped 9 times per year;
- Bintulu liquefaction plant: gas shipments of 11.84 Bcf or 30 % supply and shipped 9 times per year.

Acknowledgement
The authors would like to thank all those who have supported this research, especially the University of Indonesia through a research grant called PITTA (International Publication Indexed for Thesis of Universitas Indonesia) in 2018.

References

[1] ESDM 2015 Rencana Usaha Penyediaan Tenaga Listrik (RUPTL) PT PLN (Persero) Tahun 2016 sd 2025 (Jakarta).
[2] Garcia D and You F 2015 Supply chain design and optimization: Challenges and opportunities *Comp. Chem. Eng.* **81** 153-170.
[3] Asala H, Chebeir J, Dahi Taleghani A and Romagnoli J 2017 The application of compositional modelling to the integration of realistic re-stimulation strategies and unconventional shale gas supply chain optimization *Chem. Eng. Trans.* **57** 979-984. DOI: 10.3303/CET1757164.
[4] Gao J and You F 2018 Tackling Uncertain Performances of Multiple Stakeholders inthe Design and Optimization of Decentralized Supply Chains *Chem. Eng. Trans.* **70** 73-78. DOI: 10.3303/CET1870013.
[5] PJB 2016 Laporan akhir kajian kelayakan LNG receiving terminal PLTGU Jawa-3 (Jakarta).
[6] Sakmar S L 2013 Energy for the 21st century: Opportunities and challenges for liquefied natural gas (LNG) (Massachusetts: Edward Elgar Publishing).
[7] Nikolaou M 2010 Optimizing the logistics of compressed natural gas transportation by marine vessels *J. Nat. Gas. Sci. Eng* **2** 1–20.
[8] IGU 2017 *World LNG report: 2017 Edition* (Barcelona, International Gas Union).
[9] Lee H 2005 *Dawning of a New Era: The LNG Story, Cambridge, MA: Belfer Center for Science and International Affairs* (Massachusetts: John F. Kennedy School of Goverment).

IOP Conf. Series: Materials Science and Engineering **543** (2019) 012043 doi:10.1088/1757-899X/543/1/012043

Technology selection of LNG receiving terminals in Gresik, East Java

R Wicaksono[1] and M Sudibandriyo[1]
[1] Department of Chemical Engineering, Faculty of Engineering, Universitas Indonesia, Kampus UI Depok, Depok 16424, Indonesia

Email: msudib@che.ac.id

Abstract. Java, the centre of industry in Indonesia, needs fuel gas supplies to maintain production continuity. The addition of gas infrastructures in the form of Liquefied Natural Gas (LNG) receiving terminals can be an alternative to secure gas supplies. Such a terminal is planned to be built in Gresik, East Java. This paper aims to examine the most appropriate technology for LNG receiving terminals to be built in East Java by considering both technical and economic aspects. In this research, LNG receiving technology was selected based on the terminal location and selection of the regasification technology using the method Analytic Hierarchy Process. Moreover, economic and sensitivity analyses were also performed. Research findings suggest the land-based LNG receiving terminal with regasification technology in the form of shell and tube vaporizers as the most suitable LNG receiving terminal in Gresik, East Java. Land-based LNG receiving terminal establishment in Gresik, East Java is feasible as regasification costs 0.468 USD/MMBTU, the Net Present Value (NPV) amounts to 31943500 USD, the Internal Rate of Return (IRR) reaches 19.25%, the Benefit-Cost (BC) Ratio equals 1.23, and the Payback Period (PBP) is 3.4 years. The throughput required by the power plant generates the greatest sensitivity to IRR.

1. Introduction

The availability of natural gas for industrial fuel needs in Java as the centre for industry in Indonesia is an absolute requirement to maintain fuel gas supplies in order that production activities continue. In 2013 to 2028 there was an imbalance between natural gas supplies and demands, especially in East Java and Bali [1]. It is imperative to increase the number of gas infrastructures in the form of Liquefied Natural Gas (LNG) receiving terminals in order to maintain security of natural gas supplies.

This paper intends to examine the appropriate technology to be applied to the LNG receiving terminals to be built in East Java by taking the technical aspect and the economic aspect into account. The LNG receiving terminals to be selected are the land-based LNG receiving terminal (fixed onshore facility), the gravity-based structure LNG receiving terminal (concrete structure near the beach), and FSRU (Floating Storage Regasification Unit). As for regasification technology, those LNG receiving terminals will adopt Open Rack Vaporizer (ORV), Submerged Combustion Vaporizer (SCV), Shell and Tube Vaporizer (STV), Intermediate Fluid Vaporizer (IFV), and Ambient Air Vaporizer (AAV).

In this research, the LNG receiving technology was selected based on the location of the terminal and selection of the regasification technology using the method Analytic Hierarchy Process. The Analytic Hierarchy Process (AHP) is a Multi-Criteria Decision Making System that was developed by Thomas L. Saaty in the 1970s [2]. This procedure has been widely applied in decision making. AHP was used as a method to choose warehouse location, to choose an optimal corporate maintenance strategy and to make decisions about reverse logistics plans [3-5]. This decision support model will describe complex multi-criteria problems in a hierarchical arrangement. A hierarchy is defined as the representation of complex problems in a multi-level structure where the first level constitutes the goal, followed by the levels factors, criteria, sub-criteria, and so on up to the last level of alternatives [6].

2. Methods

The steps taken in this research were selecting technology for LNG receiving terminals as shown in Figure 1, that were comprised of: determining the case study, collecting technical data and data on the location for LNG receiving terminal construction in Gresik, selecting LNG receiving terminals based on location, selecting regasification technology, selecting scenarios and calculating sizes of the equipment, undertaking economic calculation and sensitivity assessment of the LNG receiving terminals, and drawing conclusions and suggestions.

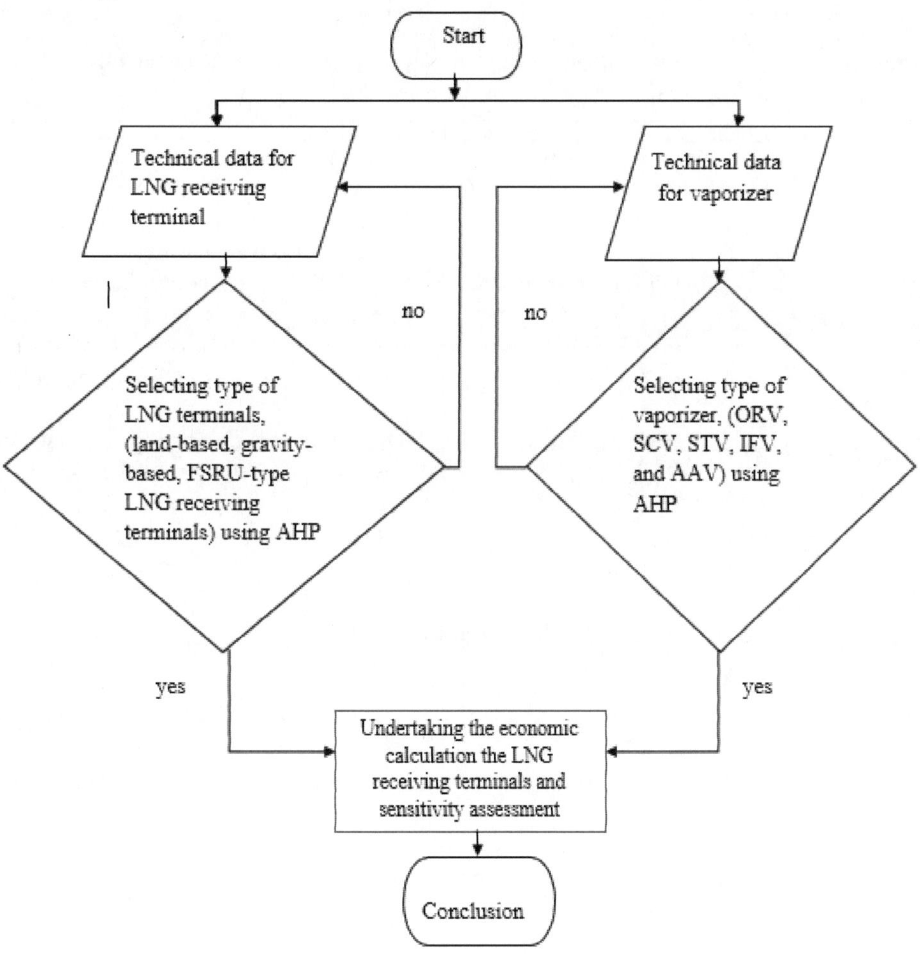

Figure 1. Research flow chart.

The steps taken in this research were selecting technology for LNG receiving terminals as shown in Figure 1, that were comprised of: determining the case study, collecting technical data and data on the location for LNG receiving terminal construction in Gresik, selecting LNG receiving terminals based on location, selecting regasification technology, selecting scenarios and calculating sizes of the equipment, undertaking economic calculation and sensitivity assessment of the LNG receiving terminals, and drawing conclusions and suggestions.

2.1. Analytic Hierarchy Process

The Analytic Hierarchy Process (AHP) method is used to select the LNG receiving terminal and regasification technology. The steps in the Analytic Hierarchy Process method are as follows: Defining the problem and the goal, Identifying the decision criteria, sub criteria and alternatives, Developing pairwise comparison matrices (the nine scale pairwise comparison is presented in Table 1) and calculating the eigenvalue and eigenvector and the consistency ratio of each pairwise comparison [6]. In the analytic hierarchy process, the construction of judgment matrix is the key in the whole analysis process, which determines the success or failure of the analysis process [7].

Table 1. Importance scale of pairwise comparisons

Intensity of Importance of an Absolute Scale	Definition	Explanation
1	Equal importance	Two activities contribute equally to the objective
3	Moderate importance	Experience and judgement slightly favour one activity over another
5	Strong importance	Experience and judgement strongly favour one activity over another
7	Very strong or demonstrated importance	An activity is favoured very strongly over another; its dominance demonstrated in practice
9	Extreme importance	The evidence favouring one activity over another is of the highest possible order of affirmation
2,4,6,8	Intermediate values between the two adjacent judgements	When compromise is needed

Table 2. Random index

n	3	4	5	6	7	8	9	10
RI	0.58	0.9	1.12	1.24	1.32	0.141	1.45	1.49

The pairwise comparison matrix with n criteria is a nxn square matrix A, whose element a_{ij} is the value of the comparison of ith over j^{th} criteria, while the value of reverse comparison is denoted by $1/a_{ij}$. Matrix W is then constructed by dividing each column of matrix A by its column sum. The principle eigenvalue of the matrix (λ_{max}) is then calculated using the following equation:

$$AW = \lambda_{max} W \tag{1}$$

AHP assesses consistency from various aspects using a consistency ratio. The consistency ratio has to be less than 10 percent, otherwise the assessment needs to be revised [8]. The consistency ratio (CR) of each pairwise comparison is then calculated as:

$$CR = CI/RI \tag{2}$$

$$CI = (\lambda_{max} - n) / (n-1) \tag{3}$$

Where CI is the consistency index and RI is the random index (see Table 2) [8]. The comparison is considered to have satisfactory consistency if $CR \leq 0.1$.

In the decision making process using the AHP method, the data to be processed consists of two types, namely quantitative data and qualitative data. The data is combined so that it can be used to calculate parameter assessments. Therefore, the quantitative data is changed to preference degree to simplify the calculation process. For all benefit attributes, the preference degree can be determined as follows:

$$P_{rk}=\frac{2(V_{rk}-V_k^{min})}{(V_k^{max}-V_k^{min})}-1 \quad (4)$$

For all cost attributes, the preference degree can be determined as follows:

$$P_{rk}=\frac{2(V_k^{max}-V_{rk})}{(V_k^{max}-V_k^{min})}-1 \quad (5)$$

Where P_{rk} is the preference degree, V_{rk} is the attribute values for the alternatives are calculated, Vk_{min} is the minimum attribute value from alternative, Vk_{max} is the maximum attribute value from alternative, r is the number of alternatives, k is the number of quantitative attributes. After obtaining a preference degree, entropy is calculated using a formula:

$$\text{Enthropy} = -\frac{1}{\ln(m)}\sum_{i=1}^{m} Yi\ln(Yi) \quad (6)$$

Where m is the number of alternatives, and Yi = preference degree value.

2.2. Analysis technical aspect

2.2.1. LNG receiving terminal. The type of LNG receiving terminal or alternative chosen is land based LNG receiving terminal, gravity based LNG receiving terminal, FSRU LNG receiving terminal. Data related to LNG receiving terminal technology consists of qualitative and quantitative data. Quantitative data can be divided into two types of attributes, namely cost attributes and benefit attributes. Cost attributes are all attributes that give effect to generating costs in the selection process. Otherwise, benefit attributes are all attributes that give a beneficial effect in the selection process. In this case the benefit attribute is the depth of the water because the greater the value the shorter the distance between the jetty and the terminal. While other parameters include the cost attribute. The quantitative data is converted into a preference degree to simplify the calculation process. Enthropy calculation is done after obtaining a preference degree.

2.2.2. Technology of regasification. Data collected related to regasification technology is capital expenditure, operational expenses, operations and maintenance, environment, safety, limited conditions. This data is obtained from literature studies. The alternative chosen is an open rack vaporizer, intermediate fluid vaporizer, submerged combustion vaporizer, shell and tube vaporizer, ambient air vaporizer. Furthermore, the selection of regasification technology is carried out using the AHP method as is done when selecting LNG receiving terminal technology.

2.3. Economic calculation

After designing equipment specifications using process modeling software, economic calculations are carried out by estimating the total value of capital investment, estimating operational costs, calculating project feasibility. In this research, Guthrie's method was used to estimate the value of total capital investment. The Guthrie method is done by determining the Free On Board (FOB) of each equipment.

FOB is the price of equipment plus the cost of transportation to the port in the country producing the equipment. The price of the equipment FOB is multiplied by the bare module. Bare module is a comparison of other costs such as piping, instrumentation, concrete, and other costs against FOB. Multiplication between FOB and total bare module factor for each tool, each added to obtain a total fixed capital investment. Land, office, working capital costs are determined based on the ratio of fixed capital investment. The sum of all fixed capital investments and other costs will be the total capital investment. Total capital investment consists of fixed capital investment and variable costs.

$$C_{TCI} = 1.18\,(C_{TBM} + C_{site} + C_{buildings} + C_{offsite}) + C_{WC} \quad (7)$$

Where C_{TCI} is total capital investment cost, C_{TBM} is total bare module cost, C_{site} is site development cost, $C_{buildings}$ is building cost, $C_{offsite}$ is 5% C_{TBM}, C_{WC} is working capital cost approximately 15% C_{TCI}.

To get the total bare module cost, it is necessary to calculate the bare module cost component. The calculation is done using the following formulas, proposed by Meyers & Kime [9] :

- Tank

$$C_T = F_M \exp[11.662 + 0.6104(\ln V) + 0.04536(\ln V)^2] \quad (8)$$

Where V is tank volume, F_m is 3.25 for stainless steel 314

- Heat Exchanger

$$C = f_d f_m f_p C_b \quad (9)$$

$$C_b = \exp[8.821\text{-}0.30863(\ln A) + 0.0681(\ln A)^2] \quad (10)$$

$$f_d = \exp[-1.1156+0.0906(\ln A)] \textit{ for fixed head} \quad (11)$$

$$f_p = 1.0305 + 0.07140(\ln A), \textit{ pressure range} = 300\text{-}600 \text{ psig} \quad (12)$$

$$f_m = 0.8603+0.23296(\ln A), \text{ SS316} \quad (13)$$

Where A is Area, ft^2

- Pump

$$C = F_M F_T C_b \quad (14)$$

$$F_M = 2 \text{ (SS 304 or 316)} \quad (15)$$

$$F_T = \exp[b_1 + b_2(\ln QH^{0.5}) + b_3(\ln QH^{0.5})^2 \quad (16)$$

$$C_b = 1.55 \exp[8.833\text{-}0.6019(\ln QH^{0.5}) + 0.0519(\ln QH^{0.5})^2] \quad (17)$$

Where Q is flow, H is head

After the investment costs are obtained, then an economic analysis calculation is performed to see whether the design of the regasification unit has been made economically feasible or not. The calculation is done by calculating the NPV value, IRR, payback period. The project will be carried out if the NPV is positive or the IRR is positive. After data on construction costs, operating and maintenance costs and income are obtained, then cash flows are made during the operation and the concession period between the government and the private sector for 20 years. Sensitivity analysis is done to see the effect of a cost component on the economic value of a factory. The analysis is done by changing the cost component to be bigger or smaller then comparing with changes in economic value. The economic value used for sensitivity analysis can be IRR, NPV or payback period. In this study, sensitivity will be made between the investment costs of the IRR and regasification capacity of the regasification costs.

3. Results and discussion

3.1. Technical aspect

3.1.1. Technology selection for LNG receiving terminal. This research was began with determining attributes or criteria for selection. Quantitative criteria includes the distance between the terminal and the power plant, regasification capacity, sea depth, land area, wave height, wind speed, currents. These data were obtained from Indonesia's Meteorology Climatology and Geophysics Council. Qualitative criteria includes investment costs, safety, environmental impact, terminal flexibility, energy efficiency, and improvement to the regasification capacity obtained from literature studies. Afterwards, a pairwise comparison matrix was formulated for each alternative technology for each criterion. Then, a synthesis and normalization were carried out in such a way that the sum of all the factors was equal to 100%. The normalized factor turned into a weight. The sum of the weights multiplied generated a weighted sum factor. The consistency vector was obtained by dividing the weighted vector by its weight. The mean of consistency vectors generated a lambda. Then, this lambda was used to calculate the consistency index.

For the criterion investment costs, the calculation was undertaken by making pairwise comparison matrices. The pairwise comparison matrices of each alternative for the criterion investment costs were developed based on the scale proposed by Thomas Saaty by comparing alternatives based on Table 1. The investment costs for gravity-based LNG receiving terminals are around 20-50% of those of the land-based LNG receiving terminals [10]. FSRU-type LNG terminals have lower capital expenditure than that of land-based LNG receiving terminals [11]. The investment costs for gravity-based LNG receiving terminals are higher than those of FSRU-type LNG receiving terminals [12].

In relation to the criterion safety, the accident most likely to occur is a gas leak that may result in an extremely dangerous fire given the location of the LNG receiving terminals that is very vulnerable to explosion. The flash fire envelope of offshore terminals reaches 4000 m while that of onshore terminals reaches 2000 m [13]. This suggests that land-based LNG receiving terminals have a flash fire envelope range that is relatively lower than those of gravity-based LNG receiving terminals and FSRU-type LNG receiving terminals.

For the criterion environmental impact, FSRU uses boil-off gases as fuels, thus eliminating the risk of flare. The whole waste undergoes treatment, including chemical waste [12]. For the criterion terminal flexibility, the land-based LNG receiving terminals do not have any flexibility to move closer to the source of supplies or demands. Conversely, the FSRU-type LNG receiving terminals can move to approach the source of supplies or demands. Similar to the FSRU-type LNG receiving terminals, the gravity-based LNG receiving terminals can also be relocated but are relatively more difficult.

In terms of the criterion improvement to the regasification capacity, it remains feasible to increase the regasification capacity of the land-based LNG receiving terminals since they have a wider area compared to the gravity-based LNG receiving terminals and the FSRU-type LNG receiving terminals. The addition of vaporizers and tanks can be done to the land-based LNG receiving terminals. For the criterion energy efficiency, the addition of power plants and use of waste heat to make LNG evaporates can improve efficiency of the overall system [14]. After comparing each criterion based on the scale in Table 1, a synthesis and normalization were then carried out to calculate the weight and consistency. The consistency ratio generated was less than 0.1, meaning that the resulting calculation can be accepted. Then, matrix values were compared for each criterion with the following description: A = investment costs, B = safety, C = environmental impact, D = location-related flexibility, E = capacity improvement, and F = energy efficiency. Afterwards, a comparison was made for each criterion to determine which criterion has the greatest weight using the method Analytic Hierarchy Process.

Quantitative criteria includes the distance between the terminal and the power plant, regasification capacity, sea depth, land area, wave height, wind speed, currents, were calculated by converting quantitative data into degrees of preference then the entropy is calculated. This entropy is summarized with the results of qualitative comparisons of paired data. Because entropy of results is zero, it does not change the results of qualitative data. The sum of the quantitative data and the qualitative data was calculated as shown in Table 3. Based on the table, it is revealed that, compared to the other alternatives, land-based LNG receiving terminals are more compatible to be applied in Gresik, East Java because the highest value is 0.4.

Table 3. Cumulative value results.

Factor	Weight	Land-Based	Gravity-Based	FSRU
A	0.16	0.11	0.26	0.63
B	0.30	0.59	0.25	0.16
C	0.30	0.25	0.16	0.59
D	0.04	0.14	0.09	0.77
E	0.07	0.70	0.18	0.11
F	0.13	0.59	0.25	0.16
Total		**0.40**	**0.21**	**0.39**

3.1.2. Technology selection for regasification technology. This research was began with determining attributes or criteria for selection. The attributes chosen included: capital expenditure, operational expenditure, operation and maintenance (O & M), environmental impact, safety, and limitation. As for

IOP Conf. Series: Materials Science and Engineering **543** (2019) 012043 doi:10.1088/1757-899X/543/1/012043

the alternatives to be chosen were: open rack vaporizer, submerged combustion vaporizer, shell and tube vaporizer, intermediate fluid vaporizer, and ambient air vaporizer. The method adopted in the selection of regasification technology was Analytic Hierarchy Process.

Submerged Combustion Vaporizer, Shell and Tube Vaporizer, and Intermediate Fluid Vaporizer have lower capital expenditure compared to Open Rack Vaporizer. While Ambient Air Vaporizer has the highest capital expenditure (Tractebel Engineering SA, 2015). For the criterion *operational expenditure*, SCV has the highest operational expenditure compared to STV, IFV, and ORV. As for AAV, it has almost no operational expenditure [15]. The ways SCV are operated are more complex than the other types of vaporizer as it requires a balance between fuels and waste heat, and coordination with the department in charge of factory operations. ORV is simpler because it only requires additional sea water pumps and filtering systems [16].

AAV does not have any significant impact on the environment except fog. While ORV, SCV, STV, and IFV affect marine life [16]. In relation to the criterion *safety*, hydrocarbon leakage in ORV and AAV is discharged into the atmosphere at the lowest level, while in SCV, STV, and IFV, it is removed through a vessel [16]. For the criterion *operating limitation*, AAV's performance depends on the weather [17]. Seawater temperature suitable for the operation of ORV is above 5° C [16]. For the criterion *resume*, environmental impact, safety, and operating limitation take priority. In the recapitulation of criteria, the value of CR <0.1 was generated, meaning that the calculation can be accepted. After collecting all the data, the sum was calculated as shown in Table 4.

Based on Table 4, it is revealed that the vaporizer technology chosen is the Shell and Tube Vaporizer that is deemed to be compatible to be applied in Gresik, East Java as it generated the highest value compared to the other alternatives. The advantage of using this vaporizer is that it is relatively more environmentally friendly and safer. In conclusion, selection of technology using Analytic Hierarchy Process chose land-based LNG receiving terminals as location-based terminal technology, and the shell and tube vaporizer as regasification technology.

Table 4. Cumulative value results.

Factors	Weight	ORV	SCV	STV	IFV	AAV
Capital Expenditure	0.10	0.12	0.41	0.21	0.21	0.05
Operational Expenditure	0.10	0.27	0.11	0.11	0.11	0.41
O&M	0.06	0.41	0.11	0.23	0.11	0.14
Environment	0.24	0.09	0.09	0.32	0.18	0.32
Safety	0.24	0.09	0.18	0.32	0.32	0.09
Limitation	0.24	0.09	0.18	0.32	0.32	0.09
		0.13	**0.17**	**0.28**	**0.24**	**0.18**

3.2. Capital expenditure and operational expenditure calculation

The method develop by Meyers & Kime can be used as an approach to capital expenditure calculation [9]. Specifications and capital expenditure are calculated for each equipment. Then, the total bare module cost is obtained. The total bare module costs generated are shown in Table 5.

If capital costs amount to 10%, life time is 20 years, and the regasification capacity is equal to 0.92 MTPA, then CRF is equal to 0.117. The capital expenditure of regasification amounts to 0.45 USD/MMBTU. The operational expenditure is equal to 4% of the capital expenditure, thus the operational and maintenance costs are 0.018 USD/MMBTU. Based on results of the calculation, it is revealed that regasification costs 0.468 USD/MMBTU. This amount is still reasonable as regasification usually costs less than 1 USD/MMBTU.

Table 5. Total bare module cost.

Equipment	Total cost (USD)
LNG storage tank	80508566.44
Shell and tube vaporizer	1949615.37
LNG pump	1790099.28
Sea water pump	1199295.31
TOTAL	85447576.40

Table 6. Total capital investment.

Equipment	Total cost (USD)
C total bare module	85447576.40
C site	5126854.58
C building	4272378.82
C offsite	4272378.82
C contingency	8554757.64
C contractor fee	1708951.53
C working capital	8554757.64
C land	17089515.28
C port	1708951.53
TOTAL	136716122.20

3.3. Economic calculation and sensitivity

The installed capacity of the power plant is 800 MW. The power plant capacity factor is 70%. Specific fuel consumption is 0.01 MSCF/KWh. The cost for investment in LNG receiving terminals amounts to 136,716,122.20 USD. Regasification costs 0.468 USD/MMBTU. Revenues are calculated by firstly calculating the throughput needed by the power plant. The throughput required by the power plant is equal to 134.4 BBTUD and therefore the revenues amount to 22,958,208 USD. In this calculation, the financing scheme is 30% for government's company and 70% for private companies. Calculating the cash flow, NPV amounting to 31.943.500 USD is obtained. Because the NPV is positive, it indicates that the project makes more money than the amount needed to pay off debts. From the cash flow calculation, IRR of 19.25% is generated. A positive IRR implies that this project is feasible.

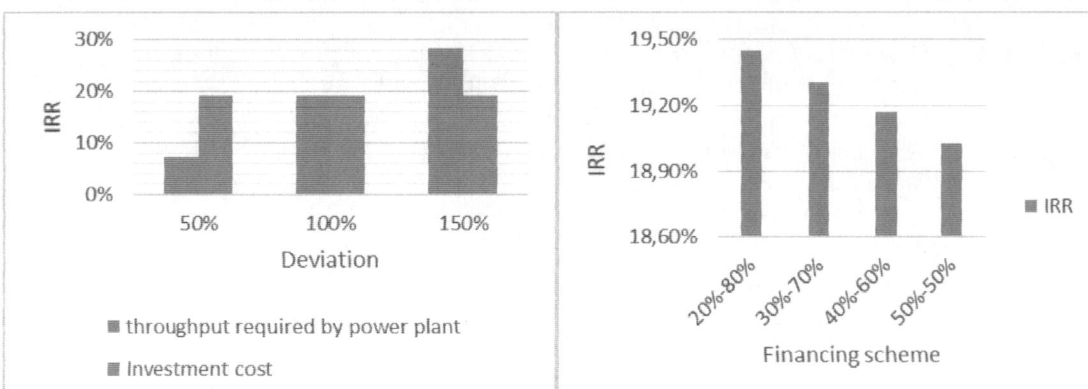

(a) (b)

Figure 2. (a) Throughput sensitivity required by power plants and sensitivity of investment costs to IRR. (b) Sensitivity of the financing scheme to IRR.

The resulting BC Ratio is equal to 1.23. The benefit is greater than the cost, making this project worth doing. Calculation results show that the payback period is 3.4 years, meaning that the whole sum of money invested will be paid back after a lapse of 3.4 years. Sensitivity calculation aims to determine the change in results if the parameters are changed. In this case, sensitivity calculation carried out was the change in the throughput required by the power plant in connection with IRR, the effect of changes in investment costs on IRR, and the effect of changes in the funding scheme on IRR. Sensitivity was measured by setting changes in the throughput required by the power plant and investment costs were set at 50%, 100%, and 150%.

Based on Figure 2.a, it is evident that the greater the throughput needed by the power plant is, the greater IRR is. Revenues increase, which in turn result in an improved ability to repay loans. Moreover, changes in investment costs do not significantly affect IRR. Sensitivity measurement was performed by making adjustments to changes in the financing scheme, which were set to be at 20% (government): 80% (private), 30% (government): 70% (private), 40% (government): 60% (private), and 50% (government): 50% (private). From Figure 2.b, it can be seen that the smaller the portion of the financing scheme is for government's company, the relatively larger IRR is. In fact, the difference in IRR between they each is relatively small, thus leading to the conclusion that it does have an effect but the resulting effect is minor.

4. Conclusion

Research findings suggest the land-based LNG receiving terminal with regasification technology in the form of shell and tube vaporizers as the type of LNG receiving terminal most suitable to be built in Gresik, East Java. The establishment of a land-based LNG receiving terminal in Gresik, East Java is feasible as regasification costs 0.468 USD/MMBTU, NPV amounts to 31943500 USD, IRR reaches 19.25%, the BC Ratio is equal to 1.23, and PBP is 3.4 years. The throughput required by the power plant generates the greatest sensitivity to IRR.

5. Acknowledgements

The authors' thank is due to any parties that have contributed to this research, especially Universitas Indonesia through its research grant, which is PITTA (Indexed International Publication for UI Student Final Projects) in 2018.

6. References

[1] Ministry of Energy and Mineral Resources 2014 National Natural Gas Policy Roadmap 2014-2030 (Jakarta: Ministry of Energy and Mineral Resources) p 53.
[2] Saaty R W 1987 *Int. Math. Mod.* 9 161–176.
[3] Hakim R T and Kusumastuti R D 2018 *J. of Tech.* 7 1405–1414.
[4] Salmasnia A, Ghasemi E and Mokhtari H 2016 *Int. J. of Ind. Eng. and Prod. Res.* 27 337–351.
[5] Chinda T and Ammarapala V 2016 Songklanakarin *J. Sci. Tech.* 38 7–14.
[6] Saaty T L 2008 *Int. J. Serv. Sci.* 1 83–98.
[7] Liang Q and He Y 2017 *Chem. Eng. Trans.* 62 1435–1440.
[8] Prasad V S and Kousalya P 2017 *Ind. J. of Sci. and Tech.* 10 1–5.
[9] Walas S M 1990 Chemical Process and Equipment: Selection and Design (Massachusetts:
[10] Butterworth-Heinemann Series in Chemical Engineering) pp 663-669
[11] Kjersem G L 2015 *Singapore Maritim Technology Conf.* (Singapore: IBC Asia)
[12] International Gas Union 2017 IGU World Report (Barcelona: International Gas Union)
[13] Magdalina M, Artana K B and Soegiono 2012 Study Of Selection LNG Receiving Terminal
[14] Technology For Bali Island Study Case LNG Supply From LNG Plant Tangguh To Bali (Surabaya: Institut Teknologi Surabaya)
[15] Ramos M, Droguett E L, Martins M R and Souza H 2014 *Int. J. of Mod. and Sim. for the Petrol. Ind.* 8 40–48
[16] Norrgard J 2018 LNG Terminals Land-Based Vs. Floating Storage and Regasification

Technology (Helsinki: Wartsila)
[17] Tractebel Engineering SA 2015 Mini/Micro LNG for Commercialization of Small Volumes of Associated Gas (Washington:World Bank Group)
[18] Patel D, Mak J, Rivera D and Angtuaco J 2013 30th *Annual Conf.* (Edinburgh: Gas Processors Association Europe)
[19] Gavelli F 2010 J. of *Loss Prev. in the Process Ind.* 23 773–780

IOP Conf. Series: Materials Science and Engineering **543** (2019) 012044 doi:10.1088/1757-899X/543/1/012044

Alternative of LNG cold exergy utilization for generating electrical energy at Gresik LNG receiving terminal

S B A Putra and M Sudibandriyo
Chemical Engineering Department, Faculty of Engineering, Universitas Indonesia, Depok, 16424, Indonesia

E-mail : msudib@che.ui.ac.id

Abstract. Gresik LNG receiving terminal will be built to meet the needs of combined-cycle power plant (PLTGU) with natural gas regasification rate of 60.95 MMSCFD. In conventional regasification process using open rack vaporizer (ORV), the potential of LNG cold exergy will be wasted to seawater which is used as a heat source from the environment. The commercially proven scheme of the regasification process with the utilization of LNG cold exergy to generate electrical energy such as Direct Expansion, Rankine Cycle and combined Direct Expansion+Rankine Cycle is simulated with Unisim computer software. The purpose of this study is to assess technically and economically three scheme mention above. The results show that the combined scheme has the ability to produce the largest electricity surplus of 39.80 kWh per ton LNG regasified with potential revenue from electrical energy sales of USD 1,140,935 per year. The energy and exergy analysis shows that this scheme also has highest thermal efficiency of 14.48% and highest exergy efficiency of 60.71%. However, based on the results of economic analysis found that Direct Expansion scheme has the highest NPV among the three schemes with a value of USD 695,032.

1. Introduction

The Government of Indonesia through RUPTL 2017-2026 seeks to reduce the use of coal and oil fuel by increasing the utilization of natural gas up to 26.7% by 2026 in the national electricity energy mix. Based on these projected needs, PT X plans to build an LNG receiving terminal to supply a power plants in Gresik. Total LNG to be regasified at terminal is 54.23 ton/h equivalent to 60.95 MMSCFD of natural gas. The terminal is designed with Open Rack Vaporizer using seawater as a heat source.

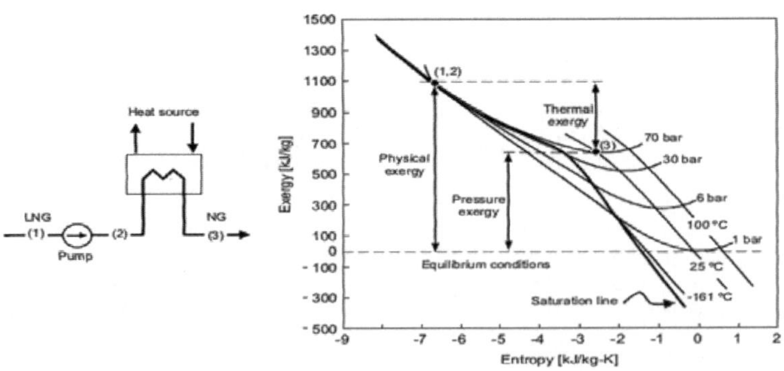

Figure 1. Change of exergy during LNG evaporation process [3]

IOP Conf. Series: Materials Science and Engineering **543** (2019) 012044 doi:10.1088/1757-899X/543/1/012044

This research focuses on the utilization of LNG physical exergy in the evaporation process as shown in Figure 1, where physical exergy consists of two components namely thermal and pressure exergy caused by the difference between LNG temperature and pressure to environment reference.

In the conventional LNG evaporation process, which mostly utilizes heat from the environment, the working fluid lost its cold temperature potential when its temperature reaches the value of the environment. If the pressure are also equal, the energy potential of pressurized gas is also lost [1]. To be able to exploit the potential of cold exergy in LNG there are several concepts that can be used, one of which is used for the generation of electrical energy. Research on this concept has been widely practiced, but only 3 power generation technologies applied commercially namely Direct Expansion (DE), Rankine Cycle (RC) and a combination of Direct Expansion+Rankine Cycle (DE+RC) [3].

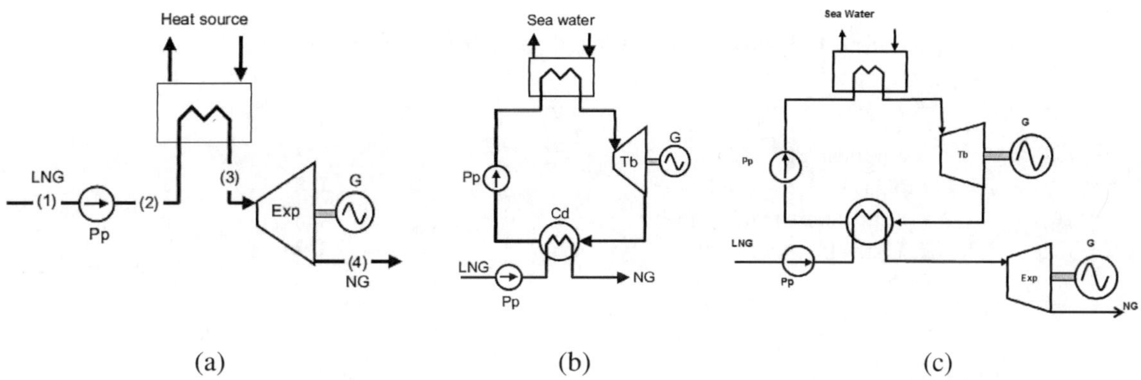

Figure 2. DE (a), RC (b) and DE+RC scheme (c) [3]

Many researchers have conducted research on modelling cold energy utilization [5] [6] [7] [8]. Research carried out by modifying the scheme, operating conditions, heat source or working fluid. A study of the economic study of cold energy utilization of LNG in South Korea has also been carried out [4]. This paper contains a technical and economic assessment of the application of LNG evaporation technology with the utilization of LNG cold exergy to generate electrical energy through Direct Expansion, Rankine Cycle and a combination scheme of Direct Expansion+Rankine Cycle at Gresik LNG receiving terminal, East Java. The working condition in Gresik is unique due to it has tropical climate condition and high send out gas pressure because it's dedicated only to supply the gas power plant, not like usual LNG receiving terminal who serve city gas with low send out gas pressure.

Energy analysis is based on the first law of thermodynamics. Energy efficiency or thermal efficiency is calculated by comparing the net work the system generates with the received heat input [5].

$$\eta_{\mathrm{th}} = \frac{\mathrm{W_{net}}}{\mathrm{Q_{in}}} \times 100\% \tag{1}$$

Where: η_{th} = Thermal efficiency of the system (%)

W_{net} = Net power production of the system (kW)

Q_{in} = Heat input rate to the system (kW)

The analysis of exergy is based on the second law of thermodynamics. The exergy efficiency is defined as exergy output divided by exergy input to the system [7] as Equation 2. The exergy input is taken as the exergy change of the heat source. The exergy output is the exergy of the net work and exergy of natural gas.

$$\eta_{ex} = \frac{W_{net} + E_{\ NG\_OUT}}{E_{\ LNG\_IN} + E_{SW}} \tag{2}$$

IOP Conf. Series: Materials Science and Engineering **543** (2019) 012044 doi:10.1088/1757-899X/543/1/012044

Where: η_{ex} = Exergy efficiency of the system (%)
W_{net} = Net power production of the system (kW)
E = Exergy on stream flow (kJ)

The economic calculations for the application of the three schemes at Gresik terminal include Fixed Capital, Operating Expenditure, Revenue and Net Present Value/NPV. Fixed Capital calculated based on purchased equipment cost data that is updated using the CEPCI cost index. The purchased cost data is multiplied by the installation factor to obtain fixed capital cost as shown in Table 1. The Operating Expenditure is assumed to be 4% of Fixed Capital and escalated by 5% per year. Revenue is calculated by multiplying the net power with electricity selling price in East Java of USD 0.0586 / kWh.

Table 1. Typical factors for estimate fixed capital cost [2]

Item	Fluids
Major equipment, total purchase cost	C_e
f_{er} Equipment erection	0.3
f_p Piping	0.8
f_i Instrumentation and control	0.3
f_{el} Electrical	0.2
f_c Civil	0.3
f_s Structures and buildings	0.2
f_l Lagging and paint	0.1
ISBL cost $C = \Sigma C_e$ x	3.3
Offsites (OS)	0.3
Design and Engineering (D&E)	0.3
Contingency (X)	0.1
Total fixed capital cost $C_{FC} = C(1+OS)(1+DE+X)$	
$= C\,x$	1.82
$= \Sigma C_e\,x$	6.00

2. Methods

System boundary of LNG cold exergy utilization to generate electrical energy at Gresik LNG receiving terminal is limited only on vaporization stage of LNG into natural gas according to Figure 3.

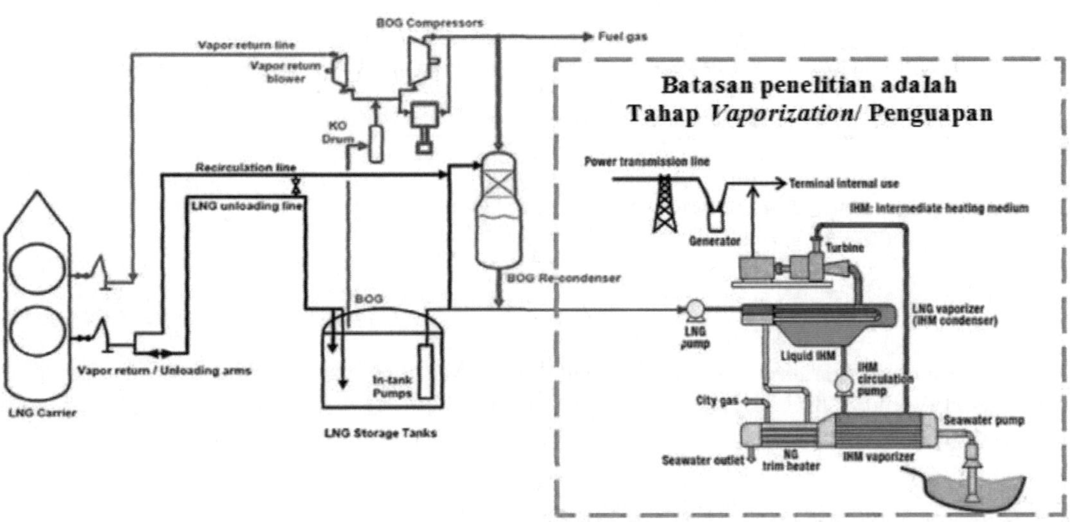

Figure 3. System boundary in research

The research methodology used in the techno-economic assessment on the alternative scheme of LNG cold exergy utilization for generating electrical energy follows the steps in Figure 4. Unisim Design R390 .1 program helps equipment design calculations of the evaporation process with the same condition of LNG inputs, technical equipment limitations and sendout gas output on all schemes in steady state conditions.

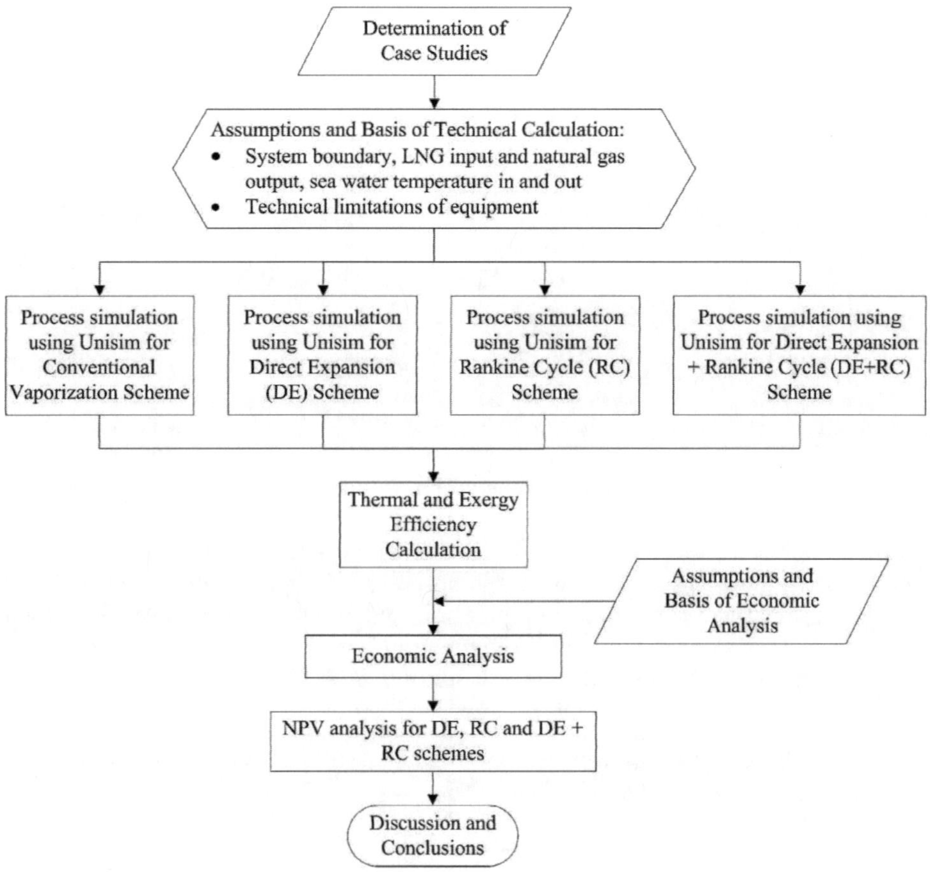

Figure 4. Research methodology

3. Results and discussion

3.1. Simulation and Efficiency Analysis Results

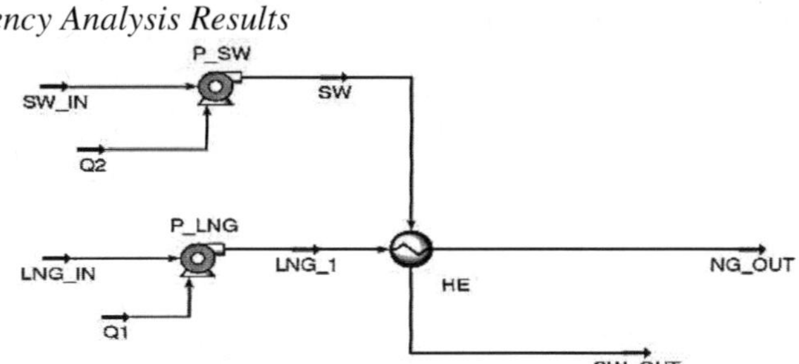

P_SW		
Energy	21.20	kW
Actual Vol. Flow	288.2	L/s
Adiabatic Efficiency	80.00	%

P_LNG		
Energy	99.20	kW
Actual Vol. Flow	32.93	L/s
Adiabatic Efficiency	80.00	%

HE		
Duty	1.252e+004	kW
UA	9.177e+005	kJ/C-h
Tube Side Delta T	177.9	K
Shell Side Delta T	-10.00	K

Figure 5. Simulation of conventional evaporation scheme

The result of the conventional evaporation scheme simulation in Figure 5 has negatif net work of 120.4 kW. This scheme required electricity of 2.22 kWh/ton LNG regasified and discards LNG cold exergy into sea water.

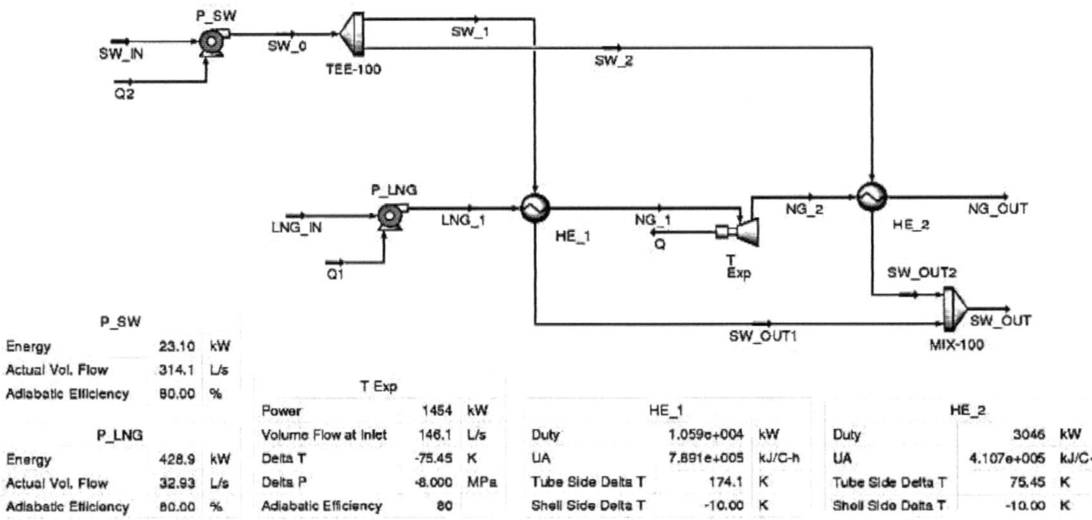

Figure 6. Simulation of DE scheme with turbine inlet pressure at 105 bar

The gas pressure entering the turbine is a key parameter in the DE scheme, because inlet pressure is directly proportional to the LNG and sea water pump power. However, the power generated by the turbine is not proportional to the turbine inlet pressure, where at a higher pressure the tendency of increasing power becomes slower. DE scheme simulation with turbine inlet pressure of 105 bar in Figure 6 have the ability to produce optimum net work of 929.04 kW and generating electrical energy of 17.13 kWh/ton LNG.

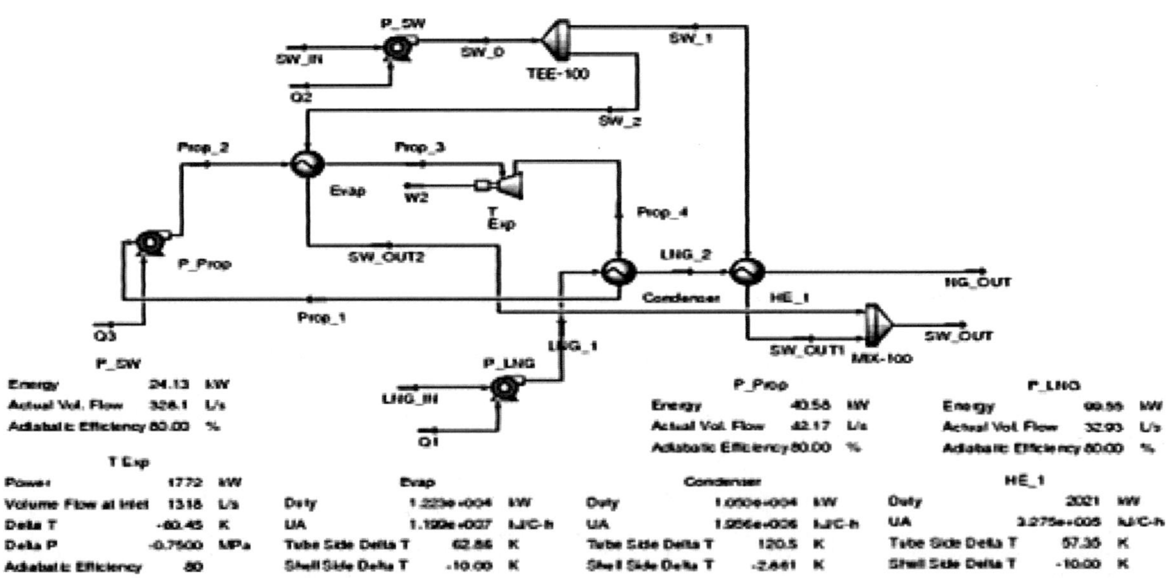

Figure 7. Simulation of RC scheme with turbine inlet pressure at 8.6 bar

The RC scheme only exploits the LNG thermal exergy to cool the condenser. The net work is affected by the turbine inlet pressure, where higher pressure means more power can be generated. However, turbine inlet pressure is limited by sea water inlet and outlet temperature of the evaporator. 8.6 bar is the optimum pressure that can be used without causing temperature cross on the evaporator. The RC scheme in Figure 7 can produce an optimum net work of 1519.30 kW and electrical energy of 28.01 kWh/ton LNG.

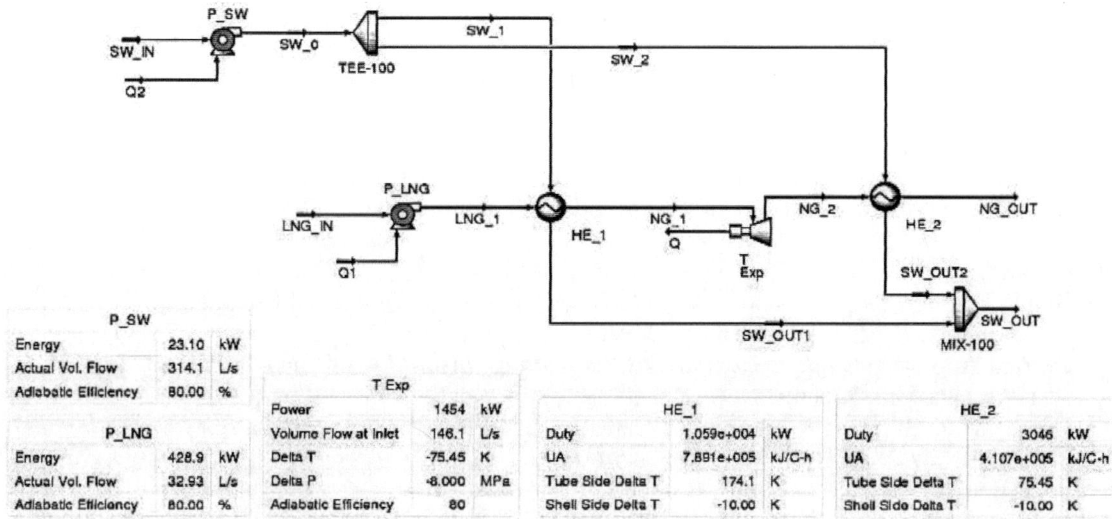

Figure 8. Simulation of DE+RC scheme with turbine 1 and turbine 2 inlet pressure at 65 and 8.6 bar

The DE+RC combination scheme can take advantage of all physical exergy in LNG. The optimum net work of 2158.38 kW and the electrical energy of 39.80 kWh/ton LNG can be obtained by maximizing the turbine inlet pressure on the RC cycle of 8.6 bar and the turbine inlet pressure at the DE cycle of 65 bar as simulated in Figure 8.

The results of the efficiency analysis in Table 2 show that the combination scheme DE+RC has the highest thermal and exergy efficiency, followed by the RC scheme and the last place is DE scheme.

Table 2. Thermal and exergy efficiency analysis results

Scheme	Wnet (kW)	kWh/ton LNG	η thermal (%)	η exergy (%)
DE	929.04	17.13	6.82	52.12
RC	1,519.30	28.01	10.68	56.24
DE+RC	2,158.38	39.80	14.48	60.71

3.2. Result of economic analysis

The economic analysis results for the application of the LNG cold exergy utilization scheme at the Gresik LNG receiving terminal in Table 3 is calculated based on the technical data of the equipment from the simulations and economic assumptions mentioned before. The DE scheme is the only scheme that has a positive NPV value of USD 695,032 while the RC and combination of DE + RC scheme has a negative NPV value. Only DE scheme is economically feasible to be applied when the electricity selling price is USD 0.0586/kWh.

Table 3. Result of economic analysis of three LNG cold exergy utilization scheme

Scheme	Fixed Cap (USD)	First Year Opex (USD/y)	Revenue (USD/y)	NPV (USD)
DE	2,412,833	96,513	491,098	695,032
RC	6,237,960	249,518	803,114	-2,201,276
DE+RC	7,959,666	318,387	1,140,937	-1,813,362

4. Conclusion

Three alternative schemes of utilizing cold LNG exergy to generate electrical energy at Gresik LNG receiving terminals have been analysed technically and economically in this paper. Based on the case study, it is known that the Direct Expansion+Rankine Cycle combination scheme has the ability to produce the highest electrical energy and the highest efficiency of energy and exergy use. However, based on the economic analysis, only Direct Expansion scheme has positive value of NPV and feasible to be applied in Gresik.

5. Acknowledgement

The authors would like to thank all those who have supported this research, especially Direktorat Riset dan Pengabdian kepada Masyarakat Universitas Indonesia through PITTA 2018 grant.

6. References

[1] Blagin E V, Uglanov D A and Dovgyallo A 2016 *About LNG Energy Utilization Efficiency Estimation Proc. Eng.* vol **152** pp 209-18

[2] Gerrard A M 2000 *Guide to Capital Cost Estimating: Institution of Chemical Engineers* (Warwickshire: IChemE)

[3] Gómez M R, Garcia R F, Gómez J R, Carril J C 2014 *Review of Thermal Cycles Exploiting the Exergy of Liquefied Natural Gas in the Regasification Process Ren. Sust. Energy. Rev.* vol **38**) pp 781-95

[4] Kim H and Hong S 2006 *Review on Economical Efficiency of LNG Cold Energy Use in South Korea (Amsterdam: 23rd World Gas Conference)*

[5] Lee Y H and Kim H K 2015 *Energy and Exergy Analyses of a Combined Power Cycle using the Organic Rankine Cycle and The Cold Energy of Liquefied Natural Gas* (Entropy) vol **17**(9) pp 6412-32

[6] Ong C W and Chen C L 2018 *Techno-Economic Analysis of Seawater Freezing Desalination using Liquefied Natural Gas Chem. Eng. Trans.* vol **70** pp. 373-8

[7] Shi X and Che D 2009 *A Combined Power Cycle Utilizing Low-temperature Waste Heat and LNG Cold Energy Energy. Conv. Manag.* volume **50(3)** pp 567-75

[8] Son H, Cho H, Gim M, Yoon S, Jang M G, Kwak D H and Kim J K 2018 *Conceptual process design and optimization of refrigeration cycles for the liquefaction of boil-off gas Chem. Eng. Trans.* vol **70** pp 169-174

IOP Conf. Series: Materials Science and Engineering **543** (2019) 012045 doi:10.1088/1757-899X/543/1/012045

Cooking ability of *Manihot esculenta crantz* as new pulp raw material with *Eucalyptus* and *Acacia mangium* wood comparison using kraft pulping method

B Irawan, V Ardina, D H Prajitno, A Roesyadi[1]
Department of Chemical Engineering, Faculty of Industrial Technology Sepuluh Nopember Institute of Technology (ITS)
Kampus ITS Sukolilo, Surabaya 60111, Indonesia

E-mail: aroesyadi@yahoo.com

Abstract. This research used *manihot esculenta crantz* of 8-10 months old, *eucalyptus* and *acacia mangium* wood of 5-year-old. The active alkali (AA) charge variations of 16% to 20% with 1% intervals. The sample position of composition of *manihot esculenta crantz* wood was 0-22 cm above ground level. The density of *manihot esculenta crantz, eucalyptus* and *acacia mangium* were 480 g/cm^3, 483 g/cm^3 and 499 g/cm^3, respectively. Base on FTIR analysis, have the same dominant functional groups. The average total yield *manihot esculenta crantz* were lower than *eucalyptus* or *acacia mangium* of 12.422% and 24.92% respectively and screen yield *manihot esculenta crantz* were lower than *eucalyptus* or *acacia mangium* of 13.678% and 26.85% respectively. Each AA charge increase of 1% in *manihot esculenta crantz, eucalyptus* and *acacia mangium* can decrease the total yield by 0.635%, 0.803% and 1.14% , the screen yield decreased by 0.543%, 0.563% and 2.196% respectively. The mean of KaNo difference of *manihot esculenta crantz* higher than *eucalyptus* or *acacia mangium* of 7.038 and 5.236 respectively. Each 1% increase in AA charge may decrease the average KaNo of *manihot esculenta crantz, eucalyptus* and *acacia mangium* were 7.693%, 3.778% and 6.628% also decreased viscosity by 3.70%, 2.99% and 0.974% respectively.

Keywords: *Manihot esculenta crantz*; Kraft Pulping; KaNo; Active Alkali; Delignification.

1. Introduction

Paper is an indipensable material in daily lie. The raw material of paper making is pulp. While the raw material of pulp is cellulose (fiber) which in this case is taken from wood. A*cacia mangium, acacia crasicarfa* and *eucalyptus* used as raw material in the pulping process [1]. The availability of raw materials used in the pulping process needs a sustainable system. About 91% of pulp and paper production in the world have been used wood as raw material, especially in developing countries [2]. Recently there have been efforts to replace wood from non-wood materials [3]. Non-wood used as pulp and paper raw material has increased by 10% over the last two decades, while the use of wood fiber has increased by only 4% [4]. To create a sustainable condition, it is necessary to look for some

[1] To whom any correspondence should be addressed.

alternative raw materials as a replacement or a mixture of raw materials that already exist. As an alternative raw material, it can be used other types of wood such as cassava wood (*manihot esculenta crantz)*. Unused cassava stems are lignocellulose waste, which is an organic component. Based on laboratory analysis, the plant stems contained 56.82% α cellulose, lignin 21,72%, ADF 21,45% and fiber length 0,05 - 0,5 cm [5]. To ensure the eligibility of *manihot esculenta crantz* wood, it is necessary to evaluate the wood cooking capability. The *manihot esculenta crantz* stems of 90% are rejected as waste, while 10% of the stems are used as seed for replanting [5]. This research used *manihot esculenta crantz* composite wood was 8 to 10 months old. *Eucalyptus* and *acacia mangium* wood 5 years old as a comparison. Pulping is performed to reduce wood or other fibrous material into a fibrous mass through chemical reactions. During pulping, lignin, cellulose and hemicelluloses are attacked and degraded by chemicals [6]. Kraft pulping is a chemical pulping method using sodium hydroxide and sodium sulfide to separate cellulose with lignin [7]. The alkali consumed in the impregnation stage affected the alkali profile of the cooking stage. The more alkali consumed in impregnation, the lower the demand in cooking. Higher alkali consumption in the impregantion stage also led to faster delignification in the cooking stage [8] Lignin will be extracted by NaOH during heating the cooking process by following the difference in NaOH concentration [9]. Delignification can be influenced by various reaction parameters such as temperature, reagent loadings and time [10]. The kraft process was chosen because it has several advantages over other mechanical or chemical pulp processes. Some of the advantages kraft process are [11].

1. Able to process various types of wood both similiar and mixed
2. Cooking time is relavely short.
3. The yield and quality of pulp produced is quite high

The phenomenon that occurs in the kraft pulping process is the consumption of alkali with events such as [12]:

1. Reaction with lignin
2. Dissolve carbohydrates
3. Reacting with some organic acids, whether derived from wood or derived from the process.
4. React with resin in wood
5. Absorption by fiber.

The effect of kraft cooking occurs at the impregnation stage takes place at 130°C for 30 minutes [11]. Kraft pulping, which cleaves the β-O-4 linkages between the C9 units in lignin by sulfide at alkaline environment and up to 170 °C [13]:

O-R1-O-R2 + OH−→O=R1OR2

O=R1OR2+HS−→O-R1S+R2Ona (fragmented)

The degree of polymerization indicates the level of cellulosic degradation during the cooking process. The higher the DP value, the stronger cellulose (fiber) in cellulose degradation events. The solubility of cellulose in soda decreases as the degree of polymerization increases [9].

2. Methodology

2.1. Material

In this research, *manihot esculenta crantz* wood, *eucalyptus* and *acacia mangium* were taken from Jambi province. Wood age *manihot esculenta crantz* was 8-10 months, while *eucalyptus* and *acacia mangium* were 5 years old. The chips size of the three woods was 10-30 mm long 3-6 mm thickness.

2.2. Methods

The purpose of this research is to know the quality of pulp from kraft pulping process for *manihot esculenta crantz* wood as new pulping raw material. The *eucalyptus* and *acacia mangium* were comparator. Wood test parameters are wood composition, FTIR and density.

The pulp test parameters are total yield, screen yield, KaNo and viscosity. The cooking process condition were variation of AA charge 16% to 20% and variation 1% , sulfidity 25%, liquid ratio 3,5, H-factor 800 and maximum temperature 165°C. The washing process was done by adding water and agitated. The aims to decipher the pulp that was formed. The screening process was done in two stages with screen 0,15 mm slot and 325 mesh.

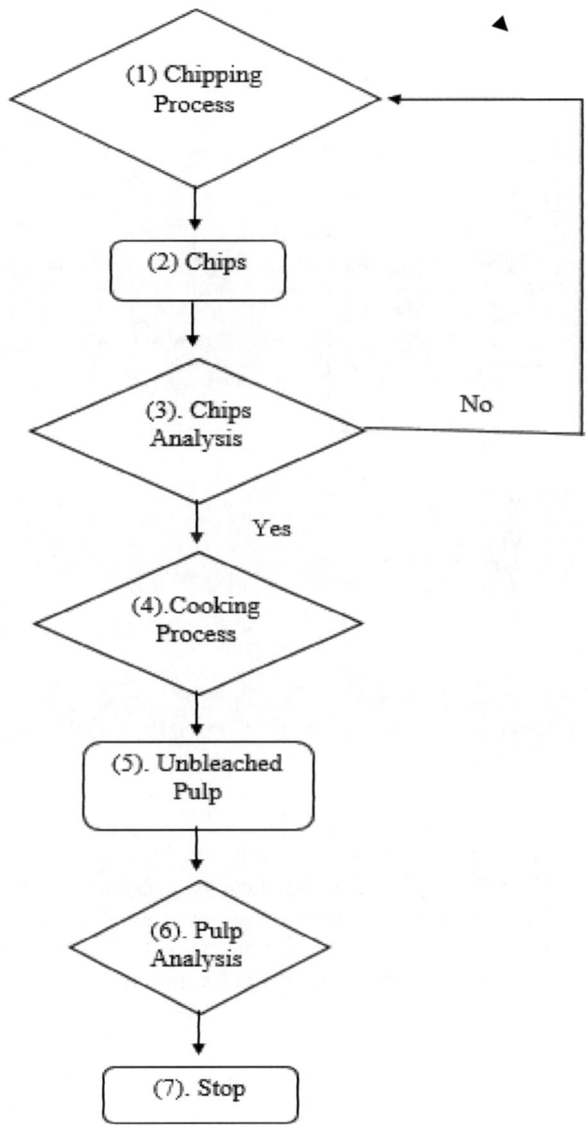

Figure 1. A series of pulping process workflow.

2.3. Instrumentation

This research the main equipment was a set of rotary digester and filter pulp with screen size of 0.15 mm slot and screen 325 mesh used sequentially as Figure 2.

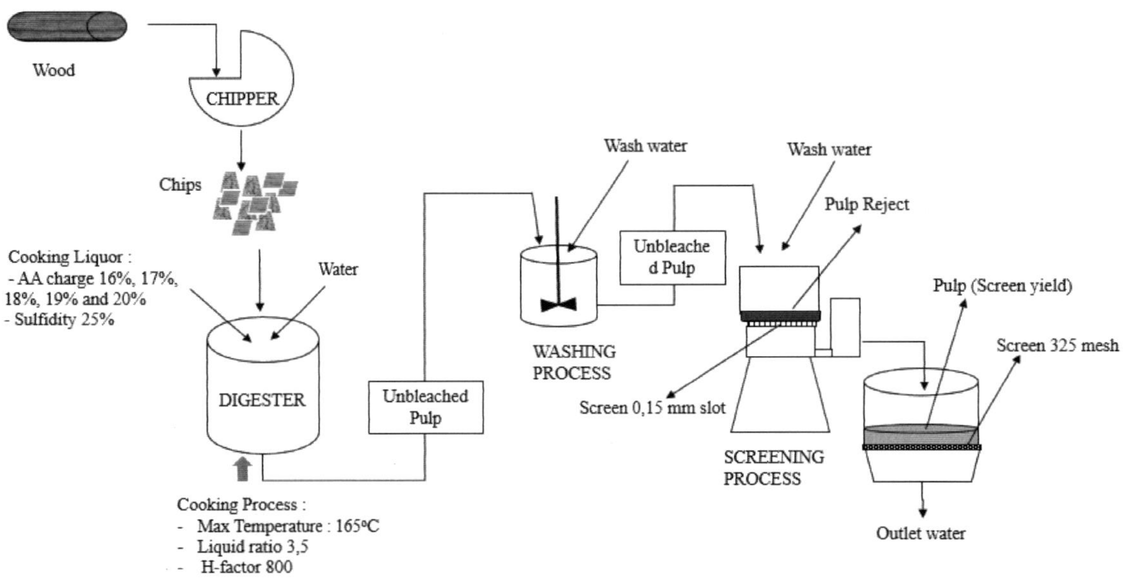

Figure 2. A series of work tools and pulping process flow including cooking, washing and screening processes.

2.4. Wood analysis

Parameter of wood gravimetry analysis in this research was wood composition in the tree, basic density of wood and FTIR analysis of wood. The wood composition in the tree was analyzed based on the proportion of dry weight of the wood component to the total weight with the sample position along the 22 cm distance from above the soil surface with a variation of 2 cm. Basic density was analyzed by using TAPPI-T258-om 2 analysis method in maximum moisture condition [14].

2.5. Cooking process

Cooking process was done by using mini digester Haato 6 chamber with maximum capacity 300 gr each of OD chips. OD (oven dry) is the condition of dry wood oven with 0% water content. Cooking conditions were maximum temperature 165°C, liquid ratio 3,5, sulfidity 25.0%, active alkali charge variation 16%, 17%, 18%, 19% and 20% and H-factor 800.

Calculation of cooking time based on the delignification rate defined as k_T / k373 where k is the delignification rate constant at T (K) and k_{373} is a constant at 373K. From the Arrhenius equation it is found

$$k = A.\ e\text{-}Ea/RT$$

IOP Conf. Series: Materials Science and Engineering **543** (2019) 012045 doi:10.1088/1757-899X/543/1/012045

Where,
Ea: Activation energy, determined with the value of 32 Kcal / mol, for kraft pulping.
T: Absolute temperature
R: Common gas permeability, 8,314 J / mol.K

Thus, k_T / k_{373} = e-Ea / R (1 / T-1/373)

The value of k is calculated starting at temperature 100°C (k = 1), from the calculation obtained the relative rate constant becomes [15]

k = kr = exp (43.2 - 16.113 / T)

H-Factor is defined as an integral of constanta relative rate (kr) with time [16].

Formula:
H = ∫ kr dt = ∫ exp (43.2 – 16.113 / T) dt
Based on the applied cooking conditions, the temperature observed during the cooking process and relative rate constanta was shown in figure 3.

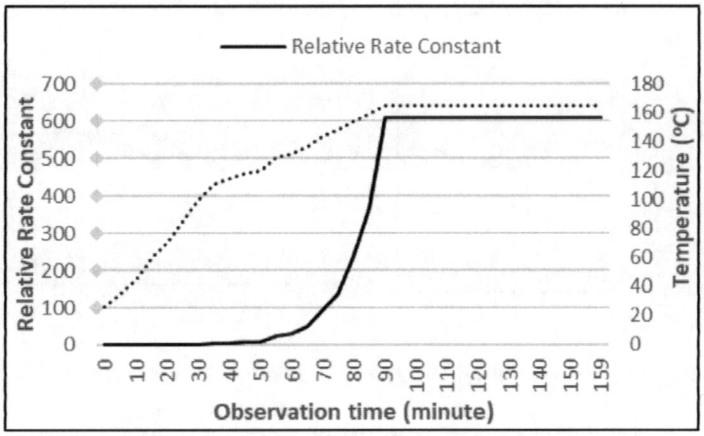

Figure 3. Observation of temperature and relative rate constanta on cooking *manihot esculenta crantz, eucalyptus* and *acacia mangium* with H-factor 800.

From Figure 3, it can be explained that the relative rate constant value can be applied at 30-minute observation with temperature of 100°C. Furthermore, the temperature continues to be observed and converted in relative rate constants to a maximum temperature of 165 ° C. At temperature of 165°C, the temperature is kept steady during cooking time. For the calculation of total cooking time to note is, the highest relative rate constant (608.3), H- Factor value and total average accumulated relative rate constant (1266.45) to be considered for calculation of total cooking time. Calculation of cooking time can be explained by the in the following steps.
Cooking Time Calculation:
H-Factor = 800
Calculation: 1266.45 x 5/60 = 105.5375
(800 – 105.5375) x 60 / 608.3 = 68.498 minutes
Cooking time = 68.498 minutes + 90 minutes = 158.498 minutes (= 2 hours, 38 minutes)

2.6. *Analysis of cooking pulp*

Cooking analyze was done among others. Total yield was difference between total pulp weight (OD) and total weight of wood sample (OD).
Screened yield was the difference between the weight of the filtered pulp (OD) by the total weight of the wood sample (OD). Kappa Number test method use TAPPI-T236-cm85 method [17] and viscosty use TAPPI-T230-om04 method [18].

3. Result and Discussion

3.1. *Result of wood analysis*

3.1.1. Gravimetric analysis of wood composition in trees.
The percentage of wood and non-wood in a tree to consider in the cooking process. Percentage of wood can be analysed by comparing between wood and non-wood components in certain positions.

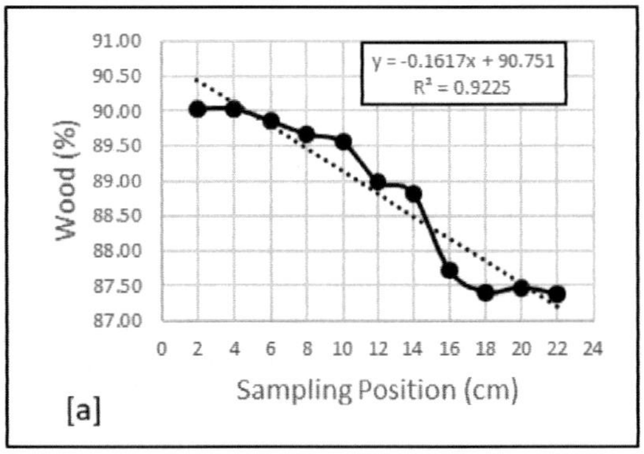

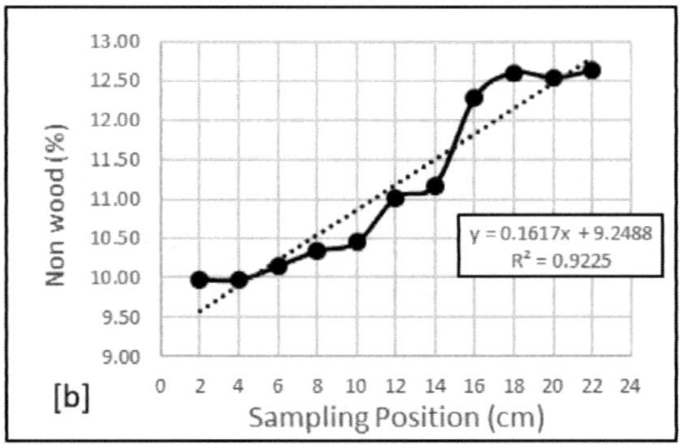

Figure 4. (a) Percentage of wood (% w) and (b) non-wood (% w) in *manihot esculenta crants.*

In Figure 4.a, it can be seen that the percentage of wood on the stem of *manihot esculenta crants* the higher the sample position, the lower the percentage of wood. The average percentage of wood at 22 cm above the ground is still above 87%. With the correlation degree between the sampling position (cm) and % w wood of 0.9225 with the correlation equation Y = -0.1617x + 90.751. The height of wood that can be recommended for the pulping process is up to 22 cm above ground level. Figure 4.b. shows the percentage of non-wood in the *manihot esculenta crants* , the sampling position the higher the non-wood percentage in the stem. The correlation degree between the height of the sampling position with percentagenof non-wood was 0.9225 with correlation equation Y = 0.1617x + 9.2488. From the graph, it can be explained that at a height of 22 cm above the soil surface of non-wood content was 12.5%.

3.1.2. Wood basic density analysis
In the pulping process, one of the main parameters to be considered in the main material is the wood density value. This wood density value shows the density of wood components, including fiber and lignin in certain volumes. The higher the density value, the higher the cellulose content in the wood.

Table 1. Wood density analysis results

Sample	Density (g/cm^3)	Average (g/cm^3)	Deviation. Std.
	0.477		
Manihot	0.476		
esculenta	0.489	0.480	0.0078
crantz	0.471		
	0.487		
	0.494		
	0.472		
Eucalyptus	0.481	0.483	0.0085
	0.480		
	0.479		
	0.510		
	0.499		
Acacia	0.489	0.499	0.0100
mangium	0.509		
	0.490		

Based on the density analysis, the *manihot esculent crantz* wood density was 0.480 g/cm^3 while the *eucalyptus* and *acacia mangium* wood density values were 0.483 and 0.499 g/cm^3 respectively. The standard deviation of *manihot esculenta crantz*, *eucalyptus* and *acacia mangium* were of 0.0078, 0.0085 and 0.0100, respectively.

3.1.3. Result of FTIR analysis of wood

This FTIR analysis aims to track the structure of organic molecules. The result of the FTIR analysis of *manihot esculenta crantz, eucalyptus* and *acacia mangium* wood can be seen in figure 5 to 7.

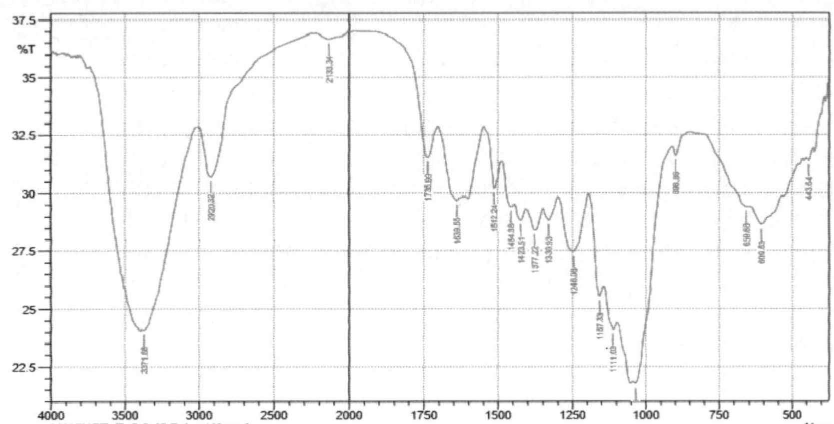

Figure 5. FTIR analysis results of *manihot esculenta crantz* wood

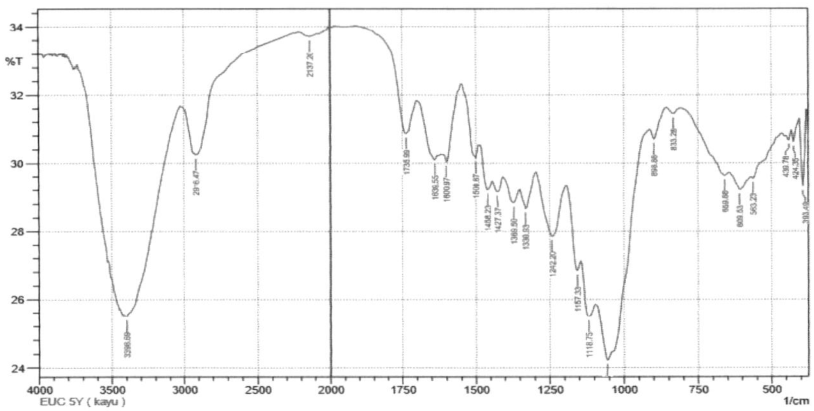

Figure 6. FTIR analysis results of *eucalyptus* wood

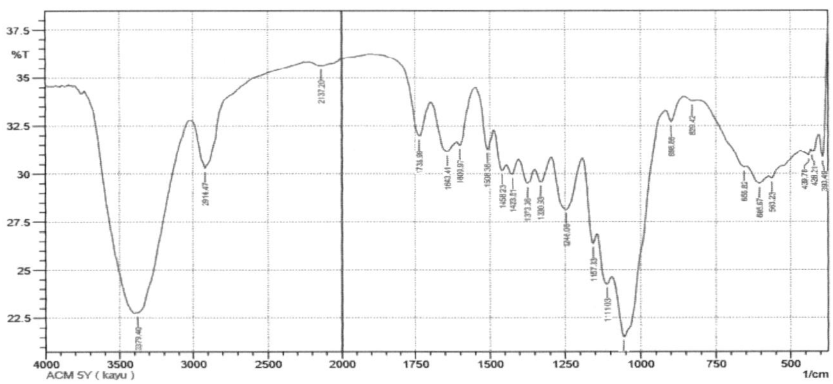

Figure 7. FTIR analysis results of *acacia mangium* wood

Base on figure 5 to 7, the comparison of FTIR analysis results can be seen that *manihot esculenta crants, eucalyptus* and *acacia mangium* have the same dominant functional groups.The functional groups were carboxylic acid (wavelengths 3500-3300 cm^{-1}), CH group (wavelengths 2970-2850 $cm^{-1)}$, alkuna (wavelengths 2260 - 2210 $cm^{-1)}$, carboxylic groups (wavelengths 1760 - 1690 $cm^{-1)}$, alkene groups (wavelengths 1650 - 1450 $cm^{-1)}$, CH3 groups (wavelengths 1470 - 1340 $cm^{-1)}$, and compounds aromatic.

3.2. Result of cooking process

In this research, total yield, secreen yield, KaNo, and viscosity parameters were analyzed. From the results of this analysis, it can be seen the differences of the quality of the products. The cooking conditions in this study based on the difference of active alkali charge of 16%, 17%, 18%, 19% and 20%. The difference of NaOH concentration based on the active alkali charge concentration.

IOP Conf. Series: Materials Science and Engineering **543** (2019) 012045 doi:10.1088/1757-899X/543/1/012045

3.2.1. Total yield

Total yield is the ratio of the weight of the pulp obtained before entering the screening stage with the weight of the chips sample in dry weight condition (OD). This total yield can illustrate the high level of wood component dissolved in the cooking liquid. Hemicellulose are more easily degraded under alkaline conditions than cellulose due to their low degree of polymerization [19]. From the result of total yield, it can be seen the total yield difference of *manihot esculenta crantz* , *eucalyptus* and *acacia mangium*, as shown in Figure 8.

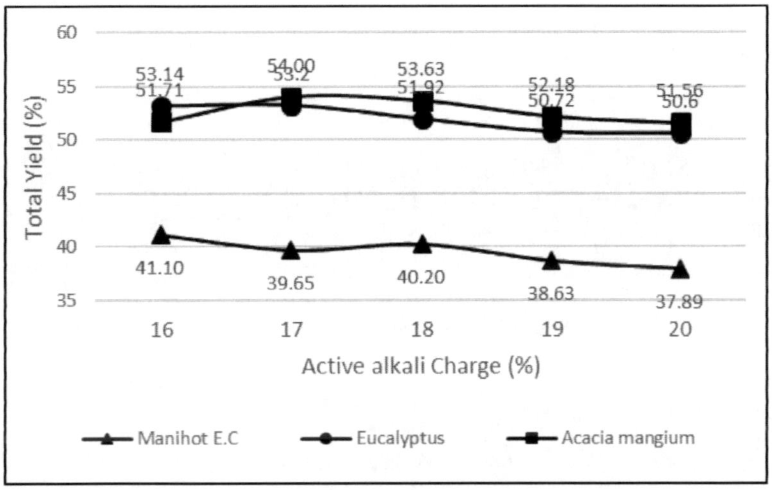

Figure 8. Effect of AA charge on total yield in cooking process of *manihot esculenta crantz, eucalyptus* and *acacia mangium.*

From Figure 8, it can be seen the difference of total yield between *manihot esculenta crantz* wood with *eucalyptus* and *acacia mangium*. Total yields of *eucalyptus* and *acacia mangium* woods were higher than that of *manihot esculenta crantz* on AA charges of 16% to 20%. The total yield difference between *manihot esculenta crantz* and *eucalyptus* at AA charge 16%, 17%, 18%, 19% and 20% were 12.04%, 13.55%, 11.70%, 12.09% and 12.71% respectively. Mean difference in total yield between *manihot esculenta crantz* and *eucalyptus* in cooking with variation of AA charge of 16%, 17%, 18%, 19% and 20% were 12,422%. The difference of total yield between *manihot esculenta crantz* and *acacia mangium* on AA charge 16%, 17%, 18%, 19% and 20% were 20.52%, 26.57%, 25.04%, 25.97% and 26.51% respectively. The average of total yield difference between *manihot esculenta crantz* and *acacia mangium* on AA charge variation of 16%, 17%, 18%, 19% and 20% was 24.92%.

Based on wood species, any increase in AA charge can affect the total yield decrease. The decrease of total yield of the *manihot esculenta crantz* , with an AA charge increase of 16% to 17%, 18% to 19% and 19% to 20% were 1.45%, 1.57% and 0.74% respectively, except for AA charge changes of 17% -18%, with slight increase of 0.55%. Any increase in AA charge of 1% in *manihot esculenta crantz* may result the reduction of the total yield of 0.803%. For *eucalyptus*, the AA charge increase of 17% to 18%, 18% to 19% and 19% to 20% may result the decrease of the total yields of 1.28%, 1.20% and 0.12% respectively, except for the AA charge of 16% to 17% with the total yield increased of 0.06%. Every 1% AA charge increase in the sample *eucalyptus* may result in the total yield decrease of 0.635%.

For *acacia mangium*, AA charge increase of 17% to 18%, 18% to 19% and 19% to 20% resulted in the decrease of total yield of 0.685%, 2.70% and 1.19% respectively. AA charge of 16% to 17% increase the total yield of 4.43%. Every 1% AA charge increase in the *acacia mangium* may result the decrease of average total yield of 1.14%.

3.2.2. Screen yield

Screen yield is calculated by comparing the weight of OD pulp that has passed through the screen process with the weight of OD chips. The AA charge influence the quality of pulp. From the screen yield results, it can be seen the screen yield difference between *manihot esculenta crantz, eucalyptus* and *acacia mangium* as shown in Figure 9.

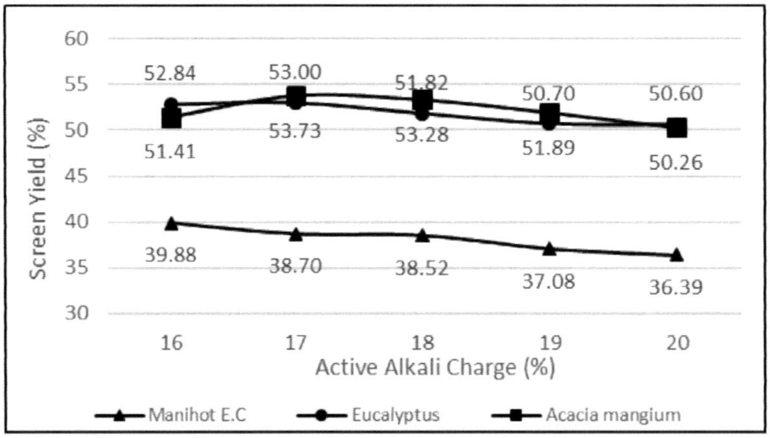

Figure 9. Effect of AA charge on screen yield (%) in cooking process of *manihot esculenta crantz, eucalyptus,* and *acacia mangium.*

Figure 9 shows, that screen yield of *manihot esculenta crantz* wood has a lower value compared with *eucalyptus* and *acacia mangium*, with AA charge of 16% to 20%. Screen yield between *manihot esculenta crantz* and *eucalyptus* with AA charge 16%, 17%, 18%, 19% and 20% were 12.96%, 14.3%, 13.3%, 13.62% and 14% respectively. The average difference of screen yield between *manihot esculenta crantz* with *eucalyptus* was 13.678%. Meanwhile, screen yield when compared with *acacia mangium,* the AA charge of 16%, 17%, 18%, 19% and 20% were 22.43%, 27.97%, 27.70%, 28.54% and 27.60 % respectively. The average difference of screen yield between *manihot esculenta crantz* with *acacia mangium* was 26.85%.

For *manihot esculenta crantz*, an AA charge increase of 16% to 17%, 17% to 18%, 18% to 19% and 19% to 20% resulted in a decrease in screen yields of 1.18%, 0.18%, 0.12%, and 0.69% respectively. Every 1% AA charge increase in *manihot esculenta crantz* may result in a 0.543% decrease in screen yield. For *eucalyptus*, the AA charge of 17% to 18%, 18% to 19% and 19% to 20% can decrease the screen yield by 1.18%, 1.12% and 0.01% respectively, except for the AA charge of 16% to 17%, with a slight increase of 0.06%. Every 1% increase in AA charge in *eucalyptus* can result in a decrease in screen yield of 0.77%. For *acacia mangium*, 16% to 17% increase in AA charge caused a 4.51% increase in screen yield, while with AA charges of 17% to 18%, 18% to 19% and 19% to 20% can decrease screen yield of 0.838%, 2.61% and 3.14% respectively. Every 1% AA charge increase in the *acacia mangium* can decrease the average screen yield by 2.196%.

3.2.3. Kappa number (KaNo)

The kappa number or KaNo is a non-unit number indicating the remaining lignin content in the pulp of a pulping process. The kappa number indicates the level of maturity and whiteness or degree of delignification of the pulp [2]. KaNo pulp values in *manihot esculenta crantz, eucalyptus* and *acacia mangium* can be seen in Figure 10.

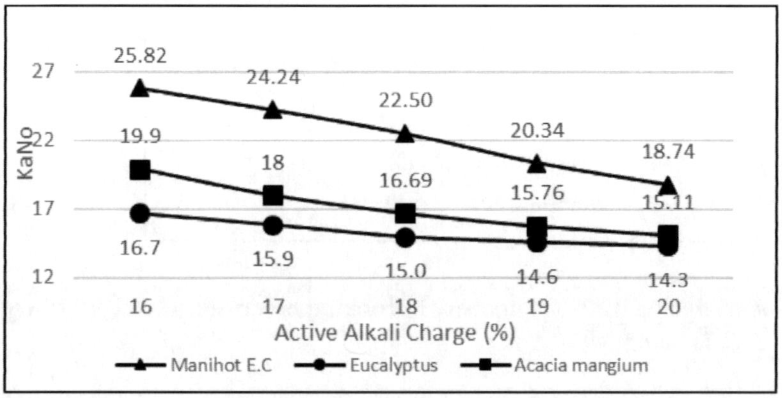

Figure 10. Effect of AA Charge on KaNo in cooking process of *manihot esculenta crantz, eucalyptus* and *acacia mangium.*

Figure 10 shows, that the kappa number of *manihot esculenta crantz* was higher than *eucalyptus* wood and *acacia mangium*. The KaNo differences between *manihot esculenta crantz* with *eucalyptus* on AA charge of 16%, 17%, 18%, 19% and 20% were 9.13, 8.37, 7.51, 5.74 and 4.44 respectively. The difference of KaNo between *manihot esculenta crantz* and *eucalyptus* was 7.038 averagely. Meanwhile, KaNo when compared with with *acacia mangium* with AA charge of 16%, 17%, 18%, 19% and 20% were 5.92, 6.24, 5.81, 4.58 and 3.63 respectively. Mean difference of KaNo was 5,236. The *manihot esculenta crantz* on AA charge increase of 16% to 17%, 17% to 18%, 18% to 19% and 19% to 20% resulted decrease KaNo were 6.12%, 7.18%, 9.60% and 7.87% respectively. Every 1% AA charge increase may result 7.693% KaNo decrease averagely. For *eucalyptus*, the AA charge increase of 16% to 17%, 17% to 18%, 18% to 19% and 19% to 20% can decrease KaNo were 4.91%, 5.55%, 2.60% and 2.05% respectively. Every 1% AA charge increase may result 3,778% KaNo decrease averagely. For *acacia mangium,* the AA charge increase of 16% to 17%, 17% to 18%, 18% to 19% and 19% to 20% can decrease KaNo were 9.55%, 7.27%, 5.57% and 4.12% respectively. Every 1% increase in AA charge may result 6.628% KaNo decreased averagely.

3.2.4. Viscosity Value

Viscosity value is a measure of fiber strength to fiber degradation process by the influence of AA charge on cooking process. Comparing the intrinsic viscosity of cellulose suspensions or solutions, before and after a chemical treatment, is a very valuable way to estimate the changes in molecular weight [20]. The higher the viscosity value indicates the higher the fiber resistance to the degradation process. The effect of AA charge on pulp viscosity in cooking process *manihot esculenta crantz, eucalyptus* and *acacia mangium* can be seen in Figure 11.

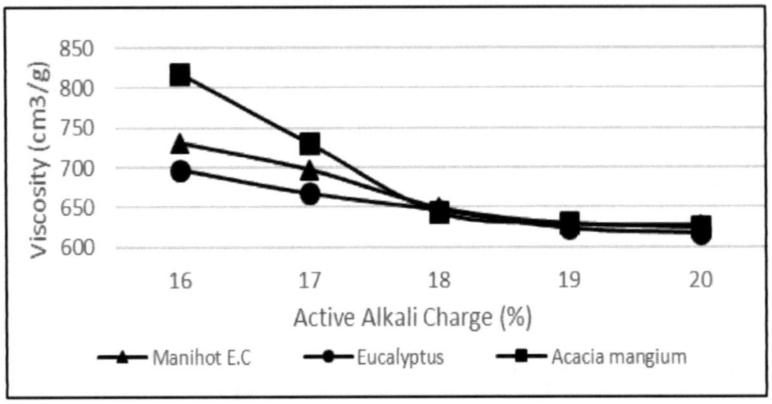

Figure 11. Effect of AA charge on the pulp viscosity in cooking process of *manihot esculenta crantz, eucalyptus* and *acacia mangium*

Figure 11, shows that the viscosity value of cooking result with variation of active alkali charge between 16% to 20% with 1% range. The active alkali charge range 16% and 17% showed viscosity significantly different when compared with AA charge 18%, 19% and 20%. The pulp viscosity value of *manihot esculenta crantz* wood was higher than *eucalyptus*. At AA charge 16%, 17%, 18%, 19% and 20% viscosity differences were 4.65%, 4.30%, 0.46%, 0.95% and 1.56% respectively. The large viscosity differences between *manihot esculenta crantz* and *eucalyptus* pulp occurred on 16% and 17% AA charges with average was 4.475%. The viscosity value of the *manihot esculenta crantz* pulp was lower than *acacia mangium* at 16% and 17% AA charge. At AA charge 16%, 17%, 18%, 19% and 20% viscosity differences were 11.9%, 4.88%, 0.77%, 0.16% and 0.32% respectively. The large viscosity difference between the *manihot esculenta crantz* and *acacia mangium* occurred of 16% and 17% AA charges was 8.39%.

For *manihot esculenta crantz*, an AA charge increase of 16% to 17%, 17% to 18%, 18% to 19% and 19% to 20% can decrease viscosity were 4.65%, 6.89%, 3.08% and 0.18% respectively. Every 1% AA charge increase in *manihot esculenta crantz* may result the reduction of the viscosity value was 3.70% averagely. For *eucalyptus*, an AA charge increase of 16% to 17%, 17% to 18%, 18% to 19% and 19% to 20% can decrease the viscosity value were 4.30%, 3.15%, 3.56%, and 0.963% respectively. Every 1% AA charge increase may result the reduction of the viscosity value was 2.99% averagely. For *acacia mangium*, AA charge increase of 16% to 17%, 17% to 18%, 18% to 19% and 19% to 20% can decrease the viscosity value 10.64%, 11.90%, 2.17% and 0.974% respectively. Eevery 1% AA charge increase may decrease viscosity value was 6.42%.

4. Conclusion

Manihot esculenta crantz is one of the raw materials that can be used for the process pulping. The proportion of wood and non-wood that on trees with a distance of 22 cm above the surface of the earth wood component is greater than 87%. Wood basic density of *manihot esculenta crantz* equivalent with *eucalyptus* wood and *acacia mangium* were for 0.480 g/cm^3, 0.483 g/cm^3 and 0.499 g/cm^3 respectively. *Manihot esculenta crants, eucalyptus* and *acacia mangium* have the same dominant functional groups were carboxylic acid, CH group, alkuna, carboxylic groups, alkene groups, CH_3 groups, and compounds aromatic. The total yield and screen yield of *manihot esculenta crantz* wood cooking process was lower than *eucalyptus* and *acacia mangium.*

The difference of total yield between *manihot esculenta crantz* with *eucalyptus* and *acacia mangium* was 12.42% and 24.92% respectively, while the difference of screen yield was 13.68% and 26.85% respectively. Lignin content was influential components in the cooking process.
The KaNo of *manihot esculenta crantz* pulp was higher than *eucalyptus* or *acacia mangium.* KaNo differences between *manihot esculenta crantz* with *eucalyptus* and *acacia mangium* were 7.038 and 5.236 respectively. However, the delignification process of *manihot esculenta crantz* more maximal than *eucalyptus* and *acacia mangium* wood. Every 1% increase of AA charge of *manihot esculenta crantz, eucalyptus* and *acacia mangium* were 7.693%, 3.778% and 6.628% respectively. Differences of pulp viscosity values between *manihot esculenta crantz* and *eucalyptus* occurred at AAcharge 16% and 17% were 4.65% and 4.30% respectively. The viscosity value of *manihot esculenta crantz* was higher than *eucalyptus*, with an average difference of 4.475%. The difference of viscosity value of *manihot esculenta crantz* with *acacia mangium* at AA charge 16% and 17% were 11.9% and 4.88% respectively. The viscosity of *manihot esculenta crantz* was lower than *acacia mangium* with an average difference 8.39%. Every 1% AA charge increase in cooking process of *manihot esculenta crantz, eucalyptus* and *acacia mangium* can decrease viscosity value were 3.70%, 2.99% and 6.42% respectively. Based on the total yield and screen yield value, indicates that the cellulose content of *manihot esculenta crantz* was lower than *eucalyptus* and *acacia mangium.* Base on the viscosity value, indicates the fiber strength of *manihot esculenta crantz* was better than *eucalyptus.* Cooking conditions, such as AA charge , maximum temperature, H-factor and liquid ratio (LR) value, were needed to increase total yield, screen yield, viscosity and to lower KaNo of *manihot esculenta crantz* pulp.

References

[1] Sugesty S, Kardiansyah T and Pratiwi T W 2015 *Jurnal Selulosa* **4** 99
[2] Fatriasari W and Risanto L 2011 *Widyariset* **14**
[3] Kamoga O L M, Byaruhanga J K and Kirabira J B 2013 *Int. J. Chem. Eng. Appl.* **4** 144
[4] Behin J and Zeyghami M 2009 *Chem. Eng. J* **152** 26.
[5] Sumada K, Tamara P E, Alqani F 2013 Jurnal *Teknik Kimia* **5** 2
[6] Aal MA 2013 *World Appl. Sci. J.* **27** 1
[7] Rahmati H, Ebrahimi P and Sedghi M. 2010 *Indian J. Chem. Techno* **17** 74
[8] Tavast D, Branval E 2017 *Nordic Pulp & Papaer Research Journal* **32** 14
[9] Ji Y, Wheeler M C and Heiningen, A 2007 *AIChE J* **53** 2681.
[10] Kumar R, Hu F, Hubbell C, Ragauskas AJ, Wyman C E 2013 *J. Biortech* **130** 372
[11] Yahya R 2016. *Molekul*, **11** 92.
[12] Gabriella A, Duarte F A S, Da Silva F G, Sacon V M 2015 *Effect of Active Alkali charge on Eucalyptus Kraft Pulping*, (Brazil: San Paulo University)
[13] Hu J, Zhang Q and Lee J D 2017 *J Biortech* 247 1181
[14] TAPPI test Method 2002 TAPPI 258-om 02 *Basic density and moisture content of pulpwood*, (Canada: TAPPI Press).
[15] Tunc M S 2003 *Relationship Between Alkaline Pulp Yield and The mass Fraction and Degree of Polymerization of Cellulose in Pulp* (Ankara: Maine University)
[16] Zou H 2012 *Effect of Kraft Pulping on Oxygen Delignification* (Ankara: Maine University)
[17] TAPPI Test Method 1999 TPPI 236-cm 85 *Kappa number of pulp* (Canada: TAPPI Press)
[18] TAPPI Test Method 2004 TAPPI 230-om 04 *Viskosity of pulp (capillary viscometer method)* (Canada: TAPPI Press)
[19] Santos R B, Jameel H, Chang H and Hart PW 2013, *Biores* **8** 158
[20] Moral A, Aguado R and Tijero A 2016 *Fibers and Polymers* **17** 857

Hydrodynamic Analysis of Hydrolysis of the Rice Husk Cellulose by Using CFD Modeling

N Novia, L Cundari, A M Jannah, M K Haphsari and M Zulfani
Chemical Engineering Department, Faculty of Engineering, Universitas Sriwijaya, Jl. Raya Indralaya–Prabumulih KM. 32 Indralaya Ogan Ilir (OI), Sumatera Selatan, Indonesia 30662

E-mail: novia@ft.unsri.ac.id

Abstract. The importance of the rule and the demand of glucose trigger the seeking of the alternative materials to produce glucose. One of the alternatives used is the utilization of biomass such as the rice husk to produce glucose. Rice husk is the agricultural residue whose availability is very abundant. In this research, rice husk contained cellulose (42.2 %), hemicellulose (17.04 %), and lignin (20.46 %). First step to produce the glucose from rice husk was pretreatment stages (Soaking in Aqueous Ammonia pretreatment and Acid pretreatment). Furthermore it was followed by the acid hydrolysis step by using dilute H_2SO_4 solution. The concentration of H_2SO_4 used in this work was 1, 2, and 3 N with the various hydrolysis times of 30, 60, 90, 120, and 150 min. The result showed that increasing of hydrolysis time slightly increased glucose concentration. The highest glucose concentration was about 21.51% at H_2SO_4 concentration of 2 N and the hydrolysis time of 150 min. CFD modeling (ANSYS Fluent 16) was performed to investigate the hydrodynamic phenomena during hydrolysis. The hydrodynamic analysis showed that the best mixing occurred at H_2SO_4 concentration of 2 N for 150 min. It is a good agreement with the experimental finding.

1. Introduction

Rice husk is agricultural residue whose availability is very abundant. It can produce glucose because it contains lignocellulose structure. The steps to produce glucose from rice husk are pretreatment and hydrolysis. Pretreatment aims to damage the lignin bond and break down a structure of cellulose crystal easily into glucose. Immersion in an ammonia solution is performed to reduce the amount of lignin and increase cellulose porosity, while the treatment using dilute sulphuric acid is done to increase cellulose porosity and help partially hydrolyse cellulose. Cellulose can be hydrolysed into glucose using aqueous media and assisted with acid or enzyme catalysts. The imperfection of the hydrolysis process can produce cellulose disaccharides. While the process of hydrolysis that runs perfectly will produce cellulose monomer in the form of glucose.

The advantage of chemical hydrolysis process is low cost. The price of chemicals is cheaper than enzymes. In addition, the enzymatically hydrolysis process takes several days while chemical hydrolysis requires relatively faster time [1]. Alternatively, dilute acids can be used to break the polymer chains of cellulose and hemicellulose. In the acid hydrolysis method, lignocellulose biomass is exposed to the acid at a given temperature and pressure for a certain time and produces a sugar monomer of cellulose and hemicellulose polymers. Sulphuric acid is the most widely studied acid and is used for acid hydrolysis [1]. Cellulose can be chemically hydrolysed using H_2SO_4 solution.

One indicator of the pretreatment effectiveness is the reduction of lignin during process. Pretreatment of the biomass results in the degradation of cellulose and hemicellulose. Effect of hydrothermal pretreatment of date palm leaflets and vine shoots on lignin was investigated by other researchers [2,3]. However they used a high temperature condition. Chemical pre-treatment using liquid ammonia and aqueous sodium hydroxide (NaOH) are recognized to change the cellulose crystalline structure and effect on enzymatic digestibility [4]. Previous authors [5], performed NaOH-dilute acid pretreatment to remove lignin in rice husk. The use of sulphuric acid is also required in the neutralization process which produces very large amounts of gypsum or lime waste and can pollute the environment [1]. Ammonia pretreatment eliminates lignin by cutting C–O–C bonds in lignin [6]. The effect of aqueous ammonia and dilute acid, on the chemical compositions, cellulose crystallinity, morphologic change, and enzymatic hydrolysis of bamboo was studied by previous researcher [7]. Therefore, this research focused on the soaking in aqueous ammonia (SAA) pretreatment and dilutes sulphuric acid of rice husk. Furthermore, hydrolysis was carried out using sulphuric acid (H_2SO_4) with various concentrations. The hydrolysis reaction between water and cellulose was catalysed by acid so that the reaction proceeded quickly and produced faster glucose.

Some studies have tried to describe the hydrodynamic phenomena in pretreatment [8] and enzymatically hydrolysis process [9–11] by using a design and CFD modelling. However this work used a different kind of reactor which is a laboratory spherical reactor. A perfect mixing process was required to obtain optimal glucose levels. Therefore, this study applied CFD modelling (ANSYS Fluent 16) to produce hydrodynamic mixing process in acid hydrolysis.

2. Materials and methods

2.1. Raw Material Preparation

The rice husk was dried under the sun for ± 1 day. Its size was reduced to 35 meshes. Then, the composition of cellulose, hemicellulose and lignin was analysed by using a Chesson method [12]. Scanning electron microscope (SEM) micrographs of cross section rice husk of were carried out by a SEM-EDX microscope Phenom Pro X.

2.2. Pretreatment Raw materials

Dried rice husk of about 40 g was put into an erlenmeyer of 300 ml and soaked in 15 % (w/v) aqueous ammonia of 160 ml at 600 °C for 5 h. The filtrate was separated from the residue by a vacuum pump. Then the residue was washed with distilled water. Pretreated sample was fed to a glass bottle, and then added 135 ml of H_2SO_4 (0.18 N). It was stirred evenly for 1 min. Then it was heated in the oven at 160 °C for 35 min. It was cooled in room temperature and filtered by vacuum pump. Pretreated sample was analysed by using a Chesson method [12].. Scanning electron microscope (SEM) micrographs of cross section pretreated rice husk were carried out by a SEM-EDX microscope Phenom Pro X.

2.3. Hydrolysis process

20 g of pretreated sample and 200 ml of H_2SO_4 (1N) were inserted into three neck flask. They were hydrolysed at a temperature of 100 °C, a speed of 200 rpm for 30 min. Repetition was performed at variable H_2SO_4 of 2 N, and 3 N for 60 to 150 min. The contents of glucose were determined using the standard procedures according to Luff-Schoorl [13].

2.4. Numerical CFD Simulation

CFD modelling of Hydrodynamic of Hydrolysis of pretreated rice husk was referred by previous authors [14,15]. ANSYS Fluent 16 with a finite volume method was used to solve the general partial differential equations that illustrate fluid flow, momentum and mass transport. The geometry of hydrolysis reactor was shown in figure 1a and 1b. The reactor had a diameter of 0.2 m. Initially, a pretreated sample was put at the reactor bottom to a level above the impeller, and then sulphuric acid was injected to the rest of the reactor. The volume fraction of 10 % of pretreated rice husk in the slurry entered to the reactor of, with the average diameter of 50 μm. The reactor had the temperature of 100 °C for 30 min and speed of 200 rpm. A reactor schematic and the initial position of slurry solution were described in figure 1c. There were three fluid zones within the domain. The first zone represented the impeller zone; the region where the cellulose slurry was initially posited; the rest of reactor where H_2SO_4 solution was fed. A calculation was started by using the initial assumption similar to the specified volume fractions. The pretreatment process was simulated for various hydrolysis time for 60 to 150 min. A pretreated rice husk entered the reactor in the form of slurry. The simulation was operated in unsteady state conditions with time step sizes equal to 0.1 s. The Eulerian-Eulerian Multiphase flow and standard k–ε model was applied in all cases for accounting the turbulence effects. Hydrodynamic analysis of the hydrolysis reactor was obtained from 5 unsteady state simulations (30, 60, 90, 120, and 150 min) at different sulphuric acid concentrations of 1N, 2N and 3 N.

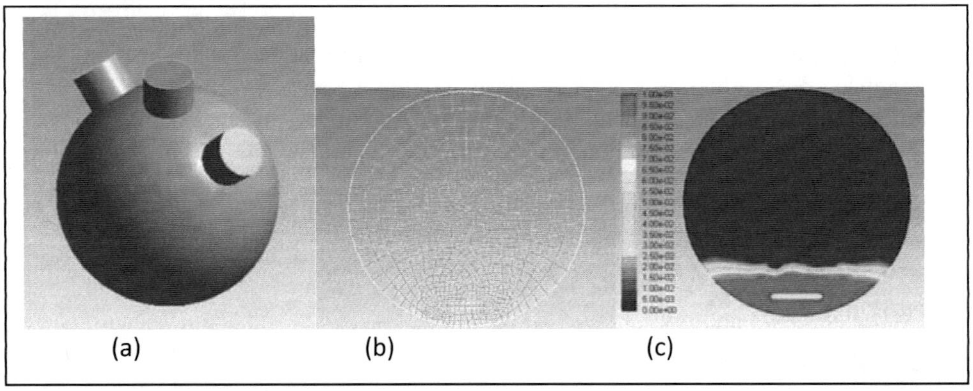

Figure 1. (a) 3-Dimensional geometry of reactor; (b) 2-Dimensional geometry of reactor; (c) initial condition of slurry.

3. Results and discussion

3.1. Morphological of the Rice Husk Structure

To ensure the successful of SAA-dilute acid pretreatment process, it is crucial to see a morphological of the rice husk structure. Chemically pretreated rice husk was analysed by SEM to investigate the change in rice husk before and after the pretreatment. Figure 2 illustrates the morphology of the rice husk structure before pretreatment (a), after SAA (b), after SAA-acid pretreatment (c). It is observed that a morphological of rice husk (figure 2a) changed to fibrous clusters (figure 2b). It is described by SEM that there was a changing in a microstructure of rice husk. This is due to more lignin was removed by SAA pretreatment. Significant lignin was also reduced after SAA-acid pretreatment (figure 2c). It can be concluded that the pretreatment of rice husk led it much more porous to be hydrolysed. The elimination of lignin and changing of its structure enhanced the degradation of cellulose and hemicellulose from biomass. In the SAA-acid pretreatment, a major part of the amorphous material (xylan and lignin) was reduced. Meanwhile, either a small part of the crystalline material (glucan) was eliminated or the crystalline structure was destroyed by swelling.

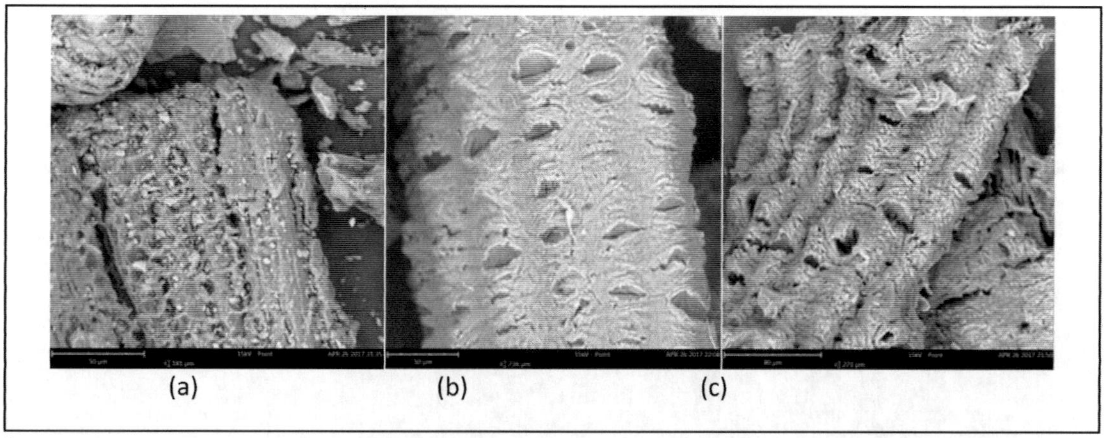

Figure 2. The morphology of the rice husk structure: (a) before pretreatment: (b) after SAA pretreatment; (c) after SAA-acid pretreatment. Magnification of images given in figure 2a, 2b and 2c are 1000X, 2000X and 3000X.

3.2. Effect of Hydrolysis Time on Glucose Levels at Various Concentrations of H_2SO_4

Figure 3 describes the effect of hydrolysis time on glucose levels at Various Concentrations of H2SO4. It is shown that the longer hydrolysis times produced the higher of glucose level. It illustrates that the length of reaction time between cellulose and water affected the amount of glucose. This is due to the longer of the reaction time caused more collisions between cellulose and water. In this study the optimum hydrolysis reaction time is 150 min. The higher concentration of acids enhanced the activation energy. The highest glucose level was about 21.51% at the H_2SO_4 concentration of 3 N for 150 min.

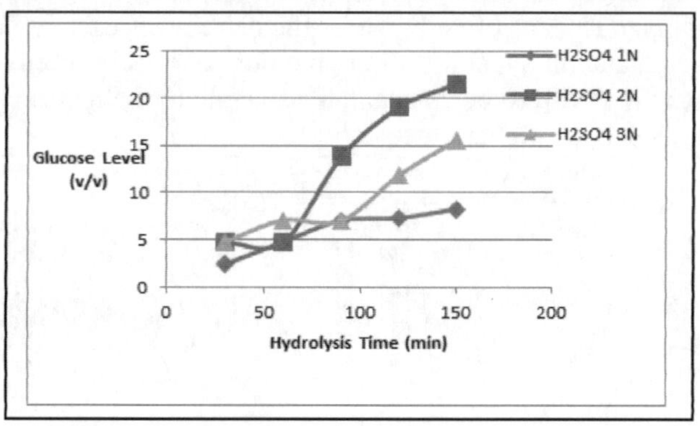

Figure 3. Effect of hydrolysis time on glucose levels at various concentrations of H_2SO_4.

3.3. Hydrodynamic Analysis of Cellulose Volume Fraction during Hydrolysis at H_2SO_4 Concentration of 1N

Figure 4 shows cellulose volume fraction during hydrolysis at H_2SO_4 concentration of 1N. It presents the volume fraction of cellulose, predicted by the CFD simulation of this research. The red colour indicates the high volume fraction of cellulose. The blue colour shows the low volume fraction of cellulose. The distribution of cellulose volume fraction was heterogeneous and mainly distributed about the middle. This non homogeneity was caused by the circular liquid flow under the solid-liquid interaction.

IOP Conf. Series: Materials Science and Engineering **543** (2019) 012046 doi:10.1088/1757-899X/543/1/012046

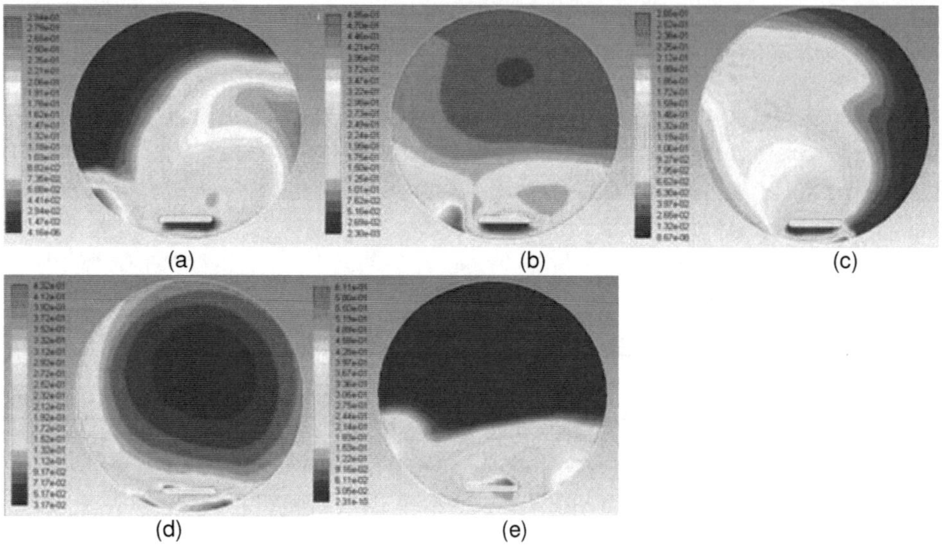

Figure 4. Hydrodynamic analysis of cellulose volume fraction during hydrolysis process at H_2SO_4 1N: (a) 30 min; (b) 60 min; (c) 90 min; (d)120 min; (e)150 min.

3.4. Hydrodynamic Analysis of Cellulose Volume Fraction during Hydrolysis at H_2SO_4 Concentration of 2N

Figure 5 describes the pattern of mixing motion in cellulose hydrolysis process at the H_2SO_4 concentration of 2 N. The predicted cellulose volume fraction distribution illustrates a higher cellulose concentration near the wall and absence of cellulose at the central of reactor. The benefit of the static mixers is that they avoid sedimentation of solids through modifications in reactor geometry. It is shown that the perfect mixing patterns tended to be constant. This is due to the mixing affected the hydrolysis process. Perfect mixing produced the higher glucose.

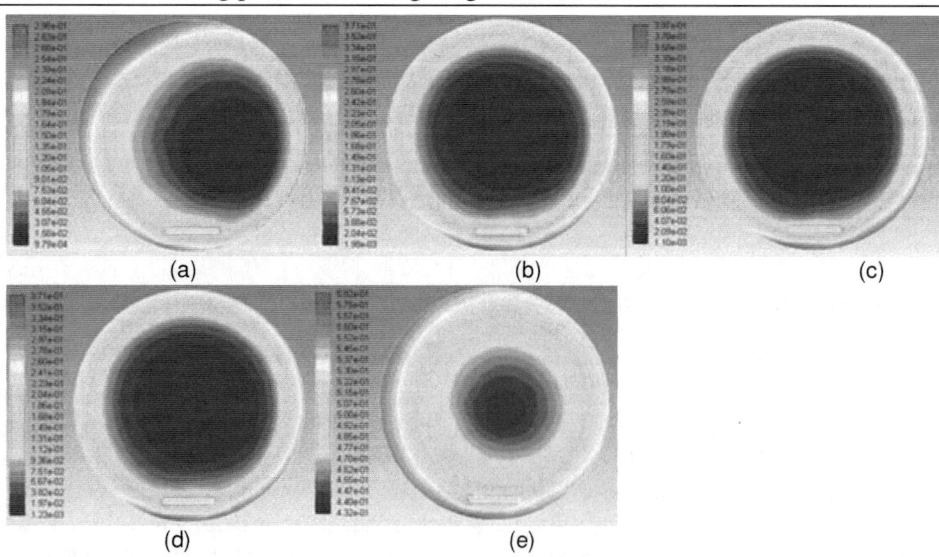

Figure 5. Hydrodynamic analysis of cellulose volume fraction during hydrolysis process at H_2SO_4 2N: (a) 30 min; (b) 60 min; (c) 90 min; (d) 120 min; (e) 150 min.

It can be seen from figure 5 that the longer time of hydrolysis time led more cellulose degrades to glucose. It is reasonable with the experimental test. It can be concluded that the optimum condition of hydrolysis process was obtained at H_2SO_4 concentration of 2N for 150 min.

3.5. Hydrodynamic Analysis of Cellulose Volume Fraction during Hydrolysis at H_2SO_4 Concentration of 3N

Figure 6 presents the flow field of cellulose volume fraction at the H_2SO_4 concentration of 3N. The distribution of cellulose volume fraction was heterogeneous and mainly distributed about the middle. This non homogeneity was caused by the circular liquid flow under the solid-liquid interaction.

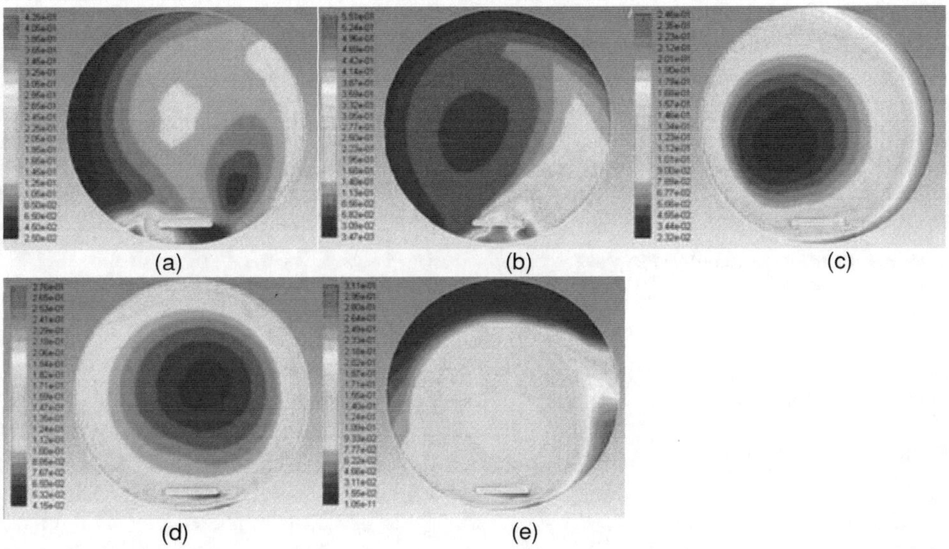

Figure 6. Hydrodynamic Analysis of Cellulose Volume Fraction during Hydrolysis Process at H_2SO_4 3N: (a) 30 min; (b) 60 min; (c) 90 min; (d) 120 min; (e) 150 min.

4. Conclusion

It can be concluded that a morphological of rice husk changed to fibrous clusters after SAA pretreatment. Significant lignin was also removed after SAA-dilute acid pretreatment. The pretreatment of rice husk led it much more porous to be hydrolyzed. The longer hydrolysis time of rice husk cellulose increased glucose level significantly. The optimum concentration of H_2SO_4 produced the highest glucose level of 21.51% at H_2SO_4 concentration of 2N for 150 min. Through hydrodynamic analysis with CFD modeling, cellulose hydrolysis with H_2SO_4 concentration of 2N for 150 min had the best mixing distribution pattern for almost every time variable. It is reasonable with the experimental test. The model and results presented in this study are valuable for the application of CFD model to calculate the flow of biomass suspensions in spherical reactors.

5. References

[1] Taherzadeh MJ, Karimi K 2007 *Bioresour Technol.* **2**(3):472–99.
[2] Dávila I, Robles E, Andrés MA, Gullón P 2017 *Chem Eng Trans.* **57**(2005):85–90.
[3] Fang C, Schmidt JE, Cybulska I, Brudecki GP, Frankaer CG, Thomsen MH 2015 *Biomed Res Int.* **2015**.
[4] Mittal A, Katahira R, Himmel ME, Johnson DK 2011 *Biotechnol Biofuels.* **4**(1):41.
[5] Novia, Pareek VK, Agustina TE 2017 *MATEC Web of Conf.* vol 101 p 1-6
[6] Salvi DA, Aita GM, Robert D, Bazan V 2010 *Appl Biochem Biotechnol.* **161**(1–8):67–74.

[7] Xin D, Yang Z, Liu F, Xu X, Zhang J 2015 *Bioresour Technol.* **175**:529–36.
[8] Martínez EL, Quiroga AG, Jardini AL, Filho RM 2009 *Comput Aided Chem Eng.* **26**:683–8.
[9] Carvajal D, Marchisio DL, Bensaid S, Fino D 2012 *Ind Eng Chem Res.* **51**(22):7518–25.
[10] Pino MS, Rodríguez-Jasso RM, Michelin M, Flores-Gallegos AC, Morales-Rodriguez R, Teixeira JA, Ruiz HA 2018 *Chem Eng J.* **347**:119–36.
[11] Hanra Sebayang A, Haji Hasan M, Hwai Chyuan O, Dharma S, Harjon Bahar A, Susan Silitonga A, Kusumo F 2017 *Chem Eng Trans.* **56**:553–8.
[12] Datta R 1981 *Biotechnol Bioeng.* **23**(9):2167–70.
[13] Egan H, Kirk R, Sawyer R 1981*The Luff Schoorl method. Sugars and preserves. In: Pearson`s chemical analysis of foods.* 8th ed. Longman Scientific and Technical: Harlow, pp. 152-153.
[14] ANSYS FLUENT 16.0 Documentation. Canonsburg, Pa, USA.: ANSYS, Inc.; 2015.
[15] Novia N, Ray S, Pareek V 2007 *Chem Prod Process Model.* **2**(2):4.

Acknowledgments
The authors gratefully acknowledge the financial supports provided by Universitas Sriwijaya for funding the research project through Anggaran DIPA Badan Layanan Umum No. 042.01.2.400953/2018 December 05, 2017 (as Research Contract: Unggulan Kompetitif Universitas Sriwijaya, No. 0007/UN9/SK.LP2M.PT/2018 June 6, 2018).

IOP Conf. Series: Materials Science and Engineering **543** (2019) 012047 doi:10.1088/1757-899X/543/1/012047

Thermal Pyrolysis of Polypropylene Plastic Waste into Liquid Fuel: Reactor Performance Evaluation

M Martynis, Mulyazmi, E Winanda and A N Harahap

Chemical Engineering Department, Bung Hatta University, Jl. Gajah Mada No. 19 Olo Nanggalo, Padang, West Sumatera, 25143, Indonesia

martynismunas@yahoo.co.id

Abstract. The amount of plastic waste is growing over years due to the vast applications of plastics in many sectors. Of the various processes that can be used to convert solid waste into fuels, the pyrolysis process has been identified as having significant potential. Pyrolysis thermally degrading long chain polymer molecules into smaller, less complex molecules, through heat and pressure with little or in absence of oxygen. In this study, a 125 dm^3 pyrolysis reactor was designed and its performance was evaluated. The feedstock for the reactor was 1 kg polypropylene type of plastic waste and conducted at temperature of 250 °C, 300 °C, 350 °C and 400 °C, each process was carried out at 30 minutes and 60 minutes of operating time. The results showed that at a temperature of 400 °C, for a period of 60 minutes, a maximum yield of 88.86% liquid fuel was achieved. Obtained pyrolysis liquid fuel is comparable with the commercial fuel set by Indonesian Ministry of Energy and Mineral Resources. Furthermore, the viscosity and calorific value of liquid fuel produced are close to those of kerosene, meanwhile the density is close to the density of gasoline.

1. Introduction

A marvel of polymer chemistry, plastics has become parts of everyday consumer life and its production and consumption have increased drastically. Over 300 million tons are globally produced every year and the demand is expected to continue to rise [1]. Plastics are non-biodegradable polymers mostly containing carbon, hydrogen, and few other elements such as chlorine, nitrogen etc. [2]. Due to its non-biodegradable nature, the plastic waste contributes significantly to the problem of municipal waste management. The disposal of plastic by landfilling and incineration will lead to environmental pollution [3]. In a long term neither the landfilling nor the incineration solve the problem of wastes, because the suitable and safe depots are expensive, and the incineration stimulates the growing emission of harmful and greenhouse gasses e.g. NO_x, SO_x, CO_x [4]. Thus, in order to reduce plastic disposal to the landfill, recycling method is considered as another alternative to manage plastic waste.

Recycling has many types of techniques including pyrolysis, gasification, hydrolysis and etc. The carbon recycling by catalytic cracking is a better alternative approach to counter the increase of polymer waste, since they are environmentally friendly compared to disposal by landfilling and incineration. Moreover, it is possible to regain energy and fuel from recycling while reducing the consumption of natural resources [5].

The most attractive recycling technique to waste minimization that has been gaining interest recently is pyrolysis. Pyrolysis is the process involves the thermal degradation of biomass or long chain polymer

IOP Conf. Series: Materials Science and Engineering **543** (2019) 012047 doi:10.1088/1757-899X/543/1/012047

molecules performed in the absence of oxygen into a wide range of products [6] and involves the removal of volatile components such as hydrocarbon gases, tars and phenols [7]. This process is a part of de-polymerization, which is in contrast to the production of plastics, which uses polymerization from syngas and crude oil [8]. The pyrolysis of plastics yields on average 45-50% of liquid fuel, 35-40% of gasses and 10-20% of char, depending on the pyrolysis technology [9]. Plastic waste mainly consists of polyethylene terephthalate (PET), high-density polyethylene (HDPE), polyvinyl chloride (PVC), low-density polyethylene (LDPE), polypropylene (PP) and polystyrene (PS) [10]. Fundamentally, different types of plastics have different compositions that normally reported in terms of their proximate analysis. Thus, different liquid fuel yield will be obtained when different types of plastic wastes undergo pyrolysis process. Table 1 summarizes the types of plastics and its application.

Table 1. Types of plastics, symbol and applications

Types of plastics	**Symbol**	**Applications**
Polyethylene terephtalate (PET)	1 PET	Beverage bottles, medicine jars, rope, clothing and carpet fiber
High-density polyethylene (HDPE)	2 HDPE	Containers for milk, motor oil, shampoos and conditioners, soap bottles, detergents, and bleaches
Polyvinyl chloride (PVC)	3 V	All kinds of pipes and tiles
Low-density polyethylene (LDPE)	4 LDPE	Cling-film, sandwich bags, squeezable bottles, and plastic grocery bags
Polypropylene (PP)	5 PP	Lunch boxes, margarine containers, yogurt pots, syrup bottles, prescription bottles, plastic bottle caps, plastic cup
Polystyrene (PS)	6 PS	Disposable coffe cups, plastic food boxes, plastic cutlery and packing foam
Polyethylene Acrylonitrile butadiene styrene (ABS) Polyamide (PA) or Nylons Polybutylene terephthalate (PBT)	7 OTHER	Baby bottles, compact discs, and medical storage containers

Different researchers have studied the effect of temperature and the type of reactor on the pyrolysis of plastic wastes. According to previous research, there are some cases where a high amount of liquid fuel yield, more than 80 wt % could be produced in the pyrolysis of individual plastic. Santaweesuk and Janyalertadun [11] studied pyrolysis process of various types of plastic found in the municipal landfill in Thailand. The results showed that PP was converted into 80% liquid fuel, LDPE to 73% liquid fuel, HDPE to 70% liquid fuel, and mixed plastic to 46% liquid fuel. Miandad et. al. [12] examined the outcomes of different types of plastic waste such as PS, PE, PP and PET on the yield from each type of

IOP Conf. Series: Materials Science and Engineering **543** (2019) 012047 doi:10.1088/1757-899X/543/1/012047

plastic and product quality from the pyrolysis process. the plastic waste of PS type showed maximum yield of liquid fuel (80.8%) with minimum amount of gases (13%) and char (6.2%) in comparison with other types of plastic. Another research was conducted to produce liquid fuel from PP type of plastic waste and successfully obtained heavy fuel with carbon number of C_4 to C_{35} which similar to fuel grade oil [13].

In 2010 Indonesia produced 46 million tons of waste with 11% being of plastic waste [14]. Thus, pyrolysis can be an alternative method to recycle this amount of plastic waste. Therefore, this research aims to obtain liquid fuels resulted from pyrolysis of plastic waste (PP type) that is safe for humans and environment with heating value and fuel quality that meet standardized-compliant.

2. Experimental section

2.1. Materials

Plastic cup wastes (polypropylene type) were collected from Bung Hatta University campus area, cleaned and dried. The dried plastic wastes were cut into small square shaped pieces (about 5 cm side). The pyrolysis setup consists of batch reactor made of stainless steel as shown in Figure 1.

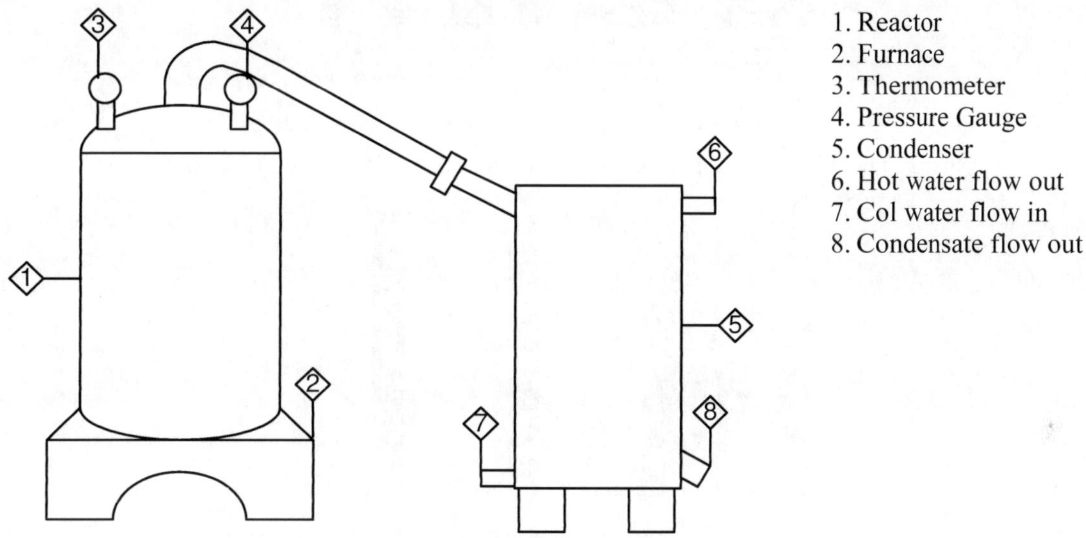

Figure 1. Pyrolysis reactor setup

The experimental setup consist of a batch reactor with capacity of 125 dm^3. The reactor was covered with 5 mm insulation jacket made of low alloy steel to avoid heat loss. It was placed on an LPG gas burner and was connected to the condenser and separating system.

2.2. Pyrolysis process

1000 g of plastic cup sample was loaded into the reactor in each pyrolysis reaction. A thermocouple was then placed in the reactor and heated with gas burner while the temperature was maintained between 250 to 400 °C according to the variable for 30 and 60 minutes. The condensable liquid products were collected through the condenser and the volume was measured. After pyrolysis, the char left was weighed and the gaseous product was calculated. The yield of liquid fuel was determined by dividing the collected material by the weight of feeding plastics.

2.3. Analysis

The calorific value of the obtained liquid fuel was performed using Bomb Calorimeter. A small container contains the pyrolysis liquid fuel sits submerged in a container of water, and ignition wires start the combustion. The bomb calorimeter will measure the heat that is released by the reaction, calculated as calorific value. Furthermore, the physical properties of pyrolysis liquid fuel, such as density and kinematic viscosity, were also measured.

3. Results

3.1. Effect of temperature on pyrolysis liquid fuel yield

The yield of liquid fuel in this experiment is shown in Figures 2 and 3.

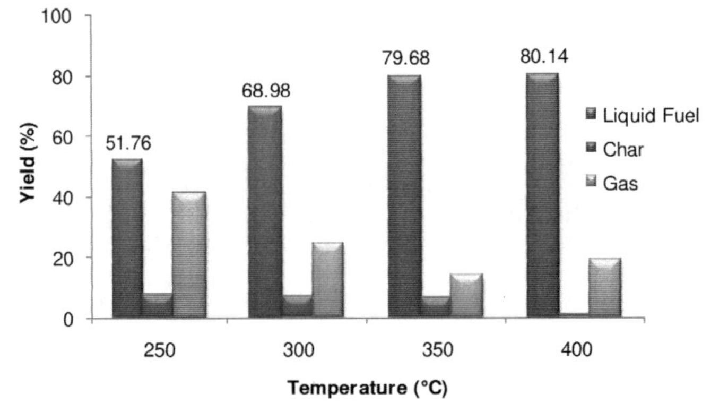

Figure 2. The effect of temperature on pyrolysis liquid fuel yield at 30 minutes

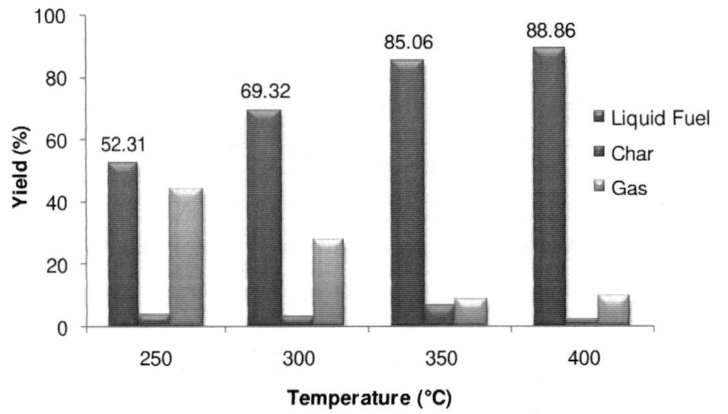

Figure 3. The effect of temperature on pyrolysis liquid fuel yield at 60 minutes

For each experiment, more amount of pyrolysis liquid fuel was acquired at higher temperature of 400 °C compared to that of lower temperatures, as shown in Figure 2 and Figure 3. The oil and gas constituted major product as compared to the char fractions. The recovery of condensable fraction (liquid fuel) increased with gradual increase of temperature. A maximum of 88.86% of the liquid fuel has been obtained at temperature of 400 °C. The maximum yield of char 5.2% was obtained at the temperature of 250 °C. With the increasing of temperature, the gas yield is decreased. Contrarily, from low to high temperature, the liquid fuel yield increases resulting in a decrease of char yield. As a result, the pyrolysis process showed an inverse ratio between solid yields (char) and liquid products (oil). It is due to the fact that the higher the pyrolysis temperature, the components in the plastic cup wastes would be decomposed sharply. As more components were decomposed, the char production is lessened and the oil produced is rising.

The effect of reaction time is also shown in Figure 2 and Figure 3. At the same temperature and same amount of polypropylene, more pyrolysis liquid fuel was observed at 60 minutes because longer reaction time resulted in further cracking of polypropylene into liquid fraction. The maximum liquid fuel obtained at temperature of 400 °C and 60 minutes of reaction time with the yield of liquid fuel, char, and gas were 88.86%, 1.84% and 9.3%, respectively. This result is comparable to those by another researcher. Sommani et. al. investigated the cracking process of used vegetable oil mixed with polypropylene plastic waste in the presence of activated carbon and able to obtained maximum of 79.69% liquid fuel at temperature of 430 °C and ratio of used vegetable oil and polypropylene of 30 : 70 (wt %) [15].

(a) (b)

Figure 4. Pyrolysis products: (a) liquid fuel, and (b) char

3.2. Pyrolysis liquid fuel physical properties
The physical properties of liquid fuel obtained in this study is summarized in table 2 and 3.

Table 2. Pyrolysis liquid fuel quality

Time (minutes)	Temperature (°C)	Density ($kg\ m^{-3}$)	Kinematic Viscosity ($mm^2\ s^{-1}$)
30	250	790.3	1.2773
	300	803.9	1.0540
	350	796.0	1.4589
	400	793.5	1.3876
60	250	781.9	1.0229
	300	780.6	0.9889
	350	795.7	1.2618
	400	796.9	1.5398

Table 3. Comparison of gasoline, diesel fuel and pyrolysis liquid fuel quality

Parameter	Unit	Pyrolysis Liquid Fuel	Gasoline (RON 88)*	Diesel Fuel*	Testing Method (ASTM)
Density	$kg\ m^{-3}$	796.9	715 - 780	815 – 860	D 1259
Kinematic viscosity	$mm^2\ s^{-1}$	1.5398	not mentioned	2.0 – 4.5	D 445
Calorific value	$MJ\ kg^{-1}$	44.95	min. of 44.56	min. of 42.7	D 240

* standard designed by Indonesian Ministry of Energy and Mineral Resources

IOP Conf. Series: Materials Science and Engineering **543** (2019) 012047 doi:10.1088/1757-899X/543/1/012047

The produced liquid fuel density after 60 minutes of pyrolysis process was increase with the increasing of temperature. According to Table 3, the density of pyrolysis liquid fuel (796.6 kg/m^3) meet the standard designed by Ministry of Energy and Mineral Resources for gasoline (RON 88) but slightly lower compared to the diesel fuel, which range between 815 and 860 kg/m^3. It should be noted that density affects the level of engine power. In terms of density, all values seem comparable with the commercial standard value for both gasoline (RON 88) and diesel fuel. Furthermore, the kinematic viscosity is another important parameter that affects the performance of diesel fuel as it represents the resistance of fuel to flow. The viscosity of pyrolysis liquid fuel obtained at 400 °C was 1.5398 mm^2/s, which is lower than the standard quality for diesel fuel.

The experimental calorific value of pyrolysis liquid fuel (44.95 MJ/kg) meet both standard for gasoline (RON 88) in addition to diesel fuel and were considered high for energy utilization. Additional analysis in characterizing the exact components of pyrolysis liquid fuel by Gas Chromatography-Mass Spectrometry analysis (GC-MS) and another fuel properties analysis (Cetane Number, Cetane Index, Octane Number, flash point etc.) were required for better understanding of pyrolysis liquid fuel properties.

4. Conclusions

The pyrolysis of polypropylene plastic waste was performed in a batch reactor at varied temperature of 250 °C, 300 °C, 350 °C, and 400 °C. The maximum pyrolysis liquid fuel of 88.86% was obtained at temperature of 400 °C after 60 minutes of reaction. The pyrolysis process showed an inverse ratio between char (solid yields) and fuel (liquid products). The densitu of pyrolysis liquid fuel produced (meet the commercial standard value for both gasoline (RON 88) and diesel fuel. However, the kinematic viscosity of pyrolysis liquid fuel obtained at 400 °C was 1.5398 mm^2/s, which is lower than the standard quality for diesel fuel. Moreover, the calorific value of pyrolysis liquid fuel meet both standard for gasoline (RON 88) as well as diesel fuel. Further analysis in characterizing the exact components of pyrolysis liquid fuel by Gas Chromatography-Mass Spectrometry analysis (GC-MS) and another fuel properties analysis (Cetane Number, Cetane Index, Octane Number, flash point etc.) were required for better understanding of pyrolysis liquid fuel properties.

References

[1] Valavanidis A 2016 Scientific Reports 1–39.

[2] Grigore M E 2017 Recy. **2** 1–11.

[3] Triassi M, Alfano R, Illario M, Nardone A, Caporale O, Montuori P 2015 *Int. J. Environ. Res* Public Health 1216–36.

[4] Lee K H and Shin D H 2003 *Korean J. Chem. Eng.* **20** 89–92.

[5] Habyarimana J B, Njiemon M, Abdulnasir R, Neksumi M, Yahaya M, Sylvester O D, Joseph I, Okoro L, Agboola B, Uche O, Jahng W J 2017 *Int. J. Sci. Eng. Res.* **8** 1193-–02.

[6] Strubinger A, Oliveros A R, Araque M A, Guerra J 2017 *Chem. Eng. Trans.* **57** 19–24.

[7] Bungay V C 2017 *Chem. Eng. Trans.* **56** 193–8.

[8] Olufemi A S and Olagboye S A 2017 *Int. J. Petrochem. Sci. Eng.* **2** 252–7.

[9] Wong S L, Ngadi N, Abdullah T A T, Inuw I M 2015 *Renew. Sust. Energ. Rev.* **50** 1167–80.

[10] Sonawane Y B, Shindikar M. R., Khaladkar M Y 2015 *Int. Res. J. Environ. Sci.* **4** 24–8.

[11] Santaweesuk C, and Janyalertadun A 2017 *Int. J. Environ. Sci. Dev.* **8** 168–173.

[12] Miandad R, Barakat M A, Aburiazaiza A S, Rehan M, Ismail I M I, Nizami A S 2016 *Int. Biodeterior. Biodegrad* 1–14.

[13] Sarker M and Rashid M M 2014 Am. *J. Environ. Energy Power Res.* **2** 1–10.

[14] Jambeck J P, Geyer R, Wilcox C, Siegler T R, Perryman M, Andrady A, Narayan R, Law K L 2015 Sci. **347** 768–771.

[15] Sommani P, Mankong N, Vitidsant T, Lothongkum A W 2013 *ASEAN Eng. J. Part B* **4** 16–24.

IOP Conf. Series: Materials Science and Engineering **543** (2019) 012048 doi:10.1088/1757-899X/543/1/012048

Testing Instrument For Water Quality and Drinking Water Using Oxidation and Electromagnetic Methods (Case Study: Local Water Company at Bangka Barat)

V Luvita, N T Eka Darmayanti, G Zaid and D Setyarini

Research Centre for Metrology, Indonesian Institute of Science, Serpong, Indonesia

Email : veny.susanto@gmail.com

Abstract. Safe clean water, especially for consumption purposes, is needed by the humans. Various regulations on water safety standards have been developed and implemented by each country, including Indonesia Regulation of the Minister of Health Number 492/MENKES/PER/IV/2010 concerning Water Quality Requirements; SNI 01-3553 2006 on Drinking Water in Packaging and Government Regulation of the Republic of Indonesia Number 82/2001 on the Management of Water Quality and Control of Water Pollution. Some parameters of physics, chemistry, biology and radioactivity are the benchmarks of water security. Raw water treatment for clean water and drinking water has been done using oxidation and electromagnetic methods. Both of these methods have been tested in local water company (PDAM) at Bangka Barat which has problems with high organic content, the amount of heavy metal content Fe and Mn that exceeds the threshold above normal and high acidity level. Test results after processing show physical and chemical parameters that meet the criteria of water quality and drinking water requirements in accordance with regulations set by the government. Other important things besides meeting the health requirements are also the quality assurance of the measuring instrument used to test the physical and chemical parameters. This measurement traceable to International System of Units through Research Centre of Metrology- Indonesian Institute of Science.

1. Introduction

Water as a vital source of life is a very important and fundamental resource for the survival of living things in this world. Cities, nations and civilization have grown near rivers. Satisfactory water supply (adequate, safe and accessible) should be available to all living creatures of humans, animals and plants. In human life, water is widely needed in all aspects such as for consumption, crop irrigation, industry, power generation and much more. Water is well known as a "universal solvent" which means that many kinds of materials can dissolve or wash away in it. This character also makes it easy for water to be polluted with things that are dangerous or harmful. The problem of polluted water is very serious and can affect to the quality life of mankind and other living things as well [1-3]. Therefore, water must be utilized wisely so that sustainability of clean water and the ecological balance can be maintained and utilized continuously by our future generations.

Safe clean water is essential to the realization of basic human rights so that an effective policy in regard to clean water supply is required. Since the diseases related to drinking water contamination

IOP Conf. Series: Materials Science and Engineering **543** (2019) 012048 doi:10.1088/1757-899X/543/1/012048

become a major burden of human health, therefore every effort should be made to achieve the highest quality of drinking water. Interventions to improve the quality of drinking water and access to safe drinking water will give real benefits to good health for the society.Access to safe drinking water is essential to develop programs at the national, regional and local levels. In some areas, it has been shown that investments in water supply and sanitation can result in net economic benefits, as the reduction of adverse health effects and health care costs are greater than the cost of intervening [4-5]. Experiences also show that interventions in improving access to clean water greatly benefit the poor, both in rural and urban areas, and can be an effective part of poverty alleviation strategies.

2. Experimental Section

Water quality testing is essential for identifying pollution problems that occur, ensuring that water is proper for intended use, ensuring that the water to be consumed is safe and also for evaluating the effectiveness of the water treatment system. In recent day, environmental issues related to environmental pollution have also been the concern of the community both in national and international scope [6]. Law enforcement is firmly enforced for companies that cause environmental pollution. It is intended to protect the environment and to preserve the balance of the ecosystems. Thus, the water quality test result data must be valid and acceptable to all interested parties, so that appropriate policy or decisions can be taken as water is a very important natural resource and is needed by all living creatures.

Metrology both in physics and chemistry are expected to play a strategic role in overcoming water problems, and become part of the clean water supply solution for mankind. It can be done by developing techniques / measurement methods and traceability needed for determining water quality, estimating the water balance and developing methods for water quality monitoring and water quality improvement.

The World Health Organization (WHO) has published international norms on water quality and human health in the form of guidelines that has been used as the basis of regulation and standard setting worldwide. In Indonesia, there are several regulation on regard to water quality including Regulation of Minister of Health of Republic Indonesia No. 492/MENKES/PER/IV/2010 concerning on Drinking Water Quality Standards; SNI 01-3553 2006 on Bottled Drinking Water and Government Regulation of the Republic of Indonesia No. 82 year 2001 on the Management of Water Quality and Control over Water Pollution. There are several of the physical, chemical, biological and radioactivity parameters that set out in most regulations as shown in the Table 1.

Table 1 Parameters that affect the quality of clean water

Parameter			
Physical	Chemical	Microbial	Radiological
Total hardness *Total dissolved solid* (TDS) Temperature Electrolytic conductivity pH	Inorganic: Arsenic, Fluoride, total Chrom, cadmium, nitrite, nitrate, cyanide, selenium Organic: Detergent, TOC, pesticide	Coliform E-coli	Gross-A Gross-B

In this study a prototype will be developed to improve water quality in a physical-chemical manner using electromagnetic resonance methods. This tool has a resonance tube which is equipped with a coil and an electric pulse generator to produce a magnetic field so that it can resonate with water molecules.

IOP Conf. Series: Materials Science and Engineering **543** (2019) 012048 doi:10.1088/1757-899X/543/1/012048

This prototype to improve water quality in a physical-chemical way works by giving resonance treatment to hydrogen atoms physically to water molecules. The magnetic resonance method in this tool works by utilizing the behavior of protons and the molecular bonds contained in water [7-8]. The behavior of protons (including those found in water molecules) due to the influence of certain static magnetic fields (on this device utilizes the earth's magnetic field) which, if then disturbed by other magnetic fields that have a certain frequency, where the direction of both magnetic fields is perpendicular or not mutually parallel, the Larmor Precession with a certain frequency (Larmor Frequency) and the duration of the resonance is also specific [9]. In water molecules mixed with other liquid or solid materials, the mixing material if it has a proton will also resonate with a certain resonance frequency and duration. Such events can be found in theories related to Nuclear Magnetic Resonance or Proton Magnetic Resonance. With this kind of treatment the molecules contained in water will become clusters, so that groups that are reactive (pure water) become larger and not blocked by mixing materials [10].

3. Result

The results of research conducted using the system created shows that there is a decrease in the value of the measure parameters. This shows the quality of treated water is getting better. In the treatment process of ex-mine lake water, it was found that all the measured mineral parameters had decreased and were degraded by the presence of oxidation and electromagnetic processes.

Table 2. The results of the Analysis of the Raw Water Measurement Laboratory in Bangka Barat PDAM before and after processed using AOPs and EWT

No.	Parameters	Unit	Experiment	
			Before	*After*
	Physics			
1	Conductivity	μS/cm	16800	5500
	Organics			
2	Ammonia (NH_3-N)	mg/L	0.265	0.312
3	Barium	mg/L	< 0.005	< 0.005
4	Iron (Fe)	mg/L	6.8	2.9
5	Chloride (Cl^-)	mg/L	3600	887.5
6	Cobalt (Co)	mg/L	< 0.004	< 0.004
7	Manganese (Mn)	mg/L	0.65	0.11
8	Nitrate (NO_3-N)	mg/L	15.8	2.15
9	Nitrite (NO_2 -, N)	mg/L	5.47	0.765
10	pH	-	1.65	7.58
11	Hardness ($CaCO_3$)	mg/L	2479.54	1050.18
12	Zinc (Zn)	mg/L	43.5	22.76
13	Copper (Cu)	mg/L	2.26	0.21
14	Cadmium (Cd)		< 0.004	< 0.004
15	Sulfate (SO_4^{-2})	mg/L	2780	720
16	Lead (Pb)	mg/L	2.92	5.01
17	Sodium (Na)	mg/L	1150.87	720.16
18	Kalium (K)	mg/L	325.71	215.48

In order to obtain analysis data, it is necessary to analyze the equipment calibration trace. Measurement traceability according to the International Vocabulary of Metrology (JCGM 200: 2012), is defined as the nature of the measurement results that can be connected to a particular reference, through a documented unbroken calibration chain, each of which contributes to measurement uncertainty. Measurement traceability activities can be considered correct if it can be proven that the measurement results are in accordance with the primary standard value. But in reality, it is not easy to directly measure primary standards. Besides because the primary standard is limited in number and may be stored in a place that is difficult to reach due to cost factors. So with consideration of cost and practicality, a standard measurement hierarchy system is created, where the primary standard position is at the top of the hierarchy.

In the concept of measurement traceability which is also known as metrological traceability, the correctness of measurement can be obtained by means of measuring activities using a measuring instrument. The measuring instrument must be calibrated with a calibration standard that has a higher accuracy, as well as the calibration standard in question which must also be calibrated with a higher one; so on until it goes to the primary standard and SI unit. It states that the existing measuring instrument can be ascertained its value in accordance with the measurement value of the primary standard [1]. So that traceability of this measurement has another purpose as a guarantee of the correctness of the value of the measuring instrument used.

4. Discussion

From this study, measurements were taken regarding the parameters that had been set. The measurement must go through a measurement hierarchy. In the concept of measurement traceability which is also known as metrological traceability, the correctness of measurement can be obtained by means of measuring activities using a measuring instrument. The measuring instrument must be calibrated with a calibration standard that has a higher accuracy, as well as the calibration standard in question which must also be calibrated with a higher one; so on until it goes to the primary standard and SI unit. It states that the existing measuring instrument can be ascertained its value in accordance with the measurement value of the primary standard. So that traceability of this measurement has another purpose as a guarantee of the correctness of the value of the measuring instrument used [11].

All measuring instruments used must be ensured that they are not interrupted by their traced chains. Moreover, the measuring instruments used are related to strategic matters, one of them is a measuring instrument that concerns the lives of many people. To guarantee the traceability of a measurement, the measuring instrument used must be calibrated. The calibration process is intended to determine and determine the correction that is owned by the measuring instrument is still within the limits of the performance of the allowed performance. This performance limit is known as tolerance or maximum permissible error (MPE). The calibration process is carried out by comparing directly to a measurement standard that already has a calibration certificate. Output released from calibration activities is in the form of a certificate and can also be a label or sticker given as a sign on a calibrated device [12].

If the process of guaranteeing traceability of measurement of a measuring instrument is problematic, then it can be ascertained that the size of the measurement that belongs to the standard is also problematic. Therefore, the task of ensuring traceability is very important, especially with regard to the primary quantity, both of which are owned by a calibration laboratory, let alone standards that are owned by national metrology institutions such as the LIPI Metrology Research Center.

5. Conclusion

All measuring instruments used must be ensured that they are not interrupted by their traced chains. Moreover, the measuring instruments used are related to strategic matters, one of them is a measuring instrument that concerns the lives of many people. To guarantee the traceability of a measurement, the measuring instrument used must be calibrated. The calibration process is intended to determine and determine the correction that is owned by the measuring instrument is still within the limits of the

IOP Conf. Series: Materials Science and Engineering **543** (2019) 012048 doi:10.1088/1757-899X/543/1/012048

performance of the allowed performance. This performance limit is known as tolerance or maximum permissible error (MPE). The calibration process is carried out by comparing directly to a measurement standard that already has a calibration certificate. Output released from calibration activities is in the form of a certificate and can also be a label or sticker given as a sign on a calibrated device.

Conflict of Interest
No conflict of interest

Acknoledgement
This research was supported in Prioritas Nasional LIPI 2018 has provided funding and support of research that has been done. Team at UPT BPI-LIPI and Mrs. Sudaryati has supported the sustainability of this research until it can be implemented in Bangka Barat.

References
[1] Langlais, B., Reckhow, D. A., Brink, D.R. 1991. *Ozone in Water Treatment: Application and Engineering*. Michigan. Lewis Publisher, Inc. **2** 369-70
[2] Munter, Rein. 2001. *Advanced Oxidation Processes – Current Status and Prospects*. Department of Chemical Engineering: Tallinn Technical University. **2** 215.
[3] Tchobanoglous, George, Burton, F.L., Stensel, H.D. 2003. *Wastewater Engineering, Treatment and Reuse*. 4th Edition. Metcalf & Eddy, Inc. Mc Graw Hill, New York. **4** 275.
[4] Fakhrurroja, Hanif. 2010. *Membuat Sumur Air di Berbagai Lahan*. Depok: Griya Kreasi. **1** 15-20.
[5] T.H.Y. Tebbutt. 2001. *Principles of water quality control*. Department of civil engineering University of Brimingham. **2** 60.
[6] Environmental Protection Agency Guidance Manual 3-42, 1999. 60.
[7] Hariyadi. Sugiono. Fakhrurroja, H. Tanu, Edi. 2011. Analisis Hasil Uji Terap Alat Penghemat BBM Electric Fuel Treatment pada Engine Diesel Genset 35 KVA dengan Beban Statis. Jurnal Teknologi Indonesia, **34** 68-76.
[8] Suradiputra, I Nyoman, Dibjo Sartono, dan Muhammad Ilman. 2005. Krisis Air di Indonesia. *Prosiding Kongres Ilmu Pengetahuan Nasinal VIII: Air untuk Pembangunan Berkelanjutan*. Jakarta: LIPI Press. 20-25.
[9] Hariyadi. 2004. Metode dan Alat untuk Meningkatkan Kinerja Bahan Bakar Minyak. No Paten ID P 0022630 Indonesia. 1 September 2004. 4-8.
[10] Howard, P., Fiona R., Metrology in Short 3rd Ed. Euramet, UK, 2008. 10-15.
[11] Drijarkara, A.P, Ghufron Z., 2005. Metrologi: Sebuah Pengantar, Pusat Penelitian Kalibrasi, Instrumentasi, dan Metrologi. 5-10.
[12] Pusat Penelitian Kimia LIPI. 2012. *Laporan Hasil Analisis: No. 0052-B/LPPK/IV/2012*. 4-7.

IOP Conf. Series: Materials Science and Engineering **543** (2019) 012049 doi:10.1088/1757-899X/543/1/012049

Process Dynamic and Control for Nonconventional Column/Rectifier Configuration with Aspen Hysys v10.0

R A Hikmadiyar[1], J P Sutikno[1], R Handogo[1], Z Azizah[1], A Hisyam[2]
[1] Department of Chemical Engineering, Institut Teknologi Sepuluh Nopember, Surabaya 60111, Indonesia
[2] Faculty of Chemical and Natural Resources, University Malaysia Pahang, Malaysia

E-mail: juwari@chem-eng.its.ac.id

Abstract. Nonconventional column/rectifier configuration is the new alternate design for the sequence multicomponent separation of three or more products. It can accomplish an efficient energy. There are two columns, the first is the rectifier and the second is the main, the vapor sidestream is taken out from the location under the feed tray of the main column then fed to the rectifier column and the liquid that from the bottom of the rectifier column is turned back to the main column. In order to completely understand the dynamic behavior of complex process of this configuration, process dynamic and control of it has to be studied intensively. The implementation of it is demethanizer and deethanizer column. The configuration has been simulated for dynamic by using Aspen Hysys v10.0. The results show that the graph of rejecting disturbance from all controllers of the main column need more attention especially for inventory control i.e. level controller for reflux drum, sump level reboiler and pressure controller for condenser. The overshoot for level controller of reflux drum of main column is 10% from set points. The better algorithm of tuning is required. The results of this studied could provide guidance for composition controller of this configuration.

1. Introduction

The sequence multicomponent separation of three or more products is usually performed in distillation column with conventional configuration [1]. Many research have studied for conventional configuration in distillation column, and the results showed that it has consumed excessive energy [2]. Increasing the need for an efficient energy, the alternate design for distillation column is required. Nonconventional column/rectifier configuration is one the alternate design for nonconventional column configuration in distillation process [3]. Thermally coupled distillation sequence was used for the basic concept for nonconventional column/rectifier configuration system. Several studies have reported it [4][5][6][7][8]. It does not use reboiler but still uses condenser. There are two columns, the first is the rectifier column and the second is the main column, the vapor sidestream is taken out from the location under the feed tray of the main column then fed to the rectifier column and the liquid that from the bottom of the main column is turned back to the main column [3]. Energy and cost saving can be accomplished by it [6]. The significant problem of nonconventional column/rectifier configuration is the recycle stream in requirement for the dynamic and control [5]. It has been proof that plant uses the recycle system is more profit in steady state condition, because it can minimize the operation cost [9]. When in dynamic system

IOP Conf. Series: Materials Science and Engineering **543** (2019) 012049 doi:10.1088/1757-899X/543/1/012049

it need more caution [5]. Process simulator such as Aspen Hysys gives good enviroments for dynamic simulation of distillation column in addition to PID controller [10]. In order to completely understand the dynamic behavior of complex process of nonconventional column/rectifier configuration, so the study about process dynamic and control of nonconventional colom/rectifier using process simulator i.e. Aspen Hysys is being important.

2. General study

2.1. Process description
The process chosen for this study is a modified demethanizer and deethanizer column [3]. Methane is mostly involved in Natural Gas coming from wells. The other important component is ethane, propane, and heavier hydrocarbons. Demethanizer and deethanizer column separate methane, ethane and propane with the heavier hydrocarbons. The top products of demethanizer are methane and small amount of ethane, the top and bottom products of deethanizer are ethane and propane with the heavier hydrocarbons [11]. There are many reasons why methane, ethane, propane and heavier hydrocarbons are separated from natural gas wells. The reasons depend greatly on the commercial products requirement. For Liquefied Natural Gas (LNG) requirement, ethane and propane with the heavier hydrocarbons have to be removed to avoid the solid formations at low temperature [3].

2.2. Process simulation
Demethanizer and deethanizer column used nonconventional column/rectifier configuration. Demethanizer was the rectifier column and deethanizer was the main column. Demethanizer had 8 stages and 31 stages for deethanizer. Feed entered the second stage of demethanizer column. The vapor sidestream was retreated from the seventh stage of deethanizer column then settled to the base stage of demethanize column and the liquid from the base of demethanizer column was fed back to seventh stage of deethanizer column [3].

Process Flow Diagram (PFD) from demethanizer and deethanizer column was simulated by Aspen Hysys v10.0. In flowsheet of Aspen Hysys, Demethanizer column used Refluxed Absorber Column model. It had a tower and condenser but no reboiler. For steady state, demethanizer column was specified 1479 kmol/hr for the distillate rate and 2950 kmol/hr for vapor sidestream rate. Deethanizer column used Distillation model. Distillation model has tower with both a condenser and reboiler. To achieve convergence simulation, deethanizer column has three specifications i.e. 507.2 kmol/hr for the distillate rate, 0.784 for reflux ratio, and 2950 kmol/hr for vapor sidestream rate. For dynamic requirement, demethanizer column needed vessel for liquid level control in the bottom of column. Pressure changers like valve, pump and compressor were added. Pump was added to transfer liquid from the base of the demethanizer column to seventh stage to deethanizer column. Compressor was given to transfer vapor from the deethanizer to demethanizer column. Valves were added in streams that will be the manipulated variables.

2.3. Control strategy
The purpose of control system was to keep the purity of methane, ethane, and propane. Nonconventional column/rectifier configurations had ten degree of freedoms i.e. Feed rate ("F") two distillates rate ("D"), one bottom product rate("B"), Vapor sidestream ("Vss"), reflux rate ("R") one reboiler duty ("Qr"), and two condensor duty ("Qc"). The stability of column was made by keeping levels in reflux drum, column bottom of the main column and top pressure of column.

This study focusses on the control configuration that will influence the dynamic behavior of nonconventional column/rectifier configuration. The control configuration for demethanizer and deethanizer column is designed with the inventory control. It will keep the production rate and mass balance [7]. Vapor sidestream flowrate is controlled. Because vapor sidestream is the recycle stream so it has to be controlled [12]. When this configuration in dynamic mode, the recycle stream that from vapor sidestream makes trouble for level controller. The recycle stream is free flow when it is not controlled. It becomes a disturbance for level controller especially for level controller system in the main column. The dynamic model uses pressure changer i.e. Compressor and pump. They need attention when temperature on the reboiler suddenly increases or decreases. When it suddenly increase, it causes the entry of vapor in pump. On the contrary if it suddenly decreases, it generates the entry of liquid in compressor. As a result the temperature controller is required. Figure 1 below shows the proposed control configuration.

All controllers that used in nonconventional column rectifier/configuration was PID controller. Flow controller was put on feed. Levels in reflux drum were made by choosing distillate rate as manipupated variables. Flow controller was put on Vapor sidesstream to control the vapor rate rate from the main column that was treated to rectifier column. Pressures in top column were controlled by making condensor duty as manipulated variables and levels in the bottom of columns were controlled by chosing bottom product rate as manipulated variables. Because the main column had reboiler so temperature controller was added to keep purity the bottom product of the main column.

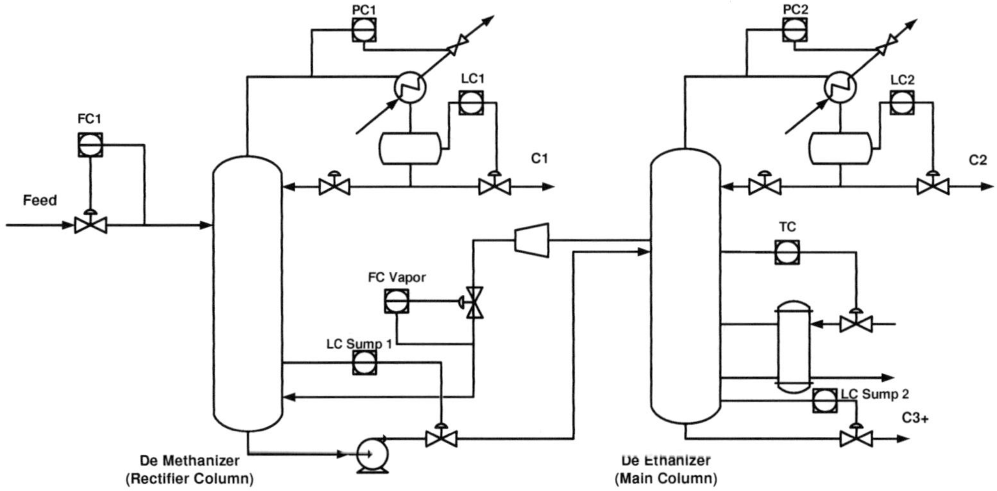

Figure 1. The proposed control configuration.

3. Results and discussion

Demethanizer and deethanizer with nonconventional column/rectifier configuration is simulated for dynamic by using Aspen Hysys v10.0. The proposed control configuration in Figure 1 is implemented in flowsheet of Aspen Hysys. All Controllers uses Proportional Integral Derivative (PID) controller. Tuning parameter for PID controller is got by Autotuner in Aspen Hysys v10.0. It uses the oscillations that will be got Ku or the ultimate gain and Pu or the ultimate periode [13]. The ultimate gain is formulated as Eq(1) below:

$$K_u = \frac{4h}{a\pi} \quad (1)$$

After getting the ultimate value, the PID tuning parameters are computed by the rule of Tyréus-Luyben as Equations (2) – (4) below:

$$K_c = \frac{K_u}{2.2} \quad (2)$$

$$\tau_I = 2.2P_u \quad (3)$$

$$\tau_D = \frac{P_u}{6.3} \tag{4}$$

For tracking set points and rejecting disturbances, Plant in Aspen Hysys v10.0 is given new set points. Disturbances is given by changing the flowrate of feed with adding +10% and-10%. Figure 2a until Figure 2h below show the results of set points and rejecting disturbances for each controller.

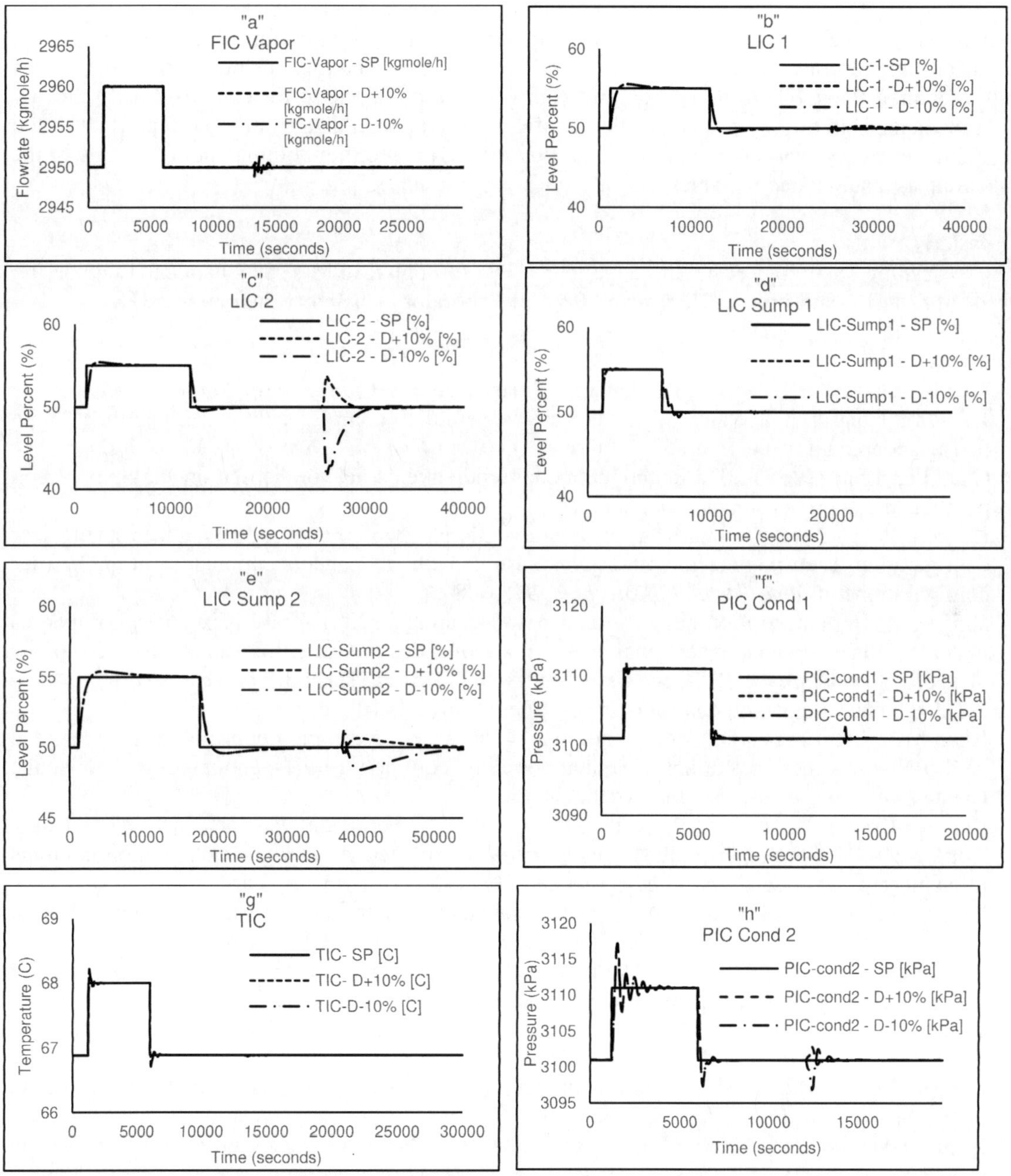

Figure 2a: Response of Flow controller for vapor sidestream. **2b**: Response of Level controller for reflux drum in the rectifier column. **2c**: Response of Level controller for reflux drum in the main

column. **2d**: Response of Level controller for sump level reboiler in the rectifier column. **2e**: Response of Level controller for sump level reboiler in the main column. **2f**: Response of Pressure controller for condenser in the rectifier column. **2g**: Response of Temperature controller in the main column. **2h**: Response of Pressure controller for condenser in the main column.

4. Conclusion

Demethanizer and deethanizer with nonconventional column/rectifier configuration has been simulated for dynamic by using Aspen Hysys v10.0. The proposed control configuration has been implemented in flowsheet Aspen Hysys v10.0. The results show that the level controller of reflux drum in the main column is most in need of attention for this proposed control configuration of Demethanizer and Deethanizer with nonconventional column/rectifier configuration. The implementation for other better tuning is required for it. The proposed control configuration can be used for composition control by creating methane from the top product of the rectifier column and ethane from the top product of the main column as process variables, and making reflux rate of the rectifier column and reflux rate of the main column as manipulated variables.

Acknowledgments

Thanks to Institut Teknologi Sepuluh Nopember (ITS) through LPPM ITS, Chemical Engineering Department FTI-ITS, and Process Design and Control Laboratory Chemical Engineering ITS.

References

[1] Tamayo-Galván V E, Segovia-Hernández J G, Hernández S, Cabrera-Ruiz J and Alcántara-Ávila J R 2008 Controllability analysis of alternate schemes to complex column arrangements with thermal coupling for the separation of ternary mixtures *Comput. Chem. Eng.* **32** 3057–66

[2] Chan L L T and Chen J 2018 Economic model predictive control of distillation startup based on probabilistic approach *Chem. Eng. Sci.* **186** 26–35

[3] Luyben W L 2016 Comparison of a conventional two-column demethanizer/deethanizer configuration requiring refrigerated condensers with a nonconventional column/rectifier configuration *J. Chem. Technol. Biotechnol.* **91** 1688–96

[4] Jiménez A, Hernández S, Montoy F A and Zavala-García M 2001 Analysis of control properties of conventional and nonconventional distillation sequences *Ind. Eng. Chem. Res.* **40** 3757–61

[5] Wu K L, Yu C C, Luyben W L and Skogestad S 2003 Reactor/separator processes with recycles-2. Design for composition control *Comput. Chem. Eng.* **27** 401–21

[6] Khan M S, Chaniago Y D, Getu M and Lee M 2014 Energy saving opportunities in integrated NGL/LNG schemes exploiting: Thermal-coupling common-utilities and process knowledge *Chem. Eng. Process. Process Intensif.* **82** 54–64

[7] Ramírez-Corona N, Mascote-Pérez D, Sánchez-Hijar A, Fernández-Pastrana M I and Jiménez-Gutiérrez A 2015 Insights on the dynamic behavior of thermally coupled distillation columns implemented on processes with recycles *Chem. Eng. Res. Des.* **93** 120–35

[8] Uwitonze H, Lee I and Hwang K S 2016 Alternatives of integrated processes for coproduction of LNG and NGLs recovery *Chem. Eng. Process. Process Intensif.* **107** 157–67

[9] Ward J D, Mellichamp D A and Doherty M F 2004 Importance of process chemistry in selecting the operating policy for plants with recycle *Ind. Eng. Chem. Res.* **43** 3957–71

[10] Ignat R and Woinaroschy A 2011 Dynamic analysis and controllability of dividing-wall distillation columns *Chem. Eng. Trans.* **25** 647–52

[11] Luyben W L 2013 Effect of natural gas composition on the design of natural gas liquid demethanizers *Ind. Eng. Chem. Res.* **52** 6513–16.

[12] Alpuche-Manrique M, Rivera-Mejía T, Ramírez-Corona N and Jiménez-Gutiérrez A 2011 Steady state analysis of snowball effects for reaction-separation-recycle systems with thermally coupled distillation sequences *Chem. Eng. Res. Des.* **89** 2207–14

[13] Åström K J and Hägglund T 1984 Automatic Tuning of Simple Regulators *IFAC Proc.* **17** 1867–72

IOP Conf. Series: Materials Science and Engineering **543** (2019) 012050 doi:10.1088/1757-899X/543/1/012050

Heavy Fraction Separation From Used Lube Oil Using Thin Film Evaporator

A J Rifliansah, R H M Utomo, J P Sutikno and R Handogo*.
Process Design and Control Laboratory, Department of Chemical Engineering, Institut Teknologi Sepuluh Nopember, Surabaya (ITS), 60111, Indonesia
*email: renanto@chem-eng.its.ac.id

Abstract. Used lube oil is a dangerous waste that must be treated before disposed on the environment. Used lube oil treatment process consists of several stages, one of them is the separation of heavy fraction. Some equipment system can be used to complete this process, and thin film evaporator (TFE) is chosen for this research. The goal of this research is to understand the effect of temperature and rotor rotation on the heat transfer that happens in the distillation process using thin film evaporator. Simulation using Aspen Plus is used and is validated by manual calculation using Matlab. Temperature is varied between 250 – 330 °C and rotor rotation are varied between 0 – 60 rotations per minute (RPM) in a vacuum pressure of 2.53 kPa. From this research, the relation between operating temperature and rotor rotation in thin film evaporator to the vapor fraction produced is directly proportional. The optimum operating condition in this research was found at a temperature of 310 °C with the agitator rotational speed of 30 RPM.

1. Introduction

Used lube oil is a valuable resource. However, in its use, used lube oil is contaminated with various components of impurities. In addition, the organic constituents in lube oils also undergo changes during its use and produce contaminants. Thus, the used lube oil can no longer be used for several reasons. Because of the basic characteristics of used lube oil and these contaminants, the used lube oil cannot be discharged directly into the environment. Therefore, it needs further treatment for this used lube oil. There are several solutions to deal with used lube oil. The first way is to use used lube oil as a fuel. For more economically and environmentally feasible to handle the used lube oil, reprocess the used oil back into the fresh lube oil which can be used for the original requirement. There are several methods for treating used lube oil, one of them is intended to separate the heavy fraction of lube oils free of the residue using Falling Film Distillation Column [1]. The second method is using agitated thin film evaporator as the processing equipment [2]. In this separation process, the thin film evaporator as one normally will distribute the feed distilled at the wall of the evaporator. The thin layer thus formed and be separated, light fraction will go to the distillate, while the heavy fraction will be heading to the residue. The heating can be done in two ways, namely by using a heating coil in the evaporator or by using heat in the beginning before it goes into the evaporator. Thin film evaporator offers the dual advantage of short residence time and the formation of a thin film which optimizes the separation process. Therefore, it is necessary to design the thin film evaporator using simulation software Aspen Plus. The goal to determine the effect of operating temperature and rotor rotation thin film evaporator. After that, simulation results are validated using MATLAB and the optimum operating conditions are determined for this separation process according to specifications feed used in this study.

IOP Conf. Series: Materials Science and Engineering **543** (2019) 012050 doi:10.1088/1757-899X/543/1/012050

2. Theory

2.1. Used Lube Oil

Used lube oil is usually composed of a mixture of various lubricating oils that have been used in motor vehicles and industry. [3]. Lubricating oils lose their effectiveness because of specific types of contaminants. These contaminants are divided into two categories, the first is a foreign contaminant and the second is the product of damaged lube oil. Foreign contaminants come from the air and metal particles from the engine. Contaminants which come from the air are sand, dirt, and humidity. The air itself can be considered as a contaminant because it may cause oil or lube oil becomes frothy. Contaminants coming from the engine are metal particles due to the use of machines, carbon particles originating from incomplete combustion, metal oxides originating from corrosion on metal, water from leaking cooling system, the water of the products of combustion, fuels or additives or by-products which may enter the engine crankcase. [3]. Used lube oil has a chemical formula of carbon chain between C21-C40. Volatile gases contained in the used lube oil are H2, CO, and CO2. The composition of the used lube oil used in this study is described in Table 1 below:

Table 1: Composition of used lube oil [4]

Component	Mass Flowrate (kg/h)	Mass Percentage (%)
Light End	0.0001	0.00
Gas Oil	0.6758	7.30
Lube Oil	5.7667	78.37
Residue	1.3273	14.33
Total	9.2623	100.00

2.2. Thin Film Evaporator

Thin film evaporator is one type of evaporator that applies a formation of the thin layer in the heat exchange process. Despite this consideration, thin film evaporator is also considered as mass transport equipment because the molecules of the liquid phase may be transferred to the gas phase during the evaporation and vice versa. Stripping is done by the upward gases and absorption is done by the downward liquid. This equipment has been applied in the production of chemicals, pharmaceuticals, and food since 1950. Illustration of vertical thin-film evaporator is given in Figure 1:

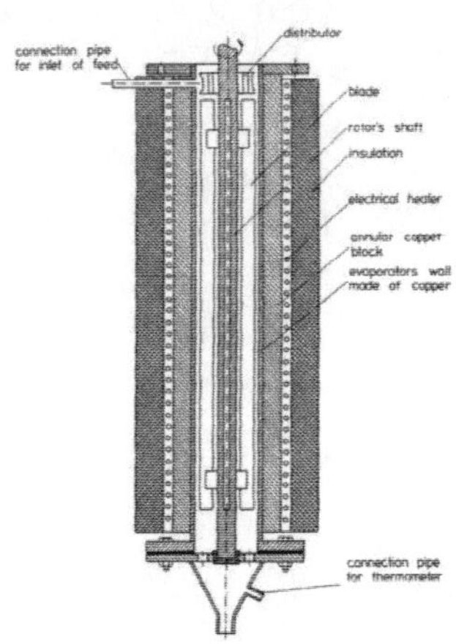

Figure 1: Vertical Thin-Film Evaporator [5]

IOP Conf. Series: Materials Science and Engineering **543** (2019) 012050 doi:10.1088/1757-899X/543/1/012050

3. Methods

Thin film evaporators are modeled on Aspen Plus using a combination of horizontal heat exchangers and vertical flash drum [6] according to Figure 2. This model is an approximation since there is no thin film evaporator as a single unit in Aspen Plus. Therefore, it needs validation using manual calculation. In this process, feed enters into the heat exchanger with temperatures of 200 °C and will be heated to 250-330 °C with increments of 20 °C. Then the value of the heat coefficient is entered which corresponds to variable rotor rotation 6-60 RPM at an operating pressure of 2.53 kPa. The liquid phase and vapor phase are then separated using a flash separator to obtain a top product and bottom product from the thin film evaporator. The value of the heat transfer coefficient (hsc) will change as the value of rotation of the rotor on the thin film evaporator changes. To get the hsc value, calculations are necessary and involve a wide range of variables with the following equation:

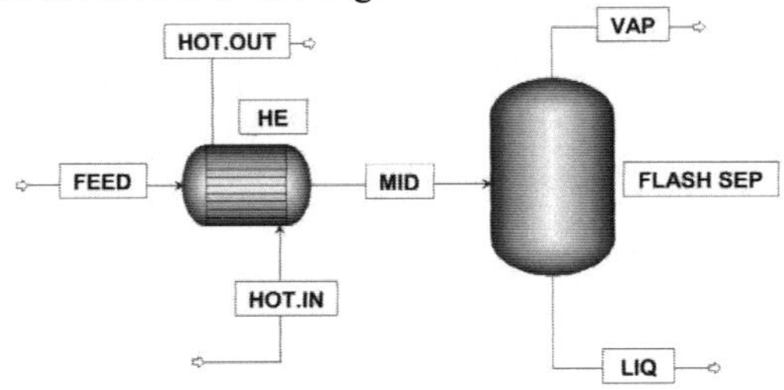

Figure 2: Thin-Film Evaporator Model

$$\%L_v = \frac{\sum_{i=1}^{k} m_{vk}}{m_z} \times 100 \qquad (1)$$

$$m_{vk} = \frac{Q_k}{\lambda_s} \qquad (2)$$

$$Q_k = h_{sc \cdot k} A_k (T_w - T_s) \qquad (3)$$

$$h_{sc \cdot k} = 2\sqrt{\frac{k_k \rho_k Cp_k}{\pi t_e}} \qquad (4)$$

$$k_k = X_{L.k-1} k_L + (1 - X_{L.k-1}) k_h \qquad (5)$$

$$\rho_k = X_{L.k-1} \rho_L + (1 - X_{L.k-1}) \rho_h \qquad (6)$$

$$Cp_k = X_{L.k-1} Cp_L + (1 - X_{L.k-1}) Cp_h \qquad (7)$$

$$t_e = \frac{1}{nB} \qquad (8)$$

IOP Conf. Series: Materials Science and Engineering **543** (2019) 012050 doi:10.1088/1757-899X/543/1/012050

$$A_k = H_I \pi D_s \quad (9)$$

$$H_I = \pi . D_s . \sin \phi \quad (10)$$

$$\phi = \tan^{-1} \frac{V_z}{V_t} \quad (11)$$

$$V_z = \frac{m_z}{A_i \rho_a} \quad (12)$$

$$V_t = \pi D_t n \quad (13)$$

$$A_i = (\frac{\pi}{4})(D_s^2 - D_r^2) \quad (14)$$

$$T_w = T_h - \frac{Q}{H_h A_h} \quad (15)$$

$$T_h = \frac{T_{hi} - T_{ho}}{2} \quad (16)$$

$$Q = m_z Cp_h (T_{ho} - T_{hi}) \quad (17)$$

$$h_h = \frac{Nu . k_h}{D_e} \quad (18)$$

$$D_e = D_{oj} - D_{ij} \quad (19)$$

$$Nu = 0{,}023 Re^{0{,}8} Pr^{0{,}4} \quad (20)$$

$$Re = \frac{D_e V_h \rho_h}{\mu_h} \quad (21)$$

$$V_h = \frac{m_h}{A_h \rho_h} \quad (22)$$

$$A_h = \frac{\pi}{4}\left(D_{oj}^2 - D_{ij}^2\right) \quad (23)$$

$$Pr = \frac{Cp_h \mu_h}{k_h} \quad (24)$$

$$A_h = \frac{\pi}{4}\left(D_{oj}^2 - D_{ij}^2\right) \quad (25)$$

IOP Conf. Series: Materials Science and Engineering **543** (2019) 012050 doi:10.1088/1757-899X/543/1/012050

With all of the equations above, the overall heat transfer coefficient can be obtained with varying the rotation of the rotor thin film evaporator (N). Heat transfer coefficient values obtained from manual calculations will be used into the simulation Aspen Plus. This overall heat transfer is also used to calculate how much product is separated using thermodynamic relation to validate the simulation results so that the error can be known. In this case the required data according to Tables 2 and 3

Table 2: Thin Film Evaporators Specification

Specification	Symbol	unit	Value
Blade Diameter	D_t	m	0.0446
Outer diameter of the annulus	D_s	m	0.0720
Inner diameter of the annulus	D_r	m	0.0625
Height	T	m	0.2200
Number of Blade	B		8

Table 3: Feed Composition

Feed composition	Symbol	Unit	Value
Inlet mass flow	m_z	Kg/s	0.0025
Feed Density	ρ_a	Kg/m^3	975
Light component fraction	X.L		0.8572
Thermal light conductivity	k.L	W/m.K	0.1210
Heavy thermal conductivity	k.H	W/m.K	0.0870
Light density	ρ.L	Kg /m^3	823
heavy density	ρ.H		975
Light heat capacity	Cp.L		2,409
Heavy heat capacity	Cp.H		2,734

4. Results and Discussion

4.1. The relation between Rotor and Rotation with Heat Transfer Coefficient (h_{sc})

The speed of rotation of the rotor can be modeled by changing the heat transfer coefficient (U) that is calculated for corresponding speed. Therefore, the manual calculation is necessary to determine the heat transfer coefficient for each variable of the rotation speed of the rotor. Results are obtained on the graph in Figure 3. Heat transfer coefficient value obtained from the calculation will be used in modeling the thin film evaporator at Aspen Plus software for each variable rotation speed of the rotor

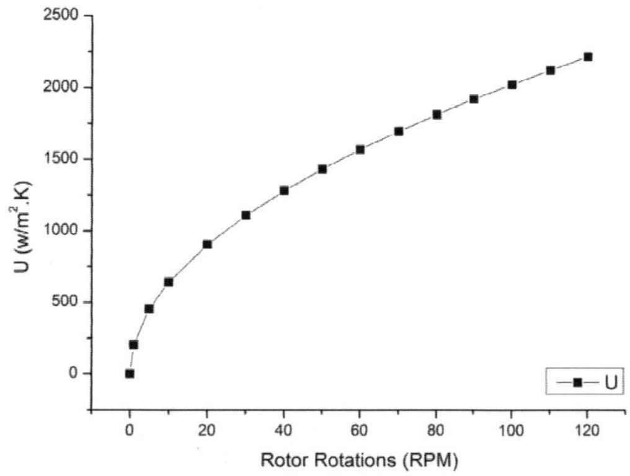

Figure 3: Comparison graph between heat transfer coefficient and the rotational speed of the rotor

4.2. Effect of Various Rotor Rotation Vapour Fraction and Vapour Mass Flowrate

Variations of rotor rotation used in this simulation were from 0 up to 60 RPM. Variations of this rotor rotation were simulated to each variable operating temperature of the thin film evaporator at a temperature of 250 °C, 270 °C, 290 °C, 310 °C and 330 °C. Parameters that would be observed are the amount of vapor fraction and composition of the vapor fraction that is carried to the top product. In this case, the operating pressure is in a vacuum of 2.53 kPa. The results of the simulation model of thin film evaporator for various rotor rotations can be seen in Figures 4 and 5. Vapor fraction and mass flow of vapor produced are proportional in general with the speed of rotation. This is true because the equation above, the overall heat transfer coefficient is proportional with rotation speed. The amount of vapor for rotor rotation increases significantly until about 6 RPM then steadily rises until 30 RPM.

4.3. Effect of Various Operating Temperature on Product Composition

Variations in temperature used in this simulation are in the range of 250 °C to 330 °C with increments of 20 °C. Rotation speed also varied in each of these variables. The percentage of lube oil and residue in distillate for each variable and mass flow of lube and residue are plotted in Figures 6, 7, 8, and 9.

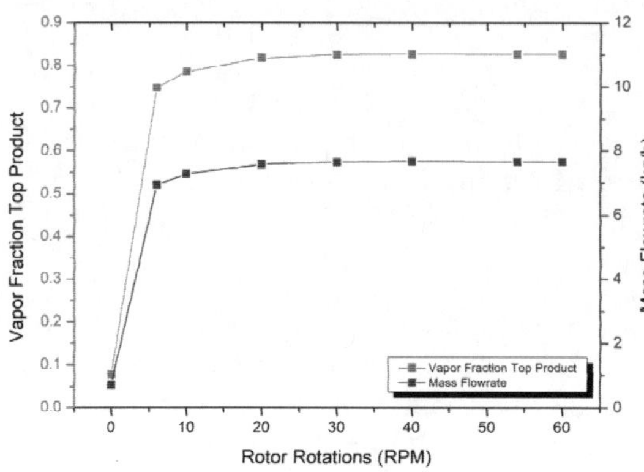

Figure 4: Graph of Vapor Fraction and Mass Flowrate Againts Rotor Rotation at 310°C

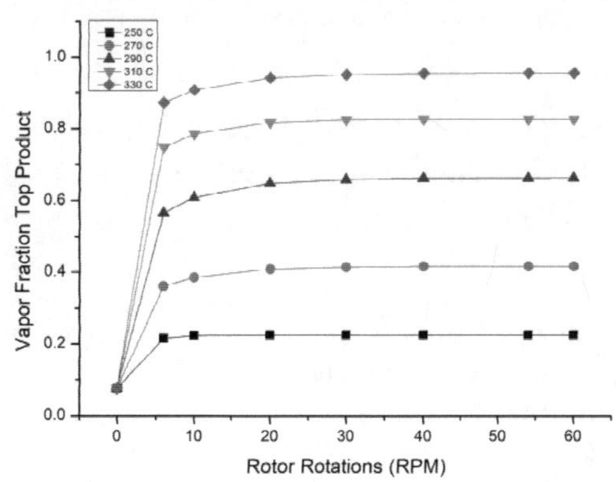

Figure 5: Graphs of Vapor Fraction Againts Rotor Rotations at Various Temperature

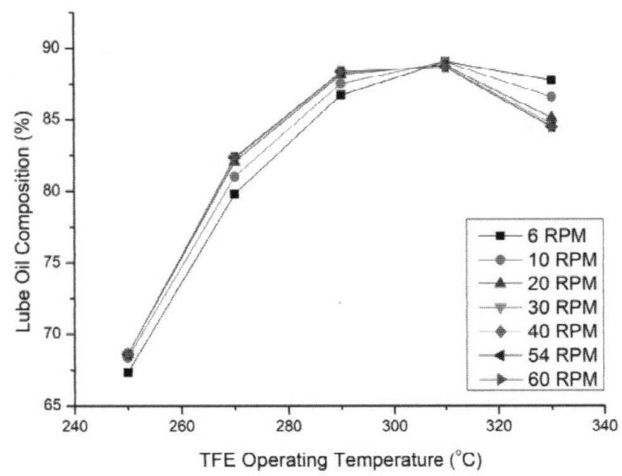

Figure 6: Graphs of Percentage Lube Oil Composition Againts Operating Temperature at Various Rotor Rotation

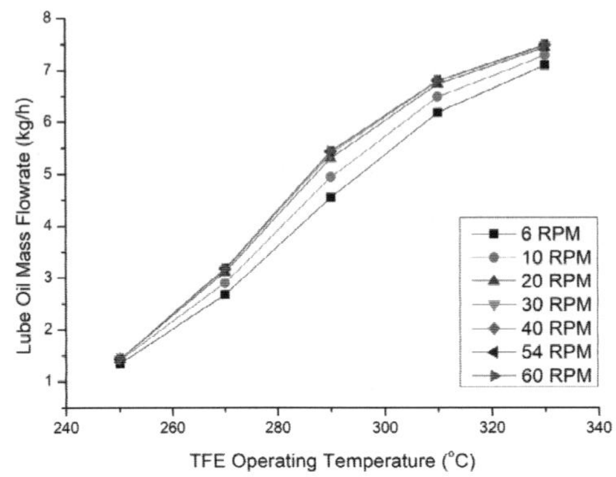

Figure 7: Graph of Percentage Residue Composition Againts Operating Temperature At Various Rotor Rotation

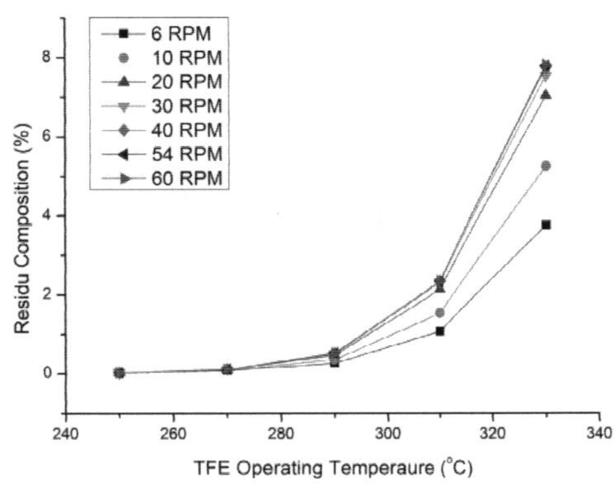

Figure 8: Graph of Lube Oil Mass Flowrate Againts Operating Temperature At Various Rotor Rotation

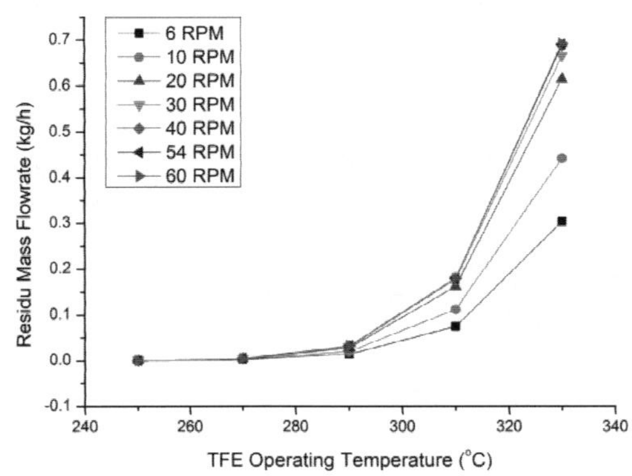

Figure 9: Graph of Residue Mass Flowrate Againts Operating Temperature At Various RotorRotation

It can be seen in Figure 6, the percentage composition of lube oil in 60 RPM is higher than in 6 RPM when the operating temperature between 250 °C - 310 ° C. Above 310 °C, the percentage composition of lube oil in 60 RPM is lower than in 6 RPM. So the highest percentage of lube oil composition is found at a temperature of 310 °C. Below 310 °C, every increase in the operating temperature of the thin film evaporator cause lube oil mass flow to increase significantly according to the Figure 8 and the flow rate of residue also increases but not significant according to the Figure 9, so that the percentage of lube oil increases. So, it can be concluded that the optimum operating conditions of this thin film evaporator at a temperature of 310 °C with the rotation of the rotor 30 RPM.

4.4. Validation Using Manual Calculation

Because of the lack of features of Aspen Plus, the tool is modeled with a combination of the heat exchanger and a flash drum. So it is necessary to validate the simulation results using manual calculation. Results of the validation are presented in Table 4. Based on the results of the calculation using MATLAB and the results of the simulation using Aspen Plus, the error is generated. The error does not exceed 5%, therefore the modeling in Aspen Plus can represent a process in Thin Film Evaporator. Percent error obtained in optimum conditions, a temperature of 310 °C and 30 RPM rotor rotation is at 2.45 %.

4.5. Production Composition of Optimum Condition

The optimum operating conditions are obtained at a temperature of 310 °C with an optimum rotor rotation is at 30 RPM. In such operating conditions, the product obtained has the following composition in Table 5.

Table 4: Comparison of Simulation Results Thin Film Evaporator Using Aspen Plus and Matlab Operating temperature 310 °C

RPM	Top Product Fraction		Error
	MATLAB	Aspen Plus	Percentage (%)
6	0.7829	0.7486	4.38
10	0.8146	0.7854	3.59
20	0.8405	0.8176	2.73
30	0.8460	0.8253	2.45
40	0.8484	0.8268	2.55
54	0.8436	0.8265	2.55
60	0.8492	0.8265	2.67
Average error			2.98

Table 5: Product Composition At Optimum Condition

Component	Mass flowrate (kg / h)	Mass Percentage (%)
Light End	0.0001	0.00
Gas Oil	0.6751	10.00
Lube Oil	5.7667	85.40
Residue	0.3110	4.61
Total	6.7529	100.00

5. Conclusion

Based on the results of the research that has been done, it can be concluded that the relation between operating temperature and rotor rotation thin film evaporator to the vapor fraction produced is directly proportional. The higher the operating temperature and rotor rotation of Thin Film Evaporator, the higher the vapor fraction generated. But for rotor rotation above 30 RPM, the increase is almost negligible. The optimum operating conditions for this process is at the rotation speed of 30 RPM and 310 °C operating temperature, which produces a vapor fraction of 0.8253 with the percentage of lube oil fractions of 85.40 % to the product and the product mass of 6.75 kg/h.

Acknowledgment
The authors acknowledged the funding from Direktorat Riset dan Pengabdian pada Masyarakat Dikti to support this work.

References

[1] Querino M V, Marangoni C and Machado R 2018 *Chem. Eng. Trans.* **69** 679-84
[2] Rossi F, Corbetta M, Geraci D, Pirola C and Manenti F 2015 *Chem. Eng. Trans.* **43** 1429-34
[3] Speight J and Exall D 2014 *Handbook of Refining Used Lube Oils, CRC Press*
[4] Carlo G L 2013 *Presentation of Study Tecnologie Progetti Srl*
[5] Dziak J 2011 *Advanced Topics in Mass Transfer* Prof. Mohamed El-Amin (Ed.), INTECH.
[6] Pawar B S, Patil R and Mujumdar U S 2011 *Drying Technol.* **29** 719-28

Water and power consumption reduction by gas conditioning tower system modification in cement industry

C W Purnomo[1,2], R B Cahyono[2], A Setiawan[3], N Amin[3], Triyono[3]
[1] Agrotechnologi Innovation Center (PIAT), Universitas Gadjah Mada, Kalitirto, Tanjungtirto, Berbah Sleman, Indonesia.
[2] Chemical Engineering Department, Universitas Gadjah Mada, Jl. Grafika No. 2 Yogyakarta 55281, Indonesia.
[3]PT Indocement Tunggal Prakarsa, Palimanan, Cirebon, West Java Indonesia.

E-mail: chandra.purnomo@ugm.ac.id

Abstract. Gas conditioning tower (GCT) is a vital equipment in cement industries. The main principle in this equipment is evaporative cooling, i.e. reducing the temperature of the hot flue gas by evaporation of cooling water. Thus, the water consumption is large enough to be able to cool the hot gas from about 400 °C to 150 °C or less. In this case study, the water consumption reduction of a GCT is evaluated before and after modification. The modification of the GCT system includes the change in operating condition and tower mechanical construction which are installation of flow deflector in the inlet pipe, the addition of two perforated plates, the reduction of the nozzle sprayers and the addition of tower length. Meanwhile, the dust removal replacement from Electrostatic Precipitator (EP) to Bag Filter (BF) system allows higher outlet gas temperature of the GCT which could reduce the water and power consumption significantly. The modification effect of the GCT system is analyzed in terms of water demand and energy savings using 100 days real plant data from PT Indocement Tunggal Prakarsa, Indonesia (ITP). It is shown that the modification can reduce the total water consumption by about 20 % and the specific power consumption by about 0.5 kWh/ton of Raw Mill product. The general equation for cooling water demand calculation based on heat balance has been modified as well for better real data approach incorporating air spray flow into the equation.

1. Introduction

Cement industries are considered as material and energy intensive industries which have a high potential contribution to air emissions. Therefore, the efforts to bring this industry to be more sustainable and greener are very essential such as the usage of alternative fuels and raw materials [1], controlling emissions [2], and lowering carbon footprints [3]. In terms of water consumption, it is considered less significant since the wet process in cement production has been replaced by dry method and now this industry is mostly dealing only with solids and gases. In a typical Material Flow Analysis (MFA) for Portland cement, water can be neglected during clinker production due to lower quantity compared with other solid raw materials [4], but it starts to be counted in the concrete and mortar processing [5]. In other words, water utilization is only counted during the cement product life cycle but not in the manufacturing processes [6]. The documented efforts of water saving in cement industry is very rare and mostly dealing with the modification of cement properties in order to reduce water demand during the application for concrete or mortar production [7] or during concrete formation [8].

Water in this industry is used only in several cooling equipment. However in the future, fresh water availability is considered as one of the major sustainability problems. Moreover most of the industries has been started to reduce water consumption as much as possible. In sugarcane industry, water demand can be reduced up to 85% theoretically by optimizing the source and sink of water in the process [9]. Meanwhile in pulp and paper industry, an optimization model can reduce the demand up to 23 % [10]. These two examples of water minimization in different type of industry can be a clue that there is always a space for resource optimization without sacrificing the main production.

Thus, cement industries worldwide must also pursue serious attempts to optimize the water consumption. Furthermore, water utilization is considered as one of the 19 initial Key Performance Indicators (KPIs) of sustainable manufacturing in cement industry [11]. Even though, it was placed in the last priority of the KPIs list in the survey results which reflected the lower priority of this indicator compared with others such as energy consumption, material cost and occupational health and safety.

A major cooling device in cement industry is a Gas Conditioning Tower (GCT). This tower is used for reducing the flue gas temperature from a suspension preheater (SP) before entering a dust removal system i.e. bag filter (BF) system or electrostatic precipitator (EP) system. The mechanism of cooling is using evaporative cooling of the cooling water flowing co-currently to the hot gas flow. Typically, a part of the hot gas flow will be diverted from entering GCT to be used for drying and transportation in Raw Mill (RM) system which is called 'compound' operation mode. Whenever the RM stop due to maintenance or troubleshooting, all gas will be flown to GCT and it is called 'direct' mode. The GCT system connecting with RM and dust removal system is represented in Figure 1.

During direct operation mode all the hot gas from SP is flown into GCT which requires higher cooling water flowrate than compound mode. The inlet gas temperature of a common GCT is about 300 to 400 ^{o}C, while outlet gas temperature depend on the dust removal system requirement. EP requires lower gas temperature than BF, thus if a BF system is used then the cooling load in a GCT will be much lower. In a typical cement plant, the cooling system (GCT) is equipped with additional water spray system outside the tower in order to further cool down and increase the moisture of the gas before mixing with the hotter gas from RM system (Figure 1). It is common in an existing cement plant when a n EP system is replaced by a new BF system, this additional spray system (water spray 2) is no longer be operated, thus, the gas cooling duty is completed inside a GCT only. Even though the main duty of GCT is cooling the gas phase by water spraying mechanism, due to a high dust content inside the inlet gas, some part of the dust is deposited in the bottom of the tower. The recovered dust will be mixed with raw mill product for clinker production.

The equipment is called gas conditioning due to the main duty of achieving the required gas properties of the adjacent equipment i.e. dust removal equipment. The different dust removal system requires different setting of the GCT operation. For instance, bag filter (BF) allows higher inlet temperature than electrostatic precipitation (EP) thus the GCT cooling duty is much lower. The consequence is the reduction of cooling water consumption. However, bag filter requires low droplet of water in the inlet gas phase thus the retention time of gas inside the tower should be sufficient to make sure that most of the sprayed water is already evaporated.

Water consumption is the main parameter than can be minimized to become sustainable cement industry. The effort in reducing the water consumption can be pursued by process modification and equipment modification for improving the contact between the cooling water droplets and the hot gas. Meanwhile, the previous literature of water utilization analysis in modern cement industry is very rare to be found. In this study, the water and power consumption are evaluated before and after GCT modification and EP to BF replacement. By observing the real plant data, it might necessary to modify general equation for calculating water demand especially for GCT equipped with not only water spray but also pressurized air spray.

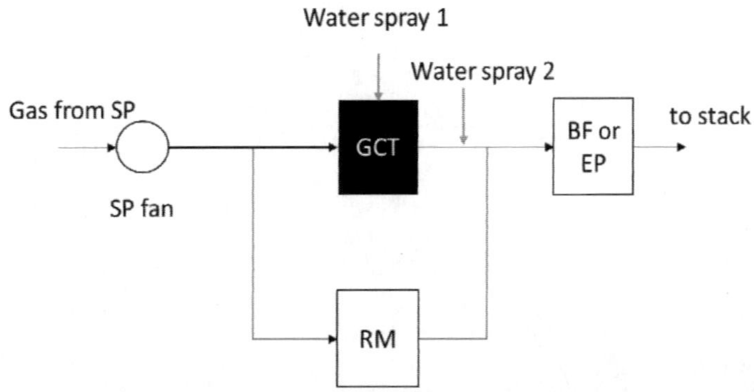

Figure 1. Flow diagram of a GCT system in compound operating mode.

2. Methods

The study data of GCT system was provided by PT Indocement Tunggal Prakarsa (ITP), Palimanan Plant line 9, Cirebon, West Java, Indonesia. The data was collected 3 months before (July – September 2015), during the modification (October 2015) and after the modification (November 2015 – January 2016)). The modification of GCT, listed in Table 1, was in the same time with the replacement of EP to BF system in this cement production line. Thus the analysis of GCT performance is extended to the dust removal system. The schematic diagram of GTC gas and water flow is represented in Figure 2.

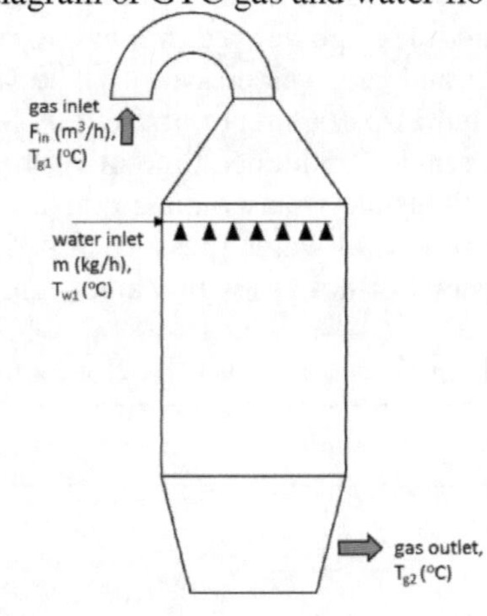

Figure 2. The GCT system diagram.

The theoretical water consumption of GCT (*mw*) can be calculated using simple energy balance represented in Eq. (1). The heat removed from the gas is equal to the total heat for increasing the water temperature from the room temperature to the boiling point, the heat for water evaporation and for increasing the vapor temperature to the temperature of outlet gas (*Tg2*). However, applying this simple heat balance equation to the real plant data may overestimate the water consumption. Thus, the efficiency of the tower (ε) evaluated by Eq. (3) may reach higher than 100%. This simple equation is suitable for design purposes during cooling water demand calculation or GCT with only water spay. For this study, the more accurate equation for equipment rating purpose is proposed in Eq. (2) for evaluation the GCT efficiency. The equation is improved by incorporating the flow of pressurized air mixture in water spray that usually being neglected. The cold pressurized air flow (*Fs*) at the same temperature of water inlet (*Tw*) has a significant contribution in gas cooling system. It can reduce significantly the cooling water demand.

IOP Conf. Series: Materials Science and Engineering **543** (2019) 012051 doi:10.1088/1757-899X/543/1/012051

$$m_w = \frac{F_{in} \cdot \rho_g \cdot C_{pg} \cdot (T_{g1} - T_{g2})}{C_{pw}(100 - T_{w1}) + \lambda + C_{pv}(T_{g2} - 100)} \quad (1)$$

$$m_{wcal} = \frac{F_{in} \cdot \rho_g \cdot C_{pg} \cdot (T_{g1} - T_{g2})}{C_{pw}(100 - T_{w1}) + \lambda + C_{pv}(T_{g2} - 100) + F_S \cdot \rho_g \cdot C_{pg} \cdot (T_{g2} - T_w)} \quad (2)$$

$$\varepsilon = \frac{m_{w\ plant\ data}}{m_{wcal}} \times 100 \quad (3)$$

F_{in} = pressure inlet (m^3/h)
F_{out} = pressure outlet (m^3/h)
Fs = cold pressurized air flow (m^3/h)
T_{g1} = temperature of inlet gas (°C)
T_{g2} = temperature of outlet gas (°C)
ρ_g = density of gas (g/cm^3)
C_{pg} = specific heat air (kJ/kg°C)
C_{pw} = specific heat water (kJ/kg°C)
C_{pv} = specific heat water vapour (kJ/kg°C)
ε = efficiency of the tower (%)
mw = the theoretical water consumption of GCT
λ = latent heat water

In this study the two GCT systems with EP or BF will be evaluated in terms of water efficiency usages. In terms of energy reduction, the reduction of power requirement to produce the final product in Raw Mill station will be compared before and after modification. The GCT modification was done for increasing the efficiency of cooling and also for incorporating the requirement of the new bag filter system. The length was added for increasing residence time of cooling water droplet to make sure all the water can be evaporated. The water droplets may damage the bag filter materials in the long run. Meanwhile, the reduction of water nozzle and water pressure aimed to reduce water demand. Then, deflector addition will create better contact of hot gas with water to improve efficiency.

Table 1. Comparison of specification of GCT before and after modification

Specification	unit	before	after
Effective Height	m	17	23
Number of spray	pc	20	12
Gas volume inlet	Nm^3/h	315,000	315,000
Temp. inlet	°C	400	400
Temp. outlet	°C	240	150
Water supply pressure	bar	33	10
Pump motor	kW	75	37
Internal accessories		n.a.	inflow deflector Screens

3. Results and discussion

The GCT modification was done for increasing the efficiency of cooling and also for incorporating the requirement of the new bag filter system. The length was added for increasing residence time of cooling water droplet to make sure all the water can be evaporated. The water droplets may damage the bag filter materials in the long run. Meanwhile, the reduction of water nozzle and water pressure aimed to reduce water demand. Then, deflector addition will create better contact of hot gas with water to improve efficiency.

The GCT tower efficiency was evaluated using Eq. (3). The calculated efficiency using plant data during three months before and after tower modification (100 days in total) is represented in Figure 3. The figure shows the general trend of efficiency reduction of the tower after modification. The average efficiency before modification is 72.1% ±16% while after modification is 58.5%±9.5%. Even tough, the cooling efficiency of the tower before modification is higher but it highly fluctuated. The larger deviation of the efficiency can be seen and in many times the efficiency are lower than after modification. Thus, the modification was able to stabilize the GCT performance with respect to inlet gas properties and operating mode changes.

The lower trend of efficiency after modification could be due to the reduction of the water spray nozzles as indicated in Table 1, but in other hand it will reduce the power consumption of the tower by reducing the power of the water spray pump. Another possible cause is due to the installation of wire mesh screen in the lower part of the conical section of the tower for improving gas distribution and avoiding flow separation. This internal part addition for correcting the gas flow could lead to the build-up of large lump of wet dust that may fall into the bottom of tower and creating blocking problem [12]. The wet lump of dust will reduce the water efficiency since it is necessary that the water injected is totally evaporated before reaching the bottom of the tower.

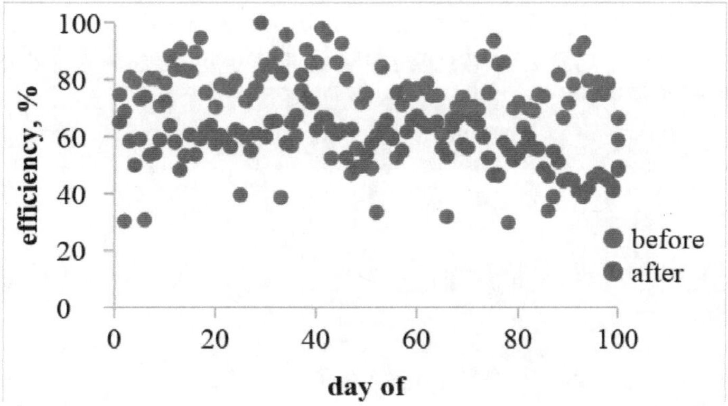

Figure 3. GCT tower efficiency before and after modification.

As expected, the data for water consumption of the tower (GCT) before modification is in general less than after modification as indicated in Figure 4. In line with the efficiency results, the water demand before modification is oscillating. The average water consumption was shifted from 9.6 m^3/h to 12.1 m^3/h after modification. This higher demand of cooling water is basically due to the termination of the additional water spray (water spray 2 in Figure 1), thus, the gas cooling mechanism is only rely on the GCT. With the new GCT configuration with less spray nozzles the higher water flow may tend to reduce the nozzle performance in distributing water droplets throughout the tower especially in the situation that one or more nozzle(s) get blocked by solid dust.

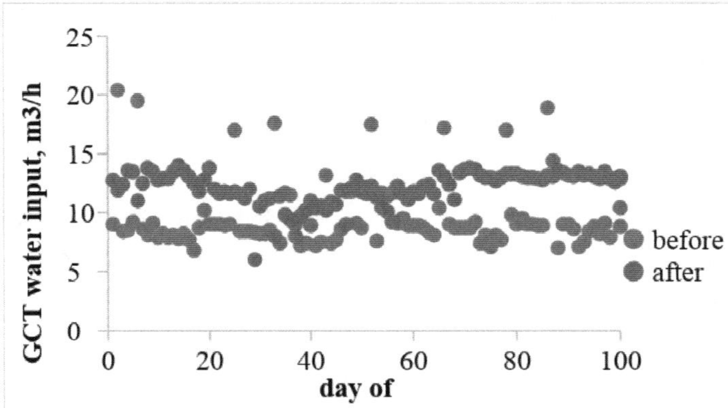

Figure 4. GCT water consumption before and after modification.

On the other hand, the total cooling water consumption is reduced as indicated by Figure 5. The reduction is caused by closing the second water spray and also the increase of GCT gas outlet temperature. The increase of outlet temperature from 150 °C before modification to 240 °C afterward will decrease the cooling water demand theoretically, even though it is not totally applied in this case. By this modification, the overall water consumption can be reduced by about 20%. There is still room for GCT performance improvement after modification. For instance, to increase the cooling efficiency the number of nozzles with a suitable nozzle type can be increased for better water distribution. Even though, the pumping power by the nozzle addition is also should be carefully evaluated.

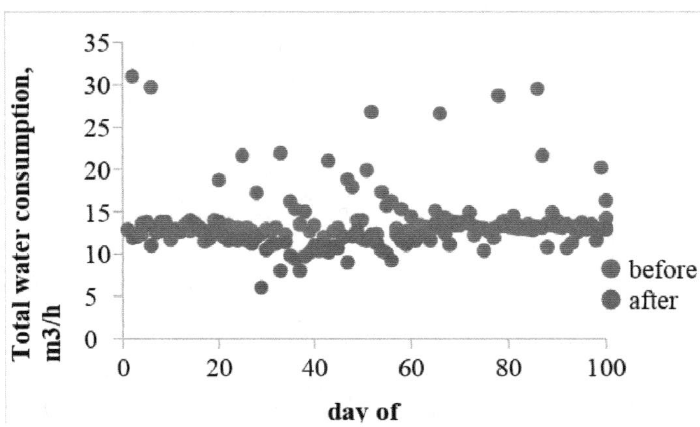

Figure 5. Total water consumption for gas cooling before and after modification.

From the power consumption point of view, in this case, the main reduction is basically due to the replacement of EP to BF since EP consumes a lot more energy due to electrostatic separation principles. The other significant power reduction is by application of single water spray in GCT only. Figure 6 shows the evolution of monthly average specific power consumption based on Raw Mill product. It can be seen that the specific power consumption (SPC) decrease about 0.5 KWh/ton after replacement (the middle of October 2015). If it is multiplied by the average Raw Mill production i.e. 6000 ton/day and also the electricity price in Indonesia i.e. 0.1 USD/KWh, the operational cost saving is estimated at about 2.7 million USD/year.

IOP Conf. Series: Materials Science and Engineering **543** (2019) 012051 doi:10.1088/1757-899X/543/1/012051

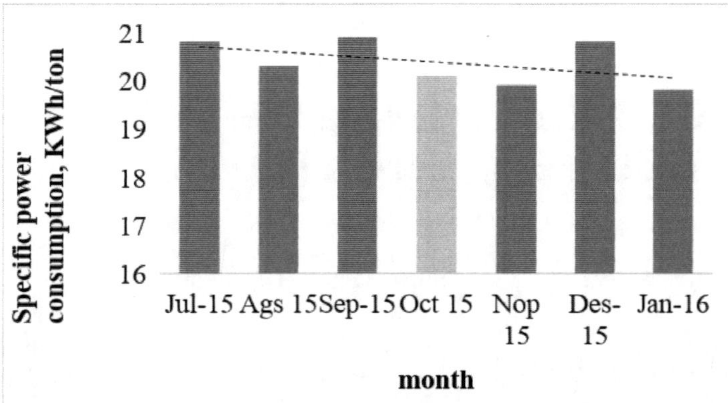

Figure 6. Specific power consumption of Raw Mill product.

4. Conclusions

The modification of GCT tower is able to increase the tower performance stability in any mode of operation and the gas feed fluctuation. The cooling water demand in the modified GTC is more stable than the previous system. However, the water cooling efficiency is lowered to compensate the stability improvement. The total water consumption by a new combination of modified GCT and BF can be reduced up to 20% and the power consumption specific to the Raw Mill product can also be reduced by 0.5 KWh/ton.

5. References

[1] Schneider M 2015 *Cement Concrete Res.* **78** 14–23

[2] Purnomo C W, Budhijanto W, Alfisyah M and Triyono 2018 IOP Conf. Ser.: Mater. Sci. Eng. **316** 012031

[3] Lu Y L, Geng J and He G Z 2015 *Adv. Climate Change Res.* **6** 202–9

[4] Wang W, Jiang D, Chen D, Chen Z, Zhou W and Zhu B 2016 *J. Clean. Prod.* **112** 787–96

[5] Gao T, Shen L, Shen M, Liu L and Chen F 2016 *J. Clean. Prod.* **112** 553–65

[6] Diego A, Raúl C, Andr D, Marx D, Jael A and Duque-rivera J 2016 *J. Clean. Prod.* **113** 114–22

[7] Proske T, Hainer S, Rezvani M and Graubne C A 2013 *Cement Concrete Res.* **51** 38–46

[8] Schneider M, Romer M, Tschudin M and Bolio H 2011 *Cement Concrete Res.* **41** 642–50

[9] Gonzalez Cortes M, Verelst H, Lam H L and Gonzalez Suarez E 2011 *Chem. Eng. Trans.* **25** 839–44

[10] Chew I M L, Foo D C Y, Lam H L, Bonhivers J C, Savulescu L E and Alva-Argaez A 2011 *Chem. Eng. Trans.* **25** 441–46

[11] Amrina E and Vilsi A L 2015 *12th Global Conf. on Sustainable Manufacturing Key* vol 26 (Berlin: Elsevier B.V) p 19–23.

[12] Nielsen N F and Lind L 2002 *Proc. 4th Int. Sym. on Computational Technologies for Fluid/Thermal/Chemical Systems and Industrial Applications (Vancouver)* vol 448-2 p 177-90

IOP Conf. Series: Materials Science and Engineering **543** (2019) 012052 doi:10.1088/1757-899X/543/1/012052

Evaluation of COD Removal using Dielectric Barrier Discharge to Produce Biogas from POME

A. Hazmi[1,*], P. Emeraldi[1], M.I. Hamid M.I[1], R. Desmiarti[2]

[1] *Department of Electrical Engineering, Andalas University, 25163, Padang, Indonesia*
[2] *Department of Chemical Engineering, Bung Hatta University, 25137, Padang, Indonesia*

*Email: *ariadi@eng.unand.ac.id*

Abstract. Evaluation of Chemical Oxygen Demand (COD) removal by a dielectric barrier discharge (DBD) system to produce biogas from palm oil mill effluent (POME) was investigated in a batch experiment. The residence time was set at 1 hour. The batch experiments were observed with voltage variation at 10, 15, 20 and 25 kV. It was found that the COD removal was measured at 22.47% to 44.94% with the applied voltage was set at 10-25 kV, respectively. The results also shown that the maximum yields of hydrogen and methane by the system were 7.94 and 1.78 mL/mL of POME, respectively that were obtained at an applied voltage of 25 kV.

1. Introduction

Palm oil industry in Indonesia is the biggest producer in the world [1]. As consequence, it also produces large quantities of POME during crude palm oil (CPO) processes. Each ton of CPO produces about 3 tons of POME [2], which is considered wastewater. POME wastewater can cause serious water pollution because it contains high levels of organic contents such as high chemical oxygen demand (COD), biochemical oxygen demand (BOD), total dissolved solids (TDS) and oil-grease. Conventional methods mostly used for the treatment of POME to produce biogas used in the palm oil industry are: open ponding, anaerobic and aerobic systems, or combinations of both. The advantage of these methods is that biological treatment is easy, but disadvantages are that they require a long hydraulic retention time (HRT) and a large area of land [3-4]. Also, POME wastewater treated in open ponds can cause great amounts of greenhouse gas emission. However, there is an opportunity to convert POME wastewater into biogases such as hydrogen and methane. POME wastewater is a very promising source for renewable energy. Each ton of POME can produce about 28 m^3 of biogas [5]. It has been reported that the treatment of POME to produce biogas requires from 1 to 10 days, respectively [6-7].

Since recently it is known that POME can be treated successfully using a dielectric barrier discharge (DBD) plasma system. This system is able to reduce the HRT for treatment of COD, BOD and oil-grease to achieve standard wastewater quality and to produce biogas simultaneously [8]. Our previous results also resulted that the maximum yields of hydrogen at 2.42 mL H_2/mL POME and methane at 1.32 mL CH_4/mL POME. Based on our previous studies [8-9], modification design of DBD plasma reactor is very important to increase the hydrogen production.

The purpose of this work is to study the performance of DBD plasma reactor to increase production of $H_2$from POME. The DBD plasma system generates a oxidize radicals (H· and ·OH) and molecules (H_2O_2, O_3, etc.) to accelerate the chemical reaction. These chemical reactants are very effective in decomposing organic compounds in wastewater [10-11].

IOP Conf. Series: Materials Science and Engineering **543** (2019) 012052 doi:10.1088/1757-899X/543/1/012052

2. Material and Methods

The characteristic of POME used in this study is shown in Table 1. The POME was taken after the fat-grease pit from a wastewater treatment system in the palm oil industry. The experimental setup for biogas production is displayed in Figure 1. This experiment used a needle-plane electrode configuration system with a separation between the needle and the POME surface of 5 mm. A stainless steel needle electrode and a plane electrode were used, with a diameter of 3 mm and a thickness of 1 mm. The upper needle electrodes and a lower plane electrode were connected to a high voltage system with a frequency of 50 Hz. The applied voltages and discharge currents were measured through a high voltage probe (P6015A Tektronix) and a current probe (P6022 Tektronix), and recorded by an oscilloscope (TDS5104 Tektronix). The applied voltages were varied at 10, 15, 20 and 25 kV. Furthermore, a container glass with a storage volume of about 3500 mL was used as the DBD reactor. The volume of the POME in the DBD reactor glass was 1000 mL. In addition, a smaller container with a storage volume of about 1500 mL was used as gas storage. The accumulated hydrogen and methane gas were detected by commercial MQ8 and MQ4 gas sensors (Hanwei), respectively. All gas sensors were connected to a data logger (ADC24 Pico) to digitize and record the conductivity change related to the electrical output signal of gas concentrations in the storage container every second for one hour, which was controlled by a personal computer (PC). Figure 2 shows the applied voltage to generate the discharge current in the DBD reactor.

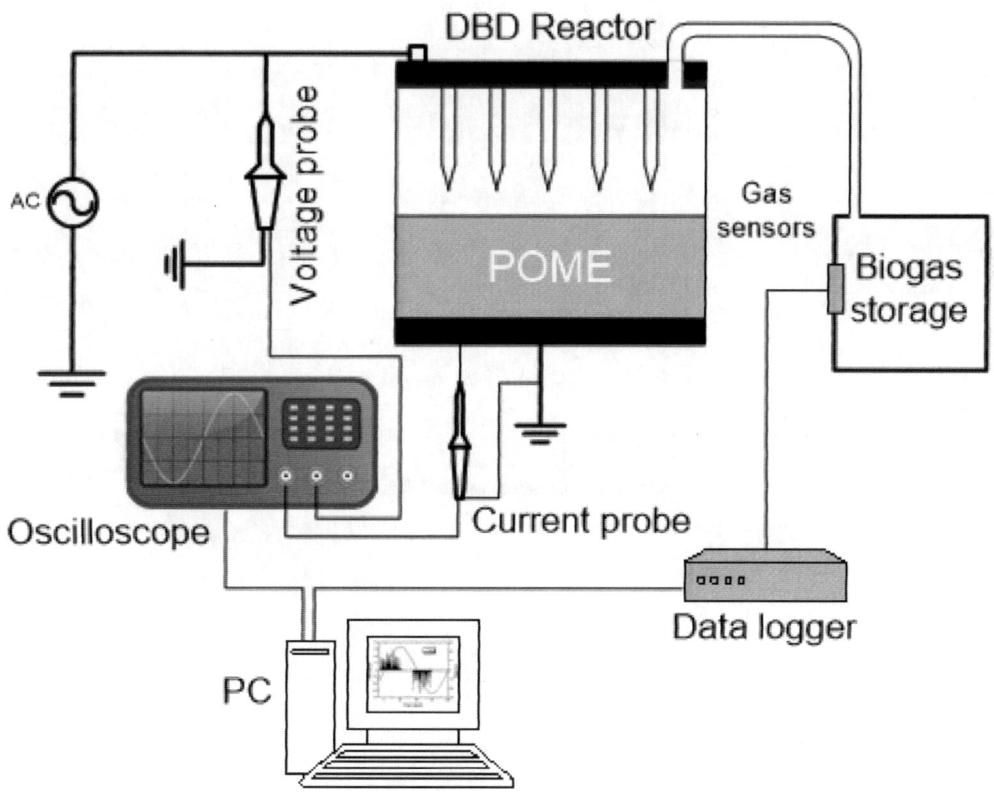

Figure 1. The experimental setup system.

Table 1. Characteristics of POME.

Parameter	Unit	Value
COD	mg/L	5520
BOD	mg/L	2000
pH		4.2

3. Results and Discussion
The experiment studied the effects of applied voltage and time on the discharge current in the DBD reactor and the product yield in the biogas storage. The applied voltage and the discharge current in the DBD reactor were measured and recorded by oscilloscope through high voltage and current probes, while the biogas product yield in the storage detected by the gas sensors was recorded by a data logger.

3.1 Discharge current characteristic
Figure 2 shows that the discharge current pulse train occurred at an applied voltage of 20 kV. The discharge current arose around the peaks of both the positive and the negative half cycles for the applied voltage as can be seen in Figure 2. The discharge current on the POME surface produced UV light andvarious active species with high oxidation potential ions and molecules such as ozone (O_3), perhydroxyl radical (□H^+), hydroxyl radical (□OH^-) and hydrogen peroxide (H_2O_2), which effectively decompose organic compounds. The decomposition process of POME produces water, inorganic salts [11] and biogases such as hydrogen (H_2), methane (CH_4) and carbon monoxide (CO).

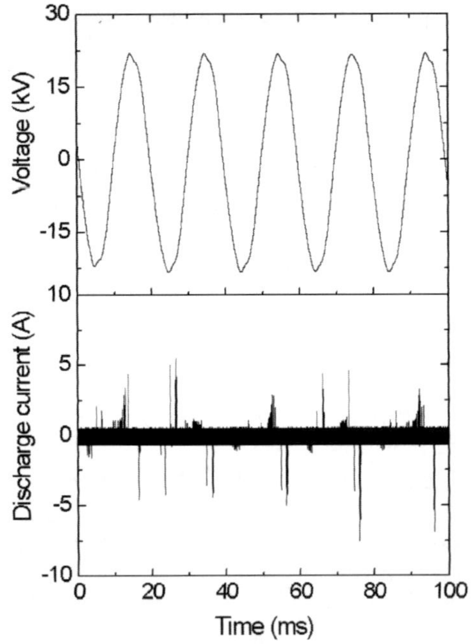

Figure 2. Typical the voltage application and the discharge current.

3.2 The effects of applied voltage on biogas production
The removal efficiency (RE) of chemical oxygen demand (COD) in the DBD system is shown in Figure 3a. The removal of COD was measured at 22.47%, 30.34%, 33.71% and 44.94% with the applied voltage was set at 10, 15, 20 and 25 kV in a batch experiment for 1 hour. These results are similar with our previous study [8] found that the removal efficiency of COD was 31% with applied voltage at 19 kV. Based on the results in [9], the decrease of COD was observed by formation of biogas such as

hydrogen and methane. It was found that the degradation rate of POME increased with an increased voltage. The effects of the applied voltage on the yield production of biogas is displayed in Figure 3b. The yield of hydrogen increased from 5.32, 6.30, 7.22 and 7.94 mL H_2/mL POME after 1 hour with the applied voltage at 10, 15, 20 and 25 kV, respectively.

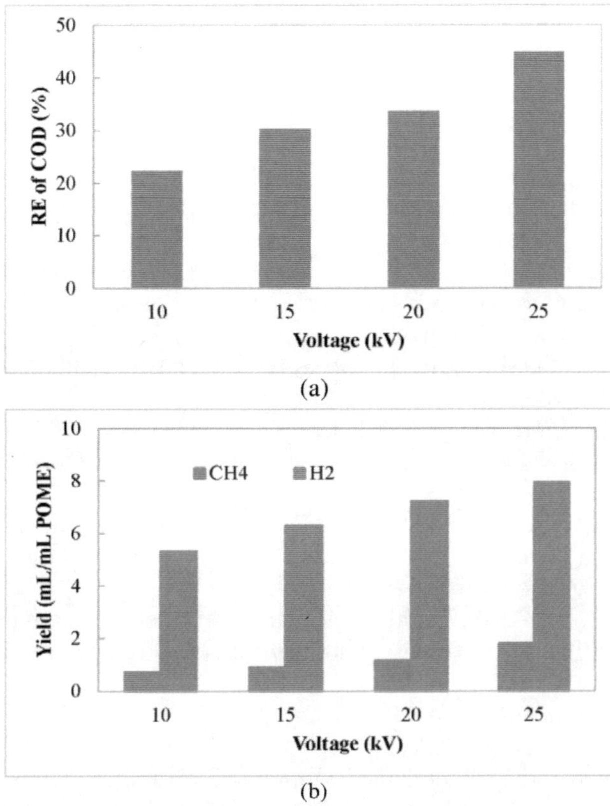

Figure 3 (a). The effects of applied voltage on removal efficiency of COD. **(b).** The effects of voltage application on yield production of hydrogen and methane.

Figure 3 also shows that the yield of methane increased from 0.71, 0.88, 1.15 and 1.78 mL CH_4/mL POME after 1 hour with the applied voltage at 10, 15, 20 and 25 kV, respectively. Norfadilah et al [7] found that the biogas production contained 5.99 mL H_2/mL POME after 10 days fermentation process. The biogas contained with 36% H_2 and 64 % CO_2. The biogas production using fermentation process depends on the substrate and microorganism concentration, pH, temperature and hydraulic retention time to produce biogas following hydrolysis, acidognesis and methanogenesis [12-13]. Latief et al [14] investigated biogas production from POME and found that when the biogas contains 63% methane with processing time at 25 days, the system should be well controlled under mesophilic condition (37 °C). The processing time in the present study was 238 times faster compared with the results using fermentation. The biogas production with DBD system consisted of 82-88% hydrogen and 12-18% methane. These results were higher than our previous studies [8-9] before modification of DBD plasma reactor.

4. Conclusion

The effects of applied voltage on biogas production from POME was studied using a DBD plasma system. It was found that increasing the applied voltage caused an increase in the yield of hydrogen and methane. The highest hydrogen and methane yields were 88% and 12%, respectively with the applied voltage at 25 kV in batch experiment for 1 hour. The processing time in the present study was 238 times faster compared with the results using fermentation. The removal of COD was calculated in the range 22.47% to 44.94% with the applied voltage from 10 kV to 25 kV.

Acknowledgement
This work was supported by PTUPT KEMRISTEKDIKTI of Indonesia with contract number 050/SP2H/LT/DPRM/2018 and 02/UN.16.17/PTUPT.EET/LPPM/2018. Thanks go to my students (Faygia and Arief).

References

[1] Ahmed Y, Yaakob Z, Akhtar P and Sopian K 2015 Production of biogas and performance evaluation of existing treatment processes in palm oil mill effluent (POME). *Renew. Sust. Energ.* Rev. **42** 1260–78
[2] Vijaya, S., A. N. Ma, Y. M. Choo, N. I. K. Meriam. 2008. Life cycle inventory of the production of crude palm oil - a gate to gate case study of 12. *J. Oil Palm Resea.* **20**: 484–94
[3] Desmiarti, R. 1999. Combination of Anaerobic Process and Membrane Filtration for Treatment Palm Oil Mill Effluent (POME), *Master Thesis* (Indonesia: Chemical Engineering Department of Institut Teknologi Bandung)
[4] Chena, C., W. Guo, H. H. Ngoa, D. J. Leeb, K. L. Tungb, P. Jinc, J. Wang, Y. Wud. 2016 Challenges in biogas production from anaerobic membrane bioreactors. *Renew. Energ.* **98**: 120-34
[5] Harsono, S.S., P. Grundmann, S. Soebronto. 2014. Anaerobic treatment of palm oil mill effluents: potential contribution to net energy yield and reduction of greenhouse gas emissions from biodiesel production. *J. Clea. Prod.* **64**: 619–27
[6] Budiman, P.M., T. Y. Wu. 2016. Ultrasonication pre-treatment of combined effluents from palm oil, pulp and paper mills for improving photofermentative biohydrogen production. *Energ. Con. and Manag.* **119**. 142-50
[7] Norfadilah, N., A. Raheem, R. Harun, F. R. Ahmadun. 2016. Bio-hydrogen production from palm oil mill effluent (POME): A preliminary study. *Int. J. Hydr. Ener.* 41(28): 11960-64
[8] Hazmi, A., R. Desmiarti, E. P. Waldi, and P. Emeraldi. 2016. Preliminary Study on Treatment of Palm Oil Mill Effluent (POME) by Sand Filtration-DBD Plasma System. *J. Eng. Technol. Sci.* **48**(1). 21-30
[9] Hazmi, A., R. Desmiarti, P. Emeraldi, M.I. Hamid, Edwardo, and E. P. Waldi. 2017. Preliminary Study on Biogas Production from POME by DBD plasma. *Telkomnika.* **15**(2). 554-9
[10] Lukes, P., M. Clupek, P. Sunka, F. Peterka, T. Sano, N. Negishi, S. Matsuzawa, and K. Takeuchi. 2005. Pulsed Corona Discharge in Combination with TiO photocatalysis, *Resear. Chem. Inter.* **31**: 285-94
[11] Sato, M., T. Soutome, S. Mii, T. Ohshima, and N. Yamada. 2007. Decomposition of Phenol in Water Using Water Surface Plasma in Wetted-Wall Reactor. *Inter. J. Plasm. Envi. Sci. & Technol.* 1: 71-5.
[12] Igoni, A., M. Abowei, M. Ayotamuno and C. Eze. 2008. Effect of total solids concentration of municipal solid waste on the biogas produced in an anaerobic continuous digester. *Agric. Eng. Inter.* CIGR E. journal, Manuscript. EE 07 010.
[13] Wang, J., and W. Wan. 2008. Effect of temperature on fermentative hydrogen production by mixed cultures. *Int. J. Hydr. Energ.* **33**: 5392–7.
[14] Lattief, F.A. 2016. A study of biogas production from date fruit wastes. *J. Clea. Prod.* **139**. 1191-5.

IOP Conf. Series: Materials Science and Engineering **543** (2019) 012053 doi:10.1088/1757-899X/543/1/012053

Biobutanol Production from Bagasse Using Ammonia Pre-treatment and Acid Hydrolysis Method

A M Jannah, D U Amalia and N S Hana

Department of Chemical Engineering, Faculty of Engineering, Sriwijaya University, South Sumatera, Indonesia

Abstract. Biobutanol is a type of alcohol that can be produced from biomass lignocellulose and used as renewable fuel. One of lignocellulose sources as raw material of making biobutanol is bagasse. Bagasse as agriculture waste was useless and did not have a potential benefit for farmers, but in this research it used as raw material because of the high cellulose contained. In this research for producing biobutanol, bagasse pre-treated by using ammonia with various concentrations (0 %, 1 %, 3 %, 5 %, 7%) and hydrolysis with 2 % sulphuric acid for 2 hours and 120 °C. Then the process continued with fermentation. In this step 10 mL *Clostridium acetobutylicum* bacteria was used and fermentation times were varied (2, 4 and 6 days). The aims of this research were analysed the content of cellulose, hemicellulose, lignin before and after pre-treatment processes and also biobutanol content after fermentation process. The results of this research showed that using 7 % ammonia was the highest lignocellulose content which had 69.68 % of cellulose, 14.39 % of hemicellulose and 9.15 % of lignin. The highest biobutanol content was 0.85 % that produced by using 5 % ammonia in the pre-treatment process and fermented for 6 days.

1. Introduction

As non-renewable, the quantity of fossil fuel decreased by the time. However, humans need the fuel every day. The decreasing of fuel quantity made many researchers were trying to find alternative fuel to substitute non-renewable energy and able to reduce the scarcity of fossil fuel. Alcohol has high potential to convert to be energy. Butanol is biofuel diversification that has high potential to be developed. Biobutanol have a great potential to partially or completely replace fossil fuels in the automotive industry due to their attractive characteristics [1]. At the first generation biobutanol used for fuel produced from organic compounds such as glucose (19.6 g/l) and corn flour (15.8 g/l) [2]. But nowadays butanol also produced from organic waste called 2nd generation biofuel. Another agriculture waste has been researches. From Plant empty fruit bunches as raw material can be produced 0.84 g/L biobutanol [3]. Biobutanol also has been researched by using city organic wastes consist of fruits and vegetables waste from traditional market in Samarinda, Indonesia. In this research, organic waste also hydrolyzed by using 0.5 % - 1 % H_2SO_4 with rasio 1:6 (g/v) at various temperature 100 – 130 °C. The result from that research produced 13.0 (g/L) biobutanol. The most glucose yield has been produced by sample with hydrolysed with 1 % H2SO4 at 110 °C for 45 min [4]. In this study, biobutanol was produced from bagasse. Bagasse as a solid waste of sugar industry is very abundant in Indonesia. Statistics Indonesia released that Indonesia sugar cane (Saccharum officinarum) in 2016 has 482,239 hectares. One sugar cane can be obtained 30 % - 40 % of bagasse by volume. Because of this reason, Indonesia has huge potential to make bagasse as biofuel and produce sustainably. Biobutanol was produced from lignocellulose biomass with main constituent compounds are cellulose, hemicellulose and lignin. The main processes of biobutanol production were in hydrolysis process and fermentation. Bagasse as the raw material in this study was gotten from waste

IOP Conf. Series: Materials Science and Engineering **543** (2019) 012053 doi:10.1088/1757-899X/543/1/012053

of sugar industry. Bagasse can be used as a raw material to produce biobutanol because it contents of 52 % cellulose [5]. Cellulose as a major compound of plant structure will be reacted with acid or alkaline solution and converted to become glucose. Then glucose will be fermented by using Clostridium bacterial in anaerobe to produce biobutanol. In the plant structure compound, lignin covers cellulose and hemicellulose bonding. Lignin inhibits cellulose to react with catalyst and convert to be glucose. Because of this reason, lignin must be eliminated. Amount of lignin in raw material was quite influential in glucose production. The process to produce biobutanol was started by decrease the lignin component of biomass and continued to the hydrolysis process and fermented the bagasse by using *Clostridium acetobutylicum* bacteria. *Clostridium acetobutylicum* bacteria is a bacteria that can help in process of conversion sugar into the butanol organic. *Clostridium acetobutylicum* is a gram-positive bacillus [6]. This bacterium is working in anaerobic condition of fermentation. Some of the newest research has investigated alternative methods to produce the industrial solvent which *Clostridium acetobutylicum* has been used for the last century to produce. In particular, butanol has received particular attention as possible alternative fuel source for automobiles.

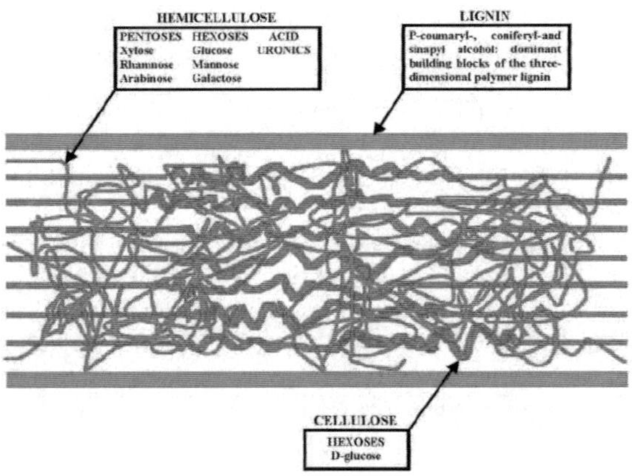

Figure 1. Lignocellulosic Materials: Composition of Major Compounds [7]

2. Processing Description

2.1. Delignification

Pre-treatment process for lignocellulose biomass aimed to degrade lignin content in the cell walls of plants. Bagasse as raw material in this research has 9.15 % by volume of lignin content. The amount of lignin has influence in the conversion process of biobutanol. Less lignin content in cell wall of bagasse was made the process of biobutanol production easier. Lignin is one of complex polymer with high weight molecule that composed of fenil propane. This component fills the space between cellulose and hemicellulose in the cell walls [8]. The main process of biobutanol production was cellulose and hemicellulose conversion into glucose and the last product was alcohol. In this case, lignin will inhibit the process of glucose formation. Because of this reason lignin has to be delignificated. The delignification process can be done by adding solvent and followed by heating. Delignification process can be done through chemical treatment, physic treatment, and chemical-physic treatment. Type of treatment used for raw material can be adjusted. Using only physic treatment for delignification was not effective to decrease lignin, therefore needed follow chemical treatment, so there will be chemical reaction to help lignin detached from the bond. Chemical delignification can be done with several treatments such as Liquid Hot Water, Acid Treatment, alkaline treatment, and Soaking in Aqueous Ammonia (SAA). Liquid Hot Water (LWH) was a biomass pre-treatment process using high temperature and high pressure water in to break the molecules. Other treatments were hydrothermolysis, hydrothermal, aqueous fractionation, solvolysis. While in Acid Treatment process was using acid in delignification. Both weak acid and strong acid can be used in this process. Using the acid in pre-treatment was effective to degrade the lignin. The use of weak acid commonly followed by high temperature and continue flow process for 5 % - 10 % by weight for substrate concentrate of solid (low solid) and low temperature and batch process for 10 % - 40 % by weight for substrate concentrate of solid (high solid). The alkaline also can be used in degradation of lignin. Alkaline solution improved the affectivity of polysaccharide beside degraded the lignin. In this research, delignification process used ammonia solution called soaking in Aqueous Ammonia (SAA). The biomass soaked in ammonia solution in 28°C – 30°C with atmospheric pressure. This solvent was effectively to decrease lignin from biomass by hydrolysis ether bonds. The using of ammonia had several advantages. It has high selectivity to lignin, maintained carbohydrate in original form, very little interaction with hemicellulose and the ammonia itself can be regenerated. SAA pre-treatment was started by washing the bagasse and well dried under the sun light. Then the bagasse chopped and milled with grinder until 60 meshes. Took 50 g of bagasse was weighed and placed using Erlenmeyer for a sample and soaked with 500 mL of ammonia solution with several concentrations (0 %, 1 %, 3 %, 5 % and 7 %) by volume for 60 minutes with 121°C and 3.5 bar using autoclaved. After heating, the samples were filtered and washed until pH 7. Then the samples dried in the oven with 105°C until well dried (8 h drying). The dried samples were ready to be analysis for lignin, hemicellulose and cellulose content.

Figure 2. Dried bagasse as raw material

2.2. *Acid Hydrolysis*

Hydrolysis is a reaction involving water to break down a bond in a molecule using water. Breaking a bond of lignocellulose can be used water, alkaline solution, acid solution and enzyme. This treatment process generated inhibitory compounds, and the detoxification was required for removing these compounds found in the hydrolysate [9]. The success of hydrolysis process in term was depending on some factors. The main factor that influenced the hydrolysis process was the content of carbohydrate in the raw material. In hydrolysis process carbohydrate will be conversed to glucose. The amount of carbohydrate content in raw material determined the amount of glucose. If in the raw material contented much carbohydrate, it decreased the molecular collisions between carbohydrate and water, so it decreased time of glucose form reaction. Bagasse as raw material in this research has 43.28 % cellulose, 16.01 % hemicellulose and 24.76 % lignin. Reaction between lignocellulose and water molecule was very slow, so needed the catalyst to make it faster. In hydrolysis cellulose and hemicellulose will be converted to become glucose by using acid solution as catalyst by activated the water, this process called acid hydrolysis. The main advantage of acid hydrolysis is that acids can prenetrate lignin without any preminary pre-treatment of biomass, thus breaking down the cellulose and hemicellulose polymers to form individual sugar molecules [10]. There were two types of acid hydrolysis, concentrated acid and dilute acid. Several types of acid, concentrated or diluted, can be used, such as sulphurous, sulphuric, hydrochloric, hydrofluoric, phosphoric, nitric and formic acid

[11], but commonly sulphuric and hydrochloric were used as catalyst for hydrolysis process [12]. The using of concentrated acid will be produced the high content of glucose (90 % glucose in theoretic). The high concentration of acid as catalyst in hydrolysis can be heated in low temperature but needed corrosion resistant equipment with high cost. It was the reasons the using concentrated acid catalyst did not common used in industry. Otherwise, the used of dilute acid was common used because this process did not need acid recovery, no acid lost in process [13] and low cost. In industry acid solutions that used were sulphuric acid (H_2SO_4) and hydrochloric acid (HCl). In this research catalyst used was dilute sulphuric acid solution. This process was started by adding 500 mL sulphuric acid with 2 % concentration by volume to the samples that has been pre-treated. Then the samples heated 121°C for 120 min by using autoclaved. After heating, the samples were washed with pure water and filtered to separate the filtrate and sediment. The samples have to be washed to neutralize the bagasse.

2.3. Fermentation

Fermentation is a process of composition of organic compounds to produce energy accompanied by conversion of substrate to new product using microbes [14]. Fermentation has been done by scientists for a long time ago; it was one of the old technic in science. They used fermentation in process of making drinking of food. There were several factors that influenced the success of fermentation. The acidity or alkalinity solution in the term has important role in fermentation. Besides that the microbes itself has important influence in fermentation. Bacteria will grow in pH 4 – 5 [15]. The microbiology applied to fermented substrate has to have optimum temperature. All bacteria have their own optimum environmental surrounding and temperatures in which they thrive the most. In order to grow, bacteria must have an energy source. Energy sourced from carbon, other required nutrients, and a permissive range of physical condition such of oxygen concentration. In this research, to ferment bagasse used *Clostridium acetobutylicum*. *Clostridium acetobutylicum* is a commercially valuable bacterium belonging to genus *Clostridium*. *Clostridium acetobutylicum* naturally produces acetone, butanol and ethanol. *Clostridium acetobutylicum* thrived in pH 5. The samples adjusted with pH 5 and added with 5 mL *Clostridium acetobutylicum* starter. The sample fermented with varying time (2, 4 and 6 days) in anaerobe condition at 37°C to produce biobutanol. The last step, product of this research has to purify to separate pure biobutanol from aqua and others impurities using distillation.

3. Result and Analysis

3.1. Cellulose, hemicellulose and lignin Analysis

In this research, the delignification bagasse used dilute Ammonia solution. This process aimed to decrease lignin content in bagasse. Cellulose, hemicellulose and lignin content were analysed not only for samples after pre-treated but also applied to the raw material. Analysis samples to get the number of cellulose, hemicellulose and lignin content used Datta Method. One gram sample (a) was added 150 mL alcohol-benzene and refluxed at 100°C for an hour with water bath. Then the product was filtered and residue was washed with 300 mL hot water. The residue dried with oven with 100°C for 8 h until the weight constant (b). After drying the sample, 150 mL sulphuric acid 1 N was added and stirred until homogeny. This solution then refluxed with water bath at 100°C for 1 h. After an hour, the sample was filtered and washed with aqua until pH 7. The residue left at the filter was dried until the weight constant with oven at 100°C for 8 h (c). Dried residue was soaked with 100 mL of 72 % sulphuric acid by volume for 4 h. Then added 150 mL of 1 N sulphuric acid and refluxed at 100°C with water bath for 1 h. The residue was filtered and washed with H_2O until the residue neutral. The neutral residue dried in the oven at 105°C until well dried (d). Then dried residue burned to ashes and weighed (e). Calculations of cellulose, hemicellulose and lignin content were used formulas below:

$$hemicellulose\ content = \frac{(b-c)}{a}\ x\ 100\ \% \quad (1)$$

$$cellulose\ content = \frac{(c-d)}{a}\ x\ 100\ \% \quad (2)$$

$$lignin\ content = \frac{(d-e)}{a}\ x\ 100\ \% \quad (3)$$

IOP Conf. Series: Materials Science and Engineering **543** (2019) 012053 doi:10.1088/1757-899X/543/1/012053

The result of ammonia pre-treatment showed in the graph below :

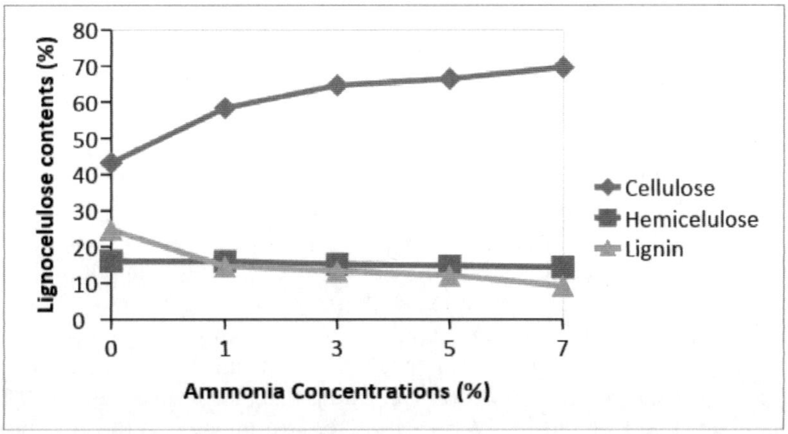

Figure 3. Cellulose, hemicellulose and lignin content after ammonia pretreatment

Pre-treatment process using ammonia aimed to break the bond between lignin and hemicellulose without this pre-treatment, hydrolysis of cellulose in biomass will be obstructed. From the figure 3, lignin content decreased significantly. The higher ammonia concentration used made more lignin depolymerized. The chemical structure of lignin can be changed naturally. Lignin will be agglomerated to be small particles and will be separated from cellulose [16]. The lignin content can be decreased until 63.05 % using 7 % of ammonia solution in pre-treatment Cellulose, hemicellulose and lignin content after ammonia pre-treatment. Bagasse pre-treatment used ammonia showed more lignin reduction compared to sodium hydroxide pre-treatment. When bagasse was soaked in 6 % of sodium hydroxide for 12 hours at room temperature can realize lignin up to 32.11 % [17], while the using of 5 % ammonia can reduce 51.21 % lignin. It showed the effectiveness of ammonia to break down lignin bond in pre-treatment process. The increasing cellulose content by increasing the concentration of ammonia solution in pre-treatment process was caused by degradation of other compounds. The other degradation compounds dissolved in black liquor decreased amount of other compounds in the substrate. Figure 3 showed amount of cellulose content increased by increasing ammonia concentrations. While hemicellulose content decreased by increasing ammonia concentration in pre-treatment. Hemicellulose and lignin were bound together by a variety of different chemical bond. Some of these attachments were easy to break but others were more difficult. When lignin dissolved in black liquor, so did amount of hemicellulose. It made hemicellulose content tended to decrease as well as lignin.

4. Biobutanol Analysis

Biobutanol analysed using gas chromatograph. But for making sure the components of lignocellulose content, the samples must be analysed refractive index. Refractometer used to analyse the refractive index for identification of cellulose contents. The refractometer works to analyse the sample using finding the angle of reflection and correlated the numbers to refractive index. Based on the research, the results showed (Table 1) that higher ammonia concentration used in hydrolysis treatment and longer fermentation time spent made refractive index increase linearly.

Table 1. Refractive Index of Samples with Various Ammonia Concentrations and fermentation Times

Ammonia	Fermentation (days)

(%)	2	4	6
0	1.33007	1.33045	1.33107
1	1.33205	1.33207	1.33209
3	1.33305	1.33307	1.33405
5	1.34407	1.34507	1.34607
7	1.34645	1.34705	1.34707

After getting refractive index, samples analysed using gas chromatograph to get the biobutanol content. The principle of gas chromatography was injection sample solution into instrument entered a gas stream which transported the sample into a separation tube called “column”. Sample was injected to GC-2010 Plus Series with condition : column type RTX-1, column temperature was 90°C with 30 m length, inside diameter 0.25 mm, film thickness 0.25 μm with nitrogen for a carrier gas, column flow 1.13 mL/min, total flow : 60.8 mL/min, linear velocity : 30.4 cm/sec, purge flow : 3.0 mL/min, split ratio : 50.0, equilibrium time : 0.5 min, hold time : 2.4 min, injection temperature : 120°C, mode : split, detector type was FID (Flame Ionization Detector), detector temperature 120°C, sample speed : 40 msec. Sample was filtered using filter Whatman-42 before injected to the column. Then the sample analysed using internal method standard with 99 % purity of butanol. The computer will be analysed the combustion process delivered by detector and showed the result with graphic. Biobutanol content calculated with the formula below:

$$Biobutanol\ content = \frac{area\ butanol\ sample}{area\ butanol\ standart}\ x\ 100\ \% \tag{4}$$

In this research the highest biobutanol content has gotten in the sample treated with 7 % of ammonia and fermented for 6 days, was 0.85 %. The optimum fermentation time was in 6 days. It showed that more ammonia concentration used for pre-treatment in this research made the most biobutanol content as product.

5. Acknowledgement

The author are grateful to Universitas Sriwijaya for funding this research by Research of Technology Science and Art No. 1013/UN9.3.1/PP/2017.

References

[1] Valentinyi N, Marton G, Toth A J, Haaz E, Andre A and Miszey P 2018 *J. Chem. Eng. Trans.* **69** 607
[2] Jin C, Yao M, Liu H, Leed C F and Ji J 2011 *J. Renewable and Sustainable Energy Reviews* **15** 4080
[3] Noomtim P, Cheirsiip B 2011 Production of Butanol from Palm Empty Bunches Hydrolyzate by Clostridium acetobutylicum *9th Eco- Energy and Materials Science and Engineering Symposium Energy Procedia* **9** 140 – 146
[4] Irawan D, Arifin Z and Irawan M 2013 Butanol Production from Waste City Using Catalyst H_2SO_4 and *Clostridium acetobutylicum Technology, Education and Science International Conference*
[5] Samsuri M, G Gozan, R Mardias., M Baiqun, H Hermansyah, A Wijanarko, B Prasetya and M Nasikin 2007 *J. Makara Teknologi* **11** 17
[6] Cato E P, George W L and Finegold S M 1986 Genius Clostridium *Baltimore* 1141-1200
[7] Kumar P, Barret D M, Delwiche M J, Stroeve P 2009 *J. Industrial & Engineering Chemistry Reseach* **48** 3713
[8] Suparjo 2008 Lignocellulose Compound Degradation by White Rod Fungi Publishing Jajo66 Wordpress.com

IOP Conf. Series: Materials Science and Engineering **543** (2019) 012053 doi:10.1088/1757-899X/543/1/012053

[9] Dussan J K, Silva D D V, Moraes E J C, Arruda P V, Felipe M G A 2014 *J. Chem. Eng. Trans* **34** 433

[10] Verardi A, Bari I D, Ricca E, Calabro V 2012 Hydrolysis of Lignocellulosic Biomass:Current Status of Processes and Technologies and Future Perspectives Rende Italy

[11] Galbe M, Zacchi G 2002 A review of the Production of Ethanol from Softwood *App. Microbiol. and Biotech.* **56** 618

[12] Lenihan P, Orozco A, O'Neil E, Ahmad N M, Rooney D W, Walker G M 2010 *J. Chemical Engineeering* **156** 395

[13] Iranmahboob J, Nadim F, Monemi S 2002 *J. Biomass and Bioenergy* **22** 401

[14] Madigan, Michael T, David P, Clarck, David S, John M, Martinko 2011 Brock Microbiology of Microorganisms, San Francisco, Benjamin Cummings publishing

[15] Winarno F G, I E Fernande 2007 Milk and Fermented Products, M-brio Press, Bogor

[16] Novikava 2000 *Appl. Biochem. Microbiol* **38** 181

[17] Gunam I B W, Wartini N M, Anggereni A M D, Suparyana P M 2011 *J. Teknologi Indonesia* **34** 24

Behavior of *Nannochloropsis oculata* In Indoor Cultivation: The Effects of Flue Gas Exposure And Illumination Period on The Growth And Lipid Content

Margono
Department of Chemical Engineering, Faculty of Engineering, Universitas Sebelas Maret, Jl. Ir. Sutami 36A Surakarta 57126, Jawa Tengah, Indonesia

E-mail : margono@ft.uns.ac.id

Abstract. Indonesian government launched an energy source substitution from fossil fuel into renewable energy such as biodiesel oil. It is known that microalgae could store high lipid content making it useful for biodiesel production. This research studied the cell growth and lipid content profile of *Nannochloropsis oculata* at various exposure periods of flue gas and illumination. The experiments were done in indoor bioreactor outfitted by a light source and flue gas absorber. Five liters of medium BG-11 was circulated through bioreactor and absorber while the exposure periods of light and flue gas were varied intermittently and/or continuously. *N. oculata* reached the highest cell density of 27.3 mg/L when it was treated at continuous illumination with the absence of flue gas exposure. The highest lipid content of 36.88% was achieved at the cell density of 14.11 mg/L in continuous illumination and 10 minutes per day of flue gas exposure. The CO_2 enrichment through flue gas exprosure showed an environmental stress toward microalgae growth and drove the lipid content as well as the periodical illumination.

1. Introduction

Indonesian goverment has developed renewable energy source and made a regulation to use 20% biodiesel in diesel blending fuels. However, the current raw material to produce biodiesel depends strongly on the production of crude palm oil (CPO) which regularly used as an edible oil. It will rise deforesting the land for plantation purpose because it always become the decision to overcome the high demand toward CPO [1]. In consequence, source diversification should be the alternative solution. The microalgae oil is a choice as a potential biodiesel feedstock due to their fatty acids similarity to vegetable oils. Microalgae has also been a good carbon biosequestriant absorbing excess carbon dioxide emission through photosynthetic process. This biological CO_2 fixation is deemed attractive as it not only reduce CO_2 emission but also it produces useful biomass [2] or starch [3]. Furthermore, harvesting microalgae oil doesn't depend on the harvesting season like the palm oil. As a result, microalgae could yield higher lipid productivity (12000 L/ha) than most of the utilized oil crop (1190 L/ha) [4].

Nannochloropsis oculata, one of marine microalgae, is widely known to utilize light and CO_2 efficiently. They stored the product of photosynthesis in the form of protein, carbohydrate and lipid. The lipid content of *N. oculata* was around 31-68% of dry cell weight. This amount is higher than oil palm seeds that only has 25% of total seed weight. Accumulation of lipid in *N. oculata* varies depending on its environmental conditions. Some of the factors are light intensity, dissolved oxygen, CO_2 concentration, N_2 concentration, temperature, pH, salinity and toxic/heavy metal content [5].

Each of microalgae species will respond differently to luminosity and photoperiod. The optimal range of luminosity for growth is between 76 to 600 mol/m^2.s, while for the lipid synthesis is 240 mol/m^2.s [6]. Wahidin et al. [7] showed that luminosity of 100 μmol m^2/s and photoperiod of 18 h light: 6 h dark resulted in 6.5 x 10^7 cells/L and lipid content of 3.31%. Optimal photoperiod provides unique cultivation technique such as dark-light technique in which the light and dark period is set alternately and periodically. The common guidance for the dark-light cycle ratio is 12:12, 14:10, 10:14 and 16:8.

Cell growth favors low level of CO_2 and the rising of its concentration is detrimental for the cell growth. It was reported that the increase of CO_2 from 6%-10% hit the population of the cell by a half. The production of the cell at lower CO_2 concentration was 0.376 g/L.d and 0.15 g/L.d at higher CO_2 concentration [8]. Lipid concentration was also affected by CO_2 concentration. Widjaya et al. [9] reported the increase of lipid content from 0.33% to 3.33% as a result of CO_2 addition to the system.

The effects of flue gas on *Chlorella* sp. related to the growth and lipid content are well explained. In achieving similar goals, the information regarding the behavior of *N. oculata* using flue gas as CO_2 source must also be perceived. However, no experimental studies have demonstrated the behavior of *N. oculata* using flue gas as CO_2 source in the culture medium. In attempt to satisfy the proposed information gap, we report the growth characterization, biomass and lipid production of *N. oculata* at medium BG-11. The medium was enriched with CO_2 by exposing flue gas to the medium at various contact time. In addition, we also extend the investigation to the effect of illumination period on the behavior of microalgae as mentioned previously.

2. Methods

2.1. Strain of N. oculata
The strain of *N. oculata* was come from Laboratory of natural feeding, Jepara, Jawa Tengah, Indonesia. It was cultured in an open pond with salinity of 2.35%, pH of 7.9 and temperature of 25-30 °C.

2.2. Medium BG-11 preparation
The enriched medium for cultivation was medium BG-11. It consisted of 1.5 g $NaNO_3$, 0.04 g $K_2HPO_4.3H_2O$, 0.2 g $KH_2PO_4.3H_2O$, 0.0004 g EDTA, 0.005 g Fe ammonium citric, 0.005 g citric acid, 0.02 g Na_2CO_3 and 1 mL trace element per 1 litre of brackish.

2.3. Starter preparation
Starter for this culture was developed in three steps. Each of them used medium BG-11 with working volume of 10, 100 and 1000 mL. The Culture was continuously aerated for 7 days. The operating temperature and pH were 30 °C and 7.3 respectively. The light intensity used was 3000 lux with 12:12 light : dark cycle.

2.4. Microalgae lipid production
The cultivation of microalgae was done at 5 litre squared photobioreactor (figure 2). The cultivation was treated with two different conditions. The first condition was photoperiod variation compromising continuous and periodical light cycle (16 hours of light and 8 hours of dark). The second variation was flue gas fixation compromising 0 minutes, 10 minutes and 30 minutes exposure. During fixation process, the reactor was continuously exposed to the light. The cultivation was lasted for 15 days for every variable tested.

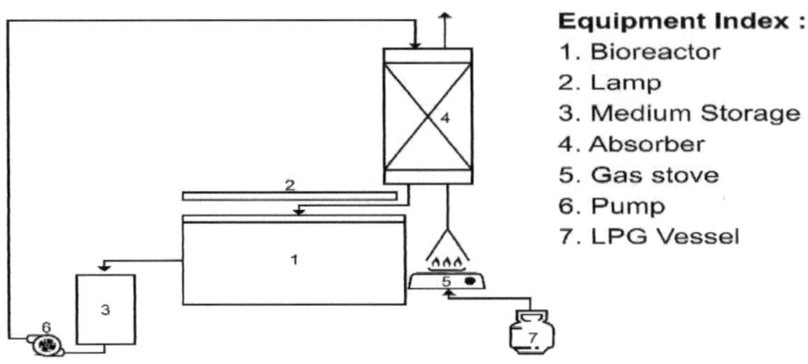

Figure 1. Photobioreactor configuration for lipid production from nannochloropsis oculata.

2.5. Analysis

Dry cell weight was measured by weighing the dried cell directly. Fifty millilitres sample taken from the reactor was centrifuged at 6000 rpm for 15 minutes. The supernatant was drained, the pellet was ovened at 80 °c for 24 hours and weighed. The lipid was extracted from the dried cell using solvent mixture of chloroform and methanol (2:1 v/v) for an hour. The extracted liquid was then centrifuged to separate the supernatant and solvent. The remaining solvent was evaporated and the remaining oil was weighed.

3. Results and discussion

The proximate analysis and lipid content of *N. oculata* before treatment were approached by analyisis results of *N. oculata* from other open pond of Situbondo, Jawa Timur, Indonesia [10]. Its moisture, volatile matter, fix carbon, ash and lipid content were 3.99%, 67.45%, 8.08%, 24.47% and 11.44% respectively.

3.1. The effect of illumination method to the growth and lipid content of cells

The effects of illumination method were investigated for 15 days and described in Figure 2. The cell grows since the initial day and reaches maximum at the eleventh day of cultivation before decreasing significantly.

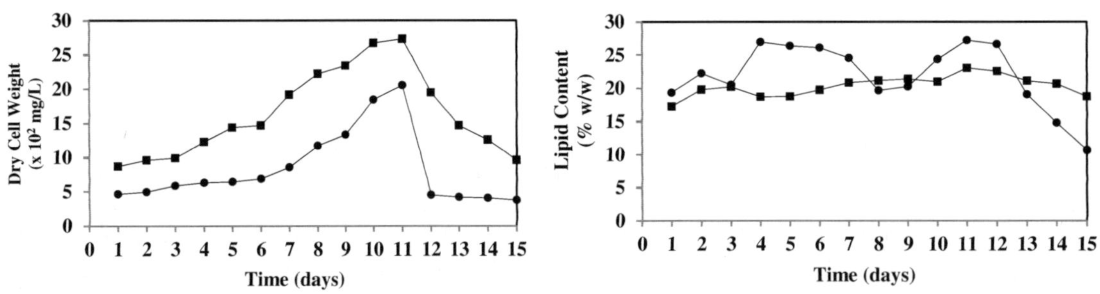

Figure 2. The cell population (a) and lipid content (b) of Nannochloropsis oculata at different illumination method. (●: (16L: 8D); ■: Continuous illumination)

N. oculata is seemed to be more comfortable growing at continuous illumination than periodical illumination. It could be observed by how long of the slow growth (lag phase) at early incubation, i.e. 6 days at periodical illumination and 3 days at continuous illumination. Prolonged lag phase may also be happened due to low and near saturation light intensity [7]. Under low intensity, photosynthesis efficiency is also low and therefore the growth is inhibited. Meanwhile, at near saturation intensity, the light induces photo inhibition destructing the devices of photosynthesis and reducing the photosynthesis efficiency.

Figure 2 shows that cell population is higher at continuous illumination than periodical illumination. The highest population is 20.5 mg/L and 27.3 mg/L for periodical and continuous illumination respectively. This result is consistent with the report from Wahidin et al. [7] using *Nannochloropsis* sp. and Amini Khoeyi et al. [11] using *Chlorella vulgaris*. They founded that an increase illumination period will have a favorable effect on biomass production so long that the light is still under light saturation limit.

The lipid accumulation in microalgae increases at stress environment such as growing under nutrient deficiency or limited light exposure. According to the Figure 2, the lipid content of *N. oculata* at continuous illumination is almost constant at 17% - 23% (w/w). In the other hand, at periodical illumination, cell accumulates more lipid than the continuous one gathering with the highest content of 27.25% (w/w). It is the evidence that periodical illumination is a form of environmental stressed to the cell body. This data is in accordance with the report from Liu et al. [12] that the total lipid content of algae reached the highest level at the dark condition. In addition, Cheirsilp and Torpee [13] reported that in the process of low light intensities, microalgae tend to use the synthesized energy to accumulate it in the lipid form.

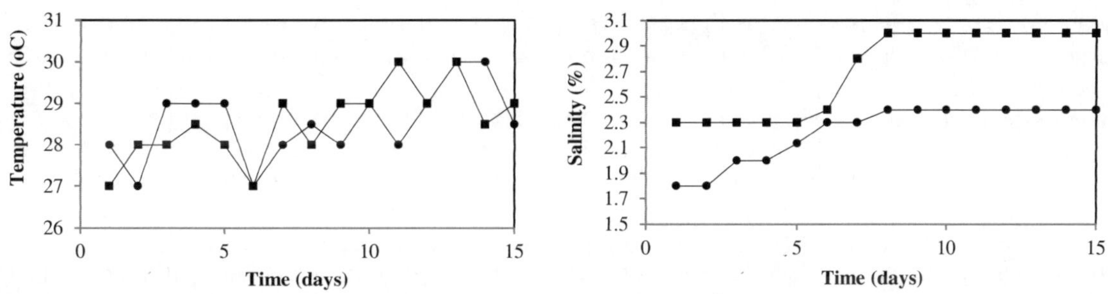

Figure 3. The profile of medium temperature (a) and medium salinity (b). (●: (16L: 8D); ■: continuous illumination)

The profiles of temperature during cultivation regardless of the photoperiod variation are unsteady (Figure 3(a)). The temperature of the medium is observed to vary from 27 °C – 30 °C and it is greatly affected by the ambient temperature. Renaud et al. [14] investigated that the optimal temperature for microalgae growth was at 25 °C – 35 °C, while Pahija et al. [15] reported similar tend to grow between temperature 22 °C and 27 °C for microalgae *Chlorococcum*. At current research, it shows that using ambient air alone to control medium temperature is sufficient to meet the appropriate temperature.
The salinity of the medium is increased continuously and starts to be steady at the eighth day (Figure 3(b)). The initial value is 2.3% and ended at 3.0% by the end of process at continuous illumination and 1.8% to 2.3% at periodical illumination. It was previously investigated that the medium salinity affects the lipid accumulation [16]. Campenni et al. [17] reported that 20 g/L NaCl produced higher lipid content than 10 g/L and 30 g/L NaCl.

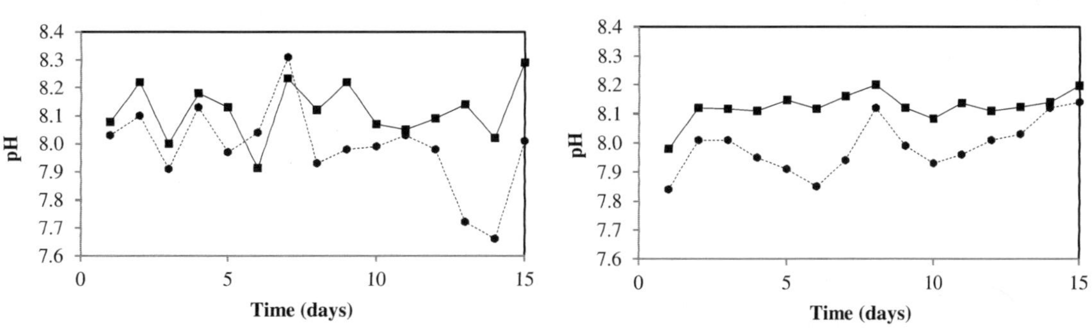

Figure 4. The profile of pH during day and night at (a) continuous illumination condition and (b) periodical illumination condition (16L: 8D). (--●--: pH during the night; —■—: pH during the day)

The pH level on the medium is affected by the availability of CO_2 for photosynthesis and the nutrient medium. It is crucial to maintain pH at appropriate values because complete culture collapse may occur due to extreme pH. At current investigation, the value of pH during cultivation is monitored (figure 4) at day and night. pH during the day is mostly higher than the night. At both illumination period, the average pH value is 8.12 and 7.99 during the day and the night respectively. According to the paper from Bartley et al. [18], the pH ±8 was the most optimum values for the growth of *Nannochloropsis* sp. Day/night pH fluctuation driven by photosynthesis and respiration rate. Photosynthesis reaction and the use of CO_2 raise pH level. In the other hands, respiration reverses this process and lowers pH level.

3.2. The effect of flue gas exposure to the growth and lipid content of cells
The photosynthetic metabolism utilizes light and CO_2 for growth and organic substance production. The aim of contacting flue gas was to enrich CO_2 in the medium and observed its effect on *N. oculata* growth and lipid content. Figure 5(a) shows the effect of flue gas exposure on the growth of microalgae. For 0 minute exposure, the highest population is 27.3 mg/L then followed by 23.55 mg/L for 30 minutes and the lowest is 14.1 mg/L for 10 minutes. It shows that CO_2 enrichment through flue gas exposure degrades the microalgae growth, while at increasing flue gas exposure from 10 to 30 minutes) seems to favor cell growth. The similar phenomenon of flue gas exposure was also found by Razzak et al. [19]. They observed the rise of total cell weight when the carbon dioxide concentration was increased from 4% to 8%.

The gas exposure gave a stress environment for *N. oculata*. This claim can be confirmed directly from Figure 5(b) describing the lipid content profile. Microalga prefers to store the energy in the form of lipid rather to utilize it to grow undér stressed condition. Based on the observation, the lowest total cell weight which is 10 minutes flue gas exposures can accumulate higher lipid content. Microalgae at 10 minutes gas exposure stored 23.3% - 36.88% lipid and at 0 minutes stored 18.69% - 23.08%. While at 30 minutes, the lipid content was 10.58% - 18.26%. de Castro Araújo and Garcia [20] reported that *Chaetoceros cf. wighamii* assimilated extra carbon mainly to protein synthesis. This phenomenon was also occurred at Phaeodactylum tricornutum [21]. However, Chu et al. [22] observed increasing in lipids and carbohydrates at protein expenses in *Nitzschia inconspicua*, when culture was enriched with 5% (v/v) of carbon dioxide.

The temperature during the process vary from 26 °C – 31 °C (Figure 6(a)). The temperature at this process is greatly affected by the environment temperature with almost no impact from metabolic activity. Although it is controlled by the ambient temperature, no significant changing is observed at lipid content and algae growth. It shows that the ambient temperature already give the appropriate temperature for the growth of *N. oculata*.

The salinity is also increased during the cultivation (Figure 6(b)). It shows that during 10 minutes gas exposure, the salinity changes from 1.8% to 2.3%, while during 30 minutes exposure, it changes from 2.1% to 2.8%. While at the absence of gas fixation the salinity ranging from 2.3% to 3.0%. The reported investigation shows salinity below normal sea water that averagely has salinity of 3.5%. It simply shows that N. oculata can live under brackish water (low salinity) even though it is categorized as marine microalgae.

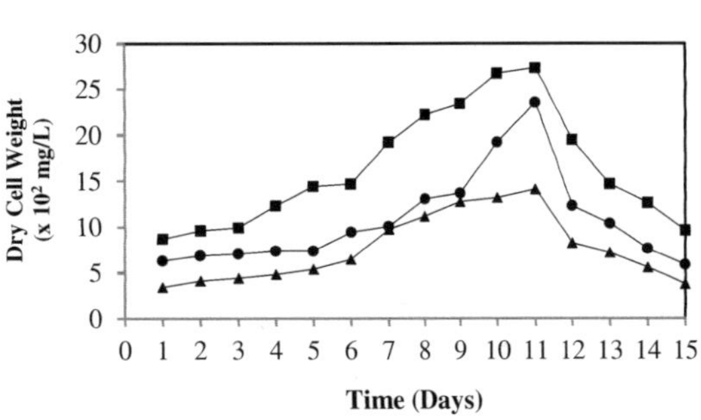

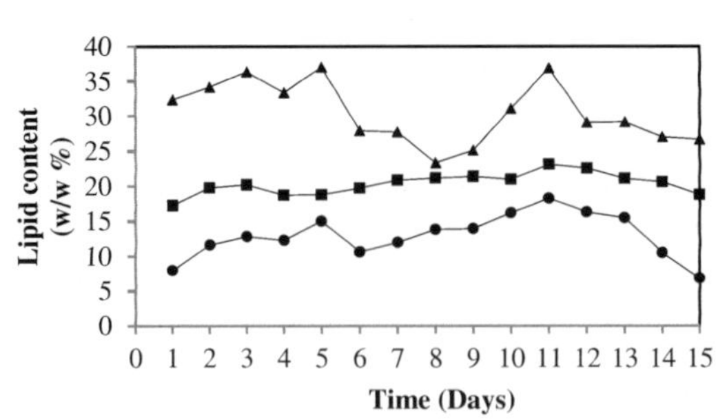

Figure 5. The cell population (a) and lipid content (b) of *Nannochloropsis oculata* in different flue gas exposure period. (●: 30 minutes exposure; ▲: 10 minutes exposure; ■: 0 minutes exposure)

IOP Conf. Series: Materials Science and Engineering **543** (2019) 012054 doi:10.1088/1757-899X/543/1/012054

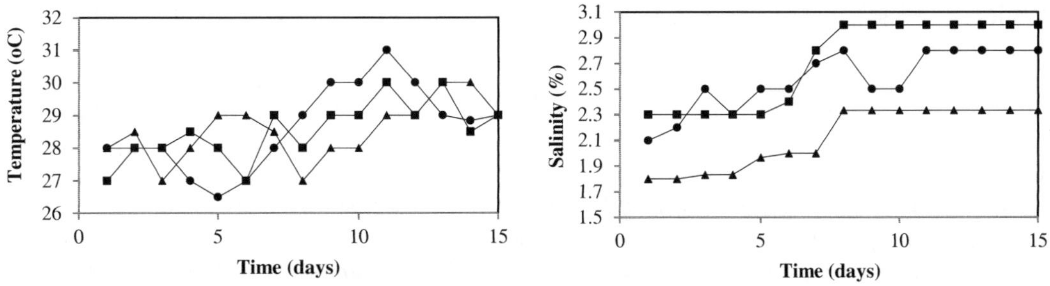

Figure 6. Temperature (a) and Salinity change (b) of medium during the process. (●: 30 minutes exposure; ▲: 10 minutes exposure; ■: 0 minutes exposure)

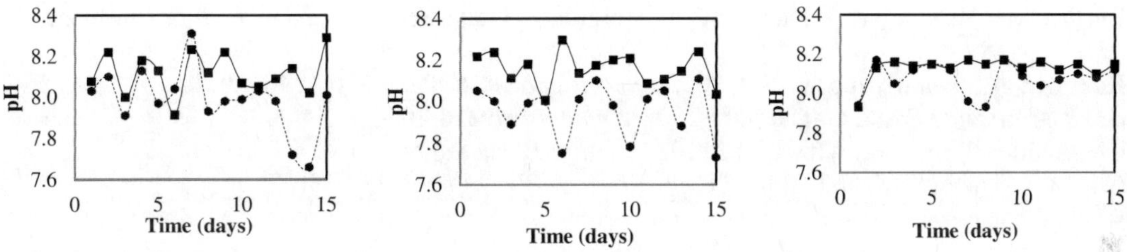

Figure 7. The profile of pH during day and night at flue gas exposure of 0 minutes (a), 10 minutes (b) and 30 minutes (c). (--●--: pH during the night; —■—: pH during the day).

The pH values of culture medium (Figure 7(a), (b) and (c)) showed a similar trend with the effect of periodical illumination. It showed that pH during the day was mostly higher than pH during the night. At variable 0 and 10 minutes exposure, the pH trend showed varied greatly through the day while the cell at 30 minutes gas exposure showed almost steady values. This condition may be occurred when the concentration of carbon dioxide in the media is enough to supply the cell to run photosynthetic continuously. It seems by exposing the media to the flue gas for 30 minutes gave enough carbon sources for the *N. oculata*. While at 0 and 10 minutes variable, it shows clear alteration between respiration and photosynthesis process. By this finding we can infer that the metabolism of *N. oculata* at phototrophic system are highly depend on the existence of nutrient and light.

4. Conclusion

The effect of illumination was investigated in lighting continuously and periodically (16L:8D). The illumination supported cell growth, while periodical illumination stimulated lipid accumulation. The cell population during continuous illumination (27.3 mg/L) was higher than that at periodical illumination (20.5 mg/L). At periodical illumination, the cell had a higher lipid content (27.25%) than at continuous illumination (17%-23%). Meanwhile, The exposure of flue gas inhibited the cell growth but increased lipid content. In this experiment, the highest cell population was acquired at 0 minutes exposure, i.e. 27.3 mg/L and the highest lipid content of 36.88% was reached by 10 minutes exposure in continuous illumination.

5. References

[1] Singh A, Nigam P S, and Murphy J D 2011 *Biores. Tech.* **102** 10–16.
[2] Huang G, Chen F, Wei D, Zhang X W, and Chen G 2010 *Appl. Energy* **87** 38–46.
[3] Gifuni I, Olivieri G, Krauss I R, D'Errico G, Pollio A, and Marzocchella A 2017 *Chem. Eng. Trans.* **57** 1423-1428.
[4] Schenk P, Thomas-Hall S, Stephens E, Marx U, Mussgnug J, and Posten C 2008 *Bio. Energy Res.* **1** 20–43.
[5] Suali E and Sarbatly R 2012 *Renew. Sust. Energy Rev.* **16** 4316-4342.
[6] Hsieh C H and Wu W T 2009 *Biores. Tech.* **100** 3921-3926.

[7] Wahidin S, Idris A and Shaleh S R M 2013 *Biores. Techn.* **129** 7–11.
[8] Doucha J, Straka F, and Livansky K 2005 *J. Appl. Phycology* **17** 403–12.
[9] Widjaja A, Chien C C, and Ju Y H 2009 *J. Taiwan. Institute. Chem. Eng.* **40** 13–20.
[10] Sukarni, Sudjito, Yanuhar U, Hamidi N, and Wardana I N G 2014 *Int. J. Energy and Envi. Eng.* **5** 279 – 290.
[11] Amini Khoeyi Z, Seyfabadi J, and Ramezanpour Z 2011 *Aquaculture Int.* **20** 41–49.
[12] Liu J, Vanormelingen P, and Vyverman W 2016 *Biores. Tech.* **200** 1080-1084.
[13] Cheirsilp B and Torpee S 2012 *Biores. Tech.* **110** 510-516.
[14] Renaud S M, Thinh L V, Lambrinidis G, and Parry D L 2002 *Aquaculture* **211** 195-214.
[15] Pahija E, Liang Y, Hui C-W 2017 *Chem. Eng. Trans.* **61** 721-726.
[16] Takagi M and Yoshida T 2006 *J. Biosci. Bioeng.* **101** 223-226.
[17] Campenni L, Nobre B P, Santos C A, Oliveira A C, Aires-Barros M R, Palavra A M, and Gouveia L 2013 *Appl. Microb. Biotech.* **97** 1383-1393.
[18] Bartley M L, Boeing W J, Dungan B N, Holguin FO, and Schaub T 2014 *J. Appl Phycology* **26** 1431-1437.
[19] Razzak S A, Ilyas M, and Ali S A 2015 *Appl. Biochem. Biotech* **176** 1290-1302.
[20] de Castro Araújo S and Garcia V M 2005 *Aquaculture* **246** 405-412.
[21] Chrismadha T and Borowitzka M A 1994 *J. Appl Phycology* **6** 67-74.
[22] Chu W L, Phang S M, and Goh S H 1996 *J. Appl Phycology* **8** 389-396.

IOP Conf. Series: Materials Science and Engineering **543** (2019) 012055 doi:10.1088/1757-899X/543/1/012055

Hydrocarbon Biofuel Production by Hydrocracking Process with Nickel-Iron Supported on HZSM-5 Catalyst

M Al-Muttaqii[1,3], F Kurniawansyah[1], D H Prajitno[2], A Roesyadi[1*]**

[1] Chemical Reaction Engineering Laboratory, Department of Chemical Engineering, Faculty of Industrial Technology, Sepuluh Nopember Institute of Technology, Surabaya, 60111, Indonesia.
[2] Department of Industrial Chemical Engineering, Vocational Faculty, Institute Technology of Sepuluh Nopember, Surabaya, 60111, Indonesia.
[3] Research Unit of Mineral Technology, Indonesian Institute of Sciences, Lampung, 35361, Indonesia

Email : aroesyadi@yahoo.com, almuttaqiimuhammad@gmail.com

Abstract. The production of hydrocarbon biofuel from coconut oil can substitute for fossil fuels to reduce the environmental issues. Hydrocracking is an effective method that can convert vegetable oil into biofuel using catalyst. In this experiment, nickel-iron was combined with HZSM-5 catalyst to produce biofuel. Hydrocracking of coconut oil was carried out in a batch reactor over Ni-Fe/HZSM-5 catalyst at a temperature of 350 and 400 °C for 2 h. Liquid hydrocarbon biofuel product was analyzed by gas chromatography-mass spectrometry (GC-MS). Based on GC-MS analysis, processing at the temperature of 350 °C produced sample containing 72.56% carboxylic acids. When the temperature increased to 400 °C, the carboxylic acids decreased rapidly from 72.56% to 24.83%. Analysis results suggested that increase of temperature promoted to the increase of desirable content, hydrocarbon compounds. At the same like, it also suppressed formation of undesirable compounds: acids, ketones, phenols, alcohols, and esters. Overall, synthesized catalyst in this study improved production of n-paraffin compounds such as (tridecane (n-C_{13}), tetradecane (n-C_{14}) and pentadecane (n-C_{15})) as high 71.85%.

Keywords: Coconut oil, Hydrocracking, Ni-Fe/HZSM-5 catalyst, n-paraffin

1. INTRODUCTION

The hydrocracking of vegetable oils (such as palm oil, jatropha oil, sunflower oil, soybean oil, coconut oil and etc) is a good promising technology for producing of biofuels to substitute the fossil fuels. The process of hydrocracking is combined of catalytic cracking process with the addition of hydrogen gas. As reported by previous study, hydrocracking with bifunctional heterogenous catalyst needs a higher temperature (300-400 °C) to crack vegetable oils into hydrocarbon compounds [1-2].

Zeolite catalyst was suitable as a catalyst for cracking of vegetable oil to produce biofuel. Some beneficial properties of zeolite are high thermal stability, high surface area, shape selectivity, and appropriate pore diameter [3]. The HZSM-5 catalyst was found to be efficient in the cracking process of vegetable oils. Metals (Ni, Mo, Co, Fe, and Zn) can be supported on ZSM-5 and have been widely used for the cracking process [4-5]. Researchers have reported the conversion of vegetable oils into hydrocarbons over a variety of catalyst lately. Iliopoulou et al. [6] was investigated catalytic cracking of biomass pyrolysis with transition metal-modified ZSM-5 (e.g., Co and Ni) and found that the transition metal-modified ZSM-5 improved the production of aromatics. When Ni/HZSM-5 catalyst was used in the catalytic hydroprocessing on FAMEs, the increasing temperature of 260-280 °C will increased n-paraffin (C_{17}-C_{18}) and isoparaffin (C_5-C_{16}) [1]. Kovács et al. [7] was investigated catalytic hydrotreating of sunflower oil with NiMoW/Al_2O_3 catalyst. They found that high paraffin (77-81%) content could be produced through decarboxylation/decarbonylation reactions took place.

Verma et al. [8] reported that hydrocracking of jatropha oil with sulphided NiMo and NiW catalyst supported on hierarchical mesoporous SAPO-11 produced kerosene range hydrocarbons containing 8 vol% of desired aromatic. Al-Muttaqii et al. [9] was reported that when Co-Ni/HZSM-5 catalyst was used in hydrocracking of Sunan candlenut oil, a slightly n-paraffins and aromatic hydrocarbon were produced at temperature of 350 °C. Leng et al. [10] reported that NiFe/Al_2O_3 improved the heating value of biofuel and the main involved reaction pathway was C-O cleavage rather than C-C cleavage during the hydrodeoxygenation process. Marlinda et al. [11] also reported that hydrocracking of Cerbera Manghas oil with Co-Ni/HZSM-5 catalyst produced pentadecane and heptadecane were the main abundant hydrocarbon compounds in gasoil-range hydrocarbon at temperature of 350 °C, 15 bar for 2 h. The author also reported that aromatic compounds (C_{11}-C_{12}) were found in this condition.

The purpose of this work is to transform of coconut oil into hydrocarbon biofuel with nickel-iron supported on HZSM-5 catalyst in a batch reactor. The influence of temperature on the hydrocracking process was also studied. The chemical composition of biofuel was analyzed by using Gas Chromatography-Mass Spectrometry (GC-MS). The ultimate goal is to produce hydrocarbon biofuel from coconut oil via hydrocracking process.

2. EXPERIMENTAL

2.1 Materials

The coconut oil was purchased from a local market in East Java Province, Indonesia. Ammonium ZSM-5 (NH_4-ZSM-5) was supplied from Zeolyst International, USA (SiO_2/Al_2O_3, Mole Ratio: 40). Nickel (II) nitrate hexahydrate ($Ni(NO_3)_2.6H_2O$) and iron (III) nitrate nonahydrate ($Fe(NO_3)_3.9H_2O$) was purchased from Merck with ≥98 % purity.

2.2 Preparation of catalyst

The Ni-Fe/HZSM-5 catalyst was prepared with procedure as reported by Al-Muttaqii et al. [9] and Marlinda et al. [12]. The first step, NH_4-ZSM-5 zeolite was calcined at 550 °C in air for 5 h to convert into HZSM-5. Then, the aqueous solution contain of $Ni(NO_3)_2.6H_2O$ and $Fe(NO_3)_3.9H_2O$ were loaded into HZSM-5 catalyst. The catalyst was dried at 120 °C for 12 h. After being dried, the catalyst was calcined in 550 °C in air for 3 h and reduced in flowing H_2 at 550 °C for 2 h to obtain metallic phase, as reported in previous study [12]. The Ni-Fe/HZSM-5 catalyst was characterized by X-Ray Diffraction (XRD) to determine crystallinity of catalyst using a PANalytical X'Pert PRO X-ray diffractometer with a Cu Kα radiation source operating at 40 kV and 30 mA and recorded on 2θ of 5 to 70°. The amount of metal impregnated on HZSM-5 of catalyst was analyzed using Scanning Electron Microscope with Energy-Dispersive X-ray (SEM-EDX) series model: EVO MA10. Surface area and pore volume of the catalyst were characterized by Brunauer-Emmett-Teller (BET) series Quantachrome NovaWin Version 11.03.

2.3 Hydrocracking test

Hydrocracking of coconut oil was carried out in a batch reactor type Parr Instrument Company 4563 with a maximum volume of 600 ml. This reactor equipped with a magnetic stirrer. For these experiments, 200 ml of coconut oil and 2 g of Ni-Fe/HZSM-5 catalyst were added into the reactor. Then it was purged with nitrogen gases for 10 min to remove the remaining air in the reactor. Then, the reactor flow with hydrogen gas for 1 h and reactor pressure of 30 bar. The reaction temperature was carried out at a temperature of 350 °C for 2 h. After processing for 2 h, the liquid biofuel products were collected and analyzed.

2.4 Product analysis

IOP Conf. Series: Materials Science and Engineering **543** (2019) 012055 doi:10.1088/1757-899X/543/1/012055

The major components in liquid biofuel product were analyzed using a Gas Chromatography-Mass Spectrometry (GC-MS, Agilent HP 6890 models 19091S-433, HP–5MS capillary column 30 m x 250 μm x 0.25 μm). The major components of hydrocarbon biofuel were identified through a National Institute of Standards and Technology (NIST) mass spectra library and literature.

3. RESULTS AND DISCUSSION

3.1 Characterization of catalyst

The XRD spectra of HZSM-5 and Ni-Fe/HZSM-5 catalyst are shown in Figure 1. As can be seen in Figure 1, the HZSM-5 catalyst shows the characteristic peaks of an MFI structure type zeolite (2θ = 7-9^0 and 23-25^0). After the impregnation metals of Ni and Fe into HZSM-5, structure of Ni-Fe/HZSM-5 catalyst was still similar to HZSM-5 catalyst. It indicates that the addition of nickel and iron on/into the HZSM-5 catalyst did not change their structures but reduce the intensity of each diffraction peaks, as reported by previous study [9, 12-13]. The result of EDAX measurement also proved the existence of nickel and iron in HZSM-5. From Table 1, the actual metal content on support HZSM-5 was 3.75 wt% of Ni and 5.43 wt% of Fe.

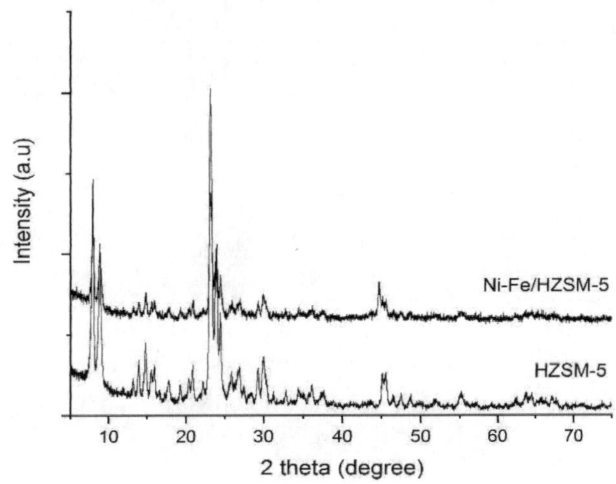

Figure 1. XRD patterns of the HZSM-5 and Ni-Fe/HZSM-5 catalyst

The surface area, average pore size, and pore volume of the Ni-Fe/HZSM-5 catalyst are listed in Table 1. The surface area of HZSM-5 catalyst was 407.386 m^2/g [14]. After the impregnation of Ni and Fe into the HZSM-5, the surface area decreased. Some of the support HZSM-5 were filled up with metal after the impregnation of Ni and Fe. These changes implied that Ni and Fe particles deposited on the surface of HZSM-5. This assumption is in accordance with the result that the total pore volume of Ni-Fe/HZSM-5 catalyst decreased compared to HZSM-5. [9, 12, 15-16].

Table 1. Properties of Ni-Fe/HZSM-5 catalyst

Catalyst	Surface area (m^2/g)	Pore Size (nm)	Total Pore Volume (cm^3/g)	Element (wt%)	
				Ni	Fe
HZSM-5*	407.386	3.12	0.32	-	-
Ni-Fe/HZSM-5	336.617	3.32	0.28	3.75	5.43

*[14]

3.2 Hydrocracking of coconut oil

3.2.1 GC-MS analysis

In this study, we used a Gas Chromatography-Mass Spectrometry (GC-MS) analysis to determine the composition of the liquid products. Figure 2 shown GC-MS chromatogram of hydrocarbon liquid biofuel product produced at temperature of 350 and 400 C using Ni-Fe/HZSM-5 catalyst. Figure 2 (a) shows n-paraffins produced from hydrocracking of coconut oil. The liquid product mainly contains tridecane, pentadecane, and heptadecane. The main component of carboxylic acids found in liquid product were octanoic acid and dodecanoic acid. According to Zhang et al. [17], triglycerides with long chain fatty acids were converted into short chain molecules by cracking, isomerization, cyclization, and aromatization reaction. It suggested that the temperatures above 350 °C were also needed to crack C=O bond through decarboxylation and/or decarbonylation reaction. Figure 2 (b) shows at temperature of 400 °C contained the highest hydrocarbon content (75.14 area%) and it contained n-paraffin, aromatic and olefin compounds. It suggested that the increasing temperature in hydrocracking reaction improve the catalytic activity and contributed to cleavage triglycerides of coconut oil into a short chain hydrocarbon molecules in liquid products as reported in previous study [9,12].

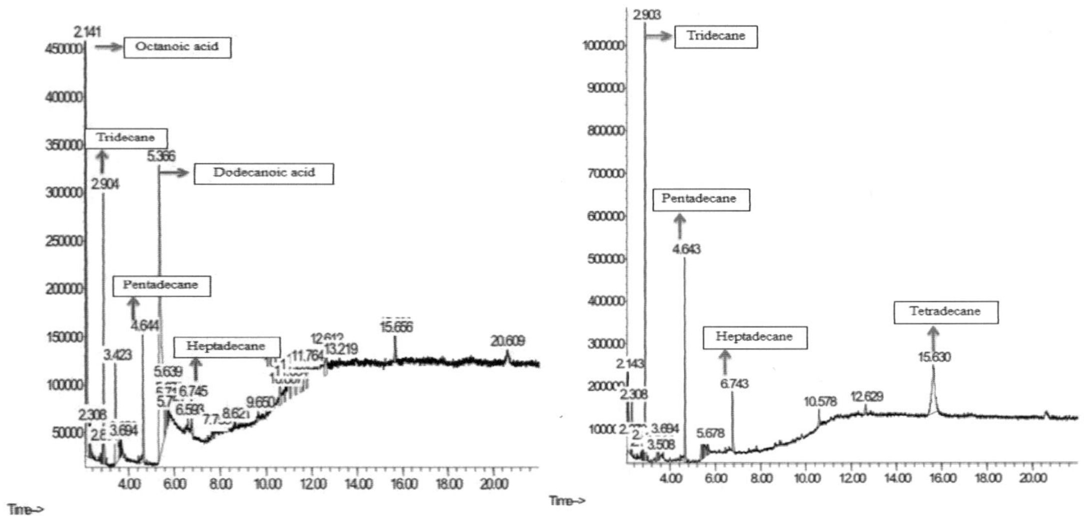

(a) (b)

Figure 2. GC-MS spectra of hydrocarbon biofuel liquid at (a) 350 °C and (b) 400 °C using Ni-Fe/HZSM-5 catalyst

Table 2 shows the major components of hydrocarbon biofuel produced at different temperature. Hydrocarbon biofuel from coconut oil mainly contained hydrocarbons (such as n-paraffin, olefin, aromatic, and cycloparaffin), oxygenated compounds and carboxylic acids. A similar observation was found from the study of Al-Muttaqii et al. [9], Marlinda et al. [11], Xu et al. [18], that the hydrocarbon biofuel produced from hydrocracking of vegetable oil contained hydrocarbon and oxygenated compounds. The total amounts of carboxylic acid and oxygenated compounds at 350 °C was 72.56 area% and 2.55 area%, respectively. These compounds are responsible for the low quality of the hydrocarbon biofuels including low heating value, comparison with conventional fuel, chemical instability, high viscosity, and acidity of hydrocarbon biofuel. Hydrocarbon biofuel produced at temperature of 400 °C contained the lowest oxygen content, suggesting that the carboxylic acid including octanoic acid and dodecanoic acid were cracked to other products such as hydrocarbons, aromatic, and etc through dehydrogenation, decarboxylation and decarbonylation reaction over the Ni-Fe/HZSM-5 catalyst [12].

As shown in Table 2, n-C_{13}, n-C_{15}, and n-C_{17} were the main products at different temperatures. The amount of smaller molecule of aromatic (i.e. hexylbenxene) that was formed when the reaction temperature reached 400 °C. The monocyclic aromatic hydrocarbons (MAH), e.g hexyl benzene was found in small amount about 0.75 area%. These compounds can be used as a lead substitute for the octane number boosting of transportation fuel. As reported by previous study [19], when Co-Ni/HZSM-5 catalyst was used in the same condition in this work, the hydrocracking of sunan candlenut oil produced a liquid product containing aromatic compounds in the form of abundant monocyclic aromatic hydrocarbons, i.e. n-hexylbenzene, n-heptylbenzene, octylbenzene, and etc. The incorporation of transition metals (Ni and Fe) on/into HZSM-5 showed a bifunctional character catalyst which included acid- and metal sites to promote oligomerization and aromatization. The addition of transition metals promoted the conversion of oxygenated compound to aliphatic (n-paraffin, olefin) and aromatics [9]. The incorporation of transition metal on/into HZSM-5 was expected to affect the deoxygenation and the hydrogen transfer during the hydrocracking process of coconut oils. Ni combined with acid site provided the ideal environment for alkenes followed by cyclization and dehydrogenation leading to small amount of aromatic formation [6].

Table 2. The Chemical composition of hydrocarbon biofuel (area% of liquid product) at temperature of 350 and 400 °C

Compounds	Molecular formula	Area% (I) 350 °C	Area% (II) 400 °C
n-paraffin			
Dodecane	$C_{12}H_{36}$	-	2.97
2,7-dimethyl undecane	$C_{13}H_{28}$	1.25	-
Tridecane	$C_{13}H_{28}$	8.99	25.25
Tetradecane	$C_{14}H_{36}$	0.37	21.57
Pentadecane	$C_{15}H_{32}$	3.91	14.95
Hexadecane	$C_{16}H_{34}$	-	1.72
Heptadecane	$C_{17}H_{36}$	1.26	5.39
2-methylhexadecane	$C_{17}H_{36}$	1.14	-
Olefin			
7-hexadecene	$C_{16}H_{32}$	1.04	-
2-tridecene	$C_{13}H_{26}$	0.68	-
5-tridecene	$C_{13}H_{26}$	-	1.52
2-tetradecene	$C_{14}H_{28}$	0.54	-
1-docosene	$C_{22}H_{44}$	2.32	-
1-octadecene	$C_{18}H_{36}$	0.09	-
3-octadecene	$C_{18}H_{36}$	0.16	-
4-dodecene	$C_{12}H_{24}$	-	1.02
Aromatic			
Hexylbenzene	$C_{12}H_{18}$	-	0.75
Cycloparaffin			
Cyclotetradecane	$C_{14}H_{28}$	0.98	-
Oxygenated compound			
1,2-Epoxyhexadecane	$C_6H_{12}O$	2.55	-
Carboxylic acid			
Octanoic acid	$C_{16}H_{32}O_2$	18.53	9.76
n-decanoic acid	$C_{10}H_{20}O_2$	7.98	2.42
Dodecanoic acid	$C_{12}H_{24}O_2$	42.17	9.6
Tetradecanoic acid	$C_{14}H_{28}O_2$	1.56	-
Hexadecanoic acid	$C_{16}H_{32}O_2$	1.28	-
Nonadecanoic acid	$C_{19}H_{38}O_2$	1.04	-
9-octadecanoic acid	$C_{18}H_{34}O_2$	-	1.13
8-heptadecanoic acid	$C_{17}H_{34}O_2$	-	1.92

The results indicate that the reaction temperature also effected the hydrocarbon product distribution. On the other hand, reaction temperature had a relatively strong influence on the ratio of n-alkanes with an even and odd carbon atom number. In this research, when the temperature rised from 350 to 400 °C, the content of n-C_{13} and n-C_{15} increased from 8.99 to 25.25% and 3.91 to 14.95%, respectively. The final product in hydrocarbon biofuel contains aliphatic compounds in the range C_{12}-C_{22}. A high n-paraffins were produced at temperature of 400 °C. Based on this results, tridecane (n-C_{13}), tetradecane (n-C_{14}), pentadecane (n-C_{15}) and heptadecane (n-C_{17}) were the most abundant hydrocarbon compounds in biofuel liquid product, as reported in previous study [12, 20]. However, the initial hydrogen pressure and reaction temperature play a crucial role in this experiment. At temperatures lower than 310 °C, the final organic liquid product contained fatty acid and triglycerides, but at higher reaction temperatures only hydrocarbons were detected, mainly C_{15}, C_{17} and C_{18} n-alkanes [21]. Based on the composition of liquid products, it can be concluded that the triglyceride is hydrogenated and broken down into various intermediate. Then the intermediates are converted into alkanes by two pathways: hydrocarbons with the same carbon number, i.e., $C_{16}H_{36}$, is the products of hydrodeoxygenation (HDO). Hydrocarbons with one carbon atoms, i.e., $C_{17}H_{36}$ and $C_{15}H_{32}$, are the products of oxygen removal from triglyceride by hydrodecarboxylation (HDC), including decarboxylation and decarbonylation. It showed that with increasing reaction temperature, hydrocracking process proceeded via HDC was the dominant reaction pathways, lead to the formation of n-alkanes with an odd number of carbon atoms chain length [14, 16].

4. Conclusion

This study present the effect of reaction temperatures on the hydrocracking process with nickel-iron supported on HZSM-5 catalyst for the transformation of coconut oil into aliphatic (n-paraffin, olefin) and aromatics compounds. The highest amounts of n-paraffin compounds produced is 71.85 area% at temperature of 400 °C. When the reaction temperature rised from 350 to 400 °C, the content of n-C_{13} and n-C_{15} increased from 8.99 to 25.25% and 3.91 to 14.95%, respectively. The increasing reaction temperature was favour to conduct further cracking of carboxylic acids into n-paraffin compounds. Hydrocracking process proceeded via HDC was the dominant reaction pathways, lead to the formation of n-alkanes with an odd number of carbon atoms chain length. The addition of transition metals (nickel and iron) promoted the conversion of oxygenated compounds to aliphatic (n-paraffin, olefin) and aromatics compounds. Ni-Fe/HZSM-5 catalyst is an effective catalyst to produce hydrocarbon biofuel by hydrocracking process of coconut oil.

References

[1] Chen L, Li H, Fu J, Miao C, Pengmei Lv, Yuan Z 2016 Catalytic hydroprocessing of fatty acid methyl esters to renewable alkane fuels over Ni/HZSM-5 catalyst. *Catal. Today* **259** 266-276.

[2] Romero M, Pizzi A, Toscano G, Casazza A A, Busca G, Bosio B, Arato E 2015 Preliminary Experimental Study on Biofuel Production by Deoxygenation of Jatropha Oil. *Fuel. Proc. Technol.* **137** 31-37

[3] Vichaphund S, Aht-ong D, Sricharoenchaikul V, Atong D, 2014 Catalytic upgrading pyrolysis vapors of Jatropha waste using metal promoted ZSM-5 catalysts: An analytical PY-GC/MS. *Renew. Energ.* **65** 70-77

[4] Abdelsayed V, Shekhawat D. Smith M W 2015 Effect of Fe and Zn promoters on Mo/HZSSM-5 catalyst for methane dehydroaromatization. *Fuel* **139** 401-410

[5] Wang H, Yan S, Salley O S, Simon Ng K Y 2012 Hydrocarbon Fuels Production from Hydrocracking of Soybean Oil Using Transition Metal Carbides and Nitrides Supported on ZSM-5. *Indust. Eng. Chem. Res.* **51** (30) 10066-10073

[6] Iliopoulou E F, Stefanidis S D, Kalogiannis K G, Delimitis A, Lappas A A, Triantafyllidis K S 2012 Catalytic upgrading of biomass pyrolysis vapors using transition metal-modified ZSM-5 zeolite. *Appl. Catal. B: Envi.* **127** 281-290

[7] Kovács S, Boda L, Leveles L, Thernesz A, Hancsók J 2010 Catalytic hydrotreating of triglycerides for the production of bioparaffin mixture. *Chem. Eng. Trans.* **21** 1321 – 1326

[8] Verma D, Rana B S, Kumar R, Sibi M G, Sinha A K 2015 Diesel and Aviation Kerosene with Desired Aromatics from Hydroprocessing of Jatropha Oil over Hydrogenation Catalysts Supported on Hierarchical Mesoporous SAPO-11. *Appl. Catal. A: General* **490** 108-116

[9] Al-Muttaqii M, Marlinda L, Roesyadi A, Danawati H P 2017 Co-Ni/HZSM-5 Catalyst for Hydrocracking of Sunan Candlenut Oil (Reutealis trisperma (Blanco) Airy Shaw) for Production of Biofuel. *J. Pure Appl. Chem. Res.* **6** (2) 84-92

[10] Leng S, Wang X, He X, Liu L, Liu Y, Zhong X, Zhuang G, Wang J G 2013 NiFe/Y-Al_2O_3: a universal catalyst for the hydrodeoxygenation of bio-oil and its model compounds. *Catal. Comm.* **41** 34-37

[11] Marlinda L, Al-Muttaqii M, Roesyadi A, Danawati H P 2016 Production of Biofuel by Hydrocracking of Cerbera Manghas Oil Using Co-Ni/HZSM-5 Catalyst : Effect of Reaction Temperature. *J. Pure. Appl. Chem. Res.* **5** (3) 189-195

[12] Marlinda L, Al-Muttaqii M, Gunardi I, Roesyadi A, Danawati H P 2017 Hydrocracking of Cerbera manghas Oil with Co-Ni/HZSM-5 as Double Promoted Catalyst. *Bull. Chem. React. Eng. Catal.* **12** (2) 167-184

[13] Vichaphund S, Aht-ong D, Sricharoenchaikul V, Atong D 2015 Production of aromatic compounds from catalytic fast pyrolysis of Jatropha residues using metal/HZSM-5 prepared by ion-exchange and impregnation methods. *Renew. Energ.* **79** 28-37

[14] Al-Muttaqii M, Kurniawansyah F, Danawati H P, Roesyadi A 2019 Bio-kerosene and Bio-gasoil from Coconut Oils via Hydrocracking Process over Ni-Fe/HZSM-5 Catalyst. *Bull. Chem. React. Eng. Catal.* **14** (2) 309

[15] Zhao X, Wei L, Cheng S, Kadis E, Cao Y, Boakye E, Gu Z, Julson J 2016 Hydroprocessing of carinata oil for hydrocarbon biofuel over Mo-Zn/Al_2O_3. *Appl. Catal. B: Envi.* **196** 41-49

[16] Al-Muttaqii M, Kurniawansyah F, Danawati H P, Roesyadi A 2019 Hydrocracking of Coconut Oil over Ni-Fe/HZSM-5 Catalyst to Produce Hydrocarbon Biofuel. *Indo. J.* Chemistry: Article in Press

[17] Zhang H, Lin H, Zheng Y 2014 The Role of Cobalt and Nickel in Deoxygenation of Vegetable Oils. *Appl. Catal. B: Envi.* **160-161** (1) 415-422

[18] Xu J, Jiang J, Sun Y, Chen J 2010 Production of hydrocarbon fuels from pyrolysis of soybean oils using a basic catalyst. *Biores. Technol.* **101** (24) 9803-9806

[19] Danawati H P, Roesyadi A, Al-Muttaqii M, Marlinda L 2017 Hydrocracking of Non-edible Vegetable Oils with Co-Ni/HZSM-5 Catalyst to Gasoil Containing Aromatics. *Bull. Chem. React. Eng. Catal.* **12** (3) 318-328

[20] Sotelo-Boyas R, Liu Y, Minowa T 2011 Renewable diesel production from the hydrotreating of rapeseed oil with Pt/Zeolite and NiMo/Al_2O_3 catalysts. *Indust. Eng. Chem. Res.* **50** 2791-2799

[21] Simacek P, Kubicka D, Sebor G, Pospisil M 2010 Fuel properties of hydroprocessed rapeseed oil. *Fuel* **89** 611-5.

IOP Conf. Series: Materials Science and Engineering **543** (2019) 012056 doi:10.1088/1757-899X/543/1/012056

Production of Bioethanol from *Colocasia esculenta (L.) Schott* (Talas Liar) by Hydrolysis Process

E Praputri and E Sundari

Chemical Engineering Department, Bung Hatta University, Jl. Gajah Mada No. 19 OloNanggalo, Padang, West Sumatera, 25143, Indonesia

Email: ertipraputri@yahoo.co.id

Abstract. Wild taro tubers (Colocasia esculenta (L.) Schott Var. Antiquorum) contain 70-80% (wt%) of starch, thus serve as potential feedstock for glucose production, providing raw materials for bioethanol production. The tubers used in this study was collected from Padang city, Indonesia. The aim of this research was to find the optimum of bioethanol yield through variation of acid catalyst (HCl, H_2SO_4 and $HClO_4$), acid concentration (0,00 N; 0,05 N; 0,10 N; 0,15 N; 0,20 N), enzyme (α-amylase, glucoamylase), enzyme volume (0.308 ml; 0.74 ml), hydrolysis time (1 hour, 2 hours, 4 hours), and fermentation time (48 hours,72 hours, 96 hours, 122 hours, 144 hours).At varied acid catalyst usage, the highest glucose content (27.54%) was obtained by using $HClO_4$ acid with a concentration of 0.10 N and the lowest glucose level of 16.64% was obtained from the usage of H_2SO_4 acid with a concentration of 0.10 N in the hydrolysis process with temperature of 120°C for 45 minutes. The highest bioethanol content as 19.10% was obtained at the time of fermentation of 96 hours with the usage of Sacharomyces cerevisiae enzyme.The highest glucose content of 20.35% was obtained by using glucoamylase enzyme of 0.308 ml.

1. Introduction

The increasing demand for fossil fuels had took place since the Industrial Revolution as the world population grew and many countries are industrialized. Enormous combustion of the fossil fuels would contrarily increase the concentration of greenhouse gas in the atmosphere resulting in greenhouse effect and global warning. In order to prevent the air pollution from burning fossil fuel and energy crisis, many countries begin to develop renewable energy source [1]. Considerable attention has been focused on biomass as an alternative energy biosource not only for them to protect the environment, but also to meet the energy demand instead of fossil fuels [2]. The interest of biofuel (biomass based fuel) production was escalated dramatically in order to replace the fossil based fuel [3]. Biofuel, such as bioethanol, biodiesel and biogas, has become a feasible and economical solution. Among them, bioethanol is by far the most widely used biofuel for transportation worldwide. Production of bioetanol from biomass is one way to reduce both consumption of fossil fuels and environmental pollution [4].
Bioethanol is the product of the sugar fermentation process from carbohydrate sources [5]. The technology of bioethanol production from biomass feedstocks consists of several steps, and varies depending on the type of raw materials used. It becomes more sophisticated as the raw materials turn from sugars and starches to cellulosic materials. Unlike starch, the specific structure of cellulose favors the ordering of the polymer chains into tightly packed, highly crystalline structures that is water-insoluble and resistant to depolymerisation [6]. However, production of bioethanol from starch and

sugars involving hydrolysis and fermentation followed by separation/purification, differ from cellulose feedstocks that require pretreatment [7].

Hydrolysis of raw materials include the processing steps that convert the carbohydrate polymers e.g. cellulose or starch into monomeric sugars. Cleavage of these polymers can be catalyzed by enzymes or chemically by acids. The reaction can be described as follows:

$$(C_5H_{10}O_5)_n + nH_2O \xrightarrow{\text{Hydrolysis}} n(C_6H_{12}O_6) \quad (1)$$

Moreover, the fermentation steps involving the chemical transformation of organic substances into simpler compounds by the action of enzymes. The fermentation reaction is caused by yeast or bacteria which feed on simple sugars [8]. The simplified reaction equation is:

$$C_6H_{12}O_6 \xrightarrow{\text{Fermentation}} 2C_2H_5OH + 2\ CO_2 \quad (2)$$

There are three types of bioethanol feedstocks: (a) sucrose-containing feedstocks (e.g. sugar beet, sweet sorghum, and sugar cane), (b) starchy materials (e.g. wheat, maize and barley), and (c) lignocellulosic biomass (e.g. wood, straw, and grasses). All of these materials can be utilised for bioethanol production [9]. In this context, some researchers have conducted studies on the use of banana and pineapple wastes, cassava and sweet potato peels, papaya waste, corn cobs, jackfruit seeds, and coffee husks [10-14].

Taro (Colocasiaesculenta (L.) Scott.), a member of Araceae family, is a plant cultivated mainly in tropical and subtropical regions such as Asia, Oceania and Africa. In Asia, taro is used to produce variety of desserts, snacks and confectionary products, leaving huge chunks of taro peel [15]. Wild taro tubers contain considerable amount of starch (70-80 g/100 g dry taro) [16]. The high starch content of taro serves as a great feedstock for glucose production, providing raw materials for bioethanol production [17]. Braide and Nwaoguikpe produced the maximum of 12.90% of bioethanol from wild taro through enzyme hydrolysis followed by *Saccharomyces uvarum* fermentation process [18]. Adegunloye and Udenze reported the effect of using *Aspergillus niger* and *Saccharomyces cerevisiae* on wild taro peels fermentation to produce bioethanol [19].

Though few works has been conducted using wild taro tubers for bioethanol production, it is yet to be determined which hydrolysis process (acid or enzyme hydrolysis) would give the highest yield of ethanol and the condition under which an optimum yield can be achieved. Thus, the knowledge would provide information required in the industrial production process design. Therefore, this research aims to establish an effective process to produce bioethanol from wild taro tubers using bread yeast (*Saccharomyces cerevisiae*) by comparing acid and enzyme hydrolysis pathway. Hydrolysis parameters (e.g. hydrolysis time, types of acid or enzyme catalysts used, acid concentration, enzyme loading amount) in addition to fermentation parameters (e.g. fermentation time) were also evaluated.

2. Experimental section

2.1. *Materials preparation*

Wild taro tubers (Colocasiaesculenta) used in this study, as shown in fig. 1, were collected from Padang city, Indonesia. Prior to use, the cleaned tubers were soaked in NaCl (1% w/v) at a ratio of 2:1. 500 g of tubers for each experiment was steamed then dried inside the oven at 60°C for 24 hours. The tubers were blended and sieved to produce taro flour.

Figure 1. Taro (Colocasia esculenta) Plant and Tuber

2.2. *Fermentation process*

IOP Conf. Series: Materials Science and Engineering **543** (2019) 012056 doi:10.1088/1757-899X/543/1/012056

For acid hydrolysis, 20 g of taro flour mixed with acid catalysts (HCl, H_2SO_4 or $HClO_4$) with various concentrations (0 N, 0.05 N, 0.10 N, 0.15 N and 0.20 N). The mixture was refluxed at 120°C for certain hydrolysis time (1 hours, 2 hours or 4 hours) then cooled and filtered. The hydrolysis solution was added with NaOH 1 N to adjust the pH into 4. The same process was repeated for enzyme hydrolysis, except for two commercial enzymes (bacterial ⟨-amylase and glucoamylase) was used instead of acid, obtained from Sigma-Aldrich, Germany. The loading amount of enzyme was also varied (0.308 mL and 0.74 mL).

Afterwards, the yeast (Saccharomyces cerevisiae) purchased from local shop in Padang city, was mixed into the hydrolysis solution at a ratio of 1:3. The mixed solution was introduced into the fermentation reactor. The fermentation process was carried out anaerobically for certain fermentation time (48 hours,72 hours, 96 hours, 122 hours, 144 hours). The fermented product was filtered and then fed into the distillation flask, the temperature is kept at 80°C. Distillation process was carried out until bioethanol stops dripping.

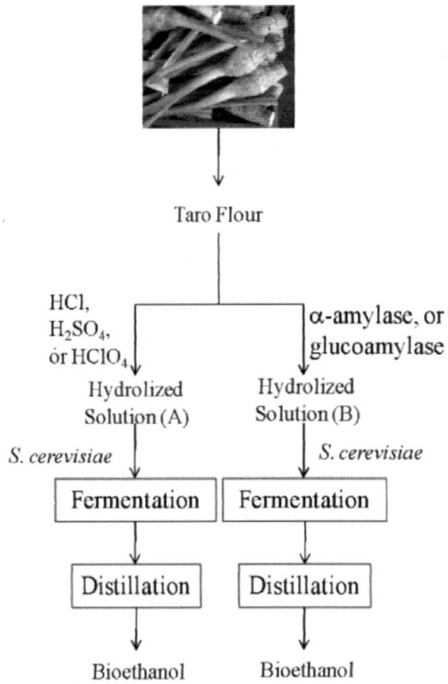

Figure 2. Flow Diagram of Bioethanol Production

2.3. *Analysis*

The starch content of taro was analyzed by using Luff-Schoorl method. 2 g of taro flour was mixed with 100 mL of HCl 3% (v/v) and heated for 3 hours. The mixture was cooled and neautralized by adding NaOH 3% (w/v) then filtered. 25 mL Luff-Schoorl solution was added to 10 mL of the filtrate followed by 15 mL of deionized water. The mixture was heated for 10 minutes and cooled immediately in cold water. 25 mL of H_2SO_4 25% (v/v) and 15 mL of KI 20% (w/v) were added slowly to prevent excessive foaming. Titrate with $Na_2S_2O_3$ 0.1 N until a dull yellow colour appeared, the starch indicator was added and the titration was completed until blue colour disappeared. The same procedure was also conducted to analyze glucose content of hydrolysis products.

3. **Results**

3.1. Hydrolysis process
3.1.1 Acid hydrolysis
The yield of glucose produced from acid hydrolysis steps is shown in figure 3.

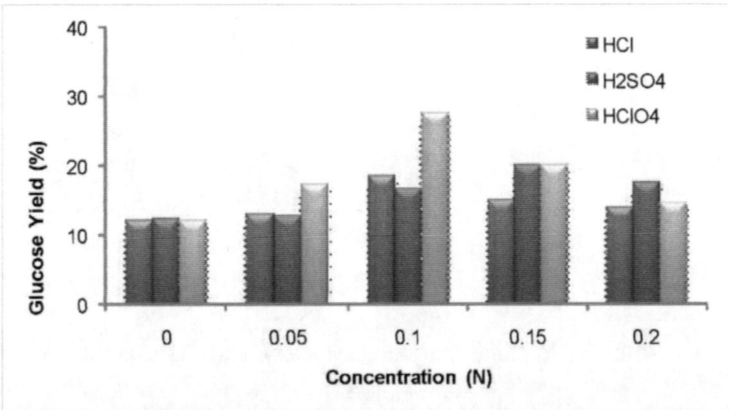

Figure 3. Glucose Yield at Various Acid Catalysts Concentration

As shown in Figure 3, concentration above 0.1 N and 0.15 N are accompanied with lower glucose yield. For H_2SO_4 and HCl, the maximum glucose yield was obtained at acid concentration of 0.15 N. However, the maximum glucose yield for $HClO_4$ catalysts was obtained at concentration of 0.1 N. Degradation of starch during hydrolysis step is a critical problem for acid processes because starch degradation also generates fermentation inhibitors that can inhibit the fermenting microorganisms during the downward fermentation process, thus jeopardizing starch to ethanol conversion yield and the overall biomass-to-ethanol conversion efficiency. The highest glucose yield of 20.15% was obtained using $HClO_4$ 0.1 N as catalysts.

3.1.2 Enzyme hydrolysis
Figure 4 shows that as enzyme loading amount increase the total glucose produced also increase. This behavior caused by higher growth rate of microorganisms at high values of inoculums concentrations which lead to higher rate of starch degradation in the process. However, glucoamylase produce glucose faster than ⟨-amylase. After 60 minutes of hydrolysis process, glucoamylase and ⟨-amylase generates 20.35% and 6.12% of glucose yield, respectively. This result caused by the nature of enzyme behavior. The ⟨-amylases are endohydrolases that randomly cleave ⟨-1,4 glycosidic bonds of linear amylase and branching amylopectin chains of the starch molecule [20]. Contrarily, glucoamylases are exoamylases that cleave both ⟨-1,4 and ⟨-1,6 glycosidic bonds from the non-reducing end of starch molecules [21], thus producing glucose faster than ⟨-amylase.

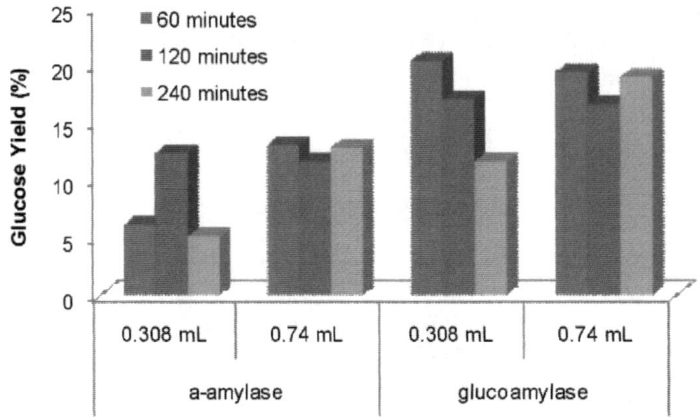

Figure 4. Glucose Yield at Various Enzyme Catalysts Loading Amount

The highest yield of glucose with 20.35% conversion was obtained from enzymatic hydrolysis as compared with 20.15% conversion from acid hydrolysis. The result suggesting that enzymatic hydrolysis is a suitable method compared to acid hydrolysis under the conditions of this study. This has been attributed to the decrease of sugar content by degradation of glucose to furfural and Hydroxy Methyl Furfural (HMF) during acid hydrolysis or conversion of glucose to levulinic, acetic or formic acids. These substances are toxic to yeast and thus can inhibit its growth by decreasing the intracellular pH of the fermenting medium, resulting in the death of microorganisms [22].

3.2. Fermentation process

The effects of fermentation time on bioethanol fermentation was determined by measuring the bioetnanol concentration after 48, 72, 96, 120 and 144 hours. The result is presented in figure 5.

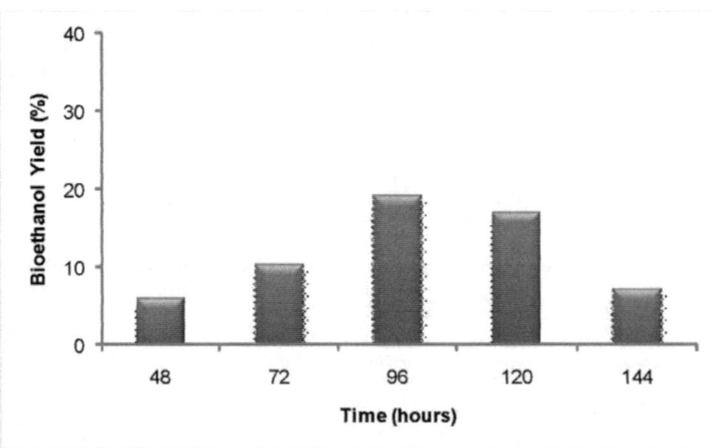

Figure 5. Bioethanol Production Over Time

The longer fermentation time, the higher bioethanol concentration obtained. However, after 96 hours the concentration decreased significantly. In the first stage of fermentation process, the bioethanol increase from 6.0% at 48 hours to 10.24% at 72 hours. The highest bioethanol concentration, which is 19.1%, obtained at 96 hours of fermentation process. This concentration decreased to 16.94% at 120 hours and after 144 hours of process the bioethanol concentration decreased to 7.24%. The results suggesting that after 96 hours the carbon source had reduced and the number of microorganisms present reduced as well, thus resulting in lower yield of bioethanol. This result is in correlation with

Ishmayana*et. al.* that after 120 and 144 hours of fermentation time the viable cell number of yeast was decreased [23].

The comparison of bioethanol production from wild taro in this study with other feedstocks can be seen in table 1. Comparative studies of bioethanol production from different feedstocks shows that wild taro tuber has higher efficiency than other biomass. The process is cost effective, thus could potentially be developed as the feedstock of bioethanol production on a large scale.

Table 1. Comparison of Bioethanol Production from Wild Taro with other Raw Materials

Raw materials	Hydrolysis process	Bioethanol yield (%)	References
Wild taro tuber	Enzyme hydrolysis	19.1	This study
Jackfruit seeds	Acid hydrolysis	57.94	[5]
Banana peel Plantain peel Pineapple peel	Enzyme hydrolysis	7.45 3.98 8.34	[10]
Cassava peel Sweet potato peel	Enzyme hydrolysis	17.6 9.3	[11]
Papaya waste	Enzyme hydrolysis	3.83 – 5.19	[12]
Corn cobs	Enzyme hydrolysis	10.08	[13]
Corns of wild taro	Corns of wild taro	12.9	[18]
Wild taro peels	Enzyme hydrolysis	5.65	[19]

4. Conclusions

Bioethanol production from wild taro tubers (Colocasia esculenta) was performed by hydrolysis followed by anaerobic fermentation. The highest glucose yield of 20.15% from acid hydrolysis was obtained using $HClO_4$ 0.1 N as catalysts. However, glucoamylase able to produce 20.35% of glucose after 60 minutes of hydrolysis process, indicating that enzymatic hydrolysis is a suitable method compared to acid hydrolysis under the conditions of this study. Moreover, the fermentation process was able to recover maximum of 19.1% bioethanol after 96 hours of fermentation time. The results showed that high carbohydrate content in taro tubers could potentially be developed as the feedstock of bioethanol production on a large scale.

References

[1] Festel G W 2008 *Chem. Eng. Tech.***31** 715–720.
[2] Cherubini F, Bird N D, Cowie A, Jungmeier G, Schlamadinger B andWoess-Gallasch S 2009 *Resour. Conserv. Recy.***53** 434–447.
[3] Ramli N A, Rahman R A, Ngadi Nand Samah R A 2017 *Chem. Eng. Trans.***56** 301–306.
[4] Balat M andBalat H, Öz C 2008 *Prog. Energy and Combust. Sci.***34** 551–573.
[5] Arif A R, Natsir H, Rohani H and Karim A 2018 *J. Phys. Conf. Ser.***979** 1–7.
[6] Mosier N, Wyman C, Dale B andLadisch M 2005 *Bioresour. Technol.***96** 673–686.
[7] Taherzadeh M J and Karimi K 2007 *Bioresources***2** 472–499.
[8] Walker G M and Stewart G G 2016 *Beverages***2** 1–12.
[9] Linoj K N V, Dhavala P, Goswami A andMaithel S 2006 *Asian Biotech. Dev. Rev.***8** 31–49.
[10] Itelima J, Onwuliri F, Onwuliri E, Onyimba I andOforji S 2013*Int. J. Environ. Sci. Dev.***4** 213–216.
[11] Oyeleke S B, Dauda B E N, Oyewole O A, Okoliegbe I N andOjebode T 2012 *Adv. Environ. Biol.***6** 241–245.
[12] Dhanaseli P B and Balasubramanian V 2014 *Biosci. Biotech. Res. Asia***11** 187–192.

[13] Itelima J, Ogbonna A, Pandukur S, Egbere J and Salami A 2013 *Int. J. Environ. Sci. Dev.***4** 239–242.
[14] Gouva B M, Torres C, Franca A S, Oliveira L S and Oliveira E S 2009 *Biotechnol. Lett.***31** 1315–19.
[15] Hsieh S C, Liu J M, Pua X H, Ting Y, Hsu R J and Cheng K C 2015 *Appl. Microbiol. Biotechnol.***100** 2629–2639.
[16] Quach M L, Melton L D, Harris P J, Burdon J N and Smith B G 2000 *J. Sci. Food Agric.***81** 311–318.
[17] Davis L, Rogers P L, Pearce J andPeiris P 2006 *Biomass Bioenergy***30** 809–814.
[18] Braide W and Nwaoguikpe R N 2011 *Int. J. Plant Physiol. Biochem.***3** 64–66.
[19] Adegunloye D V and Udenze D O 2017 *J. Adv. Microbiol.***4** 1–8.
[20] Tanyildizi M S and Durson Ö 2011 *Turkish J. Sci. Technol.***6** 47–52.
[21] Agrawal M, Pradeep S, Chandraraj K andGummadi S N 2005 *ProcessBiochem.***40** 2499–2507
[22] Jonsson J L, Alriksson B and Nits-olof N 2013 *Biotechnol. Biofuels***6** 2–10.
[23] Ishmayana S, Learmonth R P and Kennedy U J 24 to 25th Nov 2011 *2ndInt Seminar on Chem: Chemistry for a Better Future* Indonesia 379–385.

IOP Conf. Series: Materials Science and Engineering **543** (2019) 012057 doi:10.1088/1757-899X/543/1/012057

The Effects of Diethyl Carbonate in Light Naphtha Blending to Utilize New Energy Resource

R P Anugraha, R Tetrisyanda, A Altway and G Wibawa
Department of Chemical Engineering, Sepuluh Nopember Institute of Technology (ITS), Kampus ITS Sukolilo 60111 Surabaya, Indonesia

Email: gwibawa@chem-eng.its.ac.id

Abstract. Light naphtha is not reliable used as automotive fuel because of its low octane number (RON) and high vapor pressure. To utilize light naphtha as fuel, blending process may be used to obtain usable gasoline fuel. Diethyl carbonate (DEC) can be used as blending agent with light naphtha because its high octane number and low vapor pressure. Therefore in this work, the vapor pressure and octane number (RON) of DEC–light naphtha blends were measured to study the effect of DEC in light naphtha blending. The vapor pressure measurements show that vapor pressure of DEC–light naphtha blends decreased while the DEC fraction increased. The pseudo binary analysis approachment was used to correlate the vapor pressure experimental data and predict the vapor pressure of light naphtha + DEC blends. The Wilson and nonrandom two-liquid (NRTL) model are giving good correlation and prediction result with average absolute deviation (AAD) less than 1.3% meanwhile the Hildebrand & Scatchard model give quietly poor result with AAD value of 4.4%. The RON measurements show that the RON values of the blends increased while the DEC fraction increased. The DEC is successfully used as blending agent into light naphtha with 30% v/v optimum fraction.

1. Introduction

Light naphtha is a fraction of crude oil which has C5 – C6 hydrocarbons mixture and normal boiling temperature from 303.15 K to 363.15 K [1]. In general, the product of crude oil distillation gives light naphtha yield of 10% [2] and it is usually used as industrial solvent [3]. Light naphtha is not reliable if directly used as gasoline fuel without any additional treatments because of its low research octane number (RON) and high vapor pressure. Many industries have converted light naphtha become usable gasoline fuel with reforming and isomerization processes [2]. However, these processes are complicated and cost-consuming. To utilize light naphtha as fuel, blending process may be used to obtain usable gasoline fuel with an easy and low-cost process.

Oxygenate compounds are the most promising blending agent in gasoline to increase the combustion performance and increase octane number since the prohibition of tetraethyl lead [4]. Methyl tert-butyl ether (MTBE) and ethanol are widely used as blending agent in gasoline because of their high octane number [5] and high production from biomass [6]. On the other hand, MTBE and ethanol cannot be used to blend with light naphtha because these compounds are increasing the vapor pressure significantly when mixed with hydrocarbons [5]. Diethyl carbonate (DEC), an oxygenate, may be used as blending agent with light naphtha because its high octane number and low vapor pressure [7]. The DEC is also known as biochemical compound because of its production from ethanol [7]. Therefore, the blended light naphtha + DEC may be categorized as biofuel which the usage of biofuel become a concern in energy research [8]. Thus, in order to apply DEC as blending agent, the knowledge of vapor pressure and octane number of the mixture are necessary.

There are some publications about DEC as blending agent in open literature. Li et al. [9] were studying the volatility and flash point of typical aviation fuel + DMC/DEC mixture and found that DMC/DEC increase the volatility of aviation fuel. Rodrıguez et al. [10] were studying the boiling temperature of DEC + (n-hexane, n-heptane, n-octane and cyclohexane) mixture and found that DEC increase the boiling temperature of the mixture. Anugraha et al. [11] were studying the vapor pressure of DEC + (n-heptane, isooctane and toluene) mixtures and showed that DEC successfully decrease the vapor pressure of the mixtures.

Based on our knowledge, there are no vapor pressure and octane number for DEC + light naphtha published in open literature. Therefore, in this work, the vapor pressure and RON of DEC–light naphtha blends were measured. The vapor pressure data of the mixtures are used to study the DEC as vapor pressure reducer agent in light naphtha. The RON data of the mixtures are used to study the DEC as octane booster in light naphtha.

2. Experimental methods

2.1. Materials

The light naphtha used in this study was obtained from PT. Trans Pacific Petrochemical Indotama (TPPI) Tuban, Indonesia. The diethyl carbonate (DEC) was purchased from Wuhan Fortuna Chemical Co. Ltd., China. Table 1 shows the specification of the materials used in this work. All materials in this work were used directly without any additional purification process.

Table 1. Description of the materials used in this work

Component	Supplier	Molecular weight (g/mol)	Vapor pressure at 303.15 K (kPa)	Purity
Light naphtha	PT. Trans Pacific Petrochemical Indotama	81.88	44.96	–
Diethyl carbonate	Wuhan Fortuna Chemical Co. Ltd.	118.13	1.95	0.995

2.2. Measurement of vapor pressure

The apparatus and procedure for measuring vapor pressure used in this work were developed by Oktavian et al. (2013). The schematic diagram of the apparatus was shown in Figure 1.

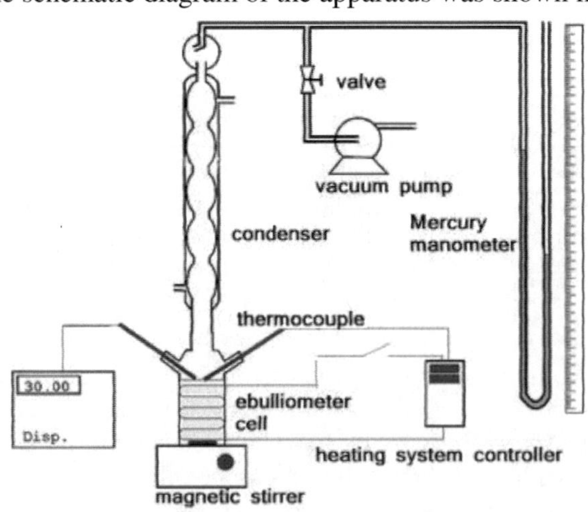

Figure 1. The schematic diagram of.the apparatus for measuring vapor pressure

The specification of main components of the apparatus was explained in our previous works [12,13]. Initially, the mixture of light naphtha and DEC with certain volume fraction was prepared using 100 mL measuring glass by measuring the required volume of each component. The prepared mixture was poured into equilibrium cell and stirred using a magnetic stirrer. Then, the vacuum pump was used to eliminate air and impurities from the equilibrium cell and the line package. The temperature condition in the ebulliometer cell was set and maintained by an electric heater equipped with a controller. The vapor pressure value of the mixture was recorded when the pressure reached a constant value at a certain temperature. The liquid composition in the ebulliometer cell was assumed to be same as initial composition because the change amount of the liquids caused by vaporization could be negligible. This procedure was repeated for other liquid mixtures and temperature conditions as listed in Table 2. The reliability of the apparatus for measuring vapor pressure of similar liquids was shown in our previous works [11,14].

2.3. Measurement of research octane number (RON)

The research octane number (RON) of a mixture was measured using portable octane analyzer KOEHLER K88600 developed by Koehler Instrument Company, Inc. This apparatus was claimed to give performance as same as ASTM D-2699 standard [15] which developed by ASTM International. Firstly, the pure isooctane was analyzed and the result was used as reference octane number. The mixture of light naphtha and DEC with certain volume fraction was prepared by measuring the required volume of each component. The prepared mixture was made with amount of 300 mL for each fraction. Then, it was inserted into the apparatus and to measure the RON value. The printed RON value was treated using reference octane number which obtained from pure isooctane to give the acceptable RON value.

3. Results and discussion

3.1. Vapor pressure data of light naphtha + DEC mixtures

The vapor pressure data of Light Naphtha + DEC mixtures were measured at temperature ranging from 303.15 K to 318.15 K as listed in Table 2. The P-T diagram of light naphtha + DEC for various compositions is presented in Figure 2. Figure 2 shows that the vapor pressure of mixtures was increasing while the temperature was increased. The temperature is increasing the vapor pressure because the vapor from the liquid is more produced due to vaporization. DEC was decreasing the vapor pressure of light naphtha mixture with the increase of DEC compositions. This is happen because the DEC has very low vapor pressure and DEC make no azeotropic phenomenon with several hydrocarbon [10,11].

Table 2. Vapor pressure experimental data of light naphtha (1) + DEC (2) mixtures at 303.15 K – 318.15 K

v_1	*P* (kPa)			
	303.15 K	308.15 K	313.15 K	318.15 K
1	44.96	54.09	63.44	74.91
0.9	44.07	53.31	62.12	73.46
0.8	42.84	51.21	60.91	72.09
0.7	41.44	49.12	59.15	70.25
0.6	39.65	46.91	56.79	66.94
0.5	36.35	43.91	51.74	65.18
0	1.95	2.61	3.43	4.46

where v_1 is volumetric fraction of light naphtha

At 30% v/v DEC composition, the vapor pressure of the mixture was reached the maximum vapor pressure specification of gasoline with value of 60 kPa at 310.95 K [16]. With 50% v/v DEC composition added to the mixture, the vapor pressure could be reduce to 20% compared with pure light naphtha.

IOP Conf. Series: Materials Science and Engineering **543** (2019) 012057 doi:10.1088/1757-899X/543/1/012057

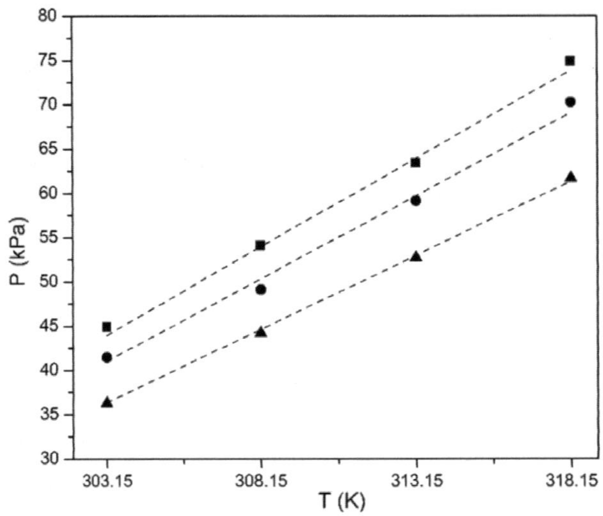

Figure 2. The P-T diagram of light naphtha + DEC for various compositions: (■) experimental data of pure light naphtha, (●) 30%v/v DEC, (▲) 50% v/v DEC, (— —) calculated by Wilson model.

Therefore, DEC is successfully to become vapor pressure reducer agent in light naphtha. Based on the vapor pressure specification of gasoline, the blended LN + DEC fuel may be used as alternative fuel resource without additional processes at the DEC composition value of 30% v/v.

3.2. Pseudo binary analysis

The experimental data of light naphtha + DEC vapor pressures were approached using pseudo binary analysis which the light naphtha was assumed as pure compound. With the pseudo binary approachment, the experimental data were treated based on isofugacity criterion and give the calculated vapor pressure of mixtures (P_{cal}) which expressed in following equation:

$$P_{cal} = \sum_{i=1}^{m} x_i \gamma_i P_i^{sat} \tag{1}$$

where m is the number of component in the mixture, x_i is the mole fraction of component i, γ_i is the activity coefficient of component i and P_i^{sat} is the pure vapor pressure of component i.

The activity coefficient of each component was correlated using Hildebrand & Scatchard [17], Wilson [18] and nonrandom two-liquid (NRTL) [19] models. The Hildebrand & Scatchard model was ever used to correlate the vapor pressure experimental data of typical fuel [9]. The Wilson and NRTL model were widely used to correlate the vapor pressure experimental data of nonelectrolyte liquid binary systems [10,11]. The parameters of the models were optimized using Barker's method [20] by minimizing the following objective function (OF):

$$OF = \sum_{i=1}^{n} \left(P_{i,cal} - P_{i,\exp}\right)^2 \tag{2}$$

where n is the number of data points and the subscript cal and exp refer to calculated and experimental values, respectively.

The parameters of each model were evaluated using average absolute deviation (AAD) between experimental and calculated values which showed in following equation:

$$AAD = \frac{1}{n} \sum_{i=1}^{n} \left| \frac{P_{cal} - P_{\exp}}{P_{\exp}} \times 100\% \right| \tag{3}$$

The best fitted parameters for each model are listed in Table 3. Figure 3 show the vapor pressure of light naphtha + DEC mixtures and the comparison of prediction each model.

Table 3. The best fitted parameters for light naphtha (1) + DEC (2) mixtures and their performance in AAD

Model	Parameter	AAD
Wilson	a_{12} = 131.50 J/mol	0.9 %
	a_{21} = 7745.00 J/mol	
NRTL	b_{12} = 8026.62 J/mol	1.2 %
	b_{21} = -1428.23 J/mol	
	α = 0.30	
Hildebrand & Scatchard	l_0 = 0.046	4.4 %
	α_0 = 6050.535	
	α_1 = 0.796	
	α_2 = 0.007	

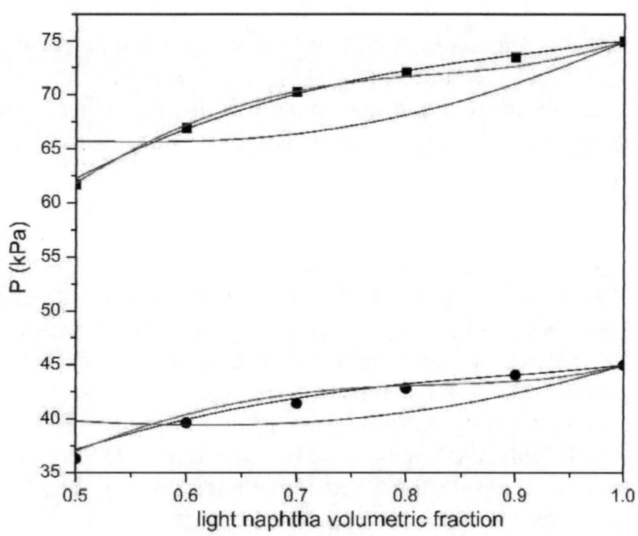

Figure 3. P vs volumetric fraction diagram of light naphtha + DEC mixture and the comparison of prediction each model: (●) experimental data at 303.15 K, (■) experimental data at 318.15 K, (—) calculated value by Wilson model, (—) by NRTL model, (—) by Hildebrand and Scatchard model.

As shown in Table 3 and Figure 3, Wilson and NRTL model are giving good correlation and prediction result with AAD less than 1.3%. The Hildebrand-Scatchard model give quietly poor result with AAD value of 4.4%. Therefore, Wilson and NRTL models may be used to predict the vapor pressure of light naphtha and DEC mixture accurately.

3.3. Research octane number (RON) of light naphtha + DEC blends

The research octane number (RON) data of light naphtha + DEC mixtures were measured as showed in Figure 4. Figure 4 is also showing the effect of DEC addition in the RON of the mixtures. From the figure, it shows that DEC is increasing the RON value in the light naphtha + DEC mixture with the increase of DEC composition. DEC could increase the RON value in light naphtha blending because DEC itself has high RON value [7].

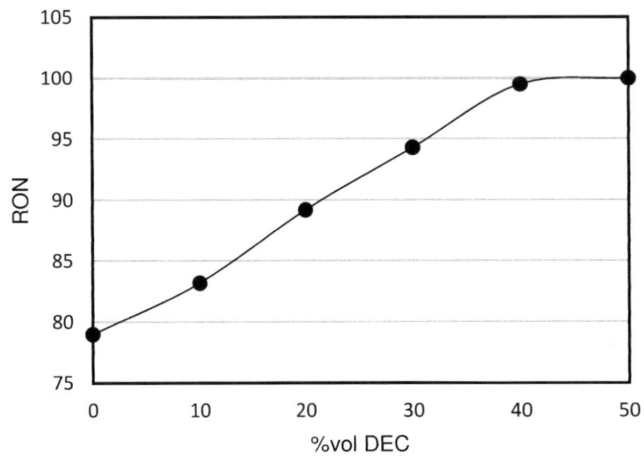

Figure 4. The research octane number (RON) data of light naphtha + DEC mixtures.

DEC, one of the oxygenated compounds, could influence the pre-flame combusting reaction with preventing undesired reaction at low temperature [5]. DEC also has branched molecule structure which give difficulties to breaking the bond and make it more resistance to pre-ignition caused by engine compression. Therefore, DEC may be used as octane booster in light naphtha blending to create the alternative fuel resources.

4. Conclusion

The vapor pressure and research octane number (RON) of DEC–light naphtha blends were measured to study the effect of DEC in light naphtha blending. The vapor pressure measurements show the vapor pressure of DEC–light naphtha blends decreased while the DEC fraction increased. The pseudo binary analysis approachment was used to correlate the vapor pressure experimental data and predict the vapor pressure of light naphtha + DEC blends. The Wilson and NRTL model are giving good correlation and prediction result with AAD less than 1.3% meanwhile the Hildebrand & Scatchard model give quietly poor result with AAD value of 4.4%. The RON measurements show that the RON values of the blends increased while the DEC fraction increased. The DEC is successfully used as blending agent into light naphtha with 30% v/v optimum fraction.

Acknowledgment

This research was funded by Ministry of Research, Technology and Higher Education, Republic of Indonesia under contract no. 128/SP2H/PTNBH/DRPM/2018. The authors would like to give credit to Dr. Susianto for his advices and guidance. The authors also give appreciation to Komang A. R. Gautama and Joy Deryl Lu for their efforts in experimental works.

References

[1] Antos G J, Aitani A M and Parera J M 2004 *Catalytic Naphtha Reforming 2nd edition* (New York: Marcel Dekker, Inc.)
[2] Jones D S J and Pujado P R 2006 *Handbook of Petroleum Processing* (The Netherlands: Springer)
[3] IHS Markit 2017 *Light and Heavy Naphtha: International Market Analysis* (London)
[4] Nadim F, Zack P, Hoag G E and Liu S 2001 *Energy Policy* **29** 1–5
[5] Silva R, Cataluna R, Menezes E W, Samios D and Piatnicki C M S 2005 *Fuel* **84** 951–959
[6] Hossain N, Zaini J H and Mahlia T M I 2017 *Int. J. Technol.* **8** 5–18
[7] Pacheco M A and Marshall C L 1997 *Energy Fuels* **11** 2–29
[8] Majid Z A, Mohsin R and Nasri N S 2016 *Int. J. Technol.* **7** 972–980
[9] Li D, Fang W, Xing Y, Guo Y and Lin R 2009 **161** 1193–1201
[10] Rodrıguez A, Canosa J, Domınguez A and Tojo J 2002 *J. Chem. Eng. Data* **47** 1098–1102
[11] Anugraha R P, Altway A and Wibawa G 2017 *J. Chem. Eng. Data* **62** 2362–2366

[12] Oktavian R, Amidelsi V, Rasmito A and Wibawa G 2013 *Fuel* **107** 47–51
[13] Wibawa G, Mustain A, Akbarina M F and Ruslim R M 2015 *J. Chem. Eng. Data* **60** 955–959
[14] Anugraha R P, Wiguno A, Altway A and Wibawa G 2018 *J. Mol. Liq.* **264** 32–37
[15] ASTM 2018 *Standard Test Method for Research Octane Number of Spark-Ignition Engine Fuel, Standard D-2699-18* (Pennsylvania: ASTM International)
[16] Ministry of Energy and Mineral Resources of Indonesia 2006 *Standar dan Mutu (Spesifikasi) Bahan Bakar Minyak Jenis Bensin yang Dipasarkan Di Dalam Negeri* Letter no. 3674 K/24/DJM/2006 (Jakarta)
[17] Hildebrand J H and Scott R L 1962 *Regular Solutions* (New Jersey: Prentice-Hall)
[18] Wilson G M 1964 *J. Am. Chem. Soc.* **86** 127–130
[19] Renon H and Prausnitz J M 1968 *AIChE J.* **14** 135–144
[20] Barker J A 1953 *Aust. J. Chem.* **6** 207–210

IOP Conf. Series: Materials Science and Engineering **543** (2019) 012058 doi:10.1088/1757-899X/543/1/012058

Production of Diethyl Ether Over Cr-Co/γ-Al_2O_3 Catalyst

M J Marbun[1,], F Kurniawansyah[1], D H Prajitno[2], A Roesyadi[1*]

[1] Chemical Reaction Engineering Laboratory, Department of Chemical Engineering, Faculty of Industrial Technology, Sepuluh Nopember Institute of Technology, Surabaya, 60111, Indonesia.
[2] Department of Industrial Chemical Engineering, Vocational Faculty, Institute Technology of Sepuluh Nopember, Surabaya, 60111, Indonesia.

Email: aroesyadi@yahoo.com

Abstract. Diethyl ether (DEE) is an important chemical in practical application. This substances actually very interesting to be object of study because its can be substitute the fuel oil such gasoline from the fossil. Its mean that DEE is one of alternative sources of energy. In this study the production of DEE was based on ethanol dehydration over a granule Cr-Co/γ-Al_2O_3 catalyst. The reaction was conducted in a fixed bed reactor at range temperature of 100 to 200^0 C and atmospheric pressure. Un nitrogen gas is fed in reactor to push the ethanol vapor. Sampling the product of reaction is taken after the condition is steady state. The catalyst was analyzed by the BET and XRD method and the result indicated that the prepared catalyst has surface area of 162,840 m^2/g with pore diameter of 0,3178 cc/g. The ethanol conversion under the variable condition attaints of 93.122% of and DEE yield is 0.34%.

1. INTRODUCTION

Ethanol is one of the organic chemicals can be used a solvent for diethyl ether production. Diethyl ether can be produced using a heterogeneous catalyst [1]. Diethyl ether (DEE) is one of the most important chemicals in industrial application. Diethyl ether was commonly used to dissolve oils, gums, resins, microselulose, parfumes, alkynid, and butadiene. Other than that application DEE is also used as anesthetic [2]. In the field of renewable energy DEE can also be utilized mixture or substitue fuel, mixed with gasoline to overcome the shortage of fuel [3]. The mixed of DEE with gasoline more compatible than ethanol- gasoline because the different of their boiling point.and also has the advantage of being a renewable energy source because very low carbon monoxide emissions [4].

DEE can be utilized to increase the octane number in the use of gasoline fuel. DEE compound began to utilize as octane level on the material diesel or biodiesel because has a high cetane number. The use of heterogeneous catalyst is very developed in the industrial chemicals. The most important factor in a heterogeneous catalyst is a acidity, surface area, porosity, mechanical properties, and stability. One of the developments in this field catalyst sectors is the preparation of nano-catalyst to improve their performance to accelerate the reaction. Nano-catalyst is a catalyst which has size of 1-100 nm. Dimension reduction of catalyst into nano-sized catalyst would increase the catalyst surface area and consequently increase the activity of catalysts in certain reactions. DEE is compound colors with distinctive aroma small and very low boiling point. Diethyl ether generally is made though the dehydration process of ethanol (Barbet process) using sulfuric acid catalyst (homogeneous catalyst).

IOP Conf. Series: Materials Science and Engineering **543** (2019) 012058 doi:10.1088/1757-899X/543/1/012058

This process has weaknesses of difficulty of catalyst separation after the reaction. To avoid this difficulty homogeneous catalysts are replaced by heterogeneous catalysts. The heterogeneous catalyst generally used for dehydration of ethanol to DEE is Al_2O_3 with promotor Cr-Co. γ-Al_2O_3 has been known for having a large surface area with acidic and alkaline active sides. Butt et al., (1962), and De boer et al., (1967). have studied reaction kinetics of dehydration of ethanol to DEE. The results shown that reaction is similar than mechanism theory formed Langmuir Hinshelwood reaction mechanics simultaneously. An excessive effort is very necessary to produce diethyl ether as an better energy resources. This research used granulated Cr-Co/ γ-Al_2O_3 catalyst prepared by impregnation method. Then, to test the activity of prepared Cr-Co/ γ-Al_2O_3 nano-catalyst. Ethanol dehydration reaction was carried out in a fixed bed reactor.

2. EXPERIMENTAL

2.1 Materials

The chemicals were used in this experiment are: $Al(NO_3)_3.9H_2O$, NH_4OH, HCl and 99% ethanol. This study was used the Al_2O_3 catalyst purchased at the Merck company, Aluminium nitrate nonahydrate (Merck), $Cr(NO_3)_3.9H_2O$, $Co(NO_3)_2.6H_2O$, molecular sieve, nitrogen gas (PT. Aneka Gas Industry) were also employed.

The starting materials, fructose (D-(-)-type, Sigma-Aldrich, 99%), silver nitrate ($AgNO_3$, AR, QRëC™), titanium tetraisopropoxide (TTIP, Sigma-Aldrich, 97%) and absolute ethanol (EtOH, HmbG® Chemicals) were used as the precursor for carbon microspheres, silver, and TiO_2 microspheres, respectively.

2.2 Preparation of catalyst

The first Al_2O_3 was synthesized using the precipitation method. Alumina nitrate monohydrate was prepared with different concentration different (0.2, 0.3, 0.4, 0.5, and 1 M) was dissolved in a solution of HCl (0.5 M). the precipitation was carried out by adding ammonium nitrate into the solution. The precipitate formed was stirred with variation of time between (3.4.5 and 6 hours) at 60, 70, 80, 90, 100°C. Furthermore, the precipitate was filtered and washed by using distilled water and ethanol and then dried at 75° C for 24 hours. The solid precipitate calcined for 6 hours at 600° C with air flow to produce catalyst γ-Al_2O_3. γ-Al_2O_3 in synthesis using XRD and BET. Size of crystal calculated by using Scherrer formula based on the data of XRD analysis. Scherrer formula used is:

$$D = \frac{\kappa \lambda}{\beta \cos \theta} \quad (1)$$

Where D is the diameter of the crystal, κ is the shape factor with a value of 0.9-1.4 (generally used ~ 0.9), λ is the wavelength of X-ray (0.15406 nm), θ is the Bragg angle (radians), and β is the FWHM (Full Width Half maximum) of the highest peak (radians).

2.3 Preparation of Cr-Co/γ-Al_2O_3 Catalysis

γ-Al_2O_3 as a support was impregnated with $Cr(NO_3)_3.9H_2O$ and $Co(NO_3)_2.6H_2O$ metal salt solution using the Dry Impregnation method with loading 10%. Phase stage as follows: The impregnation $Cr(NO_3)_3.9H_2O$) on γ-Al_2O_3 is done by spraying metal solution to γ-Al_2O_3. In stirring using magnetic stirrer at temperature 80 ^{0}C for 3 hours. After which the catalyst is dried at temperature of 110 ^{0}C for 12 hours. $Co(NO_3)_3.9H_2O$) is impregnated in same manner. and then dry catalyst was calcined with N_2 at 550 ^{0}C for 3 hour and followed by a reaction reduction with H_2 at a temperature of 600 ^{0}C for 6 hour and characterized.

2.4 The production process of diethyl ether (DEE)

The catalytic production of diethyl ether was conducted in fixed bed reactor using 3 g of catalyst. This temperature operation was of 100 – 200 ^{0}C, and in atmospheric pressure in the reactor. The ethanol and liquid products were analyzed using gas chromatography with a FID detector and using helium with a flow rate of 27,4 mL/min, temperature operation of 125-250 ^{0}C. Arrangement of equipment used for the production process DEE is shown in Figure 1 below.

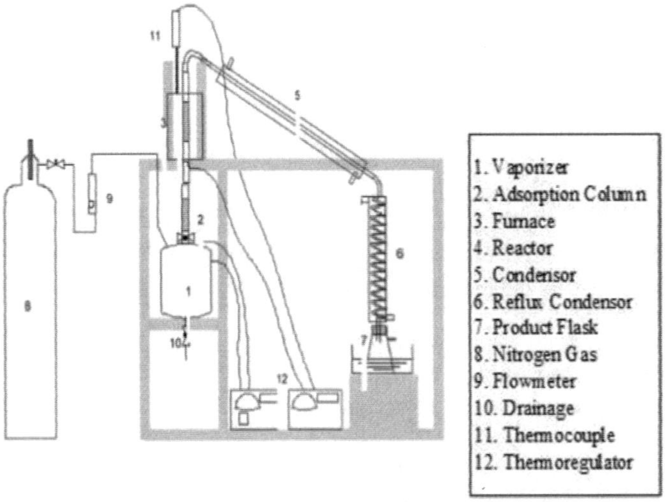

Figure 1. Diethyl Ether Production Equipment

3. RESULTS AND DISCUSSION

3.1 *Surface area and pore volume analysis of catalyst by BET*

The specific surface area, average pore diameter, and pore volumes of the γ-Al_2O_3 and Cr-Co/γ-Al_2O_3 catalyst shown in Table 1. The specific surface area for the γ-Al_2O_3 catalyst is 162.84 m^2/g. After impregnation of Cr and Co on to the γ-Al_2O_3 catalyst reduced the surface area, is 133.499 m2/g. the results indicated that the impregnation of Cr and Co particle blocked of the γ-Al_2O_3 catalyst [5].

Table 1. BET catalyst analysis results of Cr-Co/ γ-Al_2O_3

Catalyst	Specific Surface Area (m^2/g)	Average Pore Diameter (nm)	Pore volume (cm^3/g)
γ-Al_2O_3	162.840	0.3178	3.4276
Cr-Co/ γ-Al_2O_3	133.499	0.2288	3.9027

3.2 The production process of diethyl ether (DEE)

3.2.1 GC-MS analysis

Figure 2 show the chromatogram results of the DEE production process at a gas rate of 600 mL/min. At temperature of 150 ^{0}C the ethanol component was detected at retention time of 3.985 minute, while DEE detected at retention time of 5.367 minutes. From chromatogram, diethyl ether compounds was detected

IOP Conf. Series: Materials Science and Engineering **543** (2019) 012058 doi:10.1088/1757-899X/543/1/012058

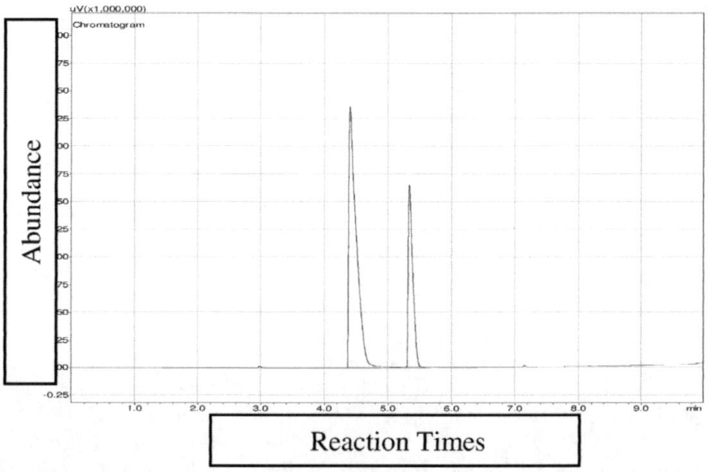

Figure 2. Chromatogram of liquid product in DEE production process using catalyst Cr-Co/ γ-Al_2O_3 of reaction temperature

Figure 2 show that at nitrogen rate 200 mL/min, the conversion of ethanol more significant value compared with 400 mL/min nitrogen gas rate. The largest conversion produced at rate of 200 mL/min is 93.12% at temperature of 100^0 C. For the conversion of 400 mL/min nitrogen gas rate 93.03% at 100 ^{0}C and at conversion rate of 600 mL/min nitrogen gas was 84.40% at temperature of 100 ^{0}C. This indicate that the greater the rate of carrier gas caused then the smaller the contact time between ethanol with surface area of the catalyst [6]. Increased temperatures can increase the conversion of ethanol according to the Arrhenius equation. In this equation, the increase of temperature will increase the reaction rate constants that will increase the rate of reaction [7].

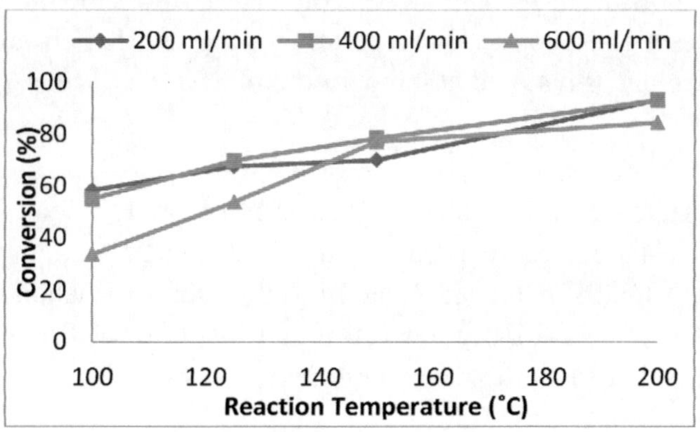

Figure 3. Ethanol conversion vs reaction temperature of dehydration (^{0}C).

Figure 3 shows that the nitrogen gas rate of 600 mL/min produce diethyl ether greater compared which rate of nitrogen gas 200 mL/min and 400 mL/min. The higher yield of DEE in this experiment attaint 0.3 % with N_2 gas rate 600 mL/min at temperature of 200 °C. As shown in Figure 4, the Increasing reaction temperature can increase the yield of DEE. Using of catalyst generally produce dehydration reaction cenvertion hight. This is happened because it has been know that the catalyst has greater surface area. Greater surface area will increase the convertion of ethanol because it can increase the opportunity for the convertion process.

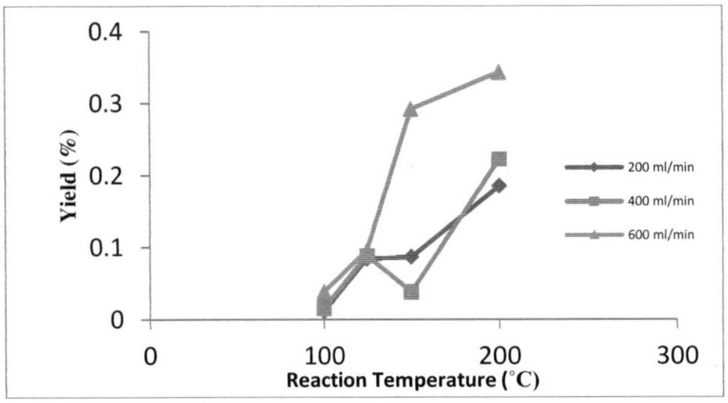

Figure. 4 Relationship between yield DEE % with temperature (^{0}C)

Figure 4 show that the nitrogen gas rate of 600 mL/min produce diethyl ether greater compared which rate of nitrogen gas 200 mL/min and 400 mL/min. The higher yield of DEE in this experiment attaint 0.3 % with N_2 gas rate 600 mL/min at temperature of 200 °C. As shown in Figure 5, the Increasing reaction temperature can increase the yield of DEE. due to more minimum (more negative) value of the Gibbs energy change (ΔGreaction) make the reaction more easy to be occurred [8].

4. CONCLUSION

Based on this research, it may be concluded that diethyl ether can be produced by using. The ethanol dehydration with γ-Al_2O_3 as a catalyst and Cr-Co as metal. The catalyst prepared in the granular state has surface area from 162.84 m^2/g to 133.499 m^2/g, and the catalyst pore volume from 3.4276 cm^3/g to 3.9027 cm^3/g. The higher conversion of ethanol attain 92,12% and yield 0.3 % of DEE yield was biggest conversion reaction has been 93.12% from temperature of 100^0 C. The biggest DEE yield was resulted at temperature of 200 ^{0}C.

ACKNOWLEDGEMENT

The authors would like to thank Ministry of Research Technology and Direktorat Riset dan Pengabdian Masyarakat–Direktorat Jenderal Penguatan Riset dan Pengembangan–Kementerian Riset, Teknologi, dan Pendidikan Tinggi Republik Indonesia for financial support under grant PTUPT Scheme for the year of 2018 granted through Institute for Research and Community Services Sepuluh Nopember Institute of Technology.

REFERENCES

[1] Cárdenas-Guerra J C, Figueroa-Gerstenmaier S, Hernandez S and Reyes-Aguilera J A 2018 Simulation study of a reactive distillation process for the ethanol production *Chem. Eng. Trans.* **69** 613-18
[2] DA Pacas, Kito-Borsa T, Selim S and Cowley S W 1998 Properties of an ethanol-diethyl ether-water fuel mixture for cold-start assistance of an ethanol-fueled vehicle *J. Ind. Eng. Chem.* **37** 3366-74
[3] Bailey B J, Eberhardt S, Erwin J and Goguen 1997 *SAE Trans.* **972** 78
[4] Sunaardi W and Yulia I 2007 Study of 2- propanol catalyst conversion reaction using catalyst and catalyst support γ-Al2O3. *Bull. Chem. React. Eng. Catal.* **2** 56-61
[5] Abdullah B, Dai-Viet V N, Sow Lock I M and Sow Lock S M 2017 Influence of palladium on Ni-

based catalysts for hydrogen production via thermo-catalytic methane decomposition *Chem. Eng. Trans.* **57** 343-48.

[6] Akira Tand Wei C X 1989 The Effects of temperature and residence time on the secondary reactions of volatiles from coal pyrolysis *Amsterdam Elsevier Science Publishers*

[7] Doğu T and Varişli D 2007 Alcohols as alternatives to petroleum for environmentally clean fuels and petrochemicals *Turkish J. Chem.* **31** 5551–67

[8] Abbott H C, Smith J M and Van Ness. 2001 Introduction to Chemical Engineering Thermodynamics 6th Edition: *New York McGraw Hill*

IOP Conf. Series: Materials Science and Engineering **543** (2019) 012059 doi:10.1088/1757-899X/543/1/012059

A Sustainability assessment of biogas plant based on fruit waste in Indonesia: case study of Biogas Plant Gamping, Yogyakarta

F Marendra[1], A Prasetya[1,2], R B Cahyono[1,2], T Ariyanto[1,2*]

[1]Waste Refinery Center, Faculty of Engineering, Universitas Gadjah Mada, Grafika Street No. 2, Bulaksumur, 55281, Yogyakarta, Indonesia
[2]Department of Chemical Engineering, Faculty of Engineering, Universitas Gadjah Mada, Grafika Street No. 2, Bulaksumur, 55281, Yogyakarta, Indonesia

E-mail: teguh.ariyanto@ugm.ac.id

Abstract. The production of biogas by anaerobic digestion (AD) from fruit waste and then converting biogas to electricity is an innovative approach in the development of renewable energy. To study the sustainability of the process, it important to analysis the environ-socio-economic benefits of the technology. In this study, the life cycle sustainability assessment (LCSA) were carried out in a fruit waste biogas plant system, which located in the Gemah Ripah Central Fruit Market in Indonesia. The AD system was designed for 4000 kg/day feed and supplies 148.5 kWh/day electricity. The results of assessment of environmental, economic and social aspects are 5.00 (very good), 3.00 (moderate), and 4.00 (good). This study indicates that biogas plant from fruit waste has a significant advantage for environment, economic, and society. Overall, the sustainability of BPG is in the category of very good (grade I) with a value 4.11 from maximum value of 5.

1. Introduction

Waste fruits and vegetables is important component of household solid waste because high quantities production of agricultural activities, supermarkets and wholesale markets [1]. The number of fruit and vegetable production in Indonesia during 2016 reached 17.5 million tons [2]. With an average depreciation of 10%, the potential fruit waste in 2016 reached 1.75 million tons/year. The amount of fruit and vegetable waste in line with the amount of population and economic activity. Therefore, it is predicted that the number of food waste will increase [3]. If not processed properly, the high number of waste will give negative impacts on the environment and humans.
Waste fruits and vegetables can be processed using anaerobic digestion technology. This process benefits from energy conservation, waste management and environmentally safe products [4], Today, anaerobic digestion is considered as the most powerful technology to convert energy from biodegradable waste [5]. Therefore, in-depth study regarding the use of anaerobic digestion to convert energy from organic waste [6] and the environmental impact of this technology [7] is needed. The AD technology has declared as the most powerful technology to convert waste to energy from biodegradable waste [8]. However, sustainability assessment includes an analysis of the environmental, economic and social of anaerobic digestion technology has not been comprehensively performed, especially in Indonesia. Also,

the previous studies mostly focused on a particular aspect, i.e. the environmental analysis performed by [9] or economic and environmental assessment as practiced by [10].

This study aimed to conduct a comprehensive sustainability evaluation for anaerobic digestion technology in the biogas power plant. The study includes analysis of the environmental, economic and social. As the case study, Biogas Power Plant Gemah Ripah (BPG) generating 148.5 kWh/day electricity with fruit waste feed, located in Gamping, D.I. Yogyakarta was chosen.

2. Methods

2.1. Description of biogas from fruit waste-generated electricity (biogas power plant)

Biogas power plant (BPG) which was analyzed in this article has a capacity of 4 tons/day, working at a pressure of 150 cmH_2O, mesophilic conditions and operating pH between 6.2 and 6.5. It is built in Gamping, Sleman, D.I. Yogyakarta, Indonesia.

The production of electricity from biogas is described as follows. Fruit waste is sorted and inserted into a temporary shelter for weighing. Fruit waste is crushed with addition of water until liquid-form known as slurry. Slurry is pumped to the two digester, and under anaerobic condition, it will produce biogas, that supplies to the power generation unit. While digestate is accommodated in the digestate storage. The unit produces biogas of 162 Nm^3/day volume which generates electricity of 148.5 kWh/day and 5.7 ton/day digestate. The electricity produced is used to lighting the plant and surrounding area. The processes in the biogas power plant is described in Figure 1.

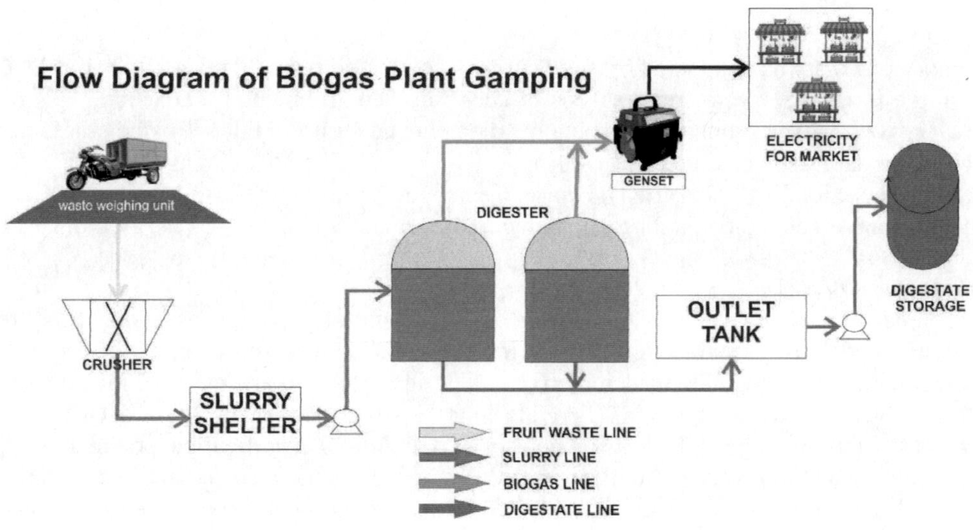

Figure 1. Process flow diagram of BPG

2.2. System boundary and basic framework of the evaluation models

System boundary and basic framework of LCSA process on BPG system is illustrated in Figure 1. The environmental assessment did not account how to obtain fruit waste but limited in start feeding fruit waste to generate electricity and digestate storage for organic fertilizer. Economic assessment is limited to the analysis of costs, benefits and the percentage of profit to the cost. Social studies was limited to the scope of the three stakeholders in BPG i.e; Worker (operator), Consumer (owners of market stalls), Local Community (private sector, organized the Gemah Ripah Fruit Market).The functional unit used in this study is a ton of processed fruit waste.

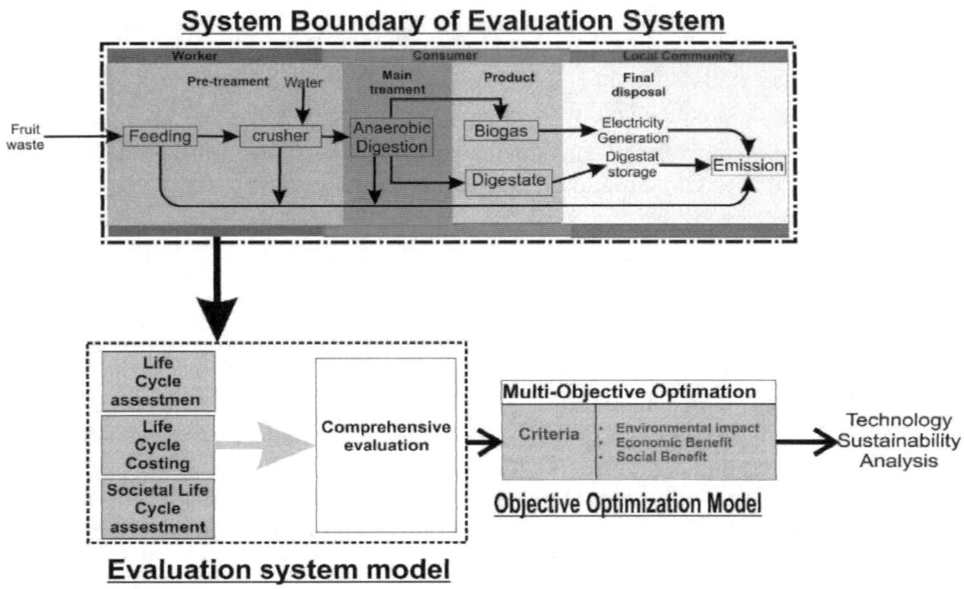

Figure 2. System boundary and basic framework of the evaluation models

2.3. Data inventory

Data inventory process was conducted by field surveys conducted from 2017-2018 year at BPG. Data collected in the form of relevant data analysis of economic, environmental and social. This data was then confirmed to the parties concerned to obtain valid data and verified. For life cycle assessment, data collected from database of Intergovernmental Panel on Climate Change (IPCC), National Environmental Research Institute (NERI), Danish Center for Environmental and Energy (DCE), Baltic Forum for Innovative Technologies for Sustainable Management and Outlook Energy Indonesia 2014.

2.4. Single factor evaluation

Data Environmental analysis was conducted using the midpoint of-life cycle assessment that refers to the operational guidelines prepared by [11]. During the process, the prediction of the environmental impact generated by BPG was divided into five categories: global warming potential (GWP, in kg CO_{2eq}), fresh water ecotoxicity potential (FAETP, in kg 1.4 dichlorobenzen (kg 1.4-DCB_{eq})), human toxicity potential (HTP, in kg 1.4 dichlorobenzen (1.4-DCB_{eq})), acidification potential (AP, in kg SO_{2eq}), and euthrophication potential (EP, in kg $PO_4^{-3}{}_{-eq}$. Each of these parameters emissions in each group was multiplied by the impact factor compiled by [12-14] to obtain the value of the operational parameters generated from BPG.

The economic analysis was performed by the method of cost-benefit analysis which refers to the way in which the [15]. In brief, the economic evaluation was divided into three categories i.e.: costs, benefits and the percentage of profit to the cost. Each category of used units per ton of processed fruit waste.

Social analysis was conducted using the product of social impact assessment refers to the operational guidelines by [16]. In this method, social impacts were divided into three categories of workers (with the highest and lowest values +24 and -24), consumer (with the highest and lowest values +4 and -4) and local communities (with the highest and lowest values +10 & -10). The question topics conducted on social assessment are as follows. For workers, topics are health and safety, wages, social benefits, working hours, child labor, force labor, discrimination, freedom of association and collective bargaining, employment relationship, training and education, work-life balance, job satisfaction and engagement. Health, safety, and experienced well-being are for consumers. Finally, health and safety, access to tangible resources, the local capacity building, community engagement and employment are for the local community.

2.5. Grades of the evaluation index

Criteria and grade of this evaluation index are very important in the study of the sustainability of the biogas power plant. The study is using grades that arranged in five grades of very good (grade I), good (grade II), moderate (grade III), poor (grade IV) and very poor (grade V). Each grade has a value in a sequence: 5,4,3,2 and 1. Each grade has a characteristic that describing per criteria on the Table 1. Environmental impact and economic benefits criteria refers to [17] while for social benefits refers to the criteria drawn up by the [18].

Table 1. Grade of evaluation index

Category		Grade I	Grade II	Grade III	Grade IV	Grade V	Ref
Level 1	Level 2						
Environmental impact	GWP (kg CO_{2eq})	≤0	246.24	402.48	738.71	984.95	
	FAETP (kg 1,4-DCB_{eq})	≤0	2.92	5.86	8.77	11.69	
	HTP (kg 1,4-DCB_{eq})	≤0	0.15	0.31	0.46	0.61	[19]
	AP (kg SO_{2eq})	≤0	0.12	0.25	0.37	0.49	
	EP (kg $PO_4{}^{3-}{}_{eq}$)	≤0	0.96	1.93	2.89	3.85	
Economic Benefit	Cost (million rupiah)	0	0.22	0.44	0.66	0.88	
	Benefit (million rupiah)	4.03	3.02	2.01	1.01	0	[20]
	The ratio profit to cost	100%	75%	50%	25%	0%	
Social benefit	worker	+24	+12	0	-12	-24	[21]
	consumer	+4	+2	0	-2	-4	
	Local community	+10	+5	0	-5	-10	

2.6. Criteria of the evaluation index

The analytic hierarchy process (AHP) method was employed to determine the weight of the evaluation index. AHP is a decision analysis method of multi-purpose labor system quantifies expert opinion using a combination of quantitative and qualitative analysis where guide the implementation process can be obtained through the article by [22].

2.7. The life cycle sustainability assessment (LCSA)

Results of single factor evaluation were employed to calculate the life cycle sustainability assessment (LCSA). First, the value of each criteria in the level 2 (*V2*) is multiplied by its weighting factor (*WL2*), hence obtaining the value of assessment level 1 (*V1*).

$$V1 = \sum (V2\,x\,WL2) \quad (1)$$

The assessment level 1 include environmental impact, economic benefit and social benefit. Total index grade *(TI)* of BPG is then calculated with following equation:

$$TI = \sum (V1\,x\,WL1) \quad (2)$$

Where *WL1* is the weighting factor of level 1. The leveling and weighting factor are shown in Table 2.

Table 2. Terminology, leveling and weighting factors of assessment levels

Category		Analysis Result	Sustainability result
Level 1	Level 2		
Environmental impact	GWP (kg CO_{2eq})	0.425	
	FAETP (kg 1.4-DCB_{eq})	0.046	
	HTP (kg 1.4-DCB_{eq})	0.172	0.395
	AP (kg SO_{2eq})	0.274	
	EP (kg $PO_4^{3-}{}_{eq}$)	0.083	
Economic Benefit	Cost (million rupiah)	0.249	
	Benefit (million rupiah)	0.157	0.291
	The ratio profit to cost	0.594	
Social benefit	Worker	0.632	
	Consumer	0.105	0.314
	Local community	0.263	

2.8. The life cycle sustainability assessment (LCSA)
This process results interpretation and recommendation which can then be applied directly to improve the biogas power plant system. This relates to the development and improvement of products/processes, strategy and planning, determination policies etc. to support sustainability in the environmental, economic and social existence of BPG.

3. Result and discussion

3.1. Result of single-factor evaluation

3.1.1. Environmental impact
Based on the assessment of environmental impact, BPG produces a low environmental impact. It can be proved by a value of 5 in the evaluation of potential emission i.e. global warming potential (GWP), fresh water ecotoxicity potential (FAETP), human toxicity potential (HTP), acidification potential (AP) and eutrophication potential (EP) as shown in Table 3. This means that the BPG system is at grade I (very good) (see criteria in Table 1). This is consistent with the results of previous research conducted by [23]. The most prominent emission value is GWP, which reaches 81.95 kg CO_{2eq}. This value is still much smaller than that of the direct landfill (984.85 kg CO_{2eq}) as described by Chen et al. (2017). Therefore, anaerobic digestion exhibits as a proven technology which could reduce CO_2 released into the environment, when of course the methane is utilized. GWP is contributed by all subsystems BPG starting from the pretreatment, the anaerobic digestion process, electricity generation and digestate storage. For the AP, EP, HTP and FAETP, the values are low. Hence, based on this assessment, BPG shows minor contribution to fresh water ecotoxicity, human toxicity, acidification and eutrophication.

3.1.2. Economic impact
The economic analysis is shown in Table 3. The results showed that the cost required for the establishment and operation of BPG is very high due to the high instalment cost. BPG included in the grade V (very poor) with a value of 1 (see criteria in Table 1). Further analysis in advantages of BPG gives different insight. Benefit category entered in the grade 2 (good), while the amount of profit to cost

ratio reached 43%. It is shown that BPG with anaerobic digestion technology for waste treatment and energy generation gives significant economic benefits. This is consistent with the result of research by [24]

3.1.3. Social Benefit

Social benefits assessment shows the important of BPG. The social benefits have a value of 4, indicating that the BPG entered at grade II (good) (see criteria in Table 1). This means that the presence of BPG provides good social benefits for workers, consumers and local communities. The workers, consumers and local communities feel the positive impact of the presence of BPG. BPG has implemented environmental health and safety regulations with the provision of relatively comprehensive health and safety information, health and safety coverage covered by BPG managers. BPG managers also have provided first aid facilities in accidents and fire extinguishers and perform prompt and appropriate handling in the event of an accident. Consumer and local community stated that the presence of BPG provides many benefits like support their trading business and they do not feel a burden due to BPG. Furthermore, consumers are also very happy that their waste can be used as raw material of BPG. The existence of this positive impact is indispensable for the sustainability of BPG considering the social aspect as a vital aspect in the development of waste management and renewable energy in Indonesia.

Table 3. Sustainability result of single-factor evaluation

Category		Analysis Result	Sustainability result
Level 1	Level 2		
Environmental impact	GWP (kg CO_{2eq})	81.95	5
	FAETP (kg 1.4-DCB_{eq})	0.0098	5
	HTP (kg 1.4-DCB_{eq})	0.0109	5
	AP (kg SO_{2eq})	0.07	5
	EP (kg $PO_4{}^{3-}{}_{eq}$)	0.02	5
Economic Benefit	Cost (million rupiah)	1.89	1
	Benefit (million rupiah)	2.64	4
	The ratio profit to cost	43%	3
Social benefit	Worker	+7	4
	Consumer	+1	4
	Local community	+5	4

3.2. The life cycle sustainability assessment (LCSA)

Kloepffer [25] mentioned that life cycle sustainability assessment (LCSA) is the sum of the life cycle assessment, life cycle costing and societal life cycle assessment. The results of LCSA of BPG system is shown in Table 4. Overall, the technology used by BPG is categorized as a very good technology (score 4.11, grade I). The environmental benefits of BPG categorized as excellent (score 5, grade I); social benefits in the category of good (score 4, grade II); and economic benefits in the category of moderate (score 3, level III). Environmental impact assessment scores are greater than the social benefits and economic benefits. The results showed that overall the AD technology has a high sustainability, which also agrees to [26]. From the result of LSCA, it is necessary to increase economic benefits of anaerobic technology for electricity production as it is the lowest criteria. Innovative approaches are needed to be introduced for development and improvement of products/processes. For the instalment cost, for example, providing better compact digestion plant is needed, thus lowering the cost for land area. Product diversities are also required e.g. digester slurry effluent for fertilizer and a direct use of gases for cooking. Another is to introduce biogas plant as education tourism.

Table 4. Comprehensive evaluation result of sustainability assessment

Category		Sustainability results (Level 2 grade)	Level 2 weighting factor	Level 1 weighting factor	Level 1 sustainability grade	Total sustainability grade
Level 1	Level 2					
Environmental Impact	GWP (kg CO_{2eq})	5	0.425			
	FAETP (kg 1. 4-DCB_{eq})	5	0.046			
	HTP (kg 1.4-DCB_{eq})	5	0.172	0.395	5	
	AP (kg SO_{2eq})	5	0.274			
	EP (kg $PO_4{}^{-3}{}_{eq}$)	5	0.083			4.11
Economic Benefit	Cost (million rupiah)	1	0.249			(grade I/very
	Benefit (million rupiah)	4	0.157	0.291	3	good)
	The ratio profit to cost	3	0.594			
Social benefit	worker	4	0.632			
	consumer	4	0.105	0.314	4	
	Local community	4	0.263			

4. Conclution

The life cycle sustainable assessment was conducted for biogas power plant. The steps of the comprehensive evaluation methods include i) determination of the goal and scope of the study, ii) the analysis of inventory of data (environmental, economic, and social), iii) assessment of the impacts (environmental, economic, social), iv) determination of the criteria and grade of index evaluation, v) the determination of the weight of the evaluation index system, vi) the calculation of the value of sustainability and vii) interpretation of obtained data. Results of the assessment of environmental benefits, economic and social are 5 (grade I/very good), 3 (grade III/moderate), and 4 ((grade II/good)), respectively. In overall, the sustainability of BPG grade is in grade I with a value 4.11 (very good category). This shows that BPG can be considered as an eco-friendly program with significant environ-socio-economic benefits.

5. Acknowledgement

The authors gratefully acknowledge the funding of the Ministry of Research, Technology and Higher Education of the Republic of Indonesia (MENRISTEKDIKTI) through Penelitian Terapan Unggulan Perguruan Tinggi (Grant No. 1870/UN1/DITLIT/DIT-LIT/LT/2018). The authors thanks to Anggun Rahmada for fruitful comments and discussion

6. References

[1] Pavi S, Kramer L E, Gomes L P, and Miranda L A S 2017 *Bioresour. Technol.* **228**, 362–7.
[2] Statistics Indonesia 2017 *Statistics of Indonesia's Annual Fruit and Vegetable Plants 2016. (Subdirektorat Statistik Hortikultura, Ed.)* (Jakarta : Statistics Indonesia)
[3] Chen T, Shen D, Jin Y, Li H, Yu Z, Feng H, Long Y, Yin J 2017 *Appl. Energy* **208** 666–77.
[4] Forster-Carneiro T, Romero LI, and Perez M 2008 *Bioresource Technology* **99** 6994–7002.
[5] Bacenetti J, Sala C, Fusi A, and Fiala, M, 2016 *Appl. Energy* **179** 669–86.
[6] Streitwieser, Almeida D, and Cadena A I 2018 *Chem. Eng. Trans* **65**
[7] Van Fan Y, Lee C T, Klemeš J J 2017 *Chem. Eng. Trans.* **37** 7-12

[8] Bacenetti J, Sala C, Fusi A, and Fiala, M, 2016 *Appl. Energy* **179** 669–86
[9] De Vries J W, Groenestein C M, and De Boer I J M 2012 *J. Environ. Manage.* **102** 173–83.
[10] Kwee L M, Hashim H, Siong H C, Shin H W, and Shiun L J 2017 *Chem. Eng. Trans.* **56** 715–20.
[11] Guinée JB, Gorrée M, Heijungs R, Huppes G, Kleijn R, De Koning A, Van Oers L, Wegener Sleeswijk, Suh S, Udo de Haes H, Bruijn H, Van Duin R, Huijbregts M A J, Lindejer E, Roorda A A H, Van Der Ven B L, Bo P, Wedema 2002 *Handbook on life cycle assessment. Operational guide to the ISO standards. I: LCA in perspective. IIa: guide. IIb: operational annex. III: scientific background*, (Dordrecht: Kluwer Academic Publishers)
[12] Hauschild M, and Wenzel H 1998 *Environmental assessment of products: scientific background (1st ed.)*, (New York : Springer US)
[13] Houghton JT, Filho L G M, Callander B A, Harris N, Kattenburg A, Maskell K 1996 *Climate Change 1995 : The Science of Climate Change (1996th ed.)* (Cambridge : Cambridge University Press)
[14] Huijbregts M 1999 *Priority Assessment of Toxic substance in the frame of LCA-Development and application of the multi-media fate, exposure and effect model USES-LCA*, (Amsterdam : University of Amsterdam)
[15] Chen T, Shen D, Jin Y, Li H, Yu Z, Feng H, Long Y, Yin J 2017 *Appl. Energy* **208** 666–77.
[16] Fontes J, Gaasbeek A, Goedkop M, Contreras S, Evitts S, Alvarado C, Saling P, Van Gelder R, Traverso M, Tame P, Das Gupta J, Morris D, Woodyard D, Bell L, Van der Merwe R, Duclaux C, Slaoui Y, Gupta A, Kimm N, Santamaria C, Laubscher M, Jacobs M, Zinck S, Stermann R, Carteron E, Nilsson S, Bolhuis A, Bogaers K, Culley H, and Challis D 2016 *Product Social Impact Assessment* (Amersfoort : PRé Sustainability)
[17] Chen T, Shen D, Jin Y, Li H, Yu Z, Feng H, Long Y, Yin J 2017 *Appl. Energy* **208** 666–77..
[18] Fontes J, Gaasbeek A, Goedkop M, Contreras S, Evitts S, Alvarado C, Saling P, Van Gelder R, Traverso M, Tame P, Das Gupta J, Morris D, Woodyard D, Bell L, Van der Merwe R, Duclaux C, Slaoui Y, Gupta A, Kimm N, Santamaria C, Laubscher M, Jacobs M, Zinck S, Stermann R, Carteron E, Nilsson S, Bolhuis A, Bogaers K, Culley H, and Challis D 2016 *Product Social Impact Assessment* (Amersfoort : PRé Sustainability)
[19] Chen T, Shen D, Jin Y, Li H, Yu Z, Feng H, Long Y, Yin J 2017 *Appl. Energy* **208** 666–77.
[20] Chen T, Shen D, Jin Y, Li H, Yu Z, Feng H, Long Y, Yin J 2017 *Appl. Energy* **208** 666–77.
[21] Fontes J, Gaasbeek A, Goedkop M, Contreras S, Evitts S, Alvarado C, Saling P, Van Gelder R, Traverso M, Tame P, Das Gupta J, Morris D, Woodyard D, Bell L, Van der Merwe R, Duclaux C, Slaoui Y, Gupta A, Kimm N, Santamaria C, Laubscher M, Jacobs M, Zinck S, Stermann R, Carteron E, Nilsson S, Bolhuis A, Bogaers K, Culley H, and Challis D 2016 *Product Social Impact Assessment* (Amersfoort : PRé Sustainability)
[22] Deng X, Hu Y, Deng Y, Mahadevan S 2014 *Expert Syst. Appl.* **41(1)** 156–67.
[23] Chen T, Shen D, Jin Y, Li H, Yu Z, Feng H, Long Y, Yin J 2017 *Appl. Energy* **208** 666–77.
[24] Jin Y, Chen T, Chen X, Yu Z 2015 *Appl. Energy* **151** 227–36.
[25] Kloepffer W 2008, **13(2)**89–94.
[26] Chen T, Shen D, Jin Y, Li H, Yu Z, Feng H, Long Y, Yin J 2017 *Appl. Energy* **208** 666–77.

IOP Conf. Series: Materials Science and Engineering **543** (2019) 012060 doi:10.1088/1757-899X/543/1/012060

Techno-economic analysis of Dimethyl Ether production using Oil Palm Empty Fruit Bunches as feedstock – a case study for Riau

H S Parbowo, A Ardy and H Susanto

Department of Chemical Engineering, Institut Teknologi Bandung, Bandung 40132, Indonesia

E-mail: herdadi.supriyop@gmail.com

Abstract. The increase of Liquid Petroleum Gas (LPG) import in Indonesia made the government promote Dimethyl Ether (DME) as partial substitution of LPG as written in the General Planning of National Energy (RUEN) 2017. DME can be produced from biomass which abundantly exists in Indonesia. Such biomass is especially oil palm empty fruit bunches (EFB) that consist of 49.07% C, 6.48% H, 0.70% N, 38.29% O, and 5.36% Ash. Techno-economic analysis of this bio-DME production plant was conducted based on literature data. The direct synthesis of DME was applied in this study, instead of using methanol as an intermediate product. Result calculated from mass and energy balance simulation was 4.1 EFB/DME. Thus, 1,090 ton per day of DME can be produced from EFB with a collecting radius of 100 km in Riau Province. Assuming 20 years plant operation with DME selling price of IDR 145,000/12-kg-LPG-equivalent and EFB price of 300 IDR/kg, economic evaluation showed that this business was economically feasible with 10.08% IRR generated. The CAPEX of this plant was 990.4 million USD and the OPEX was IDR 7,143 IDR/kg-LPG-equivalent. DME production from EFB will give the positive economics and environmental effects. Therefore, support and synergy from government, industries, and academics are needed to realize this business.

1. Introduction

The use of DME is one of government scenarios to reduce LPG import, as already written in RUEN 2017. LPG's partial substitution with DME can be implemented because of the similar characteristic between LPG and DME. The main identical characteristics of theirs are (1) both substances identify as non-toxic, (2) having colored fire, and (3) DME vapor pressure lies within the range of LPG vapor pressure. Those similar characteristics between them will minimize the change of infrastructure. The results of testing conducted by BPPT and Lemigas concluded that a mixture of maximum 20% DME in LPG can be applied in household stoves without modification. Pertamina has also conducted the market trial for the utilization of DME as household fuel in North Jakarta with satisfying result.

DME can be produced from natural gas, methanol, coal, and biomass. The utilization of biomass as feedstock is suitable to be applied in Indonesia, having the abundance of biomass. Biomass potential energy is approximately 30,000 MW_e, while only 800 MW_e of it that has already been utilized [1] whereas biomass has a potential to produce energy that economically feasible [2]. From that biomass potential, the largest proportion is oil palm biomass (39.3 %).

Oil palm biomass consists of palm trunk, palm frond, empty fruit bunch (EFB), palm kernel, and palm fiber. EFB is chosen as raw material for DME production because of its abundance in Indonesia, it not being widely utilized yet, and its localized existence in palm oil plant.

This research has been carried out with aiming to provide a preliminary techno-economical data and to propose a business plan of DME production from EFB. The results of this research are expected to be an initial reference to promote the development of DME production from EFB in Indonesia.

2. Methods

This research consisted of four major steps: determination of EFB potential, mass and energy balance calculation, economical evaluation, and business plan preparation. Overall, the steps of this research were performed based on literature study and computer aided simulation.

2.1 Determination of EFB Potential

Determination of EFB potential was implemented by creating an imaginary circle in biomass collecting area with radius of 100 km from proposed DME plant in Riau Province. The imaginary circle of biomass collecting area is illustrated in Fig. 1. The center of biomass collection is the selected plant location (District of Kampar, Riau Province). The EFB potential from this collection area was referred as the production scale of EFB to DME plant in this research.

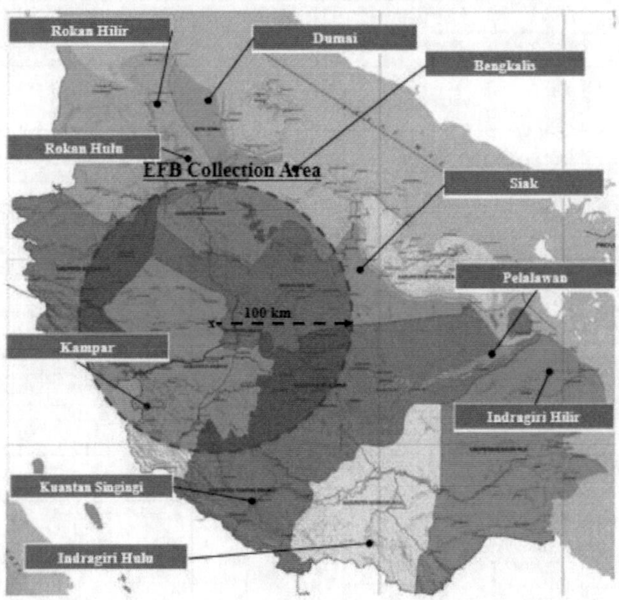

Figure 1. Selected EFB collecting area in this research

2.2 Mass and Energy Balance

The main reactions of this process are shown on Reaction (1), (2), and (3). This process uses the direct synthesis of DME instead of producing methanol and then dehydrated it on the separate reactor. The main purpose of this selection is to minimize the energy consumption. The catalysts for conducting these reactions are dual catalysts with Cu and γ-Al_2O_3 based [3].

$$4\,CO + 2\,H_2 \leftrightarrow 2\,CH_3OH \qquad \Delta H = -181.6\text{ kJ} \qquad (1)$$
$$2\,CH_3OH \leftrightarrow CH_3OCH_3 + H_2O \qquad \Delta H = -23.4\text{ kJ} \qquad (2)$$
$$CO + H_2O \leftrightarrow CO_2 + H_2 \qquad \Delta H = -41.0\text{ kJ} \qquad (3)$$

IOP Conf. Series: Materials Science and Engineering **543** (2019) 012060 doi:10.1088/1757-899X/543/1/012060

Simplified process diagram of the biomass conversion to DME is presented in Fig. 2. Calculation with Aspen Plus V8.8 and Aspen Hysys V8.8 were used to solve energy and mass balance problem. The main process parameter used in the mass and energy balance is presented in Table 1.

Table 1. Process design parameter used in this research

Feedstock	EFB with dry composition (%-mass) [5]: 49.07% C, 6.48% H, 0.70% N, 0.10% S, 38.29% O, 5.36% Ash, HHV 17.08 MJ/kg (dry basis), and 9%-mass of moisture content (wet basis).
Gasification	Equilibrium reaction at P 4 bar and T = 900°C. Heat loss is 5% of thermal input (dry basis).
Air Separation Unit	Electricity consumption 1 MJ/kg-O_2 [4].
DME Synthesis Reactor	Equilibrium reaction for reactions (1), (2), and (3) with approach to equilibrium 15°C for reaction (1), (2), and 100 °C for reaction (3) [5]. Operating P = 50 bar and T = 250°C.
Rotating equipments	Isentropic efficiency = 80%
Pinch temperature	10°C
Export/import heat or electricity	Zero
Fluid package	PRSV (Peng-Robinson Stryjek Vera)

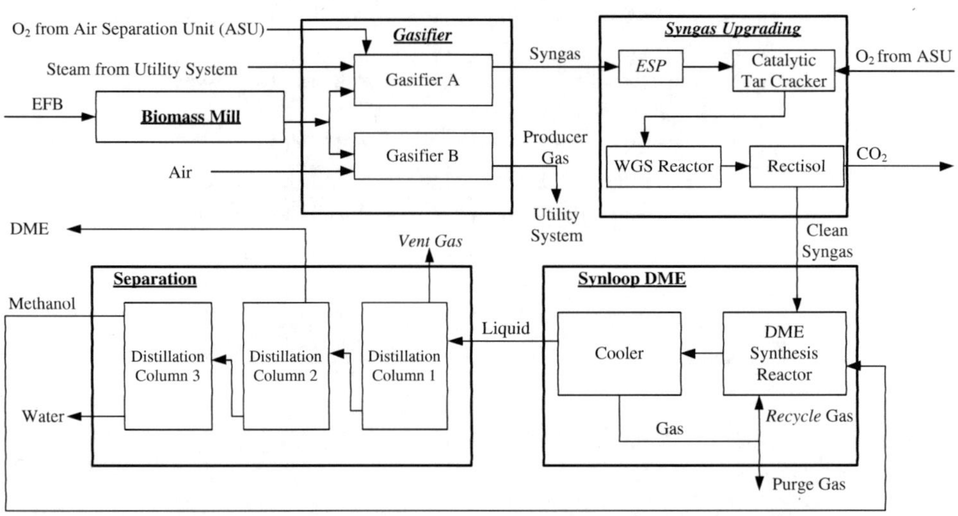

Figure 2. Simplified block diagram of EFB to DME Plant

2.3 Economic Evaluation

In order to estimate CAPEX, the assistance of APEA (Aspen Process Economic Analyzer) was employed. Values from APEA, based on 2015 data, were extrapolated using CEPCI (Chemical Engineering Plant Cost Index) to predict CAPEX of the plant in 2018. Major assumptions used in economic evaluation are presented in Table 2.

IOP Conf. Series: Materials Science and Engineering **543** (2019) 012060 doi:10.1088/1757-899X/543/1/012060

2.4 Business Plan

The result of techno-economic evaluation was then used to develop the business plan. The author's experience of interaction with some of the biomass industry practitioners had also been considered to complete the proposed business plan.

Table 2. Economic evaluation assumptions

Plant's economic age	20 years
Duration of construction	3 years
EFB cost	IDR 300/kg (plant gate)
DME selling price	IDR 145,000/12-kg-LPG-equivalent (retail)
Capital Structure	40% equity and 60% debt
Interest rate	5% per year
Depreciation Model	Linear with salvage value 0
Tax	15% from earning
Exchange Rate	13,500 IDR/USD
Internal Rate of Return (IRR) Criteria	≥8%

3. Result and Discussion

3.1 EFB Potential in Biomass Collecting Area

EFB potential collected from selected collection area mentioned in Fig. 1 was obtained using following approximation and the result is presented in Table 3.

$$P = \sum_{i=1}^{6} \left(\frac{C_i}{A_i} P_i . t . eff . fr \right)$$

P = EFB potential in the selected collection area shown in Fig. 1 (ton per day)
C_i = collection area covered in district i (m^2)
A_i = area of district i (m^2)
P_i = palm oil mill installed capacity in district i (ton per day) [6]
t = average operation time factor of palm oil mill, assumed 0,69 [6]
eff = collection efficiency of EFB, assumed 80%
fr = mass fraction of EFB in palm fruit, assumed 21%-mass
i = six districts in Riau covered by biomass collection area

Table 3. EFB potential in the selected collection area

No	District	EFB potential on selected collection area (ton/day)	
		As received (41.8% moisture content)	Air dried (9% moisture content)
1	Kampar	3,802	2,433
2	Rokan Hulu	1,158	741
3	Siak	1,808	1,157
4	Bengkalis	856	548
5	Rokan Hilir	263	168
6	Dumai	17	11
Total EFB potential in collection area		**7,904**	**5,058**

With a yield of 4.1 ton-biomass-air-dried/ton-DME based on our calculation on mass and heat balance (see Table 4), the estimated production capacity of this plant was 1,090 ton/day-DME. Thus, DME production from EFB with collecting radius of 100 km in Riau Province could replace about 7% of LPG import in Indonesia.

Selection of Kampar District in Riau Province as plant location had several advantages. The first, EFB cost could be minimized since abundant EFB was available on its surrounding and road infrastructure was available for EFB transportation. Furthermore, Kampar River was available as utility source and also enabling the ship transportation of DME to ocean.

3.2 The Mass and Heat Balance

The mass and energy balance was calculated by taking a heat integration using the pinch technology into account. So, minimum energy consumption could be expected. The summary result of heat and mass balance simulation is presented in Table 4. The process simulation has included heat integration to ensure energy utilization within the process.

Table 4. Summary result of process simulation

No	Parameter	Value	Unit
1	**Production rate**	1,090	ton-DME/day
2	**Overall thermal efficiency**	45.5	% (LHV basis)
3	**Energy consumption**		
3a	As feed	46.7	MMBTU/ton-DME (LHV basis)
3b	As fuel	13.1	
3c	Total	59.8	
4	**Feedstocks and utilities**		
4a	EFB (MC 9%)	4.1	ton/ton-DME
4b	Water	7.6	ton/ton-DME
4c	Air	17.4	ton/ton-DME
5	**Side products**		
5a	Nitrogen	4.7	ton/ton-DME
5b	Argon	0.1	ton/ton-DME
6	**Waste**		
6a	Ash	0.1	ton/ton-DME
6b	Waste Water	0.3	ton/ton-DME

3.3 Economic Evaluation

Contributions of CAPEX of each unit process and operation are presented in Table 5. It is understandable that the investment cost of the gasification unit contributed up to 17%. Moreover, to produce the synthesis gas, this process required O_2 which in turn it costed the Air Separation Unit. A twin bed gasifier without requiring O_2 is being developed for production of synthesis gas.

To validate our assessment, comparisons to other study are presented in Table 6. Result of our feasibility study was with in the range of other studies. It was surprising that the study by Clausen (2011) gave an extremely low investment cost [4]. It was because Clausen (2011) applied the biggest scale compared to others hence the advantage from the plant economies of scale can be obtained. Furthermore, Clausen (2011) used semi-processed biomass as feedstock that can reduce the equipment needed inside the plant.

Levelized cost calculation of DME production was calculated with assumption in Table 2 and the result is shown in Table 7. Operating cost needed to produce DME was 7,143 IDR/kg-LPG-equivalent. With DME selling price similar with LPG (12,083 IDR/kg-LPG-equivalent or 145,000 IDR/12-kg-LPG-equivalent), an IRR of 10.08% would be obtained, this was above IRR criterion of 8%.

Table 5. Summary of CAPEX calculation

Parameter	**Cost (million USD_{2018})**	
Gasification Island: biomass mill, gasifier, and cyclone.	169.5	17.1%
ASU	116.5	11.8%
Syngas Upgrading: ESP, tar cracker, WGS reactor, CO_2 removal.	128.9	13.0%
Synloop DME	121.2	12.2%
Power Plant: Gas Turbine, HRSG, Steam Turbine, etc.	159.0	16.1%
Separation and Purification	35.1	3.5%
Miscellaneous Utility	25.1	2.5%
Total Installed Cost (TIC)	**755.3**	
Land development (4% TIC)	30.2	3.0%
Total Direct Cost (TDC)	**785.5**	
Indirect Cost (20% TDC)	157.1	15.9%
Fixed Capital Investment (FCI)	**942.6**	
Land	0.62	0.1%
Working Capital (5% FCI)	47.1	4.8%
Total CAPEX	**990.4**	**100%**
Specific CAPEX (per MW_{th} product)	**2.74**	
Specific CAPEX (per ton/day product)	**0.93**	

Table 6. Techno-economic result comparison of biomass to DME process from various study

Parameter	**[4]**	**[7]**	**[8]**	**[9]**	**This study**
Feedstock	Torrefied woodchips	Switchgrass	Switchgrass	Woody biomass	EFB
Input Rate (MW_{th} LHV)	2302	893	601	430	795
CAPEX Specific (mil USD_{2018}/MW_{th} product)	1.17	2.91	2.97	2.68	2.74

CAPEX and the exchange rate of dollar were the most sensitive variable. On the other hand, if N_2 and Ar as side products can be sold, the minimum economic selling price might be reduced significantly. Therefore, in the feasibility study of this project in the future, the possibility of side products selling should be investigated further.

Table 7. Levelized cost calculation of DME production from EFB

No	Cost Component	Cost (IDR/kg-LPG-eq)	
1	Capital charge	2,814	23.3%
2	EFB	2,026	16.8%
3	Fixed manufacturing cost	1,876	15.5%
4	Others feedstock, waste, catalyst	206	1.7%
5	Tax	221	1.8%
	OPEX	**7,143**	
6	Distribution and selling	540	4.5%
7	Internal return minimum*)	3,688	30.5%
	Minimum economic selling price	**11,371**	
8	Additional internal return**)	712	5.9%
	Selling price	**12,083**	**100%**

*) resulting minimum IRR (8%)
**) resulting the same selling price with LPG

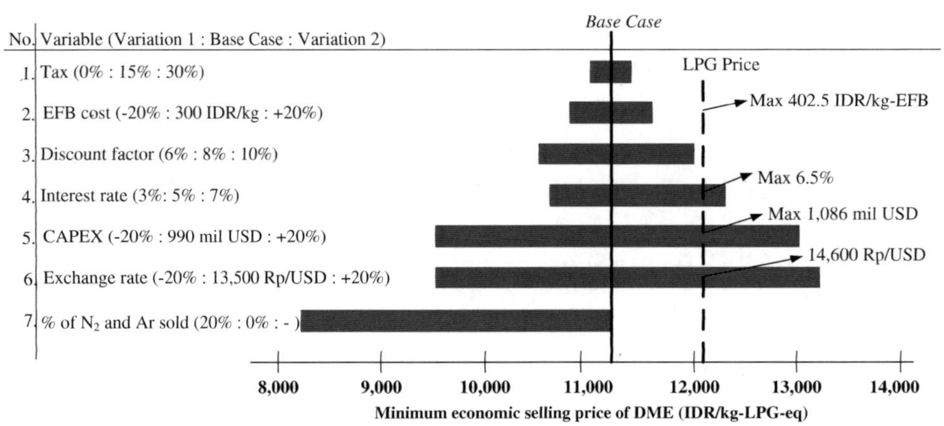

Figure 3. Sensitivity analysis result

3.4 Business Plan

The comprehensive business plan with several key points for this business is shown in Fig. 4. The first key point, EFB transportation shall be integrated with existing fresh fruit bunch (FFB) transportation to keep the EFB price on the allowable level. The integration can be performed by optimizing the FFB trucks, so that after finishing the FFB delivery to palm oil mill, the truck is then filled with EFB. The truck then returned to the plantation with twisting the route slightly to deliver EFB to the DME plant.

The second key point, the possibility to sell N_2 and Argon as side products, shall be explored during the feasibility study of this plant since it can increase the business competitiveness dramatically as shown in Fig. 3. Argon is an inert gas that is commonly used for titanium manufacturing, healthcare industries, lighting industries, and food industries. Similar with Argon, Nitrogen is also commonly used for food packaging, lighting industries, and also electricity industries. Thus, this DME plant may initiate an integrated business with these industries in order to improve each business performance.

The third key point, local content shall be maximized on this project to reduce CAPEX, decrease the linkage with the fluctuation of exchange rate, and maximize the positive domino effect for national economic growth. To maximize the local content, synergies between academics, practitioners, and government are needed to produce patents related to DME process technology from EFB and or patents related process equipment especially static equipment. Gasifier is a static equipment that is very likely to be designed and produced locally. If the gasifier can be produced locally, the CAPEX will decrease significantly since investment in the gasification island is the biggest portion of CAPEX.

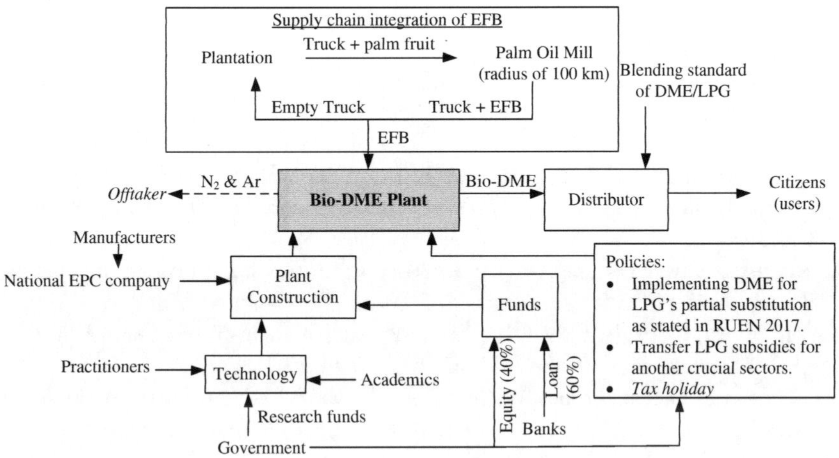

Figure 4. Proposed business plan of DME plant from EFB in Indonesia

The fourth key point, government support is needed by issuing policies that assure and regulate the DME utilization as LPG's partial substitution. If the policies to encourage DME application are fixed, the investment will get more trust from banks and any other investors. Government's support is also needed by providing the fund for the related research and development.

Implementing bio-DME as LPG's partial substitution may give positive effects. For instance, it can save the Indonesia foreign exchange and also increase the national energy security by reducing the dependency with LPG import. It will reduce GHG (Green House Gas) emission due to the use of fossil fuel and increase the renewable portion on Indonesia energy mix. Last but not least, it will solve the environmental problem caused by EFB pile in the surrounding of palm oil mills and also give the added value for EFB.

4. Conclusion

DME production from EFB was theoretically feasible with yield of 4.1 ton-EFB/ton-DME and total thermal efficiency of 45.5% (LHV basis). Commercialization of DME production from EFB will be more feasible if it is sold with the same price to LPG. DME production from EFB will give the positive economic and environmental effects. This result gave an intuition that DME production process from EFB was decent to be developed further starting from pilot scale to commercial scale. Support and synergy from government, practitioner, and academics are needed to realize this business.

Acknowledgments

This study was a part of Master's Program Thesis at Department of Chemical Engineering, ITB. The authors would like to thank to *the Indonesian Palm Oil Estate Fund* (BPDPKS) for funding our research on *Biomass to Methanol/DME*, in Fiscal Year of 2018.

References

[1] Tajalli A 2015 Panduan Penilaian Potensi Biomassa Sebagai Sumber Energi Alternatif di Indonesia (Jakarta: *Penabulu Alliance)* p16

[2] Susanto H, Suria T and Pranolo S H 2018 Economic Analysis of Biomass Gasification for Generating Electricity in Rural Area in Indonesia *IOP Conf. Ser.: Mater. Sci. Eng*, 334 012012

[3] Ardy A, Laniwati M, Rizqiana J and Susanto H 2018 Synthesis of Dimethyl Ether Using Dual Catalyst Fixed Bed for Methanol Synthesis and Its Dehydration *25th Regional Symposium on Chemical Engineering* Manila

[4] Claussen R A 2011 Design of Novel DME/Methanol Synthesis Plant Based on Gasification of Biomass *DCAMM Special Report No S123* Technical University of Denmark. 41, 55, 86, 97, 253

[5] Ginting, A S, Tambunan A H and Setiawan R P A 2014 Karakteristik Gas-Gas Hasil Pirolisis Tandan Kosong Kelapa Sawit *Jurnal Teknologi Industri Pertanian* **25**(2) Institut Pertanian Bogor 158-163.

[6] Syahza, A, Rosnita, Suwondo and Nasrul B 2013 Potential Oil Palm Industry Development in Riau *Int. Res. J. Business. Studies Vol.* ***6*** *No.2.* 142.

[7] Larson E D, Jin H and Celik F E 2009 Large-scale Gasification-based Coproduction of Fuels and Electricity from Switchgrass. *Biofuels, Biorprod. Bioref.****3.***174–194.

[8] Kreutz T G, Larson E D, Liu G and Williams R H 2008 Fischer-Tropsch Fuels from Coal and Biomass (New Jersey: *Princeton Environmental Institute*) Princeton University.

[9] National Renewable Energy Laboratory. Process Design and Economics for The Conversion of Lignocellulosic Biomass to Hydrocarbons via Indirect Liquefaction – Thermochemical Research Pathway to High-Octane Gasoline Blendstock Through Methanol/Dimethyl Ether Intermediates. 2015.

Hydrothermal treatment, pelletization and characterization of oil palm empty fruit bunches as solid fuel

Y Ahda [1,2], T Prakoso[1], C B Rasrendra[1], and H Susanto[1]
[1] Department of Chemical Engineering, Institut Teknologi Bandung, Bandung 40132, Indonesia
[2] Balai Besar Teknologi Konversi Energi, Agency for The Assesment and Application of Technology, Serpong 15314, Indonesia

E-mail: herri@che.itb.ac.id; yusuf.ahda@bppt.go.id

Abstract. Empty fruit bunch (EFB) is not yet utilized as a solid fuel for boiler due to their physical characteristics such as very bulky, high moisture and potassium contents. The hydrothermal treatment (HT) has been conducted in our research to overcome those problems. HT experiments were conducted in a 2 L digester, with 100 grams of EFB. The liquid to solid ratio was 5:1 mL/g. HT were conducted at temperature of 120, 150, 180, 200, and 220°C, and holding time of 60 minutes. The results indicated that the properties on solid products could be improved using HT, particularly its higher heating value (HHV) of 20.1 MJ/kg (obtained from HT at 180°C) compared to feedstock of 19.68 MJ/kg and commercial pellet of EFB of 13.03 MJ/kg. Pellets made of solid product from HT at 150°C and pelletized at 200 bar were considered to meet the characteristic of commercial pellets.

1. Introduction

Biomass from palm oil agro-industry is very prospective for new and renewable energy resource. In a crude palm oil mill, one ton of fresh fruit bunch produces approximately: 23%-mass empty fruit bunch (EFB), 6.5% palm kernel shell (PKS), and 13% fibre. PKS and fibre have been used internally in the CPO mill as fuel in boiler fuel. The use of EFB as a fuel for combustion and gasification is not maximum yet, due to some poor characteristics of EFB, such as high moisture content, high ash content, high alkali content in ash, low calorific value and low bulk density. Particularly, high ash and alkali contents in EFB may cause ash agglomeration, slagging and fouling in heat transfer tubes and other parts of boiler [1]. Therefore, EFB pre-treatment is needed to get better characteristic for solid fuel.

Hydrothermal Treatment (HT) is suitable for processing biomass with high moisture content without prior drying. Moreover, HT may reduce volatile matter in biomass and some alkali content such as K, Na, Mg, Ca, Mn and Fe [1-4]. HT process can be carried out for treating biomass waste order to reduce unutilized EFB waste and increased the added value of EFB. Many studies have been done to investigate HT process for various application.

The objective of this experimental study was improving the characteristic of EFB for its utilization as a solid fuel for gasification and combustion. The proposed improvement was the combination of HT and pelletization. During HT, potassium is expected to dissolve significantly into water, while pelletization is expected to increase the bulk density, and to make uniform particle size. Hopefully this HT and pelletization technique will make storage and transportation of EFB become more practical. In addition,

IOP Conf. Series: Materials Science and Engineering **543** (2019) 012061 doi:10.1088/1757-899X/543/1/012061

the hydrolysate (liquid product from HT) may content potassium which may be recovered as a source of fertilizer.

2. Experimental setup

2.1. Sample Preparation

EFB was obtained from a CPO mill in Lampung. Experiments consisted of preparation of sample, hydrothermal treatment process, making pellets and characterization of the pellet. EFB was firstly shredded into fibers having a length of about 4 cm. Then it was dried in the oven for five hours in an oven of 105°C. After drying the sample was kept the sample in a sealed plastic bag.

2.2. Hydrothermal Treatment

Hydrothermal treatment (HT) was conducted in a 2 L autoclave heated electrically. Temperatures of HT were set as an independent variable at 120, 150, 180, 200 and 220°C. The heating rate was about 4 °C/min, holding time at the desired temperature was 1 hour. The liquid and solid ratio were 5:1. One series of experiments were carried out with 500 mL water and 100 g EFB, and two series of experiments were carried out with 1000 mL water and 200 g EFB. After the HT process completed, the solid residue and the liquid separated using a filter. The solid residues were dried in an oven of 105°C for five hours, and this so-called biochar from HT was kept in sealed plastic bags for the proximate and ultimate analysis, and mineral content in ash. The liquid products of HT were kept in a sealed sample bottle for analysis of its potassium content.

2.3. Pelletization

Pellets were made manually in a mold with four holes. A pressure of 250 or 500 kg/cm^2 was applied using a hand hydraulic press. Each pellet had a diameter of 1 cm, the height of about 0.5 cm, and a weight of about 0.4 grams. This dimension of the pellet was similar to commercial EFB pellets obtained from Medan.

2.4. Pellet Characterization

Proximate analysis and ultimate analysis of solid residue were carried out in a certified laboratory, Research, and Development Center of a Mineral and Coal, the Ministry of Energy and Mineral Resources Indonesia (tekMIRA). HHV of products was also estimated using correlations based on the proximate and ultimate analysis. The correlations are presented in Table 1.

Table 1. HHV correlation formulas [1]

No	Name of author	Correlation (HHV, MJ/kg)
Based on proximate analysis*		
1	Jimenez et al, 1991	HHV = –10.81408 + 0.3133 (VM+FC)
2	Demirbas et al, 1997	HHV = 0.312*FC+0.1534*VM
3	Cordero et al, 2001	HHV = 0.3543*FC + 0.1708*VM
4	Sheng et al, 2005	HHV = –3.0368 + 0.2218VM + 0.2601*FC
Based on ultimate analysis*		
5	Dulong	HHV = 33.95 C + 144.2 (H – (O/8)) + 9.4 S
6	Boei, 1987	HHV = 0.3516 C + 1.16225 H - 0.1109 O + 0.0628 N + 0.10465 S
7	Demibars, 1997	HHV = 0.335 C + 1.423 H – 0.154 O – 0.145 N
8	Channiwala, 2002	HHV = 0.3491 C + 1.1783 H + 0.1005 S – 0.1034 O – 0.0151N – 0.0211 ash
9	Sheng, 2005	HHV = –1.3675 + 0.3137 C + 0.7009 H + 0.0318 O

*Biomass composition, VM, FC, ash, C, H, O, N, S are in the weight percent on a dry basis

IOP Conf. Series: Materials Science and Engineering **543** (2019) 012061 doi:10.1088/1757-899X/543/1/012061

Minerals content in ash were measured using AAS and analyzed in *Research and Development Center of Mineral and Coal*, the Ministry of Energy and Mineral Resources Indonesia (tekMIRA). These minerals content in ash were used to estimate the potency of fouling and slagging of ash (formula for these indices are presented in Table 2).

Table 2. Fouling and slagging properties [6]

Index	Correlation, in weight %	Value and tendencies	
Base/acid ratio (B/A)	$\frac{(Fe_2O_3+CaO+MgO+Na_2O+K_2O)}{(SiO_2+Al_2O_3+TiO_2)}$	<0.206 0.206-0.4 >0.4	Low Medium High
Fouling index (FI)	$\left(\frac{B}{A}\right) \times (Na_2O+K_2O)$	<0.6 0.6-40 >40	Low Medium High
Slagging Index (SI)	$SI=\left(\frac{B}{A}\right) \times Sd$ Sd = % of S from solid fuel	<0.6 0.6-2.0 2.0-2.6 >2.6	Low Medium High Very High
Alkali index (AI)	$AI=\left(\frac{1000}{HHV\left[\frac{MJ}{kg\text{-}dry}\right]}\right) \times ash \times (Na_2O+K_2O)$	0.17-0.34 >0.34	Probable High to Occurs

3. Result and Discussion

The HT temperature clearly influenced yields of bio-char (see Figure 1). Three series of experiments with different amount of feed gave similar trends in the decrease in yields of bio-char with increasing temperature. The decrease in yields of bio-char might be contributed by two factors, i.e. the dissolution of minerals in ash and thermal degradation of biomass. Thermal degradation of biomass could also be observed from significant changes in volatile matters and fixed carbon at HT temperatures of 200 and 220ºC (Table 3). This range of temperature had been reported in many literatures as the initial temperature for thermal degradation of hemicelluloses. At the higher temperature of HT, cellulose and lignin in EFB might start to decompose to some extent. As a result, bio-char from HT at 200 and 220ºC were more brittle and dark-black in color.

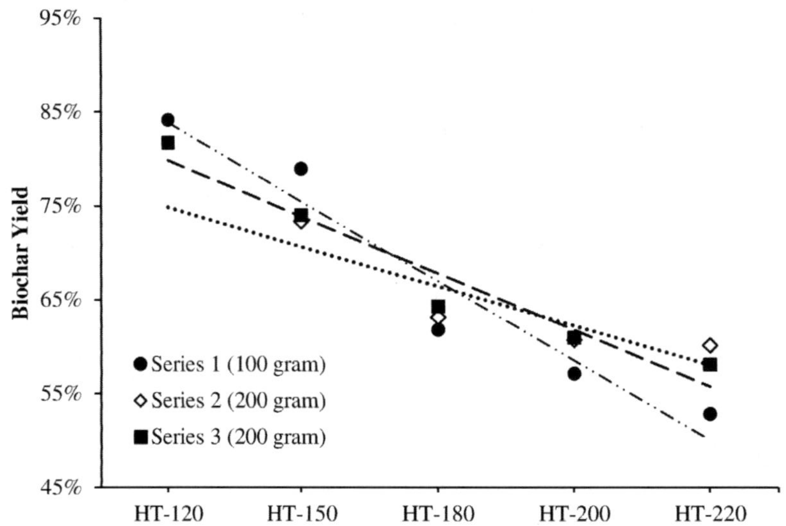

Figure 1. Biochar yield during hydrothermal treatment

In line with the proximate analysis, the ultimate analysis of bio-char also changed with the temperature of HT, although less significant (Table 3). Presumably, the elemental compositions of volatile matters were almost the same as those of raw EFB. An exception was observed again in bio-char obtained from HT at 220°C, i.e. a drastic decrease in carbon and oxygen contents.

Table 3. Proximate and ultimate analysis of HT solid products

Parameters	Previous Work		Present work					
				Hydrothermal Treatment				
	EFB [1]	EFB [10]	EFB	120°C HT-120	150°C HT-150	180°C HT-180	200°C HT-200	220°C HT-220
Proximate Analysis (%-wt dry basis)								
Volatile Matter	78.70	78.34	77.22	80.24	79.42	74.22	71.85	67.03
Fixed Carbon	15.30	18.46	18.31	17.08	19.63	20.97	24.13	29.23
Ash	5.90	3.20	4.47	2.68	2.85	4.82	4.02	3.74
Ultimate Analysis (%-wt dry basis)								
Carbon	41.81	46.77	45.87	44.78	48.14	49.06	51.17	57.50
Hydrogen	5.73	6.84	6.40	6.39	6.47	6.44	6.16	6.21
Oxygen	45.71	45.38	46.94	48.27	44.64	43.62	41.84	35.52
Nitrogen	0.84	0.71	0.78	0.56	0.75	0.88	0.83	0.97
HHV, MJ/kg	18.40	19.54						
HHV, MJ/kg *)			19.68	19.76	19.84	20.10	20.82	21.81
Potassium content in hydrolysate, mg/L				1,534	1,794	1,376	1,668	1,950

*) calculated using Cordero formula (see Table 1)

Correlation proposed by Cordero, 2001 [5] was considered as the most optimum one since this formula used the low-cost proximate analysis and gave least errors [10]. Since the HHV of bio-chars (Table 3) were calculated based on both proximate and ultimate analysis, their values had the same tendency with the changes of this analysis as the above discussion.

Table 4. Ash content and its mineral composition of EFB and HT products

No	Ash	Previous Work		Present Work					
		EFB [2]	EFB [11]	EFB	HT-120	HT-150	HT-180	HT-200	HT-220
Mineral composition in ash (%-wt of ash)									
1	SiO_2	19.00	39.46	50.60	49.30	54.20	59.30	55.10	62.65
2	$Al2O_3$	2.49	0.94	2.07	2.28	2.07	2.14	2.26	1.95
3	Fe_2O_3	5.48	1.74	2.67	8.20	5.08	2.00	1.70	2.82
4	K_2O	48.94	21.75	17.09	16.36	15.24	16.15	16.30	9.24
5	Na_2O	3.89	4.24	1.08	1.42	1.24	0.90	1.22	0.70
6	CaO	7.92	11.08	6.40	7.23	7.76	7.09	7.75	6.55
7	MgO	4.35	8.68	7.68	5.84	5.23	4.45	5.52	3.91
8	TiO_2	0.74	0.11	0.17	0.23	0.08	0.13	0.56	0.33
9	MnO	4.35	0.07	0.07	0.09	0.06	0.05	0.06	0.05
10	P_2O_5	2.96	2.21	4.57	3.54	3.77	3.11	3.41	4.31
Ash melting characteristic indices (see formula in Table 2)									
1	Base/acid ratio (B/A)	3.17 high	1.17 high	0.66 high	0.75 high	0.61 high	0.49 high	0.56 high	0.36 medium
2	Fouling index (FI)	1.68 medium	0.30 low	0.12 low	0.14 low	0.10 low	0.09 low	0.10 low	0.04 low
3	Slagging Index (SI)	0.32 low	0.34 low	0.11 low	0.11 low	0.09 low	0.08 low	0.09 low	0.06 low
4	Alkali index (AI)	1.52 high	0.39 high	0.43 high	0.25 probable	0.24 probable	0.41 high	0.35 high	0.18 probable

The most interesting result in HT process was the significant change in the ash content in solid residue from HT (Table 4). Potassium (presented as K_2O) content in ash decreased, especially in HT with temperature of 220 ^oC. On the other hand, the concentration of SiO_2 in ash increased, This was probably due to the effects of large dissolution of potassium and sodium from EFB into water, while there was only slight dissolution of SiO_2.

Liquid products of HT contained a considerable amount of potassium (lower part of Table 3). Potassium is one of an important nutrient in oil palm plantation, so recovery of potassium from palm biomass will be very valuable. A research on recovery potassium from this dilute solution is being done.

Following the change in the minerals content in ash, ash melting properties change as well (see Table 4). It was surprising, that the tendencies of fouling and slagging of raw EFB used in these experiments were already low. These tendencies were similar to our previous work using different EFB from Riau [10], but they were different from data reported by [1] probably because of different palm species. Again, change in ash properties were observed significantly in HT with temperatures of 200 and 220 °C.

As expected, HT could change the physical properties of EFB, from originally tough in raw EFB becoming brittle after HT. The higher temperature, HT will produce more brittle and darker color bio-char. Comparison of bio-char pellets from our experiment and some commercial pellets are presented in Table 5. Commercial pellets were probably made of char from nearly complete pyrolysis, so it contained a low volatile matter. While our pellets still contained a large amount of volatile matter. In this respect, our pellets would probably have a lower ignition temperature.

Table 5. Combustion characteristic of pellets

Pellet Parameters	Wood Pellets		Charcoal Briquette		Present Work					
	SNI 8021-2014	DIN 51731	SNI 01-6235-2000	Commercial EFB pellet	EFB	HT-120	HT-150	HT-180	HT-200	HT-220

Diameter, mm	≤12	4-10	<8	9	10	10	10	10	10	10
MC, %-wt	<14.00	12.00	<15.00	8.43	7.76	8.97	6.12	6.37	6.17	5.86
VM, %-wt	≤12.00	-	<8.00	51.99	71.23	73.04	74.56	69.49	67.42	63.1
FC, %-wt	≥14.00	-	-	14.49	16.89	15.55	16.64	19.63	22.64	27.52
Ash, %-wt	≤1.50	<1.5	-	25.09	4.12	2.44	2.68	4.51	3.77	3.52
HHV, MJ/kg	16.74	17.48-19.49	20.92	13.03	19.68	19.76*)	19.84*)	20.10*)	20.82*)	21.81*)

*) calculated using Cordero formula (see Table 1)

The mechanical strength of EFB and HT bio-char pellets were tested using an apparatus for catalyst test having a maximum load of 15 kg/cm^2. Raw EFB and bio-char pellets made using a pressure of 250 bar were always stronger than those using 200 bar (Table 6). As raw EFB was tough and elastic, pellets of this material were very weak. This pellet broke down at a compressive pressure of 7.5 kg/cm^2. The higher temperature of HT, the stronger pellets made from its bio-char. Uniformity particle size of feedstock affected the compactness and hence the compression strength of pellet [11, 12]. English bio-pellet has a standard compressive strength of 12.7 kg/cm^2 [11]. Based on the compressive strength, bio-char from HT at a temperature of 150°C was considered suitable for pellets production: (a) feed made in HT at a relatively low temperature (vapour pressure of about 5 bar-abs), and (b) pellet made using a compression of 200 bar.

4. Conclusions

HT and pelletization could improve the characteristic of EFB as a solid fuel by increasing fixed carbon, heating value, and bulk density. The decrease in minerals content in ash content could be expected to reduce the potencies of fouling and slagging during the use of EFB pellets in gasification or combustion process. EFB pellets made in this present work were better than commercial EFB even without using adhesive subtances. Our Hydrotreated EFB and EFB pellets also met standards of other commercial biomass pellets.

Acknowledgements

Experimental works presented in this paper were a part of Master Thesis at the Department of Chemical Engineering, Faculty of Industrial Technology, Institut Teknologi Bandung. This research was funded by the *Hibah Tim Pascasarjana* (No. 0323h/II.C06/PL/2018).

References

[1] Novianti S and Biddinika M K 2014 *Proc. Env. Sci.* **20** 46-54
[2] Paltseva J and Searlw S 2016 Potential for Advanced Biofuel Production from Palm Residues in Indonesia *White Paper the International Council on Clean Transportation*
[3] Y. Zhu Y and Niu Y 2014 *Frontiers in Energy Research* **2** 1-10
[4] Kongpanya J and Hussaro K 2014 *American J. Environ. Sci.* **10** 324-35
[5] Sheng C and Azevedo JLT 2005 *Biomass Bioenergy* **28** 499-07
[6] Liu Z and Quek A 2013 *Fuel* **103** 943-49
[7] Zhao Z and Shen Y 2014 *Appl. Energy* **131** 345-67
[8] Perry RH, *Perry's Chemical Engineering Handbook 6th* (McGraw-Hill New York 1984)
[9] Tekin K and Karagoz S 2014 *Renew Sustain Energy Rev.* **40** 673-87
[10] D. Hantoko D, Prabowo B, Yan M and Susanto H 2018 Preparation of Empty Fruit Bunch as a Feedstock for Gasification Process by Employing Hydrothermal Treatment. *Applied Energy Symposium and Forum : Low Carbon Cities and Urban Energy System*, Shanghai, China
[11] Wibowo S 2017 *Jurnal Penelitian Hasil Hutan* **35** 73-82
[12] Hendra D 2012 *Jurnal Penelitian Hasil Hutan* **30** 144-54

IOP Conf. Series: Materials Science and Engineering **543** (2019) 012062 doi:10.1088/1757-899X/543/1/012062

A Superstructure Based Enviro-Economic Optimization for Production Strategy of Oil Palm Derivatives

S A Taqwa and W W Purwanto

Department of Chemical Engineering, Faculty of Engineering, Universitas Indonesia, Depok 16424, Indonesia

E-mail: widodo@che.ui.ac.id

Abstract. Indonesia is one of the largest palm oil producers in the world. Palm oil has a wide range of derivative products that have higher values than in the upstream oil palm products. Indonesia still exports mostly crude palm oil rather than its derivatives. The objective of this research is to obtain the best strategy of developing downstream palm oil industry by considering the total price and greenhouse gas emission. Economic objective function is the total selling price of all products and environmental objective measured by the total greenhouse gas emissions. Multi-objective optimization is based on State-Task Network superstructure with fixed variable of product selling price, emission factor, conversion factor, national demand and global demand of the products. Multi-objective optimization is carried out using GAMS with Cplex 12.6.3 solver. The total selling price obtained amounted to 51.67 billion USD and total GHG emissions generated were 88.05 million tons CO_2e. The selected production pathway is the production of 54 palm oil derivatives products to meet domestic needs and 21 of them can be exported (1 FFB derivative product, 4 CPO derivative products, 1 POME derivative product, 4 EFB derivative products, 2 PKS derivative products, and 9 Palm Kernel derivative products).

1. Introduction

Indonesia produces more than a half of world Crude Palm Oil (CPO) demand, correspond to 31.8 million ton of CPO [1]. 25.1 million tons of it is exported, consisting of 7 million ton of CPO; 15.57 million ton of refined CPO; and 2.43 million tons of lauric acid, biodiesel and oleo chemicals. Almost 80% of total palm oil produced are exported in the form of CPO and refined CPO, which are low value-added products. The prices of crude products are consistently the most volatile rather than finished goods. To sustain the long-term development of the palm oil industry, the development strategy should move from production of low to high value added of products by processing CPO and unused palm oil waste to the derivative products [2]. Thus, Indonesia needs to accelerate the development of downstream palm oil industry by considering simultaneously economic and environmental aspects. In a palm oil processing, only 25-30% of the product is yielded, the remaining 70-75% is residue from waste of the processing such as: empty fruit bunch (EFB), palm oil mill effluent (POME), palm mesocarp fiber (PMF), palm kernel shell (PKS), and palm fatty acid distillates (PFAD) [3]. Currently, palm oil waste is only used as fertilizer or end up in landfills. To maximize palm oil value chain, the wastes can be reused to yield higher selling value of commodities [4]. The development of downstream palm oil industry is also affected by external factors, such as EU anti-dumping tariff for Indonesia's biodiesel, and Renewable Energy Directive (RED) to ban the use of

IOP Conf. Series: Materials Science and Engineering **543** (2019) 012062 doi:10.1088/1757-899X/543/1/012062

vegetable oils and CPO from Indonesia as a feedstock of biofuels by 2020 because of deforestation issues [5]. It can disrupt exports of 3.2 million tons per year of palm oil product to the EU.

Several studies related to this research has been conducted. Egieya et al. has developed biogas supply-chain optimization with Mixed-Integer Linear Programming (MILP) [6]. Tian and You and Ng et al. has proposed a hierarchical approach optimization [7, 8]. Andiappan et al. has proposed multi-objective optimization of integrated biorefinery with various biomass conversion system [9]. Abdulrazik et al. and Murillo-Alvarado et al. has done economic and environmental objective optimization using superstructure to obtain the most optimal production pathway [10, 11]. Abdulrazik et al. proposed superstructure optimization for only the derivative products of oil palm empty fruit bunches in Malaysia, but the environmental aspects are quantified into emission costs so optimization is done only with single objective function of maximizing profits [10]. Murillo-Alvarado et al. carried out a multi-objective optimization by maximizing net profit and minimizing environmental impact to find the best production pathway of bio-refinery in Mexico with a choice of various biomass, process and products [11].

With regard the need of innovative strategy for developing Indonesia's downstream palm oil industry, the aim of this study is to develop the strategies for production of downstream oil palm industry in supporting the decision maker's objectives that is a trade-off between maximizing selling prices of products and minimizing greenhouse gas emissions.

2. Methods

The research method consists of 2 main steps, the first one is the formation of superstructure, and the second one is the multi-objective optimization calculation. The main assumptions for this research are for number process and raw material variation, a raw material can only produce one type of product with one specific process; technical consideration is limited to conversion factor; economical consideration is selling price of products; environmental consideration is greenhouse gas (GHG) emission of CO_2, N_2O, CH_4, HFCs, PFCs and SF_6 taken from Life Cycle Assessment (LCA) method; the current installed capacity of palm oil products in Indonesia will not be considered on analysis but only as a bench mark; the amount of fresh fruit bunch or fresh oil palm fruit (FFB) is estimated from CPO production.

2.1. Formation of Superstructure

Nuh and Nasir has provided ways to specify and connect all potential raw materials and products across several series of tasks, i.e., operating units and or process units such as converting raw materials into products [12]. In this study, the superstructure is built in the State-Task-Network superstructure representation approach. Quaglia et al. has presented the way to design a superstructure representation by alternative collection method [13]. Pham and El-Halwagi (2012) has developed the merging process with forward and backward synthesis [14]. The superstructure made starting from FFB then generated CPO, POME, EFB, palm kernel shell and palm kernel. After that, superstructure developed for each of these raw feedstocks.

There are no variation of raw materials and production processes, in the sense that one type of product is produced from one type of raw material with one type of process. The superstructure obtained is shown on Figure 1. The oval shape describes the process for producing a product from a previous biomass. The colour difference denotes the difference in raw materials (FFB) (*g*), raw feedstock (*i*), intermediate 1 (*k*), intermediate 2 (*m*) and end products (*o*). Green indicates the main raw material, light yellow indicates raw feedstock, orange indicates intermediate 1, blue denotes intermediate 2, and red signifies the final product. (*h*) denotes the crude feedstock production process, (*j*) denotes the intermediate production process 1, (*l*) denotes the intermediate production process 2 and (*n*) signifies the production process of the final product. Each product is given a vertical notation of numbers to distinguish one product from another on the same column.

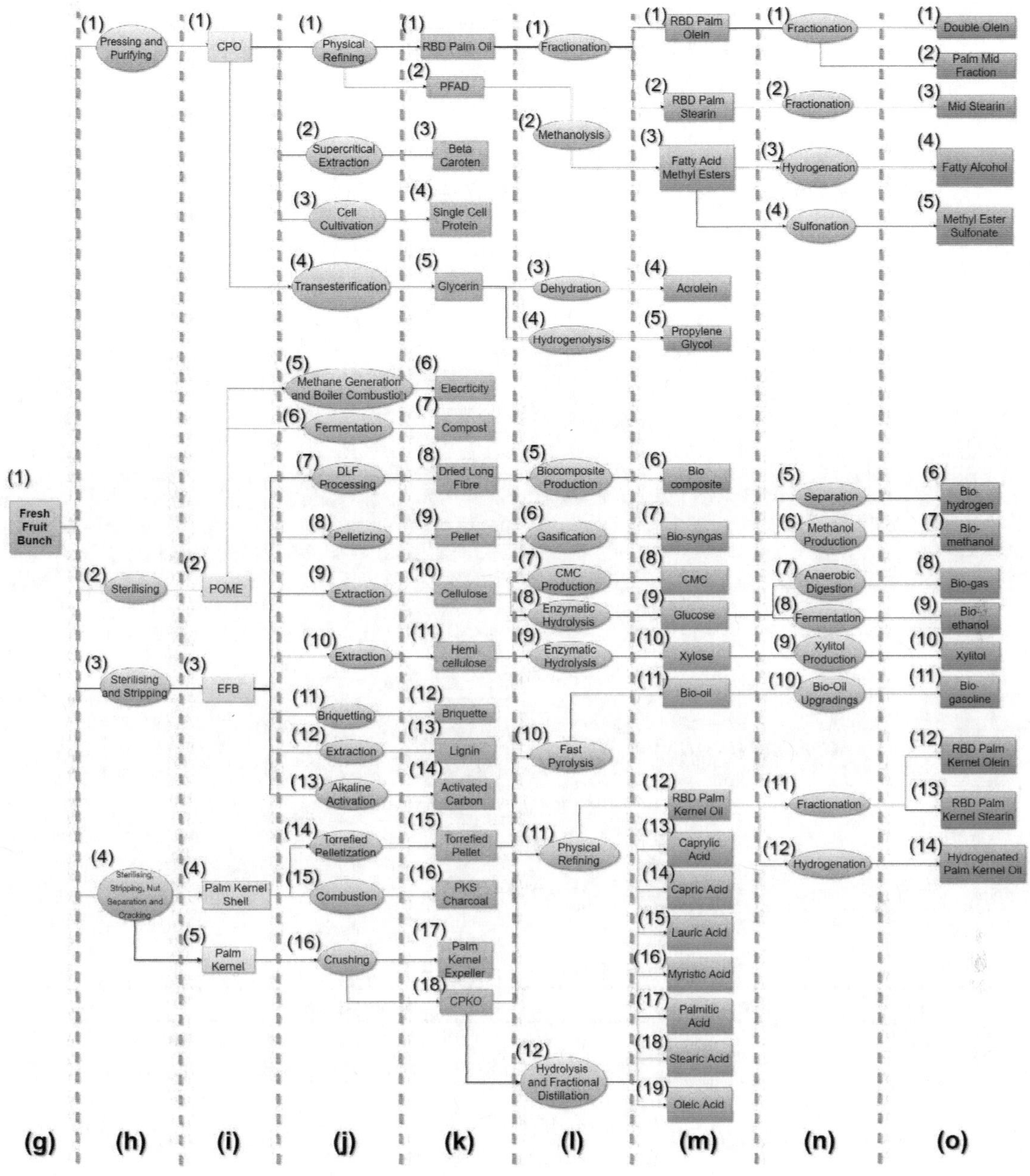

Figure 1. Generated Oil Palm Derivatives Superstructure

2.2. Mathematical Model

The objective functions of this research are to maximize the total selling price of the product and minimize total GHG emissions. Sales of the products times the products price will result to total selling price. A sale of the products is number of produced products reduced by number of products to be processed further. The number of products produced equals to amount of raw material times conversion factor. Total emission as the other objective function equals to amount of produced products times emission factor. Amount of FFB to be processed must be less or equal to amount of FFB available in Indonesia. The amount of FFB available in Indonesia is 155.3 million tons. Amount of sold products must be in between national demand and the global demand of the products. Here are the mathematical models. Table 1 describes the description of the terms.

$$Maximizing\ Total\ Selling\ Price\ =\ Max\ (TOTALPRICE) \tag{1}$$

$$Minimizing\ Total\ Emission\ =\ Min\ (TOTALEMISSION) \tag{2}$$

$$GD \geq \sum_{i=1}^{5} S_i + \sum_{k=1}^{18} S_k + \sum_{m=1}^{19} S_m + \sum_{o=1}^{14} S_o \geq ND \tag{3}$$

$$TOTALPRICE = \sum_{i=1}^{5} S_i . Pr_i + \sum_{k=1}^{18} S_k . Pr_k + \sum_{m=1}^{19} S_m . Pr_m + \sum_{o=1}^{14} S_o . Pr_o \tag{4}$$

$$S_i = P_i - \sum_{k=1}^{18} R_{i,k} \tag{5}$$

$$S_k = P_k - \sum_{m=1}^{19} R_{k,m} \tag{6}$$

$$S_m = P_m - \sum_{o=1}^{14} R_{m,o} \tag{7}$$

$$S_o = P_o \tag{8}$$

$$\sum_{i=1}^{5} R_{g,i} \leq O_g \tag{9}$$

$$P_i = R_{g,i} . CV_{g,i} \tag{10}$$

$$P_k = R_{i,k} . CV_{i,k} \tag{11}$$

$$P_m = R_{k,m} . CV_{k,m} \tag{12}$$

$$P_o = R_{m,o} . CV_{m,o} \tag{13}$$

$$TOTALEMISSION = \begin{bmatrix} \left(\sum_{i=1}^{5} PE_i\right) + \left(\sum_{k=1}^{18} PE_k\right) + \\ \left(\sum_{m=1}^{19} PE_m\right) + \left(\sum_{o=1}^{14} PE_o\right) \end{bmatrix} \tag{14}$$

$$PE_i = P_i . EF_{g,i} \tag{15}$$

$$PE_k = P_k . EF_{i,k} \tag{16}$$

$$PE_m = P_m . EF_{k,m} \tag{17}$$

$$PE_o = P_o . EF_{m,o} \tag{18}$$

Table 1. Description of the Terms

Term	Category	Description
TOTALPRICE	Variable	The total selling price of all products per year in USD
TOTALEMISSION	Variable	The total amount of greenhouse gas emissions produced in tons of CO_2e
O_g	Parameter	Number of Fresh Fruit Bunch in Indonesia that can be further processed in tons per year
S_i	Variable	Amount of the (*i*) products sold directly in tons per year
S_k	Variable	Amount of the (*k*) products sold directly in tons per year
S_m	Variable	Amount of the (*m*) products sold directly in tons per year
S_o	Variable	Amount of the (*o*) products sold directly in tons per year
P_i	Variable	Amount of (*i*) products generated in tons per year
P_k	Variable	Amount of (*k*) products generated in tons per year
P_m	Variable	Amount of (*m*) products generated in tons per year
P_o	Variable	Amount of (*o*) products generated in tons per year
PE_i	Variable	The amount of emissions generated from the production of (*i*) at tons of CO_2 is equivalent / year
PE_k	Variable	The amount of emissions generated from the production of (*k*) at tons of CO_2 is equivalent / year
PE_m	Variable	The amount of emissions generated from the production of

IOP Conf. Series: Materials Science and Engineering **543** (2019) 012062 doi:10.1088/1757-899X/543/1/012062

		(m) at tons of CO_2 is equivalent / year
PE_o	Variable	The amount of emissions generated from the production of (o) at tons of CO_2 is equivalent / year
$R_{g,i}$	Variable	The amount of product (g) to be processed further for production of (i) in tons per year
$R_{i,k}$	Variable	The amount of product (i) to be processed further for production of (k) in tons per year
$R_{k,m}$	Variable	The amount of product (k) to be processed further for production of (m) in tons per year
$R_{m,o}$	Variable	The amount of product (m) to be processed further for production of (o) in tons per year
ND	Parameter	Domestic demand of products (i), (k), (m), and (o) in tons
GD	Parameter	Global demand of products (i), (k), (m), and (o) in tons
$CV_{g,i}$	Parameter	The conversion factor for generating (i) from (g)
$CV_{i,k}$	Parameter	The conversion factor for generating (k) from (i)
$CV_{k,m}$	Parameter	The conversion factor for generating (m) from (k)
$CV_{m,o}$	Parameter	The conversion factor for generating (o) from (m)
$EF_{g,i}$	Parameter	The CO_2 emission factor for production (i) of (g) in ton CO_2e
$EF_{i,k}$	Parameter	The CO_2 emission factor for production (k) of (i) in ton CO_2e
$EF_{k,m}$	Parameter	The CO_2 emission factor for production (m) of (k) in ton CO_2e
$EF_{m,o}$	Parameter	The CO_2 emission factor for production (o) of (m) in ton CO_2e
Pr_i	Parameter	The selling price of the product (i) in USD per ton
Pr_k	Parameter	The selling price of the product (k) in USD per ton
Pr_m	Parameter	The selling price of the product (m) in USD per ton
Pr_o	Parameter	The selling price of the product (o) in USD per ton

Multi objective optimization is done by ε-constraint method. The objective function that will be maintained is maximizing the total selling price, so the objective function of minimizing total emissions will be modified into constraints using the following linear equations:

$$f_{2\,\text{limit}} = (f_1 - f_{1,\max f1})\frac{w_1(f_{2,\max f1} - f_{2,\min f2})}{w_2(f_{1,\min f2} - f_{1,\max f1})} + f_{2,\min f2} \qquad (19)$$

$f_{1,\max f1}$ shows f_1 that has been obtained with maximizing f_1 when only maximizing the total selling price. $f_{1,\min f2}$ shows f_1 that has been obtained with minimizing f_2 when only minimizing total emissions.

The mathematical models then are performed by GAMS with Cplex 12.6.3 solver for MILP problems. The analysis is started on analysing the production pathway when maximum total price is obtained, and then analyse the production pathway when minimum total emission is obtained. The generated Pareto curve and the production pathway of its trade-off point will be analysed. Sensitivity analysis based on the recent global issue also will be conducted.

3. Results and Discussions

3.1. Maximizing Total Price

In case of maximizing the selling price as an objective, the total selling price obtained is 55.6 billion USD and total emission of 91.1 million tons of CO_2e. The number of FFBs available in Indonesia is assumed to be 155.3 million tons, resulting from Indonesia's CPO production figures in 2017 of 31.07 million tons for the yield of CPO from FFB which is 0.2. The amount of FFB processed is as much as FFB available in Indonesia. It shows that it takes all available FFBs to get the maximum selling price possible. Figure 2 describes the potential production pathways of palm oil derivatives for case of maximizing of the total selling price. The products are differentiated in categories products that have already been produced in Indonesia from palm oil as raw materials which are marked in yellow, and which have not been produced in Indonesia in red. There are two types of production pathways, namely the selected path based on the optimization with a straight line, and which are not selected with the dashed line. The products exported are marked with blue box.

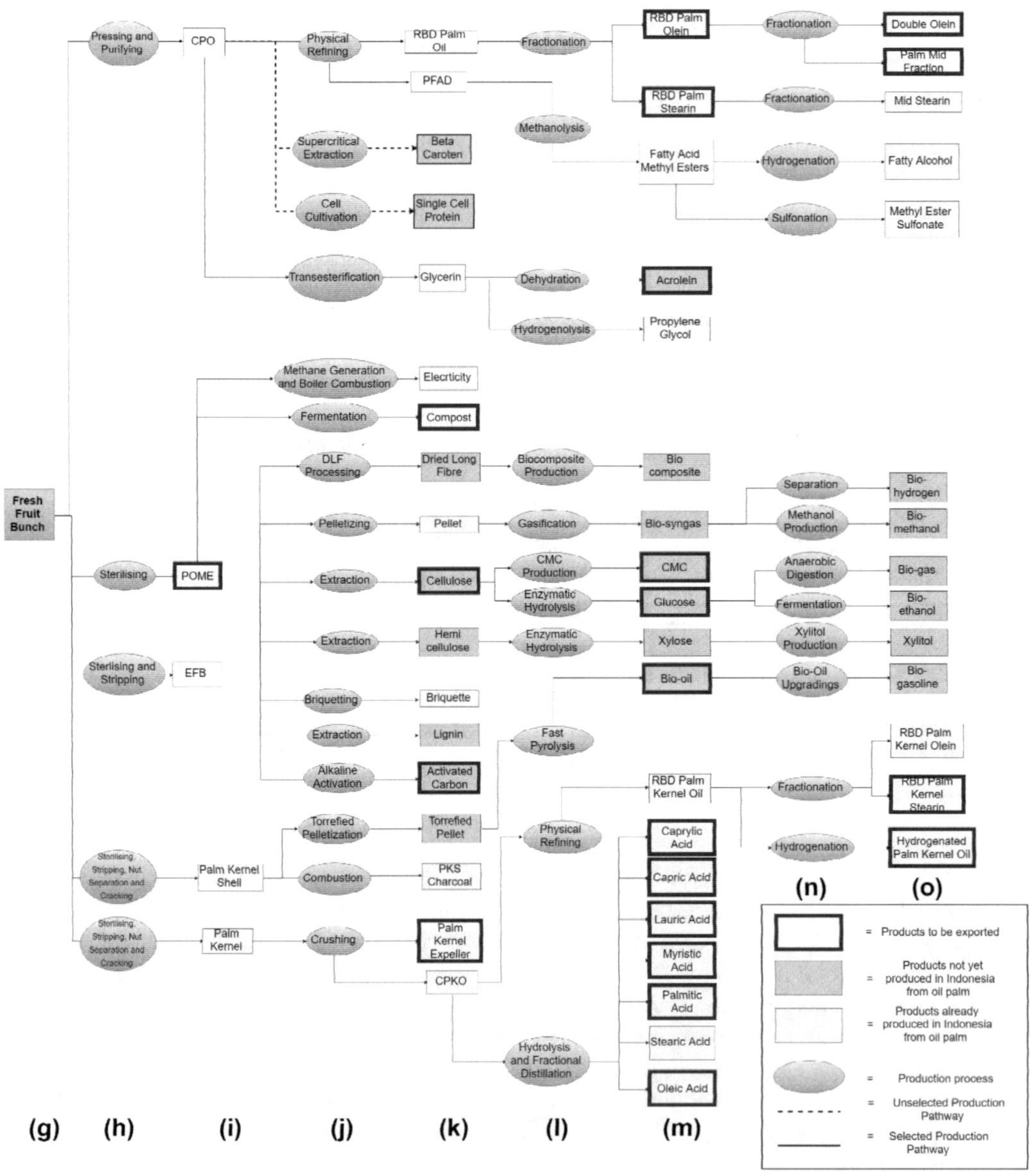

Figure 2. Production Pathway of Oil Palm Derivatives for Maximizing Total Price

All of the derivative products are produced, except beta carotene and single cell protein. So, there are 54 derivative products that are produced to fulfil domestic demands. There are 21 products exported. Most of the products that are exported are the third (*m*) and fourth (*o*) derivative products of FFB. 1 FFB derivative product (POME), 1 POME derivative product (compost), 2 EFB derivative products (cellulose and activated carbon), 1 palm kernel derivative product (Palm Kernel Expeller or PKE), 2 Refined Bleached Deodorized Palm Oil (RBDPO) derivative products (Refined Bleached Deodorized Palm Oil Olein/RBDPOL and Refined Bleached Deodorized Palm Oil Stearin/RBDPS), 1 glycerine derivative product (acrolein), 2 cellulose derivative products (Carboxymethyl cellulose/CMC and glucose), 1 torrefied pellet derivative product (bio-oil), 6 CPKO derivative products (all of the fatty acids except stearic acid), 2 RBDPOL derivative products (double olein and Palm Mid Fraction/PMF) and 2 Refined Bleached Deodorized Palm Kernel Oil (RBDPKO) derivative products (Refined Bleached Deodorized Palm Kernel Stearin/RBDPKS and Hydrogenated Palm Kernel Oil/HPKO). This is because third and fourth derivative products provide maximum value increase compared to first or second derivative products as they go through more processes.

In *i* products production, only POME is exported. In production of *k* products, beta carotene did not get any supply of raw materials, due to the very small conversion factor (0.0001) and therefore it is not optimal to produce. Same goes for Single Cell Protein (SCP), because other product demand is much higher than SCP and the selling price of SCP is not too high (650 USD per ton). PKE is produced at 4.4 million tons and should all be sold because it does not have any derivative products. Activated carbon and cellulose are produced for directly sold and meet global demand. It indicates that these products can provide the maximum value increase compared to production of other *k* products. The other products give more maximum selling price if it becomes raw material for production of next derivative products. In production of m products such as, RBDPOL, RBDPS, acrolein, CMC, bio-oil and almost all fatty acids are exported. RBDPOL and RBDPS are mostly sold directly, although they have derivative products that can increase the selling price. RBDPOL and RBDPS are produced 18.4 million tons and 4.6 million tons, respectively. Both have high global demand (53.4 million for RBDPOL, and 13.4 million for RBDPS). Double olein, PMF and mid stearin as their derivatives have higher selling prices, but global demand for these three products does not exceed 130 thousand tons each. CMC and acrolein are exported, because of having a much higher selling price than its raw material. Glucose is exported because its derivative products have a lower selling price than glucose. Only stearic acid is not exported because stearic acid has the lowest conversion factor compared to other fatty acids in the same process. Production of *o* products is to meet the national demand. Double olein, PMF and HPKO are exported to provide an optimal selling price.

3.2. Minimizing Total Emission

For the least total emission, the total selling price reached is 30.35 billion USD and total emission of 87.3 million tons CO_2e. The number of FFBs processed in the system is 155.2 million tons. In this case of the production meet only for domestic demand. It is seen that the number of FFBs processed is about 99% of the total FFB available in Indonesia. Figure 3 describes the potential production pathways of palm oil derivatives for case of minimizing total emission.

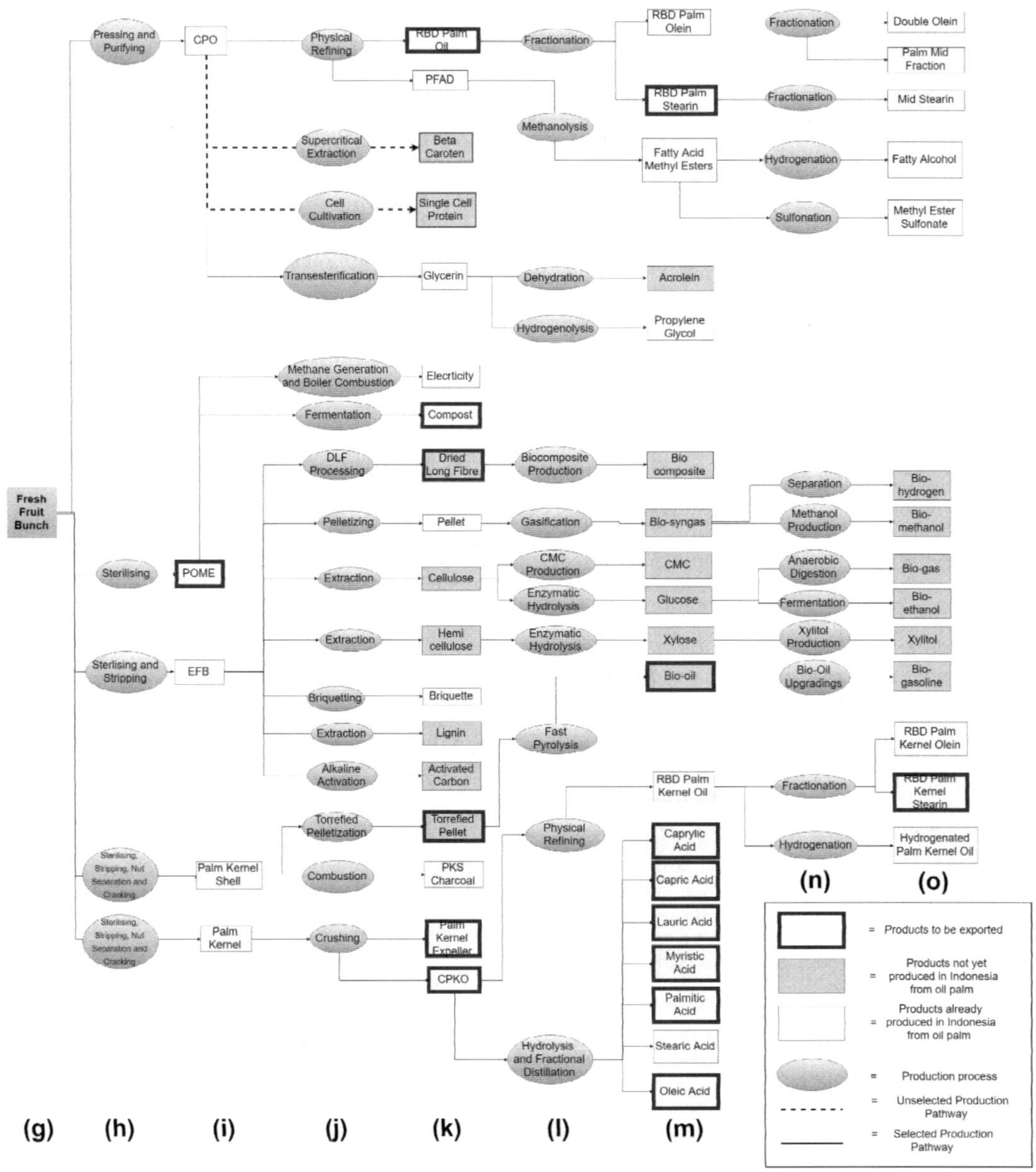

Figure 3. Production Pathway of Oil Palm Derivatives for Minimizing Total Emission

All of the derivative products except beta carotene and single cell protein (54 derivative products) are produced to fulfil national demands. The number of products that are exported is 16 products. It consists of 7 products of first (*i*) and second (*k*) palm oil derivatives, and 8 products of the third (*m*) and fourth (*o*) derivatives. 1 FFB derivative product (POME), 1 CPO derivative product (RBDPO), 1 POME derivative product (compost), 1 EFB derivative product (Dried Long Fibre/DLF), 1 PKS derivative product (torrefied pellet), 2 palm kernel derivative products (PKE and CPKO), 1 RBDPO derivative product (RBDPS), 1 torrefied pellet derivative product (bio-oil), 6 CPKO derivative products (all of the fatty acids except stearic acid), and 1 RBDPKO derivative product (RBDPKS).

The production of *i* products, POME is converted to the compost and electricity that required 913.11 million tons of POME for total POME production of 93.12 million tons, so POME should be sold to meet global needs. The exported *k* products are RBDPO, compost, torrefied pellet, CPKO and PKE. RBDPO is exported due to its high production. Compost is exported to utilize the available

POME. DLF is exported as it is produced in large quantities because of its lowest emission factor compared to the other EFB's derivatives with 0.0041. Same goes for torrefied pellet. CPKO and PKE are exported to minimize the emission of further processing it. Exported m products are RBDPS, bio-oil and all fatty acids except stearic acid. Bio-oil produces less emission than further processed for bio-gasoline production which has a production emission factor of 13 tons CO_2e/ton. On *o* product production, all are produced to meet domestic needs, except for RBDPKS that produced in the same process as RBDPKOL.

3.3. Pareto Curve Analysis

Figure 4 shows the Pareto-optimal curves generated from multi-objective optimization with weighted variations between total selling prices and total emissions.

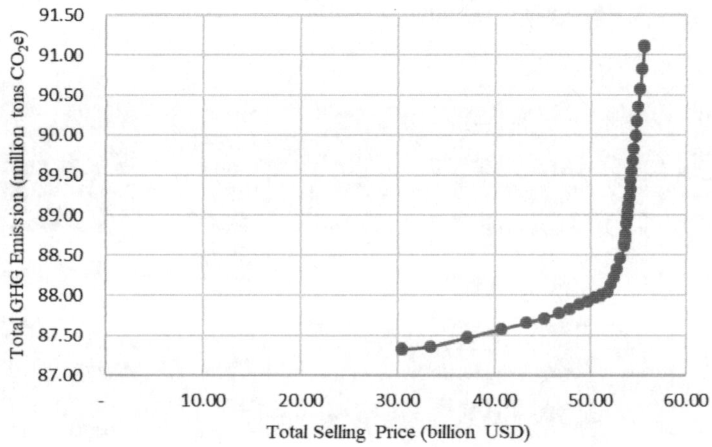

Figure 4. Pareto curve of multi-objective optimization

It can be seen that there is a trade-off point where the weighting between the emissions and the total selling price is at its most optimal. The trade-off point occurs in the weighting of 55% of total selling price and 45% of total emissions. The total selling price generated is 51.68 billion USD and total emissions generated are 88.05 million tons CO_2e at the trade-off point.

Emissions generated from production on the minimum total emission calculation and at the point of trade-off are relatively insignificant, only rising around 600 thousand tons of CO_2e to increase the total selling price to nearly 20 billion USD. The increase in emissions from both points is largely due to the emissions generated in RBDPOL production which increased from 35 thousand tons of CO_2e to 152 thousand tons of CO_2e. The significant increase in selling prices was influenced by cellulose sales which increased from 100 thousand USD to 11.5 billion USD, hemicellulose from 236 thousand USD to 1.9 billion USD, activated carbon from 26 thousand USD to 3 billion USD and RBDPOL of 3.2 billion USD to 13.6 billion USD.

The main emissions from production on the maximum total selling price calculation come from bio-oil production with emissions of 2.5 million tons CO_2e, rising considerably from the trade-off point with 52.5 thousand tons CO_2e emissions. In addition, emissions from glucose production increased from 22 tons CO_2e to 147 thousand tons of CO_2e and HPKO from 185 tons CO_2e to 186 thousand tons CO_2e. The increase in the selling price of the products is not significant. At the point of selling price optimization, the increase occurred for the sale of Bio-oil from 63 million USD to 3.1 billion USD, glucose from 86 thousand USD to 2.9 billion USD and HPKO from 2 million USD to 2.1 billion USD. However, the reduction also occurred significantly on CPKO sales which decreased from 3.5 billion USD to only 282 million USD, torrefied pellet from 1 billion USD to only 88 thousand USD and hemicellulose from 1.9 billion USD to only 240 thousand USD.

Figure 5 shows the production pathways of palm oil derivatives with multi-objective optimization.

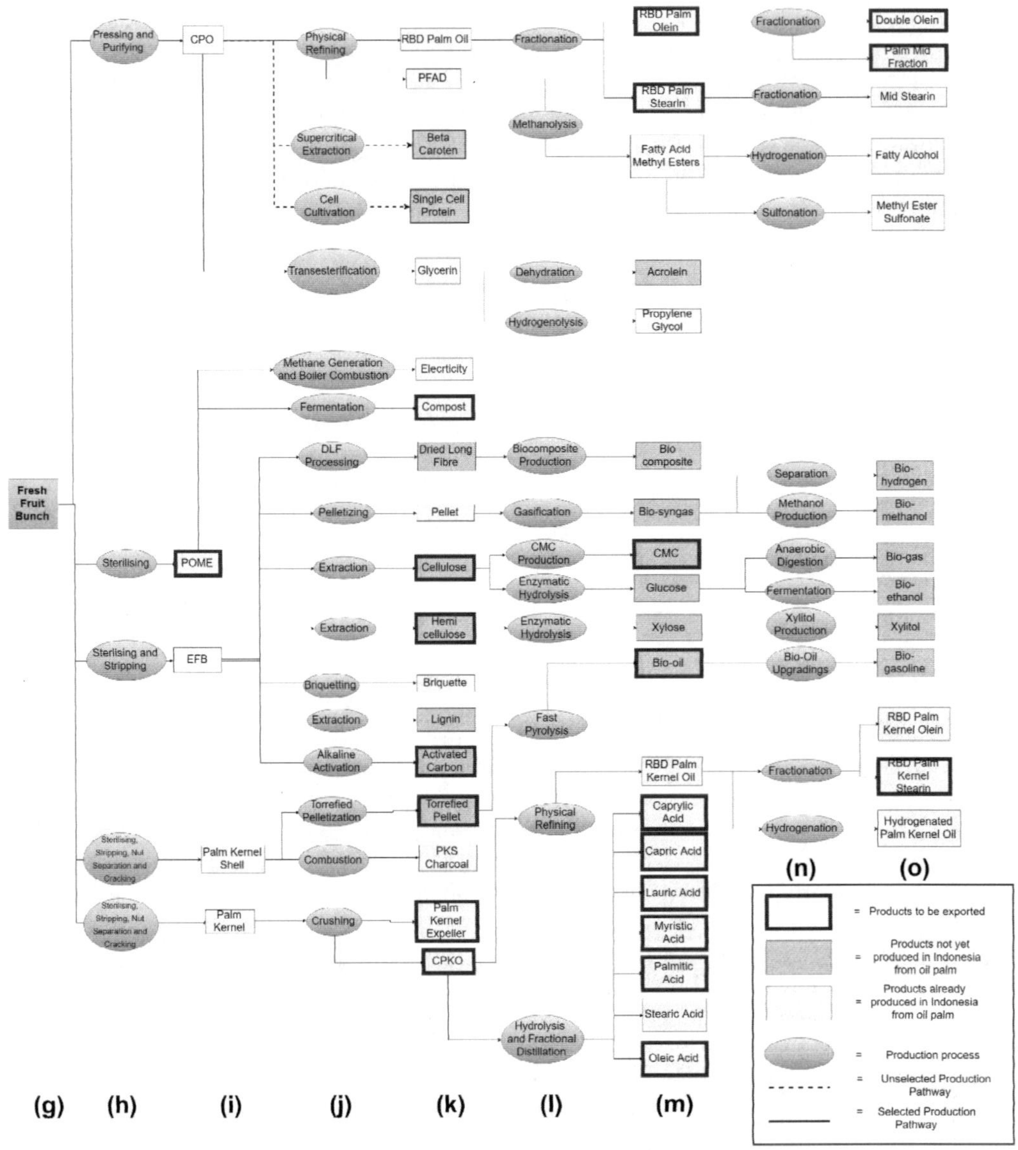

Figure 5. Production Pathway of Oil Palm Derivative Products for Multi-Objective

At the trade-off, the available EFB is not allocated to produce DLF. EFB is allocated to produce products that have higher selling prices such as cellulose and hemicellulose which have a selling price of 1.98 thousand USD and 1.8 thousand USD per ton, much higher than DLF with 134.9 USD per ton. Activated carbon is also produced more at the trade-off calculation than the minimum total emission calculation because it has a higher selling price than DLF but produces higher emissions. At the trade-off, RBDPOL and RBDPS are produced in higher quantities as more RBDPO is passed on as raw materials, RBDPO is not exported. Part of RBDPOL is continued to produce double olein and PMF to meet global needs.

Torrefied pellet, PKE and CPKO are exported and the only derivative of RBDPKO that is exported is RBDPKS. CMC, bio-oil and almost all of the fatty acids are exported. CPO, fatty alcohol and biodiesel, sold only in quantities to meet only domestic needs.

The sales figures of each product resulted from multi-objective optimization is mentioned in Table 2.

Table 2. Product Sales Figures of Multi-Objective Optimization

Product	Sales (ton)
CPO	2,900,000
POME	2,030
EFB	-
PKS	-
Palm Kernel	-
PFAD	-
RBDPO	2,610,000
SCP	-
Beta Caroten	-
Glycerin	37,963
Compost	400,000
Electricity	353,242
DLF	744
Briquette	261
Pellet	322
Torrefied Pellet	7,000,000
Lignin	5
Cellulose	5,810,000
Hemicellulose	1,053,495
CPKO	2,705,923
PKE	4,434,542
PKS Charcoal	30
Activated Carbon	1,900,000
RBDPOL	17,323,227
RBDPS	4,603,860
FAME	1,032,000
Acrolein	2,944
Propylene Glycol	28,988
RBDPKO	730,800
Caprylic Acid	715
Capric Acid	738
Lauric Acid	10,451
Myristic Acid	3,514
Palmitic Acid	1,820
Stearic Acid	542
Oleic Acid	3,318
Bio-composite	8
CMC	897,958
Glucose	51
Xylose	131
Bio-syngas	4,025,267
Bio-oil	90,474
Double Olein	556,837
PMF	556,837
Mid Stearin	3,219
Fatty Alcohol	126,890
MES	3,740
RBDPKOL	168,360
RBDPKS	112,240
HPKO	1,149

IOP Conf. Series: Materials Science and Engineering **543** (2019) 012062 doi:10.1088/1757-899X/543/1/012062

Product	Sales (ton)
Bio-gas	78
Bio-ethanol	31
Xylitol	0.02
Bio-methanol	3
Bio-hydrogen	3,272
Bio-gasoline	

3.4. Sensitivity Analysis
Sensitivity analysis is done to assess recent issues related to Indonesian CPO exports to the EU due to the Renewable Energy Directive policy that would disrupt Indonesia's CPO exports to EU countries correspond to 3.2 million tons annually. In 2008, the EU has imposed anti-dumping tariff penalties on the export of biodiesel and Fatty Alcohol from Indonesia to the European Union. It also adds scenarios where biodiesel is also produced from CPO to see its effect on total selling price and total GHG emissions.

Table 3 shows the selling prices and emissions generated from various scenarios.

Table 3. Sensitivity Analysis Results

Scenario	**Total selling price (billion USD)**	**Emission (million-ton CO_2e)**
Eliminate all exports of CPO, fatty alcohol and biodiesel.	51.68	88.05
CPO as a source of biodiesel production and eliminated all exports of CPO, fatty alcohol and biodiesel.	51.30	89.38
Eliminates Indonesia's CPO exports to the EU by 3.2 million tons	50.93	87.23
Base case (Exports of CPO by 15.39 million tons)	50.51	87.16

The three scenarios (base case, CPO export abolition to EU, and CPO-based biodiesel addition) are resulted in different values of total selling prices emissions from the trade-off. All of these scenarios resulted lower selling prices and only CPO-based biodiesel addition scenario which would have higher total GHG emissions compared to the trade-off. It can be seen that the decrease of selling price and emission of base case scenario and CPO export abolition to EU scenario are influenced by decreasing amount of CPO as the intermediate products for RBDPO, RBDPOL, RBDPS and biodiesel production. The emissions obtained from the CPO-based biodiesel addition scenario are higher than the trade-off due to its production pathway, one of which is directed to produce biodiesel and bio-oil. CPO based biodiesel produces higher emissions than those RBDPO, PFAD and their derivatives. Even though the emissions from production of RBDPO, PFAD and its derivatives is smaller than in the trade-off. This is coupled with higher bio-oil production which is one of the products with high emission factor.

4. Conclusions
Based on multi-objective optimization, selected production pathways are the production of 54 palm oil derivatives products to meet domestic needs and 21 of them can be exported comprised of 1 FFB derivative products (POME), 4 CPO derivatives products (RBDPOL, RBDPS, double olein and PMF), 1 POME derivative products (compost), 4 derivative products of EFB (cellulose, hemicellulose, activated carbon and CMC), 2 PKS derivative products (torrefied pellet and bio-oil), and 9 derived products of palm kernel (PKE, CPKO, caprylic acid, capric acid, lauric acid, myristic acid, palmitic acid, oleic acid and RBDPKS). The elimination of CPO exports, fatty alcohols and biodiesel does not

affect the selling prices and total emissions at the trade-off and higher total selling prices and emissions compared to base case.

Acknowledgments
This work has been carried out by the financial support of the Universitas Indonesia under the project "Hibah Publikasi Internasional Terindeks Untuk Tugas Akhir Mahasiswa" (PITTA) Fiscal Year 2018 number: 2557/UN2.R3.1/HKP.05.00/2018

5. References

[1] S Mielke and T Mielke 2017 Oil World Annual (Hamburg GFR ISTA-Mielke GMBH)
[2] M A Arip, L S Yee and T S Feng 2013 Assessing the competitiveness of Malaysia and Indonesia palm oil related industry *World Rev. Bus. Res.* **3** 138-45
[3] R Krismawati, R Ahdia and D Soetrisnanto 2013 Processing of facultative anaerobic pond effluent IPAL of oil palm industry in facultative anaerobic-phytoremediation as a pre-treatment of algae growing media *J. Chem. Ind. Tech.* **9**
[4] W Elbersen, K Meesters and R Bakker 2013 Valorization of palm oil (mill) residues: Identifying and solving challenges (Wageningen: WUR Food & Biobased Research)
[5] R A Dhanni 2017 Effects of World Trade Organization Policy on Developing Countries Against European-Union Anti-Dumping Policies on Indonesian Biodiesel Exports (Pasundan University: Bachelor Thesis)
[6] J M Egieya, L Cucek, K Zirngast, A J Isafiade, B Pahor and Z Kravanja 2018 Biogas supply chain optimization considering different multi-period scenarios *Chem. Eng. Trans.* **70** 985-90
[7] X Tian and F You 2018 Superstructure optimization of geothermal binary combined cooling, heating and power system using life cycle optimization *Chem. Eng. Trans.* **70** 1717-22
[8] RT Ng and D K Ng 2013 Systematic approach for synthesis of integrated palm oil processing complex. Part 1: Single owner *Ind. Eng. Chem. Res.* **52**(30) 10206-20
[9] V Andiappan, A S Ko, V W Lau, L Y Ng, R T Ng, N G Chemmangattuvalappil and D K Ng 2015 Synthesis of sustainable integrated biorefinery via reaction pathway synthesis: economic, incremental enviromental burden and energy assessment with multiobjective optimization *AIChE J.* **61**(1) 132-46
[10] A Abdulrazik, M Elsholkami, A Elkamel, and L Simon 2017 Multi-products productions from Malaysian oil palm empty fruit bunch (EFB): Analyzing economic potentials from the optimal biomass supply chain *J. Cleaner Prod.* **168** 131-48
[11] V Pham and M El-Halwagi 2012 Process synthesis and optimization of biorefinery configurations *AIChE J.* **58**(4) 1212-21
[12] M Z Nuh and N F Nasir 2017 *IOP Conf. Ser.: Mater. Sci. Eng.* **226** 012001
[13] A Quaglia, C L Gargalo, S Chairakwongsa, G Sin, and R Gani 2015 Systematic network synthesis and design: problem formulation, superstructure generation, data management and solution *Comput. Chem. Eng.* **72** 68-86
[14] P E Murillo-Alvarado, J M Ponce-Ortega, M Serna-Gonzalez, A J Castro-Montoya, and M M El-Halwagi 2013 Optimization of pathways for biorefineries involving the selection of feedstocks, products, and processing steps *Ind. Eng. Chem. Res.* **52** 5177-90

IOP Conf. Series: Materials Science and Engineering **543** (2019) 012063 doi:10.1088/1757-899X/543/1/012063

Preparation of *Catalyst Cu-ZnO-MgO-Al₂O₃* for Direct Synthesis of DME

T Swastika[1], A Ardy[1], and H Susanto [1]
[1] Department of Chemical Engineering, Faculty of Industrial Technology, Institut Teknologi Bandung, Jalan Ganesha 10 Bandung 40132, Indonesia

E-mail: herri@che.itb.ac.id; titisswstika@gmail.com

Abstract. Direct synthesis of DME catalyst was prepared using the co-precipitation method. The catalyst contained $CuO/ZnO/Al_2O_3$ about 40/27/33 (%-wt). Two types of catalyst were prepared, i.e. CZMA0 (Mg 0%-wt) and CZMA20 (Mg 20%-wt). Both of catalysts were activated using reducing gas (5% H_2 and N_2). The activity test was conducted by syngas model which contained (%-mole) 65% H_2, 28% CO and 7% N_2. The reaction was carried out in a fixed bed reactor at 5 bar, and temperature of 240, 250 and 260°C. Synthesis shown CZMA20 catalyst gave the highest catalytic activity with 73% of CO conversion and 66% of H_2 conversion (5 bar and 260°C). Direct synthesis of DME was also carried out using a dual-catalyst, i.e. catalyst for methanol synthesis (M151, the commercial catalyst of methanol synthesis) and methanol dehydration (γ-Al_2O_3-ITB). The activity of dual catalyst shown 93% of CO conversion and 91% of H_2 conversion. This activity was more stable than two other catalysts (CZMA0 and CZMA20).

1. Introduction

In 2017, the Indonesian government has decided to used DME (dimethyl ether) as partial substitution of LPG. DME has been commercially made by dehydrating methanol, including DME plant in Indonesia. DME can be used as household fuel like LPG, propellant, the substitution of diesel and intermediate of chemicals. Application of DME has been a particular concern around the world. Large production capacity of DME will be needed in order to fulfill future demand [1].

DME is generally produced through two steps process. First, natural gas, hydrocarbons, coal, or biomass would be converted to syngas that contains H_2 and CO. Furthermore, syngas would be synthesized into DME. This two steps process can be expressed by the following reactions:

a. synthesis of methanol

$$CO + 2H_2 \leftrightarrows CH_3OH \quad (1)$$

b. dehydration of methanol to DME

$$2CH_3OH \leftrightarrows CH_3OCH_3 + H_2O \quad (2)$$

c. water-gas shift reaction

$$CO + H_2O \leftrightarrows CO_2 + H_2 \quad (3)$$

The third reaction may influence the extent of the other two reactions. Many researchers are interested to develop the direct synthesis of DME. Direct synthesis of DME process can be expressed by the following reaction:

$$2CO + 4H_2 \leftrightarrows CH_3OCH_3 + H_2O \quad (4)$$

An important factor in direct synthesis of DME is the catalyst. Direct synthesis of DME (reaction 4) can be conducted by using Cu-based catalyst as reported in [1]. Unfortunately, this type of catalyst has low thermal stability, so development is being done to improve the catalyst resistance. Catalyst for direct

IOP Conf. Series: Materials Science and Engineering **543** (2019) 012063 doi:10.1088/1757-899X/543/1/012063

synthesis of DME is considered as a bifunctional catalyst: for methanol synthesis and dehydration of methanol. Metal components of bifunctional catalyst generally contain metal oxides such as CuO, ZnO, MgO, and Al_2O_3. Dehydration of methanol to DME was conducted by using solid-acid catalysts such as γ-Al_2O_3, modified silica alumina, TiO_2-ZrO_2, bohemite (AlOOH) and zeolites (H-ZSM-5, HY, mordenite, SAPO, MCM, ferrierite, chabazite, and H-beta).

There are many opportunities to develop Cu-based catalysts with various types of promoters and composition of catalysts to make a bifunctional catalyst of direct synthesis of DME. Addition 20%-wt of MgO has been proved the increases of activity of catalyst compared to those 0%-wt of MgO promoters, such as increases in CO conversion from 19% to 37%, and DME selectivity from 36% to 83% [1]. Therefore, Cu-ZnO-MgO-Al_2O_3 catalyst is interesting to be developed further.

The objective of this research was the preparation of Cu-ZnO-MgO-Al_2O_3 catalyst for the direct synthesis of DME. Activity catalyst was evaluated based on CO and H_2 conversions [1].

2. Experimental section

2.1 Catalyst preparation and characterization

The raw catalyst of CuO-ZnO-MgO-Al_2O_3 was prepared using the coprecipitation method [2]. At first, solutions of $Cu(NO_3)_2$, $Zn(NO_3)_2$, $Al(NO_3)_3$, $Mg(NO_3)_2$ were mixed with a certain ratio in order to obtain the target of catalyst composition (see Figure 1). The mixed solution of Cu, Zn, Mg, and Al would be precipitated by the solution of Na_2CO_3 simultaneously. Throughout the process temperature and pH were constant at 60°C and 7. The mixture was stirred at 700 rpm to get a homogeneous mixture and assist in the formation of the catalyst crystals. After aging, the catalyst will be wash with demineralized water and drying at 105°C. The raw catalyst that contains CuO, ZnO, MgO, and Al_2O_3 was calcinated and ready to used as a catalyst of direct synthesis of DME.

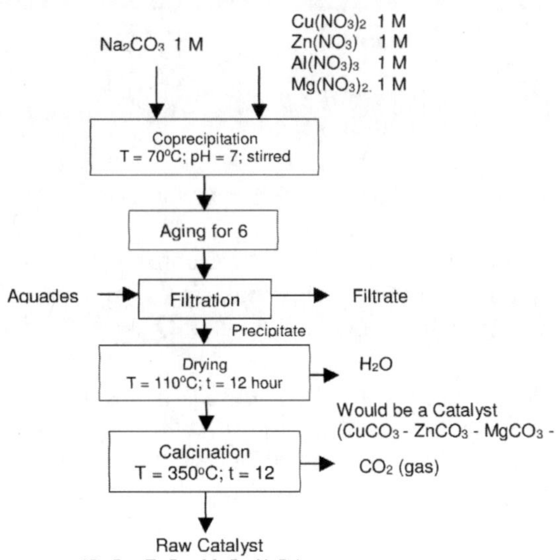

Figure 1. Preparation of catalyst by coprecipitation method

2.2 Activation of catalyst, or reduction of CuO catalyst to Cu

Pore property of raw catalyst was carried out using *Quantochrome Instruments Nova 300e*, in Laboratory of Analytical Instrument, Department of Chemical Engineering ITB. Activation of raw catalyst must be conducted to convert CuO to Cu in the same reactor before being used for the synthesis of DME. This reduction of CuO was done using 5%H_2 and N_2. The reduction process was carried out at 1.5 bar, and the temperature was increased step by step.

IOP Conf. Series: Materials Science and Engineering **543** (2019) 012063 doi:10.1088/1757-899X/543/1/012063

a. Purging reactor using N_2 at 80 mL/min (measured at 27°C), 1.5 bar and 150°C for 2 hours.
b. Gas reduction (5% H_2 and N_2) into the reactor at 80 mL/min, 1.5 bar and 150°C for 1 hour.
c. Increasing reactor temperature up to 180°C, and hold for 1 h.
d. Increasing reactor temperature up to 210°C, and hold for 1 h.
e. Increasing reactor temperature up to 240°C, and hold for 1 h.
f. Increasing reactor temperature up to 270°C, and hold for 3 h.

Started from the reactor temperature of 180°C, the progress of reduction from CuO to Cu was observed by measuring the consumption of H_2. The activation process or reduction of CuO was stopped when the H_2 concentration in the reactor effluent was the same as influent. This process was very critical related to the catalyst activity in the synthesis process. The activation and activity test of catalyst was carried out in a fixed bed tubular reactor with 1.3 cm diameter and 38 cm length. This reactor was heated using an electric heater and provided with a temperature controller. Schematic diagram of activation and activity test of catalyst are presented in Figure 2.

2.3 Catalyst activity test of direct synthesis of DME process in a fixed bed reactor.
Without opening the reactor, the reactor temperature was decreased to the desired temperature of synthesis, which is: 240, 250, and 260°C. The gas reduction flow was changed with the syngas (65% H_2, 28% CO and 7% N_2). Operating pressure 5 bar and temperature 240, 250, and 260°C. The flow rate of the syngas was 50 mL/min. In this study, the synthesis process was carried out for 10 hours.

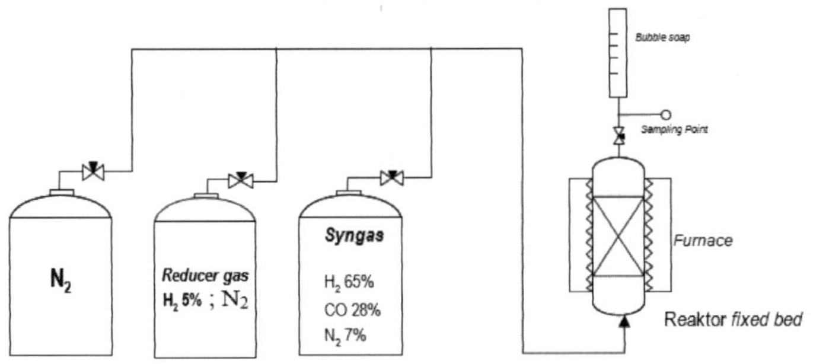

Figure 2. The eperimental device of direct synthesis of DME process

The gas compositions were analyzed using two types of gas chromatography:
(a) Shimadzu GC-14B, with separation column of Porapak-Q: CH_3OH and CH_3OCH_3 analysis
(b) Shimadzu GC-2014, with separation column of Porapak-Q and Molsieve: H_2, CO, CO_2, and N_2 analysis

Inlet gas compositions were analyzed at the beginning of experiments on catalyst activation and activity test. Outlet gas compositions during activation of catalyst were analyzed every 15 min. This data was necessary to evaluated CuO reduction. During activity tests, outlet gas sample was taken every 15 or 30 min. The samples were analyzed for gaseous composition, methanol and DME compositions.

3. Result And Discussion

The scopes of this study were catalyst characterization, activation of raw catalyst, the effect of temperature and pressure on CO and H_2 conversion from CZMA0, CZMA20, and dual catalyst.

3.1 Catalyst Characterization
The surface area of catalyst was considered as the most important catalyst character. This pore surface area was measured on the raw catalyst (before activation or reduction of CuO). Catalyst CZMA0, Cu-ZnO-Al_2O_3 with 0%-wt MgO, had a surface area of 285 m^2/g. While catalyst CZMA20 with 20%-wt MgO had a surface area of 125 m^2/g. Decreasing the surface area of the catalyst occurs because of the

increase in Mg concentration in the catalyst. This will cause changes in activities of the catalyst but increase the selectivity of DME products.

3.2 Activation of Raw Catalyst

CZMA0 and CZMA20 catalyst had similar progress in CuO reduction that indicated of H_2 conversion (Figure 3). Based on the progress of H_2 conversion, H_2 consumption of activation of CZMA20 was clearly more than CZMA0, as indicated by a higher H_2 conversion for CZMA20. The activation process was done completely after 400 min.

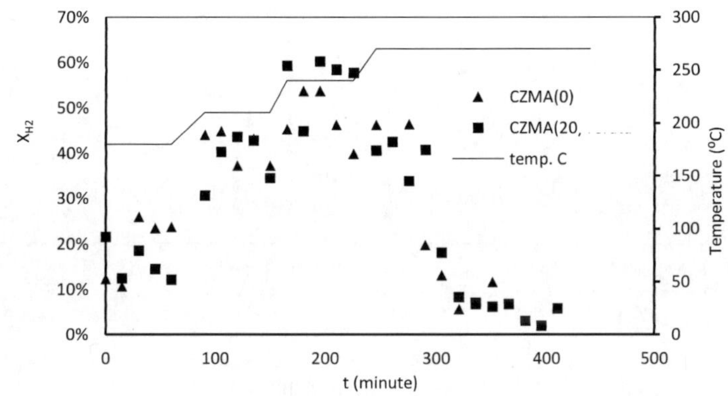

Figure 3. H_2 conversion during activation process

3.3 Effect of Temperature

The fresh catalyst was used in each condition in the activity tests. CO conversion of CZMA0 increased until 300 min, at three different reaction temperatures (Figure 4.a). The activity of CZMA0 indicate the active site of catalyst works well but after that decreased because of deactivation of the catalyst. For CZMA0 at 240°C, its activity decreased significantly after 200 min. The best activity of CZMA0 catalyst was shown 51% of CO conversion at 260°C and there was no indication of deactivation of the catalyst. Meanwhile, catalytic activity CZMA20 catalyst was lower than CZMA0. The activity of CZMA0 at 250°C was the best until 500 min, even it dropped drastically afterward. Thus, the quality of this catalyst was not homogeneous or the reproducibility in preparation of catalyst was not good. Discussion of this aspect was also applicable for catalyst CZMA20.

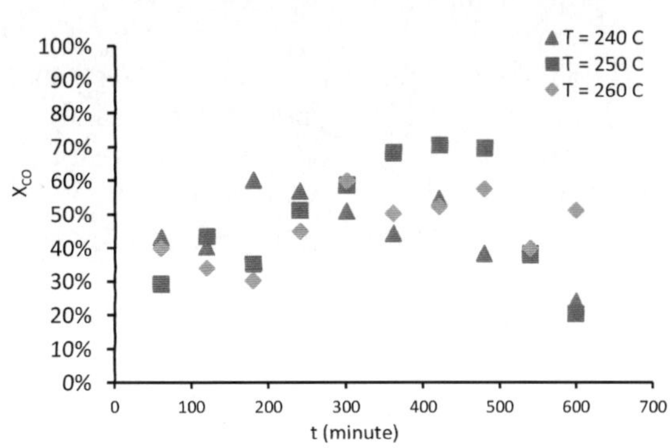

(a) CZMA0

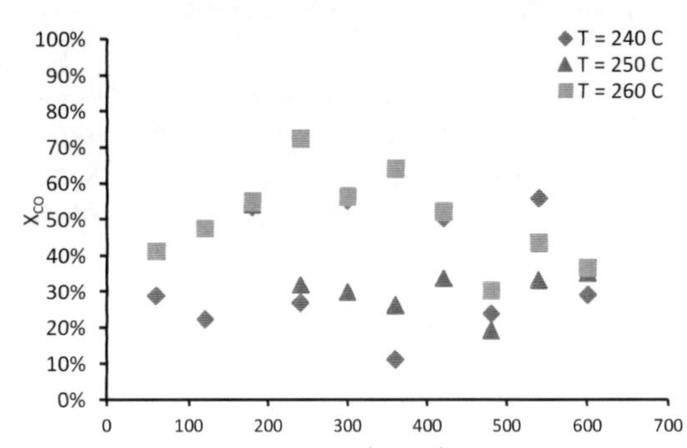

(b) CZMA20

Figure 4. CO conversion in during activity test of direct synthesis of DME
P = 5 bar; Q = 50 mL/min

Despite data fluctuation as shown in Figure 4, CO conversions on this study (60%) were relatively higher than reported by Asthana et.al (35%). with a pressure of 30 bar [1]. According to the thermodynamic of reaction equilibrium (calculated using *Aspen*), CO conversion might be as high as 90% at those reaction temperatures. Thus, there was a large chance to improve the catalyst.

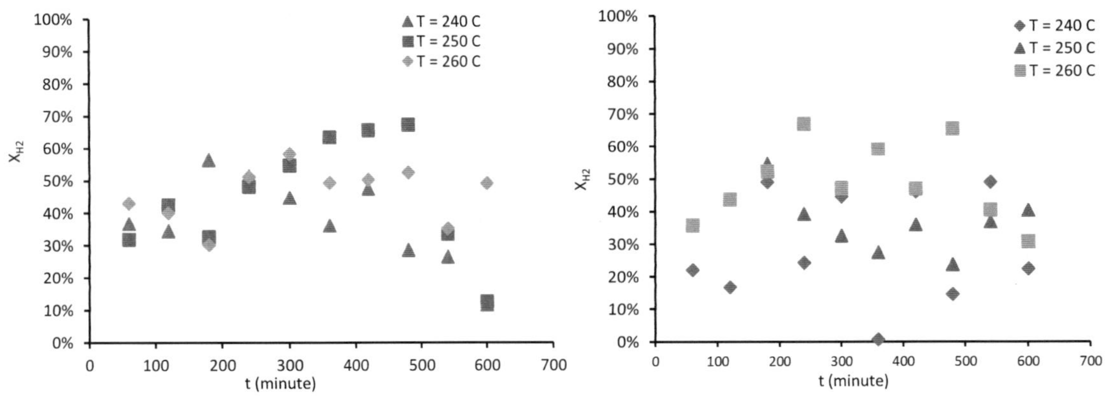

(a) CZMA0 (b) CZMA20

Figure 5. H_2 conversion in during activity test of catalyst
P = 5 bar; Q = 50 mL/min

Based on H_2 conversion (Figure 5), the stability of the catalyst was apparently very low. Progress in H_2 conversions was quite the same as CO conversion. The highest H_2 conversion was about 60%, but surprisingly it was higher than the thermodynamic equilibrium of 40% for direct synthesis of DME (equation 4). Probably, there were other reactions consuming H_2 beside direct synthesis of DME according to equation (4).

$$3H_2 + 3CO \leftrightarrows CH_3OCH_3 + CO_2 \quad (5)$$

$$H_2 + CO_2 \leftrightarrows CO + H_2O$$

$$CuO + H_2 \leftrightarrows Cu + H_2O \quad (6)$$

Reaction (5) was the overall reaction of methanol synthesis (equation 1), methanol dehydration (equation 2) and homogeneous water shift (equation 6). Reaction (6) might be possible if: (a) the catalyst activation step was not yet completed or (b) catalyst was exposed to air so there was re-oxidation of Cu to CuO.

A series of experiments using two bed of catalysts was done by a commercial catalyst of methanol synthesis (Topsøe M151) and a methanol dehydration catalyst (γ-Al_2O_3). The ratio of those catalysts was about 1:4 respectively. The reaction was carried at 240^{0}C and 5 bar. As expected, CO and H_2 conversions were significantly higher than CZMA0 and CZMA20 catalysts. A more stable activity of the dual catalyst was also observed (Figure 6). Unfortunately, the experiments were not yet carried out for sufficiently long time.

IOP Conf. Series: Materials Science and Engineering **543** (2019) 012063 doi:10.1088/1757-899X/543/1/012063

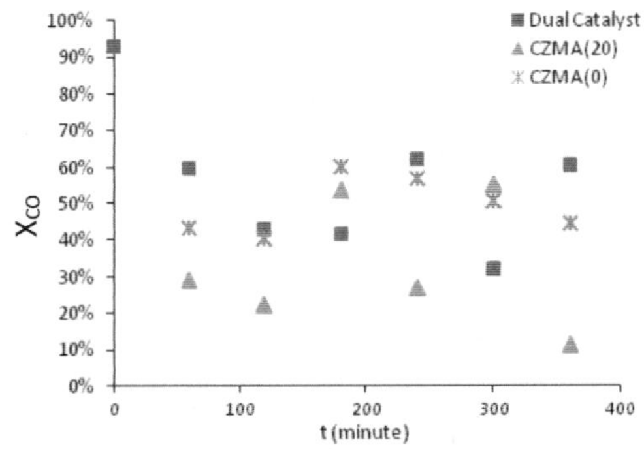

(a) CO conversion

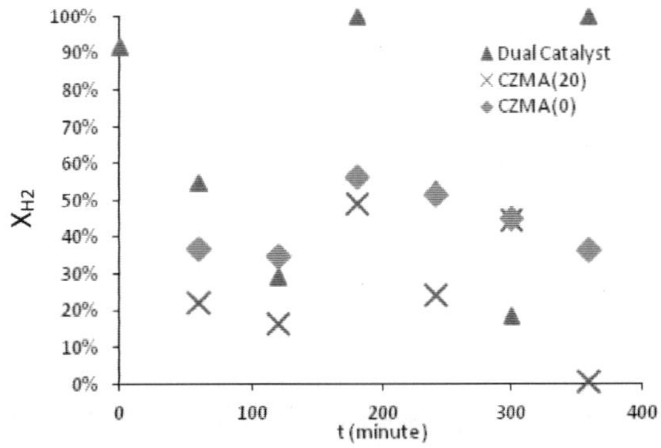

(b) H_2 conversion.

Figure 6. Comparison of the activity of CZMZ(0) CZMZ(20) and dual catalyst

3.4 Effect of Pressure

Based on equation (4), a higher pressure would give a higher conversion. Indeed, our experiments with a pressure of 7 bar resulted in higher CO conversion (Figure 7), as well as H_2 conversion. Surprisingly, the catalyst activity at a reaction pressure of 7 bar appeared more stable than that of 5 bar, for both CZMA0 and CZMA20. Furthermore, the activity of both catalysts also looked the same, based on CO conversion.

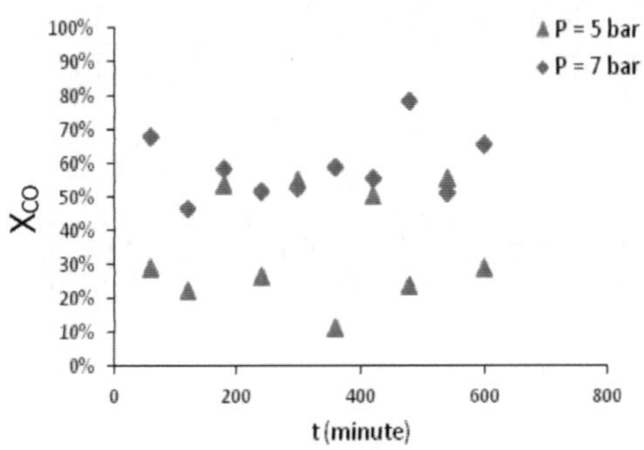

(a) CZMA0

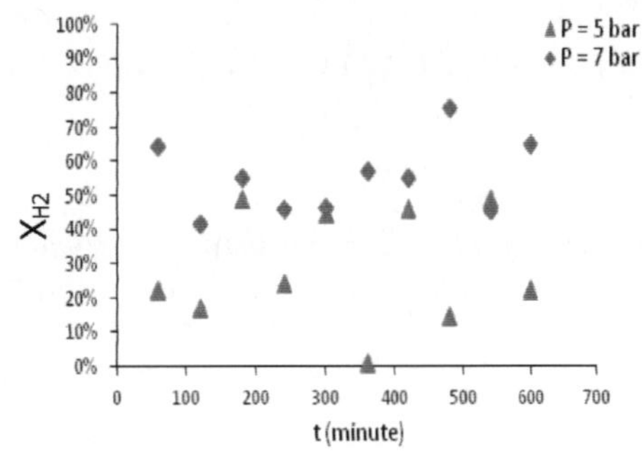

(b) CZMA20

Figure 7. Effect of pressure on CO conversion T = 240°C

Effect of pressure would be more significant on methanol synthesis (equation 1) than on methanol dehydration reaction (reaction 2). Thus, the formation of methanol would be more than that of DME. As a result, increased reaction pressure would give a higher concentration in the product. This phenomenon was reported by Prasad, et.al. [6]. So, in a dual catalyst system, the amount of catalyst for dehydration should be larger than the amount of catalyst for methanol synthesis.

3.5 Yield of DME

The yield of DME was defined as a molar fraction of inlet syngas (CO and H_2) converted into DME. Absolutely, the yield of DME was not good enough and still require a lot of improvement. In this study, the yield of DME was less than 0.01%. The yield of DME from the experiment using dual catalyst was also very low. The yields of methanol as an intermediate product of synthesis using CZMA0, CZMA20 and dual catalyst were also very low. It was suspected that the gas chromatography was not yet calibrated properly.

IOP Conf. Series: Materials Science and Engineering **543** (2019) 012063 doi:10.1088/1757-899X/543/1/012063

4. Conclusion

The catalyst for direct synthesis of DME has been prepared using the co-precipitation method several times successfully. CZMA0 without MgO had a good pore surface area of 285 m^2/g, while that of CZMA20 with 20% MgO had relatively low pore surface area of 125 m^2/g. CZMA20 consumed more H_2 and took a long time for activation. The advantage of adding MgO on the main catalyst Cu-ZnO-Al_2O_3 was not yet observed in this experimental study.

CO and H_2 conversions were found significantly high, even higher than those reported in the literature. But they were still below the thermodynamic equilibrium conversions of direct synthesis of DME. Effect of reaction pressure was also studied experimentally at 5 and 7 bar. At a higher reaction pressure, CO and H_2 conversion increased accordingly. At a higher reaction pressure of 7 bar, the activity of the catalyst was more constant during the activity test. Dual bed catalyst of direct synthesis of DME consisted catalyst for methanol synthesis and dehydration gave a slightly higher CO and H_2 conversions. This dual catalyst also gave a more stable activity of the catalyst.

Acknowledgments

This study was a part of Master Thesis at the Department of Chemical Engineering, ITB. The authors would like to thank *the Indonesian Palm Oil Estate Fund* (BPDPKS) for funding our research on *Biomass to Methanol/DME*, in the Fiscal Year of 2018.

References

[1] Asthana S, Samantha C, and Bhaumik A 2016 *J. Catal.* **334** p 89-101.
[2] Hermansyah 2010 **ISSN: 1411-4216**. p A14.1-A14.8
[3] Huang M H, Lee H M, Liang K C, Tzeng C C, Chen W H 2015 *Int. J. Hydrogen Energy.* **40** p 13583-13593.
[4] Moradi G R, Ahmadpour J and Yaripour F 2007 *Catal. Commun.* **8** p 598-606
[5] Ng K L D Chadwick and B A Toseland 1999 *Chem. Eng. Sci.* **54** p 3587-3592
[6] Prasad P S, Bae J W, Kang S H, Lee Y J and Jun K W 2008 *Fuel Proces. Technol.* **89** p 1281-1286

IOP Conf. Series: Materials Science and Engineering **543** (2019) 012064 doi:10.1088/1757-899X/543/1/012064

Activation of Coconut Shell - Randu Wood Biochar and Its Use as Heterogeneous Catalyst Support for Biodiesel Production

Y S Pradana[1], N Rochim[1], H S Mukaffa[1], H B Satriawan[1], A Hidayat[2], A Budiman[1,3,*]

[1]Chemical Engineering Department, Faculty of Engineering, Universitas Gadjah Mada, Jalan Grafika No. 2, Yogyakarta 55281, Indonesia

[2]Chemical Engineering Department, Faculty of Industrial Technology, University of Islam Indonesia, Jalan Kaliurang KM 14,5, 55584, Yogyakarta, Indonesia

[3]Master Program of Systems Engineering, Universitas Gadjah Mada, Jalan Teknika Utara 3, Yogyakarta, Indonesia

*Email: abudiman@ugm.ac.id

Abstract. In the last decade, biodiesel has been considered as one of attractive biofuels in function of replacing fossil-based diesel fuel and concerning on reduction of greenhouse gas release. For biodiesel production, the use of heterogeneous catalyst is a challenge to minimize problems related to the use of homogeneous catalyst. One of potential heterogeneous catalyst is activated carbon-based catalyst. This study investigated the effect of activating agents (hydrochloric acid and sulfuric acid), its concentration (0.5, 1.0, 1.5 and 2.0 N) and activation time (1.5, 3.0, 4.5 and 6.0 h) on iodine number of activated biochar (from coconut shell and Randu wood). The results showed that the highest iodine number (749.14 mg/g) was obtained from coconut shell biochar activated by sulfuric acid at concentration of 2 N and activation time of 6 h. Moreover, BET analysis of the best activated biochar resulted 30.88 m^2/gcat of surface area, 3.26 x 10^{-2} cm^3/g of total pore volume and 2.11 nm of average pore radius. The performance test of this potassium-impregnated activated biochar in batch transesterification reaction (palm oil-methanol) presented 29.44 % of biodiesel conversion.

1. Introduction

In the last decade, biodiesel is one of attractive biofuels in function of replacing fossil fuels which has been growing down in production [1]. In addition, the environmental concerns caused by the increasing level of CO_2 in the atmosphere has led to speed up the development of biodiesel production [2]. Biodiesel is known as a mixture of fatty acid methyl esters (FAMEs) generated from the reaction of fatty acid in renewable natural feedstocks and short-chain alcohol in the presence of catalyst [3]. It has several advantages, i.e. renewable, biodegradable, less toxic and environmentally friendly [4]. Currently, biodiesel is derived from some oil crops, such as palm oil [5], sunflower oil [6], jatropha oil [7] and soybean oil [8], because of their availability. For improving the economic value, biodiesel is also produced from low-cost feedstocks [9], such as used cooking oil [10] and palm fatty acid distillate [11]. Moreover, several studies have started to exploit microorganism as biodiesel feedstock, such as microalgae [12]. By the fact, a lot of lipid-rich feedstocks in the world makes biodiesel potential to be developed [13].

Biodiesel synthesis requires catalyst to accelerate the reaction. The common catalyst used is homogeneous alkaline catalyst such as NaOH and KOH, which have high catalytic activity and affordable in price. Nevertheless, this condition results many problems: (i) corrosive to the equipment; (ii) difficulty in separation step; (iii) more wastewater production [14]. In order to minimize these problems, the use of heterogeneous alkaline catalyst is proposed. Besides, heterogeneous catalyst can be regenerated to be used again as catalyst after impregnation. However, the active site of catalyst has possibility to be leached seriously and perform as homogeneous catalyst [15]. It results problems in reducing the heterogeneous catalytic activity and removing leached catalyst from biodiesel [16].

Among the heterogeneous alkaline catalyst, activated carbon has been shown as one of potential catalyst support for accelerating the transesterification reaction. Activated carbon has been widely used as catalyst supporter in various industrial and environmental applications. Activated carbon has characteristics: (i) high surface area; (ii) durable in high pressure and temperature; (iii) chemically stable [17]. Many researchers have been conducted for studying several carbon-based heterogeneous alkaline catalysts. Fadhil et al. [18] reported methanolysis of bitter almond oil using potassium acetate loaded on activated carbon. Baroutian et al. [19] studied potassium hydroxide catalyst loaded on palm shell activated carbon for producing biodiesel from palm oil. Wan and Hameed [20] used activated carbon supported calcium oxide catalyst to transform palm oil into biodiesel.

There are two steps for producing activated carbon: carbonization and activation. Carbonization is a process for synthesizing carbon in the form of solid, such as pyrolysis by using thermal in air limitation or no air condition [21]. Solid carbon will be formed by heating hydrocarbon materials at temperature of 400 – 600 ^{o}C. After formation of solid carbon, activation treatment, physical and/or chemical, is needed to enlarge pores by breaking hydrocarbon bond or oxidizing surface molecules so the characteristics will change. Physical activation can be applied by using heat in the presence of vapor and CO_2 gas flow. Whereas chemical activation is carried out by using chemicals, which is called activator, such as HCl and H_2SO_4.

This work presents the activation comparison of biochar synthesized from coconut shell and Randu wood as heterogeneous catalyst support in biodiesel production by using iodine number analysis. The effect of activating agents (hydrochloric acid and sulfuric acid), its concentration (0.5, 1.0, 1.5 and 2.0 N) and activation time (1.5, 3.0, 4.5 and 6.0 h) on iodine number of activated biochar (from coconut shell and Randu wood) were investigated. Moreover, the BET surface area analysis was conducted on the best activated biochar (represented by the highest iodine number).

2. Experimental

2.1. Materials preparation

Materials used in this research were coconut shell and Randu wood biochar obtained from Yogyakarta, Indonesia. These biochar were reduced in size and then sieved at 20-30 mesh. Hydrochloric acid (37 %), sulfuric acid (≥98 %), potassium iodide (≥ 99.5 %) and sodium thiosulfate pentahydrate (≥ 99.5 %) were produced by Merck. Meanwhile, distilled water was manufactured by UD. Jaya Santosa, Yogyakarta, Indonesia.

2.2. Biochar activation

The sieved biochar was then activated by using acid solution (HCl and H_2SO_4) in batch reactor with stirring rate of 300 rpm. The biochar was put into 100 ml of acid solution at desired concentration (0.5, 1, 1.5 and 2 N) for various residence times (90, 180, 270 and 360 minutes).

2.3. Analysis of iodine number

Iodine number is defined as milligram amount of iodine which is adsorbed by one gram activated carbon. Highly ability of activated carbon to adsorb iodine means it has larger surface area. After activation was

IOP Conf. Series: Materials Science and Engineering **543** (2019) 012064 doi:10.1088/1757-899X/543/1/012064

finished, activated biochar was separated from the acid solution using filter paper and was then dried at 110 °C overnight. 0.5 g of dried activated biochar was put into 30 ml of 0.1 N potassium iodide solution and then stirred at stirring rate of 300 rpm for 15 min. The solution was separated between activated biochar and filtrate using centrifuge. 10 mL of filtrate was diluted with 30 mL of distilled water and then titrated with standardized sodium thiosulfate solution. Hereafter, the iodine number was calculated using Equation 1.

$$Iodine\ number = \frac{126.9 \times (V_1 - V_2) \times N}{W} \quad (1)$$

where, V_1 : volume of standardized sodium thiosulfate needed for titrating distilled water
V_2 : volume of standardized sodium thiosulfate needed for titrating diluted filtrate
N : normality of $Na_2S_2O_3$ solution
W : mass of activated carbon

2.4. Analysis of surface area by Brunauer–Emmett–Teller (BET)

Based on the analysis of iodine number, the best activated biochar was then analyzed the specific surface area of by Brunauer–Emmett–Teller (BET) method. This method determined the surface area of solid materials by nitrogen adsorption-desorption. Moreover, BET analysis also provided representation of pore size and pore volume of activated biochar.

2.5. Performance test

Furthermore, the best activated biochar was performed to produce biodiesel in batch stirred reactor. Selected activated biochar was firstly prepared to be basic catalyst by impregnating potassium. The impregnation procedures of potassium on activated biochar was presented in previous paper by Pradana et al. [22]. Before tested in transesterification process, the solid or heterogeneous catalyst was characterized using BET and Atomic Absorption Spectroscopy (AAS) method. The prepared catalyst was then used to drive transesterification of palm oil and methanol. The steps of transesterification and separation process, as conducted by Pradana et al. [22], were carried out at 60 °C of temperature, 6:1 of methanol-oil molar ratio, and 700 rpm of stirring rate for 4 h. At the end, the conversion was determined using Equation 2 by evaluating the bounded glycerol in biodiesel sample. G_T was presented as total glycerol content and G_F was presented as free glycerol content.

$$X = \frac{\left[(G_T - G_F)_{feed} - (G_T - G_F)_{sample}\right]}{(G_T - G_F)_{feed}} \quad (2)$$

3. Results and Discussion

3.1. Effect of hydrochloric acid as activating agent

Figure 1(a) and (b) show the relation between iodine number of activated carbon from coconut shell and Randu wood, respectively, at various concentrations of hydrochloric acid (HCl) and residence times of activation step. Generally, the increase of HCl concentration and activation time were followed by iodine number increase. The highest iodine number of coconut shell activated carbon was 632.15 mg/g, obtained at 2 N of HCl concentration and 360 minutes of residence time. At the same condition, the highest iodine number of Randu wood activated carbon was reached (544.99 mg/g). This phenomena occurred because the higher concentration of HCl and the longer residence time of activation removed more impurities and opened the pores, resulting higher surface area (represented by higher iodine number).

IOP Conf. Series: Materials Science and Engineering **543** (2019) 012064 doi:10.1088/1757-899X/543/1/012064

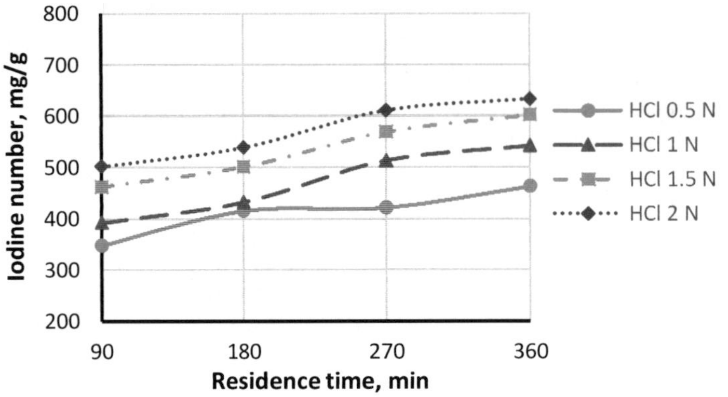

(a)

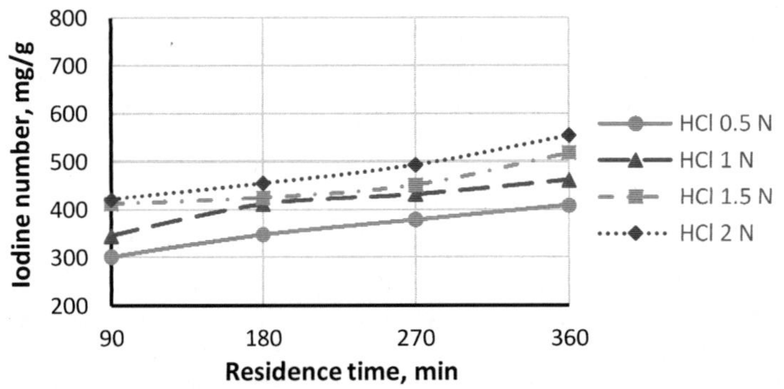

(b)

Figure 1. Effect of HCl concentration and residence time on iodine number of activated carbon from: **(a)** coconut shell [23]; and **(b)** Randu wood [24]

3.2. Effect of sulfuric acid as activating agent

Figure 2(a) and (b) show the relation between iodine number of activated carbon from coconut shell and Randu wood, respectively, at various concentrations of sulfuric acid (H_2SO_4) and residence times of activation time. These trends of H_2SO_4 performance as activating agent were similar with HCl, in which the increase of iodine number was proportionally affected by the more concentration of H_2SO_4 and activation time. Based on the results, the highest iodine number (749.14 mg/g) was obtained at 2 N of H_2SO_4 concentration and 6 hours of residence time. Meanwhile, the highest iodine number of Randu wood biochar was 556.64 mg/g. The higher concentration of sulfuric acid, refers to higher acid concentration, would remove tar and other impurities remained in the biochar and closing the pores [25]. As results, the more opened pores of biochar were generated and the adsorption of iodine was increase. Moreover, the longer residence time of activation positively affected on the iodine number. It indicated the higher activated surface area of biochar and the more pores formed.

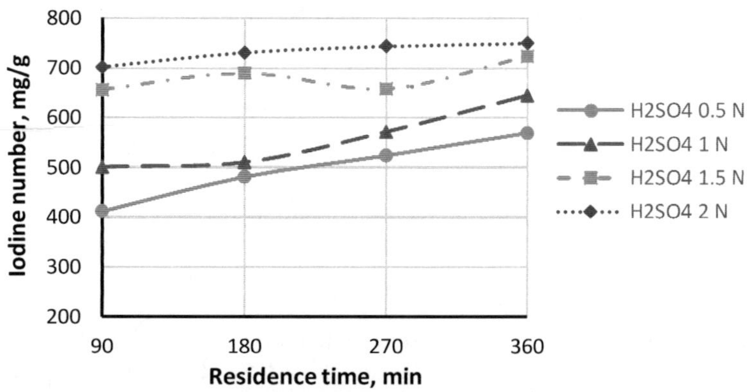

(a)

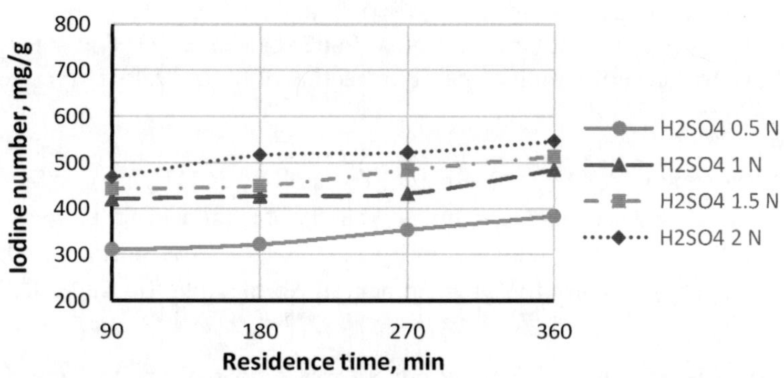

(b)

Figure 2. Effect of H_2SO_4 concentration and residence time on iodine number of activated carbon from: **(a)** coconut shell [23]; and **(b)** Randu wood [24]

3.3. Brunauer–Emmett–Teller (BET) Surface Area

The highest iodine number analysis was 749.14 mg/g, resulted from coconut shell biochar activated by sulfuric acid at concentration of 2 N and residence time of 6 h. This activated biochar sample was then analyzed the specific surface area using Brunauer–Emmett–Teller (BET) method. The analysis result showed that the surface area, total pore volume and average pore radius were 30.88 m^2/gcat, 3.26 x 10^{-2} cm^3/g and 2.11 nm, respectively. Based on this result, the activated carbon was classified as very low surface area. However, these results were very possible and prospective to be developed further because of the heterogenic functions of activated carbon in all fields [21].

3.4. Performance Test

After potassium impregnation process, the prepared heterogeneous catalyst was characterized to compare with activated carbon, as reported in Table 1. It showed slight effect of potassium impregnation on the pore size enlargement, surface area and total pore volume addition compared to activated carbon. The impregnation was proven to give additional potassium (5.17 mg/gcat) on activated carbon. Furthermore, the performance test of potassium-impregnated activated biochar resulted 29.44% of biodiesel conversion. Its transesterification showed in low conversion and catalyst activity. This results were still far from the expectation, thus the development of higher activity and durable heterogeneous catalyst was always in the research program.

Table 1. Properties of K/AC catalyst.

Characteristics	K/AC Catalyst	Unit
BET surface area*)	34.4660	m^2/gcat
Total pore volume*)	4.09×10^{-2}	cm^3/gcat
Average pore size*)	2.3740	nm
Potassium addition	5.1700	mg/gcat

4. Conclusions

The higher concentration of acid (HCl & H_2SO_4) and the longer residence time of activation removed more impurities and opened the pores, resulting higher surface area (represented by higher iodine number). The results showed that the highest iodine number (749.14 mg/g) was obtained from coconut shell biochar activated by sulfuric acid at concentration of 2 N and activation time of 6 h. Moreover, BET analysis of the best activated biochar reported 30.88 m^2/gcat of surface area, 3.26×10^{-2} cm^3/g of total pore volume and 2.11 nm of average pore radius. In performance test, this potassium-impregnated activated biochar drove transesterification of palm oil-methanol to be 29.44% of biodiesel conversion.

Acknowledgements

The authors would like to thank Ministry of Research, Technology and Higher Education of the Republic of Indonesia and Chemical Engineering Department, Faculty of Engineering, Universitas Gadjah Mada, Indonesia for financial support. We also gratefully acknowledge the partial funding from USAID through the SHERA program – Centre for Development of Sustainable Region (CDSR).

References

[1] Sulaiman S, Shah B and Jamal P 2017 Production of biodiesel from palm oil using chemically treated fish bone catalyst *Chem. Eng. Trans.* **56** 1525-1530

[2] Agarwal A K 2007 Biofiels (alcohols and biodiesel) applications as fuels for internal combustion engines *Prog. Energ. Combust.* **33** 233-271

[3] Vicente G, Martinez M, Aracil J and Esteban A 2005 Kinetics of Sunflower Oil Methanolysis *Ind. Eng. Chem. Res.* **44**(15) 5447–5454

[4] Jimenez S F, Torres J G, Correa D B and Ruiz Y P 2015 Evaluation of biodiesel production process from palm oil (*Eleais guineensis*) using exergy analysis methodology *Chem. Eng. Trans.* **43** 529-534

[5] Pradana Y S, Hidayat A, Prasetya A and Budiman A 2017 Biodiesel production in a reactive distillation column catalyzed by heterogeneous potassium catalyst *Enrgy. Proced.* **143** 742-747

[6] Reyero I, Arzamendi G, Zabala S and Gandia L M 2015 Kinetics of the NaOH-catalyzed transesterification of sunflower oil with ethanol to produce biodiesel *Fuel Process. Technol.* **129** 147-155

[7] Kusumaningtyas R D, Aji I N, Hadiyanto H and Budiman A 2016 Application of tin (II) chloride catalyst for high FFA jatropha oil esterification in continuous reactive distillation column *Bulletin of Chemical Reaction Engineering and Catalysis* **11**(1) 66-74

[8] Georgogianni K G, Katsoulidis A P, Pomonis P J and Kontominas M G 2009 Transesterification of soybean frying oil to biodiesel using heterogeneous catalysts *Fuel Process. Technol.* **90** 671 – 676

[9] Hidayat A, Rochmadi, Wijaya K and Budiman A 2016 Removal of free fatty acid in Palm Fatty Acid Distillate using sulfonated carbon catalyst derived from biomass waste for biodiesel production *IOP Conference Series: Materials Science and Engineering* **105** 012026

[10] Guan G, Kusakabe K and Yamasaki S 2009 Tri-potassium phosphate as a solid catalyst for biodiesel production from waste cooking oil *Fuel Process. Technol.* **90** 520 – 524

[11] Sudibyo H, Prasakti L, Budiman A, Fahrurrozi M and Rochmadi 2018 Continuous monoglyceride production from palm fatty acid distillate and glycerol using vacuum reactive distillation column *International Journal on Advanced Science, Engineering and Information Technology* **8**(1) 108-114

[12] Pradana Y S, Masruri W, Azmi F A, Suyono E A and Rochmadi 2018 Extractive-transesterification of microalgae *Arthrospira sp.* using methanol-hexane mixture as solvent *Int. J. Renew. Energy Res.* **8**(3) 1499-1507

[13] Herrera K C, Carmona S L, Correa D B and Ruiz Y P 2015 Evaluation of biodiesel production process from sapium tree oil sebiferum using exergy analysis methodology *Chem. Eng. Trans.* **43** 463-468

[14] Sawitri D R, Sutijan and Budiman A 2016 Kinetics study of free fatty acids esterification for biodiesel production from palm fatty acid distillate catalysed by sulfated zirconia *ARPN Journal of Engineering and Applied Sciences* **11**(16) 9951-9957

[15] Xie W and Zhao L 2013 Production of biodiesel by transesterification of soybean oil using calcium supported tin oxides as heterogeneous catalysts *Energ. Convers. Manage.* **76** 55-62

[16] Noiroj K, Intarapong P, Luengnaruemitchai A and Jai-In S 2009 A comparative study of KOH/Al_2O_3 and KOH/NaY catalysts for biodiesel production via transesterification from palm oil *Renew. Energ.* **34** 145-150

[17] Hidayat A, Rochmadi, Wijaya K, Nurdiawati A, Kurniawan W, Hinode H, Yoshikawa K and Budiman A 2015 Esterification of palm fatty acid distillate with high amount of free fatty acids using coconut shell char based catalyst *Enrgy. Proced.* **75** 969-974

[18] Fadhil A B, Aziz A M and Altamer M H 2016 Potassium acetate supported on activated carbon for transesterification of new non-edible oil, bitter almond oil *Fuel* **170** 130-140

[19] Baroutian S, Aroua M K, Raman A A A and Sulaiman N M N 2010 Potassium hydroxide catalyst supported on palm shell activated carbon for transesterification of palm oil *Fuel Process. Technol.* **91** 1378-1385

[20] Wan Z and Hameed B H 2011 Transesterification of palm oil to methyl ester on activated carbon supported calcium oxide catalyst *Bioresource Technol.* **102** 2659-2664

[21] Pradana Y S and Prasetya A 2017 Performance evaluation of household pyrolytic stove: Effect of outer air holes condition *AIP Conf. Proc.* **1823** 020069

[22] Pradana Y S, Hidayat A, Prasetya A and Budiman A 2018 Application of coconut-shell activated carbon as heterogeneous solid catalyst for biodiesel synthesis *Defect Diffus. Forum* **382** 280-285

[23] Rochim N 2012 *Coconut shell biochar activation as carrier in solid catalyst synthesis for biodiesel production - Aktivasi arang tempurung kelapa sebagai bahan penyangga pembuatan katalis padat pada proses produksi biodiesel* (Bachelor Thesis, Universitas Gadjah Mada, Yogyakarta, Indonesia)

[24] Mukaffa H S 2012 *Wood biochar activation as carrier in solid catalyst synthesis for biodiesel production - Aktivasi arang kayu sebagai bahan penyangga pembuatan katalis padat pada proses produksi biodiesel* (Bachelor Thesis, Universitas Gadjah Mada, Yogyakarta, Indonesia)

[25] Nam H, Choi W, Genuino D A and Capareda S C 2018 Development of rice straw activated carbon and its utilizations *J. Environ. Chem. Eng.* **6**(4) 5221-5229

IOP Conf. Series: Materials Science and Engineering **543** (2019) 012065 doi:10.1088/1757-899X/543/1/012065

Microwave Assisted Hydrolysis of Oil Palm Empty Fruit Bunch: Effect of Wave Distance Sources, Power, Time and Acid Concentration

I Madinah[1] and M Sarah[1,2,*]

[1]Departement of Chemical Engineering, Universitas Sumatera Utara, Padang Bulan Medan, Indonesia
[2]Sustainable Energy and Biomaterial Centre of Excellent, Universitas Sumatera Utara
*Corresponding Author

Email : Maya3@usu.ac.id

Abstract. Hydrolysis of EFB's fiber was carried out using sulfuric acid in two series reactors each operates under microwave irradiation. First reactor operates at various microwave power level (medium, medium high and high), irradiation time (5, 7.5 and 10 minutes) and acid concentration (0.5, 1 and 1.5%) while the second reactor operates at constant time using similar combination of power and acid concentration. This study placed reactor A next to the waveguide, while the reactor B on the opposite position. Microwave energy emmited from magnetron into cavity through the waveguide. This study evaluated effect of wave distance source on temperature and glucose yield in both reactors. Effect of microwave power, irradiation time and acid concentration and their relationship with glucose and temperature were investigated. Overall, increment of time and power increased yield of glucose at first hydrolysis using sulfuric acid of 0.5%. Glucose yielded from reactor A usually higher as compared to reactor B. Increment acid concentration reduced the yield except the one with high power irradiation. This study observed that second stage hydrolysis had no significant contribution to add the yield.

1. Introduction

Bioethanol is an alternative fuel that reduce greenhouse gas emissions by up to 75%. The most crucial issue for bioethanol production is selection of raw materials with respect to economic and environmental consideration. Biomass from agricultural and industrial waste can be used for bioethanol production because the source is abundant, cheap, and does not interfere with the availability of food [1, 2]. One of potential industry as a source of biomass is palm oil mill.

Production of palm fruit into palm oil contributes to several pollution problems namely palm oil mill effluent (POME), gas emission and biomass from steam batch sterilization, boiler operation, threshing, bunch digestion and pressing respectively. Biomass of palm oil mill comprises of empty fruit bunch (EFB), fiber and frond in abundant amount. They all categorized as lignocellulosic materials containing 50% cellulose, 25% hemicellulose and 25% lignin in their cell wall [3, 4]. Even though palm oil mill is implementing best practice waste management and some of those typical wastes can be reused and recycled such as utilization of POME for land application in oil palm plantation and EFB for composting. EFB is lignocellulosic material comprises of 37.1% glucose, 20.7% hemicellulose, 22.2% lignin, 1.3% ash, and 3.5% extractive substances [5]. Problem arising from composting of EFB in palm oil mill is composting performs very slow. Digestion rate requires 3 to 4 weeks to complete the fermentation while palm oil production rate is very high. This situation results abundant pile of EFB presence around the mill. With respect to those huge quantity of EFB and effort to improve economic value of EFB conversion, investigation to evaluate EFB potency to produce bioethanol is conducted. This conversion carried out through hydrolysis of EFB and glucose fermentation for bioethanol production.

Unlike hydrolysis with dilute acid to treat lignocellulosic materials that previously reported elsewhere [6], this study investigates microwave assisted hydrolysis of EFB to produce glucose. Glucose will be fermented further to produce bioethanol. Several studies reported utilization of microwave irradiation as an effective heating method either alone or in combination with other chemicals due to high heating efficiency, uniformity, and short processing times [7, 8, 9, 10, 11]. Microwaves enhances lignin removal and facilitates cellulosic and hemicellulose ultrastructure conversion thereby increasing cellulosic accessibility [12, 13]. When the treatment provides a more effective hydrolysis process in converting cellulose to glucose for further ethanol production, less cellulose is available to produce nanocellulose or other components [14]. During hydrolysis process, microwave irradiation facilitates the loosen of cellulose bonds and release lignin from lignocellulosic biomass. Microwaves increase the temperature and structure of cellulose through the expansion of water molecules, decrease cellulose crystallinity and increase the efficiency of hydrolysis [15]. The principle of heating occurs because microwaves induce substances in lignocellulose biomass which cause resonance and vibration in high frequencies. Very fast friction causes the temperature to increase rapidly in just minutes [16]. Sarah et al. (2017) reported hydrolysis of EFB using sulfuric acid assisted by microwave irradiation. Increment of microwave power from medium to high level during exposure and prolong heating from 5 to 10 min elevated final temperature of hydrolysis process approximately 97 to 105°C [17]. In this study, investigation on the effect of microwave irradiation on hydrolysis of EFB using sulfuric acid was carried out to determine glucose concentration as raw material for bioethanol production.

2. Materials and Methods

The materials in this study were oil palm empty fruit bunch (EFB), sulfuric acid, and DNS (3.5-Dinitrosalicylic acid). Prior hydrolysis process, EFB was washed, dried under the sun, cut into fibers, and stored in a dry container. Hydrolysis of EFB with sulfuric acid performed in two stages each assisted by microwave irradiation to obtain temperature for hydrolysis as shown in Figure 1. Temperature during hydrolysis measured by using thermocouple K (Krupp and Closs size diameter of 3 x 300 mm (Mineral Insulated) C/w Cable 2 m) connected with thermo controller (Shimaden). First stage hydrolysis carried out at various acid concentration of 0.5 to 1.5%, various level of microwave power (medium, medium high, and high) and irradiation times from 5 to 10 min. Liquid from first stage hydrolysis process was separated from their residue and was taken for glucose determination using DNS method and UV-VIS Spectrophotometer at a wavelength of 550 nm, while residue fed into second stage hydrolysis process. The second reactor operates at constant acid concentration using similar combination of power and time with experimental design of first hydrolysis. Similar glucose procedure determination was repeated for liquid obtained from second hydrolysis. In this study, reactor was located near (A) and far (B) from the waveguide as source of microwave and evaluated effect of wave distance source on temperature and glucose yield. Experimental rigs and reactors position in the microwave cavity during irradiation are shown in Figure 2.

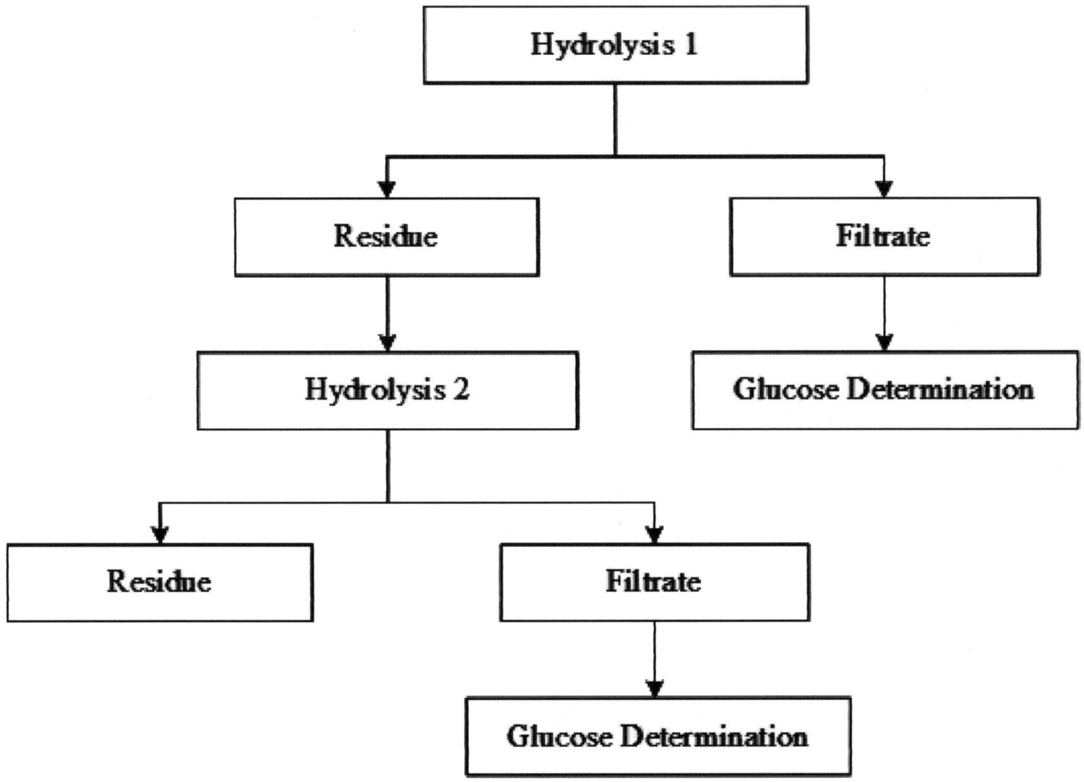

Figure 1. Scheme of experimental procedure

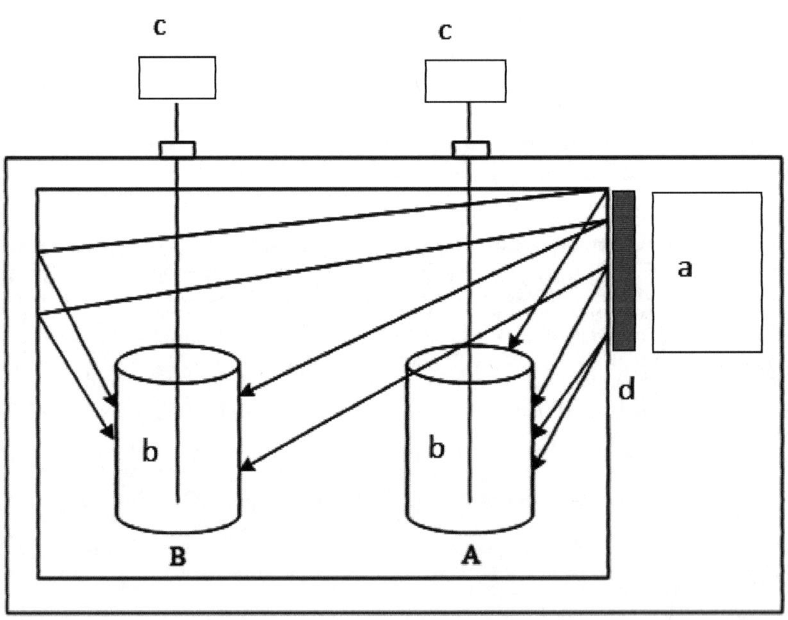

Figure 2. Experimental rigs consists of (a) magnetron (b) reactor with thermocouple (c) thermo controller (d) waveguide

3. Results and Discussions

3.1 Effect of wave distance source, power, time and acid concentration during 1st hydrolysis on glucose concentration

Effect of time, microwave power and acid concentration on glucose concentration inside reactor located near and far from the waveguide for first hydrolysis are shown in Figure 3. Overall, increment of time and power increased the glucose concentration, excluded glucose concentration from treatment at reactor B after exposure for 7.5 min. This study observed glucose obtained from first hydrolysis at medium power as depicts in Figure 3(a) increased from 15.36% and 26.22% as time extended from 5 to 7.5 min respectively. On the contrary, prolong hydrolysis period from 7.5 to 10 min at constant microwave power decreased glucose concentration fall to 19.55%. Similar phenomena observed also for time extension reactor B treatment as depict in Figure 3(b). Overall, increment of power from medium to medium high and high increased concentration of glucose along hydrolysis period exclude glucose obtained from treatment combination of irradiation using medium power for 7.5 min. The resulting glucose in reactor A increased from 29.82% to 30.56% and from 28.35% to 42.19% respectively as time for hydrolysis extended from 5 to 10 min. It was observed that glucose concentration obtained from hydrolysis at reactor A higher as compared to reactor B. It indicates wave distance source affected glucose concentration due many waves penetrated directly into the reactor A in very short time while wave in reactor B required more time to penetrate into EFB because some of the waves had to be reflected from cavity wall across the waveguide.

The effect of wave distance source to glucose concentration for irradiation using sulfuric acid concentration of 1 and 1.5% are shown in Figure 4. Glucose concentration as depicted in Figure 4(a) and 4(b) was influenced by wave distance source. EFB hydrolysis in reactor A yielded glucose higher than reactor B if hydrolysis performed for maximum 7.5 min at power of medium and medium high level. Similar phenomena observed from Figure 4(c) and 4(d) with respect to wave distance source aspect. Increment of acid concentration from 0.5% as depicted in Figure 3 to acid concentration of 1 and 1.5% as shown in Figure 4 reduced the concentration of glucose except the one with power of medium treatment. Figure 4(c) and 4(d) shows glucose only obtained from treatment with power of high level. This indicated concentration of sulfuric acid more than 0.5% should performed with combination of high power and high concentration of acid to attain temperature of hydrolysis. Lower power failed to facilitate hydrolysis of EFB with higher acid concentration.

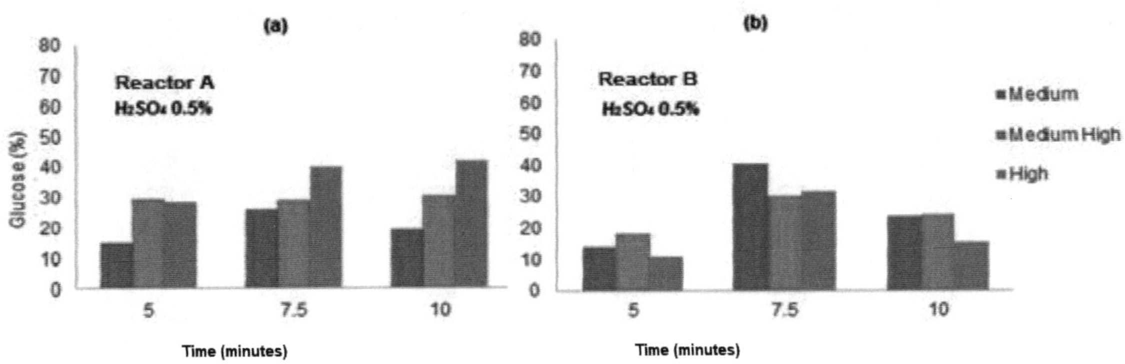

Figure 3. Resulting glucose obtained from 1st hydrolysis using 0.5% sulfuric acid in reactor A and B

3.2 Effect of wave distance source, time and power during 2nd hydrolysis on glucose concentration

This study carried out second hydrolysis to investigate remaining glucose in the residue from first hydrolysis. The 2nd hydrolysis aimed to evaluate the effects of wave distance sources, time and microwave power to glucose concentration. Figure 5 shows glucose concentration obtained after hydrolyzed residue from 1st hydrolysis in reactor A and B respectively at constant acid concentration (0.5%). The Figure 5 illustrates possible indication to conduct two stage hydrolysis of EFB with the aims to increase the yield total glucose but difficult to make any conclusion on this observation data due to inconsistency trend of resulting glucose obtained in the 2nd hydrolysis.

Figure 5(a1), Figure 5(a2) and Figure 5(a3) show performance of samples originated from residue obtained after hydrolysis for 5 min in the previous stages. Increment of time from 5 to 7.5 and 10 min as depict in Figure 5(a1), Figure 5(a2) and Figure 5(a3) have not significant effect to resulting glucose at elevated microwave power. This study observed similar phenomena also occurred during hydrolysis of samples originally came out from hydrolysis for 7.5 min in the 1st stages as depict in Figure 5(b1), Figure 5(b2) and Figure 5(b3). Excluded hydrolysis for 7.5 min, this study observed every treatment combination resulted glucose. Some of those treatment resulted glucose concentration higher than their previous hydrolysis. This explain residue from 1st stage hydrolysis still contained residual cellulosic and hemicellulose that converted further into glucose contents in 2nd stage hydrolysis. Figure 5(c1) to Figure 5(c3) show effect of time to glucose concentration for some treatments such as treatment using microwave power of medium high. Time increment for this particularly from 5 to 10 min increased glucose concentration from 10 to 33%. This study observed increment of acid concentration influenced hydrolysis of 2nd stage. Overall concentration increment from 0.5 to 1% increased the concentration of glucose and decreased at concentration of 1.5%.

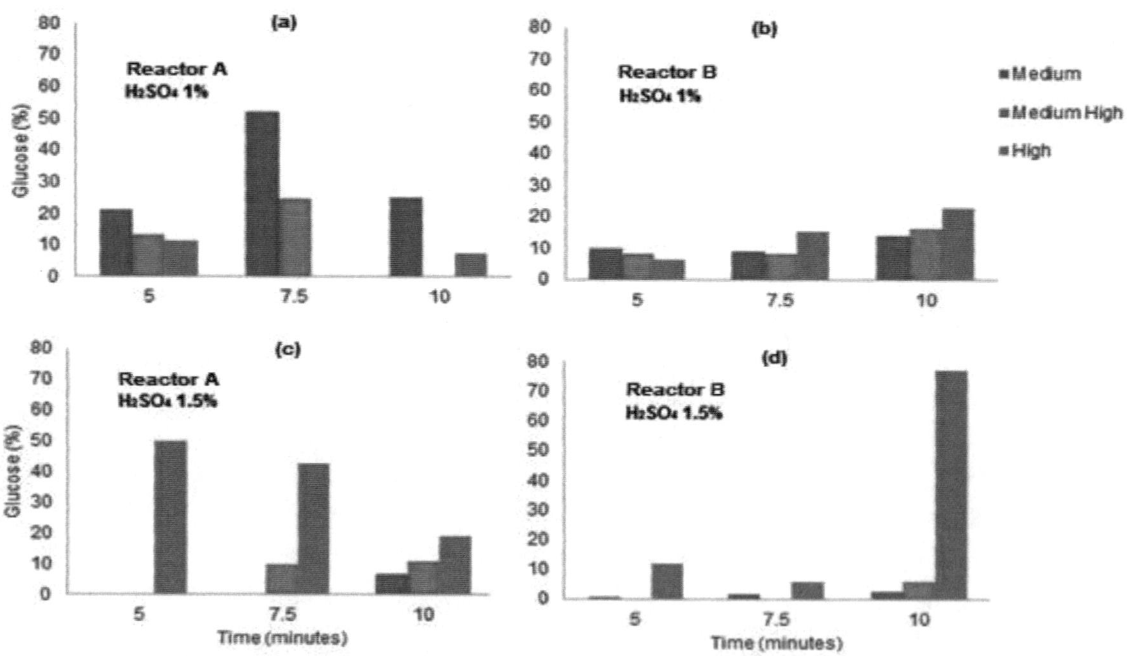

Figure 4. Resulting glucose obtained from 1st hydrolysis using sulfuric acid of 1 and 1.5% in reactor A (a and c) and reactor B (b and d)

Comparing glucose obtained in reactor A and reactor B as depict in Figure 5, hydrolysis performed in reactor A resulted higher glucose concentration as compared to reactor B's results. This study

IOP Conf. Series: Materials Science and Engineering **543** (2019) 012065 doi:10.1088/1757-899X/543/1/012065

proposed intensity of microwave energy headed for reactor B lower than to reactor A during their penetration into material in reactor B and A respectively. Reactor located near wave sources received more microwave energy and dielectric materials (residues and acid solutions) inside the reactor absorbed microwave energy and converted it into thermal energy. Overall final glucose concentration influenced by wave distance source. Reactor on the position near the waveguide (reactor A) will obtained higher glucose than reactor B that located quite far from the waveguide.

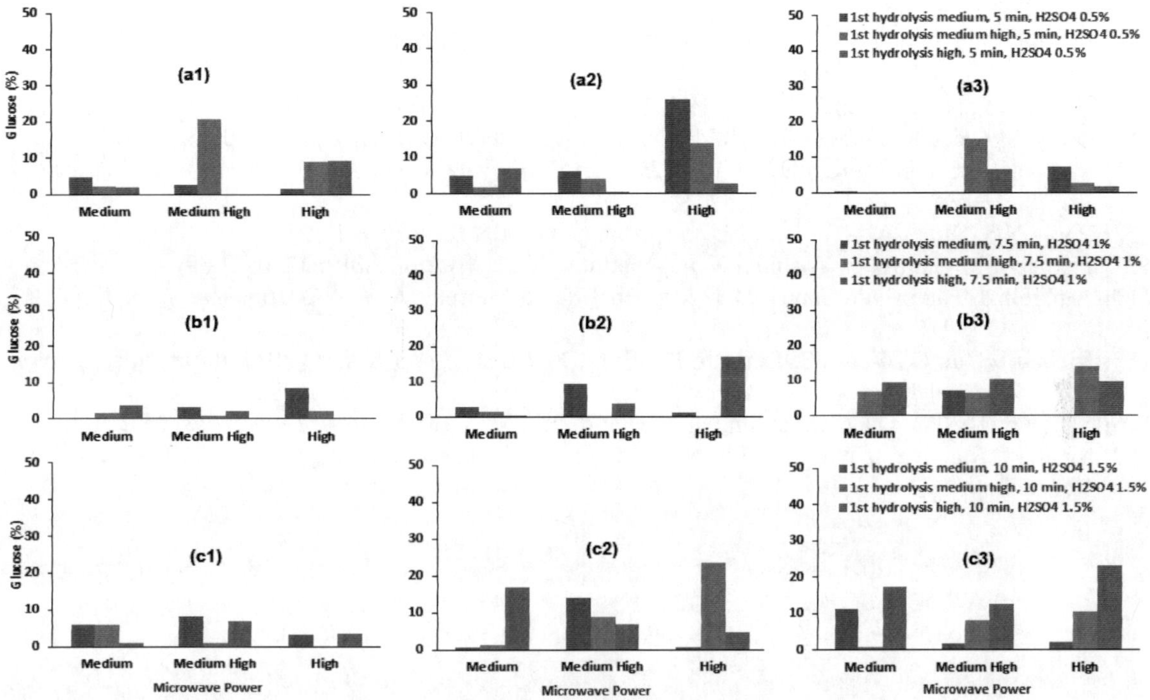

Figure 5. Resulting glucose obtained from 2nd hydrolysis using sulfuric acid of 0.5% in reactor A for 5 min (a1, b1, c1); 7.5 min (a2, b2, c2) and 10 min (a3, b3, c3)

4. Conclussion

Microwave energy can be used to assist 2 stages hydrolysis of palm oil EFB. Overall yield of glucose obtained from first hydrolysis increase at elevated time and power after irradiated by microwave energy in sulfuric acid of 0.5% solution. Reactor A usually resulted higher glucose as compared to reactor B, but treatment at elevated acid concentration reduced the yield except the one with high power irradiation. No significant contribution of second stage hydrolysis to overall glucose in this study. The maximum resulting glucose from first hydrolysis was 77% obtained from hydrolysis performed at high microwave power using sulfuric acid of 1.5% for about 10 minutes. Irradiation proceeded as microwave emitted from waveguide into the reactor that placed far from the waveguide. On the contrary, second hydrolysis yielded glucose of 43% after irradiated fiber residue for 5 minutes using sulfuric acid of 0.5% and microwave power of medium high.

Acknowledgement

This research sponsored by Directorate of Research and Community Service, Directorate General of Research Development and Reinforcement, Ministry of Research, Technology and Higher Education of Republic of Indonesia (DRPM RISTEK DIKTI) under Financial Agreement of Research and Community Service fiscal year of 2019.

References

[1] Kumar V B, Pulidindi N and Gedanken A 2016 *J. Mol. Catal. A: Chem.* **417** 163-67
[2] Sanchez C, Vidal J P, Lienemann C P and Todoli J L 2018 *Fuel Proces. Technol.* **173** 1-10
[3] Or K H, Putra A and Selamat M Z 2017 *Appl. Acoustics* **119** 9-16
[4] Eryati D, Rahmath A, Hartinie M and Mohd Khalizan S 2018 *Renew. Energ.* **129** 285-98
[5] Yang J, Kim J E, Kim H E, Yu J H, Cha Y L and Kim K H 2017 *Process Biochem.* **58** 211-16
[6] Taherzadeh M J, Millati R and Niklasson C 2001 *J. Appl. Biochem. Biotechnol.* **95** 45-57
[7] Sukaribin N and Khalid K 2009 *Ind. Crops Products* **30**179-183
[8] Sarah M and Taib M R 2017 *Proc. Int. Conf. on Chemical Materials and Process* vol 1879 (Beijing:China/AIP) p 05000 1-7
[9] Cheng S F, Nor L M and Chuah C H 2011 *Ind. Crops Products* **34** 967-71
[10] Chow M C and Ma A N 2007 *J. Microwave Power Electromagnetic Energ.* **40** 165-173
[11] Metaxas A C and Meredith R J 1988 *Industrial Microwave Heating* (London: Peter Peregrinus Ltd.)
[12] Pooja N S, Sajeev M S, Jeeva M L and Padmaja G 2018 *Biotech.* **8** 1-12
[13] Pakkapol K, Gaurav T, Yoshiaki K and Matthew J R 2016 *BioResour.* **11** 1672-89
[14] Albarelli J, Paidosh A, Santos D T, Marechal F and Meireles M A A 2016 *Chem. Eng. Trans.* **47** 49-54
[15] Richen L, Jun C, Wenlu S, Lingkan D, Binfei X, Junhu Z and Kefa C 2015 *Bioresour. Technol.* **182** 1-7
[16] Sangian H F, Pangau J R, Tamuntuan G H and Zulnazri Z 2018 *Chem. Eng. Trans.* **65** 229-34
[17] Sarah M, Maulina S, Widyastuti S, Ningsih D S, Madinah I and Salamah S 2017 *J. Eng. Appl. Sci.* **12** 5655-59.

IOP Conf. Series: Materials Science and Engineering **543** (2019) 012066 doi:10.1088/1757-899X/543/1/012066

Effect of Biomass Composition and Co-pyrolysis Retention Time on Reduction of Non-neutral CO_2 Emission from Combustion of Hybrid Coal in Power Plant

S Handoko, W Wulandari, J Rizkiana, M A Ridha, H A Prasetyo and D Sasongko
Department of Chemical Engineering, Faculty of Industrial Technology, Institut Teknologi Bandung

E-mail: slamethan@gmail.com; sasongko@che.itb.ac.id

Abstract. Indonesia's lignite coal and sawdust production are 131,05 and 11,82 million tonnes, respectively, which are potential to become raw material of hybrid coal through co-pyrolysis. Hybrid coal utilization will reduce non-neutral CO_2 emission, a part of CO_2 emission released from combustion of biomass. The aim of this study is to determine the effects on biomass composition and co-pyrolysis retention time on reduction of non-neutral CO_2 emission from combustion of hybrid coal in power plant. Co-pyrolysis was conducted in a vertical tubular furnace under an inert condition in atmospheric pressure and temperature 300° C. Biomass composition and co-pyrolysis retention time was varied from 20 to 40 %-mass and from 30 to 90 minutes, respectively. Hybrid coal was characterized with proximate, ultimate and calorific value analysis. The result of analysis was used as an input on simulation to determine reduction of non-neutral CO_2 emission on hybrid coal combustion in power plant. An increase in co-pyrolysis retention time (CRT) from 30 to 90 minutes increases the CO_2 emission from 13.25 to 13.30 %-vol. While increase in biomass composition from 20 to 40 %-mass reduce CO_2 emission from 13.18 to 12.81 %-vol. Non-neutral CO_2 emission reduction rises from 24.22 to 26.86 % along with the increased of biomass composition from 20 to 40 %-mass. Average of non-neutral CO_2 emission reduction is about 26.30 % in all variation. CO_2 emission of hybrid coal as fuel for power plant increased along with the increased of co-pyrolysis retention time (CRT), which was about 780-830 kg CO_2/MWh. The highest CO_2 emission was achieved from co-pyrolysis product hybrid coal utilization with 30 %-mass of biomass composition.

1. Introduction

Greenhouse gases emissions contain about 77 % CO_2 which make CO_2 gas as the biggest greenhouse gas in the atmosphere and give big impact in global warming [1]. Since the First Industrial Revolution took place in 1750, the raising demand for electricity due to economic growth has been secured mostly using fossil fuels, such as coal, oil and nuclear power. This adds enormous amounts of greenhouse gases (GHG) such as CO_2 to the atmosphere, which is the most significant driver of climate change [2]. In Indonesia, about 35 % of CO_2 emission in energy sector come from power plant which 67 % of them (0,203 Gt CO_2) is come from coal power plant. Indonesia's lignite coal and saw dust production are 131.05 and 11.82 million tonnes, respectively, which are potential to become raw material of hybrid coal through co-pyrolysis. Hybrid coal utilization will reduce non-neutral CO_2 emission, a part of CO_2 emission released from combustion of biomass. Hybrid coal is coal co-pyrolysis product between coal and biomass. Carbon print estimation used to calculate CO_2 emission reduction from coal and biomass

IOP Conf. Series: Materials Science and Engineering **543** (2019) 012066 doi:10.1088/1757-899X/543/1/012066

fuel (hybrid coal) and the result was a reduction of CO_2 emission about 21.2 % compare to pure coal fuel [3]. Different from previous researches [4, 5] which used coal and torrefied biomass on calculation and simulation in power plant, this research use hybrid coal and biomass in calculation and simulation of power plant. Laboratory test and simulation were conducted in this study to determine the effect of %-mass of biomass and co-pyrolysis retention time to the reduction of power plant emission.

2. Material and methods

2.1. Materials
Materials used in this study were Indonesian sub-bituminous coal and mahogany sawdust obtained from a furniture factory in Ciamis, Indonesia.

2.2. Methods

2.2.1. Co-pyrolysis experiment. Co-pyrolysis was conducted in a vertical tubular furnace under an inert condition in atmospheric pressure and temperature 300° C. Both of sawdust and coal were crushed and sieved in order to get smaller particle size, 50-100 mesh for sawdust and 30-100 mesh for coal. Process parameters which were investigated on this experiment were sawdust composition on the mixture and retention time of co-pyrolysis process. Sawdust composition were varied at 20, 30, and 40 %-mass from total mixture of sawdust and coal while co-pyrolysis retention time were varied at 30, 60, and 90 minutes. Hybrid coal characteristics were investigated by the result of proximate analysis, ultimate analysis and calorific value measurement.

2.2.2. Calculation simulation for CO_2 emission in power plant. In this study, CO_2 emission from hybrid coal combustion generated from co-pyrolysis experiment in power plant was estimated by modelling and simulation method. Model as shown in Figure 1 was used. Power plant configuration consist of boiler, steam turbine, condenser and generator. Developed model simulation was established. Simulation calculation was done by using Microsoft excel. The Gibbs minimization method (Pheng-Robinson) was used for all calculation. It is worth noting that ash and N_2 are considered as the inert material and pressure drop and mass transfer limitation are considered to have a minor effect on the process. In addition, an adiabatic condition is assumed in the simulation [6]. The composition and the temperature of flue gas was estimated using the Gibbs free energy minimization method. Given the expected operating temperature is possible to facilitate the reactions to reach the equilibrium [7]. Characterization result of six hybrid coals from co-pyrolysis experiment was used as an input data for the simulation by using 1 kg/s as flow rate base. Several assumptions which used in this simulation were 79 % N_2 and 21 % O_2 air composition, 40 % excess air, 20 bar operational pressure and 400° C temperature. Boiler, steam turbine and generator efficiency were 75 %, 85 % and 90 %, respectively. Simulation was established to model mass balanced, energy balanced and thermodynamics system of power plant.

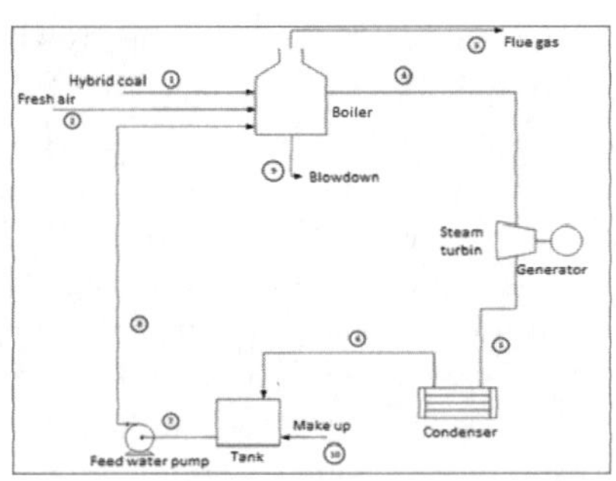

Figure 1. Process diagram of the power plant configuration

3. Result and discussion

3.1. Effect of sawdust %mass

Table 1 shows characterization data of six hybrid coals obtained from co-pyrolysis experiment. Code xxyy means xx%-mass sawdust in the feedstock pyrolized for yy minutes. Data in Table 1 shows that the biggest average calorific value was obtained from co-pyrolysis of 20 %-mass sawdust. It happened because the highest fixed carbon also obtained at 20 %-mass sawdust. This tendency was caused by synergy factor effect between coal and sawdust from co-pyrolysis at 300° C. Temperature range in which the occurrence of significant synergy is 200-500° C which has been previously reported in other study [8]. The presence of indigenous potassium species in raw biomass was a direct consequence of the synergy during co-gasification of coal and biomass [9]. Moreover, the synergy degree greatly depended on the coal and biomass origins [9].

Table 1. Characterization data of hybrid coal

Hybid coal	Proximate analysis (% mass)				Ultimate analysis (% mass)						Calorific value (cal/gr)
	FC	MC	VM	Ash	C	H	O	N	S	Ash	
2030	45.22	4.46	47.69	2.63	63.75	4.89	27.90	0.73	0.10	2.63	5895
2060	47.96	4.52	45.01	2.51	65.24	4.79	26.60	0.77	0.09	2.51	5947
2090	49.92	4.34	43.19	2.55	66.02	4.71	25.86	0.77	0.09	2.55	6025
3030	45.25	2.84	49.77	2.14	65.07	4.96	27.04	0.70	0.09	2.14	5987
3060	49.38	2.99	45.31	2.32	67.14	4.71	24.95	0.79	0.09	2.32	6150
3090	49.75	2.80	44.98	2.47	67.47	4.76	24.44	0.78	0.08	2.47	6224
4030	41.43	4.41	52.28	1.88	61.84	5.08	30.56	0.57	0.07	1.88	5659
4060	46.71	4.51	46.53	2.25	64.36	4.87	27.81	0.64	0.07	2.25	5907
4090	49.36	4.08	44.31	2.25	65.74	4.76	26.53	0.65	0.07	2.25	6031

CO_2 emission in %-mass sawdust and CRT variation is shown in Figure 2. CO_2 emission tends to decline with the increasing of %-mass sawdust in most of co-pyrolysis retention time (CRT) variation because fixed carbon content decreasing along with increasing %-mass sawdust as shown in Table 1. Fixed carbon participate in CO_2 compound formation. Therefore decreasing in fixed carbon will decrease CO_2 emission. It is notable that all biomass samples have higher CO_2 emission factors than those of coals burned herein because of their lower heating values [10]. Higher biomass composition resulted higher CO_2 emission. But for 30 %-mass sawdust composition, CO_2 emission is lower compare to 20 % and 40 %-mass sawdust. It happened because 30 %-mass sawdust has higher heating value which caused by synergistic factor that occurred during co-pyrolysis process [11]. Amount of CO_2 emission at 30, 60, and 90 minutes CRT at 20 and 30 %-mass of sawdust variation are similar, 13.2 %-vol. While at 40 %-mass sawdust, CO_2 emission have the same value, 12.81 %-vol, at 60 and 90 minutes CRT.

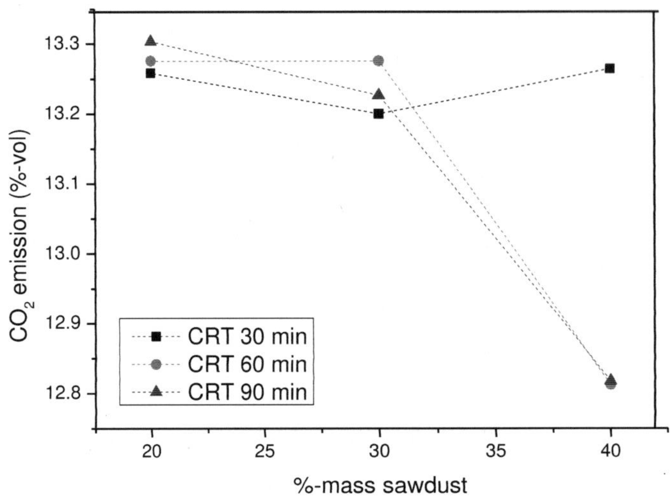

Figure 2. CO_2 emission as function of %-mass sawdust and CRT

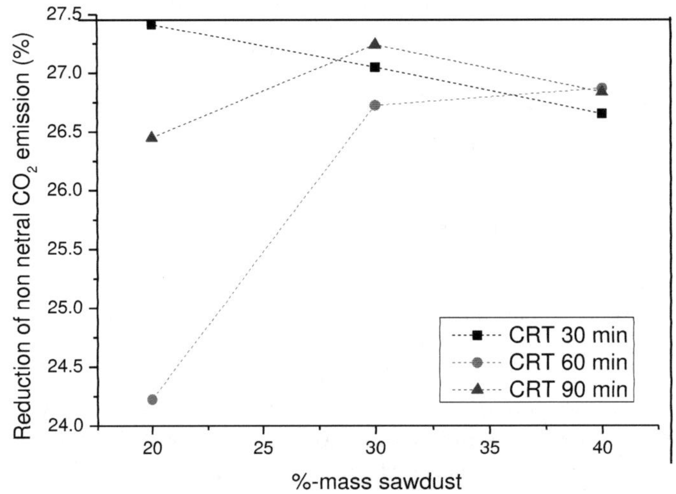

Figure 3. Reduction of non-neutral CO_2 emission as function of %-mass sawdust and CRT

Figure 3 shows effect of %-mass sawdust to non-neutral CO_2 emission reduction. One of the benefits of co-firing biomass is to reduce CO_2 emission because CO_2 from the combustion of biomass fuel source is not assumed to increase the net atmospheric CO_2 levels [9]. Biomass carbon neutrality is a term used to differentiate biomass-derived carbon from fossil-fuel derived carbon with respect to its role in the global carbon cycle. It reflects the fact that biomass-derived carbon is part of a relatively rapid natural cycle that, when in balance, neither adds nor subtracts carbon to/from the atmosphere, whereas fossil fuel-derived carbon is not part of such a rapid cycle [12].

Figure 3 shows that in 30 min and 60 min CRT, the increase of 20 % to 40 %-mass sawdust has reduced the reduction of non-neutral CO_2 emission because the increase in %-mass sawdust will increase donor C element from biomass in combustion process of power plant [9]. While in 90 min CRT, the increase of %-mass sawdust from 20-30 % has increased the reduction of non-neutral CO_2 emission but the increase of %-mass sawdust from 30-40 % has reduced the reduction of non-neutral CO_2 emission. It caused by synergy factor occured at combustion temperature between C element from coal and sawdust. Non-neutral CO_2 emission reduction which was obtained from calculation was lied between 24.22 % and 27.42 %. Average value of non-neutral CO_2 emission reduction is about 26.30 % in all variation.
Figure 4 shows the effect of sawdust %-mass to output power. At every CRT variation, the increase of sawdust from 20 to 30 % cause the reduction of output power. But the increase of sawdust %-masss from 30 to 40 % make output power rises. From ultimate analysis result, C element increased at sawdust 20-30 %-mass and decease at synergy factor between coal and biomass at 30 %-mass sawdust therefore the highest output power was obtained [11]. The thermal decomposition of the hybrid coal and the formation of volatiles are influenced by the thermal interaction between the components at a mass load of straw up to 30 %-mass, thus enhancing the release of the volatiles and improving the combustion conditions with the positive effect on the main combustion characteristics (heat output from the device and product composition). At higher straw mass loads, the thermal decomposition of the mixture and the formation of volatiles predominately are influenced by the linier decrease of the mixture HHV, hydrogen and carbon content with the correlating liniar decrease of the biomass weight loss rate, volume fractios of volatiles at the cinbustor inlet determining thus negative effect of straw combustion characteristics [13]. Output power calculated from nine hybrid coals was lied at 2870 and 3225 MW. The biggest power was gained at 30 %-mass sawdust in 60 minutes CRT.

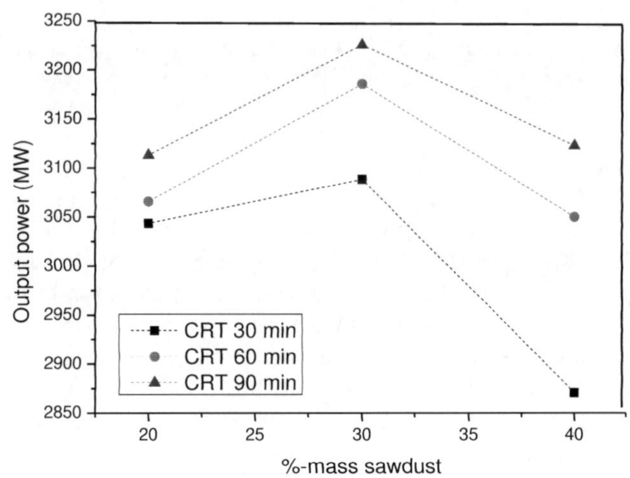

Figure 4. Output power generated as function of %-mass sawdust and CRT

Figure 4 shows that in 30 min, 60 min and 90 min retention time, the increased of sawdust %-mass from 20-30 % has increased tonnes CO_2/MW but the increased of sawdust %-mass from 30-40 % has reduced tonnes CO_2/MW. It happened because the measurements of the elemental composition and heating values of the mixtures have shown that an increase of the biomass load results in linear decrease of the

carbon and hydrogen content in the mixture with the correlating decrease of the heating value of the mixture [13].
At 30 %-mass sawdust, heating value rises so that CO_2 emission obtained per output power generated is the smallest among the three sawdust composition. It is caused by synergy effect of coal and sawdust during co-pyrolysis process at temperature 300° C [11]. Average amount of tonnes CO_2/MW as model prediction result is 0,798 kg CO_2/MW. This number is less than tonnes CO_2/MW obtained from 100 % coal. Typically, a subcritical coal-fired utility plant produces 900 kg CO_2/MWh, and 750 kg CO_2/MWh for a supercritical plant [4]. Other study shows number of 846 kg CO_2/MWh could be recaptured when 100 % fuel switched from coal to torrefied biomass [5].

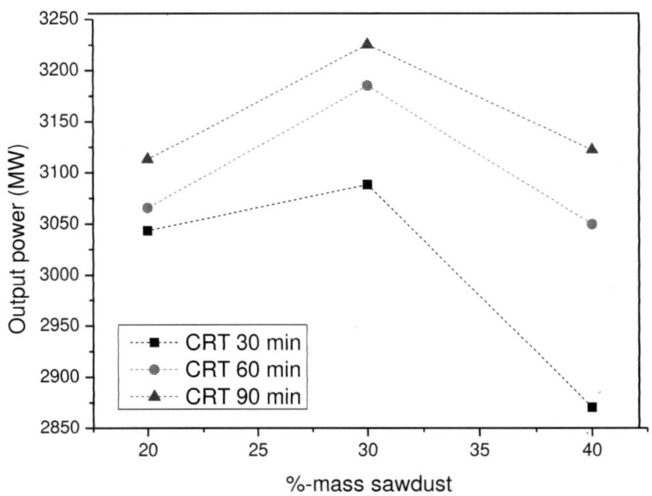

Figure 5. Amount of CO_2 generated per 1 MW power as function of %-mass sawdust and CRT

3.2. Effect of co-pyrolysis retention time (CRT)
For CRT 60 and 90 minutes, CO_2 emission generated from hybrid coal combustion tend to have the same trend. For every sawdust %-mass, 90 minutes CRT generated the smallest CO_2 emission. It happened because long CRT cause more H and O element released so that O/C and O/H ratio at hybrid coal become smaller which affect in increase in calorific value. For non-neutral CO_2 emission, the same trend also happen in every sawdust % mass. Only at CRT 30 minutes and 20 %-mass sawdust the trend is different. It is because CRT not have reached the maximum time to accommodate the release process of moisture content and volatile matter from coal and sawdust feed. Output power from power plant with hybrid coal fuel, CRT 90 minutes generated the biggest power because of it highest calorific value as a result of maximum co-pyrolysis process. CRT 30 minutes generated different trend in number of CO_2/MW because co-pyrolysis time was not sufficient to accommodate the release process of moisture content and volatile matter. Hybrid coal derived from co-pyrolysis with 90 minutes retention time and 30%-mass biomass composition reaches the highest output power and gives the lowest CO_2 emission. Summary of the result is shown in Table 2.

Table 2. Summary result

Hybrid coal	Reduction of non neutral CO_2 emission (%)	Output power (MW)	Tonnes CO_2/MWh
2030	27.41	3043.4	788.8
2060	24.22	3065.7	800.4
2090	26.45	3113.1	797.9
3030	27.05	3088.2	789.5
3060	26.72	3185.3	791.2
3090	27.24	3225.3	786.5
4030	26.65	2870.5	835.5
4060	26.87	3049.6	791.6
4090	26.84	3122.4	789.8

4. Conclusion

An increase in co-pyrolysis retention time from 30 to 90 minutes has increased the CO_2 emission from 13.25 to 13.30 %-vol. While increased in biomass composition from 20 to 40 %-mass has reduced CO_2 emission from 13.18 to 12.81 %-vol. Non-neutral CO_2 emission reduction has rised from 24.22 to 26.86 % along with the increased of biomass composition from 20 to 40 %-mass. Average of non-neutral CO_2 emission reduction was about 26.30 % in all variation. CO_2 emission of hybrid coal as fuel for power plant increased along with co-pyrolysis retention time about 780-830 kg CO_2/MWh. Hybrid coal derived from co-pyrolysis with 90 minutes retention time and 30%-mass biomass composition reaches the highest output power and gives the lowest CO_2 emission.

Acknowledgments

The authors would like to acknowledge the financial support from Kementerian Riset, Teknologi dan Pendidikan Tinggi through Penelitian Berbasis Kompetensi (PBK) 2018 Research Grant.

5. References

[1] Abdul Rahman F, Aziz Md M A, Saidur R, Abu Bakar W A W, Hainin M R, Putrajaya R, Abdul Hassan N 2017 *Renew. Sust. Energ. Rev.* **71** 112-26.
[2] Muneno and Otsuka 2018 *Chem. Eng. Tran.* **63** 43-48.
[3] Lee D W, Bae J S, Lee Y J, Park S R, Hong J C and Lee B H 2013 *Environ. Sci. Technol.* **37** 1704-10.
[4] IEA Coal Industry Advisory Board 2010 OECD/IEA, Paris, France.
[5] Li J, Zhang X, Pawlak-Kruczek H, Yang W, Kruczek P and Blasiak W 2014 *Energ. Convers. Manag.* **84** 503-11.
[6] Adnan M A and Muraza O 2017 *Energ. Fuel* **31** (7) 7471-81.
[7] Adnan M A and Susanto H 2017 *Int. J. Hydrogen Energy* **42** 10971-85.
[8] Sonobe T, Worasuwannarak N and Pipatmamomai S 2008 *Fuel Process. Technol.* **89**(12) 1371-78.
[9] Zhang Y, Geng P and Liu R 2017 *Bioresour. Technol. Part A* **245** 225-33.
[10] Rokni E, Ren X, Panahi A and Levendis Y A 2018 *Fuel* **211** 363-74
[11] Rizkiana J, Handoko S, Wulandari W, Ridha M A, Prasetyo H A and Sasongko D 2018 *AJChE* **18** (1) 68-75.
[12] EPA 2009 *Renewable fuels standard* (74 Fed. Reg. 24904).
[13] Barmina I, Kolmickovs A, Valdmanis R, Zake M, Kalis H and Strautins U 2018 *Chem. Eng. Tran.* **70.**

IOP Conf. Series: Materials Science and Engineering **543** (2019) 012067 doi:10.1088/1757-899X/543/1/012067

Pretreatment of Coal by Acid Leaching As Feedstock Preparation For Co-Gasification With Biomass

J Rizkiana, H N Firdausi, R Pranata, G A N Aulia, S F Maulana, W Wulandari and D Sasongko

Chemical Engineering Department, Institut Teknologi Bandung, West Java (40132) Indonesia

Abstract. Co-gasification process of coal and biomass is believed to be one of the promising way to produce hydrogen. The mixing of coal and biomass can solve the problems that usually happened during gasification of pure coal or pure biomass. Furthermore, the presence of synergistic effect between coal and biomass may increase the hydrogen production due to Alkali and Alkaline Earth Metal (AAEM) content in the biomass. However, the synergistic effect may be hindered by the presence of silica in the coal. The present study focuses on the reduction of silica content of coal by acid leaching. Based on the result, it is found that the use of hydrofluoric acid may significantly reduce the silica content shown by the reduction of ash content in the coal. Furthermore, acid leaching process also causes the increase of surface area of the coal which may increase the possibility of AAEM attachment to the surface and thus the gas production may increase as well.

1. Introduction

Provide an adequate background, context of the problems based on the literature review. State Gasification is thermal conversion of carbonaceous materials into synthesis gas in limited oxygen condition with gasifying agent [1]. It can be used to utilized coal cleanly as Indonesia has abundant reserve of coal. However, 92% of Indonesia coal is in low-rank state, such as subbituminous and lignite [2]. Gasification of low-rank coal results in low gas yield as it contains higher amount of volatile matter [3]. To increase gas yield, catalyst can be added to the sample to increase its reactivity. It is known that alkaline and alkaline earth metal (AAEM) can be used as the gasification catalyst as it can increase the carbon reactivity toward gasification reaction, especially in producing hydrogen [4]. AAEM can be found naturally in biomass.

Co-gasification of coal and biomass could increase carbon conversion and gas yield as well, especially hydrogen [5]. Thus, not only decreasing the non-neutral CO2 emission, co-gasification process may also increase the overall gasification efficiency [6]. Catalytic effect in co-gasification process occurs due to the existence of AAEM in biomass [7] and they are volatilized during gasification and may attach to the coal surface so that the reactivity of the coal may increase as well. This synergistic effect makes co-gasification attractive to be applied. However, silica content in coal may hinder the synergistic effects. Silica deactivates AAEM by converting them to form metal silicate which has no catalytic effect [8]. Therefore, silica content in coal must be reduced before it is used as the gasification feedstock.

Mineral content, including silica, in coal can be removed by demineralization process, such as acid leaching method [9-13]. When coal is rinsed with strong acid solution, some of the mineral (including silica) is expected to be leached out so that the mineral content in coal reduced to some extent. It is expected that by reducing the silica content, the gas production from co-gasification process of coal and biomass increase since the catalytic activity of AAEM will not be hindered. Furthermore,

Giudicianni et al. (2014) stated that demineralization of the feedstock may promote the char porosity [14]. The present study is done to determine the effect of acid solution concentration, temperature, reaction time, and particle size during acid leaching process of coal. Two main parameters are examined; mineral content and surface area of the leached coal. Low mineral content in coal makes it more reactive towards catalytic gasification and higher surface area means more volatilized AAEM attached to the coal surface and it may increase the coal reactivity as well.

2. Methods

2.1 Raw Material Preparation

Two types of low-rank coal are used in this experiment. The proximate and ultimate analysis of the coal samples were done at TEKMIRA, Bandung, and the results are shown in Table 1. The coal has particle size of 14-48 mesh and -48 mesh. Coal is dried at 105-110°C overnight.

Table 1. Proximate and ultimate analysis of sub-bituminous and lignite coal.

Parameter	Sub-bituminous	Lignite	Unit	Basis*
Proximate				
Moisture	13.54	13.81	%	adb
Ash	5.86	8.58	%	adb
Volatile matter	42.62	41.14	%	adb
Fixed carbon	37.98	36.47	%	adb
Ultimate				
Total sulfur	0.99	0.86	%	adb
Carbon	54.41	51.57	%	adb
Hydrogen	5.19	4.92	%	adb
Nitrogen	1.15	0.97	%	adb
Oxygen	32.40	33.10	%	adb

*adb = air dried basis

2.2 Demineralization of Coal

Demineralization process was done by washing coal with acid solution. 250 ml of hydrofluoric acid (HF) with concentration of 1 M, 2 M, and 3 M were used to leach 10 gram of coal samples. Coal was dispersed in HF solution in a teflon beaker and stirred by a magnetic stirrer for 2, 3, or 4 hours. Liquid was then separated by using a filter funnel, while the solid (coal) is then washed with aquadest until the pH of water is neutral. Demineralized coal was dried in the oven at temperature of 105-110°C before further analysis.

2.3 ORF Primary Frames

Demineralized coal was analyzed to measure its ash content, mineral composition, and surface area. Ash content were determined by gravimetry method. Certain amount of coal sample was weighted and then calcined at 700°C for 2 hours to remove all the burnable matter. The remain unburned component was weighted and it is determined as the ash content. Surface area of demineralized coal was analyzed by Surface Area Analyzer (QUANTACHROME NOVA e-4200) performed at Laboratorium Instrumen Analisis Labtek X ITB. Since the ash sample was limited mineral composition of coal ash was analyzed by Scanning Electron Microscopy with Energy Dispersive X-Ray (SEM-EDX, HITACHI SU3500) done at Pusat Penelitian Nanosains dan Nanoteknologi (PPNN) ITB.

3. Results and Discussion

IOP Conf. Series: Materials Science and Engineering **543** (2019) 012067 doi:10.1088/1757-899X/543/1/012067

3.1 Decreasing of Ash Content

For the demineralization process, the effect of acid solution concentration, temperature, and reaction time towards decreasing of ash content is determined. Figure 1 and figure 2 show ash content of demineralized coal which indicated by yellow colour as the highest ash content (10%) and dark blue as the lowest ash content (1%) for both types of coal. In subbituminous coal, the decreasing of ash content is 2.79-4.89%. The similar results obtained by Bolat et. al. (1998) and Steel (2001) for ash content of coal demineralized by HF which achieved the degree of demineralization up to 79%.

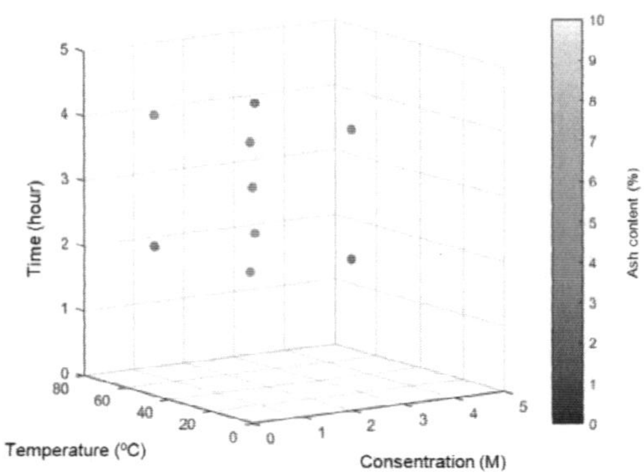

Figure 1. Ash content of demineralized subbituminous coal.

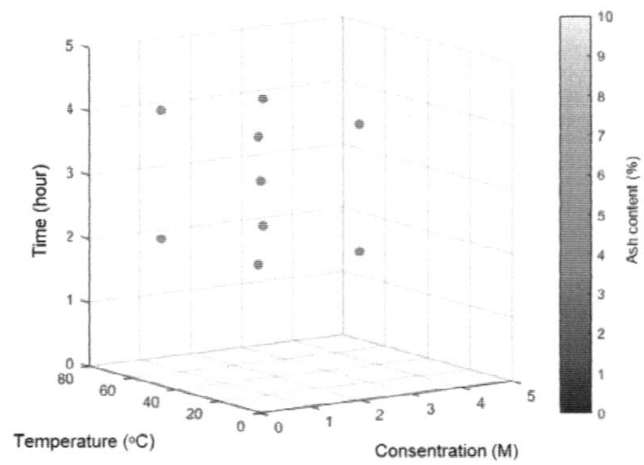

Figure 2. Ash content of demineralized lignit coal.

The present experiment utilized full factorial design (FFD) with center point which is repeated three times. For demineralization of subbituminous coal, analysis of variance (ANOVA) for factorial design is shown in Table 2. It can be seen that the main factor (i.e. concentration, temperature, and time) and interaction factor are considered significant as indicated by P-value which is smaller than α which is set to 5%. Furthermore, the R^2 value is 99.68%, indicating that the deviation of equation model is close enough with the experimental data. Thus, the model used in this case is quite accurate. Regression result of all parameters and experiment data for demineralization of subbituminous coal solution is shown in equation 1. This equation is using coded units with 1 as higher value, -1 as lower value and 0 as middle value.

$$Content_{ash} = 0.037033 - 0.002311C - 0.001127T - 0.000868R + 0.003807CT - 0.002311C\,R - 0.006149\,C\,T\,R + 0.003898\,C_t\,P_t \quad (1)$$

Table 2a. ANOVA of demineralization of subbituminous coal with HF solution.

Source	DF	Adj SS	Adj MS	F-Value	P-Value
Model	7	5.5E-04	7.9E-05	135.03	0.001
Linear	3	5.9E-05	2.0E-05	33.54	0.008
Concentration (C)	1	4.3E-05	4.3E-05	72.99	0.003
Temperature (T)	1	1.0E-05	1.0E-05	17.35	0.025
Time (R)	1	6.0E-06	6.0E-06	10.29	0.049
2-Way Interactions	2	1.6E-04	7.9E-05	135.55	0.001
C*T	1	1.2E-04	1.2E-04	198.06	0.001

Table 2b. ANOVA of demineralization of subbituminous coal with HF solution.

Source	DF	Adj SS	Adj MS	F-Value	P-Value
C*R	1	4.3E-05	4.3E-05	73.03	0.003
3-Way Interactions	1	3.0E-04	3.0E-04	516.86	0.000
C*T*R	1	3.0E-04	3.0E-04	516.86	0.000
Curvature	1	3.3E-05	3.3E-05	56.64	0.005
Error	3	2.0E-06	1.0E-06		
Total	10	5.6E-04			

3.2 Molecular Dynamics Results

Demineralization process using acid solution is expected decrease the silica content in coal ashes. The relative decrease of silica content is shown in Table 3. In demineralization of coal with HF solution, the relative decrease of silica content in subbituminous and lignite coal are 75.24% and 92.57%. This result is corresponding to some literatures state that mixing coal with HF solution could decreases the silica content of coal ash [9, 15, 16]. Previous researchers have proven that demineralization can decrease the silica content in coal ash [9, 17, 16]. The decreasing of silica content can improve the co-gasification of coal and biomass. Rizkiana et al. (2014a) stated that silica content can deactivate catalyst in co-gasification process [4]. Using demineralization process, the catalytic activity during co-gasification process is expected to work more intensive.

Table 3. Decreasing of silica content in coal ash after demineralization process.

Coal type	Acid solution	Differential relative
Subbituminous	HF	75.24%
Lignite	HF	92.57%

$$SiO_2 + 4\,HF \rightarrow SiF_4 + 2\,H_2O \qquad \Delta G = -3.2\ \text{kJ/mol} \tag{2}$$

$$C + 4\,HF \rightarrow CF_4 + H_2 \tag{3}$$

Hypothetic reactions of HF with carbon (C) are also analyzed as written in equation 3. The reaction of acid solution with carbon could affect the demineralization process because mineral matters are dispersed in carbon matrix of coal. The main composition of coal is carbon, thus reaction between acid solution and carbon played a significant role in decreasing of mineral content. The reaction is analyzed by using Gibbs free energy as well and it is found that calculated ΔG is also negative indicating that the reaction between HF and C are also likely to happen during demineralization process.

Furthermore, the effect of coal particle size in demineralization process is also examined in this experiment. The ash contents of demineralized coal with particle size variation is shown in Figure 3 for subbituminous coal, and Figure 4 for lignite coal. The effect of particle size in demineralization process using HF is not significant. Based on the Gibbs free energy calculation, HF could react spontaneously with carbon. Therefore, the mineral contents in coal matrix can to react with HF regardless the mineral are exposed to the surface of coal or not. HF coal can penetrate to the inner core of coal since it can react with the carbon matrix. Hence, in this case, the particle size does not affect demineralization process.

3.3 Surface area

In demineralization process, some reactions occur between mineral content in coal ash and acid solution as written in equation 2 and 3. Mineral content of coal is present in the coal matrix. The reaction between the mineral content and the acid solution causes the coal matrix to open to form pores in the coal. It causes the surface area of coal to increase. Data of surface area of raw subbituminous and lignite coal were 0.803 m^2/g and 8.552 m^2/g respectively. Both in subbituminous and lignite coals, the demineralization process using acid solution could increase the surface area of coal. Coal surface area data before and after demineralization process is shown in Table 4.

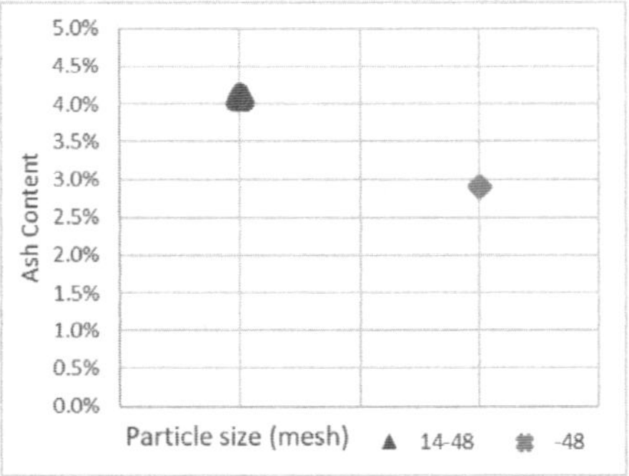

Figure 3. Effect of particle size towards the ash content of demineralized subbituminous coal.

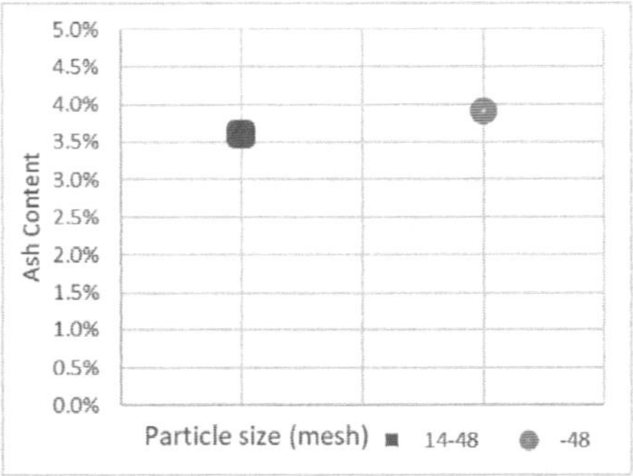

Figure 4. Effect of particle size towards the ash content of demineralized lignite coal.

Table 4. Coal surface area before and after demineralization process.

Sample no-	Coal type	Acid solution	Surface area (m^2/g)
1	Subbituminous	-	0.803*
2		HF	131.046
3	Lignite	-	8.552*
4		HF	80.516

*raw coal

The increase of coal surface area may increase gas production in co-gasification process of coal and biomass [4]. The catalysis process occurring in co-gasification begins with the release of alkali and alkaline earth metal (AAEM) contents of the biomass. Then, the AAEM content is attached to the coal surface so as to increase the conversion in the co-gasification process [4]. Increasing the surface area of coal may increase the amount of AAEM that can attach to the coal surface. Therefore, an increase in the area of the coal surface can improve the reactivity of coal during the co-gasification process.

4. Conclusion

Demineralization of coal has been performed by using hydrofluoric acid. By measuring the mineral content of the treated coal, it can be concluded that the use of acid leaching successfully decreases the mineral content of the coal to some extent. The most decreasing mineral contents occurred for Si component as it can be removed up to 92% from its original content. Furthermore, acid leaching can also increase the coal surface area quite significantly. The total surface area increases from 0.8 to 131 m^2/g for subbituminous coal sample. It indicates that coal pretreatment by acid leaching can increase the coal properties so that it is more reactive when it is used as the feedstock for gasification process.

IOP Conf. Series: Materials Science and Engineering **543** (2019) 012067 doi:10.1088/1757-899X/543/1/012067

5. References

[1] Basu P 2010 *Biomass Gasification and Pyrolysis Practical Design and Theory* (Cambridge: Academic Press)

[2] Badan Pengkajian dan Penerapan Teknologi (BPPT) 2016 *Outlook Energi Indonesia 2016* (Jakarta: Perpustakaan Nasional RI)

[3] Duan W, Yu Q, Liu J, Wu T, Yang F and Qin Q 2016 *J. Exp. And Kinetic Study Of Steam Gasification Of Low-Rank Coal In Molten Blast Furnace Slag Energy* **111** 859-868.

[4] Rizkiana J, Guan G, Widayatno W B, Hao X, Huang W, Tsutsumi A and Abudula A 2014 *Effect Of Biomass Type On The Performance Of Co-Gasification Of Low Rank Coal With Biomass At Relatively Low Temperatures* **134** 414–419.

[5] Rizkiana J, Guan G, Widayatno WB, Hao X, Li X, Huang W, and Abudula A 2014 *Promoting Effect Of Various Biomass Ashes On The Steam Gasification Of Low-Rank Coal* **133** 282–288.

[6] Ali D A, Gadalla M A, Abdelaziz O Y and Ashour F H 2016 *Modelling of Coal-Biomass Blends Gasification and Power Plant Revamp Alternatives in Egypt's Natural Gas Sector* **52** 49-54.

[7] Iwaki H, Ye S, Katagiri S and Kitagawa K 2004 *Wastepaper gasification with CO_2 or steam using catalysts of molten carbonates* **270** 237-243.

[8] Risnes H, Fjellerup J, Henriksen U, Moilanen A, Norby P and Papadakis K 2003 *Calcium Addition In Straw Gasification* **82** 641-651.

[9] Steel K M, Besida J, O'Donnel T A and Wood D G 2001 *Production Of Ultra Clean Coal Part I—Dissolution Behaviour Of Mineral Matter In Black Coal Toward Hydrochloric And Hydrofluoric Acids* **70** 171-192.

[10] Mukherjee S, Borthakur P C 2001 *Chemical Demineralization/Desulphurization Of High Sulphur Coal Using Sodium Hydroxide And Acid Solutions* **80** 2037-2040.

[11] Behera S K, Chakraborty S and Meikap B C 2017 *Chemical Demineralization Of High Ash Indian Coal By Using Alkali And Acid Solutions* **196** 102-109.

[12] Ratanakandilok S, Ngamprasertsith S and Prasassarakich P 2001 *Coal Desulfrization With Methanol/Water And Methanol/KOH* **80** 1937-1942.

[13] Wu Z and Steel K M 2007 *Demineralization Of A UK Bituminous Coal Using HF And Ferric Ions*, **86** 2194-2200.

[14] Giudicianni P, Cardone G, Cavaliere A and Ragucci R. 2014 *Effect of Feedstock Demineralization on Physico-Chemical Characteristics of Arundo Donax Derived Biochar* **37** 85-90.

[15] Meshram P, Purohit B K, Sinha M K, Sahu S.K and Pandey B D 2015 *Demineralization Of Low Grade Coal* **41** 745-761.

[16] Wijaya N and Zhang L 2011 *A Critical Review Of Coal Demineralization And Its Implication On Understanding The Speciation Of Organically Bound Metals And Submicrometer Mineral Grains In Coal* **25** 1-16.

[17] Bolat E, Saglam S and Piskin S 1998 *Chemical Demineralization Of A Turkish High Ash Bituminous Coal* **57** 93-99.

An improvement study of biodiesel production from rice bran via non-catalytic in-situ transesterification using a subcritical water-methanol mixture

M Yasmin, F D Alfaty, H S Pradipta, M Rachimoellah and S Zullaikah[1]
Deparment of Chemical Engineering, Institut Teknologi Sepuluh Nopember (ITS), Kampus ITS Keputih Sukolilo, Surabaya 60111 Indonesia

Email : szulle@chem-eng.its.ac.id

Abstract. A non-catalytic in-situ transesterification of biodiesel production from rice bran using subcritical water-methanol mixture was found to be unaffected by initial moisture and free fatty acids (FFA) contents. The method has been known as environmentally friendly conversion since no catalyst was used. Acid catalyst has an important role in in-situ transesterification due to their ability to accelerate the oil extraction from the bran and the reaction rate. In this study, CO_2 as pressurizing gas was added into subcritical water methanol mixture to increased yield and content of biodiesel. Effect of different operation pressure (P = 40-100 bar) at subcritical water methanol mixture (T = 200°C) and reaction time of 180 min on the yield and content of biodiesel were investigated. Rice bran with initial FFA content of 40.42 % has been subjected into hydrothermal reactor together with water methanol mixture to produce rice bran oil (RBO)-based biodiesel. The more CO_2 as pressurizing gas was added into the hydrothermal reactor, the more yield and content of biodiesel were obtained. The highest yield and content of biodiesel of 98.00±1.41% and 83.93%, respectively, were produced under P = 100 bar.

1. Introduction

Fossil fuels are still the most engage fuels to fulfill the energy demands in both industrial and transportation sectors. The negative effect to environment and depletion of fossil resource increase a demand of alternative fuels to replace fossil fuels utilization. Biodiesel, a promising and potential alternative fuel, is considered to replace petroleum diesel because of its non-toxicity, renewable, and environmentally friendly fuel [1-3]. Furthermore, biodiesel can also be blended with petroleum diesel or directly used in diesel engine [1,3]. Vegetable oils come from soybean, palm, rapeseed, and others are the most common feedstock used in biodiesel production on industrial scale. However, those feedstocks can drive high price of production cost [2,4]. Feedstocks which are cheaper, abundant, and sustainable can be an alternative to reduce the production cost. Rice bran, by-product of rice milling, becomes a potential feedstock for biodiesel production. Since it is abundant, cheap, and containing high oil content (10-26%) [4]. However, high free fatty acids (FFA) content in rice bran due to hydrolysis of triglycerides by lipase enzyme becomes an obstacle and cause a saponification which is reaction between oil and base catalyst. The saponification reaction can lower yield of biodiesel and hinder the separation and purification of biodiesel [5]. Various methods have been proposed to overcome those obstacles. Zullaikah et al [6] conducted two-step acid-catalyzed methanolysis to produce biodiesel from dewaxed/degummed rice bran oil (RBO) under low temperature for 8 hours. The conversion of 98% fatty acid methyl esters (FAME) was obtained under optimum condition. Ozgul-Yucel and Turkay [7] used simultaneous extraction and alkyl transesterification with acid catalyst at low temperature to convert rice bran oil into FAME. Shiu et al [2] investigated rice bran conversion using two-step in-situ transesterification from rice bran by acid and base catalyst simultaneously. While, Lai et al [8] used Novozym 435 and IM 60 as enzyme catalyst to accelerate reaction transesterification in biodiesel conversion. 98% of FAME and 81% of FAME was converted from RBO. Catalytic conversion technology is the common conversion method to convert biodiesel but the complicated separation and purification, lower rate of reaction, and longer reaction time become the drawbacks of those methods. Alternatively, non-catalytic conversion technology under sub/supercritical condition becomes an interest in rice bran-based biodiesel production due to insensitivity of FFA and water content of feedstock [4]. Kasim et al. [9] reported that 51.3% of FAME can be achieved using supercritical methanol at temperature 300 °C and pressure 30 MPa with CO_2 as co-solvent. Supercritical CO_2 can lower viscosity and have selectivity property [10]. While, Zullaikah et al. [4] proposed a non-catalytic in-situ conversion of rice bran using subcritical water-methanol mixture. The results showed that the higher yield of FAME (67.4 %) can be achieved under CO_2 atmosphere with temperature 200°C, pressure 4 MPa, and reaction time for 3 h. However, high temperature of supercritical condition was a drawback since its charred the rice bran and lower yield of biodiesel [9]. Non-catalytic conversion under subcritical condition is a potential conversion method to produce biodiesel from rice bran because it obtained high biodiesel yield in a shorter reaction time [4]. Furthermore, Parvez et al [11] stated that subcritical water extraction can be considered as less energy intensive and environmentally friendly than other systems. Zullaikah et al [4] stated that yield and content of biodiesel produced under CO_2 atmosphere was higher than those under nitrogen (N_2) atmosphere. CO_2 becomes determined factor in non-catalytic conversion method. CO_2 gas as co-solvent gave some positive effect in biodiesel conversion such as lower the operating condition, improve solubility, and increase efficiency of extraction [12]. Go et al [12] also stated that addition of CO_2 gas as a co-solvent can improve the solubility and extraction efficiency of biomass, therefore higher yield of FAME can be achieved. They proposed in-situ transesterification method of J. curcas L. seed kernel under subcritical methanol, acetic acid, and water mixture to produce biodiesel using CO_2 as pressurizing gas. Yield of FAME (98 %) based on extractable lipids of dry kernels can be achieved using this method. Miyazawa et al [13] also used CO_2 in the form of dry ice to increase polysaccharide hydrolyses under hydrothermal condition. The result showed that yield of glucose increased with the increasing of the amount of loaded CO_2 in the system. The effect of CO_2 addition to increase the product yield can be considered as one of important and determined factor to improve performance in biodiesel production specifically in non-catalytic in-situ transesterification. In this work, the effect of pressurized CO_2 injected in the in-situ transesterification of rice bran under subcritical water-methanol mixture on yield and content of biodiesel was investigated.

2. Experimental section

2.1. Materials

Rice bran (IR 64 variety) was obtained from local rice mill located in Banyuwangi, Indonesia. n-hexane and methanol were obtained from ANHUI FULLTIME (Anhui, China). CO_2 gas as pressurizing gas was obtained from Aneka Gas (Indonesia) and other chemicals used was obtained from commercial sources.

2.2. Rice bran oil extraction

Initially, rice bran was separated from its impurities by sieving method. Then, rice bran (50 g) was extracted by n-hexane as solvent using a soxhlet extractor at 70 °C for 3 h. The mixture of crude rice bran oil (CRBO) and n-hexane was separated by a rotary vacuum evaporator (Yamato RE-46). The crude

RBO obtained was weighed and analyzed to know the yield of RBO and FFA content. FFA analysis used modified titration method by AOCS official method Ca 5a-40 according to Rukunudin et al [14]. The yield of CRBO was determined as the following equation:

$$\text{CRBO yield} = \frac{mass\ of\ CRBO\ (g)}{mass\ of\ rice\ bran\ (g)} x100\% \quad (1)$$

Crude Rice bran oil yield obtained was 11.78% with the initial FFA content of 40.42%.

2.3. Biodiesel production through in-situ transesterification under subcritical water-methanol mixture

In situ transesterification process was carried out under hydrothermal reactor (V = 86 mL) made of seamless 316 stainless steel tubing (SS-T12-S-083-6ME, Swagelok, Solo, USA) which was able to handle pressure and temperature up to 338 bar and 605°C, respectively. The in-situ transesterification which is simultaneous process of extraction and (trans)esterification process was carried out at 200°C for 3 hours with rice bran/methanol/water ratio 1/4/4 (gr/ml/ml) according to Zullaikah and Rahkadima [15] and Zullaikah et al [4]. The operating pressure was varying from 40 bar to 100 bar with introducing pressurizing gas (CO_2) to increase the pressure in the reactor. The operating condition used for each varied pressure can be seen at Table 1. After completing reaction, the mixture was quenched to the ambient temperature. The mixture containing solid (defatted rice bran), water, and oil were washed by n-hexane to extract the oil phase. Separation of the oil phase from n-hexane was carry out using a rotary vacuum evaporator (Yamato RE-46) at 35°C for 45 min. The oil phase obtained was rich of biodiesel (FAME) and this product called crude biodiesel (crude FAME). High Temperature Gas chromatography (HTGC) was used to analyze the crude biodiesel composition. HTGC was performed on a *Shimadzu* GC-17A (Kyoto, Japan) gas chromatography equipped with flame ionization detector. Separations were carried out on a DB-5HT (5%-Phenyl)-methylpolysiloxane nonpolar column (15 m x 0.32 mm; *Agilent Tech.* Palo Alto, California). The temperature of injector and detector were set at 380 °C while temperature column was held at 80°C for 0.2 min, increased at 15 °C/min up to 380 °C and then maintained at 380 °C for 10 min. The split ratio was 1:50 and carrier gas used was nitrogen.

Table 1. Operating condition in hydrothermal reactor

Temperature	CO2 gas addition (bar)	Initial pressure (bar)	Operating condition (bar)
200°C	23	17	40
	37	23	60
	60	20	80
	65	25	100

The yield of FAME was determined as the weight ratio of FAME to CRBO as follows equation:

$$\text{Yield of FAME (\%)} = \frac{\{crude\ FAME\ (g) x\ FAME\ Content\ (\%)\}}{mass\ of\ CRBO\ (g)} \text{ x } 100\% \quad (2)$$

3. Result and discussion

3.1. Effect of operating pressure to acidity (pH)

Operating pressure clearly affected the acidity of the system. Figure 1 shows that pH of the system slightly decreases with increasing loaded of CO_2 into the reactor. High amount of CO_2 loaded increased the operating pressure of the system. Peng et al [16] stated that pH of the CO_2 - water system decreased with pressure increment. Increasing solubility of CO_2 in water affected the acidity reduction and decreased the pH system. However, at high pressure, the reducing pH of CO_2 - water system is slowly with increasing of pressure. At pressure of 40 bar, calculated pH in the system was 3.82 and decrease to 3.78 at pressure of 100 bar. The decreasing of pH from pressure of 40 bar to 100 bar was only 1 % even though the amount of CO_2 was increased by 65 %. Furthermore, high operating pressure does not affect the reducing of acidity system significantly.

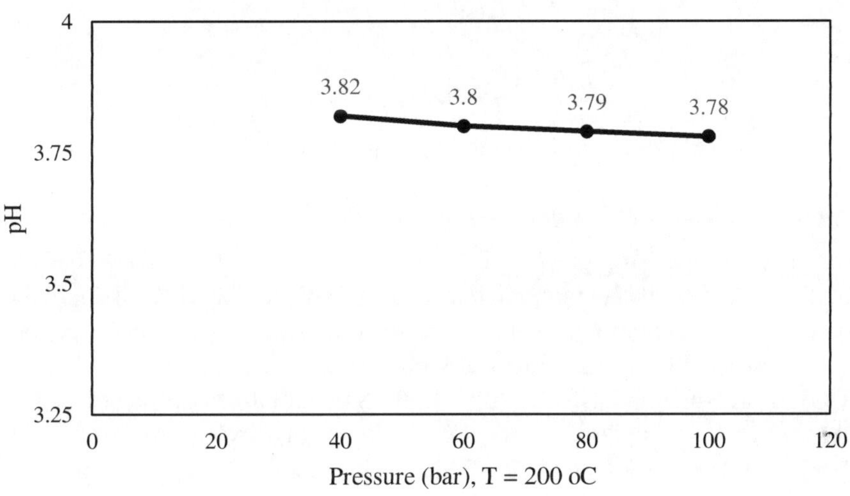

Figure 1. Effect of operating pressure to pH in the CO_2-water system at T = 200 °C

3.2. Effect of operating pressure on yield and purity of biodiesel (FAME)

The effects of increasing operating pressure on yield and content of FAME are shown in Figure 2. Yield of FAME increased significantly with increasing of operating pressure from 40 bar to 100 bar. At pressure 40 bar, yield of FAME was 35.00 ± 0.99% and kept increasing with the operating pressure. The highest yield of FAME (98.00 ± 1.41 %) was achieved at pressure 100 bar. The increment yield of FAME is caused by increment of pressure. When operating pressure in the system increase, the solubility of CO_2 in water system also increase [16]. The increasing of dissolved CO_2 in water can increase the acidity of the system due to the reaction of soluble CO_2 in water. The reaction can produce H^+ and makes the system more acidic [4,13]. In acidic condition, oil is more soluble therefore; the rate of oil extraction will be faster. Furthermore, the reaction rate of transesterification and hydrolysis of acylglycerols and esterification of FFA can also be faster. Figure 1 shows that the higher amount of loaded CO_2 introduced to the system, the lower acidity of the system. At pressure 100 bar, calculated pH in the system was the lowest and yielded the highest FAME. The highest percentage of increasing yield of FAME is obtained from pressure 80 bar to 100 bar which is 48%.

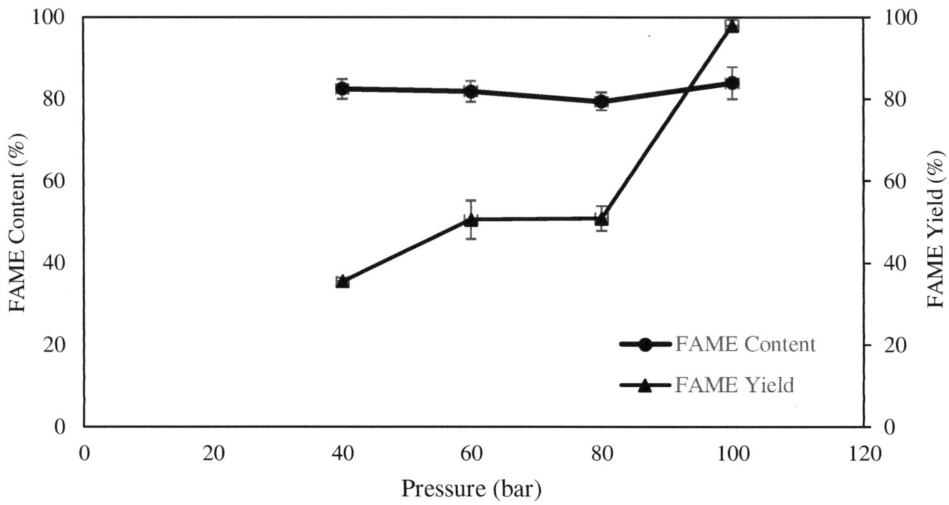

Figure 2. Effect of Operating Pressure on Yield and Content of FAME [T= 200 °C, reaction time = 3 h, rice bran/methanol/water ratio 1/4/4 (g/ml/ml)]

3.3. Effect of operating pressure on composition in crude biodiesel

The effects of increasing operating pressure on content of unreacted oils and other components are shown in Figure 3. Unreacted oils contain free fatty acids (FFA), triglycerides (TG), diglycerides (DG), monoglycerides (MG). Generally, the content of unreacted oils and other components in biodiesel was below of 10 %. It represented that the content of biodiesel produced under subcritical water-methanol mixture with CO_2 as pressurizing gas is mostly FAME. The initial FFA content in RBO was 40.42% and it lowered into below 4% under subcritical condition with varying pressure. However, the high amount of CO_2 added to the system did not affect the decreasing of FFA. The reaction pathway of FFA conversion is esterification in which FFA reacts with alcohol to form alkyl ester and water. However, the reaction pathway of acylglycerols conversion is transesterification with parallel reaction mechanism. In supercritical condition which needs high pressure and temperature, the reaction mechanism of transesterification is similar to the acid-catalyzed path. However, the reaction is much faster and completion of triglycerides to esters is more rapid than acid-catalyzed transesterification because the chemicals kinetics accelerate rapidly under supercritical conditions [17]. Ngamprasertsith and Sawangkeaw [17] proposed the mechanism reaction by the analogues between hydrolysis esters in supercritical water and transesterification of TG in supercritical alcohol. The assumption of mechanism reaction is initiated when the molecules of alcohol reacts with carbonyl carbon of TG due to the low hydrogen bond energy of alcohol and form alkyl esters and DG. Then, DG reacts with other alcohol molecules in a similar way to form alkyl esters and MG. In the last step, MG reacts with alcohol to produce alkyl esters and glycerol. The reaction mechanism proposed can be seen in Figure 4. The same reaction path also proposed by Asri et al [18] for thermal transesterification. The proposed reaction path of transesterification in supercritical condition can also be applied to subcritical condition. It is caused that those conditions need high temperature and pressure.

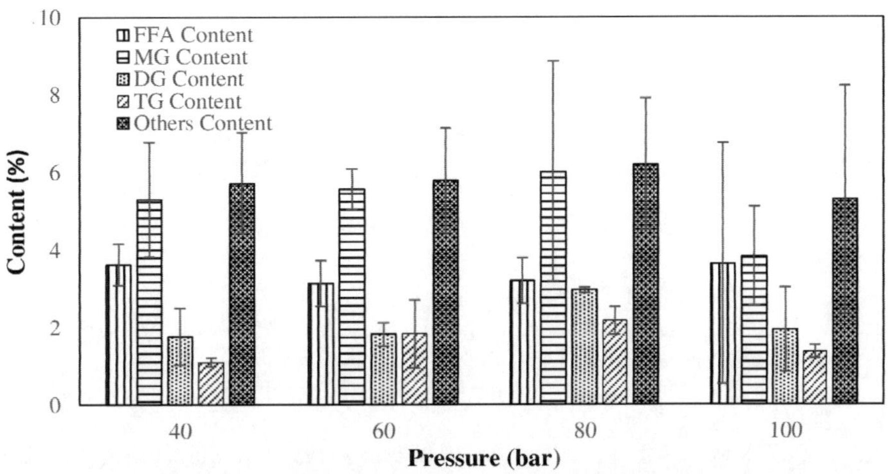

Figure 3. Effect of Operating Pressure on Composition in Crude Biodiesel [T= 200 °C, reaction time = 3 h, rice bran/methanol/water ratio 1/4/4 (g/ml/ml)]

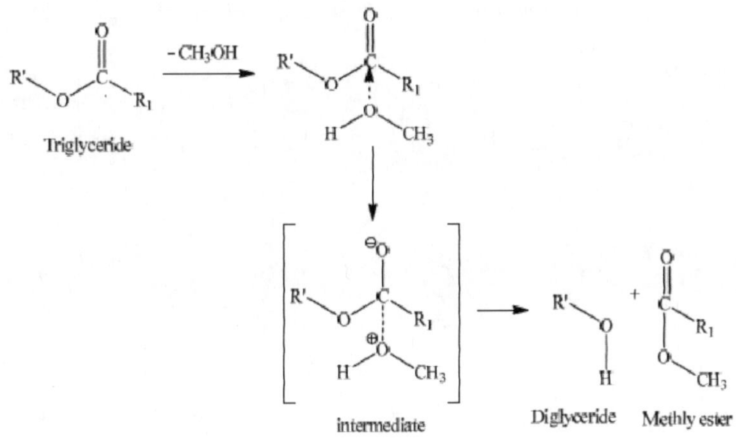

Figure 4. The proposed reaction mechanism of transesterification under thermal conditions [17]

4. Conclusions

Carbon dioxide (CO_2) as pressurizing gas can substitute the absence of catalyst in non-catalytic conversion technology. The increasing of operating pressure due to the increment amount of introduced CO_2 to the system increased the yield of FAME and slightly increased the FAME content. The highest FAME yield and content was obtained under pressure 100 bar due to the lowering acidity in the system. The increasing of pressure does not affect unreacted oils and other components in biodiesel significantly.

References

[1] Gunawan S, Maulana S, Anwar S and Widjaja T 2011 *Ind. Crops. Prod.* **33** 624-28
[2] Shiu P J, Gunawan S, Hsieh W H, Kasim N S, and Ju Y H 2010 *Bioresour. Technol.* **101** 984-89
[3] Niawanti H and Zullaikah S 2017 *Chem. Eng. Trans.* **56** 1513-18
[4] Zullaikah S, Rahkadima Y T and Ju Y H 2017 *Renew. Energ.* **111** 764-70
[5] Ju Y H and Vali S R 2005 *J. Sci. Ind. Res.* **64** 866-82

[6] Zullaikah S, Lai C C, Vali S R and Ju Y H 2005 *Bioresour. Technol.* **96** 1889-96
[7] Özgül-Yücel S and Türkay S 2002 *J. Am. Oil. Chem. Soc.* **79** 611-14
[8] Lai C C, Zullaikah S, Vali S R and Ju Y H 2005 *J. Chem. Technol. Biotechnol.* **80** 331-37
[9] Kasim N S, Tsai T H, Gunawan S and Ju Y H 2009 *Bioresour. Technol.* **100** 2399-403
[10] Hamid M A, Bakar N A Park C S, Ramli F and Wan W R 2018 *Chem. Eng. Trans.* **63** 577-82
[11] Parvez A, Megat Mohd Noor M G, Motoo U, Goto M, Muzahidul Islam A K M and Akter A 2017 *Chem. Eng. Trans.* **56** 271-76
[12] Go A W, Sutanto S, Liu Y T, Tran Nguyen P L, Ismadji S and Ju Y H 2014 *J. Taiwan. Inst. Chem. Eng.* **45** 1516-22
[13] Miyazawa T and Funazukuri T 2005 *Bioresour. Technol.* **21** 1782-85
[14] Rukunudin I, Whitea P, Bernb C and Baileyc T 1998 *J. Am. Oil. Chem. Soc.* **75** 563-68
[15] Zullaikah S and Rahkadima Y T 2015 *AIP Conf. Proc.* p 1699
[16] Peng C, Crawshaw J P, Maitland G C, Martin Trusler J P and Vega-Maza D 2013 J. *Supercrit. Fluids* **82** 129-37
[17] Ngamprasertsith S and Sawangkeaw R 2011 *Biodiesel-Feedstocks and Processing Technologies, InTech* ed Chapter In: Dr.Margarita Stoytcheva p 247-68
[18] Asri N P, Machmudah S, Wahyudiono W, Suprapto S, Budikarjono K, Roesyadi A and Goto M 2013 *Bull. Chem. React. Eng. Catal.* **7** 215-23

Acknowledgments
This research was financially supported by Ministry of Research, Technology and Higher Education and Institut Teknologi Sepuluh Nopember (ITS).

Production and Characterization of The Basic Soaps Obtained by Double Decomposition and Direct Process of Palm Stearin Oils

G F Neonufa[2], M Pratiwi[1], R Purwadi[1], T Prakoso[1], T H Soerawidjaja[1] and L Elizabeth[1]

[1] Chemical Engineering, Institut Teknologi Bandung, West Java, Indonesia
[2] Department of Agricultural Product Technology, Universitas Kristen Artha Wacana, Kupang, NTT, Indonesia

E-mail: godliefneonufa@gmail.com

Abstract. The direct and double decomposition process of palm stearin oil were evaluated for the production of basic soaps as a decarboxylation feedstock to drop-in fuel produce. The metals proposed for the saponification reaction was magnesium which has a high-basicity and is a low-cost metal. The Fourier transform infrared spectroscopy profile of the Mg(OH)-stearin basic soaps obtained by both direct and double decomposition processes showed hydrocarbon groups such as alkanes and alkene, without the oxygenate groups. The basic soap products generated by the double decomposition process showed better basicity level which was determined based on the IR spectrum intensity, especially of the –OH group. The type of saponification process used in the resulted basic soaps can provide a different effect on the generated basic soaps characteristic.

1. Introduction

Since the past few decades, excessive use of fossil resources (mainly liquid fuel) as mono-energy in transportation and industries has been leading to both energy crisis and many environmental concern [1]. To solve this problem, the renewable liquid fuel from vegetable oils has been promoted as the alternative approach to replace fossil fuel. Biofuels have been seen as the nearest-term answer to the need for alternatives to petroleum fuels in the transportation sector [2]. Biodiesel is one type of biofuel that has been established as a renewable fuel. However, biodiesel is a type of oxygenated fuel [3], therefore, its utilization is very limited (approximately 15-20 volume%) as a blended agent for petroleum fuel. Green diesel refers to liquid hydrocarbons fuel that do not contain oxygen in their molecular structure. Therefore, green diesel's properties are similar to petroleum fuel that can be used in greater quantities, even up to 100 volume% [4]. As raw materials, vegetable oils have the capacity to storage large amounts of energy. It is directly related to their chemical structure that contains similar carbon chains to the common fuels such as gasoline and diesel [5]. The main component of vegetable oil and fat is triglyceride, i.e. an ester of glycerol and three fatty acids.

The fatty acids are carboxylic acid with long aliphatic chain. It can be considered as carboxylic acids contaminated by carbon dioxide (RCOOH). Fatty acids are essential components of lipid fraction of living organisms and are common constituents of ancient and recent sediments [6]. The basic idea of hydrodeoxygenation or decarboxylation is to produce liquid hydrocarbon fuel by removing carbon dioxide from the molecule chain of fatty acids. Both hydrodeoxygenation and decarboxylation processes can be applied directly to oils and fats or fatty acids. The disadvantage of the direct application of both methods mentioned above to remove the oxygen in the molecular chains of oils and fats or fatty acids is that it required a lot of hydrogen and noble metal catalysts such as Pd and Pt [7]. In addition, biohydrocarbon products still contain many unexpected free fatty acids in the fuel [8]. Metal basic soaps are derivatives of the oils and fats of vegetable which have been proven to be excellent feedstock for decarboxylation process to produce green diesel [9]. In 1947, Chang and Wan have reported that decarboxylation of metal soaps from Tung oil can produced liquid hydrocarbon fuel [10]. The green diesel product resulting from metal basic soaps decarboxylation contains no more free fatty acids [11]. There are three main processes applied to the metal basic soaps manufacture which are double decomposition, fusion, and metal-acid reaction. The first process involves two reactions; first, aqueous caustic and fatty acids are mixed to form a sodium soap, subsequently, the sodium soap and an inorganic metal (or mix metals) salt are reacted to form the metal basic soap. The second process involves the fatty acid and metal oxide reaction. The third process involves the fatty acid and free metal powders to form polyvalent metal soap [12]. The first process or double decomposition process is the most widely used in commercial scale. In the fusion process, the main concern is that the feedstock (fatty acids) must be pure grade. The metal-acid reaction process is the least used of the three commercial processes [13]. Another method to prepare the metal basis soaps has been introduced by Blachford [14], it was the direct method. Through this method, the metal basic soaps were produced directly by reacting palm stearin and magnesium hydroxide. Each method produces metal basic soaps with different characters that tend to produce liquid products with different quality profiles as well [15]. The purpose of this study is to investigate the production and characterization of magnesium basic soaps manufactured by direct and double decomposition (metathesis) methods. In addition, it is also geared to find the best production process of metal soaps most effectively used as basic soaps feed of decarboxylation processes to generate biohydrocarbon fuel.

2. Method

Palm stearin used in this research was supplied by local oil refinery in West Java, Indonesia. The saponification and acid value of palm stearin are 190.7 mg KOH/g and 0.6 mg KOH/g, respectively. The composition of palm stearin were mainly of 68.24 mole% of palmitic acid (C16:0), 5 mole% of stearic acid (C18:0), and 20.63 mole% of oleic acid (C18:1). Commercial-grade magnesium hydroxide and zinc hydroxide were supplied by Sigma-Aldrich Corporation of Merck KGaA in Singapore. Magnesium-zinc (mixture) stearin has been obtained from direct reaction of palm stearin and magnesium-zinc hydroxides [14].

2.1 The preparation of magnesium basic soaps

Direct (Blachford) method: Magnesium basic soaps are produced from a reaction mixture which initially comprises of Mg-hydroxide, water and a palm stearin (modified of Blachford). First, the reaction mixture is agitated; next, the reactants are poured into the agitated mixture to produce a metallic soap and glycerol. The saponification reaction was carried out at 140 to 180 °C and around 10 bars pressure for 3 hours in a mini-stainless steel autoclave in the presence of excess water to dissolve the glycerol formed in the reaction mixture; as a result, reactions between by-product glycerol and the product of magnesium basic soap is substantially hindered. The amount of palm stearin (triglyceride) and $Mg(OH)_2$ were approximately the same with their stoichiometric reaction (see reaction 1 and 2). Two types of metal soaps were produced by this method, Mg-stearin ($Mg(RCOO)_2$) and Mg-stearin basic soaps (Mg(RCOO)(OH)). The chemistry of saponification reaction by direct method may be expressed briefly as follows:

$$2H_2C(OOCR)CH(OOCR)CH_2(OOCR) + 3Mg(OH)_2 \rightarrow 2CH_2OH.CH(OH).CH_2OH + 3Mg(RCOO)_2 \quad (1)$$

The saponification reaction to produce Mg-basic soaps is as follows :

IOP Conf. Series: Materials Science and Engineering **543** (2019) 012069 doi:10.1088/1757-899X/543/1/012069

$$H_2C(OOCR)CH(OOCR)CH_2(OOCR)+3Mg(OH)_2 \rightarrow CH_2OH.CH(OH).CH_2OH+3Mg(RCOO)(OH) \quad (2)$$

The reaction mixture is allowed to separate into an aqueous layer and a molten layer of product magnesium basic soap. Under a pressure like that, the aqueous layer is essentially quiescent and the molten metallic soap layer is dissociated from the aqueous layer. The molten layer was removed from the reactor, and then washed with hot water and filtered with Buchner filter. The obtained slurry was dried overnight at 60 °C in the oven, and crushed into Mg basic soaps powder.

Double decomposition (metathesis) method: Metallic basic soaps are commonly prepared by a double decomposition process. Double decomposition process involves two basic reaction [16]: the first, reaction of aqueous caustic (in excess) and the fatty acid (stearin) to form a sodium soap dissolve in an aqueous phase, and the second, reaction of the sodium soap with an inorganic metal salt (Mg salt) to form the metal basic soap (see reaction 3 and 4). The chemistry of its process may be expressed briefly as follows:

$$NaOH\ (excess) + RCOOH \rightarrow RCOONa + H_2O \quad (3)$$

$$RCOONa + Mg(OH)(NO_3) \rightarrow (RCOO)Mg(OH) + Na(NO_3) \quad (4)$$

The metal basic soap ((RCOO)Mg(OH)) has been used as feedstock to decarboxylation process to produce renewable fuels.

2.2 Magnesium basic soap characterization

The conversion of metallic basic soaps was determined by weighing the wash water before and after the purification of the soaps. The structure of the soaps solid phase was obtained by X-ray diffraction techniques. The X-ray diffraction patterns were recorded with a Rigaku Geigerflex RBRU 200. The CuKα radiations were used over the diffraction angle range $2\theta = 3^{\circ}$-65°. The topography surface of the metallic basic soaps were obtained with Scanning Electron Microscope (SEM). Fourier Transform Infrared (FTIR) were used to determine the metallic basic soaps which was based on the hydroxide ion content in the basic soaps. The infrared spectra were recorded at room temperature with a PerkinElmer FTIR Spectrometer, Model 1750 (PerkinElmer Corporation, Norwalk, CT), equipped with a deuterated triglycine sulfate detector and controlled by a PerkinElmer 7300 PC. The software used for FTIR data collection was the Infrared Data Management system. The instruments were maintained at constant humidity to minimize water vapor interference.

3. Results and Discussion

3.1 Fourier Transform Infrared Analysis of soap

Results should be clear and concise. Fourier Transform Infrared (FT-IR) spectra of Mg(OH)-stearin basic soap powders produced in the direct process and double decomposition process are shown in Figure 1a and 1b, respectively. The IR spectrum of the Mg(OH)-stearin basic soaps sample shown in Figure 1a and 1b possessed an intense broadband between approximately 3700 and 500 cm^{-1} which may be represented as a superimposition of deformational vibrations of physically adsorbed water [17], vibration of structural OH- groups [18], and of characteristic stretching vibration of the Mg^{2+}-OH^{-} bond in Mg, Al-hydroxy carbonate [18]. The wavenumbers of important absorption in the IR spectra of Mg(OH)-stearin basic soaps were compared with those of corresponding fatty acids, i.e. myristic (14:0) palmitic (16:0) and stearic (18:0).

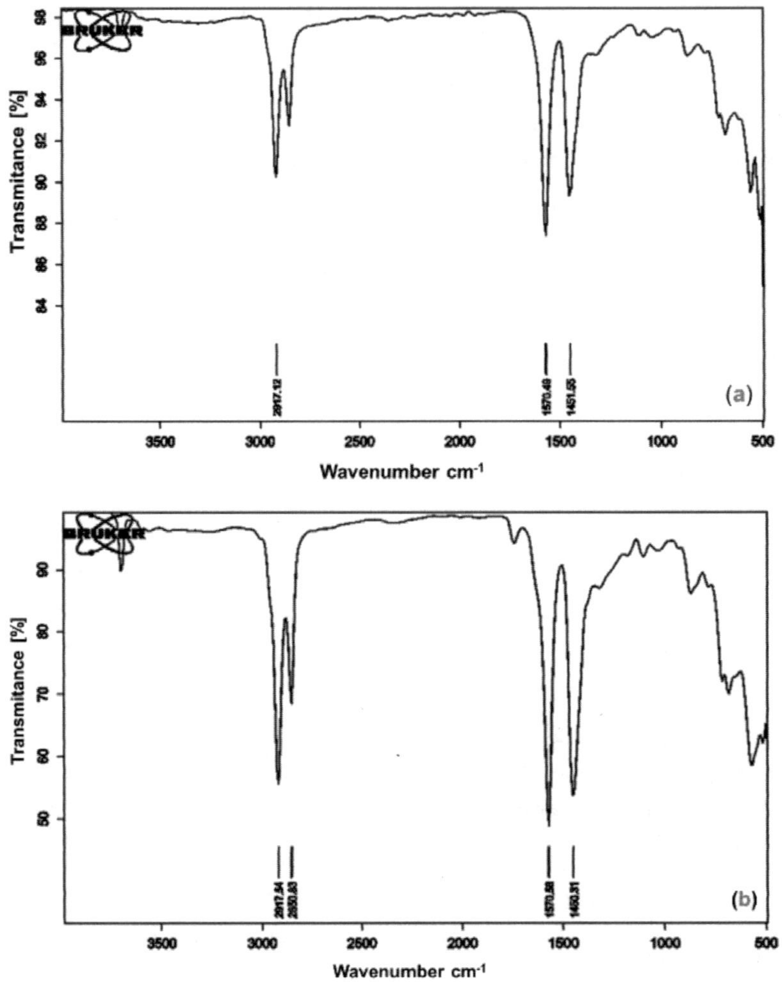

Figure 1.IR spectrum of dried Mg(OH)-stearin basic soaps based on direct method (a) and double decomposition method (b)

Infra-red analysis showed that both saponification methods had similar wavenumber characteristic. The absorption maxima near 3460-3475 cm^{-1} in the spectra of fatty acid indicates the presence of a hydroxyl group. Magnesium basic soaps based on double decomposition method had higher hydroxyl group intensity compared to the direct method. Possibly, it is due to the fact that manufacture of basic soap by double decomposition reaction is more complete than the direct method. The absorption bands were observed near 2922.54 -2855.27 cm^{-1} according to C-H stretching in the fatty acid spectrum. The intensity of C-H stretching was observed to be higher in the spectrum of Mg-stearin basic soaps based on double decomposition method than direct method. Similarly, the absorption band observed near 1570.43-1450.31 cm^{-1} was corresponding to the carboxyl group of the fatty acid. The spectrum with wavenumbers as mentioned above were observed in the basic soap that has been manufactured by direct method and double decomposition method. Even so, the spectrum intensity observed from basic soaps based on double decomposition methods was still higher than direct method. Therefore, in the case of the strength of the hydroxyl group spectrum intensity that reflects the basicity levels of magnesium basic soap, the double decomposition method is the best to be applied to produce the metallic basic soap. However, it should be noted that the drawback of the soaps manufacture by the double decomposition method is the requirement of higher material cost and the production of a lot of unexpected waste [13].

3.2. X-Ray Diffraction Analysis

The thermal decomposition of Mg(OH)-stearin basic soap was studied by X-ray diffraction. Diffractogram recorders for the Mg(OH)-stearin basic soaps resulted from both direct and double decomposition process were calcined at 900 °C for 12 hours. X-ray diffraction pattern of Mg(OH)-stearin basic soap, given in Figure 2 for direct and double decomposition process indicate that these basic soaps were sharp and had symmetrical peaks which gave clear indication that both sample were well crystallised and the peaks corresponding to 2θ values near 12°, 23° and 39° were the characteristic of magnesium soaps structures [19].

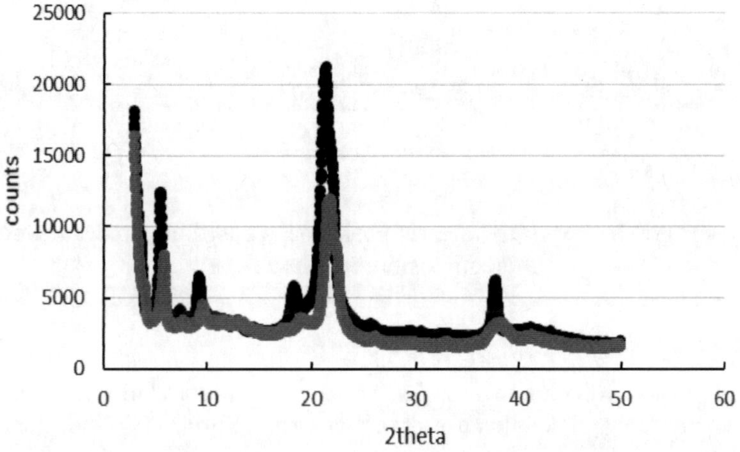

Figure 2. XRD patterns of dried Mg(OH) basic soaps based on direct (●) and double decomposition (●) methods

The diffractogram in Figure 2 showed that the basicity of Mg(OH)-stearin resulted by direct method more stable than double decomposition method at 900 °C. This can be seen from the sharp diffraction lines with low intensity displayed by diffractogram of Mg(OH)-stearin basic soaps based on double decomposition method. XRD pattern shows that all peaks at a certain 2θ for Mg(OH)-stearin basic soap based on double decomposition was lower than the direct method. It also showed that Mg(OH)-stearin basic soap resulted by direct method stabilized at 900 °C during 12 hours was more crystalline than that obtained by double decomposition.

3.3 Scanning Electron Microscope (SEM) Analysis

The particle morphology of Mg(OH)-stearin basic soaps powder produced by the direct process and double decomposition process is given as SEM microphotograph in Figure 3a and 3b, respectively. From the micrograps of Scanning Electron Microscopy (SEM), it was seen that Mg(OH)-stearin powder of both direct and double decomposition process had amorphous particle morphology structure and the average particle size was 500 μm. There appears to be no significant morphological particle difference between Mg (OH) -stearin basic soap manufactured by the direct process and the double decomposition process. Accordingly, differences in the method of manufacturing soap with magnesium metal did not have a significant effect on the morphological particle of the magnesium basic soap produced.

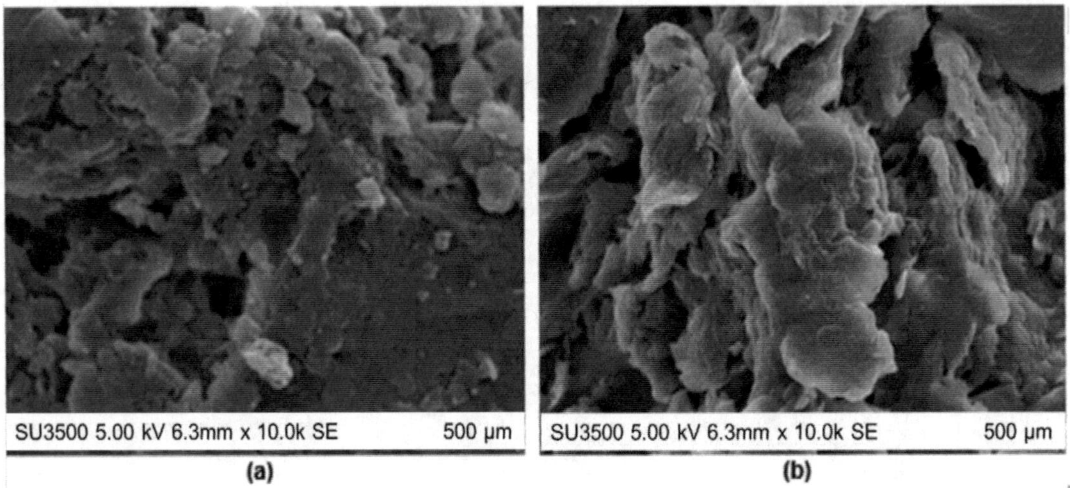

Figure 3. SEM micrograph of Mg(OH)-stearin basic soaps based on direct method (a) and double decomposition method (b)

3.4 Fourier Transform Infrared (FT-IR) Analysis of Fuel

FTIR spectra of liquid bio-hydrocarbon fuel generated by decarboxylation of Mg(OH)-stearin basic soaps based on direct process and double decomposition process are shown in Figure 4.

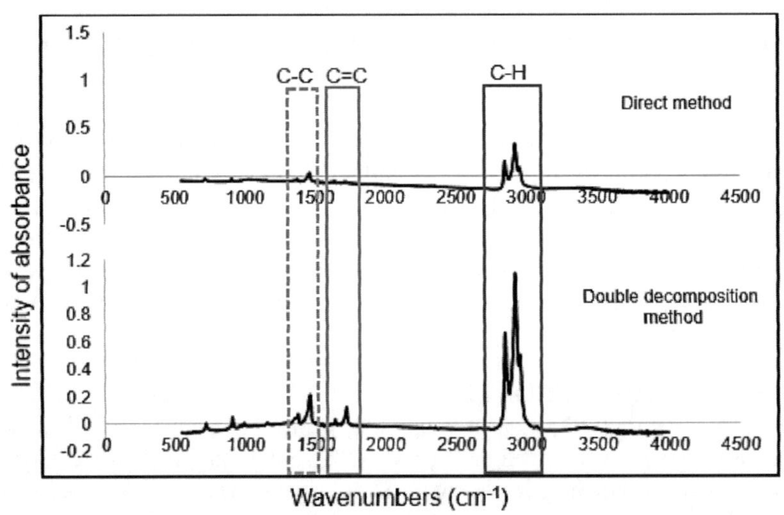

Figure 4. IR spectrum of bio-hydrocarbon fuels generated by decarboxylation of Mg(OH)-stearin basic soaps

Figure 4 shows that liquid biohydrocarbons produced by decarboxylation of Mg(OH)-stearin basic soaps, were identified to only had C-C, C-H and C = C groups without oxygenate groups. It shows that the methods of manufacturing basic soaps based on direct and double decomposition process have resulted in good basic soaps as expected for decarboxylation feedstock to produce bio-hydrocarbon fuel.

IOP Conf. Series: Materials Science and Engineering **543** (2019) 012069 doi:10.1088/1757-899X/543/1/012069

4. Conclusion

The direct method and the double decomposition method are both applicable for the manufacture of Mg(OH)-stearin basic soap. The double decomposition method resulted in Mg(OH)-stearin basic soap that has a better basicity level which is determined based on the IR spectrum intensity, especially of the –OH group. Mg(OH)-stearin basic soap resulted by direct method is more stabilized at 900 °C during 12 hours and is more crystalline than that obtained by double decomposition. There are no significant morphological particle difference between the Mg (OH)-stearin basic soap manufactured by the direct process and the double decomposition process. Liquid bio-hydrocarbons produced from the Mg(OH)-stearin soap decarboxylation process do not contain oxygen functional groups.

Acknowledgments

This work was supported by Indonesian Oil Palm Estate Fund Agency (BPDPKS) under GRS-16 Grant.

References

[1] Sinthupinyo P, Hayashi T, Habaki H and Egashira R 2009 *Chem. Eng. Trans.* **18** 623-28.
[2] Demirbas A 2008 *J.Energy Conversion and Management.* **49** 2106–16.
[3] Kubickova I, Snare M, Eranen K, Maki-Arvela P and Murzin D Y 2005 *J. Catal. Today.* **106** 197-200.
[4] Neonufa G F, 2018 *Drop-in Fuel Production Technology of Green Diesel and Jet Fuel Types via Catalytic Thermal Decarboxylation of Basic Soap Based on Magnesium and Transition Metal Combination.* PhD Thesis (Bandung: Institut Teknologi Bandung)
[5] Romero M, Pizzi A, Toscano G, Casazza A A, Busca G, Bosio B and Arato E 2018 J.*Chem. Eng. Trans.* **64** 121-26.
[6] Kallai L H, Aizensthat Z and Milosslavski I 1984 J.*Clay Minerals.* **19** 779-88.
[7] Fu J, Lu X and Savage P E 2010 *J.Energy Env. Sci.* **3** 311-17.
[8] Fu J, Lu X and Savage P E 2011 *J.Chem. Sustain. Chem.* **4** 481-86.
[9] Neonufa G F, Soerawidjaja T H, Prakoso T, 2017 *J. Eng.Tech Sci.* **49** 575-86.
[10] Chang C C and Wan S W 1947 *J.Ind.and Eng. Chem.* **39** 1543-48.
[11] Neonufa G F, Soerawidjaja T H, Indarto A and Prakoso T 2017b *J.Ambient Energy.* **40** 374-80.
[12] Elliot S B 1946 *The Alkaline Earth and Heavy Metal Soaps* (New York: Reinhold Publishing Corporation)
[13] Pratiwi M, Ylitervo P, Pettersson A, Prakoso T and Soerawidjaja T H 2016 *IOP Con. Mat. Sci. Eng.* **206** 1-7.
[14] Blachford J 1982 *U.S. Patent No 4,316,852.*
[15] Akanni M S, Okoh E K, Burrows H D and Ellis H A 1992 *A Review. Thermoc. Acta.* **208** 1-41.
[16] Rogers Jr R H and Blew Jr W R 1959 *U. S. Patent No. 2,890,232.*
[17] Allegra G and Ronca G 1978 *Acta Crystallograph.* **34** 1006-13.
[18] Roy D, Roy R, Osborn E, 1953 *American J. of Sci.* **251** 337-61.
[19] Jablonska M., Chmielarz L, Wegrzyn A, Guzik K, Piwowarsa Z, Witkowski S, Walton R I, Dunne P W and Kovanda F 2013 *J.Therm Anal Calorim.* **114** 731-47.

IOP Conf. Series: Materials Science and Engineering **543** (2019) 012070 doi:10.1088/1757-899X/543/1/012070

Development of integrated renewable energy system model for rural productivity zone in East Nusa Tenggara

N P Salsabila, W W Purwanto and M R Fuad

Department of Chemical Engineering, Faculty of Engineering, Universitas Indonesia, Depok 16424, Indonesia

E-mail: widodo@che.ui.ac.id

Abstract. East Nusa Tenggara (NTT) is a province with low electrification ratio (78.3%) and high number of underdeveloped villages (74.8%), despite its abundant renewable energy resources potential. The objective of this study is to obtain technically-economically feasible village model with integrated renewable energy system for developing productive zone, *Prukades*-postharvest cocoa processing, in Wewaria, NTT. The methodology used is techno-economic analysis, in which technical analysis include optimization for hybrid power generation system (HPGS), done by using HOMER Pro. Several financing schemes are proposed to evaluate the economic feasibility. Environmental comparative analysis of mitigated GHG emission is also evaluated. It is found that the HPGS consists of 95 kW photovoltaic, 78 kW wind turbine, and 200 kW diesel generator, equipped with battery storage, and the *Prukades* system has annual capacity of 250 tons of wet cocoa beans/year with 32.9% yield. *Prukades* system is economically feasible with all proposed schemes, and potential to increase annual income of cocoa farmers. On the other hand, HPGS is only economically feasible with FS-5 scheme with 100% grant as financial intervention. The economic attractiveness of both systems is relatively higher by using integrated scenario, both owned by the same business entity, resulting in 3 other schemes to be feasible.

1. Introduction

East Nusa Tenggara (NTT) is a province with the highest percentage of underdeveloped villages in Indonesia (74.8%) and low electrification ratio (78.29%) as per PLN data [1], despite its abundant renewable energy resources potential. This is disadvantageous since energy is the 'fuel' to economic growth and social development. Based on prior facts and challenges of increasing electrification rate and economic productivity in remote areas, the aim of this study is to obtain technically and economically feasible village model (named as RPZ Village), which consists of integrated renewable energy system and development of productive zone in Wewaria, Ende, NTT. The uniqueness of this research lies in the utilization of renewable energy for productive zone, as most international research regarding integrated renewable energy system, done by Barzola, et al. [2], Hassan [3], Chauhan and Saini [4], Sen and Bhattacharyya [5], Matin, et al. [6], and Balachander and Vijayakumar [7] still focus on meeting the energy demand of residential zone only, and have not yet emphasized on developing productive zone based on local potential, as discussed by Basu and Marett [8]. By developing productive zone, it is expected that access to energy will not be taken for granted for consumptive or non-productive use, since productive activities from income-generating are encouraged. Project of integrated renewable energy system for developing productive zone is found in other countries, e.g., done by United Nations Development Program in 2012 in Nepal, or by World

Bank and Global Environment Facility in 2003 in Sri Lanka, but has not been widely found in Indonesia. In this study, the selected productive zone to be developed is postharvest processing of cocoa. In addition, the study also proposes new integrated business scenarios between the hybrid power generation and the productive zone entity.

2. Methods

The proposed idea of developing an integrated renewable energy system and rural productivity zone as a sustainable development strategy for RPZ Village is illustrated in Figure 1.

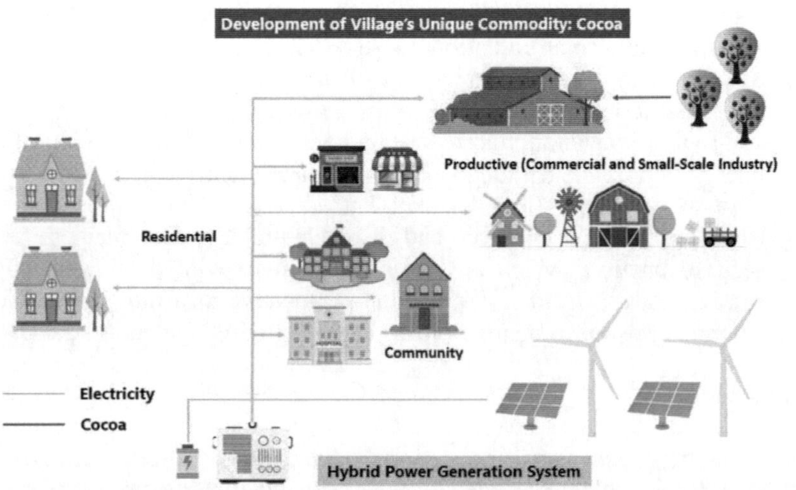

Figure 1. Proposed hybrid power generation system (HPGS) using renewable energy sources for rural productivity zone

Based on availability and reliability of potential resources in designated area, it is decided to propose hybrid power generation system for the RPZ Village. Hybrid system is selected since all this time, power generation in remote areas relies solely on diesel, in which operating and fuel cost are expensive and they are not environmentally friendly. Hybrid option is considered as the most feasible and cost-effective for remote areas where national grid extension is considered impossible and not economical. The power generation system is used to meet the electricity demand of residential, community, and productive (commercial and small-scale industry) zone. In the productive zone, there is a small-scale industry (*Prukades* or *Pengembangan Produk Unggulan Kawasan Pedesaan* or Development of Village's Unique Commodity), one of Kemendesa's strategy [9]. It is a centralized postharvest processing unit of cocoa that utilizes pod breaking machine to break pod and obtain wet beans, fermentation box to ferment the wet beans, hybrid dryer powered by solar thermal and biomass to dry the wet beans, and beans sorting machine to sort the dry beans, which could increase the added-value of cocoa. To date, most farmers do not process the harvested cocoa, resulting in low quality cocoa beans, thus low selling price. The product (dry cocoa beans) is expected to meet cocoa beans specification.

2.1. Technical analysis: Prukades system

Prukades system is designed by adhering to heuristic approach (good practices) for each postharvest processing unit and machine, based on Mulato [10], Mulato [11], and Mulato, et al. [12]. Output of system design is capacity and sizing for each equipment, also electrical and thermal energy demand of *Prukades* system. Before-after comparative analysis from environmental perspective regarding mitigated greenhouse gas mitigation from proposed design, compared to conventional design, is also evaluated as kg CO_2-equivalent, based on Shen, et al. [13].

IOP Conf. Series: Materials Science and Engineering **543** (2019) 012070 doi:10.1088/1757-899X/543/1/012070

2.2. Technical analysis: hybrid power generation system

Electricity demand profile of each zone, both for primary and deferred load, are estimated based on typical load pattern and power consumption, ownership factor, number of users, and usage duration per day from Blum, et al. [14]. Then, optimization for mini-grid of hybrid power generation system is done by using HOMER Pro (Hybrid Optimization of Multiple Energy Resources) software as explained by Lambert, et al. [15]. There are 4 essential technical input variables for the optimization: electricity demand estimation; selection of system components based on rationale and their technical and economical parameters: Caterpillar 200 kW (250 kVA and 50 Hz) diesel generator, CanadianSolar MaxPower CS6U-330P polycrystalline photovoltaic solar panels, AWS HC 1.5 kW horizontal wind turbines, lead acid Rolls Surrette 6CS25P battery, and generic converter model; RPZ Village energy resources, i.e. horizontal daily solar radiation (5.76 kWh/m^2/day), clearness index (0.5835), and average wind speed (5.03 m/s), as well as property of diesel fuel and its price; also constraints and other parameters, i.e. maximum annual capacity shortage (MACS), operating reserves, discount rate, and project lifetime. From optimization, hybrid system configuration (combination of components and optimization variables), along with detailed output parameters for each component, and its supply profile, levelized cost of electricity (LCOE), Net Present Cost (NPC), excess energy production, renewable energy fraction, fuel consumption, and dispatch strategy, are obtained. Optimum solution that meets the objective function (which is least cost, solution with the lowest NPC), is selected. Before-after comparative analysis from environmental perspective regarding mitigated greenhouse gas mitigation from optimum solution, compared to base case, is also evaluated as kg CO_2-equivalent.

2.3. Economic analysis

Economic analysis is conducted by estimating capital cost and operating and maintenance cost of both systems. Several financing schemes are proposed to evaluate the economic feasibility of both systems using Microsoft Excel to calculate the cash flow for financial performances (Net Present Value, Payback Period, and Internal Rate of Return). There are two scenarios to be discussed, i.e. if the *Prukades* system and hybrid power generation system are owned by different business entity (scenario 1), and if both systems are owned by the same entity (scenario 2).

There are 4 financing schemes to be considered for *Prukades* system (scenario 1). Base case (business-as-usual) for this system is cooperative business model (cocoa pod from farmers is purchased first as pre-financing, and dry cocoa beans is directly sold to buyers with international selling price, i.e. Rp 45,000; once the beans are sold and generate profit, the profit is shared for farmers), with 100% loan as financing source, 13% loan interest with 5 years tenure, and no grant from government. The rest of financing schemes are varied in terms of loan interest and tenure (7% within 10 year besides 13% within 5 years), also availability of government grants (with and without grant).

On the other hand, there are 10 financing schemes for hybrid power generation system (scenario 1) and integrated system (scenario 2). Base case for this system is 30% of financing source comes from equity with 12% cost of equity, and 70% of financing source comes from debt with 5.8% cost of debt, with no grant and no tax exemption (normal tax rate is applied). The hybrid power plant is owned by Independent Power Producer (IPP)/Special Purpose Vehicle (SPV) who sells electricity to PLN at maximum Power Purchase Agreement (PPA) selling price, i.e. 85% of the local electricity generation cost. For Ende, the local electricity generation cost is 17.72 cents USD/kWh, thus the maximum selling price of electricity is approximately 15.1 cents USD/kWh (Rp2,039/kWh). The rest of financing schemes are varied in terms of cost of debt (financing source intervention), availability of grant and its percentage (financial intervention), and availability of tax exemption (fiscal intervention). FS-2 represents financing scheme with fiscal intervention only (with tax exemption, not subjected to 25% income tax); FS-3 (50% grant), FS-4 (75% grant), and FS-5 (100% grant) represents financing scheme with financial intervention only (with no tax exemption); FS-6 (50% grant) and FS-7 (75% grant) represents financing scheme with combination of financial and fiscal intervention (with tax exemption); FS-8 represents financing scheme with financing source and fiscal intervention (3.77% cost of debt with no grant); FS-9 (50% grant) and FS-10 (75% grant) represents financing scheme with

IOP Conf. Series: Materials Science and Engineering **543** (2019) 012070 doi:10.1088/1757-899X/543/1/012070

combination of financing source, financial, and fiscal intervention (with 3.77% cost of debt and tax exemption).

In scenario 2, purchase of cocoa pods as raw material is not needed and selling of dry cocoa beans is not taken into account as revenue because the revenue comes from imposing cocoa processing tariff to farmers as services. In addition, there is no electricity cost because the electricity generated by hybrid power plant can be used directly for own use, and the joint cost concept is used. There are 2 types of cocoa processing tariff to be considered: scenario 2a and 2b. In scenario 2a, tariff that is applied generates farmers' income similar to scenario 1, which is equivalent to the processing cost, i.e. Rp 1,300 (9.6 cents USD) per kg of wet cocoa beans. While in scenario 2b, maximum tariff that could be charged to farmers by assuming the maximum willingness-to-pay is when Return on Investment by processing cocoa is no lower than 50% is applied. Based on the selling price of wet and dry cocoa beans, the cocoa processing tariff is set at Rp1,790 (13.3 cents USD) per kg of wet cocoa beans.

3. Results and Discussions

3.1. Technical analysis: hybrid power generation system

For Prukades system, it is obtained that the capacity of processing unit is 5 tons of wet cocoa beans per batch, equivalent to 250 tons of wet cocoa beans per year (each batch is 7 days). Dryer is selected to fully utilize the processing capacity, therefore in dry season and during daytime, thermal energy demand is met by solar, while in rainy season and during nighttime, thermal energy demand is met by biomass (firewood). From the mass balance calculation, the yield of sorted cocoa beans from wet cocoa beans reaches 32.9%. From design process, it is obtained that cocoa pod breaking machine will have capacity of 10,000 pods/hour with electricity requirement of 0.746 kW. Fermentation boxes will have capacity of 625 kg/box with arrangement of 8 rows box for 2-stage fermentation and dimension of 1.5 x 1.2 x 0.4 m. The dryer will be operated for 48 hours drying time at 55°C temperature with dimension of 5 x 3 x 0.3 m. Solar air heater will be made from galvanized iron sheet and corrugated polycarbonate sheet with 40% thermal and optical efficiency and is modelled as flat type with 269.65 m^2 area and 25° inclination to receive 194.15 kW energy incident. Backup heater is a biomass furnace equipped with heat exchanger, with capacity of burning 25 kg biomass/hour and 55% thermal efficiency. The fans installed will be axial type with 30% efficiency and 0.74 kW electricity requirement. The blower chosen is centrifugal type with 20%-30% efficiency and 0.3 kW electricity requirement. The cocoa beans sorting machine will be able to sort 1 ton/hour cocoa beans with 0.746 kW electricity requirement. The processed cocoa beans will be stored in a sack with capacity of 60 kg of beans per pieces on wooden pallet with dimension of 1.3 x 1 x 0.14 m. The proposed Prukades system (hybrid dryer powered by solar thermal and biomass) could reduce dependence on firewood by 96,800 kg per year compared to the conventional system (biomass-fired dryer) and mitigate greenhouse gas emissions by 38,096 kg CO_2-eq/year (from 165,296 kg CO_2-eq/year to 127,199 kg CO_2-eq/year), generating emissions that is 23% lower than conventional system emissions. In addition, the proposed system is relatively economical and profitable compared to the conventional system.

Meanwhile, optimization result of different configuration of system components generates 11 solutions for hybrid power generation system, in which the most optimum solution is hybrid configuration of 95 kW of photovoltaic (contributing to 44.1% of total electricity production), 78 kW of wind turbine (contributing to 29.9% of total electricity production), and 200 kW of diesel generator (contributing to 26% of total electricity production), equipped with battery (64 in quantity) and converter (38.1 kW), while base case (diesel-fired power plant) ranks as the 2nd least optimum solution. The dispatch strategy used is combined dispatch, which combines the advantages of load following and cycle charging strategy. The optimum design has its own unique yearly and daily supply profile that describes when exactly each component is operating to meet the load (presented as time series). For instance, in July 22, at 12.00 PM, the load is met by electricity generated by photovoltaic and wind turbine, and the excess energy production is stored in battery. At 06.00 PM, most of the load is met by electricity generated by diesel generator, and the rest of it is met by electricity generated by wind turbine. Information of details of levelized cost of electricity of each component can also be inferred from optimization results. The selected optimum solution of hybrid power generation system could reduce dependence on diesel by 130,163 L per year compared to base case and mitigate greenhouse gas emissions by 1,246,816 kg CO_2-eq/year (from 1,560,862 kg CO_2-eq/year to 314,047 kg CO_2-eq/year), generating emissions that is 80% lower than base case emissions, providing opportunity for the project to get access to carbon financing, which could increase the economic attractiveness of the project and lower the electricity generation cost.

3.2. Economic analysis

Cost estimation. In scenario 1, Prukades system has capital cost of Rp 234,392,800 and operating and maintenance cost of Rp 2,860,155,841. Whereas, hybrid power generation system has capital cost of USD 329,727.79, replacement cost of USD 54,636.24, and operating and maintenance cost of USD 34,510. Projected annual revenue comes from selling of 250 tons of dry cocoa beans and selling of 287,597 kWh of electricity. In scenario 2, Prukades system has capital cost of Rp 359,248,500 and operating and maintenance cost of Rp 269,446,500. Whereas, hybrid power generation system has capital cost of USD 320,479.01, replacement cost of USD 54,636.24, and operating and maintenance cost of USD 34,510. Projected annual revenue comes from selling of 250 tons of dry cocoa beans and selling of 279,530 kWh of electricity.

Impact of financing scheme. For Prukades system, all financing schemes are economically feasible comparing to diesel-fired power plant tariff. This indicates that financial intervention, neither government grants nor low-interest loan, significantly affect the economic feasibility, since the base case financing scheme is already profitable. All financing schemes are potential to generate relatively high annual income per farmer. The economic attractiveness of the Prukades system is high, regardless of the financing scheme used, since the gross profit margin of cocoa is fixed and high, i.e. Rp 4,475 per kg of processed cocoa beans. For hybrid power generation system (scenario 1), it can be concluded that hybrid system with high renewable energy fraction, has high initial capital costs (USD 329,728), making it difficult to obtain access to financing, but its operating and maintenance costs (USD 41,205) and fuel costs (USD 285,506) are low. In contrast, conventional energy system has low initial capital costs (USD 50,000), but has high operating and maintenance costs (USD 134,329) and fuel costs (USD 1,419,315), causing NPC and LCOE of base case to be much higher than that of optimum solution.

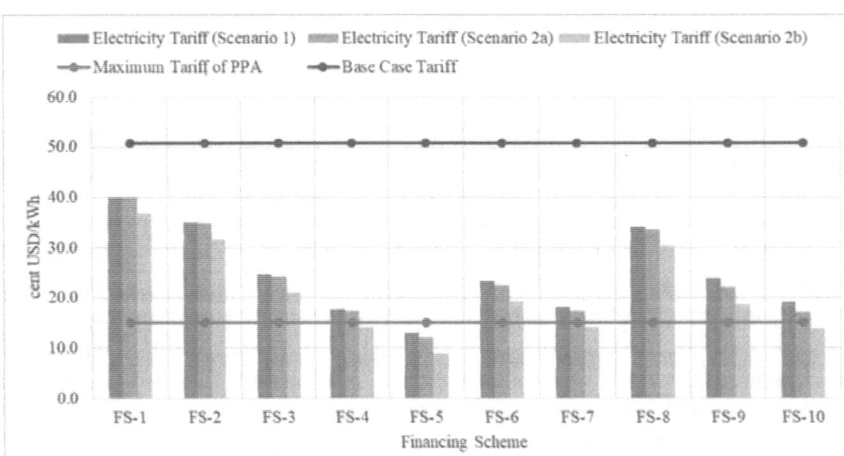

Figure 2. Comparison of electricity tariff of each financing scheme (scenario 1, 2a, and 2b)

Figure 2 compares tariff of each financing schemes in scenarios 1, 2a, and 2b maximum tariff of PPA as baseline, as well as base case tariff as benchmark tariff. It can be concluded that for all scenario and all financing schemes are economically feasibel if base case tariff as reference and for scenario 1, almost all financing scheme are not economically feasible except FS-5. Financial intervention in form of grants and low-interest loan makes schemes to be economically feasible if the value is significant

(100% grant or very-low-interest loan with long-term payment tenure that is close to project lifetime). Meanwhile, fiscal intervention in form of tax exemption does not significantly affect the economic feasibility of the project. However, compared to base case, the cost of hybrid power generation is more economical and profitable. Each year, the government can save around USD 108,654 on electricity subsidies if it prefers the optimum solution (the subsidy needed is around USD 29,047) to base case (the subsidy needed is around USD 137,701). Providing base case subsidy annually to cover the difference between electricity fees paid by user and the cost of generation generates much higher NPC (USD 1,499,351) than giving 100% grant to cover the initial capital cost of hybrid power generation project (USD 382,446). Overall, it can be concluded that for RPZ Village, hybrid solution is the optimum solution in the long term (over the project lifetime) rather than base case option. When compared to the savings gained, the decision to set maximum tariff of PPA for renewable energy generation is actually a bit unfair if the local electricity generation cost is used as reference, since electricity generation cost of conventional power plant is certainly much lower than that of hybrid power generation system. To attract more investors, the government may reconsider the PPA capping price policy (85% of local electricity generation cost). The government may also consider implementing the Feed-in-Premium, where tariff above conventional market price with cap and floor system is applied (which is slightly different from fixed Feed-in-Tariff). The government may also charge carbon tax for fossil fuel power plants or divert fossil fuel subsidies, therefore revenues gained can be used as financial incentives to finance project in renewable energy. In addition, the government may review Build-Own-Operate-Transfer (BOOT) scheme because it is considered disadvantageous to the developers, where in nonrecourse project financing, bank do not accept the power plant as collateral since at the end of project lifetime, the power plant will be transferred to the government. This scheme limits the role of small developers who do not have assets, and only accommodates large developers who can provide guarantee. It also negates the salvage value. Consequently, the government may undertake advanced legal and techno-economic assessments, as well as take many factors into account before setting tariff, scheme, or regulation by accommodating interest of various stakeholders to make effective long-term win-win solution that can increase investment certainty, with minimum amendments and revocation.

If scenario 2a is used, the result is pretty similar to scenario 1. Integration of both systems has little impact on project economics, since additional revenue stream (from Prukades system) is not high enough to compensate for the costs incurred and to contribute significantly to lower the electricity generation cost. Meanwhile, if scenario 2b is used, there are 3 additional financing schemes which were economically feasible, i.e. FS-4, FS-7, and FS-10. This indicates that imposing slightly higher processing tariff could have considerable impact on the project economics, since additional revenue stream is high enough. However, it should be noted that the willingness-to-pay of cocoa farmers is sensitive to tariff. On the other hand, the average additional income per farmer decreased from Rp 2,666,370 in scenario 1 and scenario 2a to Rp 2,255,610 in scenario 2b, illustrated in Figure 8, as processing cost per farmer increased from Rp 1,092,840 in scenario 1 and scenario 2a to Rp 1,503,600 in scenario 2b.

In general, the integrated system could increase the economic attractiveness of the project. Prukades has much lower capital and operating and maintenance costs than hybrid power generation system, yet the share of revenue contribution from Prukades is significant. To make this integration strategy more profitable, it is recommended to design processing units for commodities with higher gross profit margins or to be more flexible for multiple product (not specific to one type only), to upscale the processing unit capacity, i.e. by expanding the processing center from village to sub-district level, and to choose the right productive zone that correlates with the occupation of majority.

4. Conclusions

From the results and discussion section, it can be concluded that:

- The design of Prukades system is proposed to be a centralized hybrid (solar thermal-biomass) postharvest processing unit. The optimum solution of hybrid power generation is solar PV (95 kW)-wind (78 kW)-diesel (200 kW) equipped with battery, with potential savings of electricity subsidy up to USD 108,654 compared to base case of diesel-fired power plant.

- For diesel-fired electricity tariff as reference, all scenario and financing schemes are economically feasible and the proposed hybrid system design is an economical and environmentally-friendly option, which increases annual income per farmer.

- Based on maximum tariff of PPA, using scenario 1 and 2a, almost all financing schemes are not economically feasible except FS-5. Meanwhile, using scenario 2b, there are 3 additional financing schemes which were economically, i.e. FS-4, FS-7, and FS-10. The integration of Prukades system with hybrid power generation system could enhance the economic attractiveness and lower the electricity generation cost.

- The selected optimum solution of hybrid power generation system could reduce dependence on diesel by 130,163 L per year and generate emissions that is 80% lower than the base case (diesel-fired power plant). Meanwhile, the proposed Prukades system could reduce dependence on firewood by 96,800 kg per year and generate emissions that is 23% lower than the conventional system (biomass-fired dryer).

Acknowledgments
This work is financially supported by Universitas Indonesia under the *Hibah Publikasi Internasional Terindeks untuk Tugas Akhir Mahasiswa* (PITTA) project Fiscal Year 2018 number: 2557/UN2.R3.1/HKP.05.00/2018.

5. References
[1] Perusahaan Listrik Negara 2018 *Rencana Usaha Penyediaan Tenaga Listrik PLN 2018-2027* (Jakarta)
[2] Barzola J 2016 Analysis of hybrid solar/wind/diesel renewable energy system for off-grid rural electrification *Int. J. Renew. Energ. Res.* **6** 1146–52
[3] Hassan Q 2016 Optimization of PV/wind/diesel hybrid system in HOMER for rural electrification *J. Phys. Conf. Ser.* **745** 1–8
[4] Chauhan A and Saini R P 2016 Techno-economic feasibility study on integrated renewable energy system for an isolated community of India *Renew. Sust. Energ. Rev.* **59** 388–405
[5] Sen R and Bhattacharyya S C 2014 Off-grid electricity generation with renewable energy technologies in India: an application of HOMER *Renew. Energ.* **62** 388–98
[6] Matin M A 2013 Optimum planning of hybrid energy system using HOMER for rural electrification *Int. J. Comput. Appl.* **66** 45-52
[7] Balachander K and Vijayakumar P 2012 Optimization, simulation and modeling of renewable electric energy system with HOMER *Int. J. Appl. Eng. Res.* **7** 245–54
[8] Basu A and Marett D 2014 *Integrated sustainable rural development: renewable energy electrification & productivity zones* (Denmark: United Nations Development Programme)
[9] Kementerian Desa, Pembangunan Daerah Tertinggal, dan Transmigrasi 2017 *Percepatan Pembangunan Desa* (Jakarta: Forum Merdeka)
[10] Mulato S 2001 *Development and Evaluation of a Solar Cocoa Processing for Cooperative Use in Indonesia* (Universität Hohenheim: Dissertation)
[11] Mulato S 2010 *Cocoa Processing and Machinery Development in Indonesia* (Jember: Indonesian Coffee and Cocoa Research Institute)
[12] Mulato S 2010 *Pengolahan Produk Primer dan Sekunder Kakao* (Jember: Pusat Penelitian Kopi dan Kakao Indonesia)
[13] Shen X 2015 The comparative life cycle assessment of power generation from lignocellulosic biomass *Sustainbility* **7** 12974–87

[14] Blum N U 2013 Rural electrification through village grids–assessing the cost competitiveness of isolated renewable energy technologies *Renew. Sust. Energ. Rev.* **22** 482–96
[15] Lambert T 2006 Chapter 15: Micropower system modeling with HOMER *Integration of Alternative Sources of Energy* (New Jersey: John Wiley & Sons, Inc)

IOP Conf. Series: Materials Science and Engineering **543** (2019) 012071 doi:10.1088/1757-899X/543/1/012071

Utilization of Emission Carbon Dioxide Gas into High Economic Value Chemicals: Diethyl Carbonate

D Aslya[1], S Febriani[1], F Adian[1], R Tetrisyanda[2], A Wiguno[2], S Gunawan[2], F Kurniawansyah[2] and G Wibawa[2,*]

[1] Research Technology Center, PT. Pertamina (Persero), Jakarta, 13920, Indonesia
[2] Department of Chemical Engineering, Institut Teknologi Sepuluh Nopember, Surabaya, 60111,Indonesia

*Email: gwibawa@chem-eng.its.ac.id

Abstract. Utilizing fossil fuel is always has a deficiency in the environmental aspect. The processing of fossil fuel produce high carbon dioxide (CO_2) emissions. It is required an effort to utilize CO_2 emissions into feed of a high-economic value product of such as diethyl carbonate (DEC). The objective of this study was to develop the process of DEC synthesis from CO_2 using ethanol and ethylene oxide as reactant and potassium iodide – based catalyst. The catalyst used in the experiment were binary catalyst potassium iodide (KI) – cerium oxide (CeO_2) and KI – sodium ethoxide (EtoNa). The experiment conducted in cylindrical reactor equipped with stirrer and heating jacket. Reactant and catalyst were added into reactor and setting operation condition at initial pressure of 35 bar and temperature of 170 °C within 3 hours. The ratio of ethylene oxide to ethanol was investigated as well. Subsequently, the products of this synthesis were analyzed qualitatively and quantitatively to determine the reaction products and DEC yield. From the experiment, DEC was successfully obtained and the highest DEC yield was found to be 7.1 % by using KI-EtoNa as catalyst at mole ratio of ethylene oxide to ethanol of 1:15 within 3 hours of reaction time.

1. Introduction

Fossil-based energy sources (petroleum, natural gas and coal) are the main energy sources used by most of the Indonesian people. In its development, consumption of fossil-based energy sources will increase in the next few years. However, fossil-based energy resources are known as the main carbon emitters. Currently the petroleum, natural gas and coal industries have real responsibilities and challenges in meeting future energy needs by facing environmental impacts and issues.

The utilization of carbon dioxide as the main raw material in the process of making an economic value product is an alternative process that can be applied to reduce gas emissions from the processing and utilization of fossil-based energy. Carbon dioxide (CO_2) emissions not generated only from fuel production but also from mobiles vehicles contribute to global warming [1]. The addition of additives such as ethanol to the fuel becomes an important aspect given by the increasing levels of pollution in the environment. Ethanol is renewable energy which can be obtained from lignocellulosic materials [2]. However, the use of ethanol has the disadvantage of increasing the vapour pressure of gasoline [3-4].

IOP Conf. Series: Materials Science and Engineering **543** (2019) 012071 doi:10.1088/1757-899X/543/1/012071

The use of ethanol as an additive in a mixture of fuels besides having high RON (table 1) is also based on its non-toxic properties and abundant availability of biomass. However, the use of ethanol has the disadvantage of being able to increase vapor pressure gasoline [5]. The addition of ethanol to the isoocatane (the main component of gasoline) increases the pressure of the gasoline mixture. This increase in vapor pressure causes the gasoline mixture emissions to increase. In addition to this, the property of ethanol which is hygroscopic causes water to dissolve in gasoline to increase so that the quality of gasoline decreases [6]. So that additional additives are needed to overcome the problem. These additional additives are not expected to also have characteristics that can improve combustion performance.

Table 1. The octane value of various additives.

Additive	RON
Methyl tert-butyl ether	105-123
Ethanol	120-135
Diethyl carbonate	111
Dimethyl carbonate	101

One product that can be produced using carbon dioxide is diethyl carbonate (DEC). DEC can be used as an additive in diesel and gasoline fuel to reduce pollutant emissions because DEC has a high oxygen content [7-8]. The addition of DEC in alcohol-gasoline blends can reduce the mixtures vapour pressure [9-10]). The use of carbon dioxide emission gas into DEC is a very effective and efficient alternative in tackling emissions in the environment by integrating DEC production units with CO_2 emitting units.

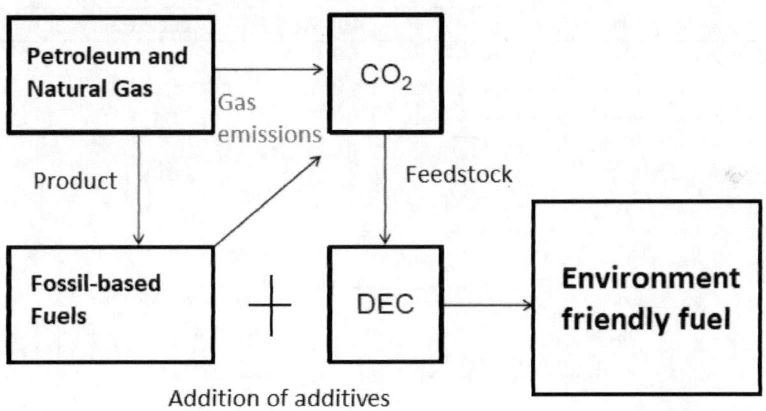

Figure 1. Integration scheme of gas emission utilization from the energy industry.

From the explanation above, the need for an environmentally friendly energy industry can be realized if there is integration between the energy producing industry and the emission gas utilization unit as shown in figure 1. Gas emissions produced from processing oil and natural gas are captured to become raw material for DEC production. After that, the use of fossil-based fuel products formed will also produce dangerous gas emissions. So adding DEC not only makes burning fuel more perfect but also makes the energy industry more environmentally friendly.

The method of synthesis of DEC has been developing for a long time. Some synthesis routes are the phosgene-ethanol method, DEC formation of ethanol into ethyl nitrite [11], oxidative carbonylation of ethanol [12], gas-phase oxidative carbonylation of alcohol [13] and activation of CO_2 [14]. DEC synthesis using carbon dioxide has been developed by many studies. The formation of DEC is done by reacting carbon dioxide with ethanol, which will produce DEC and water [15]. In the synthesis reaction of DEC through one pot reaction with ethanol and CO_2 reactants, the following equation is shown in equation (1).

$$\textit{Ethanol} + \textit{Carbon Dioxide} \rightarrow \textit{Diethyl Carbonate} + \textit{Water} \quad (1)$$

In general, the DEC synthesis reaction requires dehydration agents to shift equilibrium towards DEC products. Therefore, ethylene oxide is needed to increase DEC production. The reaction mechanism of DEC synthesis from carbon dioxide using ethylene oxide was proposed by previous work [16]. The catalyst was very influential on the DEC formation process. Therefore, the aim of this work is to develop DEC synthesis technology using heterogeneous and homogeneous catalysts. In addition, the effect of using the mole ratio of reactants between ethanol and ethylene oxide was also studied in this work.

2. Experimental Section

2.1. Description of Materials

The materials used in this work were carbon dioxide, ethanol, and ethylene oxide as reactants. Ethanol were used using further purified by molecular sieve to remove the water contaminant and other materials were used without further purification. The reactant specifications are given in table 2. The binary catalyst of potassium iodide (KI) and cerium oxide (CeO2) or sodium ethoxide (EtoNa) were used in the DEC synthesis. The binary catalyst mass ratio was 1:3.

Table 2. Properties and specifications of reactant.

Component	Source	Purity (%)	Molecular Weight ($g.mol^{-1}$)
Carbon dioxide	Samator, Indonesia	99.8	44.01
Ethanol	Merck, Germany	99.5	46.07
Ethylene oxide	Aneka Gas, Indonesia	90	44.05

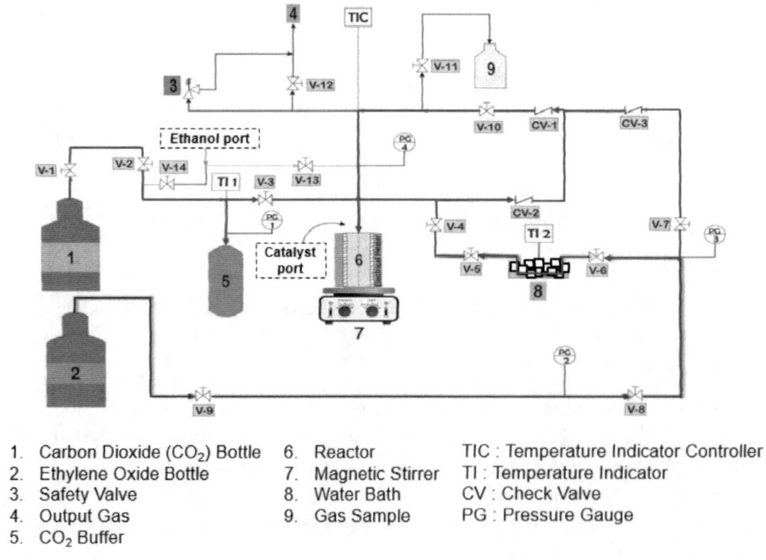

Figure 2. Experimental apparatus.

IOP Conf. Series: Materials Science and Engineering **543** (2019) 012071 doi:10.1088/1757-899X/543/1/012071

2.2. Experimental Procedures

The synthesis of DEC was conducted in 345 ml cylindrical stainless steel reactor equipped with stirrer and heating jacket. The reactor is also equipped with thermocouples and pressure indicators. Pressure control were done manually through a valve connected to the reactor. Thermocouples was used to monitor and control the temperature of the reactor using the TC-4S Autonic. The experimental apparatus was shown in figure 2.

The study was started by weighing ethanol and catalyst on the OHAUS analytic balance with an accuracy of ± 0.0001 grams. The experiment was carried out by mixing ethanol with 0.2 gram of catalyst into the reactor and ensure no air contamination in the reactor. After that, inject 0.023 mole ethylene oxide into the reactor and followed by CO2 up to initial pressure 30 bar. The operating conditions at the reactor was kept constant at 443K during the reaction time. The mole ratio between ethylene oxide and ethanol was studied between 1:5, 1:10 and 1:15 and the reaction time was 3 hours. After the specified time, the reactor temperature was reduced to room temperature then the pressure on the reactor was slowly released to the air by release valve. The liquid phase product is separated from its solid catalyst and analysed for DEC yield in the product by using gas chromatography.

3. Results and Discussion

Synthesis of diethyl carbonate (DEC) was carried out by reacting CO_2, ethanol with ethylene oxide using KI –CeO_2 and KI – EtoNa catalysts. From the GC-MS qualitative analysis it is shown that DEC synthesis has been successfully demonstrated in the three reactions that occur in accordance with the reference where four compounds were formed in this synthesis, including 2-ethoxyethanol, ethylene glycol, diethyl carbonate, and ethylene carbonate. The catalytic mechanism produces the reaction of ethylene oxide (EO) and CO_2 in the first cycle which occurs to form ethylene carbonate (EC) through the cycloaddition reaction following equation (2). At this step, the reaction runs very effectively. For the second cycle, EC transesterification with ethanol occurs to produce DEC and Ethylene glycol (EG) according to equation (3). Side reactions that occur are 2-ethoxyethanol formation from EO and ethanol reactants following equation (4).

Ethylene Oxide + Carbon Dioxide → Ethylene Carbonate (2)

Ethylene Carbonate + Ethanol → Diethyl Carbonate + Ethylene Glycol (3)

Ethylene Oxide + Ethanol → 2-ethoxyethanol (4)

The results of the quantitative analysis of each experiment are shown in table 3. DEC yield is a percentage of the number of DEC moles formed on the number of moles of ethylene oxide used in the reactor.

Table 3. DEC synthesis results using KI – CeO2 catalyst.

Catalyst	Ethylene oxide : Ethanol (Mole ratio)	DEC Yield (%)
KI – CeO_2	1:5	0.365
	1:10	1.064
	1:15	1.109
KI - EtoNa	1:5	2.158
	1:10	2.54
	1:15	7.097

Catalyst is very important parameter in DEC synthesis. The effect of homogen and heterogen catalyst was observed in this study. From figure 3, it is shown that the homogen catalyst KI – EtoNa give better DEC yield than heterogen catalyst KI – CeO_2. KI catalyst has an important role in the cycloaddition reaction and sodium ethoxide has an important role in the transesterification reaction. The mechanism of transesterification with ethanol and ethylene carbonate catalysed by sodium ethoxide is a classic addition-elimination process (Zhang et al., 2014). KI and EtoNa are homogeneous catalysts, which are soluble in ethanol, so that the catalyst can more effective in DEC formation. Thus, the homogeneous catalyst is superior in terms of activity and selectivity. Moreover, EtoNa is a base type catalyst which suitable for the reaction of DEC synthesis. CeO_2 catalyst is a heterogeneous catalyst that does not dissolve in ethanol, so that small active surface can come into contact with the reactants during the reaction. However, this heterogeneous catalyst is superior in terms of separation from the product mixture [17].

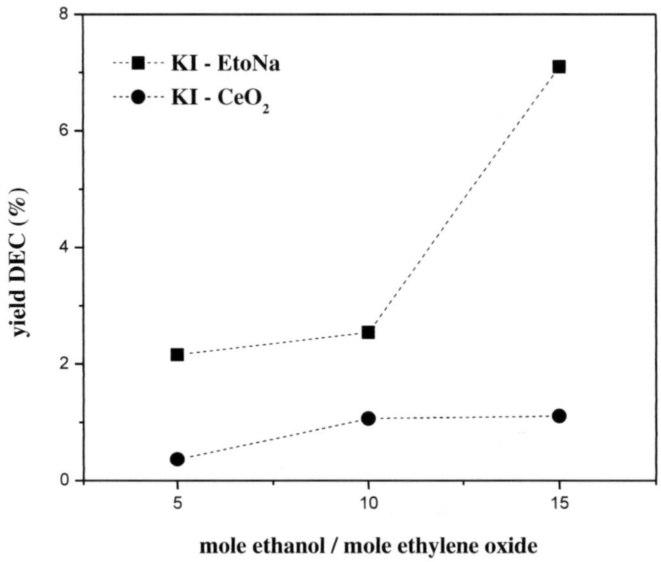

Figure 3. Effect of ethanol to ethylene oxide mole ration on DEC yield.

In DEC synthesis, dehydrating agent is one of the important compounds to be used. Ethylene oxide was used as dehydrating agent by removing thermodynamic limitations by shifting equilibrium towards the carbonate production. From figure 3, it shown that with the increasing ratio of ethanol to the ethylene oxide, the number of DEC compounds formed significantly increases with the best mole ratio 1:15. In this synthesis, ethanol is an excess reactant. With the increase in the mole ratio between ethanol and ethylene oxide it will also have an impact on the mole ratio between H_2O and ethylene oxide, which will increase the rate of hydrolysis reaction of the ethylene oxide itself, so that there will be a reaction that rivals the main reaction in DEC synthesis.

4. Conclusions

Environmental emission gas carbon dioxide has potential to be used as feed material in the process diethyl carbonate sythesis. Diethyl carbonate has high economic value and it can be used as gasoline additives. This potential can be used as an alternative to mitigate the gas emission waste of the energy industry so that it can become more environmentally friendly industry. From the results of this study, it was found that the processing of carbon dioxide to diethyl carbonate produced the best yield of 7.1% using homogeneous KI-EtoNa catalyst as catalyst at mole ratio of ethylene oxide to ethanol of 1:15 within 3 hours of reaction time. The use of homogeneous catalysts is superior to heterogeneous catalyst.

5. References

[1] Hanpattanakit P, Pimonsree L, Jamnongchob A and Boonpoke A. 2018 *Chem. Eng. Trans.* **63** 37-42
[2] Da Silva A R G, Errico M and Rong B G 2017 *Chem. Eng. Trans.* **57** 1-6

[3] Oktavian R, Amidelsi V, Rasmito A and Wibawa G 2013 *Fuel* **107** 47-51
[4] Anugraha R P, Altway A and Wibawa G 2017 *J. Chem. Eng. Data* **62** 2362-66
[5] Aulich T R, He X, Grisanti A A and Knudson C L 1994 *J. Air Waste Manag. Assoc.* **44** 1004-09
[6] Peng C, Lewis K C and Stein F P 1996 *Fluid Phase Equilibr.* **116** 437-44
[7] Pacheco M A and Marshall C L 1997 *Energy Fuels* **11** 2-29
[8] Dunn B C, Guenneau C, Hilton S A, Pahnke J and Eyring E M 2002 *Energy Fuels* **16** 177-81
[9] Anugraha R P, Picunang Z A A, Wiguno A, Tetrisyanda R, Kuswandi K and Wibawa G 2017 *Indones. J. Chem.* **17** 500-8
[10] Anugraha R P, Wiguno A, Altway A and Wibawa G 2018 *J. Mol. Liq.* **264** 32-7
[11] Matsuzaki T and Nakamura A 1997 *Catal. Surv. Asia* **1** 77-88
[12] Zhang Z, Ma X, Zhang J, He F and Wang S 2005 *J. Mol. Catal A-Chem.* **221** 141-6
[13] Yanji W, Xinqiang Z, Baoguo Y, Bingchang Z and Jinsheng C 1998 *Appl. Catal. A Gen.* **171** 255-60
[14] Tomishige K, Sakaihori T, Ikeda Y and Fujimoto K 1999 *Catal. Lett.* **58** 225-9
[15] Leino E, Maki-Arvela P, Eranen K, Tehno M, Murzin D Y, Salmi T and Mikkola J P 2011 *Chem. Eng. J.* **176-177** 124-133
[16] Wang L, Li H, Xin S, He P, Cao Y, Li F and Hou X 2014 *Appl. Catal. A Gen.* **471** 19-27
[17] Van Klink G P M, Dijkstra H P and Van Koten G 2003 *C. R. Chim.* **6** 1079-85

IOP Conf. Series: Materials Science and Engineering **543** (2019) 012072 doi:10.1088/1757-899X/543/1/012072

Biodiesel production from *Calophyllum inophyllum* L oil using Microwave with Calcium Carbonate catalyst

L Qadariyah, A Ansori, S A Wibowo, R S C Muchammad, D S Bhuana, and M Mahfud
Chemical Engineering Department, FTI-ITS, Keputih, Sukolilo, Surabaya, Jawa Timur 60111, Indonesia
Email: lailatul_2008@yahoo.com

Abstract. Biodiesel has become a promising renewable energy resource in recent years due mainly to the fluctuating global oil prices. *Calophyllum inophyllum* L seed has fairly high oil content of about 40-73 % by weight, thus, making it a great potential as raw material for producing biodiesel. This paper aims to study the effect of the microwave irradiation for biodiesel production from *Calophyllum inophyllum* L oil when solid $CaCO_3$ catalysts made from waste eggshells and synthetic were employed. Furthermore, the effects of operating conditions, including reaction time, microwave power, and amount of catalyst loading were also investigated. The initial step to produce *Calophyllum inophyllum* L biodiesel was degumming process, followed by the esterification and trans-esterification, respectively. The catalysts used are $CaCO_3$ synthetic and eggshell. The results of production biodiesel from *Calophyllum inophyllum* L oil give the highest yield of 65.36 % when performed with the microwave power of 300 Watt, 1 % eggshell catalyst and the reaction time of 10 minutes. Therefore, *Calophyllum inophyllum* L shown as potential biodiesel feedstock.

1. Introduction

Energy consumption in Indonesia mainly comes from fossil energy with crude oil (41.46 %), coal (35.81 %), and natural gas (19.36 %) [1]. However, due to the declining fossil fuel sources in the next few years and the increase of environmental issues caused by fossil energy, this has led to examine the prospect of using alternative energy resources such as biodiesel. Biodiesel is defined as the mono-alkyl esters of long chain fatty acids (FAME) derived from renewable lipids. Biodiesel is a biodegradable, nontoxic, low on emission, high cetane number, higher flash point, and give better lubricity on engine type of fuel [2]. thus, making it an ideal fuel for the future.

Biodiesel has a wide range of available feedstock with non-edible oils and therefore, regarded as second-generation biodiesel feedstocks. Biodiesel can be obtained from virgin vegetable oil like palm oil, waste frying oil, and palm oil mixtures [4]. One of the promising non-edible plants for biodiesel resource is *Calophyllum inophyllum* L, whose yield is known to be relatively higher than other plants (*Calophyllum inophyllum* L 40-73 %, Palm oil 45-70 %, Jatropha curcas 40-60 %) [5]. The composition of *Calophyllum inophyllum* L is shown on Table 1.With the Free Fatty Acid (FFA) content of 5.1%, crude *Calophyllum inophyllum* L oil needs to undergo pre-treatment process prior to transesterification, which consist of degumming and esterification. The purpose of degumming and esterification are to eliminate phosphatide gum from crude oil and to reduce the free fatty acid (FFA) content to less than 2%. [6]. High FFA in oil can result in saponification reaction instead of transesterification in the next process, therefore, the FFA content needs to be reduced by converting the FFA into methyl ester in an esterification process [7].

Microwave irradiation is the best method of accelerating and enhancing chemical reactions because it delivers the energy directly to the reactant. Therefore, heat transfer is more effective than in conventional heating and the reaction can be completed in a much shorter time and higher yields in the production of biodiesel can be obtained. It enhances the speed of the reaction and makes the separation process easier in comparison with conventional heating [8].

IOP Conf. Series: Materials Science and Engineering **543** (2019) 012072 doi:10.1088/1757-899X/543/1/012072

Table 1. *Calophylum inophyllum* L oil fatty acid composition [3]

No.	Fatty Acid Composition	Percentage %	No.	Fatty Acid Composition	Percentage %
1.	Saturated fatty acid	35.83	2.	Unsaturated fatty acid	64.17
	-Myristic acid (C14:0)	0.09		-Oleic acid (C18:1)	37.57
	-Palmitic acid (C16:0)	14.6		-Linoleic acid (C18:2)	26.33
	-Stearic acid (C18:0)	19.96		-Linolenic acid (C18:3)	0.27
	Arachidic acid (C20:0)	0.94			
	Erucic acid (C22:1)	0.24			

Catalyst selection is important for transesterification process. Heterogeneous base catalysts are more effective than heterogeneous acid catalysts and enzymes. Heterogeneous catalyst could be operated in continuous processes, could give high quality products, reusable, and eco-friendly [9]. Calcium is common in the environment and is available at very low costs compared to other catalytic metals. The calcium carbonate catalyst is cheaper and is considerably more robust than calcium oxide particles. Traces of calcium carbonate released into the environment generally slightly affect the environment [10]. One of promising $CaCO_3$ source comes from waste chicken eggshell with 99% $CaCO_3$ [11]. Therefore an effort to achieve reduction in the biodiesel production cost there is needed to find an alternative low cost catalyst was derived from waste chicken eggshell and accelerating chemical reaction using microwave

The aim of this paper is to study the effect of the microwave irradiation for biodiesel production from *Calophyllum inophyllum* L oil when a solid $CaCO_3$ catalyst from waste eggshells and synthetic was employed. Furthermore, the effects of operating conditions, including reaction time, microwave power, and amount of catalyst loading are also investigated.

2. Methods

Calophyllum inophyllum L oil was commercially available from Cilacap area of Central Java and delivered in liquid form extracted by mechanical press. The oil that came out of the press machine was dark green with high viscosity. One neck flask was used as reactor with an inner volume of 1000 ml and reflux condenser as a cooling tower. Modified Microwave (Electrolux Co., model EMM 2308X) was used as energy source microwave equipped with magnetic stirrer. [12].

2.1. Catalyst Preparation

The waste eggshells were washed by tap water to remove dust and impurities, and then dried at 110 °C and crushed to 40 mesh size. The catalyst characterization was carried out in the form of X-Ray Diffraction (XRD).

2.2. Degumming

100 ml of *Calophyllum inophyllum* L oil were loaded and heated at a temperature of 80°C then followed by the addition of 85% phosphoric acid by 5 % (v/v). *Calophyllum inophyllum* L oil were mixed for 15 minutes using magnetic stirrer. The collected product was washed three times, each using warm aquadest (60°C), in a separation funnel. The upper oil phase was withdrawn and dried using an oven at 110°C to reduce the water content in the oil and result from degumming is analyzed accordingly.

IOP Conf. Series: Materials Science and Engineering **543** (2019) 012072 doi:10.1088/1757-899X/543/1/012072

2.3. Esterification

100 ml of post-degumming oil were loaded to reactor. Methanol was added to reactor alongside with the oil, the mole ratio of oil-methanol 1:40 and 98 % H_2SO_4 catalyst by 10% (w/w). The reactions were carried out with constant stirring (~300 rpm) at 450 W microwave oven for 20 minutes. After the radiation process, separation between methanol, oil and catalyst was using separating funnel, the top layer of methanol which can be recovered and the bottom layer was a mixture of oil and methyl ester (crude biodiesel). The solutions were cleansed, washed seven times, each using warm aquadest (60°C). The crude biodiesel product was dried in an oven at 110 °C and the result from esterification was analyzed

2.4. Transesterification

Mixed Crude biodiesel (30 mL) were loaded in a methanol (mole ratio of oil-methanol 1:9) and heterogeneous base catalyst was added in reactor. The reaction was carried out with constant stirring using magnetic stirrer under constant radiation in a microwave oven. The solution was washed ten times, each with warm aquadest (60°C) using a separator funnel. The oil was dried in an oven at 110°C to reduce the water content. The variables used were heterogeneous base catalyst type: $CaCO_3$ from eggshell and $CaCO_3$ synthetic; concentration (% v/v oil): 1 %; 2 %; 3 %, 4 %, and 5 %; microwave power (W): 150, 300, 450, and 600; as well as radiation time (minute): 2, 6, 10, 20, and 30. The biodiesel yield was then analyzed with gas chromatography at UBAYA University.

3. Results and discussion

3.1. Material Characteristic

A number of analysis were conducted to determine the characteristics of crude *Calophyllum inophyllum* L oil, including density, viscosity, % FFA, color and GC test (chemical component). The results are presented in Table 2.

Table 2. Characteristic of crude *Calophyllum inophyllum* L oil

Parameter	Results
Density (g/mL)	0,9387
Viscosity (cSt)	53,6259
FFA (%)	27,9011
color	blackish green

Tabel 3. GC Analysis of crude *Calophyllum inophyllum* L oil

Components	(%)
Free Fatty Acid (FFA)	15,7568
Monoglycyrol (MAG)	12,2459
Diglycyrol (DAG)	4,6605
Triglycerol (TAG)	63,9121
Others	3,4247
Total	100

Table 3 represents the type of composition and purity of crude *Calophyllum inophyllum* L oil. Since the crude *Calophyllum inophyllum* L oil used in this study was extracted from the *Calophyllum inophyllum*

IOP Conf. Series: Materials Science and Engineering **543** (2019) 012072 doi:10.1088/1757-899X/543/1/012072

L seeds using mechanical extraction method, the oil contains relatively a lot of impurities, both from the skin and chemical compounds such as phosphatida, chlorophyll, karetenoid and others

3.2. Characteristic Catalyst

In order to characterize the crystallinity patterns of synthetic and egg shell calcium carbonate, XRD analysis was performed for both samples. The diffractograms in Figure 1 shows that diffraction patterns of synthetic and eggshell calcium carbonate have similar peaks. The patterns can be put upon being the differences only found in the intensity of the peaks. Both samples depicted same trends with one sharp peak at 2θ around 30° and weak peaks at 2Θ around 40°-50°. The sharp peaks of both samples are characteristic of the high crystallinity of the samples.

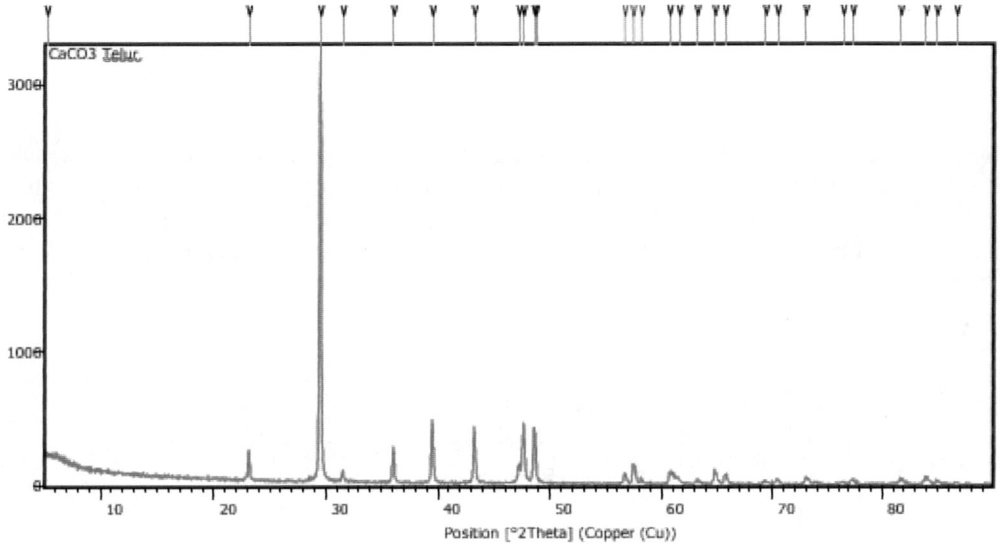

(a)

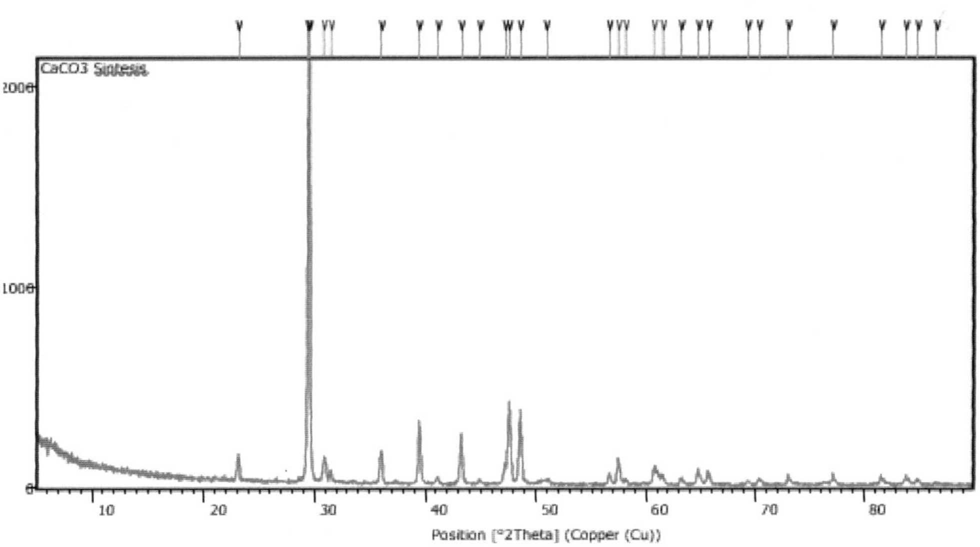

(b)

Figure 1. (a) XRD pattern of $CaCO_3$ eggshell ; (b)XRD pattern of $CaCO_3$ synthetic

3.3 Degumming

The degumming process is a process to remove impurities in the oil. The impurities that typically exist in the form of gum are phosphatida compounds which are carried from the previous process of oil pressing. The comparison between the parameters before and after degumming is presented in **Table 4**. As can be seen, there are some changes recorded on the density, viscosity, and color of the oil after the degumming. More importantly, there is also a quite significant decrease in the FFA content, i.e. from 27.9011% to 19.3069%.

Table 4. Characteristic of degummed *Calophyllum inophyllum* L oil

	Before Degumming	After Degumming
Density (gr/mL)	0.9387	0.9132
Viscosity (cSt)	53.6259	40.4428
Colour	Dark green	Reddish brown

3.4. Esterification

During the esterification stage, the triglycerides in the oil did not seem to react to produce methyl esters based on the fact that no formation of glycerol was identified. This shows that in oils with high FFA concentrations when present in acidic conditions, the reactions that occur tend to be between FFA and methanol (esterification) rather than between triglycerides and methanol (transesterification). The comparison of the oil parameters before and after the esterification process is shown in **Table 5**.

Table 5. Characteristic of esterified *Calophyllum inophyllum* L oil

	Before Esterification	After Esterification
Density (g/mL)	0.9132	0.9030
Viscosity (cSt)	40.4428	22.3905
FFA (%)	19.3074	0.4076

3.5 Transesterification

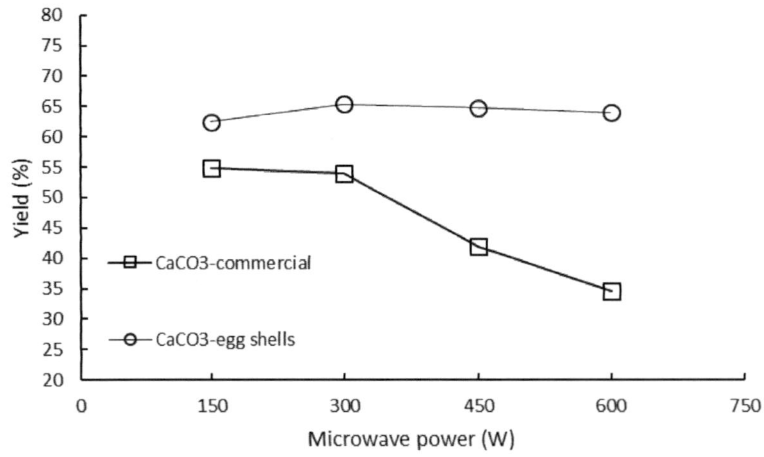

Figure 2. Effect of microwave power on biodiesel yield

Transesterification is the main process of biodiesel production, in which triglyceride contained in the oil is reacted with alcohol to form fatty acid methyl ester (FAME). **Figure 2**. shows the effect of microwave power on biodiesel yield on each $CaCO_3$ synthetic and eggshell catalyst. With increasing power from 150 to 300 W, the eggshell catalyst exhibits the increase of the biodiesel yield by 4.71 % (from 62.4 % to 65.36 %), however, for $CaCO_3$ synthetic catalyst indicates a decrease by 1.58% in the biodiesel yield (from 54.76 % to 53.89 %). Furthermore, with the increasing microwave power, the biodiesel yields are decreasing for both catalysts. The final biodiesel yields at 600 Watt are 63.93 % for eggshell catalyst and 34.5 % for $CaCO_3$ synthetic catalyst. The highest biodiesel yield was obtained at 300 Watt for both catalysts. a research done by D. Y. C. Leung, X. Wu, and M. K. H. Leung carried out a biodiesel production from *Calophylllum inophyllum* Lwith dolomite rock and spiritus at 60 °C for operation condition and examined that base catalyst at high temperature was found to decrease the biodiesel yield due to the unwanted reaction in the form of saponification[13] .

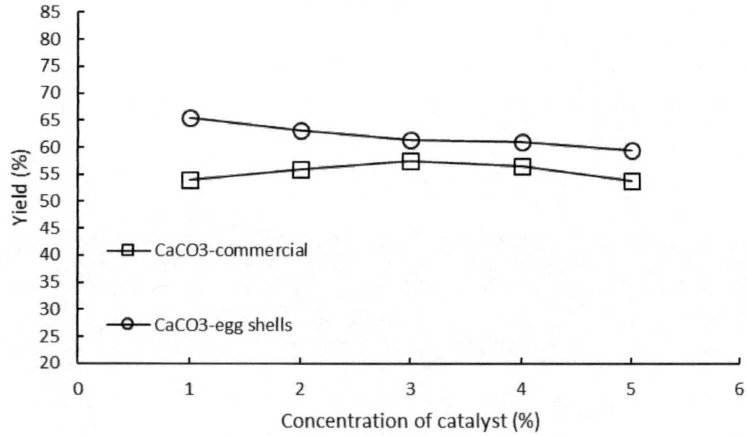

Figure 3. Effect of concentration catalyst on biodiesel yield

Catalysts have important roles in forming methyl ester to accelerate the transesterification reaction. To study the effect of catalyst concentration on biodiesel yield, a microwave power of 300 Watt was set for 10 minutes. As shown in **Figure 3.** when catalyst concentration is increased, the biodiesel yield increases and at one point the yield is decreasing for $CaCO_3$ synthetic catalyst. Yield increase from catalyst concentration 1% to 3% but decrease at concentration 3 % to 5 %. The optimum concentration catalyst (3 %) give the highest yield 57.44 % is for $CaCO_3$ synthetic. As for eggshell catalyst, the biodiesel yield is decreasing with increasing catalyst concentration. The highest yield (65.36 %) is obtained at 1 % concentration catalyst

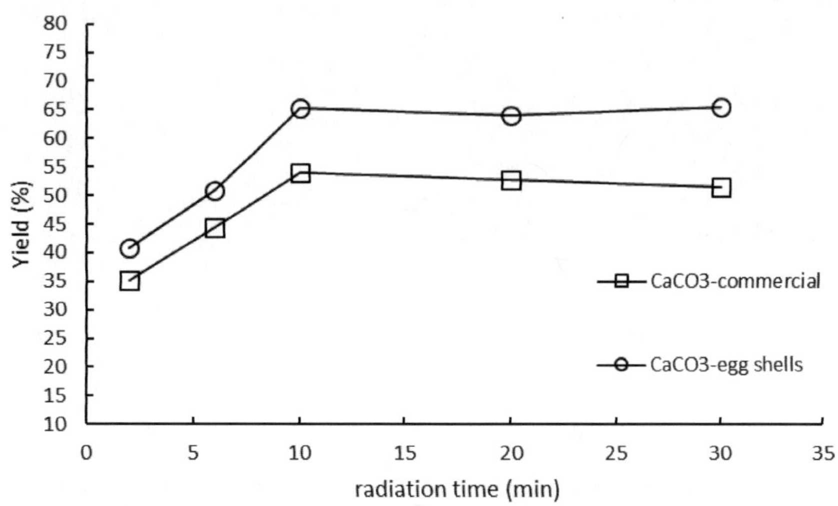

Figure 4. Effect of transesterification time on biodiesel yield

In this study, the effect of radiation time was investigated under the microwave power of 300 Watt with 1 % (w/w) catalyst concentration. For both catalysts the biodiesel yields increase over time until 10 minutes before decreasing over the next minutes. For $CaCO_3$ synthetic from 2 to 6 minutes, biodiesel yield increases by 25.81 % (from 35.20 % to 44.29 %). From 6 to 10 minutes, biodiesel yield increases by 21.69 % (from 44.29 % to 53.89 %). However from 10 to 20 minutes, the biodiesel yield decreases by 2.00 % (from 53.89 % to 52.82 %). From 20 to 30 minute, biodiesel yield decreases by 2.66 % (from 52.82 % to 51.41%). For eggshell catalyst, from 2 to 6 minutes, biodiesel yield increases by 24.72 % (from 40.81 % to 50.90 %). From 6 to 10 minutes, biodiesel yield increases by 28.41 % (from 50.90 % to 65.36 %). However from 10 to 20 minutes, biodiesel yield decreases by 2.10 % (from 65.36 % to 63.99 %) and from 20 to 30 minute, biodiesel yield increases by 2.51 % (from 63.99 % to 65.60 %). The highest yield for $CaCO_3$ synthetic is 53.89 % at 10 minutes and for eggshell catalyst is 65.36 % at 10 minutes. When transesterification reaction is continued further over time, the biodiesel yield decreases (between the 10th and 30th minutes), possibly due to the reversible reaction.

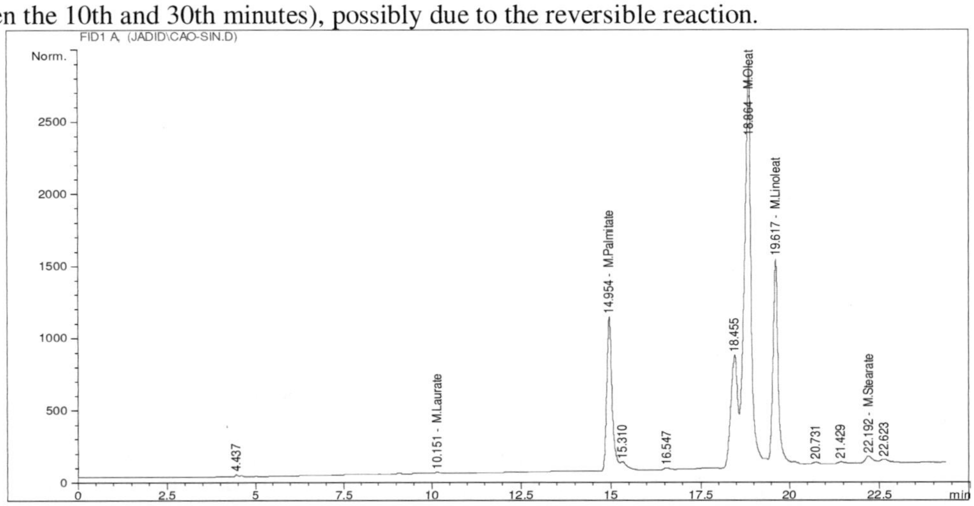

(a)

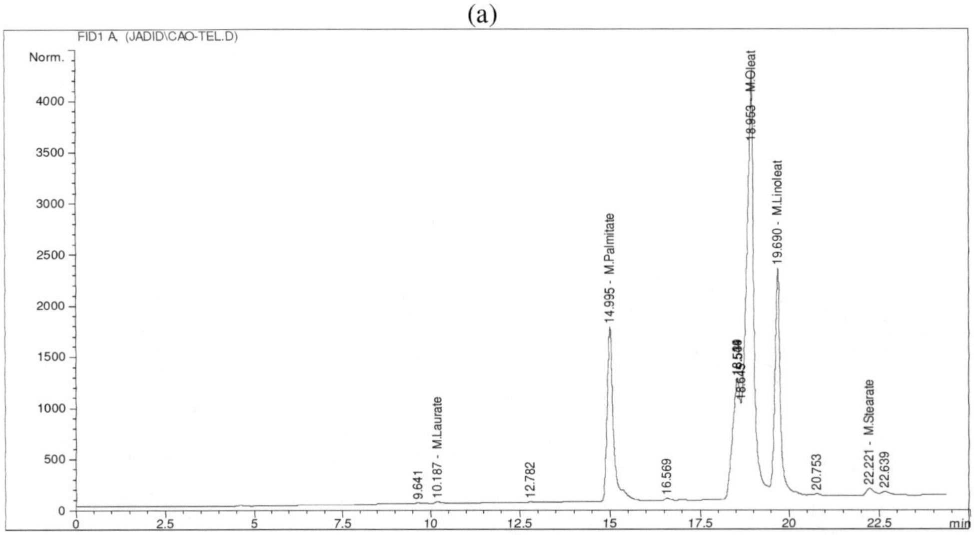

(b)

Figure 5. (a) Chromatogram of biodiesel with 300 W, 1 % $CaCO_3$ synthetic catalyst, 10 minutes reaction; (b) Chromatogram of biodiesel with 300 W, 1 % eggshell catalyst, 10 minutes reaction

The gas chromatography (GC) test result indicates there is FAME formed. From **Figure 5**(a) the composition for FAME with $CaCO_3$ synthesis are laurate methyl 0.0483%, palmitic methyl 19.4538 %, oleic methyl 575571 %, linoleic methyl 22.1287 %, and stearic methyl 0.8121 %. As for the FAME composition from biodiesel with eggshell catalyst can be seen from **Figure 5**(b), the results are hexanoat methyl 0.0167 %, laurate methyl 0.0463 %, palmitic methyl 18.3351 %, oleic methyl 58.1299 %, linoleic methyl 22.5343 %, and stearic methyl 0.9377 %.

Table 6. Density and viscosity of biodiesel yield

Catalyst	Density at 40°C (g/mL)	Viscosity at 40°C (cSt)
$CaCO_3$ synthetic	0,8980 – 0,9041	8,6867 – 12,1441
Eggshell	0,8955 – 0,9025	8,0788 – 9,7932

Unfortunately the density and viscosity of biodiesel yield are out of SNI standard range. The SNI standard for biodiesel density is 0.850 - 0.890 g/mL and for viscosity is 2.3 - 6 cSt. The closest density to SNI standard is 0.8980 g/mL for biodiesel with $CaCO_3$ synthetic catalyst with operation condition 450 W, reaction time 30 minutes, and catalyst concentration 3 %. While the closest density for eggshell is 0.8955 g/mL with operation condition 450 W, reaction time 20 minutes, and catalyst concentration 1%. A research done by J. Goli and O. Sahu [14] using chicken eggshell catalyst was measured and meet all the standards using 7% wt catalys while in this research using maximum 5% wt catalyst. The closest viscosity to SNI standard is 8.6867 cSt for biodiesel with $CaCO_3$ synthetic catalyst with operation condition 300 W, reaction time 30 minutes, and catalyst concentration 1 %. While the closest viscosity for eggshell is 8.0788 cSt with operation condition 300 W, reaction time 30 minutes, and catalyst concentration 1%

4. Conclusion

It is concluded from this study that $CaCO_3$ synthetic and eggshell catalysts can be used for biodiesel production despite the fact that the yield is relatively low. The lowest yield is 35.20% using microwave power 300 W with 1 % $CaCO_3$ synthesis as catalyst and the reaction time of 2 minutes. The highest yield is 65.36 % using microwave power 300 W with 1 % eggshell as catalyst and the reaction time of 10 minutes. $CaCO_3$ can be calcined into CaO, which is another catalyst for biodiesel.

Acknowledgments

This work is financially supported by the Ministry of Research, Technology, and Higher Education of the Republic of Indonesia (KEMENRISTEKDIKTI) under the scheme of Penelitian Dasar Unggulan Perguruan Tinggi (PDUPT) 2018.

References

[1] British Petroleum 2007 BP Statistical Review of World Energy *2017*, **66**.
[2] Rashid U, Anwar F, Moser B R and Knothe G 2008 Moringa oleifera oil: A possible source of biodiesel *Bioresour. Technol.* **99** 8175–79.
[3] Kurniati S, Soeparman S, Yuwono S S and Hakim L 2018 Characteristics and Potential of Nyamplung (*Calophyllum inophyllum* L .) Seed Oil from Kebumen , Central Java , as a Biodiesel Feedstock *Int.Research. J. Adv. Eng. Sci.* **3** 148–52.
[4] Alarcón R, Malagón-romero D and Ladino A 2017 Biodiesel production from Waste Frying Oil and Palm Oil Mixtures *J. Chem. Eng. Trans.* **57** 571–76.
[5] Atabani A E and César S 2014 *Calophyllum inophyllum* L. – A prospective non-edible biodiesel feedstock. Study of biodiesel production, properties, fatty acid composition, blending and engine performance *Renew. Sustain. Energy Rev.* **37** 644–55.
[6] Hathurusingha S, Ashwath N and Midmore D 2011 Periodic variation in kernel oil content and fatty acid profiles of *Calophyllum innophyllum* L .: A potential biodiesel feedstock in Australia *Biomass and Bioenergy* **35** 3448–52.

IOP Conf. Series: Materials Science and Engineering **543** (2019) 012072 doi:10.1088/1757-899X/543/1/012072

[7] Rodríguez D, Riesco J and Malagon-romero D 2017 Production of Biodiesel from Waste Cooking Oil and Castor Oil Blends *J. Chem. Eng. Trans.* **57** 679–84.

[8] Motasemi F and Ani F N 2012 A review on microwave-assisted production of biodiesel *Renew. Sustain. Energy Rev.* **16** 4719–33.

[9] Birla A, Singh B, Upadhyay S N and Sharma Y C 2012 Kinetics studies of synthesis of biodiesel from waste frying oil using a heterogeneous catalyst derived from snail shell *Bioresour. Technol.* **106** 95–100.

[10] Suppes G J, Bockwinkel K, Lucas S, Botts J B, Mason M H and Heppert J A 2001 Calcium Carbonate Catalyzed Alcoholysis of Fats and Oils *JAOCS* **i** 139–46.

[11] Cho Y B and Seo G 2010 High activity of acid-treated quail eggshell catalysts in the transesterification of palm oil with methanol *Bioresour. Technol.* **101** 8515–19.

[12] Duz M Z, Saydut A and Ozturk G 2011 Alkali catalyzed transesterification of safflower seed oil assisted by microwave irradiation *Fuel Process. Technol.* **92** 308–13.

[13] Leung D Y C, Wu X and Leung M K H 2010 A review on biodiesel production using catalyzed transesterification *Appl. Energy* **87** 1083–95.

[14] Goli J and Sahu O 2018 Development of heterogeneous alkali catalyst from waste chicken eggshell for biodiesel production *Renew. Energy* https://doi.org/10.1016/j.renene.2018.05.048

IOP Conf. Series: Materials Science and Engineering **543** (2019) 012073 doi:10.1088/1757-899X/543/1/012073

Biodiesel Production from Microalgae with Trans-esterification Method Using Microwave

L Qadariyah, F Mujaddid, D S Bhuana and M Mahfud
Department of Chemical Engineering, Sepuluh Nopember Institute of Technology Surabaya.

Email: lailatul_2008@yahoo.com

Abstract. *Microalgae* has a highly potential to be biodiesel fuel due to its high lipid content. The research used *Nannochloropsis oculata* microalgae refer to high lipid content (approximately 68%). The objectives of this research are to study edible oil extraction process from *Nannochloropsis oculata,* to study the effect of microwave on biodiesel yield using KOH catalyst, time reaction as well as catalyst percentage. The trans-esterification process was carried out on microwave with lipid molar-methanol comparison of 1:10, microwave power, reaction time and catalyst composition were also varied. It was found that the addition of co-solvent (addition n-hexane) seems to give the best result in terms of yield, i.e. 54.19%, which was obtained at a 600-watt microwave power, reaction time of 40 min, and 2.5% catalyst concentration. The fatty acid consists of several major constituents, such as palmitic acid 84.81%, oleic acid 12.41% stearic acid 1.87% and linoleat 0.88%. It can be concluded from the study that the in-situ trans-esterification process is such an efficient method due to the short reaction time, less energy consumption and relatively efficient owing to the combined extraction and trans-esterification steps.

1. Introduction

Biodesel has several generations, the first generation of edible generation (food), among others, including palm oil and soybeans. The second generation is a non-edible biodiesel, among others, jatropha and nyamplung are some widely known examples. The third generation is microalgae. Harvesting of microalgae can be done daily. Microalgae are the fastest organisms in photosynthesis and contain high lipid content. This can produce up to 200 times more yields than other plants [1]. Microalgae have recently become promising candidates as alternative third generation biotechnology to produce large amount of biomass energy [2]. The triglycerides contained in the microalgae can easily be extracted and esterified to obtain biodiesel. This research uses *Nannochloropsis oculata* microalgae as the main ingredient. *Nannochloropsis oculata* contains 68% fat on dry weight. *Nannochloropsis oculata* has more lipid than *Chlorella* and *Dunaliella*. *Chlorella* has 32% fat compared to *Dunaliella*'s 23% [3]. The conventional process of biodiesel making, the transesterification process is carried out after the extraction and purification process of the oil [4]. Therefore, it is necessary to develop biodiesel manufacturing process that is simple, efficient, energy efficient and can produce high quality biodiesel through in situ transesterification process. In situ method is one of the methods applied in the process of making biodiesel by doing direct extraction on the source of raw materials containing oil or fat. In the in situ process the raw material used is a solid material containing oil. This process is known as transesterification in situ. Microwave irradiation is one of the new technologies in extracting lipid microalgae. It has been reported that microwave is a simple and effective technology for extracting lipid microalgae. Warming up with a microwave can help speed up the extraction and simultaneously turn it into FAME. Other experiments confirmed that the addition of co-solvent is also considered as a way to increase the yield on in situ transesterification because co-solvent can increase the solubility of alcohol and speed up the in situ transesterification [5]. Therefore, the aim of this research is to produce a high yield of biodesel from microalgae as potential resource of biodiesel using microwave as source energy by in situ transesterification method.

2. Methods

2.1. Materials
The materials used include powder of microalgae *Nannochloropsis oculata,* which was collected from BBAP Jepara. Another material such as Methanol 96%, KOH and ethyl acetate 96 % was used from MERCK products meanwhile for n-hexane 96 % was used from FULLTIME products).

2.2. Apparatus
In situ transesterification was performed in EMM2308X Electrolux microwave unit. The apparatus of biodiesel from microalgae comprised of a vacuum pump, a condenser, magnetic stirrer and reactor pyrex. The temperature is controlled using a thermocouple.

2.3. Extraction of microalgae oil using soxhlet
The experiment was carried out in round neck flask. The loaded amount of microalgae powder was 5 gram for microalgae oil extraction with methanol were 50 ml. Microalgae powder and solvent were mixed in round neck flask. The extraction process was done at the soxhlet. the reaction time of 12 hours. After the extraction was completed, the product was cooled down and separated by filter vacuum. the filtrate and residu were obtained. The residu was dissolved in the mixture of hexane and methanol to get the remaining oil. The filtrate was put into a separatory funnel and washed using 40 ^{0}C of aquadest. After that, the filtrate was distilled to separate the solvent and microalgae oil. The microalgae oil was analyzed by gas chromatography (GC).

2.4. Biodiesel Production from Microalgae Nannochloropsis oculata by In-Situ Microwave-Assisted Transesterifiaction
The in-situ transesterification takes place in a round neck flask, insert a 5 gram microalgae powder and methanol is 50 ml for all experiments. Microalgae and methanol powder were mixed in a round neck flask. In situ transesterification was carried out at 300, 450, 600 watt microwave power, with reaction time of 10,20,30,40 minutes and catalyst concentration of 1%, 1.5%, 2%, and 2.5%, respectively. The reactant was then cooled to room temperature and separated with a vacuum filter so that the filtrate and the residue were formed. The residue was dissolved in a hexane and methanol mixture to obtain the remaining FAME. Ethyl acetate was added to the filtrate to form three layers. The upper layer was heated to a temperature of 77.1 ^{0}C. The crude biodiesel was then analyzed by mass-gas chromatography (GC).

3. Results and discussion

3.1. Characterization of Microalgae Oil
The experiments were performed inside a round neck flask under microwave irradiation. In-situ transesterification was carried out with and without the addition of co-solvent (n-hexane) using KOH as an alkaline catalyst. The use of KOH in this process will provide some more reactive advantages than acid catalysts. The biodiesel yield is defined as the mass of crude biodiesel divided by the mass of the microalgae oil. Based on the experiment, the yield of microalgae oil extraction was 12.934 % (based on its dry weight). The lipid composition of *Nannochloropsis oculata* is shown by Table 1. Nannochloropsis oculata contains 16.43% FFA and 5.7302 % triglycerides makes *Nannochloropsis oculata* potentially become biodiesel.

Table 1. Lipid composititition From *Nannochloropsis oculata*

Lipid	Composition (%)
Free fatty acid	16.43
monoglycerides	2.69
diglycerides	11.492
triglycerides	5.7302
Total	36.342

3.2. The Effect of Reaction Time and Microwave Power Without- Solvent and Co-Solvent

The co-solvent method gave the highest yield of 54.19% at 40 minutes on 2.5% catalyst. Figure 1 shows the effect of reaction time on biodiesel yield on various catalysts in solution without solvent and co-solvent. At 10 minutes, 20 minutes, 30 minutes, and 40 minutes, the biodiesel yield gradually increases with increasing reaction time. In this condition, the longer the reaction time the contact between the reacting substances will be greater so that the results obtained will also be greater. The use of microwaves that provide a large heat effect so that time will increase the temperature on the media and one of the high solubility factors is the increase in temperature so that the reaction time is longer. The temperature will increase so that the contact time will increase the yield of biodiesel. The results of this study are supported by previous research which states that time variables affect the resulting results [6].

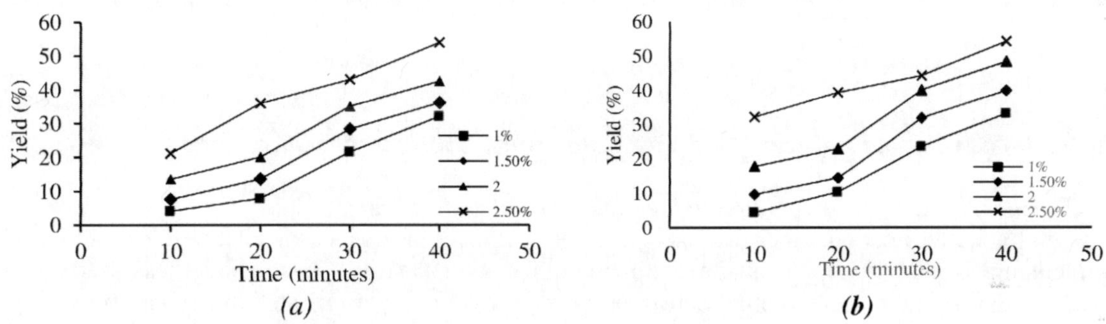

Figure 1. The effect of reaction time on yield at 600 watt without solvent (a) co-solvent (b)

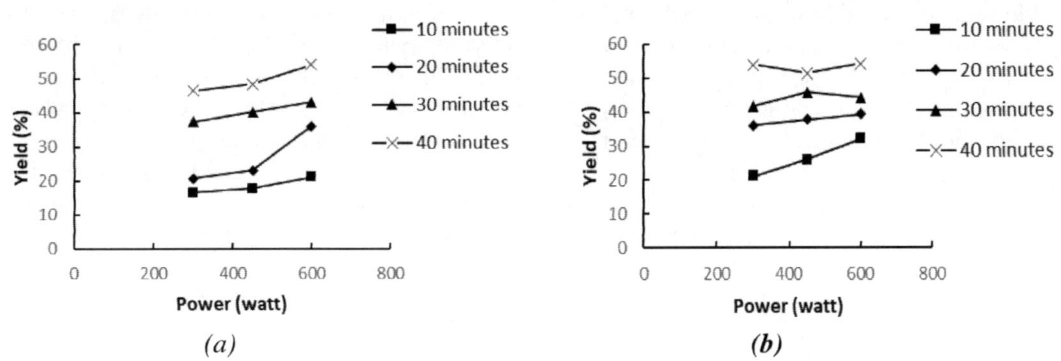

Figure 2. The effect of power on yield at 600 watt without solvent (a) co-solvent (b)

Figure 2 shows the effect of power on biodiesel yield at various reaction time. At power of 300,450 and 600 watt biodiesel yield gradually increased with power. The highest yield of biodiesel were 54.19 % at

40 minutes at catalyst 2.5 %. This is mainly because high temperatures can help break down cell walls and make contact between methanol and lipids more easily and produce higher yields [7].

3.3 The Effect of Catalyst Concentration Without Solvent and Co-Solvent

The co-solvent method gave the highest yield of 54.19% at 40 minutes on 2.5% catalyst and 600 watt. Figure 3 shows the effect of catalyst on biodiesel yield at various reaction time. The yield increases with the increase of catalyst concentration in the range 1.0 - 2.5%. The catalyst will react with methanol to form methoxide ions. This ion will react with lipids to produce methyl esters (biodiesel). The more catalysts used the greater percentage of yield product. This happens because the function of the catalyst is to decrease the activation energy. In accordance with Arrhenius law which states that the addition of catalysts can activate the reagents so that the activation energy (Ea) becomes smaller and the value of the reaction rate constant becomes greater, so as to increase the rate of reaction because the reaction rate constant is directly proportional to the reaction rate causing the formed product more abundant so that the yield percentage of the product becomes greater [8]

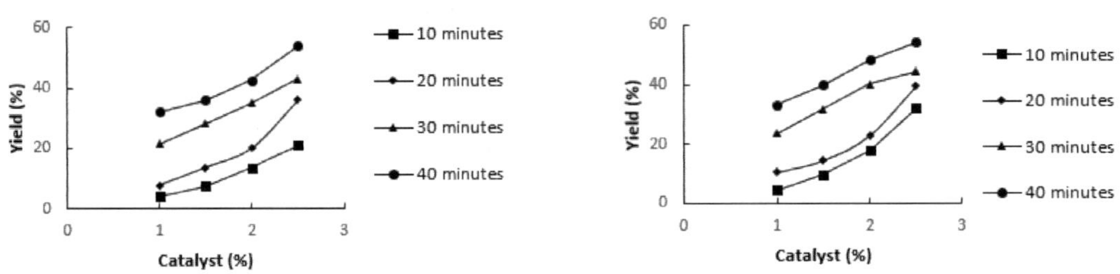

(a) ***(b)***

Figure 3. The effect of catalyst on yield at 600 watt without solvent (a) and co-solvent (b)

3.4. The Effect of Co-Solvent Addition to the Yield of Biodiesel Products

In in-situ transesterification, other than as a reactant, methanol also acts as a solvent in extracting oil. But methanol is not a good solvent for extraction, whereas hexane is. Thus, hexane was used as co-solvent in this study to assist the extraction process in in-situ transesterification. As can be seen on Figure 4, the transesterification process with the addition of co-solvent can increase the yield of crude biodiesel when compared to the transesterification process without co-solvent. The highest yield obtained in the transesterification process with co-solvent at 40 minutes was 54.19%. In this study, the in-situ transesterification process with the addition of 40 ml of co-solvent (hexane) produced a one-phase solution. Basically, hexane and methanol will not coalesce, however, with a considerable volume of methanol a one-phase solution is produced. As a co-solvent, hexane can increase the solubility of the oil in methanol. With the addition of co-solvent, the interfacial surface tension of the oil and methanol will decrease and the liquid will form one phase. In other words, the reaction will increase because the contact between the reactants is better. It is in the presence of this one-phase solution which causes the in situ transesterification process with the addition of co-solvent to increase the yield of crude biodiesel. In the in situ transesterification process. the ongoing reaction can occur in two possibilities. First, the process occurs simultaneously (extraction and transesterification take place at the same time) [9]. Second. the process begins with lipid extraction and is then simultaneously converted to FAME.Based on this research, it can be concluded that the in-situ transesterification process begins with lipid / oil extraction and then followed by the production of fatty acid methyl ester (FAME) from the lipid/oil. This statement is supported by the results of the research provided that transesterification with the addition of co-solvent (hexane) will increase the yield of crude biodiesel. it was supported by another research the addition of n-hexane give high oil extraction from microalgae. The addition of the co-solvent itself provides an important role in the initial process of extraction in in situ transesterification so that the resulting yield also increases [10]. The results obtained maintains that the addition of hexane as a co-solvent in in situ transesterification might increase the yield of biodiesel [5].

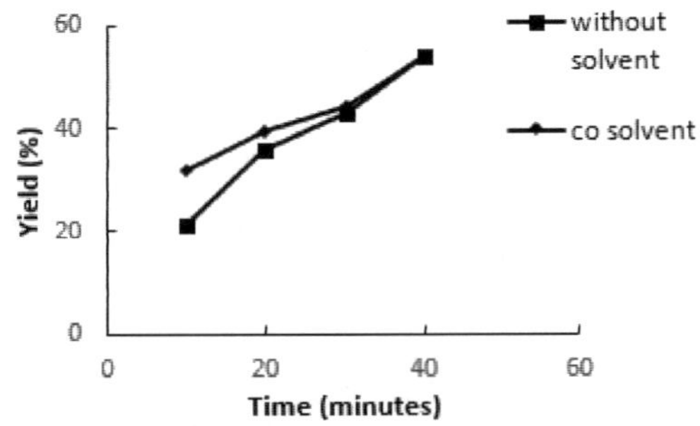

Figure 4. The effect of co-solvent on yield at 600 watt

3.5 FAME compounds of Nannochloropsis oculata

From Table 2, the main compounds of methyl ester of biodiesel were palmitic acid, stearic acid, oleic acid and linoleic acid. The results obtained are in accordance with the GC yield of raw materials with lipid composition *Nannochloropsis oculata* where the dominant fatty acids in the lipid is palmitic acid of 84.8173%. This analysis fits reporting that all plants oil which is used for biodiesel production must contain palmitic and linoleic of fatty acids. From the investigated methyl ester compounds tested with the GC, it can be concluded that the production of biodiesel from *Nannochloropsis oculata* microalgae using in situ transesterification method with microwave wave was successfully carried out.

Table 2. Methyl Ester Biodiesel Composition From *Nannochloropsis oculata*

Methyl Ester	Composition (%)
Palmitic acid	84.8173
Oleic acid	12.4169
Linoleic acid	0.8869
Stearic acid	1.8787

4. Conclusion

The in-situ transesterification was studied by microwave irradiation at microwave power of 300, 450, and 600 watt, reaction time of 10, 20, 30 and 40 minutes, and KOH catalyst concentration of 1 %, 1.5%, 2% and 2.5%, respectively. Whereas the in-situ transesterification with co solvent (addition of n-hexane) was studied by microwave irradiation at microwave power of 300, 450, 600 watt, reaction time of 10,20,30 and 40 minutes, KOH catalyst concentration of 1%, 1.5%, 2% and 2.5%. The most effective method in terms of the biodiesel amount was the one by using co-solvent (addition of n-hexane), i.e. with the highest yield of 54.19% which was obtained at 600 watt microwave power, reaction time of 40 min, and 2.5% catalyst concentration. Biodiesel yield is found to be affected by microwave power, reaction time, acid catalyst concentration. The longer the reaction times would give higher biodiesel yields.

Acknowledgments

This work is financially supported by the Ministry of Research, Technology, and Higher Education of the Republic of Indonesia (KEMENRISTEKDIKTI) under the scheme of Penelitian Dasar Unggulan Perguruan Tinggi (PDUPT) 2018.

References

[1] Yusuf C 2008 Biodiesel from microalgae beats bioethanol *Trends Biotechnol.* **26** 126–31.
[2] Yeimmy P R, Luis G O and Ángel G D 2018 Design of biodiesel and bioethanol production process from microalgae biomass using exergy analysis methodology *Chem. Eng. Trans.* **70** 1045–50.

[3] Susilaningsih D, Djohan A C and Widyaningrum D N 2009 Biodiesel from indigenous Indonesian marine microalgae *J. Biotech Res* **2** 1–4.

[4] Yeily A R B, Ismael E G O, John H S G, Zuorro A, Barajas-Solano A F and Urbina-Suarez N A 2018 The effect of temperature and enzyme concentration in the transesterification process of synthetic microalgae oil *Chem. Eng. Trans.* **64** 331–36.

[5] Dianursanti, Sistiafi A G, and Putri D N 2018 Biodiesel synthesis from nannochloropsis oculata and chlorella vulgaris through transesterification process using NaOH/zeolite heterogeneous catalyst *IOP Conf. Ser. Earth Environ. Sci.* **105** 1-5.

[6] Armando T Q, Takahisa. K, Mitsuru S and Motonobu G 2013 Microwave-hydrothermal extraction and degradation of fucoidan from supercritical carbon dioxide deoiled Undaria pinnatifida *Ind. Eng. Chem. Res.* **52** 7940–46.

[7] Lailatul Q, Mahfud, Renova P, Fatkhulloh M, and Ummu K. 2018 Biodiesel production from dry microalga biomass by microwave-assisted in-situ transesterification *Matec* **156** 1-5.

[8] Jincheng D, Benqiao H, and Jianxin L 2011 Biodiesel production from acidified oils via supercritical methanol E*nergies* **4** 2212–23.

[9] Georgogianni K G, Kontominas M G, Tegou E, Avlonitis D and Gergis V 2007 Biodiesel production : reaction and process parameters of alkali-catalyzed transesterification of waste frying oils *Energy Fuels.* **21** 3023–27.

[10] Antonio M, Juri R, Patrizia C, Antonietta C, Vincenzo L, Despina K, Evangelos H, Simeone C and Dino M 2018 Microalgae valorisation via accelerated solvent extraction: optimization of the operative conditions *Chem. Eng. Trans.* **65** 835–40.

IOP Conf. Series: Materials Science and Engineering **543** (2019) 012074 doi:10.1088/1757-899X/543/1/012074

Extraction of Natural Dye from *Ketapang* Leaf (*Terminalia catappa*) for Coloring Textile Materials

Faisal RM and A Chafidz*

Chemical Engineering Department, Universitas Islam Indonesia, Yogyakarta 55584, Indonesia
E-mail: achmad.chafidz@uii.ac.id

Abstract. The objective of the present work is to extract natural dye substances from *Ketapang* leaf. This leaf contains tannin substance that can be used as natural dye for textile materials. This natural dye is one of potential alternatives to the synthetic dyes because it is environmentally friendly and also *Ketapang* leaf is abundantly available, especially in Indonesia country. The brief detail of *Ketapang* leaf extraction process can be explained as follow: *Ketapang* leaf was soaked/immersed in distilled water for several periods (i.e. 4, 6, and 8 days) by using maceration method. From the experimental results, the optimal immersion time was reached at 6 days. The tannin content in the extract was also determined by using UV-Vis spectrophotometer. To test the performance of this natural dye, textile material made of 100% cotton was immersed in tannin solution for 60 min, squeezed, and the color was raised with three different fixer solutions, i.e. *alum* ($Al_2(SO_4)_3$, *kapur tohor* (CaO), and *tunjung* ($FeSO_4$) 50 g/L for 60 min. The fabrics were then tested for color fastness using detergent washing method and rubbing method (with crockmeter test). The test results showed that the fabric with *alum* as the fixer exhibited the optimum result for fastness test.

1. Introduction

For esthetic and additional value purpose, textile materials (i.e. fabrics) are coloured. In ancient times, these fabrics were coloured using natural dye [1]. However, in 1856, synthetic coloring agents became important and used in many various industries, such as textile industry. Most of these synthetic dyes are derived from petrochemicals, which used hazardous chemical process that can affect human's health and environment [2]. Nowdays, most of the modern textile industries uses synthetic dyes due to low price, readily available, many variant of colors, easy to use, good fastness properties, etc. [3, 4]. However, the use of synthetic dyes of these textile industries has increased water pollution caused by the wastewater discharge. Therefore, synthetic dyes began to be issued by people and people began to focus back on natural dyes. The use of natural dye for textile dyeing has attracted great interests of both researchers and textile industries due to the use of sustainable raw materials, biodegradability and thus environmentally friendly, non-toxic, non-allergic [4, 5].

Natural dyes can be extracted from various parts of plant, like leaves, flowers, fruits, wood or stem, and roots [6]. Various studies have been carried out by people, including research into the manufacture of natural dyes from mangosteen peel, betel leaves, and *indigofera* leaves. In general, natural dyes have some limitations, including limited variation of colors, must be treated prior to use as coloring agent, and less fastness. For extraction process, there are four major types of methods that can be done to extract natural dyes from their raw materials, namely: 1) Aqueous extraction; 2) Extraction of alcohol solvents / organic solvents; 3) Extraction with the help of ultrasonic; 4) Microwave-assisted extraction [7]. Another method, which is considered a very simple and conventional method is called maceration.

Maceration is a traditional technique used for many years in the extraction of many natural products. This method extracts natural dye from plants by soaking the grinded plants into a suitable solvent in a container for hours while shaked or stirred. This is the most simple method since does not require special equipments [8].
Indonesia is a country with a large diversity of plants, which is an extraordinary potential. However, there are still many plants that can be considered as source for natural dye's raw materials in Indonesian that have not been fully researched and utilized properly. Therefore, numerous research are needed to explore the raw materials of natural dyes, how to extract them, and use or apply them as coloring agent substituting synthetic dyes in textile industries. In this study, we have investigated *Ketapang* leaf as a potential source of natural dyes. In this study, natural dye (i.e. tannin) was extracted from *Ketapang* leaf using a maceration process. The extracted natural dye was then used for coloring cotton and tested for color fastness against washing and rubbing.

2. Experimental

Ketapang leaf was cut into small pieces and grinded until become fine sizes. Afterward, the extraction of natural dye was carried out using a maceration method for 48 hours with a weight ratio of material: solvent (1: 4). The extract of *Ketapang* leaf was then filtered using a filter paper and the filtrate was collected. Afterward, 1.5 ml solution of potassium ferrisianide and ammonia were added into 300 ml of *Ketapang* leaf extract. The solution was then shaked and left for 3 minutes until the color changed. Then, the extract was evaporated using a rotary-evaporator for 30 minutes. Afterward, the weight of the extract was measured by weighing the empty beaker glass and pouring the extract into an empty beaker glass. Afterward, the beaker was put in the oven and heated at 60°C until the ethanol completely evaporated. Then, the mass of beaker glass and *Ketapang* leaf extract was measured. The analysis of tannin content was carried out using with a UV-VIS spectrophotometer. For coloring cotton fabric, a vegetable oil type mordant, Turkey Red Oil (TRO) was used. TRO is a mordant mainly used for dyeing of red colour [3]. In this study, a TRO solution was prepared with a mass ratio of 2 gram of TRO and 1 L of distilled water. Then, the white cotton cloths were soaked into the TRO solution for 12 hours. Afterward, the cotton cloths were dried without being exposed to sunlight. Dip the cotton cloths that have been dried into *Ketapang* leaf extract solution. Then, the colored cotton fabrics were dried without being exposed to the sunlight. Afterward, three fixer solutions, namely *alum* ($Al_2(SO_4)_3$, *kapur tohor* (CaO), and *tunjung* ($FeSO_4$) were used with a ratio of 50 g/1 L for each fixer. The cotton cloths that have been coloured were dipped into each of fixer solutions (i.e. *tawas, kapur tohor, tunjung*) for 60 minutes (each) and then dried the cloths. The colored cloths were then tested for the color fastness via detergent washing and rubbing methods, and also their color strength by using UV-Vis spectrophotometer.

3. Results and discussion

The natural dye extracted from the *Ketapang* leaf gave brownish color, which indicates tannin content in the natural dye. A rotary evaporator was used to separate the *Ketapang* leaf extract from the solvent (i.e. ethanol). Figures 1a and 1b show the appearance of *Ketapang* leaf extract solution and the extract after evaporation of the solvent using a rotary evaporator. The working principle of this tool is the process of separating the extract from the solvent by heating and the process is accelerated by rotating the flask. The solvent can evaporate 5-10ºC below the boiling point of the solvent due to a pressure drop. By using this apparatus the extract's compounds will not be damaged by the temperature.
Tannin is polyphenol compound that has high molecular weight and has hydroxyl groups and other groups (e.g. carboxyl) so that it can form complexes with other proteins and macromolecules under certain environmental conditions. This compound contains pigments which cause brownish color. To identify the tannin content of *Ketapang* leaf, a color test was conducted quantitatively by adding folin solution and then allowed it for 3 minutes, afterward Na_2CO_3 solution was dripped, then shaked and let it idle. Blue changes that occur indicate tannin content. Additionally, Table 1 shows a quantitative result of *Ketapang* leaf extract with two different ethanol concentrations (i.e. 70% and 96%). The weight of *Ketapang* leaf extract was 0.73 and 0.83 grams for solvent (i.e. ethanol) concentrations of 70% and 96%,

respectively. The difference between the two solvent concentrations was 0.1 grams (i.e. 10%). This result indicated that the higher the solvent concentration, the greater the amount of extract produced because higher ethanol concentration will increase the extraction power.

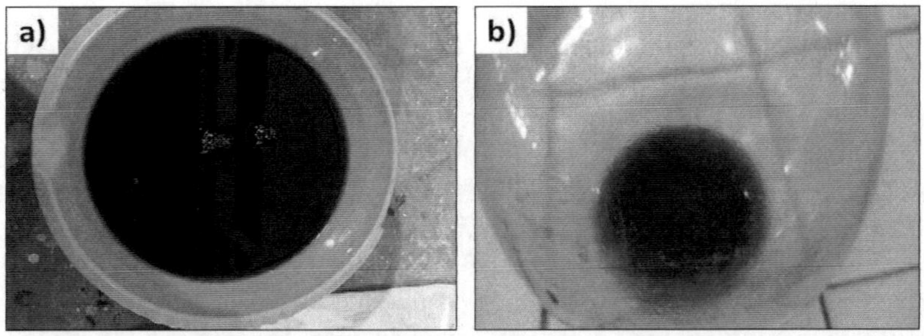

Figure 1. Photograph of a) *Ketapang* leaf extract solution; b) Ethanol evaporation from *Ketapang* leaf extract solution.

Table 1. The result of *Ketapang* leaf extract with two different concentrations of solvent (i.e. ethanol).

Ethanol concentration (v/v)	Weight of *Ketapang* leaf extract (gram)
90%	0.83
70%	0.73

Additionally, a UV-Vis spectrophotometer was also used to measure the level of tannin contained in the *Ketapang* leaf extract. By using the UV-Vis spectrophotometer, the absorbance of each extract solution was measured to determine the response of pigments/dyes to the light in the UV to Visible light spectrum with a wavelength of 405 nm [9]. The higher the absorbance value, the higher the tannin content. Figure 2 shows the effect of ethanol concentration and duration of Maceration process on the tannin level contained in *Ketapang* leaf extract. As seen in the figure, the higher tannin content was obtained from the longer extraction time. This result is in accordance with the study reported by the literature [10]. They reported that the absorbance was proportional to the tannin content contained in the extract and the longer the extraction time the greater the concentration of dye in the solvent.

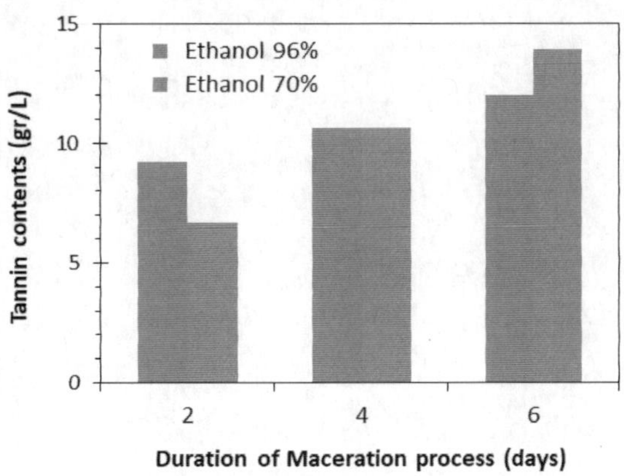

Figure 2. Tannin level in *Ketapang* leaf extract with as function of ethanol concentration and duration of Maceration process

Handayani, et al. [11] also reported that extraction time was an important factor that must be considered during tannin extraction process because it can affect the quality of extraction result. Lestari et al. [10] also reported that the longer the extraction time, the longer the contact between the solvent and the material (solute). Nevertheless, each material has an optimum limit, if the time exceeds the optimum limit, further extraction process will have no effect, instead the compound will decompose. Aside from the time, the type of solvent may also affect the maceration process. The solvent used must be non-polar solvent. From the experimental results, the highest tannin content was obtained when using solvent with 70% ethanol instead of 96% ethanol. The possible explanation for this result is as follows.

One of the ethanol works in damaging bacterial cells is by denaturing the proteins. This work will be effective if there is water in it, which can inhibits the growth of mold that will damage the compounds contained in the raw material during the maceration extraction process. Solvent with 70% ethanol is a mixture of 70% ethanol and 30% water. This solvent consists more water than the solvent with 96% ethanol. Therefore, the optimum tannin results were obtained at 6 days of maceration with 70% ethanol solvent. The dyeing of white cotton cloths using *Ketapang* leaf extract solution with different fixer solutions have resulted in several colors change of the cotton fabrics. The use of *alum* fixer solution produced a light yellow, while *kapur tohor* and *tunjung* produced a brownish yellow and a gray near black, respectively. The resulting color difference is caused by different fixer solutions used. Therefore, the color produced will vary according to the use of the fixer solution, which are *alum* will produce color, *kapur tohor* will produce older colors, and *tunjung* will produce stronger and thicker color. Figure 3 shows fabric colors before and after dyeing and fixation.

Additionally, color fastness test was also carried out to find out the resistance of color that has been bound to the white cotton fabrics to fade or bleed. The color fastness test can be done using a soap washing method, which includes slacking and desecration and rubbing method. The principle of the method of washing soap is to see the color strength attached to the fabric after the washing process when the pores of the fabric are opened by a detergent solution. The soap washing method measurement uses grey scale to see the color quality after washing. Table 2 shows the color fastness test using detergent washing method. As seen in the Table 2, the color fastness test results obtained were varied based on the type of fixer. The fastness test score for the *alum* fixer was averagely 4-5, which means that the color resistance is classified as "good". Whereas, for the *kapur tohor* fixer, the fastness test score was averagely 3, meaning that the color resistance of the fabric is categorized as "quite". Additionally, score 3-4 (in average) was obtained for the *tunjung* fixer, which can be classified as “quite good”. From the results in Table 2, it can be concluded that the fabric with the best color resistance when washed with detergent is the cloths treated by the *alum* fixer solution.

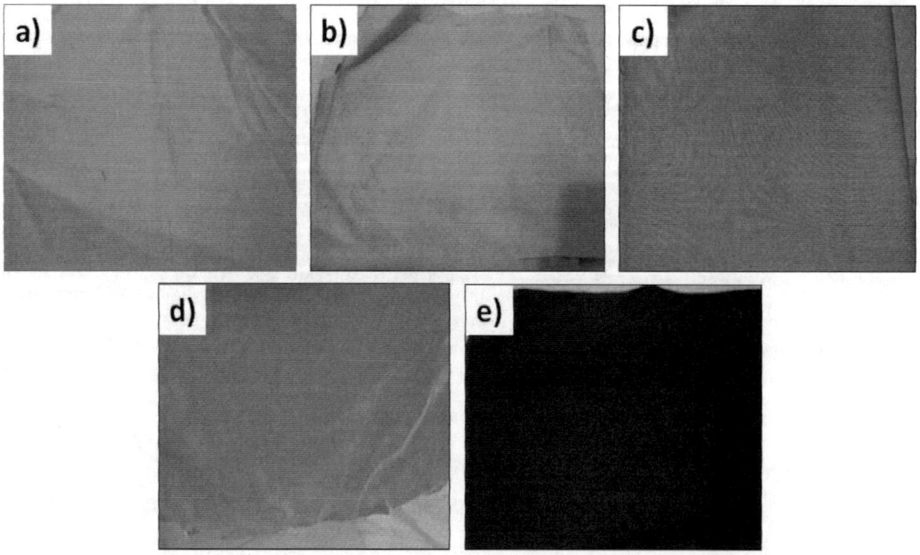

Figure 3. Colors of the cotton cloth: a) before dyeing; b) after dyeing (without fixation); c) after dyeing and fixation (with tawas); d) after dyeing and fixation (with *kapur tohor*); e) after dyeing and fixation (with *Tunjung*)

IOP Conf. Series: Materials Science and Engineering **543** (2019) 012074 doi:10.1088/1757-899X/543/1/012074

Table 2. Results of color fastness test using detergent washing method

Fixation reagent	Sample No.	Color fastness score
Tawas $(Al_2SO_4)_3$	1	4-5 (Good)
	2	4-5 (Good)
	3	4-5 (Good)
Mean		4-5 (Good)
Kapur tohor (CaO)	1	3 (Quite)
	2	3 (Quite)
	3	3 (Quite)
Mean		3 (Quite)
Tunjung (Fe_2SO_4)	1	3-4 (Quite Good)
	2	3-4 (Quite Good)
	3	3 (Quite)
Mean		3-4 (Quite Good)

Additionally, the other method to evaluate the color fastness of a fabric is rubbing method. Rubbing method uses a Crockmeter to determine the level of color resistance of the fabric sample to be absorbed into the white cotton cloth by rubbing a white cotton cloth against the fabric sample. The measurement scale in the rubbing method is staining scale. This rubbing method is divided into 2 types, i.e. dry rubbing test and wet rub test. The difference lies in the treatment of white cotton fabric. In the dry rubbing test, the white cotton cloth is completely dry, while in a wet rubbing test, the white cotton cloth is rinsed first with water. Table 3 shows the color fastness test results by using fabric rubbing method. The test results show the same scores between the wet rubbing test and dry rub test for all the fixers, i.e. *alum, kapur tohor*, and *tunjung*, which were on 4-5 scale. The number 4 indicates the color quality of the fabric is "good", while the number 5 shows the "excellent" quality. From this test, it can be concluded that the color fastness against dry or wet cloth rubbing did not differ on the type of fixer.

Table 3. Color fastness test results using fabric rubbing method

Fixation reagent	Sample No.	Dry rubbing test value	Wet rubbing test value
Tawas $(Al_2SO_4)_3$	1	4-5 (Good)	4-5 (Good)
	2	4-5 (Good)	4-5 (Good)
	3	4-5 (Good)	4-5 (Good)
Mean		4-5 (Good)	4-5 (Good)
Kapur tohor (CaO)	1	4-5 (Good)	4-5 (Good)
	2	4-5 (Good)	4-5 (Good)
	3	4-5 (Good)	4-5 (Good)
Mean		4-5 (Good)	4-5 (Good)
Tunjung (Fe_2SO_4)	1	4-5 (Good)	4-5 (Good)
	2	4-5 (Good)	4-5 (Good)
	3	4-5 (Good)	4-5 (Good)
Mean		4-5 (Good)	4-5 (Good)

The color quality of a fabric can also be measured through a color strength test using a UV-Vis specrophotometer. The working principle is to determine the percentage of reflectance (R%) of the fabric sample against the reference (white cotton fabric). The higher the R% of the fabric sample, the lower the strength of the color. This is because the measured blank fabric has a R% value of 100, so that when the fabric sample approaches R% value of 100, meaning that the color strength of the fabric sample is very close white. Table 4 shows that test results of color strength of the dye (in R%). The R% value obtained in the dyed fabric sample with no fixation were as follow: with a 60 minute of dyeing process time the R% was 96%, at 90 minutes the R% was 92%, and at 120 minutes was 81%. It can be concluded that the color of the mixture of *Ketapang* leaf extract with white cotton cloth that is dyed longer will be thicker (old) compared to just dipped briefly. The longer the dyeing process will increase the extraction capacity of the *Ketapang* leaf which causes the extract concentration to be higher and more concentrated so that the color created will be thicker.

Table 4. Test results of color strength of the dye (R%)

Fixation reagent	Sample No.	Color strength of the dye (R%)
Cotton cloth without fixation	1	103.36
Tawas ($Al_2SO_4)_3$	1	81.02
	2	73.04
	3	69.10
Mean		74.39
Kapur tohor (CaO)	1	48.84
	2	39.77
	3	41.33
Mean		43.31
Tunjung (Fe_2SO_4)	1	12.23
	2	7.32
	3	7.91
Mean		9.15

4. Conclusion

Natural dye rich of tannin was successfully extracted from *Ketapang* leaf via maceration process. Based on the experimental results, it can be concluded that the optimal tannin content was 13.9 g/L, which was obtained in the use of 70% ethanol solvent with a maceration time of 6 days. The natural dye was used for coloring cotton cloths with the help of different fixer solutions (i.e. *tawas, kapur tohor, tunjung*). The resulted colors were light yellow, brownish yellow, and gray near black for the use of fixer tawas, *kapur tohor*, and tunjung, respectively. The colored cotton cloths were then tested for their color fastness using detergent washing method and rubbing method. Additionally, the colored cloths were also tested for their color strength. For the color fastness test via detergent washing method, the most optimal results was obtained from the cloth that used fixer *alum* solution with an average fastness test score of 4.5 (i.e. "good"). This result indicated that the use of *alum* fixer gave better results in term of color resistance. Whereas, from the rubbing test method, all of the cloths with three different fixer solution gave similar or even same results on the color fastness test. Furthermore, the color strength of the fabric was also measured by using UV-Vis spectrophotometer. From the measurement results, it can be concluded that the cloth with *tunjung* fixer had the strongest color/dye level.

References

[1] Samanta A K and Konar A 2011 *Dyeing of textiles with natural dyes* (United Kingdom: InTech)
[2] Kasiri M and Khataee A 2011 Photooxidative decolorization of two organic dyes with different chemical structures by UV/H_2O_2 process: experimental design *Desalination* **270** 151-9.
[3] Samanta A K and Agarwal P 2009 Application of natural dyes on textiles *Indian J. Fibre Text. Res.* **34** 384-99.
[4] Zarkogianni M, Mikropoulou E, Varella E, Tsatsaroni E 2011 Colour and fastness of natural dyes: revival of traditional dyeing techniques *Color. Technol.* **127** 18-27.
[5] Shahid M., Islam S-ul, Mohammad F 2013 Recent advancements in natural dye applications: a review, *J. Clean. Prod.* **53** 310-31.
[6] Aberoumand A 2011 A Review Article on Edible Pigments Properties and Sources as Natural Biocolorants in Foodstuff and Food Industry *World J. Dairy Food Sci.* **6** 71-8.
[7] Kasiri M B and Safapour S 2015 Exploring and exploiting plants extracts as the natural dyes/antimicrobials in textiles processing *Prog. Color Colorants Coat.* **8** 87-114.

[8] Nwonye N U and Ezema P N 2017 Extraction and Utilization of Natural Dye Extract from Guinea Corn Leaf *Int. J. Dev. Strategies Humani. Manag. Soc. Sci* **7** 40-51.
[9] Rosita N, Satya A, Putro P, Aji M P 2006 Sintesis Pigmen Alami Daun Tanaman Andong (*Cordyline Fruticosa L.*) Sebagai Pewarna Batik Dan Analisis Sifat Optiknya *J. Fisika* **4** 88–91.
[10] Lestari P and Wijana S 2010 Ekstraksi Tanin Dari Daun Alpukat (*Persea Americana Mill.*) Sebagai Pewarna Alami (Kajian Proporsi Pelarut Dan Waktu Ekstraksi), *Universitas Brawijaya.*
[11] Handayani P A and Maulana I 2014 Pewarna alami batik dari kulit soga tingi (*Ceriops tagal*) dengan metode ekstraksi *J. Bahan Alam Terbarukan* **2** 1-6.

IOP Conf. Series: Materials Science and Engineering **543** (2019) 012075 doi:10.1088/1757-899X/543/1/012075

Production of Liquid Smoke From the Process of Carbonization of Durian Skin Biomass, Coconut Shell and Palm Shell for Preservation of Tilapia Fish

E Sari [1], U Khatab[2], Burmawi[1], E D Rahman[1], F Afriza [1], A Maulidita [1], and V Desti[1]

[1]Department of Chemical Engineering, Universitas Bung Hatta, Padang-West Sumatera.
[2]Depertment of Civil Engineering,College of Technology Payakumbuh, West Sumatera.

E-mail : sariellyta@yahoo.com

Abstract. The Liquid smoke can be use for food preservation, be obtained from pyrolysis of materials containing cellulose, hemicellulose, and lignin. The form of liquid smoke that has the ability to preserve,for their phenolic compounds,acids and carbonyl. The purpose of this research is to know the quality of liquid smoke from process of charcoal the biomass obtained from rotary pyrolysis carbonisator as fish preservative, the effect of liquid smoke on the chemical and organoleptic properties of fish, and the resilience of fish after being given smoke liquid. The metode of experiments with skin biomass of durian, coconut shell , and palm shell. Biomass is dried, then is burned in rotary carbonisator pyrolysis. The results ofthe process are namely liquid, tar and charcoal smoke. Next, Liquid smoke is precipitated and then it is distilled twice, namely ordinary distillation and vacuum distillation. Results combustion produces liquid smoke with a yield of 19% on the skin of durian, 23.6% on coconut shell, and 20.8% on palm shells. Organoleptic test results on the fish with the addition of liquid smoke from the skin of durian with a concentration of 5 % is most preferred by the panellists in terms of color, aroma, flavor, and texture.

1. Introduction

Preservatives of food and beverage products have become in separable in food industry. The food industry provides solutions by adding hazardous ingredients in food processing, such as the use of dyes, flavors, and preservatives. These ingredients can cause some dangerous diseases, if consumed in a long period of time. One solution that can be utilized is using natural preservatives that can be obtained from organic materials. One of them is by utilizing liquid smoke in the process of carbonization pyrolysis biomass as biobriquette raw material [12]. Pyrolysis of this waste results in the formation of three final products: gas, oil and charcoal, all of which havepotential for use in various applications [8]. Oil here can be liquid or tar, and this depends on the raw material that is pyrolysis. The smoke produced from burning of biomass is condensed and generate liquid smoke. Smoke can also be obtained from biobriket carbonization process where all this time it has been removed without re-use it. Therefore, research needs to be done by using smoke as preservative. Liquid smoke as pyrolysis results depend on basic materials and pyrolysis temperature [3]. Liquid smoke can be used as fish preservative.In the community, fish preservation is generally performed traditionally with curing method, by contacting the smoke directly into fish. The resulting fish smells of stinging smoke and has

IOP Conf. Series: Materials Science and Engineering **543** (2019) 012075 doi:10.1088/1757-899X/543/1/012075

blackish brown color. Therefore, we use organic liquid smoke from biomass smoke as preservative that is not too smelly. From the case, that it is necessary to diversify smoker materials, thus affecting the increase of production and the quality of liquid smoke. The quality and quantity of liquid smoke is also influenced by the condition of production process, including the ignition, pressure, temperature, and combustion duration. In liquid smoke, there are still tar compounds containing Polycyclic Aromatic Hydrocarbon (PAH) compound. This compound is an aromatic chain that causes cancer. To reduce the amount of PAH contained in the liquid smoke, it can be purified like purification of liquid smoke by distillation [6] and purification by using active charcoal [11].As a preliminary study to obtain the curing temperature using pyrolysis spinning carbonisator [12], biomass used durian skin biomass, coconut shell, and palm shell at temperature of 200°C, 300°C, and 400°C [1] .We founded that optimum temperature in durian peel biomass pyrolysis is 140°C to 175°C. The smoke obtained can be used as liquid smoke. If optimum temperature is exceeded, pyrolysis will produce black tar.This can be shown from the results of the study pyrolysis in brown algae Saccharina japonica can be produced bio-oil and bio-char at a temperature of 450°C [2]. The purpose of this research is to find out whether liquid smoke can be used as a preservative of fish, then to know the composition that exist in liquid smoke and the quality of liquid smoke as a preservative of fish from biomass durian skin, coconut shell, and palm shells and to know the resilience of fish preserved by using liquid smoke from biomass.

2. Experimental section

2.1. Material

Tools used are series of spinning pyrolysis carbonization reactor, series of distillation tools,GC-MS Tool. Material: Durian skin biomass, palm shell, coconut shell from Padang

2.2. Research Parameters

Research fixed parameters ie pyrolysis carbonization temperature: 140 to 175 °C, carbonization time of 120 minutes; parameter variables: biomass type: durian skin, coconut shell and palm shell; Liquid Smoke Concentration1%;2.5%;5%. Test parameters: Tilapia fish, Output Parameters: yield, pH, Density, Composition of liquid smoke compound, Color and flavor of liquid smoke, Durability of fish preservation, fish organoleptic (color, scent, texture, flavor).

2.3. Working Procedure

Capture of Liquid Smoke; Biomass (durian shell, palm shell, coconut shell) is reduced in size and dried about 4-5 days under sunlight, then biomass is incorporated in pyrolysis carbonisator and fire heated if for 120 minutes at 175 °C temperature, the liquid smoke coming out is accommodated in a trophy cup, the result of pyrolysis process are three products i.e. liquid smoke, tar, and charcoal. Liquid smoke is obtained from smoke condensation at pyrolysis process and precipitated for a week.

Liquid Smoke Distillation Stage; Liquid smoke after precipitation (liquid smoke Grade III) is distilled by vacuum distillation until liquid smoke comes out from the condenser end. Temperature along the process is recorded. The distillation produce Grade II liquid smoke, then distilled again to obtain Grade I liquid smoke that can be applied as food preservative.

Analysis stages; Analysis stage performed on liquid water are rendemen, GC-MS, pH and density.

Testing Stage of Fish Preservation Durability; Tilapia is weeded by throwing the scales, split the fish into a butterfly shape, then discarded gills, stomach contents, dirt and black abdominal wall layers. Then liquid smoke solution is prepared, made with concentrations of 1%,2.5% and 5%. To make liquid smoke with concentration of 1% is by taking 5 ml of liquid smoke, then added clean water of 500 ml. The fish is soaked in liquid smoke with a determined concentration for 30 minutes, then it drained until the surface dries. It was then dried at 40 °C for 30 minutes, 60 °C for 30 minutes, and 80 °C for 8 hours in oven. Then the fish is left cold in open air and then packed. Smoked Tilapia is packed with plastic and stored in storage room for 15 days.

Organoleptic Testing Stage; Each sample is placed on a white container or plate to be able to see the color difference clearly. The panelist tastes the test sample, before and after it, he should drink to wash the mouth. The panelist is required to state his / her level of satisfaction to the sample presented, by assigning a number of 1, 2, 3, 4, and 5 in each sample column considered to be in accordance with the panelist's preference level.

3. Results and Discussion

3.1. Pyrolysis Carbonization Process

Liquid smoke as a result of biomass pyrolysis, conducted using spinning carbonisator pyrolysis reactor, has black colored result. This type of liquid smoke is called liquid smoke Grade III. This liquid can not be used for food because it contains tar, but only used for wood preservation. In order for the liquid smoke pyrolysis product to be used for the preservation of fish, the liquid must be purified. The yield is shown in Table 1

Table 1. Acquisition of liquid smoke in pyrolysis carbonization process

Type of Biomass	Product form at Temperature (°C)			Liquid Smoke Yield Grade III (%)
	T= ±140	T= ±170	T= ±175	
Durian Shell	Liquid Smoke	Tar	Tar	19,0
Coconut Shell	Smoke	Liquid Smoke	Tar	23,6
Palm shell	Smoke	Smoke	Liquid Smoke	20,8

3.2. Liquid Smoke Purification

The next step to obtain liquid smoke for fish preservation is purification by using distillation 2 times. The first distillation with pyrolysis liquid smoke (liquid smoke Grade III) will produce second Grade liquid smoke. Class II liquid smoke is for fast food preservation, such as meatballs and instant noodles[14]. Liquid smoke Class II has brownish yellow color. To get liquid smoke for fish preservation, the Grade II liquid smoke is distilled again to obtain Grade I liquid smoke. The Grade I liquid smoke is clear yellow. The result and optimum temperature for each Grade of liquid smoke at the purification stage can be seen in Table 2.

Table 2. Result of Grade II and Grade I Liquid Smoke, Optimum Temperature and Distillation Time

Type of Biomass	Grade	Temperature (°C)	Time (hour)	Yield (%)
Durian Shell	Grade II	84	1,35	18
	Grade I	94	1,15	14,8
Coconut Shell	Grade II	83	1,00	22,6
	Grade I	81	1,07	18,8
Palm Shell	Grade II	84	1,20	19,5
	Grade I	88	1,25	16,3

From Table 2, it can be seen that temperature of Grade I and II is optimum temperature. This is because at the time of purification, the constant temperature is at that specific degree and it does not increase, because the purification is conducted by vacuum distillation. The temperature will increase when liquid smoke in boiling flask is exhausted and only tar is left in biomass liquid smoke. The yield of liquid smoke on pyrolysis carbonization process of each sample was 19% of durian shell biomass, coconut shell biomass 23.6%, and 20.8% shellfish biomass (Table 1). The yield in Grade II and Grade I decreased because the tar contained has begun to decrease, the yield is still very small, and can be caused by several factors are heating temperature, moisture content and the particle size of biomass used. To obtain liquid smoke with high yield, the heating should be carried out to 400°C while in this study the maximum temperature is only 175°C, resulting in very small yield. Form of Grade III liquid smoke as pyrolysis results and Grade I and II as results of vacuum distillation can be seen in Figure 1 and Figure 2

Figure 1. Liquid smoke as Pyrolysis Grade III, II, I

Figure 2. Liquid smoke as Pyrolysis Grade I

Liquid smoke biomass from durian skin, coconut shell and palm shell have clear color difference. In Figure 1 b. it can be seen that distillation of liquid smoke of Grade I palm shell has a slightly darker brown color than liquid smoke durian skin, while coconut shell smoke has the clearest result. This is because the content in the biomass varies, and when purified by distillation there is compound content that lost from the liquid smoke.

3.3. Liquid Smoke Test

3.3.1. GC-MS Test for Determining Liquid Smoke Chemical Composition
The quality of liquid smoke depends on the composition of the chemical compounds contained in the liquid smoke. The compounds contained in the liquid smoke are affected by the pyrolysis conditions and the raw materials used. Below is the result of GC-MS liquid smoke analysis from purification result with redistillation process can be seen in Table 3. The components identified in liquid smoke, primarily derived from thermal degradation of wood carbohydrates [4], such as ketones, carbonyls, acids, furans and pyran derivatives. In this study, the most common components found in durian leather and coconut shell smoke are alcohols 16.74 and 35.57%, whereas in the palm oil liquid shell the most abundant component is carbonyl 12.24%. The difference of chemical composition in liquid smoke is due to the difference of raw materials used and also influenced by pyrolysis temperature used to produce liquid smoke.In this study the pyrolysis temperature used on durian skin is 140°C, coconut shell 170°C, and palm shell 175°C. This is because at the time of research with temperatures exceeding the temperature range, liquid smoke has produced tar, an undesirable component in the resulting liquid smoke. The most influential components in liquid smoke are carbonyl, acids and phenols.

Table 3. Major Compound Content on Liquid Smoke Distillation Result of Durian Skin, Coconut and Palm shell

No	**Compound name**	**% Area**		
		Durian skin	**Coconut shell**	**Palm shell**
1	Carbonil = (AldehidedanKeton)	12.31	0.62	12.24
2	Acid and Derivatives	11.43	0.07	5.13
3	Furan and Pyran Derivatives	14.79	0.11	4.1
4	Phenol	8.68	2.71	3.06
5	Alcohol	16.74	35.57	5.11
6	Metil, etil, eter	13.22	18.32	8.39

According to Hamm and Potthast in Girard [5], maximum levels of phenol, carbonyl and acid compounds are 600°C.Each component has different effects. Carbonyl has a major influence on color, whereas its effect on the taste of asphalt product is less prominent due to interaction between carbonyl and amino groups. Acids derived from liquid smoke may affect flavor, pH, and shelf life [7]. While

IOP Conf. Series: Materials Science and Engineering **543** (2019) 012075 doi:10.1088/1757-899X/543/1/012075

phenol contributes in dye smoke product. Liquid smoke containing phenol, carbonyl, and acid compounds can simultaneously act as antioxidants and antimicrobials and influence the color and distinctive flavors of food products[8]. The analysis results shows that Policyclic Aromatic Hydrocarbon (PAH) compounds including benzopyren are not found in this liquid smoke. The most influential factor in the formation of PAH compounds is pyrolysis temperature.

3.3.2. Physical property Test of Liquid smoke (pH and Density)
The physical test of Grade I liquid smoke is shown in Table 4 and Japanese vinegar standard is used as standard comparator [14].

Table 4. Result of Liquid Smoke Grade II and I, Optimum Temperature and Distillation Time

Parameter	Physical properties of liquid smoke Grade I			Standard Wood Vinegar Japan [14]
	Durian skin	**Coconut Shell**	**Palm Shell**	
Density	1,002	1,0107	1,0015	>1,005
pH	3,83	2,4	2,32	1,5-3,7
Color	clear yellow	clear yellow	clear yellow	Yellow Brown
Transparency	Transparent	Transparent	Transparent	Transparent (impureless)

From Table 4, it can be seen that the density of liquid smoke from variety of biomass ranges from 1.002 to 1.015 and this is in accordance with Japanese standards. pH of liquid smoke from various purification processes is strongly influenced by acetic acid or phenol contained in liquid smoke. Acetic acid compounds may affect pH of liquid smoke, the taste and shelf life of smoke product. In addition, phenol levels also affect pH of liquid smoke because phenol has acid properties that are the effect of aromatic rings [13]. The test results showed that pH value of each liquid smoke is low. This means that the resulting liquid smoke has a good quality, especially as an anti-bacterial. Based on the pH above, liquid smoke that has a good quality is liquid smoke derived from palm shell biomass. This is because at low pH, microbes or bacteria tend not to live and breed well [9].

3.3.3. Organoleptic Test Results
The organoleptic test used in this study was sensory test. Sensory test is one assessment of a food product performed by humans as a measure using its five senses. The senses that act in this sensory test are vision (color), smell, touch (texture), and taste (flavor). The sensory properties proposed to 5 fixed panelists were color, smell, texture, and taste, treatment with 1%, 2.5%, and 5%. Panelist assessment was determined by choosing a range of 1-5 scores with a score of 1 being the least preferred and score 5 was the most favored. The results of panelist assessment can be seen in Table 5.

Table 5. Results of Panelists' Assessment on Fish Organoleptic Test

Sample	Kode sample	ConcentrationLiquid Smoke (%)	Average analysis (score)			
			flavor	**Taste**	**Color**	**Texture**
Fish + liquid smoke Durian Skin	A	1	4.4	4.2	4.1	4.2
		2.5	4.8	3.6	4.4	4.5
		5	3.8	5	5	5
Fish +liquid smoke Coconut Shel	B	1	4.4	4.2	4.6	4.2
		2.5	4.2	4.6	5	4.6
		5	4	4	4.2	4
Fish + liquid smoke Palm Shell	C	1	4	4.6	5	4.6
		2.5	3	4.6	4.7	4.6
		5	4.6	3.8	4.5	4.4
Non-Smoked Fish	D		1.6	3.4	2.4	3.4
Traditional Smoked Fish	E		3	3.3	2	3

IOP Conf. Series: Materials Science and Engineering **543** (2019) 012075 doi:10.1088/1757-899X/543/1/012075

Explanation : 1=Very Dislike, 2=Dislike, 3=Average, 4=Like, 5= Really like

From Table 5, it can be seen that the result of statistical analysis with Kruskal-Wallis Test indicates that panelists prefer fish color at sample A,B,C compared to sample D and E. This is because fish with liquid smoke biomass have a lighter color than others. While fish without liquid smoke has a slightly darker color compared to fish with liquid smoke biomass and traditional smoked fish is very dark and less attractive to panellists. Sample A at concentration of 5% preferably with a score of 5. In terms of flavor, sample A with a concentration of 2.5% with score is 4.8. This is shows that flavor of product is very preferred, does not cause the smell too stingy, and does not cause fishy smells. While the least favorable scent is the smokeless fish. is score only 1.6. This is because the scent on the product still smells fishy. Texture is a sensation that can be used by mouth or by touching the fingers. Texture analysis from the panelist showed that the most preferred texture are sample A with 5% concentration and score 5. While the most unlikable texture is traditional smoked fish obtained score 3 because it is still has soft texture and not tasteful. Flavor can also be used as indicator of damage on the product. The results of panelists' assessment on the most favoured flavor are sample A with concentration of 5% and score is 5, which indicates the taste in the product is well liked, and has a distinctive flavor. While the least favored taste is the sample E, with score is 3.3. This is because the panelists feel that the fish taste of smoke is too strong, that make its original taste reduced.

3.3.4. Fish Endurance Test

According to Wibowo [9], fish curing can be done by cold or hot curing. Cold curing is a method of curing carried out at low temperatures, ie at room temperature and not higher than 33°C or about 15-33°C. The use of low temperatures is intended to make fish meat is not cooked or proteins inside are not coagulated. As a result, the resulting smoked fish is still considered to be half cooked. While hot curing is conducted at high temperatures that is 80°C-90°C. Hot curing basically consists of three stages. The first stage is initial drying stage above room temperature at 30-40°C for 30-60 minutes. The second stage, is the first maturation stage where the temperature is slowly raised to 50-60°C for 30-45 minutes. The third stage is final maturation stage, where the temperature is raised to about 80°C for 3-8 hours. In this study, the fish resilience test was conducted by comparing the dipped tilapia fish and then baked, with the tilapia dipped in liquid smoke and left at room temperature. Analysis results can be seen in Table 6.

Table 6. Results of Panelists' Assessment on Fish Organoleptic Test

Sample	**Average analysis (score)**			
	Flavor	**Color**	**Texture**	**Durability (day)**
Fish + Liquid Smoke baked	smelled of smokenot fishy	yellow Brownish	Hard	30
Fish + Liquid Smoke Room Temperature	Rotten Smell Stingy	pale white Reddish	Soft	1

From Table 6, Judging from the results of fish endurance test, fish with liquid smoke and left at room temperature is not feasible to be consumed because after a day stand, because the smell will sting and blood out. While fish with liquid smoke is roasted (hot curing), fish is worth consuming despite even though having passed 30 days because of its hard texture and not fishy. So it can be concluded that liquid smoke from biomass is not suitable for cold curing and is only suitable for hot curing (heating at temperatures of >80 °C). The results of tilapia fish were given liquid smoke, then dried with an oven can be seen Figure 3.

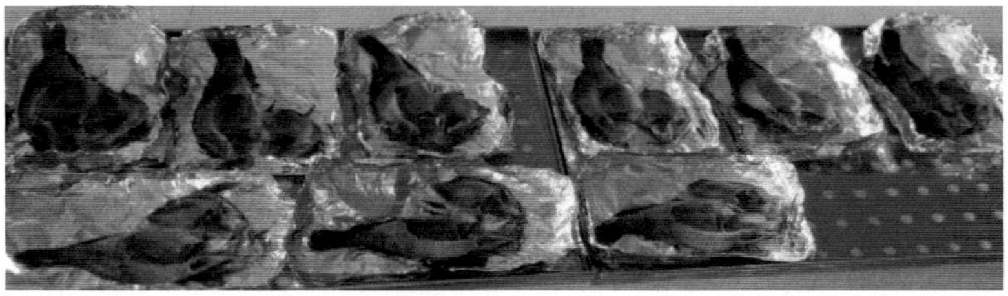

Figure 3. The results of tilapia fish were given liquid smoke and then dried with an oven.

4. Conclusion

Waste of Durian skin, coconut shell, and palm shell biomass can be utilized as an alternative of curing material to produce liquid smoke and charcoal. From the analysis using GC-MS, durian skin has carbonyl area of 12,31%, acid and derivative 11,43%, and phenol 8,68%. Coconut shell has carbonyl area of 0,62%, acid and its derivative 0,07%, and phenol 2,71%. And palm shell has carbonyl area of 12,24%, acid and its derivative 5,13%, and phenol 3,06%. Yield of liquid smoke of durian skin, coconut shell and palm shell from pyrolysis process at temperature of 140-175°Cis 19%; 23.6% and 20.8% respectively. Variations in the concentration of liquid smoke are very influential on organoleptics (color, texture, flavor, and taste) of fish. From the result of the whole organoleptic test, the most preferred by panelists are fish + liquid smoke of durian skin with a concentration of 5%.

Acknowledgement

Thank to Directorate of Research and Community Service, Directorate General for Research and Development of Ministry of Research, Technology and Higher Education and Rector and Chairman of LPPM Bung Hatta University for the opportunity and facilities provided and All parties involved in this research.

References

[1] Budaraga I K, Amim Marlida Y and Bulanin U 2016 Analysis of liquid smoke chemical components with GC MS from different raw materials varition production and pyrolysis temperatur level *Int J. of Chem. Tech. Research* **9** 06 p 694-708.

[2] Chooi J H, Woo H C and Suh D J 2014 Pyrolysis of seaweeds for io-oil and bio-char production *Chem. Eng. Trans* **37** p 121-126

[3] Darmadji P 1996 Smoke that produces various kinds of agricultural waste *J.Agritechno.* **16**4 p19-22

[4] Guillen M D, Ibargoita M I 1996 Relationship between the maximum temperature reached in the smoke generation process from Vitis viniera L. Shoot sawdust and composition of the aquaeus smoke flavoring preparation obtained *J. Agricul. and Food Chem.* **44** p 1302-1307.

[5] Girard J P 1992 Technology of meat products *New York Ellias Howard Ltd*

[6] Luditama C 2006 A Purification of liquid made from shell and coconut fiber directly and distillation *Chooi J. Undergraduate thesis of IPB* Bogor

[7] Pszczola DE 1995 Tour highlights production and uses of smoke base flavours *Food Technol* **49** p 70-74

[8] Pinto F, Miranda M, Costa P, 2015 Co-Pyrolysis of wastes mixtured obtained from rice production, upgrading of produced liquids *Chem. Eng. Trans* **43** p 2053-2058

[9] Sari N M 2015 Rice husk charcoal and the smoke quality of rice husk *J. of Tropical Forests***3** (3) p 2337-7771

[10] Rahmalinda 2014 Comparative study of the characteristics of pyrolysis liquids from durian, midrib and empty palm punches with perfect distillation purification *Undergraduete thesis of Universitas Riau*

[11] Rinaldi S, Alimuddin and Panggabean A S 2015 Purification of liquid from durian skin using activated charcoal *Samarinda Industrial Research and Standardization Center*

[12] Sari E, Pasymi, Khatab U, Desmiarti R, Ariansyah R, Hariadi and Maisaroh S 2018 Studies of carbonization process on the production of durian peel biobriquette with mixed biomass coconut and palm shells IOP *Conf. Series: Mat Sci and Eng.* **316** 012021

[13] Wijaya M, Noor E, Irawadi T T, and Pari G 2008 Changes in pyrolysis temperature to the liquid chemical structure of pine wood sawdust *J. of Forest Products***I** (2) p 73-77

[14] Yatagai M, Nishimoto M, Hori K and Ohira T 2002 Termiticidal activity of wood vinegar, its components and their homologue *J.of Wood Science* **48**(4) p 338-342

[15] Ratna Y 2008 Liquid smoke monograph as a natural preservative in meat and fish products *Undergraduate thesis of UPN Veteran JawaTimur, Surabaya*

Fabrication of Interlayer-free P123 Caronised Template Silica Membranes for Water Desalination: Conventional Versus Rapid Thermal Processing (CTP vs RTP) Techniques

M Elma[1], H Setyawan[2], A Rahma[1], A E Pratiwi[1], E L A Rampun[1]
[1] Chemical Engineering Department, University of Lambung Mangkurat, Banjarbaru, South Kalimantan 70714, Indonesia
[2] Chemical Engineering Department, Institute of Technology Sepuluh November, Jalan Raya ITS, Sukolilo, Surabaya, East Java 60111, Indonesia

E-mail: melma@ulm.ac.id

Abstract. This work demonstrates the comparison techniques between conventional and rapid thermal processing (CTP Vs RTP). It is to fabricate the interlayer-free P123 carbonised template silica membranes. The aim of this work is to understand the performance of fabricated membranes for water desalination. The silica sols pH 6 were prepared from a two acid-base catalyzed sol gel method by applying tetraethyl orthosilicate (TEOS) as the precursor. Triblock copolymer of Pluronic P123 was used as the templating agent and mixed homogenously with silica sols. The mixtures were calcined in two method called CTP and RTP. CTP is a slow calcination and takes 4 hours with 1 °C min-1 for dwelling time. On the other hand, RTP is rapid technique calcined in air (1 h without dwelling time). CTP confirms higher surface area (50 % higher), porosity (34 % higher) and thinner (60 %) than RTP. RTP technique is supposed to save time, energy and cost during membranes fabrication.

1. Introduction

Main problems which mostly facing in developed country especially is water scarcity. Portable water would be hard to get due to high population of citizen. Membrane technology can be alternative option to desalination application. It have some advantages such as does not need chemical addition, high selectivity and save more cost production [1].There many types of membrane material widely used for desalination, such as organic by PSf (Polysulfone) [2, 3], and inorganic membrane using zeolite [4] and silica [5-7].

Lately, most common material which often get attracted in many studies is silica. The silica membranes give a good performance [8, 9], thermal stability [10] and good molecular sieving properties that contribute to obtained high performances in desalination. However there is drawback of silica because the hydrophilic reacts with water and can resulting poor performance [11].

Some studies have been reported to improve silica ability material for water treatment including carbon template [11, 12], hybrid organic-inorganic [7] and metal oxide embedded [13].

Recently, for the first time Elma, Wang [5] are prepared TEOS-P123 templated (Tetraethyl orthosilicate-Triblock Copolymer of Pluronic) without interlayer. Whereas other works introduce smooth support with required interlayer coupled with many thin films at the top layer. Indeed, interlayer-free can reduce membrane thickness and offered higher water flux. But these process spend long of time production as known conventional thermal processing (CTP) with slow calcination and 1 °C min^{-1} for dwelling time [11]. Hence, rapid thermal processing (RTP) could be an option in terms of saving time because during calcination, there is no need dwelling time [14]. In this work we demonstrate the comparison between CTP and RTP method to produce interlayer-free P123 carbonised template silica membrane.

2. Methods

2.1. Materials, Xerogel Preparation and Characterisation

The synthesized of silica sols by a dual acid-base sol gel method have been done as to previous research of Elma, Wang [5], Elma, Wang [15]. Tetraethyl orthosilicate (TEOS, 99.0 %, Sigma-Aldrich) as precursor of silica was added drop wise ethanol (EtOH, 99 %) and stirred for 5 min in cold condition at 0 °C to avoid hydrolysis reaction followed by addition of (0.00078 N HNO_3, Merck) and refluxed for 1 h at 50 °C. Diluted of (0.0003 N NH_3, Merck) was added drop wise into the silica sols and stirred for 2 h at the same condition. The pH sol was adjusted to pH 6 ± 0.1 by dual catalyst. After that, the templating agent that used is P123 (triblock copolymer Pluronic, Sigma-Aldrich) was added into silica sol. The silica-P123 sol was conducted with the final molar ratios is the TEOS:EtOH:H_2O:NH_3:HNO_3:P123 sol were calculated to be 1:38:5:0.00008:0.0003:x, where x was varied of pectin concentration such as reported in [5]. In this research we choose the optimum molar ratio of P123 from our previous research [5] of P123-35 (35 %) for calcined at 350 °C in vacuum with CTP technique. Then, for the RTP technique with the same concentration that calcined in air at the same temperature. The final molar ratio is the TEOS:EtOH: HNO_3:H_2O:NH_3:P123-35 sol were calculated to be 1:38:0.0008:5:0.0003:0.00024.

The silica-P123 sol was dried into oven at 60 °C for 24 h to obtain the dried gel then grounded into powder and calcined in air condition at 450 °C for 4 h (CTP technique) with 1 °C min^{-1} of ramping and cooling rates in a furnace and 1 h at 350 °C (RTP Technique) with no ramping rates. The vibrational bands and peak wave of silanol, siloxane and carbon was measured by FTIR at wavelength range 400-4000 cm^{-1}, to performing of FTIR spectra deconvolution by using the Fityk computer program with Gaussian peak fitting as the preferred curve. The specific surface area also measured and was determined from Brunauer, Emmett and Teller (BET) method.

2.2. Membrane Fabrication and Characterisation

Silica-P123 thin film were coated four times directly onto macroporous alumina substrates (α-Al2O3 tubular support (Ceramic Oxide Fabricators, Australia)) via dip-coating process with dwelling time of 2 min and dipping and withdrawl rate of 10 and 5 cm min^{-1}, respectively. Then the membrane layer was dried into an oven then calcined in a furnace for 4h per layer (for CTP) and 1 h per layer (for RTP) at 350 °C. the total layers of membrane is 4 layers. The membrane morphology and thickness were characterized by SEM (Scanning Electron Microscopy), FESEM JOEL 7001.

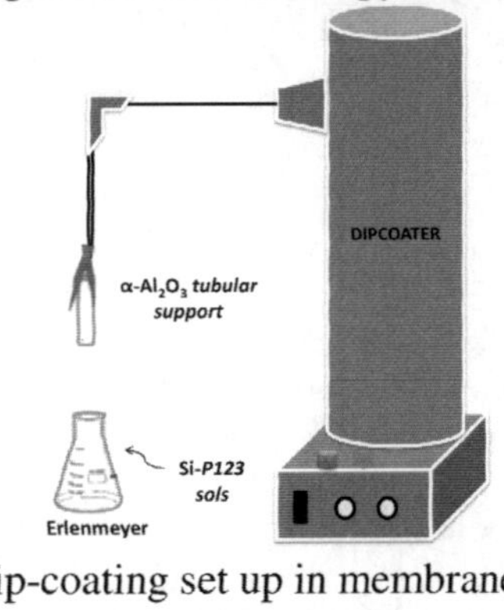

Figure 1. Dip-coating set up in membrane fabrication

3. Results and Discussion

3.1. Xerogel Characterisation

As shown at Figure 2 (a) shows that silica-P123 calcined by RTP technique was higher vibrational bands at wavelength 600-1400 cm^{-1} than with CTP technique. The peak of siloxane bridges (Si-O-Si) was assigned at 1070 cm^{-1}, silanol groups at 960 cm^{-1} and then Si-C at 800 cm^{-1}. The copolymer P123 here has a function as carbon templated that changes the Si-O-H bonds to Si-C. A carbon that contains in P123 is replacing a hydrogen that bonds with silica. As can be seen at Figure 2 (b) the peak area ratio between Si-OH against Si-O-Si shows the lowest one is P123 with RTP techniques. The highest ratio of Si-OH against Si-O-Si indicate the silanol contain in network was higher than siloxane and vice versa. The lowest peak area ratio could be explained that the Si-OH was replaced as Si-O-Si bridges or new bonding of Si-C [16]. So, it suggested that the silica-P123 using RTP technique is more hydrophobic rather than silica-P123 via CTP. The similar results were reported by Yang, Elma [7] that fabricated silica-carbon membrane from TEOS-TEVS-P123, then Saputra, Astuti [16] that synthesized silica thin film from inorganic-organic materials of TEOS and DMDMS. Both of that research obtain the hydrophobic materials.

Meanwhile, if we compare both of membranes that calcined via RTP and CTP, the membrane with RTP technique is higher although the concentration of P123 was same. That happened because the different condition of calcination included of calcine temperature and method of carbonisation. P123-CTP was calcined in vacuum condition at 450 °C of calcine temperature, whereas the P123-RTP was calcined in air at 350 °C. These condition was affected to the bonding in silica-P123 matrices and of course on the membrane surface. Ibrahim, Nagasawa [17] reported that calcination under air could improve the hydrophilicity. However, from different literature reported the peak ratio of Si-H groups was lower when membrane calcined under air atmosphere. Especially at lower temperature that caused the oxygenation of Si-H groups were started coincide with the forming of Si-OH groups [18]. That's why the membrane calcined under air condition via RTP contains more Si-O-Si bridges and Si-C bonds.

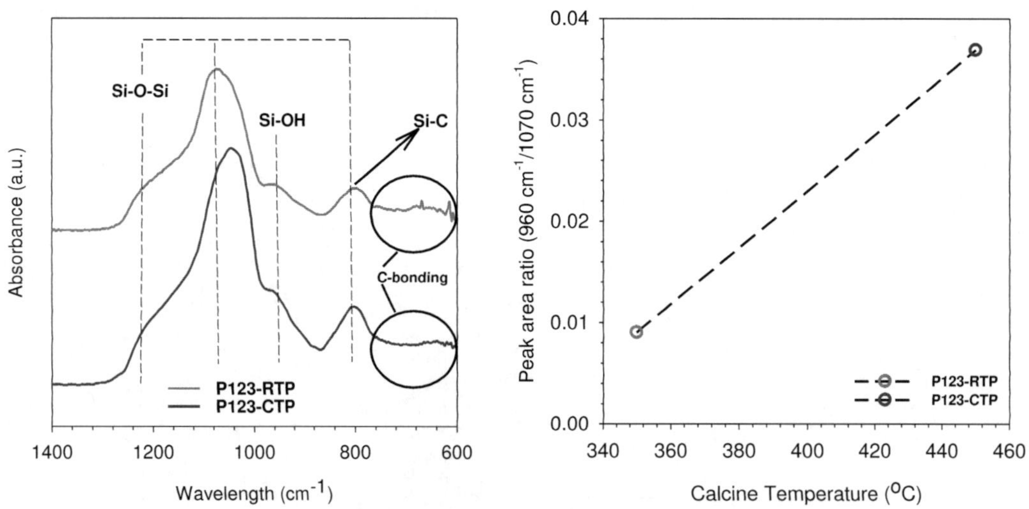

Figure 2. (a) FTIR spectra of xerogels calcined in air by CTP and RTP technique (b) peak area rations between silanol (960 cm^{-1}) and siloxane (1070 cm^{-1})

Furthermore, the surface area and pore properties were analyzed by N2 sorption as can be seen at Figure 3 that shows the differences hysteresis of both silica-P123 membrane. The P123-CTP sample started hysteresis at 0.40 p/p^{o}. While the P123-RTP the hysteresis occurred at relative pressure approximately 0.45 p/p°. The results categorized all membrane as mesoporous structures as synchronized to pore size as shown at Table 1. If compare of both xerogels, the sample that prepared via CTP techniques has a higher surface area, although the pore volume and pore size for pure silica were similar [19]. This could be explained due to the CTP techniques can lead to forming the close till dense network in silica matrices as the function of time of calcination. The samples were calcined under vacuum for 4 h per layer then via RTP techniques is faster only spend 1 h per layer. A long time of calcination influence the carbonization of carbon chains and the evaporation will done well and avoid to thermal stress. However, this work contrary to Wang, Wang [14], Wang, Wang [20] that reported the higher surface area can achieved via RTP. This could be happened due to the different of water molar ratio used in this work and different material of silica precursor that they used ES40. However, the dual catalyst using in this work makes the mesoporous structure that able to membrane fabricated for desalination.

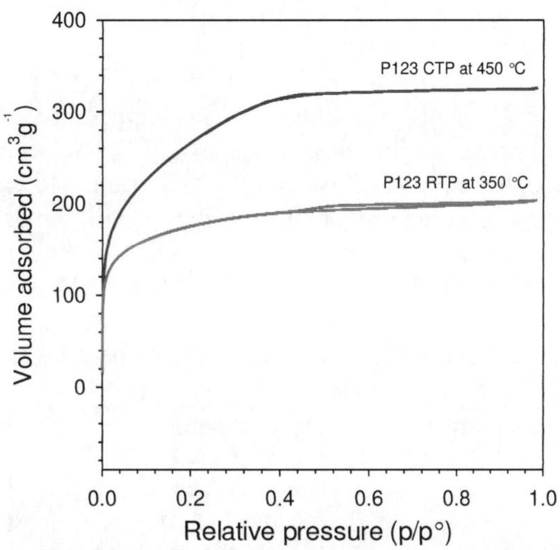

Figure 3. N_2 sorption of xerogels in air condition by CTP and RTP techniques

Table 1. Surface properties of bulk xerogels calcined in air

Membrane type	BET surface area ($m^2.h^{-1}$)	Pore volume ($cm^3.g^{-1}$)	Average pore size (nm)	Thickness (nm)	References
Silica-P123 calcined in air at 350 °C by RTP	572	0.31	2.21	~1000	This work
Silica-P123 calcined under vacuum at 450 °C by CTP	965	0.50	2.31	~400	[5]

3.2. Membrane Morphology

The effect of different condition of vacuum and air calcination must be concerned. The xerogel sample calcined under vacuum condition that led to slight densification of silica structures. Elma, Wang [5] prove that the highest decomposed of carbon was obtained in calcination under air than under vacuum or N_2 atmosphere particularly at high temperature. The carbon templated of P123 is retained in the networks by calcine under N_2 and vacuum calcination. The materials calcined under vacuum condition were more hydrophobic due to the higher carbon retained in the carbon–silica matrices [7], opposite to calcination under air. As shown at Figure 4 the surface and cross section of membrane calcined via CTP and under vacuum condition there is no defect films formation on the top of the substrate that suggested the sol-gel infiltrated into porous substrate. The vacuum pressure opposed the capillary stresses during the calcination process that caused to dense pore of film matrices. On the other hand,

IOP Conf. Series: Materials Science and Engineering **543** (2019) 012076 doi:10.1088/1757-899X/543/1/012076

RTP SEM image in air condition led to wide spread of cracks on top film like shown on Figure 4. It is because RTP technique calcined in air without dwelling and cooling rate time, so it suggested cause of thermal stressing.

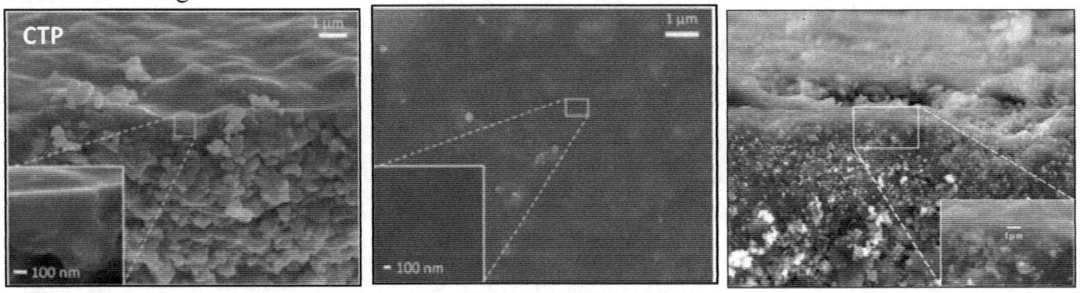

Figure 4. P123 CTP technique at 450 °C (vacuum) and P123 RTP technique at 350 °C (air) SEM image respective

Li, Chen [21] was reported the calcination under air in high temperature can enhanced the shrinkage. The decomposition some materials such as carbon can be happened with the significant effect if calcine in air (with the oxygen presence). Ibrahim, Nagasawa [17] also reported in BTESE membrane that calcine under N_2 gas atmosphere showed low weight loss compared calcine in air that explained the calcine method with N_2 was slow decomposition of the organic bonds and makes the structure more rigid. **Figure 5** shows the illustration of carbon chains bonding in silica matrices. There's different carbon chains consist in multiple techniques (CTP and RTP). The membrane that calcined via CTP has more carbon chains rather than RTP techniques. The higher carbon chains in silica network could improve mechanical properties that makes these material more strengthen [5]. So that gained best desalination performance of membrane.

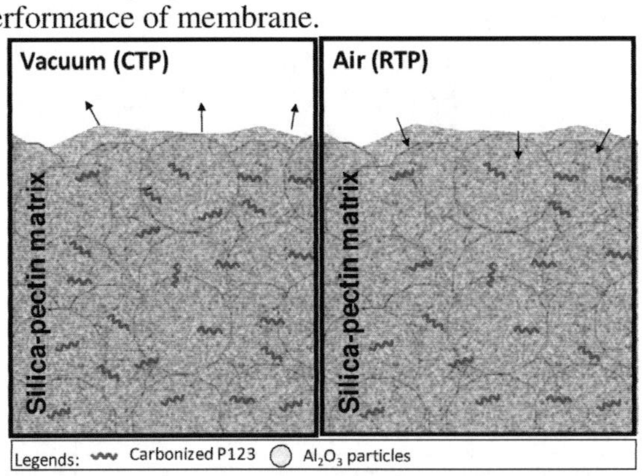

Figure 5. Illustration of carbon chains connected to Al_2O_3 particles of mesoporous alumina support for silica-P123 membrane calcined via CTP and RTP

4. Conclusion

The different of membrane fabrication technique have crucial affect toward of membrane characterisation obtained. Comparison technique between conventional and rapid thermal processing (CTP Vs RTP) for fabricate the interlayer-free P123 carbonised template silica membranes was conducted in this work. RTP results in great reducing time fabrication less than 1 day over CTP technique. Whereas CTP is a slow calcination and takes 4 hours with 1 °C min^{-1} for dwelling time. In other hand, CTP confirms higher surface area (50 % higher), porosity (34 % higher) and thinner (60 %) than RTP. RTP technique is supposed to save time, energy and cost during membranes fabrication.

Acknowledgement
The authors would like to acknowledge the financial support research grant of Postgraduates Studies (PNBP 2018) and the facilities of Process Technology Laboratory, Engineering Faculty, Lambung Mangkurat University. Muthia Elma thanks to Post-Doctoral Grant 2017-2018, Directorate of Research and Community Service, the Ministry of Research, Technology and Higher Education Republic of Indonesia and Australian Indonesian Project (ANU) Indonesian Project and SMERU Research Institute Research Grants 2016-2017 as well as the 2018 Alumni Grant Scheme (AGS) Round 2 Funding, Australia Awards in Indonesia.

References
[1] Ahmad SA. Hybrid Process (Pervaporation- Distillation): A Review 2012.
[2] Ba-Abbad MM, Mohammad AW, Takriff MS, Rohani R, Mahmoudi E, Faneer KA, et al. Synthesis of Iron Oxide Nanoparticles to Enhance Polysulfone Ultrafiltration Membrane performance for Salt Rejection. *Chem. Eng. Trans.* 2017;**56**:1699-704.
[3] Ahmad A, Yunos MZ, Harun Z, Hassan MF, Adzila S, Arifin AMT, et al. Effect of Polyaniline on Surface Properties of Polysulfone Membrane. *Chem. Eng. Trans*. 2017;**56**:691-6.
[4] Drobek M, Yacou C, Motuzas J, Julbe A, Ding L, Diniz da Costa JC. Long term pervaporation desalination of tubular MFI zeolite membranes. *J. of Mem. Sci*. 2012;**415-416**:816-23.
[5] Elma M, Wang DK, Yacou C, Costa JCDd. Interlayer-Free P123 Carbonised Template Silica Membranes for Desalination with Reduced Salt Concentration Polarisation. *J. of Mem. Sci.*. 2015;**475**:376-83.
[6] Elma M, Hairullah, Assyaifi ZL. Desalination Process via Pervaporation of Wetland Saline Water. *IOP Conf. Ser.: Earth and Environmental Science*. 2018;**175**(1):012009.
[7] Yang H, Elma M, Wang DK, Motuzas J, Diniz da Costa JC. Interlayer-free hybrid carbon-silica membranes for processing brackish to brine salt solutions by pervaporation. *J. of Mem. Sci.*. 2017;**523**:197-204.
[8] Duke MC, Mee S, da Costa JCD. Performance of porous inorganic membranes in non-osmotic desalination. *Wat. Res*. 2007;**41**(17):3998-4004.
[9] Elma M, Yacou C, Wang DK, Smart S, Diniz da Costa JC. Microporous Silica Based Membranes for Desalination. *Water*. 2012;**4**(3).
[10] El-Feky HH, Briceño K, Jardim EdO, Silvestre-Albero J, Gumí T. Novel silica membrane material for molecular sieve applications. *Microporous and Mesoporous Materials*. 2013;**179**:22-9.
[11] Wijaya S, Duke MC, Diniz da Costa JC. Carbonised template silica membranes for desalination. *Desal*. 2009;**236**(1):291-8.
[12] Elma M, Wang DK, Yacou C, Diniz da Costa JC. Interlayer-free P123 carbonised template silica membranes for desalination with reduced salt concentration polarisation. *J. of Mem. Sci.*. 2015;**475**(Supplement C):376-83.
[13] Elma M, Wang DK, Yacou C, Motuzas J, Diniz da Costa JC. High performance interlayer-free mesoporous cobalt oxide silica membranes for desalination applications. *Desal*. 2015;**365**:308-15.
[14] Wang S, Wang DK, Smart S, Diniz da Costa JC. Improved stability of ethyl silicate interlayer-free membranes by the rapid thermal processing (RTP) for desalination. *Desal*. 2017;**402**:25-32.
[15] Elma M, Wang DK, Yacou C, Motuzas J, Costa JCDd. High performance interlayer-free mesoporous cobalt oxide silica membranes for desalination applications. *Desal*. 2015(365):308-15.
[16] Saputra RE, Astuti Y, Darmawan A. Hydrophobicity of silica thin films: The deconvolution and interpretation by Fourier-transform infrared spectroscopy. *Spectr. Acta Part A: Mol. and Biomol.r Spect*. 2018;**199**:12-20.
[17] Ibrahim SM, Nagasawa H, Kanezashi M, Tsuru T. Organosilica bis(triethoxysilyl)ethane (BTESE) membranes for gas permeation (GS) and reverse osmosis (RO): The effect of

preparation conditions on structure, and the correlation between gas and liquid permeation properties. *J. of Mem. Sci.* 2017;**526**:242-51.

[18] Kanezashi M, Sazaki H, Nagasawa H, Yoshioka T, Tsuru T. Preparation and gas permeation properties of thermally stable organosilica membranes derived by hydrosilylation. *J. of Mat. Chem. A*. 2014;**2**(3):672-80.

[19] Elma M, Yacou C, Costa CJDd, Wang DK. Performance and Long Term Stability of Mesoporous Silica Membranes for Desalination. *Membranes*. 2013;**3**(3).

[20] Wang S, Wang DK, Motuzas J, Smart S, Diniz da Costa JC. Rapid thermal treatment of interlayer-free ethyl silicate 40 derived membranes for desalination. *J. of Mem. Sci.* 2016;**516**:94-103.

[21] Li L, Chen M, Dong Y, Dong X, Cerneaux S, Hampshire S, et al. A low-cost alumina-mullite composite hollow fiber ceramic membrane fabricated via phase-inversion and sintering method. *J. of the Euro. Cer. Soc.* 2016;**36**(8):2057-66.

IOP Conf. Series: Materials Science and Engineering **543** (2019) 012077 doi:10.1088/1757-899X/543/1/012077

CO_2 Freezing Area Concept for Improved Cryogenic Distillation of Natural Gas

T P Adhi[1), E P Putra[1), R B Haristyawan[2)
[1]Department of Chemical Engineering, Institut Teknologi Bandung
[2]Upstream Technology Center, PT. Pertamina (Persero)

E-mail : tpadhi@che.itb.ac.id

Abstract. Sweetening process for natural gas with high content of sour gas ($\geq$ 40% mol of CO_2 & H_2S) is difficult to be profitable. A lot of gas fields in Indonesia has high content of sour gas. New improved cryogenic distillation process of natural gas applying the concept of CO_2 freezing in the middle of the column is promising. To study the feasibility of this technology, comparison with other two methods (Absorption and Membrane process) in technical and economic aspects were elaborated. Process simulation of improved cryogenic distillation steady-state model was developed using Aspen HYSYS. CAPEX of all methods were estimated using Aspen Process Economic Analyzer. Compared to the other methods, this improved cryogenic distillation process showed tendency to have lower CAPEX and OPEX, but have some uncovered operational problems.

1. Introduction

Energy supply is one of many challenging problems many countries face recently. Natural gas is one of important energy source. More than 24% of world energy consumption in 2016 is supplied by natural gas [1]. However, many of gas fields have high content of acid gas. Gas field is classified to have high content of sour gas if that field has more than 40%-v/v total content of CO_2 and H_2S. These impurities need to be removed from natural gas to meet sales gas or pipelines specifications. One of important sales specification is acid gas content, which is < 2% v/v for CO_2 and < 4 ppm v/v for H_2S [2].

Indonesia has lot of field that has high content of acid gas (e.g. Natuna field with 71%-v/v of acid gases, West Java gas fields containing 45% - 75%-v/v of acid gas, and East Java gas fields with 40%-v/v of acid gases) [3]. Most of this field development raise a lot of questions to be profitable. Minimizing CAPEX and OPEX of separation process for this kind of natural gas is the key to make this development profitable. A new improved cryogenic distillation process has been developed by applying the concept of CO_2 freezing in the middle of the column. Feasibility study of this new improved cryogenic distillation process is needed to substantiate advantages of this method.

2. Development Of Gas Field Containing Acid Gases

Separation process of acid gas from natural gas commonly called as sweetening process. Various technologies were commercially available for this sweetening process. They are categorized based on the underlying governing phenomena, such as adsorption, physical absorption, chemical absorption, membrane, and cryogenic distillation. Natural gas sweetening technologies can be classified as shown in Table 1.

Table 1. List of Gas Sweetening Technologies [10].

<table>
<tr><th>CO_2 Removal Mechanism</th><th>Process Type</th><th>Technology</th></tr>
<tr><td rowspan="3">Chemical Absorption</td><td rowspan="2">Regenerative, Continuous</td><td>Amines</td></tr>
<tr><td>Potassium carbonate</td></tr>
<tr><td>Non regenative, Continuous (usual arrangement: lead/lag)</td><td>Sodium hydroxide</td></tr>
<tr><td>Physical Absorption</td><td>Regenerative, Continuous</td><td>Physical Solvents</td></tr>
<tr><td>Physical-Chemical Absorption</td><td>Regenerative, Continuous</td><td>Physical - Chemical Solvents</td></tr>
<tr><td>Physcial Adsorption</td><td>Regenerative, Continuous (Adsorption/Desorption sequence)</td><td>Molecular Sieve</td></tr>
<tr><td>Permeation</td><td>Continuous</td><td>Membrane</td></tr>
<tr><td>Distillation</td><td>Continuous</td><td>Cyrogenic Distilation</td></tr>
</table>

Design of an efficient and competitive processing scheme is an important stage in the development of gas fields. Table 2 shows a representative gas stream composition and condition as a basis of feasibility study exercised in this paper.

The first challenge is to select the technology that best fits the project needs. Compilation of best practice technologies in the form of graphics are most commonly used to predict the most proper technology for sweetening natural gas [4]. Important factor that affecting the technology selection are acid gas content in feed, acid gas content in sweet gas product, and gas flowrate.

New improved cryogenic distillation for natural gas was developed, particularly for gas field that has high content of acid gas. The underlying phenomena of this new improved process is the concept of CO_2 freezing in the middle of the column [5]. Applying freezing of CO_2 will improve separation of CO_2 and methane. As comparison to the selected technology, other methods are chosen based on traditional selection guidelines. Based on graphs in Figure 1, two technologies met the feed criteria (feed gas acid gas content = 40%-v/v, outlet sweet gas acid gas content = 2%-v/v, and gas flowrate=189 MMSCFD). Those are technologies based on physical-chemical absorption and hybrid method (membrane + physical-chemical absorption).

3. Comparison Methodology

Comparison of the three technologies was carried out in four steps: flowsheet simulation, preliminary equipment sizing, estimation of CAPEX, and estimation of OPEX.

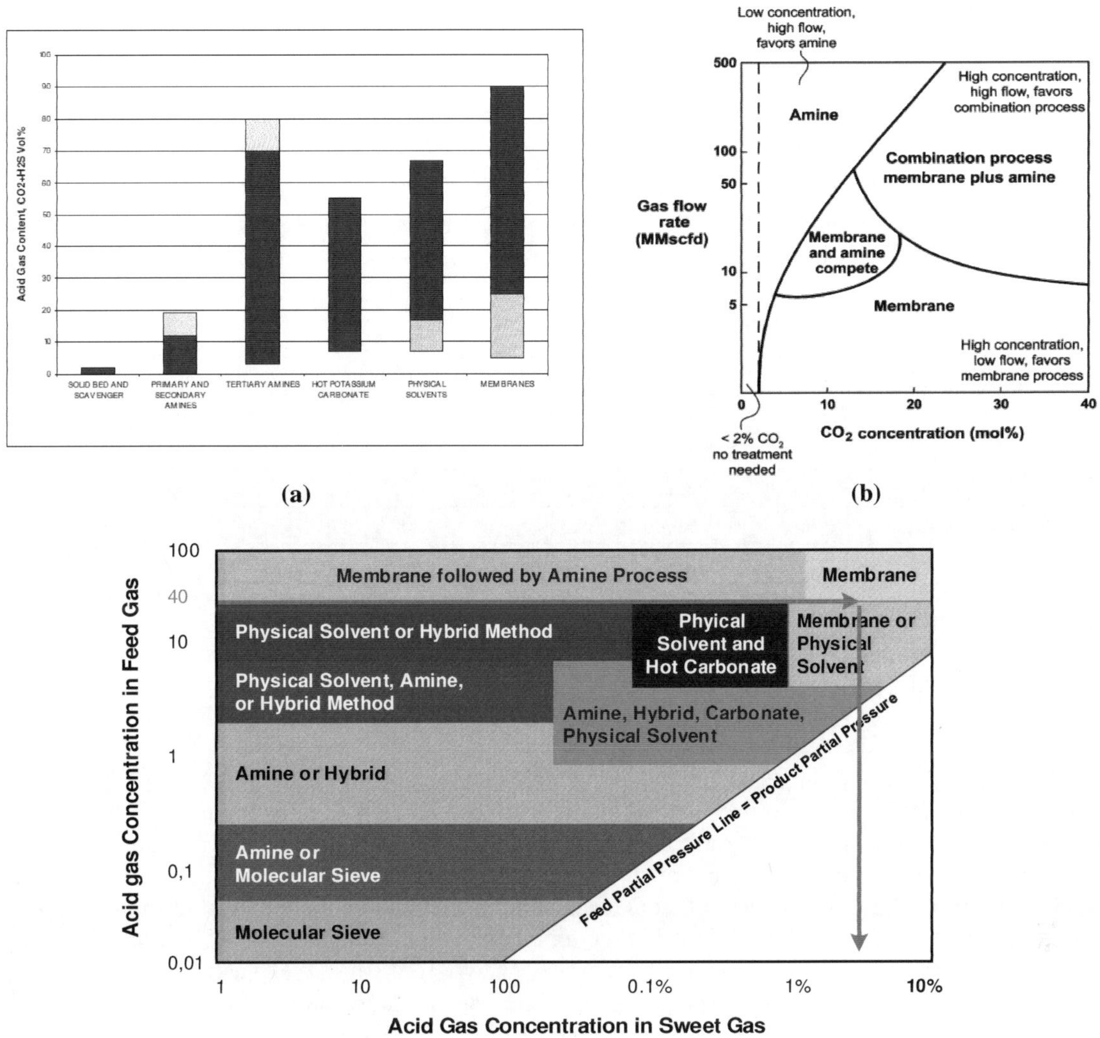

(c)

FIGURE 1. Technology Selection of Gas Sweetening Process from Different Sources (a) Abdel-Aal, et.all, 2003 [6] (b) Cameron, 2013 [7] (c) Bergel, M., [8]

3.1. Flowsheet Simulation Development

Steady-state simulation models of separation process were developed using advanced process simulation software, Aspen HYSYS v8.8. The software is appropriate to model any process equipment contained in the three technologies evaluated. Figure 2 shows the simulation flowsheet for the new improved cryogenic distillation. This process consists of four sub-processes: condensate extraction and stabilization, main fractionation, CO_2 liquid injection, and refrigeration. The heavy hydrocarbon is extracted from the feed gas by cooling utilizing the cold sweet gas stream. Using cascade propane-ethylene refrigeration unit, the lean feed gas is then further cooled down to just above the freezing temperature of CO_2 and fed to the main fractionation column at the bottom of the controlled freezing section. The liquid CO_2 is taken off the bottom of the column and then injected to underground reservoir. The cold sweet gas is drawn from the top of the column and utilized as cooling medium for the condensate extraction sub-process.

Table 2. Gas Field Composition and Operating Condition

Operating Condition		
1	Volume Flow	189 MMSCFD
2	Pressure	700 psig
3	Temperature	170 oF
Composition		
No.	**Component**	**Composition (%-mol)**
1	Methane	55,8
2	Ethane	1,92
3	Propane	0,64
4	i-Butane	0,14
5	n-Butane	0,18
6	i-Pentane	0,07
7	n-Pentane	0,06
8	C6+	0,50
9	N_2	0,32
10	CO_2	39
11	H_2S	1
12	Organic Sulfur	0,04
13	BTX	0,33

The physical-chemical absorption process has been considered as reported in Figure 3. This process consists of four sub-processes: dew point control unit (DPCU), main acid gas removal unit (AGRU), acid gas injection by compression, and refrigeration. The lean feed gas from DPCU is fed to the absorption column, where it is contacted counter-currently with the lean MDEA solution containing sulfolane. The sweet gas stream is obtained at the top of the absorber and a rich liquid stream is taken from the bottom, containing contaminants to be removed. The rich stream is then flashed to low pressure, heated in the intermediate heat exchanger, and sent to the regeneration column, where acidic gases are stripped from the solvent and obtained as gas at the top, while the lean regenerated solvent is recovered at the bottom of the regeneration column and recycled back to the absorber after being cooled at the intermediate and second heat exchangers. Make up of water, amine, and sulfolane is needed due to leakages during solvent regeneration. The acid gases is finally compressed and injected to the underground storage.

Shown in Figure 4, simulation flowsheet for the hybrid technology is similar to the previous process with the addition of two-stage membrane unit at the upstream side of the physical-chemical absorption sub-process. divided into five sub-processes: dew point control, membrane unit, main absorption, acid gas injection, and refrigeration. The acid gases from both the membrane and absorption sub-processes are finally compressed and injected to the underground storage.

Multiple assumption and process parameter have been chosen and adjusted during development and execution of the simulation models. All feed gas parameter, such as composition and flow rate, and product specification, have been kept at the same values for all technologies evaluated to make sure the results are comparable. Feed stream was saturated by water first before being further simulated. For validation and comparison of simulation results, data was collected from several references with selected parameters as follows.

IOP Conf. Series: Materials Science and Engineering **543** (2019) 012077 doi:10.1088/1757-899X/543/1/012077

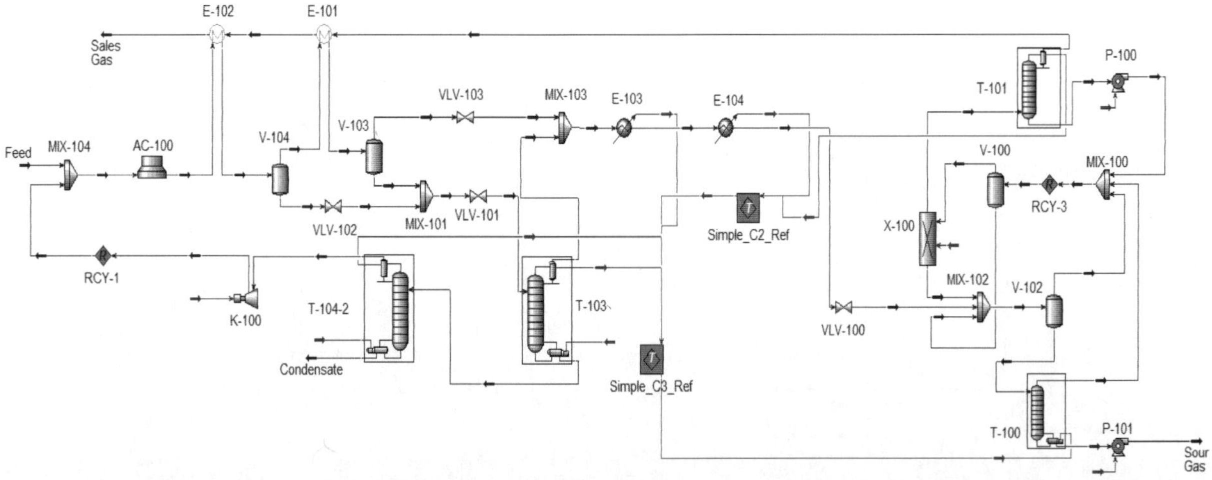

Figure 2. Simulation flowsheet of the improved cryogenic distillation process.

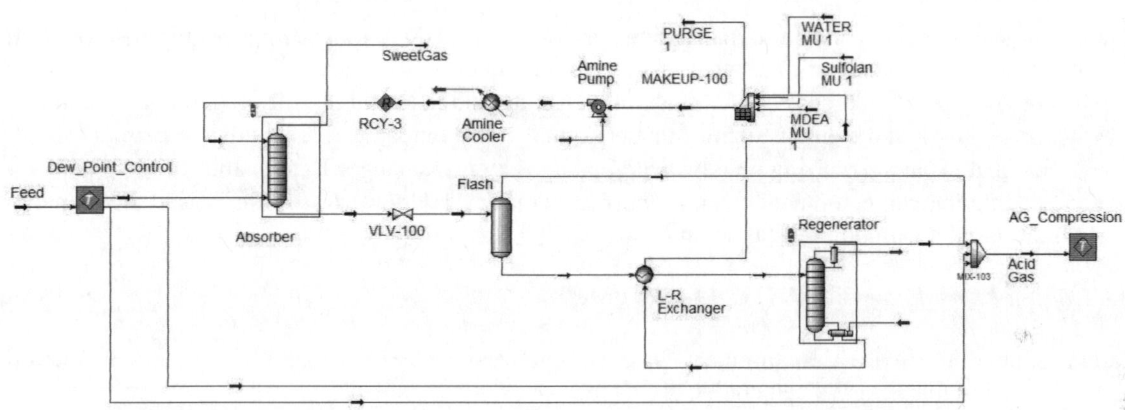

Figure 3. Simulation flowsheet of the physical-chemical absorption process.

- For all three process technologies:
  - Polytropic efficiencies for compressors and pumps are assumed 75%.
  - Cooling medium is assumed to be able to cool down the stream to 35 °C.
  - Minimum temperature approach in heat exchanger is greater than 2 °C.
  - Pressure drop of heat exchanger is 5 psi at tube side, and 2 psi at shell side.
- For Improved Cryogenic Distillation :
  - Heavier hydrocarbons have to be removed by cooling from the feed gas stream to prevent the occurrence of premature hydrocarbon freezing before entering the main fractionation column.
- For Chemical + Physical Absorption:
  - Aqueous absorbent with sulfolane content of 10.52 %-w/w and MDEA of 40.48 %-w/w.
  - Rich loading for solvent is limited in the range of 0.40 – 0.42 mole/mole.

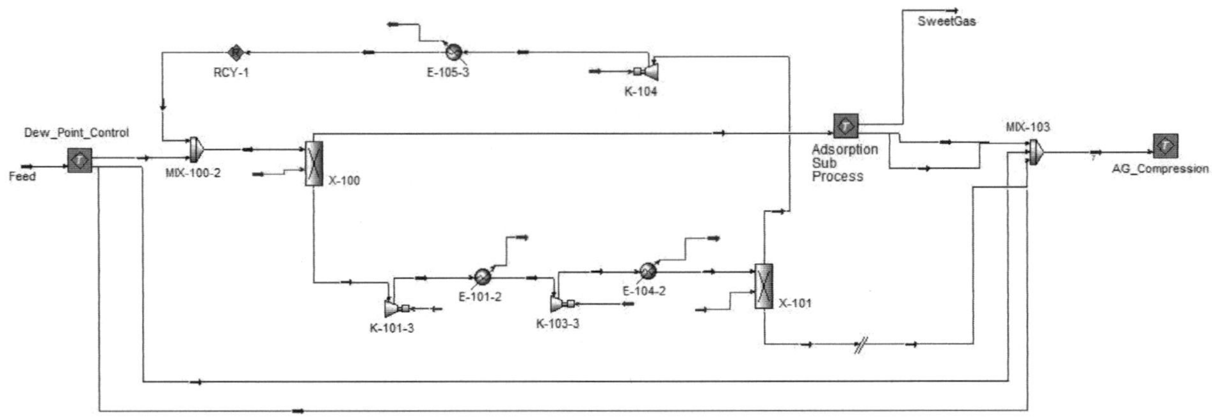

Figure 4. Simulation flowsheet of the hybrid technology.

3.2. Equipment Sizing Method

Equipment sizing was carried out to determine variables required for estimation of equipment price. In Aspen HYSYS, calculated variables differ for every equipment as follows.

- For compressor and pump, process variable is duty and taken directly from simulation results.
- For separator and column, sizing was performed using available vessel and tray sizing tools.
- For heat exchanger, sizing was estimated using Aspen Exchanger Design and Rating tool.
- For membrane equipment, sizing (surface area calculation) was done based on capacity comparison with other installed units.

3.3. Capital Expenditure (CAPEX) Calculation Method

Capital expenditure for main equipment was estimated using Aspen Process Economic Evaluation. Methods used in this CAPEX calculation are as follows.

- The cost for each equipment is calculated based on three factors: bare equipment, material (for piping, instrument, electrical, insulation, and civil), and labor. If equipment is quoted from other references, this price will be inserted as packaged unit. The calculated total cost is known as Total Direct Cost (TDC).
- Indirect cost was added as 46.52% of TDC. Indirect cost covers various components: engineering, site office, permit & license, insurances, taxes, transportation, and overhead.
- Contractor profit and contingency cost are assumed to be 5% and 25% of TDC respectively.
- The final capex was estimated with some adjustment to incorporate the conditions in Indonesia.

3.4. Operating Expenditure (OPEX) Calculation Method

OPEX calculation was based on the consumption figures of supporting materials, electricity, cooling water, hot oil, and hydrocarbon loss. Prices of the components and basis used in OPEX calculations were quoted from price list commonly used in industry. Indicators used in total OPEX calculation can be seen in Table 3.

IOP Conf. Series: Materials Science and Engineering **543** (2019) 012077 doi:10.1088/1757-899X/543/1/012077

4. Result and Discussion

Each of simulation flowsheet shown in Figure 2 – Figure 4 was executed for the same evaluation basis as listed in Table 2. The quantitative results of this flow-sheeting calculation are concisely reported as block flow diagram for each evaluated technology as shown in Figure 5 – Figure 7. Each diagram contains complete information to be used for equipment sizing, estimation of CAPEX, and estimation of OPEX.

Table 3. Variable OPEX Indicators [9].

Cost Item	Typical Range of Multiplying Factors	Value Used
1. Direct Manufacturing Cost		
a. Raw Material	C_{RM}	C_{RM}
b. Waste Treatment	C_{WT}	C_{WT}
c. Utilities	C_{UT}	C_{UT}
d. Operating Labor	C_{OL}	C_{OL}
e. Direct supervisory and Clerical Labor	(0.1 - 0.25) C_{OL}	0.18 C_{OL}
f. Maintenance and Repairs	(0.02-0.1) FCI	0.06 FCI
g. Operating Supplies	(0.1 - 0.2) Line 1.F	0.009 FCI
h. Laboratory Charges	(0.1- 0.2) C_{OL}	0.15 C_{OL}
i. Patents and Royalty	(0 - 0.06) COM	0.03 COM
Total Direct Manufacturing Cost	**C_{RM} + C_{WT} + C_{UT} + 1.33 C_{OL} + 0.03 COM + 0.069 FCI**	
2. Fixed Manufacturing Cost		
a. Depreciation	0.1 FCI	0.1 FCI
b. Local taxes and insurance	(0.014 - 0.05) FCI	
c. Plant Overhead Costs	0.15 (Line 1D + Line 1E + Line 1F)	0.177 C_{OL} + 0.009 FCI
Total Fixed Manufacturing Cost	**0.708 COL + 0.068 FCI+ depreciation**	
3. General Manufacturing Expenses		
a. Administration Costs	0.15 (Line 1D + Line 1E + line 1F)	0.177 C_{OL} + 0.009 FCI
b. Distribution and Selling Costs	(0.02 - 0.2) COM	0.11 COM
c. Research and Development	0.05 COM	0.05 COM
Total General Manufacturing Costs	**0.177 COL + 0.009 FCI + 0.16 COM**	
Total Costs	**CRM + CWT + CUT + 2.215 COL + 0.19 COM + 0.146 FCI + depreciation**	

CAPEX of the improved cryogenic distillation technology is estimated about \$265 Million. This is smaller than CAPEX of the chemical-physical absorption technology (about \$371 Million) and Hybrid (Membrane +chemical-physical absorption in series) technology (about \$319 Million). Detail of CAPEX estimation for each process section of the improved cryogenic distillation process, chemical-physical absorption technology, and hybrid technology are shown in Table 4 respectively.

IOP Conf. Series: Materials Science and Engineering **543** (2019) 012077 doi:10.1088/1757-899X/543/1/012077

CAPEX distribution for each equipment for improved cryogenic distillation, chemical-physical absorption technology, and hybrid technology are shown in Figure 8 respectively. In improve cryogenic distillation technology, CAPEX is dominated by equipment price for column + refrigeration system (63%), while the other two technologies require small cost for column (less than 5%).

Power, hot oil, propane and ethane are used as utilities in improved cryogenic distillation technology. Propane and ethane are used as refrigeration working fluids. In Chemical-Physical Absorption and Hybrid technology, make-up solvent is used as additional utility beside previous mentioned utilities. This utility consumption is used as a basis for OPEX calculation of each technology. OPEX of improved cryogenic distillation, chemical-physical absorption, and hybrid technology is shown in Table 5, Table 6, and Table 7 respectively.

Comparison for three methods based on economical aspect and hydrocarbon losses is shown in Figure 9. Improved Cryogenic Distillation is the lowest cost of all three technologies, followed by hybrid technology and chemical-physical absorption process. Improved cryogenic distillation also shows the lowest hydrocarbon losses compared to two other technologies. This shows that this new improved cryogenic distillation process is promising in economical aspect and hydrocarbon losses.

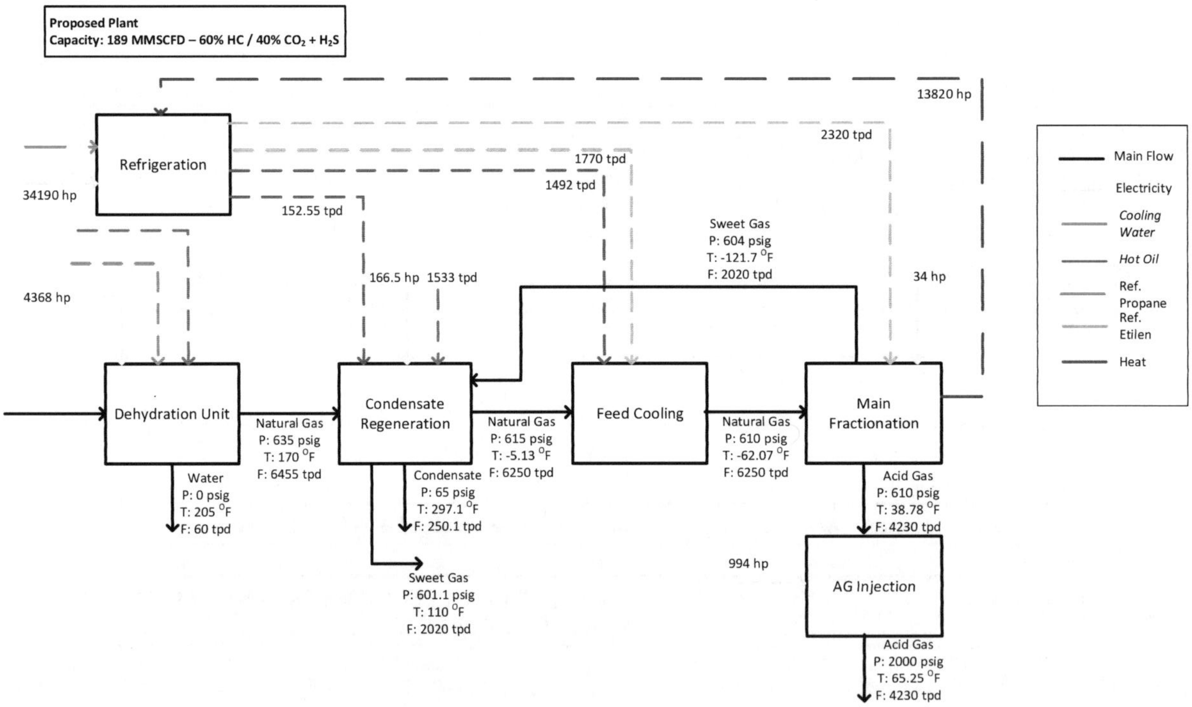

Figure 5. Block flow diagram of the improved cryogenic distillation process.

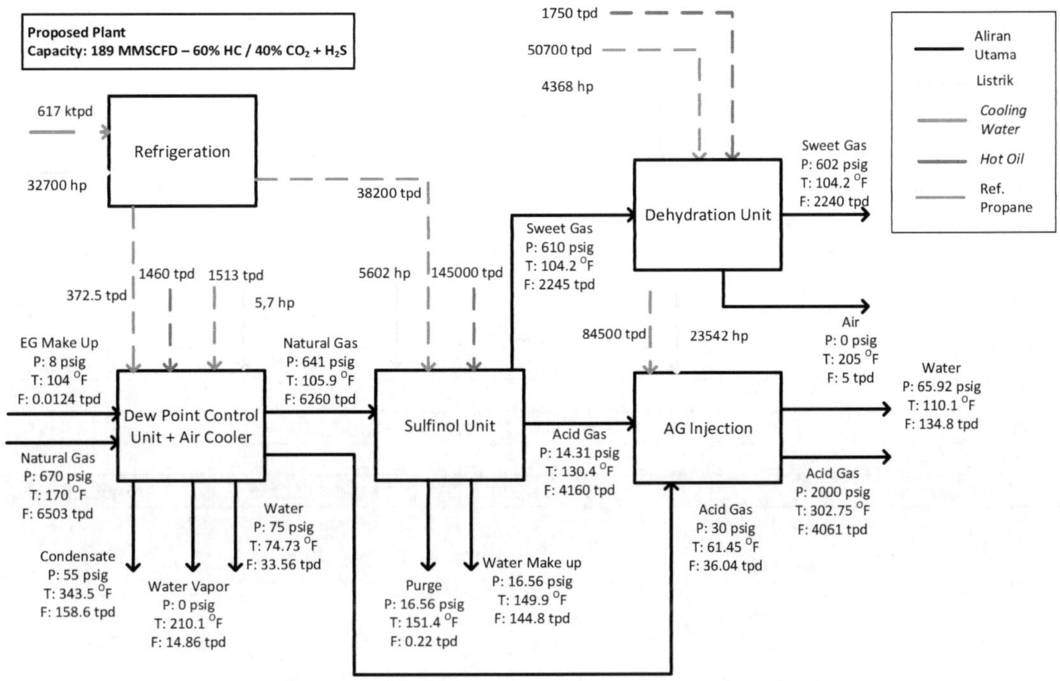

Figure 6. Block flow diagram of the physical-chemical absorption process.

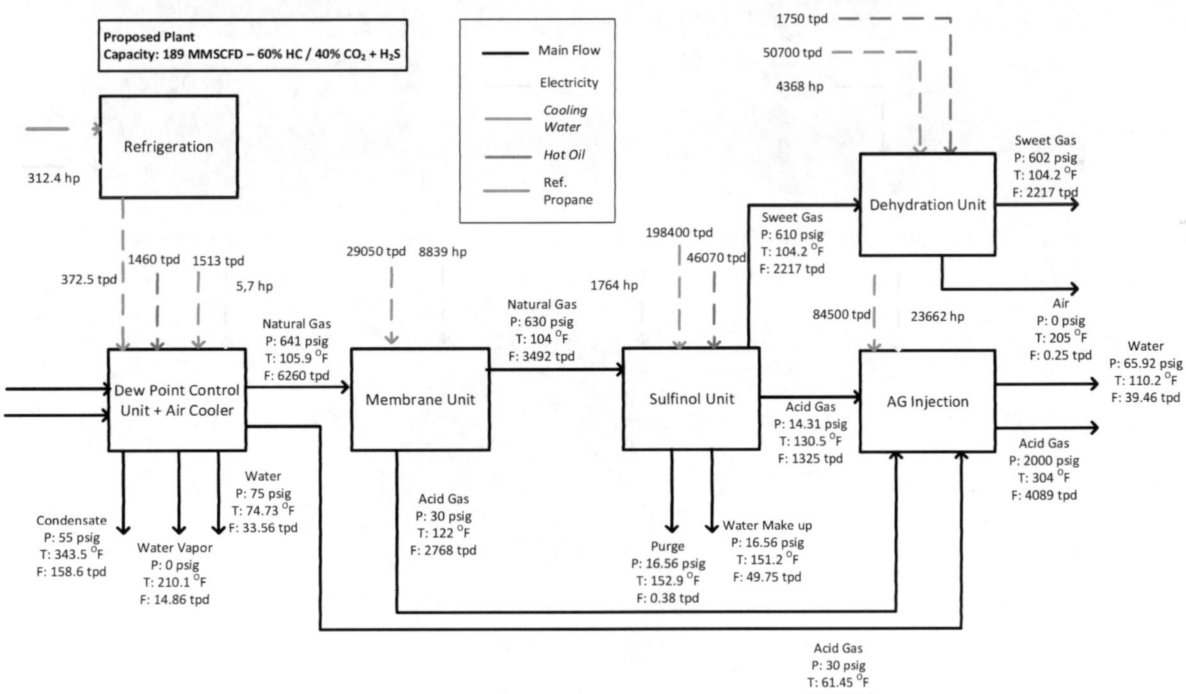

Figure 7. Block flow diagram of the hybrid technology.

Table 4. Capital Expenditure Comparison for the Three Technologies Evaluated

Process Section	Improved Cryogenic Distillation	Chemical-Physical Absorption	Hybrid Technology
Acid Gas Compression Unit	$ -	$ 60,065,850.69	$ 93,918,325.96
Acid Gas Injection	$ 2,386,653.92	$ -	$ -
Condensate Regeneration	$ 33,835,049.58	$ -	$ -
Cooling Water Handling	$ 4,892,749.93	$ 3,154,966.98	$ 7,958,199.91
Dehydration Unit	$ 15,402,525.77	$ 13,269,137.43	$ 13,256,450.12
Dew Point Control Unit	$ -	$ 13,814,073.36	$ 13,800,865.00
Hot Oil Handling	$ 928,124.33	$ 21,063,993.38	$ 7,650,742.44
Main Absorption	$ -	$ 85,135,039.32	$ 95,184,790.93
Main Fractionation	$ 85,985,178.87	$ -	$ -
Membrane Unit	$ -	$ -	$ 42,668,343.28
Power Generation	$ -	$ -	$ 37,550,981.27
Refrigeration Unit	$ 79,408,466.91	$ 115,199,696.85	$ 7,062,301.08
Turbine Generation	$ 42,513,250.68	$ 59,961,241.99	$ -
Total CAPEX	**$ 265,352,000.00**	**$ 371,664,000.00**	**$ 319,051,000.00**

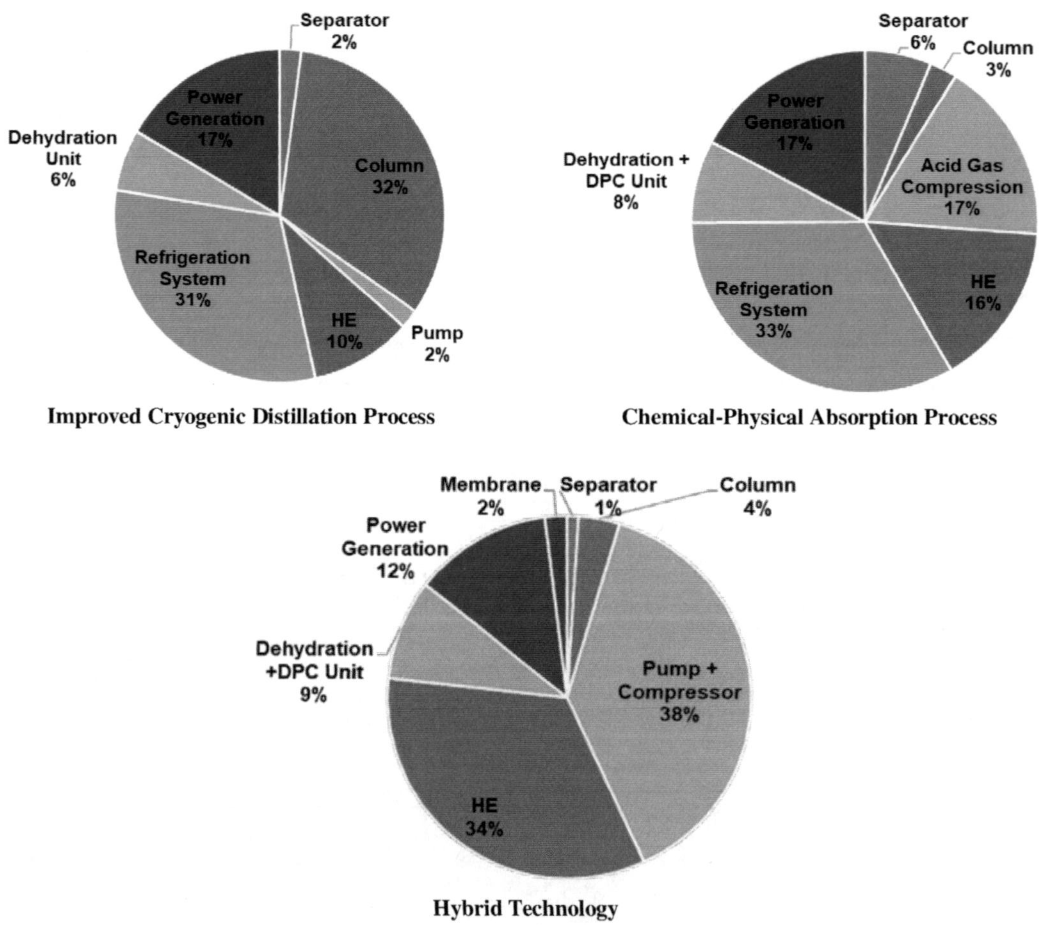

Figure 8. Proportion of Equipment Cost for the three technologies evaluated.

Table 5. OPEX Calculation of Improved Cryogenic Distillation

	Consumption		Price		Sub Total ($/yr)
Raw Materials (CRM)					
Make-up Propane	20.3	ton/h	16	$/ton	2,575,076
Make-up Ethylene	11.1	ton/h	30	$/ton	2,627,267
TEG	1.0	ton/h	167	$/ton	1,354,320
Product for Fuel	6.7	MMSCFD	10	$/MMBTU	13,597,793
Utility (CUT)					
Cooling Water	178.7	m3/h	1.32	$/m3	1,863,328
Makeup Hot Oil	0.7	ton/h	250	$/ton	1,353,513
Operating Labor (COL)	30	person(s)	4200	$/person/yr	126,000
Cost of HC Loss (CHL)	21	MMSCFD	10	$/MMBTU	42,948,905
Fixed Capital Investment (FCI)	$ 265,352,000				
Total OPEX Without CHL	0.18FCI + 2.73COL + 1.23(CRM+CUT)				76,854,036
Total OPEX With CHL	0.18FCI + 2.73COL + 1.23(CRM+CUT) + CHL				119,802,941

Table 6. OPEX Calculation of Chemical-Physical Absorption

	Consumption		Price		Sub Total ($/yr)
Raw Material (CRM)					
MDEA + Sulfolane	1.82E-02	ton/h	12000	$/ton	1,725,840
TEG	1.0	ton/h	167	$/ton	1,354,320
Make-up Propane	79.6	ton/h	16	$/ton	10,084,966
Ethylene Glycol	1.87E-02	ton/h	1200	$/ton	5,378
Product for Fuel	10.7	MMSCFD	10	$/MMBTU	21,713,219
Utility (CUL)					
Cooling Water	113.9	m3/h	1.32	$/m3	1,187,406
Hot Oil	31	ton/h	250	$/ton	61,018,079
Operating Labor (COL)	30	person(s)	4200	$/person/yr	126,000
Cost of HC Loss (CHL)	18	MMSCFD	10	$/MMBTU	37,288,484
Fixed Capital Investment (FCI)	$ 371,664,000				
Total OPEX Without CHL	0.18FCI + 2.73COL + 1.23(CRM+CUT)				186,663,226
Total OPEX With CHL	0.18FCI + 2.73COL + 1.23(CRM+CUT) + CHL				223,951,710

Table 7. OPEX Calculation of Hybrid Technology

	Consumption		Price		Sub Total ($/yr)
Raw Materials (CRM)					
MDEA + Piperazin	4.29E-03	ton/h	6000	$/ton	204,000
TEG	1.0	ton/h	167	$/ton	1,354,320
Product for Fuel	6.2	MMSCFD	10	$/MMBTU	12,673,802
Utility (CUL)					
Air Pendingin	293.5	m3/h	1.32	$/m3	3,059,420
Hot Oil	10.3	ton/h	250	$/ton	20,306,969
Operating Labor (COL)	31	person(s)	4200	$/person/yr	130,200
Cost of HC Loss (HCL)	11	MMSCFD	10	$/MMBTU	23,114,549
Fixed Capital Investment (FCI)	$ 319,051,000				
Total OPEX Without CHL	0.18FCI + 2.73COL + 1.23(CRM+CUT)				104,030,794
Total OPEX With CHL	0.18FCI + 2.73COL + 1.23(CRM+CUT) + CHL				127,145,342

5. Conclusion

Gas field with high content of acid gases requires preliminary processing technology capable of cost-effectively eliminating high acid gas content and large amounts of gas. Whether the processed acid gas will be re-injected or released into the air after the sulfur content is taken, the three process technologies studied have the potential to be applied, including new improved cryogenic distillation that applying freezing of acid gas in column.

Quantitative technical evaluation of total hydrocarbon consumption shows that Chemical-Physical Absorption technology will require the largest operating cost (OPEX), followed by Membrane Hybrid, and then new improved cryogenic distillation. The same sequence is also obtained when an evaluation is made on the required capital cost (CAPEX).

However, this new improved cryogenic distillation has concern about operability, particularly during start-up and shut down. This can be mitigated by a more specialized review by visiting a demonstration facility to discuss in more detail the operational issues.

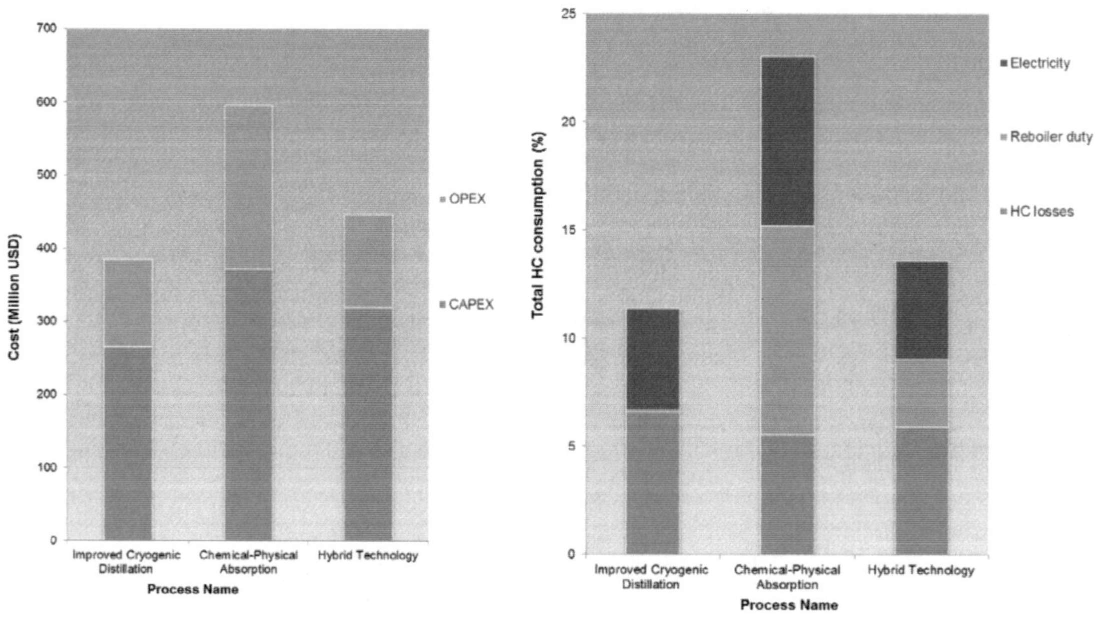

(a) (b)

FIGURE 9. Comparison of the three technologies evaluated for (a) CAPEX and OPEX (b) Hydrocarbon Losses

References

[1] British Petroleum 2017 *BP Statistical Review of World Energy 66th Edition.*

[2] Stewart, M and Arnold, K (2011) *Gas Sweetening and Processing Field Manual*. Waltham: Gulf Professional Publishing.

[3] Suarsana, P 2009 *Workshop of Natural Gas Resources with High CO_2 and Carbon Storage (CCS) in CCOP* at Bali.

[4] Valencia, J A, 2009 *The Controlled Freeze Zone Technology for the Development of Sour Gas Resources and Geosequestration of Acid Gases*, Sour Gas Workshop, s.l. : Gas Arabian.

[5] Northrop, P S and Valencia, J A 2009 *The CFZ™ process: A cryogenic method for handling high-CO_2 and H_2S gas reserves and facilitating geosequestration of CO_2 and acid gases*, Energy Procedia **1** 171-77.

[6] Abdel-Aal, H K, Anggour, M, Fahim, M A, 2003 *Petroleum and Gas Field Processing* Vol. 256.

[7] Cameron, *Cynara CO_2 Membrane Separation Solutions*, 2013.

[8] Bergel. M., Tierno, I. *Sweetening Processes - A look at the Whole Picture*. In: 24th World Gas Conference. International Gas Union, Buenos Aires, Argentina, 2009.

[9] Turton R, Bailie, R C, Whiting W B, and Shaeiwitz J A 2008 *Analysis, Synthesis, and Design of Chemical Processes.* (Westford, Massachusetts: Prentice-Hall Pearson Education)

IOP Conf. Series: Materials Science and Engineering **543** (2019) 012078 doi:10.1088/1757-899X/543/1/012078

Wet Extraction of γ-Oryzanol From Rice Bran

S Zullaikah*, Y A Pujiastuti , G Prihandini , and M Rachimoellah
Department of chemical engineering, Institut Teknologi Sepuluh November (ITS), Kampus ITS Keputih Sukolilo, Surabaya 60111, Indonesia

E-mail: szulle@chem-eng.its.ac.id

Abstract. γ-oryzanol is unique to rice bran oil (RBO) and it is known to be a potent hypocholesterolemic agent. Separation and isolation of γ-oryzanol from crude RBO is difficult due to complexity of components in crude RBO. Converting acylglycerides and FFA in RBO into FAME will concentrate bioactive compounds in the residue after FAME was removed from the RBO-based biodiesel. In this study, wet-extraction of bioactive compound from rice bran, γ-oryzanol, has been carried out using subcritical water-methanol mixture. The effects of extraction temperature, extraction time, water-methanol volume ratio and pressurizing gas type on the content and recovery of γ-oryzanol in crude RBO-based biodiesel have been investigated. It was found that the content and recovery of γ-oryzanol in RBO-based biodiesel increased with higher extraction temperature, longer extraction time and lower water concentration. The content of γ-oryzanol in RBO-based biodiesel was higher under N_2 atmosphere since CO_2 can acidify the solvent leading to hydrolysis of γ-oryzanol. The maximum γ-oryzanol content in RBO-based biodiesel was 2.16 % (corresponding to 2750 mg γ-oryzanol/kg rice bran) obtained after extraction for 7 h at 200 °C and 40 bar with water concentration of 50 % (v/v) under N_2 atmosphere.

1. Introduction

Indonesia is the third largest rice producer after China and India. Rice bran is produced during the milling of husked rice as a by-product and rice bran contains about 15 - 23 % oil [1-2]. Rice bran oil (RBO) is one of the most nutritious oil because of its favorable fatty acid composition and a unique combination of naturally occurring biologically active and antioxidant compounds. Crude RBO is extremely rich in micronutrients accounting for 4 - 6 % of the oil [1 , 3]. Among the micronutrients, γ-oryzanol, a mixture of ferulic acid esters and triterpenoid alcohols, accounts for 1 - 3 % of the total oil, and is unique to RBO [3 – 8]. The amount of γ-oryzanol in rice bran is ten times higher than that of vitamin E, therefore, γ-oryzanol is the main antioxidant in rice bran [9]. γ-oryzanol in RBO has been shown as an antioxidant against oxidation of linoleic acid in a certain molar ratios. γ-oryzanol can be degraded at a lower rate than that of α-tochoperol at elevated temperatures; its make γ-oryzanol is a promising antioxidant to be applied at high temperatures [3]. γ-oryzanol in RBO was responsible as a hypocholesterolemic agent. RBO containing γ-oryzanol was added to the rat diets was shown to lower plasma cholesterol levels compared to rats given a diet containing RBO devoid of γ-oryzanol [10 – 11]. Other beneficial effect of γ- oryzanol is their effectiveness in reducing menopause symptoms [12]. Das 1998 and Narayan 2005 have problems encountered during the extraction of γ-oryzanol from RBO soapstock, and the problem was mainly arises from variations in the compositions of RBO soapstock. Xu and Godber 2000 was investigated the extraction of γ-oryzanol from rice bran using a supercritical fluid at low temperature (T = 50 °C) and high pressure (P = 680 atm) for 25 min. They found that the yield of γ-oryzanol was four times higher than that obtained using solvent extraction. Zullaikah 2009 reported the isolation of γ-oryzanol from RBO by a two-step crystallization process. In the first crystallization step, triglycerides (TG) and steryl esters were mainly removed, while γ-oryzanol remained in the liquid phase along with free fatty acids (FFA), diglycerides (DG), monoglycerides (MG), squalene, tocols, and phytosterols. In the second crystallization step, γ-oryzanol was precipitated by adding n-hexane as anti-solvent to the liquid phase (γ-oryzanol-rich product) and keeping the mixture at 5 ± 1 °C for 48 h. This two-step crystallization process was able to produce 93 - 95 % pure γ-oryzanol, but the total recovery rate was only 59 %. Due to the difficulty on separation of γ-oryzanol from crude RBO or RBO soapstock, converting neutral oils into fatty acid methyl esters (FAME) will be concentrated γ-oryzanol in the bottom product after purification of RBO-based biodiesel through distillation. Purification of RBO-based biodiesel (initial FAME content of 89.05 %) to fulfill the standard biodiesel specification with FAME content more than 96.5 % was successfully carried out using green solvent (deep eutectic solvents, DES) [13]. In their work, FAME was immiscible in DES, while γ-oryzanol in RBO-based biodiesel was extracted into DES. Separation of extracted compounds which is rich of γ-oryzanol from DES will make the further process in the separation and purification of γ-oryzanol will be easier. Therefore, in this study, wet extraction of bioactive compound from rice bran, γ-oryzanol, has been carried out using subcritical water-methanol mixture. The effects of extraction temperature, extraction time, water-methanol volume ratio and pressurizing gas type on the content and recovery of γ-oryzanol in crude RBO-based biodiesel have been investigated.

2. Methods

2.1. Materials
Rice bran IR-64 type was obtained from Lamongan, East Java, Indonesia. Rice bran was filtered to remove any impurities and then kept in a refrigerator to minimize further formation of FFA. Standard of γ-oryzanol was obtained from Wako, Japan. Analytical grade solvents and pressurizing gases (CO_2 and N_2) were purchased from commercial source in Indonesia.

2.2. Soxhlet Extraction of RBO
Rice bran (20 g) was extracted using n-hexane as solvent in a soxhlet extractor for 12 h. In this work, it was assumed that soxhlet extraction can recover all RBO in rice bran. Yield of RBO (17.71 %) with high FFA content (37.71 %) was achieved using soxhlet extraction method, however the γ-oryzanol content in the RBO was quite low (1.33 %). The yield is expressed as percentage of RBO weight/rice bran weight.

2.3. Formatting author affiliations
The experiment setup for wet extraction of γ-oryzanol from rice bran extraction using subcritical water-methanol mixture was shown in Fig. 1. Rice bran (5 g) was mixed with 40 ml solvent (water-methanol mixture at different volume ratio) in a high pressure reactor (volume = 86 ml). The reactor was made of stainless steel and all fittings were purchased from Swagelok, Taiwan. The reactor was equipped with an external electrical heater and a thermocouple was inserted into the reactor to monitor the temperature. The extraction time was counted after a predetermined temperature was reached. The extraction process was investigated at different temperatures (T = 160 – 215 °C), extraction times (t = 1 – 8 h) and water-methanol volume ratios (5/35 – 35/5, ml/ml). The pressure was kept constant at 40 bar by using either CO_2 or N_2 as compressing gas. After a pre determined extraction time, the reactor was rapidly quenched and, as a consequence, the extraction and reaction inside the reactor was stopped. Product in the reactor were collected and mixed with n-hexane (50 ml). The mixture was stirred at 300 rpm for thirty minutes. The upper phase containing oil (RBO-based biodiesel) was collected and the bottom phase was mixed again with n-hexane (50 ml). This procedure was repeated three times. n- Hexane was then removed by using a rotary evaporator. The mass of the RBO-based biodiesel was determined and the content of γ-oryzanol was analysed

2.4. Soxhlet Extraction of RBO

γ-oryzanol content in the sample was determined by UV-vis Spectrophotometer V-550. The calibration curve was obtained by using standard of γ-oryzanol with a concentration range from 0 to 97 ppm. A straight line passing through the origin was obtained at a wavelength of 316 nm (R^2 = 0.99). To determine γ-oryzanol content, the sample was diluted in n-hexane in a 1 cm cuvet and the spectrophotometer was operated at 1 nm bandwith and 1 nm data pitch.

3. Results and Discussion

The process of wet extraction of γ-oryzanol from rice bran was carried out under subcritical condition of water-methanol mixture. Subcritical water was widely applied in the extraction, hydrolysis and wet oxidation to allow γ-oryzanol extracted in the oil phase [14]. The addition of methanol may increase the acquisition of γ-oryzanol [15]. Since γ-oryzanol is more easily dissolved into polar solvents such as isopropanol, ethyl acetate, and methanol. Dissolved CO_2 in subcritical water will form of carbonic acid, in other words CO_2 can be used as an acid catalyst [16]. Under subcritical condition, water may act as an efficient acid-base catalyst because it has a high dissociation constant. Subcritical water can also be utilized as solvent extraction of non-polar compounds. This is because the dielectric constant of water drops to 27 when in a subcritical state. This dielectric constant value is almost the same as methanol in ambient conditions [17].

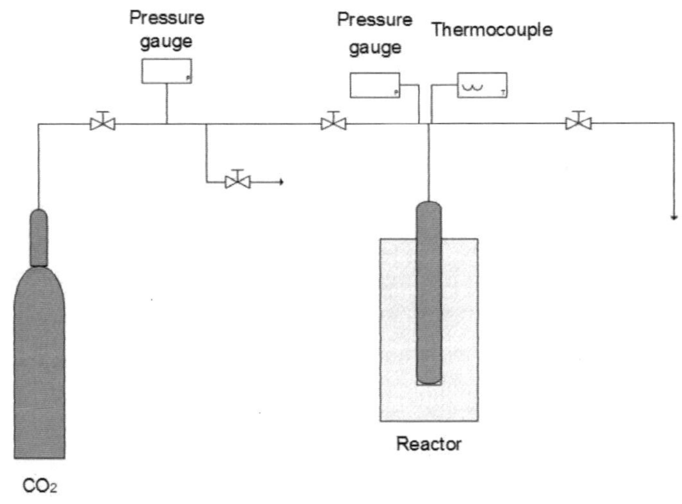

Figure 1. Experimental setup for extraction of *γ-oryzanol* using subcritical water-methanol mixture

3.1. Effect of Extraction Time

γ-oryzanol began to degrade at temperatures over 120 °C for 10 h [18]. While according to Nystrom 2007, γ-oryzanol begin degraded at 180 °C for 6 hours. In this study γ-oryzanol was slightly degraded at 200 °C for 8 h. From figure 2, γ-oryzanol content increased for both types of pressurizing gases used (N_2 and CO_2) along with increasing of extraction time. Longer extraction time was required to extract overall γ-oryzanol. The use of N_2 as a pressurizing gas obtained a higher yield of γ-oryzanol than that of CO_2, because γ-oryzanol is easily degraded under acidic conditions [19]. N_2 was an inert gas; therefore γ-oryzanol in RBO-based biodiesel obtained using N_2 as a pressurizing gas was quite stable and tend to increase [3]. The highest γ-oryzanol in RBO-based biodiesel using CO_2 as a pressurizing gas for 7 h of extraction time was 2.03 % ± 0.19, and then decreased after 8 h of extraction time (1.54 % ± 0.29). Both CO_2 and N_2 gases had shown the best conditions at 200 °C for 7 h of extraction time to extract γ-oryzanol of about 2 %. The decrease in yield and recovery is due to a greater degradation rate compared to the

IOP Conf. Series: Materials Science and Engineering **543** (2019) 012078 doi:10.1088/1757-899X/543/1/012078

extraction rate. the degradation rate may actually have occurred at 1 - 5 h extraction time, but the extraction rate is greater than the degradation rate. Zullaikah 2013 in the reaserch said that "Decreased of yield and recovery of γ-oryzanol after 8 h extraction time, it's because γ-oryzanol start degraded to phytosterol".

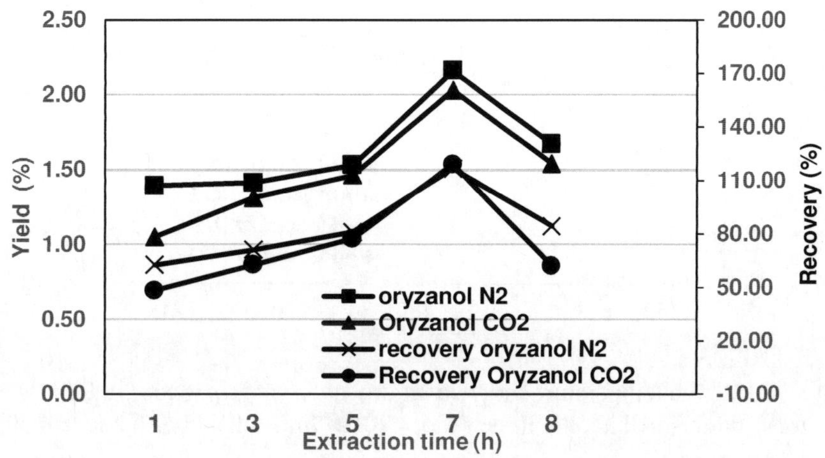

Figure 2. Effect of extraction time on yield and recovery of *γ-oryzanol* in RBO-based biodiesel. Operation conditions: T= 200 ^{0}C, water-methanol volume ratio = 20/20 (mL/mL) and P = 40 bar

3.2. Effect of Extraction Time

Rising temperature will increase the rate of diffusivity and decreased the surface tension and viscosity. The higher water temperature accordingly lowers of water dielectric constant. As the dielectric constant increases, the solubility of non-polar compounds in water will be greater [20]. This makes the water as a solvent of non-polar compounds, and γ-oryzanol compounds are quite resistant to high temperatures when compared to other antioxidants in RBO such as tocopherols and tocotrienols [21]. Figure 3 show that under the influence of CO_2 atmosphere, γ-oryzanol content increased from 1.44 % ± 0.01 to 2.02 % ± 0.18 when the temperature was increased from 160^{0}C to 200 ^{0}C. If extraction temperature was raised to 215 ^{0}C the content of γ-oryzanol decreased dramatically to 1.41 % ± 0.05. Using N_2 as a pressurizing gas, γ-oryzanol content increased from 1.27 % ± 0.05 to 2.15 % ± 0.11 when the temperature was increased from 160 ^{0}C to 200 ^{0}C. If extraction temperature was raised to 215 ^{0}C the content of γ-oryzanol decreased slightly to 2.12 % ± 0.11. Both CO_2 and N_2 gases obtain the best conditions at 200 ^{0}C for 7 hours to reach γ-oryzanol content of about 2 %. Increased extraction temperature has a positive effect on the content of extracted γ-oryzanol. At high temperatures, the physical properties of the bran are more easily penetrated by solvents thereby causing γ-oryzanol to be easily released from the matrix [22].

IOP Conf. Series: Materials Science and Engineering **543** (2019) 012078 doi:10.1088/1757-899X/543/1/012078

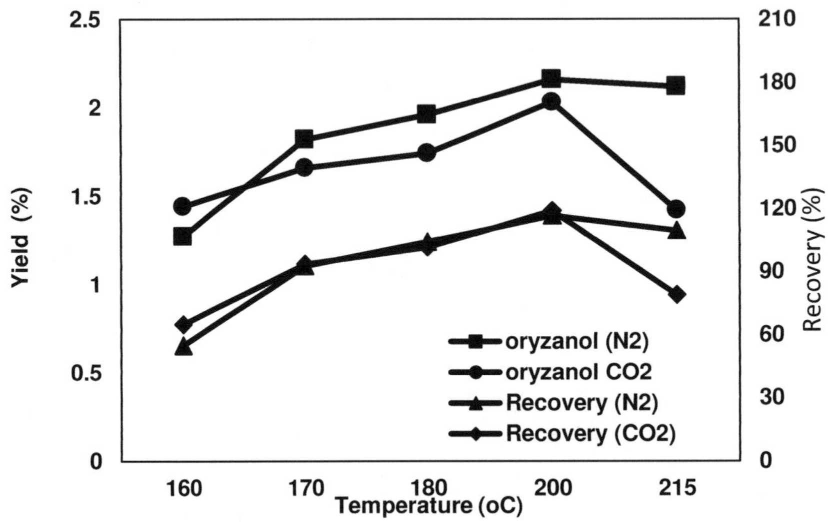

Figure 3. Effect of extraction temperature on yield and recovery of *γ-oryzanol* in RBO-based biodiesel. Operation conditions: Water-methanol volume ratio = 20/20 (mL/mL), t = 7 h and P = 40 bar

3.3. Effect of Water-Methanol Volume Ratio

Decreased water-methanol volume ratio has a positive effect on the extraction of γ-oryzanol. γ-oryzanol components was more dissolved in polar solvents such as isopropanol, ethyl acetate, and methanol [22]. Chen and Chiu, 2011 also used methanol as solvent to obtain 10 times more γ-oryzanol yields than that of using ethyl acetate. However, the addition of excess methanol does not increase the percentage of γ-oryzanol content. The use of methanol over the optimum amount will cause a reduction in the product, because methanol tends to extract more polar components such as carbohydrates. From the figure 4, it shown that N_2 as a pressurizing gas gives a positive effect on the content of extracted γ-oryzanol. The content of γ-oryzanol increased from 1.78 % ± 0.02 (methanol-water volume ratio = 5/35, mL/mL) to 2.15 % ± 0.11 (methanol-water volume ratio = 20/20, mL/mL). Then finally going down using methanol-water volume ratio of 35/5 (mL/mL) to 1.9 % ± 0.03. The use of CO_2 as a pressurizing gas also increased γ-oryzanol content from 1.6 % ± 0.08 to 2.03 % ± 0.19 when the methanol-water volume ratio was increased from 5/35 (mL/mL) to 20/20 (mL/mL) before finally decreasing to 1.65 % ± 0.07 at a methanol-water volume ratio of 35/5 (mL/mL). The presence of CO_2 as a pressurizing gas gives the acid state on the medium; therefore, γ-oryzanol was prone to degrade under CO_2 atmosphere [3 , 19]. γ-oryzanol obtained in this study was 2.15 % or 2.75 mg of γ-oryzanol / g bran under N_2 atmosphere and 2.03 % or 2.5 mg γ-oryzanol / g bran under CO_2 atmosphere, and those results obtained were better in comparison with Xu and Godber's resulted in 2000 with Supercritical Fluid Extraction (SFE) at 680 atm, extraction time for 15-20 min at 50 °C resulting in γ-oryzanol of 1.11 ± 0.07 mg / g bran.

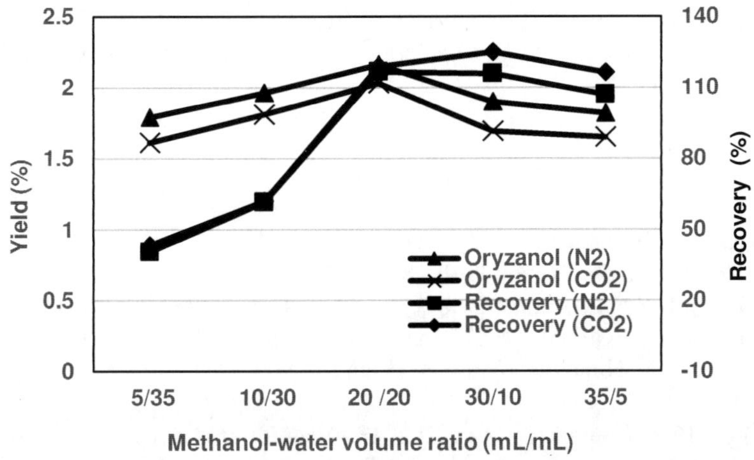

Figure 4. Effect of methanol-water volume ratio on yield and recovery of *γ-oryzanol* in RBO-based biodiesel. Operation conditions: T = 200 °C, t =7 h and P = 40 bar

4. Conclusion
Wet extraction of γ-oryzanol from rice bran was found that the content and recovery of γ-oryzanol in RBO-based biodiesel increased with higher extraction temperature, longer extraction time and lower water concentration. The content of γ-oryzanol in RBO-based biodiesel was higher under N_2 atmosphere since CO_2 can acidify the solvent leading to hydrolysis of γ-oryzanol. The maximum γ-oryzanol content in RBO-based biodiesel was 2.16 % (corresponding to 2750 mg γ-oryzanol / kg rice bran) obtained after extraction for 7 h at 200 °C and 40 bar with water concentration of 50 % (v/v) under N_2 atmosphere. And the use of subcritical technology is far more effective and efficient for extracting γ-oryzanol (7 h) comparing with soxhlet extraction (12 h). The content of γ-oryzanol in RBO-based biodiesel was higher under N_2 atmosphere since CO_2 can acidify the solvent leading to hydrolysis of γ-oryzanol

5. References
[1] Zullaikah S, Lai C C, Vali S R, Ju Y H 2005 *Biores Tech* **96** 1889-1896
[2] Zullaikah S, Rahkadima Y T, Ju Y H 2017 *Renew E* **111** 764-770
[3] Ju Y H, Zullaikah S 2013 *J Taiwan Inst Chem Eng* **44** 924-928
[4] Kasim N S, Chen H, Ju Y H 2007 *J Chinese Inst Chem Eng* **38** 229-234
[5] Lai C C, Zullaikah S, Vali S R, Ju Y H 2005 *Chem Tech and Biotech* **80** 331–337
[6] Saska M, Rossiter G J 1998 *JAOCS* **75** 1421–1427
[7] Seetharamaiah G S, Prabhakar J V 1986 *J. Food Sci Tech* **23** 270
[8] Zullaikah S, Melwita E, Ju H S 2009 *Biores Tech* **100** 299-302
[9] Xu Z, Godber J S 2001 *JAOCS* **78** 465-469
[10] Seetharamaiah G S, Chandrasekhara N 1988 *Nutr Rep Int* **38** 927–935
[11] Tabassum S, Aggarwal S, Ali S M, Beg Z H, Khan A S, Afzal K 2005 *Indian J Nephrol* **15** 10 – 13
[12] Michelle S G 2003 Metabolic aspects of oryzanol in rats M S Thesis Louisiana State University Lousiana USA
[13] Niawanti H, Zullaikah S 2017 *CET* **56** 1513-1518
[14] Toor S S, Rosendahl L, Rudolf A 2011 *J Energy* **36** 2328-2342
[15] Carr A G, Mammucari R, Foster N R 2011 *J Chem Eng* **172** 1-17
[16] Rogalinski T, Liu K, Albrecht T, Brunner G 2008 *J Supercritic F* **46** 335-341
[17] Tsigie Y A, Huynh L H, Ismadji S, Engida A M, Ju Y H 2012 *J Chem Eng* **213** 104-108

IOP Conf. Series: Materials Science and Engineering **543** (2019) 012078 doi:10.1088/1757-899X/543/1/012078

[18] Khuwijitjaru P, Taengtieng N, Changprasit S 2012 *J Silpakorn Univ Int* **4** 154-165

[19] Goufo P, Pereira J, Figueireido N, Oleivera M B P P, Carranca C, Rossa E A S, Trindade H, 2014 *J Cereal Sci* **59** 15-24

[20] Nystrom L, Achrenius T, Lampi A M, Moreau R A, Piironen V 2007 *J Food Chem* **101** 947-954

[21] Xu Z, Godber J S 2000 *JAOCS* **77** 547-551

[22] Goufo P, Pereira J, Figueireido N, Oleivera M B P P, Carranca C, Rossa E A S, Trindade H, 2014 *J Cereal Sci* **59** 15-24

[23] Baldino L, De Luca R, Reverchon E, 2018 *CET* **64** 31

[24] Chu BS, Baharin B S, Che Man Y B, Quek S Y 2004 *J food Eng* **62** 97-103

[25] Das P K, Chauduri A, Kaimal T N B, Bhalerao U T 1998 *J Agri Food Chem* 3073-3080

[26] Du Z, Deng S, Bei Y, Huang Q, Wang B, Huang J, Yu G 2014 *J Hazardous Mat* **274** 443-454

[27] Ju Y H, Huynh L H, Tsigie Y A, Ho Q P 2013 *Fuel* **105** 266-271

[28] Narayan A V, Barhate R S, Raghavarao K S M S 2006 *JAOCS* **83** 663-670

[29] Patel M, Naik S N 2004 *J Sci Indust Research* **63** 569-578

IOP Conf. Series: Materials Science and Engineering **543** (2019) 012079 doi:10.1088/1757-899X/543/1/012079

Microfiltration of Oily Waste Water: A Study of Flux Decline and Feed Types

P D Sutrisna*, J Candrawan and W W Tangguh

Department of Chemical Engineering, University of Surabaya, Jl Raya Kalirungkut, Surabaya, 60293, Indonesia

* Email : pudod@staff.ubaya.ac.id

Abstract. In recent years, membrane separation techniques gain an increasing interest to be applied in many industrial separation processes, such as the separation of produced water in petroleum industry and the treatment of used oil and used cooking oil. Used oil and used cooking oil consist of suspended solids in addition to emulsion. Both ultrafiltration and microfiltration have been used to separate oil from water. The challenges of oil-in-water emulsion separation include the possibility of big oil drops deforming through the slot or pore of membrane thus reduce the oil rejection. In addition, concentration polarisation and fouling phenomena are considered as negative attributes for membrane performances. This research applied flat sheet microfiltration membranes to treat used oil and used cooking oil. The performances of membranes in terms of flux reductions were evaluated and analysed using filtration model. Experimental results showed the decline of permeate flux with time indicating the formation of polarisation layer on the surface of the membrane. In addition, complete blocking and deep bed filtration explained the flux behavior of used oil and used cooking oil filtration, respectively.

1. Introduction

The treatment of waste water plays an important role in chemical industries in terms of operating cost and environmental protection issue. The successful treatment of waste water will reduce the operating cost and also improve the environment quality around the industries. Examples of waste that is produced by chemical industry are oil-in-water (O/W) emulsion and water-in-oil (W/O) emulsion [1-5]. These kinds of emulsions cannot be discharged directly into the body of water because the emulsions can reduce the amount of dissolved oxygen. In addition to chemical industries, O/W and W/O emulsions can also be produced by local or household activities, such as kitchen frying. Hence, it is considered very important to treat the emulsions before they can be discharged into the body of water.

Until recently, several techniques, such as chemical demulsification, gravity or centrifugal settling, and electrostatic demulsification, have been utilised on the treatment of emulsion [1]. In addition to those techniques, membrane separation technique also gains an increasing interest to be applied to treat emulsion. Microfiltration and ultrafiltration are used more widely than other types of membranes to treat emulsions. During the treatment of emulsions using membrane filter, there emerges the possibility of the deformation of particles or drops through the pores of membranes. Such deformation causes the decrease in the quality of permeates thus reduce the efficiency of separation [1]. Several studies have

been conducted to utilize microfiltration and ultrafiltration on the treatment of emulsion waste. Hlavacek [6] analyzed the demulsification process through microfiltration membrane. It was found that the demulsification process could produce water with only 30 ppm of oil, which was much lower than the concentration of oil in feed emulsion. Other study suggested to employ slotted surface filters that can produce very high permeate flux at very low trans membrane pressure [7,8]. This type of membrane was specially design to treat emulsions and it can reduce the possibility of membrane pore plugging and fouling during the treatment of emulsion.
Despite of the successful utilisation of microfiltration and ultrafiltration on the treatment of emulsion on laboratory scale, there are still limited successes of the realisation of these techniques on industrial scale. It is considered very important to find suitable model from laboratory scale that can be employed to predict the behavior of membranes when the membranes are applied to treat real emulsion on industrial scale. Research on the investigation of flux behavior during O/W and W/O emulsions have been conducted recently using synthetic emulsion. The flux decline during microfiltration of O/W emulsion using porous glass tube membranes was investigated by Ohya et al [9]. The flux decline can be explained by two stages or models, i.e. blocking filtration and cake filtration. On another line of research, Tien and Ramarao [10] evaluated four filtration models, i.e. cake filtration, deep bed filtration complete blocking, and intermediate blocking. These models were first developed by Hermans and Bredèe in 1936 [11]. In the proposed models, there are four cases that might appear during the filtration process. The cases include cake filtration, deep bed filtration, complete blocking, and intermediate blocking as described in Figure 1.

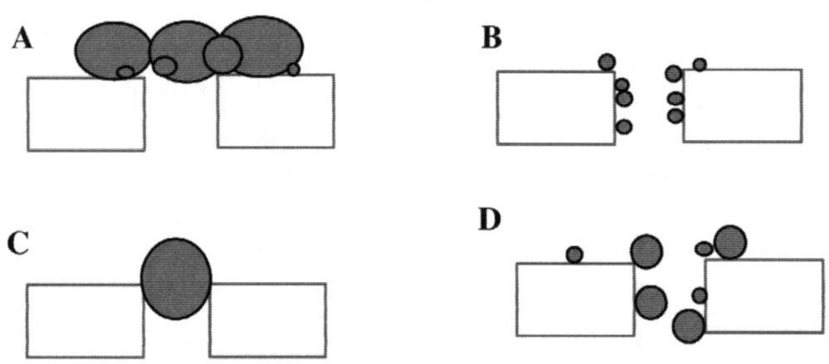

Figure 1. Schematic representation of filtration model: (A) cake filtration, (B) deep bed filtration, (C) complete blocking, and (D) intermediate blocking

In case A, the separation process is conducted with 40 wt% slurry that can form cake layer on the surface of filter. Case B explains the separation process using filter with relatively big pore sizes thus many particles will pass through to the permeate side. In case C, the filtration is done by using filter with small pore sizes, while case D using filter with pore sizes in between the pore sizes used in case 3 and case 1 or 2. All cases can also be explained by mathematical expressions as follows [10]:
Case A, cake filtration:
Filtration rate:

$$\frac{dV}{dt} = \frac{P}{\mu(AV+B)} \quad (1)$$

The correlation between V and t:

$$t = \frac{\mu}{P}\left(\frac{AV^2}{2} + BV\right) \quad (2)$$

Where V is volume of permeate, t is time, μ is viscosity, and P is trans membrane pressure. A and B are constants.

Case B, deep bed filtration:
Filtration rate:

$$\frac{dV}{dt} = \left(\frac{dV}{dt}\right)_o (1 - aV)^2 = \frac{(dV/dt)_o t}{(1+a(dV/dt)_o t)^2} \quad (3)$$

The correlation between V and t:

$$V = \frac{(dV/dt)_o t}{(1+a(dV/dt)_o t)} \quad (4)$$

Case C, complete blocking:
Filtration rate:

$$\frac{dV}{dt} = \left(\frac{dV}{dt}\right)_o (1 - aV) \quad (5)$$

The correlation between V and t:

$$V = \frac{1}{a}\left[1 - exp\left\{-a\left(\frac{dV}{dt}\right)_o t\right\}\right] \quad (6)$$

Case D, intermediate blocking:
Filtration rate:

$$\frac{dV}{dt} = \frac{(dV/dt)_o}{(1+At)} \quad (7)$$

The correlation between V and t:

$$V = \frac{\left(\frac{dV}{dt}\right)_o}{A} ln(1 + At) \quad (8)$$

The correlations or equations (1-8) will be employed in this research to explain and predict the flux decline behavior during microfiltration of used oil and used cooking oil.

2. Materials and methods

Used oils in this research were collected from local workshop in Surabaya, Indonesia. Used cooking oils were gathered from local food stall in Surabaya, Indonesia. Before employed for filtration process, the viscosities of oils were analysed using Brookfield RVT. The viscosity of used oil was around 200 cP, while the viscosity of used cooking oil was around 125 cP. Whatman Cellulose nitrate microfiltration membranes were kindly supplied by Sigma Aldrich with nominal pore sizes around 0.2 ſm. The flat sheet membranes were soaked in aquadest before filtration process to remove some preservatives.
The filtration processes were conducted in a dead-end filtration set up that is presented in Figure 2.

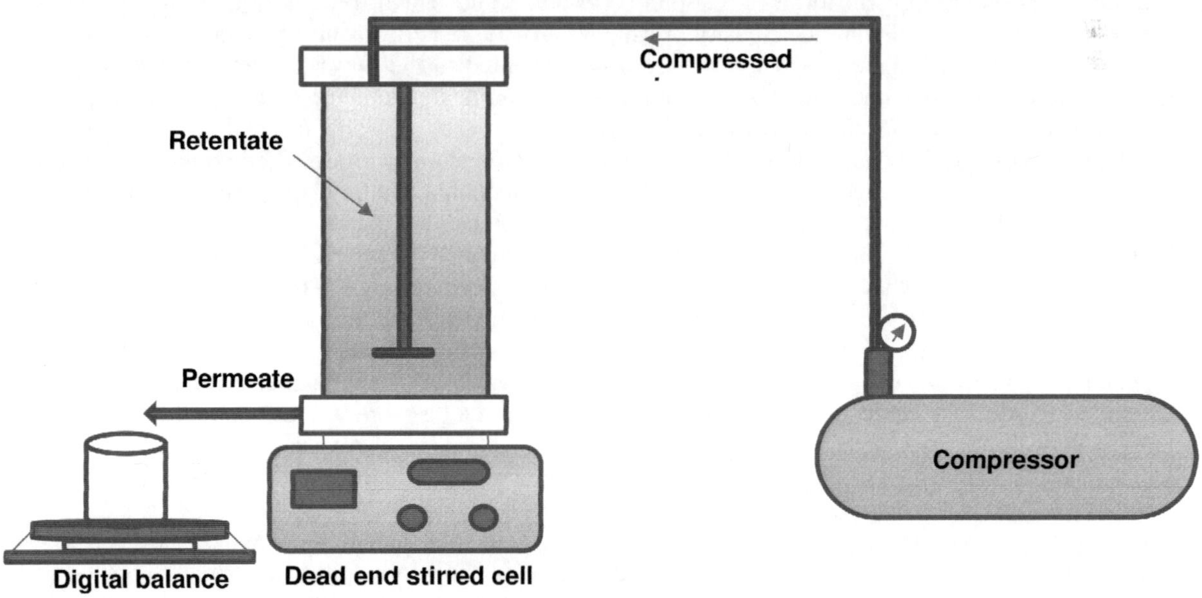

Figure 2. Experimental setup

The flat sheet membrane was placed inside the dead end stirred cell before it was contacted with feed emulsion. The emulsion was then poured into the cell and pressed with compressed air to induce feed pressure around 1 to 2 bar. During the filtration process, the permeate was collected inside a beaker glass and the flow rates of permeate were recorded. The flux of permeate through the membrane was then analysed and models based on equation (1-8) were generated. After each experiment, the membrane was dismantled and new membrane was then placed for the next experiment.

3. Results and discussions

3.1. *Flux – Time Profiles of Oil Emulsions Filtration*

In order to examine the influence of filtration time to the permeate flux, the relations of permeate flux and time for used oil and used cooking oil are presented in Figure 3.

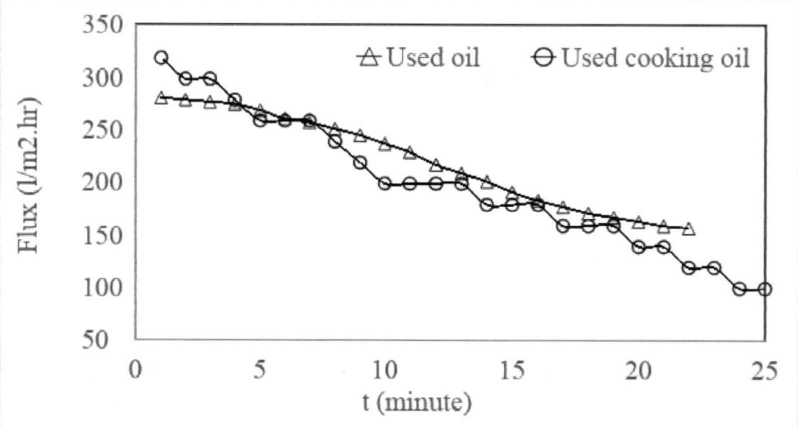

Figure 3. Permeate flux-time behavior during filtration of used oil and used cooking oil at feed pressure 1.2 bar.

The flux-time profile of used oil showed a relatively lower initial flux (~ 280 $l.m^{-2}.hr^{-1}$) compared to used cooking oil indicating the influence of solution viscosity and also the polarisation phenomenon. As mentioned in Section 2, the initial viscosity of used oil around 200 cP was higher than the viscosity of used cooking oil. This higher viscosity of used oil contributed to the lower movement of permeates through the membrane at the same trans membrane pressure. In terms of the rate of flux decline, the used cooking oil showed faster decrease in flux compared to used oil. This faster decline implies an accelerated polarisation concentration phenomenon during the filtration of used cooking oil. In short period after the experiment was conducted, there was non-blocking condition, and where the movement of the dispersed phase was only affected by the shear force generated on the surface of the membrane by the stirrer. Then, the flux behavior through the membranes will be determined by the formation of secondary membrane (dynamic membrane) on the surface of the membrane. This secondary membrane is always formed, whether automatically or by design from constituents in the feed. The formation of the secondary membrane will alter the performance of the primary membrane, in terms of flux and rejection. The formation of secondary membrane can be regarded as a beneficial effect, but this type of filtration behavior is usually associated with increasing filtration pressure, or decreasing filtration flux rates. Under these circumstances, the filtration is determined by the characteristics of the secondary membrane formed on the surface of the original membranes and, to some extent, the nature of the original membranes becomes less relevant.

Polarisation concentration will happen more severe if the feed solution contains more visible particles that can accumulate on the surface of the membranes. This phenomenon will hamper the utilisation of membrane on industrial scale because it will determine the life time of membranes. To extend the life time of membranes, polarisation concentration phenomenon needs to be handled and reduced. However, to select the most appropriate technique to reduce the polarisation concentration on the surface of the membrane, we need to examine the mechanism of pore blocking and cake formation on the surface of the membrane. Compared to literature values, the permeate flux from this research showed comparable value as presented in Table 1. The concentration of oil used in this research was much higher than other studies indicating the potentiality of the membranes in this research for O/W emulsion separation.

Table 1. The comparison of permeate flux from several studies

Membrane material	Oil concentration (ppm)	Permeate flux ($l.m^{-2}.hr^{-1}$)	Ref.
Polyacrylonitrile (PAN)	1,000	100 – 2,000	(12)
2 microns true surface filter	1,000	200	(13)
7.5 microns slotted pore filter	1,000	1,820	(8)
0.2 microns cellulose nitrate	> 1,000	280 - 320	This work

3.2. *Mathematical Modelling of Flux Decline During Filtration*

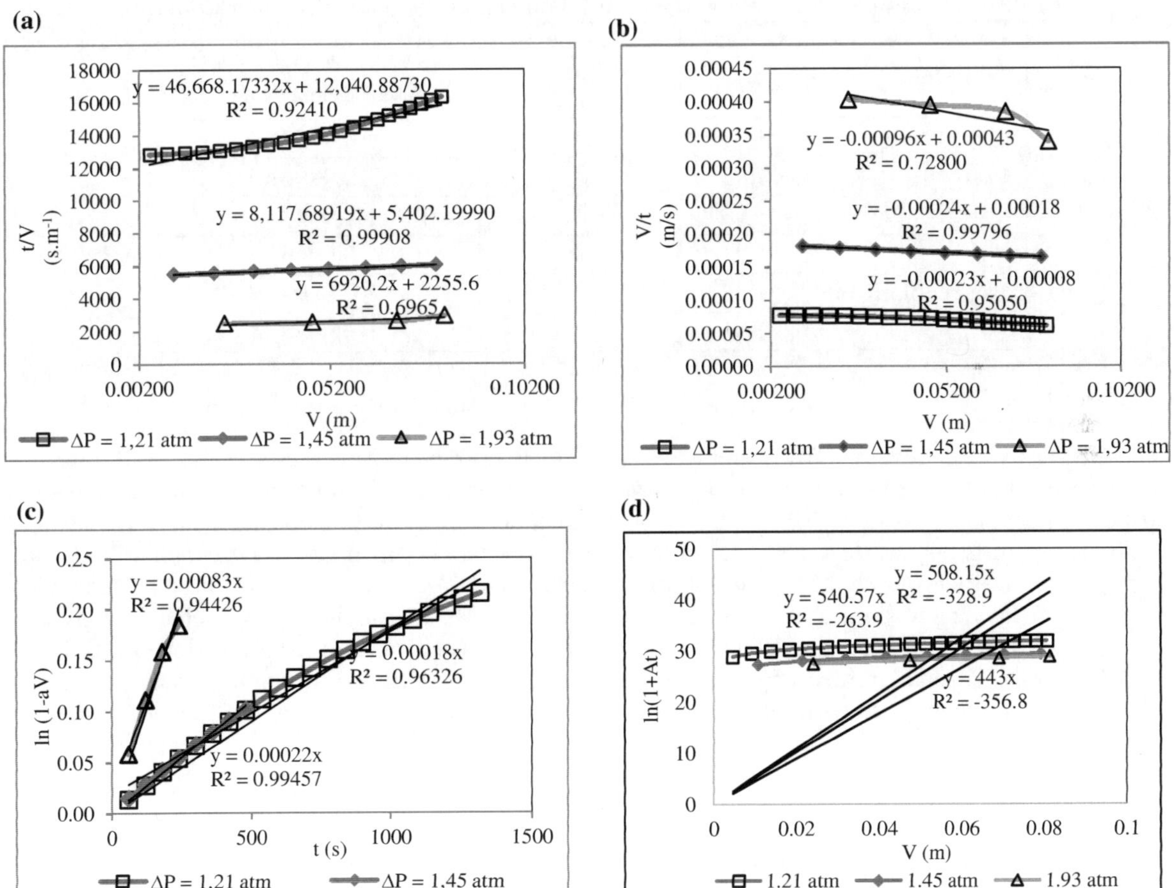

Figure 4. The results of mathematical modelling of (a) cake filtration, (b) deep bed filtration, (c) complete blocking, and (d) intermediate blocking during used oil filtration test

The correlations shown in Section 1 were used to explain the flux decline during the filtration of used oil and used cooking oil. Equation (2), (4), (6), and (8) were used and the results of the model for used oil are depicted in Figure 4.
According to the model, used oil filtration followed type 3 or complete blocking filtration model because the coefficient correlation (R^2) of that model showed the highest value at all feed pressure. Type 3 filtration model explains the blockage of membrane pores by large particles as shown in Figure 1C. Hence, the behavior of flux during used oil filtration was influenced by the presence of these large particles. The complete blocking of the pores by particles or dispersed phase could possibly decrease the flux through the membranes and the membrane can suffer from fouling phenomenon. In terms of feed pressure, the models suggest no influence of feed pressure on the flux behavior. As can be seen from Figure 4 the range of pressure used during the experiments need to be increased if one need to observe the influence of feed pressure on the flux behavior.
On the other hand, for used cooking oil, the flux behavior during experiment showed case 2 or deep bed filtration model as presented in Table 2. Deep bed filtration model explains the strong interaction between the wall of membrane pores and small particles, thus small particles will attach into the wall of membrane pores. However, this phenomenon also results in the decontamination of permeate by some small particles or dispersed phase.

Table 2. The results of modelling during used oil and used cooking oil microfiltration

Type of emulsion	⊗P (atm)	The type of filtration model	Correlation coefficient ,R^2
Used oil	1,21	3	0,9633
	1,45	3	0,9996
	1,93	3	0,9443
Used cooking oil	1,21	2	0,9966
	1,45	2	0,9902
	1,93	2	0,9638

4. Conclusions

The efforts to treat O/W and W/O emulsions play an important role to preserve the environment. Techniques such as membrane technology have been employed to separate the continuous and the dispersed phases of emulsions. Microfiltration with small pore size has been considered as a promising technique to treat oily waste water. However, its real application still needs to be explored and one particular concern is polarisation concentration phenomenon. In order to tackle this problem, the behavior of flux decline during filtration needs to be studied. This research was conducted to analysed and proposed filtration modeling of oily waste water using used oil and used cooking oil as feed emulsions. Experimental results suggested complete blocking and deep bed filtration behaviors for used oil and used cooking oil, respectively. The results of this research could bring an information to select suitable technique to reduce polarisation concentration in order to realise the implementation of the treatment of oily waste water by using microfiltration.

References

[1] Holdich R G, Cumming I W and Smith I D 1998 *J. Memb. Sci.* **143** (1-2), 263-74.
[2] Darvishzadeh T, Bhattarai B and Priezjev N V 2018 *J. Memb. Sci.* **563**, 610-16.
[3] Yang Y, Raza A, Banat F and Wang K 2018 *J. Water Proc. Eng.* **25,** 113-17.
[4] Tummons E N, Tarabara V V, Chew J W and Fane A G 2016 *J. Memb. Sci.* **500**, 211-24.
[5] C-Tepole V, Fuente E B and Sanchez B T 2017 *Chem. Eng. Res. Des.* **126**, 286-96.
[6] Hlavacek, M. 1995 *J. Memb. Sci.* **102**, 1-7.

[7] Kosvintsev S R, Sutrisna P D, Cumming I W, Holdich R G and Mason G 2007 *Chem. Eng. Res. Des.: Part A* **64**,32-44.
[8] Sutrisna P D, Holdich R G, Kosvintsev S R and Cumming I W 2005 *J. Appl. Memb. Sci. Tech.* **2**, 15-28.
[9] Ohya H, Kim J J, Chinen A, Aihara M, Semenova S I, Negishi Y, Mori O and Yasuda M 1998 *J. Memb. Sci.* **145** (1), 1-14.
[10] Tien C and Ramarao B V 2011 *J. Memb. Sci.* **383**, 17-25.
[11] Hermans P H and Bredée H L1936 *J. Soc. Chem. Ind.* **T55, 1**.
[12] Scharnagl N and Buschatz H 2001 *Desalination* **139** (1-3), 191-198.
[13] Holdich R G, Cumming I W, Kosvintsev S, Bromley A J and Stefonini G 2003 *Miner. Eng.* **16**, 121-128.

IOP Conf. Series: Materials Science and Engineering **543** (2019) 012080 doi:10.1088/1757-899X/543/1/012080

Effect of Activated Alkanolamine for CO_2 Absorption using Hollow Fiber Membrane Contactor

Y. Rahmawati*, S. Nurkhamidah, S. Susianto, N.I. Listiyana, M.A. Rahmatullah, and I.M. Hardian

Department of Chemical Engineering, Institut Teknologi Sepuluh Nopember, Kampus ITS Sukolilo, Surabaya 60111, Indonesia

*E-mail: rifqah_18des@chem-eng.its.ac.id

Abstract. Hollow Fiber Membrane Contactor (HFMC) technology has been widely developed as an alternative technology to separate CO_2 gas. HFMC technology combines gas absorption process with absorbent and membrane technology. Membrane contactor technology has advantages than conventional technology such as larger gas-liquid contact area, smaller equipment, modularity, and gas-liquid flow rates adjusted independently. In this research, the CO_2 absorption experiments was conducted for absorption process at 30°C temperature and various gas flow rate from 200 - 600 mL/min, feed gas of CO_2 20%vol. flowed into shell side and the absorbents flowed in tube side of membrane contactor. Piperazine (PZ) and monosodium glutamate (MSG) as activators with 1%w/w concentration, added into methyldiethanolamine (MDEA) 30% w/w solution to form aqueous solutions of activated MDEA. The membrane material used in this experiment was hydrophobic polypropylene membrane. The effect of liquid flow rate and various activators used to get CO_2 absorption flux and CO_2 removal in HFMC. The results showed that the best of CO_2 absorption process using activated alkanolamine was MDEA-PZ, where the highest flux value was 5.5×10^{-4} mole/m^2.s and CO_2 removal reached 96.9% at a gas flow rate of 600mL/min.

1. Introduction

Indonesia is a developing country with high population, the world's fourth most populous nation. With a large population, there are many jobs will be needed to support Indonesia's economic growth. The industry has been the fastest growing sector of the Indonesian economy since the late sixties. As the industry grows in Indonesia, it turns out give a bad impact on the environment. Waste gas from industry can cause air pollution. The content in industrial waste gas is CO_2, SOx, NOx, CO, and hydrocarbon with percentage composition as follows: 30% of CO_2 gas, 27% of CO gas, 25% of hydrocarbon gas, 10% of NOx gas, 9% of SOx gas, and 8% of particulate dust.

In recent years, CO_2 has attracted attentions because of increasing the emission of CO_2 as a determining greenhouse gas into the atmosphere. Fossil fuels, including petroleum, coal and natural gas are all non-renewable resources and contain high percentages of carbon. They are still the major sources of energy throughout the world. The emission of CO_2 into the atmosphere causing as the main reason for climate change effects including global warming, changes in sea levels, extreme hot summers and cold winters, and agricultural problems [1].

Several technologies has been developed to remove CO_2 from the gas streams, such as physical and chemical absorption with liquids especially aqueous amines solutions, solid adsorption, cryogenic distillation, membrane technology, and most recently membrane contactors [2]. However, in conventional absorption using packed column, CO_2 absorption is not efficient enough due to the flow rate of gas and absorbent has not yet separately each other so that can generate entrainment, flooding, loading and foaming [3]. On the other hand, cryogenic distillation needs a very huge installation and a very expensive operational cost. While adsorption process has a poor affectivities in CO_2 separation by using solid adsorbent [2].

In the recent time, conventional column absorption still becomes the best equipment despite needs great energy and big installation of equipment that depends on the other operational units. These shortages encourage some experiments towards the new contactor technology was expected to resolve the above problems.

Membrane contactor is a new method in contacting absorbent to CO_2 by using hollow fiber membrane contactor. This method has several advantages including wider liquid gas contact space, smaller tool, its modular characteristic which is easy to scale up , and the flow rate of gas and liquid which independently easy to be arranged; thus it can solve the operational issue [4]. Research on removing CO_2 by membrane contactors from natural gas has been carried out since 1980. Qi and Cussler [5] fabricated the first membrane for gas absorption process to remove CO_2 from a gas stream.

Typical membranes nowadays are prepared from polyethylene (PE), polypropylene (PP), polyvinyldenefluoride (PVDF), polytetrafluorethylene (PTFE) and polysulfone (PS) in gas-liquid contacting process. Currently hydrophobic membranes are widely used due to a larger contact area than the hydrophilic membranes. Among these, PTFE together with amine-based solutions are of high hydrophobicity, good mechanical properties and chemical stability [6]. Proper choice of absorbent plays an important role in determining the performance of gas liquid membrane contactor for CO_2 absorption. Various absorbent such as pure water, aqueous solution of NaOH, KOH, alkanolamines and ammonia have been studied experimentally. In this work PP membrane was used because it is more affordable and already have tested high temperature stability up to 100^0 C. Besides, the PP membrane already has tested for wetting phenomenon using alkanolamines solvents.

Many studies of CO_2 absorption-desorption with membrane contactors still use a single solvent without any mix of activators, such as Fang, et al. [7], Khaisri et al. [8] and Lv, et al. [9] who used MEA as a solvent. Then Mansourizadeh and Ismail [10], Karoor and Sirkar [11], Rahmawati et al. [12] used water, and several other studies using NaOH, MDEA, or DEA.

Recently, the blended alkanolamines or addition of activators have been done by a number of researches in order to improve the performance of CO_2 absorption. One of the previous studies, Yeon, et al. [13] was absorption using membrane contactor for absorber and desorber columns for regeneration of absorbents by mixing PZ and triethanolamine (TEA) as absorbents. The results showed that TEA with low absorption capacity can be increased by adding piperazine. TEA is a tertiary amine with high volatility that can help hollow fiber membranes to be maintained from the phenomenon of wetting.

In this work, PZ and mono sodium glutamate (MSG) as activators used into methyldiethanolamine (MDEA) solution to form aqueous solutions of activated MDEA. The activated mechanism were presented to explain the activation phenomenon. The experiment of CO_2 absorption was carried out in polypropylene hollow fiber membrane contactor (HFMC). The use of such amine solvent mixture going to increase CO_2 separation efficiency and reduce the cost of solvent usage so that it is more economical.

2. Materials and Methods

2.1. Materials

In this experiment, the material of membrane is a hydrophobic polypropylene membrane purchased from GDP Filter Indonesia membrane industry with its characteristics shown in Table 1.

Table 1 Specification of polypropylene hollow fiber membrane

Parameter	Specification
Inside Diameter (mm)	0.35
Outside Diameter (mm)	0.5
Pore Diameter (μm)	0.2
Fiber Length (mm)	300
Number of fibers	6500
Membrane porosity (%)	65
Membrane area (m^2)	1.3943

Absorbent used was MDEA, CO_2 gas cylinder with 20% volume and 80% volume N_2 balance, and activator used Piperazine (PZ) and Monosodium glutamate (MSG). In this study, concentration of absorbent used was 30% wt. with 1% wt. activator.

2.2. Apparatus and procedure

The experimental set-up for CO_2 absorption shown in Figure 1. On the absorption process, feed gas with 20% of CO_2 (*balance* N_2) was flowed with various flow rates for about 200-600 mL/min through the shell side of membrane contactor which measured using gas flow meter**.** The solvents was flowed using a diaphragm pump into the tube side of membrane contactor at flow rate of 100 mL/min which measured using liquid flow meter. Flow rate of gas and liquid was counter-current. The outlet gas from the membrane shell also called sales gas, while the outlet absorbent from the membrane tube also called rich amine.

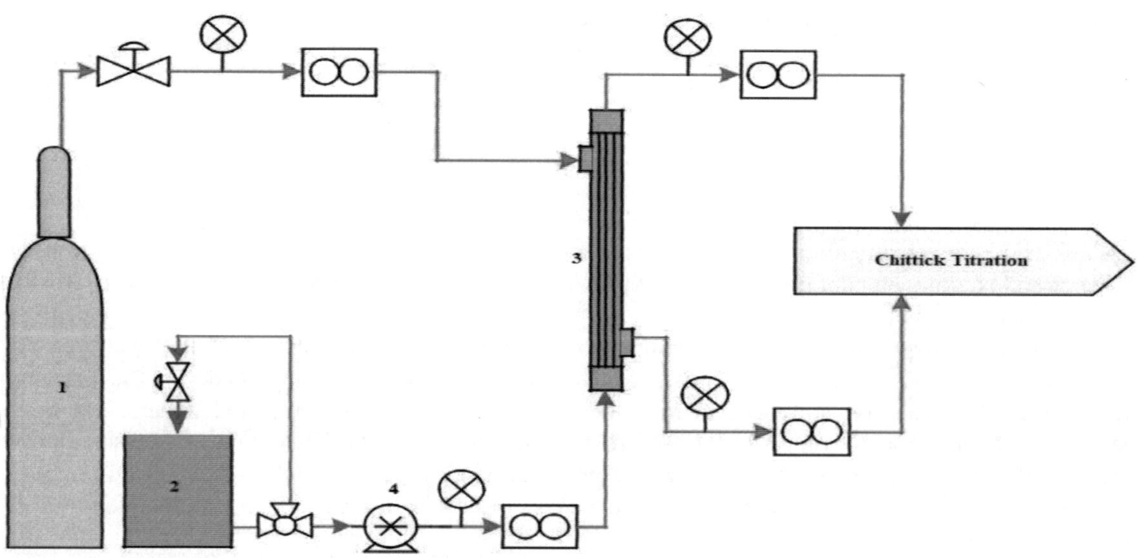

Figure 1. Experimental setup for CO_2 absorption. 1) Feed gas tube; 2) solvent tank; 3) HFMC; 4) liquid pump

CO_2 concentration in the solvent was analysed using chittick titration with 1 M HCl and indicator of methyl orange. While sales gas sample was taken and being bubbled into 1 M NaOH. Flow rate of sales

IOP Conf. Series: Materials Science and Engineering **543** (2019) 012080 doi:10.1088/1757-899X/543/1/012080

gas bubbled into NaOH was depend on the value stated in the flow meter of sales gas and it was bubbled for 10 minutes. NaOH contained CO_2 titrated using 1 M HCl and assisted with pp and methyl orange indicators.

2.3. Data analysis

CO_2 concentration in the gas and liquid phase at inlet and outlet of membrane contactor module measured by chittick titration. Amount of CO_2 in gas phase use to calculated flux absorption and CO_2 removal with equation as follow:

$$J_{CO_2} = \frac{Q_{im}\ x\ C_{in} - Q_{out}\ x\ C_{out}}{A} \tag{1}$$

Where J_{CO2} is the CO_2 flux (mole/m^2.s), Q_{in} and Q_{out} are input and output of gas flow rate (mL/min), C_{in} and C_{out} are CO_2 concentration in feed gas and sales gas (mole/m^3), and A is inner surface area of the hollow fiber membranes (m^2). And the CO_2 removal rate is an important measure for absorption performance of the system and absorbents is calculated as follow:

$$\eta = \frac{Q_{in} \times C_{in} - Q_{out} \times C_{out}}{Q_{in}\ x\ C_{in}} x100\% \tag{2}$$

CO2 concentration in the absorbent as CO_2 loading it is determined by equation from Zhang, et al. [1]:

$$\alpha = \frac{mol(CO_2)}{mol(solvent)} = \left[\frac{\dfrac{(V_{gas} - V_{HCl})(P)(273K)}{(101325\,Pa)(T)(22{,}4L/mol)}}{C_1 V_1} \right] \tag{3}$$

V_{gas} is volume change of saturated NaCl which is shown by measuring burette (mL), V_{HCl} is HCl volume that is required until the solvent sample changes in color (mL) ; from yellow to red, P is operational pressure (Pa), T is operational temperature (K), C1 is solvent concentrate (mole/L), and V_1 is the volume of solvent sample (mL).

3. Results and Discussion

3.1. Effect of feed gas flowrate on CO_2 absorption flux

The effect of feed gas flowrate on CO_2 absorption flux shown in Figure 2. The result shows that increasing of feed gas flowrate could enhance the CO_2 absorption flux for three kind of absorbents used. This can help explain to the reduction in gas boundary layer thickness due to the increase in gas flowrate. CO_2 mass transfer resistance will decrease due to reduction of gas boundary layer thickness and make it easier for CO_2 gas to diffuse from the bulk gas phase to the liquid phase [12]. Absorption occurs because of concentration difference between the gas phase and liquid phase. The feed of gas has higher CO_2 concentration will diffuse through the membrane pore and then dissolved into absorbent [14].

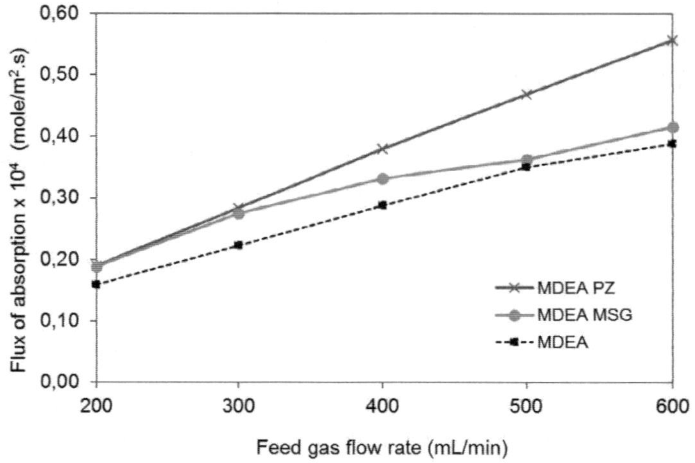

Figure 2. Effect of feed gas flowrate on CO_2 absorption flux in various solvent (CO_2: 20% vol, Q_{liq}= 100 mL/min)

Figure 2 shows that for the same operation conditions, the flux absorption with the MDEA, MDEA–MSG and MDEA–PZ solvents could arrive at the highest values of 3.8 x 10^{-5} mole.m^{-2}.s^{-1}; 4.1 x 10^{-5} mole.m^{-2}.s^{-1} and 5.5 x 10^{-5} mole.m^{-2}.s^{-1} respectively. The absorption flux of the MDEA-activated solutions are evidently better than that of the non-activated MDEA solution. Activators PZ and MSG, despite a little amount in the solutions, can effectively enhance absorption flux of membrane gas absorption. The activation of PZ is higher than that of MSG. It reveals that PZ is an efficient activator in membrane gas absorption for CO_2. PZ has a special molecular structure in which a symmetrical diamino cyclic structure exists. The symmetrical structure make PZ can easily binding the CO_2 molecule from absorbent.

3.2. Effect of Feed Gas Flowrate on CO_2 Removal

The effect of feed gas flowrate on CO_2 removal shown in Figure 3. Figure 3 shows that CO_2 removal decreased with the increase of feed gas flowrate for any given absorbent. At higher gas flowrate, the residence time of feed gas in the membrane contactor was reduced. Short residence time of feed gas in the membrane module will accelerate contact time between the gas and membrane surface and reduce the amount of diffused CO_2 from gas phase through the membrane pore to the liquid phase [15].

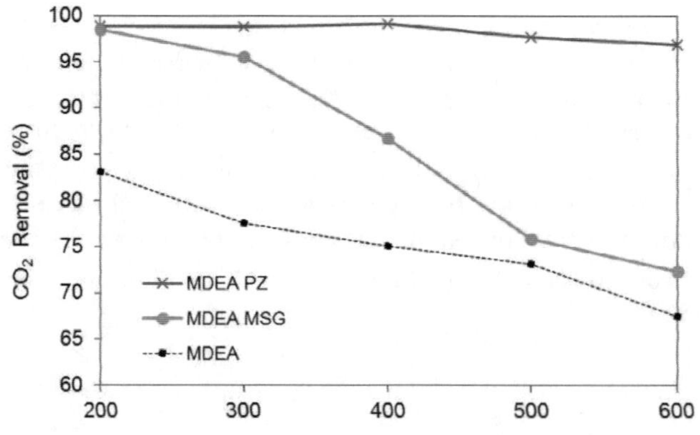

Figure 3. Effect of feed gas flow rate on CO_2 removal in various solvent (CO_2: 20% vol, Q_{liq}= 100 mL/min)

For the same operation conditions, the CO_2 removal with the MDEA, MDEA–MSG and MDEA–PZ solvents could arrive at the maximal values of 67%; 72%; and 96% respectively. PZ and MSG activators even in smaller amount can effectively increase CO_2 absorption performance in the membrane contactor.

3.3. Effect of Feed Gas Flowrate on CO_2 Loading

CO_2 *loading* defines as amount of mole CO_2 that is absorbed per mole absorbent. Figure 4 shows the effect of feed gas flowrate towards CO_2 *loading* on various kind of absorbent. The higher flowrate of the feed gas, the CO_2 *loading* value will be increase. The amount of CO_2 loading affected by the value of CO_2 absorption flux, the higher rate of CO_2 mass transfer from the surface of the membrane into the absorbent will shows the amount of CO_2 absorbed into the absorbent [9].

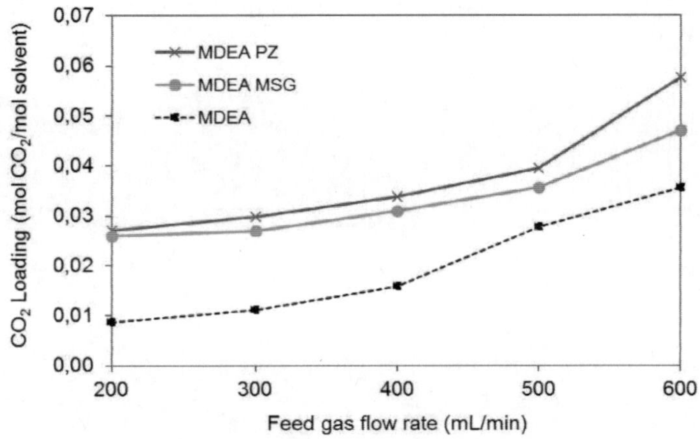

Figure 4. Effect of feed gas flow rate on CO_2 loading (CO_2: 20% vol, Q_{liq}= 100 mL/min)

Figure 4 shows that for the same operation conditions, the CO_2 loading with the MDEA, MDEA–MSG and MDEA–PZ solvents could arrive at the highest values of 0.036; 0.047; and 0.058 respectively. This value compared with the value of absorption flux of three various absorbent that MDEA-activated solutions are evidently better than that of the non-activated MDEA solution.

4. Conclusion

The best absorbent in the absorption process of CO_2 is activated MDEA. MDEA-PZ become the best absorbent for CO_2 absorption with comparison of separation efficiency 1.4 higher than unactivated MDEA absorbent. The highest absorption flux was obtained 5.5 x 10^{-5} mole.m-^{2}.s^{-1} with MDEA-PZ as absorbent and operation condition of gas flowrate 600 mL.min^{-1} and liquid flowrate 100 mL.min^{-1}. The highest CO_2 removal was 96% by utilizing MDEA-PZ solvent with operation condition of gas flowrate 200 mL.min^{-1} and liquid flowrate 100 mL.min^{-1}.

Acknowledgments

Research Grant sponsored by Directorate of Research and Community Services, Directorate General of Strengthening Research and Development, Ministry of Research Technology and Higher Education, Republic Indonesia (No. 128/SP2H/PTNBH/DRPM/2018) is gratefully acknowledged.

References

[1] Zhang Z E., Yan Y F, Zhang L, and Ju S X 2014. *Global NEST J.* **16**(2) 354-373
[2] Barbieri G, Brunetti A, Scura F, and Drioli E 2011. *Chem. Eng. Trans.* **24** 775 - 780
[3] Mansourizadeh A, and Ismail A F 2009 *J. Haz. Mat.* **171**(1–3) 38–53
[4] Kosaraju P, Kovvali S, Korikov and Sirkar K K 2005 *Ind. Eng. Chem. Res.* **44**(5) 1250–1258

[5] Qi Z, and Cussler, E L 1985 *J. Memb. Sci*, **23**(3) 321–332
[6] Clarizia G 2009 *Chem. Eng. Trans.* **17** 1675 - 1680
[7] Fang M, Wang Z, Yan S, Cen Q, and Luo Z 2012 *Int. J. Greenhouse Gas Cont.* **9** 507–521
[8] Khaisri S, deMontigny D, Tontiwachwuthikul P, and Jiraratananon R 2011 *J. Memb. Sci*, **376**, 110-118
[9] Lv Y, Yu X, Tung Tu S, Yan J, and Dahlquist E 2012 *Appl. Energ.* **97**: 283–88
[10] Mansourizadeh A, and Ismail A F 2011 *Desal.* **273** 386–390
[11] Karoor S and Sirkar K K 1993 *Ind. Eng. Chem. Res.* **32** 674–847
[12] Rahmawati Y, Nurkhamidah S, Susianto, Listiyana N I, and Putricahyani W 2017 *Amer. Inst. Phys. Conf. Proc.* **1840** 100009
[13] Yeon S H, Sea B, Park Y I., Lee K S and Lee K H 2004 *Separ. Sci. Technol.* **39**(14) 3281–3300
[14] Rahim N A, Ghasem N, and Al-Marzouqi M 2015 *J. Nat. Gas Sci. Eng.* **26** 108–117
[15] Listiyana N I, Rahmawati Y, Nurkhamidah S, Syahnur H R, and Zaelana Y 2018 *MATEC Web Conf.* **156** 08012

IOP Conf. Series: Materials Science and Engineering **543** (2019) 012081 doi:10.1088/1757-899X/543/1/012081

Lignocellulosic Properties of Coffee Pulp Waste After Alkaline Hydrogen Peroxide Treatment

L Pudjiastuti[1,*], T Iswanto[2], A Altway[2], E O Ningrum[1], T Widjaja[2]

[1] Chemical Industrial Engineering Department, Institut Teknologi Sepuluh Nopember, Sukolilo, Surabaya 60111, Indonesia
[2] Chemical Engineering Department, Institut Teknologi Sepuluh Nopember, Sukolilo, Surabaya 60111, Indonesia
lily_p@chem-eng.its.ac.id

Abstract. The effect of alkaline hydrogen peroxide (AHP) treatment on reducing lignin content in coffee pulp waste (CPW) was investigated to increase the digestibility and accessibility of cellulose during production of valuable products by enzymatic and biological process. Hydrogen peroxide (H_2O_2) at various concentrations (2.5, 5, and 7.5% (w/w of substrate)) was mixed into 10% (w/v) of CPW-NaOH solution at 35 °C for 24 h with pH 11.5. The concentration of lignocellulosic content was investigated. The characteristic of crystallinity index, functional groups, and surface morphology of CPW at before and after the treatment have been observed using XRD, FTIR, and SEM analysis, respectively. The results showed that the highest lignin removal was achieved at the addition of 7.5% H_2O_2 which decreased from 16.85% to 8.5%. The concentration of cellulose and hemicellulose in CPW increased from 53% to 64% and 11% to 24%, respectively. The crystallinity index of AHP-treated CPW has increased from 30.83% to 42.08%. A surface morphological destruction on the CPW was getting worse as along with the increase of H_2O_2 concentration. This would be useful to increase the porosity of biomass to give cellulose accessibility by enzyme for further processes.

1. Introduction

In Indonesia, CPW is reported to have big potential to be converted as valuable products due to its abundant availability with the potential amount of 30-50% of total weight of coffee production by 686 x 10^6 kg each year [1]. Moreover, it contains a high cellulose compound, about 50-63%, which can be utilized as the feedstock of reducing sugar and its derivative products, such as ethanol, butanol, methane, etc. Its utilization is important because the abundant untreated CPW in the environment will release the xenobiotics, such as caffeine, tannins, and free polyphenols, as major pollutant agent on water and soil that toxic to saprophytic microorganisms which take important biotransformation within the environment. This causes disorderliness in environmental stability [2]. Furthermore, uncontrolled emission of CH_4, N_2O, and CO_2 from the open composting area will increase the negative effect of global warming [3].

Lignin in CPW needs to be removed before a further process to increase the accessibility of cellulose by cellulosic enzyme during the enzymatic or biological process. Therefore, delignification takes an important role to obtain the higher converted product. In this case, chemical treatment using alkaline has been a favored technique for lignin degradation than acid, ammonia, liquid hot water, and pH controlled hot water because it can be applied in lower pressures and temperatures compared to other ones. Alkaline treatment is generally more effective even if the treatment is carried out at ambient conditions [4]. Recently, AHP treatment, which combine the advantages of sodium hydroxide and hydrogen peroxidase, has been widely applied in reducing lignin of some biomass such as bamboo [5], black liquor residue of poplar wood soda pulp [6], banana pseudostem [7], Douglas fir [8], corncob [9], empty palm fruit bunch [4], rice straw [10] etc. Reference [11] reported that AHP treatment could selectively remove lignin and deconstructs the cell walls which is useful for significant digestibility enhancement of biomass. Besides, hydrogen peroxide is environmental friendly for a cleaner process [12].

Although AHP treatment had been applied for much lignocellulosic feedstock, but not yet for coffee pulp waste. In this present study, the effects of H_2O_2 concentration on lignin and other lignocellulosic contents in CPW after the treatment have been investigated. As well as comparing the change in the characteristic of CPW involve cellulose crystallinity index, its functional groups and surface morphology induced by treatment.

2. Material and Methods

2.1. Materials

Coffee pulp waste was collected from Malang, Indonesia. It was dried under sun radiation for several days until a weight loss of coffee pulp about 50% and milled in a disk mill to pass through a 100-mesh screen. Then, it was dried in the oven at 60 °C until constant weight. The initial lignocellulosic composition of the CPW was observed. It contained 53% (dry basis) of cellulose, 11% of hemicellulose, 16.85% of lignin, 5% of moisture content, and 14.15% of other chemicals.

NaOH and 30% H_2O_2, glacial acetic acid, sodium bicarbonate, and concentrated sulfuric acid were purchased from Merck. Sodium chlorite ($NaClO_2$) was purchased from Sigma Aldrich. The other chemicals used were of analytical grades.

2.2. Treatment Procedure

Ten grams of coffee pulp waste was introduced into Erlenmeyer flask containing 90 ml of 1% (w/w) NaOH solution that made the solution volume approximately 100 ml. Then, 30% H_2O_2 with the following dosage; 2.5, 5, or 7.5% (w/w of substrate) was added to that flask. The pH of the solution was adjusted to 11.5 using NaOH 6 M. Then, they were placed in an incubator with a rotary shaker at 125 rpm at 35 °C for 24 h. Liquid and solid fraction from the Erlenmeyer flask were drawn at the certain interval of treatment time and separated using a Whatmann filter paper no. 41 on a vacuum filtration. The solid samples were obtained from the solid fraction after washing by distilled water until neutral pH and drying at 60°C in an oven. The treatments were performed in duplicates.

2.3. Analytical Methods

Lignocellulosic content in CPW (before and after treatment) was analyzed. Insoluble lignin was determined by TAPPI T222 om-02 method. Determination of polysaccharides in CPW solid samples was carried out based on the isolation of holocellulose which contains cellulose and hemicellulose using sodium chlorite ($NaClO_2$), glacial acetic acid and acetate buffer. To determine the cellulose content, the obtained holocellulose was hydrolyzed with dilute hydrochloric acid to remove hemicellulose at boiling temperature using a reflux condenser as described by [13].

Cellulose crystallinity index (CrI) of the samples was scanned using X-ray Powder Diffraction (XRD, PANalytical X'pert Pro) from 5° to 50° of 2θ and calculated by the equation (1) [14]:

$$CrI = \frac{(I_{002} - I_{am})}{I_{002}} \times 100\% \tag{1}$$

where I_{002} is the maximum intensity of the diffraction peak of 002 in 2θ between values of 22-23, and I_{am} is a minimum intensity of the peak between the values of 18-19 [15]. FTIR spectra were recorded to characterize the major functional groups of CPW's samples. Surface morphology of CPW samples was captured using a scanning electron microscope (SEM) after they were coated using Pb-AU. That analysis was applied to the sample at before and after the treatment process.

IOP Conf. Series: Materials Science and Engineering **543** (2019) 012081 doi:10.1088/1757-899X/543/1/012081

3. Results and Discussion

3.1. Lignocellulosic Content of CPW after AHP Treatment

Lignin can become a physical barrier and restricting the enzyme access to cellulose due to undertaking non-productive binding with the lignin, which could decrease the efficiency of enzyme performance. [4] have reported that AHP treatment could give a significant reduction in lignin content as compared to the CPW without treatment. According to the results (see Figure 1a), a similar trend of lignin degradation has been achieved in CPW after treatment which decreases along with the higher concentration of H_2O_2 and longer duration time. The different delignification rate was obtained in each concentration and being slower after 6 h and finally, exhibit a significant gap in lignin removal at 24 h of treatment time. As revealed in Figure 1a, the highest lignin removal is successfully achieved at 7.5% H_2O_2 for 24 h, decreasing from 16.85% to 8.5%.

In Figure 1b, the concentration of cellulose and hemicellulose in CPW increase during AHP treatment due to the lignin content reduction in the solid fraction sample. The same result was obtained by [7] who found that the lignin content in treated banana pseudostem using alkaline and peroxide have decreased the lignin content from 17.26% to 7.65% and 7.17%, respectively and increased cellulose content from 60.84 to 75.48 and 74.37%, respectively. In this study, the highest increment of cellulose and hemicellulose content were retrieved from treated CPW using 7.5% H_2O_2 for 24 h which increased from 53% to 64% and 11% to 24%, respectively. The increase in hemicellulose in AHP-treated samples was probably caused by relatively unassociated hemicellulose with lignin and relatively unchanged "macromolecular" structure of the hemicellulose itself [10].

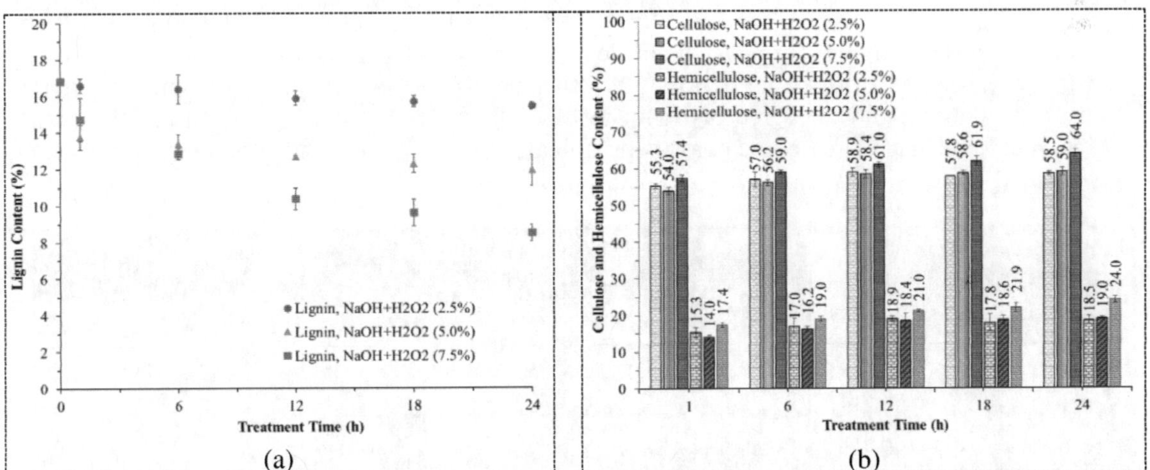

Figure 1. (a) Lignin content, (b) Cellulose and hemicellulose content, in AHP-treated CPW for 24 h

3.2. Characterization of CPW Samples after AHP Treatment

3.2.1. FTIR Analysis of CPW Treated by AHP

The X-ray diffraction of the CPW before and after AHP treatment has been investigated and the crystallinity index was calculated by equation (1), as shown in Table 1.

In Table 1, the crystallinity index value of AHP-treated solid samples is higher than CPW without treatment. AHP-treated samples for 24 h have a significant increase of the CrI from 30.83% to 31-43.68% after treatment. The increase of CrI was mainly due to the removal of amorphous part of lignocellulose which was indicated by weight loss of total solid of CPW after AHP treatment in an average of 4 g per 10 g of initial solid sample. This result was in line with [16] who reported that AHP-treated bamboo has relatively higher crystallinity, due to the dissolution of lignin and in part of hemicellulose. He found that the acid-soluble lignin, organic solvent extractives, and ash were effectively removed by AHP, except the acid-insoluble lignin and holocellulose. Moreover, [9] also

IOP Conf. Series: Materials Science and Engineering **543** (2019) 012081 doi:10.1088/1757-899X/543/1/012081

found that the CrI of APH-pretreated corncob was higher than that of the raw material due to the growth in cellulose content in that solid sample.

Table 1. Crystallinity index of CPW without and with treatment by alkaline hydrogen peroxide

Variable	CrI (%)
Untreated CPW	30.83
Treated CPW by $NaOH+H_2O_2$ 2.5%, 6 h	31.22
Treated CPW by $NaOH+H_2O_2$ 5.0%, 6 h	36.16
Treated CPW by $NaOH+H_2O_2$ 7.5%, 6 h	30.95
Treated CPW by $NaOH+H_2O_2$ 2.5%, 24 h	31.49
Treated CPW by $NaOH+H_2O_2$ 5.0%, 24 h	43.68
Treated CPW by $NaOH+H_2O_2$ 7.5%, 24 h	42.08

3.2.2. Surface Morphology of CPW

Figure 2 shows the surface morphology of AHP-treated CPW after 24 h compared to untreated CPW. At glance, there is a significantly different between each other where untreated CPW is looked more compact than others which are destroyed. The destruction on the CPW surface is getting worse as more H_2O_2 concentrations were added. This reveals by Figure 2b which has wrinkles in its surface while Figure 2b and 2c are fully ripped. Theoretically, NaOH and hydrogen peroxide (under proper pH, 11.5-12) do the synergistic activities during the lignin degradation. NaOH breaks the ester bonds cross-linking lignin [4], while hydrogen peroxide destroys the ether bonds between lignin and hemicellulose and decreases the polymerization degree of lignin in the cell wall. That condition leads to the partial dissolution of lignin and increases the porosity of biomass [17]. The increasing of porosity would useful to give cellulose accessibility by enzyme or chemicals for further processes.

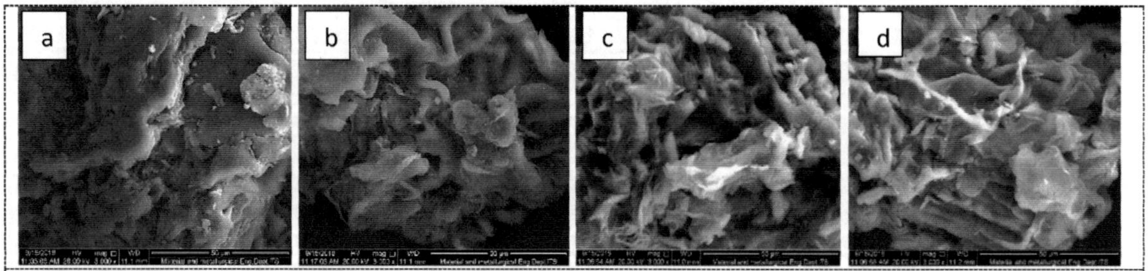

Figure 2. Surface Morphology of CPW, a) without treatment, b) treatment $NaOH+H_2O_2$ 2.5%, c) treatment $NaOH+H_2O_2$ 5%, and d) treatment $NaOH+H_2O_2$ 7.5%.

3.2.3. FTIR Analysis of CPW Treated by AHP

The FT-IR spectra of functional groups in solid CPW before and after treatment using various concentration of H_2O_2 for 6 and 24 h of pretreatment time are shown in Figure 3. The peak at 1025-1027 cm^{-1} related to methoxy groups. The signal at 1158-1163 and 890-898 cm^{-1} are the representation of C-O-C stretching at the b-(1–4) glycosidic bounds [18]. The absorption at 1235-1246 cm^{-1} and 1410-1422 cm^{-1} indicated the C=C and C=O stretching vibrations and C-H in-plane deformation, respectively. The peaks at 1315 cm^{-1} are due to symmetric CH_2. The signals at 1515 cm^{-1} have resulted from the aromatic skeletal vibration from lignin. An aromatic ring vibration is observed at 1615-1624 cm^{-1}. The peak at 1726-1729 cm^{-1} 1 is relevant to ester group of hemicellulose and lignin. The peaks at 2850 and 2919 cm^{-1} are from C-H stretching vibrations in lignin. The peaks at 3289-3321 cm^{-1} are due to the stretching of OH groups, as the representation of phenolic hydroxyl groups [6]. These results show that the AHP treatment under-investigated conditions could not eliminate most functional groups in the lignin of CPW.

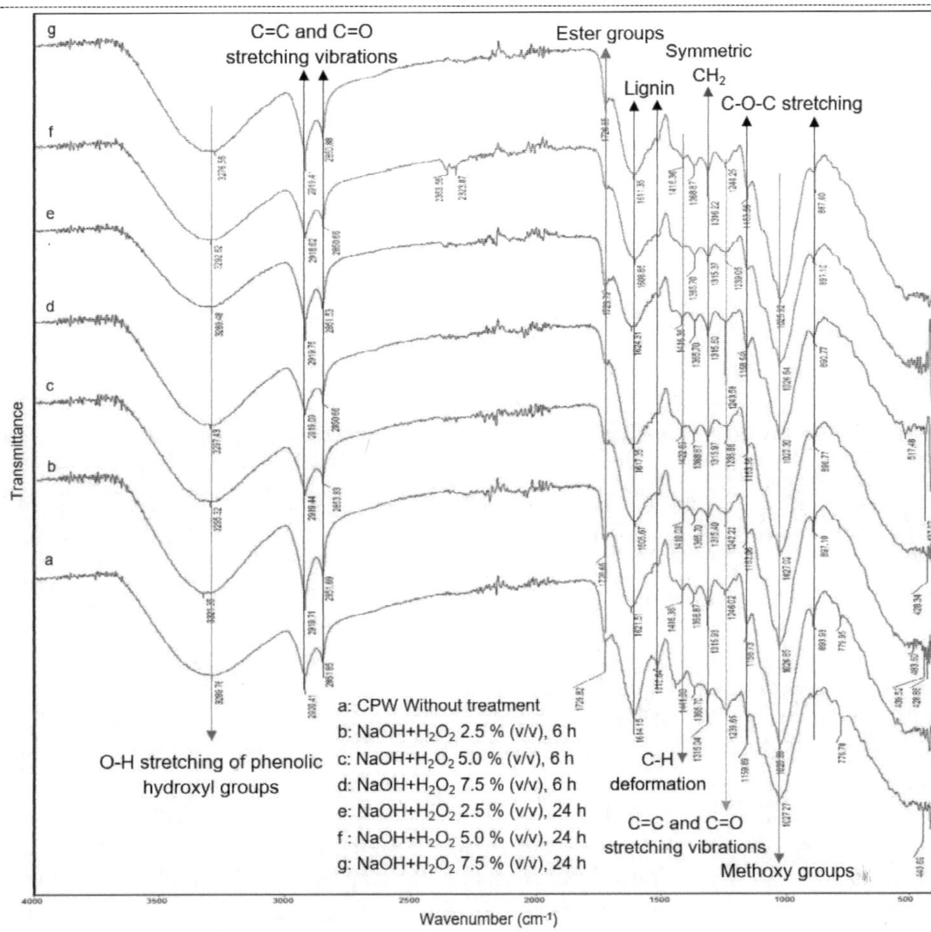

Figure 3. FTIR spectra of solid CPW before and after AHP treatment for 6 and 24 h

4. Conclusion

The effect of AHP treatment on reducing lignin content of coffee pulp waste (CPW) was conducted using the different concentrations of H_2O_2 (2.5, 5, and 7.5% (w/w of substrate)) which was mixed into 10% (w/w) of CPW solution at 35 °C for 24 h. The pH of the solution was 11.5, adjusted by NaOH. After AHP treatment for 24 h, the highest lignin removal is achieved at 7.5% H_2O_2 which decreased from 16.83% to 8.5%. In contrast with lignin removal, the concentration of cellulose and hemicellulose in CPW increased during the treatment. The highest increment of cellulose and hemicellulose content were obtained from treated CPW using 7.5% H_2O_2 for 24 h which increased from 53% to 64% and 11% to 24%, respectively. There was a significant increase in the crystallinity index of treated CPW. FTIR analysis has shown that AHP treatment could not completely eliminate most functional groups in the lignin of CPW. The captured surface morphology by SEM revealed that the destruction on the CPW's surface was getting worse along with more H_2O_2 concentrations addition into the system. This would be useful to increase the porosity of biomass to give cellulose accessibility by enzyme or chemicals for further processes. For further application, hopefully, this study can provide an overview of the process improvement in CPW pretreatment using AHP, especially in reducing the inhibitory content in CPW.

Acknowledgments
Authors would like to thank the Research and Community Service Foundation of Institut Teknologi Sepuluh Nopember (LPPM-ITS) for the financial support by Local Research Fund of ITS with contract number of 133/PKS/ITS/2018.

References
[1] Widjaja T, Iswanto T, Altway A, Shovitri M and Sri J R 2017 Methane production from coffee pulp by microorganism of rumen fluid and cow dung in co-digestion *Chem. Eng. Trans.* **56** 1465–70
[2] Ibrahim S, Shukor M Y, Syed M A, Rahman N A A, Khalil K A, Khalid A and Ahmad A 2014 Bacterial degradation of caffeine: a review *Asian J. Plant Biol.* **2** 18–27
[3] Corro G, Pal U and Cebada S 2014 Enhanced biogas production from coffee pulp through deligninocellulosic photocatalytic pretreatment *Energy Sci. Eng.* **2** 177–87
[4] Misson M, Haron R, Ahmad M F, Aishah N and Amin S 2009 Pretreatment of empty palm fruit bunch for lignin degradation *J. Teknol.* **50** 89–98
[5] Yuan Z, Wen Y and Kapu N. S 2018 Ethanol production from bamboo using mild alkaline pre-extraction followed by alkaline hydrogen peroxide pretreatment *Bioresour. Technol.* **247** 242–49
[6] Cheng Y, Zhao P X, Alma M H, Sun D F, Li R and Jiang J X 2016 Improvement of direct liquefaction of technical alkaline lignin pretreated by alkaline hydrogen peroxide *J. Anal. Appl. Pyrolysis* **122** 277–81
[7] Shimizu F. L and *et al.* 2018 Acid, alkali and peroxide pretreatments increase the cellulose accessibility and glucose yield of banana pseudostem *Ind. Crops Prod.* **115** 62–68
[8] Alvarez-Vasco C and Zhang X 2013 Alkaline hydrogen peroxide pretreatment of softwood: hemicellulose degradation pathways *Bioresour. Technol.* **150** 321–27
[9] Su Y, Du R, Guo H, Cao M, Wu Q, Su R, Qi W and He Z 2015 Fractional pretreatment of lignocellulose by alkaline hydrogen peroxide: characterization of its major components *Food Bioprod. Process.* **94** 322–30
[10] Sun R, Tomkinson J, Mao F C and Sun X F 2001 Physicochemical characterization of lignins from rice straw by hydrogen peroxide treatment *J. Appl. Polym. Sci.* **79** 719–32
[11] Banerjee G, Car S, Scott-Craig J S, Hodge D B and Walton J D 2011 Alkaline peroxide pretreatment of corn stover: effects of biomass, peroxide, and enzyme loading and composition on yields of glucose and xylose *Biotechnol. Biofuels* **4** 16
[12] Fidalgo J L G *et al.* 2018 Neural network modeling for prediction of oxidized lignin content by delignification of sugarcane bagasse through hydrogen peroxide with RAMAN spectroscopy data *Chem. Eng. Trans.* **65** 517–22
[13] Ioelovich M 2015 Methods for determination of chemical composition of plant biomass *J. Sci. Isr. Advantages* **17** 208–14
[14] Segal L, Creely J J, Martin A E and Conrad C M 1962 An empirical method for estimating the degree of crystallinity of native cellulose using the x-ray diffractometer *Tex. Res. J.* **29** 1962.
[15] Barakat A, Monlau F, Solhy A and Carrere H 2015 Mechanical dissociation and fragmentation of lignocellulosic biomass: effect of initial moisture, biochemical and structural proprieties on energy requirement *Appl. Energy* **142** 240–46
[16] Song X, Jiang Y, Rong X, Wei W, Wang S and Nie S 2016 Surface characterization and chemical analysis of bamboo substrates pretreated by alkali hydrogen peroxide *Bioresour. Technol.* **216** 1098–01
[17] Silverstein R A, Chen Y, Sharma-Shivappa R R, Boyette M D and Osborne J A 2007 comparison of chemical pretreatment methods for improving saccharification of cotton stalks *Bioresour. Technol.* **98** 3000–11
[18] Li M F, Yu P, Li S X, Wu X F, Xiao X and Bian J 2017 Sequential two-step fractionation of lignocellulose with formic acid organosolv followed by alkaline hydrogen peroxide under mild conditions to prepare easily saccharified cellulose and value-added lignin *Energy Convers. Manag.* **148** 1426–37

IOP Conf. Series: Materials Science and Engineering **543** (2019) 012082 doi:10.1088/1757-899X/543/1/012082

The Extraction of Natural Dyes from Henna Leaves (Lawsonia Inermis L.) by Ultrasound-assisted Method

L Qadariyah, N Azizah, A Q Syafa'atullah, D S Bhuana, M Mahfud
Chemical Engineering Department, FTI-ITS, Keputih, Sukolilo, Surabaya, Jawa Timur 60111, Indonesia

Email: lailatul_2008@yahoo.com

Abstract. One alternative to reduce the effect of water pollution from the textile industry is to replace the use of synthetic dye with the natural one. Henna leaves are one of the natural dye sources which has lawsone, a unique substance that produces orange colour. Some commonly used conventional extractions are soxhletation and maceration. One extraction method that is tried to be developed in this study is the Ultrasound-assisted Extraction. The effects of feed ratio to solvent, time, temperature and solvent types were investigated and discussed in this study. The highest yield from the extraction of Henna leaves was found to be 17.96% which was achieved at pH 1 with the extraction time of 10 minutes, feed/solvent ratio of 0.02, and aquadest solvent. Based on the both qualitative and quantitative analysis of the extracted colour, the ultrasound assisted method was found to be more effective and efficient than the soxhletation method.

1. Introduction

Textile demand increases with increasing appetite for colour. Most of the colouring process for the textile industries use synthetic dyes. The advantages of synthetic dyes include wider colour variations, more practical, easier to use, as well as more economical. However, synthetic dyes also have disadvantages, one of which is the waste generated after the staining process. The synthetic dye waste is harmful to aquatic ecosystems because it may contain chemicals or even heavy metals [1]. In September 2012, there was a terrible pollution in Pekalongan. The laboratory test samples from the river that passes through the Karangjompo Village, Tirto, Pekalongan showed that it contained high ammonia. It can cause water hardness that can severely affect health. Therefore, the need for alternative method to reduce the impact of textile industry waste has arisen, and replacing synthetic dye with the natural one is seems to be one of the best approaches [2]. Henna leaves contain lawsone dye that can be extracted as yellow and orange crystals. This dye can be used as a dye for skin, nail, hair, silk, and wool. Extracting the lawsone from the Henna leaves can be carried out by the mass transfer of dyestuff from solid to liquid phase (solvent) [3]. This method of extraction is commonly called as solid-liquid extraction (leaching). Some conventional methods that are commonly used for extraction include maceration, soxhletation, and reflux methods. These conventional methods have several weaknesses, such as abundant requirement of solvent, lower in yield and longer extraction time. It is necessary to using an extraction technique that safe for chemical compounds and solvent with high efficient energy such as ultrasound extraction (De Oliveira et al., 2018). Ultrasound assisted extraction is simple method and eco-friendly than conventional methods [4]. Therefore, in this study the extraction method with ultrasonic waves or Ultrasound Assisted Extraction (UAE) is to be developed in order to optimize the extraction process.

2. Methods

2.1. Material and Equipment

The materials used in this experiment include Henna in powder form, aquadest, ethanol, CH_3COOH, and NaOH. The Henna was imported from Saudi Arabia. The equipments used in this experiment are as follow: three neck flask with flatbase, beaker glass, analytical balance, pH meter, thermometer, water jet pump, filter paper and ultrasonic cleaner

2.2. Optimation of extraction condition

Various trials were carried out for the extraction of natural dye from Henna leaves in aquadest solutions, ethanol 20%, ethanol 40%, ethanol 60%, ethanol 80%, and ethanol 96%. The ratios of feed to solvent were 0.01, 0.02, 0.04, 0.06, 0.08, and 0.1. The extractions were carried out at various temperature and time durations as well, i.e. at 30°C, 40°C, 50°C, and 60°C for 5 min, 10 min, 20 min, 30 min, and 60 min. As far as to obtain the optimum pH, the extractions were carried out at pH 1, pH 4, pH 8, and pH 10.

2.3. Ultrasound assisted extraction

100 ml of solvent with Henna powder in it was placed in the ultrasonic bath and was extracted by the ultrasound for 10 minutes. The extraction time was set up and the frequency of the ultrasonic bath was set up at 40 kHz. Then solvent was then separated from the extract using water jet pump and a filter paper on top of it. The solid was separated and sticked on the filter paper. The solvent became clear from solid and the color of the solvent was orange. CH_3COOH or NaOH was then added into it to control the pH of this extract solution. The lawsone contained in the solvent made the solvent turn orange and was separated with the solvent by heating. It was heated at 80°C using waterbath until only the dye extract left.

2.4. Soxhelation

500 ml of aquadest and ethanol 60% were prepared and placed in a one neck flask roundbed flat of 1000 ml. Ten grams of Henna powder was prepared and put in the filter paper which was placed in the soxhlet accordingly. It was heated at 80°C for ethanol 60 % and at 110°C for aquadest. The extraction occured until the solvent in the soxhlet become clear or colorless. The soxhletation extraction was counted to be one cycle when the solvent filled in soxhlet and then turned back to the one neck flask roundbed

3. Results and discussion

3.1. Ethanol concentration effect

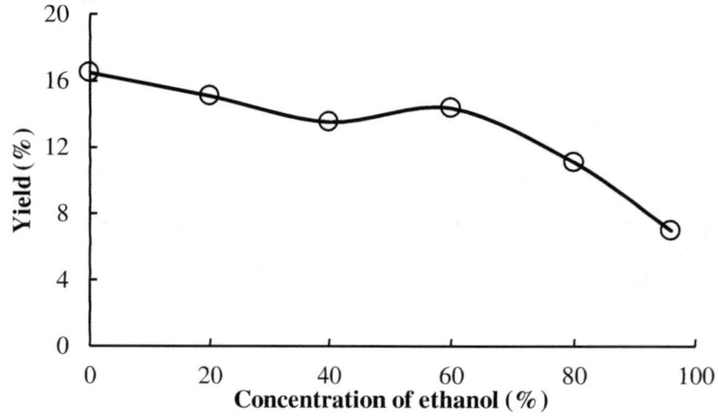

Figure 1. The effect of ethanol concentration (w/w) on the yield of henna extract (w/w) at 10 minutes and at temperature solvent 30°C

The extraction with the UAE method at various concentrations of ethanol is shown in Figure 1. showed optimum results when aquadest was used as a solvent. When higher ethanol concentration was used, the obtained extract was not as abundant. However, at a concentration of 60% ethanol there was an increase in the extraction yield. In other words, 60% ethanol solvent gave higher yield compared to that with 40% ethanol solvent. it was found using 70 % ethanol as solvent give highest rate of yield extracting colorant from sorghum husk [5]. Furthermore, if the ethanol contains more water then the solvent polarity index becomes higher and increases the efficiency of the extraction process [6]. The water polarity index is 9.0 whereas ethanol is 5.2. So it can be said that water is more polar than ethanol [7]. The aquadest solvent was found to give 16.47% yield, which was higher than 60% ethanol's 14.36% yield

3.2. Effect of Extraction Time

Figure 2 shows the effect of extraction time on the yield of Henna extract (w/w) at ratio feed to solvent 0.02 and at temperature 30°C. The extraction with UAE method using 60% ethanol solvent and aquadest gave the optimum yield at the extraction time of 10 minutes. At the extraction time between 5 minutes and 10 minutes, the yield has greatly increased but after the extraction time of 10 minutes the extract yield started to decrease and finally started to stay constant from 30 minutes onwards. This point indicates that the solvent binding capacity of the extract reaches maximum (saturated). The water content of the solvent causes the swelling of the material so that the contact area between the plant matrix and the solvent becomes larger, thus, causing the extraction results to become larger accordingly [8].

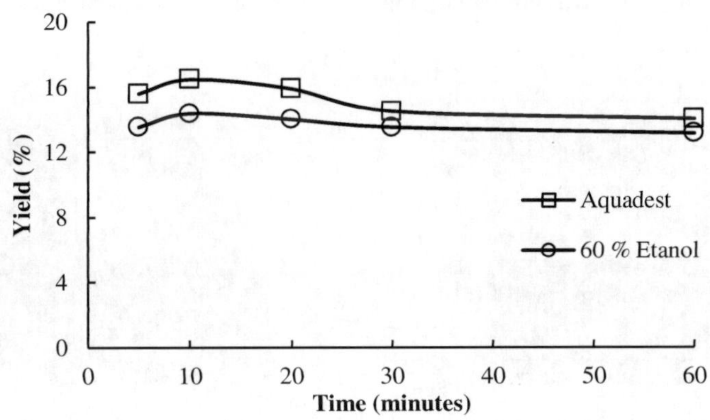

Figure 2. The effect of extraction time on the yield of henna extract (w/w) at ratio feed to solvent 0.02 and at temperature 30°C

3.3 Effect of of Ratio Feed to Solvent

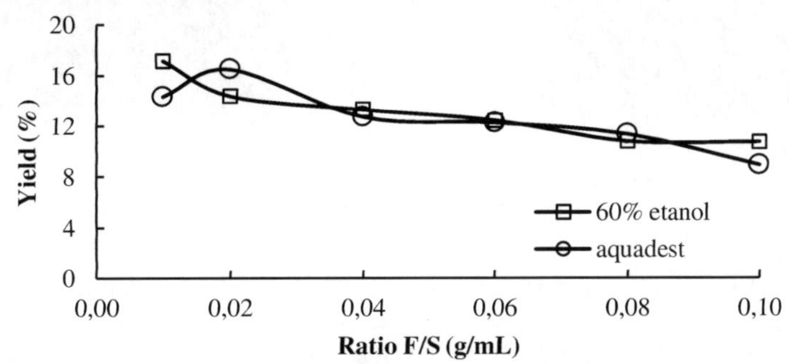

Figure 3. The effect of ratio extraction (F/S) on the yield of henna extract (w/w) at 10 min and at temperature 30°C

3.4. Effect of Temperature

The effect of temperature was investigated with a feed-solvent ratio of 0.02 and 10 minutes extraction time is shown by Figure 4. The optimum yield at room temperature was found to be 16.47% in water solvent. At higher temperatures, the yields of the resulting extracts decrease and began to rise back up at 60°C. This decrease is due to the fact that the lawsone substance in the Henna leaves is vulnerable to heat. The higher the temperature of the extraction can cause more damage in the lawsone. Thus, the optimum temperature of the solvent in Henna extraction was 30°C.

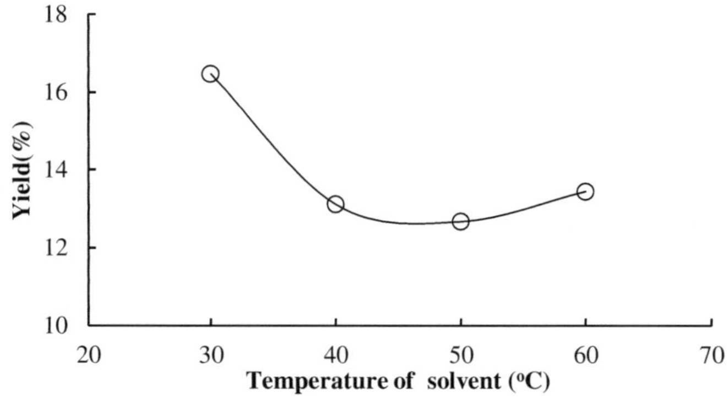

Figure 4. The Effect of temperature of solvent on the yield of henna extract (w/w) with at ratio (F/S) 0.02, time 10 min and solvent of aquadest

3.5 Effect of pH

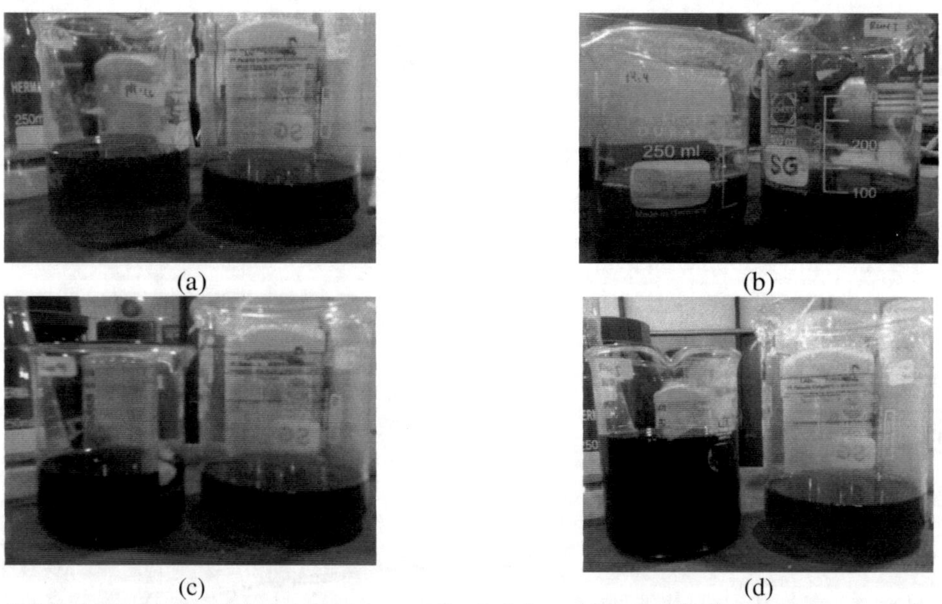

Figure 5. The effect of pH in color change. The left is acidic or alkaline solution and the right is solution at original pH, (a) at pH 1.08, (b) at pH 4, (c) at pH 8, and (d) at pH 10

Figures 5 and 6 show the effect of pH on colour change and Henna extract yield. Various pH level used in this experiment was 1.09; 4.0; 6.49; 8.0; and 9.94. The pH started to be controlled after the extraction with ultrasonic. HCl was added to the extract solution to make acidic solution and NaOH to make alkaline solution, respectively. In both acidic and alkaline conditions, no significant yield changes were recorded. Rather, it only affects the colour of the extract solution. When the extract solution was on acidic condition, the colour became fade. When the extract solution was on alkaline condition, the colour became dark.

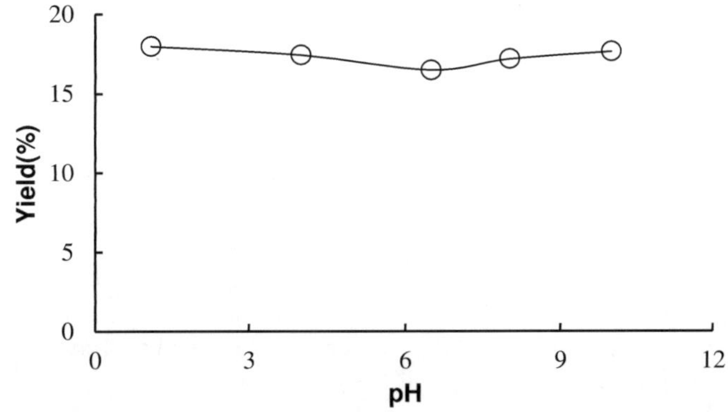

Figure 6. The effect of pH on the yield of henna extract (w/w) at ratio (F/S) 0.02, time 10 min, temperature of 30°C and solvent of aquadest

3.6 Soxhlet Extraction

Based on the experiment results, the highest yield for the UAE method with aquadest solvent was found to be 16.47% whereas 60% ethanol solvent gave 17.13% yield with 10 minutes extraction time. For soxhlet extraction method, the yield with aquadest solvent was 15.98% while 60% ethanol solvent gave 12.06% yield. The duration for soxhletation per cycle was 5-6 hours. When aquadest solvent was used, it required 10 cycles whereas 60% ethanol solvent required 12 cycles. Based on this, the soxhlet extraction method is less effective to extract dye from Henna as far as the time needed for extraction is concerned

3.7 UV-Visible Spectrophotometric Analysis of Extract Natural Dye Henna

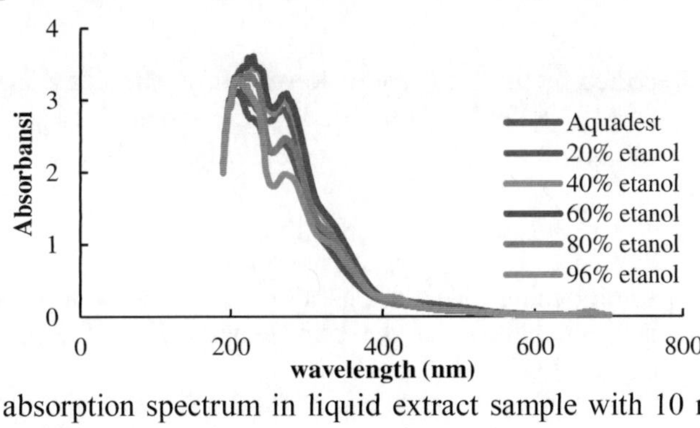

Figure 7. UV-Visible absorption spectrum in liquid extract sample with 10 min extraction time and ratio to solvent 0.02

Figures 7 and 8 show UV-Visible absorption spectrum at liquid and powder sample. The absorbance test was carried out to find out what kind of dye were present in the extracted material. The absorbance test was performed by using Thermo Scientific UV-Visible Spectrophotometry by comparing the graph of the relationship between absorbance and wavelength (nm). Aquadest and 60% ethanol solution were used as solvents for the Henna leaf extract. The light absorbed by a substance is different from the light captured by the human eye. The maximum absorption of the coloured solution occurs in the opposite colour areas. Based on the results of the UV-Vis spectrophotometric test that the maximum absorbance level on 60% ethanol solvent was 3.6 in wavelength 230 nm and 3.09 at 274 nm wavelength for the result of Henna leaf extract in liquid form, while absorbance level maximum on a powder form of 3.03 at wavelengths of 206 nm and 2 at a wavelength of 270 nm. In the previous research, it was explained that Lawson had the followings absorbance in ethanol solvents: 246, 277, and 333 nm, respectively. In wavelength of 246 nm shows that the presence of intramolecular hydrogen bonds from benzoquinone, in wavelength of 277 nm indicates that the electron delocalization of benzoquinone becomes a quinone ring, whereas in wavelength of 333 nm indicates that the presence of naphthoquinone in the form of hydrogen bonds with carbonyl [9].

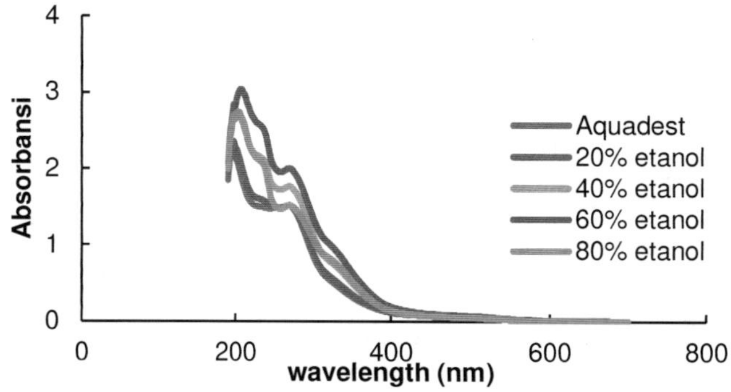

Figure 8. UV-Visible absorption spectrum in powder extract sample with 10 min extraction time and ratio to solvent 0.02

4. Conclusion

The Henna extract yield were strongly influenced by concentration of solvent, ratio of feed to solvent, pH and temperature, while time extraction has no significant effect on Henna extract yield. The highest yield was 16.47% for the UAE method with aquadest solvent whereas for 60% ethanol solvent gave 17.13% yield with 10 minutes extraction time and feed to solvent ratio of 0.02, temperature of 30°C. The soxhlet extraction method is less effective to extract dye from Henna as far as the time needed for extraction is concerned.

Acknowledgments

This work is financially supported by the Ministry of Research, Technology, and Higher Education of the Republic of Indonesia (KEMENRISTEKDIKTI) under the scheme of Penelitian Dasar Unggulan Perguruan Tinggi (PDUPT) 2018.

References

[1] Gonzalez V, Wood R, Lee J, Taylor S and Bussemaker M J 2018 Ultrasound-enhanced hair dye application for natural dyeing formulations *Ultrason. Sonochem* **52** 294-304.

[2] Anastasia F, Karolina S, Sukmann T, Iŕen J J and Andrea E 2019 Application of natural dyes on diverse textile materials *Optik* **181** 215–19.

[3] Bhuiyan M A R, Islam A, Ali A, and Islam M N 2018 Color and chemical constitution of natural dye henna (Lawsonia inermis L) and its application in the coloration of textiles *J. Clean Prod.* **167** 14–22.

[4] Yingcheng W and Wei W 2015 Optimization of Ultrasound-assisted Extraction of PhysalisalkekengiL var francheti seed oil *Chem. Eng. Tran.s* **46** 1387–92.

[5] Yiqi Y, Lu W, Boumou M, Xiuliang H, Jakpa W and Yongying T 2018 Ultrasound-microwave assisted extraction of natural colorants from sorghum husk with different solvents *Ind. Crops. Prod.* **120** 203–13.

[6] Chandrasekhar J, Madhusudhan M C and Raghavarao K S M S 2012 Extraction of anthocyanins from red cabbage and purification using adsorption *Food Bioprod. Process.* **90** 615–23.

[7] Muhamad N, Muhmed S A, Yusoff M M and Gimbun J 2014 Influence of solvent polarity and conditions on extraction of antioxidant, flavonoids and phenolic content from Averrhoa bilimbi

J. Food Sci. Eng. **4** 255–60.

[8] Wen H, An X, Hai N, Zhen J and Jiawen W 2009 Optimised ultrasonic-assisted extraction of flavonoids from Folium eucommiae and evaluation of antioxidant activity in multi-test systems in vitro *Food. Chem.* **114** 1147–54.

[9] Chandrakalavathi T, Sudha V, Sindhuja M, Harinipriya S and Jeyalakshmi R 2018 Photosonoelectrochemical analysis of Lawsonia inermis (henna) and artificial dye used in tattoo and dye industry *J. Photochem. Photobiol. A: Chem.* **360** 44–57.

IOP Conf. Series: Materials Science and Engineering **543** (2019) 012083 doi:10.1088/1757-899X/543/1/012083

Comparison of Performance Characteristic Prediction of Sieve Tray with and without Down comer

S Gondosurohardjo[1], A Altway[1]*, F Taufany[1], and S Nurkhamidah[1]

[1]Department of Chemical Engineering, Institut Teknologi Sepuluh Nopember, Kampus ITS Sukolilo, Surabaya 60111, Indonesia

*E-mail: alimohad.chem@gmail.com

Abstract. A sieve tray is a type of tray that is widely used in separation in industry. While sieve tray without down comer, frequently called dual flow sieve tray is commonly used to increase the effective area, so it can increase tray capacity. Dual-flow sieve tray operations can be classified into two categories, it is froth and transition based on gas/vapour velocity, so that there are parts of holes that are only flown by gas/vapour and at the same time there are parts of holes that are only flushed with liquid. Performance characteristics studied include clear-liquid height, dry and wet pressure drop, efficiency, and entrainment. The study aims to compare performance characteristic of sieve tray with and without down comer. The sieve tray without down comers showed almost 50% higher capacity compared to the sieve tray with down comers. However, sieve tray without down comers has 40% lower pressure drop compared to that with down comers. The results can ensure that the trays performance prediction for 298 mm diameter with open area in the range of 77-80%, 5 mm diameter holes, 12.5 mm hole spacing and 35 cm tray spacing satisfying and optimistic can be utilized in the distillation process.

1. Introduction

Separation equipment is needed in chemical industries to separate and purify the required product. Distillation column is one type of separation equipment based on the difference in volatility. The application of this equipment continues to increase in the petro-chemical industry, oil and gas industries and in the sustainable sector with renewable resources and energy [1]. Tray column is commonly used for high capacity operation. There are three types of tray commercially used in industries: sieve tray, bubble cap tray and valve tray, sieve tray is the lowest cost. Mcallister [2] has studied extensively performance of sieve tray. Sieve tray can be designed with and without down comer. In general sieve tray with downcomer has effective area 60-70% of column cross sectional area [3] due to the presence of downcomer. Outili [4] investigate effect of downcomer size on sieve tray performance. Sieve tray without downcomer has many advantages, it has larger bubbling/effective area so it can increase 20% to 30% capacity, lower pressure drop, but lower efficiency compared to sieve tray with downcomer [5-9] In general, contactor to carry out separation using sieve tray without downcomer is called dual flow tray. Efficiency of sieve tray is affected greatly by bubble movement phenomena through liquid. The contact of gas/vapour-liquid in sieve tray is similar with contact gas-liquid in stirred tank and research in this area will give important information concerning bubble size distribution, and affecting tray efficiency [10]. This phenomenon has been studied intensively by Buchanan et al [11] either for still liquid or vibrated liquid. Buchanan et al [12] also has studied the application of these phenomena on oxygen absorption operation. Colburn [13] has studied effect of entrainment on plate efficiency. The mechanism of vapour-liquid contact in sieve tray without downcomer is very different with that with downcomer. In distillation column without downcomer, vapour and liquid flow through different holes, then wave flow of agitated mixed vapour-liquid will be developed constantly in every tray [5]. With development of wavy flow, clear liquid pool with certain height is occurred, simultaneously followed by agitated mass wave above it. Xu et al [7] reported that efficiency of tray without downcomer was affected by hole area fraction, while Zhang et al [8] proposed a new parameter, fraction of holes passing gas, affecting tray efficiency. No previous studies reported comparing comprehensively performance of tray column with and without downcomer. The aim of this work is to compare the hydrodynamic performance with and without downcomer based on the published correlations in literatures [3, 14-17].

2. Methodology

This study will investigate the comparison between performance of sieve tray with and without down comer based on the published correlation in literatures. The parameters investigated are pressure drop, entrainment, aerated liquid height in tray, and tray efficiency. The tray column considered in this study has 29.8 cm diameter, and 35 cm tray spacing. The tray has 5 mm diameter holes arranged in triangular pitch with hole spacing of 12.5 mm. Liquid flow rate is fixed with constant value of 3.59 gpm. For tray with downcomer the clear liquid height is set 4 cm. For tray with and without downcomer, equation (1) and (2) respectively are used to predict entrainment,

$$e_w = 0.22\left(\frac{73}{\sigma}\right)\left(\frac{v_c}{s'}\right)^{3.2} \tag{1}$$

$$e_w = 6.31 \times 10^{-7} \times F_s^{4.57} \tag{2}$$

Where, F_s, velocity factor, is defined as,

$$F_s = v_0\sqrt{\rho_v} \tag{3}$$

The total pressure drop across the tray is predicted as follows for tray with down comer,

$$h_t = h_{dt} + h_e \tag{4}$$

Where h_e , effective head is obtained as a function of h_{sl} from Fig.8.130 in [3]. The static liquid seal, h_{sl} is determined from the following equation,

$$h_{sl} = h_w + h_{ow} \tag{5}$$

Where height of liquid above the weir, h_{ow} is determined from equation (6),

$$h_{ow} = 0.092F_w\left(\frac{L}{l_w}\right)^{2/3} \tag{6}$$

Where the correction factor F_w is obtained from Fig.8.105 in [3] as function of $\frac{L}{l_w{}^{2.5}}$ and $\frac{l_w}{D}$. The dry tray pressure drop h_{dt} from equation (4) is obtained from equation (7)

$$h_{dt} = 0.003[v_0{}^2\rho_v]\left[\frac{\rho_{water}}{\rho_L}\right][1-\beta^2]/[C_0{}^2] \tag{7}$$

Liquid height in downcomer H_d can be obtained from equation (8),

$$H_d = h_t + h_w + h_{ow} \tag{8}$$

For tray without downcomer, height of aerated liquid on tray is important parameter and can be obtained from equation (9),

$$h_{al} = 1.25\,F_c + 0.0005\,L + \frac{0.54}{\beta} - 2.45 \tag{9}$$

Equation (10) through (9) are obtained from [3]. The aerated liquid height in the tray for tray without downcomer can also obtained from [14] as follows.

$$h_f = \frac{h_L}{1-\epsilon} \tag{10}$$

Where liquid hold up, h_L is predicted using correlation by Xu et al [7] as shown in equation (11)

$$h_L = 0.01728 \frac{(LM_L)^n \left[U_s \left(\frac{\rho_G}{\rho_{Air}} \right)^{0.5} \right]^1}{\rho_L \left(\frac{A_h}{A_a} \right)^{1.5} \left(\frac{t_t}{d_h} \right)^{0.42}} \tag{11}$$

Where constant n is obtained as equation (12),

$$n = 4.3 \left(\frac{A_h}{A_a} \right)^{1.5} \tag{12}$$

Porosity in tray determined by Xu et al [7] is shown in equation (13),

$$\in = 1 - 0.0946 \left[\frac{U_s^2 \rho_G}{g h_L \rho_L} \right]^{-0.2} \tag{13}$$

Point efficiency in this study was obtained using equation (14) reported by Bennett et al [17] and adopted by Al-Hemiri and Hassan [15]

$$E = 1 - exp \left[\frac{-0.0029}{1 + m \frac{\rho_{MV}}{\rho_{ML}} \sqrt{\frac{D_V (1-\emptyset_e)}{D_L \left(\frac{A_h}{A_a} \right)}}} Re^{0.4236} \left(\frac{h_L}{d_H} \right)^{0.6074} \left(\frac{A_h}{A_a} \right)^{0.3195} \right] \tag{14}$$

3. Results and Discussion

Figure 1 shows the effect of vapour rate on total pressure drop, for tray with and without downcomer. It is appeared that the total pressure drop increase significantly with increasing vapour rate for tray with down comer. However, for tray without downcomer the total pressure drop increase very slowly with increasing vapour rate. The total pressure drop for tray without downcomer is much lower than that for tray with down comer especially for high vapour rate. For vapour rate of 8.68 cuft/s, the value of total pressure drop are 4.88 and 1.62 inch for tray with downcomer and without downcomer respectively. This is the advantage of tray without down comer having much lower pressure drop compared to that for tray with downcomer.

Figure 2 shows the effect of vapour rate on liquid height in downcomer (for tray with down comer) and aerated liquid height in the tray (for tray without downcomer). These parameters are important because flooding limit is determined by the value of these parameters. Flooding occurs if the value of these parameter exceeding one halve of tray spacing. The figure depicts that flooding limit is 8.33 cuft/second and 12.15 cuft/second for tray with downcomer and tray without downcomer. Then, the capacity of tray without down comer is almost 50 % larger than the capacity with downcomer.

Figure 3 shows the effect of vapour rate on entrainment for tray with and without downcomer. It is obvious that tray without downcomer gives lower entrainment compared to that with downcomer. At vapour rate of 8.68 cuft/sec, the entrainment is 0.18 and 0.04 for tray with and without downcomer respectively. This is the other advantage of tray without downcomer compared to tray with downcomer.

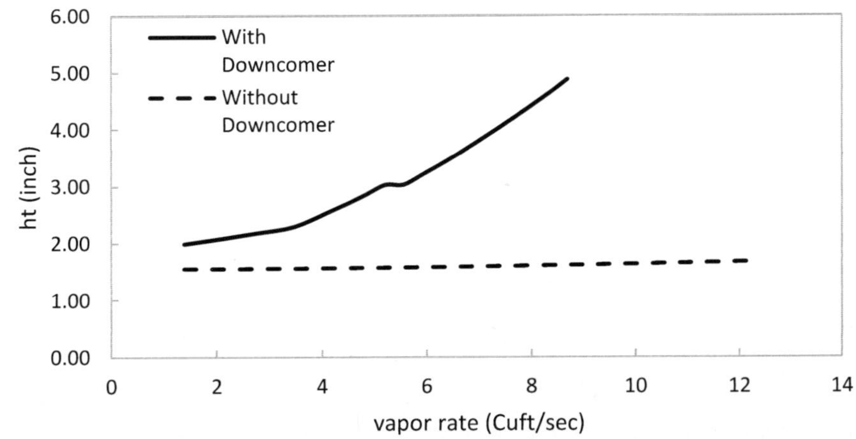

Figure 1 Effect of vapour rate on total pressure drop across the tray. Comparison between tray with and without downcomer. Hole spacing=12.5 mm

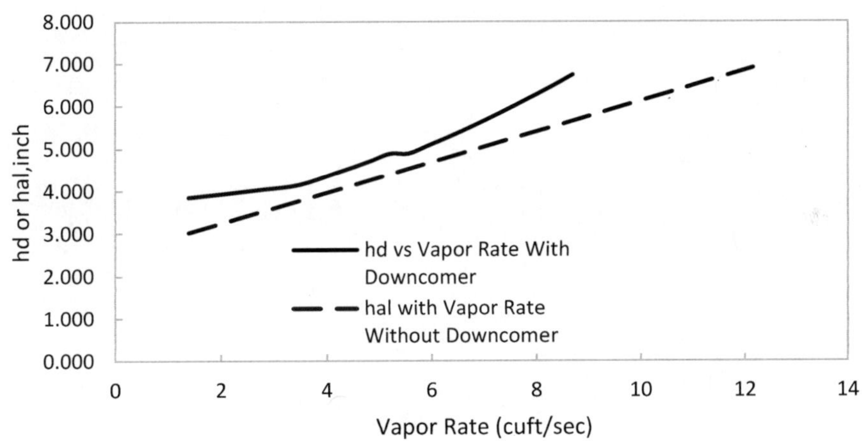

Figure 2 Effect of vapour rate on H_d (for tray with down comer) or h_{al} (for tray without down comer). Comparison between tray with and without downcomer. Hole spacing=12.5 mm

IOP Conf. Series: Materials Science and Engineering **543** (2019) 012083 doi:10.1088/1757-899X/543/1/012083

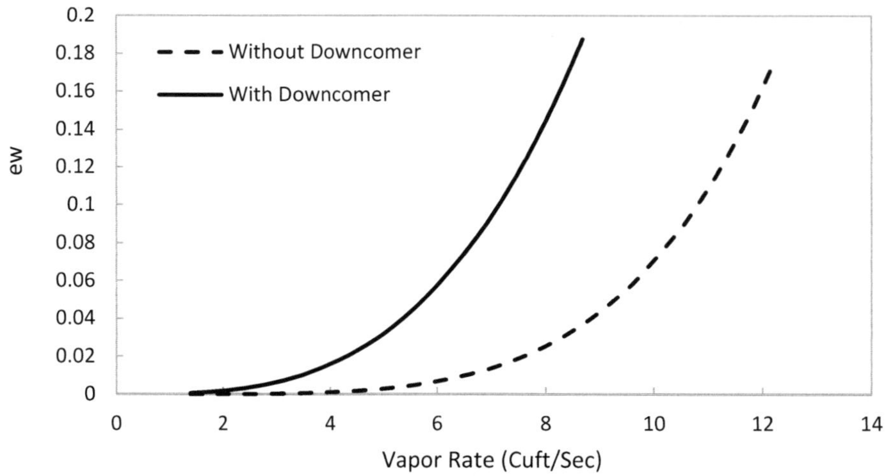

Figure 3 Effect of vapour rate on entrainment. Comparison between tray with and without downcomer. Hole spacing=12.5 mm

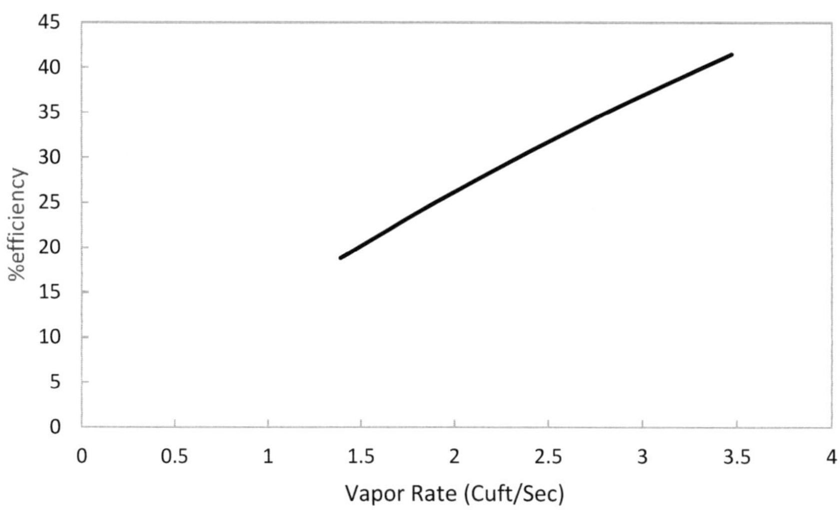

Figure 4 Effect of vapour rate on point tray efficiency for tray without down comer. Hole spacing=12.5 mm

Figure 4 shows effect of vapour rate on point tray efficiency. It can be seen that the efficiency increases with increasing vapour rate. Increasing vapour rate makes gas-liquid contact more effective. This figure also shows that the point efficiency of tray without downcomer is small, smaller than the efficiency for tray with downcomer having value more than 60 %. Further research have to be done to modify geometry of tray without downcomer to increase it's efficiency.

Figure 5 shows comparison prediction value of aerated liquid height in the tray without downcomer using Ludwig's formula and Garcia and Fair's formula. It is appeared that for small vapour rate, Ludwig's prediction is larger than Garcia's prediction. However, for large vapour rate Garcia's prediction is larger than Ludwig's prediction.

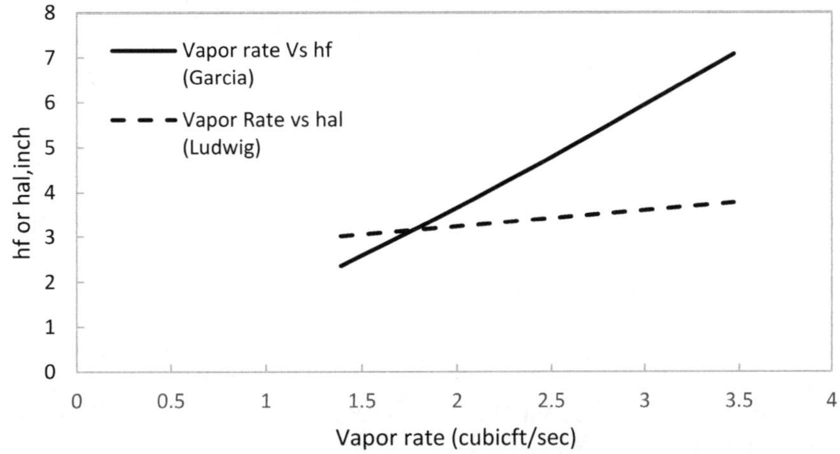

Figure 5 Effect of vapour rate on aerated liquid height in tray for tray without downcomer.

4. Conclusions

The sieve tray without downcomer showed 50% higher capacity compared to the sieve tray with downcomer. Meanwhile, sieve tray without downcomer has 40% lower pressure drop compared to that with downcomer. However, sieve tray without downcomer shows smaller efficiency compared to that with down comer. The results can ensure that the trays performance prediction for 298 mm diameter with open area in the range of 77-80%, 5 mm diameter holes, 12.5 mm hole spacing and 35 cm tray spacing satisfying and optimistic can be utilized in the distillation process. Further research work has to be done to increase efficiency of sieve tray without downcomer by modifying its geometry

Acknowledgments

The authors would like to thank to Retno Dwi Nyamiati and Arnesya Ramadhani for their support in this research.

References

[1] Osuolale F N and Zhang J 2014 Energy efficiency control and optimisation of distillation column using artificial neural network *Chem. Eng. Trans.* **39** 37-42
[2] Mcallister R A, Mcginnis P H JR and Plank C A 1958 Perforated -plate performance, *Chem. Eng. Sci.* **9** 25 - 35
[3] Ludwig E E 1997 Applied process design for chemical and petrochemical plants, **Volume 2**, Third Edition, Gulf Professional Publishing.
[4] Outili N, Chegga N, Elbahi K and Meniai A H 2013 Effect of downcomers sizes on tray column performance, *Chem. Eng. Trans.* **32** 1981-86
[5] Swamy K and Meikap B C 2017 Hydrodynamic characteristics of a three-stage dual-flow sieve plate scrubber, *S.Afr.J.Chem.Eng.* **23** 91-97
[6] Flávio D, Mayer, Feris L A,Marcilio N R, Baldo V and Hoffmann R 2014 Review of hydraulics correlations for sieve trays without downcomers *Ind.Eng.Chem.Res*, **53 (20)**, 8323–31
[7] Xu Z P, Afacan A and Chuang K T 1994 Efficiency of dual flow trays in distillation, *Can.J.Chem.Eng.* **72** 607-13
[8] Zhang J, Wang Y, Yu G, Mao X and Wang F 2015 Experimental study of two phase flowcharacteristics on the dual-flow tray *Chem.Eng.Res.Des.* **102** 90–99
[9] Takahashi T, Miyahara T, Tano T and Akagi Y 1986 Hydrodynamic characteristic of sieve trays

having large free areas without down comers *J.Chem.Eng.Jpn.* **19(4)** 339-41
[10] Garcia J A and Fair J R 2000 A Fundamental model for the prediction of distillation sieve tray efficiency. 1. Database development *Ind. Eng. Chem. Res.* **39** 1809- 17
[11] Buchanan R H, Jameson G and Djoeriaman O 1962 Cyclic migration of bubbles in vertically vibrating liquid columns, *Ind.Eng.Chem.Fund.* **1(2)**, 82-86
[12] Buchanan R H, Teplitzky D R and Djoeriaman O 1963 Oxygen absorption in low-frequency vertically vibrating liquid column *Ind.Eng.Chem.Proc.Des.Dev.* **2(3)** 173- 77
[13] Colburn A P 1936 Effect of entrainment on plate efficiency in distillation *Ind.Eng.Chem.* **28(5)** 526-30
[14] Garcia J A and Fair J R 2002 Distillation sieve trays without downcomers: prediction of performance characteristics, *Ind.Eng.Chem. Res.* **41(6)** 1632-40
[15] Al-Hemiri A A and Hassan M A R 2006 Prediction of the efficiency of sieve tray using air water system, *J.Eng*, **13(4)** 941-56
[16] Bennett D L, Agrawal R and Cook P J 1983 New pressure drop correlation for sieve tray distillation columns *AIChE J* **29(3)** 434– 42.
[17] Bennett D L,Watson D N and Wiescinski M A 1997 New correlation for sieve-tray point efficiency, entrainment, and section efficiency *AIChE J.* **43(6)** 1611– 26
[18] Bennett D L and Grimm H J 1991 Eddy diffusivity for distillation sieve trays *AIChE J.* **37(4)** 589–96

IOP Conf. Series: Materials Science and Engineering **543** (2019) 012084 doi:10.1088/1757-899X/543/1/012084

Hydrate formation calculation in the natural gas purification unit

J A Prajaka, J Himawan, S A Affandy, J P Sutikno and R Handogo*.
Chemical Engineering Department, Institut Teknologi Sepuluh Nopember (ITS), Surabaya 60111, Indonesia.

*email : renanto@chem-eng.its.ac.id

Abstract. Common problem in natural gas purification is Hydrate Formation. Hydrate is a solid – ice – like crystals in which gas molecules are trapped in the structure of solid water and causing problems especially clogging in the pipe line. Hydrate Formation usually happens at Dew Point Control Unit (DPCU) where the temperature is lowered to separate natural gas with heavy fraction by using Joule – Thomson effect. This paper aims to predict the Hydrate Formation Temperature at certain pressure by using some of HFT calculation methods. Kartz, Towler, Bahadori, and New Correlation are used to predict the Hydrate Formation Temperature with the pressure range of 700 – 2200 kPa. The obtained temperatures forming the gas at pressure of 700 kPa using Kartz, Towler, New Correlation, and Bahadori Calculation respectively are 274.26 K, 274.92 K, 275.02 K, and 275.2 K with Dew T at pressure of 700 kPa is 279.06 K, that means the Dew Point temperature is higher than the HFT, while at pressure 2200 kPa the temperature of Hydrate Formation respectively are 283.15 K, 283.46 K, 283.54 K, and 283.7 K with Dew T at pressure of 2200 kPa is 299.75 K, that means the Dew Point temperature is higher than the HFT.

1. Introduction

Nowadays, natural gas is widely used as main energy source besides oil and coal [1]. The sector that directlu drives the increase in demand for natural gas is the electric power plant sector where natural gas has replaced coal and oil as the main fuel [2]. Natural gas produces fewer CO2, NOx, sulfur, dust and harmful substance emissions than coal and oil [3]. Before it can be stored and used as fuel, natural gas must undergo a process of purification. In this process one of the problems that often occurs is hydrate formation. Hydrate of gas is a solid – ice – like crystals in which gas molecules (mainly Methane) are trapped in the structure of solid water, which occurs when water molecules from a cage-like structure around smaller guest molecules [4]. The most common guest molecules are methane, ethane, propane, isobutane, normal butane, nitrogen, carbon dioxide, and hydrogen sulfide. Natural gas generally consist of propane, butane, and a little part of ethane and penthane [5]. It should be noted that normal butane does form a hydrate, but is very usable. While many factors influence hydrate formation are (1) the gas being at the appropriate temperature and pressure, and (2) the gas being at or below its water dew point, other factors that affect hydrate formation include mixing, kinetics, type of physical site, surface for crystal formation, agglomeration and the salinity of the system. Hydrate is first discovered by Davy in 1810. Villard in 1888 presented performance data on hydrocarbon hydrates. Although hydrates are known for over 200 years, their importance to the oil and gas industry was not realized until the 1930's. In 1984, Hammerschmidt determined that hydrates were the cause of plugged natural gas pipelines, thereby leading to the regulation of gas water content, and the development of improved methods of prevention of hydrate plugs, including the injection of methanol and other inhibitors into the gas stream.

IOP Conf. Series: Materials Science and Engineering **543** (2019) 012084 doi:10.1088/1757-899X/543/1/012084

He developed the correlation that describes hydrate formation temperature as a function of only pressure. His simple correlation was modified by Ghiasi [6]. Formation behaviours of mixed gas hydrates including olefin compounds were addressed by Lee and Kang [7]. Some workers such as Jozian and Vafajoo [8] succeeded to build a mathematical modeling of the gas hydrate formation utilizing CFD technique.

2. Methods

In this research, Hydrate Formation Temperature was calculated using several methods, such as Gas Gravity Method [9], Towler & Mokhatab Calculation [10], Bahadori & Vuthaluru Method [11], and New Correlation which was done by Safamirazaei [12]. From several methods, some variables were changed such as (a) pressure on natural gas assumed to use ideal pressure on gas wells in Indonesia that is between 700-222 kPa, (b) Natural Gas molecular weight. In addition to predict the Hydrate Formation Temperature, Dew Point calculations were also made on natural gas based on the pressure influence between 700 - 2200 kPa pressure using Aspen Hysys Simulation Application and using 3 different fluid packages Peng - Robinson, Redlich - Kwong, and Lee - Kesler.

2.1 Gas Gravity Method

In 1945, Katz generated the gas gravity chart from a limited amount of experimental data and a more substantial amount of calculations based on the K-value method [9]. This plot relates the hydrate formation pressure and temperature with gas gravity defined as the apparent molecular weight of a gas mixture divided by that of air. To use this method, the gas should be primarily methane and other light hydrocarbons (ethane, propane, butane and normal pentane). Sloan showed in a statistical analysis report that this method is not accurate and the calculated pressure for the same gas gravity with different mixtures may result in 50% error. This method estimated the same hydrate formation pressure or temperature for the components having equal molecular weights, such as butane and isobutane, despite the fact that they should be different in reality. To avoid tedious calculations, a regression analysis was used to fit the curve to predict the hydrate developed for gases where specific gravity was known. Some available correlations for gas gravity method to calculate the hydrate formation conditions are Motiee correlation [13], Towler and Mokhatab correlation [10], and Bahadori and Vuthaluru correlations [11]. In which Bahadori & Vuthaluru correlation is the most accurate and reliable equation.

2.2 Towler and Mokhatab

In 2005, Towler and Mokhatab recommended a relatively simple correlation for predicting HFT of natural gas mixtures. A modified form of this correlation is shown in Eq. 1

$$T = 13.47 \ln P + 34.27 \ln \gamma - 1.675 \ln P \ln \gamma - 20.35 \quad (1)$$

where the temperature, T, is in R, the pressure, P, is in psi, and γ is the gas gravity, dimensionless.

2.3 Bahadori & Vuthaluru Correlation

Equations (2) and (3) represent Bahadori & Vuthaluru correlation [11] in which four coefficients are used to correlate hydrate formation pressure and temperature of natural gases. The coefficients shown in Equations (2) and (3) are correlated as a function of molecular weight (M)in Equations (4) to (7).

$$\ln T = a + b\left(\frac{1}{P}\right) + c\left(\frac{1}{P}\right)^2 + d\left(\frac{1}{P}\right)^3 \quad (2)$$

$$\ln P = a + b\left(\frac{1}{T}\right) + c\left(\frac{1}{T}\right)^2 + d\left(\frac{1}{T}\right)^3 \quad (3)$$

where,

$$a = A_1 + B_1 M + C_1 M^2 + D_1 M^3 \quad (4)$$

$$b = A_2 + B_2 M + C_2 M^2 + D_2 M^3 \quad (5)$$

$$c = A_3 + B_3 M + C_3 M^2 + D_3 M^3 \quad (6)$$

IOP Conf. Series: Materials Science and Engineering **543** (2019) 012084 doi:10.1088/1757-899X/543/1/012084

$d = A_4 + B_4M + C_4M^2 + D_4M^3$ (7)

This method helps to cover data from Katz gravity chart [13] in temperature variation of 260 K to 298 K as well as the pressures and gas molecular weights within the range of 1200 kPa to 40000 kPa and 16 g/mol to 29 g/mol, respectively. Although this correlation is accurate, its application is especially grueling and could be tedious. In this paper, a simple, easy to use and simultaneously accurate model for estimation of HTF is presented. Finally estimation capability of the obtained correlation and Bahadori & Vuthaluru [11] method is compared by some statistical parameters. The required data to develop this correlation includes the reported data for hydrate formation condition of natural gases as well as temperature and pressure. In this work, the hydrate formation condition is predicted rapidly by proposing a novel correlation, which has been developed, based on newly proposed numerical model. The following methodology has been applied to develop this correlation. Firstly, hydrate formation temperatures are correlated as a function of pressure for various molecular weights. Then, the calculated coefficients for these polynomials are correlated as a function of molecular weight. The derived polynomials are applied to calculate new coefficients for Equations (1) and (2) to predict the hydrate formation condition of natural gases as a function of pressure and temperature respectively. Table 1 shows the tuned coefficients for Equations (3) to (6).

2.4 New Correlation

Twenty-two experimental data points used by Bahadori and Vuthaluru [11], with specific gravities from 0.55 to 1, have been implemented to develop the correlation. Eq. 9 is suitable for estimating the HFT, especially for natural gas mixtures:

$T(K) = A\gamma^B(\ln P)^c$ (8)

This equation is the equation of the development of previous scientists, where by processing some data obtained by the determination of A B C as described above. Based on what M. Safamirzaei [12] does, according to him, New Correlation is the most effective method, because based on the % ARD value it calculates, New Correlation has the lowest %ARD with the least number of Adjustable Parameters compared to other calculation methods.

2.5 Dew Point Temperature

In some cases, the Hydrate Formation temperature is most commonly found in the Dew Point Control Unit where it separates natural gas from the heavy hydrocarbon component before entering the next treatment unit. The process used to reduce the temperature of natural gas to reach the dew point of the natural gas, that's when the possibility of Hydrate formed, therefore it is necessary to review the relationship between Dew Point with temperature hydrate by calculating the dew point value of the natural gas. Dew point calculation method could be done in various ways, Dew Point is not influenced by the composition of the natural gas, dew point is only influenced by the pressure of natural gas.

3. Result and Discussion

Using different calculation methods, the results showed not too much different between each other, but the difference was quite significant when viewed from the difference between molecular weight with each other. The results of some calculations can be seen in Table 2. As we can see in the Figures 1 to 3, all of the results show a value that is not too much different, where the higher the pressure of natural gas Hydrate Formation Temperature will also increase, then between the pressure and temperature is directly proportional, it indicates that the four methods used to calculate the hydrate formation temperature is quite accurate, only need to be compared between the calculation data and actual conditions in the field.

From all of the calculation methods, the Bahadori Method always on the highest score overall, based on research conducted by Bahadori & Vuthaluru [11], they compared their results with actual data on the same operating conditions (Pressure and gas composition) which is not much different from each value (Calculated & Actual) with % Error between 0.01 - 0.2%.

Table 1. Tuned coefficients use in Equation (3) to (6) to Predict hydrate formation temperature in K (given pressure in kPa) by Equation (1)

Coefficient	Natural Gas with molecular weight more than 23 & Pressure 1200 kPa < P < 40000 kPa	Natural Gas with molecular weight less than 23 & Pressure 1200 kPa < P < 5000 kPa
A_1	6.418507	-4.181213
B_1	-8.801710×10^{-2}	1.472639
C_1	3.557342×10^{-3}	-7.27458×10^{-2}
D_1	-4.749984×10^{-5}	1.18979×10^{-3}
A_2	-8.642628×10^{3}	4.528497×10^{4}
B_2	1.024330×10^{3}	-6.862812×10^{3}
C_2	-4.096639×10^{1}	3.424072×10^{2}
D_2	5.445005×10^{-1}	-5.642533
A_3	1.159643×10^{7}	-8.317075×10^{7}
B_3	-1.385902×10^{6}	1.260481×10^{7}
C_3	5.535314×10^{4}	-6.301857×10^{5}
D_3	-7.339994×10^{2}	1.040884×10^{4}
A_4	-4.020095×10^{9}	5.858977×10^{9}
B_4	4.791331×10^{8}	-9.663496×10^{8}
C_4	-1.903632×10^{7}	5.134731×10^{7}
D_4	2.511329×10^{5}	-8.878185×10^{5}

It can indicate that the accuracy of Bahadori & Vuthaluru Method [11] is quite high. As for New Correlation method conducted by Safamirzaei [12] has advantages in the number of variables, it only needs 2 variables, and the variables are Pressure and Natural Gas molecular weight. It certainly simplifies the users of this method even though the accuracy of the method is still under Bahadori & Vuthaluru [11] method, with% error of 0.4%. As we can see in the Figures 1 to 3, all of the results show a value that is not too much different, where the higher the pressure of natural gas Hydrate Formation Temperature will also increase, then between the pressure and temperature is directly proportional, it indicates that the four methods used to calculate the hydrate formation temperature is quite accurate, only need to be compared between the calculation data and actual conditions in the field.

From all of the calculation methods, the Bahadori Method always on the highest score overall, based on research conducted by Bahadori & Vuthaluru [11], they compared their results with actual data on the same operating conditions (Pressure and gas composition) which is not much different from each value (Calculated & Actual) with % Error between 0.01 - 0.2%, it can indicate that the accuracy of Bahadori & Vuthaluru Method [11] is quite high. As for New Correlation method conducted by Safamirzaei [12] has advantages in the number of variables, it only needs 2 variables, and the variables are Pressure and Natural Gas molecular weight. It certainly simplifies the users of this method even though the accuracy of the method is still under Bahadori & Vuthaluru [11] method, with% error of 0.4%.

Table 2. Predicted Hydrate Formation Temperature Results

P kPa	*Hydrate Formation Temperature (K)*											
	22.8435				23.8473				20.454			
	Kartz	Towler	New Cor.	Bahadori	Kartz	Towler	New Cor.	Bahadori	Kartz	Towler	New Cor.	Bahadori
700	274.6	274.9	275.0	275.2	275.3	275.6	275.8	275.7	273.5	273.5	273.7	273.4
850	276.2	276.5	276.5	276.9	276.8	277.1	277.3	277.2	275.3	274.9	275.5	275.6
1000	277.0	277.4	277.8	277.3	278.1	278.3	278.6	278.5	276.6	276.2	276.7	276.9
1150	278.4	278.5	278.8	278.3	279.2	279.4	279.6	279.7	277.7	277.3	277.8	277.9
1300	279.4	279.4	279.8	279.7	280.1	280.3	280.5	280.7	278.6	278.2	278.7	278.8
1450	280.1	280.2	280.6	280.9	280.9	281.2	281.4	281.4	279.4	279.1	279.5	279.8
1600	281.2	281.0	281.3	281.7	281.6	281.9	282.1	282.2	280.2	279.8	280.2	280.4
1750	281.8	281.7	281.9	281.9	282.4	282.6	282.7	283.0	280.9	280.5	280.9	281.5
1900	282.3	282.3	282.5	282.5	283.1	283.2	283.3	283.9	281.6	281.1	281.4	282.1
2200	283.0	283.5	283.5	283.7	284.1	284.3	284.3	284.8	282.6	282.3	282.5	283.0

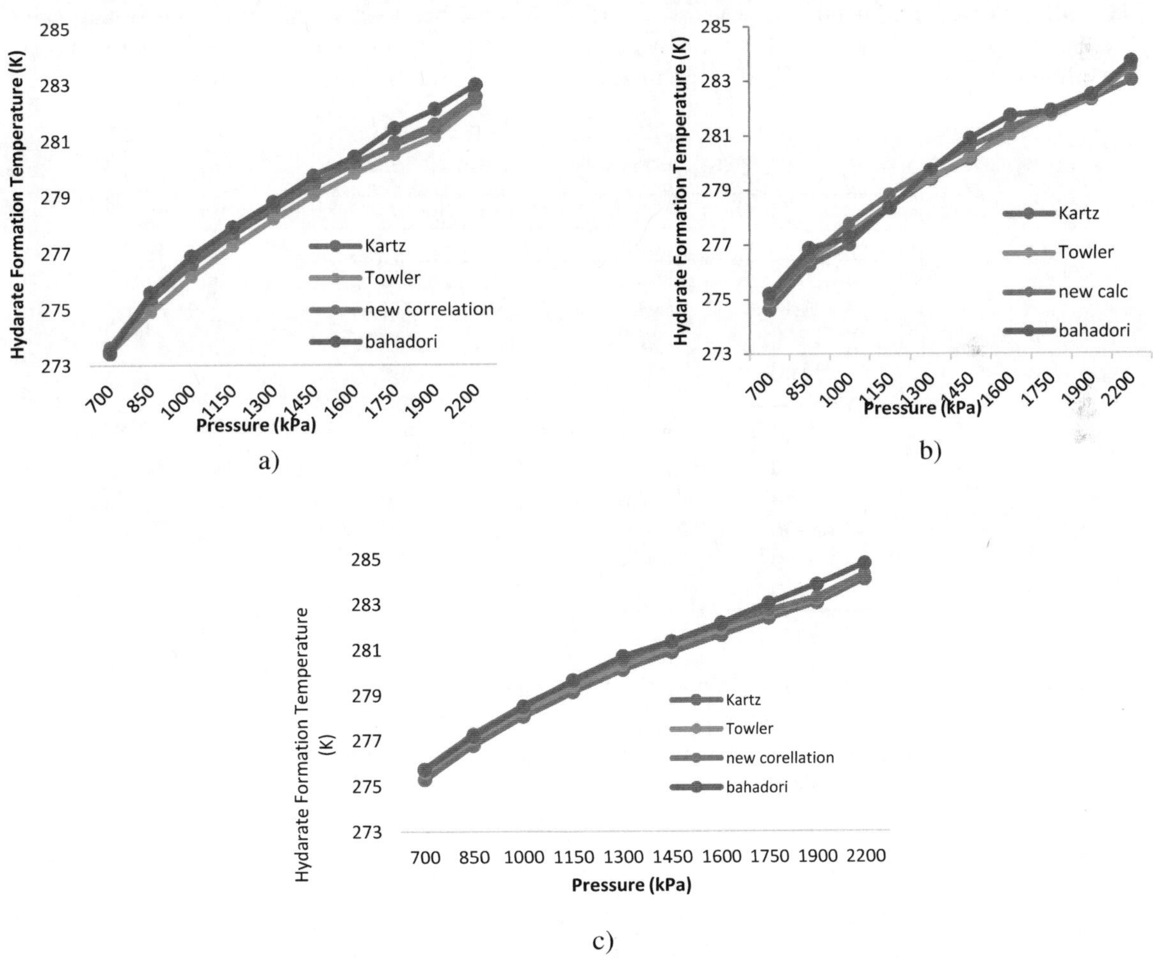

Figure 1. Comparison of HET between each method, natural gas molecular weight : a) 20.454 b) 22.2875 c) 23.8437

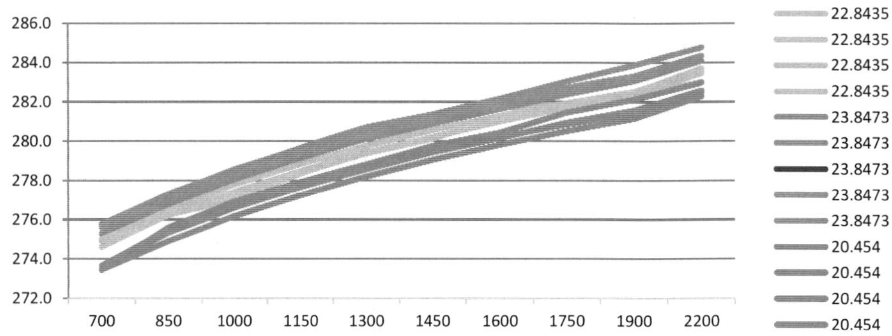

Figure 2. Comparison of HFT between each Natural Gas Molecular Weight

As seen in Figure 2, the comparison of HFT values with molecular weight differences shows that the highest average HFT value occurs in Natural Gas with a molecular weight of 23.8473 and the lowest average HFT value occurs in natural gas with a molecular weight of 20.4541, indicating that the formation of hydrate gas is influenced by the molecular weight of natural gas, where HFT will increase as the molecular weight of natural gas increases or it can be said that both are directly proportional. The molecular weight of the natural gas is influenced by the composition of the natural gas, as the more composition of the heavy hydrocarbon in the natural gas, the molecular weight of natural gas will also be greater.

Table 3. Dew Point Results

P	Dew Point (K)		
	Peng	SRK	Lee
700	280.2	281.2	280.9
850	283.5	284.9	284.7
1000	287.0	288.0	287.9
1150	289.7	290.9	290.7
1300	292.0	293.0	293.2
1450	294.1	295.1	295.4
1600	295.9	296.9	297.4
1750	297.6	298.6	299.2
1900	299.0	300.1	300.8
2050	300.4	301.5	302.3
2200	301.6	302.7	303.7

As one can see in Table. 3, results of Dew Point Calculation with several methods shows that when the pressure increases the dew point also increases. In Figure 3. shows that the results between one and other methods are quite similar, that means those calculation methods are valid. It is known, that the dew point is not affected by the composition of the components. It just depends on the pressure and the components. Figure 4 shows the comparison between Hydrate Formation Temperature and the average of dew point temperature.

As one can see the dew point values are above the Hydrate Formation Temperature, it means that the possibility of the hydrate to formed is low, it could happen because before natural gas reaches the Hydrate Formation Temperature, it will reach the Dew point temperature first which makes the natural gas will turn into the liquid phase.

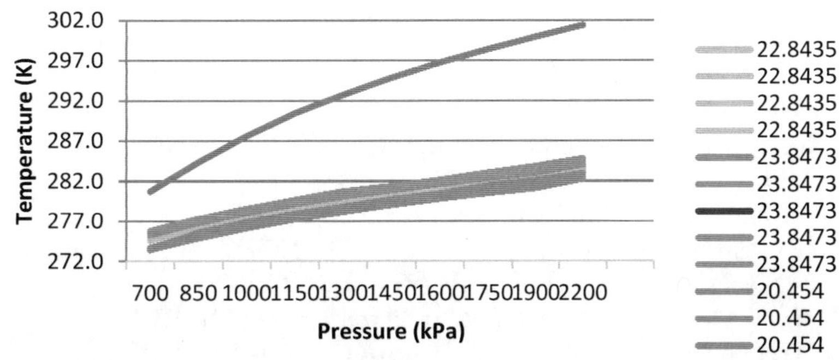

Figure 3. Comparison of Dew Point Temperature with Different Calculation Method

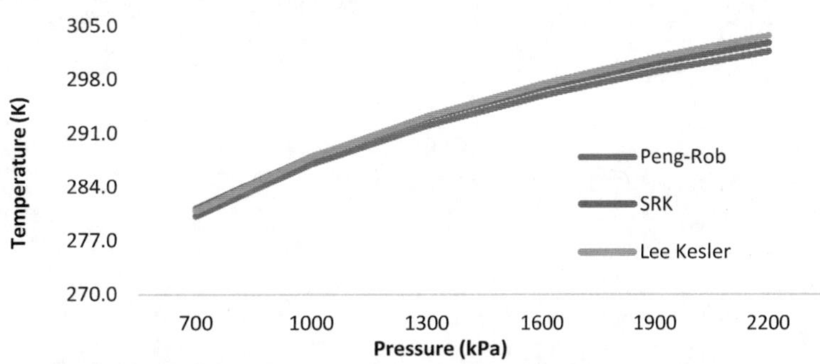

Figure 4. Comparison between Dew Point and Hydrate Formation Temperature

This process sometimes happened in the Dew Point Control Unit in Natural Gas Purification. Hydrate Formation could be avoided as long as we keep the operation temperature between the Dew Point temperature and Hydrate Formation Temperature, in this case with Natural Gas Molecular Weight between 20-23, hydrate formation could be avoided by keeping the operation Temperature between 270°C – 280°C at Pressure 700 kPa – 2200 kPa. In other cases, the hydrate formation can be avoided by injecting chemicals such as DEG (Diethylene Glycol), MEG (Methyl Ethylene Glycol), and TEG (Triethylene Glycol).

4. Conclusion

Based on the results of the research that has been done, it can be concluded that the relation between Hydrate Formation Temperature and pressure of the natural gas is directly proportional. The higher the pressure of the natural gas, the higher the hydrate formation temperature is. The relation between Hydrate Formation Temperature and the natural gas molecular weight is directly proportional. The higher the natural gas molecular weight, the higher the hydrate formation temperature is. The relation between Hydrate Temperature Formation and Dew Point temperature is if the dew point temperature is above the hydrate formation temperature, then the posibillity of the hydrate formed is low.

IOP Conf. Series: Materials Science and Engineering **543** (2019) 012084 doi:10.1088/1757-899X/543/1/012084

Acknowledgement
The authors announce that this research was funded DIKTI through PMDSU Project No. 042554/IT2.11/PN.02/2016.

Reference

[1] Anna D, Yuli W, Yi G and Mingzhi F 2018 *J. Appl. Thermal Eng.* **142** 666-73
[2] Bekhet H A and Yasmin T 2016 *IOP Conf. Ser.: Earth Environ. Sci.* **32** 012-051
[3] Song F, Xu Y, Song X and Gao H 2019 *IOP Conf. Ser.: Earth Environ. Sci.* **242** 022-029
[4] Barkan E S and Sheinin D A 1992 *J. Fluid Phase Equil.* **86** 111-36
[5] Setiyo M, Soeparman S, Hamidi N, Wahyudi S 2017 *Int. J. Tech.* **1** 114-23
[6] Ghiasi M M 2012 *J. Natural Gas Chem.* **21** 508-12
[7] Lee J W and Kang S P. 2013. *Chem. Eng. Trans.* **32** 1921-26
[8] Jozian S and Vafajoo L 2018 *Chem. Eng. Trans.* **70** 2167-72
[9] Katz D L 1945 *J. Trans. American Inst. Mining Eng.* **160** 140-49
[10] Towler B F and Mokhtab S 2005 *J. Hydrocarbon Process.* **84** 61-62
[11] Vuthaluru H B and Bahadori A 2009 *J. Natural Gas Chem.* **18** 453-57
[12] Safamirzaei M *Gas Processing & LNG Gulf Publishing Holdings LLC*
[13] Motiee M 1991 *J. Hydrocarbon Process* **70** 98-99

IOP Conf. Series: Materials Science and Engineering **543** (2019) 012085 doi:10.1088/1757-899X/543/1/012085

Ultrasound Pre-treatment for Intensification of Hydrothermal Process in Reducing Sugar Production from Cassava Starch

Sumarno*, A Bramantyo, P Febriyati, I Gunardi and P N Trisanti

Departemen of Chemical Engineering, Faculty of Industrial Technologi, Institut Teknologi Sepuluh Nopember, Surabaya, Indonesia

E-mail : onramus@chem-eng.its.ac.id

Abstract. Cassava is an abundant natural resource in Indonesia and mostly used as food. Starch, as the main component of cassava, can be converted to many of its derivative product such as glucose, fructose, and other reducing sugar. Conventionally, reducing sugar is produced from starch by acid and enzyme hydrolysis. A hydrothermal process is sophisticated biomass processing method which makes it possible to hydrolyze the starch by water only. However, the reducing sugar productivity of the hydrothermal process still not as high as the conventional process. By ultrasound pre-treatment, a hydrothermal process can be intensified, so that increase its productivity. This study set out to find the effect ultrasound pre-treatment to intensify production of reducing sugar from the hydrothermal process of cassava starch. The experiment was conducted by treating cassava starch suspension using ultrasonic irradiation with various operation time (0-120 minutes) and under a temperature of 40°C. For combination process, the starch suspension is processed under 15 minutes ultrasonic irradiation at 40°C, then continued to the hydrothermal process at 100°C and 100 bar using carbon dioxide as the pressurizing gas. The research found that the combination process can produce 0.693 mg/mL of reducing sugar

1. Introduction

Cassava is an abundant natural resource in Indonesia and mostly used as food. Indonesia produced about 21.8 million tons of cassava per year [26]. Starch, as the main component of cassava, can be converted to many of its derivative product such as glucose, fructose, and other reducing sugar. Reducing sugar is a kind of sugar that has the potential to reduce other compounds. This potential characterized by a hemiacetal bond in the Haworth structure of sugar. This bond can result in an open ring form which has open aldehyde or ketone functional group. Reducing sugar have good fermentability so that making it has more value for pharmaceutical, food and fermentation industries [6].

Cassava starch is composed by amylose (15-30%) and amylopectin (85-70%) [27]. Amylopectin is a branched chain that linked at α-(1,4) and α-(1,6) glycosidic bond. Amylopectin is a branched polymer which builds the amorphous structure, but it has strong hydrogen bonding among its molecule in native starch granule. This made its structure forms double helical chains and tends to build a semicrystalline structure of starch. Amylose is a linear chain linked at α-(1,4) glycosidic bond. It tangled at amorphous regions of the semicrystalline structure of starch. Therefore, their existence in starch constructs tight granule structure which is difficult for water to penetrate inside [5; 12; 21]. These become the inhibition factor for its degradation process.

Strong structure of the starch makes the conventional method need a higher concentration of acid, to optimize the whole process [15]. The application strong acid in degradation that allows high solubilization of starch in mostly solvent (mainly in water) lead to low in processing time. But, this process is to give serious problems in product purification, cause a negative effect on the apparatus because of the corrosion dangerous, and the difficulty in wastewater treatment handling.

Acid and enzyme hydrolysis have been accepted as a common degradation method for starch conversion to reducing sugar process [8; 13; 14]. Hydrolysis of starch has been modeled by Hoover. An interaction between hydronium ion with a glycosidic bond of starch can induce electron transfer between them. The electron from glycosidic bond transferred to hydronium ion, leaving unstable oxygen atom of the glycosidic bond. Unstable oxygen atom gets easily release its bond with D-glucose monomer and associated with OH ion. This glycosidic bond cleavage resulting in lower molecular weight polysaccharide.

For the industrial scale, the conventional acid method need a huge amount of acid to set pH around 1.65. This method also needs high-temperature solubilization around 120°C to ensure starch dispersion. Otherwise, the enzymatic method can overcome the acid and heat demand, but this method operates much longer operating batch time [15]. This encourages the development of non-conventional methods in reducing sugar production for recent years.

It has been an interesting topic to optimize its degradation process through pre-treating and processing with non-conventional methods such as hydrothermal, ultrasound, microwave, and other [2; 3; 17; 25]. Each method has several advantage and limitation. Hence, the combination among non-conventional has the potential to wield more advantage in the starch degradation process. For example, hydrothermal has been combined with several methods to yield more reducing sugar. Hydrothermal has been combined with acid and enzyme. This method can produce more reducing sugar yield than the conventional acid and enzyme process [10].

Ultrasound is one of the non-conventional methods of starch degradation which utilize ultrasonic irradiation. Its wave transmits in bulk liquid resulting acoustic pressure with a sinusoidal cycle. It induces cavitation and produces microbubbles. The microbubble compressed and decompressed, resulting in a hot spot with high temperature and high pressure inside it. The microbubble can be collapsed, then producing high local temperature and high local pressure. It generates radical ion that can help glycosidic bond cleavage then producing oligosaccharide and monosaccharide. The explosion of the microbubble also produces microjet effect which can vigorously impact the starch granule surface [18; 22; 30].

Ultrasonic irradiation can increase granule porosity, decrease the size and disarray crystalline region of starch [7]. These will make water easier to penetrate inside granule. In the hydrothermal process, the reducing sugar production rate increased because more ion product reacted with starch in one time. Therefore, it can increase its reducing sugar production with lower operating temperature as in the hydrothermal process. By lowering operating temperature this also decreases the potential of reducing sugar to degrade further into furfural or 5-HMF [25].

Hydrothermal has become one of the sophisticated methods for polysaccharides degradation since it has the potential to substitute acid and enzyme [1; 2; 20; 23]. It utilizes water at high temperature and keeps it in a liquid phase by operating at high pressure. Higher temperature exhibit increasing of molar volume, then it will weaken hydrogen bonding among water molecule. By weaker hydrogen bonding, water can produce more ionic product concentration, enhancing its ability to degrade polysaccharides. This also decreases polarity of the water so that make water become less polar solvent [9; 16]. By hydrothermal process, water properties can be tuned to a suitable value to achieve the application need. Modification of the water as a solvent can also conduct to increase solvent power properties and able to become a specific reaction media.

One of the possible methods to increase the production of reducing sugar in the hydrothermal process is increasing its process temperature. However, this increases its energy demand and risk to trigger further degradation of reducing sugar into 5-HMF [20] Temperature rises not only increase the production of reducing sugar but also increase its further degradation rate. Consequently, the total reducing sugar

IOP Conf. Series: Materials Science and Engineering **543** (2019) 012085 doi:10.1088/1757-899X/543/1/012085

concentration decreases at a specific time and temperature. Controlling the reaction rate at high temperatures will save from the occurrence of further records, and also avoid unwanted products.
The aim of this work was set to perform degradation characteristics study of ultrasonic pre-treatment to intensify hydrothermal process in reducing sugar production from starch. Both the ultrasonic process only and the combination process was carried at various process time (15-120 minutes). For the combination process, 15 minutes of the ultrasonic process was applied before the hydrothermal process. The characterization was divided into granule degradation and reducing sugar production. The product was separated into a solid product and liquid product. The solid product was freeze dried and characterized using SEM and XRD to know granule degradation. The liquid product was analyzed by the DNS method to measure total reducing sugar concentration.

2. Methods

2.1. Material
The native cassava starch prepared from fresh cassava. Fresh cassava obtained from Mojokerto Regency, East Java, Indonesia. It shredded to reduce its size. Then it was separated from other residues by soaking in distilled water then it was squeezed. The separation process did three times. The obtained starch then was dried and sorted to 40 mesh.
Sodium hydroxide (NaOH), distilled water was purchased from SAP Chemical, Indonesia. dinitrosalicylic acid (DNS), D-Glucose, and potassium sodium tartrate (PST) was purchased from Merck, Germany. Ultra high purity carbon dioxide gas (CO2) was purchased from Samator, Indonesia.

2.2. Ultrasound and Hydrothermal Treatment of Starch

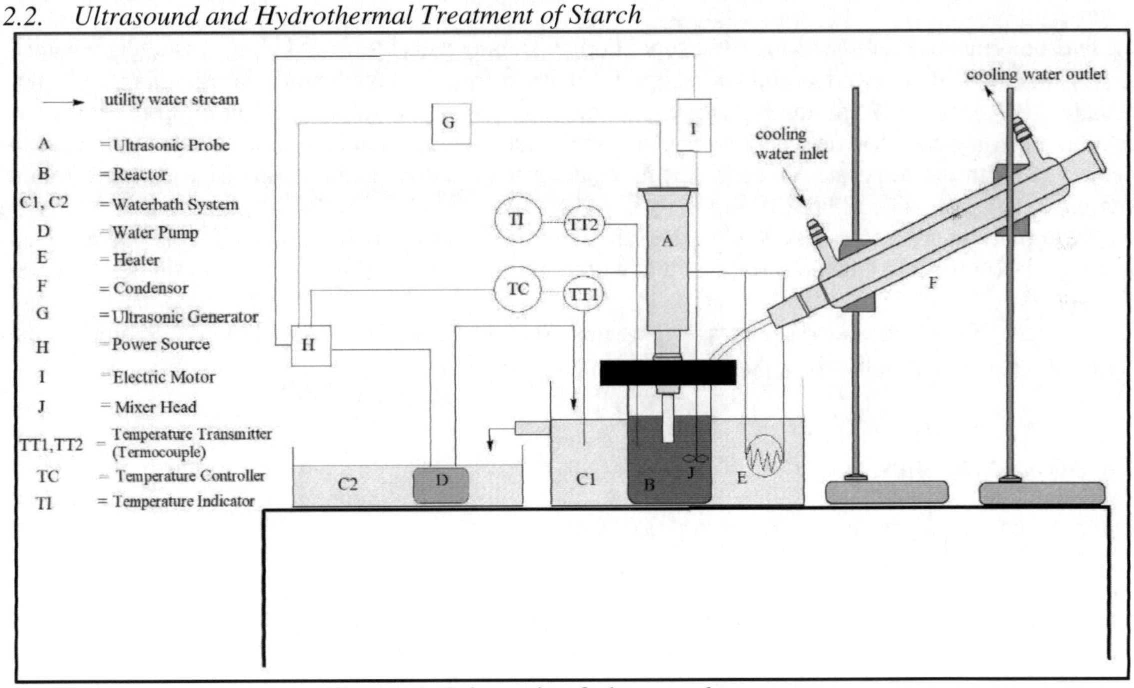

Figure 1. Schematic of ultrasound treatment

Direct ultrasonic probe VCX500 (Sonics & Materials, Inc.) with a frequency of 20 kHz and power of 500 watts was used in this experiment. Reactor capacity was 1000 mL. Ultrasound equipment arranged in figure 1. The initial starch suspension used was 1/20 (g/mL) with a volume of 500 mL. Ultrasound treatment operated at 50% amplitude and initial temperature of 40°C. Impeller was used to assist suspension mixing. The condenser was also used to avoid a concentration change effect by vaporization.

2.3. *Hydrothermal Treatment*

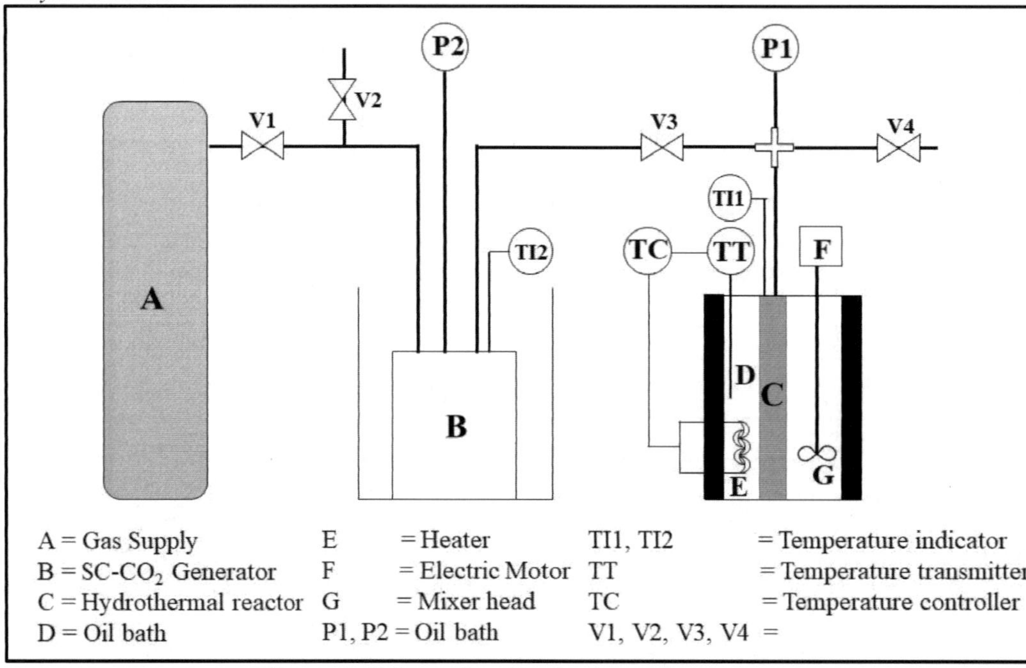

Figure 2. Schematic of hydrothermal treatment

The hydrothermal reactor was made from super duplex stainless steel tubing. Tubing system is obtained from Swagelok and arranged as shown in figure 2. Dimensions of the reactor, among others, the outer diameter 1.270 cm, 2.387 cm. Inner diameter, 29 cm in length with a volume of 20 mL. In the reactor is installed K type thermocouple with a size 0.159 cm as a sensor for the temperature inside the reactor in accordance with the setpoint. As a pressure indicator used pressure gauge (Hanyo) with a maximum pressure reading of 343.233 bar (350 kg / cm^2).

The hydrothermal process was done after the sample ultrasonicated for 30 minutes. It was operated at a pressure of 100 bar and temperature of 100°C with the hydrothermal time of 15-120 minutes (interval 15 minutes). The operating temperature condition was set by the heater and the oil bath system. Supercritical CO_2 was used for the pressurizing gas. Supercritical CO_2 was generated by compression and adiabatic decompression in supercritical CO_2 generator (B).

2.4. *Scanning Electron Microscopy (SEM)*

SEM was conducted by Zeiss Evo MA-10. The sample was frozen by nitrogen liquid before it was snapped to get represent the image of granule morphology. The sample coated by PdAu before characterized. The magnification factor of 2000 was used.

2.5. *X-Ray Diffraction (XRD)*

XRD was performed by X'pert Pro PANalytical. The signal was processed by OriginPro 2018. Diffraction angle 2θ used at a range between 12° to 28°. The data obtained smoothed by Savitsky-Golay method, 11 points of window and 2nd-degree polynomial order. Relative Crystalline Index (RCI) is calculated according to the previous studies (Wu et al., 2016). It calculated as the ratio between the crystalline area (A_C) and the total area of the crystalline and amorphous region (A_T).

$$RCI = \frac{A_C}{A_T} \times 100\% \tag{1}$$

A_C was calculated by peak integration with manually select 7 base marker at relatively same diffraction angle on the curve and 5th-degree polynomial as the fitting method. By the same manner, A_T was determined by minimum constant baseline as the fitting method.

2.6. DNS method for total reducing sugar

Dinitrosalicylic acid (DNS) method was adapted from the experiment performed by Miller. Briefly, it utilizes a reduction-oxidation reaction between reducing sugar and DNS. Reduction of DNS form 3-amino, nitrosalycylic acid resulted in orange color. Rochelle salt or PST was used to prevent oxygen dissolving by DNS [19]. The fresh reagent was made by adding 1 gr DNS, 1.6 gr NaOH, PST 30 gr was mixed in 100 mL distilled water. Then 0.2 mL sample diluted into 1 mL solution, then this considered as dilution factor of 5. The standard curve solution was made by 0.05-0.5 mg/mL of glucose for each day of the experiment. After that, 3 mL reagent was added. The solution was heated in a 100°C water bath for 10 min, then it was drenched in cool water. Subsequently, it was diluted by adding 2 mL of distilled water. Afterward, the spectrophotometry conducted by Genesys 10S UV-VIS Spectrophotometer at 502 nm wavelength.

3. Result and Analysis

3.1. Granule breakage effect

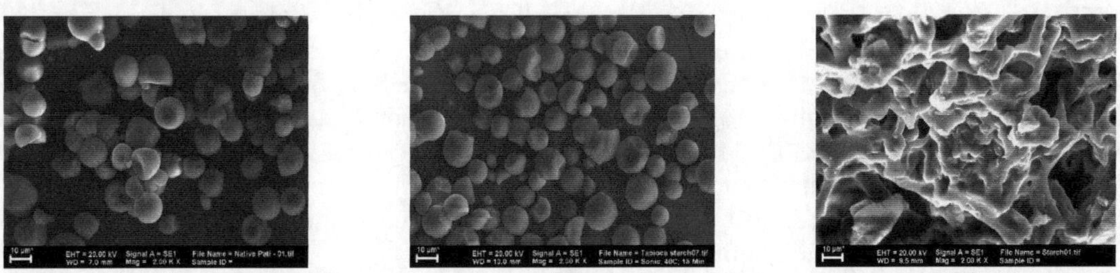

(a) (b) (c)

Figure 3. SEM result of starch: native or before processed (a), ultrasonicated for 15 minutes (b) and ultrasonicated for 15 minutes continued with hydrothermal process for 15 min (c)

Starch granule consists of amylopectin and amylose which formed as a semicrystalline biopolymer. Native starch has granules which close each other as shown as SEM result in figure 3(a). It forms a tight structure that can prevent the penetration of water inside the granule.

After ultrasonicated as shown in figure 3(b), it exhibited a small portion of surface destruction. No significant granule size reduction can be seen from the figure. However, ultrasonic treatment at 50% Amplitude and 40°C dispersed the granule arrangement. This result was similar to experiment using 170 W ultrasonic treatment of potato starch that only shows some fissure over its external surface without significant granule size reduction [28; 29].

Figure 3(c) show that the granule was severely disrupted. Heat effect conducted by 100°C hydrothermal inducing swelling of the starch granule. As porosity of starch granule grew larger, more water penetrates inside it. Then, at some point granule was not able to maintain its shape. It disrupted. This process often called starch gelatination. This process can extract most part of starch into liquid, then lead to starch solubilization. Solubilization will make glycosidic cleavage easier. By enhancing this process, conversion of polysaccharide into oligosaccharide or monosaccharide will be also increased. Unconverted polysaccharides were reassociated each other and resulting gel-like image like Figure 3. This phenomenon called retrogradation [31].

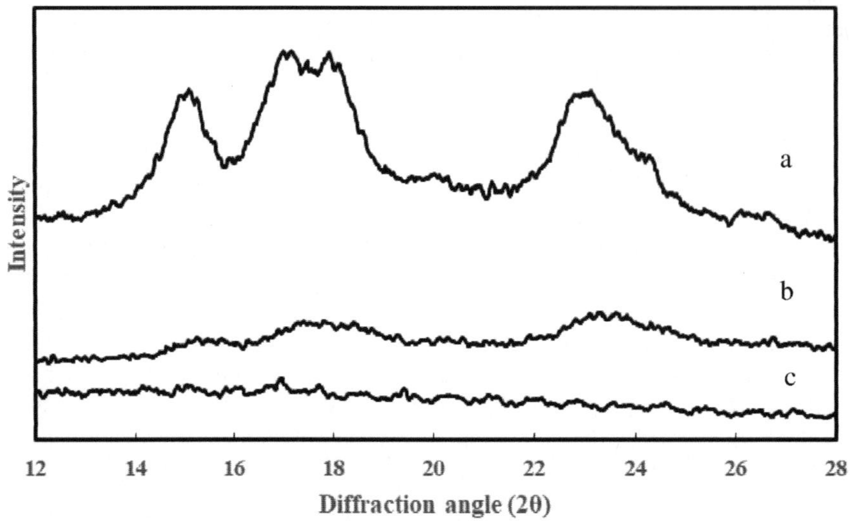

Figure 4. XRD result of starch: native (a), ultrasonicated for 15 min (b) and ultrasonicated for 15 min continued with hydrothermal process for 15 min (c)

From XRD result in figure 4(a), it has four characteristics peak which located diffraction angle (2θ) of 13°, 17°,18° and 23°. Those peaks were formed as four sloppy peaks which represented semicrystalline areas of the starch granule. After ultrasonicated, as represented in figure 4(b), the peak reduced significantly. There was crystalline region degradation by ultrasonic although there is no significant destruction as shown in figure 3(b). Microjet effect pushed water inside the granule. Cavitation might be generated inside the granule, then degrading the whole granule from inside. Some part of the crystalline region was disarrayed. Lost of the crystalline region could decrease starch thermal resistance, then it might increase degradation in the hydrothermal process.

Combination process resulted in the XRD curve as shown in figure 3(c). The process was followed by the disorientation of the main peak of the starch granule (13°, 17°, 18° and 23°). It extracted amylose from the whole structure and disrupted the semicrystalline structure of amylopectin. Disrupted of amylopectin might be made it turned into amorph form as figure 3(c).

Table 1. Relative crystallinty index calculation result for each variable

Variable	RCI
Native (a)	37.47%
Ultrasonicated for 15 min (b)	29.61%
Ultrasonicated for 15 min continued with Hydrothermal Process 15 min (c)	6.41%

The relative crystallinity index (RCI) can be calculated to represent the percentage area occupied as the crystalline area. Commercial cassava used in the experiment has an initial RCI of 37.47% as shown in Table 1. This result quite agreed with other experiment using ACS standard cassava starch which shows result relative crystallinity of 37% [11].

Starch relative crystallinity index slightly decreases to 29.61%. This supported by the XRD result in figure 4(b). The main peak of starch intensity was considerably decreased indicating some crystalline part of starch was broken and converted into amorphous form.

Combination process relative crystallinity index become 6.41% as shown in Table 1. This value represent 82% of crystallinity reduction. These crystallinty reduction show more satisfactory than ordinary Heat Moisture Treatment (HMT) at 100 °C which experimented by Gunaratne and Hoover. HMT only provided 2% of crystallinity reduction [11]. For the combination process, starch suspension treated with 30 minutes ultrasonic irradiation at 40°C. By ultrasonication, the starch granule had more pore and lower crystallinity index. It became easier to have water penetrate inside. After that, the sample was put in a hydrothermal reactor and processed at 100°C and 100 bar by supercritical CO_2 as the pressurizing gas. The starch granule excessively disrupted and converted most parts of the semicrystalline region become amorphous regions. This was agreed in figure 4(c) which show the intensity of its peak significantly decreased. The combination ultrasound treatment and hydrothermal can greatly decrease the degradation process.

3.2. Reducing sugar production effect

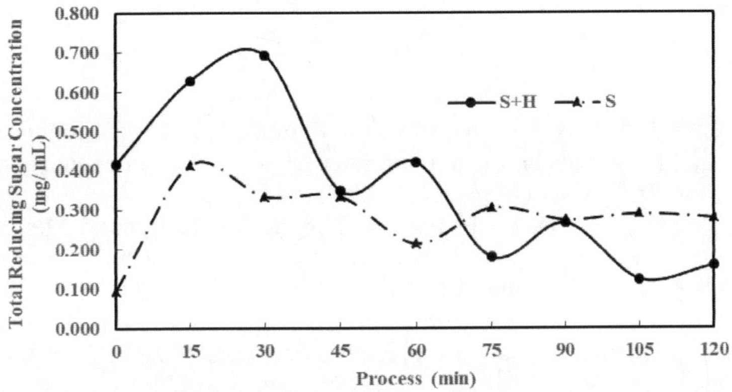

Figure 5. Total reducing sugar concentration analysis result on ultrasonicated (S) and ultrasonicated continued with hydrothermal process (S+H)

In figure 5, total reducing sugar production by ultrasound treatment was increased up to 0.415 mg/mL for 15 minutes of the process. Afterward, its total reducing sugar concentration tended to stable in around 0.291 mg/mL. From previous studies, ultrasound has a limitation for degrading polymer and tend to reach a stable point [24]. It also has possibilities to have similar phenomena at starch degradation. Physical degradation by microjet effect extracted some part of amylose and amylopectin. Then, radical ion product generated from ultrasonic irradiation cleaved the glycosidic bond similar mechanism to conventional hydrolysis.

From DNS (figure 5) and XRD (figure 4 and Table 1) analysis result, it shows that the degradation process through pre-treated hydrothermal by ultrasound (S) reducing sugar concentration reached higher maximum concentration than ultrasound only (S+H) for early process time. It possible since the S+H process utilize better crystalline region reduction that S process. Better crystalline region reduction can make starch more viable to water penetration inside granule. The increasing of water viability of starch enhanced gelatinization process brought by the hydrothermal process. The ionic product generated by water dissociation at hydrothermal process cleaved glycosidic bond like conventional hydrolysis. The decreasing of dielectric constant at the hydrothermal process also converting water into a less polar solvent. Water becomes less polar and became more suitable for starch which less polar compound. Both properties assist water to degrade starch become reducing sugar without the presence of an additional solvent.

Therefore it can increase the number of starch reacted at the same time and increase its total reducing sugar concentration to 0.693 mg/ mL. After that, the total reducing sugar concentration tends to decrease over time. This further degradation phenomenon has identical to previous research [20]. It may the hydrothermal process has transferred enough energy so that the produced reducing sugar degrade further into furfural and 5-HMF. This result had much reducing sugar than experiment conducted by Ayoola et.al. using 1-hour acid process using 1 M H_2SO_4 which resulting about 0.06 mg/ mL of reducing sugar [4].

4. Conclusion

Ultrasonic treatment of 50% Amplitude and 40°C can disperse the granule arrangement and produce reducing sugar up to 0.291 mg/mL. Granule dispersion was followed by reducing the crystalline area without resulting in significant granule size. Combination ultrasonication as pretreatment of hydrothermal at 100°C and 100 bar can completely disrupt starch granule, reducing the crystalline area and increase the produced total sugar concentration up to 0.693 mg/mL.

5. Acknowledgement

This study was supported by a grant (1341/PKS/ITS/2018) from Post Graduate Research of Local Funding Program, Institut Teknologi Sepuluh Nopember, Indonesia.

References

[1] Albarelli J, Paidosh A, Santos D, Marechal F and Meireles M 2016 Environmental, energetic and economic evaluation of implementing a supercritical fluid-based nanocelulose production process in a sugarcane biorefinery *Chem. Eng. Trans.* **47** 49-54.

[2] Alcázar-Alay S C, Osorio-Tobón J F, Forster-Carneiro T, Steel C and Meireles M A A 2017 Polymer modification from semi-defatted annatto seeds using hot pressurized water and supercritical CO_2 *J. Supercrit. Fluids* **129** 48–55.

[3] Audrey H V 2013 Catalytic depolymerisation of starch-based industrial waste: use of non-conventional activation methods and novel reaction media PhD Thesis, University of Oulu, Oulu, Finland.

[4] Ayoola A, Adeeyo O, Efeovbokhan V and Olasimbo D A 2013 Optimum hydrolysis conditions of cassava starch for glucose production *J. Adv. Res. IT Eng.* **2(1)** 93–101.

[5] BeMiller J and Roy W 2015 Starch Chemistry and Technology, 3rd Edition, Academic Press, Oxford, UK.

[6] Bertolini A., 2010, Starches: Characterization, Properties, and Applications, CRC Press: United States of America.

[7] Boufi S, Bel Haaj S, Magnin A, Pignon F, Impéror-Clerc M and Mortha G 2018 Ultrasonic assisted production of starch nanoparticles: structural characterization and mechanism of disintegration *Ultrason. Sonochem.* **41** 327–36.

[8] Cardenas-toro F P, Alcazar-alay S C, Forster-carneiro T and Meireles M A 2014 Obtaining oligo- and monosaccharides from agroindustrial and agricultural residues using hydrothermal treatments *Food Public Health* **4** 123–39.

[9] Cocero M J, Cabeza Á, Abad N, Adamovic T, Vaquerizo L, Martínez C M and Pazo-Cepeda M V 2018. Understanding biomass fractionation in subcritical & supercritical water *J. Supercrit Fluids* **133** 550-65.

[10] Gaewchingduang S and Pengthemkeerati P 2010 Enhancing efficiency for reducing sugar from cassava bagasse by pretreatment *World Acad Sci Eng Technol* **46** 727–30.

[11] Gunaratne A and Hoover R 2002 Effect of heat-moisture treatment on the structure and physicochemical properties of tuber and root starches *Carbohyd Polym* **49** 425–37.

[12] Hancock R D and Tarbet B J 2000 The other double helix—The fascinating chemistry of starch *J. Chem Educ* **77** 988-92.

[13] Hernández I P, Pérez-Pimienta J A, Messina S and Saldaña D C E 2012 Dilute sulfuric acid hydrolysis of tropical region biomass *J. Renew Sustain Ener* **4** 021201.

[14] Hoover R 2007 Acid-treated starches *Food Rev Int* **16:3**, 369-92

[15] Hull P 2010 Glucose syrups: technology and applications United Kingdom: John Wiley and Sons

[16] Kruse A and Dahmen N 2015 Water-A magic solvent for biomass conversion *J. Supercrit Fluids* **96** 36–45.

[17] Martína C, Wei M, Xiong S and Jönsson L J J 2017 Enhancing saccharification of cassava stems by starch hydrolysis prior to pretreatment *Ind Crop Prod* **97** 21–31
[18] McDonnell C and Tiwari B K 2017 *Comprehensive Analytical Chemistry*, ed E. Ibanez and A. Cifuentes (Amsterdam, Elsevier) p 111-26
[19] Miller G L 1959 Use of dinitrosalicylic acid reagent for determination of reducing sugar *Anal. Chem.* **31** 426–28.
[20] Nagamori M and Funazukuri T 2004 Glucose production by hydrolysis of starch under hydrothermal conditions *J. Chem. Technol. and Biotechnol.* **79** 229–33.
[21] Pérez S and Bertoft E 2010 The molecular structures of starch components and their contribution to the architecture of starch granules: A comprehensive review *Starch/Staerke* **62** 389–20.
[22] Pingret D, Durand G, Fabiano-Tixier A S, Rockenbauer A, Ginies C and Chemat F 2012 Degradation of edible oil during food processing by ultrasound: Electron paramagnetic resonance, physicochemical, and sensory appreciation *J. Agric. Food Chem.* **60** 7761–68.
[23] Posmanik R, Cantero D A, Malkani A, Sills D L and Tester J W 2017 Biomass conversion to bio-oil using sub-critical water: Study of model compounds for food processing waste *J. Supercrit. Fluids* **119** 26–35.
[24] Price G J and Smith P F 1993 Ultrasonic degradation of polymer-solutions: 2. the effect of temperature, ultrasound intensity and dissolved-gases on polystyrene in toluene *Polymer* **34** 4111–17.
[25] Rogalinski T, Liu K, Albrecht T and Brunner G 2008 Hydrolysis kinetics of biopolymers in subcritical water *J. Supercrit. Fluids* **46** 335–41.
[26] Statistics Indonesia 2015 *Production of food crops* Subdirectorate Food Crop Stat Indonesia, Jakarta, Indonesia
[27] Srichuwong S, Sunarti T C, Mishima T, Isono N and Hisamatsu M 2005 Starches from different botanical sources I: Contribution of amylopectin fine structure to thermal properties and enzyme digestibility *Carbohydr. Polym.* **60(4)** 529–38.
[28] Sujka M 2017 Ultrasonic modification of starch–Impact on granules porosity *Ultrason. Sonochem.* **37** 424–29.
[29] Sujka M and Jamroz J 2013 Ultrasound-treated starch: SEM and TEM imaging, and functional behaviour *Food Hydrocoll.* **31** 413–19.
[30] Suslick K S, Didenko Y, Fang M M, Hyeon T, Kolbeck K J, McNamara III W B, Mdleleni M M and Wong M 1999 Acoustic cavitation and its consequences Philos. *T. R. Soc. A* **357(1927)** 335–53.
[31] Wang S, Li C, Copeland L, Niu Q and Wang S 2015 Starch retrogradation: A Comprehensive Review. *Compr. Rev. Food Sci. Food Saf.* **14(5)** 568–85.
[32] Xue Z, Liu L and Zhang M 2017 Study on optimization of reducing sugar yield from corn stalk based on microwave-assisted acid pretreatment process *Chem. Eng. Trans.* **62** 1315-20

IOP Conf. Series: Materials Science and Engineering **543** (2019) 012086 doi:10.1088/1757-899X/543/1/012086

Ternary liquid-liquid equilibria for β-caryophyllene(1) + 1-propanol(2) + water(3) system at temperatures 303.15, 313.15, and 323.15 K

K Kuswandi*, G Wibawa, I Hidayatulloh, C Mahardhikawati and A D Pratiwi
Department of Chemical Engineering, Institut Teknologi Sepuluh Nopember, Kampus ITS Sukolilo, Surabaya 60111, Indonesia

*E-mail: kuswandi@chem-eng.its.ac.id

Abstract. The experimental liquid–liquid equilibrium data of the system β–caryophyllene(1) + 1-propanol(2) + water(3) at temperatures 303.15, 313.15, and 323.15 K are presented. The mixtures of β–caryophyllene + 1-propanol + water were stirred for 4 hours and settled for at least 20 hours at constant temperature. The samples taken from both phases were analyzed by a gas chromatography (GC). The experimental equilibrium data were correlated by NRTL and UNIQUAC models. The compatibility of the models is based on the value of root mean square deviation (RMSD). The liquid-liquid phase diagram of the system shows type I phase behavior, where two binary subsystems are miscible and only one binary subsystem is partially miscible. The temperature effect of this system on immiscible zone is very small. Based on the experiment and the calculation results, the ternary equilibrium data have been successfully represented using both the NRTL and the UNIQUAC models.

UNIQUAC models. The compatibility of the models is based on the value of root mean square deviation (RMSD). The liquid-liquid phase diagram of the system shows type I phase behavior, where two binary subsystems are miscible and only one binary subsystem is partially miscible. The temperature effect of this system on immiscible zone is very small. Based on the experiment and the calculation results, the ternary equilibrium data have been successfully represented using both the NRTL and the UNIQUAC models.

1. Introduction

Indonesia is a potential country as a source of essential oils. Indonesia produces 40 of 70 types of essential oils that were traded on the world market. From that amount 15 types have been traded on the world market, one of which is clove oil. About 60% of the clove oil produced in Indonesia is exported to many countries in the world.

Clove oil (Syzygium aromaticum) consists of hydrocarbon compounds such as terpene, sesquiterpene, etc; oxygenated hydrocarbon compounds such as alcohol, ester, ether, ketone, laktone, phenol, phenol ether, etc; and a small amount of non-volatile residue such as paraffin. Eugenol, one of the oxygenated hydrocarbon compounds, is a main compound of clove oil with a content of 78-98%. There are three types of clove oil produced from leaves, stems and flower-buds of S. Aromaticum that contain eugenol

82-88%, 90-95% and 60-90% respectively [1]. Eugenol is a colorless liquid, aromatic, and spicy that can be used in many fragrance, flavor, and pharmacy industry [2,3]. Oxygenated hydrocarbon compound (non-terpene) is a main cause of aromatic in the clove oil, while terpene and sesquiterpene are easily oxidized under the effect of light and air or on poor storage conditions that affect the fragrance and decrease the solubility of the oil in alcohol. Therefore, the removal of terpenes will increase the quality of clove oil. One of the terpene compounds is β-caryophyllene with a content of 10% in the clove oil [2].

According to the USP (United States Pharmacopeia), to be traded in the world market, eugenol should have a purity more than 98%. The purity of eugenol determines the selling price of clove oil. The clove oil with eugenol content of more than 98% is traded at prices 6-8 times higher than clove oil with eugenol content of less than 90%.

To satisfy these standards, it is necessary to process the separation of β-caryophyllene from oil. The separation process is called deterpenation which is the process to remove a part or all of terpene hydrocarbon compounds to reduce concentrations of terpenes which affect flavor and fragrance, increase solubility in alcohols and fragrance oils [4]. In the deterpenation process, the characteristic of a solvent should be considered such as polar, non-polar, volatility, and toxicity. The use of alcohol as a solvent has several advantages [4,5]: (1) They have a high solubility in aqueous solutions; (2) They increase the aromatic strength of the mixture; (3) Since the presence of the alcohol, oxidation reactions can be reduced; (4) They are non-toxic; (5) An aqueous alcohol is selective to dissolve eugenol, but cannot dissolve terpenes.

The knowledge of phase equilibrium properties of the fluid mixture is an important issue in various separation processes in the industry [6], but the availability of data is inadequate, both for the type of fluid system and operating range. In the previous studies, the equilibrium data of clove oil have been presented using various solvents [5,7-9]. In this work, liquid-liquid equilibrium data were investigated for the system β-caryophyllene + 1-propanol + water at temperatures 303.15, 313.15, and 323.15 K. The experimental data obtained were correlated by the NRTL [10] and the UNIQUAC [11] equations.

2. Experimental

2.1. Chemicals

β-caryophyllene 99.9 % was supplied by PT. Indesso Niagatama. 1-Propanol 99.8 % was supplied by MERCK and bi-distilled water was supplied by PT. Ika Pharmindo Putramas. The purity of chemicals was checked by gas chromatography.

2.2. Procedure

The LLE measurements were conducted for the β-caryophyllene + 1-propanol + water system at 303.15, 313.15, and 323.15 K. About 30 cm^3 of the mixture were stirred intensely in a thermostat equilibrium cell. The equilibrium cell was equipped by a magnetic stirrer and a water-jacket to circulate water as a heater to maintain a constant temperature. Thermocouple Anly AT-502 and Tianshi XMT-7000 with accuracy ±0.1°C was used. The mixtures were stirred by the magnetic stirrer for 4 hours and settled for at least 20 hours at constant temperature to ensure that the equilibrium is going to be reached. After phase equilibrium has been reached, samples of two layers were analyzed by gas chromatography (GC) Shimadzu 2010 plus.

2.3. Analysis

The composition of the sampled liquid phase was determined using gas chromatography Shimadzu 2010 plus with Thermal Conductivity Detector (TCD) and 30 m, 0.32 mm ID, 0,25 μm RTX-5 column. The carrier gas (helium) flowrate was maintained at 3 cm^3/min, purge flow at 3 cm^3/min, column flow at 0.59 cm^3 /min, and split ratio 50:1. The temperatures of detector and injector were set at 573.15 K. The oven temperature was set at 343.15 K hold time 5 min and increased to 473.15 K at a rate of 40 K/min, hold time 5 min. About 1 μL sample was injected to gas chromatography with maximum capacity and accuracy of Microliter Syringe Shimadzu were 10 μL and ± 0.1 μL. The peak areas of components were calibrated with gravimetrically prepared mixtures and converted to mole fraction. The results in area fraction were substituted by binary calibration equation (water + 1-propanol) and (1-propanol + β-caryophyllene) to get mole fraction each components.

3. Results and discussion

3.1. Experimental data

The experimental data of the LLE for ternary system β-caryophyllene + 1-propanol + water are presented in table 1. All concentrations are expressed in mole fractions. The ternary diagram of the experimental results can be seen in figures 1-3 for all temperatures studied.

Table 1. Experimental liquid-liquid equilibrium data of β-caryophyllene (1) + 1-propanol (2) + water (3) system at 303.15, 313.15, and 323.15 K.

Temperature	Organic phase			Aqueous phase		
(K)	x_1	x_2	x_3	x_1	x_2	x_3
303.15	0.4683	0.4046	0.1271	0.0890	0.5712	0.3397
	0.6018	0.2956	0.1026	0.0266	0.5188	0.4546
	0.5514	0.3429	0.1057	0.0437	0.5294	0.4269
	0.6758	0.2291	0.0951	0.0157	0.4260	0.5583
	0.7071	0.2006	0.0923	0.0034	0.4108	0.5857
	0.7502	0.1711	0.0788	0.0054	0.2846	0.7100
	0.7554	0.1713	0.0733	0.0043	0.1821	0.8136
	0.8111	0.1220	0.0669	0.0040	0.1399	0.8561
313.15	0.4850	0.4088	0.1063	0.0950	0.5467	0.3583
	0.5962	0.2894	0.1145	0.0268	0.5033	0.4699
	0.5591	0.3308	0.1100	0.0391	0.5258	0.4351
	0.6243	0.2639	0.1118	0.0078	0.4329	0.5594
	0.6843	0.2124	0.1033	0.0063	0.3870	0.6066
	0.7402	0.1689	0.0909	0.0073	0.2591	0.7336
	0.8238	0.0997	0.0765	0.0034	0.1446	0.8519
	0.7709	0.1457	0.0834	0.0066	0.1721	0.8213
323.15	0.4226	0.4276	0.1498	0.1273	0.5334	0.3393
	0.5988	0.2929	0.1083	0.0359	0.4982	0.4660
	0.5557	0.3229	0.1214	0.0572	0.5156	0.4272
	0.6175	0.2702	0.1124	0.0222	0.4173	0.5606
	0.7006	0.2019	0.0975	0.0145	0.3725	0.6130
	0.7423	0.1711	0.0866	0.0041	0.2591	0.7368
	0.8215	0.1059	0.0726	0.0095	0.1110	0.8795
	0.7512	0.1633	0.0855	0.0032	0.1477	0.8491

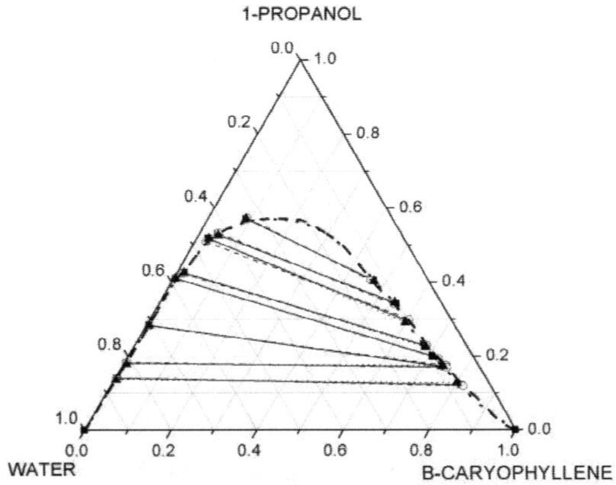

Figure 1. LLE of β-caryophyllene + 1-propanol + water system at 303.15 K; (■) experimental data; (○) correlated by NRTL; (▲) correlated by UNIQUAC; (—) experimental tie lines; (- - -) NRTL tie lines; (····) UNIQUAC tie lines; binodal curve calculated using: (- - -) NRTL Model; (····) UNIQUAC Model.

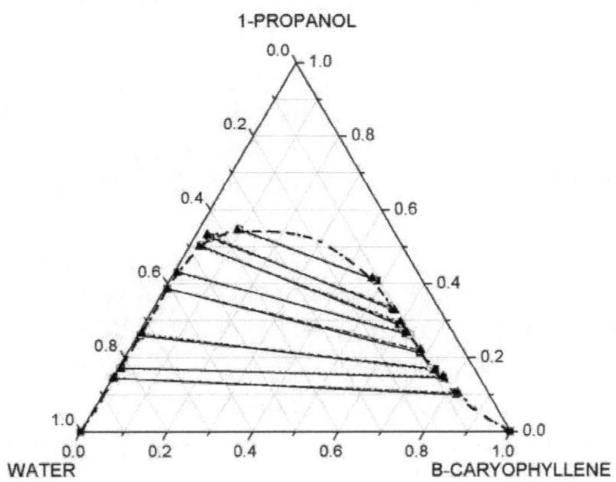

Figure 2. LLE of β-caryophyllene + 1-propanol + water system at 313.15 K; (■) experimental data; (○) correlated by NRTL; (▲) correlated by UNIQUAC; (—) experimental tie lines; (- - -) NRTL tie lines; (····) UNIQUAC tie lines; binodal curve calculated using: (- - -) NRTL Model; (····) UNIQUAC Model.

IOP Conf. Series: Materials Science and Engineering **543** (2019) 012086 doi:10.1088/1757-899X/543/1/012086

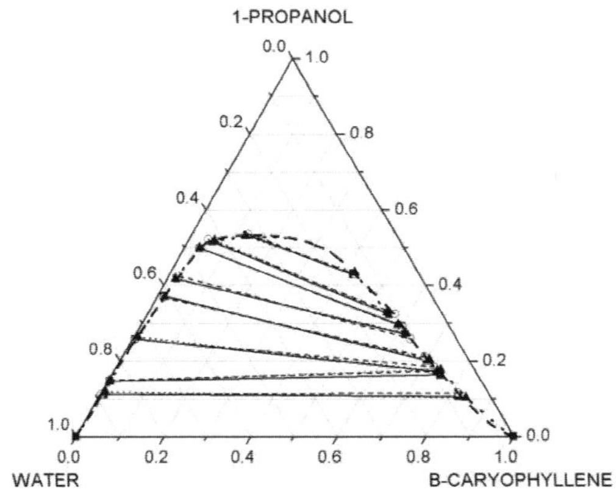

Figure 3. LLE of β-caryophyllene + 1-propanol + water system at 323.15 K; (■) experimental data; (○) correlated by NRTL; (▲) correlated by UNIQUAC; (—) experimental tie lines; (- - -) NRTL tie lines; (····) UNIQUAC tie lines; binodal curve calculated using: (- - -) NRTL Model; (····) UNIQUAC Model.

As can be observed on these figures, liquid-liquid equilibria of the ternary mixtures exhibit type I [12] that has one pair of partially miscible components (water + β-caryophyllene) and two pairs of miscible components (β-caryophyllene + 1-propanol and 1-propanol + water). The organic phase is dominated by β-caryophyllene meanwhile water is dominating in aqueous phase. 1-Propanol is more distributed in aqueous phase. It can also be observed that the temperature effect for this system is not significant with a relatively small change in the tie lines for all temperatures.

3.2. Data correlation

The experimental data for all temperatures were correlated using the NRTL and the UNIQUAC equations to obtain binary interaction parameters of the models. These parameters were important for the calculation data. In the NRTL correlation, the non-randomness (α) was set to 0,2 based on chemical characteristic group [10, 13]. This value was used for all binary pairs. For the UNIQUAC model, the structural parameters of the pure components (ri and qi) were calculated using the group contribution method and presented in table 2. The correlation results for the binary parameters using the NRTL and the UNIQUAC models are shown in table 3. Figures 1-3 show the equilibrium diagrams for the correlation data along with the experimental data at temperatures 303.15, 313.15, and 323.15 K.

Table 2. Structural parameters r_i and q_i used in UNIQUAC correlation.

Components	r_i	q_i
β-caryophyllene	9.1945	7.364
1-propanol	3.2499	3.128
water	0.92	1.40

Table 3. Binary interaction parameters of NRTL and UNIQUAC models for β-caryophyllene(1) + 1-propanol(2) + water(3) system.

T (K)	Comp (ij)	NRTL parameters			RMSD	UNIQUAC parameters		RMSD
		α	a_{ij} (J/mol)	a_{ji} (J/mol)	(%)	a_{ij} (J/mol)	a_{ji} (J/mol)	(%)
303.15	1-2	0.2	1450.19	1475.98	0.48	368.19	-65.79	0.36
	1-3		1606.29	-852.31		640.85	13.49	
	2-3		1338.72	413.23		3389.02	-53.79	
313.15	1-2	0.2	1279.97	1522.86	0.46	-13.33	152.54	0.48
	1-3		1663.61	-888.46		881.40	-68.54	
	2-3		1340.58	296.18		124.86	-96.38	
323.15	1-2	0.2	437.37	1176.78	0.68	119.87	22.28	0.61
	1-3		1718.97	-1028.39		881.48	-49.50	
	2-3		1149.59	-73.24		176.95	-35.47	

The binary parameters for each component were determined by minimizing the differences between the experimental and calculated mole fractions for each component over all the tie lines. The objective function (OF) used as equation (1):

$$OF = \sum_{k=1}^{M}\sum_{j=1}^{2}\sum_{i=1}^{3}\left[x_{ijk} - x_{ijk}^{cal}\right]^2 \tag{1}$$

where M is the number of tie lines, x indicates the experimental mole fraction, x^{cal} is the calculated mole fraction, and subscripts i, j, k denote component, phase, and tie line respectively. Optimization of the parameters can be evaluated based on the root mean square deviation (RMSD) for both phases. The RMSD between experimental and calculated data are based on equation (2).

$$RMSD = 100\,\mathrm{x}\left(\sum_{k=1}^{M}\sum_{j=1}^{2}\sum_{i=1}^{3}\frac{\left[x_{ijk} - x_{ijk}^{cal}\right]^2}{6M}\right)^{\frac{1}{2}} \tag{2}$$

The calculation of the binary interaction parameters a_{ij} and a_{ji} along with RMSD values given in table 3 was performed separately at each temperature. It can be seen that the UNIQUAC model provides a slightly better correlation than the NRTL model. Base on the RMSD values, both the NRTL and the UNIQUAC models give a good representation for the system β-caryophyllene + 1-propanol + water at all temperatures. The average values of the RMSD calculated by the NRTL and the UNIQUAC models are 0.54 and 0.48 % respectively.

4. Conclusions

Liquid-liquid equilibrium data of β-caryophyllene + 1-propanol + water system at temperatures 303.15, 313.15, and 323.15 K have been measured. The temperature effect of this system is very small. Correlation data using the NRTL and the UNIQUAC models fit the experimental data satisfactorily with the RMSD values are 0,54 and 0,48 % respectively. The UNIQUAC model gives slightly better results than the NRTL model for this system.

Acknowledgments
This research was supported by PT. Indesso Niagatama and the Thermodynamics Laboratory of Chemical Engineering, Institut Teknologi Sepuluh Nopember.

References

[1] Lawless J 1995 *The Illustrated Encyclopaedia of Essential Oils* (Northampton: Barnes and Noble Inc.)
[2] Alma M H, Ertas M, Nitz S and Kollmannsberger H 2007 *Bio Resources* **2** 265-69
[3] Wenqiang G, Shufen L, Ruixiang Y, Shaokun T and Can Q 2007 *Food Chem.* **4** 1558-64
[4] Chafer A, Torre J, Munoz R and Burguet M C 2005 *Fluid Phase Equilibr.* **238** 72-76
[5] Kuswandi K and Naryono E 2011 *Int. Rev. Chem. Eng.* **3** 663-67
[6] Pokki J, Mannisto M and Alopaeus V 2018 *Chem. Eng. Trans.* **69** 19-24
[7] Tetrisyanda R, Kuswandi K and Wibawa G 2017 *AIP Conf. Proc.* **1788** 0300781-6
[8] Parameswari Z D, Kurniawan A M, Wibawa G and Kuswandi K 2017 *AIP Conf. Proc.* **1840** 1000121-9
[9] Sucipto R K H, Kuswandi K and Wibawa G 2017 *AIP Conf. Proc.* **1840** 1000131-6
[10] Renon H and Prausnitz J M 1968 *AIChE J.* **14** 135-44
[11] Abrams D S and Prausnitz J M 1975 *AIChE J.* **21** 116-28
[12] Sorensen J M and Arlt W 1980 *Liquid-Liquid Equilibrium Data Collection Ternary Systems Part 2 Dechema Chemistry Data Series* (Frankfurt)
[13] Chafer A, Munoz R, Burguet M C and Berna A 2004 *Fluid Phase Equilibr.* **224** 251-56

IOP Conf. Series: Materials Science and Engineering **543** (2019) 012087 doi:10.1088/1757-899X/543/1/012087

Preparation of Activated Carbon from Mangrove Waste by KOH Chemical Activation

Paryanto[1], W A Wibowo[1], D Hantoko[1,2] and M E Saputro[1]
[1] Department of Chemical Engineering, Faculty of Engineering, Universitas Sebelas Maret Jl. Ir. Sutami No. 36A, Surakarta, Central Java Province, 57126, Indonesia
[2] Institute of Energy and Power Engineering, Zhejiang University of Technology Hangzhou, Zhejiang Province, 310014, People's Republic of China

E-mail : paryanto.uns@gmail.com

Abstract. In this study, Mangrove (*Rhizophora mucronata*) fruit waste which was generated from water-based extraction of natural dyes from this fruit was used as raw material in activated carbon preparation. Carbonization of chopped Mangrove fruit waste was conducted in a batch pyrolyzer for slow pyrolysis at temperature range of 300°C to 335°C and atmospheric pressure. The produced carbon was then soaked with potassium hydroxide (KOH) solution for 24 hours before activation process in an electrical furnace at 400°C for an hour. Effect of KOH concentration on physical characteristics of activated carbon was investigated in terms of moisture content, ash content, surface area, pore diameter, iodine number, and surface morphology. It was observed that the surface area and pore diameter increased with increasing the KOH concentration. Soaking the carbon in 2.5 M KOH provided the highest surface area and pore diameter of activated carbon, which were 1,920.6 m^2/g and 5.52 μm, respectively. Also, the iodine number of 1,540.13 mg/g was obtained in correspond to the adsorption capacity.

1. Introduction

Activated carbon has been well known as good absorbent for wide applications including wastewater treatment, air pollution control, and gas purifying due to high adsorption capacity, large surface area, and low cost [1]. Typically, agricultural waste such as rice husk, sawdust, bamboo, coconut shell [2], palm shell [3], and also urban organic waste [4] are used as raw material of activated carbon production.

In Indonesia, transplantation of Mangrove especially Rhizophora mucronata are used to reduce the land erosion by sea water waves [5]. The mangrove forest has been long utilized by society for construction materials, firewood, and charcoal [4]. Nowadays, the fruit of this mangrove species is used as natural dyes resources due to high tannin content (15–36%.wt) that produce reddish brown dyes [6]. Unfortunately, the significant amount of solid residue is generated in water-based tannin extraction of natural dyes production that it is only disposed without proper treatments. The production of one kg of concentrated natural dyes solution from this chopped fruit will generate 0.5 kg of fruit waste. Due to high density and carbon content of this solid residue, it is possibly to be used as a precursor in preparation of activated carbon in order to produce more valuable product.

Basically, the activated carbon is possibly be synthesized by physical and/or chemical activation process. Physical activation involves the carbonization under the inert atmosphere or using either steam, air, and CO_2 as activating agent [3, 7, 8, 9]. While, chemical activation involves the presence of chemicals such as; alkaline metal [10], zinc chloride [11], phosphoric acid [12] in an inert condition.

Due to higher carbon yield, larger surface area and lower operating cost, the chemical activation is preferred after physical activation [2, 13]. Many researchers have been observed that activated carbon produced by KOH activation from carbonaceous materials has a high surface area, good porosity, and well-developed of pore distribution [2, 10, 11, 14]. However, sufficient literatures are not available about the utilization of mangrove fruit waste as an activated carbon precursor followed by KOH solution activation.

In this study, activated carbon was prepared from mangrove fruit waste using KOH solution as activating agent. The process includes carbonization followed by impregnation of the carbon (char) and activation process were employed to develop high-quality of activated carbon. The yield and specific energy consumption were investigated in order to evaluate the performance of carbonization process. Furthermore, the physical characteristics of activated carbon were analyzed including moisture content, ash content, surface area, pore diameter, iodine number, and surface morphology.

2. Experimental section

2.1. Materials

In this study, Mangrove (*Rhizophora mucronata*, see figure 1) fruit was obtained from Bontang district, Kalimantan Timur province, Indonesia. The fruit was then chopped into 2 cm of length and then boiled in water by ratio 10:1 (L/kg) at 100°C for an hour. The solid was then filtered and separated from the solution of natural dye. This solid (magrove fruit waste) was then dried at 105°C for 24 hours before carbonization process. The proximate and ultimate analysis of mangrove fruit waste is shown in Table 1. The high-purity (more than 85%.wt) of potassium hydroxide (KOH) was used in this study. This solution was then mixed with certain quantities of aquadest to make some KOH solution which then act as activating agents.

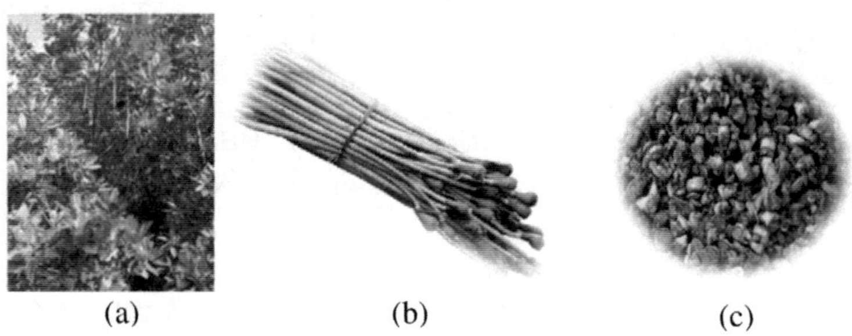

Figure 1. (a) Mangrove (*Rizhophora mucronata*); (b) magrove fruit; (c) chopped mangrove fruit

Table 1. The proximate and ultimate analysis of mangrove fruit waste

Proximate analysis (dry basis-%.wt)		Ultimate analysis (dry basis -%.wt)	
Fixed carbon	48.56%	Carbon, C	62.00%
Volatile matter	41.91%	Hydrogen, H	4.79%
Ash	3.05%	Oxygen, O	26.36%
		Nitrogen, N	1.20%
		Sulfur, S	0.03%
		LHV, (MJ/kg)	5,523

2.2. Methods

The precursor or carbon (char) was produced by carbonization process in a batch pyrolyzer with one kg of capacity. The carbonization was conducted by slow pyrolysis at temperature range of 300°C to 335°C for an hour. The heat requirement of carbonization was supplied by the combustion of liquid petroleum gas fuel. After carbonization, the precursor was crushed and sieved into 0.01 mm, and then cleaned by using deionized water. The pre-dried char was then fully soaked with KOH solution for 24 hours. The KOH solution was varied at different concentration of 0.5 M, 1 M, 1.5 M, 2 M, and 2.5 M. Samples were then filtered and dried in electric oven at 110°C for two hours. Subsequently, samples were then spread separately into a glass plate as a thin layer for the activation process. The activation process was carried out in an electrical furnace at temperature of 400°C for an hour. In order to remove the remaining activator and impurities, the produced activated carbon was then washed by using deionized water until neutral pH reached.

2.3. Analytical Methods

In order to evaluate carbonization process, the yield of char and specific energy consumption are defined as follows:

$$\text{Yield}(\%) = \frac{m_{char}}{m_{mw}} \times 100\% \qquad (1)$$

$$\text{Spesific energy consumption (MJ/kg)} = \frac{m_{gas\ fuel} \times LHV_{gas\ fuel}}{m_{char}} \qquad (2)$$

Where, m_{char} is the weight of produced char (kg) and m_{mw} is the weight of dried mangrove fruit waste (kg). The specific energy consumption is defined as the ratio of energy requirement of carbonization to the quantity of produced char. The energy requirement is calculated by multiplying the consumed fuel gas (kg) and its heating value (MJ/kg).

In this study, the physical characteristics of activated carbon are defined as moisture content, ash content, surface area, pore diameter, iodine number, and surface morphology. According to the Standard National Indonesia (SNI 06-3730-1995), the maximum moisture content and ash content are 15%.wt and 10%.wt, respectively. The Brunauer Emmet, and Teller (BET) method is used to measure the pore characteristics of the activated carbon by using Quantchrome Instrument Nova 3200e. Iodine number was analyzed in order to evaluate the adsorption capacity of activated carbon. The iodine number represents the porosity of the activated carbon which is defined as the milligrams of iodine absorbed per gram of carbon. The pore diameter and surface morphology of activated carbon was analyzed by using Scanning Electron Microscope (SEM) JEOL 650 apparatus.

3. Results and Discussion

3.1. The preparation of char by carbonization process

The effect of carbonization temperature was investigated in order to evaluate the yield of char and the specific energy consumption. As shown in Figure , the yield of char decrease with increasing of carbonization temperature. The higher temperature promotes gases formation while the char yield becomes decrease. The volatile matter in the biomass is released during the carbonization process [15]. The carbonization of hemicellulose and cellulose is quickly occurred at a temperature range of 220°C-315°C and 315°C-400°C, respectively. However, the decomposition of lignin is more difficult which is happened in the range of 160°C-900°C [16]. Char yield were found about 30%–45%.wt of dried mangrove fruit waste. Due to the endothermic reaction of carbonization, the heat requirement should be considered. The specific energy consumption of carbonization increase with increasing temperature and it's ranged from 5.99×10^{-5} to 14.27×10^{-5} MJ/kg in correspond to fuel gas consumption during combustion.

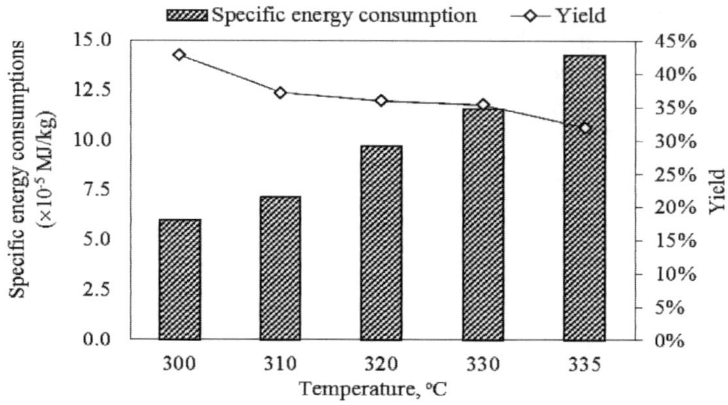

Figure 2. The yield of carbonization (%) and specific energy consumption (MJ/kg)

3.2. The effect of KOH concentration on the activated carbon properties

In this study, activated carbon was prepared from char of mangrove fruit waste by using several concentrations of KOH solution as activating agent. Activation temperature in furnace was kept constant at 400°C. The previous studies have been reported that optimum activation temperature is in the range of 400°C to 500°C [1, 12, 17]. The effect of different concentration of KOH on the surface area of activated carbon and iodine number is shown in figure 3a. An increase of surface area of activated carbon is observed with increase in the concentration of KOH. The surface area of activated carbon with KOH 2.5 M is obviously larger than without KOH, with the value of 1,920.6 m^2/g and 1,200.7 m^2/g, respectively. In general, the larger surface area of activated carbon will promote good adsorption performance.

Iodine number is used to estimate the surface area of activated carbon. Iodine number increase with increasing surface area of activated carbon, this is closely related to the increase in adsorption capacity. It was found to attain 1.540,13 mg/g at 2.5 M of KOH concentration. In addition, the pore diameter is also an important property of activated carbon. It can be seen in figure 3b that the pore diameter significantly increased from 0.93 nm to 5.52 nm with the increase of KOH concentration. It was observed that KOH activation led to a porous formation of the activated carbon which has microporous and mesoporous structures. Similar trends were also reported by Ma [11] and Kawano [14] in their studies on the preparation of activated carbon by chemical activation.

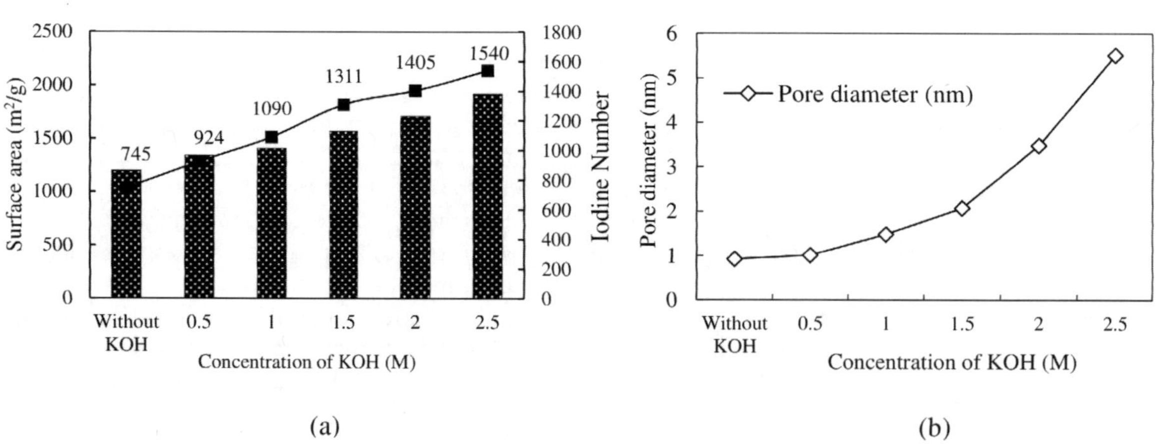

Figure 3. The effect of KOH concentration on: (a) surface area of activated carbon and iodine number, and (b) pore diameter

IOP Conf. Series: Materials Science and Engineering **543** (2019) 012087 doi:10.1088/1757-899X/543/1/012087

The surface morphology of activated carbon are shown in figure 4. The increasing of concentration of KOH encouraged the alteration of pores, indicating that the volatile matter was released during the activation process. The releasing of volatile matter enhanced the pore structures of activated carbon [18]. Moreover, a higher concentration of KOH not only produce pores but also dramatically destroys the carbon structure as reported by Budi et.al [2]. During the activation process, the reaction mechanisms are given as follows [10, 11].

$$C + KOH \rightarrow 2K + CO_2 + H_2 \quad (3)$$
$$C + KOH \rightarrow 2K + H_2O + CO \quad (4)$$
$$CO_2 + 2KOH \rightarrow K_2CO_3 + H_2O \quad (5)$$

The mesoporous and microporous are formed in correspond to the consumption of carbon by KOH, that is acts as a dehydrating agent. Furthermore, the use of KOH promote the formation of microporous and transformation of lignin [2]. The pores distribution has to gain considerable attention due to activated carbon applications, but no analysis was made in this study. And also, due to the use of KOH as an activating agent, the point of zero charge (PZC) should be determined due to its next applications, but no measurement was made in this point. Based on the pore diameter range of activated carbon, the possible application of this activated carbon is as adsorbent of synthetics dyes in wastewater treatment.

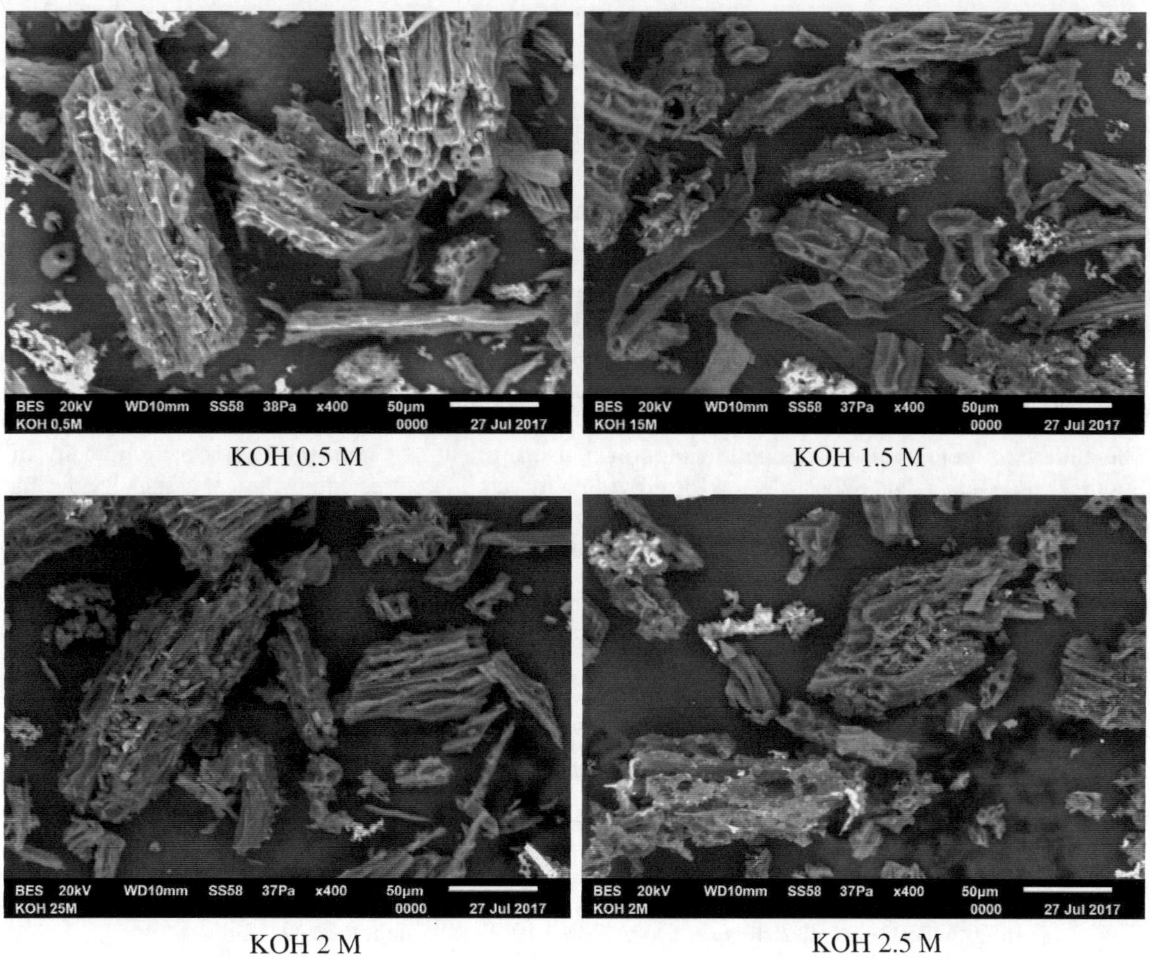

Figure 4. The SEM images of activated carbon

According to Standard National Indonesia (SNI), the moisture content of activated carbon and ash content must lower than 15% and 10%, respectively. As shown in Table , both moisture content and ash content of produced activated carbon meet the requirements of SNI 06-3730-1995. All the biomasses contain ash minerals which are formed in organic or inorganic compounds [19]. In this case, the higher content of ash may reduce the adsorption capacity of activated carbon due to the presence of minerals such as potassium, calcium, magnesium, sodium in the pores.

Table 2. Moisture content and ash content of produced activated carbon

Concentration of KOH (M)	Activation temperature (°C)	Moisture content (%.wt)	Ash content (%.wt)
0	-	6.6	8.1
0.5	400	3.2	7.9
1	400	4.6	7.8
1.5	400	6.1	7.7
2	400	7.7	7.5
2.5	400	8.5	7.1

4. Conclusion

The activated carbon was prepared from Mangrove (*Rhizophora mucronata*) fruit waste that was generated from water-based extraction of natural dyes from this fruit by KOH activation. The effects of KOH concentration were investigated in order to evaluate the physical characteristic of produced activated carbon. The specific energy consumption of carbonization process were obtained from 5.99×10^{-5} to 14.27×10^{-5} MJ/kg with the yield of char from 30%.wt to 45%.wt. Increasing the KOH concentration in activation process improved the specific surface area, pore diameter and iodine adsorption capacity of produced activated carbon. The highest specific surface area and pore diameter were obtained at KOH concentration of 2.5 M, which are 1,920.6 m^2/g and 5.52 μm, respectively, and the adsorption capacity was obtained at 1,540.13 mg/g of iodine number.

Acknowledgement

The authors gratefully acknowledge the support of Department of Chemical Engineering, Faculty of Engineering, Universitas Sebelas Maret for providing research facilities and technical assistance during the experiments. Also, the authors thankful to Ministry of Research, Technology and Higher Education for the research funds in the scheme of Penelitian Unggulan Perguruan Tinggi 2017.

References

[1] Shafeeyan M S, Daud W M W A, Houshmand A and Shamiri A 2010 *J. Anal. Appl. Pyrolysis* **89** **(2)** 143-151.
[2] Budi E, Umiatin, Nasbey H, Bintoro R A, Wulandari F and Erlina 2016 *AIP Conference Proceedings 1712* (American Institute of Physics, Melville, NY) p 1-6.
[3] Herawan S G, Hadi M S, Ayob M R and Putra A 2013 *Sci. World J.* **6**
[4] Ahmed I, Jangsawang W and Gupta A K 2012 *Appl. Energy* **91 (1)** 173-179.
[5] Nguyen T P, Luom T T and Parnell K E 2017 *Ocean. Coast. Manage.* **149**(Supplement C) 12-21.
[6] Punrattanasin N, Nakpathom M, Somboon B, Narumol N, Rungruangkitkrai N and Mongkholrattanasit R 2013 *Ind. Crop. Prod.* **49** (Supplement C) 122-129.
[7] Hoseinzadeh Hesas R, Arami-Niya A, Daud W M A W and Sahu J N 2013 *Chem. Eng. Res. Des. J.* **91** (12) 2447-2456.

[8] Tan I A W, Ahmad A L and Hameed B H 2008 *Chem. Eng. J.* **137** (3) 462-470.
[9] Yang K, Peng J, Srinivasakannan C, Zhang L, Xia H and Duan X 2010 *Bioresour. Technol.* **101** (15) 6163-6169.
[10] Wang J and Kaskel S 2012 *J. Mater. Chem.* **22** (45) 23710-23725.
[11] Ma Y 2017 *Waste Biomass Valori.* **8** (3) 549-559.
[12] Srinivasakannan C and Abu Bakar M Z 2004 *Biomass Bioenerg* **27** (1) 89-96.
[13] Ahmadpour A and Do D D 1996 *Carbon* **34** (4) 471-479.
[14] Kawano T, Kubota M, Onyango M S, Watanabe F and Matsuda H 2008 *Appl. Therm. Eng.* **28** (8) 865-871.
[15] Chun Y N, Ji D W and Yoshikawa K 2013 *J. Mec. Sci. Technol.* **27** (1) 263-272.
[16] Yang H, Yan R, Chen H, Lee D H and Zheng C 2007 *Fuel* **86** (12–13) 1781-1788.
[17] Alam M Z, Muyibi S A, Mansor M F and Wahid R 2007 *J. Environ. Sci.* **19** (1) 103-108.
[18] Budi E, Nasbey H, Yuniarti B D P, Nurmayatri Y, Fahdiana J and Budi A S 2014 *AIP Conference Proceedings 1617* (American Institute of Physics, Melville, NY) p 130-133.
[19] Werther J, Saenger M, Hartge E U, Ogada T and Siagi Z 2000 *Prog. Energ. Combust. Sci.* **26** (1) 1-27.

IOP Conf. Series: Materials Science and Engineering **543** (2019) 012088 doi:10.1088/1757-899X/543/1/012088

Adsorption Capacity and Isotherm of Methylene Blue Removal in Aqueous Solution onto Regenerated Activated Carbon

L Cundari[1], **A N Utari, M Septikarini**
Chemical Engineering Department, Faculty of Engineering, Universitas Sriwijaya, Palembang 30139, Indonesia

E-mail: liacundari@ft.unsri.ac.id

Abstract. Adsorption is one of the potential ways to removal of methylene blue in aqueous solution. The adsorbent used in the adsorption process is the activated carbon of betel nuts that produced from regeneration. The purpose of this study was to determine the effect of regeneration method to capacity and isotherm adsorption. The research occurs in a batch with three various regeneration conditions that is chemical, physical, and chemical-physical. In chemical, activated carbon is regenerated with HCl with various concentrations of 5 N, 6 N, and 7N. In physical, activated carbon is regenerated by heating it at temperatures of 160°C, 175°C, and 190°C. In chemical-physical regeneration, the condition is combination of the two. The concentration of methylene blue was analyzed using ultraviolet-visible spectrophotometer (UV-visible). The result showed the best regeneration conditions occur in chemical-physical regeneration with HCl concentration of 7 N, temperature of 190°C, and contact time of 25 minutes with an adsorption capacity on the surface is 3.57 mg/g and efficiency of 98.98%. Adsorption isotherm of removal methylene blue onto regenerated of betel nuts activated carbon follows the Langmuir model with regression value 1 ($R^2 = 1$).

1. Introduction

Dyestuffs and pigments are very widely used in several industries such as textile, paper, plastic, leather, food, and cosmetics. Industrial wastewater that contains dyes or pigments will pollute the aquatic environment [1]. One of the dyes found in aquatic is methylene blue. Methylene blue causes irritation to the digestive tract by ingested, cyanosis by inhalation, and irritation to the skin [2]. Methylene blue is an aromatic hydrocarbon compound that is toxic and includes cationic dyes with very strong adsorption power. In general, methylene blue is used as silk dyes, wool, textiles, papers, equipments, and cosmetics. Methylene blue is one type of base color that is widely used in the textile industry [3]. This compound is a dark green crystal. Dissolved methylene blue in water or alcohol will produce a blue solution.

The textile industry is one of the fastest growing industries in Indonesia. According to the regulation of the Minister of Environment No.51/MENLH/10/1995 concerning the standard quality for liquid waste, the maximum concentration of permissible methylene blue is 5-10 mg/L [4]. The method that can be used to reduce methylene blue as an aquatic environmental pollutant is adsorption. Adsorption is the best alternative that is done to overcome dye pollution [1].

One of the adsorbents that can be used in the adsorption process is activated carbon. Activated carbon is an amorphous compound that is produced from materials containing carbon or charcoal that are

[1] To whom any correspondence should be addressed.

specifically needed to obtain high adsorption power. Activated carbon adsorb gas and certain chemical compounds or selective adsorption properties, depending on the size or volume of pores and surface area [5]. Raw material of the activated carbon that developed is betel nuts [6-7]. Betel nuts contain 32.56% water, 3.35% protein, 2.17% fat, 1.06% ash, and 60.86% carbohydrate [7]. Betel nuts also have a cellulose content of 70.2% which shows more carbon content in betel nuts [8]. The removal direct dye onto betel nuts activated carbon was 76.4% adsorption [7]. The activated carbon to be used is the activated carbon of betel nuts produced from regeneration. Regeneration is usually cheaper than replacing adsorbents [9].

Sanada et al [10] showed that removal of methylene blue with KATK (Coconut Shell Activated Carbon) and KABB (Coal Activated Carbon) were respectively 48.54% and 66.79%. Kow et al [11] used chemical regeneration to the activated carbon. The results showed that optimal regeneration conditions were produced at a volume of 15 ml of 6 N NaOH solution per gram of carbon with a contact time of 30 minutes, with a regeneration efficiency of 98% in the adsorption of methylene blue [11]. Bello et al [12] conducted research on the adsorption of methylene blue using treated sawdust by physical regeneration method. The results showed that the regeneration efficiency was 90% for 6 hours of contact time.

Hassan et al [13] conducted a study on adsorption capacity and isotherm in batches using commercial activated carbon from regeneration in the removal of synthetic dyes. The results showed that the adsorption capacity using physical methods was 61.73 mg/g and the chemical method was 53.72 mg/g. Physical methods have a higher regeneration efficiency value than chemical methods, 96.30% and 83.80% respectively.

The use of Freundlich and Langmuir isotherms is carried out with the aim of obtaining an equilibrium equation that can be used to determine the mass of the adsorbate absorbed by the adsorbent (adsorption capacity). Liquid-solid phase adsorption usually adheres to the Freundlich and Langmuir isotherm types [14]. The Freundlich isotherm is used with the assumption that multilayer bonds between the adsorbent and the adsorbate occur due to Van der Walls style, so the bond is not too strong. The Langmuir isotherm assumes that the formed layer is a monolayer which has a strong adsorbate because of chemical bonds [15]. Based on the previous studies, the development study of the effect of regeneration method of betel nuts activated carbon to the capacity and isotherm adsorption in removal of methylene blue was obtained.

2. Research methodology

In this study, the activated carbon made from betel nuts is a waste produced from the adsorption process of Jumputan wastewater. The activated carbon was saturated. This adsorbent obtained in three regeneration method, namely chemical, physical, and chemical-physical regeneration. Chemically method used 5 N, 6 N, and 7 N HCl for 30 minutes immersion. Physically method occurs at a temperature of 160°C, 175°C, and 190°C for 1 hours heating. Chemically-physically regenerated carried out by using 5 N, 6 N, and 7 N HCl for 30 minutes immersion, then heating at 160°C, 175°C, and 190°C for 1 hour. The regenerated activated carbon is dried in an oven 105°C to remove the water content. The quality of regenerated activated carbon analyzed by the water content.

The process of making a standard curve is done by making a standard solution of methylene blue with concentrations of 20, 40, 60, 80, and 100 mg/L. Each standard of methylene blue solution is measured for absorbance using UV-Visible Spectrophotometer at a wavelength of 560.5 nm. The experiment carried out by mixing 6 grams of regenerated activated carbon into 200 ml of methylene blue solution with a stirring speed of 150 rpm. Samples were filtered and analyzed using a UV-visible spectrophotometer.

The adsorption efficiency is determined using the equation in (1). Adsorption efficiency is a comparison between the initial concentration (Co) (mg/L) and the final concentration (Ce) with the initial concentration.

$$\text{Adsorption efficiency} = \frac{C_0 - Ce}{C_0} \times 100\% \tag{1}$$

Langmuir adsorption isotherm can be based on several assumptions, adsorption only occurs in a single layer (monolayer), adsorption heat does not depend on surface closure, and all sites and surfaces. The Langmuir equation assumes that the surface of the adsorbent is homogeneous, and a site can only be occupied by one pollutant molecule [16]. q_m and K are the maximum theoretical adsorption capacity (mg/g) and Langmuir constant (L/mg). The constant associated with the Langmuir isotherm model is calculated by the slope and intercept of the plot, obtained from C_e/q versus C_e. From this concept, the equation can be derived as follows (equation 2):

$$\frac{C_e}{q_e} = \frac{1}{q_m.K} + \frac{C_e}{q_m} \quad (2)$$

The approach to the adsorption isotherm can be explained by the Freundlich equation. If the concentration of the solution in equilibrium is plotted as ordinate and the concentration of adsorbate in the adsorbent as abscissa in logarithmic coordinates, gradients n and intercept will be obtained [16]. The Freundlich isotherm model shows that the adsorbate layer formed on the surface of the adsorbent is multilayer [16]. q_e is the number of adsorbates adsorbed (mg/g), K_F is the Freundlich constant (L/mg), and n is the adsorption constant. These model constants are calculated by the slope and intersection of the plot obtained from log q_e versus log C_e. Where the Freundlich equation is written as follows (equation 3):

$$log\, q_e = log k_F + \frac{1}{n} log\, C_e \quad (3)$$

Based on the calculation of concentration dyestuff, it can be found that the adsorption capacity (q_e) using the equation (4), where C_0 and C_e are initial and final concentration (mg/L), V is volume of solution treated (L), and m is the dosage of the adsorbent used (grams).

$$q_e = (C_0 - Ce) x \frac{V}{m} \quad (4)$$

3. Result and discussion

The quality of the regenerated activated carbon tested based on the water content. The fresh activated carbon contains 0.05% water and waste activated carbon (before regeneration) of 0.059%. The best water content results are obtained on chemical-physical regeneration with 7 N and a temperature of 190°C, which is equal to 0.02%. The regenerated one has lower water content because the HCl concentration used greater that HCl used in the activation process, and the addition of heating process. The low water content is because the surface of the activated carbon contains fewer polar functional groups [17]. The water content analysis results obtained from this study meet the activated carbon quality standards in accordance with SNI 06-3730-1995 [18]. The result showed that regenerated activated carbon has a good performance as an adsorbent and can be used to adsorption process.

3.1. Effect of chemical regeneration method on methylene blue adsorbed

The adsorption capacity in various HCl concentration was in range 2.37 to 3.28 mg/g. At HCl concentration 5 N, q_e increases through the first 15 minutes rapidly, then continue to decline slowly. The 6 N HCl concentration has the same trend, q_e increases rapidly to the first 20 minutes adsorption. Regeneration using HCL 7 N shows slowly increases by the time, with initial adsorption capacity lower than 5 N and 6 N. Up to the 35th minute, the adsorption capacity has increased over the 5 N and 6 N.

Figure 1 shows that the adsorption of methylene blue on regenerated activated carbon has increased with increasing HCl concentration. At a concentration of 7 N and a contact time of 35 minutes, the ability adsorbent to adsorb the methylene blue was 3.28 mg/g. The higher concentration will produce greater the surface area available, more pores that can be passed, so contact of dyes with activated carbon wider [19]. Figure 1 is also shown that at a concentration of 7 N HCl the amount of mass of methylene blue adsorbed increases with increasing contact time. Kow [11] reported that at a concentration HCl of 8 N the adsorbed mass decreases. This shows that if the concentration of regenerator used more than 7 N will damage the activated carbon structure that effect its ability to adsorb the dye. In this study, longer contact time is needed to see a decrease in the mass of adsorbed, because longer contact makes the

IOP Conf. Series: Materials Science and Engineering **543** (2019) 012088 doi:10.1088/1757-899X/543/1/012088

methylene blue pass through the pores longer. Regeneration efficiency of 98% in the adsorption of methylene blue onto activated carbon achieved at 30 minutes contact time [11]. The highest adsorption efficiency of methylene blue achieved by regenerated of betel nuts activated carbon was 89.87%. This efficiency is greater than activated carbon based on coconut and coal [10].

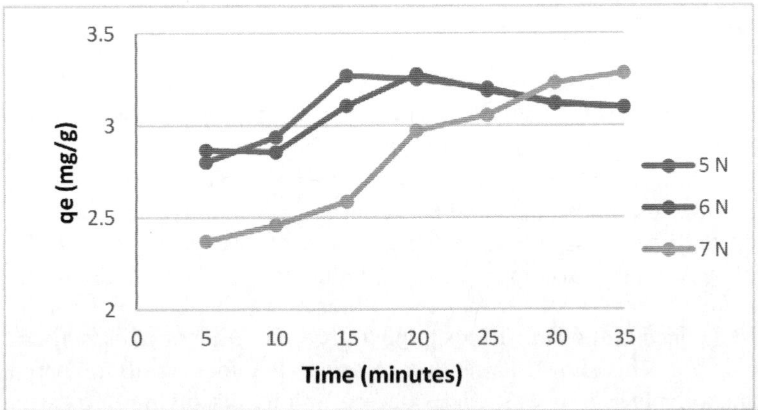

Figure 1. Effect of Chemical Regeneration Method on Methylene Blue Adsorbed

3.2. Effect of physical regeneration methods on methylene blue adsorbed
Physical regeneration method occurs by heating the waste activated carbon at 160°C, 175°C, and 190°C. The adsorption capacity is 3.13 – 3.51 mg/g. Overall data shows that qe go up rapidly to first 10 minutes adsorption then fluctuate up and down slowly. The adsorption process occurs in two roles, the initial adsorption occurs fast and then the releases of substances slowly because the amount of the adsorbed substance exceeds the maximum adsorption capacity. Langmuir explained that on the surface of the adsorbent there are active sites whose numbers are proportional to the surface area of the adsorbent [20]. If the active site has been saturated, the addition of the contact time will not affect the increasing of adsorption ability, the q_e tends to decrease.

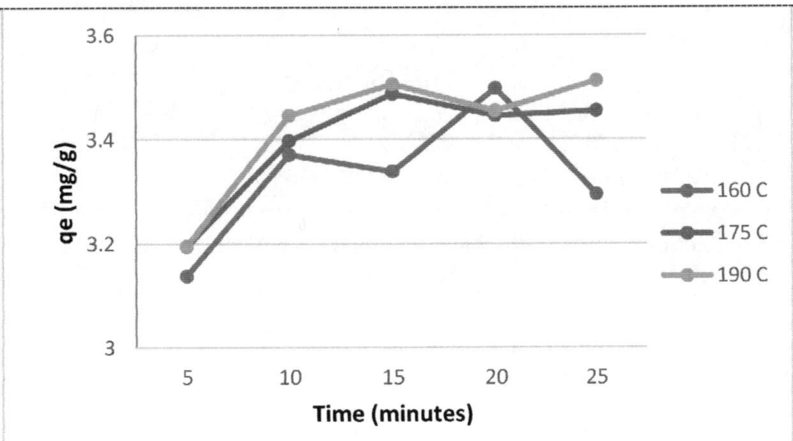

Figure 2. Effect of Physical Regeneration Method on Methylene Blue Adsorbed

Figure 2 describes the largest mass of methylene blue was obtained at a temperature of 190°C for 25 minutes of contact time, which is 3.51 mg/g. At a temperature of 160°C and 175°C, the water and organic matter content evaporated are less than at 190°C. The higher the temperature and the length of contact time, the more water and organic matter evaporates and makes the pore surface area even bigger. Increasing the pore surface area will affect the adsorption capacity of an adsorbent [21]. The amount of q_e in the three various temperature does not show a significant difference.

The highest adsorption efficiency of methylene blue with the physical regeneration method is 96.16%, that achieved at 190°C and 25 minutes contact time. This amount is greater than efficiency methylene blue removal onto treated sawdust [12]. The efficiency of adsorption of physical regeneration is higher than chemical, same report as Hassan et al [13].

3.3. Effect of chemical-physical regeneration method on methylene blue adsorbed

Chemical-physical regeneration method obtains 2.01-3.62 mg/g of adsorption capacity. The combination of chemical-physical regeneration will make the pore surface of activated carbon greater than just chemical or physical regeneration. Larger pores will certainly facilitate the process of adsorption of methylene blue substances to be removed.

Most of the q_e on the chemical-physical regeneration with a concentration HCl of 5 N for various temperature is below 3 mg/g. Variation of 5 N HCl and 190°C show the smallest q_e all the time. This shows that active site, water and volatile matter on pores of waste adsorbent has not opened completely. At 6 N and 7 N HCl concentration at various temperature, the amount of q_e is almost the same, most of in range 3.60-3.62 mg/g. This shows that with increasing HCl concentration, there is an increase in the ability to clean the pore surface of waste activated carbon by dissolving into cations and heating. The ion H^+ in this activator has a higher electronegativity, with the addition of high temperature and contact time, so the ability of the regenerated adsorbent to adsorb of methylene blue will increase.

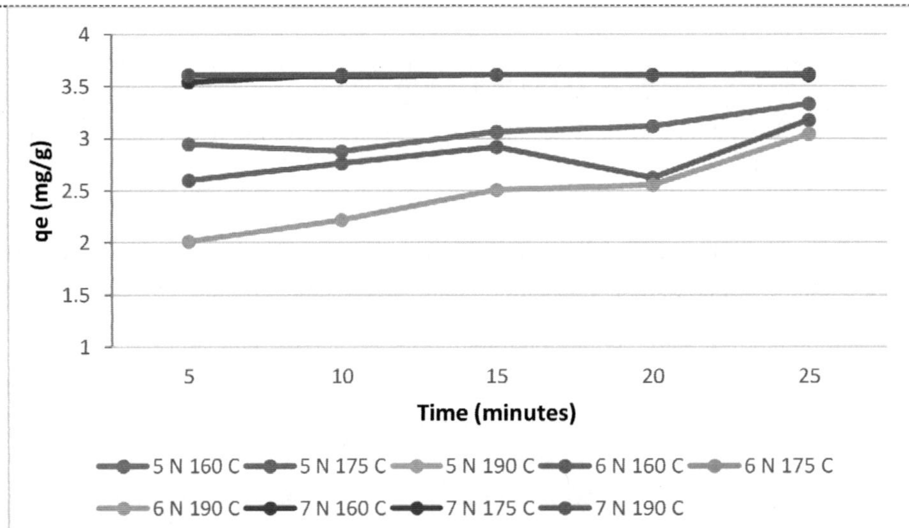

Figure 3. Effect of Chemical-Physical Regeneration Method on Methylene Blue Adsorbed

Compare with the previous studies, the q_e of methylene blue onto fresh coconut and coal activated carbon was 1.18 mg/g and 1.81 mg/g respectively [10]. This shows that the use of waste activated carbon made from betel nuts that regenerated is superior compared to coconut shell and coal activated carbon with maximum mass absorbed of 3.62 mg/g.

On the chemical-physical regeneration method, the highest adsorption efficiency of methylene blue is achieved under conditions of 7 N HCl concentration, temperature 190°C and contact time 25 minutes, which is equal to 98.98%. Larger pore surface area is due to chemical reactivation and followed by physical so that the activator solution can easily enter the pores and can release the adsorbate or impurities that trapped inside. The presence of high concentrations and temperatures will increase the pore size and push the formation of new pores. The removal of methylene blue onto regenerated activated carbon made from betel nuts is more effective and efficient than coconut, coal, commercial activated carbon and sawdust [10-13].

3.4. Adsorption isotherm of activated carbon with chemical regeneration method

Based on Figures 4 and 5, the isotherm adsorption which has a value of R close to 1 is the model that best matches the adsorption process on regenerated carbon. The comparison of R^2 values from the Langmuir and Freundlich isotherm adsorption shows that the adsorption process to remove methylene blue in aqueous solution was fitted to the Langmuir isotherm.

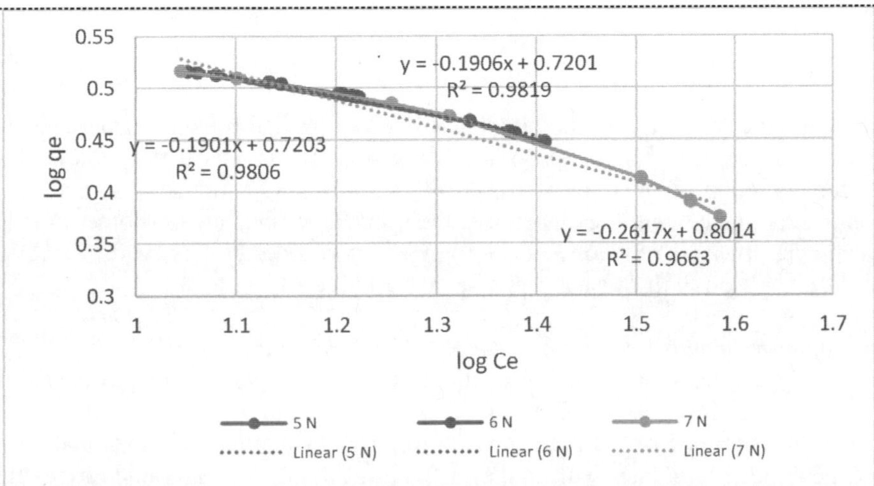

Figure 4. Freundlich Isotherm of Methylene Blue in Chemical Regeneration Method

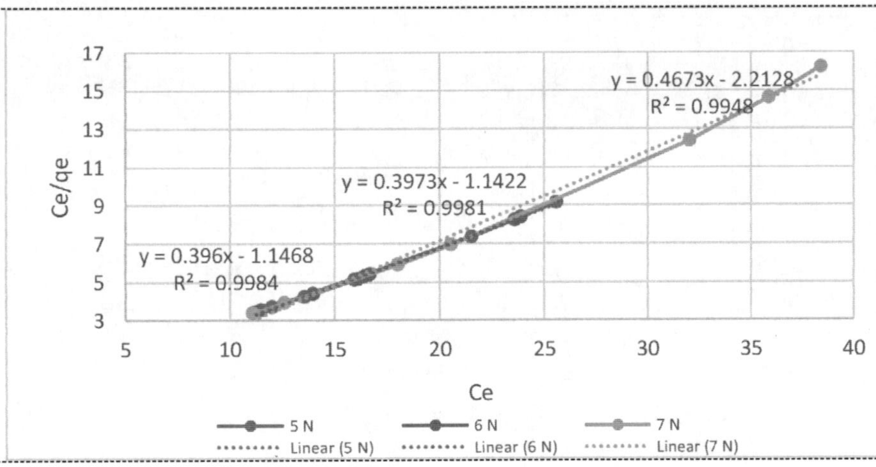

Figure 5. Langmuir Isotherm of Methylene Blue in Chemical Regeneration Method

Table 1 describes the function values of the Freundlich and Langmuir isotherm on the adsorption of methylene blue onto chemically regenerated carbon. Adsorption capacity in Freundlich was adapted from K_F, the greater is the best [23]. The slope value of 1/n with a range of 0-1 is the adsorption intensity or the heterogeneity of the adsorbent surface, increasingly heterogeneous if 1/n approaches to 0 [12, 22]. The cooperative and weak adsorption bond indicates with n value less than 1 [23-24]. As shown in Table 1, the n value of the process is 3.82-5.26, it means favorable sorption process occurred, where n refers to $1<n<10$ [24].

Table 1. Adsorption Isotherms of Methylene Blue in Chemical Regeneration Method

Isotherm Model	Chemical Regeneration	R^2	Constant			
Freundlich	5 N HCl	0.981	K_F	5.24 L/mg	n	5.24
	6 N HCl	0.980	K_F	5.25 L/mg	n	5.26
	7 N HCl	0.966	K_F	6.32 L/mg	n	3.82
Langmuir	5 N HCl	0.998	q_m	2.52 mg/g	K	2.87 L/mg
	6 N HCl	0.998	q_m	2.52 mg/g	K	2.89 L/mg
	7 N HCl	0.994	q_m	2.13 mg/g	K	4.73 L/mg

Theoretical adsorption capacity (q_m) of the regenerated adsorbent that produces from chemical method is 2.13-2.52 mg/g. Value q_m at a concentration of 5 N and 6 N HCl is greater than 7 N HCl. It indicates that the higher concentration of activator affected to less methylene blue that adsorbed (q_m) [25]. The data obtained is in accordance with the theory that the specific surface area becomes smaller, because with the increasing pH (low activator concentration) more active sides will be available while at low pH (high activator concentration) will cause a decrease in the active site.

3.5. Adsorption isotherm of activated carbon with physical regeneration method

Regression value of the Freundlich and Langmuir adsorption isotherm onto physically regenerated carbon shows that the Langmuir has a value of R^2 closes to 1. It can be concluded that the isotherm of adsorption is fitted to Langmuir. Freundlich constant in physical smaller than chemical, in range 3.97-4.04 L/mg, but the n value is greater with range 11.27-12.28 (over the favorable range). The q_m is in range 2.99-3.05 mg/g, greater than physical method, that shows more methylene blue can be adsorbed by the carbon.

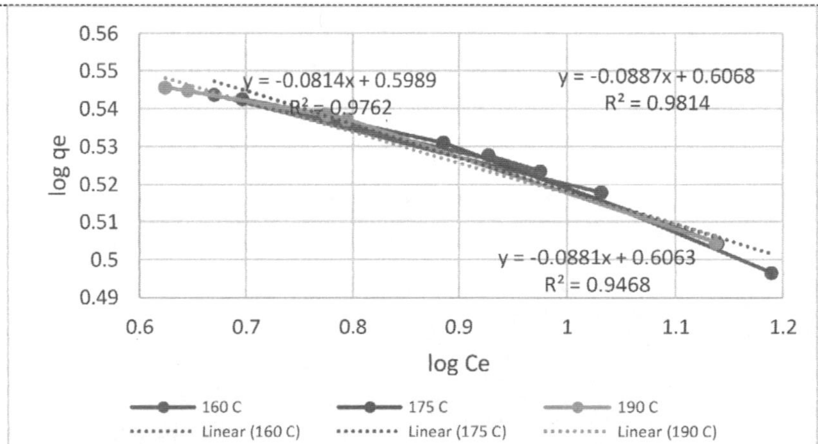

Figure 6. Freundlich Isotherm of Methylene Blue in Physical Regeneration Method

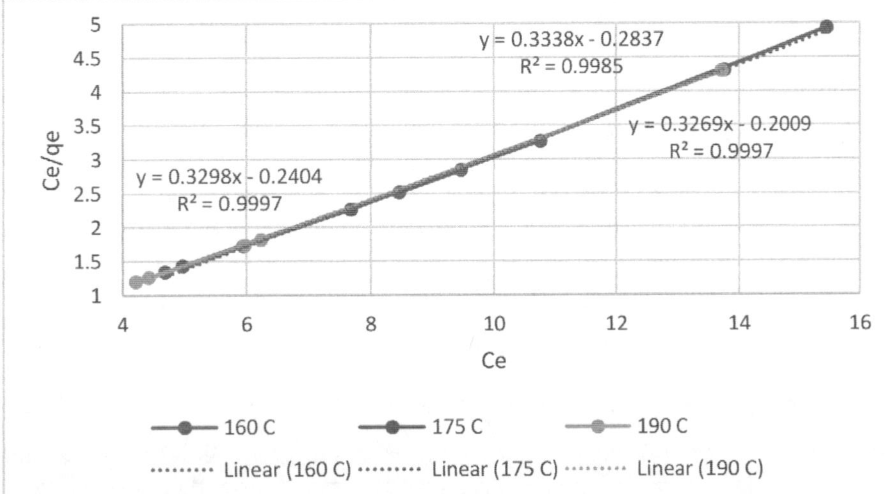

Figure 7. Langmuir Isotherm of Methylene Blue in Physical Regeneration Method

Table 2. Adsorption Isotherm of Methylene Blue in Physical Regeneration Method

Isotherm Model	Physical Regeneration	R^2	Constant			
Freundlich	160°C	0.946	K_F	4.03 L/mg	n	11.35
	175°C	0.981	K_F	4.04 L/mg	n	11.27
	190°C	0.976	K_F	3.97 L/mg	n	12.28
Langmuir	160°C	0.998	q_m	2.99 mg/g	K	0.84 L/mg
	175°C	0.999	q_m	3.03 mg/g	K	0.72 L/mg
	190°C	0.999	q_m	3.05 mg/g	K	0.61 L/mg

3.6. Adsorption isotherm of activated carbon with chemical-physical regeneration method

Table 3 describes the function values of the Freundlich and Langmuir isotherm on the adsorption of methylene blue onto the chemical-physical regenerated adsorbent. From all the data, the regression value of the two approaches to 1, but Langmuir isotherm shows the best results.

Based on the q_m and K_F values in Table 3, it can be seen that the greatest adsorption capacity occurs in waste activated carbon which regenerates chemical-physically with 5 N HCl concentrations and 190°C, which is equal to 9.77 L/mg and 10.11 L/mg respectively. The Langmuir isotherm model has the highest adsorbent capacity is 3.57 mg/g.

From Figure 1-3 and Table 1-3, adsorption capacity experimental and Langmuir is adjacent, on average 3.09-3.25 mg/g. Determination of equilibrium models depends on the determinant coefficient (R^2) at the highest value [26]. Based on the results of the calculation of R^2 from the three regeneration methods is closer to 1 by the Langmuir isotherm. The results obtained were also in accordance with the other results [27-30].

Table 3. Data of Adsorption Isotherm Methylene Blue in Chemical-Physical Regeneration Method

Isotherm Model	Chemical-Physical Regeneration	R^2	Constant			
Freundlich	5 N 160°C	0.9764	K_F	4.86 L/mg	n	6.09
	5 N 175°C	0.9812	K_F	6.28 L/mg	n	3.95
	5 N 190°C	0.9506	K_F	10.11 L/mg	n	2.48
	6 N 160°C	0.9994	K_F	3.62 L/mg	n	82.64
	6 N 175°C	0.9999	K_F	3.62 L/mg	n	83.33
	6 N 190°C	0.9994	K_F	3.62 L/mg	n	89.28
	7 N 160°C	0.9974	K_F	3.62 L/mg	n	50.50
	7 N 175°C	0.9986	K_F	3.62 L/mg	n	70.92
	7 N 190°C	0.9992	K_F	3.62 L/mg	n	88.49
Langmuir	5 N 160°C	1	q_m	3.56 mg/g	K	0.01 L/mg
	5 N 175°C	1	q_m	3.56 mg/g	K	0.01 L/mg
	5 N 190°C	0.9812	q_m	1.68 mg/g	K	9.77 L/mg
	6 N 160°C	1	q_m	3.56 mg/g	K	0.01 L/mg
	6 N 175°C	1	q_m	3.56 mg/g	K	0.01 L/mg
	6 N 190°C	1	q_m	3.57 mg/g	K	0.01 L/mg
	7 N 160°C	1	q_m	3.50 mg/g	K	0.03 L/mg
	7 N 175°C	1	q_m	3.55 mg/g	K	0.02 L/mg
	7 N 190°C	1	q_m	3.57 mg/g	K	0.01 L/mg

4. Conclusion

The best regeneration conditions of waste activated carbon made from betel nuts in the process of adsorption of methylene blue are by chemical-physical regeneration methods at HCl concentration of 7 N and temperature of 190°C. Adsorption capacity on the surface and the highest adsorption efficiency in the chemical-physical regeneration are 3.57 mg/g and 98.98%. Adsorption isotherm of removal methylene blue onto regenerated of betel nuts activated carbon follows the Langmuir model with regression value 1 ($R^2 = 1$).

Acknowledgments

This research was supported by the Separation and Purification Laboratory of Engineering Faculty of Universitas Sriwijaya. The authors thank Universitas Sriwijaya for funding the research under SATEKS Scheme.

References

[1] Tanasale M, Killay A and Laratmase M 2012 *J. of Indonesian Natur* **14** 2 165-171
[2] Handayani L, Riwayati I and Ratnani R 2015 *Momentum J.* **11** 1 19-23
[3] Derakhshan Z, Baghapour M A and Faramarzian M 2013 *Int. Quarterly J.* **2** 3 136-144
[4] Minister of the Environment 1995 *Regulation of Minister of Environment No. 51/MENLH/10/1995: Liquid Waste Quality Standards* (Jakarta)
[5] Hendra D and Darmawan S 2007 *Forrest Product Research* **8** 6 1-18
[6] Cundari L, Sari K F, Anggraini L 2018 *J. Phys.: Conf. Ser.* **1095** 012018
[7] Cundari L, Sari K F, Anggraini L 2018 *IOP Conf. Ser.: Mater. Sci. Eng.* **345** 012041
[8] Syauqi M R, Bali S and Itnawita 2016 *Repository unri* (Riau: University of Riau) p 1-10
[9] Lu P J, Lin H C, Yu W T and Chern J M 2011 *J. of the Taiwan Institute of Che. Engineers* **42** 305-11
[10] Sanada R, Moersidik S and Suwartha N 2014 *Repository ui* (Depok: University of Indonesia)
[11] Kow S H, Fahmi M R, Abidin C A, Ong S A and Ibrahim N 2016 *J. Desalination and Water Treatment* **57** 60 1-6
[12] Bello O S M, Adelaide M A, Hammed O A M and Popoola 2010 *Macedonian. J. Che. and Che. Eng.* **29** 1 77-85
[13] Hassan L G, Ajana B N, Umar K J, Sahabi D M, Itodo A U and Uba A 2012 *Nigeria J. of Basic and App. Sci.* **20** 20 93-104
[14] Atkins P W 1999 *Physical Chemistry* (Jakarta: Erlangga)
[15] Reyra A S, David S and Yenti S R 2017 *FTEKNIK J.* **4** 2 1-9
[16] Jaafari J, Ghhozikali M G, Azari A, Delkhosh M B, Javid A B, Mohammadi A A, Agarwal S, Gupta V K, Sillanpaa M, Tkachev A G and Burakov A E 2018 *J. of Ind. and Eng. Chemistry.* **57** 1 396-404
[17] Qin C, Yao C and Jian M G 2014 *J. of Materials Letters* **13** 5 123-6
[18] Muna A N 2011 *Final report* (Semarang: University of Negeri Semarang)
[19] Wirosoedarmo R, Haji A T S and Hidayati E A 2018 *J. of Natural Resources and Env.* **2** 1 170-6
[20] Handayani M and Sulistiyono E 2009 *Proc. of the National Seminar on Science and Nuclear Technology* (Bandung: ITB)
[21] Idrus R, Lapanporo B P and Putra Y S 2013 *Prism of Physic J.* **1** 1 50-55
[22] Kumar A, Chaundry P and Verma P 2013 *J. of Che. Eng.* **1** 2 111-6
[23] Ghazi M B H I, Weng C N, Rahman H A and Zakaria N A 2013 *J. Geogr. Earth Sci.* **1** 1 1–8
[24] Dada A O, Olalekan A P, Olatunya A M and DADA O 2012 *IOSR J. Appl. Chem.* **3** 1 38–45
[25] Ilmi M M, Khoiroh N, Firmansyah T B and Santoso E 2017 *J. of Mechanical Eng.* **6** 1 69-76
[26] Gultom E M and Lubis M T 2014 *J. of Che. Eng. Dep. University of North Sumatera* **3** 1 5-10
[27] Kanawade S M and Gaikwad R W 2011 *Int. J. of Che. Eng. and App.* **2** 1 317-19
[28] Priya E S and Selvan S 2014 *Arabian J. of Chemistry* **1** 1 25-31
[29] Satabas O, Tejer M and Konuk Z *Global NEST J.* **14** 10 24-31
[30] Ghanizadeh G and Asgari G 2011 *J. of Reaction Kinetic* **10** 2 127-142

IOP Conf. Series: Materials Science and Engineering **543** (2019) 012089 doi:10.1088/1757-899X/543/1/012089

Influence of *Shewanella oneidensis* MR-1 Bacterial Metabolism Process On Waste Treatment Of Cr And Mn Metals In Reactor Microbial Fuel Cell (MFC)

S R Juliastuti, R Darmawan, N Hendrianie, G A Prakoso, T A Bachtiar
Department of Chemical Engineering, Faculty of Industrial Technology, Institut Teknologi Sepuluh Nopember, Campus Keputih-Sukolilo, Surabaya 60111,Indonesia

Email : julizainul@gmail.com

Abstract. Chromium (Cr) and manganese (Mn) are heavy metals which have high oxidation potential and cause mutagenesis and carcinogenic to living organisms. Microbial Fuel Cell (MFC) appears as one of the applications in the processing of Cr and Mn metals. MFC is bioreactor which converts chemical energy of organic compounds into electrical energy through the catalytic reaction of microorganisms in anaerobic conditions. The aim of this research are to find out the role of *Shewanella oneidensis* MR-1 bacteria on manufacture electrical energy by utilizing organic wastes, and to find out the concentration of *Shewanella oneidensis* MR-1 on the most influential anode side for the reduction of metallic Cr, Mn, and its mixed metal on the cathode side by using a MFC. The research method used molasses with bacteria concentration of 10%; 12.5%; 15% (v/v) on the anode side, while on the cathode side was used Cr(VI), Mn(VII) solutions, and mixtures of Cr and Mn solutions with the ratio of 1:0 ; 0:1 ; 1:1 (w/w). 15% bacteria concentration on mixture of metal solution resulted the highest electricity production, which was 46.33 (mW/m^2), and the best reduction of Cr^{6+} to Cr^{3+} 82.15%, while for the reduction of Mn^{7+} to Mn^2 was 61.17%.

1. Introduction

The concentration of Chromium (Cr) and Manganese (Mn) in the environment is increasing as the number of industries developing. Environmental pollution by heavy metals has the potential causing ecosystem damage if the concentration exceeds the threshold. One way of generating alternative energy is bio-based fuel cell based generator. Microbial Fuel Cell is a bioreactor that converts the chemical energy of organic compounds into electrical energy through the catalytic reaction of microorganisms in anaerobic conditions [1]. Microbial Fuel Cell generate electricity by oxidizing anaerobic organic matter through the aid of bacteria [2]. The electricity generated by MFC is relatively low and often power output is fluctuated due to unstable internal resistance presence in the MFC environment [3]. Microbial metabolic activity could as well as be hindered by the macro-environment created by the electrolyte and complexity of the substrate used as the source of fuel [4]. The catalytic activity and proton transfer are carried out using enzymes or bacteria. Microbial Fuel Cell has several advantages that can generate electricity from organic waste and renewable biomass [2]. The electricity generation was then investigated in an air cathode microbial fuel cell, using pure cultures of *E. coli* and *Shewanella sp.* and their mixtures as well [5,6]. Bacteria are able to catalyze and adapt well to different organic materials present in environmental waste to produce electrons. In the cathode compartment, there is a conductive electrolyte solution. Microbial Fuel Cell working principle is the using of bacteria that perform metabolism of the medium at the anode to catalyze the converter of organic matter into electrical energy by transferring electrons from the anode through a cable and generating current to the cathode. The electron transfer from anode received by free electron at the cathode. Protons transported to the cathode through proton exchange membrane (salt bridge) [7]. In the cathode electrons were received by Cr(VI) and Mn(IV). Then the power density was generated by the electron movement from anodic to the cathode and Cr(VI) and Mn(IV) reduction occurred [8]. In a previous study, pH 4 circumstances Chromium conductive electrolyte solution in cathode chamber were the best for reducing chromium concentration and also generate power density. Previous study, pH 4 circumstances Chromium conductive electrolyte solution in cathode chamber were the best for reducing chromium concentration and also generate power density. The present study has successfully verified that a promising MFC technology can achieve high output power in removing Cr(VI), revealing the enormous potential of scale-up MFCs for heavy metal treatment [9]. MFCs may represent a promising in bioremediation unit by using an anaerobic digestor plants for simultaneous nitrogen removal and electricity generation using digestate as substrate [10]. Using hydrogen peroxide as electron acceptor in the cathode compartment improved the stability of performances and the relation between the time of incubation and the power density of the MFC was showed [11]. The current and voltage output of a *Saccharomyces cerevisiae* based MFC are directly correlated to the cells on the micro-environment of the electrode and to the presence of an electron mediator such as methylene blue (MB). It suggested that reduced methylene blue was entrapped into the cells, enhancing the electron transfer on the graphite electrode as an internal mediator of the cellular metabolism, probably linked to the oxidative electron cascade [12].

2. Experimental Section

2.1. Microbial Fuel Cell Design Instructions

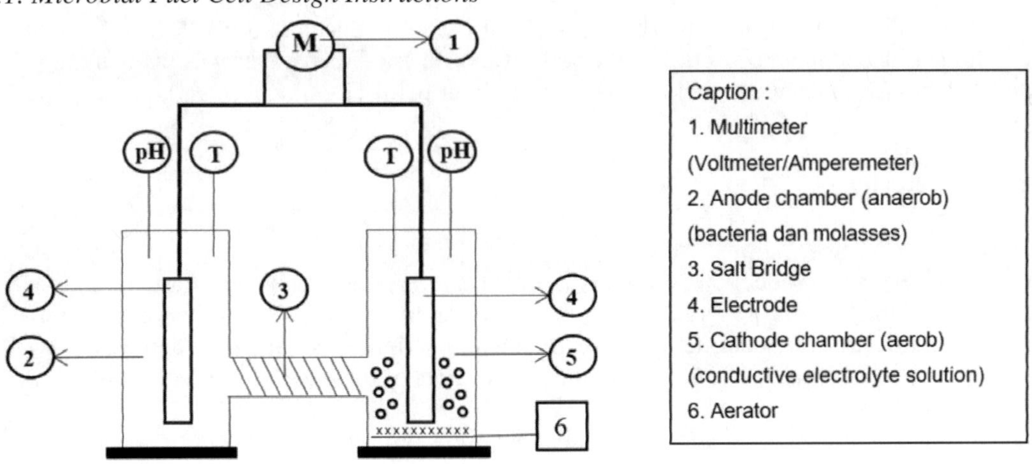

Figure 1. Dual Chamber Microbial Fuel Cell Design

2.2. Electrode Activating preparation

Graphite electrodes need to be submerged into HCl 1 M solution for 1 day, later flush with aquades. After that, submerged again into NaOH 1 M solution for 1 day, then flush with aquades. Electrodes submerged in aquades until the electrodes will be installed

IOP Conf. Series: Materials Science and Engineering **543** (2019) 012089 doi:10.1088/1757-899X/543/1/012089

2.3. Buffer solution preparation

For cathode chamber using citrate buffer pH 4 solution, the formula is preparing 0.1 M citric acid solution (21.04 gr of citric acid in 1 L aquades) and 0.1 M sodium citrate solution (29.41 g Sodium citrate / sodium citrate in 1L aquades). Then mixed (33 ml of citric acid solution 0.1 M) + (17 mL sodium citrate solution 0.1 M) and dissolved aquadest until the volume reached 100 mL. For anode chamber using phosphate buffer pH 7 solution, the formula is weighing NaH_2PO_4 0.4063 grams and Na_2HPO_4 0.9247 grams. Then dissolve all in 500 mL aquadest.

2.4. Anode and cathode solution preparation

In cathode chamber was filled with 400 mL conductive electrolyte solution and 400 mL pH 7 buffer mix both solution of them. In anode chamber was inoculated with the bacteria and sugar waste with 5% concentration. phosphate. The conductive electrolyte solution is solution 400 mL $K_2Cr_2O_7$ of 8 mg/L, or solution 400 mL $KMnO_4$ of 8 mg/L, or 400 mL

2.5 *Procedure and measurement*

Digital multimeter was connected on both the electrodes, the postive end at a cathodic chamber and the negative end at a anodic chamber. Recorded power density generated every 24 hour calculated by using equation 1

$$\text{Power density } (mW/m^2) = (I\ (mA) \times V(Volt))/A(m^2) \quad (1)$$

Where P is the power density in milli watts (mW), V is the voltage in milli volts (mV), I is the current density in milli ampere (mA) and A is the surface area of electrode. Voltage and current measurement is carried out by using a multimeter with a system OCV (Open Circuit Voltage) by getting the voltage and the current, then it can be calculated that the resulting power density.

The variable of bacteria concentration of this experiment is 10%, 12.5%, 15% (v/v) with volume of chamber is 800 ml. For example, to obtain the variable of bacteria concentration 10% is measured 10 ml for the volume of bacteria and 790 ml for the wastewater. In anodic chamber was maintain at pH 7 circumtances, otherwise in cathodic chamber was maintain at pH 4 circumtances and measure the Cr and Mn reduction.

3. Result and discussion

3.1. Power density generated

The highest power density production in this experiment at 15% bacteria concentration variable on mixture of metal solution which was 46.33 (mW/m^2). Power density was affected by voltage, electric current and the surface area of electrode. In this experiment, maximum voltage recorded was 1.037 V; an electric current recorded was 0.1 mA; and for the surface area of the electrodes was 22.844 cm^2.

The amount of electrical energy in the Microbial Fuel Cell system was influenced by the rate of metabolism carried out by bacteria. The presence of bacteria in this process becomes very important because the electrons were generated by the bacterial metabolism process. the more microorganisms it would be more electrons were released. The power production were shown in figure 2.

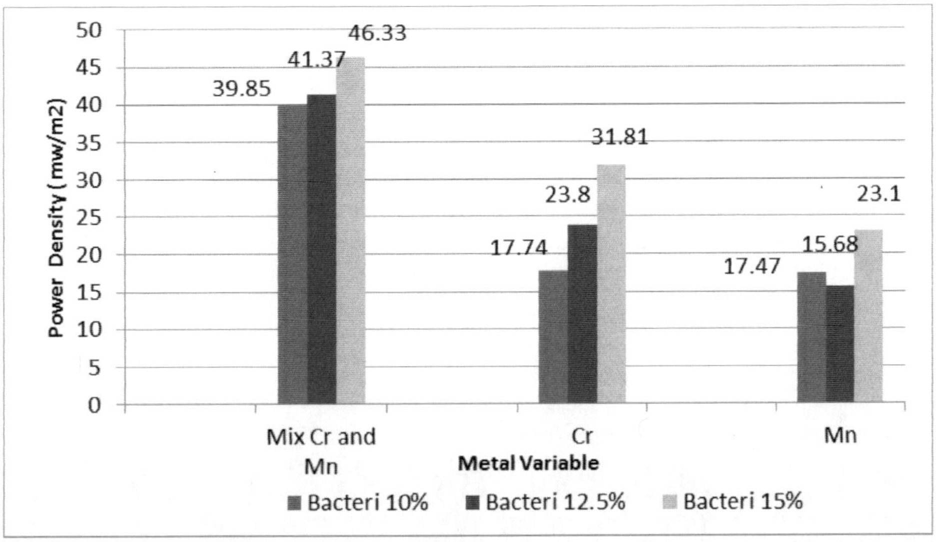

Figure 2. The highest power production relation curve generated with the bacterial population for all variables

3.2 Reduction of metal Cr, Mn and both mix Concentrations

The best reduction of heavy metal Cr^{6+} to Cr^{3+} in mixed metal solution with initial concentration of 8.01 mg/L to 1.43 mg/L and removal percentage of Cr^{6+} was 82.15%. While for the reduction of heavy metal Mn^{7+} to Mn^{2+}, the best result was obtained on mixed metal solution with initial reduction concentration from 8.01 mg/L to 2.95 mg/L and removal percentage of Mn^{7+} was 61.17%.

The reduction of Mn metal concentration was smaller than Cr metal, because the position of Mn metal was on the left of Cr. According to the voltaic series, the more right the position of a metal in the series, the less reactive the metal (it was harder to remove the electrons) and the stronger the oxidant (the easier to reduce).

Oxidation and reduction precisely refered to changes in oxidation numbers due to electron transfer. So that oxidation was better defined as an increase in oxidation number, and reduction as a decrease in oxidation number. The redox reaction could be illustrated in the figure 3 :

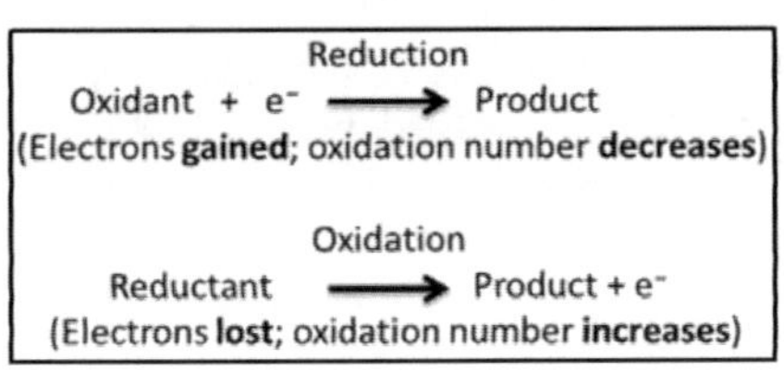

Figure 3: Reduction and oxidation reactions

In cathode chamber, reduction reaction occured, where Cr^{6+} would turn into Cr^{3+}, also Mn^{7+} turned into Mn^{2+} with the help of the electrons coming from the anode as in equation below :

$$O_2 + 4H^+ + 4e^- \rightarrow 2H_2O \quad E0 = 1.29\ V \tag{2}$$

$$Cr_2O_7{}^{2-} + 14H^+ + 6e^- \rightarrow 2Cr^{3+} + 7H_2O \quad E0 = 1.33\ V \tag{3}$$

$$MnO_4^- + 4H^+ + 3e^- \rightarrow MnO_2^- + 2H_2O \quad E0 = 1.70\ V \tag{4}$$

$MnO_2^- + 4H^+ + 2e^- \rightarrow Mn^{2+} + 2H_2O$ E0 = 1.23 V (5)

In addition to the largest population of bacteria, it also affected the production of electricity generated because in the mixed variable of Cr and Mn metals, the electrode potential (E0) by Mn metal would propose the formation of electrical energy. In the cathode compartment where Cr^{6+} and Mn^{7+} would be transformed into Cr^{3+} and Mn^{2+} with the aid from anode compartment in the form of electron supply through copper wires reacting with protons passing through the salt bridge, and forming water molecules. The longer the cathode side was rich in protons so that the reduction reaction of Cr, Mn, and mixed metal concentrations moved and the concentration decreases.

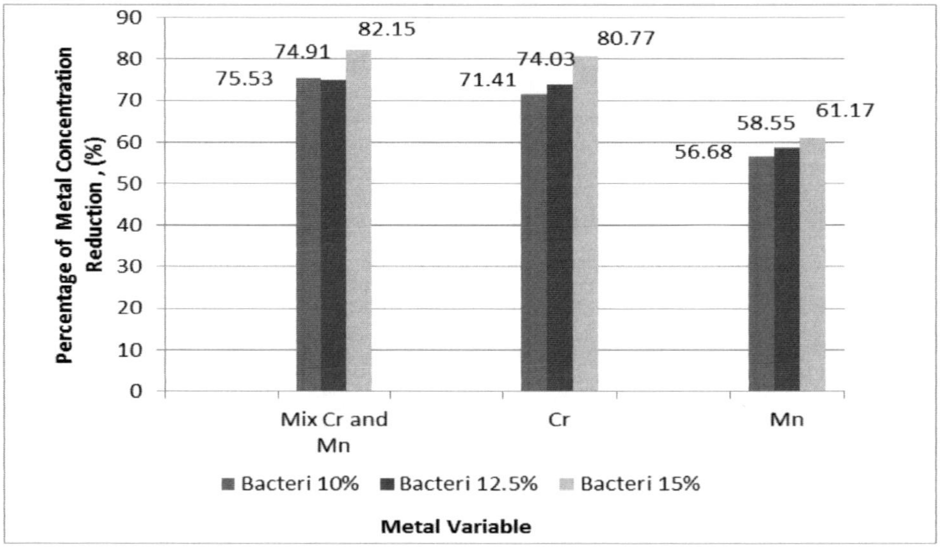

Figure 4. Curve correlation relationship of metal concentration with bacterial population for all variables

4. Conclusion

15% bacteria concentration variable on mixture of metal solution resulted the highest electricity production, which was 46.33 (mW/m^2), and the best reduction of heavy metal Cr^{6+} to Cr^{3+} in mixed metal solution with initial concentration of 8.01 mg/L to 1.43 mg/L and removal percentage of Cr^{6+} was 82.15%. While for the reduction of heavy metal Mn^{7+} to Mn^{2+}, the best result was obtained on mixed metal solution with initial reduction concentration from 8.01 mg/L to 2.95 mg/L and removal percentage of Mn^{7+} was 61,17%.

Acknowledgement
We would like thank to Wastewater Treatment Laboratory, Chemical Engineering Department, Institute of Technology Sepuluh Nopember, for the facilities provided to accomplish the research.

References

[1] Logan B E, Hamelers B, Rozendal R, Schröder U, Keller J, Freguia S, Aelterman P, Verstraete W, Rabaey K 2006 *Env. Sci. & Tech.* **40 (17)** 5181-92.

IOP Conf. Series: Materials Science and Engineering **543** (2019) 012089 doi:10.1088/1757-899X/543/1/012089

[2] Lovley, D R 2008 The microbe electric conversion of organic matter to electricity *Current Opinion Biotechnol.* **19** 564-71.
[3] Wang X, Zhang E, Yuc E H and Scott K 2016 *Chem. Eng. Trans.* **51** 37– 42
[4] Mahmood N A N, Khairul A A W, Amirul A A N, Ahmad A M A, Nazlee F G and Eraricar M S 2018 *Chem. Eng. Trans.* **63** 607–12.
[5] Miroliaei M R, Samimi A, Mohebbi-Kalhori D and Khorram M 2014 *J. Ind. Eng.Chem.* **25**, 42-50
[6] Miroliaei M R, Abdolreza Samimi and Davod Mohebbi-Kalhori 2015 *Env. Prog. Sust. Energy* **34 (4)**, 1097-05.
[7] Ramya N, Renganathan K, Barathi S and Venkatraman K 2013 *Int. J. Adv. Res. Technol.* **2 (5)** 326-30
[8] Carmalin Shopia A and Saikanth S 2016 *J. Env. Chem. Eng.* **4** 2402–09.
[9] Meng L, Shaoqi Z, Fuzhen M, Liangliang Z and Xuan Z 2018 *Chem. Eng. J.* **334** 1621-29.
[10] Enea G D D, Gianluca P, Daniele M, Alberto G, Luca D P and Fiorentina A 2015 *J. Biomed. Biotech.* DOI: 10.1155/2015/351014
[11] Ruggero R and Leonardo S 2016 *Env. Eng. Manage J.* **16** (9), 2011-18.
[12] Ruggero R ,Matia C and Leonardo S 2016 *Chem. Eng. Trans.* **49** 559-64

Utilization of plastic waste with mix plastic softening agregate method as performance of stability and quality asphalt concrete

N Hendrianie [1], **S R Juliastuti**[1], **R Darmawan**[1], **Widjonarko**[2], **F I N Aini**[1] **and M L Hakim**[1]

[1]Department of Chemical Engineering, Institut Teknologi Sepuluh Nopember, Surabaya 60111, Indonesia.

[2]Department of Civil Engineering, Institut Teknologi Sepuluh Nopember, Surabaya 60111, Indonesia.

E-mail: nuniek@chem-eng.its.ac.id

Abstract. To increase sidewalks in Indonesia there are several things that must be considered, such as aggregate quality, implementation methods and asphalt quality. The amount of waste plastic waste in Indonesia at this time and in the years to come, can be reprocessed, one of them is by using it as an additive in the technique of making asphalt concrete. It is expected that a complex problem related to the amount of plastic waste piles can be turned into an opportunity to build the acceleration and quality of concrete asphalt roads in Indonesia. The method used for this research is the wet process, which is the direct method of mixing plastic material into asphalt which is heated at temperature according to Tait's parameters, then added aggregate. The optimum asphalt used in this study was 5.7%. The variables used are 5%, 7%, and 9% of the plastic for substitution at the optimum level of asphalt used. Plastic variable added 100% high density polyethylene (HDPE) plastic, 100% polypropylene (PP), HDPE: PP and mixture (50:50), then Marshall Test. The best results were obtained by adding 9% plastic mixture variable with HDPE: PP ratio (1:1) resulting in an increase of 179% MQ then normal asphalt.

1. Introduction

The higher production of plastic waste generated will cause the necessity of processing of plastic waste. Based on data from the Ministry of Environment in 2012, the amount of garbage piling up in Indonesia has reached 175.000 ton / day or equal to 64 million tons / year. This issue needs to be handled seriously; the government has been reviewing many national solutions and strategies for handling plastic waste. People in general often burn plastic waste to reduce the amount of plastic waste in the environment when burned plastic waste will produce hydrogen sulphide gas (H_2S) that can release toxic to the environment. One of the most dominant plastic waste is HDPE (High Density Polyethylene) plastic type, because this plastic has a nature that tend to be harder and resistant to high temperature or around 120 °C [1]. Plastic waste can also be converted into gas and naphtha fractions [2]. Various waste materials, plastic waste and municipal waste are of great concern in its management, among others, by finding ways to properly process plastic waste [3]. The process of removing plastic waste used by consumers, for example incineration or soil, is not fully accepted according to international standards. On the other hand, road traffic is increasing. Traffic intensity increases. Carrying capacity of road loads must be increased. The use of PCA (Plastic Coated Asphalt) for asphalt pavement helps for the reuse of plastics

waste and for the improvement of road strength. This is a new innovative technique developed by the author. Plastics, a versatile packing material and a close to common man, become a problem to the environment after its use. Most used materials are bags, cups, films and foams, made up of PE, PP or PS [4].
According to the Director of Coastal and Marine Control and Damage Control of the Ministry of Environment and Forestry, until the end of 2016, Indonesia recorded as the second largest plastic waste contributor in the world. Every year Indonesia contributes 3.2 million tons of plastic waste annually. While China, the world's largest plastic waste contributor, produces 8.8 million tons of plastic waste per year.
The large amount of plastic waste in Indonesia today and in the coming years, can be recycled one of them by utilizing it to be an additive on the technique of making concrete asphalt. Plastics have the same carbon chain derivative characteristics as the asphalt but have different properties so as to have a positive impact on asphalt. It is expected that from a complex problem related to the amount of plastic waste pile can be transformed into an opportunity to build the acceleration and quality of asphalt concrete roads in Indonesia. In this study, additive materials added to asphalt concrete will come from two plastic variables i.e HDPE (High Density Polyethylene) and PP (Polypropylene). From the two variables will be done the process of grinding and mixed then identified the positive characteristics properties that produced on mixing of asphalt concrete, where if founded the most similar characteristics with asphalt and has a large advantage character then this research is expected will be used as an additive on the manufacture of asphalt concrete. Furthermore, the results of this research are concrete asphalt mixed with plastics will be tested its quality in the form of density test, stability, flow and Marshall Quotient (MQ) in laboratory.

2. Methodology

This research will be conducted at Industrial Waste Processing Laboratory, Chemical Engineering Department, Industrial Technology Faculty and Road Material Laboratory, Civil Infrastructure Engineering Department, Vocational Engineering Faculty, Institut Teknologi Sepuluh Nopember Surabaya from June to February 2018. This method used wet process to reduce the amount of asphalt by replacing it with plastic. Type of plastic used is Polypropylene (PP), and High Density Polyethylene (HDPE).
Polypropylene (PP) is a thermoplastic type polymer with properties between LDPE plastic and HDPE made from a combination of propylene monomers. Polypropylene (PP) In some parts for packaging, food packaging, textile materials, laboratory equipment, loudspeakers, automotive components, book binders and clothes hangers [5,6].
HDPE is a thermoplastic material formed from carbon and hydrogen atoms that unite and form high molecular weight products. Then, with the application of heat and pressure, polyethylene is formed [5, 6].

2.1. Mix plastics with bitumen used wet process

To improve the quality of asphalt pavement, stone aggregates are used with specific characteristics based on SNI standard and AASHTO. Bitumen was mixed with waste plastic material by the following process. The waste plastics shredded to the required size of 2.5 mm – 5 mm. The bitumen is heated to 170 °C and aggregate also heated to 210 °C. The shredded waste plastic was sprayed over the hot bitumen (wet process) then plastics got softened and mixed dissolved with bitumen. After that, the aggregate should be fast mixed completely with it. The hot plastics coated aggregates was compacted into a block using compacting machine operated hydraulically and cooled. Then the block was subjected to a compressive test using universal testing machine [4].

2.2. Determine optimum asphalt content

Determination of Optimum Asphalt Content (OAC) is one of the first steps to find out how the best amount of asphalt can be added. The determination of OAC can be estimated theoretically by calculation, and then the result is made a range of values adding the asphalt. The materials used for asphalt concrete mixture in this study consisted of 60/70 AC-WC, coarse aggregate, medium aggregate, fine aggregate and cement as filler.

2.3. Calculation range of Tait parameter
Determination of the melting temperature range of plastics, carried out by calculation approach of density and plastic viscosity at temperature above melting point and atmospheric pressure. The Tait equation is used to determine plastics density range.

$$Volume\ Molar\ (V(P,T)) = \mathrm{V(0,T)}\ \{\ 1 - C \ln\left[1 + \frac{P}{B(T)}\right]\} \quad (1)$$

Where: C is the common polymer constant (0.0894), V(0,T) is the isotherm condition with no pressure (cm3/g), B(T) is virial coefficient (bar), T is temperature (°C), then from equation 1, we determine the value of density by equation 2

$$Density\ (P) = \frac{1}{V(P,T)} \quad (2)$$

The variables used were 5%, 7%, and 9% (wt. %) of plastic in the substitution on the value of the optimum levels of the asphalt which is used, with 100% High Density Polyethylene (HDPE) plastic, 100% Polyprophylene (PP), HDPE: PP and mixtures (50:50), then conducted Marshall test. From the results of tests performed include stability, Flow, void in the mixed (VIM), void filled with asphalt (VFA), VMA (Voids in Mineral Aggregate), up to the value of the Marshall Quoetion (MQ) [7].
From the calculation results will be a reference in determining the value of plastic viscosity. The range of values of plastic viscosity at high temperatures and atmospheric pressure can be known through comparison of plastic density data with comparative density data in this case kerosene, as explained. But to find the flow time of plastic at high temperatures is very difficult, because of the limitations of the tool. From the viscosity equation, it is known that the viscosity is directly proportional to the density, it can be estimated that of all types of plastic to be used, the more temperature increases the lower the viscosity value, seen from the density value which also decreases with increasing temperature. Of the three types of plastic used plastic with the lowest viscosity is PP. These two physical properties may affect the physical properties of concrete asphalt, which will affect one of the tests carried out.

3. Results and Discussion

3.1. Result of Planning of Combined Aggregate Proportion
Mixed planning using the method established by [8] begins with the effective bitumen content that remains in accordance with the provisions. The available aggregate mixing is made into several variations to fulfill the requirements based on Marshall Parameters. In this study used merging 4 aggregate fractions by analytical way that is determination of composition based on trial and error. The aggregate proportion, including coarse aggregate, medium aggregate, fine aggregate, and filler. The following table 1 about combined aggregate proportion fraction [8].

$$P = aA + bB + cC + dD \quad (3)$$

Where; P = Per cent pass the desired size (mm) filter, A= Per cent pass filter coarse aggregate fraction size d (mm), B = Per cent pass filter medium aggregate size d (mm), C= Per cent passes through the fine aggregate fraction filter size d (mm), D = Per cent pass through filter filler aggregate fraction size d (mm)
For percentage values of a, b, c obtained from graphical calculations in accordance with the specifications set forth in the Bina Marga (2010) regulation concerning aggregate limits of AC WC mixtures. The following Table 1 is concerning the percentage of aggregate aggregation

The results obtained based on the combined aggregate calculation in table 1 that the aggregate boundary fits within the AC-WC specification with a 12% percentage for rough aggregates, 48% of medium aggregate, 38% fine aggregate, and 2% Filler.

Table1. Combined aggregate 4 fractions

Filler type		CA 10 - 20 mm		CA 10 - 10 mm		MA 5 - 10 mm		Cement (FA) 0-5 mm		Filler		Results	Specification
No	Size	% pass	0	% pass	12	% pass	48	% pass	38	% pass	2		AC-WC
10	3/4"	0.00	0	100.00	12.00	100.00	48.00	100.00	38.00	100	2	100	100
9	1/2"	0.00	0	28.57	3.42	99.58	47.79	100.00	38.00	100	2	91	90-100
8	3/8"	0.00	0	10.18	1.22	98.28	47.17	99.55	37.82	100	2	88	77-90
7	No.4	0.00	0	7.65	0.91	47.59	22.84	97.65	37.10	100	2	63	53-69
6	No.8	0.00	0	0.00	0.00	8.67	4.16	81.40	30.93	100	2	37	33-53
5	No.16	0.00	0	0.00	0.00	6.76	3.25	42.09	15.99	100	2	21	21-40
4	No.30	0.00	0	0.00	0.00	6.52	3.13	28.42	10.80	100	2	16	14-30
3	No.50	0.00	0	0.00	0.00	0.00	0.00	21.60	8.21	100	2	10	9-22
2	No.100	0.00	0	0.00	0.00	0.00	0.00	11.12	4.23	100	2	6	6-15
1	No.200	0.00	0	0.00	0.00	0.00	0.00	7.75	2.96	100	2	5	4-9

3.2. Determination of Optimum / Theoretical Asphalt Level

To determine the level of asphalt which will be used as a basic plan, the calculation of the amount of proportion for each asphalt sample test material is first carried out. Such as asphalt weight and aggregate type with a total weight of 1200 grams for each sample. The determination of optimum asphalt content must fulfill the asphalt characteristic value according to the general specification requirements of Bina Marga in 2010.

$$Pb = 0.035\ (\%CA) + 0.045\ (\%FA) + 0.18(\%) + constants \tag{4}$$

Where; Pb = Optimum Asphalt Level (%), CA, FA, FF = the fraction of coarse, medium and fine aggregates (%)

From the calculation results it was found that the theoretical asphalt content was 5.7%, so we tried to see the actual optimum levels by making a range of 1% below and above 5.7%, namely 4.7%, 5.2%, 5.7%, 6.2% and 6.7%. Then the Marshall test was carried out, to find out the best mix.

Marshall Test on hot asphalt concrete mixture is stability, VMA (voids in mineral aggregate), VFA (voids filled with asphalt), VIM (voids in the mixture), flow (Mars) and Marshall Quotient (MQ) on the test specimen of each bitumen contents of one object. Table 2 Data Test Results Marshall Parameters [8].

Table 2. Data test results marshall parameters

Marshall Parameter	Specification	**Asphalt Level** (%)				
		4,7	5,2	5,7	6,2	6,7
Stabilisation	>1000 Kg	1316	1356	1593	1524	1731
Flow	> 3 mm	3.80	4.25	4.32	4.30	4.10
VIM	3.0% - 5.5%	2.70	3.01	3.17	2.78	2.31
VFA	> 65%	79.82	79.46	79.98	83.16	86.50
VMA	> 15%	13.31	14.64	15.81	16.50	17.10
MQ	>300 kg/mm	346.3	319.4	364.7	354.3	422.3

Of these, the optimum bitumen content that we can use is 5.7% because it reaches all parameters of Marshall Test. The addition of 5.7% asphalt will be a reference for the mixed composition plastic substitution. Total quantity of the asphalt sample used is 1200 grams, with a composition of 38% fine aggregate, 48% of medium aggregate, 12% coarse aggregate, 2% filler and 5.7% bitumen. The weight calculation of the asphalt added in each sample was 68.40 grams, thus it can be calculated the amount of addition of plastic (gram) substituted to asphalt for variables 5%, 7%, and 9% plastic addition.

Table 3. Test results marshall parameters

Marshall Parameter	Specification	Range	Optimum Asphalt Level (OAL %)
Stabilitation	>1000 Kg	4.7 – 6.7	
FLOW	> 3 mm	4.7 – 6.7	
VIM	3.0 - 5.5%	4.7 – 6.7	
VMF	> 65%	4.7 – 6.7	
VMA	> 15%	5.2 – 6.7	
MQ	>250 kg/mm	4.7 – 6.2	
Asphalt Level Plan (weight%)			4.7 5.2 5.7 6.2 6.7
Optimum Asphalt Level (%)			5.7

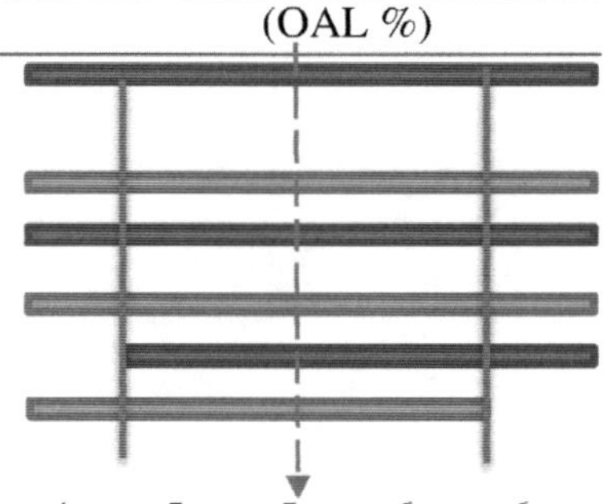

Table 4. Number of plastic substitutes against asphalt weight (grams)

Levels of Plastic Substitution (%)	Plastic weight (gram)	Asphalt Weight (gram)	Total (gram)
5	3.420	64.980	68.40
7	4.788	63.612	68.40
9	6.156	62.244	68.40

3.3. Marshall Parameter Test Results Plastic Mixing Compared with Asphalt

Wherever As described previously, the asphalt has a standard specification following the general standard provisions of the Bina Marga 2010. The result of the concrete asphalt with the addition of plastic is then tested with Marshall Parameters and it is obtained with mixed characteristics, namely the value, of stability, flow, VIM, VFA, VMA and Marshall Quotient.

Marshall Quotient (MQ) value is a mixture of stiffness properties. If the MQ value is too high, the mixture will tend to be too stiff and easy to crack. Conversely, if the MQ value is too low, the pavement becomes too flexible and tends to be less stable. The MQ value is obtained from the ratio between the stability value and the (melt) flow. The value of MQ will be greater if the stability value gets bigger and the value of the flow gets smaller.

Marshall Parameter tests have been carried out at Road Material Laboratory of Department of Infrastructure of Civil Engineering ITS Surabaya. Marshall Parameter test results can be seen in table 5. The test results show the MQ values were qualified by MQ specification for all plastic mixtures with additive except for variable with addition of 5% HDPE PP mixture and 7% HDPE PP mixture, i.e. more than 300 kg / mm. The value of Marshall Quotient (MQ) on the asphalt test results with the addition of HDPE plastics 5%, 7% and 9% respectively MQ value experienced an increasing trend of 316.80 kg / mm, 326.25 kg / mm, and 556.10 kg / mm. In the mixing of asphalt with the addition of PP plastic 5%, 7% and 9% are also successively MQ value has an upward trend of 444.88 kg / mm, 412 kg / mm, 570

kg / mm although at concentration 7% MQ. In mixing the asphalt with the addition of mixture of HDPE and PP plastic 5%, 7% and 9% also the MQ value has an increasing the trend of 204.37 kg / mm, 234.47 kg / mm, 652.16 kg / mm.
From the results of calculations by comparing the results of asphalt MQ values that use additional plastic with 100% pure asphalt, it shows that the percentage increase in asphalt concrete quality occurs at the addition of 9% HDPE, PP and, mixture of Hof stability DPE and PP, 152%, 156%, respectively and 179%. This shows that the addition of plastic with a concentration of 9% can significantly improve the quality of concrete asphalt the main text (11 points).

Table 5. Results Marshall parameters of combain plastic and asphalt

Variable combination	Stabilisation (kg)	Flow (mm)	VIM (%)	VMA (%)	VFA (%)	MQ (kg/mm)
Asphalt concrete 100%	1593.00	4.3	3.17	15.81	79.98	364.70
Add 5% HDPE	1501.48	4.6	3.70	16.31	77.10	316.80
Add 5% PP	1807.33	3.9	3.90	16.45	76.31	444.88
Add 5% HDPE and PP	1362.45	6.4	4.95	17.36	71.50	204.37
Add 7% HDPE	1529.28	4.5	4.67	17.12	72.73	326.25
Add 7% PP	2015.88	4.7	4.26	16.77	74.57	412.00
Add 7% HDPE and PP	1807.34	7.4	3.53	16.13	78.13	234.47
Add 9% HDPE	2085.39	3.6	4.26	16.77	74.57	556.10
Add 9% PP	2850.03	4.8	3.85	16.41	76.52	570.00
Add 9% HDPE and PP	2989.06	4.4	4.47	16.94	73.64	652.16

4. Conclusion
The most significant Marshal Question (MQ) test results were seen in samples from mixed asphalt with the addition of 9% plastic. The percentage increase in MQ value from the addition of 9% HDPE, PP, and HDPE: PP mixture, compared to the use of pure asphalt (100% asphalt) was 152%, 156% and 179%. This shows that the addition of plastic with a concentration of 9% can significantly improve the quality of asphalt concrete.

References
[1] Purnamasari E P and Suryaman F 2010 *Indones. Nt. Civ. Eng. Conf. Proc.* **I**-397-04
[2] Akah A, Hernandez-Martinez J, Rallan C and Garforth A A 2015 *Chem. Eng. Trans.* **43** 2395-00.
[3] Bungay V C 2017 *Chem. Eng. Trans.* **56** 193-98
[4] Vasudevan R, Sekar R C, Sundarakannan B and Velkenedy R 2012 *Constr. Build. Mater.* **28** 311-20.
[5] Achilias D S, Roupakias C, Megalokonomos P, Lappas A A and Antonakou E V 2007 *J. Hazard. Mater.* **149** 536–42
[6] Christensen T H 2011 *Solid Waste Technology & Management* (Lyngby: John Wiley and Sons Ltd)
[7] Rodgers P A 1993 *Appl. Polym. Sci* **48** 1061-80.
[8] Bina Marga 2010 *Gen. Spec. Rev.* ***3***.

IOP Conf. Series: Materials Science and Engineering **543** (2019) 012091 doi:10.1088/1757-899X/543/1/012091

The Role of Calcium Compound on Fluorine Leaching Concentration

F F Hanum[1], E R Desfitri, Y Hayakawa and S Kambara
Environmental and Renewable Energy System, Graduate school of Engineering, Gifu University, 1-1 Yanagido, Gifu Japan 501-1193

E-mail: farrah_fh@yahoo.co.id

Abstract. The performance of calcium compound in leaching mechanisms has been known through the addition of high calcium compound material. The addition of those materials was proven that the additives could simultaneously inhibit the leaching of several trace elements such as arsenic, selenium, boron and fluorine. This study will be focused on the role of calcium compound onto fluorine leaching concentration. Fluorine is one of trace elements consisting in most of coals that has effect in environmental pollution. The application of mixture additives in leaching process was given a promising effect in inhibit the leaching of fluorine until less the environmental limit (0.8 mg/L). This researched will present the information about calcium compound related with fluorine leaching mechanism after the addition of mixture additives (Ca(OH)2, PS ash 8, and BF cement) into 14 different coal fly ashes (FA A, B, C, B, E, F, G, H, I, J, K, L, M, and N). This information will be important in the coal fly ash utilization to minimize the effect of fluorine leaching concentration into the environment.

1. Introduction

Coal has been predicted as the largest source of power generation in 2040 with a share almost 30 percent [1]. It is means that the production of coal fly ashes will still affected the environment because of the hazardous element containing in the coal fly ash [2]. One of the toxic trace elements consisting in coal fly ash is fluorine. Fluorine is the most reactive of all the chemical elements which including into hazardous trace elements in coal, and study was reported that about 10-40% of fluorine in coal entered into coal fly ash through the combustion process [3-5]. Therefore, the utilization of coal fly ash in various purposes needs to be awaring about the fluorine effect into the environment. Fluoride (the anion of fluorine) has been listed as one of the contamination in water by the WHO because could cause many problems in human health [6]. In plants, fluorine could be absorbed by plants under natural condition and it is bound into insoluble compounds such as CaF_2 [7].

In the development of additives for inhibit the leaching of the trace element into the environment, previous study was found that calcium has positive effect in decreasing trace element leaching concentration [8]. Then, the application of mixture additives was given promising effect in inhibits the leaching of several trace elements, including the leaching of fluorine [9]. This study was purposed to investigate about the calcium compound which plays important role during the fluorine leaching mechanisms. This information will require in controlling the effect of fluorine into the environment.

[1] To whom any correspondence should be addressed.

IOP Conf. Series: Materials Science and Engineering **543** (2019) 012091 doi:10.1088/1757-899X/543/1/012091

2. Methods

2.1. Coal Fly Ashes

Coal fly ashes sample used in this research is obtained from different coal-fired power plants in Japan. The chemical composition of these fourteen coal fly ashes were determined by X-ray fluorescence analysis (WDXRF S8 TIGER, Bruker AXS). Based on the XRF analysis, the coal fly ashes samples consisting of different calcium content from the lowest (FA F, 0.66%) to the highest calcium content (FA B, 10.80%).

Table 1. Chemical composition for coal fly ashes by XRF analysis

Fly Ash (FA)			FA A	FA B	FA C	FA D	FA E	FA F	FA G	FA H	FA I	FA J	FA K	FA L	FA M	FA N
Power station			Unit 2	Unit 2	Unit 2	Unit 2	Unit 1	Unit 1	Unit 1	Unit 1	Unit 1	Unit 1	Unit 1	Unit 1	Unit 1	Unit 1
Chamber of electrostatic precipitator			EP1	EP1	EP1	EP1	EP1	EP1	EP1	EP1	EP1	EP1	EP1	EP1	EP1	EP1
chemical composition	SiO_2	[%]	52.61	51.42	64.34	74.78	55.30	66.99	66.21	59.25	59.21	65.07	56.82	57.32	64.84	63.08
	Al_2O_3		31.35	22.39	22.79	16.68	30.84	26.40	26.65	25.63	26.32	21.83	21.06	20.68	23.28	22.73
	TiO_2		2.08	2.16	2.27	1.14	1.94	2.01	1.77	1.99	1.77	1.11	1.06	0.96	1.12	1.24
	Fe_2O_3		7.06	7.01	3.71	3.93	5.75	2.15	2.57	7.49	6.79	7.17	7.23	7.17	5.81	5.65
	CaO		3.26	10.80	2.71	0.45	2.28	0.66	0.81	2.05	2.07	1.43	8.86	8.79	1.39	2.67
	MgO		0.58	1.05	0.85	0.22	0.98	0.51	0.54	0.79	0.91	0.52	0.96	1.02	0.82	1.12
	Na_2O		0.38	1.23	1.20	0.38	1.14	0.27	0.29	0.60	0.74	0.44	0.75	0.80	0.52	1.25
	K_2O		0.96	1.19	0.80	1.43	1.20	0.58	0.54	1.56	1.44	1.76	1.97	1.97	1.67	1.44
	P_2O_5		0.61	0.22	0.07	0.10	0.19	0.04	0.06	0.18	0.18	0.16	0.28	0.27	0.13	0.20
	MnO		0.09	0.18	0.06	0.04	-	0.12	0.11	-	0.10	-	-	-	0.00	0.07
	V_2O_5		0.00	0.00	0.00	0.00	0.09	0.01	0.02	0.03	0.03	0.15	0.17	0.17	0.08	0.04
	SO_3		0.27	1.22	0.35	0.28	0.28	0.26	0.44	0.42	0.50	0.35	0.84	0.87	0.35	0.53

2.2. Additives

An additive is used to elevate the calcium content in the coal fly ash sample. There are two kinds of mixture additives used in this researched: (1) The mixture of PS ash 8 and BF Cement (2) The mixture of $Ca(OH)_2$, PS ash 8 and BF cement. The previous researched presented that these two mixtures were given promising effect in the leaching of several trace elements from coal fly ash with low calcium content (FA C and FA H). The calcium containing in PS ash 8 and BF cement respectively are 51.22% and 48.35% (Table 2). The coal fly ash samples mixed with the additives in the certain ratio before the leaching process. The ratio of the additives based on the total sample is 3% of calcium hydroxide ($Ca(OH)_2$), 10% of PS Ash 8, and 10% of BF Cement. Then, 1:10 of coal fly ashes and distilled water prepared for the 6 hours leaching in the room temperature.

2.3. Analysis and Instrumentations

Ion chromatography analysis (ION ANALYZER IA-300) was performed in order to determine the fluorine leaching concentration from the 14 different coal fly ashes for before and after the application of the additives. Then, the data of leachate alkalinity was provided by pH/ION METER D-53, HORIBA in order to probe the effect of pH into fluorine leaching concentration. In addition the data for thermal gravimetric analysis (TG/DTA6300 SII EXSTAR 6000, HITACHI), ethylene glycol analysis, ICP-AES (ULTIMA2, HORIBA Ltd) and XPS analysis was used to explain the role of calcium onto the fluorine leaching mechanisms.

Table 2. Chemical composition for additives by XRF analysis

Chemical composition		Additives	
		PS ash (8)	BF Cement
SiO_2	[%]	28.76	31.03
Al_2O_3		15.41	13.32
TiO_2		0.35	0.19
Fe_2O_3		0.91	0.44
CaO		51.22	48.35
MgO		2.76	3.77
Na_2O		0.02	0.08
K_2O		0.15	0.36
P_2O_5		0.10	0.00
MnO		0.04	0.05
V_2O_5		0.02	0.02
SO_3		0.27	2.39

2.4. Analysis and Instrumentations
Ion chromatography analysis (ION ANALYZER IA-300) was performed in order to determine the fluorine leaching concentration from the 14 different coal fly ashes for before and after the application of the additives. Then, the data of leachate alkalinity was provided by pH/ION METER D-53, HORIBA in order to probe the effect of pH into fluorine leaching concentration. In addition the data for thermal gravimetric analysis (TG/DTA6300 SII EXSTAR 6000, HITACHI), ethylene glycol analysis, ICP-AES (ULTIMA2, HORIBA Ltd) and XPS analysis was used to explain the role of calcium onto the fluorine leaching mechanisms.

3. Results and Discussions

3.1. Effect of pH in fluorine leaching process
The different calcium compound on coal fly ashes also affected pH of the leachate. The pH of the leachate from the fourteen coal fly ashes before the application of mixture additives is linier with the calcium content on each coal fly ashes based on the XRF analysis (Table 1). After the application of the mixture additives, the pH of the leachate is become higher. The leachate pH from the mixture additives PS Ash 8 and BF cement is slightly lower than mixture additives $Ca(OH)_2$, PS Ash 8 and BF cement (pH around 12, Figure 1). This is proven the study that explained alkalinity is one of the factors which affected the trace element leaching concentration into the environment and the high alkaline pH is giving positive effect in decreasing the trace element leaching concentration [10-12].

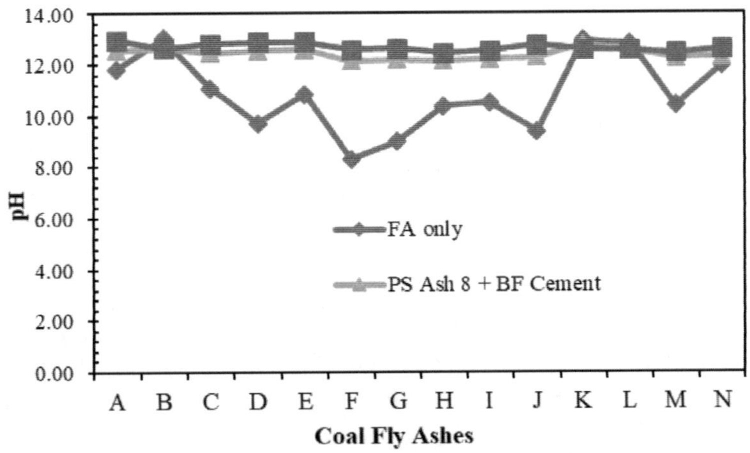

Figure 1. Effect of mixture additives into the leachate pH from different coal fly ashes on fluorine leaching process

3.2. Effect of additives into fluorine leaching concentration

Based on the pH analysis result, both of the mixture additives have pH in around pH 12. Piekos and Palombo et al explained in their study that higher pH on the leaching process could lowered the leaching of the fluorine leaching. As the application of the mixture additives, the pH of the leachate was become higher, so that the additives might have promising effect on the fluorine leaching process. There are fourteen different coal fly ashes were tested in order to study the effect of the additives mixture into fluorine leaching concentration. Figure 2 showed that the leaching of fluorine from coal fly ash B, C, D, G, I, J, L, M, and N without the additives are above the environmental limit. The others have fluorine leaching concentration under the environmental limit. Based on the Ministry of the environment of Japan, the environmental limit of fluorine in Japan is 8 mg/L for other than sea area.

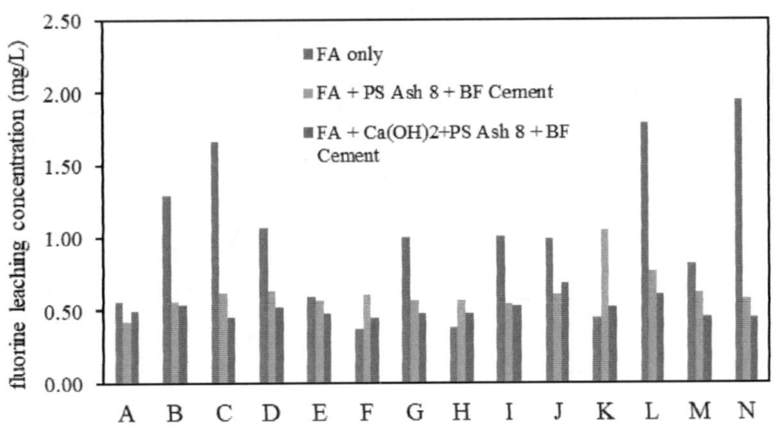

Figure 2. Effect of the mixture additives into different coal fly ashes on fluorine leaching concentration

In addition, amongst the coal fly ashes which have fluorine leaching higher than the standard, the calcium content containing in those coal fly ashes is very various including the lowest (0.81%) and the highest (10.80%) once. This is proven that the amount of calcium consisting in the coal fly ash sample is not linier with the fluorine leaching concentration. Furthermore, this result also stated that calcium compound containing in the coal fly ashes are diverse and there is a specific calcium compound which affected the fluorine leaching concentration. Later, this study will inform the calcium compound which plays an important role during the fluorine leaching concentration.
The both of mixture additives performed a positive decreasing in fluorine leaching concentration into the coal fly ashes which have higher leaching concentration than the environmental standard (Figure 2). The mixture of $Ca(OH)_2$, PS Ash 8 and BF cement could decrease the amount of fluorine in the leachate higher, except on the coal fly ash J (FA J; Ca 1.43%). This result means that the mixture of $Ca(OH)_2$, PS Ash 8 and BF cement is applicable in fluorine leaching mechanisms. In addition, this mixture also containing the calcium compounds which important in inhibit the fluorine leaching concentration, which might be the same with calcium compound that affected the fluorine leaching concentration on the coal fly ashes mentioned before.

3.3. The role of calcium in fluorine leaching process

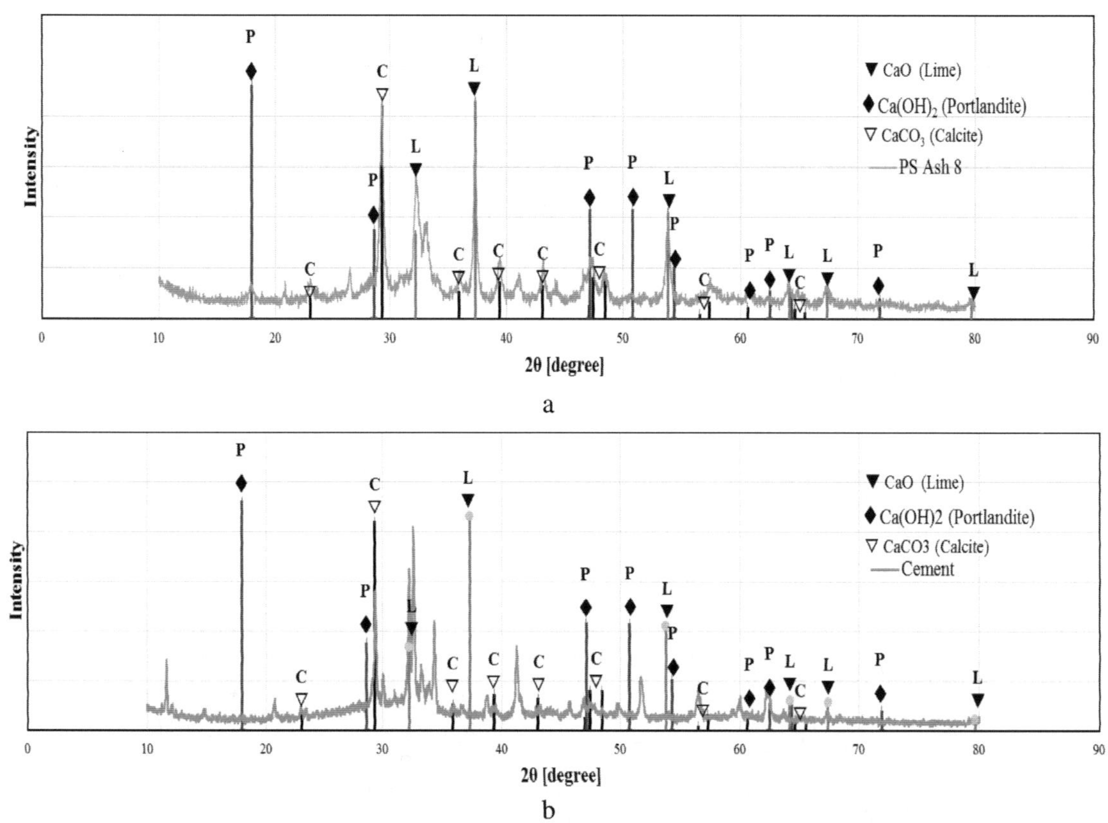

Figure 3. XRD patterns of (a) PS ash 8 and (b) Blast Furnace Cement (BFC)

The both of the mixture of the additives presented the promising effect in minimizing the effect of the fluorine leaching into the environment. Calcium compound consisting in the coal fly ashes and in the additives were believed to take important role during the leaching of the fluorine. Figure 3 showed the XRD analysis result of the PS ash 8 and BF Cement. This figure explained that CaO, $Ca(OH)_2$ and $CaCO_3$ were the main calcium containing on these additives.

Meanwhile, thermal gravimetric analysis (TGA) and ethylene glycol analysis were carried out in order to find the amount of those calcium compounds in the both of PS ash 8 and BF cement (Table 3). This data explained that the most calcium compound which might affected the fluorine leaching concentration is calcium oxide (CaO) and calcium carbonate ($CaCO_3$). These calcium compounds will be reacting with other chemical composition which will stabilized the hazardous heavy metals consisting in coal fly ashes.

Table 3. Calcium compound analysis consisting in the additives materials

Additives material	CaO (%)	$Ca(OH)_2$ (%)	$CaCO_3$ (%)
PS ash 8	11.1	1.44	19.32
BF Cement	1.62	0.31	6.45

Several studies have been researched the calcium and fluorine bearing compound consisting in coal fly ashes which might be produced during the combustion process [13-15]. The study said that mainly fluorine compound in coal is insoluble fluorine such as CaF_2, MgF_2, FeF_3, and AlF_3. These fluorine compounds are the main occurrence state after coal combustion process and difficult to break down even at high temperature. Wang, et al explained that several oxide compounds inside the coal including CaO in the fly ashes were took part in the formation of insoluble fluorides. This is confirmed the previous result that CaO is one of the main calcium compound which plays role in fluorine leaching process.

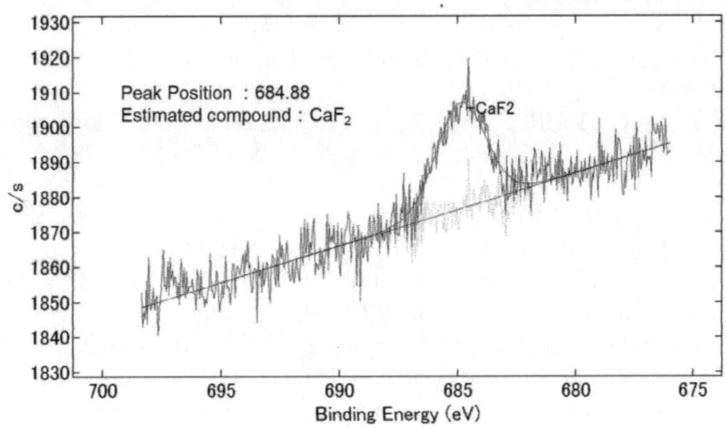

Figure 4. XPS analysis result on the mixed coal fly ash sample and additive (FA C and $Ca(OH)_2$)

Other than that, based on the XRD analysis data and TG/EG analysis data, the amount of calcium hydroxide is a few. But, calcium hydroxide is native calcium compound to enrich the calcium content on the coal fly ash samples. Even though the amount that little amount of $Ca(OH)_2$, this compound have ability to decrease the fluorine leaching. The XPS analysis was did into one of the coal fly ash sample (FA C) which already mixed with $Ca(OH)_2$. The peak showed that after the mixing process, the estimated calcium compound consisting in the samples is CaF_2. This is only a qualitative analysis, so that the exact amount of CaF_2 which might be produced was not well known yet. However, the leaching process might be affected the increasing amount of CaF_2 or other insoluble compound of fluorine which may cause by the chemical stabilization during the leaching process.

4. Conclusion

The application of mixture additive in leaching process has shown promising effect into fluorine leaching concentration. The various kinds of coal fly ashes was treated with the mixture of $Ca(OH)_2$, PS Ash 8 and BF cement as the additives, then the results confirmed that these additive mixtures is appropriate to apply in fluorine leaching concentration. CaO and $CaCO_3$ was mainly the main calcium compound consist in the additives. These calcium compounds was took part in the chemical stabilization of fluorine during the leaching concentration to produce the insoluble fluoride compounds. CaF_2 believed to be the one of fluoride compound consisting in the mixture of coal fly ashes and additives which difficult to break down even with the high temperature.

Acknowledgement

The author would like to thank Akihiro Takeyama for his valuable contributions in data analysis.

References

1. BP Energy outlook. 2018. BP Statistical Review of World Energy 67th Edition. Available online at: https://www.bp.com/content/dam/bp/en/corporate/pdf/energy-economics/statistical-review/bp-stats-review-2018-full-report.pdf, Accessed: July 9, 2018.
2. Bchtrong P, Nguyentanth L and Pham T H 2017. Effect of fly ash from thermal power plant on the dielectric properties of polymer composites materials based on matrix epoxy DER 331. *Chem. Eng. Trans.* **56** 1207-12
3. Robertson J D, Wong A S and Hower, J C 1998 Fluorine in coal and coal by products Reported by : Departement of Chemistry and Center for Applied Energy Research, pp 500-03.
4. Machado P M, Mores S, Pereira, Ederson E, Welz B, Carasek, E and de Andrade J B 2015 Fluorine determination in coal using high-resolution graphite furnace molecular absorption spectrometry and direct solid sample analysis. *Spec. Acta. Part B* **105** 18-24.
5. Deng S, Shu Y, Li S, Tiang G, Huang J, and Zhang F 2016 Chemical forms of the fluorine, chlorine, oxygen and carbon in coal fly ash and their correlation with mercury retention *Jour. Of Haz. Mat.* **301** 400-06
6. He J, Li Yong, Xue X., Ru H, Huang, and Yang H 2017 Leaching of fluorine and rare earths from bastnaesite calcined with alumium hydroxide and the recovery of fluorine as cryolite *RSC Adv.* **7** 14053-59
7. Szostek R and Ciecko Z 2017 Effect of soil contamination with fluorine on the yield and content of nitrogen froms in the biomass crops *Env. Sci. Pol. Res.* **24** 8588-01
8. Hartuti S, Hanum F F, Takeyama A, and S Kambara 2017 Effect of additives on arsenic, boron, and selenium leaching from coal fly ash *Min.* **99** 1-19.
9. Hanum F F, Erda Rahmilaila D, Yukio H and S Kambara 2018 Preliminary study on additives for controlling As, Se, B and F leaching concentration from coal fly ash *Min.* **11** 493 - 03
10. Piekos S and Paslawska S 1998 Leaching characteristics of fluoride from coal fly ash *Res. Report: Fluoride* **4** 188-92
11. Wang T, Wang J, Tang Y, Tang Y, Shi H and Ladwig K. 2009 Leaching characteristics of arsenic and selenium from coal fly ash: role of calcium *Energy Fuel* **23** 2959-66
12. Palumbo Anthony V, Tarver Jana R, Fagan, Lisa A, McNeilly, Meghan S, Ruther R, Fisher L S, Amonette and James E 2007 Comparing metal leaching and toxicity from high pH, low pH, and high ammonia fly ash *Fuel* **86** 1623-30
13. Wang G, Luo Z, Zhang J and Zhao Y 2015 Modes of occurence of fluorine by extraction and SEM method in a coal-fired power plant from Inner Mongolia China *Min.* **5** 863-69
14. Guo S, Yang J and Liu Z 2006 The fate of fluorine and chlorine during thermal treatment of coals *Env. Sci.* **40** 7886-89
15. Xie Pan Pan, Guo W, Yan X. and Zheng X. 2017 Fluorine in Lopingian superhigh-organic-sulfur coals from the Lalang Coal Mine, Guangxie, Southern China *Fuel* **208** 483-90

IOP Conf. Series: Materials Science and Engineering **543** (2019) 012092 doi:10.1088/1757-899X/543/1/012092

Calcium Performance in Paper Sludge Ash as Suppressing Material

E R Desfitri, F F Hanum, Y Hayakawa and S Kambara
Environmental and Renewable Energy Systems Division, Gifu University, 1-1 Yanagido, Gifu, Gifu Prefecture 501-1193, Japan

Email: rahmilaila01@gmail.com

Abstract. Paper sludge (PS) ash is a waste material collected from the paper industry, which containing calcium compound. In the previous research, several paper sludge ashes have been tried into coal fly ashes in order to suppress the leaching of trace elements such as As, Se, and B into the environment. The result showed that paper sludge ashes has potential effect in inhibit the leaching of these trace elements because of the calcium contains in the paper sludge ashes. Therefore, this research provide information about the amount of calcium in paper sludge ash as suppressing material. Nine kinds of paper sludge ashes tested into coal fly ash C (FA C) in order to know the ability of each paper sludge ash in suppressing the leaching of trace elements. Inductively coupled plasma (ICP-AES) used to determine the leaching concentration of As, Se, and B. Then, the kind of calcium in paper sludge ashes were analysed by X-ray diffraction (XRD). Thermal gravimetric analysis (TG) used to determinate $Ca(OH)_2$ and $CaCO_3$. The percentage of CaO in paper sludge ashes analysed by combination ethylene glycol, ICP-AES and TG. This research found the kind of PS which most influential for controlling leached out of trace elements.

1. Introduction

Fly ash, a waste product from the thermal power plants, is increasing day by day [1]. So, in the next few years, coal will be still an important source of electricity, not only in Japan but also in the world. As the coal burns, toxic pollutants such as arsenic (As), selenium (Se), boron (B) are transferred into the wastes [2]. Because of this, the study about how to minimize the effect of trace element into the environment still needs to be researched. In order to decrease the effect of trace element into the environment, many researchers already found that calcium is affected trace elements leaching mechanism said that calcium compound could control the leaching of trace element from coal fly ash [3]. Another research also states that CaO dominates in coal fly ash sample directly affects the mobility of trace elements by determining pH of the leaching medium [4].

The role of calcium through the addition of additives during the leaching of trace elements has been studied in the previous work [5]. The study explained that paper sludge as the additives is one of the promising solution in minimize the hazardous effect of trace element into the environment. However, the study also showed that the higher calcium in paper sludge ashes did not always linier into a better effect during mechanisms. Paper sludge ash is waste which lately was utilized by some industry, especially as the cement kiln feed and the cement blending because after incinerating paper sludge at approximately 800°C, the resultant fly ash may contain reactive silica and alumina (in the form of metakaolin) as well as lime (CaO). The composition of the ash not only depend on the specific fuel, but also on the combustion technology and conditions [6].

There are specific characteristic of calcium which may affect the leaching mechanisms. Therefore, this research will study the performance of several kinds of paper sludge ashes based on the calcium content. Nine kinds of the paper sludge ashes will be tried into coal fly ash C, in order to analyse their

IOP Conf. Series: Materials Science and Engineering **543** (2019) 012092 doi:10.1088/1757-899X/543/1/012092

performance during the leaching mechanisms. In addition, the performance of calcium through the addition of paper sludge ashes will be study by the analysis by inductively coupled plasma- Atomic Emission Spectroscopy (ICP-AES) analysis, X-ray Diffraction (XRD) analysis, thermal gravimetric (TG) analysis and ethylene glycol (EG) analysis. The result will provide information about the paper sludge ash which effective in decreasing the As, Se and B leaching concentration simultaneously.

2. Methods

2.1 Materials

Fly ash sample which used in this research was Fly ash C (FA C). It derived from a pulverized coal-fired power plant unit 2 electrostatic precipitators number 1. This coal fly ash containing As 26.39 µg/L, Se 186.51 µg/L and B 7.85 mg/L based on ICP-AES analysis.

Paper sludge ash as the suppressing material has been tested for stabilized As, Se, and B in FA C. Inorganic Chemical composition of fly ash C and paper sludge ashes could be shown in Table 1.

Table 1. Chemical composition in FA C and paper sludge ashes based on XRF analysis

Samples		FA C	PS Ash 3	PS Ash 4	PS Ash 5	PS Ash 6	PS Ash 7	PS Ash 8	PS Ash 9	PS Ash 10	PS Ash 11
	SiO_2	64.3	31.47	44.21	42.36	26.03	32.39	28.76	37.82	40.6	26.2
	$Al2O_3$	22.8	12.40	22.23	19.8	18.31	15.94	15.41	19.39	19.57	15.04
	TiO_2	2.27	0.38	2.56	2.11	3.47	0.62	0.35	3.01	2.27	0.42
	Fe_2O_3	3.71	5.13	2.63	5.56	1.58	0.98	0.91	6.11	3.13	-
Ash	CaO	2.71	46.31	18.77	19.51	41.45	44.24	51.22	23.27	24.29	54.76
composition	MgO	0.85	3.28	3.42	3.3	2.09	2.72	2.76	3.05	2.88	2.56
[%]	Na_2O	1.20	0.24	0.95	0.41	-	0.26	0.02	0.41	0.71	0.16
	K_2O	0.80	0.20	2.09	1.99	0.93	0.53	0.15	1.87	2.34	0.07
	P_2O_5	0.07	0.18	1.75	1.54	0.94	0.43	0.10	1.32	2.00	0.16
	MnO	0.06	0.03	0.05	0.06	0.04	0.05	0.04	0.06	0.06	0.03
	V_2O_5	-	0.01	0.02	0.02	0.02	0.02	0.02	0.02	0.02	0.02
	SO_3	0.35	0.36	1.31	3.33	5.15	1.83	0.27	3.67	2.16	0.56

2.2 Leaching Test

2.2.1 Leaching test for trace elements

Leaching test for this experiment based on the procedure of standard leaching test for fly ash notification No.13 by the Environmental Agency of Japan.

2.2.2 Ethylene glycol leaching method

CaO needs a high temperature (>1000°C) to decompose, so it is difficult to measure with thermal gravimetric directly. In this research, determination of CaO in paper sludge ash with the combination of TG and ethylene glycol have been developed. At the beginning of this method 1 mg PS ash was weighed and added 25 ml warm ethylene glycol (T=80°C). Then stirred at 200 rpm (temperature was keep at 80°C) for one hour. After one hour, sample was filtered by 0.45 µm cellulose acetate membrane filter. The concentration of Ca in solution measured by inductively coupled plasma atomic emission spectrometry (ICP-AES).

2.3 Instrumentation

Some instruments have been used to analysis Calcium compound in paper sludge ashes. X-ray Diffraction (XRD) labX XRD6100, Shimazu, Kyoto, Japan has been used for qualitative analysis. The sample in powder form was pressed into a sample holder so that have a smooth flat surface. The sample was irradiated with Cu Kα-radiation between 10° and 80° (2θ). Scanning speed is 2°/min, at an acceleration voltage of 40kV and tube hazard 30 mA. Software module "DDView and Sleve," database PDF − II release 2013 from the International Center for Diffraction Data (ICCD) was used to analysis diffraction pattern after measurement. PDF II means database which contains full information on a particular phase including cell parameters. TG/DTA6300 SII EXSTAR 6000, Hitachi, Hongkong has been used to measured $Ca(OH)_2$ and $CaCO_3$. Paper sludge ashes sample (9-12mg) was heated with a measurement temperature from 30 to 1000°C at a heating rate of 10°C/min under a nitrogen atmosphere at a flow rate of 200ml/min. This instrument work based one measure of weight/mass change (loss weight) as a function of temperature, time and atmosphere. Percentage of CaO measured by combining ICP-TG through ethylene glycol leaching process because CaO needs the high temperature (>1000°C) to decompose, so it is difficult to measure with thermal gravimetric directly. However, not only CaO can dissolve by the ethylene Glycol but also $Ca(OH)_2$. Therefore this method has to combine with TG analysis. The content of CaO in a paper sludge is calculated from the amount of calcium determined by the EG method and $Ca(OH)_2$ obtained by TG.

3. Results and Discussion

3.1 Effect of Suppressing Material

In order to know the effect of paper sludge ashes into trace element leaching concentration each paper sludge ash have been tested into FA C. The result shown that the addition of paper sludge ash can increase the pH value of leached and decreasing leaching concentration of trace elements in FA C as demonstrated in figure 1. It consentient with statement [4] that CaO in coal fly ash sample directly affects the mobility of trace elements by determining pH of the leaching medium. Figure 1(a) shows effect each paper sludge ashes in controlling leaching Arsenic (As). Arsenic leaching concentration in FA C without additive material was 17.55 μg/L, but after added paper sludge ashes it could be decreased. All of paper sludge ashes could be decreased As leaching concentration under environmental standard, but the most effective to decreasing Arsenic concentration in FA C was PS ash 11 followed by PS ash 3 and PS ash 6 (0.01 μg/L, 1.04 μg/L and 1.65 μg/L respectively). Then, pH also one of the most important parameters that control the leaching process [8]. Paper sludge ashes could increase the alkalinity of FA C from 8.6 increases until 11.7 pH value by PS ash 11 as shown in Figure 1. Selenium (Se) leaching concentration in FA C without additive material was 186.51 μg/L. Paper sludge ashes also could keep stabilization selenium in coal fly ash. It proves by figure 1b Selenium leaching concentration decreasing until 0.615 μg/L, 14.65 μg/L and 17.02 μg/L by PS ash 11, PS ash 5 and PS ash 9 respectively but only PS ash 11 addition could decrease until under environmental limit as display in figure 1(b). Boron leaching concentration also keep stable after added paper sludge ashes. Different kinds of paper sludge ashes give different effect stabilization to Boron leaching concentration. Boron leaching concentration without paper sludge ashes was 3.88 mg/L. It could be decreased until 0.38 mg/L, 0.39 mg/L and 0.64 mg/L by PS ash 11, PS ash 7 and PS ash 8 shown in figure 1(c). PS ash 6 good at controlling Arsenic, but no in controlling Selenium and Boron. PS ash 5 and 9 good in selenium but no in controlling Boron. PS ash 7 good at controlling Boron, but no in arsenic and selenium. Different with PS ash 11, it good in three of trace elements: arsenic, selenium, and boron. PS ash 3 and PS ash 8 also keep stable to decreasing leaching concentration of three kinds trace elements in FA C. In order to know why some paper sludge ashes give good effect but the others no, calcium qualitative and quantitative analysis in FA C have been done.

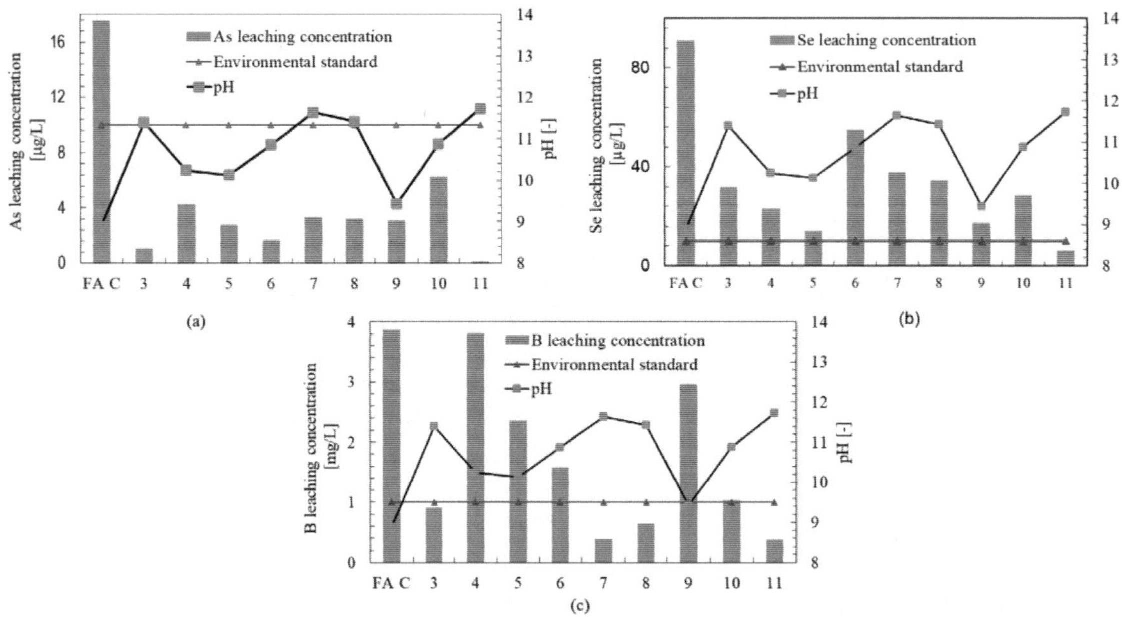

Figure 1. The effect of additive material in controlling trace elements leaching concentration (a) Arsenic, (b) Selenium, and (c) Boron

3.2 Qualitative Analysis by XRD

X-ray diffraction analysis was performed to analysis calcium compound in the paper sludge ashes. Calcium oxide (CaO), Calcium Hydroxide ($Ca(OH)_2$) and Calcium carbonate ($CaCO_3$) pure have been used as a standard to analysis calcium compound in paper sludge ashes. As can be seen from Figure 2 raw peak of each paper sludge ash have different intensity. Figure 2(a) indicate that peak of diffraction pattern from PS ash 3, 8 and 11 was detected as CaO, $Ca(OH)_2$ and $CaCO_3$. Based on figure 2(a), PS ash 3, 8 and 11 predicted have high calcium compound. Different with figure 2(b) the higher intensity of sample there are on different intensity of calcium standard. Only a small part of the diffraction peak detected as calcium compound. It predicted that PS ash 4, 5, 6, 7, 9 and 10 have low calcium compound. In order to make sure this hypothesis quantitative analysis has been done.

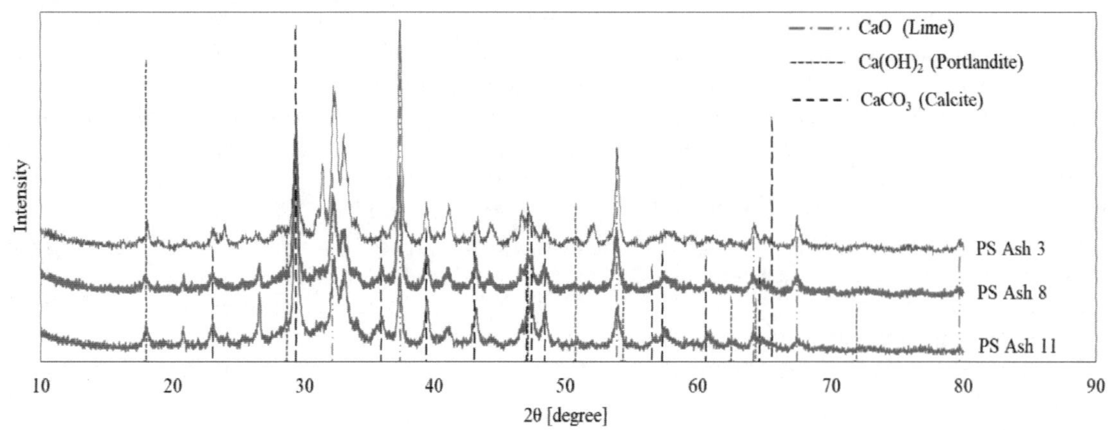

IOP Conf. Series: Materials Science and Engineering **543** (2019) 012092 doi:10.1088/1757-899X/543/1/012092

(a)

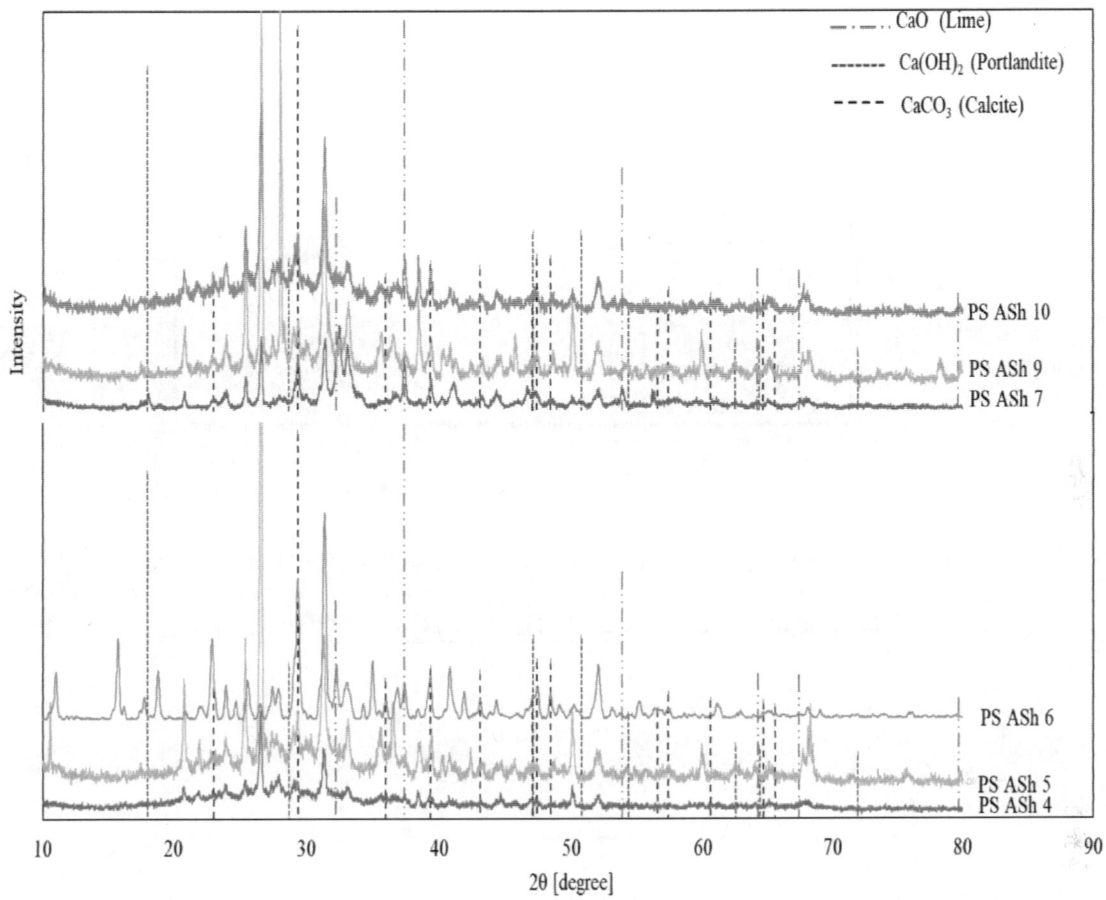

(b)

Figure 2. Combination pattern of paper sludge ashes and three of calcium standard; (a) PS Ashes pattern which have generally same peaks with calcium standard, (b) PS Ashes pattern which slightly consist of calcium standard.

3.3 Quantitative Analysis by TG and Ethylene Glycol Method

TG analysis was carried out on paper sludge ashes to confirm percentage of $Ca(OH)_2$ and $CaCO_3$ based on weight losses in TG curve. $Ca(OH)_2$ starts to lose weight in temperature range 350-450°C decomposed into CaO and H_2O. $CaCO_3$ decomposed into CaO and CO_2 at the temperature around 600-790°C. Both of them have been used as a standard to determine calcium compound in paper sludge ash. Figure 3 shows decomposition of PS ash 3, a weight loss from the temperature around 355-445°C and 600-780°C for the $Ca(OH)_2$ and the $CaCO_3$ respectively. Percentage of each calcium obtained from calculation of weight loss, molecule weight and total weight. Since the decomposition temperature of CaO is above 1000°C and cannot detected by TG only, the analysis was carried out by combining ethylene glycol extraction/ICP-AES. Based on the previous research [8] has been known that CaO and $Ca(OH)_2$ are dissolved in Ethylene glycol well but $CaCO_3$ dissolved hardly. As shown in the Table 2, it became clear that CaO and $Ca(OH)_2$ can be completely extracted by the Ethylene Glycol method. Calcium compound extracted from paper sludge ash by the ethylene method was converted into CaO and the CaO content in the paper sludge ash was determined by subtracting the $Ca(OH)_2$ amount.

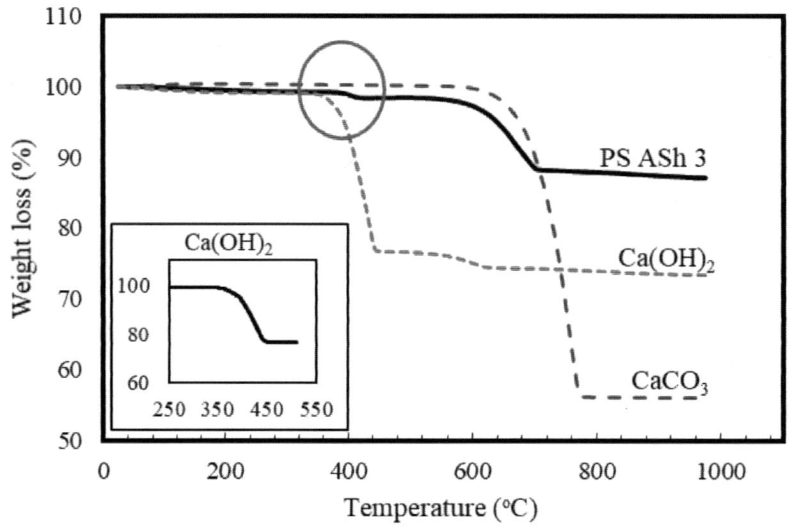

Figure 3. Thermal decomposition curves of PS ash 3, $Ca(OH)_2$ and $CaCO_3$

Table 2. Solubility of CaO, $Ca(OH)_2$ and $CaCO_3$ in ethylene glycol

Calcium compound	Dissolve amount (%)
CaO	100.5 ± 0.5
$Ca(OH)_2$	96.7 ± 0.6
$CaCO_3$	0.1 ± 0.0

3.4 Calcium Performance in Paper Sludge Ashes

It has been known that each paper sludge ashes give different effect to control leaching concentration of As, Se, and Boron in coal FA C. In aims to know the kind of calcium compound TG analysis and EG method has been done. Figure 4 shows the concentration of calcium compounds which contained in paper sludge ashes.

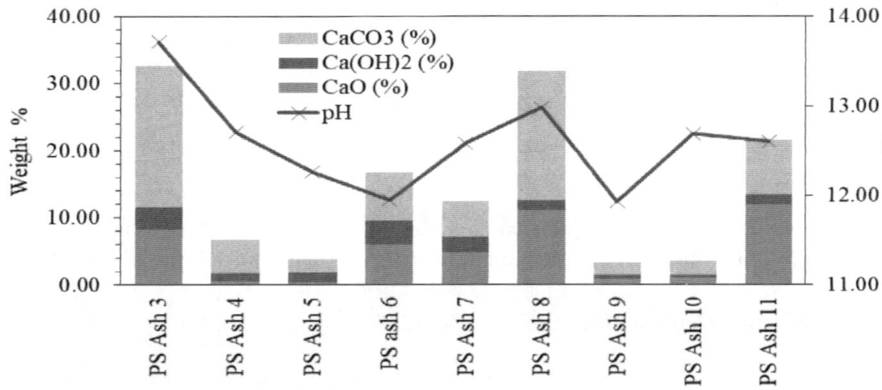

Figure 4. Calcium compounds content of paper sludge ashes and it relation with the alkalinity

$Ca(OH)_2$ and $CaCO_3$ were result from TG and CaO was result from EG method. The lowest calcium concentration was found on PS ash 9, 10, and 5. Otherwise, the highest calcium compound contained on PS ash 3, 8, 11 and 6 with the total 32.69%, 31.86%, 21.53%, and 16.81% respectively. The highest 21.12% $CaCO_3$, 3.55% $Ca(OH)_2$ and 12.03% CaO consecutively consist on PS ash 3, 6 and 11. The alkalinity of paper sludge ashes which could increasing pH value in leaching process of coal fly ashes also could be seen in figure 4. The highest alkalinity found in PS ash 3, 8 and 11 (13.72, 12.98 and 12.61 respectively).

Related to the result of additive material effect in controlling trace elements leaching concentration arsenic, selenium and boron can be known that the highest calcium compound not give the best effect in controlling leaching concentration simultaneously. Figure 5 shows which one is the best PS ashes in controlling As, Se and B simultaneously.

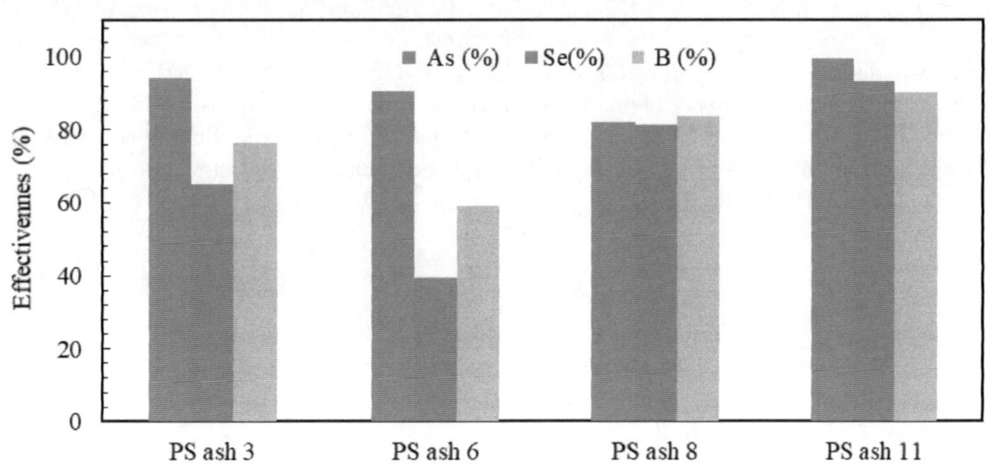

Figure 5. The capability of paper sludge ashes in decreasing trace elements

Percentage of effectiveness shows the capability of PS ashes which contain the highest calcium in decreasing leaching of trace elements. PS ash 3 which contains the highest calcium compound and consist of the highest $CaCO_3$ only can control leaching concentration As, Se and B were 94.07%, 65.13% and 76.42% respectively. PS ash 6 which contains the highest $Ca(OH)_2$ can control leaching concentration of 90.60% As, 39.48% Se, and 59.28% B. PS ash 8 which contains the second highest of calcium compound can control around 80% of As and B but only around 60% of Se and B. The most effective to control leaching three of As, Se and B is found in PS ash 11. PS ash which containing the highest CaO can control As, Se and B were 99.43%, 93.22% and 90.34% respectively. It means the kind of calcium which most affected is CaO and the most effective to control leaching concentration of As, Se, and B is PS ash 11.

4. Conclusion
Based on this research has been known that each PS ash give different effect in controlling leach out of trace elements in FA-C suitable with kind and total calcium which contained in each PS ash. In this research found that PS ash 11 is most effective for stabilization of trace elements in coal fly ash because it can control trace elements leaching more than 90% simultaneously.

References
[1] Bachtrong P, Nguyenthanha L and Pham T H 2017 Effect of fly ash from thermal power plant on the dielectric properties of polymer composites materials based on matrix epoxy der 331 *Chem. Eng. Trans.* **56** 1207-12.

[2] Alper B, Gulbin G and Fatma S 2010 Leaching characteristic of fly ash from fluidized bed combustion thermal power plant *Fuel Proc. Tech.* **91** 1073-80.

[3] Facun J, Longlong Z, Norihiro, Atsushi S and Yoshihiko 2011 Role of calcium compounds on reduction of arsenic and selenium during fluidized bed coal combustion *Meas. Cont. Serv. Eng.* **7** 11-23.

[4] Akar G, Polat M, Galecki G and Ipekoglu U 2012 Leaching behavior of selected trace elements in coal fly ash samples from Yenikoy *Fuel Proc. Tech.* **104** 50–56.

[5] Hartuti S, Hanum F F, Takeyama A and Kambara S 2017 Effect of additives on arsenic, boron and selenium leaching from coal fly ash *Minerals* **7** 99-18.

[6] Rummel L, Neshumayev D and Konist A 2018 Power plant ash composition transformations during load cycling *Chem. Eng. Trans.* **70** 655-60.

[7] Komonweeraket K, Cetin B, Aydilek A H., Craig H B and Tuncer B E 2015 Effects of pH on the leaching mechanisms of elements from fly ash mixed soils *Fuel* **140** 788–02.

[8] Kato M, Tatsuya H, Saito S and Masami S 2014 Determination of free lime in steelmaking slags by use of ethylene glycol extraction/ICP-AES and Thermogravimetry *Japan Sci. and Tech. Agency* **100** 340-45.

Study of Nanoparticle Distribution in Water Treated with Combined Filtration-Inductively Coupled Plasma System

Y Trianda[1], R Desmiarti[1,*], A Hazmi[2], M Martynis[1], and F Li[3]

[1]Chemical Engineering Department, Universitas Bung Hatta, 25147, Padang, Indonesia
[2]Electrical Engineering Department, Andalas University, 25166, Padang, Indonesia
[3]River Basin Research Center, Gifu University, 1-1 Yanagido, Gifu 501-1193, Japan
Email : renidesmiarti@gmail.com; renitk@bunghatta.ac.id

Abstract. Nanoparticles represent a specific type of organic or inorganic matter with at least one dimension less than 100 nm and possess unique physical and chemical properties. The presence of nanoparticles in drinking water may pose a direct human health threat or an indirect risk through ingestion. Thus, the removal of nanoparticles from drinking water is needed. This research was carried out to study the distribution of nanoparticles in water treated with a combined filtration-inductively coupled plasma system by continuous processing. The flow rate was set at 100 and 200 mL/minute. The results showed that after 180 minutes of treatment, the amount and volume of nanoparticles in the treated water produced were different from those contained in raw water. The nanoparticles were within the diameter ranges from 2.70 to 6.50 nm and 0.62 to 712 nm for flow rate at 100 and 150 mL/minute, respectively. The particles within these diameter ranges belong to the group of proteins, small molecules and atoms. The measured zeta potential of the treated water had also decreased, indicating that the stability of the nanoparticles was reduced and removal by the combined filtration-inductively coupled plasma system had occurred. Further research is required to find the optimum flow rate and determine the exact components of the groups of nanoparticles.

1. Introduction

The demand for safe drinking water increases day by day due to the rapid growth of the world population and the insufficient capabilities of conventional treatment methods. Various conventional and contemporary purification techniques are available, such as basic filtration and adsorption as well as advanced technologies, such as membrane filtration and advanced oxidation processes (AOPs). In general, the aim of drinking water treatment is to provide potable water that is free from toxic matters (organic, inorganic and biological) [1]. Among the advanced technologies, AOPs are recognized as highly effective in producing safe drinking water that is free of organics, inorganics and microbes [2].

AOPs are one of the advanced drinking water treatment technologies that involve the production of hydroxyl radicals (OH•) in sufficient quantities to induce water purification at standard temperature and pressure [3]. Hydroxyl radicals are extraordinarily reactive species with an oxidation potential of 2.80 V, which is higher than that of ozone of 2.07 V [4]. This causes the hydroxyl radicals to easily react with and decompose toxins in water by a series of chemical reactions, which ultimately leads to the conversion of organic compounds in solution into carbon dioxide and water along with mineral acids and salts [5]. Thus, AOPs are considered an environmental-friendly drinking water technology.

The aforesaid physical processes leading to hydroxyl radical formation in water occur when plasma is contacted with water. Plasma is generated by introducing electrical energy into a reaction zone, which leads to various physical and chemical effects, such as the formation of oxidizing species, including radicals (H•, O•, and OH•, etc.) and molecules (H_2O_2, O_3, etc.), shockwaves, ultraviolet light, and electrohydraulic cavitation [6]. The formed reactive species can drive decomposition and ultimately destruction of organic contaminants in solution, including microorganisms.

Numerous reports have been published about plasma application in water purification. Sun et al. [7] have reported the using of a pulsed-streamer corona discharge for organic compound removal with 83% of the total organic carbon removed without oxygen bubbling and 86% removed with oxygen bubbling. Desmiarti et al. [8] reported that the removal efficiencies of fecal coliforms and total coliforms by radio frequency plasma treatment were 95% and 93%, respectively. Recent research by Desmiarti et al. [9] showed that the inductively coupled plasma system (ICPS) was able to kill microorganisms in water with removal efficiencies of fecal coliforms and total coliforms at 100% and 74%, respectively. The effect of flow rate on removal of fecal coliforms, total coliforms, other coliforms [10] and Salmonella [11] by means of radio frequency ICPS was also investigated. However, there are very few reports regarding the removal of nanoparticles in drinking water treatment.

Nanoparticles are particles that have at least one dimension at the nanoscale (approximately 1 to 100 nm), a feature that alters their behavior. They are commercially applied in sunscreens, toothpastes, sanitary ware coatings and even food products [12]. Little has been reported regarding the use of engineered nanomaterials for water treatment. Chiavola et al. [13] have reported the application of iron-based nanoparticles as adsorbents for arsenic contamination removal from groundwater. In spite of the benefits of nanomaterials having been widely publicized, their potential health and safety risks pose a concern for drinking water utilities because these systems are designed to deal with microparticles and not with nanoparticles [14]. Consumption of drinking water contain nanoparticles such as normal organic matter and virus caused a direct or indirect risk human health including DNA damage [15].

This research aimed to study the distribution of nanoparticles in water by a combined filtration-inductively coupled plasma system (ICPS). The effect of flow rate was studied and the nanoparticle distribution was considered to study the number and volume distribution. The nanoparticle was measured by Zetasizer Nano. To expect the stability of the nanoparticles in the water, the zeta potential was also investigated.

2. Materials and Methods

2.1. Materials

A water sample was collected from the Kuranji River, a river in Padang City, Indonesia, and used as water source for a drinking water treatment plant. Prior to use, the sample was filtered to remove relatively large organic matters. The initial water properties were analyzed, of which the results are shown in Table 1.

Table 1. Kuranji river water properties

Parameters	Unit	Value
pH		10-10.8
Total dissolved solids (TDS)	mg l^{-1}	49-65
Dissolved organic carbon (DOC)	mg l^{-1}	28.75
Fecal coliforms	CFU/100 ml	3,800
Total coliforms	CFU/100 ml	16,400
Average particle size (diameter)	nm	1.12-2.70

2.2. Experiments

The experimental set-up of the combined filtration-ICPS was adapted from our previous study [9]. The designed system consisted of a plasma reactor and a plasma generator. The plasma reactor was made of 2 mm thick glass, with a length of 30 cm and a diameter of 1 cm, wrapped in 2 mm copper wire. The plasma generator was equipped with a high-voltage probe (Tektronix P6015A, Tektronix Inc., USA), a current probe (Tektronix P6022, Tektronix Inc., USA). To record the applied voltage and discharged currents, a Picoscope 4424 (Picotech, United Kingdom) with four channels was used. The frequency was set at 1.8-1.9 MHz and concurrently the flow rate was controlled at 100 and 150 ml/minute using a peristaltic pump.

2.3. Analysis

All water samples were filtrated using a 0.2 μm cellulose acetate membrane filter (ADVANTEC Corporation) before being measured to a zeta potential meter (Zetasizer Nano, manufactured by Malvern). The data were recorded as number distribution and volume distribution. The zeta potential and the electrophoretic mobility were also measured to expect the stability of the nanoparticles in the treated water.

3. Results and Discussion

3.1 Nanoparticles size distribution

The nanoparticle size distribution is displayed in Figures 1 and 2. From Figures 1 and 2 it can be seen that after 60 minutes of running process, particles in the ranges of 1.74 to 4.85 nm and 0.71 to 6.5 nm categorized the majority of nanoparticles occupying the volume for flow rates of 100 ml/minute and 150 ml/minute, respectively. After 180 minutes of running process, there was a shift in the size distribution. The average size of the nanoparticles found in the treated water produced at a flow rate of 100 ml/minute was increased, as larger sizes of nanoparticles occupying the treated water were identified. This shows that during the treatment removal of nanoparticles from the water sample occurred, resulting in only larger particles remaining in the treated water. After the treatment, the particle sizes ranged from 2.70 to 6.50 nm. The particles within this diameter range belong to the group of proteins, small molecules and atoms [16].

However, this phenomenon was not found in the treated water produced at a flow rate of 150 ml/minute. After 180 minutes of treatment, nanoparticles with smaller sizes still remained in the treated water, as the particle size ranged from 0.62 to 712 nm. This result indicates that the flow rate is a very important parameter. Further research to investigate the optimum flow rate in the combined filtration-ICPS to increase the removal efficiency with respect to nanoparticles can be conducted.

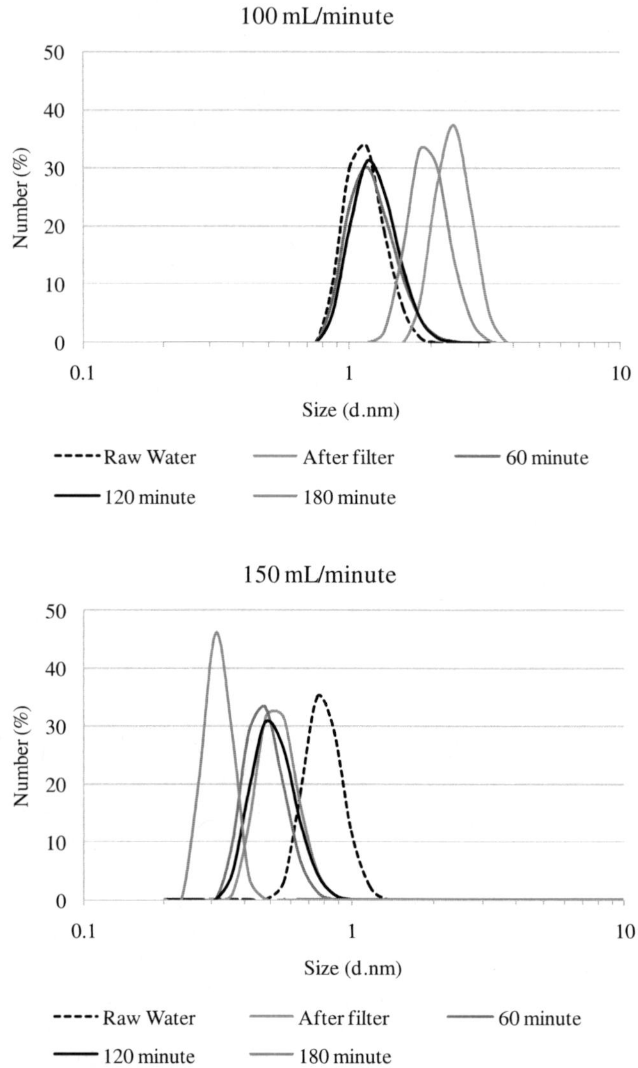

Figure 1. Number size distribution of nanoparticles.

IOP Conf. Series: Materials Science and Engineering **543** (2019) 012093 doi:10.1088/1757-899X/543/1/012093

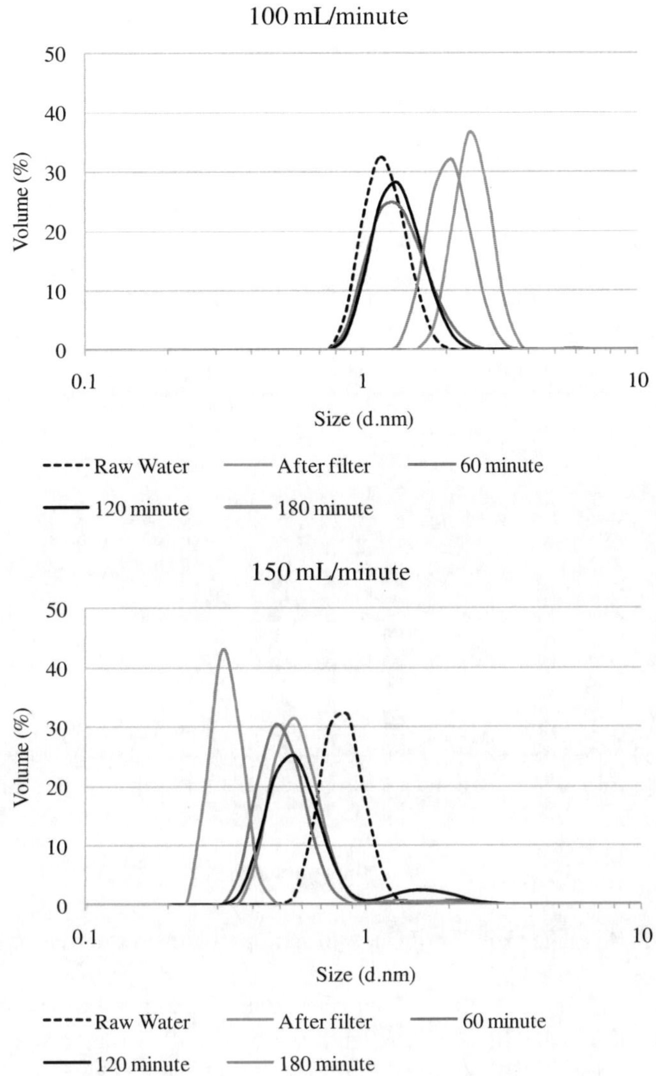

Figure 2. Volume size distribution of nanoparticles.

The average size distribution of nanoparticles occupying the treated water during the treatment process is shown in Figure 3. It can be seen that at flow rates of 100 ml/minute and 150 ml/minute a trend of gradual decrease of nanoparticles with smaller size was detected. These results suggest that the combined filtration-ICPS seems to be a feasible and efficient prospective method for nanoparticles removal in drinking water treatment plants.

3.2 Zeta potential and electrophoretic mobility

Table 2 summarizes the zeta potential and electrophoretic mobility values obtained for the raw water and the treated water produced with the combined filtration-inductively coupled plasma system. The table also consist of the conductivity of treated water values.

Table 2. Zeta potential and electrophoretic mobility values of raw and treated water produced by combined filtration-inductively coupled plasma system.

Sample	Flow rate (ml/minute)	Time (minutes)	Zeta potential (mV)	Electrophoretic mobility ($m^2Vs^{-1} \times 10^{-8}$)	Conductivity ($mS\ cm^{-1}$)
Raw water			-19.3	-1.5110	0.114
Treated water	100	60	-11.4	-0.8938	0.111
		120	-10.6	-0.8274	0.108
		180	-8.18	-0.6411	0.109
	150	60	-12.4	-0.9743	0.143
		120	-10.3	-0.8055	0.141
		180	-18.4	-1.4390	0.144

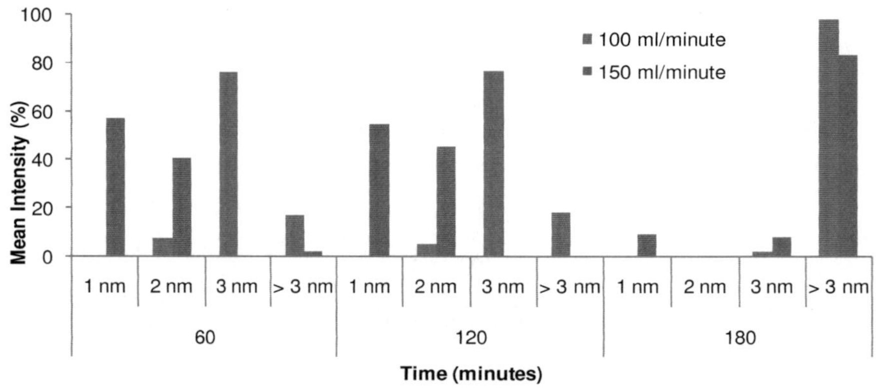

Figure 3. Average size distribution of nanoparticles in the treated water.

In the results as shown in Figure 3, the values of measured mobility after 60 minutes of running experiment were increased for each flow rate. After 180 minutes of running process, the mobility values became less negative, i.e. the magnitude of mobility decreased. This shows an increase in the average size of the nanoparticles present in the treated water, resulting in obscuration of light transmission. This result supports the previous result, where the average particle size had increased after treatment. Furthermore, the measured zeta potential values were reduced after 180 minutes of running process for both flow rates. Solutions with low zeta potential tend to have poor stability, suggesting that the removal of nanoparticles by the combined filtration- inductively coupled plasma system had occurred.

4. Conclusion

Nanoparticles are particles with at least one dimension at the nanoscale (approximately 1 to 100 nm) and might pose a direct human health threat or an indirect risk when consumed. Thus far, conventional and advanced treatment processes have been designed for dealing with microparticles and not nanoparticles. The combined performance of the filtration-inductively coupled plasma system with respect to nanoparticles were investigated in this study. The effect of flow rate on nanoparticle size distribution were also studied. The results showed that the size distribution shifted after treatment. The average particle sizes ranged from 2.70 to 6.50 nm and 0.62 to 712 nm at a flow rate of 100 ml/minute and 150 ml/minutes, respectively. The measured mobility values became less negative and the zeta potential was decreased after 180 minutes of treatment, indicating that the removal of nanoparticles by the combined filtration-inductively coupled plasma system had occurred. These results suggest that the combined filtration-inductively coupled plasma system seems to be a feasible and efficient prospective method for nanoparticle removal in drinking water treatment plants. Further research is required to find the optimum flow rate and determine the exact components of the groups of nanoparticles.

References

[1]. Mishra N S, Reddy R, Kuila A, Rani A, Mukherjee P, Nawaz A, Pichiah S 2017 *Curr World Environ* **12** 470-490
[2]. Bin A K, Sobera-Madej S 2012 *Ozone:Sci* **34** 136-139
[3]. Andreozzi R, Caprio V, Insola A, Marotta R 1999 *Catal Today* **53** 51-59
[4]. Kasih T P 2017 *J Ecol Eng* **18** 1-6
[5]. Foster J E 2017 *Phys Plasmas* **24** 055501
[6]. Jiang B, Zheng J, Qiu S, Wu M, Zhang Q, Yan Z, Xue Q 2014 *Chem Eng J* **236** 348-368
[7]. Sun B, Sato M, Clements J S 1999 *J Phys D Appl Phys* **32** 1908-1915
[8]. Desmiarti R, Hazmi A, Trianda Y, Sari E 2015 *Res J Pharm Biol Chem Sci* **6(1)** 889-897
[9]. Desmiarti R, Hazmi A, Emeraldi, Martynis M, Trianda Y, Sutopo U M 2018[a] *MATEC Web Conf* **156** 1-5
[10]. Desmiarti R, Hazmi A, Martynis M, Sutopo U. M, Li F 2018[b] *ICIMECE2018 AIP Conf Proc* 1931 030012
[11]. Desmiarti R, Hazmi A, Li F 2018 *J Eng Technol Sci* **49** 737-749
[12]. Hoet P H M, Brüske-Hohlfeld I, Salata O V 2004 *J Nanobiotechnology* **2** 1-15
[13]. Chiavola A, D'Amato E, Stoller M, Chianase A, Boni M R 2016 *Chem Eng Trans* **47** 325-330
[14]. Good K D, Bergman L E, Klara S S, Leitch M E, VanBriesen J M 2016 *J Am Water Works Assoc* **108(1)** E1-E17
[15]. Singh N, Nelson B C, Scanlan L D, Coskun E, Jaruga P, Doak S H 2017 *Int J Mol Sci* **18** 1-15
[16]. Erickson H P 2009 *Biol Proced Online* **11** 32-51 E1-E17

Acknowledgments
The Ministry of Research, Technology and Higher Education, Republic of Indonesia supported this work with research contract number: 001/research contract /K10/KM/2018. The authors are also thankful to the Water Quality Laboratory and the River Basin

IOP Conf. Series: Materials Science and Engineering **543** (2019) 012094 doi:10.1088/1757-899X/543/1/012094

Alkali modified jackfruit wood sawdust as bio adsorbent for removal of Pb(II) ions from wastewaters

T Mutiara*, L Setyaningsih, A Chafidz, B S Panandita and R Raharjo
Department of Chemical Engineering, Universitas Islam Indonesia, Jl. Kaliurang Km.14.5 Yogyakarta 55584 Indonesia
E-mail : tintin.mutiara@uii.ac.id

Abstract. This work was conducted to study the adsorption behaviour of bio adsorbent of alkali modified jackfruit wood sawdust and to investigate its possibility as adsorbent to remove heavy metals from wastewaters. The effects of sodium hydroxide concentration on the bio adsorbent adsorption capacity were investigated. The bio adsorbent prepared through chemical modification with solution of 2% sodium hydroxide showed the highest adsorption capacity with the value 3.32 mg/g at initial Pb(II) concentration 18.69 mg/L and percent removal Pb(II) ions was 88.86%. This study proved that adsorption process of Pb(II) ions onto bio adsorbent was perfectly illustrated by second order kinetic equation and adsorption isotherm of bio adsorbent was well represent by Freundlich isotherm model.

1. Introduction

The world is facing major problem related to human health called water scarcity. Almost 1.1 billion people worldwide have problem to access water. Even though water is covering up 70% of the earth, ironically fresh water is very scarce. Only 3% of world's water is fresh and 2/3 of it is in frozen glaciers form and cannot be used. With the growth of technology and industry, environmental health and hygiene are also increasingly threatened, especially the clean water. Hazardous contaminants in wastewaters include chemicals, heavy metal ions, dyestuffs, pathogenic bacteria, agricultural waste and household waste.

Industrial processes such as mining, painting, smelting, batteries manufacture, and others, release various types of side product streams containing high amount of residual metal ions dispose to wastewaters [1]. Several processes have been applied and developed to remove heavy metals from wastewaters, for example adsorption, chemical precipitation, ion exchange, electrochemical, biological operation, membrane processes [2] and many others. However, some of these methods are not very effective, expensive and require large energy input. Adsorption becomes popular in the last decades because of its advantages, such as low operation cost, simple in design and easy to operate [3]. Usually wastewaters contain low concentration of heavy metals. Adsorption has been proven as a proper method to eliminate heavy metals in wastewaters, since it is able to treat wastewaters containing heavy metals as low to 1 ppm [4].

Many kinds of bio-based adsorbent have been studied as low cost adsorbent to treat wastewaters. The advantages of using bio adsorbents are low cost, biodegradability and no disposal problems [5]. For example, white pine sawdust [6], Beech wood sawdust [7], salvinianatans biomass [1], coffee grounds [5], and many more. In the previous work, it was proven that untreated jackfruit wood sawdust has higher adsorption capacity compared to the ones which treated by acetic acid, citric acid and tartaric acid [8]. According to Sciban et al. [9] alkali modification of softwood sawdust resulted higher

IOP Conf. Series: Materials Science and Engineering **543** (2019) 012094 doi:10.1088/1757-899X/543/1/012094

adsorption capacities compared to unmodified ones. Sodium carbonate solution can be used to modify, but this alkaline solution is less efficient than sodium hydroxide solution. This experiment also established that the ion exchange is not only adsorption mechanism, but also precipitation of metal-hydroxide in the liquid.

In this work, jackfruit wood sawdust was modified by alkali solution sodium hydroxide to investigate the effect of chemical modification on jackfruit wood sawdust adsorption capacity. Adsorption experiments performed to determine the adsorption capacity of alkali treated jackfruit sawdust, adsorption kinetics and adsorption isotherm.

2. Materials and methods

2.1. Materials

Jackfruit wood sawdust used in this study obtained from local sawmill industries around Yogyakarta, Indonesia. The sawdust was sieved to 80 mesh size and rinsed several times with distilled water. After clean sawdust was obtained, then the sawdust oven dried at 50 ℃ for 48 h. All chemicals used throughout this study were analytical grade and distilled water used to prepare all solutions.

2.2. Methods

Bio Adsorbent Preparation: clean jackfruit wood sawdust was soaked in sodium hydroxide (NaOH) solution with concentration 1%, 2% and 3%. The ratio between sawdust and alkali solution was 1 gram to 10 ml. The mixture sawdust in alkali solutions constantly mixed for 6 hours at room temperature. Afterwards, the sawdust was separated from solution by filtration using filter paper and washed with distilled water for several times until the pH of washing water was constant. After that, the sawdust put in the oven at 50 ℃ until it dry.

Adsorption Experiments: batch Adsorption was performed to determine the adsorption capacity of bio adsorbent prepared with various concentration of alkali solution. 0.5 gram bio adsorbent was equilibrated with 100 ml of Pb(II) solution with known concentration. The mixture was stirred at 150 rpm for a day. A 10 ml sample was taken and analysed using Atomic Absorption Spectroscopy (AAS) to determine the final concentration of solution after adsorption process.

The amount of Pb(II) ions adsorbed in every gram of bio adsorbent was calculated by the following equation:

$$q_e = \frac{(c_o - c_e)V}{W} \tag{1}$$

The percentage removal of Pb(II) ions was calculated by the following equation:

$$R\% = \left(\frac{c_o - c_e}{c_o} \right) x100\% \tag{2}$$

Where q_e (mg/g) is the mass of Pb(II) ions adsorbed in every gram of bio adsorbent at equilibrium, c_o (mg/l) and c_e (mg/l) are concentration of Pb(II) ions at initial condition and at equilibrium, respectively. V (liter) is the volume of Pb(II) solution and W(gram) is the mass of bio adsorbent.

To investigate the kinetics adsorption behaviour of Pb(II) ions onto bio adsorbent, time varied adsorption experiment was conducted. 2.5 gram bio adsorbent soaked and mixed in 500 ml Pb(II) solution with known concentration. Sample taken every certain time and the concentration analysed using AAS. The kinetics of adsorption Pb(II) ions onto bio adsorbent was tested using Lagergren first order and second order equation.

The Lagergren first order equation is in the following form:

$$\frac{dq_t}{dt} = k_1 (q_e - q_t) \tag{3}$$

With the boundary condition $q_t = 0$ at $t = 0$ and $q_t = q_t$ at $t = t$, equation 3 becomes the following:

$$\ln(q_e - q_t) = \ln q_e - k_1 t \tag{4}$$

Where q_t (mg/g) is the amount of Pb(II) ion adsorbed in a gram of bio adsorbent at t time (minute) and k_1 is the rate constant of Lagergren first order (1/min). Rate constant k_1 can be obtained by plotting ln(q_e – q_t) versus t [1011].

The second order equation is in the following form:

$$\frac{dq_t}{dt} = k_2\left(q_e - q_t\right)^2 \tag{5}$$

With the boundary condition $q_t = 0$ at $t = 0$ and $q_t = q_t$ at $t = t$, equation 5 becomes the following:

$$\frac{t}{q_t} = \frac{1}{k_2 q_e^2} + \frac{1}{q_e} t \tag{6}$$

Where k_2 is the rate constant of second order equation (g/mg/min), can be obtained by plotting the value of t/q_t versus t [12, 13, 14].

Adsorption isotherm of bio adsorbent was tested by Langmuir and Freundlich isotherm equation. Langmuir isotherm equation is used to ensure whether the adsorption is caused by chemical forces metal ions to the surface of adsorbent and Freundlich isotherm equation is applied to non-specific adsorption on heterogeneous solid surfaces [15]. The Langmuir isotherm equation [16] and Freundlich isotherm equation [17] are in the following form:

$$q_e = \frac{q_o K_L c_e}{1 + K_L c_e} \tag{7}$$

$$q_e = K_F c_e^{1/n} \tag{8}$$

Where q_o and K_L are Langmuir equilibrium coefficients, K_F and 1/n are Freundlich equilibrium coefficients.

3. Results and discussion

This experiments were conducted to study the adsorption behaviour of jackfruit wood sawdust which chemically modified by alkali to remove Pb(II) ions from aqueous solutions. The effects of alkali concentration on the adsorption capacity of bio adsorbent, the adsorption kinetics and the adsorption isotherm were investigated.

3.1. Effect of alkali concentration on adsorption capacity

This experiment was performed to investigate the effect of alkali concentration on the adsorption capacity of bio adsorbent. Sodium hydroxide was used to modify jackfruit wood sawdust to enhance the adsorption capacity of sawdust and to prevent plant materials release soluble organic compounds during water treatment. The effect of alkali concentration on adsorption profile of bio adsorbent is shown in Figure 1. At low concentration of NaOH, the increase of alkali concentration follows by the increase of adsorption capacity of sawdust. At Pb(II) initial concentration 18.69 mg/L, maximum adsorption capacity has been achieved by bio adsorbent which modified by 2% NaOH solution with adsorption capacity at equilibrium was 3.32 mg/g and percent removal Pb(II) ions was 88.86%. This probably happened because of wood modification with a mild alkaline solution causes the liberation of new adsorption sites on the sawdust surface [9].

In this study, the adsorption capacity decrease at bio adsorbent modified by 3% NaOH solution. It needs further detail study to investigate the optimum NaOH concentration within 1% to 3%, to treat jackfruit wood sawdust into low cost bio adsorbent.

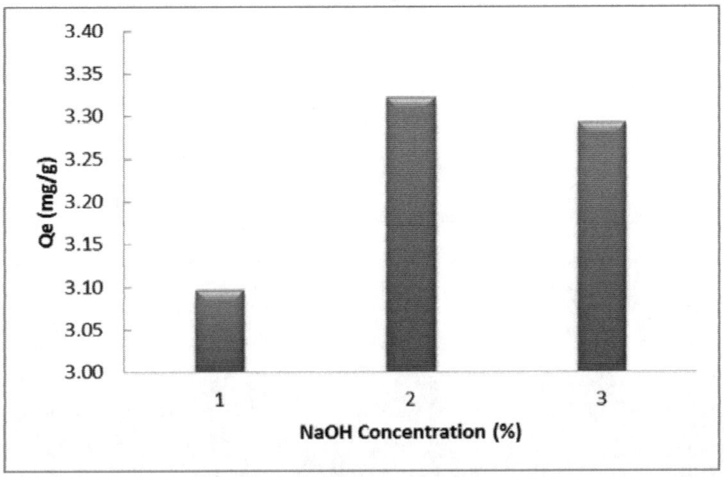

Figure 1. The adsorption capacity of modified jackfruit wood sawdust at various NaOH concentration. Mass of sawdust was 0.5 gram, Pb(II) initial concentration was 18.69 mg/L and volume of solution was 100 ml

3.2. The adsorption kinetics

The kinetics of adsorption Pb(II) ions onto bio adsorbent was analysed using Lagergren first order and second order equation. The amount of Pb(II) ions adsorbed in every gram of bio adsorbent at equilibrium q_e, the rate constants of first order and second order equation k_1, k_2, and Sum of Squared of Errors (SSE) are shown in Table 1.

Table 1. Lagergren first order and second order rate parameters for Pb(II) ions adsorption on bio adsorbent

Kinetics equation	q_e (mg/g)	k	SSE
First order	0.5890	0.0027 min^{-1}	0.6566
Second order	0.5890	0.0885 g/mg/min	0.0680

Both first order and second order equation have relative small SSE, but the SSE value for second order is smaller, so the second order equation is more suitable for Pb(II) ions adsorption on bio adsorbent from jackfruit wood sawdust modified by NaOH. Adsorption kinetics equation of Pb(II) ions onto bio adsorbent is in the following form:

$$\frac{dq_t}{dt} = 0.0885\left(0.5890 - q_t\right)^2 \tag{9}$$

The visual comparison of both kinetics models and experiment data is expressed in Figure 2.

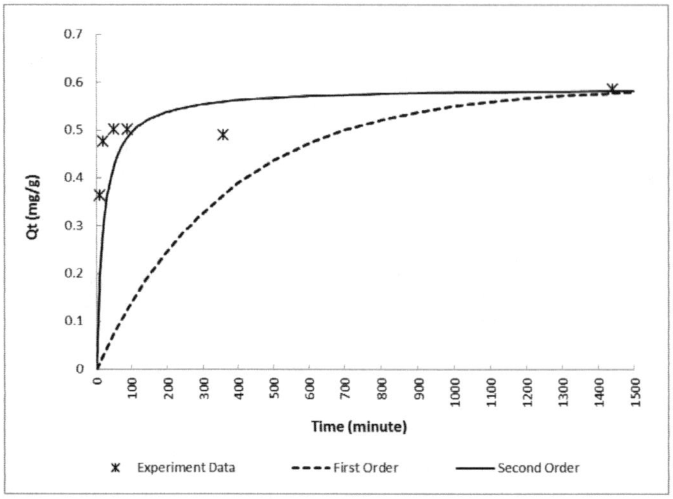

Figure 2. The mathematics equation of adsorption kinetic of bio Adsorbent. Mass of sawdust was 2.5 gram, Pb(II) initial concentration was 3.5 mg/L, and volume of solution was 500 ml.

3.3. The adsorption isotherm

This experiment was performed to determine data for adsorption isotherm of bio adsorbent at 30 ℃. The experiment data were tested with Langmuir and Freundlich isotherm model. The Langmuir equilibrium coefficients q_o, K_L, Freundlich equilibrium coefficients K_F, 1/n, SSE value for both models are shown in Table 2 and Figure 3 shows the adsorption isotherm of Pb(II) ions on bio adsorbent.

Table 2. Langmuir and Freundlich adsorption isotherm parameters of bio adsorbent at 30 °C

Langmuir Isotherm	q_e (mg/g)	K_L	SSE
	-26.9542	-0.0058	0.0147
Freundlich Isotherm	1/n	K_F	SSE
	1.0895	0.1365	0.0053

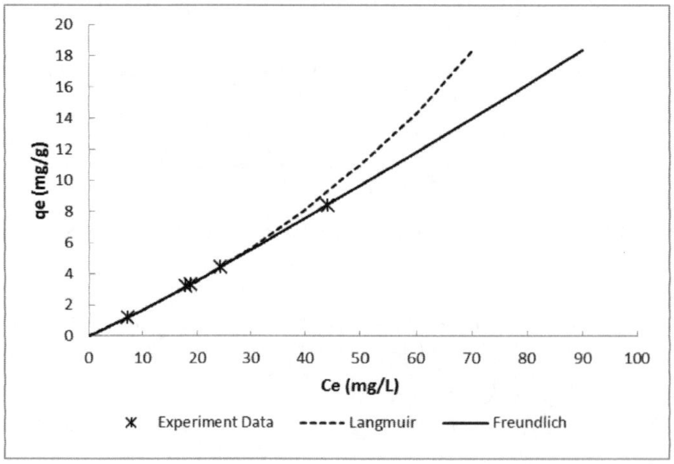

Figure 3. The adsorption isotherm of Pb(II) ions on bio Adsorbent at 30 ℃. Mass of sawdust was 2.5 gram and volume of solution was 500 ml.

The relative small value of SSE and good linearity of the graph mean that the plot is well represent experiment data to both Langmuir and Freundlich isotherm model. Based on Figure 3 and value of SSE can be concluded that the adsorption of Pb(II) ions on bio adsorbent follows the Freundlich isotherm. This model is suitable to non-specific adsorption on heterogeneous solid surfaces [15]. Adsorption isotherm equation of bio adsorbent is in the following form:

$$q_e = 0.1365 c_e^{1.0895} \tag{10}$$

Table 3 summarizes adsorption capacity of this work to previous works. Based on Table 3, there is a significant progress in adsorption capacity of jackfruit wood sawdust treated by NaOH compared to untreated sawdust and other chemical treatment.

Table 3. Adsorption capacity of bio adsorbent from jackfruit wood sawdust

Jackfruit wood sawdust	Heavy metal	Initial concentration (mg/L)	Adsorption capacity (mg/g)	References
Untreated	Fe(II)	24	1.05	[8]
Treated by acetic acid (CH_3COOH)	Fe(II)	24	0.81	[8]
Treated by phosphoric acid (H_3PO_4)	Pb(II)	25	1.40	[18]
Treated by sodium hydroxide (NaOH)	Pb(II)	24	4.43	This work

4. Conclusion

Bio adsorbent from jackfruit wood sawdust modified by alkali NaOH was succeed to remove Pb(II) ions from aqueous solutions. The optimum concentration of NaOH to treat jackfruit wood sawdust was 2%, with the value of adsorption capacity 3.32 mg/g at initial Pb(II) concentration 18.69 mg/L and the removal percentage of Pb(II) ions was 88.86%. The kinetics of adsorption Pb(II) ions onto bio adsorbent was analyzed using Lagergren first order and second order equation. The second order equation resulted lower SSE compared to first order equation, this indicates that kinetics adsorption of Pb(II) ions onto bio adsorbent correlated well with second order kinetic equation. The data for adsorption isotherm of bio adsorbent were tested with Langmuir and Freundlich isotherm. Freundlich isotherm model is well represent the experiment data with smaller SSE value compared to Langmuir isotherm model.

5. Acknowledgement

Financial support from Department of Chemical Engineering, Faculty of Industrial Technology, Universitas Islam Indonesia is gratefully acknowledged.

6. References

[1] Lima L K S, Silva J F L, Da Silva M G C and Vieira M G A 2014 Lead biosorption by salvinianatans biomass: equilibrium study *Chem. Eng. Trans.* **38** 97-102
[2] Reynolds T D and Richards P A 1996 *Unit Operations and Processes in Environmental Engineering* 2nd ed. (Boston: PWS Publishing Company) 166–754
[3] Jusoha, A, Lam S S, Noraaini A and Noor M J M M 2007 A simulation study of the removal efficiency of granular activated carbon on cadmium and lead *Desalination* **206** 9–16
[4] Tripathi A and Ranjan M R 2015 Heavy Metal Removal from Wastewater Using Low Cost Adsorbents *J. Bioremed. & Biodegr.* **6** 1000315

[5] Patterer S, Bavasso I, Sambeth J and Medici F 2017 Cadmium removal from aqueous solution by adsorption on spent coffee grounds *Chem. Eng. Trans.* **60** 157-162
[6] Salazar-Rabago J J and Leyva-Ramos R 2016 Novel biosorbent with high adsorption capacity prepared by chemical modification of white pine (Pinusdurangensis) sawdust. Adsorption of Pb(II) from aqueous solutions. *J. Env. Manag.* **169** 303-312
[7] Bozic D, Gorgievski M, Stankovic V, Strbac N, Serbula S and Petrovic N 2013 Adsorption of heavy metal ions by beech sawdust – Kinetics, mechanism and equilibrium of the process. *Ecolog. Eng.* **58** 202–206
[8] Mutiara T, Karisa P C and Mujahidah I 2018 Acid modified jackfruit wood sawdust as biosorbent for the removal of Fe(II) from aqueous solutions. *MATEC Web of Conferences* **154** 01019
[9] Sciban M, Klasnja M and Skrbic B 2006 Modified softwood sawdust as adsorbent of heavy metal ions from water *J. Haz. Matl.* **B136** 266–271
[10] Namasivayam C and Kavitha D 2003 Adsorptive removal of 2-chlorophenol by low-cost coirpitch. *J. Haz. Matl.* **B98** 257–274
[11] Chen C Y and Zhuang Y Y 2003 Adsorption equilibrium and kinetic of dyes on dried activated sludge *J. Saf. Env.* **3** 46–50
[12] Otero M, Rozada F, Calvo LF, Garcia A I and Mordn A 2003 Kinetic and equilibrium modelling of the methylene blue removal from solution by adsorbent materials produced from sewage sludges *Bioche. Eng. J.* **15** 59–68
[13] Zhang X J and Land H Y 2003 Synthesis and adsorption kinetics of the chelate compound of chitosan with ferrous ions *Chinese J. Appl. Chem.* **20** 749–753
[14] Lin C Y and Li L 2003 Kinetics of phenol sorption to organobentonite from water *Acta Scientiae Circumstantiae* **23** 738–741
[15] Li Q, Zhai J, Zhang W, Wang M and Zhou J 2007 Kinetic studies of adsorption of Pb(II), Cr(III) and Cu(II) from aqueous solution by sawdust and modified peanut husk *J. Haz. Matl.* **141** 163–167
[16] Langmuir I 1916 The constitution and fundamental properties of solids and liquids *J. American Chem. Soc.* **38** 2221–2295
[17] Freundlich H M F 1906 Over the adsorption in solution *The J. Phys. Chem.* **57** 385–471
[18] Mutiara T, Rofiki I and Al Ghifari M A D 2018 Bio Adsorbent from Modified Jackfruit Wood Sawdust for Removal of Lead Ions *6th Asia Conference on Mechanical and Materials Engineering* (Trans Tech Publications) **934** 159-164

Performance Evaluation of Coal Fly Ash Based Zeolite A for Heavy Metal Ions Adsorption of Wastewater

T Paramitha, W Wulandari, J Rizkiana and D Sasongko

Department of Chemical Engineering, Faculty of Industrial Technology, Institut Teknologi Bandung, Bandung, 40132, Indonesia

Email : winny@che.itb.ac.id

Abstract. One of the environmental characteristics of chemical industries is producing wastewater effluent that containing heavy metals, which are hazardous to the aquatic life and the human health. The purpose of this research is to evaluate the use of zeolite A as a sorbent for the heavy metal ions removal from wastewater. Zeolite A was synthesized from coal fly ash by fusion-hydrothermal method by the authors contains cation exchange capacity of 5.05 meq/g. This study used a series of artificial wastewater containing one and mixed heavy metal ions (Cu (II) and Zn (II)). The result showed that wastewater containing one heavy metal ion was almost fully adsorbed, for instance percentage sorption of Cu (II) and Zn (II) ions were 95.61 % and 92.15 %. On the other hand, for binary system, the percentage sorption slightly decreased becoming 91.51 % with the increase the amount of Zn (II) ion in solution. Based on kinetic data, sorption of Cu (II) and Zn (II) ions is controlled by pseudo-second order. In addition, adsorption isotherm follows Langmuir model. From this study, it is confirmed that zeolite A has a good possibility to be used as sorbent of heavy metal ions removal in wastewater.

1. Introduction

Coal fly ash is a by-product produced in large quantities from combustion of coal-based thermal power plants and is generally considered as waste product [1]. However, only 20-30% of the coal fly ash was utilized on worldwide in 2015, while the remainder was disposed of in landfills [2]. Therefore, it is necessary to develop potential alternatives in the sustainable use of coal fly ash. Several papers by other authors show coal fly ash can be applied for mixed material of cement and concrete production, mine reclamation [3], as well as there is evidence that it can also be applied directly for the removal of pollutants, such as heavy metals.

Heavy metals ions are the abundant pollutants in many industrial wastewater such as in electroplating facilities, mining operation, the production of battery, the manufacture of paints and pigments, and ceramic and glass industries [4]. Wastewater streams of these industries commonly contain Sb, Cr, Cu, Pb, Zn, Co, Ni, etc. which may be found in high concentrations even up to 500 mg l^{-1} [5]. Heavy metal ions are not biodegradable and can cause various diseases and disorders in living organisms [6].

However, the direct use of coal fly ash as sorbent of heavy metals is not effective due to the variability in physical and chemical properties of different coal fly ash [7]. In addition, the sorption capacity is relatively low [8]. Coal fly ash is suitable as precursor for synthesizing zeolite with enhanced surface area and sorption characteristics, due to contents of silica and alumina.

Zeolite is a crystalline material composed aluminosilicates with a three-dimensional network of silicon and aluminium tetrahedral linked by oxygen atoms [9]. In comparison with all types of synthetic zeolites, zeolite A has high cation exchange capacity (CEC) due to substitution of Si (IV) by Al (III) in its framework, high porosity and specific surface area, as well as high affinity towards polar molecules [10].

In this study, coal fly ash was specifically converted into zeolite A by alkaline fusion followed by hydrothermal method. Furthermore, zeolite A was used as sorbent for sorption of heavy metal ion. Cu (II) and Zn (II) ions were chosen as heavy metal ions sorbed. Moreover, the work was conducted to investigate the selectivity of mixed heavy metal ions.

2. Experimental section

Coal fly ash was obtained from PT YTL Paiton East Java, Indonesia. Before synthesis process, coal fly ash was pre-treated using hydrochloric acid technical grade. Sodium hydroxide and sodium aluminate was used in zeolite synthesis. Two heavy metal ion solutions were prepared using analytical grade of $Cu(NO_3)_2.3H_2O$ and $Zn(NO_3)_2.4H_2O$. For pH adjusting purpose, 2 % HNO_3 solution and 0.1 M NaOH solution were prepared.

The experiments were conducted in three steps including pre-treatment of coal fly ash, zeolite synthesis, and batch sorption. In pre-treatment step, coal fly ash was heated at 850˚C for 2 h. Then, it was mixed with 10% of hydrochloric acid solution (ratio of 1 g : 25 ml) in a 500 ml three-neck flask. The mixture was heated at 80˚C for 1 h and stirred constantly at 300 rpm. After that, pre-treated coal fly ash was recovered by filtration, repeatedly washed with distilled water, and dried at 100˚C for 24 h. The zeolite synthesis process consists of fusion, aging, and hydrothermal. In fusion process, 10 g of pre-treated coal fly ash was mixed with 12 g of sodium hydroxide. The mixture was put into crucible nickel and left to fuse at 550˚C for 1 h. The sintered granules of fused coal fly ash were cooled down, grinded, and added sodium aluminate solution. Furthermore, the mixture was shaken at 30˚C for 16 h in incubator shaker for aging process and subsequently heated to temperature of 100˚C during 1 h for hydrothermal process. After hydrothermal step was complete, the suspension was filtered and zeolite was washed repeatedly until pH of solution of 9-10. Then, zeolite was dried at 105˚C for 16 h.

In batch sorption step, Cu (II) and Zn (II) metal ion solutions were separately prepared. The sorption experiments were performed under following conditions: temperature of 30 ˚C, stirring of 150 rpm, synthetic zeolite A mass/solution ratio of 0.125 g/50 ml. Kinetic experiments were conducted at constant metal ions concentration of 200 mg l^{-1} with different time intervals (from 5 to 480 min). Furthermore, sorption capacity experiments were conducted in various initial metal ion concentrations of 200-645.26 mg l^{-1} for 480 min. For binary metal ion sorption experiments were carried out in various concentrations of Cu (II) and Zn (II) ion for 480 min. After the batch sorption step, the supernatants were recovered by filtration and final concentration of heavy metal ions in the solution was determined using Atomic Adsorption Spectrophotometer (AAS). Percentage sorption was (%) calculated as:

$$\boldsymbol{percentage\ sorption}\ (\%) = \frac{\boldsymbol{C_i - C_f}}{\boldsymbol{C_i}} \times \mathbf{100\%}$$

Where C_i and C_f are the concentrations of the heavy metal ions in the initial and final solutions, respectively.

The sorption capacity was calculated as:

$$\boldsymbol{sorption\ capacity}\ (\boldsymbol{q_e}) = \frac{\boldsymbol{C_i - C_e}}{\boldsymbol{C_i}} \times \frac{\boldsymbol{V}}{\boldsymbol{m}}$$

Where q_e is the sorption capacity at equilibrium time (mg g^{-1}); C_i is initial concentration of heavy metal ions (mg l^{-1}); C_e is equilibrium concentration of heavy metal ions (mg l^{-1}); V is volume of solution (l); and m is mass of sorbent (g).

3. Results

3.1. Characterization of zeolite A

The coal fly ash used in this study contains major components such as Fe_2O_3 40.2%, CaO 25.2%, SiO_2 18.6% and Al_2O_3 7.18%. However, significant amount of Fe_2O_3 and CaO would be inhibitor for zeolite synthesis. In this work, the pre-treatment process of coal fly ash was conducted using hydrochloric acid solution and was found successfully to reduce the amount of Fe_2O_3 and CaO.

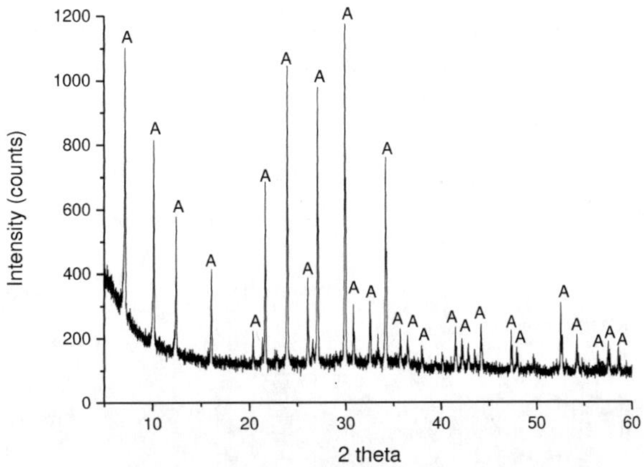

Figure 1. XRD pattern of zeolite A.

Figure 1 illustrates the XRD pattern of zeolite A obtained from pre-treated coal fly ash. According to XRD pattern obtained that there are no detected peaks of impurities such as hydroxysodalite and crystalline components of coal fly ash, indicating the formation of relatively pure zeolite A without hydroxysodalite and high conversion of coal fly ash to zeolite A. In addition, the cation exchange capacity of zeolite A is 5.05 meq/g. This cation exchange capacity is relatively high for the application of zeolite A in sorption of heavy metal ions based on ion exchange.

3.2. Effect of contact time

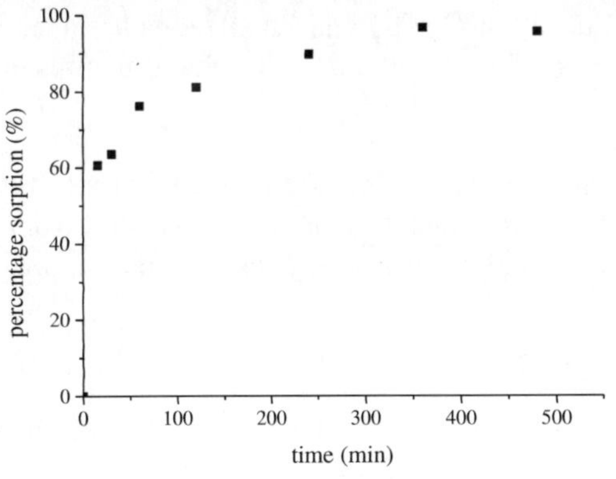

Figure 2. Percentage sorption of zeolite A for Cu (II) ion against contact time

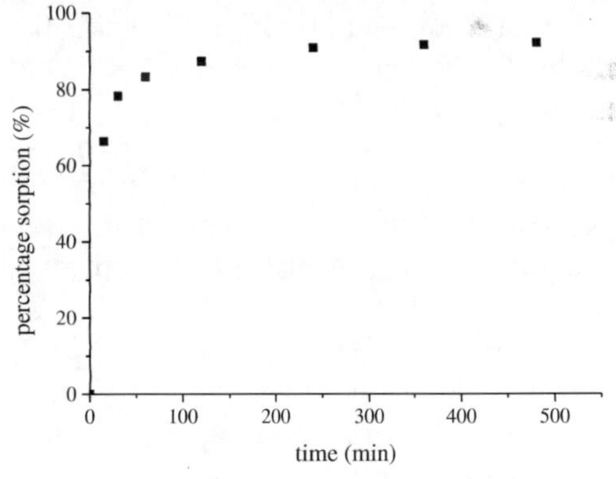

Figure 3. Percentage sorption of zeolite A for Zn (II) ion against contact time

The effect of contact time on the percentage sorption of Cu (II) and Zn (II) into zeolite A are shown in Figure 2 and Figure 3. These figure show that the sorption of Cu (II) and Zn (II) ions are quite fast initially followed by increase slightly to reach an equilibrium time of around 360 min for Cu (II) and 240 min for Zn (II). These tendency related to the fact that abundant of active sites are available in the initial process and heavy metal ions interact easily. In addition, the fast sorption initially corresponds to high driving force that can be seen from the high concentration difference between bulk solution and solid-liquid interface [6]. Furthermore, active sites become increasingly used up and the sorption rate become slower [11]. The results showed that the percentage sorption at 200 mg l^{-1} are 95.61 % for Cu (II) and 92.15 % for Zn (II).

3.3. Effect of initial metal concentration

The effect of initial metal ion concentrations on the percentage sorption of Cu (II) and Zn (II) ions by zeolite A are shown in Figure 4 and Figure 5. When the initial metal ion concentrations increase from 200 to 645.26 mg l^{-1} for Cu (II) and 200 to 540.08 mg l^{-1} for Zn (II), the percentage sorption gradually decrease. A decrease from 95.61% to 36.14% and from 92.15% to 42.13% for Cu (II) and Zn (II) respectively are obtained. These phenomenon occur due to the fixed number of active sites used in this study. Therefore, the active sites become saturated at high concentrations [12].

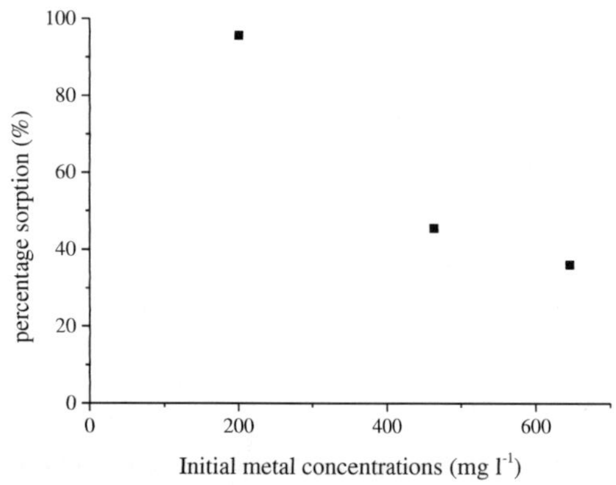

Figure 4. Percentage sorption of zeolite A for Cu (II) ion against initial metal concentration.

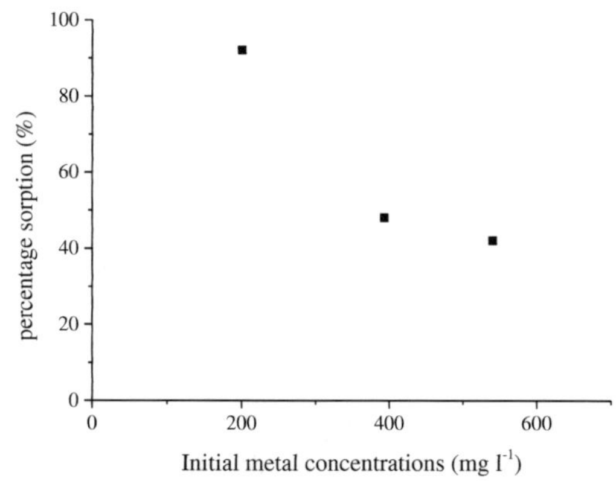

Figure 5. Percentage sorption of zeolite A for Zn (II) ion against initial metal concentration.

3.4. Kinetic sorption

In order to investigate the kinetic mechanism and rate law describing the Cu (II) and Zn (II) ions sorption with zeolite A, the data of effect of contact time on 200 mg l^{-1} of Cu (II) and Zn (II) ions are analysed using two kinetic equations, the pseudo-first order and the pseudo-second order. The equation of pseudo-first order can be expressed as:

$$ln(q_e - q_t) = ln\, q_e - k_1 t$$

Where q_e is the sorption capacity at equilibrium time (mg g^{-1}); q_t is the sorption capacity at time t (mg g^{-1}); and k_1 is the pseudo-first order rate constant (min^{-1}). Parameter k_1 can be determined from the slope of the ln (q_e - q_t) against t plots. Furthermore, the equation of pseudo-second order can be expressed as:

$$\frac{t}{q_t} = \frac{1}{k_2 q_e^2} + \frac{1}{q_e} t$$

Where k_2 is the pseudo-second order rate constant (g $mg^{-1}.min^{-1}$). Parameter k_2 can be determined from the intercept of the $\frac{t}{q_t}$ against t plots.

The values of kinetic parameters for each equation are given in Table 1. Based on the Table 1, pseudo-second order equation offers the best correlation both Cu (II) and Zn (II) ions. These results are in accordance with previous research by Hui et al. (2005) [5].

The pseudo-second order based on the assumption that sorption follows chemisorption, indicating the rate limiting step may be a chemical sorption occurred valance forces through sharing or exchange of electrons between zeolite A and heavy metal ions [13]. This assumption does not consider that diffusion as limiting step in the sorption process.

Table 1. The values of kinetic parameters of pseudo-first and second order in adsorption of Cu (II) and Zn (II).

Kinetic parameters	Cu (II)	Zn (II)
Pseudo-first order		
k_1 (min^{-1})	0.0092	0.0105
q_e calculation (mg g^{-1})	32.589	16.013
Correlation coefficient (R^2)	0.9815	0.968
Pseudo-second order		
k_2 (g mg^{-1}.min^{-1})	0.0009	0.00198
q_e calculation (mg g^{-1})	78.125	74.627
Correlation coefficient (R^2)	0.9978	1

3.5. Sorption isotherms

This section examines the values of sorption isotherm based on the Langmuir and Freundlich isotherm model. The Langmuir isotherm model can be expressed as:

$$\frac{C_e}{q_e} = \frac{1}{k_L q^*} + \frac{C_e}{q^*}$$

Where q_e is the sorption capacity at equilibrium time (mg g^{-1}); q^* is the monolayer sorption capacity (mg g^{-1}); C_e is equilibrium concentration of heavy metal ion (mg l^{-1}); and k_L is Langmuir constant. Parameter k_L and q^* can be determined from the slope and intercept of the $\frac{C_e}{q_e}$ against C_e plots. In addition, the Freundlich isotherm model can be expressed as:

$$\log q_e = \log k_f + \frac{1}{n} \log C_e$$

Where k_f is the Freundlich constant (mg g^{-1}) and n is the constant indicative of the intensity of the sorption process. Parameter k_f and n can be determined from the intercept and slope of the log q_e against log C_e plots.

Table 2. The values of adsorption isotherm parameters of Cu (II) and Zn (II).

Adsorption Isotherm Parameters	Cu (II)	Zn (II)
Langmuir		
q^* (mg g^{-1})	92.5926	90.0901
k_L (mg l^{-1})	0.1166	0.0769
Correlation coefficient (R^2)	0.9947	0.9773
Freundlich		
Correlation coefficient (R^2)	0.8384	0.4751

Based on correlation coefficient on Table 2, it is obtained that the sorption of Cu (II) and Zn (II) ions by zeolite A are more appropriate following the Langmuir isotherm model, which is indicated from the correlation coefficient close to one. It could be assumed that maximum removal corresponds to interaction between solute molecules and the zeolite surface and results in monolayer coverage on zeolite surface [14].

According to q^* parameter, Cu (II) ion is sorbed more than Zn (II) ion onto zeolite A with value of 92.5926 and 90.0901 mg g^{-1}. Therefore, it could be concluded that the zeolite A is higher affinity towards Cu (II) ion than Zn (II) due to ion electronegativity and ionic radius. Ion electronegativity of Cu (II) and

Zn (II) ions are 1.93 and 1.63, while the ionic radius of Cu (II) and Zn (II) ions are 0.072 nm and 0.083 nm [15]. The smaller ions are easier to pass through the micropores and channels of zeolite A [6].

3.6. Metal ions removal in binary systems

The metal ions removal in binary systems represented by relatively zeolite A without hydroxysodalite are shown in Table 1. The concentration of heavy metal ions are adjusted at 200 mg l^{-1}. As mentioned in 3.2, the percentage sorption of Cu (II) and Zn (II) ions at 200 mg l^{-1} are 95.61 % and 92.15 %. Compared with the single system, percentage sorption of binary system is slightly lower. The percentage sorption of binary system decrease from 94.04 % to 91.51 % with the increase of Zn (II) ion in solution. These results indicate that sorption selectivity of Cu (II) ion is higher that of Zn (II) ion and in good agreements with the results of sorption capacity.

Table 3. Heavy metal ions sorption of zeolite A in binary system.

Metal ions	c_o (mg l^{-1})	c_f (mg l^{-1})	Total percentage sorption (%)
Cu (II)	50	0.30	91.51
Zn (II)	150	16.68	
Cu (II)	100	2.66	93.25
Zn (II)	100	10.85	
Cu (II)	150	4.99	94.04
Zn (II)	50	6.93	

4. Conclusion

Kinetic sorption was analysed and was found that the rate limiting step for sorption of Cu (II) and Zn (II) ions follows chemical sorption which occurred valance forces through sharing or exchange of electrons between zeolite A and heavy metal ions. The sorption capacity was calculated from Langmuir model and was found that the sorption capacity of Cu (II) and Zn (II) ion are 92.5926 and 90.0901 mg g^{-1}. Therefore, zeolite A is higher affinity towards Cu (II) ion than Zn (II). At binary system, sorption selectivity of Cu (II) ion is higher that of Zn (II) ion. Overall, this study showed that zeolite A obtained from coal fly ash is possible to remove the heavy metal ions.

Acknowledgement

The authors would like to thank Program Penelitian, Pengabdian kepada Masyarakat, dan Inovasi (P3MI) 2017 Institut Teknologi Bandung for funding in this research.

References

[1] Izidoro J C, Fungaro D A, Abbott J E and Wang S 2013 *Fuel* **103** 827-34.
[2] Shu Y, Wei X, Fang Y, Lan B and Chen H 2015 *J. Hazard. Mater.* **286** 517-24.
[3] Ahmaruzzaman M 2010 *Prog. Energ. Combust. Sci.* **36** 327-63.
[4] Salam O E A, Reiad N A and Elshafei M M 2011 *J. Adv. Res.* **2** 297-03.
[5] Hui K S, Chao C Y H and Kot S C 2005 *J. Hazard. Mater.* **127** 89-01.
[6] Motsi T, Rawson N A and Simmons M J H 2009 *Int. J. Miner. Process.* **92** 42-48.
[7] Simate G S, Maledi N, Ochieng A, Ndlovu S, Zhang J and Walubita L F 2016 *J. Environ. Chem. Eng.* **4** 2291-12.
[8] Nascimento M, Soares P S M and Souza V P 2009 *Fuel* **88** 1714-19.
[9] Kongnoo A, Tontisirin S, Worathanakul P and Phalakornkule C 2017 *Fuel* **193** 385–94.

[10] Golbad S, Khoshnoud P and Abu-Zahra N 2017 *J. Minerals Mater. Charact. Eng.* **5** 237–51.
[11] Jha K, Matsuda M and Miyake M 2008 *J. Hazard. Mater.* **160** 148-53.
[12] Akpomie K G, Dawodu F A and Adebowale K O 2015 *Alexandria Eng. J.* **54** 757-67.
[13] Visa M 2016 *Powder Technol.* **294** 338-47.
[14] Santasnachok C, Kurniawan W and Hinode H 2015 *J. Environ. Chem. Eng.* **3** 2115-26.
[15] Putra W P, Kamri A, Yusoff S N M, Ishak C F, Mohamed A, Hashim N and Isa I M 2014 *J. Encapsulation Adsorpt. Sci.* **4** 25-35.

IOP Conf. Series: Materials Science and Engineering **543** (2019) 012096 doi:10.1088/1757-899X/543/1/012096

Activated Carbons from Palm Kernels Shells Prepared by Physical and Chemical Activation for Copper Removal from Aqueous Solution

T M Asnawi[a], H Husin[a]*, A Adisalamun[a], W Rinaldi[a], M Zaki[a], F Hasfita[b]
[a]Chemical Engineering Department, Faculty of Engineering, Syiah Kuala University, Darussalam-Banda Aceh, 23111, Indonesia
[b]Chemical Engineering Department, Faculty of Engineering, Malikus Saleh University, Lhokseumawe, 24300, Indonesia

*e-mail:husni_husin@che.unsyiah.ac.id

Abstract. Preparation and evaluation of activated carbon from the palm kernel shell (PKS), an abundant agricultural residue, by physical and chemical activation were carried out. The activation treatment of adsorbent with and without chemicals modification was evaluated. The physical activation using nitrogen gas and chemical treatment using 10 wt.% of sodium hydroxide and nitrogen gas as the activating agent. This study evaluated the activation temperature and nitrogen gas rate. The results showed that activated carbon treated with NaOH followed by heating under nitrogen gas had the highest iodine number of 1062.8 mg/g and methylene-blue number of 247.2 mg/g (heating at 750 °C, nitrogen gas flow rate of 100 mL/min). The equilibrium data obtained at various initial concentration reasonably fit well with the Langmuir adsorption isotherm. The adsorption of Cu(II) ions fit with the Langmuir isotherm pattern with adsorption capacity (Xm) of 8.9445 mg/g and kf value of 0.16802 L/mg for activated carbon without NaOH treatment. The adsorption capacity of activated carbon with NaOH and heating at 750 °C under nitrogen gas was obtained 12.0773 mg/g and kf value of 0.15076 L/mg. Those results suggested the prepared adsorbent could be considered as a promising candidate for Cu(II) ions for wastewater treatment.

1. Introduction

Recent development of industrial revolution generation 4.0 is appealing to focal concerns on environmental impacts. Waste derived from industries, hospitals, and mining can be chemicals such as pesticides, herbicides, chlorine, heavy metals and other organic wastes [1]. Some of it have the following hazardous and toxic heavy metals such as: Cu, Cr, Hg, Cd, etc. [2]. Metals can affect the ecosystem damaged or dead, illness, reduce speed, decrease the metabolism ability of biota, and cause various diseases of living organisms [3]. Heavy metals elements that enter the environment have an inherent nature and are readily absorbed by biota so that immersed into humans body and food chain [4]. Copper as heavy metal, Cu(II) is containing in water body in the form of ions and compounds. Cu(II) include essential heavy metals, even though it is toxic but small amount are needed in the human body. The certain amount of metal is required for human health, but if it exceeds the threshold of its presence in water will be toxic to humans [5]. Liquid wastes from refineries, tannery plants, electroplating and batteries, metal finishing and manufacturing are the main copper source that pollutes the ecosystems. Cu(II) naturally enters water through erosion, rock erosion, or in the atmosphere brought down by rainwater. Heavy metal ions could not be decomposed but can be accumulated in living organisms [6]. Many technologies have been used to reduce heavy metal ions in industrial wastewater, such as: filtration, sedimentation, flocculation, adsorption, and others. These methods are less effective considering the nature of water in a dissolved state, not as suspension or sediment, and adsorption could be the more cost-effective [7]. The most widely used and effective adsorbent applied for heavy metals removal is activated carbon. This is due to the heavy metals are in a dissolved state in the water body so that it is not easily removed by other methods. The commercial activated carbon is the most widely used adsorbent recently in adsorption process but its price is still comparable expensive. The efforts to explore the potential for a comparable low-cost adsorbent from the low-cost raw material are of great concern among researchers. One of the agricultural residues that attract much attention in this research is Palm Kernel Shells (PKS). PKS are available in abundance as by-product generated from Palm Oil Mills with approximately 15 tonns per year in Indonesia [8]. This biomass by-product is classified as a potential low-cost adsorbents and are also environmentally friendly materials [9]. PKS is the shell fractions left after the nut has removed after crushing in the Palm Oil mill. Kernel shells are a fibrous material and could be handled in bulk directly from the product line to the end user. This study proposed the preparation of activated carbon from PKS. The adsorption of Cu(II) ions in aqueous solutions by Palm Kernel Shells activated carbon (PKS AC) was investigated over independent variables i.e activating by chemical sodium hydroxide (NaOH) concentration, temperature of physical activation, initial adsorbate concentration, initial pH of adsorbate solution and adsorption temperature on the preparation method of PKS AC on its adsorption capacity in term of iodine and methylene blue number and adsorption capacity.

2. Materials and Methods

2.1 Materials

The PKS waste was collected from an Oil Palm Mill in Meulaboh, in the district of West Aceh, Aceh Province, Indonesia. The production of PKS as by-product of Palm Oil Mill approximately 6.6 ton/y. The 2.0 kg PKS was rinsed using tap water, and then it was dried in an oven (Memmert: NN-ST342M, Germany) at 110 °C for 24 h. The Cu(II) ions aqueous solution was made by dissolving 4 g (0.001) of CuSO4.5H2O (99% pure, from Aldrich) in 2000 mL of distilled water in the Erlenmeyer flask. The sample of 2 mL was taken, it was diluted with 10 mL distilled water in the vial for Atomic Absorption Spectrophotometer (AAS) analysis. The concentration of Cu(II) ions in the stock solution were calculated based on the dilution factor of 6. The predetermined Cu(II) ions concentration for all the adsorption experiments was prepared using the stock solution and general dilution formula. The equipment are a vertical reactor, tube furnace, shaker, spectrophotometer, vacuum filter, oven, scales, sieve, grinder, Erlenmeyer, burl 50 ml, 1 L flask, dropper, and pH meter. While the materials used are PKS, distilled water, methylene blue, NaOH 0.1 N, HCl 0.1 N, sodium thiosulfate, phenolphthalein indicator, starch, iodine solution, copper nitrate Cu (NO3)2.3H_2O, HNO_3, ethanol, and nitrogen gas.

2.2 Synthesis of activated carbon from palm kernel shells

The research consisted of three stages; PKS pyrolisis, physical activation, and chemical activation. The dried PKS was milled to the powder using a ball mill and then it was sieved to get the size in the range of 80-100 mesh. A tube furnace (TF-80/120/160, HumanLab Inc. 300-1500 °C, Korea) was used for pyrolysis of the dried PKS powder. The tube furnace (TF) heating rate was at 45 °C/min, and it took 12.75 min to get the TF final temperature of 600 °C for pyrolysis with nitrogen gas pumped at 5 mL/minwith the initial TF temperature of 30 °C. The physical activation took place at 600 °C (1 °C) for 1 h. After the TF was turned off, the PKS carbon was left in the TF overnight to gradually decrease the PKS carbon temperature. Chemical activation of the PKS carbon was conducted for 6 h in 250-mL beaker glass consisting of 200 mL NaOH solution at 0.1 M (prepared using NaOH, Merk). It was stirred at 75-rpm using a hot plate stirrer (IKA, type C-MAG HS 7) at 30 °C. The PKS activated carbon (PKS AC) was washed using distilled water and filterred out using vacuum filter for many times until the waste water reaching neutral pH of 7. The chemical activation was separately repeated to prepare the PKS AC with 0.3 M NaOH and the PKS AC with 0.5 M NaOH. The two types of PKS AC were separately dried in porcelain cups using an oven drier (Memmert, type NN-ST342M, Western Germany) at 110 °C for 2 h to remove the remaining water. The PKS AC were stored in different sealed bottles in a desiccator for adsorption experiments within 2 weeks. Cu(II) solution samples of 2 mL at the contact time of 0, 10, 20, 40, 60, 90 and 120 min (in time series) were analyzed using the AAS.

2.3 Experiments of Cu (II) adsorption

Adsorption experiments of Cu(II) ions as shown in Figure 1were conducted in batch mode. The system of the PKS AC-solution consisted of 1.0 g of the PKS AC and 100 mL of Cu(II) ions solution (Cu (NO3)2.3H2O) in 200-mL beaker glass, and it was stirred at 75-rpm and 1 atm. The Cu(II) ions adsorption capacity of the PKS AC over independent variables was investigated by varying Cu(II) initial adsorbate concentrations(5, 10, 25, 50, 75, 100, 125, 150, 175, 200, 225 and 250 mg/L, based on the AAS reading), the contact time (0–80 min), NaOH concentration for chemical activation (0.1–0.5 M), initial pH of Cu(II) ions solution (3–6) and adsorption temperature (30–60 °C).

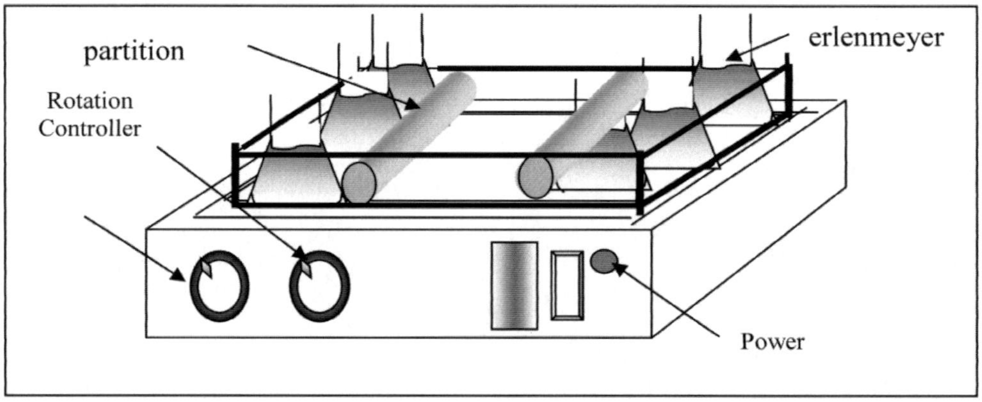

Figure 1. Experimental set up of batch mode adsorption

The optimum condition with the highest Cu(II) ions adsorption capacity of the PKS AC was obtained. Cu(II) solution samples of 2-mL at the contact time 0, 10, 20, 40, 60, 90 and 120 min (in time series) were analyzed using the AAS based on the procedure in the previous study. The Cu(II) ions adsorption isotherm studies were using the experimental data.

3. Result and Discussion

3.1. Iodine and methylene blue number

Iodine number is the amount of iodine adsorbed by 1.0 g of PKS activated carbon. The iodine number represents a pore volume measure of size ranging from 10 to 28 Å [10]. This number gives an indication of the magnitude of the adsorption capacity of the activated carbon micropore. Activated carbon that has a large percentage of pore size in the range of 10 s/d 28 Å is particularly suitable for absorbing substances that have small molecular weights from its solution. The iodine number of physically activated carbon is greater when compared with the chemically activated one method the physically activated under the same activation conditions [11]. Experimental results showed that the iodine number of physically activated carbon ranging from 875.7 mg/g to 1019 mg/g, while for the chemically activated one ranging from 1098 mg/g to 1098 mg/g, as shown in Figure 2.a.

IOP Conf. Series: Materials Science and Engineering **543** (2019) 012096 doi:10.1088/1757-899X/543/1/012096

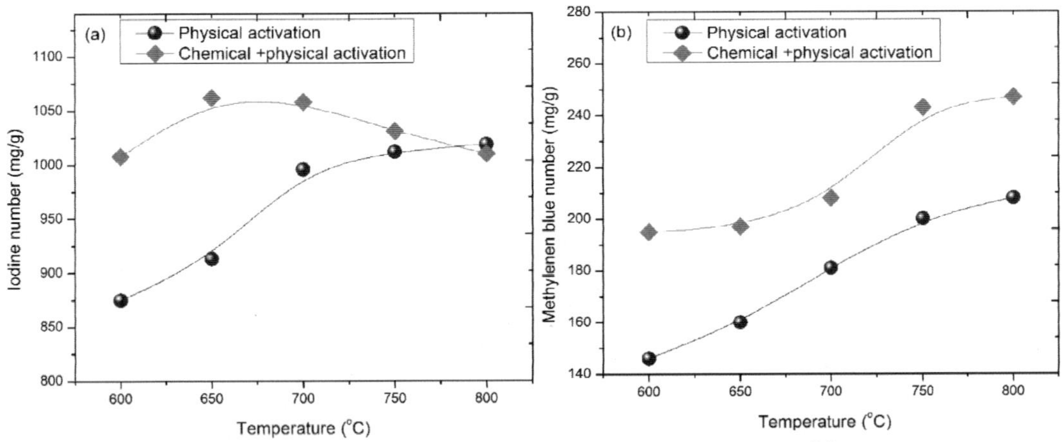

(a) (b)

Figure 2. (a) Iodine number and (b) methylene blue using activated carbon (AC) by physical and physical + chemical activation by NaOH under N_2 gas atmosphere of 100 mL/min for 1 h.

Iodine number is the amount of iodine adsorbed by 1.0 g of PKS activated carbon. The iodine number represents a pore volume measure of size ranging from 10 to 28 Å [10]. This number gives an indication of the magnitude of the adsorption capacity of the activated carbon micropore. Activated carbon that has a large percentage of pore size in the range of 10 s/d 28 Å is particularly suitable for absorbing substances that have small molecular weights from its solution. The iodine number of physically activated carbon is greater when compared with the chemically activated one method the physically activated under the same activation conditions [11]. Experimental results showed that the iodine number of physically activated carbon ranging from 875.7 mg/g to 1019 mg/g, while for the chemically activated one ranging from 1098 mg/g to 1098 mg/g, as shown in Figure 2.a.

Iodine number was increased with the increase of temperature to 750 °C for physical activation treatment. As the activation temperature increasing to 800 °C, the iodine number slowly decreased and tended to stationary. In the chemical activation the iodine number appears to increase with the increasing temperature until the activation temperature reached 650 °C, at the same N_2 gas flow rate with the physical activation. As the activation temperature raised higher, the iodine number decreases after reaching the maximum value. This phenomenon is likely due to the absolute volume of macropores decreased with the increasing of burn-off. It was revealed that the weight loss was not lead to the formation of new micropores. The decrease in absolute micropore volume at higher burn-off may affect the burning of walls between adjacent pores. This process causes the micropores changing into larger pores. The maximum iodine number obtained from the study was 1019 mg/g for physical activation adsorbent pretreated at 750 °C under the nitrogen gas flow rate of 150 mL/min, meanwhile for chemical activation one was 1062.8 mg/g under nitrogen gas flow rate 100 mL/min. This result was in line with the evidence that the chemical activation requires lower temperature for activation as compared to the physical activation [12].

The methylene blue number indicated the molecular adsorption capacity of the activated carbon that has similar molecular sizes to the methylene blue and the mesoporous presence in the carbon active [13]. Both physical and chemical activation methods show the same tendency to increase the methylene blue number. The methylene blue number increased as the activation temperature increasing. Methylene blue number obtained for physical activation methods ranging from 146 mg/g to 208.5 mg/g and for chemical activation methods range from 195.0 mg/g to 247.2 mg/g. The increasing of methylene blue number is due to the burn-off process which continues to produce new mesopores resulting from the release of volatile components contained in carbon. The formation of new mesopores can also be produced by the breakdown of micropores [14]. The higher methylene

IOP Conf. Series: Materials Science and Engineering **543** (2019) 012096 doi:10.1088/1757-899X/543/1/012096

blue number in the chemical activation method is due to the more micropores in activated carbon being split into mesopores and macropores than that of physical activation. The iodine and methylene blue numbers obtained from this experiment obtained higher than the required standards for activated carbon.

3.2. The Cu(II) Ions Adsorption Equilibrium Study

The Cu(II) ions equilibrium study was carried out on activated carbon which gave maximum absorption capacity. The physically activated carbon pretreated at temperature 750 ^{0}C under the nitrogen gas flow rate of 100 mL/min, and chemically activated one the temperature was 650 ^{0}C under the nitrogen gas flow rate 100 mL/min.

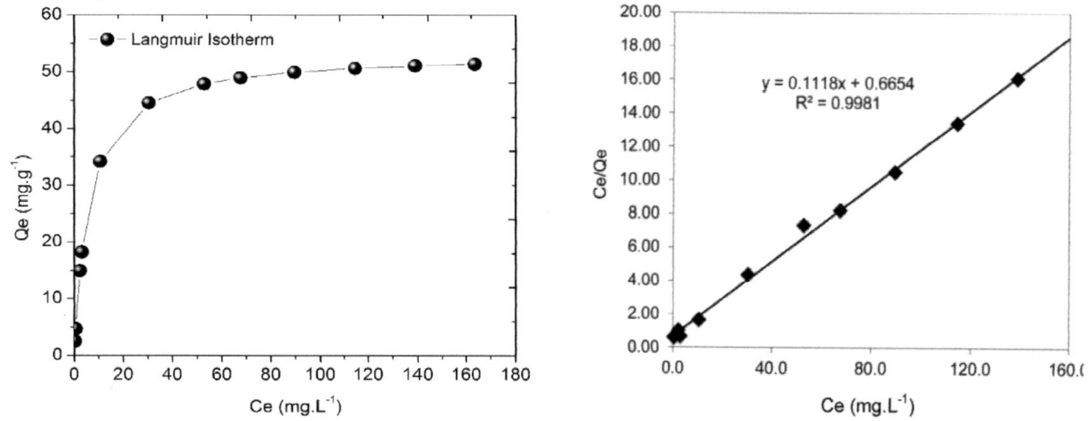

(a) (b)

Figure 3. Langmuir's Isotherm and linearized curve of Cu(II) ions adsorption on physically activated PKS AC at 750 ^{0}C under 100 mL/ min of nitrogen gas

The Cu(II) ions adsorption isotherm in the activated carbon is obtained by plotting a curve as the relationship between the equilibrium concentration in the liquid phase (Ce) with the equilibrium concentration in the solid phase (Qe). The plot of the Ce value against Qe shows the adsorption characteristics of Cu(II) ions by activated carbon fitting very well to the Langmuir isotherm, as shown in Figure 3 and 4.

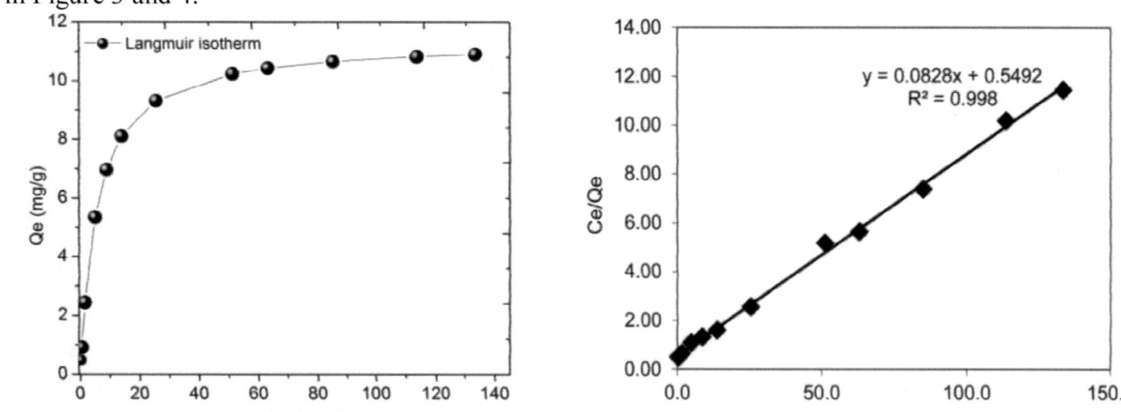

(a) (b)

Figure 4. Langmuir's Isotherm and linearized curve of Cu(II) ions adsorption on physical + chemical activated with NaOH PKS AC at 650 ^{0}C under 100 mL/ min of nitrogen gas

The adsorption isotherms for Cu(II) onto PKS AC and their Langmuir and Freundlich adsorption isotherm plots are shown in Figs. 2 and 3 it can be seen that the absorption of Cu(II) ions follows Langmuir's isotherm with maximum adsorption capacity (Xm) and Langmuir (K) constant value and obtained by plotting Ce/Qe to Ce as follows:

a) Physical activation

$$Y = 0.1118X + 0.6654 \quad (1)$$

Where; X_m = 8.9445 mg/g dan K_f = 0.16802 L/mg

b) Chemical activation

$$Y = 0.00828X + 0.5492 \quad (2)$$

Where; X_m = 12.0773 mg/g and K_f = 0.15076 L/mg

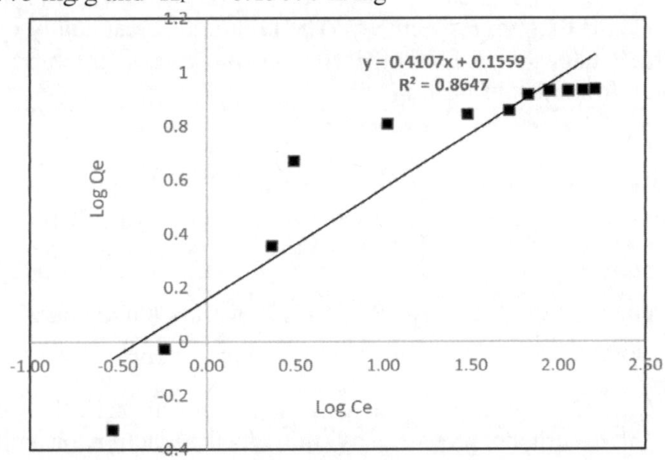

Figure 5. Freundlich Isotherm and linearized curve of Cu(II) ions adsorption on physically activated PKS AC at 750 0C under 100 mL/ min nitrogen gas

The following comparison shows the graph of Friendlich Isotherm on the absorption of Cu(II) ions using chemically activated with NaOH PKS AC at 650 ^{0}C under 100 mL/ min of nitrogen gas. From Figure 4 and 5, it can be seen that the corresponding values of the constants appearing in the Langmuir and the Freundlich equations. The correlation coefficients (R^2) obtained from the Langmuir and Freundlich model are 0.998 and 0.8647 for Cu(II) onto PKS AC, respectively. These results show that the Langmuir model gives a better fit of the experimental data than that of the Freundlich model for the adsorbent. It would be indicated that the surface of the physical activated PKS AC is energetically homogeneous, the adsorbed species are independent of each other (no associative adsorption) and they form a monolayer. The Langmuir monolayer adsorption capacity value toward Cu(II) adsorption onto PKS AC of chemical and physical activated obtained by the linear method of analysis was Xm = 12.0773 mg/g; Kf = 0.1451 L/mg and Xm = 8.9445 mg/g and Kf = 0.16802 L/mg, respectively. The higher adsorption capacity of chemical activated PKS AC compared to the physical one can be explained by the enhancement in the specific surface area and metallic affinity due to the small size and chemical structure of surface enriched with hydroxyl functional groups during chemical activation with NaOH on the surface of the adsorbent. This functional groups expected to develop metallic affinity enhancement. It can be concluded that chemical activation with NaOH plays an important role in the enhancement of Cu(II) removal efficiency.

4. Conclusion

The adsorption of Cu(II) ions onto activated carbon prepared from palm kernel shell (PKS AC) was investigated in batch mode over independent variables of activating chemicals (NaOH) concentration, the temperature of physical activation, initial adsorbate concentration, initial pH of adsorbate solution and adsorption temperature. Chemical activation treatment of PKS AC provided a better adsorption characteristics on Cu(II) ions in solution compared to physical activation. The PKS AC iodine numbers activated by the physical and chemical activation reached 1019.1 mg/g and 1062.8 mg/g, by activation conditions temperature 750 oC and 650 °C under nitrogen gas flow rate 100 mL/min, respectively. Methylene blue number of the PKS AC was obtained for physical and chemical activation was reached to 208.5 mg/g and 247.2 mg/g, respectively. Methylene blue number increased with the increasing activation temperature and nitrogen gas flow rate. The equilibrium model for the absorption of Cu(II) ions onto activated carbon prepared from palm kernel shell (PKS AC) obtained fitting with the Langmuir isotherm model. The Cu(II) adsorption capacity by physical activated and chemically activated of PKS AC obtained was Xm = 8.9445 mg/g with K = 0.1713 L/mg, and Xm = 12.0773 mg/g with K = 0.1451 L/mg, respectively. The high uptake capability of PKS AC adsorbent makes it potentially attractive for the removal of Cu(II) ions in wastewater treatment, especially for the treatment of toxic effluent from metal finishing process.

Acknowledgements
The authors would like to appreciate the Syiah Kuala University, Indonesia, for providing financial support under Universitas Syiah Kuala, Laboratories Research Grant year 2018.

References

[1] Nethaji S., Sivasamy A. and Mandal A.B., 2013, Preparation and characterization of corn cob activated carbon coated with nano-sized magnetite particles for the removal of Cr(VI), *Biore. Tech.*, **134**, 94-100.

[2] Zhang J., Fu H., Lv X., Jie T. and Xu X., 2011, Removal of Cu(II) from aqueous solution using the rice husk carbons prepared by the physical activation process, Biomass and Bioenergy, **35**, 464-72.

[3] Sun Y., Yue Q., Gao B., Gao Y., Qian L. and Wang Y., 2013, Adsorption of hexavalent chromium on Arundo donax Linn activated carbon amine-crosslinked copolymer, *Chem. Eng. J.*, **217**, 240-47.

[4] Song J., Zou W., Bian Y., Su F. and Han R., 2011, Adsorption characteristics of methylene blue by peanut husk in batch and column modes, *Desalination*, **265**, 119-25.

[5] Samarghandi M.R., Al-Musawi T.J., Mohseni-Bandpi A. and Zarrabi M., 2015, Adsorption of cephalexin from aqueous solution using natural zeolite and zeolite coated with manganese oxide nanoparticles, *J. of Molec. Liquids*, *211*, 431-41.

[6] Olaniran A.O., Balgobind A. and Pillay B., 2017, Bioavailability of heavy metals in soil: impact on microbial biodegradation of organic compounds and possible improvement strategies, *Int. J. of Molec. Sci.*, **2013**, 14, 10197-28.

[7] Othman N.H., Alias N.H., Shahruddin M.Z., Abu Bakar N.F., Nik Him N.R. and Lau W.J., 2018, Adsorption kinetics of methylene blue dyes onto magnetic graphene oxide, *J. of Environ. Chem. Eng.*, **6** , 2803-11.

[8] Hambali E and Rivai M., 2017, The Potential of Palm Oil Waste Biomass in Indonesia in 2020 and 2030, *IOP Conf. Series: Earth and Environ. Sci.*, **65**, 012050.

[9] El-Aziz A., Kamal M.E., Ali K.H., Abdel-Aziz K.A., and Kamel M.S., 2018, Biodegradable grafting cellulose/clay composites for metal ions removal, *Int. J. of Biol. Macromol.*, Volume 118(Pt B), pp.2256-2264.

[10] Jow J.J., Guo Z.-S., Chen H.-R., Wu M.-S. and Ling T.-R., 2010, Determination of the iodine adsorption number of carbon black by using a direct cathodic reduction method, *Electrochem.*

Comm., **12**, 1605-08.

[11] Kadirvelu K., Thamaraiselvi K. and Namasivayam C., 2001, Removal of heavy metals from industrial wastewaters by adsorption onto activated carbon prepared from an agricultural solid waste, *Biores. Tech.* **76**, 63-65

[12] Gebru K.A. and Das C., 2017, Removal of Pb (II) and Cu (II) ions from wastewater using composite electrospun cellulose acetate/titanium oxide (TiO_2) adsorbent, *J. of Wat. Proc.*, **16**, ,1-13.

[13] Narvekar A.A., Fernandes J.B. and Tilve S.G., 2018, Adsorption behavior of methylene blue on glycerol based carbon materials, *J.of Environ. Chem. Eng.*, **6** ,1714-25.

[14] Prauchner, M. J. and Rodríguez-Reinoso, F., 2012, Chemical versus physical activation of coconut shell: A comparative study, *Microp. and Mesop. Mat.*, **152** 163–71

IOP Conf. Series: Materials Science and Engineering **543** (2019) 012097 doi:10.1088/1757-899X/543/1/012097

Biogas Production from Wastes of Tofu Industry with Effects of Water Hyacinth and Cow Manure Additions

S Sa'diah and M D Putra[1]

Department of Chemical Engineering, Lambung Mangkurat University, South Kalimantan, Indonesia

E-mail: mdputra@ulm.ac.id

Abstract. Biogas is one of the potential sources of energy to cope with the limitations of fossil energy. On the other hand, liquid waste of tofu industry, water hyacinth, and cow manure are poorly managed that can potentially pollute the environment. The type of those wastes was potential source for biogas production through anaerobic digestion due to high content of protein and nutrient. This study aims to determine the effect of addition of liquid waste of tofu industry (WT) and water hyacinth (WH) to cow manure (CM) on the biogas production. The amount of water hyacinth and cow manure were varied at 0-4 kg, while the liquid waste of tofu industry was fixed to 8 kg. The anaerobic digestion was conducted in biodigester container equipped with thermometer and biogas shelter. The biogas production was then analyzed using Gas chromatography–mass spectrometry (GC-MS). The optimum biogas was achieved at raw material ratio of 4:2:2 (WT:WH:CM) for 21 d. At this condition, 60 ppm of CH_4 and 10,744 ppm of CO_2 were obtained. Thus, the utilization of liquid waste of tofu industry, water hyacinth and cow manure as raw material of biogas production can minimize the environment problem and supply energy.

Keywords: biogas; tofu; water hyacinth; cow manure; anaerobic digestion

1. Introduction

The increased energy demand is triggered by the increased transportation uses, industrial development, and human growth. The existence of unrenewable energy sources such as natural gas, coal and petroleum will ultimately run out. Renewable energy utilization is an alternative that can be used to replace the use of fuel oil or natural gas as fossil fuels, such as biogas [1]. Biogas is a process produced from organic waste that can be decomposed by certain microorganism in anaerobic environment [2]. So far, many researchers still utilize various resource of organic wastes and also combine them to acquire the optimum results [3-5]. However, these renewable energy sources should be further explored.

Most tofu factories still use simple technology which causes relatively high waste production. Tofu production is a process involving soybean milling, boiling, filtration, protein coagulation, preservation, and packaging [6]. During the process, up to 30 % of soybeans become waste. However, only a small portion of waste is used as feed nutrition, while most of it is burned and or reclaimed as industrial waste, which contributes to serious pollution problems [7].

The abundance of liquid waste in the production process if it is disposed of directly can pollute the environment, especially to the quantity and quality of water [8, 9]. The liquid waste is a pollutant waste, with COD values ranging from 4000-6000 mg/L and BOD content ranging from 3000-4000 mg/L. Tofu industry wastewater is biodegradable or easily biologically degraded [10, 11].

Tofu liquid waste contains high levels of organic compounds and nutrients consisting of the content of N, P, K and Mg, protein, fat, crude fiber and ash which have the potential to produce biogas through anaerobic digestion process [12]. In general, biogas contains methane, CO_2, H_2S and a little water, which can be used as a substitute for kerosene or liquid petroleum gas. By converting tofu wastewater into biogas, tofu factory owners not only contribute to protecting the environment but also increase their income by reducing fuel consumption in the tofu-making process [7, 13].

Cow manure is known as waste that can produce biogas due to the high content of cellulose and hemicellulose [14]. On the other hand, water hyacinth also contains about 40-60 % cellulose and hemicellulose and small content of lignin [15]. The contents lead to a potential value for biogas production [16], because the little amount of lignin makes an easier process to break down structural linkages at the initial conditions [17].

Conventional energy that is scarce and the high production of tofu liquid waste that always pollutes the environment, encourages the creation of researches related to alternative energy that is environmentally friendly as a solution to these problems. This study utilizes tofu liquid waste combined with water hyacinth as a substrate which is expected to increase the heating value of biogas. In addition, it also utilizes the addition of bio-starter cow dung which is useful to accelerate the process of biogas production from tofu industrial wastewater.

2. Materials and methods

2.1. Raw material preparation

Water hyacinth was taken from the Sipai river in Martapura, South Kalimantan. It was then cut to a size of 2 cm and then dried for ± 6 h in sunny weather. Liquid waste from tofu factory was taken from industry of Tahu Sekumpul. Raw materials for tofu wastewater about 8 l were placed in biodigester and then 4 kg dry water hyacinth was added. The mixture was then added with a starter of 2 kg cow manure. The substrate composition for anaerobic fermentation was shown in Table 1.

Table 1. The composition of fermentation substrates.

Digester (Tofu waste: hyacinth: cow manure)	Waste of tofu (kg)	Hyacinth (kg)	Cow manure (kg)
4:2:1	8	4	2
4:0:1	8	0	2
4:1:1	8	2	2
4:2:0.5	8	4	1
4:2:2	8	4	4

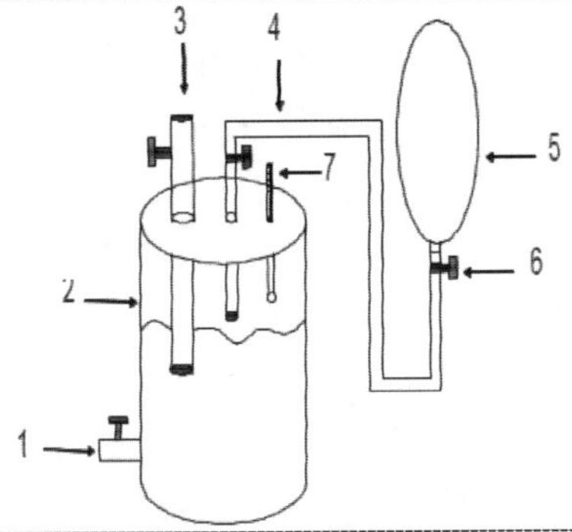

Figure 1. Schematic process of biodigester in laboratory scale: 1. Outlet; 2. Biodigester; 3. Inlet of raw materials; 4. Biogas hose; 5. Biogas shelter; 6. Control valve; and 7. Thermometer.

2.2. Anaerobic process

The mixture was reacted with the anaerobic digestion method for 28 d, measured by pH and temperature every day during the process. During the 28-day process, the biogas content, the amount of heat and the liquid compost were analyzed in accordance with a predetermined time range. Biogas produced by each digester was taken by injection of 10 mL per sample to analyze CH_4 and CO_2 content every 7, 14, 21, and 28 d using GC-MS. As for the analysis of liquid compost fertilizer, each sample was taken with a measuring cup of 50 ml per sample to analyze the amount of N, P, K content using the Kdejahl method. The schematic process of anaerobic process in biodigester was shown in Figure 1.

3. Results and discussion

3.1. Kinetic profile of methane and carbon dioxide productions

Figure 2 shows the production of methane and carbon dioxide in the process of breaking down raw materials for tofu waste, water hyacinth, and cow dung. The formation of methane and carbon dioxide on the seventh day was caused by the degradation of organic matter in the biodigester as the hydrolysis process. Hydrolysis is the decomposition of complex compounds or long chain compounds into simple compounds. At this stage, organic materials such as proteins, lipids, and fats are degraded into short-chain compounds, such as peptides, amino acids, and simple sugars. O'Reilly et al. [18] revealed that species such as Methanosaeta were predominantly present in raw materials from low-strength psychophilic synthetic waste bioreactors to treat. Cellulose hydrolysis bacteria group was found to be dominant in biogas reactors using raw materials with rich in lignocellulose such as pig manure and stalk [19, 20]. The significant production of methane was found at 28 d after pass through the lag phase as commonly occured in the fermentation process [21].

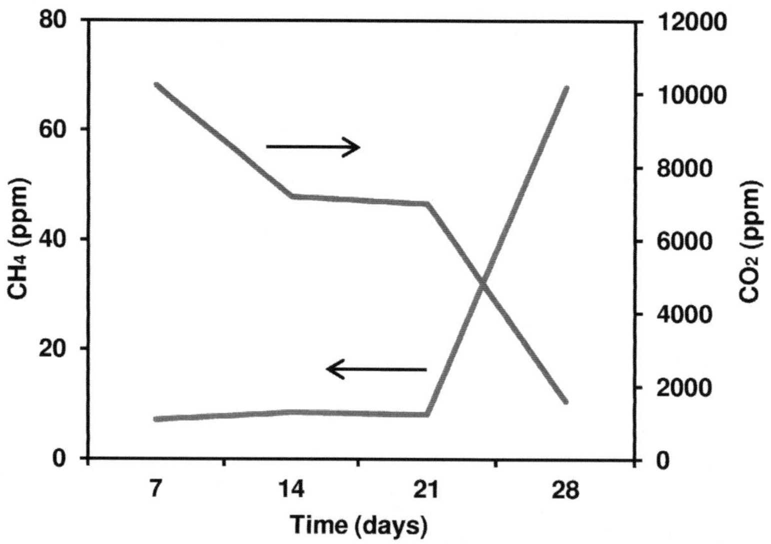

Figure 2. Profile of methane and carbon dioxide productions from waste materials of tofu waste (8 kg), hyacinth (4 kg) and cow manure (2 kg). Blue and red lines represent CH_4 and CO_2 contents, respectively.

3.2. Effect of hyacinth on biogas production

Figure 3 shows the production of methane and carbon dioxide at variation of water hyacinth with the content of tofu waste and cow manure of 8 kg and 2 kg, respectively. The greater the amount of water hyacinth led to the smaller the amount of methane and the greater carbon dioxide. The difference in results is obtained in research without waste tofu [22], where there is an increase in biogas production by increasing the amount of water hyacinth in a certain amount; however, carbon dioxide production also increases. Decrease in methane levels and increase in carbon dioxide levels in this study due to substrate conditions. Large substrate size can inhibit the growth of microorganisms in the anaerobic fermentation process because microorganisms must degrade the particle size of the substrate first so that it takes longer for the microorganism to produce methane. The structure of the lignocellulose complex results in anaerobic degradation process running very slowly, especially in the hydrolysis process. On the other hand, the presence of tightly bound recalcitrant lignin, crystalline cellulose and hemicellulose polymers, can delay production and reduce biogas production during anaerobic digestion [23, 24].

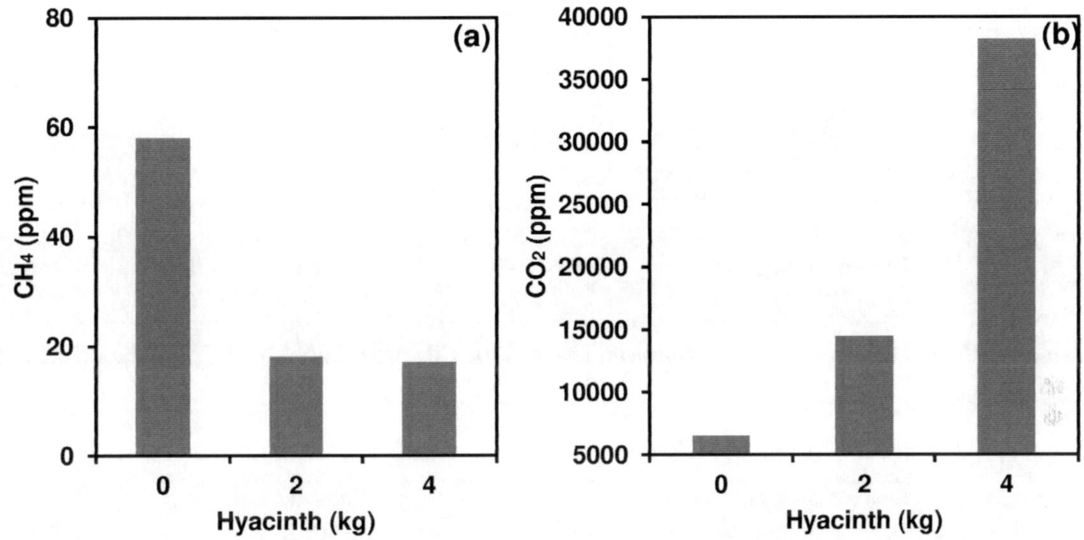

Figure 3. Effect of hyacinth on methane and carbon dioxide productions analyzed at experimental time of 21 d

3.3. Effect of cow manure on biogas production

Figure 4 presents the production of methane and carbon dioxide at variation of cow manure with the content of tofu waste and water hyacinth of 8 kg and 4 kg, respectively. The greater amount of cow dung produced greater methane [25, 26]. Cow manure is a habitat for various microorganisms that serve to accelerate the degradation of components in the digester. At this stage, there is degradation of organic matter in the biodigester, namely the hydrolysis process, where complex organic polymers will be converted into short organic components [27]. However, the amount of carbon dioxide experienced a significant decrease in the addition of 4 kg of cow dung; this probably due to a slight increase in alkalinity thus limiting the hydrolysis process from organic matter to produce carbon dioxide [26]. The methane content was undetected for the substrates without cow manure indicating the role of cow manure in the fermentation.

3.4. Content in liquid products and calorific value

Table 2 shows the content of N, P and K from liquid products from the results of the biodigester and the calorific value of biogas. The nutritional content of the liquid product shows the potential material to be used as fertilizer for plants. Percentage of N content was higher when the water hyacinth was added, while P and K levels decreased. This was likely due to the amount of N content in water hyacinth in the degradation process [28]. However, the difficulty of degradation in the material with the addition of water hyacinth because of the recalcitrant lignin presence of tightly bound, crystalline cellulose and hemicellulose polymers, resulted in low heating values produced [23].

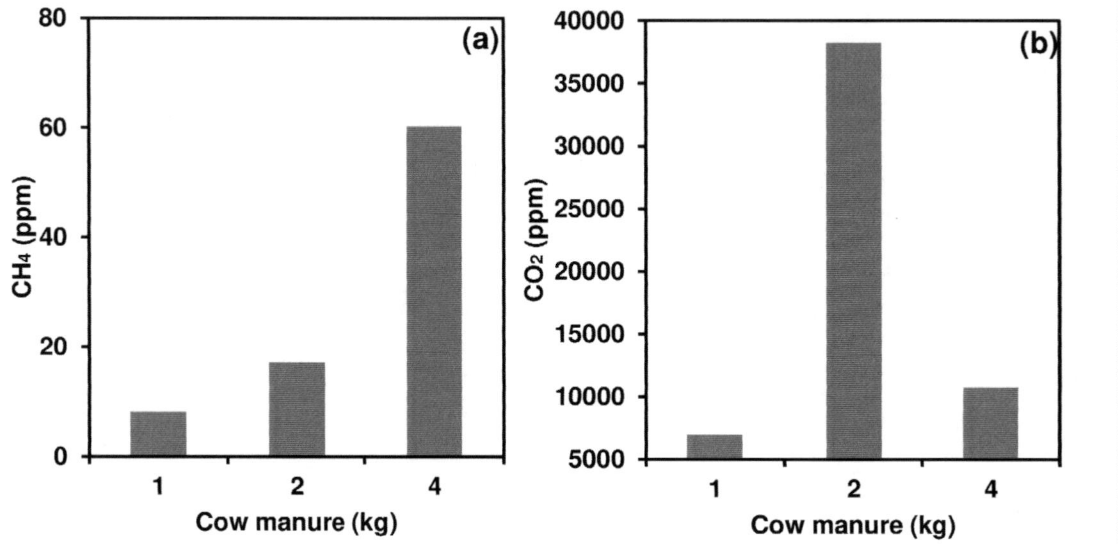

Figure 4. Effect of cow manure on methane and carbon dioxide productions analyzed at experimental time of 21 d

Table 2. Content of N, P, and K in liquid products and calorific value.

Hyacinth Content, kg	N, %	P, %	K, %	Calorific value, J/cm3
0	3.50	0.33	1.30	0.39
4	6.03	0.12	0.88	0.15

4. Conclusion

Tofu waste led to the potential material to be processed for biogas production. The optimum biogas was achieved at raw material ratio of waste of tofu to water hyacinth to cow manure (4:2:2) for 21 d. At this condition, 60 ppm of CH_4 and 10,744 ppm of CO_2 were obtained. Thus, the utilization of liquid waste of tofu industry, water hyacinth and cow manure as raw material of biogas production can minimize the environment problem and supply energy.

5. Acknowledgments

The authors thank the Lambung Mangkurat University for supporting the work.

6. References

[1] Sakhmetova, G, A Brener, and R Shinibekova Scale-up of the installations for the biogas production and purification. 2017 *Chem. Eng. Trans.* **61** 1453-1458.
[2] Wang, H, J Xu, L Sheng, and X Liu Effect of addition of biogas slurry for anaerobic fermentation of deer manure on biogas production. 2018 *Energy* **165** 411-418.
[3] Khayum, N, S Anbarasu, and S Murugan Biogas potential from spent tea waste: A laboratory scale investigation of co-digestion with cow manure. 2018 *Energy* **165** 760-768.
[4] Okonkwo, UC, E Onokpite, and AO Onokwai Comparative study of the optimal ratio of biogas production from various organic wastes and weeds for digester/restarted digester. 2018 *J. King Saud Univ. Eng. Sci.* **30** 123-129.

[5] Xiao, B, W Zhang, H Yi, Y Qin, J Wu, J Liu, and Y-Y Li Biogas production by two-stage thermophilic anaerobic co-digestion of food waste and paper waste: Effect of paper waste ratio. 2019 *Renew. Energy* **132** 1301-1309.

[6] Chai, X, Y Mi, P-L Yue, and G Chen Bean curd wastewater treatment by membrane separation. 1999 *Sep. Purif. Technol.* **15** 175-180.

[7] Kim, M-S and D-Y Lee Fermentative hydrogen production from tofu-processing waste and anaerobic digester sludge using microbial consortium. 2010 *Bioresour. Technol.* **101** S48-S52.

[8] An, G-H, D-Y Lee, and H-J Ahn Tofu-derived carbon framework with embedded ultrasmall tin nanocrystals for high-performance energy storage devices. 2017 *J. Alloys Compd.* **722** 60-68.

[9] Fei, Y, L Liu, D Liu, L Chen, B Tan, L Fu, and L Li Investigation on the safety of Lactobacillus amylolyticus L6 and its fermentation properties of tofu whey. 2017 *LWT* **84** 314-322.

[10] Chua, J-Y, Y Lu, and S-Q Liu Evaluation of five commercial non-Saccharomyces yeasts in fermentation of soy (tofu) whey into an alcoholic beverage. 2018 *Food Microbiol.* **76** 533-542.

[11] Kim, D-H, D-Y Lee, and M-S Kim Enhanced biohydrogen production from tofu residue by acid/base pretreatment and sewage sludge addition. 2011 *Int. J. Hydrogen Energy* **36** 13922-13927.

[12] Barber, ST, J Yin, K Draper, and TA Trabold Closing nutrient cycles with biochar- from filtration to fertilizer. 2018 *J. Cleaner Prod.* **197** 1597-1606.

[13] Huang, X, J Zhao, Q Xu, X Li, D Wang, Q Yang, Y Liu, and Z Tao Enhanced volatile fatty acids production from waste activated sludge anaerobic fermentation by adding tofu residue. 2019 *Bioresour. Technol.* **274** 430-438.

[14] Ren, NQ, JF Xu, LF Gao, J Qiu, and DX Su Fermentative bio-hydrogen production from cellulose by cow dung compost enriched cultures. 2010 *Int. J. Hydrogen Energy* **35** 2742-2746.

[15] Thi, BTN, LK Ong, DTN Thi, and YH Ju Effect of subcritical water pretreatment on cellulose recovery of water hyacinth (*Eichhornia crassipe*). 2017 *J. Taiwan Inst. Chem. Eng.* **71** 55-61.

[16] Jayaweera, MW, JAT Dilhani, RKA Kularatne, and SLJ Wijeyekoon Biogas production from water hyacinth (*Eichhornia crassipes* (Mart.) Solms) grown under different nitrogen concentrations. 2007 *J. Environ. Sci. Health, Pt. A: Toxic/Hazard. Subst. Environ. Eng.* **42** 925-932.

[17] Harun, MY, ABD Radiah, ZZ Abidin, and R Yunus Effect of physical pretreatment on dilute acid hydrolysis of water hyacinth (*Eichhornia crassipes*). 2011 *Bioresour. Technol.* **102** 5193-5199.

[18] O'Reilly, J, C Lee, G Collins, F Chinalia, T Mahony, and V O'Flaherty Quantitative and qualitative analysis of methanogenic communities in mesophilically and psychrophilically cultivated anaerobic granular biofilims. 2009 *Water Res.* **43** 3365-3374.

[19] Tian, G, et al. The effect of temperature on the microbial communities of peak biogas production in batch biogas reactors. 2018 *Renew. Energy* **123** 15-25.

[20] Zheng, Z, J Liu, X Yuan, X Wang, W Zhu, F Yang, and Z Cui Effect of dairy manure to switchgrass co-digestion ratio on methane production and the bacterial community in batch anaerobic digestion. 2015 *Applied Energy* **151** 249-257.

[21] Putra, MD, AE Abasaeed, HK Atiyeh, SM Al-Zahrani, MH Gaily, AK Sulieman, and MA Zeinelabdeen Kinetic modeling and enhanced production of fructose and ethanol from date fruit extract. 2015 *Chem. Eng. Commun.* **202** 1618-1627.

[22] Tasnim, F, SA Iqbal, and AR Chowdhury Biogas production from anaerobic co-digestion of cow manure with kitchen waste and Water Hyacinth. 2017 *Renew. Energy* **109** 434-439.

[23] Barua, VB, VV Goud, and AS Kalamdhad Microbial pretreatment of water hyacinth for enhanced hydrolysis followed by biogas production. 2018 *Renew. Energy* **126** 21-29.

[24] Mosier, N, C Wyman, B Dale, R Elander, YY Lee, M Holtzapple, and M Ladisch Features of promising technologies for pretreatment of lignocellulosic biomass. 2005 *Bioresour. Technol.* **96** 673-686.

[25] Achinas, S, Y Li, V Achinas, and GJ Willem Euverink Influence of sheep manure addition on biogas potential and methanogenic communities during cow dung digestion under mesophilic conditions. 2018 *Sust. Environ. Res.* **28** 240-246.

[26] Zhao, Y, F Sun, J Yu, Y Cai, X Luo, Z Cui, Y Hu, and X Wang Co-digestion of oat straw and cow manure during anaerobic digestion: Stimulative and inhibitory effects on fermentation. 2018 *Bioresour. Technol.*

[27] Amirta, R, E Herawati, W Suwinarti, and T Watanabe Two-steps utilization of shorea wood waste biomass for the production of oyster mushroom and biogas – A zero waste approach. 2016 *Agric. Agric. Sci. Proc.* **9** 202-208.

[28] Qin, H, Z Zhang, M Liu, Y Wang, X Wen, S Yan, Y Zhang, and H Liu Efficient assimilation of cyanobacterial nitrogen by water hyacinth. 2017 *Bioresour. Technol.* **241** 1197-1200.

IOP Conf. Series: Materials Science and Engineering **543** (2019) 012098 doi:10.1088/1757-899X/543/1/012098

Treatment of Oily Wastewaters from Tanjung Perak Port by Electrocoagulation Using Aluminum Electrodes

S Soeprijanto, D F Nury, L Pudjiastuti

Department of Industrial Chemical Engineering, Faculty of Vocational Studies, Institut Teknologi Sepuluh Nopember, Kampus ITS Sukolilo, Surabaya, 60111, Indonesia

E-mail: s.soeprijanto@gmail.com

Abstract. Electrocoagulation is electrochemical water and wastewater treatment technology that is the easiest technology using an electrochemical cell supplied by DC power to the electrodes. The application of the electrocoagulation process was purposed to treat oily bilge water. The results indicated that the oil removal efficiency was achieved approximately 99.62% at an initial oil concentration of 60 g/l in batch processes. The current density of 62.5 mA/cm^2 applied the process produced oil removal efficiency of 99.95%.In a continuous mode with the initial water only present in the reactor, the oil removal efficiency was achieved approximately 100% with a hydraulic residence time (HRT-1) of 6 h for 60 min of operations. However, the initial oil/water emulsion at a concentration of 63.89 g/l present in the reactor, the oil removal efficiency was approximately 73.33% with HRT-1 of 6 h for 90 h of operations.

1. Introduction

Oily wastewaters are mostly disposed of by many sources, such as cooking oil industries, petroleum refineries, petrochemical, food, bilge water, and other general industrial sources. The wastewaters in the form of bilge water are generated by Ships and commercial vessels. Oil containing bilge water is derived from oily fluids, lubricants, cleaning fluids that accumulate in the lowest part of a vessel. The content of metals in bilge and ballast water is not able to be removed through treatment. These metals are able to cause human health problems for example lead poisoning when they are ingested. This oily wastewater generation is one of the problems when disposed of to the environment. Some attempts have been made to develop an improve oily wastewater treatment processes. This development of the treatments includes coagulation [1], adsorption [2], biological processes [3,4], coupling coagulation-flocculation [5], photocatalytic [6], microwave-assisted catalytic wet air oxidation [7], and electrochemical technology [8-13]. Recently, treatment of wastewater in the oil industry is mostly conducted by several methods for example electrocoagulation, ozonization, Fenton oxidation and UV working effectively. All of the methods, electrocoagulation, and electrooxidation work more effectively [14, 15]. Soeprijanto et al. [15] studied on the treatment of oily wastewater emulsion with aluminum electrodes in a batch mode. Most of the electrocoagulation were also applied to the treatment of synthesis of oil/water emulsion [5, 6,16], petroleum refinery wastewaters and strain Industry oily Wastewater [9, 19, 20], biodiesel wastewater production [21], palm oil mill effluent [22,23], pollutant from restaurant wastewaters [24,25], and pollutant from agro-industry wastewaters [26]. Also, the Electrocoagulation technologies have been widely used for the treatment of urban wastewaters [9] and the removal of heavy metal from metal plating wastewaters [28]. Electrocoagulation is a process involving chemistry and physics process using sacrificial electrodes for the generation of coagulants. The most commonly materials used as anode electrocoagulation are aluminum and ion. An electrical current was supplied between the electrodes anode-cathode with the generation of hydroxyl ions and hydrogen gas simultaneously happening at the cathode. The hydrogen production will also help the float flocculation of particles separate them from

the water. Electrochemical reaction of the aluminum or iron metals as the anode materials can be seen as follows [29]:
For the aluminum anode:

$$Al_{(s)} \rightarrow Al^{3+}_{(aq)} + 3e \quad (1)$$

With the iron electrodes:

$$Fe_{(s)} \rightarrow Fe^{2+}_{(aq)} + 2e \quad (2)$$

Oxidation of Fe^{2+} to Fe^{3+} by anodic oxidation can be considered as follows:

$$Fe^{2+}_{(aq)} = Fe^{3+}_{(aq)} + e \quad (3)$$

$$2\,Fe^{2+}_{(aq)} + 5\,H_2O + \tfrac{1}{2}\,O_{2(g)} \rightarrow 2\,Fe(OH)_{3(s)} + 4\,H^+ \quad (4)$$

And reaction occurring at the cathode can be represented by:

$$2\,H_2O + 2e \rightarrow H_{2(g)} + 2OH^- \quad (5)$$

The purpose of the study was to apply electrocoagulation process using aluminum electrodes for the treatment of oily bilge water from a Port PT. Pelindo Marine Service Tanjung Perak, Surabaya, Indonesia.

2. Materials and Experimental Set-up

2.1. Materials

Bilgewater was collected from a Tanjung Perak Port in Surabaya, Indonesia. Sodium hydroxide and Hydrochloric acid were used to adjust the pH. Salts were used for the electrolyte. All of the chemicals were purchased at shops Chemistry in Surabaya, Indonesia. Aluminum plates were purchased at a flea market in Surabaya, Indonesia.

2.2 Experimental Set-up

The batch and continuous electrochemical reactor were constructed in a Laboratory scale (Figure 1).

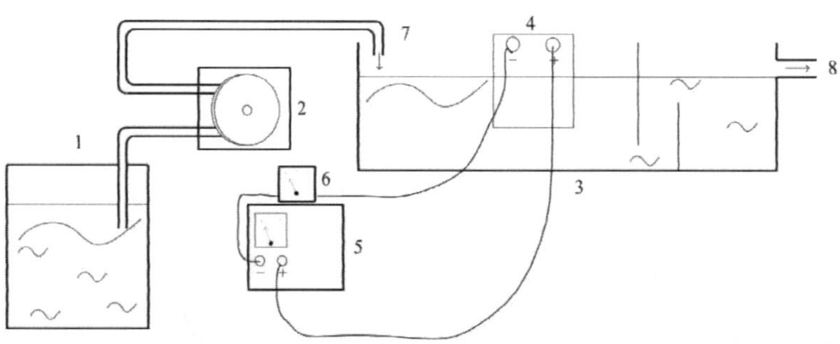

Figure 1. Apparatus of electrocoagulation process;(1) Bilge water tank storage; (2) Peristaltic pump; (3) Electrocoagulation reactor; (4) Aluminum Electrodes; (5) Voltmeter and power DC; (6) Ampere meter; (7) Influent of fluid; (8) effluent of fluid.

A volume of the glass reactor was 15 l, and the effective volume was 10 l. Aluminum electrodes with dimensions of 20 cm × 8 cm × 0.2 cm were used and vertically mounted with a distance of 4 cm and immersed at a depth of 8 cm to the solutions. The electrodes were then connected to DC power supply with an electric current approximately 10 A used as a source of direct current. The total area of effective work electrodes was 160 cm^2. The process was kept constant at around 15-90 min. Sampling was conducted for analyzing oil content every 15 min. The oil concentration was measured gravimetrically. The efficiency of oil removal during the process was measured using Equation (6) [30,31]:

$$Eff= \frac{C_0 - C}{C_0} \times 100\% \quad (6)$$

Where Co and C is the concentration of oily wastewater before and after electrocoagulation process in mg/l, respectively.

3. Results and Discussions

3.1 Batch Process

3.1.1 The Effect of pH on Electrocoagulation Process

In a batch process of electrocoagulation was carried out using Aluminum electrodes for 60 min. Table 1 shows the correlation between pH and time. The results indicated that the pH changed during the process. The pH increased from 6.41 to 7.43 with a time of 15 to 30 min. However, after 30 min the pH decreased slowly up to 60 min and achieved 7.08. It indicated that during the process Al^{3+} ions were released in solutions came from a sacrificial anode and water electrolysis process that generated gas and hydroxide ions as well as the release of electrons with time according to ionic reactions. The increase in hydroxide ions in a solution might cause an increase in the pH that caused alkaline conditions. The decreasing in pH after 30 min was because of the reaction between hydroxide and Al^{3+} ions that were released from the anodes which generated precipitation of $Al(OH)_3$.

Table 1. pH measurements

Time (min)	pH
15	6.43
30	7.43
45	7.31
60	7.08

3.1.2 Effect of Initial Concentration

A variety of initial concentration of oily wastewater from 60-145 g/l was conducted in a batch process. Figure (2) shows the correlation between the efficiency of oil removal and time at different oil content. The results show that the initial oil removal was rapid approximately for 15 min, then afterward, the removal was relatively constant. It was obtained that the removal was achieved by 99.62%, 99.50%, 98.12% and 96.25% for 15-60 min, respectively, at the initial concentration of 60 g/l, 75 g/l, 90 g/l, and 145 g/l, respectively. However, these results contradict the results of some authors. Sekman et al. [32] showed the electrocoagulation time for removal of these pollutant determined for 20 min, the removal efficiency of 81% at the initial oil & grease concentration of 600 mg/l. Ulucan et al. [33] studied that oil & grease removal obtained was 81.7% in approximately 13 min at the initial oil & grease content of 338 mg/l. Also, Chavalparit and Ongwandee [21] studied that removal of oil & grease was 98.42% for 25 min operation at an initial concentration of 6,020 mg/l.

IOP Conf. Series: Materials Science and Engineering **543** (2019) 012098 doi:10.1088/1757-899X/543/1/012098

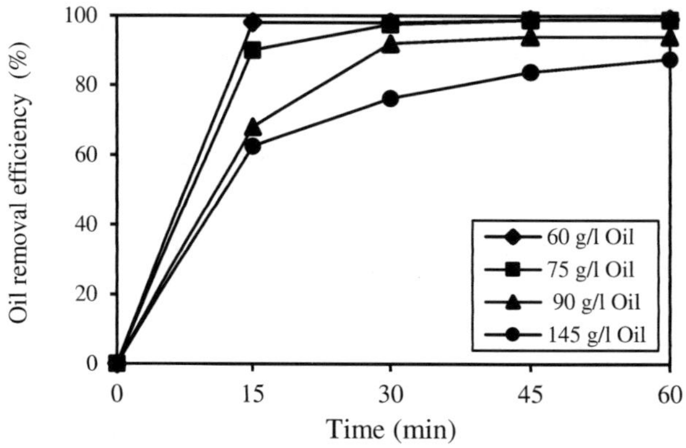

Figure 2. The influence of the initial oil concentrations on removal efficiency in a batch mode.

3.1.3. Effect of Current Density

Figure (3) shows the correlation between oil removal efficiency and current density. The removal of oil was affected by the current density. The results showed that an increase in current density might cause an increase in oil removal. The applied current density of 62.5 mA/cm^2 in the process achieved a maximum oil removal efficiency 99.95% and the use of a current density of 36.39 mA/cm^2 minimum removal was found to be 68.33%. This was due to the fact that an increase in current densities would cause an increase in the amount of the Aluminum anodes dissolved in the solutions, therefore an increase in the formation of metal hydroxide flocs. Thus, the formation of complexed aluminum hydroxide could lead oil removal. Therefore, the efficiency of oil removal increased with increasing current density. However, this result contradicts the results of some authors. Ulucan et al. [34] found to be oil &grease removal efficiency of 81.7% under a current density of 9.87 mA/cm^2. Chavalparit and Ongwandee [21] found that oil & grease removal efficiency was approximately 98.42% at a current density of 20.8 mA/cm^2.

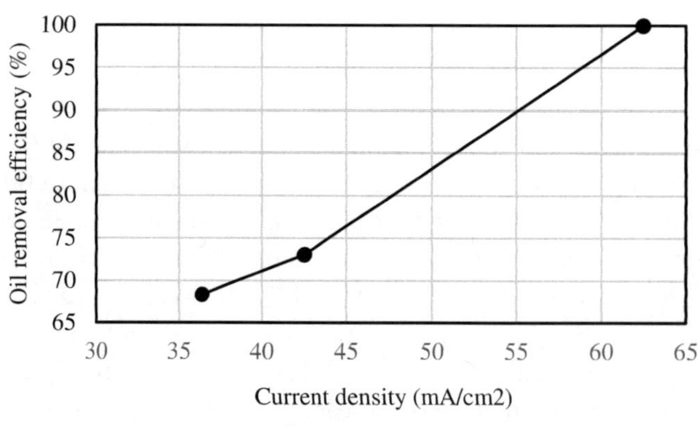

Figure 3: Effect of current density on oil removal.

3.2 Continuous Mode

3.2.1. Effect of Oil Removal on the water present in a Basin

These experiments were conducted in order to obtain time steady state of oil removal in a continuous mode. At starting-up of the experiment, an electrocoagulation reactor was filled with only water, then oily bilge water was introduced continuously at oil concentration of 63.89 g/l with a variety of hydraulic residence time (HRT-1 = 6 h; HRT-2 = 5 h; HRT-3 = 4.5 h; HRT-4 = 4 h), respectively. The HRT might cause the effect of oil removal during the process. The results of the experiments are shown in Figure (4) indicating that the oil removal was obtained 100% with HRT-1 at steady state conditions of approximately 5 min and followed by 94%, 93.33%, and 87.5%, respectively, with the HRT-2, HRT-3, HRT-4 and time of approximately 60 min, respectively.

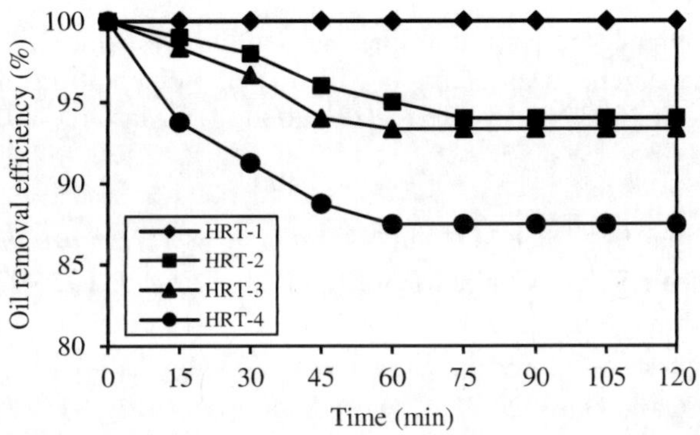

Figure 4. The influence of HRT on oil removal efficiency at the initial water present in a basin.

3.2.2. Effect of Oil Removal on the Oil Present in a Basin

The effect of oil removal efficiency is shown in Figure (5). At Starting-up of the experiments, oil/water emulsion synthetic was filled in the reactor at a concentration of approximately 63.89 g/l. A variety of HRT was introduced to the reactor, and they showed influences on oil removal. The results indicated that an increase in the HRT might cause a decrease in oil removal. At steady state, all of the experiments was achieved for 90 min. The maximum efficiency of oil removal was obtained by 73.33% with HRT-1 and followed by the efficiency of oil removal of 60%, 50% and 48% with HRT of HRT-2, HRT-3, and HRT-4, respectively.

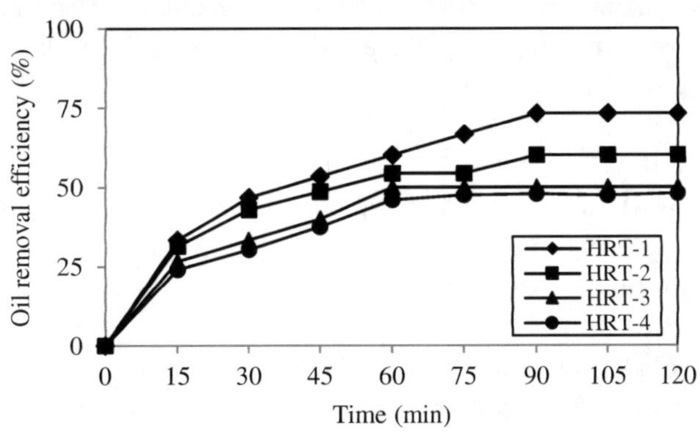

Figure 5. The influence of HRT on oil removal efficiency at the initial oil present in a basin.

Conclusions

The experiments could be concluded that electrocoagulation process had a potential for the treatment of oily bilge water. The treatment of oily wastewater using electrocoagulation process achieved a high removal efficiency in a short time. The maximum efficiency of oil removal was around 99.62% and the minimum was 96.25% at an initial oil concentration of 60 g/l and 145 g/l in batch processes, respectively. The maximum oil removal of 99.95% and a minimum of 68.33% with the current density of 62.5 and 36.39 mA/cm^2, respectively. In a continuous process with the initial water only present in the basin, the maximum oil removal efficiency was achieved approximately 100% with a hydraulic residence time (HRT-1) of 6 h for 60 min of operations. In a continuous process with the initial oil/water emulsion at a concentration of 63.89 g/l present in the basin, the maximum oil removal was achieved approximately 73.33% with HRT-1 of 6 h for 90 h of operations.

Acknowledgment

The authors would like to thank the Institut Teknologi Sepuluh Nopember, Surabaya, Indonesia for financially supporting this research, in association with the implementation of University Excellent Research in the year of 2015 No. 003246,150/ IT2.11/ PN.08/ 2015 March 5, 2015.

References

[1] Bensadok K, Benammar S, Lapicque F and Nezzal G 2008 *J. Hazard.Mat.* **152** 423–30.
[2] El-Naas M H, Al-Zuhair S and Alhaija M A 2010 *Chem. Eng. J.* **162** 997–05.
[3] Kuyukina M S, Ivshina I R, Serebrennikova M K, Krivorutchko A B, Podorozhko E A, Ivanov R V and Lozinsky V I 2009 *Int. Biodeterior. Biodegrad.* **63** 427–32.
[4] Lu W, Qu L, Henry K and Dai L 2009 *J. Power Sources* **189** 1270-1277.
[5] Santo C E, Vilar V J P, Botelho C MS, Bhatnagar A, Kumar E and Boaventura R A R, 2012 *Chem. Eng. J.* **183** 117–23.
[6] Shahrezaei F, Mansouri Y, Zinatizadeh A A L and Akhbari A 2012 *Powder Technol.* **221** 203–12.
[7] Sun Y, Zhang Y and Quan X 2008 *Sep. Purif. Technol.* **62** 565-70.
[8] Abdelwahab O, Amin N K and El-Ashtoukhy E S Z 2009 *J. Hazard. Mat.* **163** 11–16.
[9] El-Naas M H, Al-Zuhair S, Al-Lobaney A and Makhlouf S 2009 *J. Environ. Manag.* **91** 180–85.
[10] Yang C L 2007 *Sep. Purif. Technol.* **54** 388-95.
[11] Yavuz Y, Koparal A S, Öğütveren Ü B 2010 *Desalination* **258** 201–205.
[12] Wei L, Guo S, Yan G, Chen C, and Jiang X 2010 *Electrochim. Acta* **55** 8615-20.
[13] Yan L, Ma H Z, Wang B, Wang Y F and Chen Y S 2011 *Desalination* **276** 397–02.
[14] Asselin M, Drogui P, Brar S K, Benmoussa H, Blais J F 2008 *J. Hazard. Mat.* **151** 446–55.
[15] Soeprijanto S, Perdani A D, Nury D F and Pudjiastuti L 2017 110015. http://dx.doi.org/10.1063/1.4982345.
[16] Canizares P, Martinez F, Jimenez C, Saez C and Rodrigo M A 2008 *J. Hazard. Mat.* **151** 44–51.
[17] Carmona M, Khemis M, Leclerc J P and Lapicque F 2006 *Chem. Eng. Sci.* **61** 1237–46.
[18] Fouad Y O A, Konsowa A H, Farag H A and Sedahmed G H 2009 *Chem. Eng. J.* **145** 436–40.
[19] Martínez-Delgadillo S A, Morales-Mora M A and Barceló-Quintal I D 2010 *Sust. Environ. Res.* **20**, 227-31.
[20] Ozyonar F 2016 *I. J. Electrochem. Sci.* **11** 1456 – 71.
[21] Chavalparit O and Ongwandee M 2009 *J. Environ. Sci.* **21** 1491–96.
[22] Daud Z, Latiff A A A, Azis N A A and Awang H 2013 *Aust. J. Basic Appl. Sci.* **7** 457-63.

[23] Un U T, Koparal A S and Ogutveren U B 2009 *J. Environ. Manag.* **90** 428–33.

[24] Chen X, Chen G and Yue P L 2000 *Sep. Purif. Technol.***19** 65–76.

[25] Xu X and Zhu X 2004 Treatment of refectory oily wastewater by electrocoagulation process, *Chemosphere*, **56** 889–94.

[26] Drogui P, Asselin M, Brar S K, Benmoussa H and Blais J F 2008 *Sep. Purif. Technol.* **61** 301–10.

[27] Cotillas S, Cañizares P, Martin De Vidales M J, Sáez C, Rodrigo M A and Llanos J 2014 *Chem. Eng. Trans.* **41** 133-38.

[28] Rehman A, Kimb M, Reverberi A P and Fabiano B 2015 *Chem.Eng. Trans.* **43** 2251-56.

[29] Balasubramanian N and Srinivasakannan C 2010 *Nova Science Publishers*, Inc. NY. The USA, 19.

[30] Daneshvar N, Oladegaragoze A and Djafarzadeh N 2006 *J. Hazard. Mat.* **B129** 116–22.

[31] Daneshvar N, Ashassi-Sorkhabi H and Tizpar A 2003 *Sep. Purif. Technol.* **31** 153–62.

[32] Sekman E, Top S, Uslu E, Varank G and Bilgili M S 2011 *I. J. Environ. Res.* **5** 1079-86.

[33] Ulucan K, Kabuk H A, Ilhan F, Kurt U2014 *Int. J. Electrochem. Sci*. **9**, 2316 – 26.

IOP Conf. Series: Materials Science and Engineering **543** (2019) 012099 doi:10.1088/1757-899X/543/1/012099

Liquid Organic Fertilizer from Waste of Coconut Porridge Manufacturer and Molasses using Various Microorganisms

R Darmawan*, F Hidayat, S R Juliastuti, N Hendrianie, S Gunawan, and H W Aparamarta

Department of Chemical Engineering, Institut Teknologi Sepuluh Nopember, Kampus ITS Sukolilo, Surabaya 60111, Indonesia

*E-mail: rdarmawan@chem-eng.its.ac.id

Abstract. Excessive use of inorganic fertilizers can aggravate soil conditions. The development of organic fertilizer is urgently required. The fertilizers are used to supply organic materials to improve the physical, chemical, and biological properties of the soil. Meanwhile, many potential waste from coconut porridge household industry in Ponorogo Distric of Indonesia have not been utilized properly yet. The aim of this research are to improve the quality of liquid organic fertilizers by using various microorganisms with addition of liquid materials such as coconut water waste, water of immersed coconut fiber and molasses; and to observe the growth of chilies, tomatoes, and eggplants which were treated using the organic fertilizer. Liquid organic fertilizer was made by mixing the liquid materials using various microorganisms, such as Aspergillus niger, Pseudomonas putida, Bacillus muchilaginosis, Azotobacter chrococum and bioactivator EM4. The liquid fertilizer was carried out with a mixed batch reactor equipped of aeration 4 L/ min for 10 days. From this research was conducted observations on the development of height and fruit on the assesed plants using the produced fertilizer.

1. Introduction

Coconut/ *Cocos nucifera* is a strategic commodity that has a social, cultural and economic role in the Indonesian people lives. Indonesia belongs one of the countries that has the largest coconut plantations in the world with an area of 3.54 million hectares and producing coconut 2.87 million tons [1]. Ponorogo is one of regency in East Java Province, Indonesia. According to the BPS of Ponorogo Regency in 2015 [2], the number of coconuts that can be produced is 6.170,09 tons. The benefits of coconut plants are not only in the flesh of the fruit, but all parts of the coconut plant have great benefits, starting from tree trunks, husks, shells to coconut water. Ponorogo district has a potential product from coconut especially coconut porridge in household industrial scale. There are some coconut porridge manufacturers to make the traditional food. The manufacturer release coconut water for coconut porridge production which is not utilized properly. The use of waste of coconut water is considered as a good chance to generate useful thing to the community. This effort is also one of promising ways by producing an organic liquid fertilizer from waste water of coconut. The fertilizer based on organic can be used to nourish plants and supply organic materials to improve nutrients and water uptake, growth and tolerance of plant to abiotic and biotic factors. Furthermore, the bio-fertilizers have been as eco-friendly and cost effective inputs for the farmers and have a key play role for production and improving the sustainability physical, chemical, and biological properties of the soil [3]. On the other hand, the existence of molasses as an abundance side product from sugar factory is also limited using in Indonesia. Furthermore, the existence of Lapindo mud in Sidoarjo – Indonesia as a result of the spurt of material in the earth to the surface which has been determined as a national disaster has not been handled well and still leaves many problems, besides the potential possessed of the potential content of microorganisms, this experiment was conducted. This study has aims to utilize the waste of coconut water and molase to produce the organic liquid fertilizer by supplementing various microorganisms.

2. Materials and Methods

2.1 Microorganisms
There are five microorganisms and the mixed culture used in this study. Four types were ordered from the Microbiology Laboratory of Chemical Engineering Department, Faculty of Industrial Technology, Institut Teknologi Sepuluh Nopember (ITS) - Indonesia, namely Aspergillus niger, Azetobacter chrococum, Pseudomonas putida and Bacillus mucilagenosus. Effective Microorganism 4 (EM4) [4] was as bioactivators also added to the process. Meanwhile, the mixed cultured was isolated from Lapindo mud, Sidoarjo - Indonesia by applying enrichment culture for three weeks.

2.2. Medium
Minimum medium was prepared by adding some chemical such as 1 gram of $NHNO_3$, 1 gram of KH_2PO_4, 1.3 gram of $K_2HPO_4.3H_2O$, 0.1 gram of $MgSO_4.7H_2O$, 0.1 gram of $FeCl_3.6H_2O$, 0.008 grams of $CaCl_2$ which were are dissolved in 1000 ml of sterile distilled water for making solution in enrichment culture stage [5]. NB (nutrient broth) of 8 grams was dissolved within 1 liter and sterilized as a bacterial medium.

2.3. Liquid Material
Waste of coconut water and molasses were purchased from coconut porridge manufacturer in Ponorogo and a sugar factory from Mojokerto – Indonesia, respectively.

2.4. Enrichment culture
Medium was prepared first by making minimum medium. K_2HPO_4 50 ppm and UREA 50 ppm were added into medium. Then it was sterilized using autoclave for 15 minutes. The medium including mud was incubated for three weeks where every week, 1 ml of liquid culture transferred into new enrichment medium.

2.5. Process treatment
Liquid fertilizer of organic was made by mixing 3 l of coconut water waste, 6 ml of molasses waste and microorganisms into the bioreactor (Figure 1).

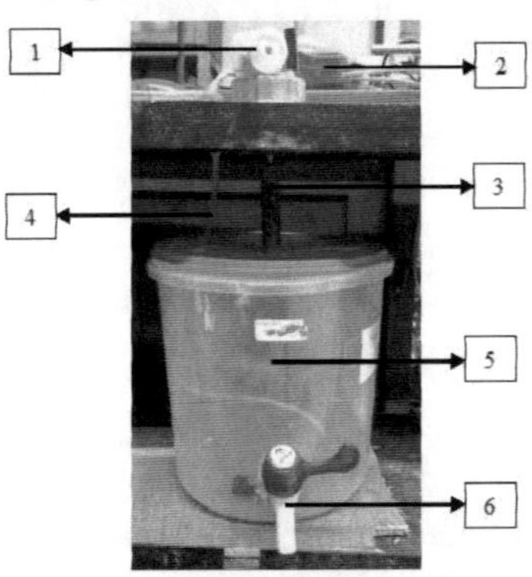

Figure 1. Reactor for biological process; with explanation no 1, 2, 3, 4, 5 and 6 are dynamo agitator, aerator, agitator air hose, reactor, and output valve, respectively

Various microorganisms such as Aspergillus niger, Azetobacter chrococum, Pseudomonas putida, Bacillus mucilagenosus, Effective Microorganism 4 (EM4) bioactivators, and mixed culture of Lapindo mud were added by considering the log phases of every microbe. The number of microorganisms was calculated using counting chamber method [6]. There were seven variables, they were 5 microbes (10^8 cells /mL), 4 microbes without EM4 (10^8 cells /ml), 5 microbes added mixed culture (10^8 cell /ml), 4 microbes without EM4 added mixed culture and only mixed culture (10^8 cell /ml). The bio-fertilizer was conducted in the mixed batch reactor with air aeration for 10 days. The result of fertilizer was analyzed N, P, K and C-organic content at t = 0 days, t = 5 days, and t = 10 days, then tested by manuring the bio-fertilizer to chili, tomato and eggplant.

3. Results and Discussion

There were five microorganism groups applied as variables for producing the organic liquid fertilizer. They were 5 Microbes (Aspergillus niger, Azetobacter chrococum, Pseudomonas putida, Bacillus mucilagenosus and EM4); 5 Microbes with Mixed Culture; 4 Microbes (Aspergillus niger, Azetobacter chrococum, Pseudomonas putida and Bacillus mucilagenosus); 4 Microbes with Mixed Culture and only Mixed Culture from Lapindo Mud. The resulting of the biofertilizer was analyzed by N, P, K and C-organic content at t = 0 days, t = 5 days, and t = 10 days, then tested on pepper, tomato and eggplant.

3.1 Effect of incubation process and microorganisms used on N, P, K and C elements

Many factors or parameters affected to the number of N, P, K and C elements on bio-fertilizer, such as the initial element and concentration material sources from coconut water and molasses, respectively, microbes used and duration of incubation. Based on the analysis of the bio-fertilizer product, it was found that the best result on N, P, K and C elements were obtained at t = 5 days with concentrations of 0.059%, 113.18 mg /l, 401.63 mg /l and 1.1%, respectively. Meanwhile, the percentage enhancing of macro elements were presented in the figures below. From Figure 2, 3, 4 and 5, it can be seen percentage enhancing of N, P, K and C elements, respectively during process incubation period.

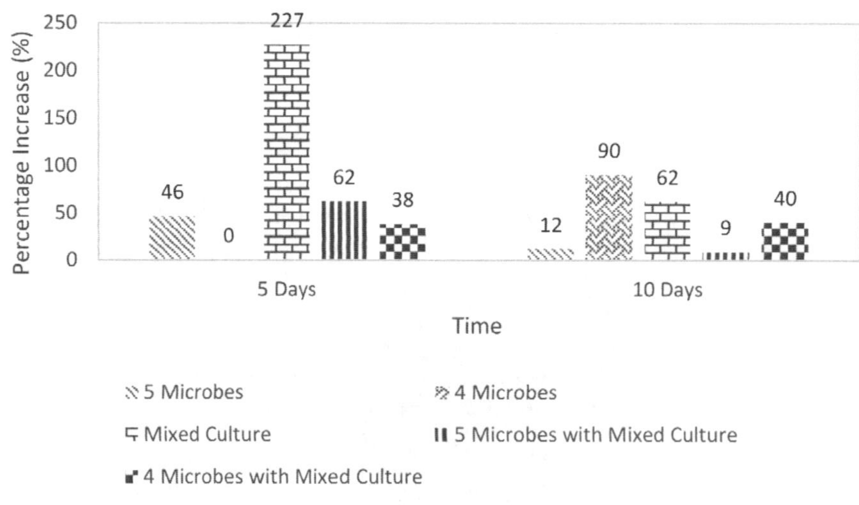

Figure 2. Percentage enhancing of N during process incubation period

Figure 2 shows the results of percentage increasing of N element for 10 days incubation period, where this figure indicates the highest enhance of N (227 %) was achieved by mixed culture at 5 days.

As shown in Fig. 2, the mixture culture isolated from Lapindo Mud reported that it had the highest increase of nitrogen compound. These culture from LapindoMud believed can upgrade and have a good role as the nitrogen contributors on the plants [7]. These microbes might be belonging to diazotroph bacteria where the isolates have a critical role to fix nitrogen. The capability of the isolates might be the existence a nitrogenase enzyme which have ability for combining nitrogen and hydrogen [8, 9].

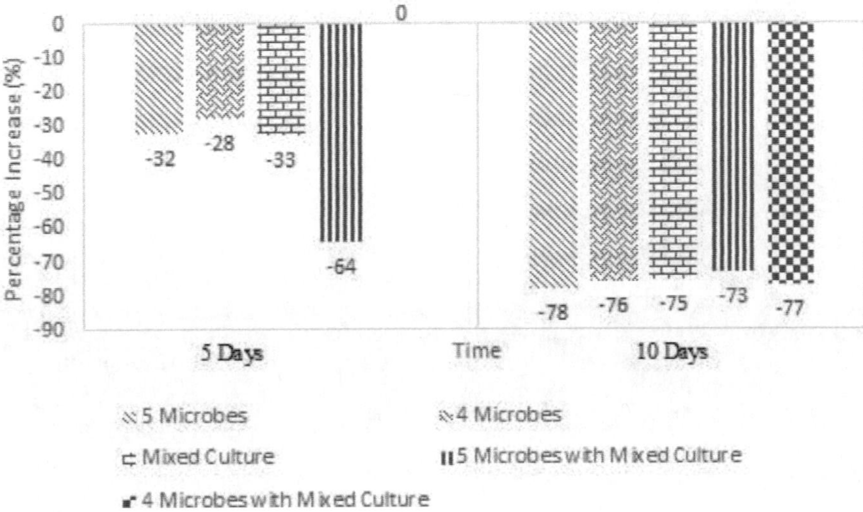

Figure 3. Percentage enhancing of P during process incubation period

Figure 3 shows the results of percentage decreasing of P during process incubation period. With increasing incubation period, the enhancing of phosphor elements were decreased. After 5 days incubation, the highest content can be achieved (-28 %) then the element composition was decreased (-73 %) for 10 days incubation period. The best result of P element enhancing was obtained by 4 microbes (-28 %) variable at 5 days incubation. From the result indicates that the combination of microbes (Aspergillus niger, Azetobacter chrococum, Pseudomonas putida and Bacillus mucilagenosus)
4 microbes) were not suitable to enhance element of phosphor [7].

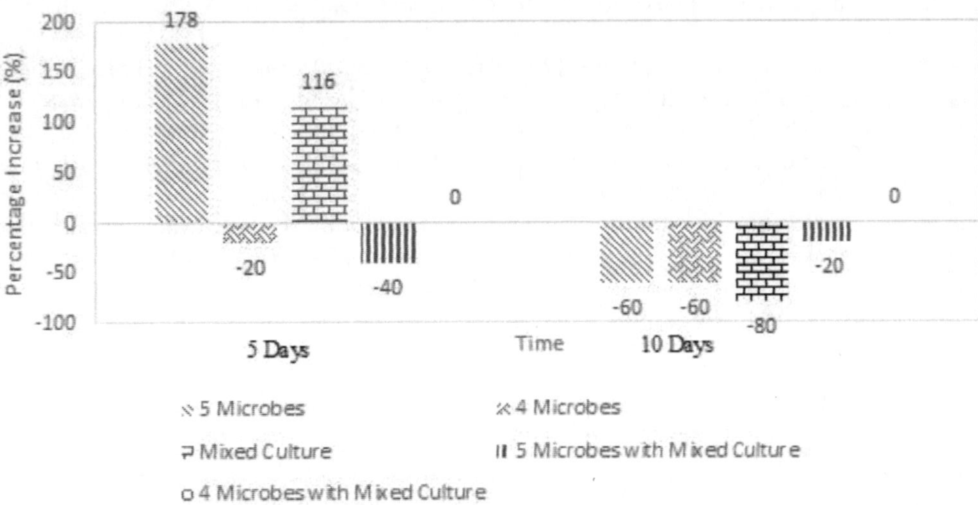

Figure 4. Percentage enhancing of K during process incubation period

Potassium element from several variables were increased along with the incubation experiment conducted after time 5 days then for all variables were decreased after 10th days period. From Figure 4, it can be seen that highest percentage enhancing of K during process incubation period was obtained by 5 microbes at 5 days. The enhancing of potassium component might be caused by the hydrolysis and proteolysis reactions so that it lead to release K^+ from the organic matter.

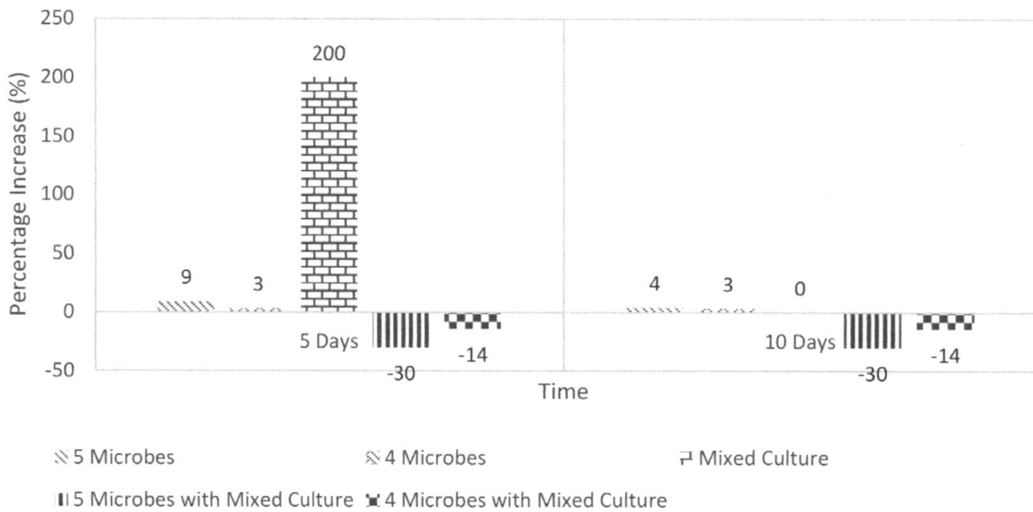

Figure 5. Percentage enhancing of C during process incubation period

Figure 5 exhibits that best percentage decreasing of C element during process incubation period was obtained by 5 microbes with mixed culture at 5 days. In the conducted experiment, the carbon elements from coconut water and molasses were degraded into a simpler compound with more stable. All variables represent the decreased with enhancing the incubation period. The condition occurs might be caused by absence of the added microorganisms that consume C-organic within coconut water and molasses as a carbon source to enhance the microorganisms metabolism.

3.2 Observation of organic liquid fertilizer application to test plants

After bio-fertilizer was obtained, the observation of fertilizer application on the test plants for 45 days. There were three test plants, i.e. chili, tomato and eggplant. The growth observation was conducted including difference high of plants before and after of manuring, the average number of fruit and leaf. Each fertilizer experiment variable on the test plant used three plants each, so that for the height, fruit and number of leaves were calculated on average. Table 1, 2 and 3 show about observation of plant growth on tested plants such as chilies, tomatoes and eggplants. From these tables, it can be seen the highest growth on chilies, tomatoes and eggplants were achieved by 4 types of microbes with mixed culture, 4 types of microbes with mixed culture, and mixed culture variables after 45th day with high plants of 56.20, 141,23 and 62.67 cm, respectively.

Table 1. Observation data of plant growth on chilies

No	Variable	Δh (cm)			
		0 day	15 day	30 day	45 day
1	5 types of microbes	37.67	40.90	41.20	42.90
2	5 types of microbes with *mixed culture*	35.67	38.40	38.97	40.90

3	4 types of microbes without EM4	36.67	40.03	41.50	48.23
4	**4 types of microbes with *mixed culture***	42.00	50.20	51.90	**56.20**
5	*Mixed culture*	36.67	39.90	41.23	43.57
6	Only coconut water	40.67	44.23	44.87	46.27
7	Only water	39.67	37.57	38.23	39.57

Table 2. Observation data of plant growth on tomatoes

No	Variable	Δh (cm)			
		0 day	15 day	30 day	45 day
1	5 types of microbes	32.67	57.2	67.6	89.23
2	5 types of microbes with *mixed culture*	38.33	65.2	92.63	97.75
3	4 types of microbes without EM4	33.67	52.97	71.67	85
4	**4 types of microbes with *mixed culture***	40.67	71.67	101.6	**141.23**
5	*Mixed culture*	34.67	65.3	89.63	106.85
6	Only coconut water	40.67	70.9	81.97	97.23
7	Only water	39.33	75.53	82.57	86.35

Table 3. Observation data of plant growth on eggplants

No	Variable	Δh (cm)			
		0 day	15 day	30 day	45 day
1	5 types of microbes	30.33	34.93	35.93	38.73
2	5 types of microbes with *mixed culture*	26.67	31.27	32.6	36.3
3	4 types of microbes without EM4	34.33	43.9	46.6	51.7
4	4 types of microbes with *mixed culture*	31.67	41.83	47.23	56.3
5	***Mixed culture***	36	47.93	50.93	**62.67**
6	Only coconut water	30.33	34.53	35.97	38.83
7	Only water	30.67	33.63	35.87	38.03

Meanwhile, Table 4, 5 and 6 show about observation of the number of fruit on tested plants. From these tables, it can be seen the highest fruit number on chilies, tomatoes and eggplants were all variables achieved by 4 types of microbes with mixed culture, after 45th day with the average number of fruit of 1.67, 7.67 and 1.67, respectively.

Table 4. Data on observing the average number of fruit on chilies

No	Variable	Mean of fruit number			
		0 day	15 day	30 day	45 day
1	5 types of microbes	-	-	0.33	0.33
2	5 types of microbes with *mixed culture*	-	-	-	-
3	4 types of microbes without EM4	-	0.67	1.33	0.67
4	**4 types of microbes with *mixed culture***	-	-	0.67	**1.67**
5	*Mixed culture*	-	-	-	1.33

6	Only coconut water	-	-	-	1
7	Only water	-	-	-	-

Table 5. Data on observing the average number of fruit on tomatoes

No	Variable	Mean of fruit number			
		0 day	15 day	30 day	45 day
1	5 types of microbes	-	-	-	5
2	5 types of microbes with *mixed culture*	-	-	1	6
3	4 types of microbes without EM4	-	-	-	5.67
4	**4 types of microbes with *mixed culture***	-	-	0.67	**7.67**
5	*Mixed culture*	-	-	0.67	3.67
6	Only coconut water	-	-	-	3.67
7	Only water	-	-	0.67	4

Table 6. Data on observing the average number of fruit on eggplants

No	Variable	Mean of fruit number			
		0 day	15 day	30 day	45 day
1	5 types of microbes	-	-	-	-
2	5 types of microbes with *mixed culture*	-	-	-	-
3	4 types of microbes without EM4	-	-	0.33	0.33
4	**4 types of microbes with *mixed culture***	-	-	0.67	**1.67**
5	*Mixed culture*	-	-	0.67	1
6	Only coconut water	-	-	-	-
7	Only water	-	-	-	-

Furthermore, Table 7, 8 and 9 show about observation of the number of leaves on tested plants. From these tables, it can be seen the highest leaves number on chilies, tomatoes and eggplants were all variables also achieved by 4 types of microbes with mixed culture, after 45th day with the average number of leaves of 48.33, 24.67 and 24.67, respectively.

Table 7. Data on observing the average number of leaves on chilies

No	Variable	Mean of leaf number			
		0 day	15 day	30 day	45 day
1	5 types of microbes	11.33	14.67	15	18.67
2	5 types of microbes with *mixed culture*	11	15	21	27
3	4 types of microbes without EM4	16	21.67	40.67	45.67
4	**4 types of microbes with *mixed culture***	12.33	23.33	37	**48.33**
5	*Mixed culture*	7.33	14.67	18	25
6	Only coconut water	12.33	14.67	18.67	19
7	Only water	10.33	13.67	18.67	18.33

Tabel 8. Data on observing the average number of leaves on tomatoes

No	Variable	Mean of leaf number			
		0 day	15 day	30 day	45 day
1	5 types of microbes	7.67	9	11.33	17.33
2	5 types of microbes with *mixed culture*	7.67	9.33	13.67	23.67
3	4 types of microbes without EM4	6.67	8.67	11.67	20.33
4	**4 types of microbes with *mixed culture***	7.33	10.33	15.67	**24.67**
5	*Mixed culture*	7.67	9	12	21.33
6	Only coconut water	8	9.67	11.67	17.33
7	Only water	7.33	8.67	11.33	17

Table 9. Data on observing the average number of leaves on eggplant

No	Variable	Mean of leaf number			
		0 day	15 day	30 day	45 day
1	5 types of microbes	9.33	8	11	12.67
2	5 types of microbes with *mixed culture*	6.33	7.67	9.67	13
3	4 types of microbes without EM4	6.67	9.67	12.67	16.67
4	**4 types of microbes with *mixed culture***	7.67	10.67	15.33	**24.67**
5	*Mixed culture*	7.67	12.67	19	19.67
6	Only coconut water	6	7	8.67	10.33
7	Only water	5	4.33	6	7.67

Based on three parameters observation results including the height, fruit and number of leaves show that variable of 4 types of microbes with mixed culture was the best combination of microorganism to give good effect of chilies, tomatoes and eggplants. This condition might be beside 4 microorganisms such as Aspergillus niger, Azetobacter chrococum, Pseudomonas putida and Bacillus mucilagenosus, the mixed culture isolated from Lapindo mud have a key play role on the plants. This occurs because the mixed culture of Lapindo mud may contain good microorganisms and can increase soil so that plants can thrive and bear fruit well. In the future, it is necessary to conduct bacterial community analysis in order to know the name of bacteria contained in Lapindo mud.

4. Conclusion

The coconut waste water supplemented by molasses can be used as the potential material sources for generating liquid biofertilizer. A combination between Aspergillus niger, Pseudomonas putida, Bacillus mucilagenosus, Azetobacter chrococum and mixed culture isolated from Lapindo Mud have been found as a promising combination cultures to make the organic liquid fertilizer for manuring to chilies, tomatoes and eggplants.

5. Acknowledgments

The authors would like to thank the Institut for Research and Community Services (LPPM) of Institut Teknologi Sepuluh Nopember (ITS) on the Program (*Pengabdian Kepada Masyarakat Berbasis Penelitian*) for financial supporting with agreement document No: 1474/PKS/ITS/2018.

6. References

[1] Directorate General of Estate Crops 2018 Tree Crop Estate Statistics Of Indonesia, Coconut 2015-2017
[2] BPS Kabupaten Ponorogo, 2015.
[3] Itelima J, Bang WJ, Sila, Onyimba IA and Egbere OJ 2018 *J Microbiol Biotechnol Rep.* **2**(1) 22-28
[4] Higa T, Parr J F 1994 International Nature Farming Research Center Atami, Japan
[5] Darmawan R, Nakata H, Ohta H, Niidome T, Takikawa K, Morimura S 2015 *Bioremed. & Biodegrad.* **6**, 3, doi:10.4172/2155-6199.1000283
[6] Williams R E O 1952 *J. gen. Microbiol.* **7**, 89-97
[7] Marbun, Posma, Hardy G and Putri J A 2013 Universitas Sumatera Utara
[8] Yin T T, Pin U L, and Ghazali A H A 2015 *Trop. Life Scie*. Res. **26** 1
[9] Issar S and Gaur R K 2013 *Intern.J. of Adv. Res.* **1**, 8

Institute of Physics
Dirac House, Temple Back
Bristol BS1 6BE UK

ISSN: 1757-8981
ISBN 978-1-5108-8922-4

9 781510 889224

CURRAN ASSOCIATES INC.
proceedings.com